Intellectual Property Management in Health and Agricultural Innovation

a handbook of best practices

Intellectual Property Management in Health and Agricultural Innovation

a handbook of best practices

VOLUME TWO

EDITED BY

ANATOLE KRATTIGER

RICHARD T. MAHONEY

LITA NELSEN

JENNIFER A. THOMSON

ALAN B. BENNETT

KANIKARAM SATYANARAYANA

GREGORY D. GRAFF

CARLOS FERNANDEZ

STANLEY P. KOWALSKI

MIHR
A STEP NEARER EQUITY

PIPRA
public intellectual property resource for agriculture

Ministério da Saúde
FIOCRUZ
Fundação Oswaldo Cruz

bioDevelopments
INTERNATIONAL INSTITUTE

Published by

MIHR (Centre for the Management of Intellectual Property in Health Research and Development), Oxford Centre for Innovation, Mill Street, Oxford, OX2 0JX, U.K. www.mihr.org, mihr@bioDevelopments.org

PIPRA (Public Intellectual Property Resource for Agriculture), University of California, Davis, Plant Reproductive Biology Bldg., Extension Center Circle, Davis, CA, 95616-8780, U.S.A. www.pipra.org, info@pipra.org

Oswaldo Cruz Foundation (Fiocruz), Av. Brasil 4365, Rio de Janeiro, RJ 21040-900, Brazil. www.fiocruz.br, ccs@fiocruz.br

*bio*Developments-International Institute, Cornell Business and Technology Park, PO Box 4235, Ithaca, NY 14852, U.S.A. www.bioDevelopments.org, info@bioDevelopments.org

www.ipHandbook.org, info@ipHandbook.org

Manuscript Editors
Paula Douglass (Dryden, New York, U.S.A.)
with Jacqueline Stuhmiller (Ithaca, New York, U.S.A.) and Katy Dixon (Champagne, Illinois, U.S.A.)

Layout and Design
Linette Lao (Invisible Engines, Ypsilanti, Michigan, U.S.A.)

Copy Editor
Barry Hall (OmniGlyph, Ypsilanti, Michigan, U.S.A.)

Online version
Dynamic Diagram (Providence, Rhode Island, U.S.A.)

Citation
Krattiger A, RT Mahoney, L Nelsen, JA Thomson, AB Bennett, K Satyanarayana, GD Graff, C Fernandez, and SP Kowalski (eds). 2007. *Intellectual Property Management in Health and Agricultural Innovation: A Handbook of Best Practices.* MIHR: Oxford, U.K., and PIPRA: Davis, California, U.S.A. Available online at www.ipHandbook.org.

Translations
Requests for permission to reproduce the book in its entirety or for translations,
whether for sale or for non-commercial distribution, should be addressed to MIHR or PIPRA.

Library of Congress Preassigned Control Number
2007925343

ISBN 978-1-4243-2027-1

Distributed free of charge to low- and middle-income countries
(subject to availability of funding and support for distribution).
www.ipHandbook.org

green press INITIATIVE

Bio Developments International Institute is committed to preserving ancient forests and natural resources. We elected to print *Intellectual Property Management Volume Two* on 30% post consumer recycled paper, processed chlorine free. As a result, for this printing, we have saved:

29 Trees (40' tall and 6-8" diameter)
12,331 Gallons of Wastewater
4,959 Kilowatt Hours of Electricity
1,359 Pounds of Solid Waste
2,670 Pounds of Greenhouse Gases

Bio Developments International Institute made this paper choice because our printer, Thomson-Shore, Inc., is a member of Green Press Initiative, a nonprofit program dedicated to supporting authors, publishers, and suppliers in their efforts to reduce their use of fiber obtained from endangered forests.

For more information, visit www.greenpressinitiative.org

PRINTED WITH SOY INK

Table of Contents

Section 2: Specific Strategies and Mechanisms for Facilitating Access to Innovation

Section 3: The Policy and Legal Environment for Innovation

SECTION 4: THE IP TOOLBOX

Section 5: Institutional Policies and Strategies

Section 6: Establishing and Operating Technology Transfer Offices

Section 7: Contracts and Agreements to Support Partnerships

Section 8: Inventors and Inventions

Section 9: Evaluation and Valuation of Technologies

Section 10: Patents and Patenting: Balancing Protection with the Public Domain

VOLUME TWO

Section 11: Technology and Product Licensing

Section 12: Dealmaking and Marketing Technology to Product-Development Partners

SECTION 13: THE PUBLIC SECTOR AND ENTREPRENEURSHIP

SECTION 14: FREEDOM TO OPERATE AND RISK MANAGEMENT

SECTION 15: MONITORING, ENFORCEMENT, AND RESOLVING DISPUTES

Section 16: Bioprospecting, Traditional Knowledge, and Benefit Sharing

Section 17: Putting Intellectual Property to Work: Experiences from around the World

SECTION 11

Technology and Product Licensing

CHAPTER 11.1

Licensing Biotechnology Inventions

JOHN W. FREEMAN, *Principal, Fish & Richardson P.C., U.S.A.*

ABSTRACT

After providing an overview of licensing in the field of biotechnology, the chapter carefully examines the key components of a license agreement, particularly in relation to the field's unique concerns. The chapter raises a number of issues that licensors and licensees should consider when negotiating patent license agreements. It offers precise definitions of key terms, points out areas of the agreement that merit special attention (including the relative merits of exclusive and nonexclusive licensing), considers the difficult question of how to determine a patent's value (especially when the patent is being used for screening purposes), and gives much-needed attention to the complexities of confidentiality agreements, especially those involving academic research institutions. To make negotiations easier and more realistic, the incentives for licensors and licensees are discussed, as are some of the finer points of development collaboration. In addition, the author offers some advice about how to define patent misuse, offering some helpful suggestions about what to do should things go bad. The goal of this chapter, however, is to ensure that agreements succeed.

1. BIOTECH LICENSING OVERVIEW

The issues raised in licensing patents are similar to those raised when prosecuting and enforcing biotech patents. In the case of licensing, however, the process is somewhat of an art, and the characteristics of the biotech industry are the artist's tools. No other industry requires so much time and so much money to market a product. Indeed, biotech patent applications typically are filed, and biotech patent licenses typically are executed, well before commercial goals are even in sight. This is particularly true for inventions with important medical applications that involve a drug or a diagnostic that will travel an extraordinarily long road before being manufactured commercially and used clinically. Even for inventions that are not related to medicine, extraordinary amounts of money are likely to change hands long before commercial goals are reached, if they ever are. Often, patent licenses play a key role in the development of biotech inventions.

Indeed, the likelihood of successfully commercializing any medical application embodied in a patent is a battle against the odds. According to an article by Henry Grabowski, professor of economics at Duke University, less than 1% of compounds examined in preclinical studies makes it into human testing, and only 20% of the compounds entering clinical trials survives and gains marketing approval.[1] Thus, less than one-fourth of 1% of newly developed compounds makes it to market. Once the product achieves marketing approval the task does not get much easier. The product will face enormous pressures from competition and will have significant difficulties establishing an infrastructure to manufacture and commercialize the drug product.

This is not to say that a biotech patent license needs to address all of these issues in detail. That would be impossible. These issues are raised to

Freeman JW. 2007. Licensing Biotechnology Inventions. In *Intellectual Property Management in Health and Agricultural Innovation: A Handbook of Best Practices* (eds. A Krattiger, RT Mahoney, L Nelsen, et al.). MIHR: Oxford, U.K., and PIPRA: Davis, U.S.A. Available online at www.ipHandbook.org.

suggest some of the ways biotech patent licenses differ from patent licenses in other industries. Moreover, knowing that a biotech invention is unlikely to succeed should heighten the license drafter's sensitivity to the kinds of reasons permitted for terminating the agreement, as well as what the impact of that termination would be. Other industry characteristics that the license drafter should keep in mind include:

- long, costly lead times to market that can result in limited patent life remaining after commercialization
- process of discovery, proof, and development into a product that requires a synergy of complex operations
- very high risks combined with high (often deferred) reward

2. KEY COMPONENTS OF THE LICENSE

Given the inherent complexities, in terms of business and science, of biotech patent licensing, it is easy to forget that a biotech patent license is merely a contract. All of the basic principles of contract law apply. The license drafter must take a step back from business terms and scientific subject matter to consider how the document will stand up to questions of enforceability, breach, and so forth.

A patent license, like other contracts, is enforceable in a legal action seeking either (a) damages for the aggrieved party in an amount corresponding to the benefit of the bargain that was breached; or (b) equitable (injunctive) relief giving the aggrieved party the benefit of its bargain. To withstand the scrutiny that a license will face, particularly if there is legal action for breach of contract or patent infringement, the licensing document should be precise and written in complete, clear sentences without errors in grammar, use, or syntax that could make interpretation difficult. Above all, the license should use terminology consistently (as is true for a patent claim) and avoid using different words for the same thing or using the same word to indicate different things.

Completeness and clarity are important goals, but some ambiguity is unavoidable. The parties need to use good judgment in tolerating ambiguities that cannot be resolved at the contracting stage.

The license document governs the parties' rights over a substantial period of time during which unforeseen events very likely will occur. The license cannot address explicitly all of the possibilities.

During the negotiations, it is important to consider, along with their consequences, events that are unlikely to occur. However, attention to these unlikely events can easily consume a disproportionate amount of time and effort and can sidetrack progress toward agreement on core issues. Thus, care should be taken to devote an amount of attention that is proportional to the potential cost or benefit associated with such an unlikely event. Keep in mind that alternate ways of mitigating the risks may be equally appropriate. For example, excessively negotiating over the division of risks and liabilities, and trying to structure the language of the agreement accordingly, may be less efficient than agreeing on insurance coverage to address those risks. This chapter will review some key components of the license to identify issues that recur during the negotiation and enforcement of biotech patent license rights.

2.1 *Background*

The background section of a license agreement identifies the factual predicates (or basis) for the license, including the parties, the effective date, and the parties' motivations and expectations. Definitions of critical terms may also appear in the background section.

Certain types of problems commonly arise when drafting this section. One type involves the identification of participants. Because corporate structure can be extremely fluid in the biotech industry—companies are acquired and spun off, and they frequently collaborate—and because small companies may have key personnel whose participation in product development is more important than the other assets of the licensee, careful attention must be paid to the identification of the party who is obligated to perform under the contract. The parties would be wise to consider the following questions:

- Does the obligation carry over to *affiliates*?

- Is the term *affiliates* defined in a way that meets expectations about who the other party should be? Does the term include a well-capitalized corporation that can be expected to survive other less well-capitalized affiliates?
- Could a competitor be defined as party to the license through its affiliation with another company that is more directly involved in your negotiation? (For example, does the definition of *parties* include companies that could sell to your customers or to the customers of your affiliates?)
- Should the flow of confidential information be restricted to certain affiliates in the family?
- Is a competitor company a shareholder in the licensor?
- Does a competitor company have a right of refusal in the commercialization of certain technologies or in certain territories based on previous agreements?

Terms such as net sales, net profits, and licensed product will likely appear and need to be defined in the background section. The following list presents a few of those terms and some notes on how they are likely to be treated:

- **Net Sales.** Includes deductions from gross sales before figuring royalty. Typical exclusions from net sales can include transportation costs, returns, bad debt, actual trade, quantity or cash discounts, broker's/agent's commissions, credits or allowances made or given on account of rejects or returns, and so on.
- **Net Profits.** Can be used instead of net sales but can be problematic as a basis for calculating royalty because profit figures can vary tremendously depending on accounting practices.
- **Licensed Product(s).** Identifies the product(s) whose sales constitute the royalty base. Include(s) any product covered by the licensed patents, or any product made by a method covered by the licensed patents. The scope of licensed products should be limited by field in accordance with the license grant.
- **Licensed Patent.** Usually includes particular patents identified by number. Problems may arise over patents issuing on applications that are continuations, divisionals, foreign counterparts, reissues, reexaminations, and continuations-in-part of known patents. Another issue is whether the license covers all of the licensor's patents that could ever be used in conjunction with the technology of the licensed patent. For example, the licensee may want to license "*all patents covering a licensed product.*" Such a definition is unclear, because the applicability of other licensor patents would depend entirely on what embodiment(s) the licensee chose to practice. For an academic institution with wide-ranging patent positions in many fields, this type of open-ended license is likely to raise problems and should be avoided. An even worse definition would sweep in "*all patents necessary to practice the licensed invention.*" In addition to the problem of not knowing exactly what embodiments the licensee will practice (and therefore not knowing which patents are being licensed), this definition is circular when combined with the standard definition of *licensed products*: products licensed are those covered by the licensed patents, and the licensed patents are those necessary to make, use, or sell the licensed product. Further, this definition is problematic because it implies a license to patents belonging to third parties. Finally, a license to "improvements" can raise problems (see also sections 2.5 and 5. below).

Seemingly innocuous definitions in the background section of the license agreement may decide key issues, including the scope of the license and the nature of the parties.

2.2 *Grant*

The grant section of a license establishes whether the license is exclusive to the licensee or whether others (including the licensor) may practice the invention. The grant section establishes limitations on the grant, such as restrictions on the technical or commercial fields or on the geographical areas within which the license may be practiced. The grant section may set out rights to sublicense or assign, or it may say that there are no such rights under the license.

The right to allow sublicensing or a prohibition on sublicensing should be explicit, as should be a right to assign or a prohibition on assigning. A party can retain some level of control on future events by using provisions allowing for the assignment of the license only with the consent of that party. The licensor should be aware that withholding the right to sublicense or even to assign does not guarantee that the nature and character of the licensee will remain constant. In one case, a very large player in HIV diagnostics purchased controlling stock in a relatively minor player that had a license under a key patent from a third party licensor, with no right to assign or sublicense. The licensor's intent in making the license personal to the minor company was to avoid competition from a large competitor. By purchasing controlling stock in the small licensee, the large competitor frustrated the licensor's purpose (*Institut Pasteur v. Cambridge Biotech Corp.*).[2]

In certain cases, it may be desirable to allow an assignment of interests without consent when a significant change in control occurs (for example, a merger or acquisition of a party) provided that the surviving entity assumes all of the obligations and benefits of the merged/acquired party. This can be advantageous to a corporate entity considering merger or spinout scenarios because it can simplify such transactions. This may be acceptable when a licensor is more concerned about income and less concerned about who is paying (and getting access to the license) and what future research/development interactions may arise with a partner.

Biotech licenses frequently are limited to specific medical indications, treatment modalities (for example, route of administration) or diagnostic formats (for example, screening versus confirmatory diagnosis). One reason for this might be that the technology is in a very early stage and substantial resources are needed to commercialize the technology, even in one limited field. Many biotech inventions feature basic ideas or technologies that may be used for a number of different medical indications, and the licensor may seek to increase its chances of success by establishing different licensees in different fields, particularly if no one licensee is likely to have the resources or interest to give top priority to all fields. Examples of such basic or *platform* technologies include viral constructs to deliver genes to a patient for gene therapy, diagnostic formats, and methods of screening.

Another reason the parties may prefer to negotiate a license with a limited field of use is to tailor the field of use to the strength of the licensee. Even large pharmaceutical companies generally specialize to some degree in certain medical indications. One may have made a strategic decision to invest in cystic fibrosis therapies; another may favor clotting disorders. A company with ongoing research projects related to both indications may decide to prove the technology in one area first before trying it in a second.

For these and many other good reasons, the licensor may want to license a number of companies exclusively, but in different fields. Some cautions are appropriate. Some biotech patent claims define the invention functionally (for example, by molecular mechanism). While claim language that relies heavily on functional limitations should generally be avoided, if possible, or supplemented with narrower claims that avoid descriptions of events at the molecular level (for reasons explained elsewhere in these materials), such functional language does have a place in patent claims when there is no other way to broadly express the inventive contribution. That does not mean that similar functional expressions are suitable to define license fields. No matter how certain scientists are about the molecular mechanisms, nature has a way of foiling neat pigeonholes. Functional limitations in patent claims can cause problems for patent claim interpretation and validity.[3] When it comes to licensing, functional descriptions in fields of use can be the seeds of a major disaster, in effect granting the same rights to multiple licensees, each of which was thought to have a distinct field. For example, it might seem safe to license a broad patent on administration of substance X exclusively in each of two fields (say, protection of central nervous system neurons and relaxation of blood vessels) thought to be distinct when the two licenses were executed. Should the data indicate that the substance helps glaucoma patients both by relaxing blood vessels to reduce intraocular pressure and by protecting the retinal ganglion from damage due to hypoxia, then which licensee is authorized to treat glaucoma may become a hot topic of dispute. The point is simply that fields of use typically should be defined according to medical indications so that licensees are less likely to trip over each other.

One problem with licenses limited to treating certain medical indications concerns so-called off-label uses. If the license is limited to a particular one of several uses of a patented drug, the licensee will want to consider procedures that can be put in place in the contract to prevent, or at least limit, the extent of overlapping sales by the products of other licensees. The licensee should also consider ways to avoid a possible charge of infringement if it allows its products to be sold for other uses. Even careful labeling of the drug for use in the licensed field does not ensure that doctors will not prescribe it for off-label uses, or that the product from the licensee will not be used outside the licensee's field.

Licensors may also grant multiple exclusive licenses based on geographic territory. The advantages to the licensor include: having access to multiple research and development partners, (thus tapping additional expertise as well as ameliorating the risk of a single development partner), allowing the selection of a partner with particular sales/marketing expertise in that geographic area, and allowing the selection of a partner with regulatory agency experience in a particular territory.

A note of caution about the decision to grant multiple licenses, whether exclusive in a field or nonexclusive: it is important to establish a financial incentive for at least one party to defend the patent. A licensor who is not prepared or able to spend the money and effort to defend its patent is well advised not to establish a nonexclusive licensing program. Nonexclusive licensees rarely, if ever, have an incentive to defend the patent, which leaves enforcement solely to the licensor. If the licensor lacks the resources, or will be unwilling to enforce the patent for some other reason, its licensing program may stall at the starting gate. Believing the patent will not be enforced, potential licensees may have no incentive to accept fair license terms.

Indeed, situations justifying nonexclusive licenses as a purposeful strategy from the outset (as opposed to a basis for settling legal actions) are rare. One such exceptional situation was a license to a family of the early patents on manipulating genetic material—Stanford University's so-called Cohen/Boyer patents on gene splicing. Stanford sought to make this technology available throughout the industry under nonexclusive licenses. This strategy was highly successful, in part because the license fee was fixed very low, but perhaps also because it was the first of its kind. Companies were willing to accept the first such license, but they soon drew the line and refused to spend money for nonexclusive licenses to later patents from other licensors, complaining that their fragile commercial beginnings would be substantially jeopardized by the multiple royalty burdens imposed by licenses for such broad-based patents. Of course, when dealing with federally funded or co-owned inventions, political considerations may rule out exclusive licensing, even if exclusive licensing represents the best business strategy.

2.3 *Fixed payments, royalties, or both?*

Nearly every license negotiation involves a tradeoff between risks taken for a large sum in the future (for example, getting a percentage of sales) and the more-certain enjoyment of a smaller, upfront sum. This choice is particularly significant in biotechnology, where both the upside potential and the risk are enormous. Licensees may wish to save the upside for themselves and not share it. On the other hand, they face substantial expenditures for commercializing the technology, and they may not want to add to their cash-flow burden in the near term, particularly in view of the low probability that a marketable product will result from the technology. From the licensor's standpoint it may be hard to accept the idea that someone else stands to realize more from developing and commercializing an idea than those who originated it and obtained patents.

Royalties are typically calculated as a percentage of a royalty base (such as net sales). Where the license is exclusive (and therefore the licensor gives up the opportunity to commercialize the invention itself or through other parties) the agreement typically provides minimum annual royalties, or at least reversion to nonexclusivity if a minimum royalty is not paid in a given period of time. The problem with the latter provision is that the licensor can no longer grant an exclusive license to another party, so long as the original licensee retains any license rights. Thus, diligence provisions, coupled with a complete reversion right for failure to meet those provisions, are desirable to ensure that a technology moves through the development stage, either with another partner or alone.

In return for an exclusive license, the licensor should place contractual requirements to ensure that the licensee exerts sufficient efforts to commercialize the invention. In addition to rather vague efforts requirements, such as "reasonable efforts" or similar language, the licensor should consider easily measurable requirements, such as minimum sales amounts or clinical achievement milestones. Conversely, if the licensor requires a minimum annual payment, the licensee may want to specify that the minimum annual fee is in lieu of best (or other) efforts, so the licensee retains the exclusive rights by paying the annual minimum fee, even if it sits on the technology and develops a competing product.

Milestones at which additional fixed payments may be due from the licensee (for example, selection of a clinical candidate, initiation of a clinical trial, completion of a satisfactory clinical trial, and filing of a nondisclosure agreement) provide a convenient middle ground for the risk/reward trade-off. The licensee with commercialization rights should be able to obtain additional financing at that milestone. Moreover, some of the risk of project failure at the clinical-trial stage is shifted to the licensor, justifying higher payments than would have been due at the license signing date. Other common milestones that indicate progress in accordance with the business plan and that are likely to bring funds to the licensee include U.S. Food and Drug Administration (FDA) marketing approval, the execution of an agreement with a marketing partner or some other collaborator, the first commercial sale, and/or the creation of a joint venture.

A common licensee complaint in the biotech field is royalty stacking, which is the need to pay royalties to multiple parties for commercializing a single product. For instance, a pharmaceutical company that screens a combinatorial chemistry library for compounds that bind to and block a particular neuronal receptor might owe royalties to the various owners of patents covering the library, the general screening assay, the isolated receptor, a cDNA encoding the receptor, and an expressed sequence tag (EST) derived from the cDNA (if the EST patent claim is written in open-ended *"comprising"* language). Stanford University met with success in its Cohen/Boyer patent license program, in part because Stanford University was the first with a broad biotech patent. Afterward, biotech companies were heard increasingly to say that they would not pay multiple royalties for a single product.

One compromise on stacking is to permit an offset to royalties up to but not more than some percentage (say, .5%) of the nominal royalty, if the accumulated nominal royalties add up to more than a set percentage of sales. In effect, the licensor is funding one-half of the cost of obtaining licenses under additional patents.

2.4 *Confidentiality*

Depending on the extent to which the parties exchange confidential information and biological materials, confidentiality provisions can be extremely important in the agreement. In some cases, patent protection may be narrowly limited to biological material that is not reproducible, and that alone is important confidential information, at least until the patent issues.[4] In such cases, the applicant may decide to abandon allowed but extremely narrow claims instead of making available the key biological deposits required for those claims to be issued.

Nucleic acid and amino acid sequence information is another type of confidential information. With modern sequencing technologies, however, such information arguably becomes nonconfidential when materials become available in a form pure enough to sequence easily.

In any confidentiality provision, it is important to spell out how long each type of information and materials remains confidential under the agreement, the disposition of written information and materials when no longer needed, and ownership of inventions made when the recipient makes authorized use of the materials and information internally.

One particularly important implication of confidentiality provisions is that they hinder a party's freedom to look for another partner should the collaboration fail. Having been "contaminated" by the first partner's confidential information, a licensor or licensee may be unattractive to future partners who are risk averse and do not want to have to deal with the possibility of a legal action for "misappropriation" of that information.

One solution is to limit the time period of confidentiality and to provide (in a sort of prenuptial agreement) an understanding that if certain milestones are not reached, the parties may collaborate with others on the same subject matter. Of course, such an understanding does not amount to a license under improvements that one or both parties may have made during the collaboration using confidential information. If the agreement does not specify who owns such improvements, there may need to be inordinate emphasis on murky and contentious ownership and inventorship issues related to improvements that are made after the license is executed.

2.5 *Enforcement against infringers*

As with payment terms, the decision about which party shoulders the burdens and realizes the benefits from enforcing the licensed patent against infringers often involves allocating the risks and rewards of the overall success of the venture. The party standing to make the most money from the operation typically wants (and should have) the right to enforce the patent against infringers. Litigation strategy (particularly settlement) of expensive and protracted patent infringement actions should be guided by proper business incentives and not by an entity on the financial sidelines of the litigation. For example, it is undesirable to have a licensee who can maintain unreasonable positions in patent enforcement litigation when the licensor is paying for the litigation, directly or indirectly (for example, with an offset to royalties that is carried forward to future years when it exceeds current-year royalties due). To the extent that the license provides a total offset to royalties, the licensor is, in effect, partially financing litigation it doesn't control, which is a very frustrating position to be in. Even deferral (as opposed to permanent offset) of guaranteed minimum royalties increases the licensor's risk, because if the patent is struck down or narrowed, those deferred royalties probably will never get paid.

One solution is to allow the licensee commercializing the invention to control litigation and to defer some portion (not all) of the royalties due each year, down to some minimum amount that is due no matter what legal expenses the licensee incurs. The offset ceases when the licensee's legal expenses in a given royalty period fall below a certain level. A variation on this theme allows the licensee to deduct a certain percentage of legal expenses due in a given year. If the total royalties owed in the year are less than the amount of that deduction, the question is whether any legal expenses from that year can be carried forward to reduce royalties in future years. While the fact patterns and license provisions vary tremendously, it is generally a good idea to set up the license so that the licensee will experience at least some significant nonrecoverable legal expenses and thus will have an appropriate economic incentive (litigation cost) to conduct and/or settle the litigation efficiently.

On the other side of the table, the licensor who wants to reduce or eliminate any risk of litigation expense should understand that its valuable patent property is at risk. It may make sense for the licensor to at least partially fund and fully control the litigation, as a strategy for avoiding an inept defense of the patent by the licensee. This is particularly true if the patent represents an important asset for the licensor in the form of income from other sources, such as royalties from other licensees or increased licensor profits due to the licensor's enhanced market position under the patent outside the licensee's field. Moreover, to give a licensee responsibility to fund and control litigation, with no offset or deferral of royalty payments, may deprive the licensee of the resources and incentive to defend the patent properly.

One important incentive for the licensee is exclusivity under the patent, at least in one important field. In general, only an exclusive licensee has a strong interest in maintaining the patent. A nonexclusive licensee is likely to face competition with or without the patent. Moreover, as far as the nonexclusive licensee is concerned, a royalty is owed so long as the patent is valid, yet the validity of the patent does not give the licensee a significantly better market position. In some cases, the nonexclusive licensee may have a substantial incentive to invalidate the patent, so it is unwise to place such a licensee in control of patent enforcement. Indeed, nonexclusive licensees lack standing to enforce the licensed patent, so even if the parties want the nonexclusive licensee to enforce the patent, the infringement action will probably be brought in the licensor's name (*Ortho Pharmaceutical Corp. v. Genetics Institute*).[5] Moreover, even when the licensee is the enforcing party, the licensor may be a necessary party under the Federal Rules of Civil Procedure, so the accused infringer can force the licensor to be joined in the action.

In sum, when negotiating the terms of patent enforcement, one should keep an eye on the business incentives that are created. Obviously, these questions depend on the context of a given license, such as the relative financial strength of the parties and their relative interest in maintaining the patent.

2.6 *Term and termination*

As with most licenses, the biotechnology license will often have a term that coincides with the patent term. Also, the right to premature termination for material breach typically includes a grace period for correcting the breach after notice.

One common provision is that a bankruptcy filing by either party constitutes termination. It is unlikely, however, that courts will uphold such provisions when the licensee declares bankruptcy under chapter 11. This is because the license is viewed as an executory contract under 11 U.S.C. § 365, with substantial performance remaining due on both sides (*Institut Pasteur v. Cambridge Biotech Corp*).[6] Therefore, the trustee in bankruptcy has the option to assume the rights and obligations under the license.

3. INCENTIVES FOR LICENSING

As with any contract negotiation, it is important to know how the deal will benefit both parties. Without knowing both parties' incentives, it is difficult to negotiate effectively.

Biotech patent owners grant licenses for a number of reasons:

- to trade long-term risk and the possibility of substantial income for the certainty of a, perhaps more modest, short-term payoff
- to obtain development and marketing assistance beyond the owner's abilities
- to obtain clinical development for applications of academic discoveries
- to obtain funding for further research
- to exploit areas that would not be developed in-house by the patent owner
- to enhance reputation in a field by collaborating with a well-known company

In granting licenses, the owner is exposed to several risks:

- adding a competitor if the product is in an area the licensor already exploits
- having to depend on the choice of the licensee to realize the value of the discovery (if the licensee fails, the opportunity may be lost)
- having to share profit in the long run if the invention succeeds
- losing control over information that could be kept secret if development were done in-house

The licensee takes a license for any of several reasons, such as:

- to ensure freedom to use a product line
- to obtain exclusivity for a product line
- to become current quickly without the cost of internal research

- to gain access to technology from a leader
- to gain access to trained personnel

In exchange, of course, the licensee:

- adds to costs and reduces profit margin
- undertakes potential liabilities associated with long-term confidentiality agreements
- undertakes a long-term obligation to share internal financial information with the licensee

Understanding the balance of pros and cons in a given situation is critical for assessing how much the opposite party will be willing to pay and what other terms are critical for them. Not surprisingly, the balance the parties strike will be different in different licensing contexts.

4. DEVELOPMENT COLLABORATION

Usually a great deal of work with uncertain success remains to be done between the time the license is signed and the date that the biotech product reaches the market. Unless that work is carried out entirely by the licensor or handed off entirely to another entity, collaboration will be necessary. The licensor has made the initial discoveries and knows their nature and promise best. The licensee, however, generally is best equipped to develop those discoveries further to the point of marketability. The synergies achieved by combining these disparate strengths are the rationale for the collaboration of licensee and licensor, at least in theory. Such collaborations, however, often raise additional licensing issues.

4.1 *Confidentiality in the context of collaboration*

We have already discussed some of the confidentiality issues raised in nearly all biotech-licensing situations. Where there is a genuine collaboration, in which employees of each company share ideas and information, confidentiality provisions become even more important.

Confidentiality provisions in a collaborative license should address several points. First, they should forbid any use or disclosure of confidential information by the recipient for any purpose other than the furtherance of duties under the collaboration. Second, if each party brings existing expertise (and confidential information) to the collaboration, the agreement should be two way, with each party disclosing and receiving information solely pursuant to confidentiality provisions. Third, it is important not to give either party an excuse to create a confidentiality obligation for information that was never intended to be confidential. To avoid doing so, it helps to identify in the background section of the license agreement the technical expertise of each party and the technical nature of each party's expected contribution. This information may also be helpful for sorting out inventorship.

While the following points apply generally to confidentiality agreements, they take on particular significance when the information at issue is disclosed as part of a long-term mutual exchange of information and skill. In effect, nonemployees are given the type of information and access to information usually reserved for employees. These long-term exchanges make the confidentiality issues extremely important.

4.1.1 *The nature of confidential information*

Put simply, any information that gives a commercial advantage over those not possessing the information can be a trade secret. The authors know of no meaningful distinctions between *trade secret* versus *proprietary* versus *confidential* information. Regardless of the label used, information that is valuable and obtained as part of a confidential relationship is in theory protectable. The ability to recreate information by combining numerous public sources does not necessarily establish that the information was readily available to those outside the confidential relationship. The standard for considering information confidential is not nearly so high as it is for nonobviousness, and analysis akin to a patent obviousness test has no place in determining whether something is confidential. Items of commercial value, such as customer and vendor lists, price lists, and selection of certain specific combinations of steps out of a large number of known alternative ways of approaching each step, may in some cases be protected. Typical exceptions to confidentiality include information that has been

- published
- independently developed by the recipient of the information (sometimes limited to information developed before receipt of the confidential information)
- independently learned by the recipient from a third party not obligated to the disclosing party
- ordered to be disclosed by a judicial- or regulatory-body process (subject to notice and best efforts to oppose such a process)

It makes sense to put the burden on the recipient of the information for invoking one of these exceptions. They should document the factual basis for the exception and notify the disclosing party before the recipient's disclosure or use of the information. The key is to avoid letting these exceptions become after-the-fact justification for improper disclosure or use.

4.1.2 *Duration of obligation from time of disclosure*

What is, or will be, the value of the lifetime of the information? Information that is about to be published will be confidential for only a short time. On the other hand, biological materials that cannot be duplicated may retain value indefinitely. It is important to be realistic about the length of time, so as not to provide a wide-open opportunity for a dispute on this subject.

Of course, there should be no obligation to maintain confidence for information that has been published or otherwise made public. This principle is easily stated, but not easily applied, because the typical fact pattern does not involve a wholesale publication of all information on a given topic. Instead, the information may dribble out over time in many publications, and a unified knowledge of the entire process, from start to finish, may continue to be valuable business information that is not generally available to competitors or other members of the public without a great deal of work.

4.1.3 *Survival of obligation*

Parties may be bound to maintain confidence for at least some period after the collaboration ends (so long as the information still qualifies as confidential information), and this obligation may affect the parties' ability to work on the subject matter alone or with others. The confidentiality obligation therefore creates a disincentive to terminate the collaboration because the parties' freedom to develop the technology separately is in doubt. This doesn't mean one has to avoid post-collaboration confidentiality obligations. In fact, the client may want such obligations to protect its own information.

4.1.4 *Recordkeeping for confidentiality*

Often the agreement requires the disclosing party to label information as confidential, if that party wishes it to be treated as such. Because of the proof issues raised about the content of the information disclosed, information disclosed orally with no written record before or after the disclosure generally is not treated as confidential. In this situation, the one making oral disclosures of confidential information has the burden of following up with a written disclosure. That procedure may seem unnecessarily cumbersome, but the alternative is to seek protection of orally disclosed information, which entails the burden of proving in detail the nature and full content of the information disclosed (along with the confidentiality of that information). Thus, sound business practice dictates making a record of the disclosure. A requirement to put a legend on the written disclosures is useful, but it should not apply when the nature of the information and the context of the disclosure make clear that the parties' understanding is that the information is confidential.

4.2 *Ownership of inventions resulting from collaboration*

Deciding who owns inventions is the hardest part of any collaboration negotiation. Without a contractual arrangement, ownership will depend on inventorship. Inventorship decisions can be contentious, and the law can be difficult to apply to individual facts. Therefore, consider avoiding the standard solution, for which each side owns its inventions and joint inventions are jointly owned. One option is to put ownership of all inventions in the field of the collaboration in a single party, with the other party having exclusivity in its field. Alternatively, ownership can be divided by field or geography. The parties' inability to agree on these issues may indicate that they want to keep open their option to compete and that the collaboration is not really a long-term arrangement. The inability to agree on ownership issues may reflect an inability to decide at an early stage about the relative sharing of risk and reward that is implicit in every license. A party may want to share in the ultimate success of the venture, even though the party's near-term contributions (capital plus IP plus commitment to use resources) are not commensurate with the other party's contribution.

Finally, ownership of an invention at the time the invention was made can determine whether commonly owned patents or inventions are prior art under 35 U.S.C. § 102(e), (f) and (g) as those sections are applied through § 103. A well-thought-out collaboration agreement should address ownership in a way that will minimize or avoid serious prior-art problems arising from inventions and patent applications that the parties bring to the collaboration. This issue had been quite a thorn in the side of biotech-patent license drafters for many years. Fortunately, however, with the passage of the Cooperative Research and Technology Enhancement (CREATE) Act in December 2004, the scope of common ownership was expanded. The existence of prior art under 35 U.S.C. § 102(e), (f) and (g) does not preclude patentability where the related inventions were made pursuant to a joint research agreement (in addition to the already existing safe harbors under 35 U.S.C. § 103[c]). New terms in the amendment, such as *joint-research agreement,* are certain to go through some interpretive growing pains. Still, it is interesting to note that the CREATE Act was pushed in large part by the biotech industry. This change recognizes the realities of collaborative practices in the biotech industry.

4.3 *Collaborators' rights to practice and sublicense*

An exclusive license is presumed to prevent even the licensor from practicing the invention. If the licensor intends to practice the invention, even in a narrow field, the license must explicitly reserve or grant that right.

In the United States, each joint owner may practice the invention without authorization from the other owner(s), and the licensor/owner need not account to other owners (35 U.S.C. § 262). In the absence of an agreement, therefore, joint owners can compete with each other. Indeed, a prospective licensee may force the owners to compete each other. Also, by definition, neither joint owner can unilaterally grant an exclusive license, because the other owner and the other owner's licensees are free to practice the invention.

Japan and Europe also permit each owner to practice the invention, but the countries differ from the United States when it comes to licensing. A licensee of a European or Japanese patent position must have authorization from all owners in order to practice the invention. If your business plan calls for licensing overseas, and your co-owner's plan calls for practicing the invention on his or her own, you should obtain the co-owner's agreement that you can license for both parties.

5. LICENSING FROM ACADEMIC INSTITUTIONS

Academic institutions pose special licensing issues. Part of the academic mission is to make worthwhile technology available to the public, particularly medical technology. Of course, money helps to do that, but other factors are equally, if not more, important. The licensee's stability, competence, incentive, and willingness to use its resources, technical expertise, and business skill to achieve this end are critical to the academic licensor's goal of bringing the invention to the public. Another factor in achieving this goal is the relationship between the licensee and the investigator. Cooperation between the parties increases the chances that the licensee will be able to develop clinical applications of the invention.

Many academic research institutions depend heavily on federal government funding. In comparison, licensing revenue is relatively minor. Under the terms of most government research grants, the licensing of inventions made with grant funding is controlled to some degree by the government. The key tool for control is legislation known as the Bayh-Dole Act.[7] The terms of the research grant typically follow that legislation, providing that the recipient of the grant (usually the academic institution as the grantee under the grant) must retain title, so that the government can regain title if certain conditions are not met. These conditions include a requirement that the academic institution or its licensee make reasonable progress toward commercialization of inventions resulting from funded research. Also, the government must have advance notice of the abandonment of patent applications in time to take over ownership and prosecution of those applications. In either case (failure to make progress or abandonment of the application), the government may take over. The government also has a royalty free, paid-up license to practice the invention—for example, to use such medical inventions as vaccines for military personnel.

In addition to the government's residual rights, certain other provisions are generally essential in an academic license. First and foremost, the inventors must retain the right to publish, although the licensee often is given the right to review manuscripts to identify potential inventions prior to submission or publication of the manuscript. In addition, the academic institution will require indemnification and insurance covering legal actions (for example, workers' compensation, commercial general liability, umbrella liability, product liability, or personal injury) growing out of development activities, sometimes naming the licensor as an insured party. There should, however, be flexibility in the insurance requirements depending on local regulations and customary business practices in the territory.

Many academic inventions are early stage and based on work that will be or has been published. Thus, confidential information generally is not a long-term asset. In an academic context, the value of the license to the licensee lies in the patents, and the value of the patents depends on:

- the likelihood of getting broad coverage from early-stage patent applications that will dominate later improvements
- the likelihood of getting patents on narrow improvements after the original work has been published
- recognition that the licensee is free to use unpatented, published work without a license
- the licensee's ability to obtain an option to license improvements under reasonable terms

6. PATENT MISUSE

Patent misuse is a defense to patent infringement. In asserting this defense, the accused infringer takes the position that the patent owner has misused its government-granted monopoly, thereby forfeiting the right to enforce that monopoly in a patent infringement action (*C.R. Bard, Inc. v. M3 Systems, Inc.*[8]). A body of case law has evolved to address the application of this doctrine to patent licensing practices, and in 1988, the Patent Misuse Reform Act[9] was enacted to amend 35 U.S.C. § 271 (d) regarding certain aspects of patent misuse.

Unenforceability due to misuse does not call into question the inventor's entitlement to a patent under the provisions of 35 U.S.C. It is distinguished from a defense of invalidity, which would require proof that the U.S. Patent and Trademark Office (PTO) was not empowered to grant the patent because the invention application did not meet the statutory requirements for patentability.

Most often, resolution of misuse issues involves a balancing of the inherent tension between patent law and antitrust law. To establish a claim of patent misuse, it must be shown that the patent owner misused its government-granted right, or in other words, used the patent to improperly extend its power in the marketplace. Patent-misuse analysis is acknowledged to be somewhat convoluted, due in part to its close interplay with antitrust analysis, which makes it susceptible to contemporary societal/regulatory pressures at the moment of analysis, and also in that often such analyses are particularly fact specific, leading to narrowly applicable analyses. Historically, certain activities were considered *per se* patent misuse. Other activities, such as those governed by 35 U.S.C. § 271(d), were evaluated under a "rule of reason" analysis similar to that in antitrust analysis. (*Virginia Panel Corp. v. MacPanel Co.*[10]).

It is now abundantly clear that the mere existence of a patent right does not establish market power in the antitrust sense and that certain licensing provisions that were once thought to unfairly extend the patent monopoly do not constitute patent misuse, per se. Rather, the courts require a factual analysis (a rule of reason) of whether the patent owner possessed market power, and the patent is simply one factor in that analysis. The Supreme Court dealt with an allegation that a patentee misused its patent by tying sales of a patented printhead and ink container to sales of unpatented ink in *Illinois Tool Works, Inc. et al. v. Independent Ink, Inc.*[11] The court held that a patent does not necessarily confer market power upon the patentee in every case involving a tying arrangement. The plaintiff seeking a finding of illegal tying and monopolization in violation of the Sherman Act must prove that the patentee has market power in the tying product.

The United States Court of Appeals for the Federal Circuit relied on the *Illinois Tool Works* decision when it recently held that various Monsanto marketing practices for sales of seeds resistant to its Roundup® pesticide did not constitute patent misuse (*Monsanto Co. v. Scruggs et al.*[12]). The facts in that case involved a complex marketing scheme that included flexibility to react to FDA approval of competitive products.

CSU, LLC, et al. v. Xerox Corporation[13] raised the basic issue of whether a refusal to license is anticompetitive activity under § 2 of the Sherman Act (15 U.S.C. § 2). CSU brought an antitrust action charging that Xerox had engaged in anticompetitive behavior when it tried to monopolize markets for sales and service of Xerox high-volume copiers and printers. Xerox counterclaimed for patent infringement, and CSU raised a misuse defense. The Kansas District Court denied Xerox's motions for summary judgment, in part based on the conclusion that CSU may have a valid defense of misuse (*In re Indep. Serv. Orgs. Antitrust Litig.*[14] and *In re Indep. Serv. Orgs. Antitrust Litig.*[15]). The Federal Circuit, however, ultimately supported the notion that although a patentee's right to exclude is not without limits, a unilateral refusal to sell or license a patent does not exceed the scope of the patent grant and does not rise to patent misuse (*CSU LLC, et al. v. Xerox*[16]).

A per se rule on whether refusal to license always (or never) amounts to misuse seems unlikely. Such a rule would eviscerate the patent system and exceed judicial authority to compel patent owners to license in all situations. On the other hand, it seems artificial to ignore a patent owner's licensing activities (or lack of them) when viewing the overall picture of monopolization. The practitioner is left to exercise judgment in the vast middle ground.

One interesting aspect of the CSU case involves the accused monopolist's state of mind (*"intent"*). In concluding that it must take evidence on the misuse issue, the Kansas District Court expressly declined to follow the Federal Circuit's subjective intent standard for evaluating misuse. The Kansas District Court also refused to adopt a per se rule on the ground that refusal to license violates the Sherman Act. This trend away from per se rules has been going on for a long time (*Eastman Kodak Co. v. Goodyear Tire & Rubber Co.*[17]).

Another example of potential patent misuse is a license requiring royalty payments after expiration of the patent of the licensed technology. Case law that has not been explicitly overruled holds that such license agreements are illegal and unenforceable and are per se misuse (*Brulotte v. Thys Co.*[18]; *Scheiber v. Dolby Laboratories, Inc.*[19]). Conditioning a license grant upon the payment of royalties on unpatented products has also been found to be a per se wrong (*Zenith Radio Corp. v. Hazeltine Research, Inc.*[20]). Another example is charging royalties twice (*PSC v. Symbol Tech.*[21]). This example was analyzed under a rule-of-reason analysis. It is open to question whether any such license arrangement will be misuse, per se (that is, without an analysis of market power).

A federal district court addressed the issue of whether a license requiring reach through royalties to products (for example, drugs), discovered using patented screening tools, constitutes patent misuse in *Bayer A.G. v. Housey Pharmaceuticals, Inc.*,[22] affirmed on other grounds,[23] further proceedings on other grounds,[24] affirmed by the Court of Appeals of the Federal Circuit.[25] Bayer first alleged that misuse arose because the license contemplated royalties on products and activities not covered in the licensed patents by claims relating to screening. As Housey offered alternative compensation structures to licensees, for example, lum-sum payment, royalty based on discovered-product sales, or royalty based on licensee's total R&D expenditure (the selection of which was explicitly stated in the agreement as the "most appropriate" and "convenient" approach), the district court found that Housey did not "condition" the license on products/activities outside the patent, and therefore there was no misuse. Bayer next alleged that misuse arose because the agreement imposed a requirement of royalty payments beyond the term of the patent, which was a per se misuse under Brulotte. The district court, also finding no misuse by Housey on this issue, held that collection of royalties after expiration of a patent was not per se misuse. The district court reasoned that a patentee can charge a royalty for practicing an invention prior to the expiration of the patent covering the invention and that payment for such can be postponed beyond the expiration date of that patent. Whether the payment is for pre- versus post-patent expiration use appeared to be determinative to the district court. Thus, agreement language explicitly delineating that payment is "time-shifted" for the convenience of the parties, and is not for post-patent expiration use, seems to be an important factor in this district court's analysis of patent misuse.

In sum, it remains risky for a patentee that has external (nonpatent) market power to engage in the above licensing practices, but it is likely that the rule-of-reason analysis will be required to find misuse.

7. SPONSORED RESEARCH

Sponsored research, for example, at an academic institution, should not be viewed as a typical collaboration but as a special case. The sponsor will nearly always want exclusivity over the fruits of the research, regardless of inventorship. Also, disputes about confidential information may arise should the sponsor want to establish a competitive advantage by maintaining confidence, at least until a patent application is filed, and maybe for some time thereafter. The researcher will want freedom to obtain future funding from others, given that current funding will be limited in amount and duration. If the researcher is an academic, he or she will want the freedom to publish without interference, though he or she may be willing to delay publication for a short period to give the sponsor an opportunity to prepare and file a patent application. In a highly competitive field, however, even a month can give another laboratory a chance to scoop the researcher in print. The researcher is unlikely to cede any control over the content of his or her publication, with the exception of information that originated with the sponsor.

The extent to which the issues discussed above will present serious problems for any given sponsored research arrangement depends on specific circumstances, particularly the extent and duration of the funding. A researcher whose entire operation is funded to a substantial extent by a single sponsor obviously will have fewer problems with such issues as the right to collaborate with other companies. Ideally, a sponsor desires a representation and warrant from the researcher that no confidential information of a third party or proprietary material or process of a third party is utilized in the sponsored research. In reality, particularly with the multiple funding scenarios from both institutional and government sources, such representation and warrants cannot be made.

Maintaining the confidentiality of sponsors' confidential information can also be a challenge. Some institutions may not allow some of their researchers to be a party to confidentiality agreements. In such instances, it is necessary to identify the specific researchers (in addition to the principal investigator) and what their exposure to confidential information will likely be. Mechanisms for protecting information should be carefully considered. Representations and warrants that the materials will not be used other than as agreed and that the materials will specifically not be analyzed or reverse engineered, may also be appropriate.

One common problem when drafting a sponsored research agreement in an academic setting is the "mobility of funding" culture. Typically, a principal investigator has the freedom to move his or her operation, funding and all, to another institution. If the sponsor wants to remain with a particular investigator should the investigator move from one institution to another, the agreement must be clear on this point. Otherwise, if the principal investigator moves, the sponsor could be left in the position of being obligated to fund other researchers at the original institution. One solution is to clearly state that the sponsor's funding obligation terminates if certain named individuals (usually just the principal investigator and perhaps one or two others) cease employment. The sponsor then has the freedom to decide whether to continue funding the project elsewhere.

Another problem arises from the culture of authorship and even ownership of technology as discretionary privileges to be controlled by the principal investigator. It is common for a principal investigator to assume that he or she has the right to determine the inventorship and content of a patent application, just as he or she has the power to control content and authorship of journal publications. Obviously, these decisions must instead be controlled by inventorship law, patent prosecution strategy, and the sponsored research contract. For these reasons, the sponsor may want to control the prosecution of patent applications arising from the research.

A similar problem arises from multiple grants for a single laboratory. Investigators are used to deciding to some degree how grant funds will be allocated among a number of projects. Here again, the agreement should contain a carefully drafted statement of the work and the field of the research, coupled with clear entitlement to exclusivity in the investigator's work in the field.

8. LICENSING TOOLS FOR DRUG SCREENING AND DEVELOPMENT

Even biotech discoveries that are too fundamental to support a patent claiming a clinical therapeutic or diagnostic use may support a patent on screening. Driven by the rapid increase in knowledge about molecular (including DNA) bases for diseases, coupled with automated equipment for synthesis, screening, and analysis, the interest in rational drug design and screening has exploded. Indeed, licensing inventions featuring drug screening and development are all the rage.

8.1 *The computer software component*

The computer software developed in connection with rational drug design and screening can be protected by patent, copyright, and/or trade secret. The particular form of protection will depend upon the ability to reverse engineer the software, and/or the effect upon the company of making the software public, as will happen in connection with patent protection. No matter what form(s) of protection are selected, the license agreement will include several elements that are unique to the software environment.

For example, various limitations upon the use of the software, and the availability of the software (in source code or object code form) need be addressed. Further, will the licensee, if he or she is able to obtain source code, be permitted to modify and improve the software, and if so, which of the improvements, if any, will flow back to the licensor? Will the use of the software be limited to a particular database, CPU, physical location, number of users, simultaneous users, and/or application?

If the license is for object code only, will the licensee insist, as well he or she might, that the source code be placed in escrow in case computer software bugs develop that are not corrected by the licensor? (The nature of the escrow agreement, and who shall hold the escrow, is typically the subject of yet another agreement.)

If software is provided, will it be subject to a maintenance agreement, that is, an agreement by which the licensor submits to providing improvements, fixing problems if they develop in the software code, and in return receiving an annual maintenance fee? If maintenance is provided but not taken by the licensee, will the licensor disclaim all responsibility for operation of the software after a fixed period of time, for example, one year?

If the software being provided is experimental software and there is a software bug, the licensor will likely limit his or her liability to either a return of any monies paid or to using reasonable efforts to correct the code. On the other hand, most academic institutions provide software code "as is," without any obligation on the institution's part to provide any further help. (As a result, there is often a consulting arrangement with the developer of the code to aid in fixing problems or improving the code, if improvements are allowed under the license agreement.)

One should also consider the distinction between providing the software code, the technology, and the license to develop similar functionality under a patent license. With respect to the latter, no technology may be transferred at all, only the license to use the technology as covered by the patent claims. The provision of technology invokes many of the elements noted above with regard to protecting the technology being transferred.

8.2 *Controlling the reagents used to screen*

The reagents used for screening typically are protectable trade secrets. For example, monoclonal antibodies, specific peptide fragments or DNA fragments, and cellular components that are used in a screen may not be publicly known or available. When licensing others to perform the screen, the agreement should be clear that the license is limited (for example, in time or in the number of compounds that can be screened) and that the materials are to be returned when that license has run out. At least, the license should provide (as do software licenses) that the reagents can only be used in limited ways (for example, on the premises in certain types of screen formats) and can be duplicated only to provide a secure backup in case the primary reagent is lost or damaged. The reagents (or their derivatives) should not be duplicated and used in additional screens at other sites or by other companies. In cases where the PTO is unlikely to grant broad protection, this type of contractual protection may be the only meaningful protection available.

8.3 *Valuation of screening patents*

Assessing the value of screening patents poses special issues. Because screening patents specifically focus on research activities and do not cover commercial products or manufacturing processes, and, indeed, by their nature are practiced before any product is identified—much less ready to market—traditional valuation techniques (discounted stream of sales over time) may be inappropriate.

One way to evaluate screening patents is to estimate the amount of research expense saved by licensing the screen from outside rather than engaging in an in-house project. Another way is to consider the screen in view of its proportion to the total R&D budget or to the appropriate program or screening budget. As discussed below, however, other factors come into play.

8.3.1 *Concerns about screening preissuance*

Since in the United States there can be no infringement until the patent issues, screening preissuance cannot give rise to damages absent an issued patent having claims covering the screening.[26] However, the American Inventors Protection Act[27] provides provisional rights. If the application is published, a resulting patent will include the right to a reasonable royalty for the period between the date of publication and the date of grant, if: (1) notice of the published application is provided, and (2) the patent claims are substantially identical to the claims of the published application. Given the ordinary course of at least two years pendency for biotech patent applications, the potential licensee should evaluate the likely duration of its screening project to determine how long, if at all, screening will continue after patent issuance.

8.3.2 *Damages for unlicensed use*

For screening that is likely to be conducted after issuance, the question remains of how much to pay for a license. Of course, the licensor would like to have a percentage of sales of drugs discovered using the screen, but there is no reason to believe that measure is common in the industry, or that it would be used by a court in fixing "reasonable" royalty damages for infringement. More typically, screening assays will produce a royalty based on the length and intensity of use and the noninfringing alternative screens available. Thus, a screen used occasionally to confirm results of a noninfringing screen would be compensated at a much lower rate than a screen so well accepted that it is effectively required to get approval for human clinical trials.

Finally, use of a screen to generate data for submission to the FDA may not constitute infringement at all. It may be difficult for many reasons to obtain suitable value when licensing screening technologies.

8.3.3 *Compositions used for screening*

In general, licenses of patents covering compositions used for screening are subject to the same considerations as those discussed above. To take into account the situation in which the reagents may have some other, more valuable use, the license should restrict use of the reagents to screening (for example, as a field of use) and should explicitly exclude clinical uses.

9. CONCLUSION

Licensing of biotech inventions requires special considerations and specialized license drafting with clear provisions that unambiguously detail the obligations of the licensors and licensees. In large part, this attention is needed because of the nature of biotech inventions and the risks and uncertainty that are integral to the biotech business. For example, development of an invention into a product requires a synergy of complex operations. Hence, the biotech invention may be unlikely to succeed, or may entail long, costly lead times to market, resulting in limited patent life remaining after commercialization. Such high risks are combined with high (often deferred) rewards. Therefore, licenses are structured to reflect this risk/reward reality of the biotech business. Key considerations include: fees and royalties, royalty stack ceilings, fields of use, setting milestones, mergers and acquisitions, exclusivity of licenses, patent maintenance, patent enforcement, confidentiality, patent misuse, and issues relating to collaborations. Notwithstanding this rather daunting list of considerations, there are many incentives that drive successful licensing of biotech inventions.

For the ***licensor***, incentives include obtaining:

- development and marketing assistance beyond the owner's abilities
- clinical development for applications of academic discoveries
- funding for further research
- assistance in areas that would otherwise not be developed

For the ***licensee***, incentives include:

- ensuring freedom to use a product line
- obtaining exclusivity for a product line
- becoming current quickly without the cost of internal research
- gaining access to technology from a leader and accessing or developing trained personnel

Hence, by balancing the inherent risks and potential rewards, properly structured biotech licenses serve to coherently actualize the incentives of licensors and licensees, such that all parties are winners, and biotech R&D advances toward commercialization for the benefit of all. ■

JOHN W. FREEMAN, *Principal, Fish & Richardson P.C., 225 Franklin Street, Boston, MA, 02110-2804, U.S.A. Freeman@fr.com*

---

1 Gradowski H. 2002. Patents, Innovation and Access to New Pharmaceuticals, July. *J. Int. Econ. Law* 5(4):849-860.

2 104 F.3d 489 (1st Cir. 1987)

3 See also, in this *Handbook*, chapter 4.2 by C Nottenburg.

4 If those in the art cannot duplicate the material based on a written description and public starting materials, then patent law requires that the material be deposited with a public depository. Absent publication of a foreign patent or issuance of the U.S. patent, however, this deposited material is probably not available to the public, and it remains valuable confidential information.

5 52 F.3d 1026, 34 USPQ2d 1444 (Fed. Cir. 1995).

6 104 F.3d 489, 41 USPQ2d 1503 (1st Cir. 1987).

7 PL96-517 (1980); see 35 U.S.C. § 200 and following.

8 157 F.3d 1340, 1372 (Fed. Cir. 1998).

9 (PL 100-73, 102 Stat. 4674 (H.R. 4972).

10 133 F.3d 860 (Fed. Cir 1997).

11 126 S. Ct. 1281, 547 (US ____ 2006).

12 (Fed Cir. Slip op., August 16, 2006; 01-1532; 05-1120; 05-1121).

13 203 F.3d 1322 (Fed. Cir. 2000).

14 964 F. Supp. 1454 (D. Kan. 1997).

15 964 F. Supp. 1469 (D. Kan. 1997).

16 203 F.3d 1322, 1328 (Fed. Cir. 2000).

17 114 F.3d 1547, 42 USPQ 2d 1737 (Fed. Cir. 1997).

18 379 U.S. 29 (1964).

19 63 USPQ 2d 1404 (7th Cir. 2002).

20 395 U.S. 100 (1969).

21 26 F. Supp. 2d 505 (WDNY1998).

22 228 F. Supp. 2d 467 (D. Del. 2002).

23 340 F.3d 1367 (Fed. Cir. 2003).

24 386 F. Supp. 2d. 578, 582 (D. Del. 2005).

25 (Fed. Cir., August 4, 2006) (Slip op.)

26 We ignore for the moment the question of whether use of a screen patented in the United States to identify a compound renders the importation or use of the compound in the United States infringement of the screening patent under 35 U.S.C. § 271(g).

27 PL 106-113. Enacted 29 November 1999.

CHAPTER 11.2

Licensing Agreements in Agricultural Biotechnology

RICHARD S. CAHOON, *Executive Director, Cornell Center for Technology, Enterprise & Commercialization and Senior Vice President, Cornell Research Foundation, U.S.A.*

ABSTRACT

Though similar in many ways to other kinds of license agreements, agri-biotech licenses have some unique elements that require special attention. Considering first the similarities, this chapter looks closely at the typical boilerplate language that all license agreements share and outlines the basic structures and concerns of all such agreements. The chapter then turns to the singularities of agri-biotech licenses, focusing on such issues as multiple property types that often cover a single technology and/or product, freedom to operate issues that drive anti-royalty-stacking provisions, philanthropic- and humanitarian use clauses, and stewardship obligations.

1. INTRODUCTION

"Agricultural biotechnology" is a relatively broad term that can include cell culture, fermentations, bioprocessing, breeding and animal husbandry, diagnostic methods and apparatus, and biocontrol of plant disease and pests. An important, challenging area of IP management and licensing in agricultural biotechnology relates to the *genetic engineering* of plants and animals through applied nucleic acid chemistry and related technologies. These technologies include methods and materials for isolating functional pieces of DNA (for example, genes and promoters), creating *genetic constructs* (that is, functional packages of DNA sequences), and stably inserting genes into plants and animals. This chapter focuses on these issues (the terms *agricultural biotechnology* and *agri-biotech* will be used synonymously to describe this area of genetic engineering). Since the largest amount of genetic engineering activity in agriculture to date has involved plants, the discussion focuses on plant-related technology. But many of the principles of intellectual and biological property-based management and licensing in plant-based agri-biotech apply equally to animals and microbes.

This chapter's topic is license agreements. It explores the basic nature and purpose of a license agreement: the definition and transfer of certain property rights between two or more parties under a specified sharing of rights and obligations between those parties. A license is distinguished from a "sale" in that ownership of the property does not transfer but remains with the original owner. In a license, the owner, called the *licensor*, transfers certain rights of possession and use (but not ownership) to the recipient of those rights (the *licensee*).

As in any area, the process of creating a license agreement in agri-biotech involves the precise definition of the property of interest, an articulation of the exact rights of the licensor and licensee in the property after the agreement is signed, and the ongoing rights and obligations of each party. The elements of this process are defined below, and the attendant issues in agri-biotech licensing are described. Preferred licensing methods are also suggested.

Cahoon RS. 2007. Licensing Agreements in Agricultural Biotechnology. In *Intellectual Property Management in Health and Agricultural Innovation: A Handbook of Best Practices* (eds. A Krattiger, RT Mahoney, L Nelsen, et al.). MIHR: Oxford, U.K., and PIPRA: Davis, U.S.A. Available online at www.ipHandbook.org.

2. BACKGROUND ISSUES IN AGRI-BIOTECH LICENSING

A decision about whether to license an agri-biotech invention is typically based on a few important background issues:

- the significant cost to create, develop, and commercialize agri-biotech products
- the critical role of government regulations in testing and commercializing products
- the importance of public perception and acceptance of agri-biotech products
- the necessity of using numerous, different (and often proprietary) technologies to create agri-biotech products

This last issue leads to the following related problems:

- the "tragedy of the anticommons" problem, which creates different technology owners with respect to a single product
- the challenge of obtaining freedom to operate (FTO) for agri-biotech technologies and products
- the royalty-stacking problem, in which each owner of a proprietary technology expects a significant royalty on sales
- the existence of multiple forms of property that can exist simultaneously in any one technology or product, namely:
  - utility patents
  - plant patents
  - plant breeder's rights (for example, plant variety protection based on the UPOV Convention)
  - trade secret
  - trademark
  - tangible biological property
- the unique attributes of the agricultural industry, that is:
  - low profit margins
  - commodity economics
  - national food security issues
  - humanitarian concerns over hunger and malnutrition

3. OVERVIEW OF AGRI-BIOTECH LICENSES

The factors described above combine to configure and constrain agri-biotech license terms and conditions. For example, the multifaceted aspects of possible property instruments in agri-biotech require the type and scope of property rights contained in the license to be carefully described. Does the license include a patent and a plant variety protection certificate on a new plant variety? Does the license include limited rights of possession of tangible materials such as seeds, vegetative cuttings, or tissue cultures?

Similarly, the precise nature of the rights granted to the licensee must be clearly stated. Is the grant limited to a nonexclusive, freedom to operate for testing only or an exclusive right to make, use, and sell? Does the grant include rights in improvements to the technology or product and to related future inventions (for example, does the right to make, use, and sell a transgenic plant include rights to all crosses made with that plant using traditional breeding techniques)? And does the grant of rights permit ownership of further developments by the licensee? For example, does the grant of rights to a transgenic plant include the right to use individual components of the genetic construct (individually or in combination) in other constructs and "transgenic plant events" made by the licensee? Agri-biotech licenses should also define the precise rights of sublicensing granted to the licensee. For example, is sublicensing limited to specific transgenic events or to genetic components? Finally, what is the geographical scope of these rights? Are certain rights granted in one country but not in another? Breeding rights, for example, could be limited to one country and sales to another.

The low profit margins typical of commodity agriculture naturally depress the royalty rates that a technology owner can expect. For similar reasons, the large up-front license fees more typical of pharmaceuticals are unlikely.

The flipside of rights is obligations, and several sections of the license will define the obligations of the licensee. The most obvious are the financial obligations. Licensee payments will be defined, which may include license fees, royalty on product sales, milestone payments, and IP expenses. Such obligations can be defined in many different structures, schedules, and unique terms. In agri-biotech licenses, milestones may include the achievement of successful field tests, regulatory approval, and first product sale. Other obligations of the licensee are likely to include adherence to applicable laws, assumption of business risk, and product quality assurance. The license may also include licensee obligations for mandatory sublicensing, diligence in commercial development, labeling requirements, trademark use, confidentiality, and requirements for certain philanthropic and humanitarian uses, especially in developing countries.

The license is also likely to contain obligations for the licensor. For example, the licensor may be obligated to provide a specified amount of biological material over a certain time period. Similarly, the licensor may be required to provide know-how, and/or access to proprietary data, documents, and related information. On occasion, licensors will be obligated to perform certain tests or laboratory work or to provide access to future inventions and improvements. Almost certainly, the licensor will be obligated to guarantee its ownership rights and perhaps also product performance, noninfringement of licensed IP, and so on.

Of course, the parties to the license will be obligated to adhere to a set of legal requirements that are standards of contracts, such as formal notifications, protocols for contract amendment, dispute resolution, use of names, and the delineation of legal remedies and venues. Although each part of a contract has importance, one of these sections of legal boilerplate, warrants and representations, is especially critical. This language exactly defines the commitments being made by the parties and must always be scrutinized carefully.

The important sections of an agri-biotech license are described in more detail below, and some of the implications unique to licensing in this area of technology are discussed.

4. IMPORTANT SECTIONS OF AGRI-BIOTECH LICENSES

4.1 *The preface*

The preface sections, which precisely define the parties and provide background and context for the agreement, are not unique to agri-biotech licenses. Like any license, the WHEREAS clauses of an agri-biotech license provide a good background to the terms and conditions of the agreement—when they are written well.

4.2 *Definition of property rights*

It is particularly important in agri-biotech licensing to precisely define the property rights contained in and transferred by the agreement. Biological materials should be described precisely. For example, complete lists of named plant-breeding lines, cell type sand lines, plasmids, and the like should be attached to the agreement. All patents, patent applications, and plant protection certificates should be listed in an attachment that includes serial numbers and their applicable countries. It should also be clear what derivates of patents and applications are to be included in the grant of rights, including continuations, continuations-in-part, divisionals, and reexaminations.

4.3 *Grant of rights*

This section of the license agreement precisely defines the rights conveyed by the owner-licensor to the licensee. In agri-biotech, there will likely be a mix of such rights granted. For example, the licensee may receive an exclusive right to sell a specific line of transgenic plant but not to make variants of the line. The grant of commercial exclusivity to a transgenic plant line will very likely not include the right to make, use, or sell any of the components of the genetic construct alone or in combination, but only as an inextricably linked part of the specific transgenic plant.

The grant of rights should also define any territorial limitations. As with any IP, agri-biotech patents are country-specific. But in agri-biotech this might include limits on export from countries where the right to make and sell has been granted. In addition, licensors in agri-biotech will frequently provide incentives for licensees to sublicense, especially when the sublicense will cover markets in which the licensee may not be strong or even have a presence. The grant of sublicensing rights and its scope, therefore, is often an important issue.

It is particularly important in agri-biotech to define whether the licensee may use the technology to create new variants. For example, will the licensee have the right to make crosses of the exclusively licensed plant line with its own proprietary germplasm? If so, will this affect other license terms, such as the royalty rate owed?

The grant of rights will define the nature of rights exclusivity and whether there are any time limits to the exclusivity. For example, some exclusive licenses provide only an exclusive lead-time of five years or so, after which the license reverts. Nonexclusive licenses are common in agri-biotech licensing, but sole, exclusive, and co-exclusive licenses are also often granted.

Finally, agri-biotech licenses are relatively unique with regard to the scope of rights concept field-of-use. In agri-biotech licenses, field-of-use typically refers to a crop type that may be broadly or narrowly defined. For example, the grant of rights may broadly include the right to make, use, and sell all monocots and dicots created using the technology. Or, the field-of-use might grant only monocots, or only corn. The field-of-use grant is particularly prevalent in the licensing of agri-biotech genetic construct components, such as genes, selectable markers, translation enhancers, or promoters. This is due to the technologies' frequently broad applicability.

4.4 *Consideration*

The consideration section of the agreement is one of the most familiar. It is common to all licenses, including agri-biotech. What did the license cost? How valuable is the license? These are standard issues dealt with in the consideration. This section is designed to deal with the *opportunity cost* to the licensor and to account for the potential value, cost to develop, and market potential of the licensed rights. Agri-biotech licenses may provide for exchanges of germplasm and access to other technology owned by the licensor. For example, the licensee may provide the licensor of a genetic construct access to the licensee's valuable germplasm for future transformations. As mentioned above, agri-biotech licenses have typically lower license fees and are often characterized by milestone payments at critical commercial development stages.

4.5 *Royalty payments*

Like most licenses, agri-biotech agreements contain provisions for a royalty payment linked to sales volume. Frequently, this link is a percentage of net sales. Due to low profit margins in agriculture, this percentage is almost always much less than 10%. In fact, royalties of between 1% and 5% are common.

A relatively unique aspect of agri-biotech royalty rate setting is the important problem of *royalty stacking*. This problem arises when several different owners of intellectual or tangible property components in an agri-biotech product all expect a reasonable royalty on each sale. All of the owners will then "stack" their royalty expectation on the sale of each product. While this may be relatively manageable for two or three separate stacked royalties, it is wholly unmanageable when there are several and/or when any one of the component owners expects a royalty that is too large. For example, it is common for each of four or five different owners of different proprietary technical components to request half of the profit margin. Obviously, that kind of royalty stacking makes commercializing an agri-biotech product economically unfeasible. The royalty stacking provisions of agri-biotech licenses are designed to mitigate this problem. Although such provisions can be difficult to negotiate, when implemented they can provide a pro rata sharing protocol that self-adjusts as the technology-property-ownership mosaic changes over time.

Other popular royalty mechanisms include fixed-fee payments based on some type of added-value calculation. For example, in the United States, royalty on the sale of transgenic corn with lepidopteran and/or herbicide resistance (that is, Bt corn or Roundup Ready® Corn) has been based on a fixed *tech fee* on each bag of seed. Rebates, trademark use, incentives, and other mechanisms act to modify the fixed-fee amount.

4.6 *Minimum royalty payment*

Minimum royalty payment obligations are not unique to agri-biotech licensing. They are common in all exclusive licenses. In agri-biotech licenses, such payments are often linked to the scope of rights granted, particularly territory and field-of-use rights. For example, the licensor may use increased or decreased minimum payments as an incentive (or disincentive) for the licensee to pursue commercialization in certain crop types or countries.

4.7 *Philanthropic and humanitarian use*

There is often pressure to establish philanthropic- or humanitarian-use provisions in agri-biotech licenses, particularly if the crops are important food staples (for example, rice or wheat) in developing countries. Such provisions are designed to establish clear boundaries between the commercial sphere and uses that directly impact a country's poor population. Although there are a variety of ways to define these boundaries, they are often based on the scale of production and the scope of commercial activity. Such definitions depend on the crop, the country, and the particular socio-economic situation. For example, growing three avocado trees would very likely be defined as philanthropic use in Bangladesh. Growing twenty-five trees there may or may not be philanthropic; a plantation of 500 hectares would most certainly be considered commercial. However, if the production of these 500 hectares was used by a nonprofit organization to feed the poor, it would likely be considered philanthropic use. Carefully designing and implementing philanthropic-use boundaries is essential, as is ongoing monitoring for compliance. Philanthropic use should always be considered when staple crops in developing countries are involved. However, such provisions should not be used to disguise commercial-scale use.

Philanthropic- or humanitarian-use provisions of a commercial agri-biotech license will often identify a third party responsible for implementing the noncommercial provisions. The license may also define certain protocols for the interaction of the commercial licensee and the philanthropic-use licensee. A separate philanthropic-use license will be in place between the technology owner and the noncommercial partner. Such licenses usually would contain royalty or other payment obligations. However, stringent obligations for controlling and monitoring the technology and products may be imposed on the licensee to ensure the achievement of philanthropic and commercial goals. Despite the licensor's waiver of royalty payments for philanthropic use, nominal fees may be required by the philanthropic licensees to support dissemination of the technology. Both commercial and philanthropic-use licenses must be designed to enhance—and not hinder—the respective purposes of each agreement.

4.8 *Stewardship of technology*

The issue of stewardship arises frequently in agri-biotech licensing. Although precise definitions vary, *stewardship* generally refers to the ongoing oversight and guidance of the commercial development and dissemination of the new technology. It typically refers to the importance of maintaining a licensor's overall interests in sustaining the long-term use of transgenic crops. Stewardship clauses in agri-biotech licenses have been particularly concerned with smooth regulatory approvals, good government relations, effective management of public relations, and mitigation of the loss of product efficacy caused by inappropriate or less-than-optimal implementation. For example, stewardship clauses in an agri-biotech license will most certainly obligate the licensee to actions that will not harm regulatory approvals or relations between relevant government officials, the licensee, and/or the licensor. These clauses may also prescribe rights and obligations of the licensor and licensee that are designed to allow the licensor to maintain effective control over public relations efforts. Finally, on the technical side, stewardship clauses have been used to avoid the development of pest resistance in transgenic crops by mandating certain crop management techniques, such as rotations, buffers, and pest reservoirs.

4.9 *Enforcement and litigation*

Successful agri-biotech products have a history of significant patent-infringement litigation. For example, large agri-biotech companies such as Monsanto, Syngenta, Bayer, and Pioneer Hi-Bred International Inc. (now a division of DuPont) have engaged in numerous, complex patent infringement actions against each other and their sublicensees. Although litigation can be viewed as generally undesirable, it may be unavoidable. Therefore, agri-biotech licenses should contain enforcement and litigation provisions that are designed with this eventuality in mind.

5. PRACTICAL EXAMPLES

Cornell University's long history of licensing its agricultural intellectual property (IP) began with veterinary vaccines. Cornell patented and licensed these animal vaccines in the early 1930s after establishing its patent and licensing subsidiary, Cornell Research Foundation (CRF). Years before this, Cornell had an informal technology transfer process through which it delivered new crop varieties to New York farmers. Using this informal process, Cornell transferred new seed varieties to the commercial sector (farmers) through the New York Seed Improvement Program (NYSIP), a function of the New York Agricultural Experiment Station within Cornell's College of Agricultural and Life Sciences. Although not a licensing process per se, NYSIP provided farmers with Cornell-developed seed under a long-held tradition in which farmers paid a nominal fee to NYSIP in exchange for the seed. And, following a practice that characterizes Cornell's IP technology transfer today, NYSIP transferred these seeds from the University to the private sector nonexclusively.

Nonexclusive licensing reflects Cornell's public mission and its fundamental desire to see Cornell technology widely disseminated.

Given the long history of the NYSIP seed-distribution program, it's not surprising that after vet vaccines, the next significant effort of Cornell's patenting and licensing in agriculture was a program to transfer new varieties of tree, vine, and other fruits through nonexclusive licenses. In the early 1980s, Cornell began a program to patent and license new raspberry and strawberry varieties. This activity was driven, in part, by the arrival of a new generation of plant breeders who saw patents and licensing as an important part of the mission for plant breeding at a land-grant university. More-traditional breeders at Cornell, responsible for Cornell's apples and other tree fruits, were resistant to the notion of such using intellectual property to control dissemination of new varieties. They preferred the traditional route of placing new-fruit varieties in the public domain, involving no intellectual property, no controls over distribution, and no financial return to Cornell or its breeding program.

This traditional view of public domain releases began to change with the release of Cornell's "Jonagold" apple variety. Although this variety was a modest success in the United States (often labeled as other, more common apples), Jonagold was hugely popular throughout Europe. For many years, it was the most popular European apple. But, because Cornell had not sought protection for the variety, there was no intellectual property in place, and this marketplace popularity did not translate into financial benefit for Cornell. This fact, coupled with a decline in state and federal support for apple breeding, changed the traditional "public domain" mind-set among certain groups at Cornell once and for all.

Since the mid-to-late 1980s, Cornell has had a comprehensive program of patenting and domestic licensing of apples, cherries, plums, grapes, apple rootstocks, raspberries, and strawberries. These licenses are nonexclusive, simple, two-page contracts that provide for a royalty to be paid to CRF on sales of plants. These licenses have no up-front fees or minimums. While these licenses have accomplished the goal of widespread use of Cornell varieties, they have also been a disappointment because nonexclusive licensees provide little or no incentive to invest in developing the market for the licensed variety. So, sales volume per licensee stays small.

In one rare instance, Cornell decided to license a raspberry variety, "Watson," exclusively, with significant license fees, minimum royalty payments, and higher royalty amounts per sale. The license proved to be a financial success for Cornell and its fruit-breeding program and one that catalyzed significant market development for Watson. But this exclusive license was a political failure. Various political constituencies at Cornell, including farmers, nursery owners, state legislators, and others, protested this license. Thus, until recently, all domestic licenses for Cornell fruit varieties have been nonexclusive. And, although the royalties gained from these nonexclusive licenses have provided significant support for Cornell's fruit breeders, one wonders if Cornell fruit varieties might have been even more successful in the market if exclusive licenses had been allowed to incentivize market development.

Despite this adherence to nonexclusive licensing in the crop sector, Cornell continued to license veterinary technology on an exclusive basis. This was in consideration of the large investment necessary by the licensee to bring the product to market, but also the lack of political resistance to exclusive licenses in the animal-health area. These conditions likewise existed in the food-process and agricultural-device fields. Throughout the seventies, eighties, and nineties, Cornell patented and exclusively licensed several food-manufacturing processes including: egg pasteurization and vegetable blanching, as well as the supercritical CO2 fluid extruder. The latter was unique in that the licensed device required a royalty payment on sales of food product made using the patented machine.

During this same period, a number of biological control technologies were patented and licensed, all exclusively. Two of these are notable because the technologies were commercialized through start-up companies. In both cases, CRF took an equity stake in the companies. One company, Bioworks, sells a patented fungal species for control of plant disease. Bioworks is privately held, and Cornell retains strong ties to this New York company. A second company, Eden Bioscience trades on NASDAQ and was responsible for one of the largest equity-liquidation events realized by Cornell for its patented inventions.

The policy decision to allow CRF to take equity in start-ups as part of a patent license was a watershed event. That decision, made in the late 1980s, was driven by one of the first and most important inventions in plant biotechnology—the "gene gun." The gene gun, which is based on a biolistics process, was invented by Cornell professors, John Sanford and Edward Wolf. CRF patented the invention but was unsuccessful in licensing it to existing agriculture-related companies. Sanford and Wolf founded a company, Biolistics, which was ultimately purchased by DuPont, that actively commercialized the device. CRF had founder's equity in Biolistics and realized significant benefits on the sale of the company to DuPont.

Although the Biolistics story was a technology transfer success in many respects, the early participants were not fully aware of certain implications of some of the intellectual property aspects of the license arrangements. In particular, Cornell failed to retain its own right to use the invention for research and technology transfer purposes and also failed to carve out certain philanthropic or humanitarian uses from the commercial license. This has presented problems for some who wish to use the technology without having to abide by constraints imposed by DuPont and its sublicensees. Cornell has been criticized for this lack of foresight and, perhaps, rightly so. However, at the time, few people understood the full implications of licensing agri-biotechnologies that were largely unproven.

There was one, very positive outcome of the gene-gun experience. After the gene gun, every invention licensed by CRF was also made available for philanthropic and humanitarian purposes. Furthermore, all licensing by CRF contained explicit conditions that would ensure diligent use of Cornell technologies for any and all crops and in any geographical region.

After the gene-gun experience, Cornell and CRF actively pursued a two-pronged approach in agri-biotech licensing: nonexclusive and exclusive. Nonexclusive licensing is more common, and when exclusive licenses are granted, they contain quite stringent requirements for diligent development in all applications, as well as carve-outs for philanthropy and orphan crops. For example, the "harpin" technology was licensed to Eden Bioscience under two different sets of terms: one for topical applications of the harpin proteins (for plant-disease control and yield enhancement), and the other for transgenic expression of the harpin genes. This provided for two sets of diligence requirements and financial terms.

A good example of Cornell's nonexclusive licensing strategy in agri-biotechnology has been the licensing of the rice actin promoter. This promoter, discovered in rice, has widespread utility in monocot crops. It has particular utility in transgenic corn and has been used in corn lines with stacked traits of herbicide and insect resistance. Use of the rice actin promoter in corn has stimulated widespread interest in licensing. Cornell's strategy of nonexclusive licensing has successfully disseminated the invention while providing reasonable compensation to Cornell. However, the licensing effort has been complicated by the varied business models of the various nonexclusive licensees. Although Cornell attempted to maintain a standard set of license terms, each successive licensee asked for variations that were tailored to their particular business models. In order to maintain fairness to all licensees, this tailoring of license terms required Cornell to adjust the balance of rights and obligations. For example, significant adjustments have been required in the sublicense provisions. Of course, no sublicensing of the promoter, per se, was allowed. However, the extent to which sublicensees could develop new crosses has been a frequent area of license negotiations.

An aspect of the nonexclusive rice actin licensing strategy has been the development of a hybrid of paid-up and royalty-bearing licenses. The agri-biotechnology industry has demanded paid-up licenses. The industry's complaint was that royalty on each sale was too much of an accounting burden. But, such terms make it difficult for the licensor to realize a significant return; unless the paid-up amount is very, very large. So, Cornell developed a hybrid for which the licensee would not pay an ongoing royalty on each sale; rather, lump-sum payments (of a predetermined amount) are owed upon reaching certain defined milestones. For example, payments are owed on signing, first successful field trial, first regulatory approval, first sale, third anniversary of first sale, and so on.

Today, Cornell uses a variety of licensing strategies to accomplish the privacy goal of assuring delivery of Cornell technology to the marketplace. This practice relies heavily on nonexclusive licenses, but exclusives are more readily accepted. Cornell continues to try new and innovative licensing strategies to satisfy its multifaceted mission.

6. CONCLUSION

Agri-biotech license agreements share many similarities with other types of intellectual-property-based technology licenses. Much of the standard, legal boilerplate will be similar to that of any other license technology agreement. However, there are unique aspects of agri-biotech that set its licenses apart. Those differences include:

- multiple property types often covering a single technology and/or product
- freedom-to-operate issues that drive anti-royalty-stacking provisions
- philanthropic- and humanitarian-use clauses
- stewardship obligations.

Common themes, structures, and contract conventions are part of this technology domain, but the complex nature of agri-biotech and its industry requires each license agreement to be unique, with special, built-in mechanisms that foster the mutual agreement of licensor and licensee. Hopefully, this overview will take us a step closer to a greater understanding of both the common and the unique aspects of agri-biotech licensing. ■

RICHARD S. CAHOON, *Executive Director, Cornell Center for Technology, Enterprise & Commercialization and Senior Vice President, Cornell Research Foundation, Cornell University, 20 Thornwood Drive, Suite 105, Ithaca, NY, 14850, U.S.A.* rsc5@cornell.edu

The In- and Out-Licensing of Plant Varieties

MALIN NILSSON, *Marketing Manager, Value Chain Cereals and Oilseeds, Svalöf Weibull AB, Sweden*

ABSTRACT

Variety licensing is a tool for plant breeding companies and institutions to commercialize their varieties and to transfer technology to farmers efficiently. As the seed industry becomes increasingly privatized, interest in in-licensing new varieties, both from national and international sources, is likely to increase. Likewise, financial pressure on public sector breeding will increase the need for the targeted commercialization of varieties through out-licensing. As the seed sector becomes more transparent, the market should see more foreign investment from companies who wish to make their varieties available through licensing. That, in turn, should promote local seed production and variety testing. The licensee and the licensor should focus primarily on the practical content of the license agreement, specifically, exclusivity to plant material and territory, plant variety protection, variety trials, national registration, royalty payment, and information transfer. The purpose of this chapter is to provide guidance for prospective licensors and licensees in the practical issues of in- and out-licensing of varieties.

1. INTRODUCTION

Variety licensing allows breeding companies or institutions to commercialize their products (plant varieties) and is also an efficient tool for technology transfer. New technology in a variety, represented by improved genetics and expressed mostly through improved agricultural performance, can be transferred to farmers by licensing out seed production and distribution rights to seed companies. The variety license itself consists of an agreement between the owner of the varieties, or an authorized representative, and a legally eligible person who wishes to commercialize the variety.

As described by Louwaars,[1] the first problem in seed policy development is the dual function of seeds. Seeds are a method of technology transfer, *and* each seed itself is a commercial commodity. These two functions are among the most important issues to address in establishing long-term success in variety in- and out-licensing. The technology embedded in the seed of a new variety is easily transferred to farmers on a large scale and can be used instantly. In many countries, public breeding has supplied varieties for use by seed producers and farmers at no cost. This free sharing of varieties makes it difficult to give recognition, in terms of royalty payments, for the variety improvement work.

Further use of the technology—and its improvements—depend on the seed's other function, that of a commercial commodity. The seed must be used in trade. Once the seed is circulating in the marketplace, a portion of the profits can be re-invested in further breeding and the development of new technology and plant varieties. This is possible because the incentive, especially for the private seed business, for continued crop development lies in the possibility of getting a return on the investment.

Nilsson M. 2007. The In- and Out-Licensing of Plant Varieties. In *Intellectual Property Management in Health and Agricultural Innovation: A Handbook of Best Practices* (eds. A Krattiger, RT Mahoney, L Nelsen, et al.). MIHR: Oxford, U.K., and PIPRA: Davis, U.S.A. Available online at www.ipHandbook.org.

Development of the private seed sector will increase competition and could speed up efforts to reach a larger part of the farming community. Small- and medium-sized seed companies need to develop their product portfolios through in-licensing of varieties (whereas public institutes could increase profitability by out-licensing their varieties). The privatization and increased transparency of the seed sector could promote foreign investment from companies wishing to make their varieties available through licensing, which in turn would promote local seed production and variety testing.

Access to new varieties requires proper handling of intellectual property (IP). This can be accomplished through variety license agreements, which also provide a strategy for developing and introducing new varieties. A variety license agreement can be divided into two main parts: first, those clauses describing the key rights and obligations of the parties and the conditions that make the framework of the license—these clauses will set the standards for cooperation and outline what the parties wish to achieve—and second, "boilerplate" clauses that are not specific to the agreement but are legally relevant (for example, processes for dealing with arbitration, relevant law, legality, assignability, warranty, and force majeure). The purpose of this chapter is to provide guidance for establishing the first part of a variety license, and the key elements have been divided into the following sections:

- exclusivity
- territory
- evaluation of the licensed material
- protection of germplasm
- national registration and plant variety protection
- royalties
- effect of termination
- reporting to licensor

In this chapter, the words *breeder* and *variety owner* will be used interchangeably, to mean a breeding company, an individual plant breeder, or a person with the legal rights of ownership to a licensed plant material.

2. THE DRIVING FORCES BEHIND LICENSING

2.1 *In-licensing*

In-licensing plant varieties can raise market share or offer competitive advantages by increasing the ability to meet customer demands. The most obvious reason for in-licensing varieties is to enhance or complete a company's variety portfolio. This applies both to companies with their own breeding programs and to companies working exclusively with in-licensed varieties. Those species for which a company has existing breeding programs—or other species that may be of interest to the market—are potentially subject to in-licensing. Demand for certain products from farmers, the processing industry, or consumers could be met by a company obtaining a license from the variety owner to supply the market with seed of that variety. These parties may demand things such as a species not available on the existing market, varieties with improved agricultural characteristics, or improved nutritional value.

In-licensing gives breeding and seed companies access to new technology (like hybrid varieties); breeding companies may profit from this new technology without obtaining a license to use the hybrid system itself in variety development. Another advantage, or, rather, side effect, is the possibility for breeders to compare their material with that of their competitors in the early stages of variety development.

2.2 *Out-licensing*

The most common reason for a company to out-license its varieties is to maximize the return on its investment by allowing others to produce and sell its varieties in markets that the company cannot reach. Small- or medium-sized breeding companies, for example, may not have the resources to establish their own sales organization either within their own country or in different countries. Thus the companies will use out-licensing to fully exploit the potential of their breeding program.

2.3 *Plant variety protection*

The importance of *plant variety protection* (PVP) legislation as a driving force for successful variety licensing cannot be stressed enough. PVP confers IP rights, known as *plant breeder's rights* (PBR), which provide an incentive to plant breeders for the development of new varieties of crops. This, in turn, fosters progress in sustainable agriculture and generally improves the economic circumstances of farmers and growers, since it gives them access to new and improved varieties. However, without the legal framework for acknowledging the ownership of the licensed varieties, the variety owner will have difficulty getting a return on investments made in variety development. Effective PVP legislation supports the interests of both the variety owner and the farmer. It will also facilitate the transfer of technology and provide incentives for further investments in the development of new plant varieties. In many countries, PVP legislation is based on the International Union for the Protection of New Varieties of Plants (UPOV) Convention, which exists in three revised versions (adopted 1961, 1978, and 1991, respectively). Currently, 61 countries[2] have ratified the UPOV Convention. This makes it the most widely adopted form of a sui generis IP protection system developed specifically for plant varieties. The latest revision of the Convention has not been ratified by all member countries; however, all new members are required to ratify the Convention of 1991.

Major differences in the conventions will affect the approach to licensing. These differences include the species and genera for which PVP provides IP protection, exemptions from PBR (that is, the plant breeder's exemption and the farmer's, or crop, exemption, also known as the "farmer's privilege"), the period of protection, and the scope of protection under PBR. The latest UPOV Convention strengthens the rights of the breeder: member states are obliged to provide protection to all botanical genera and species (Chapter II, Article 13(1–2)); the Convention also extends the duration of the breeder's right by five years (Chapter V, Article 19(2)), and extends the scope of protection to include conditioning for the purpose of propagation, export, import, and stocking (Chapter V, Article 14(1)). The farmer's privilege is an optional exemption from the PBR (Chapter V, Article 15(2)). It may limit the farmer's rights to use on-farm harvested material—obtained from a protected variety on the same farmer's holdings—as propagating material. This propagating material is commonly called *farm-saved seed* (FSS), and this exemption stems from the basic rights outlined in the 1961 and 1978 UPOV conventions (though the exemption is not optional in either and is not as clearly defined as in the 1991 version).

The PVP legislation of the UPOV members is well documented and should not pose any large problems for prospective licensors and licensees. An awareness of the differences will facilitate the development of the variety license agreement. On the other hand, it may prove more difficult to influence PVP legislation in nonmember countries, and licensors are strongly advised to gather as much information as possible about the PVP system in a new territory so that they can adapt their licensing strategy accordingly.

3. KEY ISSUES IN VARIETY LICENSING

When establishing a license agreement, whether for in- or out-licensing, it is important to discuss and agree upon those issues that will constitute the spirit of the agreement and set the foundation for good cooperation.

3.1 *Exclusivity*

The following section on exclusivity has been divided into two parts. The first section discusses the rights granted under the license. The second defines the material for which an exclusive license is granted.

Nonexclusive licenses are rare, and experience has shown that breeders grant exclusive licenses more willingly than nonexclusive ones. Exclusive licenses are preferred because breeders believe that the mutual commitment will be stronger when working exclusively. A good variety provides a competitive advantage and will thus create revenue for the company with the exclusive rights. It is in the best interest of both parties to make the variety as profitable as possible, and the commitment resulting from exclusive rights is considered to lead to the best market coverage possible. Indeed, working on a nonexclusive basis is considered to have smaller market potential.

The extent of exclusivity is defined by various factors (such as the territory for which crop or variety exclusivity is granted) that will be discussed in greater detail later.

3.1.1 *The rights granted*

The exclusive rights granted to the licensee often correspond, either in part or in whole, to the rights that can be obtained through the plant breeder's rights (PBR) protection for a variety. As defined in the UPOV Convention Act of 1991[3, 4] (Chapter V, Article 14 (1)), the following actions shall require prior authorization from the breeder:

- production or reproduction (multiplication)
- conditioning for the purpose of propagation
- offering for sale
- selling or marketing
- exporting
- importing
- stocking for any of the purposes mentioned above

These provisions are recommended as a starting point for discussions about what rights the licensee will be allowed to exercise. The most important factors in determining the type of license to grant include: former experience, seed production and distribution infrastructure accessible to the licensee, type of species to be licensed, and plant variety protection.

There are two major types of licenses. The first type is the *distribution license*, which includes the rights to market and sell the licensed material. The second is a *production license*, which in addition to these rights includes the rights to seed multiplication and production. For varieties that are easily and rapidly multiplied, such as those of species with small seeds and low sowing rates, the licensor may prefer to keep all or most of the seed production within its own control. This would limit the exclusive rights for a distribution license. For varieties of species with high sowing rates and low multiplication factors (for example, cereals), the transportation cost of the commercial seed to the licensee is likely to be high, and so a production license is usually preferred.

Breeders can partially preserve variety protection by limiting access to seed for propagating purposes. If the licensor allows only for marketing and sales, the variety is better protected because the licensor will not have to leave out early generations of seed for multiplication from its internal control system. However, under certain circumstances, the final seed generation, or the commercial seed, may be more expensive because the total seed costs increase if the seed has to be transported between countries or over long distances within the same country. Giving the licensee responsibility for seed multiplication and production will decrease margins (actual sales revenue for the seed itself) for the licensor because the income will then be based on royalties (revenues derived from licensed use, propagation, sales, and so on), as opposed to sales margins *and* royalties, that is, a more lucrative double revenue stream. Licensed production may, however, be advantageous for the licensor because risks in seed multiplication will be spread, as will the costs for handling the seed in the production chain.

High transaction costs in the chain from the breeder to the farmer can present large problems since many factors influence these costs.[5] High transaction costs result in expensive seed, which makes it difficult to realize sales on the market. This is especially true for countries using large amounts of farm-saved seed or for places that market predominantly public varieties; these countries have a hard time realizing sales because both of these seed categories are chosen for their low costs to farmers. Still, if the licensee has access to the required seed production infrastructure (basically, farm capacity for growing, harvesting, processing, storing, and transporting seed), costs can be kept low when incorporating new varieties. This will increase the value of the seed for the licensee and promote local agricultural business. Still, as stated earlier, contracting seed production to small-scale enterprises will spread the risks in seed production and lower transportation costs because the seed can be produced closer to the market.

The number of generations of seed the licensee is allowed to multiply can also be a matter of discussion. Generally, the number of generations is decided on a case-by-case basis rather than regulated through the license agreement. National legislation, as well as international rules and directions (such as the OECD Seed Schemes,[6, 7] as laid down by the Organisation for Economic Co-operation and Development [OECD][8]), should be consulted during licensing, since they regulate the number of generations that any seed may be reproduced. Because the reproduction system will influence the stability of a specific variety, the number of generations varies between cross-pollinated and self-pollinated species.

The rights of the licensee to hybrid varieties are most commonly restricted to marketing and sales of the commercial seed. Hybrid seed production is more expensive and considerably more complex than the production of line varieties. The owner control of the hybrid components may influence the possibilities for out-licensing the production of hybrid seed. Moreover, by keeping hybrid seed production within its own control, the licensor, to some degree, protects the hybrid components. In addition, in some jurisdictions (for example, the United States) inbred seed lines can be protected as trade secrets. Or, to be legally, technically accurate, the "information" embedded in the seeds is protected as a trade secret.

The licensor may wish to restrict the rights of the licensee to import seed from sources other than the licensor. It may also wish to similarly limit the export of seed from the defined territory. In contrast, the licensee may want to retain these rights, and it is not always possible to restrict seed import and export, since this may be prohibited by legislation. For example, according to the [European] Community Plant Variety Rights (Chapter III, Article 13(2)),[9, 10] authorization of the holder is required for export from the European Community (EC) and format import to the EC of a protected variety. Between EC member countries, the export and import of protected variety material can only be restricted if the material is for propagating purposes (that is, higher seed generations than certified seed).

3.1.2 *Defining the licensed material*

The second part of exclusivity deals with the definition of the licensed material. The access to varieties a licensor is prepared to give a prospective licensee depends on such factors as earlier experience, market penetration ability, the licensee's existing variety portfolio, and ongoing cooperation with other breeders. The exact size of the material must also be determined on a case-by-case basis. Exclusivity to the licensor's material may be granted on different levels:

- single varieties
- selected crops/species
- all crops/species

The most common type of exclusivity at the beginning of a partnership is likely to be *first right of refusal*, or exclusivity based on single varieties provided by the licensor. The licensor provides a few varieties of its choice, or it may allow the licensee to choose its candidates among a number of varieties for commercialization. The licensor may freely dispose of the remaining varieties through other marketing channels within the same territory. Exclusivity is maintained, for single varieties only, and the licensor has the opportunity to evaluate the licensee's ability to commercialize the licensed variety. This can also be a strategic tool to distribute varieties among a number of licensees, in the hopes of stimulating competition and obtaining a larger total market share in a particular market.

Granting a licensee exclusive rights to the whole set of crops in a breeding program occurs rarely, but this differs based on the number of crops or species within which the licensor is active. This kind of exclusive relationship between the breeder and the licensee is likely to result from strategic decisions concerning the long-term relationship between companies, a wish to strengthen connections with key partners or between mother/daughter companies, and so forth.

The other type of exclusivity is to grant exclusive rights to selected crops or species. In a country with limited participants in the seed business, participants will likely specialize in certain crops. In such cases it could be appropriate to grant exclusivity to all material from a breeding program.

In certain circumstances, exclusivity may limit the work of a company or public institute. The public sector or other external funding source might support a company's breeding program in whole or in part. These funds may come with provisions restricting the breeder's options to offer exclusivity in out-licensing. Public sector breeding may also be unable to grant exclusivity to selected licensees, because this may limit public access to the varieties.

License agreements may regulate continued access to new varieties from the same licensor. Where the license agreement is limited to a single variety, it is likely that continued access would require a request from either party and could be part of the written agreement. For collaboration based on more-extensive variety trials, it would be sensible to settle an appropriate number of new breeding lines or varieties to submit each year to the licensee, subject to availability and request from either party.

3.2 *Territory*

Territory defines the geographic area where the licensee has the right to exercise its exclusive rights. The territory is not necessarily restricted to a country; it could be a part of a country, one or more countries, continents, or even the world.

In variety licensing, however, the most common territory is that of a country. Depending on the market coverage capabilities of the licensee, it may also be suitable to instead define the territory as a group of countries or established unions, such as the European Union,[11] the African Union,[12] or the Mercosur.[13] In places such as these, the common rules for PVP, seed trade, and other relevant areas are more harmonized. Such territories have a tendency to change over time, and so it is recommended that parties in a licensing agreement consider defining a union as its member countries when the agreement is signed.

Definition of the territory may be influenced by existing PVP legislation. As discussed above, not all countries are UPOV members, and even UPOV members differ in PVP legislation depending on which version of the UPOV Convention the country has ratified. Many countries, especially developing countries, are not UPOV members. This should be taken into consideration when defining the territory and the rights that the licensee will be given by the licensor to exercise within that territory.

3.3 *Evaluation of the local adaptation of the varieties*

The aim for both parties when in- and out-licensing varieties is to select varieties for marketing that show improved agricultural performance or have other desired characteristics. Apart from the market (end-user) demand, the value of a variety is largely ascribed to its adaptation to local growing conditions. Depending on the plant species, varieties can be transferred between geographic areas and climatic zones. Introducing new varieties usually requires the local confirmation of agricultural performance, which is done for the purpose of national listing and/or marketing advantages. Either the public system of variety testing or private trials can be used to introduce the new variety.

The trial strategy and the minimum requirements for assessing local adaptation should be discussed and settled in the agreement, including any decisions about cost sharing. Commonly, the licensor will require the licensee to evaluate the value of the varieties at its own cost, with the aim of including them in the national list, recommended list, or any corresponding list of varieties officially registered for release in the territory. These trials are often referred to as VCU (*value for cultivation and use*) trials. Of course, the trial strategy can also consider whether it is necessary to have a variety officially listed in the territory or not. For example, within the European Union, varieties included on a national list in one member state or in any of the European Free Trade Association (EFTA) countries can be marketed in any other member state without any prior demand of inclusion on an additional national variety list.

Plant variety protection has to be applied for separately from the local adaptation trials. All three versions of the UPOV Convention provide the legal means to provisionally protect the variety from the date of filing an application until the grant of PBR. This gives the applicant the right to enforce the provisional rights in case of breach during the evaluation period, whether in a private or an official trial network, provided an application for PBR has been filed. If no such system for provisional protection exists, the licensor may add clauses in the license agreement that will regulate the distribution conditions of the plant material for trials.

3.3.1 *Private trials*

Private trials in this context are defined as all trials that are not part of publicly performed trials. The trials can be conducted by the licensee or any other skilled partner equipped to perform them (for example, other seed or breeding enterprises, farmers' cooperatives, universities, or agricultural extension service centers). In countries without an official trial system, the role of the private trials can be significant.

Private trials are a potential tool for the licensee to test varieties and select the best candidates for official trials. Some countries require a minimum number of *station data* for entering a variety into official trials. Collection of these data can occur either in one year from the number of stations required for the application, or on fewer stations over a period of two or more years.

Unfortunately, breeders, either through neglect, procrastination, or possibly selfish motivation, might abuse the private trial system by keeping varieties within the private trial system until they are too old for market introduction. This could either prevent competitors from including the variety in their portfolio or prevent breeding companies from entering the market with that specific variety. In order to avoid this abuse, it is necessary to limit the number of years a variety can be tested in the private trial network before it will be included in national list trials. For annual crops, a maximum of two years or two growth cycles should be sufficient for evaluation unless some unpredictable event occurs, in which case the period can be extended by one year or growth cycle.

3.3.2 *Official trials*

Official variety trials, also referred to as national or recommended list trials, are carried out to evaluate the candidate variety's value for cultivation and use. This incorporates the varieties' agricultural performance and quality characteristics. Varieties that show an improvement compared to standard control varieties qualify for inclusion in the national list, a register of varieties approved for release on the national market. A national list or register of varieties does not provide any PVP for the varieties included. Instead, it is a means of safeguarding the quality of the varieties released on the national market—they have been tested and proved valuable in cultivation and use, in comparison to the other varieties on the list.

The private sector can undertake VCU trials in countries where the public sector does not perform such trials. It is possible also to establish private trial networks that will enable new varieties to be independently evaluated.

3.4 *Germplasm protection*

It is important for a breeder to obtain protection for finished varieties and those still in trials. Due to the importance of protection, it is essential to include a section in the agreement outlining the handling and supervision of plant material before it has obtained plant breeder's rights (PBR) protection. If the production and sale of a variety is initiated before PBR has been granted, there is a risk that the variety will not be eligible for protection. It is advisable to restrict the licensee's distribution rights of the not-yet-protected material to third parties and use of the germplasm to the licensee's own breeding programs. This restriction could either be part of the license agreement or part of a separate material transfer agreement.

3.5 *Plant breeder's rights and official variety registration*

3.5.1 *Plant breeder's rights*

Plant variety protection (PVP) is important when granting access to new varieties. It provides protection of the proprietary rights of particular species in a territory. There is no blueprint solution for implementing PVP laws because the policies between countries differ greatly. Europe and the United States, both members of UPOV, are good examples of public versus private responsibility systems. Both systems provide protection for plant varieties and a legal means of enforcement of the rights, and both seek to grant PBR based on trials, usually referred to as *DUS* trials, that show that the variety is *distinct, uniform, and stable*, and have received a novelty declaration from the breeder. The European Union (E.U.) has harmonized PBR legislation, and European countries have generally adopted a system based on testing and registration that is fully controlled and performed by designated authorities. PBR can be applied for at the *community plant variety office* (CPVO) and will be valid throughout the entire union. The system in the United States is based on self-control. The plant variety protection office (PVPO) issues PBR certificates, and the applicant is responsible for carrying out the necessary trials and filing an application based on forms and guidelines from the PVPO.[14]

The PBR legislation in the defined territory will determine two matters: the strategy chosen by the licensor and the licensee to protect licensed varieties and what action to take if there is a breach of rights of the protected varieties.

In the first case, the licensor and the licensee can jointly decide on the appropriate way to protect the licensed varieties, as well as when to apply for protection. In some countries, even though there is PBR legislation in place, it may prove difficult to enforce the rights. Critics argue that, in these cases, the PVP system is a way to finance and maintain the bureaucracy rather than protect IP. Others claim that using the system, despite enforcement difficulties, is a way to ensure its improvement. At any rate, the licensor and the licensee have to decide jointly on the best approach for protecting the varieties under the current circumstances. This strategy should be clearly stated in the agreement.

The use of hybrid technology can provide additional IP protection in plants. Although F_2 seed harvested from hybrid varieties can be used as seed, the agronomic advantages from hybrid vigour and a homogenous crop cannot be maintained in the second seed generation. This provides a self-regulating kind of protection for hybrid varieties and increases profitability for the licensee and the licensor through repeated seed sales. It should be noted that national PVP legislations differ: some permit the use of farm-saved seed of the F_2 seed from hybrid varieties, others do not.

3.5.2 *Official registration of varieties*

Many countries require that new varieties undergo official trials following official registration of the approved varieties. Official registration of a variety results in its inclusion in a national list of recommended varieties approved for market release. As mentioned above, the official trial system is one method of maintaining quality control for a variety, since the listed varieties have been tested for their agricultural performance and quality. Release decisions are based either on results from independent public trials, on testing data supplied by the breeder, or on both. The appropriate trial strategy for the official registration should be jointly decided by the licensee and the licensor and included in the license agreement.

3.5.3 *Responsibility and cost sharing*

In addition to decisions concerning PBR and official registration strategies, the licensor and the licensee must agree upon who will be in charge of applying for and maintaining the PBR and national list entries. It is also important that neither party withdraw the PBR grant or the national list entry without obtaining a written confirmation from the other about the decision. Even if the licensee wishes to stop marketing a variety, continued protection may be required for other purposes (for example, if the variety is used as a hybrid component, for marketing it through another channel or to allow for continued collection of FSS royalties).

The application and maintenance of varieties for protection or official listing has associated costs. If the licensee has exclusive rights to the varieties in the territory, the licensee usually carries the costs connected to variety protection and the national list (including trials for either purpose). However, if the licensee has nonexclusive rights to the variety, the licensor will usually carry these costs. In the European Union, where it is possible to obtain either national PBR or Community PBR (valid within the entire union), the cost for maintaining national PBR protection is commonly absorbed by the licensee, whereas the licensor is responsible for the cost for community PBR.

Costs for trials, such as marketing or demonstration trials, are commonly paid by the licensee. The licensor could make other contributions (for example, providing promotional material, field signs, technical support through information material, or by attending field days, and supplying seed bags with the licensor's logotype).

3.6 *Royalties*

For the rights to commercial exploitation of the plant varieties granted under the license agreement, the licensee pays the licensor a royalty. A royalty can include not only the fee agreed to by the licensor and the licensee, but all fees connected with the use of the licensed varieties, such as fees for FSS and acreage fees.

The royalty should be at a level acceptable to the market. It must neither be so high that the farmers cannot buy the seed, nor so low that the licensor will not find it profitable. It is common practice for the licensor and the licensee to split the collected royalty. The proportions of the royalty paid to each party are a matter of negotiation. The amount depends on the structure of sharing costs related to trials, maintenance of national list entries, PBR, market support, and other factors. There is no blueprint solution: for each variety license the royalty has to be negotiated separately. Nevertheless, a few royalty-calculation principles can be used on their own or in combination: fixed royalty rate, royalties connected to the seed price, minimum royalty rate, royalty intervals and sold quantities, and multiplication acreage and end-point royalties.

3.6.1 *Fixed royalty rate*

Setting the royalty at a fixed rate is the most common remuneration system. It requires knowledge of the seed business in the territory and the farmers' ability to pay for the seed. The fixed rate is independent of the sales price and is calculated per weight unit of seed bags containing a specified quantity. One can also calculate a fixed royalty based on the units of a specified number of seeds. The latter system is used, for example, for winter oilseed rape (*Brassica napus*) in Europe, where the seed is sold in units of 1.5 or 2 million germinating seeds (hybrid and line varieties, respectively, in Germany) and 2 million seeds (hybrid varieties in France).

Royalties can also be settled centrally in negotiations between breeder and farmer representatives. This is done, for example, by GESLIVE[15] in Spain and SICASOV[16] in France. The royalties are negotiated and fixed annually for each species and seed generation—they could potentially be settled for individual varieties.

3.6.2 *Royalty connected to the seed price*

A royalty level connected to the price of the seed will instantly change as seed prices increase or decrease. The rate may be calculated as a percentage of the net sales price to the farmer, and since the actual net sales prices may be difficult for the licensor to verify, trust between the licensee and the licensor is of great importance.

3.6.3 *Minimum royalty rate*

A minimum royalty rate paid annually is a less common form of royalty and must be combined with some other royalty system. In this system, the royalty is calculated on one of the calculation principles described above, but a minimum royalty is added to it. For example, if the royalty is calculated on a fixed rate and the total royalty collected exceeds the minimum royalty, the royalty based on the fixed rate will be paid to the licensee. If the total royalty collected is below the minimum rate, the minimum rate will be paid regardless of the actual total royalty.

3.6.4 *Royalty intervals connected to sold quantity*

Royalties can also be connected to the seed quantities sold. The royalty rates per unit can be fixed at intervals of sold seed quantities. The licensee either pays the royalty rate for the highest interval achieved for all seed sold or for the royalty corresponding to each interval.

3.6.5 *Multiplication acreage and end-point royalties*

There are royalty systems that are independent of the actual seed sales. If sales volumes are difficult to control, it might be more efficient to use a royalty system calculated on the multiplication acreage with a fixed rate per surface unit.

In countries or areas where much of the agricultural produce is not used on the farm, a so-called *end-point royalty* can be successfully implemented. When the farmer delivers his or her produce, a royalty based on the delivered quantity will be charged, regardless of whether the farmer has purchased the seed or used his or her own. This royalty system can be based on variety, use of certified seed, or other criteria.

3.7 *Effect of termination*

Termination of the agreement will have both immediate and long-term effects on the licensee and the licensor. Controversy can be avoided by defining the consequences of termination on the licensed varieties and the remaining seed at termination. The varieties can be divided into three groups:

1. Marketed varieties
2. Varieties to enter the market soon
3. Varieties in trials

The varieties of the second group usually include varieties in official trials and varieties that recently have been officially listed but are not yet marketed.

If the agreement is terminated for reasons that allow for immediate termination, the licensor is likely to require that all rights to all varieties be rescinded immediately and that any seed still in the licensee's possession be retuned to the licensor.

If the agreement is terminated for other reasons, the licensor may want to treat the three variety groups differently. Usually, the agreement will continue for the lifetime of the varieties with regard to the varieties in groups (1) and (2), but will be terminated immediately with regard to the those in group (3).

3.8 *Reporting to licensor*

It is recommended that the agreement specify the information that should be transferred between the parties (usually from the licensee to the licensor) on a regular basis. This information could include anything relevant to the activities resulting from the license agreement, such as:

- marketing plans and sales targets for the season(s)
- sales reports and forecasts throughout the season
- royalty statements
- variety trialing plans
- variety trial results
- seed certification reports
- copies of documents connected to PBR and a national list, such as application forms and PBR certificates

Establishing such routines through the agreement will facilitate establishment of a transparent communication and relationship and will help both parties achieve their goals and continue to improve cooperation.

4. CONCLUSIONS

The seed sector in many developing countries is moving toward decreased funding of public sector breeding and increased privatization. This trend is leading to a decrease in new varieties entering the market on the one side and an increased opportunity for introduction of new varieties on the other. Seed companies need to in-license varieties, while private sector breeders, national and international, may need to out-license their products. The financial pressure on public sector breeding makes it difficult to maintain development of improved varieties; thus, incomes could be generated through variety out-licensing. Privatization could further attract foreign seed companies by making their varieties available for local production and sales. This would also provide local seed companies and, presumably, farmers with access to new technology. The development of new varieties—as well as good geographic coverage of the private seed sector—requires that breeders and seed companies get a return on their investment. This is achieved when farmers buy seed and a royalty is paid to the breeder. It is also important for a breeder to obtain proper protection for the IP of a new plant variety. Proper PVP legislation is also needed. Providing the legal framework for breeders to get a fair chance to profit from their breeding efforts will promote further incentives for investments in variety development.

The discussions around PVP in this chapter have dealt exclusively with PVP based on the acts under the UPOV Convention. Granting PBR is the predominant system for IP protection of plant varieties; in most countries of the world where plant varieties are not patentable, it is the only system for such protection. The major difference between PBR and patent rights lies within the breeder's exemption and the farmer's privilege of the PBR, as there are no similar exceptions from the rights in the patent.

The license agreement is a written statement of what the licensor and the licensee wish to achieve together. The principal objectives of the license must be clearly stated; otherwise, they will never be achieved. This chapter has described the key elements of variety licensing and how to approach them. The conditions of the license agreement should set out the framework and the standards for cooperation, but it is also important to recognize that a license agreement is not static. There are certain provisions to follow, but these provisions also need to be flexible. Changes in the market, seed legislation, and PVP laws should be reflected in the agreement, because it is partly built upon them.

The issues discussed in this chapter should make it possible for prospective licensors and licensees to focus on the part of a license agreement that will have the largest impact on its successful implementation. ■

MALIN NILSSON, *Marketing Manager, Value Chain Cereals and Oilseeds, Svalöf Weibull AB, SE – 268 81 Svalöv, Sweden.* Malin.Nilsson@swseed.com

1 Louwaars, NP. 2002. Seed Policy, Legislation and Law: Widening a Narrow Focus. In *Seed Policy, Legislation and Law: Widening a Narrow Focus.* (ed. NP Louwaars). The Haworth Press, Inc.: Binghamton, New York. pp. 1–14

2 www.upov.org/en/about/members/index.htm.

3 UPOV. 1991. Act of 1991. International Convention for the Protection of New Varieties of Plants.

4 www.upov.org/en/publications/conventions/index.html.

5 van der Meer, C. 2002. Challenges and Limitations of the Market. In *Seed Policy, Legislation and Law: Widening a Narrow Focus.* (ed. NP Louwaars). The Haworth Press, Inc.: Binghamton, New York. pp. 65–75.

6 OECD. 2005. OECD Seed Schemes 2005: Rules and Directions.

7 www.oecd.org/agr/seed.

8 www.oecd.org.

9 European Union. 1994. Council Regulation (EC) No 2100/94 of 27 July 1994 on Community Plant Variety Rights.

10 www.cpvo.fr/default.php?res=1&w=1024&h=583&lang=en&page=droit/legislation.htm.

11 www.europa.eu.int/.

12 www.africa-union.org/.

13 www.merco-sur.net/.

14 www.ams.usda.gov/science/PVPO/apply.htm#proof.

15 www.geslive.com.

16 www.sicasov.com.

CHAPTER 11.4

Potential Use of a Computer-Generated Contract Template System (CoGenCo) to Facilitate Licensing of Traits and Varieties

ANATOLE KRATTIGER, *Research Professor, the Biodesign Institute at Arizona State University; Chair, bioDevelopments-International Institute; and Adjunct Professor, Cornell University, U.S.A.*

JOHN DODDS, *Founder, Dodds & Associates, U.S.A.*

DONNA BOBROWICZ, *Technology Transfer Specialist, Loyola University Chicago, Office of Research Services, U.S.A.*

ABSTRACT

Licensing between companies of both traits and varieties is routine, and there is no reason that it should be anything other than routine between companies and public sector institutions, as well. Some public entities struggle to gain experience in this area. This leads companies to shun negotiations and, even, discussions. Yet opportunities for the public sector to in-license traits (in the form of well-characterized and deregulated transgenic "events") and varieties are vast and could lead to earlier access with respect to transgenic events (through backcrossing into local varieties) and to improved varieties for subsistence farmers. In order to improve the ability of the public sector to both in-license and out-license germplasm, a test version of a software program, the "Computer Generated Contract Template System" (CoGenCo), was developed. It aims to facilitate the exchange (or licensing) of commercial varieties by "walking" potential licensors and licensees though a systematic list of questions and tested parameters. CoGenCo is a pragmatic way of increasing the licensing of both finished varieties and germplasm containing transgenes for backcrossing, and its flexibility would make it especially suited for use in developing countries. This chapter explains the concept behind the software's test version and leads the reader through its use. The authors very much welcome comments and suggestions about the software and look forward to collaborating with interested parties to further develop CoGenCo into a comprehensive and widely available system.

1. INTRODUCTION

The international agricultural-development community, the crop industry, and various advocacy groups disagree about how to transfer protected varieties and biotechnological inventions to developing countries. Yet everyone agrees that access to these inventions in developing countries should be improved and accelerated, either through donations or "open-source" licensing or through a variety of other strategies. But too often this goal is made complicated by too much industry incrementalism, or by activist demagoguery. From a humanitarian perspective, such debates distract from the only focus that matters—the urgent need for farmers to access improved traits and varieties.

There is no reason that the licensing of germplasm and traits, particularly to meet the needs of resource-poor farmers in developing countries, need be more difficult than out-licensing for routine business purposes. Any plant-breeding company that does the latter—virtually all of them—considers out-licensing routine. Consider Holden's Foundation Seeds, a company now owned by Monsanto, the sole revenue of which comes from the out-licensing of its foundation seeds. In terms of developing country licensing, however, most companies are reluctant to even enter into discussions, let alone negotiations, partly because many variables are unknown or little tested, and because few companies have any experience in this area. For these reasons, a small

Krattiger A, J Dodds and D Bobrowicz. 2007. Potential Use of a Computer-Generated Contract Template System (CoGenCo) to Facilitate Licensing of Traits and Varieties. In *Intellectual Property Management in Health and Agricultural Innovation: A Handbook of Best Practices* (eds. A Krattiger, RT Mahoney, L Nelsen, et al.). MIHR: Oxford, U.K., and PIPRA: Davis, U.S.A. Available online at www.ipHandbook.org.

project was undertaken to develop a test version of a software program, the Computer Generated Contract Template System (CoGenCo).

2. WHAT IS COGENCO?

CoGenCo was designed to contribute to facilitating the exchange (or licensing) of commercial varieties by "walking" potential licensors and licensees though a systematic list of questions and tested parameters. The word *commercial* here is used because the licensor transfers commercial varieties primarily for commercialization in developing countries (following appropriate backcrossing, as necessary). Such commercialization may be in the form of donations, through national agricultural-research systems, or directly through seed companies.

CoGenCo is a concept proposed as a pragmatic way of increasing licensing of proprietary and finished varieties that may or may not incorporate proprietary technologies. Essentially, CoGenCo facilitates the awarding of out-licenses to developing country institutions, including germplasm from the Consultative Group on International Agricultural Research (CGIAR). Under the legally binding terms of these license agreements, several entities in a given country could compete against one another on price in *poor* (developing) countries but would not be allowed to compete against the patent holder in developed countries, where revenues and incentives for developing new varieties and new technologies would be undiminished. Under appropriate circumstances, the germplasm and/or traits could also be licensed royalty free. Such out-licensing separates these fundamentally different markets and promotes access to improved germplasm and technologies, all by reaffirming various statutory protections as indispensable for successful agricultural research and development.

The CoGenCo system, therefore, is aimed at establishing a certain international standard license. The more institutions use the CoGenCo template, the more the system becomes valuable. For this reason, we intend to make the CoGenCo system available for free once it is fully developed.

3. THE TEST VERSION

Based on discussions with several lawyers and licensing experts, we generated a basic license template. First, we developed a set of key variables and agreed on different options to choose from within defined ranges. A software engineer translated the concept into a "workable" software version that would provide a feel for what a finished product would look like. We selected Microsoft® Access® as the backbone of the system because it provides flexibility and easily expands into a version that can be used via a Web interface. Users around the world would thus be able to access the system without having to invest in expensive database software.

The primary objective of this test version was to see how different types of potential users would use it. The software allows for certain parameters to be adjusted. For example, for "humanitarian" licensing, a royalty of 0% could be specified, whereas for larger farmers, a sliding-scale royalty rate could be chosen. Depending on the option preferred, a different set of follow-up options will arise, such as liabilities, payment terms, auditing requirements, and so on. The software will be developed in such a way that individual users may customize the software. For example, they could include their own institutional standard language where appropriate. It could also eventually be downloadable from the online version of this *Handbook*.[1]

Figure 1 shows one of many screenshots that allow users to input various parameters and select from a range of options. For example, by selecting the tab *License*, the user is offered a screen that lists all the pertinent licensing details, including the territory (countries), and many more. The user basically walks through the different issues that should be considered in a license and is provided with one, two, or more options.

The software thus presents users with an interactive decision tree, which allows for multiple choices or user inputs. The key factors included are:

- country
- commodity/crop
- technology
- farm size

- material transfer/reach-through clauses
- farm income
- import/export matters
- cooperative farm issues
- sliding scale for royalties
- royalty stacking issues
- warranties
- liabilities
- third-party distribution issues
- farmer-seeds issues

For example, the software system will ask the user whether tangible material is being transferred under the license. If *NO* is selected, then the next options will be limited to IP licensing aspects (including patents and/or know-how and/or trademarks and/or copyrights). If *YES* is selected under tangible material, then a specific question arises as to the conditions of the transfer, primarily in terms of possible reach-through clauses. To include reach-through clauses has certain advantages and disadvantages. If the user selects *YES*, then he or she will be prompted with different language and issues to consider. Also, if the user selected *YES*, then later down the path, an alternative liability clause will be offered that is somewhat different from the scenario under which no material transfer takes place.

To illustrate, if the user clicks *YES* under material transfer, he or she will be offered options such as these:

1. Is the licensor transferring the material with certain claims of ownership on new inventions based on the transferred material?

   ☐ No, the licensor makes no claims on ownership of new inventions.

FIGURE 1: USER INTERFACE FOR THE SPECIFICATION OF LICENSEE, LICENSOR, AND LICENSE DETAILS

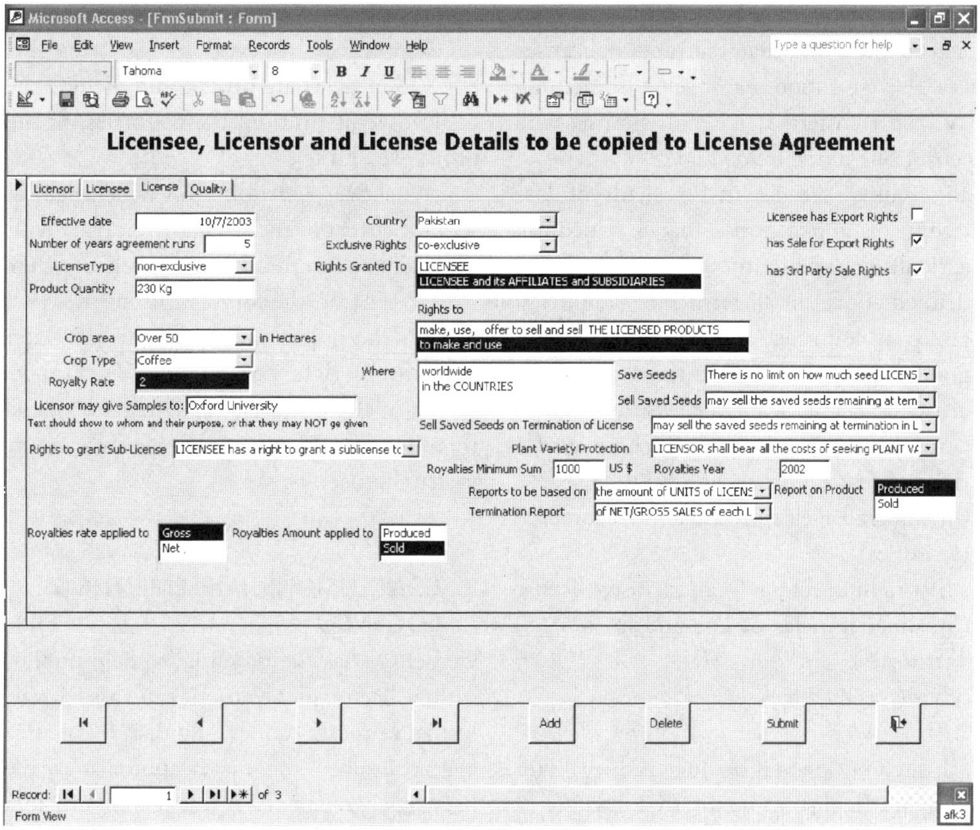

→ The software system will proceed to the next topic.

☐ Yes, the licensor does make some claim of ownership.

→ The software system will offer some of the options illustrated below:

First, the user will be offered some text about reach-through clauses, their utility, and their rationale, and information about how common such clauses are under different conditions. Basically, licensors want to ensure that if the licensee makes an improvement, the licensor is not prevented from using/licensing the improved licensed technology and benefiting from the improvements.

There are several levels of ownership a licensor may wish to exercise. Which one is chosen depends on the commercialization strategies of the licensor, including the symmetry of negotiations. Generally, three levels are typical (whether for commercial or humanitarian use). These will be listed, together with a blank field for the users to specify their own terms. For example:

Licensor gives the material, and if licensee improves the invention or invents something based on the transferred material, licensee will give licensor one of the following:

1. an exclusive license in all Fields of Use (crops or applications, that is, medical, agricultural, environmental, and so on, as defined above) in all territories (countries, group of countries, as defined above) and grant back a royalty-free nonexclusive license to licensee in Field of Use
2. a royalty-free nonexclusive license ***and*** a right of first refusal to an exclusive license (in some/all Fields of Use and in some/all territories)
3. a first right of refusal to an exclusive license (in some/all Fields of Use and in some/all Territories).
4. other (specified by user)

Each such option will be linked to legal language in plain English to be inserted into the license. For example, under 2. above, the following language would be inserted:

In consideration of Licensor's contribution of Materials (defined above), Licensee grants to Licensor a paid-up, worldwide, nonexclusive license to make, have made, use, have used, import, export, sell and have sold products and processes developed from Materials and an option to obtain a fee-bearing, worldwide, exclusive license to make, have made … (terms to exercise option to be defined; software will prompt user with a new screen on the ways in which such options can be exercised; depending on which is selected, the legal language and clause will be amended accordingly).

For number 3. above, the clause could read:

In consideration of Licensor's contribution of Materials (defined above), Licensee grants to Licensor an option, exercisable at any time up to two years after expiration or termination to obtain a royalty-bearing, worldwide, exclusive license with a right to grant sublicenses to Company affiliates and subsidiaries in the following Field of Use (defined where field of use refers to crops) in Territory (geographic region, limited or worldwide) or a combination thereof.

Other fields are diverse and include the type of licensee institution, the countries, or the type of license (Table 1).

As above, depending on which field is chosen, other text in the database template will automatically be inserted into the license agreement.

To generate the complete license in Microsoft® Word®, the user now presses the tab Submit at the bottom right corner. See Box 1 for an example of the output (see also the Appendix to this *Handbook* for a comprehensive commercial variety license).

4. CONCLUSIONS: IMPLEMENTING COGENCO

CoGenCo has the potential to help public institutions license plant varieties and associated intellectual property more easily than before. It offers a very flexible, pragmatic approach to drafting licensing agreements. A test version of CoGenCo and a preliminary user's guide, together with the draft license, are available to interested parties.

It will require running Microsoft® Access® on a Windows XP or higher system. The authors very much welcome comments and suggestions about the software and look forward to collaborating with interested parties to further develop CoGenCo into a comprehensive and widely available system. ■

ACKNOWLEDGMENTS
We are grateful to the the Rockefeller Foundation for a grant to *bio*Developments-International Institute, Interlaken, New York, that made this project possible. We would also like to thank many people (too numerous to mention) for their valuable suggestions and inputs during this initial consultation/feasibility phase.

ANATOLE KRATTIGER, *Research Professor, the Biodesign Institute at Arizona State University; Chair, bioDevelopments-International Institute; and Adjunct Professor, Cornell University, PO Box 26, Interlaken, NY, 14847, U.S.A. afk3@cornell.edu*

JOHN DODDS, *Founder, Dodds & Associates, 1707 N Street NW, Washington, DC, 20036, U.S.A. j.dodds@doddsassociates.com*

DONNA BOBROWICZ, *Technology Transfer Specialist, Loyola University Chicago, Stritch School of Medicine, 2160 S. First Avenue, Building 120, Room 400, Maywood, IL, 60153 U.S.A. dbobrowicz@lumc. edu*

1 See www.ipHandbook.org.

2 This means that the LICENSOR shall be the first party to which a worldwide exclusive license is offered. Only after the LICENSOR has refused from such a license may the LICENSEE offer the license to others.

Table 1: Options under License Type

License Type	Note
Exclusive	Exclusive license is a promise by the licensor not to practice under the licensed intellectual property and not to grant any further licenses.
Nonexclusive	Nonexclusive license ensures that the owner of the licensed intellectual property shall not sue the licensee with respect to acts done within the scope of the license. The licensor can grant several nonexclusive licenses to same intellectual property.
Coexclusive	Coexclusive license is otherwise similar to the exclusive license but the licensor retains rights to itself practice the intellectual property.

Box 1: Sample Noncommercial Variety Licensing Agreement

Underlined and bolded text means that these gaps will be filled in when completing the agreement using the software.

Italicized and bolded text means that these are one or more alternatives to be chosen depending on the parties, the circumstances, and so forth.

Bold indicates text that may not apply to given agreement.

Effective **day** of **month**, **insert year** (hereafter, the EFFECTIVE DATE) **full name of organization licensing out to the other**, having a principal place of business at **address** ________ (hereafter LICENSOR) and **full name of organization licensing in**, having a principal place of business at **address** ____________ (hereafter LICENSEE) agrees as follows:

I. PARTIES

LICENSOR being
a) a not-for-profit organization with the objective to ____________
b) a not-for-profit company in business of____________
c) a for-profit entity in business of ____________

and LICENSEE being
a) a small farmer____________
b) a farmer's association____________
c) a for-profit entity in business of ____________
d) a not-for-profit organization with the objective to ____________

have agreed to ____________***(for example, commercialize and produce seeds of the variety CCC)***

LICENSOR represents that it owns the rights to
patents
plant patents
trademarks
plant varieties
trade secrets
copyrights
in respect to which it is prepared to grant
nonexclusive
exclusive
coexclusive
license to the LICENSEE.

LICENSEE, wishes to acquire a license under selected
patents
plant patents
trademarks
plant varieties
trade secrets
copyrights
for purposes of **(for example, seed production, distribution, sale, to have sold, etc.)**

Box 1 (continued)

LICENSOR and LICENSEE are hereunder commonly referred to as PARTIES.

ARTICLE 1. DEFINITIONS

In this Agreement defined terms shall have the meanings set out below:

[optional] AFFILIATE of a PARTY means any person or legal entity that is a general licensee of such PARTY in the field of this agreement and that has a contract with such PARTY entitling it to receive continuing technical services from the PARTY, but any such person shall be deemed to be an affiliate only so long as it has such a contract and continues to be such a licensee.

COMMERCIAL SALES means the sales made by LICENSEE in the TERRITORY.

COOPERATIVE means an enterprise or organization jointly owned or managed by those who use its facilities or services.

[optional] COUNTRY means a country in which the LICENSEE makes EXPORT SALES. A list of COUNTRIES is attached as an integral part of this license agreement. Such list may be updated in writing by the parties from time to time by mutual agreement.

[optional] EXPORT SALES means the sales made by the LICENSEE in COUNTRIES.

FARMERS' ASSOCIATION means an organized body of farmers.

[optional] GROSS SALES means income at invoice values received for goods and services over a given period of time.

[optional] INVENTION means the invention, which is the subject matter of patents, PVP or any other form of Intellectual Property Protection.

INTELLECTUAL PROPERTY means the patents, copyrights, trademarks, design rights, data protection rights, plant variety rights and any other statutory rights for inventions, improvements, designs, and any other intellectual property rights in any territory of the world relating to the INVENTION.

KNOW-HOW means all information, data, results and know-how (including without limitation reports, notebooks, drawings, papers, documents, manuals and databases) but excluding MATERIAL.

LICENSED CROP means the crop or crops listed in Appendix I, initially derived from the plant variety XXXX.

LICENSED KNOW-HOW means KNOW-HOW relating to the INVENTION.

LICENSED
PATENTS
PLANT PATENTS
TRADEMARKS

(Continued on Next Page)

Box 1 (continued)

PLANT VARIETIES
TRADE SECRETS
means ______________________________________

MATERIAL means all forms of living and nonliving biological material including without limitation, strains, clones, antiserum, plants, parts of plants, cultivars, germplasm, genetic material, gene constructs, and microorganisms.

[optional] NET SALES means gross sales reduced by customer discounts, returns, freight out, and allowances

[optional] NONPROFIT CORPORATION means a corporation no part of the income of which is distributable to its members, directors, or officers. Corporation organized for other than profit-making purposes.

[optional] NONPROFIT ORGANIZATION means an organization for purposes other than generating profit, such as charitable, scientific, or literary organization.

[optional] PATENTS mean any and all patents (including but not limited to patents of implementation, improvement, or addition; utility model and appearance design patents; and inventors' certificates; as well as divisions, reissues, continuations, renewals, and extensions of any of these), applications for patent, and letters of patent that may issue on such applications.

[optional] UNIT OF PRODUCT EMBODYING THE INVENTION means **kg of seeds of the VARIETY or number of fruits of the VARIETY.**

PRODUCTS EMBODYING THE INVENTION means for example, fruit, seed or plant parts of the PLANT VARIETY.

PROPRIETARY GERMPLASM means Germplasm, which in the relevant TERRITORY or COUNTRIES is the subject of intellectual property protection owned or controlled by LICENSOR.

PRODUCTION COST means combined cost of raw material and labor incurred in producing seeds.

[optional] PVP means Plant Variety Protection; the protection of varieties as a form of exclusive ownership and use rights determined based on distinctness, uniformity, and stability of the Plant Material.

SAMPLES means any samples or copies of the MATERIAL distributed to third parties for testing purposes.

SMALL FARMER means a farmer

- *a) owning and operating a farm smaller than* the area *and growing* the crop *on at least* (percentage) % *of the area*
- *b) having yearly sell less than* $**amount**

[optional] SUBSIDIARY of a PARTY means any corporation over 50% of the voting stock of which is directly or indirectly owned or controlled by such a PARTY.

TECHNOLOGY means the INVENTION, LICENSED KNOW-HOW, LICENSED MATERIAL, and INTELLECTUAL PROPERTY.

(Continued on Next Page)

Box 1 (continued)

TERRITORY means the geographic territory of **name of the territory (for example, Uganda).**

[optional] VARIETY means plant variety as described in relevant certificate of Plant Variety Protection.

VARIETY NAME means the name.

ARTICLE 2. LICENSE GRANT

LICENSOR grants to LICENSEE an

a) exclusive/nonexclusive license to produce and use PRODUCTS EMBODYING THE INVENTION worldwide/throughout the TERRITORY and offer to sell and sell worldwide/ throughout the TERRITORY/in the COUNTRIES

and/or

b) an exclusive/nonexclusive license to produce and use worldwide/throughout the TERRITORY but not offer to sell and sell PRODUCTS EMBODYING THE INVENTION worldwide/throughout the TERRITORY

LICENSEE
has
has not
a right to export the PRODUCTS EMBODYING THE INVENTION to the COUNTRIES. LICENSEE
has
has not
a right to sell the PRODUCTS EMBODYING THE INVENTION to a third party exporting or aiming to export. LICENSEE
has
has not
a right to sell the PRODUCTS EMBODYING THE INVENTION through any third party, including any FARMER'S ASSOCIATION.

[optional] The license granted in this article is subject to a reserved nonexclusive license to the LICENSOR to produce, use, sell, offer for sale and import the PRODUCTS EMBODYING THE INVENTION.

ARTICLE 3. OBLIGATIONS OF THE PARTIES

[optional] 3.1. RIGHT TO SAVE SEEDS

a) There is no limit to how much seed LICENSEE may save
b) LICENSEE may save enough seed to plant his/her own farm holding
c) LICENSEE has no right to save seeds

[optional] If a or b was selected from above then 3.1.1. RIGHT TO SELL SAVED SEEDS

a) The saved seeds may not be sold without permission of LICENSOR
b) LICENSEE may sell the saved seed

(Continued on Next Page)

Box 1 (continued)

a) but only by VARIETY NAME
b) but only under the LICENSED TRADEMARK

When this agreement is terminated LICENSEE
a) may not sell the saved seeds/
b) may sell the saved seeds
a) but only by VARIETY NAME
b) but only under the LICENSED TRADEMARK
c) shall sell the seeds to the LICENSOR at the production cost

[optional] 3.1 RIGHT TO GIVE SAMPLES

[optional] The LICENSOR reserves a right to give SAMPLES to
a) any third party for <u>(e.g. research, testing)</u> purposes/
b) to <u>(e.g. research) institutes</u> to <u>(e.g. research)</u> purposes

3.3 RIGHT TO GRANT SUBLICENSES

a) LICENSEE has not a right to grant a sublicense to a third party/

b) LICENSEE has a right to grant a sublicense to a third party. LICENSEE has such a right only at such times, as it is not in material default with any of its obligations to LICENSOR under this agreement. Any such sublicense should be in writing and shall be accepted in writing by any such third party.

The operations of such third party shall be deemed to be the operations of LICENSEE, and LICENSEE shall account therefore and be primarily responsible for the performance by such third party of all of its obligations hereunder.

LICENSEE shall notify LICENSOR promptly in writing of any such sublicense.

Any sublicense granted by the LICENSEE shall be deemed to terminate upon termination of this Agreement terminates.

3.4. ADVERTISING, MARKETING AND PROMOTION COSTS

LICENSEE shall bear all costs associated with the advertising, marketing, and promotion of MATERIAL and TECHNOGIY covered by this license. **[optional] LICENSEE shall make reasonable efforts to share with LICENSOR details of such campaigns in advance of release.**

3.5. GOVERNMENT AND REGULATORY APPROVALS

LICENSEE shall be responsible for adhering to all laws and regulations and for obtaining and complying with all government and regulatory approvals, licenses, clearances and consents pertinent to or required to cover its activities under this agreement.

[optional] 3.5. PLANT VARIETY PROTECTION

LICENSOR shall bear all the costs of seeking PLANT VARIETY PROTECTION in TERRITORY and/or

(Continued on Next Page)

Box 1 (continued)

COUNTRIES when it is mutually agreed that the potential markets justify such costs.

3.6. INDEPENDENT ENTITIES

Each PARTY is acting as an independent entity. Nothing in this Agreement shall be construed so as to constitute a partnership or joint venture of any kind between the PARTIES hereto. This document merely serves to license MATERIAL and TECHNOLOGY from LICENSOR TO LICENSEE.

ARTICLE 4. ROYALTIES

4.1. RATE OF ROYALTIES

a) LICENSEE shall pay royalties to LICENSOR from COMMERCIAL SALES at the rate of number % of a) the gross sales/ b) net sales of the PRODUCT EMBODYING THE INVENTION and/or
b) LICENSEE shall pay royalties to LICENSOR from EXPORT SALES at the rate of number% of a) the gross sales/ b) net sales of the PRODUCT EMBODYING THE INVENTION and/or
c) LICENSEE shall pay royalties to the LICENSOR from COMMERCIAL SALES at the rate of US$ the amount per UNIT of a) PRODUCT EMBODYING THE INVENTION sold/ 2) PRODUCT EMBODYING THE INVENTION produced
d) LICENSEE shall pay royalties to the LICENSOR from EXPORT SALES at the rate of US$ the amount per UNIT of 1) PRODUCT EMBODYING THE INVENTION sold/ 2) PRODUCT EMBODYIN THE INVENTION produced.

[optional] In case the royalties paid do not aggregate a minimum of the sum US$ dollars for the year ending December 31, the year, and for each succeeding calendar year during the life of this agreement, LICENSEE will pay to LICENSOR, within thirty (30) days of the end of such year, the difference between the royalties actually paid under this Agreement for such year and such minimum sum.

4.2. REPORTING

a) LICENSEE agrees to a) report/ b) make written report to LICENSOR
- ***i.) once a year***
- ***ii) twice a year during the life of this Agreement stating in each such report the number and description of***
  - ***a. net***
  - ***b. gross sales of each PRODUCT EMBODYING THE INVENTION sold or otherwise disposed of during the preceding***
    - ***i. 12 months***
    - ***ii. 6 months***

b) LICENSEE agrees to report to LICENSOR once a year during the life of this Agreement the amount of UNITS of PRODUCTS EMBODYING THE INVENTION a) produced/b) sold.

LICENSEE agrees to make a written report to LICENSOR within thirty (30) days after the date of termination of this Agreement stating in such report the number and description
- ***a) of net/gross sales of each PRODUCT EMBODYING THE INVENTION sold or otherwise disposed***
- ***b) amount of UNITS of PRODUCTS EMBODYING THE INVENTION***
- ***c) amount of UNITS of PRODUCTS EMBODYING THE INVENTION produced***

and on which royalty is payable hereunder but that were not previously reported to LICENSOR.

(Continued on Next Page)

Box 1 (continued)

4.4. RECORD KEEPING

LICENSEE agrees to keep records showing the sales or other dispositions of the PRODUCTS EMBODYING THE INVENTION in sufficient details and further agrees to permit its books and record to be examined from time to time to the extent necessary to verify the reports provided above. Any costs of the examination of the books are due to the LICENSOR.

4.5 TERMINATION OF OBLIGATION TO PAY ROYALTIES

The obligation to pay royalties shall terminate when this Agreement terminates.

[optional] ARTICLE 5._

The PRODUCTS EMBODYING THE INVENTION and aimed to COMMERCIAL SALES or EXPORT SALES shall be
a) of high quality which is at least equal to comparable products produced and marketed by LICENSEE and in conformity with a standard SAMPLE approved by LICENSOR/
b) of the quality of certified seeds/
c) shall have germination percentage of at least ___(%)

If the quality of such PRODUCTS EMBODYING THE INVENTION falls below such quality as previously approved by LICENSOR, LICENSEE shall use its best efforts to restore such a quality. In the event that LICENSEE has not taken appropriate steps to restore such a quality within number days after notification by LICENSOR, LICENSOR shall have the right to terminate this Agreement.

Before selling PRODUCTS EMBODYING THE INVENTION, LICENSEE shall submit to LICENSOR, at no cost to LICENSOR and for approval as to quality, number sets of samples of the PRODUCTS EMBODYING THE INVENTION, which LICENSEE intends to sell and one (1) complete set of all promotional and advertising material associated therewith. Failure of LICENSOR to approve such samples within number working days after receipt hereof will be deemed approval. If LICENSOR should disapprove any SAMPLE, it shall provide specific reasons for such disapproval. Once such SAMPLES have been approved by LICENSOR, LICENSEE shall not materially depart therefrom without LICENSOR's prior express written consent that shall not be unreasonably withheld.

The LICENSEE agrees to permit LICENSOR or its representatives to inspect the facilities where the PRODUCTS EMBODYING THE INVENTION are being produced
and/
or
packaged.

[optional] ARTICLE 6. INVENTIONS

6.1. NOTIFICATION OF INVENTIONS, IMPROVEMENTS, OR DISCOVERIES

If during the term of this Agreement LICENSEE generates any INVENTION, improvement, or discovery that improves the MATERIAL or TECHNOLOGY, it shall notify LICENSOR immediately and the PARTIES shall meet to discuss the ownership and patenting of the NEW MATERIAL, TECHNOLOGY, or INVENTION, and if appropriate the TERRITORY and COUNTRIES in which such patent protection should be sought. Should such MATERIAL, TECHNOLOGY, or INVENTION be

(Continued on Next Page)

Box 1 (CONTINUED)

patentable LICENSOR will be granted a royalty-free worldwide nonexclusive commercial license thereunder including the right to sublicense for all applications and a first option[1] to negotiate worldwide exclusive access or all uses.

6.2. LICENSEE'S RIGHTS TO NEW INTELLECTUAL PROPERTY
In any event LICENSEE shall retain royalty bearing nonexclusive licenses for use
a) in the TERRITORY
b) in country
of any such intellectual property generated by LICENSEE arising during the term of this Agreement.

[optional] 6.3. LICENSEE'S OBLIGATIONS
LICENSEE shall not make or permit to be made by any employee, appointee, agent contractor, or otherwise any publication or results, or data arising under or in connection with this Agreement, nor disclose the existence or content of this Agreement without the prior written consent of LICENSOR.

ARTICLE 7. CONFIDENTIAL INFORMATION

Any information provided under this Agreement to LICENSEE, which LICENSOR considers confidential, will be provided in a written or oral form or in the form of a sample. LICENSEE agrees that it will treat such information and material confidential and will not divulge or provide such information and material to any third party. LICENSEE further agrees that it will not make any use of such information or material except as required or authorized by LICENSOR.

ARTICLE 8. TERMINATION OF THE AGREEMENT

[optional] In case royalties paid through December 31, **year** or any subsequent full calendar year do not equal or exceed minimum of **amount in letters** dollars **$amount in numbers.** LICENSOR may at its option terminate this Agreement and the license granted to LICENSEE by thirty (30) days' notice in writing to LICENSEE. Such termination shall not release LICENSEE from any liability or obligations to LICENSOR, which occurred on or prior to the date of such termination.

This Agreement may be terminated by either PARTY upon written notice to the other PARTY specifying a material breach by the other party of the provisions thereof. The nonbreaching PARTY may terminate this Agreement in the event the specified breach has not been cured within sixty (60) days after the written notice.

Unless earlier terminated, this agreement shall extend for **number of years)** years from the date of execution of this agreement.

ARTICLE 9. LIABILITIES

LICENSOR shall in no event be liable for damages, whether direct or otherwise, arising out of the use by LICENSEE or any third party of information or materials supplied hereunder.

In no event shall LICENSOR be liable for lost or prospective profits or special or consequential damages, whether or not LICENSOR has been advised of the possibility of the damages, nor for

(CONTINUED ON NEXT PAGE)

Box 1 (continued)

any claim by a third party against LICENSEE.

LICENSOR warrants that it is the sole owner of the (**describe the IP**) and that it has the right to grant licenses.

ARTICLE 10. APPLICABLE LAW

This agreement shall be governed by and construed according to the laws of **country or state.**

ARTICLE 11. VALIDITY

If any of the provisions of this Agreement are held to be invalid or unenforceable, the PARTIES will attempt to replace them with new provisions, which have the same force and effect and the remaining provisions shall not be affected.

ARTICLE 12. DISPUTE RESOLUTION

In the event a dispute shall arise between the PARTIES to this Agreement, the PARTIES agree to participate in at least four (4) hours mediation in accordance with the mediation rules of ______ ________. The PARTIES agree to share equally the costs of the mediation. In case the PARTIES are unable to resolve the dispute in mediation they agree to submit the dispute to

- ***a) final and binding arbitration under the arbitration rules of ______,*** **[optional] and the judgment upon the award rendered by the Arbitrator(s) may be entered into any court having judgment thereof.** ***The PARTIES agree to share equally the costs of the arbitration.***
- ***b) court decision***

IN WITNESS WHEREOF, the PARTIES hereto have caused this Agreement to be executed by their duly authorized representatives as of the dates below.

____________________ For	____________________ Date
____________________ For	____________________ Date

a This means that the LICENSOR shall be the first party to which a worldwide exclusive license is offered. Only after the LICENSOR has refused from such a license may the LICENSEE offer the license to others.

Trade Secrets and Trade-Secret Licensing

KARL F. JORDA, *David Rines Professor of Intellectual Property Law and Industrial Innovation and Director, Kenneth J. Germeshausen Center for the Law of Innovation and Entrepreneurship, Franklin Pierce Law Center, U.S.A.*

ABSTRACT

Exploiting the overlap between intellectual property (IP) categories, especially between patents and trade secrets, is an important facet of IP management. Patents (which require full disclosure) and trade secrets (which are kept confidential) are not incompatible. On the contrary, they can complement one another: patents protect inventions and trade secrets protect collateral know-how. Using patent and trade-secret protection together in a synergistic manner results in a potent exclusivity. Moreover, as licensing has become the preferred instrument for technology transfer, most technology licenses are hybrids, covering both patents and trade secrets. This situation has evolved because licenses that cover patents but do not allow access to collateral know-how usually do not permit patented technology to become commercialized. Despite the ease of obtaining trade-secret protection—immediate efficacy and low cost—this type of IP protection is too often neglected.

1. INTRODUCTION

The term *trade secret* refers to information that is maintained in secrecy and has commercial value. World Trade Organization (WTO) treaties (General Agreement on Trade and Tariffs [GATT] and the Agreement on Trade-Related Aspects of Intellectual Property Rights [TRIPS]), which have 150 nation-signatories, protect trade secrets. The following is an excerpt, addressing the concept of trade secrets, from the TRIPS Agreement:

> *Natural and legal persons shall have the possibility of preventing information lawfully within their control from being disclosed to, acquired by, or used by others without their consent in a manner contrary to honest commercial practices so long as such information:*
> *(a) is secret in the sense that it is not, as a body or in the precise configuration and assembly of its components, generally known among or readily accessible to persons within the circles that normally deal with the kind of information in question;*
> *(b) has commercial value because it is secret; and*
> *(c) has been subject to reasonable steps under the circumstances, by the person lawfully in control of the information, to keep it secret.*[1]

If national legislation is not already in compliance, all WTO countries must adopt this treaty provision. Although the provision eschews the actual term *trade secret,* it certainly refers to what are commonly known as trade secrets and follows the definition of the American Uniform Trade Secrets Act (UTSA) of 1985, cited below (section 2). The language of the North American Free Trade Agreement (NAFTA), binding upon the Canada, Mexico, and the United States also conforms closely with the definitions in the UTSA.

2. DEFINING TRADE SECRET

The UTSA, now in force in 45 U.S. states, defines *trade secret* as follows:

Jorda KF. 2007. Trade Secrets and Trade-Secret Licensing. In *Intellectual Property Management in Health and Agricultural Innovation: A Handbook of Best Practices* (eds. A Krattiger, RT Mahoney, L Nelsen, et al.). MIHR: Oxford, U.K., and PIPRA: Davis, U.S.A. Available online at www.ipHandbook.org.

> *A trade secret is any information, including a formula, pattern, compilation, device, method, technique, or process, that: (i) derives independent economic value, actual or potential, from not being generally known to, and not being readily ascertainable by proper means by, other persons who can obtain economic value from its disclosure or use, and (ii) is the subject of efforts that are reasonable under the circumstances to maintain its secrecy.*[2]

The most widely used definition, from 1929, of *trade secret* is found in the Restatement of Torts.[3] It reads:

> *A trade secret may consist of any formula, pattern, device or compilation of information which is used in one's business, and which gives him [or her] an opportunity to obtain an advantage over competitors who do not know or use it. It may be a formula for a chemical compound, a process of manufacturing, treating or preserving materials, a pattern for a machine or other device, or a list of customers.*[4]

In applying this 1929 definition to determine whether trade secrets exist, courts have relied on the following criteria:

- extent to which the information is known outside of the business
- extent to which it is known by employees and others involved in the business
- measures taken to guard the secrecy of the information
- value of the information to the business and to competitors
- amount of effort or money expended in developing the information
- ease or difficulty with which the information could be properly acquired or duplicated by others

The most recent and, in this author's view, the broadest and best definition of *trade secret* is set forth in Restatement (Third) of Unfair Competition:[5]

> *A trade secret is any information that can be used in the operation of a business or other enterprise and that is sufficiently valuable and secret to afford an actual or potential economic advantage over others.*

This definition most likely will eventually replace the earlier definitions. As of 1996, the Economic Espionage Act (EEA), a federal criminal trade-secret statute, includes the following definition:

> (A) *The term trade secret means all forms and types of financial, business, scientific, technical, economic, or engineering information, including patterns, plans, compilations, program devices, formulas, designs, prototypes, methods, techniques, processes, procedures, programs, or codes, whether tangible or intangible, and whether or how stored, compiled, or memorialized physically, electronically, graphically, photographically, or in writing if—*
>
> (B) *the owner thereof has taken reasonable measures to keep such information secret; and the information derives independent economic value, actual or potential, from not being generally known to, and not being readily ascertainable through proper means by, the public.*

3. WHAT IS AND WHAT IS NOT A TRADE SECRET

The definitions included above provide a fairly clear picture of what constitutes a trade secret. At the most basic level, a trade secret is simply information and knowledge. More specifically, it is any proprietary technical or business information, often embodied in inventions, know-how, and show-how. The definitions roughly agree on three requirements that must be met for enforceable trade secrets to exist. The proprietary information must be:

1. secret, in the sense that it is not generally known in the trade
2. valuable to competitors that do not possess it
3. the subject of reasonable efforts to safeguard and maintain it in secrecy

There are critical limitations on trade secrets and pitfalls in trade-secret enforcement and litigation. The requirement to maintain secrecy is a frequent pitfall. Moreover, any information that is readily ascertainable, or is derived from the personal skills of employees, cannot be considered an enforceable trade secret.

Trade secret protection applies not just to manufacturing processes, early stage inventions, and subpatentable innovations, as is sometimes believed. Patentable inventions can be considered trade secrets; this was made clear in the Supreme Court decision in *Kewanee Oil v. Bicron*, which recognized trade secrets as perfectly viable alternatives to patents.[6] In holding that state trade-secret law is not preempted by the federal patent law, the court tellingly held:

> *Certainly the patent policy of encouraging invention is not disturbed by the existence of another form of incentive to invention. In this respect, the two systems are not and never would be in conflict… . Trade secret law and patent law have coexisted in this country for over one hundred years. Each has its particular role to play, and the operation of one does not take away from the need for the other… . We conclude that the extension of trade-secret protection (even) to clearly patentable inventions does not conflict with the patent policy of disclosure.*

Since the essence of the patent system is the public disclosure of inventions, it is sometimes suggested that keeping inventions secret is wrong. This is a serious misconception. The decision in *Dunlop Holdings v. Ram Golf* made clear that the public benefits from trade secrets. Trade secrets generally do not suppress economic activity, because employees, suppliers, licensees, and others are given access to the necessary information.[7] Additionally, given the high incidence of employee mobility and inadvertent or deliberate leakage, many trade secrets dissipate within a few years. Possible reverse engineering and analysis of products are additional ways that trade secrets may dissipate or become compromised. In other words, trade secrets are secret only in a limited legal sense.

Contrary to conventional wisdom, trade-secret protection can be used in conjunction with patents to protect the tremendous volume of associated know-how that exists for any patentable invention but that cannot be disclosed in a patent specification.

It is useful, also, to specify the use of the terms *know-how* and *trade secret*. While the key requirement of a trade secret is secrecy, know-how does not necessarily require or imply secrecy, as can be seen from the following definitions:

- the knowledge and skill required to do something correctly.[8]
- information that enables one to accomplish a particular task or to operate a particular device or process.[9]
- knowledge and experience of a technical, commercial, administrative, financial or other nature, which is practically applicable in the operation of an enterprise or the practice of a profession.[10]

Know-how is not protectable as an IP right. Know-how acquires trade-secret status only if it is secret and has economic value and if measures are in place to secure its secrecy. Know-how is intellectual property, however, and is protected if it qualifies as a trade secret. Since we do not speak of "invention and patent licenses," it is likewise inappropriate to refer to "know-how and trade-secret licenses."

4. HISTORY OF TRADE SECRETS

Trade secret law is the oldest form of IP protection. In ancient Rome, trade secret laws established legal consequences for a person who induced another's employee (or slave) to divulge secrets relating to the master's commercial affairs. Trade secrecy was practiced extensively in Medieval European guilds. Modern trade-secret law, however, evolved in the early 19th century, in England, in response to the growing accumulation of technology and know-how and the increased mobility of employees. In 1868, a Massachusetts court held, in *Peabody v. Norfolk,* that a secret manufacturing process was considered property, and was protectable against misappropriation, and that a secrecy obligation for an employee outlasted the term of employment. The decision also held that a trade secret can be disclosed confidentially to others who need to practice it, and that a recipient can be enjoined from using a misappropriated trade secret. *Peabody v. Norfolk* clearly anticipated the main features of our present trade-secret system, and by the end of the 19th century the principal aspects of contemporary law were well established. [11]

5. IMPORTANCE OF TRADE SECRETS

Trade secrets are the crown jewels of corporations. Indeed, trade secrets are now even more relevant than they were a few decades ago as a tool for protecting innovation, and the stakes involved in their protection are getting higher. Injunctions are now a greater threat in trade-secret misappropriation cases than only a decade ago, and damage awards have been in the hundreds of millions of U.S. dollars in recent years. In a recent trial in Orlando, Florida, two businessmen were seeking US$1.4 billion in damages from the Walt Disney Company, accusing them of stealing trade secrets for use in a Walt Disney World sports complex. The jury awarded the businessmen US$240 million.[12] In another recent case, Cargill, Inc. was found to have misappropriated genetic-corn-seed trade secrets belonging to then Pioneer Hi-Bred International, Inc., and was forced to pay US$300 million. In another instance, Lexar won US$465.4 million in damages from Toshiba for misappropriation of controller technology that enabled a memory chip to communicate with its host device.[13]

Mark Halligan recently proclaimed, "*Trade secrets are the IP of the new millennium and can no longer be treated as a stepchild.*" James Pooley concurred, "*Forget patents, trademarks and copyrights ... trade secrets could be your company's most important and valuable assets.*"[14] Henry Perritt[15] said trade secrets are "*the oldest form of IP protection,*" and that, "*patent law was developed as a way of protecting trade secrets without requiring them to be kept secret and thereby discouraging wider use of useful information.*" This interpretation makes patents a supplement to trade secrets, rather than the other way around.

In fact, according to a 2003 survey on strategic IP management sponsored by the Intellectual Property Owners Association (IPO), patents are rarely viewed as an IP panacea, but rather as a supplement to other forms of IP protection.[16] Patents have limits, such as early publication, invent-around feasibility, and strict patentability requirements. Survey respondents did rate proprietary technology highly as a key source of competitive advantage, and a large majority of respondents (88%) cited skills and knowledge as the most important intellectual assets. Trade secrets are therefore directly implicated in the protection of proprietary skills and knowledge.

Moreover, patents are only the tips of icebergs in an ocean of trade secrets. Over 90% of all new technology is covered by trade secrets. And over 80% of all license and technology transfer agreements cover proprietary know-how (trade secrets) or are hybrid agreements covering both patents and trade secrets. Bob Sherwood, an international IP consultant, calls trade secrets the "*workhorse[s] of technology transfer.*"

Finally, and very importantly, trade-secret protection operates without delay and without undue cost, while patents are territorial, expensive to obtain, and can be acquired only in certain countries.

6. TRADE SECRET CHARACTERISTICS

From the above trade-secret definitions, we can understand the following salient characteristics of trade secrets and how they differ substantially from other types of IP rights.

For trade secrets, there is no subject matter or term limitation, registration or tangibility requirement. Furthermore, there is no strict novelty requirement, and trade-secret protection obtains as long as the subject matter is not generally known or available.

What *does* matter is secrecy—that the information is not known by outsiders. And maintaining secrecy requires reasonable affirmative measures to safeguard it. Such measures might include:

- stipulating in writing a trade-secret policy
- informing employees of the trade-secret policy
- having employees sign employment agreements with confidentiality obligations

- restricting access to trade-secrets (on a need-to-know basis)
- restricting public accessibility and escorting visitors
- locking gates and cabinets to sites that house trade secrets
- labeling trade-secret documents as proprietary and confidential
- screening the speeches and publications of employees
- using secrecy contracts in dealing with third parties
- conducting exit interviews with departing employees

It is important to consider that while sufficient economic value or competitive advantage is significant, the proper touchstone for a trade secret is not *actual use* but only *value to the owner*. This means that negative R&D results can give a competitive advantage (just as positive results can), in that the owner of the information has a greater knowledge of what are, and *what are not*, feasible and/or viable options for further commercialization. If competitors become privy to what is not feasible, by sidestepping known blind alleys, their R&D activities can accelerate, and any strategic or competitive advantage originally held by the owner will diminish.

Finally, the misappropriation of trade secrets is actionable if the secrets were acquired improperly, if a trade secret that was acquired improperly is either used or disclosed, or if an individual violates a duty to maintain confidentiality. A trade secret is acquired by *improper means* if it was obtained through theft, bribery, misrepresentation, breach or inducement of a breach of a duty to maintain secrecy, or through espionage, including electronic espionage. Remedies for misappropriation of trade secrets include actual and punitive damages, profits, reasonable royalties, and injunctions. The *proper means* of acquiring a trade secret (which do not support a claim for misappropriation) include independent discovery, reverse engineering, chemical analysis, or discovery from observing what has been allowed to enter the public domain.

7. INTEGRATION OF IP RIGHTS

Literature and presentations on IP strategies, IP valuation, and other IP topics almost always address patents and patent portfolios. This focus on patents, however, overlooks the fact that legal protection of innovations of any kind, especially in high-tech fields, requires the use of more than one IP category. This overlap assures dual or multiple protections.

Jay Dratler, in his *Intellectual Property Law: Commercial, Creative, and Industrial Property*, was the first to "*tie all the fields of IP together.*" According to Dratler, IP rights, formerly fragmented by specialties, are now a "*seamless web*" due to progress in technology and commerce.[17] Six years later in 1997, the authors of *Intellectual Property in the New Technological Age* also stressed the need to "*avoid the fragmented coverage ... by approaching IP as a unified whole*" and by concentrating on the "*interaction between different types of IP rights.*"[18] Today, we have a unified theory of IP management, a single field of law with subsets, and a significant overlap between IP fields. Several IP rights are available for the same IP or for different aspects of the same IP. Not taking advantage of the overlap misses opportunities, and, according to Dratler, amounts to a kind of "*malpractice.*"

Especially for high-tech products, trademarks and copyrights can supplement patents, trade secrets, and mask works ("blueprints" used in the R&D and production of semiconductor chips). One IP category, often patents, may be the "center of gravity" in certain instances. Other IP rights categories are then supplemental but equally valuable. The supplemental forms of IP may function to:

- cover additional subject matter
- strengthen exclusivity
- invoke additional remedies in litigation
- provide a backup if a primary IP right becomes invalid, thus providing synergy and optimal legal protection

Dratler provides the following examples:

a) Multiple protection for a data processing system can involve:
   - patented hardware and software

- patented computer architecture on circuit designs
- trade-secret production processes
- copyrighted microcode
- copyrighted operating system
- copyrighted instruction manual
- semiconductor chips protected as mask works
- consoles or keyboards protected by design patents, or as trade dress under trademark principles
- trademark registration

b) Multiple protection for a diagnostic kit involving monoclonal antibodies:
  - product patent on the test kit
  - process patent on the preparation of the antibodies
  - trade secrecy for production know-how
  - copyright for test kit's instructions
  - trademark

Even these examples are somewhat limited, because trade secrets can protect not only know-how and processes, but also large amounts of collateral data, information, and other know-how that are not found in patent specifications.

Other valuable examples:

c) Multiple protection of aesthetic designs:
  - patent
  - copyright for separable features
  - trademark for nonfunctional features
  - trade dress for overall appearance
  - utility patent for functional features
  - trade secrets for collateral and collateral know-how and data

d) Multiple protection for plants and plant parts:
  - plant patents
  - plant variety protection (PVP) certificates
  - utility patents
  - trade secrets[19]

To encapsulate the IP integration concept, numerous practitioners recommend to clients to do the following:

- exploit the overlap
- develop a fall-back position
- create a web of rights
- build an IP estate
- build a "wall"
- overprotect (multiple layers of IP rights protection)
- lay a "minefield"

The most important IP management and technology licensing strategy is to exploit the overlap between patents and trade secrets.

8. INITIAL PATENT/TRADE-SECRET EVALUATION

IP management always requires deciding during development between seeking patent protection and maintaining trade secrecy. The Initial Patent/Trade Secret Evaluation Questionnaire (Box 1) can be used to facilitate the decision and to help determine the center of gravity (often patents for products and trade secrets for processes).[20] To avoid the implications of the term *invention* and to cover the wide variety of innovations that may be addressed by this questionnaire, the term *development* is used generically.

The 11 questions are arranged by function, not importance, and roughly correspond to marketing (questions 1–4), technical (questions 5–8), and legal (questions 9–11) categories. Each question should be answered on a scale from 1 to 10. The responses are then totaled. With the current number of questions, the total would range from 11 to 110. If the sum approaches the higher end of the scale (above 75), trade-secret protection would seem favorable; a sum at the lower end (below 45) would suggest that patent protection would be more advantageous. At times, values in the middle range (45–75) will result. Such a score suggests that it doesn't really matter which approach is followed initially. For example, trade-secret protection might be appropriate for manufacturing-process technology, which competitors might find easier to re-create; patents make sense for products that can be analyzed or reverse engineered. However, there need be no prejudice about resorting to the other strategy to protect collateral aspects and improvements.

Box 1: Initial Patent/Trade Secret Evaluation Questionnaire

1) Is the development likely to be a commercial product or the subject of licensing?
1 2 3 4 5 6 7 8 9 10
Likely — Unlikely

2) How much of a competitive advantage would be provided if the company maximized exclusivity?
1 2 3 4 5 6 7 8 9 10
Very Great — Very Little

3) How much of a competitive disadvantage would it be if a competitor obtained exclusivity?
1 2 3 4 5 6 7 8 9 10
Very Great — Very Little

4) Is it likely the commercial significance of the development would be limited in time?
1 2 3 4 5 6 7 8 9 10
Yes-Limited — No

5) Is it likely one could develop alternatives ("design around")?
1 2 3 4 5 6 7 8 9 10
Unlikely — Likely

6) Can the nature of development be ascertained from commercial product (could the product be "reverse engineered")?
1 2 3 4 5 6 7 8 9 10
Likely — Unlikely

7) Would disclosure of this development require or permit access to other, unprotectable information?
1 2 3 4 5 6 7 8 9 10
No — Yes

8) Is it likely others will independently arrive at the same development?
1 2 3 4 5 6 7 8 9 10
Likely — Unlikely

9) If a patent was obtained, what are the chances of validity being upheld by a court?
1 2 3 4 5 6 7 8 9 10
High — Low

10) Is it likely that dissemination of the development from within the company would be difficult to control?
1 2 3 4 5 6 7 8 9 10
Yes-Difficult — Not Difficult

11) Would it be difficult to determine if competitors are using the development?
1 2 3 4 5 6 7 8 9 10
Not Difficult — Difficult

Total Score ______

To obtain the most-accurate results from the questionnaire, the following considerations for each question will be helpful in interpreting the survey responses.

Question 1. If the development is likely to be commercialized or licensed, patent protection would seem preferable to trade-secret protection. There might be some exceptions (such as the Coca-Cola® situation), but presumably these would be limited to situations where the nature of the product could not be easily ascertained by reverse engineering (see Question 6).

Note that Question 1 pertains to commercialization of the development itself. Thus the mere use of a process to produce a commercial product is not commercialization of the process (see Question 4, about commercial significance). The desirability of patenting the process itself would depend on the answers to Questions 2–11.

Question 2. Here the aim is to ascertain whether exclusivity on the development would be meaningful commercially. A development of marginal commercial importance might be better kept as a trade secret. One that provided a significant commercial edge, however, probably should be patented.

Question 3. This addresses the opposite of the issue in Question 2, namely the defensive value of a patent publication. Hence, while the development may be of minimum commercial advantage to the company, thereby favoring trade secrets, a patent (or publication) should be considered if a competitor's exclusivity would be disadvantageous.

Question 4. This is a difficult question. Some writers have suggested that a product with a short commercial life favors a patenting approach, while a long life favors trade secrets. In this author's view, life span is not a particularly useful criterion since it depends on factors unrelated to the development itself. Estimating the future lifespan for a product under development may also be a highly subjective matter. In some circumstances this question might not have to be considered.

Question 5. The ability to design around an invention is a function of the nature of the patent protection. If a claim is easily avoided, its value is considerably reduced. The destructive effect of trade-secret protection by publication is therefore unchanged, and the relative value of the trade-secret option is higher (because of the decreased value of patent protection).

Question 6. Counterbalancing Question five is the issue of whether, if the trade-secret route is chosen, a competitor will nevertheless be able to ascertain the nature of the development from the product. If competitors can reasonably easily ascertain the nature of the product, patent protection would be favored.

Question 7. The issue of disclosure is often overlooked. For example, the required disclosure of a culture collection-deposit number could provide competitors with access to the culture itself, and this access might greatly outweigh the value of patent protection. The impact of a disclosure of an unclaimed or intermediate process might also have a bearing on whether the final product should be patented.

Question 8. In many cases, evaluating whether others could arrive at the same development independently could be extremely difficult. If, however, it is known that others are working in the field, it would seem quite possible that they could arrive at the same development and patent it first. Consequently, one might eventually be excluded from using the product if patent protection is not sought.

Question 9. Even though patent protection might be indicated for other reasons, this could be counterbalanced by the fact that any coverage eventually obtained would be weak. A weak patent, ignored by competitors and for which the company is unwilling to sue, is as good as no patent. In fact, it may be worse, since the opportunity for trade-secret protection would have been irrevocably lost through publication.

Question 10. Ideally, the dissemination of information from within the company can be controlled. If not, however, a trade secret might be lost. If this risk exists, for example when numerous employees, visitors, and suppliers have access to the development, patent protection is more attractive. The same question arises with scientific publications.

Question 11. This question is related to question nine but goes to the issue of inherent enforceability rather than patent strength. If detecting infringement would be extremely difficult, the ultimate value of a patent would be reduced. Such reduced value must be weighed against the cost of the loss of trade-secret protection caused by patent publication. If the patent rights cannot be effectively enforced, then what ensues may become a de facto release of a trade secret.

9. THE PATENT/TRADE SECRET INTERFACE

Trade secrets are the first line of defense, but they not only come before patents but can go with patents and even follow patents (see sections 11 and 12, below). Moreover, as a practical matter, licenses under patents without access to associated or collateral know-how are often not enough for taking advantage of the patented technology commercially. This is because patents rarely disclose the ultimate scaled-up commercial embodiments. Data and know-how, therefore, are immensely important. In this regard, consider the following persuasive comments:

- *In many cases, particularly in chemical technology, the know-how is the most important part of a technology transfer agreement.*[21]
- *Acquire not just the patents but the rights to the know-how. Access to experts and records, lab notebooks, and reports on pilot-scale operations, including data on markets and potential users of the technology are crucial.*[22]
- *It is common practice in industry to seek and obtain patents on that part of a technology that is amenable to patent protection, while maintaining related technological data and other information in confidence. Some regard a patent as little more than an advertisement for the sale of accompanying know-how.*[23]
- *[In technology licensing] related patent rights generally are mentioned late in the discussion and are perceived to have 'insignificant' value relative to the know-how.*[24]
- *Trade secrets are a component of almost every technology license... [and] can increase the value of a license up to three to ten times the value of the deal if no trade secrets are involved.*[25]

A very striking case about the importance of proprietary know-how comes from Brazil. Brazilian officials learned a quick and startling lesson when they decided, some years ago, to translate important patents that issued in developed countries into Portuguese for the benefit of Brazilian industry. They believed that this was all that was necessary to enable their industries to practice these foreign inventions without paying royalties for licenses. Needless to say, without access to the necessary know-how, this scheme was an utter failure. This oversight is somewhat surprising, since Brazil, following the amazing progress and successes of the Asian tigers, had years earlier begun a project of importing technology (including know-how) from developed countries to be adapted and improved for local needs. They expected that the cost of importing the technology would be money well spent. And, in fact, importing the technologies led not only to exports of improved products, but also to exports of the resulting improved technology to developing countries in Africa, the Middle East, and the rest of Latin America. Such an importation/exportation policy is termed reverse technology transfer.[26]

To reiterate, patents and trade secrets are not mutually exclusive but actually highly complementary and mutually reinforcing. This is partly why the U.S. Supreme Court has recognized trade secrets as perfectly viable alternatives to patents: "*The extension of trade-secret protection to clearly patentable inventions does not conflict with the patent policy of disclosure.*"[27] Interestingly, in his concurring opinion in the Kewanee Oil[28] decision, Justice Marshall was *"persuaded"* that *"Congress, in enacting the patent laws, intended merely to offer inventors a limited monopoly [*sic*] in exchange for disclosure of their inventions [rather than] to exert pressure on inventors to enter into this exchange by withdrawing any alternative possibility of legal protection for their inventions.*" Thus, it is clear that patents and trade secrets can not only coexist but are also in harmony with each other. "*[T]rade-secret/patent coexistence is well-established, and the two are in harmony because they serve different economic and ethical functions.*"[29]

In fact, patents and trade secrets are inextricably intertwined, because the bulk of R&D data and results, and of associated collateral know-how for any commercially important innovation, cannot, and need not, be included in a patent application. Such information deserves, and requires, the protection that trade secrets can provide. In the past, and sometimes still today, if trade-secret maintenance is contemplated (for example, for a manufacturing process technology) the question is always phrased as a choice between patents and trade secrets. For example, titles of articles discussing the matter read, "Trade Secret vs. Patent Protection"; "To Patent or Not to Patent?"; "Trade Secret or Patent?"; and "To Patent or to Padlock?" This perspective imagines that patents and trade secrets are substantially different in terms of duration and scope of protection and have clearly perceivable advantages and disadvantages. However, as this chapter has demonstrated, the perceived differences are illusory. The life of a patent is roughly 20 years from filing, and an average trade secret may last but a few years. Nor do they differ in regard to the scope of protection, since virtually everything produced with human ingenuity is potentially patentable. And while a patent protects against independent discovery and a trade secret does not, a patent can lead competitors to attempt to design or invent around it. A properly guarded and secured trade secret, however, may withstand attempts to crack it.

10. HOW PATENTS AND TRADE SECRETS ARE COMPLEMENTARY

It is unnecessary and, in fact, shortsighted to choose one IP strategy over another. Indeed, the question is not so much whether to patent or to padlock, but rather what to patent and what to keep a trade secret. Of course, it may be best to both patent and padlock, thus integrating patents and trade secrets for the optimal, synergistic protection of innovation.

It is true that patents and trade secrets are opposed on the issue of disclosure. Information that is disclosed in a patent is no longer a trade secret. But patents and trade secrets are indeed complementary, especially under the following circumstances. In the critical R&D stage, before any patent applications are filed and before applications are published and patents issued, trade-secret law dovetails very nicely with patent law.[30] If an invention has been fully described so as to enable a person skilled in the art to make and use it, and if the best mode for carrying out the invention, if available, has been disclosed (as is required in a patent application), all associated or collateral know-how not divulged can, and should, be retained as a trade secret. All of the massive R&D data—including data pertaining to better modes developed after filing, whether or not inventive—should also be maintained as trade secrets, if the data is not disclosed in subsequent applications. Complementary patenting and padlocking is tantamount to having the best of both worlds, especially when technologies are complex and consist of many patentable inventions and volumes of associated know-how.

11. BEST MODE AND ENABLEMENT REQUIREMENTS

The conventional wisdom is that, because of best mode and enablement requirements, trade secret protection cannot coexist with patent protection. This, also, is a serious misconception. These requirements apply only at the *time of filing*, only to the *knowledge of the inventor(s)*, and only to the *claimed* invention.

Patent applications are filed early in the R&D stage to get the earliest possible filing or priority date. The patent claims tend to be narrow in order to achieve distance from prior art. Therefore, the specification normally describes rudimentary lab experiments or prototypes in only a few pages; the best mode for commercial manufacture and use are developed later. The best mode and the enablement requirements are thus no impediments to maintaining, as trade secrets, the mountains of collateral know-how developed after filing.

The recent decision in *CFMT v. Yieldup International* is particularly germane to this point: "*Enablement does not require an inventor to meet lofty standards for success in the commercial marketplace. Title 35 does not require that a patent disclosure enable one of ordinary skill in the art to make and use a perfected, commercially viable embodiment absent a claim limitation to that effect ... [T]his court gauges enablement at the date of the filing, not in light of later developments.*"[31] Such reasoning applies equally well to the best mode requirement.

In Peter Rosenberg's opinion, "*patents protect only a very small portion of the total technology involved in the commercial exploitation of an invention ... Considerable expenditure of time, effort, and capital is necessary to transform an (inventive concept) into a marketable product.*"[32] In the process, he adds, valuable know-how is generated, which, even if inventive and protectable by patents, can be maintained as trade secrets. Rosenberg asserts that there is "*nothing improper in patenting some inventions and keeping others trade secrets.*" Likewise, Tom Arnold asserts that it is "*flat wrong*" to assume, as "*many courts and even many patent lawyers seem prone*" to do, that "*because the patent statute requires a best mode disclosure, patents necessarily disclose or preempt all the trade secrets that are useful in the practice of the invention.*"[33]

Gale Peterson also emphasizes that "*the patent statute only requires a written description of the* claimed *invention and how to make and use the* claimed *invention.*" He therefore advises that, since allowed claims on a patentable system usually cover much less than the entire scope of the system, the disclosure in the application be limited to that necessary to support the claims in a 35 U.S.C. §112 sense (that is, having sufficient information to enable one to make and use the invention) and that every effort be taken to maintain the remainder of the system as a trade secret.

In short, manufacturing-process details, even if available, are not a part of the statutorily required best mode and enablement disclosure of a patent, and it is in this process area where "best modes" for scale-up toward actual production very often lie.

12. EXEMPLARY TRADE SECRET CASES

Of course, it goes without saying that technical and commercial information and collateral know-how that can be protected with trade secrets cannot include information that is generally known, readily ascertainable, or constitutes personal skill. But this exclusion still leaves masses of data and know-how that are protectable as trade secrets—and often also with additional improvement patents. For example, GE's industrial-diamond-process technology is an excellent illustration of the synergistic integration of patents and trade secrets to secure invulnerable exclusivity.

The artificial manufacture of diamonds for industrial uses was very big business for GE, and they had the best proprietary technology for making these diamonds. GE patented much of its technology, and when the patents expired, much of the technology was in the technical literature and in the public domain. But GE also kept certain distinct inventions and developments secret. The Soviet Union and a Far East country were very interested in obtaining licenses to this technology, but GE refused to license to anyone. After getting nowhere with GE, the Far East interests resorted to industrial espionage. A trusted fast-track star performer at GE, a national of that country, was enticed with million dollar payments to spirit away GE's precious trade secrets. The employee was eventually caught, tried and jailed.

Similarly, Wyeth has had an exclusive position on Premarin®, the high-selling hormone-therapy drug, since 1942. Their patents on the manufacturing process (starting with pregnant mares' urine) expired decades ago, but the company also held closely guarded trade secrets. On behalf of a pharmaceutical company that had been trying to come out with a generic form of Premarin® for 15 years, Natural Biologics stole the Wyeth trade secrets. Wyeth sued, prevailed, and got a sweeping injunction, as this was clearly an egregious case of trade-secret misappropriation.

These cases illustrate the value of trade secrets and, more importantly, the merits of marrying patents with trade secrets. Indeed, these cases show that GE and Wyeth could have the best of both worlds, patenting their inventions and still keeping their competitive advantage by maintaining production details in secrecy. Were GE's or Wyeth's policies to rely on trade secrets in this manner or was Coca Cola's decision to keep its formula a secret rather than to patent it, unwise and careless? Clearly not.

Other recent decisions, such as *C&F Packing v. IBP and Pizza Hut* and *Celeritas Technologies v. Rockwell International*, demonstrate that dual or multiple IP protection is not only possible but essential to exploit the IP overlap and provide a fallback.[34]

In the *Pizza Hut* case, for instance, Pizza Hut was made to pay US$10.9 million to C&F for misappropriation of trade secrets.[35] After many years of research, C&F had developed a process for making and freezing a precooked sausage for pizza toppings that had the characteristics of freshly cooked sausage and surpassed other precooked products in price, appearance, and taste. C&F had obtained a patent on the equipment to make the sausage and also one on the process for making the sausage. C&F improved the process after submitting its patent applications and kept its new developments as trade secrets.

Pizza Hut agreed to buy C&F's precooked sausage on the condition that C&F divulge its process to several other Pizza Hut suppliers, ostensibly to assure that backup suppliers were available to Pizza Hut. In exchange, Pizza Hut promised to purchase a large amount of precooked sausage from C&F. Accordingly, C&F disclosed the process to several Pizza Hut suppliers and entered into confidentiality agreements with them. Subsequently, Pizza Hut's other suppliers learned how to duplicate C&F's results. Pizza Hut then told C&F that it would not purchase any more of their sausage without drastic price reductions.

One of Pizza Hut's largest suppliers of meat products other than sausage was IBP. Pizza Hut furnished IBP with a specification and formulation of the sausage toppings and IBP signed a confidentiality agreement with Pizza Hut concerning this information. In addition, IBP hired a former supervisor in C&F's sausage plant as its production superintendent, but then fired this employee five months later, after it had implemented its sausage-making process and Pizza Hut was buying the precooked sausage from IBP.

C&F then brought suit against IBP and Pizza Hut for patent infringement and misappropriation of trade secrets, and the court found on summary judgment that the patents of C&F were invalid because the inventions had been on sale more than one year before the filing date. However, the court determined that C&F possessed valuable and enforceable trade secrets, which had indeed been misappropriated. What a great example of trades secrets serving as backup where patents fail to provide any protection!

In certain instances, a patent is a weak instrument indeed, given the many potential patent attrition factors, such as:

- doubtful patentability due to patent-defeating grounds
- narrow claims granted by a patent office
- the fact that "*only about 5% of a large patent portfolio*" has commercial value[36]
- the short life of a patent (average effective economic life is "*only about five years*")[37]
- enforcement of patents is daunting and expensive
- limited nature or lack of coverage in some countries

13. TRADE SECRETS AND HYBRID LICENSES

In trade-secret licensing practice, the threshold concern one encounters is the so-called black box dilemma. Two pieces of Anglo-Saxon wisdom describe it vividly. The trade-secret owner cannot "let the cat out of the bag," and the potential licensee will not want to "buy a pig in a poke." In plainer words, unrestricted disclosure of a new invention or proprietary know-how would result in the certain loss of trade-secret rights. On the other side, the potential recipient is unlikely to acquire something sight unseen. Fortunately, there is a perfect way out of this quandary. It is a secrecy agreement, also called a nondisclosure agreement, a confidential disclosure agreement, or a prenegotiation agreement. In negotiating and drafting such an agreement, the parties have different concerns that have to be addressed.

Trade secret owners will want to know:

- What mechanisms and procedures should be used to divulge the contents of the black box?

- What restrictions should be placed on recipients with respect to their use of the information in the black box, if they elect to use the information or if they decide not to use the information?
- How long and how thoroughly should recipients be permitted to examine the contents of the black box?
- How much should they charge for a peek into the black box?

On the other side, trade-secret recipients will want to know:

- What restrictions should they accept on use of the information if they want to license and use it?
- What restrictions should they accept on the future use of the information, if they do not want to license it?
- What if the information is already in the public domain?
- What if it turns out that they are already in possession of the information, or an important part of it?
- How much should they pay for a look into the black box?

A written agreement is the safest way to preserve secrecy and the best way to arrange an agreement. It should have provisions that define the area of technology with precision, establish a confidential legal relationship between the parties, furnish proprietary information for a specific purpose only, oblige the recipient to hold information in confidence, and spell out exceptions to secrecy obligations. The last could include information already in the public domain, information that later becomes public knowledge other than through the fault of the recipient, information that is already known to the recipient or that later comes into the possession of the recipient through a third party that has no secrecy obligation to the owner. Very importantly, the written agreement should limit the duration of the secrecy obligation.

Similar critical provisions should be incorporated into trade-secret licenses, technical assistance agreements, and hybrid patent/trade-secret licenses. The provisions should accompany the typical operational clauses that spell out license grants, royalty payments, indemnities, warranties, terms and termination conditions, and other miscellaneous matters.

While such hybrid agreements are very prevalent in the United States, they are quite problematic, since it is a misuse of a patent or an antitrust violation to exact royalty payments after a patent ceases to be in force.[38] This could happen, since the lives of trade secrets are potentially indefinite while patents have a finite lifetime. Hence, depending on how a license agreement is drafted, in the United States it can become impossible to agree to spread royalty payments over a specified term that extends beyond the lives of patents or trade secrets that are embodied in such an agreement. In an American hybrid licensing agreement, the obligation to pay royalties thus ends, even though valuable trade secrets are still in play. But there are solutions to this predicament:

- separate patent and trade-secret agreements
- make initial lump-sum payment(s)
- clearly differentiate between patent and trade-secret rights
- separate allocation of royalties to each of the rights
- provide for appropriate decreases in the royalty rate if patents terminate or are declared invalid or if applications do not issue
- reduce the royalty-payment period (for example to 10 years)
- grant a royalty-free license to patents
- grant a trade-secret license but no patent license

The choice would depend largely on the relative role and value of patents and trade secrets in the given technology.

14. CONCLUSION

Trade secrets are a viable mode of IP protection. They can be used instead of patents, but, more importantly, they can and should be used side-by-side with patents, so that inventions volumes of collateral know-how can be protected. Far from being irreconcilable, patents and trade secrets make for a happy marriage as equal partners: it is patents *and* trade secrets, not patents *or* trade secrets.

With patents *and* trade secrets it is clearly possible to cover additional subject matter, strengthen exclusivity, invoke different remedies in litigation, and have a backup when the first protection tool becomes invalid or unenforceable. Exploiting the overlap between patents and trade secrets for optimal protection is a practical, profitable, and rational IP management and licensing strategy.

License agreements have become the preferred instruments for technology transfer. Hybrid patent/trade-secret agreements are also prevalent, since patent disclosures generally cover only embryonic or early stage R&D results, which are insufficient for commercializing the patented technology, absent access to collateral proprietary know-how. This know-how, protectable as trade secrets, need not be included in patent applications and is usually developed after filing applications. Such hybrid agreements require clauses that not only maintain trade secrecy for the benefit of the trade-secret owner, but also provide appropriate limitations for the protection of the trade-secret licensee. ■

KARL F. JORDA, *David Rines Professor of Intellectual Property Law and Industrial Innovation and Director, Kenneth J. Germeshausen Center for the Law of Innovation and Entrepreneurship, Franklin Pierce Law Center, 2 White Street, Concord, NH, 03301, U.S.A. kjorda@piercelaw.edu*

---

1 General Agreement on Tariffs and Trade, Multilateral Trade Negotiations (Uruguay Round). Agreement on Trade-Related Aspects of Intellectual Property Rights (TRIPS), 15 April, 1994. Part II, Section 7: Protection of Undisclosed Information, Article 39 (2).

2 Uniform Trade Secrets Act § 1(4), 14 U.L.A. 372 (1985 & Supp. 1989).

3 Restatement (First) of Torts (1929).

4 Restatement (First) of Torts, § 757 Comment B (1929).

5 Restatement (Third) of Unfair Competition, § 39 (1995).

6 *Kewanee Oil Co. v. Bicron Corp.*, 416 U.S. 470 (1974).

7 *Dunlop Holdings, Ltd. v. Ram Golf Corp.*, 188 U.S.P.Q. 481 (7th Cir. 1975), *cert. denied*, 189 U.S.P.Q. 256 (1976).

8 The American Heritage® Dictionary. 2000. 4th edition. Houghton Mifflin: Boston, Mass.

9 McCarthy JT. 2004. *McCarthy's Desk Encyclopedia of Intellectual Property* (3rd ed.), p. 330.

10 International Association for the Protection of Intellectual Property. 1973. Mexican Congress Resolution. Mexico DF.

11 *Peabody v. Norfolk*, 98 Mass. 452 (Mass. 1868).

12 *All Pro Sports Camp v. Walt Disney Co.*, 727 So. 2d 363 (Fla. 5th DCA 1999).

13 Price D. 2000. Cargill Reaps Bitter Harvest in Pioneer Dispute, *Fin. & Com.*, May 17, 2000. www.finance-commerce.com/recent_articles/051700b.htm; *Lexar Media v. Toshiba Corp.*, No. 1-02-CV-812458 Santa Clara Co., Calif.(Super. Ct. 2005).

14 Halligan and Pooley are prolific authors and frequent lecturers and hence well-known experts on trade secret law and practice. Halligan also teaches two advanced courses in trade secrets at John Marshall Law School. Pooley will be President of the American Intellectual Property Law Association later in 2007. Their statements were made at conferences attended by the chapter's author.

15 Perritt H. 2006. *Trade Secrets, A Practitioner's Guide*. Practising Law Institute. New York: NY. p1-1, 3-7.

16 Intellectual Property Owners Association. 2003. Survey on Strategic Management of Intellectual Property.

17 Dratler J. 1991. *Intellectual Property Law: Commercial, Creative, and Industrial Property*. Law Journal Seminars Press: New York, NY.

18 Merges RP, PS Menell and MA Lemley. 2003. *Intellectual Property in the New Technological Age* (3rd edition). Aspen Publishers: New York, NY.

19 *Advanta USA Inc. v Pioneer Hi-Bred Internat'l Inc.*, No. 04-C-238-S, slip op. (W.D. Wisc. Oct. 27, 2004) states that the Plant Varieties Patent Act does not preempt trade secrets.

20 The Initial Patent/Trade Secret Evaluation Questionnaire was created by the author of this chapter.

21 Blair H. 1978. Understanding Patents, Trademarks, and Other Proprietary Assets and Their Role. In *Technology Transfer and Licensing: The Practical View*, Franklin Pierce Law Center: Concord, New Hampshire. p. 6.

22 R Ebish, freelance writer.

23 Rosenberg P. 2001. *Patent Law Fundamentals*, vol. 2 3.08 (2d ed.). West Group: St. Paul, Minn.

24 Michael Ward, Honeywell VP Licensing.

25 Jager M. 2002. The Critical Role of Trade Secret Law in Protecting Intellectual Property Assets. In *The LESI Guide to Licensing Best Practices* (R Goldschneider, ed.) Wiley: Hoboken, New Jersey. p. 127.

26 See presentations by Karl F. Jorda at a) Santiago University, Santiago de Comporela, Spain, Jan. 1997, b) IABA XX Conference, Atlanta, Georgia, May 1977; and c) Fifth ASIPI Congress, Rio de Janeiro, May 1977.

27 *Kewanee Oil Co. v. Bicron Corp.*, 416 U.S. 470 (U.S. 1974).

28 *Kewanee Oil Co.*, 416 U.S. 470.

29 Chisum DS and MA Jacobs. 1992. *Understanding Intellectual Property Law*. 3B(1), Matthew Bender: New York. pp. 3–7.

30 *Bonito Boats v. Thunder Craft Boats, Inc.*, 489 U.S. 141 (1989).

31 *CFMT, Inc. v. YieldUp Int'l Corp.*, 349 F.3d 1333 (Fed. Cir. 2003).

32 See *supra* note 23.

33 Goldscheider R and GJ Maier (eds.). 1988. *Licensing Law Handbook*. Clark Boardman and Company: New York. p. 37.

34 *C&F Packing Co. v. IBP, Inc.*, 224 F.3d 1296 (Fed. Cir. 2000); *Celeritas Techs. v. Rockwell Int'l Corp.*, 150 F.3d 1354 (Fed. Cir. 1998).

35 *C&F Packing Co. v. IBP, Inc.*, 224 F.3d 1296 (Fed. Cir. 2000).

36 Emmett Murtha, ex-IBM and former LES president.

37 Ibid.

38 *Brulotte v. Thys Co.*, 379 U.S. 29 (U.S. 1964).

Use of Trademarks in a Plant-Licensing Program

WILLIAM T. TUCKER, *Executive Director, Research Administration & Technology Transfer, University of California, Office of the President, U.S.A.*
GAVIN S. ROSS, *Vice President, Business Development, HortResearch (U.S.A.), U.S.A.*

ABSTRACT
The principal forms of IP rights protection for plant varieties are plant patents, plant variety protection patents (PVPs), and utility patents. However, trademarks can also provide long-lasting and significant protection for plant varieties. One advantage that trademarks have over the statutory forms of IP protection for plants (plant patents, PVPs, utility patents) is that trademarks can be protected indefinitely, as long as the product is marketed and the trademark enforced. The most important agreements dealing with international trademark registration are the Madrid system and the Madrid Protocol (of which the United States is a signatory). Licensing of a trademark can either stand alone or be combined with another form of IP rights protection, such as with a hybrid PVP/trademark license.

1. INTRODUCTION
The top ten global "brands"[1] in 2006: Coca-Cola®, Microsoft®, IBM®, GE®, Intel®, Nokia®, Toyota®, Disney®, McDonalds®, and Mercedes-Benz®—with a collective estimated brand value of a staggering US$396 billion[2]—each rely on a successful branding strategy, an important part of which is a recognizable trademark. Successful product branding can create phenomenal intangible value for companies. Intangible assets today have been estimated to account for at least 80% of the market value of publicly traded companies.[3]

The fresh-fruit-and-vegetable business sector, however, has not fully taken advantage of the value that can be created by a successful branding and trademark strategy. But that is changing, as multinational companies develop brand names (which are usually trademarked names, such as Dole®, Del Monte®, and Chiquita®) and others commercialize varieties under recognizable trademarks (for example, plums using the Sun World Black Diamond® trademark and green and gold kiwifruit using the ZESPRI® trademark).

2. WHAT IS A TRADEMARK?
A trademark is any marking, sign, or designation that, during the course of trade, indicates a connection between certain goods and services and the trademark owner. Trademarks identify goods and services, distinguish them from similar goods and services, and indicate their source or origin, thereby guiding and influencing consumers' decisions. A trademark guarantees that a certain good or service is of known and reliable quality, for example, a bottle labeled with the Coca-Cola® logo indicates to the consumer that the bottle is filled with a specific cola drink. In many jurisdictions, trademarks can be registered at the local patent and trademark office. A registered trademark (or a very similar version of it) cannot be used by anyone else in association with goods or services, and the owner of the mark can bring proceedings for trademark infringement against anyone else who attempts to use the mark. However, ownership of a registered trademark does not prevent

Tucker WT and GS Ross. 2007. Use of Trademarks in a Plant-Licensing Program. In *Intellectual Property Management in Health and Agricultural Innovation: A Handbook of Best Practices* (eds. A Krattiger, RT Mahoney, L Nelsen, et al.). MIHR: Oxford, U.K., and PIPRA: Davis, U.S.A. Available online at www.ipHandbook.org.

others from making or selling the same or a similar product under a clearly different mark.

Trademarks can come in a variety of different forms. Registrable trademarks often include distinctive, sometimes nonsense, words (for example, *Kodak*). Registered trademarks can take other forms as well: numbers, number and word combinations, slogans, designs, images, colors, sounds, pictures, labels, smells, and three-dimensional configurations (such as the triangular form of Toblerone® chocolates).

In order to be protectable, trademarks must be reasonably distinctive. They are classified according to their distinctiveness, from most protectable to nonprotectable:

1. *Fanciful marks* are the most distinctive and protectable. They are unique nonsense words. Examples include *Clorox*, *Exxon*, and *Pepsi*.
2. *Arbitrary marks* are real (not nonsense) words, but they have no readily apprehensible connection with the goods or services with which they are associated. Examples include *Apple* (computers), *Apple* (records), *Domino's* (pizza), and *Sonic* (restaurants).
3. *Suggestive marks* suggest, but do not explicitly describe, a characteristic of the goods or services. For example, the name *Holiday Inn and Suites* suggests that it is a "holiday" to stay in this guest residence.
4. *Descriptive marks* refer to the purpose, function, quality, size, geographical origin, and so on, of a good or service. In order to qualify as *distinctive*, and therefore protectable, consumers must be able to associate such marks with a particular good or service. For example, *Fried Chicken* as a descriptive mark would not qualify since it merely qualifies a chicken. *Kentucky Fried Chicken*, however, means more to consumers than simply "chicken, fried in a style that is popular in Kentucky": it indicates a place where customers can obtain a meal of known and predictable quality.
5. *Generic terms*, such as *soap*, *tomato*, or *car* cannot be registered as trademarks. Interestingly, and unfortunately for trademark owners, some trademarks have transformed from fanciful to generic over the years; examples include now-common words such as *linoleum*, *aspirin*, *kerosene*, and *escalator*. (Also see the discussion of genericide in the next section.)

3. BENEFITS, RISKS, AND OBLIGATIONS OF A TRADEMARK

A trademark has no inherent value. It only gains value when the good or service with which it is associated is accepted by consumers, who then come to rely on the brand/trademark as an indicator of consistent quality. In contrast, plant patents, plant variety protection, and utility patents on plants (together called plant variety rights or PVRs) have an immediate tradeable value that may or may not decline from the time of the patent grant to the time of the patent expiration (Figure 1).

A significant advantage of a trademark over a PVR is that, unlike other forms of IP rights protection such as patents and copyrights, trademarks can be owned indefinitely, so long as they are used appropriately, are enforced, and their registration is kept current (through renewals). Trademarks are recognizable, and therefore valuable, even after the term of a patent or PVR has expired. The pharmaceutical industry owns a number of powerful trademarks: Schering-Plough Corporation, maker of Claritin®, has managed to retain a significant market share of this antihistamine even after the patent expired and generic equivalents entered the market.

Registering a trademark is usually an inexpensive and straightforward process. Some money must be put into creating a distinctive, and therefore protectable, mark. When designing a mark for use in global commerce, it is important to research the trademark registries of countries where the product is to be sold in order to ensure that the mark, or something very similar to it, has not already been registered by another party. It is not a good idea to use different trademarks in different countries or to put the same trademark on different goods, as these practices can confuse consumers and will then reduce the mark's value. Trademark owners should be aware that a nonsense word in one language might be a real word (and perhaps

Figure 1: Relative Value of Trademarks Compared to the Value of Patents and Plant Variety Rights Over Time

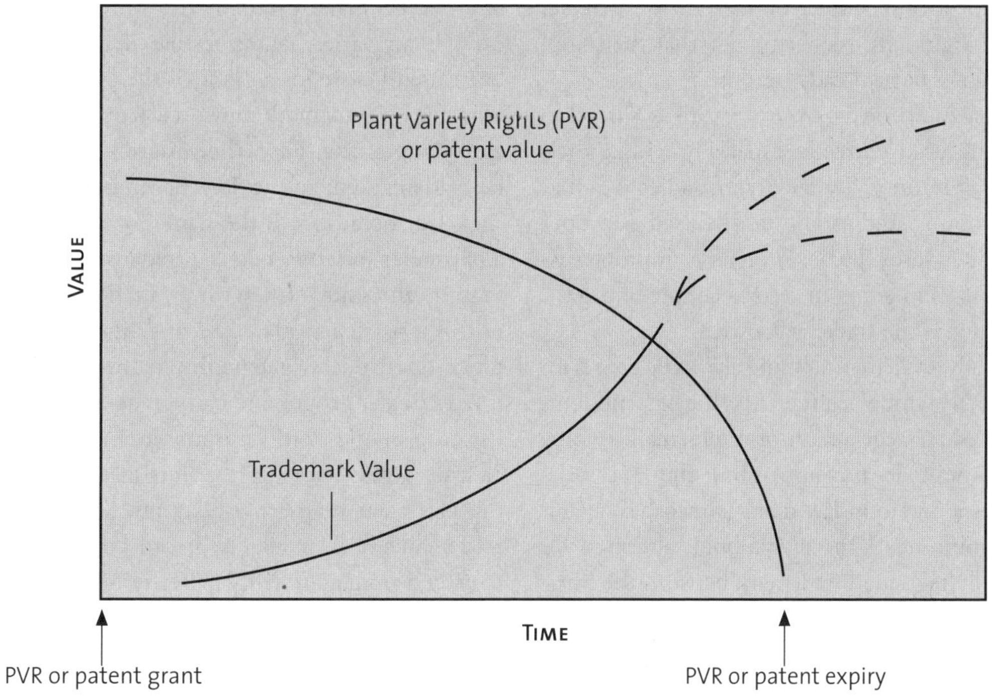

Source: Diagram kindly supplied by A MacKenzie, HortResearch.

one with a negative connotation) in another language; fanciful marks that essentially mean nothing in any language (such as *Exxon*) are usually safe.

Trademarks are a "use it or lose it" commodity. First, a trademark only has value if the good or service that it represents is of consistent quality and is continuously available; the marketplace can have a very short memory. Furthermore, and more seriously, a trademark can be invalidated if it is not used in a country for a continuous period, usually three years in most countries.

It costs considerably more to promote and develop consumer recognition of a trademark than it does to register the mark. The trademark owner will need to identify the target audience and develop promotional material tailored to that audience, a process that can become quite complex if globally marketed products are involved. It may be worthwhile to delegate these tasks to an advertising company.

The trademark owner must invest not only in establishing and maintaining a brand presence in the marketplace but also in protecting the trademark. The trademark owner will need to appoint IP managers to monitor the filing and licensing of trademarks, the policing of trademark use, and the prosecution of those who use registered trademarks illegally.

4. USING TRADEMARKS CORRECTLY

A trademark will become generic if, because of uncontrolled use, it no longer indicates that goods or services come from a particular source. Once a trademark is generic, then it is free for all to use. Such "genericide" has been the fate of many famous trademarks such as *cellophane* and *thermos*—words that are now part of the common lexicon. Though the trademarks *Xerox*® and *Kleenex*® are still protected, it has become common practice to substitute the phrase "*Xerox machine*" for a photocopier or *Kleenex* for a tissue, and the argument has been made that the trademark names have already become generic.

Trademark owners must try to ensure that marks are used correctly, especially within their own organizations. Trademarks are adjectives that qualify nouns, and should not be used as proper nouns or as verbs. For example, it is improper usage to say, "I'm going to xerox a couple of pages," even if one is the trademark owner.

Finally, trademarks should always be used with the ® or ™ symbol. In the U.S., the ® symbol indicates federal registration of a trademark (which has significant legal connotation); the ™ symbol indicates a common law mark (which has far less legal significance). The ™ symbol is also used for a federally registered trademark between the filing and registration period. Trademarks should always be used to modify a generic noun, for example, *Del Monte Gold* ™ *pineapple* or *Jazz*™ *apple*. In order to avoid violating trademark laws, breeders and growers must refer to a plant variety using the variety name and not the trademark. This can be a challenge, especially if the trademark is particularly catchy (which it should probably be in order to be successful!) or the variety name is alphanumeric.

5. TRADEMARKS IN AGRICULTURE

Trademarks have helped create value for agricultural products. One example is the Roundup Ready® trademark, which designates crops developed by Monsanto that contain transgenes that encode tolerance to the herbicide glyphosate.

Trademarks have been used to emphasize distinctive and attractive attributes of plant varieties (for example, *Pink Lady*® [apples], *Superior Seedless*®[grapes]) and *Sun-Maid*®[raisins])[4] is a branding success story: its trademark has made an otherwise pedestrian agricultural product so attractive to consumers that the owners of the mark license it for use in association with products that contain their raisins.

It is important to note that plant variety names are not the same as plant variety trademarks. Traditional plant variety names range from descriptive to fanciful, and are often chosen by the plant breeder. The only restriction on a plant variety name is that it cannot have been used before for a plant of the same species. Choosing a trademark, however, requires considerably more care. First, the variety name cannot be trademarked: the variety name is considered "generic" because it is the name for all plants of a particular variety, whereas a trademark serves to identify the source (the grower, marketer, and so on) of a particular plant. Second, the trademark office often rejects geographic names, especially if a particular geographic name is associated with the crop in question (for example, "Valencia" for citrus, "Turkey" for figs). Colors associated with the particular crop are usually not acceptable as trademarks, either. Finally, it can be difficult to register a trademark if it is already being used to refer to a related good or service, even if the good or service is different.

In order to illustrate some of the complications that may arise when attempting to trademark a product, let us take the example of the Shasta Gold® seedless mandarin, owned by the University of California. The U.S. trademark examiner objected to the use of a geographical name in the trademark, but the university argued that *Shasta* was not a region in California that is associated with citrus. The examiner objected to the use of a color in the trademark, but the university argued that *Gold* referred to the fruit's quality, not its color. Having prevailed at the U.S. Patent and Trademark Office, the university was then challenged by the Shasta Beverage Company, which claimed that the existence of the Shasta Gold® mandarin would impact sales of its own Shasta® fruit-flavored sodas. Ultimately, the parties reached a compromise out of court. Had the university simply chosen to call the variety "Shasta Gold" (without trademarking it) in the relevant U.S. Plant Patent, there would have been no conflict.

Using a trademark to cover a whole category of produce is a particularly powerful strategy. Sun World[5] uses its *Amber Crest*® trademark for various early peach varieties. These varieties are all similar in appearance and taste, but ripen at different times. Individual varieties are protected with distinct names (for example, *Supechsix*, *Supechnine*), but the consumer knows them only by their trademark name, *AmberCrest®*. This strategy has allowed Sun World to develop new varieties of early peaches while still maintaining a consistent brand image. Another strategy is to develop secondary marks or qualifying names for individual products within a brand. An example of this is the trademarked Zespri® kiwifruit from New Zealand: the yellow-fleshed kiwi is called *Zespri® GOLD* and the original fruit is called *Zespri® GREEN* (Figure 2). Because the qualifying names are common words, they cannot be trademarked.

Trademarks, if used judiciously, can add value to a single variety. The Pink Lady® apple is a good example. Whereas few consumers would recognize the variety name *Cripp's Pink*, most are familiar with the trademarked name *Pink Lady®*. Trademarks gain their value from continuous market presence and acceptance, so it may not make financial sense to create a trademark for a seasonal variety. Pink Lady® apples, however, are available year-round, so this trademark has been very successful.

Recent changes in the structure of the retail market will affect the use of trademarks in the fresh produce industry. In developed countries, the supermarket business is becoming increasingly consolidated, and these supermarkets are often expanding beyond their countries of origin. In order to keep up with the competition, supermarket chains are seeking ways to distinguish themselves from their competitors, and focusing much of the effort on the stores' produce sections

Large chains have the necessary marketing power to support trademarked produce, but the only produce varieties that are likely to provide a return on such an investment are those with unique consumer appeal: they might have an unusual or improved shape, color, texture, flavor, or other quality (such as seedlessness), or an atypical or extended market availability (such as with an early or late variety).

The growing power of supermarket chains can also work to the disadvantage of the variety's owner. The retailer may choose to reject an owner's mark in favor of its own. This is the situation in Australia, where two supermarket chains control about 80% of the fresh produce retail market. Both chains are developing their own overarching produce brands, so they are unwilling to

Figure 2: Zespri® Gold Kiwi

decrease the potential value of their trademarks by stocking and marketing products that bear other trademarks.

Because plant variety rights are not available (or particularly enforceable) in many countries, trademark protection is often stronger than, and can serve as a proxy for, variety rights protection. For example, the University of California was able to register the name *Camarosa* for a strawberry variety in certain countries where PVR was not available, and then licensed production of the Camarosa® strawberry. The central part of the license was the use of the trademark. Although third parties who were not licensed to commercialize the Camarosa strawberry could still grow them in these countries where PVR was not available, they could not sell them under the protected name of Camarosa. However, as PVR protection compliant with the International Union for the Protection of New Varieties of Plants (UPOV) becomes more common in developing countries, and if multistate protection (as exemplified by the Community Plant Variety Office [CVPO] of the European Union [E.U.]) becomes available in other regions, using trademarks as a proxy for PVR may become obsolete.

6. INTERNATIONAL TRADEMARK PROTECTION

Under the Madrid system,[6] which is administered by the World Intellectual Property Organization (WIPO), a trademark can be protected in several countries (members of the Madrid Union) if the owner files one application directly with his own national or regional trademark office. In contrast, PVR procedures are much more complicated: the variety owner must file for protection in every country (with the exception of PVRs filed in E.U. countries, which are protected throughout the European Union). The Madrid system can reduce the amount of money a trademark owner must spend on both outside lawyers' fees and filing fees.[7] The United States is not a member of the Madrid Union but is a member of the similar Madrid Protocol, adopted in 2002 and implemented in late 2003.[8]

The Madrid system has helped, in some circumstances, to curb the problem of trademark piracy and extortion, provided that the trademark owner makes use of the system and possibility to file for trademark protection in many countries at once. Consider the following scenario: a rogue entity, seeing a product on the market in one country and recognizing that it might have commercial success in another country, registers the same or a very similar mark in the second country (most countries do not require that a registered mark ever be used). When the product owner wants to enter the market in the second country, the pirate then attempts to sell the plagiarized mark to him. Taking a trademark plagiarist to court costs time and money, and the pirate relies on the probability that the trademark owner will want to settle out of court rather than engage in formal proceedings. This scenario occurred in conjunction with one of the strawberry varieties owned by the University of California: in a foreign country, a pirate registered the name of one of the university's strawberry varieties and then challenged its right to sell plant material in that country under the registered name. The ability to protect trademarks in several countries at once under the Madrid system gives product owners a useful tool for thwarting such schemes.

7. LICENSING ISSUES

A license that addresses both PVR and trademark rights, as well as when and how these rights will expire, is called a *hybrid* license. Trademarks are perpetual if the trademarked product is continuously marketed, but PVRs have a limited term. A licensee will naturally want to maintain his rights to use the trademark even after the PVR has expired and others are selling the same product. The license agreement can therefore be structured so that any given right and its associated obligations are distinct from, and can expire (or be terminated) without compromising any other rights or obligations. Box 1 provides some sample language for a licensing agreement. In addition to granting rights and specifying product marking requirements, it is important that a hybrid licensing agreement define the amount and kind of compensation to be paid for use of each right. For example, an agreement could specify a royalty for use of the PVR and a royalty for use of the trademark. In this case, after the PVR expired, the licensee would pay only the trademark royalty. Not all products may meet the quality standards required under the terms of the trademark license, so an agreement might permit the licensee to sell low-grade produce through other channels (for example, nonexport-grade products might be sold to the processing industry or local markets) without using the trademark. For these *off-grade* sales, the licensor would only collect a royalty for use of the PVR.

The licensing agreement must also cover forseeable contingencies. The quality of goods or services sold under trademark must be strictly controlled. A license agreement must require, therefore, that the licensee use the trademark only in conjunction with the licensed plant variety, and only on products that meet a prescribed quality standard (such as size/count or grade, whichever is applicable). Once a licensee has created brand equity in its own mark, it may very well terminate the license agreement and sell the licensed variety or a very similar variety under that mark; such an act would obviously be illegal, but Madrid system or not, it can be time-consuming, costly, and logistically difficult for a licensor to enforce its rights in many foreign countries. In order to avoid this kind of situation in the first place, the license can forbid the licensee to use any other trademark that could be confused with the licensed mark. Alternatively, a clause can be included in the license that requires any mark that was created and used by the licensee in association with the licensed product to revert to the licensor, should the licensee terminate the agreement.

8. LAUNCHING NEW FRUIT PRODUCTS FROM DEVELOPING COUNTRIES

Many future novel fruit products will likely come from the tropics, a region that includes many developing countries. The owners of such varieties may want to adopt a strategy that stimulates global demand for the product, while maximizing commercial returns for themselves. A global trademarking program that relies on consumer demand may be more feasible than a PVR strategy that relies on licensing for return on investment.

The developer of new branded fruit products must remember the four critical aspects of any trademarking strategy:

1. **Determine what is to be trademarked.** The owner must clearly define the registered product, as well as the standards and brand values it wishes to develop. Developing countries with variable agricultural practices may find it challenging to achieve product consistency.
2. **Register the trademark where it will be used.** The owner must have a well-developed commercialization plan with separate strategies for each country in which the fruit might be sold. The owner may need to register the trademark at the local patent and trademark office in every country or territory in which the product will be marketed.
3. **Promptly register the trademark.** Trademarks should be filed in the early stages of product conceptualization, before competitors can do so.
4. **Enforce the trademark.** The owner will need to invest money to ensure that the trademark is used appropriately, and only by those with rights to do use it. Fruit producers in developing countries may try to use a successful trademark (or a close copy) on their own products. Care must be taken to ensure that a trademark is not used so indiscriminately that it becomes a generic descriptor.

9. CONCLUSION

If chosen well and used effectively, a trademark can add substantial value to a plant variety. However, the time, effort, and up-front costs are significant, so a variety owner must be willing to make the needed investments. Moreover, an effective global trademark strategy especially requires the IP owner and its licensees to work together for mutual benefit. ■

Box 1: Example Trademark Clauses in a Master License Agreement, Where the Master Licensee Is Expected to Sublicense to Nurseries, Growers, Packers, and Distributors:

Grant Clauses:

1.1 Subject to the limitations set forth in this Agreement and the reservation of rights set forth in Paragraph XX, Licensor hereby grants to Licensee under Trademark Rights:

1.1.1 the right to use the Trademark in association with the testing and marketing of Trademark Products;

1.1.2 the exclusive right to sublicense Propagators to use the Trademark in association with the Sale of Trademark Propagator Products;

1.1.3 the exclusive right to sublicense Growers to use the Trademark in association with the Sale of Fruit;

1.1.4 the exclusive right to sublicense Packers to use the Trademark in association with the Sale of Fruit; and

1.1.5 the exclusive right to sublicense Distributors to use the Trademark in association with the Sale of Fruit.

1.2 Licensee will use the Trademarks on all promotional materials produced that refer to Licensed Products. Licensee will use the Trademarks in a featured and prominent manner. Sublicenses will require Sublicensees (a) to use the Trademark in association with, and only with (i) Trademark Products Sold or offered for Sale, and (ii) any marketing or advertising describing Trademark Products; and (b) to use the Trademarks in a featured and prominent manner. With respect to Sublicensees' Sale of Fruit, such Sublicenses will require Sublicensees to use the Trademarks with, and ONLY with, the highest grade of Fruit Sold or offered for Sale.

1.3 Neither Licensee, a Sublicensee, nor any entity which is an Affiliate, Joint Venture, or Related Party of a Licensee or a Sublicensee, will use any other trademark or name in association with Trademark Products that is confusingly similar to or, in Licensor's judgment, suggestive of, the Trademarks. Licensee and all Sublicensees will not use the Trademarks except as permitted by this Agreement.

If Licensee learns, either directly or upon notice from a Sublicensee, of any unauthorized use of the Trademarks or any colorable imitation thereof or any name or mark confusingly similar thereto, Licensee will immediately inform Licensor in writing of such unauthorized use in accordance with the provisions of Paragraph XYZ. Moreover, Sublicensor will require Sublicensees to notify Licensee (often through Sublicensor) of any unauthorized use of the Trademarks or any colorable imitation thereof or any name or mark confusingly similar thereto.

Product-marking clause:

Licensee will require all its Packers and Distributors to attach to Fruit (where commercially practicable and consistent with normal industry practice) and its cartons, boxes, pallets, or containers, sold under the terms of this Agreement, a durable and legible label or tag specifying the correct name of the Licensed Cultivar and the corresponding Trademark, if applicable.

ACKNOWLEDGMENTS
The authors would like to thank Wendy Cashmore, Michael Carriere, Andrew MacKenzie, and Clint Neagley for their constructive assistance and comments during the preparation of this chapter.

WILLIAM T. TUCKER, *Executive Director, Research Administration & Technology Transfer, Office of the President, University of California, 1111 Franklin Street, 5th Floor, Oakland, CA, 94607, U.S.A. william.tucker@ucop.edu*

GAVIN S. ROSS, *Vice President, Business Development, HortResearch (U.S.A.), 430 F Street, Suite F, Davis, CA, 95616, U.S.A, gross@hortresearch.com*

---

1 The term *brand*, as opposed to a statutorily protected trademark, encompasses a name, sign, or symbol used to identify items or services of a seller aimed at differentiating them from goods of competitors. Put differently, a brand is simply a promise, allowing customers to identify a product or service. It thus includes nonprotected assets (or liabilities) linked to a brand's name or symbol, which adds to (or subtracts from) the value provided by a product or service. It is the collection of perceptions in the mind of consumers.

2 Interbrand. 2006. *Best Global Brands: A Ranking by Brand Value*. Interbrand and Business Week. www.ourfishbowl.com/images/surveys/BGB06Report_072706.pdf.

3 Bucknell D. 2006. United States: Global Brand Strategy and the Top 100 for 2006 (the "B.R.A.N.D.I.N.G." approach). Published only on the Web. www.mondaq.com/article.asp?articleid=41882&searchresults=1.

4 www.sun-maid.com/brandlicensing.html.

5 www.sunworld.com.

6 www.wipo.int/madrid/en.

7 Information about the Madrid system is available on the Internet from patent and trademark offices in member countries (for example, www.uspto.gov; www.ipaustralia.gov.au) and from attorney firms who have trademark-law practices.

8 To learn more about the Madrid Protocol and its important differences from the Madrid system, see the U.S. Patent and Trademark Web site. www.uspto.gov/web/trademarks/madrid/madridindex.htm.

Commercialization Agreements: Practical Guidelines in Dealing with Options

MARK ANDERSON, *Solicitor (Attorney), Anderson & Company, U.K.*
SIMON KEEVEY-KOTHARI, *Barrister (Attorney), Formerly with Anderson & Company, U.K.*

ABSTRACT

An option to acquire rights in university intellectual property (IP) may be encountered in several guises: as a standalone agreement, as a clause within an agreement (for example, a sponsored research agreement or a material transfer agreement), or as a "pipeline," or IP framework, agreement for a university spinout company. Although the grant of an option may often form quite a small part of a larger agreement, the grant can raise important issues in terms of an organization's IP commercialization strategy. This is especially true of pipeline agreements that are, effectively, a specialized form of option agreement. The purpose of this chapter is threefold:

1. to provide an introduction to options, and their uses, and including legal, practical, and negotiating issues
2. to provide suggested templates along with guidelines concerning completion of the templates
3. to consider and discuss some of issues that are problematic or of particular concern to universities.

The chapter attempts to provide information that is useful for both the beginner and the experienced research-contracts or technology transfer professional. The breadth of material covered may give the mistaken impression that university contracts are wrought with legal and commercial difficulties. Usually, this is not the case. But sometimes differences of expectation, practice, or legal culture can arise between parties negotiating an agreement, particularly in international transactions.

FOREWORD

This chapter is based on one of a series of *UNICO Practical Guides*. Over recent years, the knowledge commercialization profession has grown and matured, creating a huge wealth of knowledge, experience, and best practice relating to university commercialization contracts. The *UNICO Practical Guides* have been produced specifically to share this knowledge, experience, and best practice within the profession. They are practical guidebooks on university contracts designed primarily for use by people both new and experienced in the profession that tap into the collective learning of colleagues and peers. The guides have been produced as a resource for knowledge commercialization professionals, primarily in the United Kingdom. The guides are not designed to replace or compete with existing manuals or other guides, but to provide a new and, we at UNICO believe, vitally important set of support materials to those who deal with university commercialization contracts on a daily basis. We hope that you find this document useful. (Kevin Cullen, University of Glasgow; Chair, UNICO).

1. INTRODUCTION

1.1 *What is an option?*

An option may be either an agreement or a clause within an agreement. Typically, an option gives one party to the agreement the right:

Anderson M and S Keevey-Kothari. 2007. Commercialization Agreements: Practical Guidelines in Dealing with Options. In *Intellectual Property Management in Health and Agricultural Innovation: A Handbook of Best Practices* (eds. A Krattiger, RT Mahoney, L Nelsen, et al.). MIHR: Oxford, U.K., and PIPRA: Davis, U.S.A. Available online at www.ipHandbook.org.

Editors Note: We are grateful to UNICO for having licensed one of its *UNICO Practical Guides* for inclusion as a chapter in this *Handbook*. The original version of the Guide was published by UNICO (www.unico.org.uk). The work was prepared by Anderson & Company, the Technology Law Practice™, U.K. (www.andlaw.eu), in cooperation with UNICO. The guide was edited by MIHR/PIPRA for this publication.

- to acquire a particular right (for example, a patent license) or asset (for example, a patent)
- to require another party to enter into an agreement (in a specified form) or to negotiate the terms of a further agreement
- to evaluate materials, products, or assets to determine whether to enter into further agreements (such as further research or licensing arrangements)

Usually, options are granted on an exclusive basis. Thus, where a university grants an option to acquire rights to a package of intellectual property, the option terms may require the university not to license that intellectual property to anyone else during the option period. This may be implicit in the grant of an *exclusive option*, but sometimes the parties prefer to add a clause to the option that states explicitly that the university will not license anyone else while the option continues. Sometimes, wording may go further and prohibit the university from talking to anyone else about a possible license during the option term. This type of explicit wording (when it is used) is most often requested by the grantee of the option.

The main types of agreement that an individual working in technology transfer will come across, and about which an understanding of options is useful, include the following:

- a stand-alone option agreement in which the main subject matter of the agreement is the granting of an option, such as an option to take a license to a specific patent application, and which is not part of a larger contract (See Box 1, at the end of this chapter, for a sample option agreement.)
- an option and evaluation agreement, often referred to just as an evaluation agreement, and commonplace in regard to computer software (For example, under such an agreement one party provides an item of software for a second party to evaluate, over a defined period of time, to enable the second party to ascertain whether it wants to take a license to the software. The evaluation period gives the second party an option to acquire such a license if it so wishes. See Box 2, at the end of this chapter, for a sample software evaluation agreement.)
- a research collaboration/sponsorship agreement, in which the collaborator/sponsor is sometimes given an option of acquiring rights in the intellectual property generated by the university under the research program
- a license agreement, where in addition to the licensee obtaining a license to a university's particular patents and know-how, there may be a provision for the licensee to acquire rights in *improvements* to the licensed technology (Such a provision is usually made by granting an option to such improvements and by including an appropriate definition of improvements in the agreement.)
- pipeline agreements and rights of first refusal, which are similar to options, outlined separately, and in slightly more detail, below, along with a brief explanation of how they differ from basic option agreements and clauses (See Box 3, at the end of this chapter, for a sample pipeline agreement.)

1.2 *What is a right of first refusal?*

People sometimes use the terms *option* and *right of first refusal* loosely, and interchangeably, to refer to any kind of opportunity right. (See Box 4, at the end of this chapter, for examples of options, rights of first refusal, and similar provisions.)

The authors of this guide are not aware of any official definition of these terms. However, a right of first refusal is often understood as having the following, more precise meaning, and it is considered best practice to adopt this meaning.

The key distinction between an option and a right of first refusal, involves who initiates the grant of rights. Typically, with an option, the party benefiting from the option (the grantee) is given a period of time in which to claim the prize—to notify the party granting the option (grantor) that it wishes to obtain the grant of rights (such as a license or an assignment).

By contrast, if the grantee is given a right of first refusal, it cannot initiate the grant of rights. The grantor is in control of the process. If the grantor wishes to grant the rights, it must notify the grantee and give the grantee an opportunity to accept, or refuse, those rights.

Typically, right of first refusal clauses operate at one or both of the following stages:

1. When the grantor first decides it is ready to grant the rights (or is about to start offering the rights to third parties), it must offer the rights to the grantee.
2. When the grantor is about to sign an agreement with a third party, the grantor must give the grantee an opportunity to match the terms agreed upon with the third party. If the grantee accepts this opportunity, the grantor must grant the rights to the grantee on those terms, instead of granting them to the third party.

Rights of first refusal are often encountered where the other party to an underlying agreement (for example, a research agreement) is either sponsoring the research (financially or in kind) or providing materials. Indeed, many university research agreements and material transfer agreements (MTAs) that originate from large pharmaceutical companies often incorporate a right of first refusal.

A right of first refusal can therefore cover the following situations:

- If party A negotiates with party B over certain terms (for example, a license agreement), then party A will give party C an opportunity to match those terms.
- If party A creates intellectual property from a research program or produces something (such as a prototype), then before party A offers to license it or assign it (either generally or to a specific party, B) party C will be given a first opportunity to acquire the right or product.

Depending on how rights of first refusal over intellectual property are drafted, they can present practical difficulties, particularly in the situation described in the second bulleted item, above. Negotiations over the grant of IP rights can take months to complete, and usually require a degree of confidence building with regard to the potential value of the technology and IP rights and to how the parties will work together under the agreement. A practical issue arises when one party in a negotiation must decide when to tell the other party that a third party has a right of first refusal over the same rights. If the second party is told at the outset, will it be willing to spend time and resources in negotiating terms? If the second party is told only when the third party exercises the right of first refusal, the second party may feel that it has been misled.

Universities may therefore wish to resist granting rights of first refusal that operate immediately prior to signing an agreement with a third party. Where it is commercially necessary to grant a right of first refusal, one solution the authors have found is to draft the right of first refusal so that it operates immediately before signing a nonbinding term sheet with the third party. The third party may be less likely to complain if it is trumped at this stage.

Another variation on options and rights of first refusal is termed *right of first opportunity*. This expression is used less frequently than right of first refusal and probably its meaning is more in flux. Where the authors have encountered right of first opportunity, it has tended to mean a right of the grantee to make a proposal to the grantor at some defined point in time (for example, when the grantor decides to grant rights) but with the provision that the grantor has no obligation to accept the grantee's proposal or negotiate exclusively with the grantee. Sometimes this level of right is described as having a (nonexclusive) seat at the negotiating table. As with other types of options, the precise meaning and extent of any right of first opportunity, and the procedure to be followed when exercising it, should be clearly set out in an agreement.

Sometimes one encounters heavyweight clauses that are a composite of both an option and a right of first refusal. For example, there may be an option to negotiate a further agreement, and if the parties cannot agree on terms, then the university can grant the rights elsewhere but must come back to the other party before entering into an agreement with terms that are no better for the university than those that the other party offered.

Any such clauses need to be carefully scrutinized to ensure that they are workable and do not prejudice discussions with the third party.

1.3 *What is a pipeline agreement?*

A pipeline agreement is normally encountered only in contracts involving the formation of a university spinout company. Under these circumstances, the university (or its technology transfer office) would have assigned or licensed certain intellectual property to the spinout. The intellectual property in question usually has its origins in the laboratory/department of the academics who created it. These academics usually end up being the founders of the new spinout company.

A pipeline agreement is basically a sophisticated form of option agreement, the purpose of which is to set out the rights the spinout has to future intellectual property generated in the founders department. Under such an agreement, the recipient of the option (the spinout company) is obtaining a "pipeline" to enable it to obtain rights in the intellectual property from the originating university department.

A typical pipeline agreement is therefore normally entered into by three parties:

1. The technology transfer company/office (TTO) of the academic organization
2. The spinout company
3. The original inventors/academics (often defined as the founders in company-formation agreements) involved in the creation of the invention or technology that has been assigned or licensed to the spinout company

A scenario that normally generates a pipeline agreement might include the following parts:

- The founders or their laboratory identifies or creates further intellectual property related to an original invention or technology, or, possibly, not related to the original invention or technology.
- The further intellectual property is created within a limited time span (for example, one or three years from the date of the pipeline agreement).
- The spinout company gets an option to obtain an assignment or license of the further intellectual property.

Furthermore, pipeline agreements generally include:

- a requirement for the founders to report regularly on their work and to identify any intellectual property that will be covered under the option
- a clause allowing the company to identify intellectual property suitable to be covered under the option
- clauses dealing with intellectual property created during the term of the agreement that may involve third-party rights or third-party funding, that incorporates third-party intellectual property (or technology), or that has been developed subject to third-party restrictions (for example, on assignment or licensing), or is subject to third-party licensing, assignment, or option requirements
- provisions giving the university a license back to (or reservation of rights over) any IP or technology licensed to the company under the pipeline agreement (for example, for research and/or teaching or for "noncommercial" use [setting out the parties' understanding of noncommercial] or for use outside a defined field)
- provisions imposing, on the company, an obligation to develop and commercially exploit the intellectual property and technology assigned, or licensed, to it under the pipeline agreement
- provisions stating which party is responsible for obtaining IP protection and bearing the costs of IP protection and when the protection should be sought and the costs borne

The negotiation and drafting of a good option agreement, right of first refusal agreement, and especially pipeline agreement are substantial tasks, during which consideration must be given to many issues—legal issues as well as commercial ones.

Options and similar agreements should never be taken lightly and should be clearly and comprehensively negotiated and drafted, in order to reflect fully the intentions and expectations of the parties.

2. SUMMARY OF BEST PRACTICE IN DEALING WITH OPTIONS

The practices described in this section are put forward for consideration as possible best practice (some of the practices, readers may feel, are ideal practice) with respect to the preparation of options.

Policy. Have in place an institutional policy for the different types of options, covering such matters as:

- whether to enter into them at all, and if so, which type is appropriate—that is, a basic option, a right of first refusal, or a pipeline
- what "due diligence" should be carried out to ensure that obligations under an option do not conflict with obligations under other existing agreements and to ensure that the terms of each option do not conflict with, or prejudice, an IP commercialization strategy
- use of questionnaires to be completed by the relevant researcher/department, to provide information relevant to the option and/or surrounding intellectual property
- who has authority to sign the option for the institution

Templates. Have in place templates for each type of option agreement ready for use in individual transactions.

Negotiation. Decide who has responsibility for negotiating the terms of options. Does that person have the required level of training and skill? Set out a procedure for referring difficult issues to a more specialist advisor (for example, an in-house lawyer).

Terms. Have in place clear "bottom lines" regarding terms that must, or cannot, be accepted in each type of option agreement. Possible key issues might include:

- law and jurisdiction (is it covered by relevant insurance policies?)
- duration of option
- exactly how the option is exercised
- clarification of what happens when the option is exercised (that is, there may be a need to enter into a further agreement)
- whether warranties or indemnities can be accepted in the different types of options

Monitoring. Implement procedures to monitor obligations under option agreements, including maintaining a database of options (and other agreements).

3. COMPLETING A TEMPLATE AGREEMENT

The following section provides a quick step-by-step list of points to be noted when drafting/completing a standard option agreement, or option clause comprising part of a larger agreement. The assumption, for purposes of this text, is that the basic starting point is an agreement similar to, or the same as, the templates set out in Box 1, although the comments below are generic enough to be of universal value. The issues referred to here have already been dealt with in the main text, but it seems appropriate to state them briefly again, so that one may have a one-shot view of the drafting of suitable option wording.

Signature Date. This is the date of the agreement and is usually (unless otherwise agreed) the date on which the last person/party signs. It is not advisable to backdate the agreement by merely inserting an earlier date at the beginning of the agreement; if one wishes the agreement to cover periods prior to the date of the agreement, one should insert, in the definitions section, a separate definition of a commencement date, effective date, that is, a date after which the rights and obligations under the agreement are effective.

Parties. For a university: parties must be authorized signatories. It is sometimes the case that senior members of an academic department may think they have authority to enter into legally binding agreements on behalf of the university, when they, in fact, do not.

For U.K. companies: The full address of the company should appear (this may be a registered address or business address; it must be stated which address is being provided). Consideration should be given to providing the company number.[3]

For individuals: The home address should be provided (people move from one employer to another, which can prove problematic if they need to be found to sign further documents or in the event of a dispute).

The "Recitals," or "Whereas" section. The section generally appears on the first page of the agreement, after the "Parties" section, but before the main body of the agreement (the part that usually commences with "It is agreed as follows" or similar language). Recitals are intended to give some background to the agreement, but, strictly speaking, they are not necessary.

Definitions. This may or may not be a separate clause in the agreement. Quite often definitions are found throughout the document; the standard way of providing definitions is to follow a definition with its term, with initial caps and inside parenthesis. Thereafter, throughout the agreement, the phrase Effective Date would be used in place of the actual date. If a separate clause is used for definitions, the convention generally is to place the defined term in between quotation marks. For example:

1.4 "Contract Period" shall mean the period beginning on the Effective Date and ending on the [third] anniversary of the Effective Date, subject to any earlier or later termination in accordance with Clause 8;

From a drafting, as well as a contractual interpretation point of view, both versions are very efficient approaches.

Obligations: The option agreement needs to set out clearly:

- the intellectual property covered by the agreement, or if it is future intellectual property in a pipeline agreement, it needs to be properly ring-fenced by, for example, defining it as intellectual property in a particular field, generated by a specific research group, during a limited period
- the duration period of the option
- how the option can be exercised
- what happens if it is not exercised
- what happens to any materials/software transferred under the option agreement once agreement is terminated

Jurisdiction: The law governing the agreement should as far as possible be English law, while jurisdiction should be the "Non-Exclusive Jurisdiction of the English Courts," as discussed earlier.

4. KEY NEGOTIATING ISSUES IN OPTIONS

4.1 *Key terms of a typical option agreement*

Although the detailed terms of option agreements vary, they often include provisions covering the following points:

- a description of the general subject matter of the option
- a detailed definition of "option intellectual property/pipeline intellectual property" (that may refer to existing intellectual property or future intellectual property based on some existing intellectual property)
- stating what the option is for, for example, to take an exclusive license or assignment
- in an evaluation agreement, obligations to use the intellectual property only for a defined purpose
- the option exercise period (for example, "for a period of three months from the date of the agreement"; or "within one month of the Company being informed of new intellectual property arising under a pipeline agreement")
- the method of how the option is actually exercised
- a statement of what happens after the exercise of the option, for example, obligations of the parties:
  - to execute a formal assignment of specific patents
  - to enter into a detailed license agreement on pre-agreed terms, for example, those terms set out in a schedule accompanying the option agreement
  - to negotiate the terms of further agreements, for example, a license agreement

or assignment, including any time limit for such negotiations and what would result if the parties are unable to reach agreement
- payments clause setting out the option fee, including the reimbursement of any historic patent costs
- general confidentiality obligations
- various IP-related provisions, including ownership of intellectual property, any warranties that may be given, or a provision that no warranties are given relating to any information/IP provided for evaluation (that is, the material, information or IP license is provided as is)
- in an evaluation agreement, or a research agreement containing option provisions, obligations to disclose the results of research or evaluation
- in a pipeline agreement, obligations to promptly inform the spinout company of arising intellectual property that may fall within the pipeline
- standard boilerplate provisions
- termination provisions

4.2 *What are the common areas of negotiation?*

The terms that are often negotiated in option agreements include the following:

- the extent of the intellectual property covered by the agreement, especially in pipeline situations, where the university needs to keep the pipeline narrow (defined by inventors and research groups, field, sources of funding of the research, and so on), often against the wishes of the spinout company (and their investors)
- the option fee
- the duration of the option
- the name of the party who has control over (and pays for) patenting during the option period
- the detailed terms of the "further agreement" (for example, license agreement) or, if these have not yet been agreed to at the time the option agreement is negotiated, the extent to which the parties are required to negotiate, in good faith, the terms of the further agreement, for example, the actual final license of the intellectual property and the consequences of failing to agree those terms (for example, whether the terms are settled by an expert and whether the grantee receives a right of first refusal

Sometimes, as a halfway point between items entering into a detailed license agreement and negotiating the terms of further agreements, certain key commercial terms of the future license or assignment are agreed to as part of the option agreement, for example, that there will be an exclusive license, with royalty payments. However, certain provisions, such as the actual percentage figure for royalties, may be left for agreement at a later stage (with provisions for referral to an expert where the parties cannot agree).

5. A CHECKLIST OF OPTION PROVISIONS

A checklist in Table 1 (see end of chapter) lists:

- preliminary points that may need consideration
- the main clauses usually found in an option together with the main issues that should be addressed regarding each provision

6. SPECIAL LEGAL ISSUES IN OPTIONS

Note: the following comments are based on English law, and different considerations may apply in other jurisdictions, e.g. as to the enforceability of obligations to negotiate in good faith.

The enforcement of option agreements depends on both (1) the terms of the agreement and (2) the effect of the underlying law relating to such matters as "agreements to agree" among others. The manner in which an option agreement is drafted might have a similar effect as when parties use and characterize documents as letters of intent or "heads of terms" in the course of negotiations—the document is not as much setting out all of the details of the overall transaction as it is anticipating future events (and perhaps further written agreements too) down the line.

Generally, where substantial and necessary terms of an option agreement are left open for future negotiations, a contract has not been created. Ideally (from the point of view of legal enforcement) all the terms of the further agreement (for example, license agreement) will be set out as a schedule to the option agreement, so that all the parties have to do when the option is exercised is sign the further agreement. However, the parties do not always wish to spend time negotiating detailed license terms at the time of negotiating the option agreement. An alternative is to specify that the parties will negotiate the detailed terms once the option is exercised. Unless carefully drafted (in particular, with a default mechanism stating what happens if the parties cannot reach agreement, for example, referring the terms for settlement by an independent expert), this may amount to an unenforceable agreement to agree.

Where a party intends to create a legally binding option agreement, it should refrain from merely agreeing to "agree in the future," even if future agreements will be necessary corollaries to the contract at issue. Instead, the parties should specifically describe the responsibilities and obligations of each party, clearly stating the consideration for each party's obligations. By avoiding the inclusion of uncertain terms requiring future negotiation, a party can help ensure that a binding contract has been formed.

If certain commercial terms cannot be determined at the time of the execution of the option agreement, the parties should provide a method for determining the matter. For example, in relation to any options fees or other payments to be paid at a later date, the parties can agree upon a formula that permits the calculation of fees/prices in the future, or such fees/prices will be determined by a specified independent person, that is referred to an expert. These matters should not be left for the court to decide.

7. DETAILED DISCUSSION OF COMMERCIAL ISSUES IN OPTIONS

Compared with other topics covered in the *UNICO Practical Guides*, there are relatively few detailed commercial issues to discuss, once the key drafting and negotiating issues have been resolved, that is, the scope and duration of the option and the procedure for exercising it.

7.1 *Option for license or option for assignment?*

As has already been noted, there are many different types of options and many different subject matters these options can address—for example, acquisition of shares, intellectual property, contractual rights, and income streams. In the context of technology transfer activities, and where the subject matter of the option is intellectual property, a key question is whether an option should give the grantee the ownership of the intellectual property (that is, by means of an assignment) or merely a license, with ownership remaining with the university.

From the university's perspective, the main advantage of retaining ownership (that is, licensing rather than assigning) is the degree of control (or at least influence) that ownership gives. The main areas of control may be:

- control over patenting (the licensee or assignee's interests may not always coincide with those of the university)
- control over development and commercial exploitation of the intellectual property
- recovery of rights if the company becomes insolvent

Diligence obligations can, of course, be included in an assignment agreement. However, if the grantee obtains outright ownership of the intellectual property, regaining control of the intellectual property may be more difficult (if the assignee is in breach of contract) than if only a license had been granted. A license can be terminated; an obligation to assign back intellectual property may be more difficult to enforce. If the grantee owns the intellectual property and then sells it (for example, through the grantee's liquidator, as part of a winding-up process), the new owner may be able to avoid complying with the obligations under the assignment agreement (and this is an even greater risk if the new owner were not aware of these obligations).

In the case of pipeline agreements with spinout companies, the company's investors may push hard for an assignment rather than a license of intellectual property (both in relation to the original package of intellectual property that is being acquired from the university and in relation to any further intellectual property that is acquired under a pipeline agreement). A few universities are becoming more resistant to such pressure and granting only a license, or, in some cases, granting only a license initially, but converting the license to an assignment once the company has generated a certain level of investment.

7.2 *Options as part of research agreements*

Take the example of an agreement under which a company sponsors a program of research at a university. Such an agreement will usually include provisions that determine which of the parties would own the results of the research, including any resulting intellectual property. Sometimes, the agreement will specify that the results are owned by the university and that the sponsor is granted an option to acquire a license to develop and commercialize the results. Some of the "Lambert" agreements (agreement number 2, Clause 4.6) include such option terms.[4]

This approach—the grant of an option to acquire a license to commercialize results—is just one of a number of possible ways of "carving up" any intellectual property generated from a sponsored research program. The Lambert agreements offer some alternative ways of dealing with this issue. Other possible approaches include:

- sponsor owns all the results (solely or jointly with the university)
- sponsor has an automatic license to the results (either for all purposes, including commercialization, or for research purposes only)
- sponsor gets no automatic rights to, or option over, the results

Other variations include granting rights in specific fields or territories.

7.3 *No automatic offer of license or assignment: the U.S. approach*

Although Lambert may assist U.K. universities in developing a more standardized approach to the question of intellectual property arising from research contracts, U.K. universities have not yet become as consistent in their approach as many U.S. universities are. Generally, in the United States, the policy of most universities is to only grant options to arising intellectual property that is generated under a research contract.

Although exceptions may be made in certain (rare) circumstances, U.S. universities generally retain ownership of any intellectual property that arises from the results of its own research. However, they are willing to negotiate the grant of commercial rights to a sponsor through an appropriate license, so that the sponsor may commercialize the intellectual property. This approach has evolved for two reasons—first, universities feel the need to have a certain degree of control of the discoveries made in-house (no matter who funded the research), and second, the Bayh-Dole Act prohibits universities from transferring ownership of intellectual property to a company if federal funding has helped support the work—instead, the law encourages the transfer of technologies to industry through licensing.

The Bayh-Dole Act was passed in 1980 in the United States, and the policy set down in the act encourages the utilization of inventions produced under U.S. federal funding. The policy promotes the participation of universities and small businesses in the development and commercialization process. The policy permits exclusive licensing with the transfer of an invention to the marketplace for the public good. The U.S. government enjoys royalty-free, nonexclusive licenses to use such inventions for government purposes (including for use by government contractors).

Some licenses granted by U.S. universities must be nonexclusive either because federal requirements demand it or because the research has had multiple sponsors. Under some circumstances, U.S. universities are willing to grant an exclusive license to a company. However, care is taken to ensure that, first, the field of use specified in the license is limited to the application of commercial interest to the company (so that the university researchers can continue to conduct research on other applications and develop other licensing possibilities), and second, the university will wish to ensure that the company is diligent in pursuing commercialization opportunities (a diligence clause is normally inserted into license agreements to allow the university to terminate the license if the company does not take the promised steps to develop or market the product).

In addition, licenses granted by U.S. universities normally obligate the company to pay or to reimburse the university for historic expenses associated with obtaining patents, as well as paying to the university licensing fees and/or royalties on the sale of products. If the company and the university are unable to reach agreement, or the company does not wish to obtain a license, the university is then generally free to negotiate with other parties.

In cases in which research is sponsored by a private company, a U.S. university might consider granting the sponsor a free, nonexclusive, nontransferable, royalty-free license, for internal research purposes only, to intellectual property generated by academics under the agreement. In addition, the university could, in consideration for a fixed annual fee (or royalties), grant the company the option to a nonexclusive, nontransferable, royalty-free license without the right to sublicense for the company to make products using the intellectual property.

A good example of the U.S. model is Massachusetts Institute of Technology (M.I.T.). In the majority of cases where M.I.T. research agreements involve a single sponsor, the sponsors accept M.I.T.'s standard IP clause, which gives the sponsor a number of options (including an option to an exclusive license) with regard to the licensing of patents and copyrightable materials, including software. In situations in which a sponsor wants to negotiate particular "nonstandard" IP provisions, M.I.T. is willing to enter into further negotiations. If an M.I.T. research agreement involves a consortium, the standard licensing options are limited to nonexclusive licenses.[5]

In relation to software licensing, whether intellectual property arises from sponsored research or not, companies are often willing to accept nonexclusive licenses. Also, because of the large number of patents involved in a typical electronic consumer product and because accounting for the use of each patent in such a product is onerous, many companies do not like royalty-bearing licenses in such cases. Therefore, universities might consider offering royalty-free licenses but with an upfront fee—a good example of the use of such an approach is Stanford University's EPIC (Engineering Portfolio of Inventions for Commercialization) Program, a subscription-type system with standard fees.[6] Such an approach should increase a university's chances of licensing its software technologies.

7.4 *When is an option agreement a pipeline agreement?*

An agreement will generally be described as a pipeline agreement if the party wishing to obtain rights in the intellectual property is a university spinout company and the intellectual property that is the subject of the agreement is future intellectual property that may be generated by the university (normally developed in the spinout of the department of the founding academics, or founders). Most standard option agreements, on the other hand, quite often relate to a discrete, existing item of intellectual property that a party wishes to evaluate and, possibly, obtain a license to commercially exploit.

Given that a pipeline agreement involves different pieces of (as yet unidentified) intellectual property, and also serves to set out the future relationship of the spinout and the university (and/or the university's technology transfer office), the pipeline agreement is necessarily a more complex type of agreement than a straightforward option.

Pipeline agreements usually grant an option to obtain an assignment or license of intellectual property. A pipeline agreement will usually include a definition of "pipeline IP" that will serve to define and limit the intellectual property that is to flow through the pipeline. Usually, a university will wish to limit the pipeline flow to intellectual property generated by the founders, or their laboratory, during a defined period. The university may wish to exclude from the definition any intellectual property that is subject to obligations to third parties, for example, obligations to sponsors, or to that in which any third party owns rights (for example, joint inventions made with academics employed by other universities). The method by which new intellectual property is correctly identified as pipeline IP needs to be set out in detail—that is, provisions should be set out for the submission of regular reports, by the university/founders about their relevant research, to the spinout company, in order that the company may then choose to exercise its options.

In addition, a pipeline agreement will address which of the parties is responsible for IP protection going forward, as well as certain diligence obligations on the company in relation to its commercial exploitation of the intellectual property.

7.5 *Should the university be entering into a pipeline agreement at all?*

In ascertaining whether it is really in the university's interest to grant a pipeline to a spinout company, various factors need to be taken into account. A fundamental point is whether the university spinout in question is really the best company to commercialize the intellectual property coming out of the pipeline. Often, the assumption is made that a spinout is the automatic licensee for further developments made by the university in the same field as the intellectual property on which the spinout is based (and bearing in mind that the academic inventors of the new intellectual property in question are also involved in the spinout and have a close relationship with the technology in question). However, this assumption may not always be correct. Another company may be better able to develop the new items of intellectual property, for example, because of its greater resources or because of its complementary product offerings.

Another scenario where a spinout may not be the "licensee of choice" is one in that the university may decide to grant nonexclusive licenses—for example, if several companies are possible infringers of the university intellectual property in question and may be interested in taking out a license.

7.6 *Scope, duration, and procedure for exercise*

The option agreement should be clear in relation to:

- the period of time during which the option can be exercised—the option agreement should clearly set out the relevant commencement and termination dates for exercise of the option. Options sometimes have provisions covering several different periods:
  - the period during which the grantee can decide to exercise the option, for example, during the period of a research program and for a defined period after the final report is produced
  - if the grantee exercises the option, the period during which the parties are required to negotiate the terms of a further agreement, for example, a license agreement (Sometimes, this period is vaguely specified, and there is merely an obligation on the parties to negotiate, with no clear cut-off point. From the university's point of view this approach is highly undesirable.)
  - if the option incorporates a right of first refusal, the period of that right of first refusal (For example, the clause might provide that if the parties fail to agree the terms of the further agreement within a defined period, the university is free to license to a third party, but must offer to the grantee the terms offered to the third party. Sometimes this right of first refusal will only operate for a specified period of time, for example, a year after the collapse of negotiations with the grantee.)
- what the option is exactly for, for example, whether it is a right to negotiate something or a right to acquire something, specifying exactly what the subject matter of the option is—a specific piece of technology or a specific patent, for example (Precise definitions on that subject are generally needed.)
- consequences of any failure to agree to the terms of any further agreement (The two main alternatives are: (1) the option lapses

or (2) referral to an expert who will decide the terms of the further agreement.)

7.7 *Payments*

Sometimes, options are granted without charge. This usually happens in cases in which the grantee of the option is perceived to be in a sufficiently strong bargaining position to demand a period of exclusivity prior to deciding whether to acquire rights to the asset in question.

In many situations, however, the university may take the view that the grant of an option has commercial value that should be recognized in an option fee. One possible argument for such a fee is that if an exclusive option is granted, the university is prevented from pursuing its licensing activities with other companies during the option term. The fee could be either or both of the following:

- a fee payable for the grant of the option (for example, payable on signature of an option agreement)
- a fee payable on exercise of the option

The amount that should be charged for the grant of an option is clearly a commercial, rather than a legal, issue. The authors have seen option fees of the order of tens of thousands of pounds, but much will depend on the technology, the market, the extent of rights granted, and so on. Usually, a university will wish to recover its incurred patent costs on exercise of the option, in addition to any option fee. Option fees should not be confused with initial payments under any further agreement (for example, a license agreement). Various standard techniques have been applied for the valuation (and therefore pricing) of technology generally.[7]

8. ADMINISTRATION OF OPTIONS

It is important to keep track of options—both during the review and negotiation period and once options agreements have been signed. This task is probably best administered centrally, for greater ease of checking existing options that may have already been signed with the same party, and any other agreements, for potential conflicts with the option under review. Once a party has decided to grant an option, then a number of administrative issues may need to be addressed.

8.1 *Standard operating procedure (SOP)*

It is extremely helpful to the person negotiating the option if his or her institution has an established written policy, or written standard operating procedure (SOP) for dealing with options, that includes guidelines regarding particular clauses and issues. It is particularly helpful if written guidance exists for nonnegotiable provisions as this enables the negotiator to take a more confident stance. The guidance should be updated regularly and honed in light of practical issues experienced by the negotiators on a daily basis.

In addition to aiding the negotiator, having an SOP is also in the institution's interest. By issuing clear guidelines (and emphasizing which clauses should be referred to more senior staff or legal advisers) the potential for errors or oversights is reduced. An SOP might usefully include:

- checklist of provisions that should (or should not) be included
- guidance on when to refer particular issues to more senior staff
- reminders to enter certain details of a finalized option on the relevant database and to send a copy to appropriate academics
- list of authorized signatories and the relevant procedures for holiday cover
- whether or not to have an option questionnaire for relevant academics to complete (Unlike Material Transfer Agreements, which may be quite complex and require a more structured approach in order to ensure that the university has not granted identical rights to rival sponsors or contaminated its own background, options tend to be more straightforward. In the author's view, the essential information can probably be captured in an e-mail, with a follow-up telephone conversation if necessary.)

8.2 *Getting all the essential information for a new option*

The researcher or scientist requesting or receiving the option holds the essential information that enables the negotiator to understand the relevant issues and establish a position that will best protect the interests of the institution (and the academic). Even if the organization does not use a formal questionnaire and, instead, gathers information by e-mail/phone, having a note of the relevant questions on an SOP has the advantage that (1) the negotiator does not need to rely on memory for the appropriate questions to ask and (2) it saves time.

8.3 *Deciding which information should be disclosed*

Where a suite of confidential information is concerned, it may be safest to provide only some of the confidential information to the recipient and withhold the most valuable, sensitive, and confidential parts of the information. Or, it may be prudent to disclose the most sensitive information at a later date, for example, when a further agreement has been signed or when a patent application has been filed.

Other detailed issues and best practice suggestions in relation to confidential disclosures of information are discussed in the *UNICO Practical Guide: Confidentiality Agreements*.

8.4 *Appointing a coordinator*

It may be desirable to appoint someone, for example, a senior secretary or contracts officer, to make sure that an option has been signed prior to disclosure and to oversee the disclosure and receipt of information under the option. Other duties could include:

- monitoring any deadlines (for example, the expiry date of the option)
- where appropriate, keeping a log of which employees have received the confidential information of an external party
- noting any unusual provisions or deviation of an option from one's own standard option
- sending a copy of the signed option to the relevant academic together with a covering letter highlighting any particular obligations
- recording details of the option in a contracts database and filing the original in a safe (or designated area)

8.5 *Making employees and others aware of their obligations*

It is good practice to ensure that employees are aware of their obligations with respect to options. In order to achieve this, all third-party confidential information should be clearly identified, perhaps by labeling it clearly as confidential. Any employee who receives third-party information should be informed that the information must be kept confidential and not used except as permitted under the option with the third party. In some cases it may be appropriate to provide a copy of that option to the employee.

8.6 *Contracts databases*

Many universities enter into large numbers of IP contracts, including options, with many different organizations. It can be difficult to keep track of whether, if the university wants to talk to a third party, there is already a option in place between them, and if so, whether it is in force and whether it covers the type of discussions that are contemplated. Maintaining a general contracts database (or even better, having a discrete database just for options) that includes brief details of the terms of each option, and searchable fields, can be of invaluable assistance.

8.7 *When to involve the lawyers*

Liability and indemnity provisions are probably the main areas where more-specialized legal advice is sought. It is also important to ensure that the procedures for exercising the option are unambiguously worded and do not leave the option in limbo for a prolonged period of time. However, unfamiliar phrasing within any clause is often worth checking. Some institutions may have a set policy that requires a final legal review before signature before certain nonstandard options are passed. Whether or not this is the case, a legal review of a random selection of nonstandard options at regular intervals may be useful as part of a due diligence exercise. ■

ACKNOWLEDGEMENTS

Many people and organizations contributed to the creation of the *UNICO Practical Guides* and we would like to thank all of them. In particular, we would like to recognize:

- Jeff Skinner, commercial director of University College London, who first came up with the concept of the practical guides.
- Tom Hockaday, of Isis Innovations, and Phil Clare, of Bournemouth University, who have managed the process of development and delivery on UNICO's behalf.
- The Department of Trade and Industry, London, U.K., who generously funded the production of the practical guides.
- All of the universities and individuals, inside the profession, who have contributed to and helped to review the *UNICO Practical Guides*.

UNICO is based on, and thrives upon, the sharing of ideas within the profession. We believe that the *UNICO Practical Guides*, and this chapter, are the latest tangible example of this. We thank everyone who has contributed to them, and we thank you for taking the time to read and use them.

Corresponding author:
MARK ANDERSON, *Solicitor (Attorney), Anderson & Company, 76 Wallingford Road, Shillingford, Oxfordshire OX10 7EU, U.K.* Mark@andlaw.eu

1 This chapter includes an overview and discussion of certain legal issues from the authors' perspectives as lawyers who are qualified in England and Wales. This overview and discussion is not intended to be comprehensive and does not constitute and must not be relied upon as legal advice. Readers should consult their institution's own legal advisers on any specific legal issue that may arise. UNICO members based in Scotland and Northern Ireland should be aware that, whilst some areas of law are the same throughout the United Kingdom, other areas (such as Scots contract law) differ significantly from that in England and Wales. To the fullest extent permitted by law, neither Anderson & Company nor UNICO nor any of their employees or representatives shall have any liability, whether arising in contract, tort, negligence, breach of statutory duty or otherwise, for any loss or damage (whether direct, indirect or consequential) occasioned to any person acting or omitting to act or refraining from acting upon any advice, recommendations or suggestions contained in this chapter or from using any template or clause contained in this chapter.

2 See www.unico.org.uk or write to UNICO, St John's Innovation Centre (Unit 56), Cowley Road, Cambridge CB4 0WS, U.K. info@unico.org.uk.

3 In the U.K., consider inserting the company 'number' (a company can change its name, but the original number given to it by Companies House never changes).

4 The Lambert agreements were developed in the UK by a committee consisting of university and industry representatives, and chaired by Mr Richard Lambert (now the Director General of the Confederation of British Industry (CBI)). The agreements consist of 5 alternative template agreements with different IP terms; they were designed to reduce the time spent in negotiating IP issues in university research contracts www.innovation.gov.uk/lambertagreements.

5 See also www.mit.edu.

6 otl.stanford.edu/industry/resources.html.

7 See, for example, Anderson M. 2003. *Technology Transfer: Law, Practice, and Precedents* (Second edition), ch. 3. Tottel Publishing: U.K. In this book, techniques such as net present value, benchmarking, and going rate are discussed, and a table of published royalty rates is included.

Box 1: Sample Option Agreement

THIS AGREEMENT dated the ___ day of ____________ 2007 is between:

University Technology Transfer Ltd a company incorporated in England and Wales whose registered office is at [ ] ("University Technology Transfer") and

[name of company] a [U.S. corporation incorporated in the State of] whose principal place of business is at [address] (the "Company").

WHEREAS

A. University Technology Transfer is responsible for the development and commercialization of certain technologies that have been developed at [University] ("University").
B. Either University Technology Transfer or University has filed patent application number(s) [state number(s)] in [the United Kingdom] in respect of an invention made by a University employee [name], relating to [specify invention].
C. The Company wishes to acquire an Option to obtain a license under the Patent Rights, [and is willing to fund work to establish a "proof of concept" for the said invention that, it is intended, will enable the specification and claims of the Patent Application to be improved,] and University Technology Transfer is willing to grant the Company such an Option in accordance with the provisions of this Agreement.

IT IS AGREED as follows:

1. Definitions

In this Agreement, the following words shall have the following meanings:

Commencement Date
[date]

Option
The Option described in Clause 2.1

Option Fee
The sum of [Currency]

Option Period
The period of [90] days from the Commencement Date, subject to any earlier termination of the Option under Clause 2.4

Patent Rights
The patent application(s) referred to in Recital B[, together with any continuations, continuations in part, extensions, reissues, divisions, and any patents, supplementary protection certificates and similar rights that [are based on or] derive priority from the foregoing].

2. Option

2.1 In consideration of the Option Fee, University Technology Transfer hereby grants to the Company an exclusive Option (the "Option"), during the Option Period and subject to the provisions of this Agreement, to negotiate an exclusive, worldwide license (with the

(Continued on Next Page)

Box 1 (continued)

right to sublicense) under the Patent Rights to develop, manufacture, have manufactured, market, use, and sell products [in the Field] (the "License Rights").

2.2 During the Option Period, University Technology Transfer and the Company shall negotiate in good faith the terms of a license agreement between them under which the Company would be granted the License Rights. [Any such license agreement would include, without limitation, terms based on the provisions of Schedule 2.] Upon agreement of the terms of the license agreement during the Option Period, the Parties shall forthwith execute a license agreement between them on such terms.

2.3 If the Parties are unable to agree the terms of a license agreement during the Option Period, despite negotiating in good faith, the Option will lapse.

2.4 During the Option Period, University Technology Transfer shall consult with the Company in relation to the filing and prosecution of patent applications in respect of the Patent Rights. The Company shall reimburse to University Technology Transfer all of University Technology Transfer's costs and expenses in relation to the filing and prosecution of Patent Applications, including without limitation patent agents' fees. If at any time during the continuation of this Agreement the Company notifies University Technology Transfer that it does not wish to reimburse University Technology Transfer's costs in respect of any family of patent applications, the Option shall terminate in respect of such patent applications on the date of University Technology Transfer's receipt of such notification, and the Company shall not have any responsibility for such patent costs arising after such date.

2.5 [If the Option lapses and University Technology Transfer licenses any of the Patent Rights to a third party, University Technology Transfer shall seek to recover any patenting costs paid to it by the Company in respect of such Patent Rights from the third party and reimburse such recovered costs to the Company.]

3. Payments

3.1 In consideration of the Option, the Company shall pay to University Technology Transfer the Option Fee (plus taxes, if applicable) within [30] days of the date of this Agreement.

3.2 During the continuation of the Option, the Company shall:

3.2.1 reimburse to University Technology Transfer all of University Technology Transfer's costs and expenses in relation to the drafting, filing and prosecution of the Patents, including without limitation patent agents' fees[; and]

3.2.2 [pay to University Technology Transfer the amounts described in the attached Schedule 1, on the dates stated in Schedule 1, by way of funding for the work described in that Schedule.]

3.3 For the avoidance of doubt, all intellectual property and other rights in the work referred to in Clause 3.2 above shall vest in University Technology Transfer, but if an agreement is reached pursuant to Clause 2.2, such intellectual property and rights shall be included in the license to the Company contemplated by Clause 2.2.

3.4 All amounts stated or referred to in this Agreement are exclusive of VAT, and VAT will be charged by University Technology Transfer to the Company, in addition to such amounts, if applicable and at the appropriate rate.

(Continued on Next Page)

Box 1 (continued)

4. General

4.1 This Agreement is made under English law and the parties submit to the exclusive jurisdiction of the English courts in respect of any dispute arising out of or relating to this Agreement.

4.2 Any notice to be given under this Agreement shall be in writing and shall be sent by first class mail, or by fax (confirmed by first class mail) to the address of the relevant Party set out at the head of this Agreement, or to the relevant fax number set out below, or such other address or fax number as that Party may from time to time notify to the other Party in accordance with this Clause 4.2, and marked for the attention of the representatives of the parties set out below:

4.2.1 University Technology Transfer's representative for notices—[insert name]

4.2.2 University Technology Transfer's fax number—[insert number]

4.2.3 Company's representative for notices—[insert name]

4.2.4 Company's fax number—[insert number]

4.3 Notices sent as above shall be deemed to have been received three working days after the day of posting (in the case of inland first-class mail), or on the next working day after transmission (in the case of fax messages, but only if a transmission report is generated by the sender's fax machine recording a message from the recipient's fax machine, confirming that the fax was sent to the number indicated above and confirming that all pages were successfully transmitted).

AGREED by the Parties through their authorized signatories:

For and on behalf of	For and on behalf of
University Technology Transfer Ltd	[...]
Signed ______________________	Signed ______________________
Print name ______________________	Print name ______________________
Title ______________________	Title ______________________
Date ______________________	Date ______________________

[Schedule 1]
[description of work to be done and amount and dates of payment]

[Schedule 2]
[Key points to be incorporated in license agreement]

Box 2: Sample Software Evaluation Agreement

THIS AGREEMENT is made on ____________________________ 2007 by and between:

(1) [ ] a company incorporated in [England and Wales] under company number [ ] whose registered office is at [ ] (the "Licensor"); and

(2) [ ] a company incorporated in [England and Wales] under company number [ ] whose registered office is at [ ] (the "Licensee").

WHEREAS:

A. The Licensor has developed the Software (as defined below).

B. The Licensee is interested in evaluating the Software with a view to taking a Software License (as defined below) [on [advantageous][the] terms as annexed to this Agreement) and is willing to evaluate and test the Software at its own risk subject to the provisions of this Agreement.

NOW IT IS AGREED as follows:

1. Definitions

In this Agreement, the following words shall have the following meanings:

1.1 "Documentation" shall have the meaning as described in the Software License.

1.2 "Evaluation Fee" shall mean the fee to be paid by the Licensee to The Licensor as described in Schedule 1, Part B to this Agreement.

1.3 "Evaluation Period" shall mean the period of time, commencing on the date of this Agreement, during which the Licensee is permitted to use, evaluate [and test] the Software as described in Schedule 1, Part C to this Agreement.

1.4 "Site" shall mean [ ]."

1.5 "Software" shall mean the software to be licensed under this Agreement and potentially under the Software License as described in Schedule 1, Part A to this Agreement.

1.6 "Software License" shall mean the software license annexed as Schedule 2 to this Agreement.

2. Software license

2.1 In consideration of the Licensee paying the Evaluation Fee to the Licensor, the Licensor hereby grants the Licensee the nonexclusive right to use the Software for the purpose of internal evaluation only during the Evaluation Period at the Site and in accordance with the

(Continued on Next Page)

Box 2 (continued)

provisions of the Software License, except to the extent that such terms are varied by this Agreement.

2.2 [The Licensee agrees and undertakes to use the Software and to undertake its [testing and] evaluation for the Licensor [without charge to the Licensor] for the Evaluation Period.]

2.3 Within 30 days after the end of the Evaluation Period, unless the Licensee terminates this Agreement in accordance with Clause 2.4, the Licensee may enter into the Software License subject to the financial and other terms set out in the Software License.

2.4 The Licensee may at any time during the Evaluation Period, and must at the end of the Evaluation Period if the Licensee decides not to enter into the Software License, uninstall the Software from its computer system and return to the Licensor all copies of the Software, together with all documentation for the Software and all other material containing information concerning the Software that has either been supplied to it or of which it has become aware, whereupon the Licensee's obligations under this Agreement and under the Software License shall cease, other than those under Clause 4 of this Agreement and those in the Software License that are expressed to continue to subsist after its termination.

2.5 [For the avoidance of doubt, Documentation will not be provided by the Licensor to the Licensee under this Agreement.]

3. Licensee's Obligations

3.1 During the Evaluation Period the Licensee shall:

(a) install and keep the Software installed on its computer system in its offices and [permit the Licensor to] install upgrades to the Software as soon as they become available;
(b) provide for the Software to be used at the Site by at least [ ] of its employees, being employees who would normally use such a product;
(c) produce verbal [weekly] written reports on the Software's performance (addressing quality, content, and functionality of the Software as well as its marketability), which reports shall also identify any errors, bugs, or shortcomings in the Software as well as the Licensee's comments and observations as the Licensor may from time to time reasonably request;
(d) make those of its employees who are using the Software available for meetings and discussions with the Licensor from time to time;
(e) at the request of the Licensor from time to time provide, and will procure that its staff provide, free of charge, references and information as to their practical experience of using the Software to potential and actual licensees nominated by the Licensor;
(f) comply with the terms of the Software License (except in so far as varied by this Agreement) and with the terms as to confidentiality set out in Clause 4.

4. References to Licensee's Use

The Licensor may state in any publicity and other promotional materials that the Licensee is a user of the Software during the existence of this Agreement.

(Continued on Next Page)

Box 2 (continued)

5. Confidentiality

5.1 During and after the Evaluation Period the Licensee shall treat the Software and all information concerning it that is either supplied to it or of which it becomes aware as confidential and accordingly shall not:

(a) disclose any such information to any third party; or

(b) disclose any such information to any employee who has not acknowledged in writing the confidentiality of such information; or

(c) use any such information other than for the purpose of its own internal use, testing and evaluation of the Software except to the extent that such information is or becomes public knowledge other than through any fault of the Licensor; and shall at the request of the Licensor and at its own cost take such proceedings as may be necessary to preserve the confidentiality of such information.

6. Noncompetition (It is advisable to seek legal advice before including this clause)

6.1 During the period of [ ] [months][years] from the commencement of the Evaluation Period the Licensee undertakes not supply to, and/or develop on behalf of any third party or develop or supply to any third party, any product that competes whether directly or indirectly with the Software. Any such product shall include any software that operates as a stand-alone product, or whether as part of, or integrated into, another software product, whether can only operate in conjunction with another product, whether another product is owned, licensed to or used by the Licensee.

6.2 This obligation shall not restrict the Licensee from itself undertaking internal research and development work in respect of such competing product but the Licensee shall not undertake any marketing or promotional activities in respect of the same prior to expiry of such period.

6.3 For the avoidance of doubt, the provisions of this Clause 6 shall survive the expiration of this Agreement and/or the Software License.

7. Exclusion of Warranty

Notwithstanding any warranty to be given by the Licensor in the Software License, the Licensee acknowledges that during the Evaluation Period the Software will still be under development, will be for test and evaluation purposes only, is being provided at a fee less than that normally charged by The Licensor and accordingly is provided "AS IS" without any warranty of any kind and is being tested and evaluated by the Licensee at its own risk.

8. General

8.1 The Licensee may not assign its rights and/or obligations under this Agreement.

8.2 In the event that all or any part of the terms, conditions or provisions contained in this Agreement are determined by any competent authority to be invalid, unlawful, or unenforceable to any extent such term, condition or provision shall to that extent be severed from the remaining terms, conditions, and provisions that shall continue to be valid and enforceable to the fullest extent permitted.

(Continued on Next Page)

Box 2 (continued)

8.3 This Agreement shall be governed by and construed in accordance with the laws of England and Wales to the [nonexclusive] jurisdiction of the courts of which the parties hereby submit.

8.4 This agreement does not create any right enforceable by any person not a party to it.

AGREED by the parties through their authorized signatories:

For and on behalf of [......]	For and on behalf of [......]
Signed __________	Signed __________
Print name __________	Print name __________
Title __________	Title __________
Date __________	Date __________

Schedule 1

A. Description of the Software:
B. The Evaluation Fee:
C. The Evaluation Period:

Schedule 2
The Software License

Box 3: Sample Pipeline Agreement

THIS AGREEMENT is made the ___ day of ________ 2007 by, between and among:

1. ABC LIMITED whose registered office is at [] ("the Company"); and

2. THE INDIVIDUALS DEFINED BELOW AS THE FOUNDERS ("the Founders"); and

3. UNIVERSITY TECHNOLOGY TRANSFER COMPANY LTD whose registered office is at [ ] ("Technology Transfer")

   WHEREAS:

   A. Technology Transfer is responsible for the commercialization of Pipeline IPR (as defined below) generated within the University (as defined below).

   B. The Research Group (as defined below) of the University carries out activities that include work in the Field (as defined below).

   C. The Parties envisage that some of this work will be of commercial interest to the Company.

   D. The Founders and Technology Transfer are prepared to grant the Company an opportunity to exploit Pipeline IPR generated in the course of the Research Group's work in the Field on the terms of this Agreement.

IT IS AGREED as follows:

1. Definitions

In this Agreement, the following terms shall have the following meanings:

1.1 "Affiliate" shall mean, in relation to a Party, any entity or person that controls, is controlled by, or is under common control with that Party. For the purposes of this definition, "control" shall mean direct or indirect beneficial ownership of 50% or more of the share capital, stock, or other participating interest carrying the right to vote or to distribution of profits of that entity or person, as the case may be;

1.2 This "Agreement" shall mean this pipeline agreement together with all of its schedules, annexes, and amendments;

1.3 "Candidate Technology" shall mean an invention, know-how or other IP rights that:
   (a) are generated by the Research Group in the Research Work during the Option Exercise Period;
   (b) are considered suitable and ready for commercialization and protection by the Company; and
   (c) are identified by a Party in accordance with Clauses 2.1 to 2.3;

1.4 "Contract Period" shall mean the period beginning on the Effective Date and ending on the [third] anniversary of the Effective Date, subject to any earlier or later termination in accordance with Clause 8;

(Continued on Next Page)

Box 3 (continued)

1.5 "Department" shall mean the Department of [ ], that is within the Faculty of [ ] of the University;

1.6 "Effective Date" shall mean [XXXX] [the date of this Agreement];

1.7 "Encumbered," with respect to any Pipeline IPR, shall mean that Technology Transfer is not entitled to assign such Pipeline IPR to the Company free of all liens, encumbrances and Third-Party rights and obligations, and "Encumbrance" shall be interpreted accordingly. As examples, but without limitation, Pipeline IPR may be Encumbered if:
- (a) it incorporates IP rights or materials that are owned wholly or partly by someone other than the University or Technology Transfer (for example, but without limitation, where a person who is not a University employee contributed to its development); or
- (b) it was developed under an agreement with a Third Party on terms that restricted or prevented the University's use or disclosure of such Pipeline IPR or vested rights in such Pipeline IPR in the Third Party or any other person;
- (c) it was developed in the course of a project that was funded wholly or partly by an external funding body on terms that restricted the University's ownership, use or disclosure of the results; or
- (d) in cases falling outside (a) to (c) above, it is the subject of an option, license, agreement to assign, or other commercial arrangement with a Third Party; or negotiations for the grant of commercial rights to a Third Party are continuing;

1.8 "Exclusive Commercial License" shall mean an exclusive, worldwide license to research, develop and commercialize products and services, with the right to grant sublicenses, subject to any limitations or reservations on such license stated in this Agreement;

1.9 "Expert's Decision" shall mean the procedure set out in Schedule 2;

1.10 "Field" shall mean the field of low power circuits for use in chip designs for wireless communication applications;

1.11 "Founders" shall mean Professor [ ] and [ ];

1.12 "Inventive Contribution" shall mean a contribution to an item of Pipeline IPR that, in the absence of this Agreement, would entitle the maker of the contribution, or his or her employer, to be an owner or joint owner of the Pipeline IPR as a matter of applicable IP law. In particular, it is understood that being named as a joint author of an academic paper that describes the research in which the Pipeline IPR was generated shall not, of itself, be evidence of an Inventive Contribution;

1.13 "Major Territory" shall mean any of the following territories: [United States of America, Canada, United Kingdom, Germany, France, Italy or Japan];

1.14 "Net Sales Receipts" shall mean the amount of any payment (excluding Value Added Tax), and the value of any nonmonetary receipt, received by or due to Company or its Affiliate, in any transaction or series of linked transactions that involve the sale by the Company or its Affiliates of products that incorporate technology that is the subject of any Pipeline Patents or Pipeline Trade Secrets that are assigned or licensed to the Company pursuant to this Agreement ("Relevant Transaction"), and including any of the following:
- (a) up-front, milestone (whether at the stage of development, marketing or otherwise), success, bonus, maintenance and periodic (including annual) payments, and minimum payments, received pursuant to any license or other transactions involving the Pipeline Patents or Pipeline Trade Secrets;

(Continued on Next Page)

Box 3 (continued)

(b) any receipt greater than actual incurred cost ("Incurred Costs") in respect of the funding of research or development activities ("R&D Funding") relating to the Pipeline Patents or Pipeline Trade Secrets; provided that Incurred Costs shall not include any costs that were incurred prior to the date of the agreement under which the R&D Funding was provided;

(c) any premium paid by the licensee (or its affiliate) for shares, options or other securities in the share capital of Company or its Affiliate over and above the fair market value of such shares, options or securities, pursuant to a Relevant Transaction (such fair market value to be determined on the assumption that Technology Transfer had not granted, nor agreed to grant, any rights to Company in respect of any Pipeline IPR);

(d) any loan, guarantee or other financial benefit made or given other than on normal market terms by the licensee (or its affiliate) pursuant to a Relevant Transaction; and any shares, options or other securities obtained from a third party pursuant to a Relevant Transaction;

1.15 "Net Licensing Receipts" shall mean the amount of any payment (excluding Value Added Tax), and the value of any nonmonetary receipt, received by or due to Company or its Affiliate, in any transaction or series of linked transactions that involve the grant or assignment of any rights (including the grant of any option over such rights) of any Pipeline Patents or Pipeline Trade Secrets that are assigned or licensed to the Company pursuant to this Agreement ("Relevant Transaction"), and including any of the following:

(a) up-front, milestone (whether at the stage of development, marketing or otherwise), success, bonus, maintenance, and periodic (including annual) payments, and minimum payments, received pursuant to any license or other transactions involving the Pipeline Patents or Pipeline Trade Secrets;

(b) any receipt greater than actual incurred cost ("Incurred Costs") in respect of the funding of research or development activities ("R&D Funding") relating to the Pipeline Patents or Pipeline Trade Secrets; provided that Incurred Costs shall not include any costs that were incurred prior to the date of the agreement under which the R&D Funding was provided;

(c) where any license or sublicense is to be granted under cross-licensing arrangements, the value of any third-party license obtained under such arrangements;

(d) any premium paid by the licensee (or its affiliate) for shares, options, or other securities in the share capital of Company or its Affiliate over and above the fair market value of such shares, options, or securities, pursuant to a Relevant Transaction (such fair market value to be determined on the assumption that Technology Transfer had not granted, nor agreed to grant, any rights to Company in respect of any Pipeline IPR);

(e) any loan, guarantee or other financial benefit made or given other than on normal market terms by the licensee (or its affiliate) pursuant to a Relevant Transaction; and

(f) any shares, options, or other securities obtained from a third party pursuant to a Relevant Transaction;

1.16 "Nondepartmental University Academic" shall mean a person who is employed by the University but is not part of the Research Group;

1.17 "Option Exercise Period" has the meaning given in Clause 3.1;

1.18 "Party" shall mean any of the Company, each Founder, and Technology Transfer, and "Parties" shall mean all of them;

1.19 "Patent Rights" shall mean patents and patent applications, petty patents, utility models and certificates, improvement patents and models, certificates of addition, and all foreign counterparts thereof, including any continuations, continuations in part, extensions,

(Continued on Next Page)

Box 3 (continued)

reissues, divisions, and including any patents, patent term extensions, supplementary protection certificates, and similar rights;

1.20 "Pipeline Know-How" shall mean technical information that is generated by the University in the course of the Research Work and protected under the law of confidence, and that is not Pipeline Patents or Pipeline Trade Secrets but that [relates directly to] Pipeline Patents or Pipeline Trade Secrets;

1.21 "Pipeline IPR" shall mean Pipeline Patents, Pipeline Trade Secrets, Pipeline Know-How, [and Pipeline Other Intellectual Property];

1.22 ["Pipeline Other Intellectual Property" shall mean all IP rights that are generated in the course of the Research Work by the University and are owned by the University or Technology Transfer, other than Pipeline Patents, Pipeline Trade Secrets, and Pipeline Know-how; such IP rights may include, without limitation, copyright, database right, design rights (registered and unregistered), property rights in respect of physical materials (including biological samples), and similar rights existing in any country of the world;]

1.23 "Pipeline Patents" shall mean all Patent Rights that are developed in the course of the Research Work and are owned by the University or Technology Transfer;

1.24 "Pipeline Trade Secrets" shall mean inventions and discoveries made in the course of the Research Work that the University's patent attorneys consider to be suitable to be the subject of patent applications and that, if such applications were made, would be Pipeline Patents, but that the Company elects to keep secret in accordance with the provisions of Clause 5;

1.25 "Research Group" shall mean the Founders and their postdoctoral research assistants and postgraduate students when working under any of the Founders' sole or joint, direct supervision in the Department in the Field;

1.26 "Research Work" shall mean all research carried out in the Field by the Research Group during the Contract Period; but shall exclude (unless otherwise agreed under such separate agreements) work done under:
- (a) any separate agreement(s) between (1) the Company and (2) the University and/or Technology Transfer (including without limitation research or consultancy agreements); or
- (b) any private consultancy agreement between (1) the Company and (2) any employee of the University;

1.27 "Selected Technology" shall have the meaning given in Clause 3.2;

1.28 "Software and Database Net Receipts" shall mean the amount of any payment (excluding Value Added Tax), and the value of any nonmonetary receipt, received by or due to Company or its Affiliate, in any transaction or series of linked transactions that involve the grant or assignment of any rights (including the grant of any option over such rights) of any of the Pipeline Other Intellectual Property that is assigned or licensed to the Company pursuant to this Agreement ("Relevant Transaction"), and including any of the following:
- (a) up-front, milestone (whether at the stage of development, marketing or otherwise), success, bonus, maintenance and periodic (including annual) payments, and minimum payments, received pursuant to any license or other transactions involving the Pipeline Other Intellectual Property;

(Continued on Next Page)

Box 3 (continued)

(b) any receipt greater than actual incurred cost ("Incurred Costs") in respect of the funding of research or development activities ("R&D Funding") relating to the Pipeline Other Intellectual Property; provided that Incurred Costs shall not include any costs that were incurred prior to the date of the agreement under which the R&D Funding was provided;

(c) where any license or sublicense is to be granted under cross-licensing arrangements, the value of any third-party license obtained under such arrangements;

(d) any premium paid by the licensee (or its affiliate) for shares, options, or other securities in the share capital of Company or its Affiliate over and above the fair market value of such shares, options or securities, pursuant to a Relevant Transaction (such fair market value to be determined on the assumption that Technology Transfer had not granted, nor agreed to grant, any rights to Company in respect of any Pipeline IPR);

(e) any loan, guarantee, or other financial benefit made or given other than on normal market terms by the licensee (or its affiliate) pursuant to a Relevant Transaction; and

(f) any shares, options, or other securities obtained from a third party pursuant to a Relevant Transaction.

1.29 "Third Party" shall mean any party other than the Parties, the University, and their respective employees and agents;

1.30 "Transferred Technology" has the meaning given in Clause 3.5;

1.31 "Unencumbered" shall mean, with respect to any Pipeline IPR, that it is not Encumbered; and

1.32 "University" shall mean [ ]; and every reference to a particular Clause or Schedule shall be a reference to that Clause or Schedule in or to this Agreement.

2. Identification of Candidate Technologies

2.1 Identified by Founders. Whenever the Founders identify any Candidate Technology, they shall promptly notify Technology Transfer and the Company in writing.

2.2 Quarterly reviews. Without limiting the Founders obligations under Clause 2.1, every three months during the Contract Period, the Founders shall provide Technology Transfer and the Company with a written description of the current status of the Research Work in sufficient detail to enable any resulting inventions, know-how, or other IP rights to be identified. Using this written description, the Founders, in consultation with Technology Transfer and the Company, will identify any Candidate Technologies and will jointly prepare for the Company a report specifying these Candidate Technologies, and identifying whether they are Encumbered as described in Clause 2.4.

2.3 Identified by Company. If the Company (other than pursuant to Clause 2.1 or 2.2) identifies a Candidate Technology that it wishes to attempt to protect or commercialize, it shall promptly notify the Founders and Technology Transfer in writing, and the Founders shall notify all employees or students of the University who made an inventive contribution to the Candidate Technology ("Inventors") of the Company's interest.

2.4 Encumbered Technology. When a Candidate Technology is identified pursuant to Clauses 2.1, 2.2, or 2.3, Technology Transfer shall promptly inform the Company whether or not the Candidate Technology is Encumbered. If the Candidate Technology is Encumbered, the

(Continued on Next Page)

Box 3 (continued)

Company shall only be entitled to acquire rights in the Candidate Technology under this Agreement to the extent not in conflict with such Encumbrances.

2.5 Other research contracts. For the avoidance of doubt, nothing in this Agreement shall prevent Technology Transfer or the University from entering into sponsored research contracts in the Field under which the Pipeline IPR arising from such contracts is Encumbered.

2.6 [Record-keeping. The Founders shall ensure that all members of the Research Group shall maintain laboratory notebooks in a suitable form to provide evidence of inventions in accordance with patenting practice in the United States.]

3. Grant of Option

3.1 Option Exercise Period. Where a Candidate Technology is first identified to or by the Company, the Parties shall for a period of three months beginning on the date of such identification ("the Option Exercise Period") not discuss that Candidate Technology with any Third Parties (subject to Clause 5), nor grant any rights therein, unless and until either: (a) Technology Transfer notifies the Company that the Candidate Technology is Encumbered; or (b) the Company notifies Technology Transfer during the Option Exercise Period that it does not wish to exercise the Option.

3.2 Exercise of Option. The Company shall have the Option, exercisable at any time before the termination of the Option Exercise Period, to require Technology Transfer by notice in writing to deal with the Candidate Technology in accordance with Clauses 3.4 and 3.5 ("the Option").

3.3 Expiry of Option. If the Option Exercise Period in respect of a Candidate Technology expires without Technology Transfer receiving notification that the Company wishes to exercise the Option, the Option in respect of that Candidate Technology shall lapse, and Technology Transfer shall be free to dispose of that Candidate Technology as it wishes.

3.4 Assignment of Pipeline IPR to Technology Transfer. If the Company exercises the Option during the Option Exercise Period, the Candidate Technology shall be considered Selected Technology and the procedure described in Clauses 3.4.1 to 3.4.2 shall be followed.

3.4.1 Where the Pipeline IPR in the Selected Technology vests automatically in the University, Technology Transfer shall procure that the University shall assign such Pipeline IPR to Technology Transfer.

3.4.2 If the Selected Technology does not vest automatically in the University, the Founders and Technology Transfer shall use their reasonable endeavors to obtain an express assignment to Technology Transfer of the Selected Technology.

3.5 License of Pipeline IPR to the Company. Subject to Technology Transfer successfully acquiring all Pipeline IPR in the Selected Technology (pursuant to Clauses 3.4.1 and 3.4.2), Technology Transfer shall then deal with the Selected Technology in accordance with Clauses 3.5.1 to 3.5.2. Selected Technology that is licensed to the Company pursuant to Clauses 3.5.1 or 3.5.2 is referred to in this Agreement as "Transferred Technology."

3.5.1 Generated solely within the Department. If the Selected Technology was generated solely by members of the Research Group, the Pipeline IPR therein shall be licensed to the Company on the terms set out in Schedule 1.

(Continued on Next Page)

Box 3 (continued)

3.5.2 Generated jointly with Nondepartmental University Academics. If the Selected Technology was generated jointly by members of the Research Group and Nondepartmental University Academics, then:

(a) Noninventive. If Technology Transfer is advised that the contribution of the Nondepartmental University Academic(s) to the Selected Technology was not an Inventive Contribution, the Pipeline IPR therein shall be licensed to the Company on the terms set out in Schedule 1; but

(b) Inventive. If Technology Transfer is advised that the contribution of the Nondepartmental University Academic(s) to the Selected Technology was an Inventive Contribution then [Technology Transfer shall have no obligation to license such Selected Technology to the Company and the provisions of this Agreement shall lapse with respect to such Selected Technology][, subject always to the consent of those Nondepartmental University Academic(s), Technology Transfer shall negotiate in good faith with the Company during the Option Exercise Period for the grant to the Company of a license (at the discretion of Technology Transfer) of the Pipeline IPR in such Selected Technology on terms to be agreed, taking into account Technology Transfer's policy of compensating all University researchers when Pipeline IPR that they have generated is commercially exploited].

3.6 License back. The Company hereby grants to Technology Transfer and the University a perpetual nonexclusive royalty-free license to use all Transferred Technology and Project IPR therein on the following terms:

(a) Technology Transfer and the University shall be entitled to use Pipeline Patents for the purposes of teaching and research, including use as enabling technology in research and development projects that are funded by Third Parties; and

(b) Technology Transfer and the University shall be entitled to use Pipeline Trade Secrets, Pipeline Know-How [and Pipeline Other Intellectual Property] in the Field for the purposes of teaching and research, including use as enabling technology in research and development projects ("Funded Research") that are funded by Third Parties ("Funding Parties"), and Technology Transfer and the University shall have the right to license Pipeline Trade Secrets, Pipeline Know-How and Pipeline Other Intellectual Property to Funding Parties for use in connection with the development and commercial exploitation of the results of Funded Research. Nothing in this Agreement shall restrict the rights of Technology Transfer and the University to use, license, or otherwise exploit Pipeline Trade Secrets, Pipeline Know-How, and Pipeline Other Intellectual Property outside the Field.

4. Payments

4.1 Options and Equity. In consideration for the grant of Option rights under this Agreement, the Company shall: (a) allot and issue of [relevant shares equivalent to 10% of the Company's equity as on the [Effective Date]] shares in the Company to Technology Transfer; (b) register Technology Transfer as the holder of the [relevant] shares in the Company; and (c) prepare and deliver to Technology Transfer share certificates in respect of such shares.

4.2 Licenses. In consideration for the execution of any licenses that are executed pursuant to Clause 3.5, the Company shall:

(a) upon executing any such license, pay to Technology Transfer the amount of any patenting costs that Technology Transfer incurred, prior to the date of execution, in respect of any Pipeline Patents or Pipeline Trade Secrets that are the subject of such license; and

(Continued on Next Page)

Box 3 (continued)

(b) pay to Technology Transfer the amounts and rates described in Schedule 1.

4.3 Payment terms. All sums due under this Agreement:

(a) are exclusive of Value Added Tax that where applicable will be paid by the Company to Technology Transfer in addition;

(b) shall be paid directly into Technology Transfer' bank account number [ ], sort code [ ] with [ ] Bank, [address] or such other account as Technology Transfer may specify from time to time;

(c) shall be paid in pounds sterling and, in the case of Net Sales Receipts, Net Licensing Receipts [or Software and Database Net Receipts] received by the Company in a currency other than pounds sterling, the income shall be calculated in the other currency and then converted into equivalent pounds sterling at the rate charged by the Company's U.K. bankers for converting such other currency into sterling in the Company's bank account on the last business day of the quarterly period with respect to which the payment is made;

(d) shall be made without deduction of corporation tax or other taxes charges or duties that may be imposed, except insofar as the Company is required to deduct the same to comply with applicable laws. Any and all taxes levied by a proper taxing authority required to be withheld by the Company on account of royalties or other payments accruing to Technology Transfer under this Agreement may be deducted from such payment provided that (a) such amount is paid for and on behalf of Technology Transfer to the appropriate tax authorities within the applicable payment period and (b) the Company furnishes Technology Transfer with official tax receipts or other appropriate evidence of payment issued by the appropriate tax authorities. The Parties shall cooperate and take all steps reasonably and lawfully available to them to avoid deducting such taxes and to obtain double taxation relief.

4.4 Exchange controls, etc. If at any time during the continuation of this Agreement the Company is prohibited from making any of the payments required hereunder by a governmental authority in any country, then the Company will within the prescribed period for making the said payments in the appropriate manner use its reasonable endeavors to secure from the proper authority in the relevant country permission to make the said payments and will make them within 7 days of receiving such permission. If such permission is not received within 30 (thirty) days of the Company making a request for such permission then, at the Option of Technology Transfer, the Company shall deposit the payments due in the currency of the relevant country either into a bank account designated by Technology Transfer within such country, or such payments shall be made to an associated company of Technology Transfer designated by Technology Transfer and having offices in the relevant country designated by Technology Transfer.

4.5 Statements. The Company shall send to Technology Transfer at the same time as each payment is made in accordance with Clause 4.2 a statement, where relevant, showing how any amounts paid have been calculated.

4.6 Records. The Company shall keep at its normal place of business detailed and up-to-date records and accounts showing the amount of income received by it in respect of Net Sales Receipts, Net Licensing Receipts [and Software and Database Net Receipts], on a country-by-country basis, and being sufficient to ascertain the payments due under this Agreement. The Company shall make such records and accounts available, on reasonable notice, for inspection during business hours by an independent chartered accountant nominated by Technology Transfer for the purpose of verifying the accuracy of any statement or report given by the Company to Technology Transfer under Clause 4.5, such inspection to take

(Continued on Next Page)

Box 3 (continued)

place not more than once in any calendar year (other than re-inspection of accounts where errors have been found). The accountant shall be required to keep confidential all information learned during any such inspection, and to disclose to Technology Transfer only such details as may be necessary to report on the accuracy of the Company's statement or report. Technology Transfer shall be responsible for the accountant's charges unless there is an inaccuracy of more than 5% (five percent) in any royalty statement, in which case the Company shall pay his or her charges in respect of that particular inspection. The Company shall ensure that it has the same rights as those set out in this Clause 4.6 in respect of any Affiliate or licensee (including any agent or distributor appointed by the Company, its Affiliate or licensee) of the Company that is licensed any Pipeline IPR pursuant to this Agreement.

5. Confidentiality and Publications

5.1 General obligation. Subject to Clauses 5.3 to 5.5, each Party shall maintain in confidence any information or materials provided to it directly or indirectly by the other Party under, or in contemplation of, this Agreement and shall use the same only for the purpose of exercising rights under this Agreement.

5.2 Exceptions. The obligations set out in Clause 5.1 shall not apply to any information or materials that the Party receiving the same ("Receiving Party") can prove by written records:

(a) were already the Receiving Party's property or lawfully in its possession prior to receiving it from the other Party;
(b) were already in the public domain when they were provided by the other Party;
(c) subsequently enter the public domain through no fault of the Receiving Party;
(d) are received from a Third Party who has the right to provide them to the Receiving Party without imposing obligations of confidentiality;
(e) that it has been advised by its information officer that it is required to disclose under the Freedom of Information Act 2000; or
(f) are required to be disclosed by an order of any court of competent jurisdiction or governmental authority PROVIDED that reasonable efforts shall be used by the Receiving Party to secure a protective order or equivalent over such information and PROVIDED further that the other Party shall be informed as soon as possible and be given an opportunity, if time permits, to make appropriate representations to such court or authority to attempt to secure that the information is kept confidential.

5.3 Disclosure of Selected Technology during Option Period. The Founders, the University, and Technology Transfer shall use their reasonable endeavors to prevent the publication of any information relating to a Selected Technology during the Option Exercise Period for that Selected Technology.

5.4 Postexpiry of Option Period. If the Company has not exercised the Option before the expiry of the Option Exercise Period, the University and the Inventors shall be free to publish information forming part of the Selected Technology in accordance with normal academic practice.

5.5 Postexercise of Option. If the Company exercises the Option before the expiry of the Option Exercise Period then, following the exercise of the Option, the following provisions of this Clause 5.5 shall apply:

(Continued on Next Page)

Box 3 (continued)

5.5.1 The Company acknowledges that the University is an academic research organization supported by charitable funds and that timely publication of research results is essential to the University. The University acknowledges that the Company is a commercial organization and that patent protection of inventions with commercial value is essential to the Company.

5.5.2 To allow time for review of any proposed disclosure of information that may be patentable, the University shall provide to the Company:

(a) a copy of any manuscript that discloses any Transferred Technology at least 14 days prior to submission of the manuscript for publication; and

(b) a copy of any slides to be used in an oral presentation that would disclose any Transferred Technology at least 14 days prior to making such oral presentation.

5.6 The Company shall review all material provided to it under Clause 5.5.2 promptly. If in the Company's opinion the proposed disclosure does not include patentable subject matter, the Company shall notify the University and the University shall thereafter be free to make the disclosure. If in the Company's opinion the proposed disclosure does include patentable subject matter and the Company anticipates that it may wish a patent application to be made, it will so inform the University within the said 14 day period, in which event the University shall delay such intended public disclosure for up to [30 days][three months][six months] to allow patent application(s) to be made, provided that the Parties shall seek to minimize any such delay.

6. Diligence

6.1 The Company shall diligently proceed to develop and commercially exploit Transferred Technologies to the maximum extent worldwide, or as otherwise agreed between the Company and Technology Transfer.

6.2 Without prejudice to the generality of the Company's obligations under Clause 6.1, the Company shall provide at least annually, to Technology Transfer, an updated, written development plan, showing all past, current and projected activities taken or to be taken by the Company to commercialize the products based on Transferred Technologies worldwide. Technology Transfer's receipt or approval of any such plan shall not be taken to waive or qualify the Company's obligations under Clause 6.1. Technology Transfer shall hold all development plans submitted under this Clause 6.2 in confidence, and shall disclose the same only to its own employees and to employees of University on a need-to-know basis.

6.3 If Technology Transfer considers at any time during the period of this Agreement that the Company has without legitimate reason failed to proceed diligently to develop and commercially exploit specific Transferred Technologies (the "Specific Technologies"), Technology Transfer shall notify the Company and the Parties shall use their best endeavors to resolve the situation amicably. If such a resolution is not reached within three months of Technology Transfer first notifying the Company, Technology Transfer shall be entitled to refer to an independent expert the following questions:

(a) whether the Company has acted diligently in its attempts to develop and commercially exploit the Specific Technologies; and if not

(b) what specific action the Company should have taken ("Specific Action") in order to have acted diligently.

6.4 The independent expert shall be appointed in accordance with the provisions of Schedule 2 and his or her decision shall be final and binding on the Parties.

(Continued on Next Page)

Box 3 (continued)

6.5 If the expert determines that the Company has failed to comply with its obligations under this Clause 6, and if the Company fails to take the Specific Action within six months of the expert giving his or her decision in accordance with Schedule 2, the Company shall lose all rights in and to all such Specific Technologies.

7. Patents

7.1 [Following the identification of Candidate Technology in accordance with Clauses 2.1 to 2.3, Technology Transfer shall be responsible for making any initial patent applications, at its cost and discretion, in respect of such Candidate Technology.]

7.2 Upon the Company exercising an Option under Clause 3.2 with respect to any Pipeline Patents or Pipeline Trade Secrets in respect of item of Candidate Technology, responsibility for (including paying the costs of) pursuing any Pipeline Patents shall be the responsibility of Technology Transfer. [Subject to any terms to the contrary agreed in any license granted to the Company following the exercise of the Options contained in Clause 3, Technology Transfer shall have the right, at its discretion, to discontinue patent prosecution or maintenance of any invention licensed to the Company.] It shall be the responsibility of [Technology Transfer][the Company], in consultation with [the Company][Technology Transfer], to prepare, file, and prosecute (at the Company's sole expense) such patent applications. [The Company shall consult with Technology Transfer and keep Technology Transfer informed of all developments with respect to such patent applications, and on request shall promptly supply Technology Transfer with copies of any documents relating to the prosecution thereof.]

7.3 If any of the Results are capable of being the subject of a patent application, Technology Transfer may file a patent application at its own discretion and expense or shall do so at the request and expense of the Company.

7.4 Where Technology Transfer files or has filed a patent application at the request and expense of the Company, the Company shall give Technology Transfer at least three months' written notice of the Company's intention to cease payment of any costs and expenses incurred in connection with such filing. On receipt of the Company's notice, Technology Transfer may either abandon that patent application or continue to prosecute that patent application but at Technology Transfer expense.

8. Term and Termination

8.1 Term. This Agreement shall become effective upon the Effective Date and shall continue in force for the full duration of the Contract Period unless terminated earlier in accordance with the provisions of this Clause 8.

8.2 Founders leaving. In the event that any one of the Founders ceases to be employed by the University, this Agreement shall continue in force but the definition of "the Founders" shall be automatically amended by removal of that Founder's name.

8.3 Founders joining. Any member of the academic or permanent research staff of the University who is active in the Field may become a Party to this Agreement such that this Agreement shall continue in force with the definition of "the Founders" amended to include such person, subject to the written agreement of that person, the Founders, the Company,

(Continued on Next Page)

Box 3 (continued)

[,the University], and Technology Transfer.

8.4 All Founders leaving. In the event that all the Founders cease to be employed at the University, this Agreement shall automatically terminate.

8.5 Breach or insolvency. Without prejudice to any other right or remedy it may have, either Technology Transfer or the Company may terminate this Agreement at any time by notice in writing to the other of those two Parties ("Other Party"), such notice to take effect as specified in the notice:

(a) if the Other Party is in breach of this Agreement and, in the case of a breach capable of remedy within 30 days, the breach is not remedied within 30 days of the Other Party receiving notice specifying the breach and requiring its remedy; or

(b) if the Other Party becomes insolvent, or if an order is made or a resolution is passed for the winding up of the Other Party (other than voluntarily for the purpose of solvent amalgamation or reconstruction), or if an administrator, administrative receiver or receiver is appointed in respect of the whole or any part of the Other Party's assets or business, or if the Other Party makes any composition with its creditors or takes or suffers any similar or analogous action in consequence of debt.

8.6 Consequences of termination. Termination of this Agreement by any Party for any reason shall not affect the rights and obligations of the Parties accrued prior to the effective date of termination of this Agreement. Upon any termination, all Options that have not been exercised prior to termination shall automatically lapse. No termination of this Agreement, however effected, shall affect the Parties' rights and obligations under Clauses 3 to 7 with respect to Selected Technology in respect of which the Company has exercised an Option prior to termination.

9. General

9.1 Nothing in this Agreement and no action taken by the Parties pursuant to this Agreement shall constitute or be deemed to constitute a partnership association, joint venture, or other cooperative entity between the Parties, and none of the Parties shall have any authority to bind the others in any way except as provided in this Agreement.

9.2 It is acknowledged and agreed that this Agreement relates to results of experimental research the properties and safety of which may not have been established, and that, accordingly:

(a) any results, materials, information, Candidate Technology, Selected Technology, Transferred Technology, and Pipeline IPR provided under this Agreement ("Delivered Items") are provided "as is" and without any express or implied warranties, representations or undertakings other than those set out in this agreement; and

(b) the Company shall indemnify and hold harmless the University and Technology Transfer, their Affiliates, and their respective officers, employees, consultants, agents, and representatives ("the Indemnitees") against all Third-Party Claims that may be asserted against or suffered by any of the Indemnitees and that relate to the use of any Delivered Items, or the manufacture, distribution, sale, supply or use of any products or services that incorporate any Delivered Items, by or on behalf of the Company or its licensee or subsequently by any Third Party, including without limitation claims based on product liability laws.

(Continued on Next Page)

Box 3 (continued)

9.3 None of the Parties shall without the prior written agreement of the other Parties assign or otherwise transfer the benefit and/or burden of this Agreement.

9.4 Any agreement to change the terms of this Agreement in any way shall be valid only if the change is made in writing and approved by mutual agreement of authorized representatives of the Parties.

9.5 Any notice or other communication to be given pursuant to or made under or in connection with the matters contemplated by this Agreement shall be in writing in the English language and shall be delivered by courier or sent by post using the addresses of the Parties set out above.

9.6 This Agreement shall be governed by English Law and shall be subject to the exclusive jurisdiction of the English courts.

IN WITNESS of which this Agreement has been executed as a Deed and delivered the date and year first above written.

EXECUTED AS A DEED by [ABC] LIMITED acting by:

____________________ ____________________

Director Director/Secretary

EXECUTED AS A DEED by [UNIVERSITY TECHNOLOGY TRANSFER] LIMITED acting by:

____________________ ____________________

Director Director/Secretary

SIGNED AS A DEED by PROFESSOR [ ]

____

in the presence of:

____

Witness's signature

____

Name

(Continued on Next Page)

Box 3 (continued)

Address

Schedule 1
Detailed Arrangements for Licensing of Selected Technologies

1. Pipeline Patents

Upon exercise of an Option in respect of a Pipeline Patent then, subject to the provisions of this Agreement, Technology Transfer hereby grants to the Company an Exclusive Commercial License under that Pipeline Patent in the Field.

Upon the first receipt by the Company of Net Sales Receipts in respect of a Transferred Patent, the Company shall pay to Technology Transfer a royalty on Net Sales Receipts. Such royalty will be agreed between the Company and Technology Transfer at the time of receipt of such first Net Sales Receipts on normal arm's-length commercial terms [and is anticipated to be between 4% to 8%].

Upon first receipt by the Company of Net Licensing Receipts from a license in respect of a Pipeline Patent pursuant to Clause 3.5 (the licensed Pipeline Patent being referred to below as a "Transferred Patent"), the Company shall pay to Technology Transfer a royalty on Net Licensing Receipts. Such royalty will be agreed at the time on normal arm's-length commercial terms.

2. Pipeline Trade Secrets

Upon exercise of an Option in respect of a Pipeline Trade Secret then, subject to the provisions of this Agreement, Technology Transfer hereby grants to the Company an Exclusive Commercial License under that Pipeline Trade Secret in the Field.

The Parties acknowledge that Pipeline Trade Secrets arise where the Company elects not to pursue a Pipeline Patent in respect of a Transferred Technology and instead elects to maintain the invention as a Pipeline Trade Secret. Accordingly, upon exercise of an Option in respect of a Pipeline Trade Secret, the Company shall pay to Technology Transfer the relevant amount that would have been due, under Section 1 of this Schedule, if a Pipeline Patent had been pursued.

3. Pipeline Know-How

Upon exercise of an Option in respect of Pipeline Know-How then, subject to the provisions of this Agreement, Technology Transfer hereby grants to the Company an Exclusive Commercial License under that Pipeline Know-How in the Field.

4. Pipeline Other Intellectual Property

Upon exercise of an Option in respect of an item of Pipeline Other Intellectual Property then, subject to the provisions of this Agreement:

(a) Technology Transfer hereby grants to the Company an Exclusive Commercial License under

(Continued on Next Page)

Box 3 (continued)

the Pipeline Other Intellectual Property in the Field; and

(b) The Company shall pay to Technology Transfer, with respect to each such item of Pipeline Other Intellectual Property, either (and at the Company's election made and notified to Technology Transfer on receipt of the first Software and Database Net Receipts):
   (i) A one-time fee of [currency]X on receipt of first Software & Database Net Receipts with respect to that Pipeline Other Intellectual Property; or
   (ii) A royalty of X% on all Software & Database Net Receipts received by the Company with respect to that Pipeline Other Intellectual Property.

Schedule 2
Expert's Decision

1. Any matter or dispute to be determined by an expert under this Agreement shall be referred to a person suitably qualified to determine that matter or dispute who shall be nominated jointly by the relevant Parties. Failing agreement between the Parties within 30 days of a written request by one Party to another seeking to initiate the expert's decision procedure, either of the relevant Parties may request the president for the time being of the relevant Professional Institution to nominate the expert.

2. In all cases the terms of appointment of the expert by whomsoever appointed shall include:

   2.1 a commitment by the Parties to share equally the expert's fee;

   2.2 a requirement on the expert to act fairly as between the Parties and according to the principles of natural justice;

   2.3 a requirement on the expert to hold professional indemnity insurance both then and for three years following the date of his or her determination; and

   2.4 a commitment by the Parties to supply to the expert all such assistance, documents, and information as he or she may require for the purpose of his or her determination.

3. The expert's decision shall be final and binding on the Parties (save in the case of negligence or manifest error).

4. The Parties expressly acknowledge and agree that they do not intend the reference to the expert to constitute an arbitration within the scope of any arbitration legislation. The Expert's Decision is not a quasi-judicial procedure, and the Parties shall have no right of

Box 4: Examples of Options, Rights of First Refusal (and Similar Provisions)

appeal against the Expert's Decision provided always that this shall not be construed as waiving any rights the Parties might have against the expert for breaching his or her terms of appointment or otherwise being negligent.

Note: the following examples of rights of first refusal ("ROFRs") have been included to illustrate the variety of ROFRs that are encountered. In general, universities should be cautious about giving any ROFR, and legal advice should generally be sought on the wording of the ROFR.

Example 1: Simple, Pro-University Option Clause.

(a) Subject to the provisions of this Clause [ ], the University grants to the Company an exclusive Option (the "Option") to acquire an exclusive, worldwide license (with the right to sublicense) under the Arising Intellectual Property to develop, manufacture, have manufactured, market, use, and sell products in [the Field] (the "License Rights").

(b) The Option shall be exercisable [at any time during the agreed period of the Research] [and] [up to three months following the University's submission of the final Report]. The Option shall be exercised by the Company giving notice in writing to the University ("Notice of Exercise of Option").

(c) On receipt of the Company's Notice of Exercise of Option, the Parties shall negotiate in good faith, for a period of up to 90 days from the date of such receipt, the terms of a license agreement between them under which the Company would be granted the License Rights. [Any such license agreement would include, without limitation, terms based on the provisions of the attached Schedule [x]]. Upon agreement of the terms of such license, the Parties shall forthwith execute a license agreement between them on such terms.

(d) [If the Parties fail to agree the terms of a license agreement within 90 days of the University's receipt of the Company's Notice of Exercise of Option, the Option will lapse.]

Example 2: ROFR to be tacked on to Option (fairly brief).

If LICENSEE and TTCO or UNIVERSITY, as the case may be, are unable to agree on the terms of a license agreement within 90 days of TTCO's or UNIVERSITY's (as applicable) receipt of LICENSEE's Notice of Exercise of Option, despite negotiating in good faith, the Option will lapse; provided, that TTCO or UNIVERSITY, as the case may be, may not thereafter, without first offering such terms and conditions to LICENSEE, enter into an agreement with a THIRD PARTY on terms and conditions equal to or more favorable to such THIRD PARTY than the terms and conditions negotiated between TTCO or UNIVERSITY, as the case may be, and LICENSEE.

Example 3: Strong option and ROFR to expand field; milder option to expand territory.

1.1 Expansion of Field

1.1.1 With respect to each Compound, Owner hereby grants to Licensee a first right to expand the then current Field for such Compound and all Licensed Products based on such Compound to include additional disease indications in humans and disease indications in animals. This right may be exercised by Licensee only in the event that

(Continued on Next Page)

Box 4 (continued)

Owner determines to pursue development and commercialization (whether directly or through an Affiliate or Sublicensee) of a Compound in the Territory in one or more additional disease indications in humans or in one or more disease indications in animals outside the then current Field.

1.1.2 Within a reasonable period after such determination by Owner, Owner shall provide written notice to Licensee of proposed terms for such expansion of the Field in the Territory and disclose to Licensee all information that is within Owner's control and reasonably related to such expansion of the Field. Within sixty (60) days of such written notice from Owner, Licensee shall provide written notice to Owner as to whether it is interested in such expansion of the Field. If Licensee is not interested in such expansion of the Field or if Licensee does not provide written notice within such sixty (60) day period, Owner shall be free to develop and commercialize (whether directly or through an Affiliate or Sublicensee) the Compound and all Licensed Products based on such Compound in such additional disease indications in the Territory.

1.1.3 If Licensee provides written notice indicating its interest in such expansion of the Field within such sixty (60) day period, the Parties shall negotiate in good faith to reach agreement within one hundred twenty (120) days of the written notice from Licensee.

1.1.4 If the Parties are unable to reach agreement within such one hundred twenty (120) day period (or any mutually agreed upon extension), then Owner shall be free to (i) submit the matter to arbitration for resolution pursuant to Section 14.8 or (ii) enter into an agreement with a third party during the subsequent twelve (12) month period (but not to develop or commercialize directly or through an Affiliate) to license rights to practice the Owner Patent Rights and use the Owner Know-How for such purpose in the Territory; provided, however, that Licensee is first given the right to enter into any proposed agreement reached by Owner with a third party on substantially the same financial terms and conditions as such proposed agreement reached by Owner (it being understood that Licensee shall have the right to substitute cash or Licensee equity for equity of the third party).

1.2 Expansion of Territory. With respect to each Compound, in the event that Owner is approached by a potential Sublicensee that desires to pursue development and commercialization of such Compound or Owner determines to pursue development and commercialization of such Compound through a Sublicensee, in each case, in one or more countries outside the then-current Territory for such Compound, Owner shall promptly inform Licensee. As available, Owner will advise Licensee of the structure of the proposed license (for example, the field and countries that are the subject of the potential license) and Licensee will thereupon have the nonexclusive right to negotiate for such a license from Owner.

Example 4: ROFR (very brief).

ABC agrees with XYZ that it will not sell or otherwise transfer all or any material part of its [•] business to any third party without first giving the XYZ the opportunity to purchase such business on terms identical to those offered to such third party.

(Continued on Next Page)

Box 4 (continued)

Example 5: ROFR to purchase shares.

Unless Seller otherwise agrees, Purchaser may not sell, assign, encumber, pledge, convey, grant, or otherwise transfer any of the Shares, or any interest therein (collectively and individually "Transfer"), except to an unaffiliated third-party bona fide purchaser of value, in which case Seller shall have a "Right of First Refusal" for any Shares, or any Interest In any Shares, that Purchaser desires to Transfer to the third party. In the event Purchaser desires to Transfer some or all of the Shares, Purchaser shall provide a written notice ("Transfer Notice") to Seller describing fully the proposed Transfer, including the number of Shares proposed to be Transferred, the proposed price for the Transfer, the proposed method of payment for the Shares, the name and address of the proposed transferee, and proof satisfactory to Seller that the proposed Transfer will not violate any applicable federal or state securities laws. The Transfer Notice shall be signed by both the Purchaser and proposed transferee and must constitute a binding commitment of both parties to the Transfer of the Shares. Seller shall have the right to purchase some or all of the Shares on the terms of the proposal described in the Transfer Notice (subject, however, to any change in such terms permitted under Subsection 2(b) below) by delivery of a notice of exercise of the Right of First Refusal within thirty (30) calendar days after the date Seller received the Transfer Notice. The Right of First Refusal shall be freely assignable, in whole or in part, by Seller at its sole discretion.

Example 6: ROFR to acquire royalty stream.

Transfer of other interests: If the Educational Institution, at any time on or after the Start Date [until April [ ], 2012], wishes to Transfer any other rights to any royalty stream it may own derived from intellectual property (the "Remaining Royalty Interests"), then the Educational Institution will give notice to SPONSOR of (i) its wish to Transfer such royalty stream, and (ii) the proposed consideration, payable by a named bona fide third party, for such royalty stream, and SPONSOR shall have ninety (90) days to offer to purchase such royalty stream. In the event SPONSOR does not offer to purchase such royalty stream, for equal or higher consideration than the said bona fide third-party offer, within ninety (90) days of such notice, the Educational Institution shall be free to sell such royalty stream to a third party for a consideration equal to or higher than that specified in the aforesaid notice.

Table 1: Checklist of Preliminary Issues and Considerations Commonly Found in Option Agreements

Clauses	Considerations
PRELIMINARY	
Parties	Are the parties the correct ones? For example, in a pipeline, the parties should comprise the technology transfer office/department of the university, the spinout company, and the founder academics
	Have their correct legal names and addresses been included?
Authorized signatory	Does the option need to be signed by a central part of the organization, for example, a technology transfer office?
	Do you need to remind the other party about their authorized signatory?
Are materials or software are under evaluation	Have the materials/software and intended uses been correctly identified?
	Have the materials/software been adequately described?
THE OPTION AGREEMENT	
Recitals	Is it useful/appropriate to cross-refer to a parallel agreement (for example, a research collaboration agreement, in the case of a pipeline)?
	Check the terms of the other agreement to ensure no conflicts exist.
	Is there anything in the recitals that should really be in the body of the contract? (Recitals may not be legally binding.)
CONTRACT TERMS	
Date of the agreement	This is the date when the option is signed. The official/legal date will be the date when the last party signs, and this should be the date entered onto any contracts database.
	Do the parties want to have a particular date from which the agreement is effective? If so, agree and define an "Effective Date" or "Commencement Date" to be used as the starting point of any option period. It is bad practice to try and backdate an agreement by entering a prior date in the signature block.

(Continued on Next Page)

Table 1 (continued)

Term

- Does the agreement specify a time period?
- Should it?
- Are there any obligations (for example, return of materials/software) when the term ends?
- Is there any obligation to seek to renew the option (for example, three months) prior to expiry?
- Are there any confidentiality obligations that extend beyond the term?
- Should termination provisions be included?

Meaning of the rights that would be subject to the option:

- Would all the intellectual property arising from a particular research project be covered?
- Should only certain intellectual property be covered (for example, within a defined field—or that related to materials that have been provided for evaluation)?

What exactly is the option for?

- to negotiate a further agreement
- to evaluate materials/products
- to obtain a product, material or right
- to enter into an agreement on set terms

How will payment (option fee) be handled?

- Is the option fee separate to any other payments being made under the agreement by the person being granted the option?
- If it is a separate payment, when is it to be paid? Upon signing the agreement, or upon exercise of the option?
- What is the method of payment? By check or by direct transfer?

When is the option exercised?

- a set number of days from the date of the agreement?
- on the occurrence of a particular event or result ("trigger event"), such as
  - a patent is filed
  - an invention is made or new or improved technology or intellectual property is created resulting from, for example, research work
  - a proof of concept is shown
  - software development reaches beta stage
  - another specified event
- at any time during the existence of the option agreement (or the agreement in which the option is incorporated)

(Continued on Next Page)

Table 1 (continued)

What if a trigger event occurs?
- Is the party creating the trigger event under an obligation to notify the other party?
- Within what period must the other party be notified? (for example, within 30 days of the trigger event occurring)
- How must the other party be notified? (for example, by written notice)
- Must the written notice clearly specify certain matters? (for example, describe exactly what the trigger event is, providing details)

How is the option to be exercised?
- by written notice
- by such notice being given to a specified representative of the other party

When does the option period start?
- when sent by the party exercising it
- when received by the other party

What follows exercise of the option?
- parties are to negotiate
- parties are to negotiate in good faith or using their best or other specified endeavors
- parties are to negotiate for a fixed period, for example, a period of [X] days from receipt of notice by the other party
- parties are to negotiate to achieve some achievable outcome such as entering into a further agreement

What if a further agreement is to be negotiated?
- Is there a specified set of terms to be used during the negotiations?
- Are there minimum conditions (such as milestones and payments) that must be included in any agreement?

What happens if there is a failure to agree?
- the option lapses
- the provisions of the agreement are settled by a third party
- a right of first refusal arises

Questions regarding settlement by a third party:
- Is the third party to have the final decision on the terms?
- How is the third party to be chosen? By the parties themselves, or by another third party or by a specific organization (a professional body such as the Law Society of England and Wales)?
- Are the terms that are to be settled based on an agreed minimum set of terms (such as those attached to the option agreement)?

(Continued on Next Page)

TABLE 1 (CONTINUED)

Right of first refusal:

- If the option lapses, and there is a right of first refusal, what are the circumstances that will bring the right of first refusal into play?
- What must the optionee be offered? (for example, the right to match the terms offered to the third party)
- For how long can the parties negotiate once the right of first refusal has arisen?
- When must the third party be informed about the right of first refusal?

Confidentiality provisions:

- Are there any? Should there be?
- Is it more appropriate to have a separate confidentiality agreement, which could be cross-referenced?
- What is covered by the definition of confidential information
- Does confidential information include any information generated by a party evaluating materials/software provided to it?
- For how long do any confidentiality obligations extend?

Considerations involving materials or software:

- What exactly is to be supplied and when is it to be evaluated? Are these points clearly stated in the agreement?
- What endeavors/efforts would the supplier to use to supply them?
- Is the responsibility for shipping, packaging, and insurance allocated?
- Who is responsible for the costs if materials are to be returned when the term ends?
- Are there any regulations governing materials use (for example, the regulations governing the use of genetically modified organisms)? Which party is responsible for compliance?
- If software is being evaluated, have appropriate disclaimers been included?
- Generally, should any warranties or disclaimers be given by either party?
- Does the definition of materials/software include confidential information/documents? If so, check relevant intellectual property, publication, and confidentiality clauses.
- What exactly is the receiving party to do with the materials?
    - perform (specified) experiments with the materials
    - determine whether the materials can be used for creating new products
    - prepare business, marketing, and scientific reports
    - specify how the material can be (commercially) exploited at the end of the option/evaluation period
    - inform the supplier whether the receiving party wishes to enter into a further agreement, for example, a license agreement
- Do the stated nature and purpose of the evaluation reflect the parties' understanding of what is to be carried out in relation to the materials?
- Should the receiving party have a duty to disclose information generated during the course of the evaluation?

(CONTINUED ON NEXT PAGE)

Table 1 (continued)

Liability and indemnity

- Are any warranties being given in relation to the subject matter of the option? Should liability be limited?
- Are any indemnities being given? If so, are they (1) appropriate and (2) covered by your institution's insurance policies?
- In cases in which your institution is giving an indemnity, should you insist on having control of any proceedings brought by a third party (against the other [indemnified] party)?
- Should indemnities be restricted only to third-party claims?

Law and jurisdiction

- Has the law governing the option been stated?
- Has jurisdiction also been specified (that is, which party's courts would hear any dispute)?
- Is it appropriate to specify exclusive or nonexclusive jurisdiction?
- If confidentiality provisions are important, should a right to obtain an injunction in any jurisdiction be included?

Boilerplate provisions

- entire agreement
- force majeure
- notices (may be useful if option notices should go to technology transfer office rather than address of legal entity)

Schedules

- Is a schedule appropriate for a description of the materials/software to be evaluated?
- Have the contents been agreed/checked with the relevant academic/department?
- Is it attached?
- Has the intellectual property that is the subject of the option been described in sufficient detail?

Field-of-Use Licensing

SANDRA L. SHOTWELL, *Managing Partner, Alta Biomedical Group, LLC, U.S.A.*

ABSTRACT
Field-of-use licensing provides the licensor with greater control over the use of its intellectual property, while maximizing the use and value of the technology. In order to maximize the use of a given technology, managers will have some additional work to do as they identify, negotiate with, and manage more than one licensee. Special issues related to multiple licensees in distinct or overlapping fields will have to be handled with forethought and a balancing of interests. When is field-of-use licensing worth the extra effort? When more than one company is needed to fully develop a technology's potential, when different licensees are needed to address different markets, or when field-of-use licensing has the potential to significantly increase the financial return from a technology. In all of these situations, field-of-use licensing can produce better results for everyone involved.

1. INTRODUCTION
Innovative organizations can license a technology exclusively or nonexclusively without any limitations on its commercial use. The licensee can use the technology to make soup, pharmaceuticals, or integrated circuits. Use is limited only by the obligations set out in the license agreement (and the current and future applications of the technology).

Often, however, value can be obtained from limiting the uses available to any *single* licensee. One company may not be able to develop *all* the possible uses of a technology because of its business focus or limited resources. Having multiple licensees with different fields of use may help to ensure that many uses of a technology are developed, may speed different types of products to market, and may increase the return to the licensor. Guidelines issued by agencies that fund inventions can sometimes be honored, in part, through field-of-use licensing.[1] It also can be used to focus company attention on humanitarian markets and ensure commercialization of products to serve the different needs of those markets (though this may be handled through territory limitations, rather than field of use). For any of these reasons, field-of-use licensing can be valuable. On the other hand, a restriction on field of use imposed by a potential licensor can reduce the motivation of a potential licensee, so a balance must be struck between the needs and motivations of each party to the license.

Even if a licensor sees only one possible field of use for an invention, it makes sense to limit an exclusive licensee to that field. Technology changes so rapidly that a new use for the invention would have a very good chance of developing during the life of the patent. A licensor should

Shotwell SL. 2007. Field-of-Use Licensing. In *Intellectual Property Management in Health and Agricultural Innovation: A Handbook of Best Practices* (eds. A Krattiger, RT Mahoney, L Nelsen, et al.). MIHR: Oxford, U.K., and PIPRA: Davis, U.S.A. Available online at www.ipHandbook.org.

Editors' Note: We are most grateful to the Association of University Technology Managers (AUTM) for having allowed us to update and edit this paper and include it as a chapter in this Handbook. The original paper was published in the *AUTM Technology Transfer Practice Manual* (Part IX: Chapter 4).

keep open the option of working with the best possible licensee for a new use, should one arise.

2. TECHNOLOGIES THAT ARE APPROPRIATE FOR FIELD-OF-USE LICENSING

A field-of-use license grants rights to the licensee to practice, not all uses of the licensed technology, but only a subset of those uses. The scope of the license could be limited by a general field of use (for example, digital recording or therapeutics) or a very specific field of use (for example, products for the treatment of human non-Hodgkin's lymphoma). In any case, the licensee's right to use the technology is limited in scope, leaving the licensor free to work with other companies on other uses.

Many types of technologies are appropriate for field-of-use licensing. In general, any technology that has, or may come to have, multiple, distinct uses may warrant this approach. Examples are easily found in the electrical engineering, computer, chemical, and health care areas. In the biochemistry department of a university, for instance, a new gene may be isolated and sequenced and its protein product expressed. This sounds like one technology, but it could easily lead to at least nine separate commercial uses:

1. Selling the protein product to the research reagent market
2. Making and selling antibodies directed against the protein to the research reagent market
3. Making and selling antibody-based diagnostic products
4. Making and selling DNA-based diagnostic products
5. Performing DNA-based diagnostic tests as a service
6. Making and selling the protein as a therapeutic product (this may be further focused by disease if the gene is involved in multiple disease states)
7. Using the gene and protein in-house for screening pharmaceutical drug candidates
8. Using the gene in gene therapy
9. Using the gene to develop a therapeutic based on antisense approaches

A company that sells to the research reagent market may not be in a position to make and sell therapeutic drugs (too much investment required). A company that develops therapeutics may not be interested in performing DNA-based diagnostic tests as a service (not enough return). A company that provides the DNA-based diagnostic service may not be capable of putting the protein on the research reagent market (no marketing and sales staff). Yet, each of these products is useful, further develops the technology, and is a potential source of revenue for the licensor.

What approaches can a licensor take when presented with a technology that has many distinct uses? There are at least three options:

1. License it to one company with no limitations, sit back, and hope that as the company maximizes its value from the license, all the markets will be served, and the licensor's returns also will be maximized
2. License it to one company with the requirement that it develop all uses, either directly or through sublicensing, and work closely with that company to ensure that it meets its obligations
3. License it to multiple companies with field-of-use licenses

This chapter is about the third option, a do-it-yourself approach, which entails more work, provides more control, and has a higher probability of maximizing the return for the licensor.

3. STRUCTURING THE LICENSE AGREEMENT TO LIMIT THE FIELD OF USE

Some technologies clearly have multiple uses from the outset. For other technologies the potential uses may not be so obvious, but it is worth planning for the possibility. In either case, a licensor has several approaches available for drafting agreements for distinct fields of use.

First, however, some homework must be done: one must ascertain the possible fields of use. For example, the potential licensor could ask: Is the latest product from the organic chemistry department useful as a fertilizer? A food additive? A perfume ingredient? A pharmaceutical? If it is useful as a food additive, can it be used in liquid products? Dried soups? Animal feed? If it is useful in animal feed, will it be useful in pet food? Livestock feed? Included as part of the normal market-evaluation process that most technology transfer professionals undertake, this exercise will yield essential information for developing the best field-of-use approach to take.

Once the possible fields of use are clearly defined, the next step is to market the technology to companies serving one or more of the markets those fields represent. Given a willing licensee and agreement on the scope of the license, several approaches can be evaluated for limiting the field of use in the actual license agreement.

3.1 *The grant clause*

The field of use can be limited in the grant clause by adding a phrase that delineates the field. The examples in this and the following two sections use various modifications to grant clauses from publicly available agreements to limit the field of use granted. (The original clauses and full agreements can be found on the example licensor's Web pages. Addresses can be found in endnotes.)

a. **PHS** hereby grants and **Licensee** accepts, subject to the terms and conditions of this **Agreement**, an exclusive license under the **Licensed Patent Rights** in the **Licensed Territory** to make and have made, to use and have used, to sell and have sold, to offer to sell, and to import any **Licensed Products** in the field of use of veterinary medicine and to practice and have practiced any **Licensed Processes** in the field of use of veterinary medicine.[2]

The approach in example *a* works well if the term being used to describe the field of use has a commonly accepted meaning. If it does not, or if clarification is needed, an additional (for example, exclusionary) sentence can be added to the grant, as in the following example:

b. Subject to the terms and conditions of this Agreement, Stanford grants Licensee a license under Licensed Patent to provide DNA-based diagnostic services in the Licensed Territory for providing DNA-based diagnostic services. This license specifically excludes the right to sell Licensed Product(s).[3]

In example *b*, there might be some ambiguity about whether the field of use of "providing DNA-based diagnostic services" includes selling DNA-based diagnostic products that enable others to carry out a diagnostic test. The additional sentence clarifies the limitation on the licensee: the licensee cannot sell Licensed Products. Providing diagnostic services must therefore be limited to an activity in which the licensee itself uses the Licensed Products.

In these two examples, the underlined language in the grant clause limits what otherwise would have been an unlimited license for any and all uses of the technology. Note that the language can define what is included in the field, as well as what is excluded. This approach to limiting the field of use in the grant can be taken with no other field-of-use-specific language in the license agreement, or in conjunction with related language in the Definitions section, as described below.

3.2 *Defining the field*

Perhaps the most common approach to limiting the field of use in the license agreement is to establish *Field* or *Licensed Field of Use* as a defined term in the agreement. It then can be used to limit the field in the grant clause. This approach has the advantage of simplifying the grant clause, while allowing a full definition of the field elsewhere. This is especially advantageous in a grant clause that is already lengthy or segmented, or for a field that cannot be expressed adequately in a few words. Examples of possible paired definition and grant clauses follow:

a. Field of Use. shall mean the field of research reagent products. LICENSED FIELD OF USE specifically excludes the field of human diagnostic products.

   OHSU hereby grants and Licensee accepts, subject to the terms and conditions of this Agreement, a nonexclusive license under the Licensed Patent Rights in the Licensed Territory to make and have made, to use and have used, and to sell and have sold any Licensed Products and/or Licensed Processes in the Licensed Field of Use.[4]

b. FIELD shall mean the field of human vaccines and human therapeutics for Acquired Immune Deficiency Syndrome.

Dartmouth hereby grants to Company and its Subsidiaries an exclusive, royalty-bearing license under Dartmouth Know-How and Dartmouth Patent Rights to make, have made, use, and/or sell Licensed Products in the Field in the Territory. Notwithstanding the foregoing, Dartmouth expressly reserves a nontransferable royalty-free right to use the Dartmouth Patent Rights and Dartmouth Know-How in the Field itself, including use by its faculty, staff and researchers, for educational and research purposes only. Company agrees during the period of exclusivity of this license in the United States that any Licensed Product produced for sale in the United States will be manufactured substantially in the United States.[5]

An alternative construction would include a phrase *in the Grant* to limit the license, and then define that phrase in the Definitions. As an example:

c. Human Cancer Therapeutics shall mean the treatment of human patients exhibiting malignant tumors, including but not limited to carcinomas, sarcomas and lymphomas.

Subject to the terms and conditions of this Agreement, Stanford grants Licensee a license under Licensed Patent in the field of Human Cancer Therapeutics.

Example *c* has the advantage of being custom tailored, while examples *a* and *b* have the advantage of being model documents that can be revised more simply for a new technology. The only change needed to the model document during drafting is in the Definitions; the Grant is designed to be used without modification and to be limited as to field of use by an appropriately defined term.

3.3 *Limiting rights through reference to patent claims or separate patent applications*

A third general approach to limiting the field of use of a license involves limiting the grant of rights to specific patent claims, or to a specific family of related patent applications. A well-written patent application will cover broad areas related to the technology. If the claims, however, fall into distinct groups, one could reference the claims necessary for the intended field of use or specifically exclude claims that cover uses not intended for inclusion in the license. Here are some examples of grant language that could be used in this type of approach:

a. Where an issued patent exists and is all that is referenced in the Definitions section under patent rights, the approach is straightforward. Determine the issued claims that are required for the field of use and reference them by number in the Grant. For example:

PHS hereby grants and **Licensee** accepts, subject to the terms and conditions of this **Agreement**, an exclusive license under claims 1 through 7 in the **Licensed Patent Rights** in the **Licensed Territory** to make and have made, to use and have used, to sell and have sold, to offer to sell, and to import any **Licensed Products** and to practice and have practiced any **Licensed Processes**.

b. Another reasonably straightforward situation is where a distinct invention associated with the field of use is contained within one patent application within a family of related applications that otherwise covers broader uses of the technology outside of the intended field of use. In this situation, the patent application can be the basis of the definition of licensed patents, but care must be taken not to intermingle different uses of the technology between patent applications during prosecution. The grant language would be unchanged, and the definition of the patent rights to be licensed would be limited to the appropriate patent application, as in the following example:

> ***Licensed Patent Rights*** shall mean:
>
> 1) U.S. patent application (serial number) filed (filing date), the inventions claimed therein, and to the extent that the following contain one or more claims directed to the inventions claimed in U.S. patent application (serial number), all divisions and continuations of this application, all patents issuing from such application, divisions, and continuations, and any reissues, reexaminations, and extensions of all such patents;
> 2) to the extent that the following contain one or more claims directed to the invention or inventions claimed in U.S. patent application (serial number): *i)* continuations-in-part of a) above; *ii)* all divisions and continuations of these continuations-in-part; *iii)* all patents issuing from such continuations-in-part, divisions, and continuations; and *iv)* any reissues, reexaminations, and extensions of all such patents;
> 3) to the extent that the following contain one or more claims directed to the invention or inventions claimed in U.S. patent application (serial number): all counterpart foreign applications and patents to *a* and *b* above.
>
> **Licensed Patent Rights** shall *not* include *a, b,* or *c* above to the extent that they contain one or more claims directed to new matter which is not the subject matter of a claim in U.S. patent application (serial number).

Note that this patent rights definition allows for the usual possibilities during prosecution (divisions, continuations, foreign counterparts); but where a normal descendant, a continuation-in-part, may bring in new matter, the definition limits that case's inclusion to claims related to the subject matter of the original patent application. This provides some assurance that uses of the invention beyond the intended field of use will not be wrapped into the license during the process of attempting to get a patent to issue.

It should be noted that there are some drawbacks associated with limiting the field of use solely by reference to a patent application still in prosecution. It is much cleaner to refer to an already issued claim (see section 3.3, paragraph a, above). The claims of a case still in prosecution can change through modification, deletion, or addition; in theory, they could change in ways that are not consistent with the intended field of use. Thus, when working with a patent application, as opposed to an issued patent, the approach outlined in this section can be combined with language that specifically states the field of use (see 3.2.a and 3.2.b, above). This "belt and suspenders approach" ensures that the field of use will be clearly defined, while separating out the claims to that field in a separate patent application. The additional value of having one licensee's claims in a separate patent property will become apparent in the following sections on "Reimbursing patent expenses" and "Handling patent infringement/interference issues."

4. SPECIAL ISSUES IN FIELD-OF-USE LICENSING

Several problems may be encountered if, instead of granting all rights associated with a technology to a particular company, a licensor divides those rights by field among several companies. These problems, which are described in the following three sections, arise whether or not the field-of-use licenses are exclusive; in fact, some of the problems are the same as those that occur when licensing nonexclusively without limitation as to field of use. The good news is that, with some planning, a licensor can minimize these problems.

4.1 *Overlap of rights between licenses*

In the field-of-use licensing, the licensor works to clearly define the possible fields of use for a technology. While attempts can be made to distinguish fields as much as possible with currently available information, only hindsight can be crystal clear. The licensor and licensees should be aware that overlap in fields might occur in the future. An overlap could be due to different interpretations of the rights granted under licenses or to unexpected future technical developments.

Such overlap could have significant economic impact on a licensee. For example, it could render nonexclusive a market segment that the licensee expected to hold exclusively, which could reduce a licensee's income stream in its field of use. While the economic interests under dispute affect the licensees, it is through the contract with the licensor that the situation can be resolved most effectively.

It is wise to lay the groundwork early on for resolving potential disputes related to this specific issue. A provision in each license that allows the licensor to resolve disputes may be acceptable. Alternatively, there could be a commitment to mediation, arbitration, alternative dispute resolution, or some other means short of litigation. Of course, the best course involves ongoing, constructive dialogue between the licensee and licensors, so that when problems arise, good communication and strong relationships needed to encourage negotiated solutions will already exist. If all parties enter the relationship with awareness of the potential need for dispute resolution, and if they agree, before problems arise, on a balanced way to deal with a dispute, then such problems will be easier to manage if and when they arise.

A variation on this theme is the issue of cross-prescription or cross-marketing—when the licensee sells products for use under its field, but the products are usable by the purchaser outside that field, in a field licensed to another company. Again, advance planning can help head off serious problems. For example, in the area of therapeutics, it would be worthwhile to group together fields that will use the technology in the same delivery form, and then grant a license to one company for these fields. If a therapeutic can be used intravenously, at similar concentrations, to treat both cancer and heart disease, it may be wise to license both uses to one company. There are multiple benefits to all parties in such instances. One party can handle research, development, regulatory approval, and sales more efficiently. Cross-prescription will not be a problem because proceeds flow to the same licensee. In addition, the licensee can choose independently to work with another company through sublicensing to develop one or more of the uses, staying in closer control while accessing needed resources. Grouping related uses together in a larger field provides the licensee with a larger incentive to invest in the technology and reduces problems for the licensor.

4.2 *Maintaining control of patent prosecution*

The interests of licensee and licensor do not always overlap during prosecution. This truism is amplified when a licensee has a limited field of use. The licensee may not be willing to support prosecution of certain claims or may seek to modify claim language to enhance the patent's value to the licensee at the expense of other licensees or the licensor. For this and other reasons, it is recommended that the licensor retain control over patent prosecution, while seeking to fairly distribute costs over field-of-use licensees.

4.3 *Reimbursing patent expenses*

As with any program involving multiple licensees for a technology, the field-of-use licensor must manage patent expenses creatively. With no single licensee committed to paying or reimbursing all costs, the licensor must choose another mechanism to cover patent expenses. The possibilities include the following:

a. The licensor covers patent expenses up front, reimbursing them from the royalty stream. This model results in licenses that have no patent-reimbursement language.
b. If the field-of-use licenses have been structured to relate to distinct patent applications or patents, costs can be cleanly linked to a specific license, and patent-reimbursement language as per a standard, exclusive license agreement will suffice.
c. The licensor prorates patent expenses over multiple licensees. This approach involves patent-reimbursement language in the license, with a variation on the standard theme. For example, *"On March 1 of each year during the term of this Agreement, Licensor shall provide Licensee an invoice for Patent Expenses equal to the patent costs for*

the prior calendar year divided by the number of licensees of Licensed Patents during that calendar year. Costs will be prorated for licenses that are effective for only a portion of said calendar year. Licensee shall pay this invoice within thirty days of receipt."

d. In some situations, considerable patent expenses can accrue before a technology is successfully licensed. In this scenario, if costs are to be reimbursed by the licensees, language can be used to include future licensees in that reimbursement. A fixed sum of past patent expenses can be attached to each license, or the initial licensee(s) can reimburse all the costs to make the licensor whole and then use those payments as credits as new licensees sign up. This last approach has the advantage of providing some incentive to licensees to have other companies also licensed under the technology.

4.4 *Handling patent infringement/ interference issues*

In field-of-use licensing, as with nonexclusive licensing, the lack of an all-inclusive license held by any one company reduces the licensee's motivation to protect the patent in an interference or infringement situation. The exclusive field-of-use licensee has more motivation than a straight nonexclusive licensee, because it has some exclusivity and would possibly have significantly more competition in the absence of a valid patent. Other parties (the other licensees), however, would also benefit from the patent being upheld, so that any one company may be unlikely to agree to bear the total cost of interference or litigation.

Again, there are clear advantages to designing the patent filing strategy for field-of-use licensing. If a field-of-use licensee is the only licensee of a particular patent or application in a family of related patents on a technology, the standard arrangements made with an exclusive licensee still can be used, focusing on that particular case.

If the field-of-use licensing has been undertaken in such a way that more than one licensee has an interest in a particular patent property, the simplest approach is for the licensor to carry interference and infringement costs alone, recovering them through royalties or settlements. Using this approach, the licensor retains more control. The approach also places the risk and cost on the licensor, and thus should be taken only when the potential reward justifies the resources required. Financial and legal support for these events could be obtained from other sources within the licensor's organization, supplied from a set-aside created at the beginning of the royalty stream, or covered by an insurance product carried by the licensee or licensor. Part of the planning process for field-of-use licensing (as for nonexclusive licensing), therefore, includes developing a strategy to manage the possibility of sizable future costs that might be borne solely by the licensor.

Another approach to addressing possible infringement and interference actions would be to work out a mechanism to share the costs and management of these activities with one or more licensees. For example, a licensee could be allowed or required to take the lead in litigating infringement in its field of use. The net proceeds could be treated as net sales or profits, as appropriate, for earned royalty purposes. Alternatively, both parties could share the costs and proceeds within the licensee's field, or the licensor could take the lead in litigating infringement, retaining all proceeds. These suggestions are much the same as those a licensor would select from for any exclusive license. In this case, the licensed field of use limits the infringement or interference actions that would trigger licensee responsibility.

It should be noted that the existence of more than one exclusive licensee makes it more likely that a licensor will be drawn into litigation as the only party having standing to sue. The license can require that the licensee cover any licensor legal costs, but for licensors that do not want to be named as a party to a lawsuit, a single exclusive licensee with an undivided interest that is required by the license agreement to take the lead in litigation may still be preferable.

4.5 *Diligence*

Managing diligence by the licensee is one of the issues that become simpler with field-of-use licensing. For example, if one company has responsibility for developing products for less developed countries, or for developing a human therapeutic, it is straightforward for the licensor to assess licensee performance. Having a field of use isolated from other fields removes the need to stage commercialization of products for multiple fields because of resource limitations for a single licensee with responsibility for more than one field.

5. CONCLUSIONS

The guidance provided here is intended to help licensors maximize the reach of their innovations into multiple fields, whether those fields exist at the time of the license, or arise as the innovation develops. Sometimes one licensee can develop the full potential of a technology, but often it will take multiple partners, each with its own focus, resources and expertise, to fully realize that potential. ■

SANDRA L. SHOTWELL, *Managing Partner, Alta Biomedical Group LLC, 7505 S.E. 36th Avenue, Portland, OR, 97202, U.S.A.* *shotwell@altabiomedical.com*

---

1 For example, the National Institutes of Health from time to time issues guidelines intended to ensure broad access to certain types of technologies, such as biomedical research tools, and suggests limitations on how such technologies should be licensed. (See, for example, Sharing Biomedical Research Resources: Principles and Guidelines for Recipients of NIH Research Grants and Contracts at ott.od.nih.gov/policy/rt_guide_final.html#20.) The approach some institutions have taken to follow these guidelines has been to issue nonexclusive licenses for the research reagent market and exclusive licenses for therapeutics or other fields requiring significant investment.

2 See model agreements at ott.od.nih.gov.

3 See sample documents at otl.stanford.edu/industry/resources.html#documents.

4 See sample agreements at www.ohsu.edu/tech-transfer/index.shtml.

5 See www.autm.net/aboutTT/aboutTT_policies.cfm or www.dartmouth.edu/%7Etto/standard.html.

Problems with Royalty Rates, Royalty Stacking, and Royalty Packing Issues

KEITH J. JONES, *Executive Director, Washington State University Research Foundation, U.S.A.*
MICHAEL E. WHITHAM, *President, Whitham, Curtis, Christofferson & Cook, P.C., U.S.A.*
PHILANA S. HANDLER, *Associate, Whitham, Curtis, Christofferson & Cook, P.C., U.S.A.*

ABSTRACT

Virtually all products now developed using biotechnology, genetic engineering, and chemistry are technologically complex, incorporating many different inputs. While this alone complicates R&D efforts, there is also the added complexity of potentially relevant intellectual property (IP) rights held by third parties, attached to these inputs. For example, R&D for a new vaccine might have used numerous inputs with corresponding third-party proprietary rights attached: research tools, recombinant techniques, DNA sequences, transformation vectors, cell lines, adjuvants, and delivery devices. Hence, when the vaccine is ultimately ready for use, it will likely be subject to royalty obligations to many licensors. This dilemma of multiple royalty obligations is called royalty stacking. This occurs when various licenses combine to impose aggregate royalty obligations of 6%–20% (or greater). Royalty packing, a similar situation where multiple technologies are bundled together (for example, multiple vaccine packages), is sometimes imposed by the licensor or by best practices within an industry or health ministry. The resulting aggregate-royalty problem is the same as with royalty stacking. There are several techniques to manage royalty stacking and packing: royalty ceilings, royalty floors, variable royalties, and royalty alternatives (lump-sum payments and patent pools). Royalty stacking and packing are serious licensing issues that any organization involved in IP management and technology transfer can, and must, proactively and preemptively plan for and manage.

1. INTRODUCTION

Virtually all products developed using biotechnology and chemistry are protected by one or more tools of intellectual property (IP) rights, for example, patents, material transfer agreements, and trade secrets. Royalty rates that licensees must pay on sales or use of these products can vary widely depending on how the products will be used, where they will be used, and the relative bargaining positions of the licensees and licensors at the time of drafting the license agreement for the product. In addition, most biotechnology products are made using one or more patented-research tools, each of which may have *reach through* royalty obligations; obligations to pay for sales of products made using the research tool, even though the patent holder does not have a patent on the product which is produced. This type of requirement should not be confused with patent misuse which may include a violation of antitrust laws.[1] Those royalties may be related to a product identified using a proprietary research tool and requiring the use of several different patented technologies owned by several different entities.

One example of *royalty stacking* would occur under these circumstances: a potential vaccine is identified and tested using one or more proprietary research tools that have all been licensed by different companies; the vaccine is produced using recombinant techniques and employs proprietary DNA sequences; at the same time, the vectors used for insertion and expression are owned by additional companies, while production of

Jones KJ, ME Whitham and PS Handler. 2007. Problems with Royalty Rates, Royalty Stacking, and Royalty Packing Issues. In *Intellectual Property Management in Health and Agricultural Innovation: A Handbook of Best Practices* (eds. A Krattiger, RT Mahoney, L Nelsen, et al.). MIHR: Oxford, U.K., and PIPRA: Davis, U.S.A. Available online at www.ipHandbook.org.

the vaccine employs a proprietary cell line; the vaccine itself is packaged with one or more proprietary adjuvants and is delivered to patients using a patented delivery method or device. When the vaccine is ultimately ready for use, it may be subject to royalty obligations to several different companies or licensors. The various licenses involved may ultimately impose combined royalty obligations of 6%–20%, or more, of the selling price of the product. Further complicating matters is the need for separate reporting and accounting to each of the licensors. Table 1 provides another example of royalty stacking involving a multiantigen vaccine with a proprietary adjuvant. This situation might require total royalties on the selling price of 8%, with separate reporting requirements to four different entities.

Often, a burden of 8% versus 4%, for example, can make the difference as to whether the vaccine is commercialized at all. Similar problems arise in agriculture where a genetically engineered crop might be made using proprietary varieties, proprietary vectors, proprietary gene sequences, and proprietary research tools, all owned by different companies. In one case, a published *freedom to operate* report[2] indicated that Golden Rice,[3] a line of rice genetically engineered at a university to have significant expression of pro-vitamin A, was covered by 45 patents or patent families and patent applications by more than 20 different owners in the United States. Fortunately, for the 124 million individuals severely afflicted with vitamin A deficiency (VAD) and the 500,000 cases of irreversible blindness, it was possible to obtain royalty-free licenses for use in developing countries, thanks to the strong support this project received from many companies. However, in the commercial realm, potential royalty obligations for a particular product may be too high collectively to allow for development and commercial implementation of the product. The royalty stacking problem can often be compounded in agricultural technologies. For example, a new vaccine for a pig disease will often need to be *packaged* along with vaccines for other pig diseases, if the vaccines must be administered at the same time.

Individuals that are charged with the management of IP in health and agriculture will need to deal with issues involving royalties and royalty stacking on almost every product or technology they encounter. This paper is intended to highlight some of these issues, explain the competing interests, and provide commentary on practices that can be adopted.

2. WHAT DOES THE ROYALTY APPLY TO?

2.1 *The "royalty basis"*

Clearly, one of the goals of an IP license is to allow the licensor to receive a quantifiable sum of money based on a licensee's use of a proprietary technology, or sale of products made using or incorporating the proprietary technology. The license

Table 1: Royalty Components of a Multiantigen Vaccine

Vaccine component	Royalty on sales of vaccine
Antigen A, Proprietary to Company A	2%
Antigen B, Discovered with proprietary tool of Company B	2%
Antigen C, Nonproprietary	0%
Proprietary assembly technique of Company C	2%
Proprietary adjuvant	2%

should include a provision for basic reports that identify the sales on which royalties are due and that itemize any deductions (for example, documented returns of product, damaged product, and free samples) that have been agreed upon. The licensee should keep accurate records so that sales records can be audited and reports can be verified. The records should allow the licensor to confirm that it is receiving accurate royalty revenue and that the licensee is complying with all milestones and other provisions of the license, such as the reporting of minimum sales figures.

Seemingly simple operations can be difficult in some licensing situations. Tallying up unit sales and multiplying the total by a percentage or price-per-unit royalty can become complicated when the licensee bundles a licensed product with other licensed products. A licensor may believe that its technology makes the product more valuable in combination with others, and that the licensor should be due a royalty on the selling price of the *combination or collection* product. Without a prior agreement on and consideration of such a product-combining approach, the licensee may risk patent infringement litigation. For an example, refer to *Georgia-Pacific Corp. v. United States Plywood Corp.* 318 F. Supp. 1116 (S.D.N.Y., 1970). In this case, the court sought to provide royalties based on the value of the IP, rather than the resulting combination. (Court-imposed royalty rates may be higher or lower than either party has agreed to in advance.)

In cases involving a combination or collection product, the licensee may be of the opinion that the portion of the collection covered by proprietary rights of the licensor constitutes only a small fraction of the value of the combination or collection product. Resolving the value of the proprietary product versus the value of the combination or collection product can be especially difficult if the proprietary product is not being, or has never been, sold separately by the time a dispute arises. One way of handling this type of problem is to add a valuation calculation methodology to the license agreement. However, it should be recognized that parties to a license agreement may be motivated to make the calculation work in their own favor, and disputes can arise on how calculations are made. To avoid this type of problem, the agreement may stipulate that the product be sold only as a single unit unless otherwise agreed to by the licensor. Still another way to address the issue is to specify in the agreement that royalty will be calculated based on the sale price of the proprietary product if it is sold alone, or on the sale price of the combination or collection product if the product is sold as a combination or collection.

Often, license agreements will specify that a licensed product is one that infringes *valid claims* of a licensed patent in a territory where the licensed product is made, sold, or used. This type of provision has the immediate effect of eliminating royalties on products manufactured and sold in areas where licensed patents do not exist. Further, this type of language can permit the licensee to refuse payment of royalties on the grounds that a valid patent does not exist in the territory where royalties are sought. From the licensee's perspective, there will be a concern that the licensee will have competition from unlicensed competitors in territories where patents do not exist. However, from the licensor's perspective, particularly in cases where an exclusive license is given and where data, information, and other know-how is provided in addition to rights under patents and patent applications, a licensee benefits from more than just the patent rights provided under the license and should be obligated to pay royalties on all sales of licensed products.

This issue can be addressed by designing the license agreement to address both patents and know-how.[4] Such agreements should include: (1) provisions that separate royalties from different technologies (such as royalties from patented technologies and royalties from use of trade secrets); (2) provisions that eliminate royalties from patents that expire or are invalidated (see *Brulotte v. Thys.* 379 U.S. 29, 33 (1964) and *Pitney-Bowes, Inc. v. Mestre* 517 F. Supp. 52 (S.D. Fla. 1981), which represent the view that royalties should not be due on patents upon expiration or invalidation; (3) provisions that address when a trade secret becomes known or subject to a patent; and (4) a provision that the license to know-how and/or trade secrets continues after expiration of a patent. Care must be taken to define what the obligations are for transferring know-how. For example, a university, private nonprofit, or governmental body would likely not want to be obligated to provide the same services implicated in a know-how license that commercial transaction might involve (for example, the delivery of a working prototype or a provision for a certain number of hours of instruction time).

Another way of avoiding the problems involving royalties on products manufactured and sold in areas where licensed patents do not exist is to include a provision that the licensor receives reduced royalties in territories where patents do not exist or to provide for the payment of royalties for a shortened term in territories where patents do not exist. It may be appropriate to set the royalty rate at zero in developing countries where no patent exists.

With respect to tying the royalties to valid claims covering a product produced or sold by a licensee, the technology manager at a university or within a government agency in a developing country should recognize that such a requirement favors the licensee and that the licensee may be able to benefit, for very little money, from a proprietary position on a technology (that is, prevent the licensor from licensing to others for a period of years) by commercializing a product which, according to the licensee, does not infringe the patent claims. Further, the licensee could take this position in any of several different countries or jurisdictions in the world (that is, challenge the validity of a patent in India while separately challenging the validity of a related patent in the United States). Such actions could force the licensor to attempt to prove in court that the product being produced by the licensee indeed infringes the patent claims, or attempt to license the technology to another party (in which case the value of the technology would be likely to be less because the remaining patent term would be less, obviously, than the term of the original agreement with the licensor). Neither option is very helpful to a licensor who has had its technology tied up with a company that will ultimately not commercialize the technology. The licensor could address this potential frustration by requiring the licensee to agree in advance that, regardless of any finding of patent infringement, royalties will be due on the product under development by the licensee.

Further, the license agreement might define valid claim to include any claim in any patent that has not been adjudicated, by a court of competent jurisdiction, to be invalid and from which no appeal has or can be taken. With this provision, the licensor might be able to collect royalties up until a final adjudication of patent invalidity. Of course, such a definition would not benefit the licensee in cases where prior art that is *spot on* is identified to the licensor.

2.2 *Royalty stacking*

Royalty stacking occurs when multiple patents affect a single product and thus involve multiple licenses. As noted above, a biotechnology product may require separate licenses for use of such items as research tools, gene sequences, expression vectors, cell lines, and adjuvants. Thus, from the prospective of the company making the product, the multiple royalty demands must be "stacked" together to determine the total royalty burden on producing the product. Because royalty stacking involves many IP holders, efficient exploitation of a product subject to royalty stacking may be inhibited (that is, development can be delayed or discontinued completely) and the development of future products might be impeded.

2.3 *Royalty packing*

Royalty packing occurs when there is a requirement to bundle one technology with other technologies. Such a requirement could be imposed by the licensor, but also could be imposed by best practices within an industry or by a health ministry. For example, a vaccine could be required to be administered simultaneously with one or more different vaccines that are proprietary to one or more different companies in order to reduce the cost of administration. In this situation, the royalties imposed on each of the proprietary products that are administered will be "packed" together. Royalty packing may result in the aggregate cost of the several packed products being too high.

3. TECHNIQUES TO MANAGE ROYALTY STACKING AND PACKING

A licensee may seek to impose a *ceiling* for royalties in any agreements it makes with licensors. For example, the licensee might establish a ceiling of 6% for combined royalties on product sales. In turn, if the stacked royalties exceeded 6%, each of the licensors would be agreeing to have the royalties they are to be paid reduced on a pro rata basis, so that the total royalties due to the licensors would be 6%. In this situation, the licensee may be motivated to add more technologies to its product or process because its total royalties per unit are capped. To the contrary, the licensor may dispute the need to add the additional technologies to the product and may be frustrated if its own share decreases much below the expected return. In many situations, licensors take the position that their technology is the *most important* and that their share of the royalties should not be depleted pro rata. These types of competing interests require the parties to have a good understanding of how and when reductions would apply when the agreement is made and good communications between the parties when new technologies are incorporated into a product that would affect the licensor's expected royalty stream. Also, there may be a need to differentiate some types of royalties from others. For example, some licensors may be willing to agree to a pro rata reduction in royalties when other proprietary technologies are used in the product to be commercialized. But the licensors may not be willing to agree to a reduction due to reach through licenses resulting from the licensee's use of proprietary research tools.

A licensor may seek to impose a *floor* below which its share of the royalties may not fall. For example, if additional technologies are required to exploit a product, a licensor might agree to have its royalties reduced on a pro rata basis, but not below a specified floor (for example, the license requires royalties of 5% but allows for reduction, if additional licenses are required, with the proviso that in no event will the amount due be less than 2% per unit sold). The licensor may agree to a reduction to the floor only if a license from a third party with a dominant patent position to the licensor is required to effectively use the licensor's technology. That is, a licensor may not agree to a reduction if additional technologies are *desired* by the licensee to make a better product, but not *needed* to use the invention—for example, the license agreement might specify that if an additional license to practice the invention described in the licensed patent(s) is required from a third party, the licensee may reduce its royalty payments by 50% (or by an amount equal to the amount that would have been due to the licensor, but in no event shall such reduction be more than 50%). It is not unusual to have in the same license both a ceiling on stacked royalties and a hard floor below which royalty rates could not fall. The hard floor may need to take into account other deductions from royalty payments that are allowed by the license. For example, a deduction of patent costs may be allowed, but will be limited in any year by the hard floor in royalty payments.

Licensees and licensors might agree to have *variable royalties* that depended on, for example, the importance of the technology in relation to the creation of the product. The more important the role a proprietary technology plays in a product, the higher the royalties, and vice versa (for example, the owner of proprietary antigen in a vaccine raised against the antigen would receive higher royalties than the owner of a proprietary expression system for expressing the antigen). In this situation, however, it is likely that licensors and licensees would disagree over the importance of the proprietary technology in relation to the product being developed.

Packing issues may be handled by requiring that the royalty be calculated based on the sale prices of the product if sold alone, or the sale price of the combination or collection product if the proprietary product is sold as a combination or collection.

4. OTHER MATTERS

Not every arrangement requires revenues in the form of a royalty stream. For example, a lump-sum payment for use of a research tool may be an appropriate way to disseminate and exploit a patented technology. Some technologies may best be collected in *patent pools* which allow for free use of the technologies or use of the technologies at fixed prices. A patent pool can make the licensed technology more widely available for use in different markets (for example, different products could incorporate the technology), and, further, access to a number of other different but related technologies that would be useful to a university or nonprofit organization might be available within the patent pool. Such arrangements may allow research and development using a variety of proprietary technologies without the need to negotiate licenses.

5. CONCLUSIONS

License agreements should clearly define when and how a licensor will be paid a royalty. An important part of any agreement is a clear definition of the product, such that both parties understand what royalties will be based on. Further, to avoid any disputes on royalty payments, the agreement should also clearly define when royalties are not due. Royalty stacking should be recognized and understood by those involved with managing IP in the health and agriculture fields, particularly when biotechnology products, services, and research tools are involved. Providing agreements that allow commercialization of a product that embodies the proprietary technology of several different companies, and for which royalty payments are due to each of those companies, requires recognition by the parties of the role each technology performs if royalty ceilings, floors, or other mechanisms to address stacking are to be adopted. Finally, alternatives to royalty-bearing arrangements should be considered, including the use of lump-sum payments and patent pools.[5] ■

KEITH J. JONES, *Executive Director, Washington State University Research Foundation, 1615 NE Eastgate Blvd., Pullman, WA, 99163, U.S.A. jonesk@wsu.edu*

MICHAEL E. WHITHAM, *President, Whitham, Curtis, Christofferson & Cook, P.C., 11491 Sunset Hills Road, Suite 340, Reston, VA, 20190, U.S.A. Mike@WCC-IP.com*

PHILANA S. HANDLER, *Associate, Whitham, Curtis, Christofferson & Cook, P.C., 11491 Sunset Hills Road, Suite 340, Reston, VA, 20190, U.S.A. Philana@WCC-IP.com*

---

1 Feldman RC. 2003. The Insufficiency of Antitrust Analysis for Patent Misuse. Hastings Law Journal 55:399-450. www.robinfeldman.com/Insufficiency%20pdf.pdf.

2 Kryder RD, SP Kowalski and AF Krattiger. 2000. The Intellectual and Technical Property-Components of pro-Vitamin A Rice (Golden Rice): A Preliminary Freedom-to-Operate Review. ISAAA Briefs No 20. ISAAA: Ithaca, NY. www.isaaa.org/kc/Publications/pdfs/isaaabriefs/Briefs%2020.pdf.

3 Potrykus I. 2001. Golden Rice and Beyond. *Plant Physiology* 125:1157–1161. See also childinfo.org/areas/vitamina/.

4 See, for example, Bleeker RA, BH Geissler, A Lewis and R McInnes. 2003. Certain Clauses in Know-How and Hybrid License Agreements in Several Jurisdictions. *Les Nouvelles* 38(1):10–17.

5 Additional publications of great relevance to this topic are:

Adhikari R. 2005. Patents, Royalty Stacking and Management. *World Pharmaceutical Frontiers* 25. www.worldpharmaceuticals.net/pdfs/025_WPF008.pdf.

Clark V. 2004. Pitfalls in Drafting Royalty Provisions in Patent Licences. *Bio-Science Law Review*. pharmalicensing.com/articles/disp/1087832097_40d70021d738c.

Feldman RC. 2003. The Insufficiency of Antitrust Analysis for Patent Misuse. *Hastings Law Journal* 55:399–450. www.robinfeldman.com/Insufficiency%20pdf.pdf.

Lemley M and C Shapiro. 2006. Patent Hold-Up and Royalty Stacking. Working Paper. Haas School of Business, University of California at Berkeley: Berkeley, California. http://faculty.haas.berkeley.edu/shapiro/stacking.pdf.

Mireles MS. 2004. An Examination of Patents, Licensing, Research Tools, and the Tragedy of the Anticommons in Biotechnology Innovations. *University of Michigan Journal of Law* 38:141.

Mowzoon MM. 2003. Access Versus Incentive: Balancing Policies in Genetic Patents. *Arizona State Law Journal* 28:1077.

Sutton G and E Ewing. 2002. Government Alliances with Biotechs Introduce "Royalty Stacking" Issues. www.klng.com/files/tbl_s48News/PDFUpload307/8169/Ewing_Sutton_12_2002_MHT.pdf.

CHAPTER 11.10

In-Licensing Strategies by Public-Sector Institutions in Developing Countries

KANIKARAM SATYANARAYANA, *Chief, IP Rights Unit, Indian Council of Medical Research, India*

ABSTRACT

In the past, it was possible for some countries to ignore IP (intellectual property) management while pursuing economic development and improved public health. Globalization, however, has brought the world closer and closer together, and with the advent of the Agreement on Trade-Related Aspects of Intellectual Property Rights (TRIPS), no country can afford to be isolated from the global IP system. This chapter explains how developing countries can use this new system to their advantage through in-licensing technologies (that is, bringing technology into the public sector through patent license agreements). Offering an overview of the usual requirements of a license agreement, the chapter also considers issues that are uniquely relevant to public-sector institutions in developing countries as they negotiate such licenses.

1. INTRODUCTION

Thanks to globalization, the rules governing intellectual property (IP) are changing rapidly. Many countries, such as India, that formerly stood outside the patent system have become fully compliant with the Agreement on Trade-Related Aspects of Intellectual Property Rights (TRIPS). For developing nations with strong science and technology bases, established pharmaceutical industries, and emerging biotechnology industries, adherence to TRIPS compliance and the ensuing changes have created both challenges and opportunities. Developing countries can produce health products in two ways: first, by licensing technologies developed by public-sector research and development (R&D) institutions to the pharmaceutical industry (including the biotechnology industry in general, which encompasses agricultural applications); and second, by in-licensing technologies from the pharmaceutical industry. While the public sector wants to introduce affordable health products to the marketplace, the biotechnology industry is primarily interested in optimizing its investment returns. But compromises can be made. For example, IP developed by the biotechnology industry can be transferred to the public sector for further development.

In-licensing is a well-recognized strategy for transferring technologies from companies to the public sector. In-licensing allows many parties to manufacture products, thereby creating enough competition to bring down the costs of public health products (like drugs, diagnostics, vaccines and other biologicals) and crops in agriculture. IP licensing is often complex because the parties concerned have conflicting objectives. Furthermore, the biotechnology industry, at least in developing countries, usually is not very eager to work with often-times inefficient and incompetent government officials. In any case, all parties involved in IP licensing need:

- the skill to negotiate a deal
- a strategy for negotiation

Satyanarayana K. 2007. In-Licensing Strategies by Public-Sector Institutions in Developing Countries. In *Intellectual Property Management in Health and Agricultural Innovation: A Handbook of Best Practices* (eds. A Krattiger, RT Mahoney, L Nelsen, et al.). MIHR: Oxford, U.K., and PIPRA: Davis, U.S.A. Available online at www.ipHandbook.org.

- practices that protect the interests of the public sector

2. TYPES OF AGREEMENTS

2.1 *General Requirements*

IP transfer agreements must address a number of aspects: confidentiality, material transfer, development (the licensee assumes all responsibility for further development), co-development (two parties collaborate on continued development), and distribution.

Such agreements are at least two-way because more than one public-sector institution can be involved in developing a product. For example, if the Indian Council of Medical Research, New Delhi, (ICMR) were to in-license a technology for developing a vaccine from a private company, there could be at least three parties involved in the agreement: the ICMR, which is the licensee and a public-sector institution; the licensor, which is a private company; and the Ministry of Health & Family Welfare, Government of India, which will fully or partly fund the project, conduct clinical trials, and make the vaccine available to the public. Usually, either the public-sector agency or the private company will provide the first draft of a negotiation agreement.[1] It is important that all the parties, especially the licensee, clearly understand the basic philosophy behind the deal: to provide a product to people who would not have access to it without government support. A good agreement is one that benefits all parties.

Well-drafted agreements should allow government officials to negotiate quickly, get approval from the bureaucracy, as appropriate, and come to a consensus. Since it takes several years to bring a product from the laboratory bench to the patient's bedside, mutual trust is very important during the negotiations and implementation of the project, especially if some renegotiation is needed partway through. Court battles are messy, expensive, and generally unwelcome, especially if they involve a foreign party.

Parties intending to enter a long-term working relationship with each other may either sign a series of agreements, one omnibus comprehensive agreement (with smaller specific agreements attached), or one broad, general agreement with two or more related, but separate, specific agreements. The following sections describe the kinds of agreements that can be signed by two parties engaged in jointly developing a product. The appendices provide examples of agreements that might be used by public-sector organizations.

2.2 *Confidentiality agreements*

The development of a proprietary health product usually involves the use of confidential information: research data, sources of materials, methods of production, designs of specialized proprietary equipment, and other nonscientific business information. The involved parties should therefore enter into a confidentiality/nondisclosure agreement. Such an agreement not only protects commercially useful information but also indicates the value of that information. Such agreements allow all parties to exchange sensitive information confidently.

2.3 *Materials transfer agreement*

A materials transfer agreement is drawn up whenever a potential licensee wants to evaluate a new product or process. The licensor should be willing to provide samples or information but, naturally, will want to assure that the other party does not misuse them (such as by passing on a portion of a sample to some third party or using it to generate additional material for unlicensed use). The Center for the Management of Intellectual Property in Health Research and Development (MIHR) recommends that public sector research organizations use the Uniform Biological Materials Transfer Agreement and the implementing-letter format developed by the U.S. National Institutes of Health (NIH). The wording of the agreement is uniform for all IP transfers, with only the Implementing Letter specifically tailored to each transfer.

2.4 *Co-development through collaboration*

Even after acquiring new IP from a private company, it is not always possible or feasible for a single public sector agency to carry out all stages of production and marketing. The agency may, for example, need to collaborate with other public sector laboratories in order to complete product evaluation (preclinical toxicity tests, clinical trials, and so on). Also, high-quality, good manufacturing practice (GMP) production facilities, which most public sector research organizations lack, are needed to develop products for the market. The licensee can either pay other agencies to perform some of the tasks, or, preferably, form partnerships with them. Collaborating agencies may request a share of the IP rights or a portion of the revenue generated by product sales. It is possible that the final stages of product development will require new IP.

Requests for collaboration often take the form of open tenders. In the absence of established procedures (since technology commercialization by the public sector is still an emerging area), various means have been adopted by the public sector—primarily to "protect" the public sector institution from the unlikely event of a commercial blunder—most government departments resort to what is called a "committee approach" through which a group of officials, including tech transfer professionals, administrators, finance people, and so forth, work in a transparent manner to negotiate a deal. Public communication is important because the government that is funding the initiative will expect the deal to be performed with complete transparency. Furthermore, transparency reassures partners and investors.

2.5 *Technology licensing agreement*

Technology licensing agreements allow one party to use the proprietary materials or know-how of other parties. Standard technology licensing agreements clearly define the period of time for which the license is valid, the kind of license (exclusive or nonexclusive), the territory in which the license is valid, the market in which the product will be released (public sector or open market), whether or not the product can be sublicensed, the amount of money to be paid up front, and the royalties that the licensor will receive.

2.6 *Standard elements of typical agreements*

2.6.1 *Confidentiality*

A confidentiality agreement requires all information to be carefully protected. Access to confidential information should be given only to the proven trustworthy, as improper use of confidential material can seriously erode mutual confidence between partners and even lead to litigation. Scientists, especially those in the public sector, should be especially careful because they, in other contexts, discuss science openly.

2.6.2 *Territorial exclusivity*

In a licensing agreement, the territory is the geographic region in which the licensee is permitted to sell the product. The territory could be part of a country, part of a subcontinent, several countries, or the whole world;[2] or, alternately, territory can refer to a segment of the market in a single company like public sector or private sale. Sometimes, nonexclusive licenses are awarded to licensees in order to promote competition between them. Or an exclusive license may be granted to market an expensive product within a limited market—unless such market exclusivity is guaranteed, no one may be willing to manufacture it. Commissioning a professional agency to carry out market research in order to make sure that the product is correctly priced and appropriate for the intended territory is always advisable. (Commissioning such surveys is slowly becoming routine practice due to a lack of in-house expertise and the system of government regulations.) The guiding principle for deciding whether to grant exclusive licenses of nonexclusive licenses should be that while it is most important to bring new products to market at affordable prices.

2.6.3 *Product liability*

Health-related products can lead to liabilities; especially susceptible products, such as vaccines, are tested on healthy volunteers. Often, companies are unwilling to market a product because of potential liabilities. The licensing agreement for a health-related technology must define the cases in which the investigators will, and will not, be held responsible (for example, such cases might involve bad or inferior product, improper storage and use, administration of the wrong dosage) and the licensee must take out an appropriate amount of insurance before starting trials. The clinical trial agreement should also describe how, and how much, an individual who is harmed by a health product should be compensated.

2.6.4 *Up-front fees and royalties*

Ultimately, marketability and price decide a product's fate. The licensor must decide the kind and number of licenses, how much market access, and so on, it will grant. The parties must agree on how much money the licensor will receive both up front and via royalties. These decisions will be influenced by the amount of revenue the product is expected to generate. A committee of experts, administrators, and financial advisors usually negotiates on behalf of public-sector institutions. A balance must be struck between the desires of the licensee (to pay less up front and more through royalties) and those of the licensor (to receive as much money as possible at the beginning). Factors that affect the price of the license include the expected life of the product, the duration of IP rights, the existence of a competing product, purchasing capacity, and whether or not there is a committed market (in other words, governments offering purchase commitments), and so on.

2.6.5 *Arbitration*

The licensing agreement must stipulate the terms of arbitration in case something goes wrong and there is disagreement between parties. Arbitration procedures can be relatively simple if the parties are in the same country. If governments are involved in such arbitration proceedings, such governments will often dictate the outcome. Arbitration becomes very complex when parties from different countries are involved, especially if the arbitration is conducted in a third country. Of course, all efforts should be made to settle issues amicably.

3. CONCLUSIONS AND RECOMMENDATIONS

In developing countries, it is important for the pharmaceutical industry, in general, and the biotechnology industry, in particular, to develop products (drugs, diagnostics, and vaccines) with a potential global market. This reorientation from an exclusive concentration on markets in developed countries to a product development plan that includes developing countries can be achieved through partnerships between the public and private sectors in both developed and developing countries.

Most developing countries do not have the expertise to deal with complex IP licensing issues. Public officials in developing countries often postpone making decisions in order to cover up their ignorance and lack of expertise, thereby discouraging private companies that might be interested in collaboration with them. Professional help in all areas, from product valuation to drafting IP agreements, would be useful. The following drivers are needed for developing countries to optimize their success:

- **a business strategy** that aims to balance the objectives of the public sector (to bring affordable health products to market) with those of the private sector (making profits)
- **a marketing strategy** that prices products realistically, using up-to-date marketing information (any existing products, their price structure, potential customers, the size of the potential market in private and public sectors, and so on)
- **the proper legal expertise** is usually already locally available, as many legal firms in developing countries are familiar with basic licensing procedures. Marketing and scientific experts could assist in valuating patents

Perhaps the ideal solution to the lack of know-how in developing countries is two fold: first, the establishment of a national technology transfer office; and second, the development of core team of experts drawn from diverse disciplines devoted to helping to negotiate product in-licensing. ■

KANIKARAM SATYANARAYANA, *Chief, IP Rights Unit, Indian Council of Medical Research, Ramalingaswami Bhawan, Ansari Nagar, New Delhi 110029, India. kanikaram_s@yahoo.com*

1 Some argue that in general, the public sector organization should offer the first draft of a licensing agreement. (See for example, in this *Handbook*, chapter 12.1 by RT Mahoney.) This approach is generally much easier than trying to work from a draft prepared by the private sector organization, because the draft needs to cover a number of topics of particular concern to public sector organizations, and these topics probably would not be addressed in a private sector organization's draft.

2 In India, as perhaps in other poor countries, there are states, or equivalent entities, that are rich, and politically stable, with promising markets, while other states—often those with unstable governments—have uncertain market potential. Currently, each state in India has its own drug regulator. These officials have varying expertise and, along with other factors, can determine the marketability of products in their states. Additionally, while a price can be the same over the entire country, each state has its own rates for sales tax and other taxes.

A Checklist for Negotiating License Agreements

DONNA BOBROWICZ, *Technology Transfer Specialist, Loyola University Chicago, Office of Research Services, U.S.A.*

ABSTRACT

This chapter provides a road map for licensing professionals to identify the most common terms, contractual obligations, and other provisions that are likely to be encountered in crafting a license agreement. Emphasis is placed on agricultural technology licenses. Since most people engaged in deal making are involved in multiple deals at the same time, important aspects can be forgotten or overlooked at any time and for any deal. The checklist format allows the licensing practitioner to check off each item once it has been addressed to the parties' satisfaction. While expansive, it does not necessarily fit all contexts and is therefore intended to serve as a basis from which institutions and individuals can develop their own checklists.

1. INTRODUCTION

A checklist to aid in negotiating a licensing agreement, much less to aid in actually preparing and writing the agreement itself, may sound like a simplistic tool to an experienced negotiator or contract attorney. After all, most people in such positions are well educated and used to dealing with multiple projects having many details in the scientific, legal, and business arenas, all at the same time. If they did not have the competence to deal with this type of work situation, they would not last long in the active, high-pressure licensing environment. But it is precisely because of myriad details that a checklist can be life (or deal) saving for the working licensing officer or attorney. Since most people engaged in deal making are involved in multiple deals at the same time, important aspects can be forgotten or overlooked at any time and for any deal. One of the simplest ways to make sure that a crucial or costly mistake does not happen because of an oversight is to use a tool such as the checklist presented here.

2. SPECIFIC CHECKLIST SECTIONS

This section introduces and discusses for both licensors and licensees each element of the checklist. If your work requires you to draft license

Bobrowicz D. 2007. A Checklist for Negotiating License Agreements. In *Intellectual Property Management in Health and Agricultural Innovation: A Handbook of Best Practices* (eds. A Krattiger, RT Mahoney, L Nelsen, et al.). MIHR: Oxford, U.K., and PIPRA: Davis, U.S.A. Available online at www.ipHandbook.org.

agreements, download the checklist from the online version of this *Handbook* where it is given without the annotations.

2.1 *Section 1 – The parties*

Although seemingly self-evident, having all pertinent information about the parties in one place, such as their legal names, the negotiating party's contact information, and the legal addresses is a time saver when the final agreement is being written. No more last-minute telephone calls or e-mails to get information that should have been exchanged at the first meeting.

PARTIES:

1. Licensor's Name: ______________________
   Address: ______________________
   Principal Office: ______________________
   Incorporated In: ______________ Short Title: ______________
   Contact Name: ______________________
   Contact Title: ______________________
   Contact Tel/Fax: ______________________
   Contact E-mail: ______________________

2. Licensee's Name: ______________________
   Address: ______________________
   Principal Office: ______________________
   Incorporated In: ______________ Short Title: ______________
   Contact Name: ______________________
   Contact Title: ______________________
   Contact Tel/Fax: ______________________
   Contact E-mail: ______________________

2.2 *Whereas clauses*

The following set of "whereas clauses" is offered as a guide for detailing the background of the license. Not all parties use whereas clauses; some prefer to make the background information a standard set of clauses that follow language specifying that "the following are terms of the Agreement" or similar language. Some use of background information in a contract is recommended because within a short period of time after the deal is done and the agreement signed, negotiators memories will fade and a short set of statements regarding the background of the deal may become invaluable should the contract need to be interpreted by a court or an arbitrator.

WHEREAS CLAUSES:

1. Licensor owns/controls certain Intellectual Property/Tangible Property including inventions ______, patents ______, applications ______, know-how ______, other ______ relating to ______________

2. Licensor represents that it has the right to grant a license to ________

3. Licensee owns/controls certain Intellectual Property/Tangible Property including inventions ______, patents ______, applications ______, know-how ______, other ________ relating to ______________

4. Licensee represents ____________________________

5. Licensee desires license relating to __________ in order to ______________

2.3 *Definitions*

A simple contract will not need to have a section devoted to definitions, as the definitions can be presented when special terms are first encountered. A complex document should present all definitions in one section for ease of drafting and later interpreting the contract. General terms used throughout the contract should be placed in this section, as should technical terms that are used frequently. Either an alphabetical or a hierarchical order is recommended, the latter being used when a number of terms are closely related and having them near to each other would allow the reader to more easily navigate the agreement.

Each license will have its own specific set of definitions, so a short list that includes only the most commonly used terms is presented here.

DEFINITIONS:

All other appropriate terms should be listed and defined. Clear definitions will add great clarity to a license. Care should be taken to write definitions that, in general, stand alone and are not circular in construction.

A good place to begin thinking about what to define is with a definition of the parties. If dealing with a company, is it the company and all its affiliates? All of its subsidiaries? Or only the parent company? Products/Processes licensed should be specifically defined as Licensed Products or Licensed Processes. If only certain types of inventions are covered, define the inventions here and refer to them as Inventions; include the patent number and/or patent application number that is being licensed, and specify if Know-how is included.

(Continued on Next Page)

DEFINITIONS (continued)

Licensee, sales, net sales, profit, territory, field, patents, patent rights, intellectual property, and nonprofit are examples of other relatively common terms, and there are many more. Once defined, these terms will usually appear, throughout the rest of the contract, with the first letter capitalized or in all capitals.

2.4 *The grant sections*

The following sections may seem to be overkill to the licensing professional. However, each and every section, if not handled with care and forethought, can result in a deal that is more than unsatisfactory to one or both parties.

2.4.1 *Rights granted*

The exact grant language should be specified. This includes which intellectual property rights the license is given under: patent right only or know-how right or both and exclusive right, coexclusive with the licensor, or nonexclusive. The section should also specify the term of the exclusivity and/or nonexclusivity, and whether such right is irrevocable; and if there is a right to grant sublicenses. Each organization will find that it tends to make deals in a certain way and may find that certain combinations of grant language will be used repeatedly. In that case, this section may be easily amended to the specific organization's needs.

1. RIGHTS GRANTED:

a) All substantial (statutory) rights to practice under the rights in specified Intellectual Property/Tangible Property (detail here) ______________;
b) and to make ____, have made_____, use_____, import_____, offer for sale____, and sell _____ products and processes;
c) Exclusive for ______ years and nonexclusive thereafter, or
d) Non-exclusive ______, to make (manufacture) ______, or
e) Exclusive _____ to have made for own use ______; or
f) Exclusive except as to Licensor ______, to use ______, to export ________, to make and sell in limited markets _______;
g) Irrevocable ______, to sell ______, have sold ________;
h) With right to grant sublicenses ______, to lease ______, rent ______.

2.4.2 *License restrictions*

This section deals with the field, territory, prior licensee's rights, and the commercial rights retained by the licensor. Some of what is contained in this section appears under Section 1 (the parties), and may not be needed in all situations.

2. LICENSE RESTRICTIONS:

Limited to the Field ____

Limited to Territory ____

Subject to prior Licensee (identify, if any) rights ___________________________

Subject to Licensor's right to make ______, have made ______, use ______, have used ______, export ______, import ______, sell ______, have sold ______ (as many as applicable).

2.4.3 *Reservation of rights*

This section is particularly important when the licensor is a nonprofit and must ensure that certain rights to use the intellectual property are reserved for academic, nonprofit research, or humanitarian uses in developing countries, or according to the terms of the Bayh-Dole Act (in the United States). Forgetting to include the needed reservation of rights in a license could make the license invalid and/or could lead to an expensive court fight to determine what rights are in fact owned by the licensor.

3. RESERVATION OF RIGHTS:

a) Licensor hereby reserves an irrevocable, nonexclusive right in the Technology (on behalf of itself and all other nonprofit/academic research institutions)
b) For Educational and Research uses_____, including uses in Sponsored Research ____ and nonprofit collaborations_____.
c) For Humanitarian Purposes_____, or
d) For uses in Developing or Economically Disadvantaged countries_____ (specify countries)_ ________,
e) For the U.S. government under the Bayh-Dole Act ______.

2.4.4 *Right to grant sublicenses*

The grant of a right to grant sublicenses to third parties also has a number of important choices that must be considered by parties when awarding this portion of the license. Sublicensees may be anyone or may be limited to, for example, only parties in privity with the licensee; only affiliates of the licensee; only a specified number of third parties; or only parties preapproved by the licensor.

4. LICENSEE MAY GRANT SUBLICENSES:

a) To any other party ____;
b) To limited number of parties _____;
c) To Affiliates of Licensee ____ only _____;
d) To third parties preapproved by Licensor ____;
e) To nominees of Licensor ____;
f) At specified consideration (indicate) __________________;
g) Consideration to be shared with Licensor _______________;
h) Copies of sublicense to be furnished to Licensor ________;
i) Under other conditions ___________________________________

2.4.5 *Territory*

The territory that is granted to the licensee under the license must be specifically identified.

5. TERRITORY:

a) All countries ______
b) All countries except ____________________
c) Following country/countries_________________________________
d) That portion of a specific country comprising _________________

2.4.6 *Term of the agreement*

The date the agreement begins, the effective date, should be noted, as well as the ending date of the agreement, by whatever method that is calculated. Some of the most common ways are listed below.

6. TERM OF AGREEMENT:

Effective Date is _______.
For ______ years/months/day (as agreed), until (specify date) _____; or
For the life of a specific patent or other intellectual property ________; or
Until some future event (specify) __________________________

2.5 *Improvements*

This section deals with any improvements made and/or patented (by whom and paid for by whom) during the term of the license by either the licensor or licensee and what obligations are present in the deal as to whether or not to include future technology under the present license or to have future technology fall under the reservation of rights to the licensor.

7. IMPROVEMENTS BY:

LICENSOR:	LICENSEE
Included _____	Included _______
Not included _____	Not included _________
Who will file ____________________________	Who will file ________________________
Who will pay costs _______	Who will pay costs _________
Assigned/licensed to Licensee _____.	Assigned/licensed to Licensor_________

2.6 *Consideration*

The consideration sections of the checklist is relatively involved, and can be cut back if equity is not part of the payment for the license. Royalty, milestone payments, type of currency, determining rate of exchange, and equity-ownership issues are listed here, as is the issue of minimum annual payments, particularly important in the case of an exclusive license.

8. CONSIDERATION FOR LICENSE:

Royalty free ___; or

Royalty, ___ per cent; of profits _____; of gross sales _____; of net sales _____; specific amount (specify) ______ per unit (specify) ______; other (specify) ________;

Single sum (license fee) of ________;

Milestones (what they are and amount owed) ______________________;

Payment is to be made in currency of which country __________;

At the then current rate of exchange __________________;

At the rate of ______(currency) for _______ (currency)

If exchange rate decreases or increases by ____(specify a percentage) % the payments shall decrease or increase by like amount; or exchange rate shall be that published in _______________.

Equity: Stock of Licensee (specify) ___________________________

stock of existing company _____; new company ______

value of the shares of stock shall be market value ____ at date of agreement _______ book value ______ according to Schedule ____; stock shall have full voting rights ______; nonvoting ______;

9. MINIMUM ANNUAL PAYMENT FOR LICENSE:

Amount ______ per calendar year; per 12-month period ______
Payable in advance ______
Payable at end of calendar year ______; of 12-month period ______
Credited against earned royalties, yes ______; no ______

2.7 *Reports and auditing of accounts*

Royalties based on any measure tied to a product's sales should be paid to the licensor accompanied by a report stating how the royalty was calculated. It should be decided how often and when these reports (and royalties) are due. Additionally, the right of the licensor to audit the books that generate these reports should be a part of the license.

10. STATEMENTS OF EARNED ROYALTY:

Quarterly, within ______ days of end of quarter
Annually, within ______ days of end of year
Other periods, (specify) ____________________
In writing, and certified by __(official or auditing firm) ____
With names and addresses of sublicenses ______
With copies of sublicenses ______
Together with payment of royalty accrued ______

11. INSPECTION OF LICENSEE'S ACCOUNTS:

Not permitted ______
Permitted ______
at any time during business hours ______
at specified times ______
by Licensor's authorized representatives ______
by Certified Public Accountants ______
Audit to be paid by Licensor unless underpayment is greater than ___%

2.8 *Representations/warranties*

Certain basic representations and warranties should be given by each party to the other, such as the ability to enter into this agreement, the validity of the intellectual property, and a standard warranty disclaimer. These and others are listed below.

12. REPRESENTATIONS/WARRANTIES:

A. Validity of Licensed IP
Not admitted ______
Admitted to Licensee ______
If patents held invalid, then:
Licensee may terminate:
as to invalid claims ______
entire agreement ______

B. Good title to Intellectual Property in _______ (specify countries)

C. Authority of Licensor to enter into the License _____
Authority of Licensee to enter into the License _____

D. Standard warranty disclaimer, of fitness for particular purpose
Merchantability ______; Express or Implied ______.

2.9 *Infringement*

These sections deal with how past infringement by the licensee is handled; if the IP is infringed by third parties, how such infringement will be handled, and if there is a recovery for the infringement, how that will be divided between the licensor and licensee. Indemnification by the licensor of the licensee to practice under the IP rights is also covered.

13. INFRINGEMENT:

A. INFRINGEMENT OF LICENSED INTELLECTUAL PROPERTY/TANGIBLE PROPERTY
Past infringement by Licensee
forgiven ______; not forgiven ______
forgiven for payment of ______
If infringed by others:
Who will notify ______________
Who will file suit ______________
Who is in charge of suit ________
Costs: borne by ______________
divided ______________

(CONTINUED ON NEXT PAGE)

13. INFRINGEMENT (continnued)

B. INFRINGEMENT OF OTHER'S INTELLECTUAL PROPERTY/TANGIBLE PROPERTY

No indemnity by Licensor ______

Licensor indemnifies Licensee ______

Licensee indemnifies Licensor ______

Who will notify ____________

Who will defend ____________

Who will pay costs __________

Costs: borne by ____________

divided ______________

C. RECOVERY AFTER DECREE

Retained by ______; Divided ______

Right to settle suit:

by Licensor ______; by Licensee ______

by Licensor only with consent of Licensee ______

by Licensee only with consent of Licensor ______

2.10 *Diligence*

Diligence covers the concept that the exclusive licensee will do all it can to operate under the license so that the licensor reaps a monetary benefit under the license. If this issue is not covered, then the exclusive licensee can sit on the technology and keep others from exploiting it and bringing money to the licensor.

14. DILIGENCE BY LICENSEE (Usually in absence of minimum royalty):

No obligation ______

Licensee will use its best efforts to ______

Licensee will use its reasonable best efforts ______

Licensee agrees to:

produce ______ or sell ______ specified units ______

produce ______ or sell ______ specified products ____

invest specified amount ___________________

satisfy demands of trade ______

not to refuse reasonable request for sublicense ______

Penalty for lack of diligence:

license converted to nonexclusive ______

Licensor may nominate Licensees ______

Licensor may terminate __ upon __ days' notice in writing

2.11 *IP defined*

Intellectual property (IP), and how it is paid for, must be defined in the agreement, whether it is only one patent or if it includes various reports and tangible materials. This part of the checklist may be more relevant to for-profit licensors, but nonprofit licensors may also have more than just a patent (and its family) to include in the definition of IP.

15. INTELLECTUAL AND TANGIBLE PROPERTY OF LICENSOR:

Not included, except as described in patents or applications ______
Included for products (specify) ____________________
For term of agreement ______; for specified term ______
For territory of license ______; for other territory _______

A. NATURE OF INTELLECTUAL AND TANGIBLE PROPERTY

- i. Invention records __Know-how, not confidential ___
- ii. Laboratory records ___Know-how, confidential ____
- iii. Research reports ___Employee to be bound ______
- iv. Development reports ______
- v. Laboratory notebooks ______
- vi. Construct components and design ______
- vii. Test field lay-out and design ______
- viii. Production specifications ______
- ix. Raw material specifications ______
- x. Quality controls ______; ISO 9000 procedures _______
- xi. Economic surveys ______
- xii. Market surveys ___; Producer lists __; Brokers ___
- xiii. Promotion methods ______
- xiv. Trade secrets ______
- xv. List of customers ______
- xvi. Drawings and photographs ______
- xvii. Models, tools and parts ______
- xviii. Germplasm ____________________
- xix. Other (specify) ____________________________

B. PAYMENT FOR INTELLECTUAL AND TANGIBLE PROPERTY

Included in royalty ______
Not included in royalty ______
Single payment of ________________________________
Stock in amount of ______________________________
Annual service fee of _____________________________
for term of agreement _______________________
for specified term __________________________
If Intellectual Property surrounding it is held invalid:
Know-how payment stops ______
Know-how payment continues ______

16. INTELLECTUAL AND TANGIBLE PROPERTY OF LICENSEE:

Not included, except as described ______
Included for products (specify) ____________________
For term of agreement ______; for specified term ___
For Territory ____________________
Nature of Property included: ________________

2.12 *Right of inspection; technical personnel*

If the licensee has licensed seed that is being produced by the licensor and that will include the transfer of tangible material (the seed) to the licensee, the licensee may want to have the right to inspect the licensors research data and fields during the term of the license. Whether or not licensors personnel shall be used to transfer know-how or tangible materials to the licensee, and at what cost, is also an important item to note in the contract.

17. RIGHT OF INSPECTION:

Licensee shall have the right to inspect Licensor's:
- Research laboratory ______
- Development laboratory ______
- Laboratory notebooks ______
- Test fields ______
- Production fields ______; Nurseries ______; Greenhouses _____

Number of visits permitted per year ______; Number of persons ______

Special conditions of visits ____________________________________

Licensor shall have reciprocal rights of inspection __________________

18. TECHNICAL PERSONNEL:
Licensor shall provide technical personnel to deliver Intellectual Property/Tangible Property (specify) _________:
- At Licensor's expense ______; At Licensee's expense ______
- Not more than ______ persons for not more than ______ days
- At a fee which shall be the salary, plus ______ per cent
- Travel expenses ______; living expenses ______
- borne by Licensor ______; borne by Licensee ______

(Continued on Next Page)

TECHNICAL PERSONNEL: (continued)

Number and duration of stay of technical personnel determined by:
Licensor ______; Licensee ______; mutually ______
Ownership of reports made by technical personnel ________

2.13 *Remaining sections*

The remaining sections of the checklist are what may be identified as the "boilerplate sections" of the license, even though all of these terms are subject to negotiation. In any case, confidentiality terms, provisions for export control, the non-use of each party's name by the other party, arbitration (or not), terms of breach that will cause termination of the contract and the ramifications thereof, force majeure, assignment, favored-nation clause, notices, integration, language, modifications, applicable law, and schedules should be standard items considered by every licensing professional.

2.14 *Confidentiality*

If a confidentiality, or nondisclosure, agreement has been entered into by the parties and will remain effective during the term of the license agreement, nothing else is needed. If this hasn't been done, a section dealing with terms of confidentiality may be put into the license agreement. If the previously agreed-to confidentiality agreement is weak, now is the time to bolster it and to make sure that these terms in the license agreement take precedence over earlier agreements.

19. CONFIDENCE OF CONFIDENTIAL INFORMATION:

No obligation ______; Licensee obligated ______

Both parties obligated ______

Confidence maintained for specified time ____; Without limitation as to time ______; life of agreement ______

Until published by owner ______

Existence of this agreement confidential ___; Terms and conditions of this License to be kept confidential ___

Other ______________________________

2.15 *Export regulations; use of party's name*

Export regulations are important in deals where technology is exported from the United States. All exports must comply with U.S. export control laws and regulations, and in particular, those goods and IP that may have a military use. It is a topic outside of the scope of this chapter, but as an item on the checklist, it alerts the negotiator that this is a topic to be considered. Other countries may have laws dealing with the same topic or with issues or registering the final agreement with the government. Again, this is a memory jog for the negotiator.

In some cases, either one or all of the parties will not want its/their name used in connection with any licensed products advertised or sold, as it may suggest that the licensing institution is recommending these goods. If this is the case, this should be stated in the agreement.

20. A. EXPORT CONTROL ______

B. Government registration regulations _____

21. NON-USE OF NAMES

Licensor's ______, with permission ______
Licensee's ______, with permission ______

2.16 *Arbitration*

In the case of a major disagreement about the terms of an agreement, parties may wish to take the issue to arbitration. Arbitration can be carried out in many different ways and it is easier to specify in the agreement the rules to be used for arbitration, before there is an issue to arbitrate.

22. ARBITRATION:

No right of arbitration ______
Parties will use their best efforts ______
Parties agree to arbitration by:
American Arbitration Association ______
By other body ______
By three persons, one selected by each party and a third by the selected persons ______

Appeal from arbitration decision:
Not permitted, decision final and binding ______
Permitted ____________ to ___________

2.17 *Termination*

The termination section of an agreement can be quite complicated, or it can be very simple. I have seen agreements that have been hung up on determining what to do with the rights of the parties if a material breach were to occur. Thought should be given to this area, but beware of having it take over the negotiation. Areas to consider include the right of either party to end the agreement for no reason at all; the rights of the party that has performed when confronted with a party that refuses to perform; material breach issues; and length of notification of breaching activity and time given to the breaching party to cure the breach before losing rights and/or being charged penalties. Issues dealing with the natural expiration of the license should be considered, as well. What happens to the know-how (if any) upon the expiration of all patents? And what are the confidentiality provisions?

23. TERMINATION:

A. By Licensor:

If certain person incapacitated ___ (name) ___

If certain person terminated __ (name) __

At specified time ______

Upon breach after __ days written notice if not remedied within ____ days

Other ___________________________________

B. By Licensee:

At any time upon ______ days written notice

On any anniversary date ______

At a specified time ______

Only upon payment of penalty of __________ dollars

Upon breach after ___ days written notice if not remedied within __ days

Other ___________________________________

C. Upon expiration, Licensee assigns to Licensor:

Trademarks ______

Patents ______

Copyrights __________

Sub-licenses __________

As to any specified patents or applications ______

Germplasm _________________

As to any specified country ______

Of exclusive license with right to continue as nonexclusive ______

Whenever any essential claim held invalid ______

Upon bankruptcy of either party ______

(CONTINUED ON NEXT PAGE)

D. Upon Termination, without breach, Licensor assigns to Licensee:
Trademarks ______
Patents ______
Copyrights ___________
Sublicenses ______
As to any specified patents or applications ______
Germplasm _________________
As to any specified country ______
Of exclusive license with right to continue as nonexclusive ______
Whenever any essential claim held invalid ______
Upon bankruptcy of either party ______

E. Upon Termination with breach, Licensee assigns to Licensor:
Trademarks ______
Patents ______
Copyrights _________
Sublicenses ______
As to any specified patents or applications ______
Germplasm _________________
As to any specified country ______
Of exclusive license with right to continue as nonexclusive ______
Whenever any essential claim held invalid ______
Upon bankruptcy of either party ______

F. Upon termination, with breach, Licensor assigns to Licensee:
Trademarks ______
Patents ______
Copyrights _________
Sublicenses ______
As to any specified patents or applications ______
Germplasm _________________
As to any specified country ______
Of exclusive license with right to continue as nonexclusive ______
Whenever any essential claim held invalid ______
Upon bankruptcy of either party ______

2.18 *Force majeure*

This is the "it is out of my control" reason for not performing under the license. A hurricane has just wiped out your seed crops for the year, and you have no seeds to provide or to sell; your chemical plant just went up in flames. Things happen, and this fact of life should be considered in the contract. The key is to determine what is required after the force majeure occurs to get the licensed product out the door, or the goods to the licensee as quickly as possible. Technically a French term, it literally means "greater force."

24. FORCE MAJEURE:

Licensor has right ______
Licensee has right ______
Both parties have right ______
Nature of Force Majeure:
Natural events: fire, floods, lightning, windstorm, earthquake, subsidence of soil, etc. (specify) ______________
Accidents: fire, explosion, equipment failure, other ___________
Civil events: commotion, riot, war, strike, labor disturbances, labor shortages, raw material and equipment shortages ______
Governmental: government controls, rationing, court order ______
Any cause beyond control of party ______
Time after occurrence that the exclusive license becomes nonexclusive _____months
If there are fixed payments, are they excused during FM period ___?

2.19 *Assignment provision*

A license is considered to be personal to the licensor, especially in the case of an exclusive license. The licensor hand picks the licensee, for many reasons, and rejects others for many reasons. Additionally, an exclusive licensee may be interested in taking a license from a particular licensor, and not from another. In these cases, the right to assign a license may be forbidden, or at least greatly limited to "only with the permission of the nonassigning party." Nonexclusive licenses tend to be more open to assignment, especially if there are many licensees. There may or may not be fees attached to the transfer, or assignment, of a license.

25. ASSIGNMENT OF AGREEMENT AND LICENSE:

a) Not assignable by either party ______
b) Assignable by Licensor, without consent of Licensee __; only with consent __
c) Assignable by Licensee, without consent of Licensor; only with consent __
d) By either party upon:
Merger ______
To successor of portion of business involving: license___; or only entire business ___
To any company of which a majority of stock is owned ______
To any company of which a controlling interest is owned ______

Binding upon heirs, successors and assigns ______
Fee for assigning _______ How much? ________

2.20 *Favored nation*

A licensee may demand that they pay the same royalty and/or fee as another licensee that pays the least for the same license. This can be limited, for example, to the same royalty rate, but not to up-front fees, or not take in consideration the worth that cross-licenses to IP bring to a deal. Generally, it is very tough to determine if one party has a better deal than another unless it is a straight money deal.

26. FAVORED NATION CLAUSE:

Licensee guarantees performance (and amount of return) ______
Licensor required to notify Licensee of similar license ______
Licensee has option to take term of similar license ______
License changed to terms of more-favorable license ______
Licensee may terminate if not given cheaper license ______

2.21 *Notices; integration; language; modifications; law; signatures*

You will find that clauses that involve the following issues tend to be boilerplate clauses:

- Notices. the handling of any notices, payments, and so forth, that you must make or should receive
- Integration. a statement that this is the controlling document, no matter what else was said or signed previously, unless specifically stated in the license.
- Language: deals with languages used in writing the license (Will each translation of the license be acceptable? Or only the license written in one of the languages?)
- Modifications: specifies whether amendments to the license are to be in writing (If oral changes are OK for your deal, or for portions of it, specify it here.)
- Law: specifies which country's laws will be applied to interpreting the license; what courts will hear a lawsuit; and in what country, specifically, lawsuit would be filed.
- Signature: recommended to type in the name and title of the signatory (Two years after signing, all parties to the deal may have changed, and many signatures may be illegible by then.)

27. NOTICES AND ADDRESSES:

By registered mail ______
By registered air mail (for foreign licenses) ______
By overnight mail ______
After ___ days if by FAX with confirming telephone call ___
After ____ hours if by e-mail to ____specify_____
Licensor's legal address for notice: ________________
Licensee's legal address for notice: ________________

(CONTINUED ON NEXT PAGE)

28. INTEGRATION:

This instrument is the entire agreement between parties ______
This agreement supersedes all ______ prior agreements between the parties or the agreement dated ____________________

29. LANGUAGE (for agreement with foreign language licenses):

The official language(s) shall be __specify language(s)____
Copy in _____ language shall be official ______; unofficial __

30. MODIFICATIONS AND AMENDMENTS:

This License can not be modified or amended __________
No modification effective unless written and signed by both parties __

31. APPLICABLE LAW:

To be read, construed, understood and adjudicated according to the laws of _______ in the courts located in _________.

32. SIGNATURES:

For Individual:
Witnessed by ______ witness(es)
For Corporations:
By officer ______
Title shown ______

2.22 *Schedules*

This is the place to give very specific listings of items covered in the license, background documents, and research project outlines and specific procedures. It can be easier to modify a schedule than the whole contract, should the need for changes arise. A few types of schedules are listed.

33. SCHEDULES:
    A. PATENT LIST (Give inventor, number, issue date, official title)
    B. PATENT APPLICATIONS (Give inventor, number, filing date, official title)
    C. DESCRIPTION OR COPIES of official documents, such as sublicenses, assignment, prior license, etc.
    D. ACCOUNTING PROCEDURES for determining sales, net sales, sale value of stock, or other property
    E. EXISTING LICENSES AND/OR SUBLICENSES
    F. SPECIFICS OF EQUITY ARRANGEMENTS
    G. RESEARCH PROGRAM DETAILS

3. CONCLUSION

This license checklist is a comprehensive tool useful for capturing very important concepts and terms in a complex license. Nonetheless, the checklist can and should be modified by each institution to reflect the way it does business. Having key concepts available to the negotiator and license draftsperson with a quick reading of a checklist can save much aggravation and potential misery should a deal go bad during its lifetime. It is much more cost effective to craft a sound license up front, having key terms as well-defined as possible, than it is to fix the problem through arbitration or litigation later on. ■

DONNA BOBROWICZ, *Technology Transfer Specialist, Loyola University Chicago, Stritch School of Medicine, 2160 S. First Avenue, Building 120, Room 400, Maywood, IL, 60153 U.S.A. dbobrowicz@lumc.edu*

SECTION 12

Dealmaking and Marketing Technology to Product-Development Partners

Negotiating an Agreement: Skills, Tactics, and Best Practices

RICHARD T. MAHONEY, *Director, Vaccine Access, Pediatric Dengue Vaccine Initiative, International Vaccine Institute, Republic of Korea*

ABSTRACT

License negotiations involve substantial real or potential value. They therefore should be supported by a team of experts. The essential skills and expertise needed for conducting successful negotiations include: business strategy and development for leading the negotiations, marketing for estimating commercial potential, law for evaluating IP and patents and carrying out a variety of related tasks, science and medicine for evaluating new and potential health products, manufacturing and production know-how to determine equipment and additional training needs, and finance for analyzing input from other experts on the team to combine into a comprehensive report. The strength of such a team is in its interdisciplinary composition; each of the skill areas can complement the other. From the perspective of international licensing, licensors can seek to improve the availability of health products in developing countries, possibly moving from the "traditional" approach to licensing toward one that incorporates public sector needs. The best approach for a public sector organization negotiating an agreement with a private sector entity is usually to offer initial terms that the organization would be willing to agree to if it were on the other side of the table. Negotiating a fair licensing agreement should not be seen as a process of "bargaining." Rather, a licensing agreement is establishing, in written form, the rules of operation for an ongoing relationship where mutual trust and confidence will be necessary for success.

1. INTRODUCTION

An agreement is a means of transferring *value* between two parties. Each party has something of value that the other party needs or desires. For example, one party may have a product that can potentially have a very large market, while the other party has research, manufacturing, or distribution capabilities essential to reaching that market. Therefore, the key to successful negotiation is having a clear understanding of the value each party brings to the relationship. Value has several facets. There is an objective value: represented by, for example, how many units can be sold at a certain price, yielding a certain level of profit. There are also qualitative values illustrated by these examples: (1) One company feels that a particular product, owned by a second company, would enhance or complete a particular product line. For instance, it produces hepatitis B vaccine and would like to have a hepatitis A vaccine; and (2) One company may believe that access to a certain product, owned by a second company, would allow it to develop the expertise to handle other similar products. By learning how to produce recombinant DNA hepatitis B vaccine, the first company enhances its capability to produce other recombinant health products in the future. It is important that both parties to a potential agreement think carefully about the benefits that will or could be obtained through a license agreement. Only with a clear understanding of the transfer of value can both parties intelligently and fairly negotiate an agreement.

This chapter should be of help mainly to the public sector R&D organization that is

Mahoney RT. 2007. Negotiating an Agreement: Skills, Tactics, and Best Practices. In *Intellectual Property Management in Health and Agricultural Innovation: A Handbook of Best Practices* (eds. A Krattiger, RT Mahoney, L Nelsen, et al.). MIHR: Oxford, U.K., and PIPRA: Davis, U.S.A. Available online at www.ipHandbook.org.

either in-licensing the technology it needs or out-licensing technology it has developed. The discussion applies to a technology that is quite advanced in development. Nevertheless, the information should also be of use to university technology transfer managers and others who are not necessarily directly connected with ongoing R&D programs.

We discuss the licensing process from three points of view: the skills needed, the tactics used, and the practices employed to protect the interests of the public sector.

2. SKILLS NEEDED

Because a license negotiation involves substantial real or potential value, it should be supported by a team of experts. Private sector managers commonly complain that public sector organizations are poorly prepared to undertake effective negotiations, often demand unrealistic conditions, and cannot present a convincing case about the reasonableness of their demands. Obviously, we can do better.

There may be only one or two persons conducting the negotiations, but they should be able to call upon experts in different areas. The following are essential skills for negotiations:

- business strategy or business development
- marketing
- law
- science and medicine, including regulation
- production
- finance

2.1 *Business strategy*

Often, the business strategist is the lead negotiator. With considerable experience in structuring business relationships, the strategist will use the inputs of all the other experts to assemble the negotiating package. This person needs to have a clear sense of how the particular negotiation relates to the overall goals of the organization. This is important because without this sense, the negotiations may lead to a result that will not be useful to the organization. After all, signing an agreement does not necessarily mean that negotiations were successful. The business strategist's goal is to maximize the benefits to all parties. Of particular concern is developing a strategy to be implemented by public sector organizations that helps to ensure that the resulting product is available, appropriate, adoptable, and affordable by the poor in developing countries. Such a strategy, known as a global access strategy,[1] has been the focus of much analysis recently, and the business strategist and his or her team should have prepared a global access strategy, as appropriate for their product. The negotiations of a license agreement should lead to terms that help achieve the specific goals of the strategy, which are defined in the agreement.

2.2 *Marketing*

Expertise in marketing and market analysis is essential to negotiating a good agreement. Omission is dangerous because it can lead either to an overestimation or underestimation of the market potential, which, in turn, can lead to a suboptimal agreement or a rejection of an agreement that could have been successful. Lack of marketing knowledge may also make it difficult to negotiate the best (fairest) deal. In the context of this *Handbook*, we define markets as both private sector markets and public sector health systems. For products such as a malaria vaccine, the public sector market will often be the most important, but an understanding of the travelers' market in developed countries will also be essential. A marketing specialist should ask the following questions:

- What level of sophistication is required to market the product?
- How does the new product complement or compete with existing products?
- Would the product be directed at old or new customer groups?
- If the product is to be sold in both the public and private sectors, what are the barriers to achieving a profitable market?
- What types of information would be needed to promote the product to both the government and the private sector?
- What are feasible prices and would these prices be sufficient to support the project?
- How fast would the market grow and what would be the minimum sales for sustainability/profitability?

With the answers to these questions in hand, the public sector agency will be well prepared to conduct negotiations.

2.3 *Law*

The need for legal assistance is clear.[2] The lawyer should possess IP expertise, be able to evaluate patents, and have a variety of additional skills or be able to access those skills. A party wishing to license a technology will need to be able to assess the value of the patents. This assessment will include an evaluation of the claims of other similar patents. While patent offices try to avoid granting patents with duplicate claims, it is very common to find many patents with the same or similar claims, especially for health products—a number of patents may be issued that claim different methods to produce the same health product. The lawyer will need to determine the potential for claims of patent infringement. The lawyer might also advise on the need to obtain a license from another patent holder before using the offered patent. This assessment (called a freedom to operate assessment) will help in determining the true value of a patent. Such an assessment would answer the questions: Are there other patents that actually are more important? Who owns them? A lawyer will also be needed to advise on the laws of the various countries in which work would be carried out. For example, it may be necessary to evaluate the legal aspects of various arrangements for paying up-front fees and royalties. Some countries tax royalty payments quite heavily but have low or no tax on legitimate charges for technology transfer. Other legal, country-specific matters include validity of termination conditions and validity/enforcement of milestone conditions.

2.4 *Science, medicine, and regulations*

The negotiating team should have scientists and medical experts who are knowledgeable about the products under discussion. In this age of highly sophisticated science, a lead negotiator would be ill-advised to proceed without obtaining good scientific advice about a new health product technology. Not only is it important to assess the feasibility of the new product from a scientific point of view, but it is also important to know what is going on in the field broadly. One must ask, for example, if there were several methods for production of a health product: Which is best? Which is easiest to control? What are the safety considerations of each? It is also important to understand the regulatory framework, or lack thereof, for the potential new product. What kinds of clinical trials, in how many settings, and for what length of time will be needed? In the absence of a regulatory framework for a truly innovative product, how can such a framework be created and how long will it take?

2.5 *Production*

The production staff also should be involved in the licensing negotiation. They need to contribute their knowledge about required production equipment, the needs for additional training, and facility requirements. Production experts can also provide cost estimates for establishing production and for approximating variable costs at given production volumes. (Variable cost studies help determine the extent to which cost is sensitive to production volumes.) Production staff will also be able to advise on requirements for adequate quality control. For codevelopment agreements, production experts can be indispensable for advising on production feasibility. Product developers working in the lab often are unrealistically optimistic about how easy it will be to produce a product in commercial quantities. Production staff can bring reality to the discussions. A final topic for production experts is to understand the potential costs that might be incurred in different settings (for example, developed versus developing countries). It may be desirable to seek production in a developing country to ensure the lowest costs.

2.6 *Finance*

Before negotiating, carrying out a careful financial assessment of the project is essential. The assessment will help the manager determine what new funds will be required to launch and sustain the project, which will require factoring in such variables as the cost of funds (interest payments), hard currency requirements, break-even points (the length of time it takes to recover the initial investment given certain assumptions about sales and costs), return on investment, impact of royalties and other technology acquisition fees, and opportunity costs (involving the question, could the money be used more profitably in some other way?). The financial analyst will take inputs from all the other experts and combine them to prepare a report.

It should be clear that each of the skill areas complements the others. For example, in a technology licensing agreement, it will be necessary to assess the relative capabilities of the potential licensee's production and marketing departments. A licensee might be strong in production but weak in marketing, or strong in marketing but weak in production. If the differences are too great, implementing the agreement may be difficult. In these cases, the agreement should have tangible performance obligations for activities in which the firm is weak and flexibility where the firm is strong. The marketing, finance, and production staffs will need to work together to complete these assessments.

Not all groups have direct access to a complete complement of staff resources. In those cases, expertise could be obtained through consultants or related institutions that do have the capabilities.

3. TACTICS FOR NEGOTIATING A LICENSE AGREEMENT

Once two organizations have decided to seek to conclude a licensing agreement between them, the first step is to designate the negotiating teams. Each organization should clearly indicate who the members of the negotiating team are and what their respective responsibilities are. The principle line of communication should be between the two lead negotiators. However, the two groups may need to exchange technical information. For example, it may be necessary for one organization to share scientific information with the other. In that case, the scientific staff of each organization should carry out the exchange. Or it may be necessary to go into technical detail about production issues, in which case the production staff of each organization should be involved. When there is an exchange of technical information, the discussion should be limited to the information itself, and the technical individuals should not enter into any negotiations with respect to the licensing agreement unless such involvement is requested by the lead negotiator.

In general, the public sector organization should offer the first draft of a licensing agreement. This approach is much easier than trying to work from a draft prepared by the private sector organization because the draft needs to cover a number of topics of particular concern to public sector organizations, and these topics probably would not be addressed in a private sector organization's draft. The topics of concern are jurisdiction, liability issues, ownership of IP, protection of the public sector, and others. It is much easier to start with a draft that has all of these issues clearly laid out—and is based on previous experience—than to try to insert those issues into a draft that does not include them.

The public sector organization's lead negotiator may ask for examples of the kind of agreement that the other organization feels comfortable with. The lead negotiator may extract some of the key wording in clauses from the example agreements and insert them in the prototype of the public sector organization agreement. In certain cases, primarily for in-licensing, it may be necessary to use the private sector organization's standard agreement, either because the organization requires that its agreement be used or because it has extensive experience in the kind of licensing agreement at issue, and time and energy would be saved.

One variation in developing a first draft of a license agreement is to prepare a *term sheet*. A term sheet lists the major issues that are expected to arise in the negotiations and indicates the outcome that the proposing party hopes to achieve. For example, if the agreement includes the development of a commercially viable production process, the term sheet would indicate a schedule for achieving various stages of production capability, the number of units to be produced, and the quality standards that the units would have to meet. A term sheet is a straightforward way for the parties to discuss key issues without having to wade through a long document that contains a lot of routine boilerplate. Table 1 provides an example of a term sheet for a clinical testing agreement.

The best approach for a public sector organization negotiating an agreement with a private sector entity is usually to offer initial terms that the organization would be willing to agree to if it were on the other side of the table. Negotiating a fair licensing agreement should not be seen as a process of "bargaining." This is because a licensing agreement establishes, in written form, the rules of operation for an ongoing relationship where mutual trust and confidence will be necessary for success.

At the beginning of the negotiations, it is important for each group to clearly state what it hopes to achieve from the negotiations, although, of course, there will always be confidential

TABLE 1: PROTOTYPE TERM SHEET TO FACILITATE NEGOTIATIONS

TERM SHEET	
Clinical Research Agreement	
Territory	Kenya
Phase I/II conducted by [DATE]	
Initiation	2007
Completion	2008
Subjects	250
Funding	100% paid by [DATE]
Phase III conducted by [DATE]	
Initiation	2009 or 2010
Completion	2012
Subjects	10,000
Funding	100% paid by [DATE]
Diligence	
Phase I/II initiation by [DATE]	1/1/07
Phase III initiation by [DATE]	1/1/10
Regulatory submission by [DATE]	1/1/12
Clinical trial design by [DATE]	Licensor consent
Manufacturing	Licensor or its agent
Transfer prices to [DATE]	
ncGMP (noncurrent good manufacturing practice) material for phase I/II trial	Paid by licensor
cGMP (current good manufacturing practice) material, per unit	US$10
Cost sharing for manufacturing scale-up	To be determined
Investigational New Drug (IND) preparation by licensor	$0
Quality control monitor for clinical trial	100% paid by [DATE]
Regulatory license holder	[DATE]
Indemnification	[DATE] indemnifies licensor

information that cannot be revealed. The public sector organization may be interested in working with a group that can develop a superb and economical production methodology for a new product that the public sector organization has developed. The counterpart organization may be interested in participating in the development of regulatory guidelines for a particular kind of product. By stating their primary objectives clearly at the beginning of the negotiations, it will be easier for both parties to take into account the needs of the other.

Negotiating a license agreement often takes much longer than either party would like. This can be frustrating for the technical staff of the public sector organization, who would like to resume research and development activities as rapidly as possible but have to put on hold many such activities until the license agreement is signed. There are a number of reasons why license negotiations often take longer than anticipated. The license must be approved at multiple levels in each organization and will undergo review from technical, financial, legal, and other experts with varying points of view. Often the views may differ internally, which requires internal negotiations that take some time to resolve.

4. PRACTICES TO PROTECT THE INTERESTS OF THE PUBLIC SECTOR

Table 2 illustrates how licensors can seek to improve the availability of health products in developing countries. It summarizes the "traditional" approach to licensing and then indicates a more public sector option.

Two examples of a clause pertaining to territory are provided below. The clause is for use in agricultural research and development but can be adapted to health research and development. The clause would be used in a license issued by a university to a private company.

Example 1: Public Intellectual Property Resource for Agriculture (PIPRA)[3]

Definition of Humanitarian Use:

Definitions:

"Humanitarian Purposes" means (a) the use of Invention/Germplasm for research and development purposes by any not-for-profit organization anywhere in the World that has the express purpose of developing plant materials and varieties for use in a Developing Country, and (b) the use of Invention/Germplasm for Commercial Purposes, including the use and production of Germplasm, seed, propagation materials and crops for human or animal consumption, in a Developing Country.

"Commercial Purposes" means to make, have made, propagate, have propagated, use, have used, import, or export a product, good or service for the purpose of selling or offering to sell such product, good or service."

"Developing Country" means any one of those countries identified as low-income or lower-middle-income economies by the World Bank Group at the time of the effective date of this agreement and all other countries mutually agreed to by Licensor and Licensee.

Reservation of rights

Notwithstanding other provision of rights granted under this agreement, University hereby reserves an irrevocable, nonexclusive right in the Invention/Germplasm for Humanitarian Purposes. Such Humanitarian Purposes shall expressly exclude the right for the not-for-profit organization and/or the Developing Country, or any individual or organization therein, to export or sell the Germplasm, seed, propagation materials or crops from the Developing Country into a market outside of the Developing Country where a commercial licensee has introduced or will introduce a product embodying the Invention/Germplasm. For avoidance of doubt, not-for-profit organization and/or the Developing Country, or any individual or organization therein, may export the Germplasm, seed, propagation materials or crops from the Developing Country of origin to other Developing Countries and all other countries mutually agreed to by Licensor and Licensee.

Table 2: Illustrations of Best Practices for Licensing to Meet Public Sector Goals

Topic	Basic concept	Public sector consideration
Areas of use	This clause specifies the limitations on the application of the patent in developing products. The simplest approach is to grant the licensee an exclusive right to all possible applications of the patent, including not only those specified in the patent, but others that may emerge as further research and development proceeds.	The clause could grant an exclusive license only for those products that the licensor actually wishes to pursue. Also, the clause could grant an exclusive license only for those products that were unlikely to have a significant market among the poor in developing countries.
Territory	This clause specifies the geographic areas in which the licensee has the right to exercise the patent. The simplest approach is to grant the licensee an exclusive right to all possible territories. Usually a license is valid only in the countries where a patent has been filed, but the license can give the licensee the right, at the licensee's expense, to file for patent protection in additional countries.	The clause could grant an exclusive right to a major portion of developed countries, for example, North America. The licensor could grant another exclusive limited license to countries in Europe. Finally, the licensor could grant nonexclusive licenses to both licensees for an agreed list of developing countries. Then the two primary licensees would have to compete for sales to developing countries.
Price	In most licensing agreements, there will be no conditions with respect to price. The licensor assumes the licensee will determine the best price to ensure the greatest return on investment.	The licensor can consider several options of setting a condition of the price to the public sector in developing countries. • The price could be specified, for example, US$0.30 per tablet. This is feasible only when the licensor has detailed technical knowledge of the production, marketing, and distribution costs. • The price could be set at cost of production plus a reasonable markup, for example, 15% of cost of production. This is feasible when the licensor has a reasonable expectation of being able to monitor the cost of production. • The price could be set at "no higher than the lowest price offered to any private sector buyer." This may be preferred in cases where it is expected there will be large bulk purchases by private sector buyers who are good at negotiating the very best price.

CONTINUED ON NEXT PAGE

TABLE 2 (CONTINUED)

TOPIC	BASIC CONCEPT	PUBLIC SECTOR CONSIDERATION
Labeling	In most licensing agreements, there will be no conditions about labeling. The licensor assumes the licensee will prepare labeling in conformity with national drug regulatory agency requirements.	The licensor can help ensure that the product is licensed properly, especially in developing countries where national regulatory agency requirements for labeling may not be rigorous or enforced. For example, if some of the research that led to the patent was supported by the World Health Organization (WHO), the license can specify that the name of WHO cannot be used without prior written approval of WHO. Additionally, the license could state that any claims for the use, safety, and effectiveness of the product should receive prior written approval.
White knight condition	This concept has been developed by the U.S. National Institutes of Health. It calls for the licensee to undertake some specific actions that will benefit the public sector.	The licensor can ask for a number of actions including donation of product for clinical evaluation in public sector research programs, joint efforts to develop markets in developing countries, free supply under specified condition to developing countries, and so on.

Example 2: Donald Danforth Plant Science Center

Reservation of IP Rights for Humanitarian Purposes

> COMPANY and Danforth shall diligently and in good faith negotiate the terms of the license, making provision for preserving the availability of the Intellectual Property for meeting the needs of developing countries.
>
> or
>
> Danforth shall retain the right to use Phase I Materials and Phase II Materials for both academic and commercial research purposes, which shall include the right to use such technology for the benefit of countries eligible for International Development Association funds as reported in the most recent World Bank Annual Report.

This clause has been part of the Donald Danforth Plant Science Center's IP policy since 2002.[4]

5. CONCLUSION

The negotiation of licenses is a complex undertaking that involves various tactics and a variety of skills. To meet the needs of the public sector, the negotiations should include special considerations in many clauses of the agreement. Moreover, because IP management involves matters of real or potential considerable value, it should be given the resources and personnel it needs to do the job well. No serious private sector company would enter into IP negotiations without allocating an appropriate level of resources and personnel. Because public sector research organizations are concerned with saving human life, their imperative to do the same should be no less. ■

RICHARD T. MAHONEY, *Director, Vaccine Access, Pediatric Dengue Vaccine Initiative, International Vaccine Institute, San Bongcheon-7dong, Kwanak-ku, Seoul 151-818, Republic of Korea.* rmahoney@pdvi.org

1 Mahoney RT, A Krattiger, JD Clemens and R Curtiss. 2007. The Introduction of New Vaccines into Developing Countries IV: Global Access Strategies. *Vaccine* (in press). See also Krattiger A, et al. 2006. Global Access Strategy for the live recombinant attenuated Salmonella anti-pneumococcal vaccine for newborns. Arizona State University: Tempe. www.biodesign.asu.edu/centers/idv/projects/, and Anonymous. 2006. Strategic Plan. Dengue Vaccines: The Role of the Pediatric Dengue Vaccine Initiative. Strategic Partnerships, Supportive Research & Development, Evaluation, and Access. International Vaccine Institute: Seoul. http://www.pdvi.org/PDFs/PDVI%20Strategic%20Plan.pdf.

2 See, also in this *Handbook*, chapter 6.10 by J Dodds and chapter 6.9 by M Goldman.

3 www.pipra.org/docs/HumResLanguagePIPRA.doc. See, also in this *Handbook*, chapter 2.1 by AB Bennett.

4 Beachy R. 2006. Donald Danforth Plant Science Center. St. Louis, U.S.A. Personal communications. See, also in this *Handbook*, chapter 17.9 by K Schubert.

An Introduction to Marketing Early-Stage Technologies

MARCEL D. MONGEON, *Intellectual Property Coach, Mongeon Consulting Inc., Canada*

ABSTRACT

This chapter describes marketing concepts and how to use them to create marketing plans for newly developed technologies in the health and agricultural sectors. The traditional marketing model invokes the "four Ps" of marketing: product, price, place, and promotion. This chapter, however, concentrates on the "five Ws" of marketing, which are more relevant to early-stage technologies: who? what? where? when? and why? The author then discusses the concept of the unique selling proposition (USP) and, finally, considers the marketing of technology transfer activities, or internal marketing.

1. INTRODUCTION

Because marketing is usually taught only in formal business programs, it is often not understood by scientists, technologists, and engineers. This lack of understanding can impede the transfer of technology from the laboratory to the commercial sphere.

Common misconceptions about marketing include that it is:

- only relevant to for-profit companies
- just a fancy name for advertising
- making buyers buy things they do not need
- just about one's skill in selling something to others

Let us begin with the first misconception: Marketing is only relevant to for-profit companies. Of course, a for-profit company will not be successful unless it sells products; this means that the company must understand its markets. However, the same is true for a not-for-profit or a government agency: neither can be successful without understanding the markets for its technologies. The reason for this is that the need of these types of persons to have users who will be interested in the technologies. If there are no users, the technology will not be adopted.

Next, marketing early-stage technologies has little to do with advertising. While understanding how markets become aware of technology is important (a key concern of advertisers), this is only one of many pieces of information required to understand how a market will respond to a specific early-stage technology. Advertising is only one small part of an overall marketing strategy for any product; advertising promotes awareness of a product inside potential markets. However, in the case of early-stage technologies, other aspects of marketing are more important: after all, if someone does not know where their potential markets are, advertising will likely be ill-conceived or prepared. Marketing includes identifying markets as well as the features of the technology that will be of interest to those markets.

Third, marketing is frequently characterized as a type of behavior-modification technique that alters buyers' intentions and makes them buy

Mongeon MD. 2007. An Introduction to Marketing Early-Stage Technologies. In *Intellectual Property Management in Health and Agricultural Innovation: A Handbook of Best Practices* (eds. A Krattiger, RT Mahoney, L Nelsen, et al.). MIHR: Oxford, U.K., and PIPRA: Davis, U.S.A. Available online at www.ipHandbook.org.

products they do not need based on deceptive advertising. It is true that if a product is advertised as having features or benefits it does not have, buyers will be dissatisfied. Children discover this common sense, for example, when a doll they have seen advertised on television proves not to be able to dance!

Finally, although "selling skills"—such as "cold-calling" a prospect, introducing a potential investor/licensee to the idea of an early-stage technology, and conducting a licensing negotiation—are certainly important, such skills are only one aspect of marketing.

Put simply, marketing is:

Understanding the buyer's needs and how to satisfy those needs.

Accomplishing these simple objectives, however, often requires a complex strategy.

2. "PUSH" AND "PULL"

A frequent criticism of technology transfer is that people are too concerned with "pushing" technologies into the market rather than allowing buyers' needs to "pull" those technologies in naturally. But the real problem is that, very often, buyers don't even know what their needs are!

For example, consider the Internet. Although today, most people who use it would say they can't live without it (or they *need* it), 20 years ago, the idea that all computers might be connected by some overarching network was the stuff of science fiction. However, few science-fiction writers envisioned that such a "web" might allow us to place orders for goods and services or to receive communication and information. However, once Internet technology was pushed on to consumers, a market was created. Consumer demand has pulled more and more technologies into the market ever since. The original Internet technology was created despite any study of consumers' need for it; The success of the Internet technology was not anticipated until the early-1990s other than by a few visionaries. Rather, consumers adopted it when they discovered that it satisfied their needs.

3. TECHNOLOGY TRANSFER AND EARLY-STAGE TECHNOLOGY

In this chapter, it will be important to understand two key concepts: *technology transfer* and *early-stage technology*.

Technology transfer refers broadly to any means of moving a scientific idea from a laboratory to practical use application in a production environment. Technology transfer can be formal and well-regulated: for example, assigning intellectual property ownership for a new technology, licensing the technology, and starting up a new company based on the new technology.

Some technology transfer is informal and less regulated. For example, many of the technologies that contributed to the personal computer revolution (such as the laser printer, Ethernet, WYSIWYG, and the mouse) were developed at Xerox's Palo Alto Research Center in the 1970s. Xerox did not capitalize on these technologies by actually bringing any of them to the market as products, and they were eventually transferred to other companies when the employees who had originally worked on those projects left Xerox.

Technology transfer is also generally used to refer to the process used to ensure that research findings are translated into actual use. Rather than relying on inventors to determine the practical uses of their inventions and put the appropriate structures in place to bring that use to market, an intermediate person or department (referred to as the *technology transfer office* [TTO]) takes responsibility for that work.

Early-stage technology refers to a scientific, technical, or engineering finding that is not embodied in an existing product and that does not obviously lend itself to a commercial enterprise.

Early-stage technology, for example, led to the creation of the Roundup Ready® line of genetically modified seeds sold by Monsanto. A gene that makes a plant tolerant to glyphosate had been discovered. In itself, the finding had little practical value. However, the already existing herbicide Roundup® was based on glyphosate, and researchers discovered that plants containing the new gene could be safely used with the herbicide. The herbicide kills weeds, but does not harm the Roundup Ready® crops.

Careful marketing work (usually done by the TTO) can help an organization turn an early-stage finding into a commercial product. The TTO accomplishes this by determining possible uses for the finding, identifying potential users, recognizing the features of the end product that will attract users, and then getting the resulting product to those users.

Utility is what a product allows the customer to do. Using a bicycle, for example, allows someone to get from point A to point B faster than on foot. *Marginal utility* is what a particular product does better than any other. A bicycle may have limited marginal utility since it may not be the *only* way that someone can travel a short distance, and it may not be the *best* way, either. A bicycle with square wheels, designed to roll over a roadbed comprised of inverted catenary structures (also known as a "washboard" surface)[1] would likely have a limited marginal utility for almost everyone. Such a bicycle would only appeal to people who not only want or need a bicycle but who also live near a lot of roads with surfaces that follow a very specific, very unusual structure. However, for a few people, a square-wheeled bicycle would have a very high marginal utility, since no other vehicle could travel over such roads.

Let us consider an example of an early-stage technology. A new membrane designed for the separation of proteins has a utility that is similar to many existing technologies such as filter paper or gel electrophoresis. However, if our new membrane has the additional benefit of being able to separate proteins based on their ionic charge, then the marginal utility becomes the ability to separate proteins on this basis. Those users who are interested in this feature (which is likely to be a large number) will be interested in the new product's marginal utility over the general utility of all types of filtering and separation methods.

4. THE "FIVE WS" OF MARKETING, PLUS ONE "H"

4.1 *The who of marketing*

When marketing a product, it's first important to know who will be buying it: What if we were dealing with a new drug that has been identified for hypertension (high blood pressure)? Who is the "buyer" of this drug?

We might begin by assuming that the buyer is the patient because he or she actually pays the pharmacist for the drug. Upon further reflection, however, we realize that it is the prescribing physician who makes the decision about which drug to prescribe. In fact, the patient has little input into that decision; so, in effect, the buyer may be the prescribing physician. Then again, in many jurisdictions, larger organizations—HMOs (health management organizations) or governments—decide for which drugs, and under what conditions, patients will be reimbursed. Thus, the buyer of our new drug may not be the same from jurisdiction to jurisdiction.

Now, let us turn to a different sort of product, an early-stage technology. A researcher has identified a specific genotype that makes pigs much more susceptible to porcine stress syndrome (PSS). Pigs that have PSS are significantly smaller than those without it (do not have the genotype). Farmers who raise the pigs (producers) sell the pigs to slaughterhouses, which in turn sell the carcasses to processors. Processors will pay less money to the slaughterhouses for PSS carcasses, and the slaughterhouses, in turn, pay less to the producers.

Who is likely to buy PSS-identification technology: the producers, the processors, or the slaughterhouses? Most likely, the producers: by using the technology, they can cull PSS-positive swine from their stock and save themselves the cost of raising inferior animals. In turn, the producers can sell to the slaughterhouses with the promise that their herds are PSS free.

4.2 *The what of marketing*

What do buyers want? In order to understand the market, TTO professionals must understand:

- how buyers will use the product
- what factors buyers will consider when making decisions to buy
- what product characteristics buyers find attractive

It is dangerous not to understand exactly what buyers want and what they are willing to pay for what they want. For example, the Concorde airplane was able to cut the usual trans-Atlantic flight time (approximately seven hours) by a little more than half. However, in order for the company to turn any profit at all, a Concorde flight cost more than three times the price of the average nonsupersonic flight. As you probably already know, the Concorde went out of business.

What went wrong with the Concorde's marketing concept? In all likelihood, the marketers overestimated the amount that buyers of long-distance travel were willing to pay for reduced flight times. In the 1950s, buyers of long-distance travel had certainly been willing to pay more for faster travel: they opted to pay substantially higher prices in order to travel by air rather than rail or ship. Because buyers were happy to make the trade-off, the size of the air-travel market expanded rapidly, which allowed airlines to reduce costs, leading to further market expansions (a so-called virtuous circle). Ultimately, this led to the almost complete replacement of rail and sea travel by air travel. But while travelers in the 1950s were happy to pay for improved travel technology that saved them days worth of travel, Concorde customers did not feel that the prices they were being charged were worth a mere four-hour time savings (a three-hour flight rather than a seven-hour one).

Consider another example. A new set of obstetrical forceps[2] has been devised made from a molded plastic rather than the existing standard of metal. The plastic allows a limited amount of play at the fulcrum point of the forceps. This play ensures that no more than a set amount of force will be put on the head of the baby being delivered. The new technology meets with a great deal of resistance in the marketplace. Why?

In part the answer comes from misunderstanding the *what* of marketing. Buyers (which include obstetricians) obviously consider many factors in purchasing such a device. The use of plastic rather than steel was likely perceived as a deficiency due to the perception that somehow that material is less sterile than metal, which is well known in delivery rooms. In addition, the change in material results in a significant change of weight and the perception that the plastic device is less robust than its metal counterpart. If these perceptions of buyers had been considered, alterations to the product may have resulted in an easier adoption of the technology.

4.3 *The why of marketing*

Once we have established *who* will buy our product and *what* they want to buy, we need to ask *why* someone should buy our product as opposed to someone else's. In order to answer this question, we must ask a broader one: why does a company (after all, most early-stage technologies are not sold to consumers) buy anything? To put it another way what are the drivers or forces acting on a company?

Michael Porter suggests that there are five such forces:[3]

1. Competition among businesses in the industry
2. The threat of new businesses in the industry
3. The threat of new, competing products
4. The bargaining power a company has with its suppliers
5. The bargaining power a company has with its buyers

By understanding these forces, marketers can determine what is of interest to potential users for any early-stage technology. If the technology can help the user address a company's concerns in any of these forces, it is more likely that the technology will be adopted; if there is no effect in any of these forces, there is little likelihood that the user will be interested.

If we analyze these forces for a user deciding whether or not to adopt a product derived from an early-stage technology, we find that the new product must give the user an advantage in one of the five areas. For example: does the new product:

1. **Lessen the potential competition among those already in the industry.** This could be accomplished by creating a new class of products that competitors will not be able to create for a number of years.
2. **Lessen the threat of new companies coming into an industry.** For example, increasing

the barriers to entry for new companies would make this happen.

3. **Lessen the threat to companies within the industry of new, competing products.** By ensuring that there is good IP protection around the new product, the possibility of new, competitive products is lessened.
4. **Affect reliance on existing suppliers.** By either reducing the amount required from existing suppliers or by bringing new suppliers into the picture, the new product would provide added bargaining power over suppliers.
5. **Affect the relationship between buyers.** A new product can significantly alter the relationship with buyers by, for example, providing buyers with product features that they are not able to obtain from anyone else.

It is important to articulate *which* of these forces a technology will help the business customer address—something we might call the "So what?" test. In order to answer this question, you need to consider what your product offers customers in the way of:

- features (the obvious attributes of your product)
- advantages over other, similar products
- benefits to the user

Remember that it is a f-a-b idea to make sure that your product is competitive!

Furthermore, any product or early-stage technology needs a unique selling proposition (USP): that is, something that distinguishes your product from any other (discussed in section 5).

4.4 *The where of marketing*

We have figured out *who* will buy our products. Next we must ask: *where* are products or early-stage technologies sold? After all, there is no eBay for technologies yet (although a number of technology exchanges are in the works).

Products typically move through "channels of distribution." Let us take the example of a hypothetical new technology that allows us to amplify DNA. How can we get that technology into use? What channels of distribution would exist? First, there could be use in research laboratories. Laboratory use could be subdivided into academic and for-profit (such as in a pharmaceutical company) research labs. There is also use of the amplification technology with various practical tests for patients: paternity testing and predictive genetic testing, as well as forensic crime-scene testing. Finally, the amplification technology could also be used in certain drug-production processes. Without much work we can see how one relatively simple early-stage technology may have a large number of uses.

These different uses have an intellectual property implication. Although that aspect is beyond the scope of this chapter, it is important to realize that certain types of uses for a technology may be prohibited by IP protection making it important to derive as many different uses as possible: some of these may be hindered from use by IP considerations; others may be free for use.

Consider another example: software that helps hospitals use their imaging equipment. What channels of distribution do marketers need to consider? Depending on the jurisdiction, hospitals may be free-standing private institutions, part of a health-management organization (HMO), or part of a government or quasi-government organization. In addition, a hospital may not be entirely independent: it may be associated with other hospitals or healthcare providers in a buying group.[4] In other words, the person who makes the buying decision (or even lists a software product in a catalog) may be some distance from the users of the software.

Furthermore, should the software company be separate from, or in alliance with, the companies that sell the imaging equipment? This is not a trivial decision, because it is likely to determine who the buyer is. For example, an equipment vendor is more likely to sell directly to medical staff, whereas a software vendor is more likely to sell to the computing and information services department. Not only are there a number of potential buyers within the hospital in at least two different departments, but also the hospitals or departments may buy software through a number of different channels (directly from equipment manufacturers, or through one or more buying groups). There are at least eight channels[5] of distribution!

4.5 *The when of marketing*

The last question to ask is *when* can you sell something to buyers?

4.5.1 *The long-term when*

Many technologies exist long before people become interested in them as products. For example, after the discovery of the double-helix nature of DNA in the 1950s, it took approximately 40 years (until the 1990s) before actual products depending on DNA were generally available. This long-term aspect becomes important when one considers that the term of patent protection is usually limited to twenty years. In other words, even if the original discovery of the structure of DNA had been patented, any actual revenues resulting from the discovery would only have been seen after the expiry of the relevant patents.

Another technology that took more than a century to be adopted by the public was the fax machine. It was invented in the early 1800s, but it was initially too slow: transmission took six minutes or more per page. Public interest in the fax machine only arose in the late 1980s, when digital compression technology allowed a page of data to be sent in less than one minute.

4.5.2 *The short-term, seasonal, or cyclical* when

Governments and institutions have differing equipment needs, depending on where they are in their annual budget cycles, how old or up-to-date their equipment is, and whether or not regulations have recently changed. For example, the software that allows accountants to create non-tamperable digital images of documents moves off the shelves most slowly in February, March, and April. Why? The answer is simple: at that time of year, accountants are too busy dealing with their clients' taxes to consider purchasing new tools for their own administrative needs.

4.6 *The* how *of marketing*

Thus far, we have considered the five Ws:

- *Who* is going to buy our product
- *What* product features should be emphasized
- *Why* buyers should want to buy the product
- *Where* we should sell the product and *where* along the distribution channel buyers are
- *When* buyers will be most interested in the product

So *how* should marketers use this information?

Usually, the *how* is answered with a *marketing plan*, a written document that answers each of the previous questions in detail. If the product is an early-stage technology, there are probably not going to be any concrete answers. In fact, it may be sufficient to identify *possible* answers and their ramifications. The early marketing plan can also be considered a provisional document that will be regularly revised as research and development continue.

It is essential to point out that this marketing plan is likely an important function of the technology transfer office. The creation of such a plan will be done once the office answered the questions that we have posed in this chapter and add considerable value to the early-stage technology. Value is added by identifying potential markets, products that can be sold into those markets and the features, advantages, and benefits those products will have.

Although there may be no hard answers at this point, market research will never go to waste. It may come in handy when the company considers licensing or spinouts. Also, potential buyers can be contacted early, and their responses can be useful for later market research. Moreover, people who work in early-stage technology are usually happy to cooperate with someone who is researching the market for a new technology.

5. THE UNIQUE SELLING PROPOSITION

The *unique selling proposition* is the advantage or benefit that the product offers to the buyer, *not* a description of the technology that creates that advantage or benefit. To see the difference, consider the following examples.

During the California Gold Rush in the 1870s, miners complained that their pants wore out very quickly. In response, a tailor named Levi Strauss put copper rivets at the corners of the pockets of his denim pants. Miners quickly recognized the superiority of these pants, and to this day the USP of Levi's jeans is their durability. The copper rivets are the source of that durability, but they are not the advertised feature.

USPs are also used in the automobile industry. Volvo represents the ultimate in safety, Ferrari represents the ultimate in speed, Rolls-Royce represents the ultimate in luxury, and Toyota's Prius the most environmentally friendly hybrid car. These companies advertise the concepts of safety, speed, luxury, or environmental friendliness—not the technologies that make their cars safe, fast, luxurious, or environmentally friendly. Brands such as GM and Ford, which no longer have any USP associated with their mark, are doing rather badly compared to those with clearly defined USPs.

Likewise, the computer industry uses USPs. Apple, for example, emphasizes how easy its computers are to use rather than advertising the specific technologies that make its computers user-friendly.

In order to develop an attractive USP for an early-stage technology, marketers must emphasize what buyers need over what the technology can offer. The tendency of many marketers to overemphasize the technology may explain why they are often accused of "pushing" their products into the market rather than letting them be "pulled" in by virtue of consumer demand.

Let us take, for example, a technology that allows certain vaccines to be administered using an aerosol rather than an injection. There is no question that the science may be exciting and of interest to potential users. However, the USP has nothing to do with this exciting science. Rather, the real potential which will increase user demand is to point out that aerosol vaccine delivery will allow significantly easier (and painless) delivery to the end users as well as potentially an easier storage and delivery of the vaccine prior to administration.

Sometimes a USP is bound up with the business model of the company. FedEx guarantees overnight package delivery; Domino's Pizza specializes in extremely rapid, hot, home-delivered pizza. Both of these companies have business models that allow for unusually fast delivery of products or services. With this kind of USP, of course, it is vital that delivery be as timely as promised: even small delays may send customers elsewhere.

6. CONCLUSION

Marketing is the technique of identifying markets. For early-stage technologies, it can be a difficult process given the uncertainty of what uses the technology can be put to. Nonetheless, for some early-stage technologies, the work done in the marketing phase can actually add significant value, since it identifies potential uses and buyers that may not have been considered by the original scientists. ■

MARCEL D. MONGEON, *Mongeon Consulting Inc., 301 Sunnymeade Drive, Ancaster, ON, L9G 4L2, Canada.* *marcel@mongeonconsulting.com*

---

1 Peterson I. 2004. Riding on Square Wheels. *Science News Online*, 165 (14). www.sciencenews.org/articles/20040403/mathtrek.asp.

2 See U.S. Patent No. 5,849,017.

3 Porter ME. 1979. How Competitive Forces Shape Strategy. *Harvard Business Review* 57(2): 137–145. An explanation can also be found at en.wikipedia.org/wiki/Porter_5_forces_analysis.

4 MEDBUY® is an example of a purchasing group. www.medbuy.com.

5 The actual distribution channels for such a product, defined by those who might make the decision to buy this product, include: (1) the radiologists who are ultimatelyresponsiblefortheequipment;(2)themedical administrators of the hospitals; (3) the information technology department in charge of software at the hospital; (4) the purchasing department in charge of purchasing imaging equipment; (5) a buying group that acts on behalf of an aggregate of hospitals such as an HMO; (6) a paying authority thatauthorizes any new acquisitions such as a government department or an HMO; (7) the manufacturer of the equipment looking to integrate the software; and (8) individual physicians and departments who find out about the software and are looking to acquire the tool outside of the normal channels.

CHAPTER 12.3

Technology Marketing

ROBERT S. MACWRIGHT, *Executive Director, University of Virginia Patent Foundation, U.S.A.*
JOHN F. RITTER, *Director, Office of Technology Licensing, Princeton University, U.S.A.*

ABSTRACT

Finding out how to market your technology to potential licensees can be a perplexing process. There is no common consensus about how to approach technology licensing, and workshops on the topic tend to offer a haphazard mix of tools and strategies that cannot be applied generally. This chapter emphasizes the importance of actively marketing your technology. It offers a systematic marketing approach supported by numerous models for contacting and prioritizing your contacts. The chapter also includes numerous helpful worksheets to guide and focus your approach. By following the steps laid out in this chapter, you will have learned a great deal about the market for your "merchandise," its potential licensees, and its value. You may have even found a licensee!

1. INTRODUCTION

If you ask ten seasoned licensing professionals about how they locate potential licensees, you are almost guaranteed to receive ten different answers. The truth is that technology marketing, although one of the most important and difficult aspects of technology licensing, is rarely carried out in a systematic way.

There is no consensus about the best way to approach technology licensing, and many people are not willing to share their expertise. Marketing experts in technology transfer learned the ropes just like about everyone else learns the tricks of their trade: by experimenting with hit-or-miss techniques. This haphazard approach probably explains why most training workshops on the topic offer smorgasbords of tools and strategies that one person or a few people found useful and that may or may not be useful to someone else; the workshops never offer much guidance about which tools to use, when to use them, or in what order.

The following materials suggest that it is possible to construct a marketing plan that will (1) work for both the novice and the expert in most, if not all, situations and (2) allow the licensing professional to continually refine his or her marketing strategy by systematically examining the feedback received from various sources.

2. MOVING MERCHANDISE

To fully appreciate how important technology marketing is to your licensing program, consider this simplified step-by-step plan of how technology marketing works:

1. You begin by having to market technologies that are "raw materials."

MacWright RS and JF Ritter. 2007. Technology Marketing. In *Intellectual Property Management in Health and Agricultural Innovation: A Handbook of Best Practices* (eds. A Krattiger, RT Mahoney, L Nelsen, et al.). MIHR: Oxford, U.K., and PIPRA: Davis, U.S.A. Available online at www.ipHandbook.org.

Editors' Note: We are most grateful to the Association of University Technology Managers (AUTM) for having allowed us to update and edit this paper and include it as a chapter in this *Handbook*. The original paper was published in the *AUTM Technology Transfer Practice Manual* Second Edition (Part VII: Chapter 3).

2. By investing capital in patent applications or other IP protection, you convert the raw materials into "merchandise."
3. Licensing converts your merchandise (nonliquid IP assets) into capital (liquid assets). These assets fall into two categories: recovered capital and profits.
4. Recovered capital (and, optionally, profits, as well) can be re-invested with the aim of converting more raw materials into merchandise, the licensing of which will generate more recovered capital and additional profits.
5. If the rate of licensing is slower than the rate at which raw materials are converted into merchandise, your inventory will grow. Eventually, most of your capital will be tied up in nonliquid assets, and you will go out of business.

The point is that *you must move your merchandise.*

3. HOW TO MARKET

Our approach to technology marketing makes use of the telephone extensively and requires that each call to a prospective licensee be followed up in writing.

Although direct mail communication with potential licensees is perhaps the least costly approach, the response rate to such mailings is extremely low, and there is no way to answer any questions that potential licensees might have. The same can be said for computer databases and bulletin boards, which require potential licensees to log on, search for, and find advertisements and information about your technology. The limitations of such an approach are evident.

In an ideal world, the licensing professional would personally meet with all potential licensees: much more information can be communicated in person, and the response to the presentation can be gauged more easily. But few companies have the resources to keep their marketing professionals on the road. Although conferences are an efficient way to meet many potential licensees in person, they do not happen frequently enough to be adequate as a sole source of new contacts; besides, not all companies send representatives to such meetings.

Although telephone conversations are not quite as good as face-to-face meetings, phone conversations are a close second choice. The greatest advantage of using the telephone is that you can easily and inexpensively communicate with potential customers who are geographically distant and dispersed. Follow up each phone call with a brief letter and a nonconfidential description of the technology you hope to license. This follow-up activity will remind your potential customer about your offer and allow you to offer materials that can be sent to his or her company's scientists for further consideration.

4. DISCLAIMER

Keep in mind that the ideas shared in this chapter are new and have not yet been put to the test in the "real world." However, they are based on more than 20 years of experience by licensing professionals. We believe that these are practical materials, and we hope that you will put these materials to the test. We look forward to hearing your comments and criticisms.

The strategy outlined here is meant to serve as a template. We expect each user to modify it to suit his or her own needs and personal style. Some professionals may eventually choose to abandon this strategy altogether for a more free-form approach to marketing.

Finally, we have recommended particular reference texts or databases with reluctance; some professionals in the field might feel that we are promoting the interests of certain companies. We would like to point out, however, that 1) not one of the contributors has ownership interest in any of the companies recommended here and (2) none of us has received any compensation or consideration for our recommendations. Furthermore, we acknowledge that many other services and resources may be just as good as those we have recommended, and some may be far better; many more resources exist that we have been able to personally evaluate. We therefore invite you to explore the alternatives for yourself. The Association of University Technology Managers

(AUTM) Web site contains a section on marketing resources in its business section that can help you to begin your exploration.[1]

5. SYSTEMATIC MARKETING

This systematic technology marketing approach can be divided into four major activities:

Step 1. Collect information from the inventors.

1. Attach the marketing information sheet shown in Box 1A to your disclosure form (all Boxes are at the end of this chapter). This form explains the importance of technology marketing to the inventors.[2]
2. Attach the subquestionnaire, shown in Box 1B to the disclosure form, which asks the inventors to consider a variety of marketable applications for their invention. Each inventor should fill out this portion of the questionnaire: each person is likely to have different ideas and different contacts.
3. Based on any information you have on hand (or that you can reasonably estimate) about the current situation of the market(s) into which the invention might be introduced, fill in the summary sheet shown in Box 1C. Fill out one sheet for each hypothetical product or service envisioned by you or the inventor(s). Keep this sheet updated as you collect relevant information.
4. In order to collect further information that may aid in marketing the invention, consult with the inventor(s) about the contents of the summary sheet in Box 1C, and ask them the questions on the checklist in Box 1D.
5. For each target market, prepare a tailored, single-page, nonconfidential disclosure, in accordance with the guidelines and sample text shown in Box 1E.

Step 2. Collect information about potential licensees.

1. Begin with online searches. You may decide to manually search for potential licensees, for example, using the CorpTech hard-copy directory.[3]
2. Subscribe to a service that provides an online database that you can search for potential licensees (for example, Knowledge Express Data Systems [KEDS] or another system of your choice).
3. Install the database software by following the tutorials and step-by-step instructions provided. Review any additional instructional materials that come with the database, paying particular attention to information on how to use the database.
4. Develop both a list of keywords that will help you identify potential licensees and a profile describing your ideal licensee, and also develop a CorpTech-like profile for your ideal licensee.
5. Search the databases using the parameters you have collected: your keywords, CorpTech profiles of companies that might be possible customers, and the profile you created of the ideal licensee. Identify the five companies that seem to be the best matches for your technology. If you are having trouble identifying the top five, use the worksheet in Box 2 to narrow down your list of companies.
6. If you are using KEDS, you can substantially expand the number and focus of hits by using the Knowledge Express "hypertext" function. This function allows you to quickly determine which of the many available databases have entries that match the keywords you have identified. You can then search each database individually for possible licensing prospects. The hypertext function will often find entries on advanced technologies in the CorpTech and BioScan databases (the latter is a database that focuses on biotechnology and related disciplines), *Business News* (which contains current information and lists companies that are not listed elsewhere), and SBIR (which lists awards made by the Federal Small Business Innovative Research program for small, high-tech companies).

Step 3. Review and prioritize your prospects list. Examine your list of prospects. Using the worksheet in Box 3, assign each of the top five corporate prospects a rank from 1 to 5, with 1 the highest priority and 5 the lowest priority.

Step 4. Make contact with potential clients.

1. Review the guidelines (Box 4A) for finding the right person to talk to. Write down the company's telephone number, and, if possible, make a list of names and titles of potential contacts.
2. Review the three cold-call transcripts (Box 4B) and familiarize yourself with the sorts of conversations you can expect, depending on whether your prospects are very interested, not at all interested, or somewhat interested.
3. Review the "What to Get Across to Your Contact When You Call" checklist (Box 4C), and make sure you have all of the information you will need to convey. You may want to write it down so that you do not forget any of it.
4. Make the call. Call the company with the lowest priority of the five you have selected. Box 4A explains how to find the right person to talk to.
5. During and after the call, record information about the prospective company and how your contact responded on the "Reaction Data Sheet" (Box 4D).
6. Send the prospect a follow-up letter, modeled after one of those in Box 5, along with a copy of the nonconfidential disclosure (regardless of whether or not the prospect requested one).
7. Repeat steps 4 through 6 for each of the other prospects, working from the one with the least potential to the one with the greatest potential (in other words, beginning with number 4, then number 3, and so on).
8. Next, call those prospects ranked 6, 7, 8, and so on in order of decreasing potential.
9. If you have found a licensee, congratulations! But do not stop. One prospect is fine, but two or more prospects are better: if you are planning to offer an exclusive license, more prospects will give you more bargaining power; if you are planning to offer nonexclusive licenses, each new prospect means more payoff for your marketing efforts. If, on the other hand, you have not been able to find a licensee, assess your results using the guidelines in Box 6 and decide what you want to do next: Continue looking for prospects using the same strategies? Continue looking for prospects using new strategies? Wait a year and try again? Write off, as a loss, the capital invested in IP protection for this invention?

6. CONCLUSIONS

By following these steps, you will have learned a great deal about the market for your merchandise, its potential licensees, and the value of your product. You may have even found a licensee. Build on whatever success you have found by taking the time to learn from your experience and by analyzing the feedback you have obtained from your systematic marketing approach. And share what works with others.

For further information, suggestions, or guidance regarding this marketing strategy and how it might be customized or refined, please feel free to contact the authors at the numbers shown below. We would also appreciate your feedback on how this approach has worked for you, and how you believe it might be improved. Please share with us copies of any revisions you may make to the instructions or forms. ■

ACKNOWLEDGEMENTS

We are grateful to Teri Willey, Managing Partner, ARCH Development Partners, who contributed to the original published material.

ROBERT S. MACWRIGHT, *Executive Director, University of Virginia Patent Foundation, 250 West Main Street, Suite 300, Charlottesville, VA, 22911, U.S.A.* Robert@uvapf.org

JOHN F. RITTER, *Director, Office of Technology Licensing, Princeton University, 4 New South Building. Princeton, NJ, 08540, U.S.A.* jritter@princeton.edu

1 www.autm.net (accessible to AUTM members) First select "Business," then "Marketing," then "Resources to Review."

2 See, also in this *Handbook*, chapter 8.4 by DR McGee.

3 www.corptech.com.

4 See, also in this *Handbook*, chapter 7.2 by SP Kowalski and A Krattiger.

5 See, also in this *Handbook*, chapter 11.8 by S Shotwell.

Box 1: Collecting Information from the Inside (Step 1)

A. Filling Out the Invention Questionnaire

When you complete the attached Invention Questionnaire, you will notice that it includes questions not only about the technical aspects of your invention, but also about its potential commercial market(s).

If you are like most inventors, you will probably not be very interested in thinking about how to market your invention. However, your answers to these questions are at least as important, if not more important, than your answers to the technical questions. Why? Remember that a patent is, first and foremost, an economic vehicle. It gives patent holders a monopoly on the manufacture, use, and sales of an invention for the life of the patent. The government grants such monopolies in order to provide an incentive for individuals and companies to invest the resources and effort needed to bring new products to the marketplace.

If patents were free, we could patent every invention and make profits on whichever ones reached the marketplace. Unfortunately, obtaining a patent is always costly. The application procedure for a typical U.S. patent costs between $10,000 and $20,000 from start to finish, and foreign patent applications can cost more than $100,000 for a single invention.

Therefore, we, as technology transfer specialists, have to try to determine in advance which inventions are likely to be of interest to licensees. The goal is to license each patented invention in exchange for a royalty, so that we can both recover the costs of the patent application process and generate additional revenues. If we patent inventions without first considering their licensing potential, we risk losing the money we have invested in patenting costs.

Granted, market exploration is not your job—it is ours. However, though you may not think that you know anything about marketing, experience has shown that inventors are one of the most valuable sources of market information. You know your new technology better than anyone else. You probably know how it might be used, and you might even know who would be interested in licensing it.

Now you know why we are asking you for help with marketing. Please answer the following marketing questions to the best of your ability. If you do not know the answer to a question, or are unsure whether you really understand the question, try to answer it anyway, and make your answer as comprehensive as possible. Please feel free to provide additional information that we have not specifically requested.

(Continued on Next Page)

Box 1 (continued)

B. INVENTION QUESTIONNAIRE

Docket ______________ Title ______________________________

Date ______________ Completed by ______________ Form ___of___

Please feel free to attach additional sheets if you need more room or if you want to explain your responses. In addition, please attach any materials that you think might help illustrate or supplement your answers.

PRODUCTS AND SERVICES

List as many products or services (whether actual or hypothetical) as you can think of that might benefit from your invention. Be adventurous: try to think of both broad and narrow applications, as well as applications that are outside of your own field.

1. ______________________________

2. ______________________________

[etc.]______________________________

COMPETING PRODUCTS AND SERVICES

List as many existing products or services, and the companies that provide them, as would be in competition with your new invention if it were to be used for all the functions you listed on the Products and Services form. You may wish to refer to catalogs or databases in completing this next list. Please attach any relevant product brochures or descriptions.

Product or service *Company*

1. ______________________________

2. ______________________________

[etc.]______________________________

POSSIBLE LICENSEES

List the names of companies you think would be interested in using your invention to make, use, or sell products or services. If you have a contact at any of these companies, be sure to provide a name and telephone number. (We will obtain your permission before we contact anyone.)

Company *Contact* *Phone*

1. ______________________________

2. ______________________________

[etc.]______________________________

ADVANTAGES

If we are to convince companies to invest in the commercial development of your invention, we will have to be able to explain why it is superior to alternative products, processes, or services. List all of the advantages of your invention. Attached is a list of possible advantages for you to consider and to help you generate other ideas.

1. ______________________________

2. ______________________________

[etc.]______________________________

(Continued on Next Page)

Box 1 (continued)

B. INVENTION QUESTIONNAIRE (continued)

POSSIBLE ADVANTAGES OF YOUR INVENTION

CHEAPER	The invention is cheaper to make or use than currently available products or processes.
EASIER TO USE	The product or process is less complicated, less labor intensive, or more user friendly than those of currently available products or processes.
EASIER TO MAKE	The product is less complicated to make, or its manufacturing process is less complex, than those of currently available products.
SAFER	The product or process is safer for the operator, bystanders, or animals than currently available products or processes.
MORE ECOLOGICAL	The product or process recycles materials that usually end up in landfills or is less polluting than currently available products or processes.
FASTER	The product or process works faster than currently available products or processes.
MORE PRECISE	The product or process yields a more exact result than those produced by currently available products or processes.
MORE ATTRACTIVE	The product would be attractive to a broader segment of the marketplace than those products currently on the market.
NOVEL	The product or process is novel: people would ask, "Why didn't I think of it?"
CLEAR VALUE	Other products or processes are similar enough that the value of this one will be apparent.
QUIETER	The product or process is quieter or the sound it produces is less irritating than is true of currently available products or processes.
SMELLS BETTER	The product or process produces no smell, or a more pleasant smell, than is true of currently available products or processes.
TASTES BETTER	The product (if intended to be tasted) tastes better than currently available products.
BETTER SIZE	The product is more compact, or is larger and has greater capacity, than currently available products.
BETTER WEIGHT	The product is lighter or heavier (whichever is preferable) than currently available products.
MORE DURABLE	The product is more durable than currently available products.
MORE RELIABLE	The product breaks down less frequently, or the process is more consistently successful, than currently available products or processes.
EASIER TO FIX	The product is less complicated or costly to fix or adjust than currently available products.
LARGE MARKET	There is already a large market for this product or process, or the appeal of the product or process will likely create a large market where one did not previously exist.
GROWING last MARKET	There has been steady growth in the target market for your product or processes over the several years.
LASTING MARKET	The need or demand for the product will last a very long time.
EASY FOR MANUFACTURERS TO SWITCH	The product or process is similar enough to currently available products or processes that users or manufacturers can easily switch.
HARD TO DUPLICATE	Competitors will have difficulty producing an equivalent product or process, or to solve problems without it.
HIGHER PROFIT MARGIN	The product or process is easier and cheaper to make than currently available products or processes, but can be sold for a comparable amount.

(Continued on Next Page)

Box 1 (continued)

C. MARKET SUMMARY DATA

Docket ____________	Title ____________
Date ____________	Completed by ____________ Form ___of___

Note to reader: Since this sheet is completed before any systematic research is performed, the information is likely to be both highly speculative and incomplete. You may need to fill out a separate form for each product or service that you envision for this invention. Use this form as a guide when discussing marketing issues with the inventor(s).

Product or service ____________

Market size ($ million) *Worldwide* ________ *U.S.* ________

Europe ________ *Asia* ________

Top companies ____________

____

Other companies ____________

____

Competing products or services ____________

Market cycle status ☐ *growing* ☐ *stable* ☐ *contracting*

Regulatory requirements ____________

Expected regulatory costs ($ million) ____________

Other investment needed (rough estimate) ____________

(Continued on Next Page)

Box 1 (continued)

D. QUESTIONS FOR INVENTOR INTERVIEW

Ask each of the inventors the following questions, preferably in person or by telephone, rather than in writing. Depending on the direction the conversation takes, you may decide to ask other questions that occur to you that are not on this list. You may find that the inventors are more candid if you speak to each of them privately.

1. Do you have any family members, friends, or ex-classmates who work for a company that might have an interest in your technology?
2. Do you have a company of your own? Are you interested in starting a company?
3. Do you have any consulting or other relationships with companies? Would these companies be interested in your technology?
4. When we license the technology, would you be willing to collaborate with the licensing company as a principal or as a technical advisor?
5. Do you know of anyone who might want to invest in this technology (venture capitalists or private investors, for instance)?
6. Where did you work before you started working here? Do you know anyone from your previous position(s) who might be of help?
7. Would you be willing to spend a little time calling friends and colleagues to find out what they think about your technology and its possible applications?
8. Can you give us a few names and telephone numbers of people with whom we could speak about your technology and possible licenses?
9. Would you be willing to speak to potential licensees about your technology?
10. Would you be willing to make prototypes or samples, or carry out demonstrations, in order to help us in our licensing efforts?

(Continued on Next Page)

Box 1 (continued)

E. DRAFTING THE NONCONFIDENTIAL DISCLOSURE

A nonconfidential disclosure (NCD) should be nonenabling, that is, it should not contain enough information to allow a person skilled in the field to reproduce the invention without undue experimentation. The NCD should, however, contain enough information to pique the interest of the person reading it. Only on very rare occasions should an NCD exceed one page in length. There are many possible formats for an NCD, but we recommend the following one:

1st Section. Begin with an introductory sentence such as: "A novel dengue virus vaccine has been developed by BioReplicon Corp. and is available for licensing." The remainder of this section should give a punchy, brief explanation of the field of the invention.

2nd Section. Briefly describe the state of the art before the invention, and then highlight the important advantages that the invention offers over the currently available alternatives.

Keep in mind that you can often disclose performance data without giving anything else away. For example, you can say, "Vials of one milliliter in volume, having walls 0.1 millimeter thick, were able to withstand sustained pressures measuring in excess of ten atmospheres." A reader would be able to see that the material in question is very sturdy without being able to figure out what it was or how it was made.

If at all possible, refer to and append any data (charts, tables, graphs) that show the invention's technical superiority and/or compare the technology with currently available alternatives.

3rd Section. Describe the terms of licensing and provide contact information, should the reader wish to make further inquiries.

4th Section (optional). Provide brief biographies of the inventors, especially if they are well known in their fields.

An example of an NCD follows:

New Invention

A novel method for manufacturing piezoelectric composites has been developed at Moorhead University and is available for licensing.

Piezoelectric composites are composed of two layers, an "active phase" and a "passive phase." The active phase physically deforms when an electrical current is applied, thereby producing sound waves. By improving the match between the sound impedance of the active phase and the target of the sound waves (for example, the skin), the passive phase improves the efficiency of sound transmission. Piezoelectric composites are used in medical imaging devices, hydrophones, and various sensor applications.

The industry currently uses a "dice-and-fill" method to make such composites. This method involves sawing slits into blocks of active-phase material, and then filling them with passive-phase polymer. Our new method overcomes many of the disadvantages and limitations of the dice-and-fill method:

Improved efficiency: The process takes fewer manufacturing steps to produce the same composite.

Less waste: No material is lost, because no slits have to be sawed.

Increased flexibility: The dice-and-fill method can create only two-phase composites, but our method can create multiphase composites. (See attached page for diagrams of the types of multiphase composites that are possible to make using our technology.)

Improved preformation: Our method allows for the variance of active-phase volume content, thus decreasing the out-of-plane distortions of the transmitted signal.

This new technology is available on an exclusive or nonexclusive basis.
For further information, please contact:

John Smith
Technology Licensing Associate
Office of Technology Transfer
Someplace University
Somewhereville, LA 12345
Phone +1-800-555 1212, Fax +1-800-555 1213
smight@someplace.edu

Dr. Arnold Smuthers, co-inventor of the described invention, is a world-renowned authority in the field of piezoelectrics, and holds over 30 U.S. and foreign patents.

Box 2: Collecting Information from the Outside (Step 2)

A. WORKSHEET FOR DEVELOPING A SIMPLE SEARCH STRATEGY

Docket ____________	Title ____________	
Date ____________	Completed by ____________	Form ___of___

Because you have already collected some information about the technology and its market from the inventor(s), developing a licensee search strategy should be easy. Ask yourself:

1. **In what product development areas might potential licensees be interested?** List single- and multiple-word descriptions that might be used as search identifiers. Keep in mind that you may want to find several licensees, each holding a license to make, use, and sell licensed products in a different field of use.[5]

2. **Do I already know of a few companies that might be good licensees for this technology?** Search for information on these companies, and then use that information as a guide to search for other, similar companies.

3. **Create a profile of the ideal licensee.** Imagine the ideal licensee (or describe a licensee known to you that you think would be ideal) for the technology. Complete one copy of this form for each product or service that you have envisioned for this technology. Use additional copies as necessary.

Company size	☐ *large*	☐ *medium*	☐ *small*	☐ *start-up*
Structure	☐ *private*	☐ *public*	☐ *nonprofit*	
Country	☐ *U.S.*	☐ *foreign*	☐ *multinational*	

State/province ____________

Sales per year ____________ *$ (million)*

No. of employees ____________

Products and/or services ____________

4. **If you are stuck, imagine that you are the president of a company that would be an ideal licensing partner, and ask yourself the following questions:**

   1. What is our product development focus? How does this product fit?
   2. What kind of personnel do I have? What kind of personnel would I need if I were to license this technology?
   3. What is my existing manufacturing capability? Can I manufacture this technology? Can I create the ability to manufacture it? Can I outsource its manufacture?
   4. Do I have access to complementary technology?
   5. What kind of capital resources do I have? Where will the research funds come from?
   6. What kind of marketing expertise do we have? If it is limited, can we partner with other companies that have more marketing expertise?
   7. Is it important for this technology to have international markets? Do we have the ability to develop international markets?
   8. What regulatory issues are involved? Can we handle these, given our current levels of resources and expertise?
   9. Do we have experience with this type of early-stage technology? (For example, [the applicable type of technology].)

Now, go back and re-address questions 1, 2, and 3.

Box 3: Ranking Prospects: A Worksheet (Step 3)

For each of the potential licensees identified, assign a score for each, using the criteria listed below. If you have no information, leave the space blank. Rank the companies, with the most promising prospect being the company with the highest total score. If you have more than five prospects, use additional sheets.

Write the names of the prospect companies in the spaces at the right, and on the similar spaces on the next page.					
CRITERIA					**SCORE (1-5)**
Portfolio includes products like this one					
Has large share of relevant market					
Could expand its share of that market					
Has patents on related technology					
Has personnel needed					
Has relationship with you or your office					
Has relationship with inventor					
Company not too big or small					
Company already expressed interest					
Good fit with other company products					
Located nearby					
Has known licensing experience					
Good fit with company R&D focus					
Has long history, established management					
Known for being an innovative company					
Respected by the inventors					
Has introduced new products recently					
Has membership in professional association					
Is well known, has good reputation					
Has large marketing and sales force					
Has international marketing capability					
Has successfully licensed from you in the past					
Would big part of company's business					
Can manufacture or out-source it					
Can afford necessary re-tooling					
Has product development resources					
Can afford up-front, minimum payment					
TOTAL					
RANK					

Box 4: Making Contacts (Step 4)

A. CONTACT IDENTIFICATION GUIDELINES

As you contemplate which individual in a company might be best to contact, it is worthwhile to consider how someone wishing to license to or from your organization would identify you. You hope the person would find you, but, in the end, the path between you and that person might not be direct. Furthermore, it may take a few calls before you identify the "right" person at the company you have identified as a licensing prospect.

The following guidelines should help you to make contact with the right person.

1. Utilize the knowledge of secretaries

   Receptionists and secretaries are often knowledgeable about who does what at their company. Secretaries of higher-level executives generally are the most knowledgeable about sophisticated functions such as licensing. If you are having trouble finding out who to talk to, try asking the secretary of a vice president or the president. The secretary for the legal department may also be quite helpful. Describe carefully who you are and what you need.

2. Try to look up your contact

   Regardless of the apparent size of a company, it is always worth the time to first look up the company in the LES directory and the AUTM directory. Even small companies sometimes belong to one or both of these organizations, and if the target company is listed, any one of the members included in the listing is most likely a "direct hit."

   If the company has more than one member, look up all of the members' titles before you decide who to call. If the company is fairly large, unless your technology is a revolutionary invention, you are probably better off calling the second or third most senior licensing person. He or she is more likely to spend the time to hear you out, and to take the time to follow up after the call is over.

   If the company is of substantial size, look up the company in CorpTech, Dunn & Bradstreet's, or Moody's directory, if available (you can also do this online). Look under the corporate officer's listing, and look for titles such as:

   - director of licensing
   - director of technology acquisition
   - vice president for new ventures
   - patent counsel
   - general counsel
   - director of new product development
   - vice president for new product development
   - new technology analyst
   - director of marketing
   - vice president for research and development

   The listing should give the officeholder's name. Although that person might not be the person you need to speak to, having a name and title that is at least somewhat relevant make

3. Make a call or two

   If you have found a name or at least a title that looks promising, call the company and ask for the person, or the person with that title. In all likelihood, a secretary will answer. Tell him/her your name, the organization you are from, and explain that you have a new technology that you think the company would be very interested in acquiring. Ask if the person you have called is the right person to speak to. The secretary may believe that someone else is the right person or that a different department would be better able to help; in either case, ask to be transferred. On the other hand, the secretary may not know who or which department to refer you to. If that is the case, ask to speak to the person you called. Then, give that person the same introduction and ask if he or she is the right person to speak to. If he or she is not the right person, ask to be transferred.

   Whenever you are transferred to another line, start by saying, "[name's] office thought you might help me," or "The president's office thought you might help me," for example. This will avoid the possibility of being referred back to someone you've already spoken with and will suggest to the second person that the first person thought it was worthwhile to help you, so they should, too. Introduce yourself as described above, and proceed in the same way.

   For a company that is not listed in the LES directory, the AUTM directory, CorpTech, Dunn & Bradstreet's, or Moody's, it is likely that the company is fairly small. For fairly small companies, it is sensible to start "at the top." Call and ask to speak to the president. Usually an executive secretary will screen the president's calls and will ask why you have called. Give your name and the name of your organization, and explain that you have a new technology you believe the company would be very interested in. You will likely be connected to the president, a vice president, or research director. Introduce yourself, and ask if you were properly directed. Proceed as described above.

Box 4 (continued)

B. COLD-CALL TRANSCRIPTS

The following transcripts illustrate the sorts of conversations you might encounter when talking with a prospective licensee. Keep in mind that these are examples and that you should be prepared for conversations that do not follow any of these patterns. However, we do not mean to suggest that a company of one size is a better prospect or will be more receptive to your call than a company of another size. Good licensing deals can be made with companies of all sizes.

Also, do not assume that the length of these transcripts is necessarily representative of the length of the conversations you will have with potential licensees. Conversations can be quite long and cover many subjects, especially if your contact is very interested in what you have to say. Be sure to leave plenty of time for the call, and hope that you need it.

1. The call we all want. (It really does happen this way sometimes.)

Licensor: Hello, this is Jake Sinclair, and I'm from the University of Maui. I'm calling because our Professor Mahalo has invented a new fiber-optic stethoscope that we thought your company would be interested in.

Prospect: University of Maui, huh? I went there as an undergraduate. Great school. Who did you say was the inventor?

Licensor: Professor Mahalo.

Prospect: Oh, yeah! I took a course on biomedical engineering with him about ten years ago. I'm sure anything he's invented is really good. What can you tell me about it?

Licensor: Well, it has an electronic pickup device that picks up even very faint sounds. It then converts the signal to a light beam, and transmits the beam through a fiber-optic fiber to a decoder that is about the size of a large felt-tip marker. The decoder electronically filters out background noise, then transmits the filtered sounds to a pair of headphones.

Prospect: A fiber-optic stethoscope. Pretty neat. As you know, stethoscopes are our only business here at Stethoscope Technologies.

Licensor: Yes, we know. That's why we thought of you. Also, you have an excellent reputation in this field.

Prospect: And, as it turns out, we have been looking for a high-tech product to sell to the top end of our market. But it would be very important to us that the device we sell look and handle like our other, more traditional stethoscopes.

Licensor: Dr. Mahalo feels that the pickup and decoder could be miniaturized enough for that with a little engineering work.

Prospect: Well, this certainly seems interesting. Do you have any patent protection?

Licensor: Yes, we have applied for two U.S. patents, and on one of them, we have already filed a worldwide application under the PCT [Patent Cooperation Treaty].

Prospect: Hmm. Wow, this sounds like it may be just what we have been looking for. Could you send us some detailed technical information so we can talk with our product design team about it?

Licensor: Sure. Of course, we will need to have you sign a confidentiality agreement first.

Prospect: Oh, that's no problem for us. If you would fax one to me, I'll courier it back to you tonight, and maybe you could send us a copy of the patent applications. After we've had a chance to review them, if we're still interested, we could come visit you and Dr. Mahalo next week on our way back from Japan.

Licensor: Sounds great. However, we would prefer not to show you the claims until it becomes more certain that you are interested in a license.

Prospect: That's fine.

Licensor: Well, I've really enjoyed talking to you, and I'll fax you the confidentiality agreement right away.

Prospect: Great. And tell Dr. Mahalo that I look forward to seeing him again.

Licensor: Sure will. Bye.

Box 4 (continued)

B. COLD-CALL TRANSCRIPTS

2. The "No thanks" call. Because few technologies are attractive to everyone, quite a few of your calls may be of this type.

Licensor: Hello, my name is James Sulkind and I am in charge of out licensing for the Omed Marine Corporation. I'm calling because one of our scientists has developed a radio beacon technology that is simply too high tech for our manufacturing capability, but we thought it might be right up your alley.

Prospect: Radio beacons? We make televisions and FM receivers, but we've never made marine stuff. The market's too small.

Licensor: Well, we know the market is relatively small now, but marine radio equipment is growing increasingly sophisticated, even in pleasure boats, and we thought it might be a new and growing market for you.

Prospect: Nah, we're volume producers, and that market will never be big enough for us to bother with. We even gave up the portable radio market, and that was probably ten times bigger than the one you're talking about.

Licensor: Are you sure you wouldn't be interested?

Prospect: Yes, I'm sure. But why don't you send me something anyway?

Licensor: Sure, be happy to.

Prospect: Thanks. Bye.

Interestingly, even if the person is not interested, he or she usually wants something in writing anyway. Some may circulate it to their R&D and marketing staff, just to double check that your technology is not something they want to pursue. Others may just want a nonconfidential disclosure to attach to their monthly reports in order to show their bosses that they have been actively considering new technologies. Regardless of your contact's intentions, follow up on the phone call and send the written disclosure. It may or may not get a second look, but at the least, it will encourage that individual to take your call the next time when you have another technology to offer.

3. The "Gee, I don't know" call. Another common situation is one in which the person you call has some interest in what you have to say, but really is unsure if the company would be interested or not. In this situation, it helps to have persuasive skills and to have spoken with your inventors in advance about the benefits that your technology can offer.

Licensor: Hello, my name is Beverly Houghton, and I'm a licensing associate at Ethridge University. I'm calling because Dr. Cuthbert of our computer science department thought that you would be quite interested in his new neural network approach to "just in time scheduling" for automotive parts production.

Prospect: Neural networks? We just got our computerized production scheduling system on the market last year. I don't think we are ready to make any big changes in it at this point. Coordinating all of our warehouses and car dealers was an enormous investment. Besides, our inventories are already stable and at very low levels compared to the old days.

Licensor: Well, Dr. Cuthbert is familiar with your system, and he thinks that it could really benefit from this new approach. He also thinks it could be implemented easily and quickly.

Prospect: Oh, really? What does he think would be the benefit?

Licensor: Dr. Cuthbert says he thinks that the processing time would be reduced by at least 50 percent, and that this time savings would be directly translated to increased speed at the parts department terminals.

Prospect: Well, node speed has been an issue.

Licensor: Yes, and you could increase node speed by, say, 20 percent, and then have processing time left that would allow you to receive and transmit more data in real time. This increased information transit may allow you to have even lower levels of standing inventory than you currently think possible.

Prospect: Interesting. What does Dr. Cuthbert think it will cost us to do this?

Licensor: In terms of hardware, nothing. On the software side, he already has compatible software elements that he and your programmers could easily weave in.

Prospect: But there's a catch, right? You guys aren't going to let me use this for free.

Box 4 (continued)

Licensor: You're right. But because we hope to license this technology to others, too, the cost to you should be relatively low. We would like to get something up front, plus about $100 per node per year. Of course, there would also be some costs for Dr. Cuthbert's time, and we are looking for about $50,000 per year for use of his neural-network system software.

Prospect: Well, when you add it up, that's a fair amount of money. Besides, if we tell our dealers that we're going to mess with this system again, they'll scream bloody murder.

Licensor: Only until they see what it can do.

Prospect: Well, maybe. What can you send me about this?

Licensor: For starters, I can send you a nonconfidential disclosure. If you're still interested, I can send you a copy of the patent application, and maybe have you talk to Dr. Cuthbert.

Prospect: Well, at this point, just send me the nonconfidential stuff. If the operations guys are interested, I'll call you back.

Licensor: It's on its way. If you like, maybe we could also set up a demonstration for your operations guys.

Prospect: Well, since you're right here in Detroit, maybe that isn't such a bad idea.

Licensor: How about if I have Dr. Cuthbert call you to set it up?

Prospect: Well, let's not get ahead of ourselves. I'll give you a call after we've thought about it here.

Licensor: Great. I look forward to your call. Bye. (After hanging up.) Who knows? I better make a note to call him back.

C. "WHAT TO GET ACROSS TO YOUR CONTACT WHEN YOU CALL" CHECKLIST

The following checklist should help you make sure that you cover the basics on each call. Of course, there may be something else you want to get across that is not on this checklist. Also, the person you call will likely ask questions that are listed here.

To some extent, the level of your contact's interest will determine how far down this list you get. In any event, failing to get some things across is not fatal.

In the beginning, you may want to write notes to yourself to make sure that you know exactly what you need to say. But don't sound as if you're speaking from a script. Even when you have more experience, you may still find it helpful to check off items as you cover them.

- ☐ Your name
- ☐ Your organization
- ☐ Your location
- ☐ A general overview of the technology
- ☐ Who the inventor is (if he/she is an academic or well known)
- ☐ Why you think the company should be interested in the technology
- ☐ The advantages that the new technology offers over existing products, processes, or services
- ☐ Whether prototypes or demonstrations of the technology are available
- ☐ Whether you have applied for patents, copyrights, and/or trademarks; whether there are trade secrets
- ☐ Whether you are looking for an exclusive or nonexclusive licensee (or are undecided)
- ☐ Whether other licenses have already been granted
- ☐ That you can provide written nonconfidential information about the technology
- ☐ That you would be willing to enter into a confidentiality agreement with the company
- ☐ What confidential information you could provide

Box 4 (CONTINUED)

D. REACTION DATA SHEET

Docket ______________________ Title ____

Date _________________________ Completed by ___________________________ Form ___of___

Complete a copy of this form after each call to each potential prospect. Make sure to review any prior forms before you make each call. They will help you remember what the person's personality is like and help you interpret his or her reactions.

The checklist is a general barometer of your prospect's reactions. It is a supplement to, but not a substitute for, the notes you will take during the call regarding what was said and what needs to be done.

IDENTIFYING INFORMATION

Who made the call ____

Company name ____

Company address ____

____

Contact ____

Title ____

Secretary's name ____

Telephone no. ____

Date of call ____

Company size	☐ *large*	☐ *medium*	☐ *small*	☐ *start-up*
Location	☐ *U.S.*	☐ *foreign*	☐ *multinational*	
Structure	☐ *private*	☐ *public*	☐ *nonprofit*	

CONTACT'S MOOD

☐ *calm*	☐ *hurried*	☐ *somber*	☐ *annoyed*	☐ *curious*
☐ *amused*	☐ *angry*	☐ *tired*	☐ *guarded*	☐ *happy*

CONTACT'S ATTITUDE

☐ *receptive*	☐ *enthusiastic*	☐ *sarcastic*	☐ *disinterested*	☐ *encouraging*
☐ *sincere*	☐ *mysterious*	☐ *sinister*	☐ *secretive*	☐ *confused*
☐ *aloof*	☐ *friendly*	☐ *condescending*	☐ *respectful*	☐ *nervous*

CONTACT'S COMMUNICATION STYLE

☐ *hardly spoke*	☐ *asked questions*	☐ *made suggestions*	☐ *made jokes*
☐ *made small talk*	☐ *talked about company*	☐ *conversant*	☐ *gave opinions*
☐ *talked about market*	☐ *talkative*	☐ *talked about LES*	☐ *talked about family*

Box 4: Making Contacts (Step 4) continued

CONTACT'S LEVEL OF INTEREST

- ☐ *expressed a lot of interest*
- ☐ *expressed minor interest*
- ☐ *moderately interested*
- ☐ *disinterested*
- ☐ *expressed a lot of interest*
- ☐ *bored*
- ☐ *expressed some interest*
- ☐ *expressed lack of interest*

NEGATIVE COMMENTS CONTACT MADE ABOUT THE TECHNOLOGY

- ☐ *retooling costs too high*
- ☐ *technology too complex*
- ☐ *technology too costly*
- ☐ *market too small*
- ☐ *market too committed*
- ☐ *market too unpredictable*
- ☐ *benefit not worth price*
- ☐ *benefit too small*
- ☐ *prototypes not available*
- ☐ *technology not proven*
- ☐ *licensor/inventor not known*
- ☐ *demonstrations not available*
- ☐ *similar technology flopped*
- ☐ *market in decline*
- ☐ *profit margins too low*
- ☐ *bad fit with market needs*

POSITIVE COMMENTS CONTACT MADE ABOUT THE TECHNOLOGY

- ☐ *modest retooling costs*
- ☐ *technology not too complex*
- ☐ *technology inexpensive*
- ☐ *market large*
- ☐ *market would be receptive*
- ☐ *market predictable*
- ☐ *benefit well worth price*
- ☐ *large benefit*
- ☐ *satisfies current and future market needs*
- ☐ *technology well proven*
- ☐ *high profit margins likely*
- ☐ *market expanding*

CONTACT'S REASONS FOR BEING DISINTERESTED IN THE TECHNOLOGY

- ☐ *resources are already committed to other projects*
- ☐ *technology is a bad fit with the company's other products*
- ☐ *company is not innovative*
- ☐ *working on better one*
- ☐ *company doesn't like in licensing*
- ☐ *got burned last time*
- ☐ *economy is bad*
- ☐ *technology is a bad fit with the company's goals*
- ☐ *company has no licensing experience*
- ☐ *company has a small sales/ R&D staff*

CONTACT'S REASONS FOR BEING INTERESTED IN A LICENSING DEAL

- ☐ *product is a good fit with the company's other products*
- ☐ *company prefers high-technology products*
- ☐ *ample resources available*
- ☐ *company has in-licensing experience*
- ☐ *the company is innovative*
- ☐ *product is just what they need*
- ☐ *company has strong R&D, marketing and sales capabilities*
- ☐ *licensor/inventor is known and respected*
- ☐ *working on inferior version*
- ☐ *company likes to in license*
- ☐ *economy is good*
- ☐ *product is a good fit with company goals*

CONTACT'S REASONS FOR NOT LIKING THE TERMS

- ☐ *does not understand the technology*
- ☐ *wants a different degree of exclusivity*
- ☐ *wants to limit up-front licensing costs*
- ☐ *wants to limit royalty burden*
- ☐ *does not like confidentiality agreements*
- ☐ *does not like usual license terms*

FOLLOW-UP ACTION YOU PROMISED

- ☐ *provide nonconfidential disclosure*
- ☐ *provide a confidentiality agreement*
- ☐ *provide a demonstration/sample*
- ☐ *have an inventor or scientist call*
- ☐ *send a sample*
- ☐ *arrange a demonstration*
- ☐ *call again*

Box 4 (continued)

FOLLOW-UP ACTION PROMISED BY THE CONTACT

- ☐ *ask technical staff about the technology*
- ☐ *review the technology with management*
- ☐ *provide a confidentiality agreement*
- ☐ *get in touch if interested ("don't call us, we'll call you")*
- ☐ *call back ("we'll call you, but we don't mind if you call, too")*

CONCLUSION ABOUT THE CHANCES FOR PROSPECT

SURE THING: *We have a deal in the making.*

HOT PROSPECT: *Good follow up will likely make a deal.*

LUKEWARM PROSPECT: *Hard work might make it happen.*

LONG SHOT: *Miracles can happen!*

TOTAL DEAD END: *Forget it.*

Box 5: Follow-Up Letters

Follow up with your licensing prospect by sending a letter similar to one of the following examples. Decide which letter format to use based on whether the reaction from your licensing prospect was hot, lukewarm, or cold.

In writing such a letter, keep it short and personalize it a bit: for example, mention something from the conversation to show that you were truly interested in what the person was saying. Remember, these are just examples; improvise!

1. Letter to a hot prospect

Dear Charles:

I very much enjoyed speaking with you this afternoon about our new rotary device for applying plaster casts. Although I knew that CastCorp was a major supplier of plaster for hospitals and physicians' offices, I did not know that you also made plaster-room and operating-room equipment, as well as orthopedic surgical supplies. No wonder you were so interested in our new invention.

As promised, a nonconfidential description of the rotary cast applying device is enclosed. Since you were quite interested in the technology, I have taken the liberty of sending a copy of our standard confidentiality agreement. Of course, we would be happy to discuss the agreement with you and address any concerns you might have about it. If the agreement seems reasonable to you, we can send you a copy of our patent application. Also, we would like to invite you to see a demonstration of the device.

Should you have any questions about the technology or the confidentiality agreement, please feel free to call me [phone #]. We look forward to hearing from you soon.

Sincerely,
Lawrence Muvaney
Licensing Associate

2. Letter to a lukewarm prospect

Dear Ms. Hollister:

Thank you for taking the time to speak to me today about Dr. Mortimer's new gene-therapy vector system. We are aware, as you pointed out, that there are quite a few similar systems already on the marketplace. However, Dr. Mortimer and his colleagues feel that this new system is substantially simpler and more flexible than the systems currently available.

As promised, I have enclosed a nonconfidential description of the vector system. I hope that the description encourages you and your scientists to find out more about it. If you should have any specific questions, please feel free to call me at any time at [phone #].

Sincerely,
Janice Datillio
Licensing Assistant

3. Letter to a long shot

Dear Mr. Corman:

Thank you for taking the time to speak with me today about Dr. Kaufman's new process for making microcrystalline polypropylene fibers. I understand that at this time PolyCo only manufactures bulk polypropylene. However, perhaps the enclosed nonconfidential description of our new process will encourage PolyCo to consider making specialty products in the future.

If you have any questions, please feel free to call me at [phone #].

Sincerely,
Martin Howard
Licensing Associate

Box 6: Assessing Your Results

Docket ______________ Title ______________________________

Date ______________ Completed by ______________________ Form ___of___

By this point, you have spoken to at least five companies about your new technology. Go back and look at your Reaction Data Sheets.

If you heard at least some maybes, it may be worth continuing to look for prospective licensees. At the very least, make sure you follow up with those "maybes."

If you only heard nos, ask yourself:

- Was there a pattern in the reasons people gave for saying "no"? If so, consider them carefully. They may point to a flaw in the technology or your marketing strategy.
- Did people give reasons for saying no that seemed to focus on the unsuitability of the technology for this particular company, or for the market in general? Comments in the former category suggest that you may still be able to persuade them that the technology is advantageous to them: perhaps the technology could be more effectively marketed to another type of company.

Based upon your answers to the above questions as well as your gut instincts, check off one of the boxes below. You have spent a fair amount of time with this market and this technology by now, and you are entitled to make an honest assessment. If it looks bad, go ahead and say so.

CONCLUSIONS ABOUT THE TECHNOLOGY

A SURE WINNER: We just need to find a receptive company.

A GOOD PROSPECT: A close match will likely make a deal.

AN UNCERTAIN PROSPECT: We might find a licensee with hard work, but it may not be worth it.

A LONG SHOT: Maybe someone will love it.

TOTAL DEAD END: There is no possibility and no hope.

If your technology is a sure winner or a good prospect, go back to Step 2, and find other potential licensees and contact them in order of their ranking. If it's an uncertain prospect or a long shot, you may want to revisit the technology in six to 12 months: the situations of the market and/or potential licensees might have changed, or the technology might be improved by its inventors. But if it's a total dead end, *write it off*—at least in your own mind—and focus your energy on moving your other more promising merchandise.

CHAPTER 12.4

IP Portfolio Management: Negotiating the Information Labyrinth

JEREMY BURDON, *Director of Intellectual Assets, Health Science Ventures, Arizona Technology Enterprises, LLC, U.S.A.*

ABSTRACT

The management of intellectual property is all about managing innovation with the procedures and processes that are required to turn that innovation into valuable patent rights. A truly strategic approach to IP management will span conception to product market release. Integrating IP management into the R&D, advance development, and product development cycles seamlessly provides opportunities to gain and enhance IP protection while offering the potential to reduce risk and lower costs. The following chapter discusses some of the key elements of IP portfolio management and how the combination of the right IP tools, procedural know-how, and organizational attributes and behaviors can contribute to successful implementation.

1. INTRODUCTION

The role and importance of patent professionals in IP (intellectual property) portfolio management (IPM) are increasing significantly within business, academic, and legal entities. Driven by the speed and magnitude of today's technological development, the sheer volume of patent information, and the increasingly competitive, global environment, there is a need to more effectively manage the patent process to enhance efficiency and gain a competitive edge in the marketplace. In many respects, this means deploying tools and processes that have been prevalent in the business world:

- data mining and databases for information gathering and storage
- state-of-the-art software tools and processes for data acquisition and analysis
- program management methodologies
- effective communication across technical, business, and legal teams

Couple these with effective, continuous improvement processes, and you have a recipe for efficient generation and management of intellectual property with predicted outcomes and balanced risk (see Figure 1).

2. IPM: THE WORK PRODUCT

The planning, gathering, and analysis of IP information is vital in any organization engaged in efficient competitive intelligence and strategic decision making. From the perspective of IP-portfolio management, the processes and tools that enable acquisition, analysis, and organization of IP information are usually the same, regardless of whether the final outcome is supporting a tactical or a strategic approach. However, the breadth and scope of a patent search, resultant IP analysis, and delivery of information is often quite different. Information developed to support tactical decision making may be narrower in scope and rely on a well-defined product specification within a

Burdon J. 2007. IP Portfolio Management: Negotiating the Information Labyrinth. In *Intellectual Property Management in Health and Agricultural Innovation: A Handbook of Best Practices* (eds. A Krattiger, RT Mahoney, L Nelsen, et al.). MIHR: Oxford, U.K., and PIPRA: Davis, U.S.A. Available online at www.ipHandbook.org.

known competitor landscape. Conversely, generating reliable, accurate IP information to support a strategic decision usually requires, among other things, a much broader scope of patent-information search, multiple analysis methods, and various information-delivery vehicles.

A unique blend of skills is required to manage intellectual property successfully. Portfolio managers, or an IPM team, need broad technical knowledge, business acumen, strong communication skills, and a thorough knowledge of U.S. and foreign patent laws and procedures. State-of-the art patent search and analysis tools are needed to gather and analyze patent data, while robust IP database tools maintain invention records, patent information, patent prosecution files, and associated business, licensing, and financial information.

The type and scope of IP analysis that IPM professionals are called upon to research and deliver varies immensely in complexity. Table 1 defines and describes most of the main defined IP-analysis tasks, along with their scope and complexity.

Commercially available IP databases such as Derwent,1 STN,2 Thomson,3 Delphion,4 and Micropatent5 offer comprehensive coverage and are well-suited to both simple queries and complex searches limited by patent class or extended-Boolean-technology keyword strings. Free patent searching is available at the U.S. Patent and Trademark Office (PTO),6 the European Patent Office (esp@cenet),7 and other country-specific office databases, but is currently unsuitable for detailed patent searches. Databases such as esp@cenet are useful for rapid screening of IP data that has been generated using commercial databases, providing rapid access to an individual patent publication, or an issued patent, in a convenient, user-friendly interface.8

IPM professionals are usually trained to generate complex keyword strings from the initial invention disclosure, a combination of invention disclosure, and provided references, or following a technology scan in the technology area of the invention. Synonyms of key technologies will be determined and a search will be performed using specific combinations of technology keywords, with Boolean logic deployed between main searches or search subsets. Patent classification systems are powerful tools, and intelligent use of patent classification (either alone, or in combination with other keyword searches) is extremely effective for relevant patent retrieval. The major patent classification systems are the International Patent Classification (IPC), European Patent Office Classification, and the U.S. Patent Office Classification.

Figure 1: Key Elements of Effective IP Portfolio Management

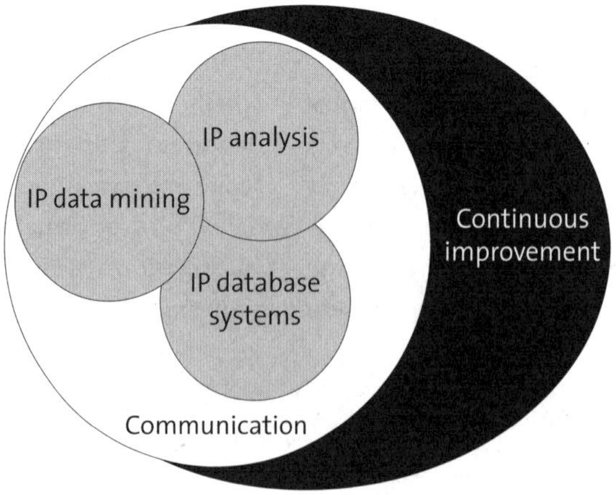

Table 1: IP Portfolio Management Task Definitions

IP Task	Definition, Scope, and Complexity
Technology Scan	High-level scan of the patent and nonpatent literature to gauge current technology status. Used prior to invention conception or may facilitate technology brainstorming
Current Awareness/IP Surveillance	Monitoring of newly published patent applications or issue patents; supports "patent intelligence"/"competitive intelligence" initiatives
Licensing/Business Development IP Support	Patent portfolio maintenance, patent-prosecution support, updating patent status information, generating reports on IP status
Patent Development/ Patentability	Targeted IP search and analysis to determine similar, overlapping, or identical technology. A search is conducted within the full specification of U.S. and foreign patent applications and issued patents
Patent Landscape	Analysis of IP in one or more specific areas of technology; integration of detailed IP analysis information into defined format such as a "landscape" enabling both high-level overviews or detailed analysis (may support patentability or claims analysis activities)
Infringement	Targeted IP claims analysis to determine if one or more patents may be infringed by a new product release to market
Validity	A search for a prior-art reference that may render a target patent or patents invalid

A brief scan of the patent and nonpatent literature is usually performed to provide a quick analysis of a particular technology area. This task may precede or facilitate technology brainstorming, or may be used to aid in and verify invention conception. With the availability and access of free online search tools for literature and patent searching, the task is often performed directly by the scientist or engineer without the need or support of an IPM professional. If the technology concept is in its early stages or is broad in nature, an IPM professional may help to focus the IP search, eliminate irrelevant search data, and help in the analysis and interpretation of the results.

IP surveillance is simply the monitoring of newly published patent applications or issue patents, usually in well-defined technology areas. This activity is usually ongoing with research, advanced development, and product-development activities and supports "patent intelligence"/"competitive intelligence." Currently available commercial patent-search tools allow the generation of sophisticated search terms with automated search frequency and delivery of the results via e-mail. The level of analysis and delivery of that analysis is user-defined. In most circumstances, it is necessary only to provide the patent number, title, and assignee (if known). Individual patent documents can be provided if the number is small, or alternatively, a list with direct hyperlinks to the patent document can be generated. Occasionally it may be necessary to provide a brief summary of the patent document, and/or provide a list of the independent claims. The IPM professional can generate this data, often, by performing a brief scan of the patent specification and claims. IP with complex specifications may require a more-extensive analysis to derive an understanding of the claimed invention. Alternatively, commercial services such as Derwent are available to provide a summary of the invention.

Licensing and business-development support activities including patent portfolio maintenance, patent-prosecution support, patent-status information updates, and generating reports on IP status are key responsibilities of IPM professionals. IP management software systems such as Inteum C/S®9 are indispensable database management tools capable of integrating patent data (invention disclosure, patent applications, issued patent information, and so forth) with current financials (licensing, fees, patent prosecution, annuity and maintenance fees, and so on). In most circumstances, data will be extracted from the IPM database and an updated patent search performed and cross-referenced to ensure the most accurate patent status? It may also be necessary to access the current prosecution status using the PTO's PAIR10 or by communicating with the prosecuting attorney to ascertain the most current status.

A patentability, or novelty, search is a search and analysis to uncover technology that may be similar, overlapping, or identical to the intellectual property for which the patent is being sought. A search is conducted within the full specification of U.S. and foreign patent applications and issued patents (in other words, it is not limited to the claims, as a patent or patent publication is potentially prior art for all that is disclosed). In most cases, a patentability search is best conducted by a patent professional. Depending on the nature of the technology and scope of the invention, the volume of search results can quickly become unmanageable. A well-structured search can greatly reduce the search time, eliminate irrelevant search data, and streamline the analysis. It is highly desirable to have completed a patentability search prior to writing claims and generating a patent application. It is often the responsibility of the IPM professional to ensure that this key step is performed, providing analysis of the results relative to the invention disclosure.

A patent "landscape," or "map" is generally an analysis of IP in one or more specific areas of technology. IP search results are analyzed and the information integrated into a defined format such as a visual landscape, or map enabling both high-level overviews or detailed analyses of specific patent documents. The level and complexity of a patent landscape are defined by the question posed. A patent landscape may be useful for providing information on potential areas of research and invention, indicating current position strength, (comparing new disclosures, prefile applications, patent applications in prosecution, and issued patents relative to competitors), or defining technology "gaps" or "white space." The IPM professional should be cautious when employing a patent landscape/map to define a technology pathway or the potential patentability of an invention, particularly if the data interpretation does not include a detailed analysis of the patent and what information has been disclosed. A technology space may seem to be extremely crowded if defined at a high level with a simple (broad) search strategy, or even somewhat complex (narrow) search strategies. Successive refinement of the landscape using additional subsearches may be required to define 'white space,' and a detailed analysis at the disclosure level for patentability should be performed to assure there are no lost opportunities. In short, it is only when the patent data is analyzed (which usually means reading each patent in the landscape search) that an accurate IP landscape can be generated.

An IPM professional may provide patent search and analysis support for an infringement, for freedom to operate (FTO), or for a validity opinion. An infringement analysis involves a search only at the claims level of a patent and has the purpose of determining whether one or more patents may be infringed by a new product release to market. A validity search is performed for a prior-art reference that may render a target patent or patents invalid. The complexity of a validity search is similar to that of a technology scan or patentability search. A search at the claims level for an infringement/FTO search is simpler, however, the data analysis will be more complex. Here the claims are analyzed in the form of a "claims chart," which allows comparisons from each element of the claim to elements or features of the potentially infringing product. The claim chart is a key tool of attorneys who are litigating patent cases.

3. INTEGRATION WITH INNOVATION MANAGEMENT

Phased-gate innovation management is a process for managing the development of new technology, widely used by mid- to large-size technology companies. The process provides a framework for evaluating a "funnel" of conceptual ideas and early-stage concepts while providing a mechanism for reducing the investment risk. Figure 2 illustrates a phase-gate development process for (A) product development and (B) research and development scenarios. At the end of each stage, numerous input and output factors are analyzed, and the risk, based on the status of the technology, the business impact, market environment, and financial status is analyzed prior to moving to the next gate.

The timely development of a robust patent position, effective patent portfolio management, and continuous monitoring of patent information for competitive analysis and infringement are all important for reducing risk.

Typically, however, IP strategy is applied only at the initial conception stages and at the later stages of product development (after product definition and prior to product release). Patent applications may be filed on early-stage concepts without regard to further modifications or improvements, and monitoring of the competitive IP position. This can leave R&D and business development groups with a false sense of security, believing that the simple act of filing provides solid IP protection.

Embedding the IP management process into the technology-development process is a key strategic approach to new technology development, IP portfolio development, and strategy. By integrating IPM continuously into the phase-gate development process—from conception through R&D—advance development, and product development, an organization may evolve a stronger patent position, optimize R&D costs, reduce patent expenses over the long haul, and minimize the potential for patent infringement and litigation risk. This approach is illustrated in Figure 3, which shows a phase-gate technology development with integrated IP management processes.

During the initial phase of project definition or concept development, the use of patent landscape or mapping methods may be useful for providing information about potential areas for research and invention, partnering, or licensing opportunities. There may be relevant disclosure in one or more patent applications already in prosecution, patent protection may already exist in a specific technology area of preliminary interest, or there may be an opportunity to license-in the technology. Discovery of prior applications or issued patents can be advantageous or detrimental depending on the breadth and scope of the invention as disclosed in relation to what may now be perceived to be new and novel. Prior disclosure may not be enabling for the new invention, however, an earlier published application or issued patent may be prior art. Given a analysis of the current IP portfolio, there may be opportunities to amend applications in process, abandon and refile, or file for reissue to gain broader protection. In-licensing may provide an opportunity to gain access to a key technology in the very early stages of product development, providing an opportunity to significantly lower the cost of development and decrease time to market. IP development will be most active during the early-concept and R&D/advanced-development stages, tapering off in the later stages of product development as the product becomes more defined. However, effective IPM processes need to be maintained in these later stages to ensure that patent prosecution is adequately supported. Provisional patent applications filed during the initial stages may at this stage be nonprovisional applications that are one or two years into prosecution, or PCT applications may be reaching the national stage. Continued advanced-development activities or product development may involve generating new inventions requiring patentability analysis and tactical or strategic positioning relative to the growing patent portfolio. Meanwhile, continuous patent monitoring may indicate that the competitor IP landscape is shifting, opening up the possibility of minor or major modifications being needed with respect to the product development strategy.

4. CONCLUSION

Technology development and IP management need to be intertwined to ensure commercial success and company viability. The increased complexity of high-technology research and development, the need to develop global-market strategies, reduction of product-life-cycles, and broadening product portfolios require an integration of IPM practices and procedures into innovation and product development. Organizations can capitalize on the integrated IPM approach by blending state-of-the art IP search and analysis tools and techniques, IP database management systems, continuous improvement processes, and seamless communication between R&D, business, and legal teams. Successful integration of

FIGURE 2: INTEGRATING THE IP MANAGEMENT PROCESS INTO THE TECHNOLOGY-DEVELOPMENT PROCESS

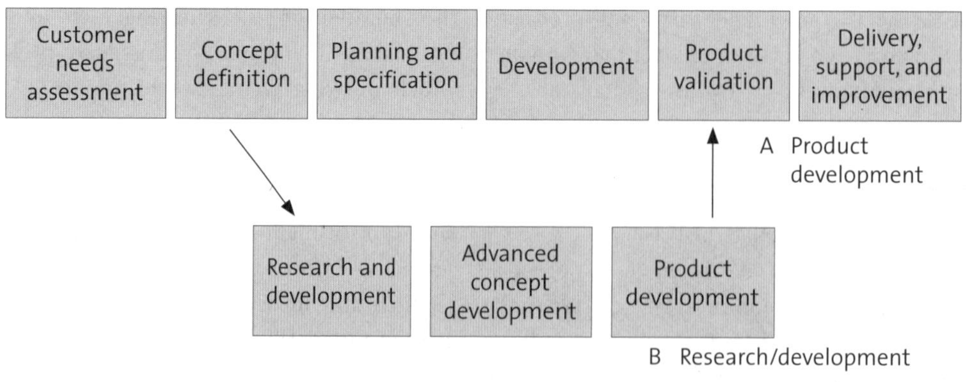

this model can enable the transformation of innovation into value, by defining strategic direction and the protection of rights based on a broad, high-quality patent portfolio. ■

JEREMY BURDON, *Director of Intellectual Assets, Health Science Ventures, Arizona Technology Enterprises, LLC, 699 South Mill Avenue, Suite 601, Tempe, AZ, 85281, U.S.A. jburdon@azte.com*

1 www.derwent.com/.

2 www.cas.org/patents/index.html.

3 www.thomson.com/content/scientific/brand_overviews/patent_store.

4 www.delphion.com/.

5 www.micropatent.com/static/index.htm.

6 www.uspto.gov/patft/index.html.

7 ep.espacenet.com/?locale=en_EP.

8 See, also in this *Handbook*, chapter 14.3 by H Thangaraj, RH Potter and A Krattiger.

9 www.inteum.com/inteum.html.

10 See portal.uspto.gov/external/portal/pair.

FIGURE 3: TECHNOLOGY DEVELOPMENT: EMBEDDING IP INTELLIGENCE

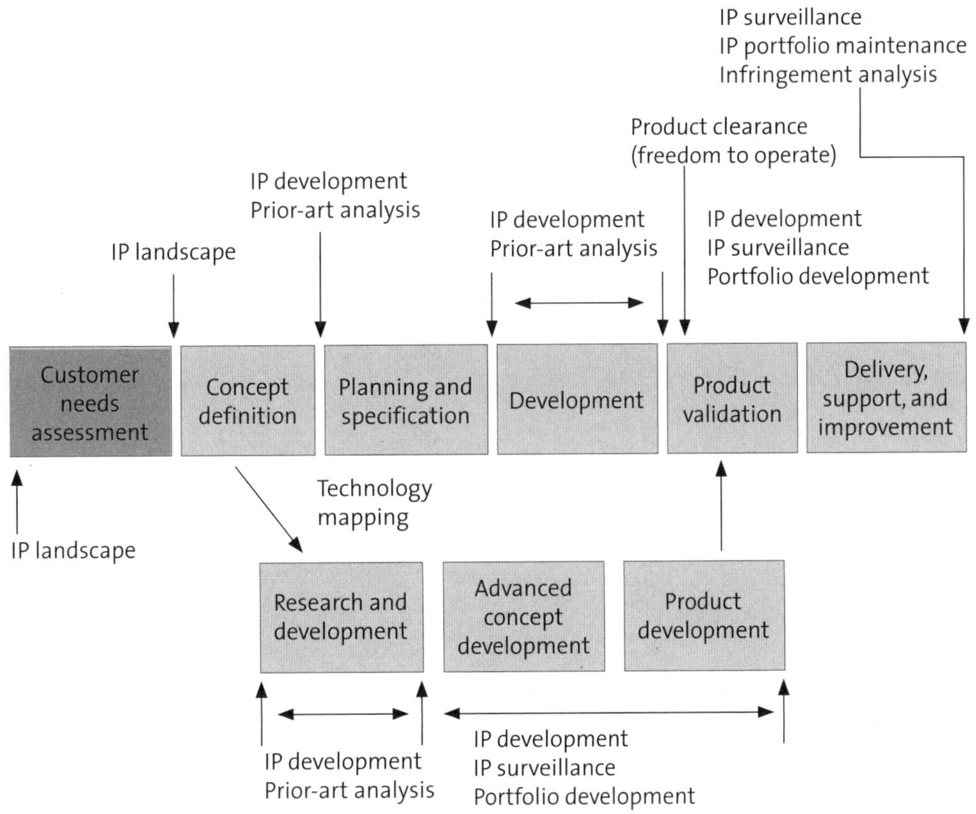

CHAPTER 12.5

The IP Sales Process

TODD S. KEILLER, *Director of Technology Transfer, University of Vermont, U.S.A.*

ABSTRACT

Marketing an institution's intellectual property (IP) is essential but challenging work. This chapter provides helpful information about how to locate potential licensees, how to determine whether or not they are qualified to manage a particular technology, and how to persuade them to begin licensing negotiations. The chapter stresses the importance of self-knowledge: having a clear sense of your institution's own IP goals, as well as the institution's strengths and weaknesses. Having this awareness makes it possible for a technology transfer office to choose wisely when it evaluates the strengths and weaknesses of potential marketing targets. Indeed, the chapter, rather than simply providing a basic overview of the marketing process, offers concrete suggestions and tough questions for those who aim to successfully market academic intellectual property.

1. INTRODUCTION

The goal of marketing IP is to bring motivated parties to a license negotiation. Technology transfer managers must locate potential licensees and make them aware of a technology's promise. A technology transfer office (TTO) can best attract licensees by placing the *right information* in the *right hands* of the *right companies* at the *right time*. Getting all of these "rights" right is a challenge for any marketing effort, but some marketing challenges are unique to marketing intellectual property. First of all, the products (university inventions) are not developed in response to market needs. Thus, a TTO must convince businesses of the marketability of potential products before businesses have recognized the usefulness of such products—and the existence of which they may have never even imagined. Of course, university inventions are early-stage technologies. Often, the technology has not been demonstrated: the buyer (the licensee) cannot "touch the merchandise," and the inventors themselves may have a hard time defining the technology's utility. In fact, no one may even be sure that it will work.

Moreover, persuading potential customers to begin license negotiations is difficult because a business takes on considerable risk when licensing intellectual property. Of course, there are license fees, but greater costs come in the form of reorienting internal resources and priorities, investing enormous sums in development, and changing company behavior (in terms of manufacturing processes, kinds of products offered, and so on). And if the invention is a "bust," it is the licensee who usually bears the financial burden.

On the other hand, everyone knows that new technologies can offer the promise of enormous value. Innovation is the engine behind any

---

Keiller TS. 2007. The IP Sales Process. In *Intellectual Property Management in Health and Agricultural Innovation: A Handbook of Best Practices* (eds. A Krattiger, RT Mahoney, L Nelsen, et al.). MIHR: Oxford, U.K., and PIPRA: Davis, U.S.A. Available online at www.ipHandbook.org.

Editors' Note: We are most grateful to the Association of University Technology Managers (AUTM) for having allowed us to update and edit this paper and include it as a chapter in this *Handbook*. The original paper was published in the *AUTM Technology Transfer Practice Manual* (Part VII: Chapter 2).

growing business. Therefore, for a marketer of an institution's intellectual property, the task is to make a licensing deal as attractive as possible by reducing the risk/promise ratio.

2. GETTING STARTED

To overcome the difficulties, one must begin at home. Indeed, when we think of "selling" an institution's intellectual property, a logical place to start is to ensure that the objectives of the TTO match those of its institution. The TTO and the institution it works for have a common goal and a common vision. This may seem rather obvious, but it is best for the institution to understand and endorse how the TTO operates (including its policies for such issues as conflict of interest, equity holdings, royalty splits, and even the direction of the research being licensed). Without this endorsement, a technology transfer manager's marketing efforts will not be supported and, in a worst-case scenario, a negotiating process that took a great deal of time and effort to achieve will be rejected by your institution. If the objectives of the TTO are not clearly in line with that of the institution, it will also be difficult to create and maintain an atmosphere of trust and cooperation between the TTO and the university—much less between the TTO and its potential customers.

A written policy—approved by the appropriate authorities and available to all investigators—will establish the ground rules for the TTO's operations. In addition to emphasizing the need to create economic benefit both for the institution and the community, this policy should reflect the philosophy of the institution. The following are sample objectives one might consider.

1. To increase research support from industry while maintaining these principles:
    - free and open communication among colleagues
    - collaborative research, as appropriate, among colleagues
    - an atmosphere of cordiality and mutual respect among scientists and clinicians
2. To provide guidelines for fairly distributing the economic benefits of academic–industry relationships and to ensure that these relationships enhance the institution's basic mission in the areas of teaching, research, and community outreach
3. To provide reliable, expeditious processes and procedures for resolving conflicts of interest in academic–industry relationships
4. To ensure that partnership companies act ethically and in a socially responsible manner, so that they diligently promote the development and dissemination of the institution's research products for the greatest possible public benefit

Publicly articulating such principles for the campus community will make the TTO's efforts more focused, transparent, and effective. This is partly because the institution will be able to get behind the TTO wholeheartedly and partly because sharing these goals with potential business partners can go a long way toward fostering mutual understanding, which is always helpful for facilitating the negotiation process.

3. TECHNOLOGY AUDITS

A common TTO complaint is that "no one has time to audit the inventory of inventions." If technology transfer managers do not know what is in the pipeline, then it will be impossible to organize a coherent sales or marketing strategy. Understanding what inventions are in the patent process, what investigators are actively working on, and whether this work matches the department chairperson's expectations is valuable, not least because such understanding lays the foundation for an effective sales strategy.

Auditing the status of each technology is such a critical starting point that it could be worth the expense to bring in an outside consultant to augment the review of the invention disclosures, understand the patent situation, evaluate the commercial potential, and recommend commercialization alternatives.

3.1 *Resource assessment*

Once a technology transfer manager knows the "inventory" of the TTO, the manager can assess the resources needed to implement a sales strategy, especially in relation to staffing. Balancing cases among available licensing professionals, for example, will allow for an even allocation of time for those cases that are close to closing. A technology transfer manager would not want to have one professional attempting to close ten cases, while another has none closing. In general, a caseload of up to 40–50 inventions in various stages of qualification per person is possible if good planning is in place.

However, realistically allocating cases among available resources may result in a shortfall. Once again, an outside consulting group may need to be brought in to handle a series of unattended cases. Moreover, it is always difficult to decide when to drop a case—the institution risks incurring unrecoverable patent expenses by carrying a case too long. Therefore, TTOs should not have cases lying dormant without having a strategy for eventually marketing them. Giving the case to a consulting group on a success-fee basis, with a small retainer to manage expenses, may be a logical action plan for cases that cannot be attended to by TTO personnel. The challenge is to ensure that the consultant's approach is fully aligned with the strategy and personality of the TTO in order to match the mission of the institution, manage the interface with the commercial targets, and make sure the investigator is feeling the technology is adequately being attended to, rather than being overlooked or pushed aside.

3.2 *Sales strategy*

Keeping up with the ongoing stream of new inventions, managing the existing portfolio of projects, and negotiating and closing the transfer of technology—all of this provides lessons in priority setting and planning. Careful preparation allows a technology transfer manager to be efficient and fair to all parties involved. After all, a scientist with a technology of little value may invent the next blockbuster royalty generator for the institution. The key to success in all of these areas is to keep up with the technology stream while building up an inventory of cases.

If building a long-term royalty stream is a goal for the institution, a manager cannot do this without closing contracts. The technology transfer manager should therefore consider creating an objective for the TTO of closing a certain number of contracts per year. Having this goal as a cornerstone of the sales strategy will create a sense of urgency, enhance office performance, and provide a sense of focus for the staff. A TTO might consider holding a monthly "to do list" meeting that realistically sets goals for the next 30 days, with the primary goal being a task related to closing a contract. Academic settings often revolve around fiscal years or semesters, while the TTO customers revolve around monthly, or at most, quarterly objectives. Having a TTO work around shorter-term priorities can potentially enhance the velocity at which the office either moves technologies "up" toward licensing, or "out" to the "abandoned" file.

4. WHO IS THE CUSTOMER?

4.1 *Identifying customers*

To develop a sales strategy, a technology transfer manager needs to thoroughly understand the customer so that he or she can ensure that the customer best matches the technology's requirements and potential. Exactly who the customer is in a technology transfer is not always evident. On the one hand, the TTO must enter into tough negotiations with research sponsors and other prospective licensees; on the other hand, the TTO serves the institution and research scientists. The bottom line is, however, that the manager needs to remember that the industrial sponsor/licensee pays the royalties. To be sure, the scientist is the producer of the package to be sold, so treating that person as the TTO's client and partner is equally important. The TTO must maintain a delicate balance.

Listening to the customer throughout the process can be a difficult challenge, but a deal could very well depend upon how well the TTO staff is listening. In particular, the manager must recognize that the technology is usually competing with other priorities in the company's development plan. Open communication will allow the manager to respond to the customer's needs and also let the TTO determine whether the customer is right for the technology.

4.2 *Finding potential licensees*

For most technologies, a list of potential partners can be easily generated. Indeed, the explosion of Web-based databases makes it simple to get a list of potential customers that may be appropriate to contact.[1] Sites like biospace.com, not only allow the technology transfer professional to "reach out" and find customers, but maintaining your own Web site, that is updated routinely, allows companies to "reach in" to the institution portfolio. A TTO may be surprised at how companies are getting more sophisticated in searching university Web sites. The Massachusetts Association of Technology Transfer Offices has gone a step further and maintains a central Web site that can search 19 institutions through the use of key words.[2] The site is updated nightly for any additions/deletions made by an individual institution. Other programs like TechEx.com also allow companies to reach in to the institutional portfolio from members worldwide who have listed their available technologies. Such lists, however, need to be sifted through before drawing up a targeted prospect list.

Another useful source of industry contacts is the team of scientists working at your institution. Scientists will often already have an industry contact for a given technology, and a scientist's relationship with a company is invaluable for initiating negotiations. In fact, AUTM data have shown that 54% of licensees were initiated due to investigator-company relationships.[3] So TTO staff must be sure to ask the scientists about their contacts. (Knowing where their graduates have gone can often provide useful leads.) When exploiting an inventor's personal contact, however, one must make sure that the technology transfer manager is serving the best interests of the technology and not limiting its possibilities by deferring to the inventor/scientist.

Other sources of contacts may come from the TTO members' industrial experience, experience from previous cases, AUTM members who have dealt with the targeted field of technology, or other members of the institution who have dealt with the company. Industry directories, professional association directories and materials, and trade publications and newsletters can all provide useful leads.

Of course, if you are a TTO manager, remember to think about your own contacts! Who do you know? Who do your friends know? Who has come to see you in the recent past? Networking begins with you.

4.3 *Qualifying potential licensees*

Evaluating companies means asking at least these four key questions:

1. Does the technology fit the company's need?
2. What is the company's time frame to develop the product?
3. Does the company have the budget to develop the product?
4. Is there any reason why the company would be unwilling to work with the institution/scientist?

It is often difficult to get accurate answers to these questions. The company contact may not be able to answer them, which may require the technology transfer manager to try to get the company to open up and explain its position. A simple tip is to ask questions beginning with the words "*who, what, when, where,* and *why*." With these types of questions, the contact cannot give a simple yes or no answer. Most importantly, the TTO manager must remember to listen after asking the question! It is pointless to ask a question and then have a colleague (or yourself) answer it instead of the customer.

5. KEY QUALIFYING QUESTIONS

5.1 *Does the technology fit the need?*

The good way to start is by asking, clearly, whether the technology field matches the company's current business development strategy. The question should be posed to the scientific contact at the company, as well as to the business contact, preferably at the executive level or at least with the top business development manager. The technology transfer manager should be on the lookout for company scientists eager to work in an area that does not match the company's overall business goals. While such scientists may have the capability to fund initial work for the technology, he or she will most likely be unable to move the technology any further.

Asking for a review of the company's business strategy is appropriate, and good customers will want to provide this—confidentially—to ensure that everyone knows where this potential partnership would fit. After all, the company's scientific efforts must be matched with its marketing endeavors for a licensed technology to be commercialized.

The company should also be able to provide a sense of the market for the proposed product. Such information should include market size, trends, participants, and contacts, as well as recent deals relevant to the market and the company's overall approach to the market. Specific questions might include the following:

- Does the product fit into an easily identified market niche?
- What is the total market potential (range)?
- How fast is the overall market growing?
- Is the market prone to frequent innovation or is it a traditional/static market?
- Is market demand stable, cyclical, or seasonal?
- How many major competitors exist?
- Is market power diffused among many participants or concentrated in a few?
- Is the market characterized by critical price constraints, (for example, regulation, industry, association, dominant price leader, and so on)?
- Are competitors generally aggressive or relatively passive in their marketing?
- Are others working on similar developments?
- What competing research/development efforts exist?
- How easy would it be to duplicate the product?
- At what stage of development are others involved in this area of technology?
- How large are barriers to entry in this industry?
- How large a market share would be required to achieve the company's objectives?
- How fast will consumers recognize and respond to this innovation when available?

Ideally, both parties come to the table with a clear idea of their needs. The TTO will have a list of the strengths of the technology, the strengths of the investigator, and the strengths of the institution, while the company will arrive with a clear definition of what it needs to accomplish its strategic goals. A close match will allow the manager to move on to the next qualifying question.

5.2 *Do time frames mesh?*

Where does the project fit in with the company's development plans? The due diligence clauses in the contract need to match the answer to this question. The technology transfer manager might have negotiated a terrific royalty on product sales, but the company may not have plans to insert the technology into its product development group until the year 2015. Reviewing the business plan would be helpful in assessing the intentions of the company.

The company needs to express its intent to commercialize the technology in an acceptable time frame in order for the negotiation to proceed. Too many TTOs have been surprised by their partners' lack of diligence, and asking this question in the beginning establishes the groundwork for moving on to the next qualifying question. Diligence can be ensured by attaching milestone payments, minimum annual royalties, or research-funding-level commitments to development activities.

5.3 *Is the company's budget adequate?*

How much money does the company have budgeted to develop this technology? The answer must match both the institution's and the company's needs. Will the scientist be comfortable with this level of funding? What research should be carried out at the company versus at the institution? The answers to these questions may reveal a flaw in the company's intentions. For example, it may desperately want this technology to round out a portfolio that would help the company raise additional funds but not really have the budget to undertake the project. The TTO might then miss the opportunity to license the technology to another party who has adequate funding available.

Typically, this question can come down to a company having any funds versus having the right funds. While having "any funds" may be acceptable, all involved need to understand this prior to entering into an agreement.

5.4 *Do prejudices exist?*

Prejudice against an institution, TTO, or scientist should not be overlooked in the qualification process. The TTO, for example, may have found the ideal company for commercializing a technology, but it turns out that the scientist is a leading consultant for the competition. Or perhaps the company has a major program in this field with another institution, and wants to avoid diluting its efforts. Perhaps previous negotiations with the company have been poorly handled, and so the company is reluctant to negotiate with the institution again.

Such prejudices need to be addressed. Any of these situations can cause negotiations to break down or even never begin. If historical prejudice involved former personnel or a situation that no longer exists, then the prejudice may be irrelevant, but there need to be assurances from the company.

6. MARKETING PACKAGE

6.1 *Tailoring to your customer*

The marketing package depends on the stage of customer qualification. Initially, when inventory is made, a short, nonconfidential abstract of the technology should be prepared. Organizing these abstracts by market segment allows the TTO to provide tailored packages to prospects. The technology transfer manager must understand that industrial business development offices receive hundreds of technology proposals. Proposals that align with the interests of such offices will have a much better chance of getting attention. Do not, however, overplay this aspect. Potential customers will reveal their level of market knowledge when they are qualified in the "technology to fit the need" questioning. It is extremely dangerous to tell a company how to conduct business in its field, even if a scientist thinks the company is approaching it incorrectly. Boxes 1 and 2 present two approaches for initiating the search for a company to license and develop a technology. Rifle-shot marketing[4] (Box 1) is most appropriate when the TTO has a handful of good partnering prospects. The shotgun-marketing approach (Box 2) provides advantages for small tech-transfer offices.[5] It is a no-frills approach that allows for a wide range of notification without a huge investment of time, but it requires careful orchestration.

An up-to-date Web site, with available technologies easily accessible, will augment your marketing approach. Make it easy for customers to navigate to a technology area and provide your nonconfidential abstracts. It could also be helpful to allow a link to pdf files of the abstract and of other publications so that the person searching can easily share the information with other internal staff. The TTO might also consider developing a list of quick pitches on video with the investigator taking 3–4 minutes to explain the technology. Technology today can produce videos relatively inexpensively, and setting a goal of adding 1–2 per month will help build the inventory without diverting too much energy from other tasks.

6.2 *Getting it (confidentially?) right*

An even more targeted approach than that of rifle-shot marketing will give the right information, to the right person, at the right time. Such precision requires a tremendous amount of effort, and managers should evaluate the opportunity cost of pursuing this approach in relation to other technologies that could be marketed using other methods. To pursue the "right-right-right-right" method,[6] be sure to offer the "right information" including:

- title
- abstract
- patent or serial number
- summaries and digests
- catalogs and lists
- patent applications
- venture summaries
- business plan outline
- inventor discussions

As far as knowing how much information to give—and the form in which to give it—be sure to emphasize the *benefits* of the invention rather than its features. Describe *what* the invention

Box 1: Rifle-Shot Marketing

1. Present to one company at a time (or at most three or four).
2. Do not spend time and money publishing lists of "available cases."
3. Present technologies handpicked for your contacts—but do not wear out your welcome.
4. Send as much nonconfidential information as you can, including published papers, if possible.
5. Do not send confidential information uninvited, but include a confidentiality agreement for easy access to more information.
6. Include the names of all the inventors; for example, "R. Jones and Albert Einstein" not "Jones, et al."
7. Send a cover letter that explains:
   - what the case is all about (one paragraph)
   - why the case might interest the company
   - what the licensing situation is
   - how to get more information
8. Don't be unnecessarily protective of information.
9. Do answer phone calls and letters promptly.

Box 2: Shotgun Marketing

Principle features of the shotgun marketing approach:

- many companies notified at once
- "cold mailings" used instead of targeted mailings
- preference to hit "more" instead of "less"
- follow-up time reduced

Special techniques for using the shotgun approach:

- provide a marketing package with a nonconfidential abstract for the invention/technology
- use letterhead, stationery, and other paper goods that clearly identify the institution
- use careful selection criteria to identify marketing targets
- maintain as much contact as possible with technology liaisons of the primary marketing targets
- explain to potential licensees why you are using this approach

does rather than *how* it does it. Compare the invention to one or more current alternatives, and highlight the invention's advantages but be prepared to knowledgeably discuss its disadvantages. Identify and evaluate the market potential, estimate production methods (and costs, if possible), and estimate the investment required to commercialize the invention. For the latter, be sure to consider what other technical, marketing, or distribution resources would be required. Also, share knowledge you may have of any regulatory, governmental, or other factors that are important to commercializing the particular technology. Finally, develop an intuitive feel for how the invention would fit in a company's strategic technical plans. As part of this attempt, try to use a title that will have marketing appeal, instead of a patent-type title. For example, turn "Synthesis of Conducting Tim Films by Nitridation of Spin-on Oxides" into "Improved Fabrication for Titanium Nitride Films Using a Sol-Gel Process." This will show that you have carefully thought not only about how the potential product would fit into the company's product portfolio but also how it might fit more generally into the market.

To get your information into the "right hands" at the "right company," you will need to have identified who the "right hands" are. Consider what company level or function is most suitable for your pitch:

- top: chief executive officer, president, general manager, vice president, director
- bottom: scientist, engineer, operations staff, marketing/sales personnel
- middle: licensing, patent counsel, tech transfer
- by function: R&D, engineering, marketing, business development

Be sure to take full advantage of alumni employees, departed inventors, and others who may still have very useful contacts and information that can help you get your materials into the right hands. Of course, before you can identify the right hands, you will need to have identified the right company. Resources for finding the right company include:

- inventors
- online services
- business directories
- trade journals
- professional and trade associations
- scientific conference attendees/speakers
- government contacts (for example, Small Business Innovation Research grantees)

To find the right time to contact the right hand at the right company with the right information, you will need to be aware of changes in government regulation, shifts in business focus, external circumstances (for example, war or macroeconomic changes), personnel changes, technical breakthroughs, and other relevant current events. Think hard, then roll the dice.

It is possible to provide even more detailed information after confidentiality agreements have been signed. But more and more companies are scrutinizing their willingness to sign such agreements, especially for devices. At any rate, in confidence, more scientific detail may be provided, including a more detailed patent-status description. Depending upon the opportunity's potential size, the TTO may go further and provide a full business plan to prospective investors.

The key to any successful information package is to find answers to as many questions as possible as to what companies would partner well with the institution, and then tailor the package to handle any objections raised by the customer. Be sure to emphasize the benefits of the invention related to the market. For example, could the invention lead to any of following?

- a product or service that performs an entirely new function
- improved performance of an existing function
- improved manufacture of an existing product
- additional functions of an existing product
- an existing product in a new market
- integration of two existing products

If the answer is yes, be sure to say so. Finally, and most importantly, *follow up and keep track of contacts.*

7. CLOSING THE TRANSACTION

7.1 *Terms*

Hopefully, the basic terms of the technology transfer will become evident after the qualification effort is complete. However, it would not be unusual for the terms to reveal the true answers to qualification questions. This is when it is critical for both sides to really understand what is expected from each party. Budget and remuneration issues should certainly be resolved at this stage and not left to the execution copy stage.

The technology transfer manager should not take a term sheet lightly. The institution attorney will caution the TTO that the term sheet could be construed as a binding document. Therefore, it should not be used for loose negotiating, but instead as a sincere effort to understand each other's responsibilities for the transaction. This includes not only the financial commitments, but the personnel, laboratory, institutional, and corporate resource commitments.

7.2 *Transaction time and negotiation process*

Transaction time, or the time taken to negotiate a contract from start to finish, is critical to the TTO if it is going to keep up versus build up its inventory. Lengthy negotiations, long meetings without agendas or outcomes, and lack of preparation all contribute to prejudices that could interfere with current and future transactions.

The technology transfer manager should keep in mind that royalties cannot begin without the completion of the transaction. A six-month delay due to a lack of focus or commitment may mean six months of lost revenue to the company and lost royalties to the institution. Moreover, competitive technologies often have a limited window of opportunity. It is a real disservice to all involved if an opportunity is missed because of an inability to work through the issues. One should always remember that, instead of languishing, it is usually better to determine quickly that a potential partner is not actually a qualified customer and then move on to another party that is more capable. The TTO has to look at such options as an opportunity cost: there are always other cases that could be moved forward but for a delayed qualification process.

This author has found it helpful as a member of a technology transfer department to review regularly the top three to six projects that are nearest to closing. Department members contribute to the process by suggesting ways to move things toward closing. The exercise also reminds the professional to spend an appropriate amount of time completing the task. In short, the TTO often needs to be the facilitator as much as the negotiator.

7.3 *Follow up*

The signatures on the execution copy of the contract are usually (1) the signal for celebration and (2) the opportunity to move on to the next case. However, the follow-up to a contract is often overlooked, and this can be a costly mistake. One must maintain contact in order to ensure that the company's original goals with respect to the technology remain the same. Be aware that the company may have been saying yes, when it really meant no, to questions during deal negotiations or during the ongoing commercialization of the institution's technology. This indecision can manifest itself when the TTO has presented a technology to a company that either does not want to, or cannot, make a decision about commercialization. The institution, for example, may be a big customer of the company's existing products, and the company does not want to upset the current relationship by passing on an opportunity to license a technology. But because the company does not know what to do, it does nothing, and the technology sits.

There is also no alternative to tracking contracts to make sure that payments are made and milestones are reached. Indeed, the diligence of all parties needs to be assured in order to eventually see a product enter the market. A database program should be used to automatically flag events, activities, and payments so that the TTO can more effectively follow up with the sponsor, collect fees, and monitor progress. By following up and measuring the success of a program, one gains useful information for future contracts. Indeed, a relationship can be built with the company that allows for more-efficient future negotiations.

8. CONCLUSION

Marketing intellectual property has unique challenges, not the least of which is trying to sell undeveloped (and, therefore, unproven) technology. The intangible and uncertain nature makes finding companies to develop such technology difficult, and yet critical to bringing the technology to market. Taking the many special considerations into account, marketing intellectual property can keep a technology transfer manager on top of IP developments at his or her institution, be an intellectually and socially stimulating part of the job, and be a successful foundational element of a TTO's overall achievements. ■

ACKNOWLEDGEMENTS

Teri Willey, Managing Partner, ARCH Development Partners, contributed to this material, which has now been further updated by the author.

TODD S. KEILLER, *Director of Technology Transfer, University of Vermont, 1 Pendulum Pass, Hopkinton, MA, 01748, U.S.A. todd.keiller@uvm.edu*

---

1. For example, AUTM lists, on its Web site, Technology Transfer Offices and companies that support technology transfer activity. www.autm.net/directory/search_org_results.cfm?searchby=all.
2. www.masstechportal.org.
3. See Jansen C and HF Dillon. 1999. Where Do the Leads for Licenses Come From? Source Data from Six Institutions. *Journal of Association of University Technology Managers* XI. www.autm.net/pubs/journal/99/leads.cfm.
4. Special thanks to Lita Nelsen for this outline.
5. Special thanks to Connie Armentrout for this information.
6. Special thanks to Richard (Dick) Cahoon for this information.

Patent Licensing for Small Agricultural Biotechnology Companies

CLINTON H. NEAGLEY, *Associate Director, Technology Transfer Services, University of California, Davis, U.S.A.*

ABSTRACT

A small agricultural biotechnology (agri-biotech) company needs to establish a strong IP portfolio. Such a portfolio provides a foundation for R&D, encourages outside investment and funding, and supports product commercialization. An important step in establishing an IP portfolio is in-licensing patent rights from third-party patent holders. Nonexclusive licenses typically give a company freedom to operate and open up the possibility of creating commercializable products. Exclusive licenses give a company an exclusive position for commercialization under the patents in question.

This chapter discusses in-licensing as it applies to small agri-biotech companies. It describes the types of technologies that may be subject to in-licensing, the procedures attendant upon in-licensing, and the terms that may be delineated by in-licenses.

1. INTRODUCTION

In order to be successful, a technology company needs to build a proprietary position in intellectual property (IP); that is, it needs to build a strong IP portfolio. The portfolio should be composed primarily of both company-developed patent rights and patent rights acquired through licensing, but it may also include know-how, trade secrets, copyrights, and trademarks. The IP portfolio should include a diverse set of IP rights that provide the company with both freedom to operate (FTO), which clears the path to commercialization, and a position of exclusivity, which provides a unique competitive position. Acquiring license arrangements and the FTO or exclusivity they provide increases a company's value, its attractiveness to funders, and its chances for acquisition or public offering.

Company-owned intellectual property is an important part of any company's portfolio, but R&D to develop IP takes time and money. In-licensing allows a company to obtain IP rights at an early stage, without having to invest in research. Nonexclusive in-licensed rights, that is, rights granted to more than one licensee (see below), provide FTO under the given patent rights. On the other hand, exclusive in-licensed rights, that is, rights that are granted to only a single licensee (see below), provide FTO under the given patent rights and assure the licensee of a commercial position of exclusivity on production, sales, or use, at least for a certain length of time.

A strong IP portfolio is key for companies based in countries with established patent systems. A strong IP portfolio can also be an asset for companies in the rest of the world: it makes them more competitive in their home countries. Moreover, a strong IP portfolio may be necessary if such a company wishes to export its products to countries with established patent systems.

Neagley CH. 2007. Patent Licensing for Small Agricultural Biotechnology Companies. In *Intellectual Property Management in Health and Agricultural Innovation: A Handbook of Best Practices* (eds. A Krattiger, RT Mahoney, L Nelsen, et al.). MIHR: Oxford, U.K., and PIPRA: Davis, U.S.A. Available online at www.ipHandbook.org.

2. NONEXCLUSIVE AND EXCLUSIVE LICENSES

An IP license (or IP license agreement) is a contract in which a holder of IP rights (the licensor) grants certain rights to another party (the licensee) in return for compensation (monetary or otherwise). The scope of a license depends on the rights that are licensed, as well as how, when, and where these rights may be used or practiced. The rights granted by a patent license include rights granted under the patent itself, but may also include trademark rights, copyrights, know-how rights, or rights over tangible material (personal property). The characterization of an IP license depends on one's perspective: the licensee considers it an *in-license* (because the licensee takes the license, as well as responsibilities and benefits thereof, *into* its IP portfolio) and the licensor considers it an *out-license* (because the licensor grants IP rights *out* of its own portfolio). In the case of a *cross-license*, parties pay for in-licenses from each other by granting out-licenses to each other.

In-licensing of patent rights may be either on a nonexclusive or an exclusive basis. Each type of licensing arrangement serves a different purpose, involves different contractual terms, and comes with a different price tag.

In general, a *nonexclusive license* gives the licensee FTO for the patented technology, but not an exclusive position. The licensor may grant licenses to others for the same technology. A nonexclusive license may contain a *nonassert* clause: that is, the licensor agrees not to assert any other patents against products developed by the licensee using the original license. It is not uncommon for small agri-biotech companies to acquire a series of nonexclusive licenses so that they have the right to develop technologies that they can eventually use to create new products.

In contrast, an *exclusive license* gives the licensee FTO for the patented technology and an exclusive position on its use; in other words, having an exclusive license to a patent is, in certain ways, like holding the patent itself. Exclusive licenses can help a new company to establish itself in a research area and to generate income for its own research activities. The trade-off is that an exclusive license typically costs more than a nonexclusive license.

"In-between" licensing positions may also be possible. For example, a company could seek a nonexclusive license with the option within a certain period of time to convert the nonexclusive license to an exclusive license. Such an option grant is normally more costly for the licensee than a nonexclusive license alone because the licensor agrees not to grant licenses to others during the specified period of time.

3. TYPES OF AGRICULTURAL TECHNOLOGIES COVERED BY LICENSING

A small agri-biotech company should develop a competitive IP portfolio that includes patents and licenses for enabling technology, trait technology, and also plant material.

Enabling technologies (in other words, research tools) are used to bioengineer new organisms. Enabling technologies include plant transformation technologies; promoters and other expression systems, including constitutive, inducible, tissue-specific, and temporal-specific promoters; markers, including selectable and screenable markers; vectors; gene-suppression technologies; leaders, transits, and signals; excision technology; and other components introduced into a bioengineered plant that are not trait- or phenotype-specific.

In-licensing is typically nonexclusive for enabling technologies. Nonexclusivity allows the licensor to grant many licenses and thus widen its revenue base; at the same time, the licensee can acquire technology and FTO at a lower cost. At times, however, in-licensing of enabling technologies may be exclusive, either for broad use or for specifically defined use, such as a defined crop area or a defined trait area. Licensing enabling technologies may involve a transfer of rights over tangible property (for example, DNA sequences) that may be regulated by material transfer agreements or bailments.[1]

Trait- or *phenotype-specific technologies* can be used to create plants with new genes that express desirable traits. The genes may be derived from any type of organism, for example, viral, bacterial, fungal, plant, or mammalian. The genes may be expressed as desirable agronomic traits, for example, biotic or abiotic resistance, or desirable consumer traits such as color, flavor, texture, or fragrance.

In-licensing is often exclusive for trait-specific technologies. A license may only authorize the licensee to work with a particular crop or group of crops. Exclusive licenses allow the licensor to be compensated for genes that it is not currently exploiting itself; at the same time, such licenses allow the licensee to hold an exclusive position with respect to the use of these technologies and to develop new commercial products with them. Licensing of trait technologies may involve a transfer of rights over tangible property, for example, genes or gene constructs, which may also be regulated by material transfer agreements or bailments.

A third type of technology is the *plant material* into which enabling technology and trait technology can be introduced. Plant material encompasses model plants, for example, *Arabidopsis*, that are used in early-stage research, as well as commercial-crop plant material (either breeding material or varietal material) that is used both in research and later-stage development or commercial work.

Plant material can be in-licensed if it is protected by patents (or plant patents) or by plant variety protection/plant breeder's rights. If the plant material is not protected by intellectual property, access may be through material transfer agreements or bailments. However, not all plant material is protected by IP laws; some is in the public domain or freely available, for example, from the U.S. Department of Agriculture.

4. LICENSING PROCEDURES

Licensing is a time-consuming and expensive procedure. Normally, each company involved in licensing has a team that includes one or more in-house technical people (and often the head of research), as well as one or more business people. In addition, in-house and outside patent specialists should be available to provide input. Patent specialists include patent counsel (in the United States, lawyers who are qualified to practice before the U.S. Patent and Trademark Office, or PTO) and patent agents (in the United States, nonlawyers with technical training who are qualified to practice before the PTO). If the company is not large enough to have in-house patent counsel, then outside counsel who understand the company's technology and budget requirements should be retained. Even when in-house patent counsel (and/or in-house patent agents) is present, outside patent counsel should still be held at the ready to assist with difficult or special situations.

The company should develop a patent plan for each R&D project it hopes to undertake. In addition to planning IP protection for company-developed inventions, the patent plan should identify the existence and status of third-party patents for which it would be useful to obtain licenses. As the research plan matures, and as the third-party patent landscape changes, the patent plan will need to be revised.

The process of identifying third-party patents is detailed elsewhere in this *Handbook*.[2] But briefly, third-party patents may be identified based on information available from a number of sources, including published patent applications, patent grants, publications, conference presentations, Web sites, Securities and Exchange Commission submissions, and the popular press. Patent applications are published by the PTO; by the World International Patent Organization, which publishes patent applications under the Patent Cooperation Treaty; and by individual foreign patent offices.

Although it is important to consult published patent applications, a few caveats are called for. First, the patent application is published 18 months after the patent is filed, so it does not contain up-to-date information. Second, the published patent application normally contains the claims as filed, not as may be amended in prosecution or as will be granted. After the patent application is published, however, the patent file is made available to the public and it will be possible to track any changes of the patent claims during the patent prosecution. Third, there is no guarantee that the patent application will issue as a patent. Fourth, it is not uncommon for more than one applicant to seek patent rights for the same invention. In countries outside the United

States, the general rule is that the first to file a patent application is entitled to the patent. In the United States, however, it is the first to invent who is entitled to the patent.

Once important third-party patents are identified, they and their file histories should be studied to determine the scope of patent claims and their applicability, or lack thereof, to the project being considered. If the patent is applicable to the project, if a license is available, and if its price is within the company's budget, the company might decide to seek the license. If the patent is applicable to the project but a license is unavailable, or not economically feasible, the project plan should be reevaluated; there may be *work-arounds*, that is, alternative ways of achieving the same results, that avoid the patent.

If the company decides to seek a license, the company should determine whether it wants nonexclusive or exclusive rights, decide what it is willing to pay for them, and decide whether it wants license rights or option rights.[3] Contact with the patent holder (the potential licensor) can be made directly or through an intermediary, such as an outside law firm. Using an intermediary may be useful if the company does not want to identify itself to the potential licensor until it is certain that a license is available. Negotiations can be direct or conducted through an intermediary and are often governed by mutually agreed-upon confidentiality agreements. During the negotiations, the licensor may ask for a business plan from the potential licensee(s) if the licensor is deciding among several potential licensees and/or in order to calculate the level and type of compensation it will request. The negotiation is normally conducted under the direction of, or at least with the input of, each company's business and legal team. Typically, discussions lead to the creation of a term sheet, which in turn is followed by negotiation of the terms and language of the license agreement.

5. TERMS OF LICENSE AGREEMENTS

The core of a patent license agreement consists of two parts: first, the rights to be granted to the licensee, and second, the compensation to be paid to the licensor. The rights granted are generally determined by the scope of the patent, though not always. The license may also delineate other rights that are to be granted, for example, tangible property rights, copyrights, know-how, trade secrets, or trademarks. The licensor receives compensation by way of a negotiated payment arrangement of fixed fees and/or royalty fees. Other key provisions of the license agreement typically include responsibility for liability; diligence requirements (defined below); the licensee's rights of participation in patent procedures; the term or duration of the agreement; and license assignability (defined below).

5.1 *Patent rights*

The rights conferred by a license, or *patent rights*, are normally based on the rights covered by one or more defined patent applications or patents, along with rights to any related filings (such as continuations, divisionals, and reissues). If the license is to be applicable in a foreign country, patent rights will also include rights under the counterpart patent(s) of that country. As noted above, the license may also confer rights under any other patents of the licensor that cover products covered by the defined patents (nonassert clause).

5.2 *Rights granted to the licensee*

According to a strict definition of an exclusive license, the licensor keeps the title to the patent but retains no other rights for itself (although, as noted below, in practice the license will often specify certain retained rights for the licensor). In a *sole* license, the licensor grants a single license while retaining full rights for itself. In a *coexclusive* license, the licensor grants licenses to a defined number of licensees (typically two).

There are several key ways that a license grant, either nonexclusive or exclusive, can be limited or defined. First, the grant can be limited *territorially*, for example, it can be restricted to certain countries, or certain geographical areas within the United States. Second, the grant can be limited in terms of *duration*, for example, it can be limited to the life of a given patent, or some other defined period of time. Third, the grant can be limited to a defined *field of use* (for example, research use, or use of certain crops or traits).

The grant, even where exclusive, may also be limited by specified *retained rights* of the licensor, that is, those rights that continue to be held by the licensor or that can be granted by the licensor to other licensees interested in a different business area, in a different territory, or for different fields of use. For instance, the Public Intellectual Property Resource for Agriculture (PIPRA) recommends that agri-biotech licensors retain rights that will allow them to license their technology to others for humanitarian purposes.[4] If a patented technology is developed using U.S. government funding, any license is subject to the rights of, and the obligations owed to, the U.S. government (Bayh-Dole Act, 35 U.S.C. § 200 et seq.).

Normally, the grant will specify whether or not the licensee has the right to grant sublicenses to affiliates, other corporate partners, or other third parties. There may also be express sublicense rights to allow others to make or sell products on behalf of the licensee. Exclusive license agreements often allow broader sublicensing rights than do nonexclusive license agreements.

In addition, the grant may also provide for release or forgiveness for past acts of infringement by, or on behalf of, the licensee. The license may also grant additional rights in the form of *most-favored-nations clauses*, in nonexclusive licenses, or in the form of *right-of-first-refusal clauses* for future licensor improvements. A most-favored-nation clause provides that, in the event the licensor grants more favorable terms in a license with another party for the same patent rights, the licensor will offer the same more favorable terms to the original licensee. A right-of-first-refusal clause provides that, in the event the licensor develops improvements of the licensed patent rights and chooses to make those improvements available for licensing, the licensor will offer to license such improvements to the licensee before offering to license them to others.

5.3 *Compensation due to the licensor*

Compensation may be a combination of fixed fees, which can be paid up-front and/or periodically, and earned royalty fees. Both the level and timing of compensation are important to the company with respect to its planning and budget. In determining what compensation it is willing to pay, the company will need to estimate the potential value of the licensed technology and assess the potential value of any commercialized products that might be developed under the license. This analysis should take into account many factors, including the product's potential market size, its likely market share, the nature of any competition, the strength of the licensor's patent rights, the scope of the license, advantages (whether monetary or otherwise) of in-licensing, projected costs of future development, and the likelihood that the product will be successfully commercialized. Previous licensing agreements for the same or similar technology are relevant to the analysis. The licensee may seek to pay less if it must obtain licenses from other licensors in order to commercialize a product covered by the license agreement (*stacking* royalties).[5]

Compensation may also take nonmonetary forms: stock in the licensee company, an exchange of license grants, or cross-license arrangement, or a *grantback* to the licensor. Grantback compensation involves the licensee granting the licensor rights to future inventions made by the licensee using rights received from the licensor.

5.4 *Liability*

The licensee may want the licensor to provide assurance of the right to license, and assurances with respect to the scope or strength of the licensed patents rights. The licensor may want the licensee to indemnify the licensor against liability resulting from licensee's activities under the license agreement. Additionally, the licensor may seek to impose insurance requirements on the licensee. Such liability-related clauses often are the subject of negotiation.

5.5 *Diligence terms*

The licensor typically wishes to ensure *diligence* on the part of the licensee in developing products and making certain that the products reach the commercial market. Diligence is particularly important for exclusive licenses, since the licensor may not receive sufficient benefit from its patent rights absent diligent licensee activity. In nonexclusive licenses, diligence on the part of the licensee may likewise be important as a means of ensuring both that the license arrangement provides some value to the licensor and that the products created by the licensed technology will enter the marketplace.

Diligence terms (or requirements), particularly in the case of exclusive license agreements, typically identify *milestones*. These are specified steps in the process of research, development, and commercialization that the licensee is required to reach by specified dates. In agri-biotech, such milestones may include the development of a model plant system, the development of a crop system, field trials, obtaining regulatory approval, initial commercialization, and commercialization at predetermined levels. If the licensee fails to achieve the specified milestones at the specified times, the licensor may terminate the license or, if the license is exclusive, reduce it to nonexclusive status. The diligence terms may include a provision for extending timelines in exchange for additional compensation. The licensee will want to protect itself against a loss of rights if unforeseen circumstances slow down the process of development and commercialization; the licensor, on the other hand, will want to make certain that it has recourse in case the licensee does not fulfill its end of the bargain.

In addition to, or occasionally in place of, the fulfillment of milestones, diligence terms may require the licensee to make periodic payments (often minimum annual payments), regardless of the licensee's level of sales under the license agreement. Such payments may be set at a fixed amount or be gradually increased according to business projections. The licensor may ask for both periodic payments and the fulfillment of milestones, in order to ensure that it will receive compensation and that the technology will enter the marketplace.

5.6 *The licensee's responsibilities vis-à-vis the patent*

In a nonexclusive license agreement, the licensee may not be required to pay patent costs, that is, the costs of filing, prosecution, and maintenance of patent filing; under such an agreement, the licensee typically will not have the right to participate in patent decisions, such as the opportunity to review and comment on patent submissions. On the other hand, a nonexclusive licensee may be asked to pay a pro rata share of patent costs; or, if it is the first licensee, it may be asked to pay all the patent costs until other licenses are granted.

In an exclusive license agreement, the licensee is often asked to pay patent costs. In return, the exclusive licensee typically has the right to participate in patent decisions. The exclusive licensee may also have the right to opt out of patent costs in the event such steps as appeals, interferences, or oppositions are undertaken, but the licensee may give up its own rights to such filings by opting out. The exclusive licensee may also have the right to control prosecution and maintenance of any licensed filings that the licensor chooses to abandon.

License agreement terms may delineate the licensee's rights in case of patent enforcement procedures, for example, if and when a licensee is entitled to participate in enforcement actions, or how or whether the licensor and licensee, or licensees, will share the costs of enforcement proceedings and any compensation that may result from them.

5.7 *License term and termination*

The term of a patent license agreement typically extends for the life of the patent. The licensee is typically allowed to terminate the agreement at any time, so long as the licensee provides adequate notice and pays any accrued fees and any applicable patent costs. In contrast, the licensor is usually only allowed to terminate the agreement if the licensee violates the license, for example, by a material breach or failure to satisfy the diligence requirements.

5.8 *Assignability*

A small company licensee will likely be concerned about the *assignability* of the license agreement by the licensee, that is, the licensee's right to transfer the license to another party in the case of corporate restructuring or acquisition of the licensee. The licensor may not wish to agree to such assignability in advance because the licensor cannot know who the successor licensee will be. In order to resolve such conflicts, various in-between terms are possible; assignability might be allowed only in certain situations, for example. The licensee, on the other hand, may want an express clause to the effect that in any assignment of the license by the licensor, the new holder of the license (new licensor) will be bound by the terms of the license agreement.

5.9 *Other provisions*

License agreements typically contain a number of other provisions, often called *boilerplate* or *standard* clauses, such as clauses for reporting of the licensee's progress; confidentiality of communications; procedures for arbitration or litigation of disputes between licensor and licensee; compliance with requirements of applicable laws and regulations; and choice of governing law.

6. CONCLUSIONS

A small agri-biotech company, whether based in a developed or developing country, can help substantially to build its patent portfolio and commercialization position through patent license agreements with third parties. The company should determine what license rights it wants to seek, whether it wants to seek these rights on a nonexclusive or exclusive basis, and under what terms it is willing to license the rights. Such license agreements can provide the company with an important complement to its company-owned intellectual property, both in terms of the company's freedom to operate and in terms of the company's exclusive proprietary position. ■

CLINTON H. NEAGLEY, *Associate Director, Technology Transfer Services, University of California, Davis, 1850 Research Park Drive, Suite 100, Davis, CA, 95618-6134. U.S.A. chneagley@ucdavis.edu*

---

1 See, also in this *Handbook*, chapter 7.3 by AB Bennett, WD Streitz and RA Gacel.

2 See, also in this *Handbook*, chapter 14.2 by SP Kowalski.

3 See, also in this *Handbook*, chapter 11.7 by M. Anderson and S Keevey-Kothari.

4 See, also in this *Handbook*, chapter 2.1 by AB Bennett.

5 See, also in this *Handbook*, chapter 11.9 by K Jones, ME Whitham and PS Handler.

CHAPTER 12.7

Business Partnerships in Agriculture and Biotechnology that Advance Early-State Technology

MARTHA DUNN, *Licensing Manager, Syngenta Biotechnology, Inc., U.S.A.*
BRETT LUND, *Licensing Manager, Syngenta Biotechnology, Inc., U.S.A.*
ERIC BARBOUR, *Head, NAFTA Seeds Licensing, Syngenta Seeds, Inc. U.S.A.*

ABSTRACT

Given the expertise of large agricultural companies with respect to product development from cutting-edge research, these companies often choose to in-license technologies from small biotechnology companies and universities rather than relying solely on in-house efforts. This chapter provides an overview of the interest of large industry players in sourcing early-stage technologies from companies, how best to communicate those opportunities to companies, and what to expect in terms of valuing the technology and structuring a licensing deal. Large companies are generally interested in creating new products or new technologies that are commercially viable and that help establish sustainable agricultural economies. But, in addition, they generally support providing products and technologies that bolster subsistence farming and humanitarian efforts, while recognizing the need to protect the company's intellectual property against unauthorized uses for commercial or other unintended purposes.

1. WHY LARGE COMPANIES LICENSE TECHNOLOGY

Not unlike most other industries, large companies in agriculture excel in the product development portion of research and development (R&D). Nevertheless, they have come to recognize that a large share of the innovative, early-stage, cutting-edge research in agriculture takes place at universities and smaller companies. Large companies have invested heavily in the infrastructure needed to develop, register, and bring products to market. While product development requires significant resources and funds, such investment is economically feasible because it has inherently less risk than investment in early-stage research. Partnerships and collaborations with other entities allow large organizations to diversify away the higher risk associated with early-stage research by creating the opportunity to access a much larger portfolio of technologies developed by thousands of different entities, as opposed to relying solely on the large organization's own internal research programs. Smaller companies and universities can focus on cutting-edge research and discovering new solutions, without carrying the burden of investing resources, and instead can realize value from their discoveries through licensing and/or partnering with larger companies for subsequent product development and commercialization. This model has been adopted by the pharmaceutical industry: in its quest to discover blockbuster drugs, most large pharmaceutical companies have chosen to in-license technologies from small biotechnology companies and universities rather than relying on in-house research alone.

2. THE AGRICULTURAL INDUSTRY

Although the agricultural and pharmaceutical industries have come to share the model of in-licensing new early-stage technologies as opposed to investing internally in higher-risk research, a number of fundamental differences with regard

Dunn M, B Lund and E Barbour. 2007. Business Partnerships in Agriculture and Biotechnology that Advance Early-State Technology. In *Intellectual Property Management in Health and Agricultural Innovation: A Handbook of Best Practices* (eds. A Krattiger, RT Mahoney, L Nelsen, et al.). MIHR: Oxford, U.K., and PIPRA: Davis, U.S.A. Available online at www.ipHandbook.org.

to the model exist between the two industries. These differences are reflected in how the pharmaceutical and agricultural companies tend to structure the relationships and agreements with their technology partners.

The length of time required to develop seed products is considerable. When using classical breeding approaches, developing a conventional seed product takes a minimum of five years, on average. When transgenic traits are involved, the time needed to develop and commercialize a new seed product, including the time needed to obtain regulatory approvals in multiple countries for the import, export and cultivation of the crop, can be seven to ten years.

There are additional reasons for the lengthy development time lines, including limited planting times, long growing cycles, and rigorous multilocational testing for efficacy and environmental impacts. From an investment perspective, an early-stage–genetic-trait technology may not begin to return a profit until ten years from the initial discovery, if it ever does.

The cost of bringing an agricultural product to market can be less than a pharmaceutical product, and the per-unit value of an agricultural product is also far less. Additionally, in the agricultural arena there are only a few major crops of interest, and within those crops a relatively small number of higher-value agronomic traits—for example, drought, insect, disease, and herbicide tolerance as well as a number of quality traits—that can justify the investments needed to develop a transgenic crop solution. This is different from the situation in the pharmaceutical industry where there are many different therapeutic areas companies can target. It should be no surprise that the few large agricultural companies investing in the development of early-stage technologies have significantly overlapping interests, making the industry extremely competitive, with a strong focus on protecting IP (intellectual property) rights. As evidence, over the last decade there has been significant consolidation, and today there remains only a handful of major competitors investing in new technologies for the agricultural industry.

Similar to companies in the pharmaceutical industry, agricultural companies vigorously protect against competitors and do so through various means including patent protection, plant variety protection, trade secrets, and trademarks. Also, unlike most small companies, which have only a regional focus, large companies look to market their products worldwide, including in developing and emerging markets.

Companies are also partnering in new ways, with foundations and public sector institutions, to support basic research, local markets, and subsistence farming in developing countries. In addition to the more immediate humanitarian and capacity-building benefits, the ultimate objective of these partnerships is to develop new, profitable and sustainable agricultural markets for local farmers and growers, ensuring a reliable and safe food supply in those countries. Companies, including Syngenta, have provided strong support and donated proprietary technologies through a number of foundations, including the Syngenta Foundation for Sustainable Agriculture. Companies are generally willing to offer their proprietary products and technologies in support of subsistence farming and humanitarian efforts, while recognizing the need to protect their intellectual property against unauthorized uses, such as for commercial or unintended purposes. This good will is often simpler to extend to places where commercial opportunities are limited.

3. MARKETING NEW TECHNOLOGIES TO LARGE COMPANIES

In contacting a company, there are generally two approaches: (1) contact a licensing or business-development individual or (2) contact a company's research organization. With respect to the first approach, it is possible to develop relationships with licensing and business-development professionals by being active in organizations, such as the Association of University Technology Managers (AUTM), Biotechnology Industry Organization (BIO), and Licensing Executives Society (LES). This way, relationships can be easily established through networking and through these contacts professionals can gain an understanding of a potential partner's interests and how well matched those interests are to a subject technology that one may be hoping to out-license. Companies have a tendency to be more responsive to people they know and with whom they have shared experiences. Also, companies are able to be more responsive when they are provided information that seeks to target their needs and interests. If no personal contact inside the company has been established, a promoter can at least visit a company's Web site and review the available information on that company's current products and research interests. Targeting specific technologies to specific companies that are likely to take an interest in the technologies usually has a much greater impact than does using mass e-mails to describe multiple technologies to potential partners. A technology that may be of interest to a company can be overlooked in a long list. Also, having an up-to-date, easy-to-navigate Web site with technologies displayed allows a company to see, on their own time, what is of interest.

When sending information to a company's licensing department, it is important to note that often such information is reviewed quickly and, only if it has some quality or aspect that fits specifically with the needs and strategic interests of the company, does it gain further review by personnel who may be able to gauge the relevance and value of the technology. Thus, it is important to include clear information on the potential uses and commercial value of the technology. Without this, depending on how quickly the information is read, something of a highly technical nature may end up being overlooked.

The second method for approaching a company is on a scientist-to-scientist basis. This typically provides a more direct route into a company, because scientists (especially those used to operating in a commercial environment) are usually uniquely situated to see the fit of a technology and determine whether it provides a solution to a real business need. Companies rely, among other things, on their researchers to scout technologies, in their respective areas of expertise, that could result in new products that further the company's business objectives.

4. WHAT COMPANIES ARE LOOKING FOR

Agricultural companies look to in-license technologies that have commercial applications, resulting in better products or more efficient methods of producing existing products. Ultimately, a technology will be reviewed in terms of its financial impact. Many technologies are interesting from a scientific point of view but do not have clear commercial applications. Licensors can make their technologies more attractive to agricultural companies by focusing on the potential commercial relevance of the technology. The commercial applications must also be financially feasible from a product development and competitive perspective.

Ultimately, every technology needs a champion within the target company, someone who has identified and believes in the scientific and commercial relevance of the technology. Champions are usually the very scientists who will ultimately develop the technology for market. Champions on both sides of a deal are critical if the deal is to be successful. Too many times, technology is in-licensed and sits on the shelf or is applied inappropriately because champions were absent or were under-resourced. Part of the due diligence for in-licensing any technology should be to ensure that the project is resourced sufficiently and that champions are identified and are able to make the project move in accordance with agreed-upon timelines.

4.1 *Risks of technology*

Most technologies from universities or small companies are at an early stage and so, by nature, carry significant risk from a product development perspective. Licensors need to recognize the significant time, resources, and money required to move a project through development to a successful launch. Costs include R&D expenditures, IP and patent costs, regulatory-approval costs, and production and marketing costs. All of these need to be taken into account when allocating the value associated with bringing the technology to market. Later-stage technology (such as one that has already been proven in a relevant crop) would of course have a higher value. How data is generated to prove a technology also needs consideration.

Studies conducted in a greenhouse or in non-elite germplasm do not always translate well into the field where the product may be exposed to the full range of environmental and other effects. Many times, a company will want to evaluate a technology over the course of two or three years in order to understand how it works, across multiple environments outside of the laboratory or greenhouse environment, before agreeing to negotiate final commercial terms. Because of the risk associated with technology, large companies often prefer to start with a research or evaluation license, with an option for a commercial license, building in key terms to the option that ensure that commercializing the product, if field trials are successful, will be economically feasible.

4.2 *Type of technology*

Different types of technologies have different applications and so have different values associated with them. An agricultural technology can generally be classified in one of two ways: (1) as an enabling technology that helps or enables a product to be created (for example, gene promoters that drive the expression of proteins or tools that enable or enhance the ability to transform a particular crop) or (2) as a technology that is itself a product or that causes a seed product to contain a characteristic or trait that provides a benefit to the grower, the manufacturer, or an end-user of the product and for which the seed company can derive additional value.

Enabling technologies are helpful for bringing products to market, but in many cases such technologies are only alternatives or improvements on other methods or technologies that accomplish similar tasks. Because a number of substitutes may exist for an enabling technology, they are usually of less value than technologies that embody products. Accordingly, large agricultural companies are likely only interested in a nonexclusive license for enabling technologies, allowing freedom to operate with the technology. The companies are likely hesitant to pay running royalties, preferring instead up-front fees, annual fees, or milestone payments. It should be noted that while enabling technologies often are used across a number of projects, the majority of these technologies and projects will not progress to market.

Product technologies, on the other hand, are those that are brought to market. For this category of technologies, agricultural companies are often interested in exclusive rights in order to obtain a strategic advantage in the marketplace. Because such technology directly translates to sales and revenues, it has an inherently higher value.

5. TECHNOLOGY VALUATION

Valuing technologies is a difficult and complex task because of all the uncertainties in getting a technology to market. Often, there is a disparity in the value attributed to a technology by the licensor and by the licensee. This is particularly true in the agricultural industry due to an asymmetry of information: one company having access to more complete information than the other for determining the cost of bringing a product to market and the potential revenue sales of the end-products would bring. In the agricultural industry there are not always comparable deals with which to compare prospective products, especially as companies embark on new market areas that involve traits outside of established traits, such as insect resistance and herbicide tolerance. Additionally, in order to sell certain traits in the market, the traits must be combined with other input or agronomic traits to which the licensor has not contributed. Value will also be influenced by the presence of competitive traits in the market. This adds additional complexity to the value-capture discussion.

The value of an early-stage technology needs to be discounted based on time to market, the time value of money, technical risk, and the risk associated with obtaining regulatory approvals. Value also must account for the amount of resources invested in commercialization. Many licensors discount or overlook these factors because they are deemed to be out of their control, but the risks remain and should influence the value-sharing discussion. Other factors that effect value sharing include whether additional licenses are needed for commercialization for ensuring that a product can be brought to market with maximum freedom to operate. If other licenses are needed to bring a technology to market, the issue of "royalty stacking" comes into play, whereby multiple royalties on a product can exceed the profit margin on the product, making it impractical to commercialize.

Traditional royalties based on net sales rarely work in agricultural licensing deals because of the issues associated with royalty stacking and the fact that many technologies—from early-stage enabling technologies to trait-related technologies—may be employed in developing the final product. Companies understandably try to avoid paying royalties to licensors on the value contributed by other technologies, whether in-licensed or developed by the company. For the same reasons, large companies also try to avoid paying product-based royalties on enabling technology because the enabling technology by itself may not drive additional revenues.

In most cases, companies can agree to a royalty based upon the value that a particular technology adds in the marketplace. Models such as a percentage of trait-related revenue or fixed-fees per unit are available to licensors.

6. TERMS OF THE LICENSE

When companies choose to in-license technologies, especially in the agricultural and biotechnology industries, the parties need to consider several issues that must be specified in the license:

- **payments:** Fees for a deal need to be balanced in accordance with the use and risk profile associated with a technology. In some instances, this balance will be achieved over the life of the license during which payments through license fees, milestones, and royalties can be paid on net sales. In other instances, for example, involving a nonexclusive license to enabling technology, this may be a one-time payment. For product technologies, payments are traditionally spread out over the life of the license, reflective of the risk factors and the development timeline, so that when there is heavy R&D spending, license costs are not excessive, and do not become disincentives, but do reflect the time frame over which revenue is actually obtained from the product.

  It is important for a licensor to maintain flexibility with regard to how payments are structured, in order to meet the needs of agricultural companies, especially as new markets are explored. Many times small start-up companies are seeking to exit within three to five years from the time they are established, usually because of the expectations of the venture-capital-investor community. This can create tension in getting a deal done because of the expectation to be paid out, while there is still significant development and product risk remaining, long before the company begins to see revenue from the investments it has made and is making.
- **exclusivity:** Every company would relish being able to exclude others from obtaining a strategic advantage in the market, but sometimes obtaining exclusivity may be neither necessary nor cost effective. Many factors will effect the need or desire for exclusivity, including financial implications, the opportunity to block or license competitors, and the opportunity to create a competitive position in the marketplace.
- **field of use:** For licenses where the licensor intends to carve out exclusivity in a field of use, the licensor will want to ensure that fields don't overlap and that fields are divided in such a way as to not destroy value for other potential licensees. Agricultural companies will many times consider specific fields of use (for example, specified crops, or specific traits of interest) as a way to obtain exclusivity in a particular market.
- **diligence:** With regard to diligence provisions, the parties need to acknowledge that these provisions and timelines should be reasonable but flexible. This is especially true for certain agricultural technologies, for example, seed products, due to the uncertainty and risks associated with it, including technical, field and environmental risks, and regulatory science-related risks. Agricultural companies recognize the desire

of the licensor in having diligence provisions, but overly restrictive provisions can put a license at risk. Most companies welcome reasonable diligence requirements as they ensure that a technology will be evaluated and developed in a commercially reasonable timeframe. The role of champions to encourage open and ongoing communication between the licensor and the licensee with regard to diligence provisions, making adjustments as necessary so that the technology develops to the benefit of both parties.

- **publication:** Licensors need to work with the large agricultural companies to ensure that publications made after the license term begins (especially for exclusive licenses) do not interfere with the opportunity to capture intellectual property and, therefore, diminish the value of the technology. Close cooperation should ensure that the right to publish is not compromised while ensuring that appropriate protections are obtained before making the publication. Mechanisms for handling publication are fairly well established between public sector institutions and industry.
- **improvements:** In order for a technology to reach its full potential, it will be in the interest of both parties to allow agricultural companies to access improvements to the underlying technology.
- **timelines:** It is important for the licensor and the licensee to be responsive when negotiating a license agreement. In instances where delays are expected, these should be communicated promptly as the business may be relying on a particular timeline to drive product development. Excessive delays can result in a loss of interest and/or a loss of funds.
- **after the deal:** Transfer of know-how or materials as provided for in the license needs to be carried out in a timely manner. The agreement should define whom the appropriate contacts are to ensure that the potential of the technology can be fully realized, especially in those instances where the company is evaluating the technology and questions may arise. Often times continued access to technology experts is expected and should be welcomed in order to realize the full benefit of the license.

7. CONCLUSION

Large agricultural companies are interested in accessing and utilizing technology that helps them gain competitive advantages in the marketplace. Universities and research institutes can, through licensing agreements, partner with these companies, which have the resources, as well as the product development and marketing capabilities to translate early-stage technologies into products that bring benefit to consumers. Furthermore, such technology partnerships can result in products or new technologies that can provide, not only humanitarian benefits in the developing world, but also can help establish sustainable agricultural economies in all countries. ■

MARTHA DUNN, *Licensing Manager, Syngenta Biotechnology, Inc, 3054 Cornwallis Rd., Research Triangle Park, NC, 27709, U.S.A.* martha.dunn@syngenta.com

BRETT LUND, *Licensing Manager, Syngenta Biotechnology, Inc., 3054 Cornwallis Rd., Research Triangle Park, NC, 27709, U.S.A.* brett.lund@syngenta.com

ERIC BARBOUR, *Head, NAFTA Seeds Licensing, Syngenta Seeds, Inc., 7500 Olson Memorial Highway, Golden Valley, MN, 55427, U.S.A.* eric.barbour@syngenta.com

Biotechnology and Pharmaceutical Commercialization Alliances: Their Structure and Implications for University Technology Transfer Offices

MARK G. EDWARDS, *Managing Director, Recombinant Capital, Inc., U.S.A.*

ABSTRACT

Understanding biotechnology and pharmaceutical commercialization alliances in the context of several evolving business models has implications for university technology transfer offices (TTOs), as well as for public policymakers intending to promote biotechnology regionally. This chapter identifies the principal structural and economic elements of biotechnology and pharmaceutical commercialization alliances and the factors that influence partner selection for a particular alliance. The four characteristics of an alliance that generally define the allocation of value between an originator and a commercialization partner include stage of development, product supply, market opportunity, and scope. The chapter explains the types of economic terms typically found in biotechnology alliances and makes an empirical analysis of the economic terms from a sample of biotechnology alliances established between 1981 and 2000. Four specific alliances entered into at different stages of development are detailed as case studies. Several recommendations are provided for university TTOs, along with guidelines for drafting commercialization alliances.

1. INTRODUCTION

Since the 1940s, the pharmaceutical industry has largely followed a vertically integrated business model. This was the period when the first antibiotics were being introduced, leading to augmented manufacturing capabilities and, soon after, to the development of sales and marketing organizations. Over the next half century, the industry was sustained by the productivity of its medicinal chemists, who isolated natural products from microorganisms, plants, and animals, designed analogs and, sometimes, stumbled upon molecules with completely unexpected activity.

The emergence of biotechnology over the past several decades has transformed the drug business and ushered in a host of new participants and several novel business models. In the early 1980s, recombinant DNA and monoclonal antibody (mAb) technologies formed the basis of the first biotechnology business model, based on intellectual property (IP) relating to the isolation and/or production of novel compounds. Strong IP positions and difficult-to-master production methods would presumably allow biotechnology start-ups to initially partner with, and then compete against, established pharmaceutical companies. Assuming a series of novel products and increasingly favorable terms from partners, this model purported to be a blueprint for becoming a fully integrated pharmaceutical company, or FIPCO. Although most of the more than 100 biotechnology companies that went public prior to 1992 adopted this model, Amgen and Genentech are the only two companies from this era to have attained FIPCO characteristics to date.

By the early 1990s, two new biotechnology business models emerged. The first of these—a technology-platform model—was based on the

Edwards MG. 2007. Biotechnology and Pharmaceutical Commercialization Alliances: Their Structure and Implications for University Technology Transfer Offices. In *Intellectual Property Management in Health and Agricultural Innovation: A Handbook of Best Practices* (eds. A Krattiger, RT Mahoney, L Nelsen, et al.). MIHR: Oxford, U.K., and PIPRA: Davis, U.S.A. Available online at www.ipHandbook.org.

use of novel techniques to discover new drugs and/or to increase the productivity of the drug discovery process. With a broad platform, a biotechnology company could perform fee-for-service research for multiple pharmaceutical partners while accumulating expertise to pursue programs for its own benefit. The earliest technology-platform companies developed novel assays for screening compounds. However, these screening companies depended on pharmaceutical partners for compounds to screen, and the terms were generally unattractive.

Other types of technology platforms soon emerged, including those using proprietary technologies to produce novel compounds from oligonucleotides (for example, antisense and gene therapy), lipids, carbohydrates, peptides, and combinatorial chemistry. With the sequencing of the human genome in the late 1990s, the technology-platform model broadened yet again to include companies that discover and validate novel drug targets. Joining them were companies making the instrumentation and software to handle the increased throughput of genomic materials, combinatorial libraries, and structural information.

These technology-platform companies had in common a fundamental reliance on corporate partners to pay for at least a portion of the platform's utilization and enhancement while adding to the biotech's infrastructure and expertise. Gilead Sciences and Vertex Pharmaceuticals are current examples of successful companies that have adopted the technology-platform business model.

A third business model to emerge in the early 1990s focused on diseases with significant unmet needs and specialized patient populations, such as cancer, dermatology, and neurodegenerative diseases. These companies sought to capture more of the value of innovative products by retaining commercial rights into clinical development—and potentially through to commercialization for selected market niches. Using this strategy, disease-focused companies attempted to create a balanced mix of discovery, development, and sometimes commercial-stage programs. However, the latter were typically less innovative products, used primarily to build a sales infrastructure and prepare the organization to eventually sell the more innovative products under development. Amylin and MedImmune are current examples of successful companies that have adopted the disease-focused business model.

By the mid-1990s, however, many of these disease-focused biotechnology companies had curtailed their drug-discovery programs owing to lack of investor interest. Similarly, technology-platform companies that had partnered their top drug-discovery programs to pharmaceutical companies came to view discovery research as an unattractive use of resources. With the consolidation of major pharmaceutical companies, these companies recognized that product-acquisition opportunities would emerge that were "flying below the radar" of ever larger drug companies. These companies turned their attention to in-licensing of approved and late-stage development compounds from pharmaceutical companies. Since most of these biotechnology companies focused on specialty markets that could be addressed with relatively small sales forces, such as cancer, anti-infectives, and dermatology, by the late 1990s investors came to view this group as a new business model, dubbed specialty pharma. Cephalon and Celgene are current examples of successful companies that have adopted the specialty-pharma business model.

The collective impact of these four biotechnology business models on the pharmaceutical industry has been to significantly enhance pharma's opportunity to obtain and divest compounds via licensing. This has eroded pharma's vertically integrated business model, to the point where most pharmaceutical companies now derive 25 to 50 percent of their product pipelines from external sources. In turn, pharmaceutical companies are the principal mode of commercialization for biotechnology products—of the 100 top-selling biotechnology drugs in 2005, 63 were partnered in development for at least some territories, as were eight of the ten top-selling biotechnology products in 2006.

Understanding biotechnology and pharmaceutical commercialization alliances in the context of these several evolving business models has implications for university technology transfer offices (TTOs), as well as for public policy-makers intending to promote biotechnology regionally.

First, under certain circumstances and with significant intellectual property and/or compounds to offer, TTOs may be in a position to play a role comparable to biotechnology companies as the licensor to a commercialization partner, whether that partner is a traditional pharmaceutical company, an emergent biotech, or a regional marketing company. Frequently, however, a TTO will be the upstream licensor of intellectual property and/or compounds that are bundled and developed by a biotechnology company before being sublicensed to a commercialization partner. In these instances, it may be important to understand, and perhaps influence, the likely terms of an eventual commercialization alliance in order to protect or augment the value contributed by the TTO's technology.

This chapter aims to identify the principal structural and economic elements of biotechnology and pharmaceutical commercialization alliances[1] and the factors that influence partner selection for a particular alliance. Section 2 describes four characteristics of an alliance that generally define the allocation of value between an originator and commercialization partner. Section 3 discusses the types of economic terms typically found in these alliances. Section 4 consists of an empirical analysis of the economic terms from a sample of biotechnology alliances established between 1981 and 2000. Section 5 describes four specific alliances entered into at different stages of development. Section 6 concludes with several recommendations to TTOs and guidelines for drafting commercialization alliances.

2. CHARACTERISTICS OF ALLIANCE-VALUE ALLOCATION

2.1 *Stages of development*

Drug development is broken into phases largely shaped by the regulatory requirements for new-drug approval. These are often referred to as discovery, lead, preclinical, investigational new drug (IND) filing, Phase I clinical trials, Phase II clinical trials, Phase III clinical trials, new drug application (NDA) filing, approval, and postapproval (Phase IV) clinical trials. Generally, the later in drug development an agreement is struck, the higher the share of consideration paid to the originator.[2] This industry practice reflects, in part, the cumulative investments of the parties to date, as well as the increased likelihood of getting the compound approved and on the market.

For example, as a compound successfully navigates various stages of drug development, there is less risk associated with the compound, and this increases the total value of the economic benefits that parties to an agreement will share. Other things being equal, a license negotiated later in a compound's development will bear a higher share of consideration paid to the originator than if the same license were negotiated earlier in the compound's development.

Conversely, a company in the early stages of developing a new compound faces substantial costs and risks as it invests in developing a new product that will probably fail. In order to have adequate incentive to take on those risks, the licensee of such a compound will demand a larger share of the expected sales or profits from the new product if it proves to be successful.

At the far end of the development spectrum, a company that has a fully developed product with a track record of increasing sales and substantial profit margins in one or more geographic markets faces relatively little risk as it attempts to expand the geographic reach of the product. All else being equal, the marketing partner of such a product will receive a much smaller share of the expected sales or profits from their efforts in expanding the geographic reach of the product.

In most instances, an originator has few nonreimbursable development obligations following the signing of a commercialization agreement at each stage of development. This reflects, in part, the commercialization partner's interest in controlling the pace and expenditures required for commercialization, as well as the originator's interest in retaining any prelaunch consideration paid for rights to the compound or technology. Exceptions occur, however, when the originator continues to have significant development obligations after signing. Such exceptions, generally associated with co-development or distribution alliances, are discussed in Section 3.2 and typically would require that a higher share of consideration be paid to an originator.

2.2 *Product supply*

While many commercialization alliances simply provide a license to intellectual property and/or know-how associated with a compound or technology, some agreements additionally provide that the originator will undertake to supply all, or a portion, of a compound through commercialization. In such instances, the originator will incur greater costs and risks than in the absence of such supply obligations. As a result, alliances involving an obligation on the part of the originator to provide at least primary or bulk manufacturing of a compound through clinical development and commercial supply will typically increase the share of consideration paid to the originator.

2.3 *Market opportunity*

The gross margins of marketed pharmaceuticals have been high historically, often in the range of 75 to 95 percent. This is due to the benefits new products often bring compared to alternative treatments and the high costs and risks of development, combined with the significant regulatory and intellectual property barriers faced by new market entrants. With high gross margins and significant economies of scale in sales and distribution, top-selling pharmaceuticals (the so-called blockbusters) drive the overall profitability of major pharmaceutical companies. As a result, competition to access compounds with the greatest potential market size is intense. By contrast, compounds having relatively small market potential, such as those intended for niche markets, attract far less interest and less-favorable terms to the originator. Typically, therefore, the more attractive the market opportunity, the higher the share of consideration paid to the originator.

2.4 *Scope*

The scope of any particular commercialization alliance refers to a broad array of nonfinancial terms that either limit or broaden the rights conveyed under the agreement. Such terms might include whether the license granted is exclusive, semiexclusive, or nonexclusive, with greater exclusivity generally yielding a premium to the originator. Similarly, the larger and more economically attractive the territory, and the longer the duration of the alliance, the higher the share of consideration paid to the originator. This is because rights and any associated economic benefit would generally revert to the originator post-termination. Other things being equal, therefore, one would expect to see higher consideration paid to an originator for a long-term alliance than for one of limited duration entered into at the same time.

Should the alliance provide that one or more additional compounds or fields of use might be included as an option for the commercialization partner, such an element would also typically increase the share of consideration paid to the originator. Such an option potentially provides a broader pipeline to the commercialization partner, while minimizing this party's expenditure and development risk for the sustenance of such a pipeline. From the originator's viewpoint, granting a multicompound or multifield option to a commercialization partner would foreclose alternative arrangements, including forward integration by the originator itself, and so would normally require a premium as compared to a more limited scope.

3. TYPES OF ECONOMIC TERMS FOUND IN ALLIANCES

3.1 *Up-front payments*

Commercialization alliances typically will include an initial (so-called up-front) payment. The up-front payment may be due upon execution of the agreement and/or staged over a period of months or several years, but in the latter instance the payment obligation is noncancelable. This is not the case with development-milestone payments (see Section 3.3), wherein the payment obligation is contingent upon the achievement of predetermined events.

The up-front payment represents a "buy-in" by the commercialization partner, reflecting all or a portion of the originator's expense and risk in bringing the compound or technology to its stage at signing. Discovery-stage deals may also entail an up-front payment, often described as a technology access fee.

For biotechnology companies, up-front payments are an important signal to investors that the partnered program is of high quality and that the commercialization alliance is being struck from a position of strength, rather than weakness. Such payments are generally nonrefundable, once paid, so their inclusion in an agreement will increase the risk-adjusted share of consideration paid to the originator.

3.2 *Reimbursement or apportionment of R&D costs after signing*

With respect to the research and development (R&D), manufacturing, and launch costs incurred during the course of bringing a pharmaceutical product to market after signing, commercialization alliances involving biotechnology companies are generally one of three types, although these types are sometimes blended or combined by product or territory.

Most biotechnology agreements are in the first category, wherein the commercialization partner takes over all costs after signing, including reimbursement of the originator's post-signing costs of continued R&D and manufacturing, as well as paying directly all other costs associated with the product's development, manufacture, regulatory approval, and launch. Such costs can be very substantial, and the risk of failure in development is largely borne by the commercialization partner.

Alliances that require reimbursement of the originator's R&D expenses after signing typically require that the originator provide a specified number of full-time equivalent scientists (FTEs) per year for one to five years, along with quarterly reimbursement at a maximum fixed rate per FTE. The originator is at risk for cost overruns, however. For example, if the FTE reimbursement rate is US$250,000 per FTE per year for ten FTEs, and the actual annual R&D expenditure by the originator is US$2.7 million, only US$2.5 million is reimbursed. Conversely, if the actual R&D expenditure by the originator is US$2.2 million, a credit of US$300,000 is carried forward to the next year's R&D reimbursement.

In the second category are alliances with regard to which both parties share costs (so-called co-development). In co-development alliances, up-front and milestone payments are generally used to adjust the parties' interests in the R&D program, and subsequent development and other costs are shared. In a typical co-development alliance, an originator may possess only a portion of the capability or resources to complete clinical development, commercial supply, and/or launch of a compound. Such alliances tend to have profit splits during the post-commercialization period, reflecting the parties' respective interests in the product. While the percentage or level of cost sharing varies by agreement, such alliances usually provide a mechanism whereby one party may reimburse excess costs incurred by the other, often at a premium.

With respect to the third category of alliances, the originator continues to incur all or substantially all development, manufacturing, and regulatory costs after signing, but the commercialization partner bears some or all launch costs and ongoing sales and marketing expense. Alliances of this third type are generally described as distribution agreements, if the originator relinquishes all sales and marketing responsibilities, or else co-promotion or co-marketing alliances, if both parties are involved in commercialization of the product.

Although a commercialization partner may commit substantial resources to a biotechnology alliance in the form of FTE reimbursements, such payments are not enriching to the originator, unlike up-front and development-milestone payments. Other things being equal, therefore, the share of consideration paid to an originator will be lowest for the type of alliance with respect to which all post-signing costs are borne by the commercialization partner, in the mid-range for co-development deals, and highest for distribution-type agreements. This industry practice reflects, in part, the total expected investments of the parties through product launch, as well as the proportion of risk borne by the commercialization partner that the compound will fail in development.

3.3 *Development-milestone payments*

Most biotechnology alliances involve contingent (so-called development milestone) payments that track the progression of the R&D program through the sequential stages of development achieved after signing of the agreement.

For an early-stage alliance, typical development milestones might be technical feasibility, patent issuance, lead compound designation, IND filing, start of Phase II clinical trials, start of Phase III clinical trials, NDA filing, and first regulatory approval. For a late-stage alliance, development milestones might track individual medical indications or market entry into major markets such as the United States, Japan, or the European Union.

Like up-front payments, development-milestone payments are generally nonrefundable once paid, so their inclusion in an alliance will increase the risk-adjusted share of consideration paid to the originator.

3.4 *Equity investments*

Approximately 15 to 20 percent of biotechnology alliances include one or more minority-equity investments by the commercialization partner in the biotechnology's equity as a component of the agreement. Such equity purchases usually involve newly issued shares, so the investment proceeds are available for use by the company. If the securities of the biotechnology company are publicly traded at the time of such an investment, the commercialization partner may purchase the shares for the fair market value (FMV) or may agree to pay a specified premium over FMV at the time of purchase. Shares purchased in nonpublic biotechnology companies, as part of an alliance, are typically purchased at a 20 to 50 percent premium over the FMV of shares sold in the most recent prior round of share issuance.

Unlike up-front and development-milestone payments, however, equity investments involve an exchange of capital for an ownership interest, so the extent of enrichment to the originator, if any, depends on the premium paid by the commercialization partner as compared to the FMV of the shares.

3.5 *Post-commercialization payments*

Post-commercialization payments usually consist of one or more of five types: (1) royalties on product sales paid by the commercialization partner to the originator; (2) payments for manufactured goods (so-called transfer prices) paid by the commercialization partner to the originator as supplier of bulk or final product; (3) one-time payments on achievement of post-commercialization milestones (so-called sales-threshold payments) paid by the commercialization partner to the originator; (4) a net profit allocation between the parties (so-called profit splits); or (5) marketing fees paid by the originator to the commercialization partner.

3.5.1 *Royalty rates*

The royalty rate paid by the commercialization partner to the product's originator commonly increases with greater product sales. For example, an alliance will specify a base royalty rate that will pertain to annual (or cumulative) product sales up to a certain sales level. Above this level, a higher royalty rate will apply until a second sales threshold is met, at which point a still higher rate will pertain, and so on, through three to five different *royalty tiers*. This practice is consistent with the industry's preference and competition for blockbusters over products for niche markets.

3.5.2 *Transfer prices*

Transfer prices for bulk or final product supplied by the originator to the commercialization partner are typically specified via one of three approaches: as cost plus a specified margin, as a specified price per unit, or as a percentage of the product's selling price. Since commercialization agreements are usually silent on the actual or anticipated cost of manufacture, it is difficult to ascertain the profit contribution from the transfer price. Of the three approaches, agreements that specify a transfer price as a percentage of the product's selling price are most informative, insofar as general industry practice is to attempt to price a new product such that the cost of manufacture is typically 5–10% of the product's selling price. This implies that a transfer price in excess of 10% of the product's selling price is usually enriching to the extent of the excess.

3.5.3 *Sales-threshold payments*

Sales-threshold payments may be paid to a product's originator as one-time events. As with development-milestone payments, sales-threshold payments are typically nonrefundable.

3.5.4 *Profit splits*

Profit splits may vary by time period, or licensed region, and may or may not be inclusive of other types of payments specified by the alliance. In co-development deals, following the buy-in payments that adjust the parties' positions for preexisting risk taken and preexisting value created, profit splits tend to track the level of each party's clinical development expenditure after signing—for example, a party paying 40 percent of development costs would be entitled to 40 percent of net profits. In such agreements, the parties precisely define the development, manufacturing, regulatory, launch, and marketing expenditures that are deemed "allowable" for purposes of reaching or adjusting the agreed-upon profit split.

3.5.5 *Marketing fees*

Marketing fees paid by the product's originator to the commercialization partner generally apply only in the event that the originator is responsible for booking the sale of the product, as is sometimes the case in distribution and co-promotion alliances. Such fees are often termed royalties, except that the originator pays them to the marketing or co-promotion partner. In such agreements, there may be a static or moving level of sales (a so-called baseline) below which the commercialization partner is not compensated, reflecting the originator's capability to sell the product in the absence of the marketing party's assistance.

4. EMPIRICAL ANALYSIS OF THE ECONOMIC TERMS OF ALLIANCES

4.1 *Sample selection*

Biotechnology companies that are publicly traded on stock exchanges in the United States are required by the U.S. Securities and Exchange Commission (SEC) to file material documents. Biotechnology companies have historically interpreted this requirement conservatively and often file their contracts involving alliances with commercialization partners, as well as upstream licenses with universities and other technology providers.

Recombinant Capital's (Recap) Alliances Database contains copies of more than 20,000 research, development, license, supply, co-development, distribution, and similar alliances established since 1973. Recap analysts collected these agreements from SEC filings, predominantly by biotechnology companies, as material disclosures. In aggregate, Recap's analysts have tracked the SEC filings of approximately 1,400 companies, the vast majority of which consist of biotechnology companies engaged in pharmaceutical discovery and development.

Companies can and usually do request confidential treatment for sensitive business information in these alliances, including royalty rates and other payments, but such grants of confidentiality are time limited. Recap's analysts first collect these SEC-filed agreements and then attempt to secure unredacted copies through use of Freedom of Information Act (FOIA) requests made to the SEC.

Figure 1 shows the number of alliances selected for inclusion in a sample of development-stage R&D alliances entered into between 1981 and 2000 by the 20 most active biotechnology and pharmaceutical commercialization partners. The "Top 20" commercialization partners were selected on the basis of their total number of biotechnology alliances over the past three decades, including alliances established by commercialization partners subsequently acquired by one of the Top 20. For example, Novartis has in aggregate more than 700 biotechnology alliances, including those entered into by Ciba-Geigy and Sandoz. Thirty-two Novartis alliances are included in the sample. These are all of the unredacted, development-stage R&D alliances involving Novartis as the commercialization partner in Recap's Alliances Database as of February 2006. A similar process was followed for the other 19 most active commercialization partners of biotechnology R&D programs, resulting in a final sample of 259 unredacted development-stage R&D alliances.

4.2 *Prelaunch payments*

Figures 2 and 3 show the average and median prelaunch payments, respectively, for biotechnology alliances established by the Top 20 commercialization partners between 1981 and 2000. The alliances are grouped by the stage of development at signing, where *mid stage* refers to alliances signed at the preclinical or Phase I clinical trials stages, and *late stage* refers to alliances signed at the stages of Phase II or III clinical trials or NDA filing.

The data in Figures 2 and 3 supports the observation that the later in drug development an agreement is struck, the higher the amount of consideration paid to the originator. For example, median prelaunch payments to originators of mid stage alliances were US$21.8 million, but US$30.7 million for late-stage deals. While median prelaunch payments for discovery-stage alliances exceed those for lead-stage deals, the largest component of such discovery-stage payments are for R&D reimbursement, and so are not enriching to the originator.

4.3 *Royalty and other post-commercialization payments*

Figures 4 and 5 show the average and median effective royalty rates (that is, rates adjusted for royalty tiers) and maximum royalty rates (which include consideration from transfer prices), respectively. This data also supports the observation that the later in drug development an agreement is struck, the higher the amount of consideration paid to the originator. For example, the data shows that the median effective royalty rate promised to a product's originator in the event of annual sales of US$500 million was seven percent for discovery-stage alliances, eight percent for lead stage, 9.6 percent for middle stage and 15 percent for late stage. Likewise, on average, the effective royalty rate increases with greater annual sales of the product.

When transfer prices and the maximum royalty rate are combined, the analysis shows that the median compensation to a product's originator increases to eight percent for discovery-stage alliances, 10 percent for lead stage, 15 percent for middle stage and 20 percent for late stage. However, none of these average or median post-commercialization payments includes the effect of the 44 alliances that involve profit splits, since this form of consideration is not directly comparable to royalties.

5. ILLUSTRATIVE INSTANCES OF ALLIANCES AT SEVERAL STAGES

5.1 *Discovery-stage alliance*

In May 1997, Eli Lilly and MegaBios (later merged to become Valentis) signed a worldwide alliance to develop gene-therapy products to treat cancer. At the time of commencement, MegaBios had a technology platform for gene therapy, but no lead compounds had yet been developed in the field of cancer.

As shown in Figure 6, the technology originator, MegaBios, received no up-front payment, but Lilly committed to US$7 million in FTE and manufacturing-process payments over two years. Lilly was responsible for all other development, clinical, manufacturing, and regulatory expenses. Development-milestone payments totaled US$27.5 million, consisting principally of amounts associated with the clinical development of compounds to treat ovarian and breast cancer. Lilly purchased US$3 million of MegaBios' equity at signing. In the post-commercialization period, Lilly committed to paying tiered royalties to MegaBios, increasing with annual net sales from six to 13 percent. Such royalties would be due for either the life of any issued patents, or the seven-year-period following product launch, whichever was longer, on a country-by-country basis, after which Lilly would retain a paid-up license.

5.2 *Lead-stage alliance*

In December 2000, Novartis and Celgene signed a worldwide alliance to develop treatments for osteoporosis. At the time of commencement,

Celgene had several lead compounds based on selective estrogen-receptor modulators (SERMs).

As shown in Figure 7, the compound originator, Celgene, received a US$10 million up-front payment, plus US$4 million in FTE payments over two years. Novartis was responsible for all development, clinical, manufacturing, and regulatory expenses. Development-milestone payments totaled US$30 million. There was no equity investment. In the post-commercialization period, Novartis committed to paying to Celgene tiered royalties that increased with annual net sales from ten to 12 percent. Such royalties would be due for either the life of any issued patents or the ten-year–period following product launch, whichever was longer, on a country-by-country basis, after which Novartis would retain a paid-up license.

5.3 *Midstage alliance*

In November 1997, Eli Lilly and Ligand Pharmaceuticals signed a co-development, license, and co-promotion alliance for worldwide rights to RXR retinoids for the treatment of diabetes. At the time the parties entered into the alliance, several of Ligand's RXR compounds were undergoing preclinical testing.

As shown in Figure 8, the compound originator, Ligand, received a US$12.5 million up-front payment. There were US$49 million in FTE payments over five years, and Lilly was responsible for all development, clinical, manufacturing, and regulatory expenses. Development-milestone payments totaled US$73 million, divided among six separate types of compounds and ranging from US$6.5 million to US$14 million per compound. There was no equity investment. In the post-commercialization period, Lilly committed to pay tiered royalties to Ligand, increasing with annual net sales and varying by type of compound from five to 12 percent of net sales. Such royalties would be due for either the life of any issued patents or the ten-year–period following product launch, whichever was longer, on a country-by-country basis, after which Lilly would retain a paid-up license.

5.4 *Late-stage alliance*

In December 1993, Burroughs Wellcome (later acquired by GlaxoSmithKline) and Centocor (later acquired by Johnson & Johnson) signed a co-development, license, distribution, and supply alliance for rights outside of Asia to Panorex, a monoclonal antibody for use as adjuvant therapy for the treatment of colon and colorectal cancers. When the parties entered into the alliance, Panorex was undergoing Phase III clinical trials.

As shown in Figure 9, the compound originator, Centocor, received US$19 million in up-front payments, US$10 million on signing, plus an additional US$9 million when the territory was expanded to include Asia in 1994. There were no FTE payments, and Centocor was responsible for the completion of Phase III trials. Development-milestone payments totaled US$47.5 million. Wellcome purchased US$23.5 million of Centocor's equity—US$20 million on signing plus an additional US$3.5 million when the territory was expanded. In the postcommercialization period, Centocor committed to paying a transfer price of 50 percent on the first US$200 million in annual net sales, then 40 percent on the next US$200 million, then 35 percent on net sales greater than US$400 million. The term of the agreement would be for the duration of product supply by Centocor.

6. RECOMMENDATIONS AND CONCLUSIONS

Although lacking vendor booths or trading floors, a robust marketplace exists for the exchange of discoveries, intellectual property, and services related to the development and commercialization of products in the life sciences. After several decades of trial and error, biotechnology and pharmaceutical companies have settled upon the principal structural and economic elements in the identification, creation, and sharing of value in this marketplace.

As the authors have noted in previous publications,[3] the economic stakes of university TTOs, primarily in the United States and Great Britain, as upstream licensors and enablers in this marketplace are also well established.

New entrants to this marketplace, especially university TTOs representing institutions in territories other than the United States, Great Britain and, to a lesser extent, Canada, Germany, and France, have an opportunity to join this marketplace with knowledge of its inner workings. At a minimum, new entrants should be in a position to undertake programs of technology or compound development with the knowledge that downstream events that would be likely to be perceived as value creating. Conversely, should these institutions be able to assemble significant intellectual property and/or compounds to offer, such TTOs may choose to supplant biotechnology companies and take it upon themselves to deal directly with prospective commercialization partners, be they traditional pharmaceutical companies or regional marketing firms.

This chapter has attempted to identify the principal structural and economic elements of biotechnology alliances and the factors that influence their selection. In the interest of brevity, only the most important structural terms have been discussed. Other provisions that are usually addressed in these alliances are noted in Box 1. ■

MARK G. EDWARDS, *Managing Director, Recombinant Capital, Inc., 2033 N. Main St., Suite 1050, Walnut Creek, CA, 94596 U.S.A.* medwards@recap.com

1 Since this chapter is principally concerned with development-stage biotechnology R&D programs, the term *alliance* is used to describe generally the relationship between the parties. Such relationships typically involve a license and/or sublicense, as well as other rights and responsibilities of the parties. Except where specifically noted, the terms alliance, agreement, deal, partnership and license are used interchangeably in this chapter.

2 In this chapter the term *originator* refers to one who licenses (a licensor) a compound or technology to a commercialization partner. When the originator is a biotechnology company, the conveyed intellectual property may include one or more sublicenses of university-derived intellectual property.

3 Edwards M, F Murray and R Yu. 2003. *Value creation and sharing among universities, biotechnology and pharma. Nat. Biotechnol.* 21: 618–24. Also Edwards M, F Murray and R Yu. 2006. *Gold in the ivory tower: equity rewards of outlicensing. Nat. Biotechnol.* 24: 509–15.

Box 1: Guidelines for Drafting Licensing Deals

I. Research & Development:

A. **Scope of Agreement**
- Effective date
- Nature of the collaboration
- Field of research
- Method of joint development
- Identify key research terms

B. **Research Period**
- Term of sponsored research program (if any)
- Note possible extensions

C. **Reimbursement Basis or Cost Sharing**
- R&D payments (amount and type)
- FTE (full time equivalent) reimbursement rates

D. **Upfront Payment**
- Payment(s) upon signing (or calendar based)
- Technology access fees
- Credit given for option payments received prior to signing?

E. **Benchmark Amounts**
- Pre-commercial milestones (i.e., IND, NDA)
- Sales-based milestones
- Creditable against royalties? Credit limitations

F. **Technology Acquisition Fees**
Applicable for asset purchases & assignments

G. **Payment Schedule**
i.e., quarterly

H. **Budgets**
- Approved in advance?
- Are budgets appended to agreement?

I. **Reimbursement Start Date**
- Typically on signing

J. **Regulatory Filings**
- Who controls and pays for regulatory filings?
- Do responsibilities vary by stage, territory or product?

K. **Specific Capital Requirements**
- Capital equipment paid for by licensee
- If special equipment is purchased, who keeps it upon termination?
- Transfer of materials

L. **Patent Ownership**
- Know how, patents, IP, material ownership
- Who owns the patent rights?
- Joint inventions

M. **Patent Filing Costs**
Who pays filing, prosecution, maintenance costs?

N. **Patent Defense Costs**
- Who has first right to sue third-party infringers?
- Who pays for the patent defense costs?
- Allocation of recovery from such action

O. **Third-Party Patents**
- Who has first right to respond to 3rd party suits for infringement?
- If royalties due to third-party, typically 50% of such payments are creditable against 50% of amounts due to licensor

P. **Non-compete Provision**
Each party can or cannot compete in the Field

Q. **Publications**
- Approval procedure
- Licensee may request delay for patent prosecution

R. **Core Technology**
- Who owns core technologies?
- Visiting scientists, retained rights, etc.

S. **Cancellation Amounts**
- Any amount due in the event of termination?
- May include wind-down of sponsored R&D

T. **Termination**
Termination rights include (i) mutual, (ii) licensor, (iii) licensee.

(Continued on Next Page)

Box 1 (continued)

U. Product Reversion
- Who keeps product rights after termination?
- Royalties due to the non-terminating party?

V. Change in Control
- Typically "not assignable without the prior written consent of the other party"
- Are co-promotion and/or supply rights lost in the event of change in control?

W. Options/Other
- Additional research options (i.e., added fields, products)
- Right of first refusal (ROFR) to other research

II. Product License

A. License Holder/Type
- License grant(s), including make, have made
- Exclusive, nonexclusive or semiexclusive (note limitations)
- Commercialization rights (right to sublicense?)
- Is know-how included?

B. Product Field of Use
- Define product field of use
- Does IP have utility beyond scope of license?

C. Territory Splits
- Define territory; what are major markets?
- Are there territory options for inclusion/exclusion?

D. Royalty Rate
- Royalty rates and/or profit splits
- Adjustments under certain conditions (type of IP protection, gross margins, competition)
- Note limitations to royalty offsets for third party patents and/or credits for prior payments

E. Right to Sublicense
- Is prior consent required?
- Impact on royalty rates
- Pass-through payments to upstream licensor

F. Term/Patent Life
- How long does license agreement last?
- Term of royalty obligations ("life of license") ("continue until the last to expire patent....")
- What happens to exclusivity upon expiration of royalty obligations?
- Note any rights of licensee to sell product after expiration (subject to royalty?)

G. License Maintenance and Diligence
- Annual license maintanence fees and/or minimum royalties
- Due diligence (e.g., IND, Phase I, NDA filing by certain dates, "use reasonable efforts to develop," etc.)
- Terminate or non-exclusive for non-performance

H. Royalty Accounting
- Define "net sales" or equivalent
- Other defined terms for royalty calculations?
- Audit provisions
- Late-payment fees, penalties, interest

I. Patent-Royalty Tie-In
- Are royalty rates tied to the granting of patents?
- Step-down rates for know-how only
- Treatment of pending patents by country if product launched prior to patent issuance

J. Options/Other
- Co-promotion rights, if any
- Commercialization options for related products

(Continued on Next Page)

Box 1 (continued)

III. Manufacturing & Supply:

A. Right Holder/Type
- Who has the right to manufacture?
- ID on packaging
- What about second source or back-up supply?

B. Bulk/Dosage Form
- Bulk or final form
- Does this change by stage of development or scale?

C. Territory
Supply territory

D. Reimbursement Basis
- Define basis of payment (e.g., fixed price per unit, manufacturing cost plus markup, percentage of net sales)
- If transfer price, inclusive/exclusive of royalty?

E. Process Development Terms
- Terms with respect to manufacturing process development
- Who is responsible for manufacturing program?
- Timing of orders and delivery commitments
- Ownership of production equipment

F. Clinical Use Manufacturing
- Who supplies compound for clinical trials?
- Reimbursement basis for clinical supplies

G. Shipment Terms
- FOB (freight on board) place of shipment
- Standard cost for bulk?
- Terms for replacement of non-spec shipments

H. Financing
- Is licensee providing financial arrangements for Licensor to meet supply obligations?

I. Escape Clause
- If Licensor cannot satisfy supply requirements, right of licensee to make or have made such quantities
- Trigger event(s) of default
- Temporary or permanent?
- Product/territory specific?

J. Product Liability
- Indemnification, including standard and limitations
- Insurance requirements

K. Options/Other
- Supply options
- Options to repurchase product

IV. Collaboration Management:

A. Representation
- Governance of program
- Committees established between the parties
- Make-up of committee, mandates

B. Quorum
Any specific quorum?

C. Basis of Actions
Unanimous vote or majority rule?

D. Meetings
How often does the committee meet?

E. Disagreements
- Dispute resolution (escalation procedure)
- Arbitration or mediation and applicable rules
- Appeal?

F. Buyout/Windup
- Applicable for JV arrangements
- Purchase option(s) in the event of termination/ expiration of the JV

G. Options/Other
- Any other terms relating to the governance of collaboration

Box 1 (continued)

V. Equity Investment:

A. Type of Security
Number and type of shares purchased

B. Pricing
Price paid

C. Board Seat
- Board seats granted?
- Specific individual or named by party when relinquished

D. Research Tie-Ins
If proceeds must be used for R&D

E. Options & Rights
- Additional equity purchases
- Convertible loans
- Rights/obligations of purchaser:
  - registration rights
  - anti-dilution protection
  - sales restrictions
  - standstill
  - market standoff
  - right of first refusal

VI. Signatories:

A. For University or Biotech Co. (R&D Co.)
Name, title, company

B. For Biotech or Drug Co. (Client Co.)
Name, title, company

Figure 1: Unredacted Biotech Alliances of the Top 20 Pharmas (259 Alliances Signed between 1981 and 2000, by Stage at Signing)

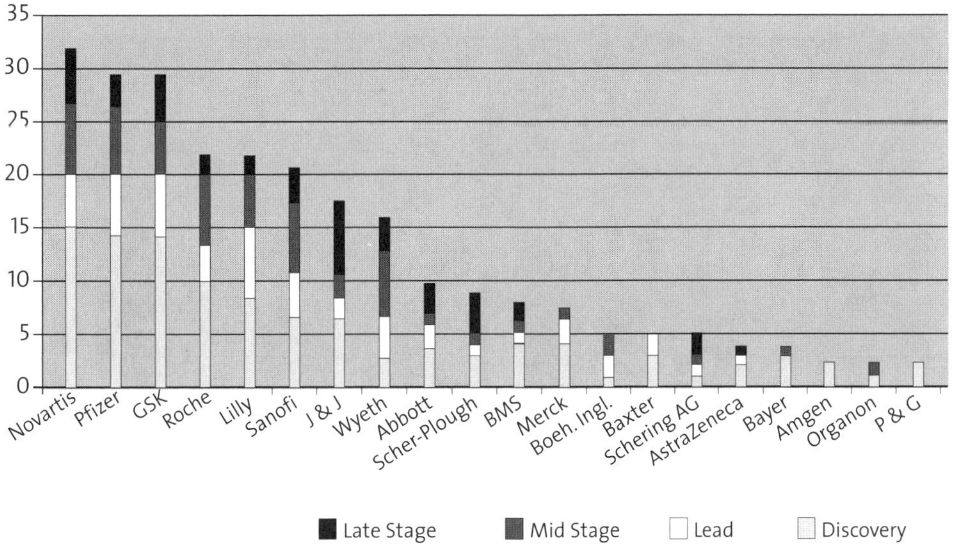

Source: Recombinant Capital www.recap.com

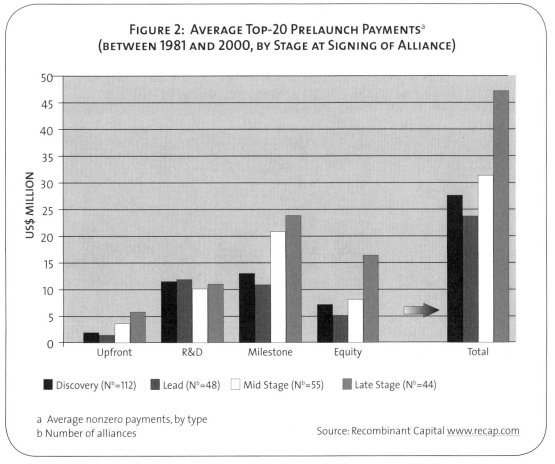

Figure 2: Average Top-20 Prelaunch Payments[a] (between 1981 and 2000, by Stage at Signing of Alliance)

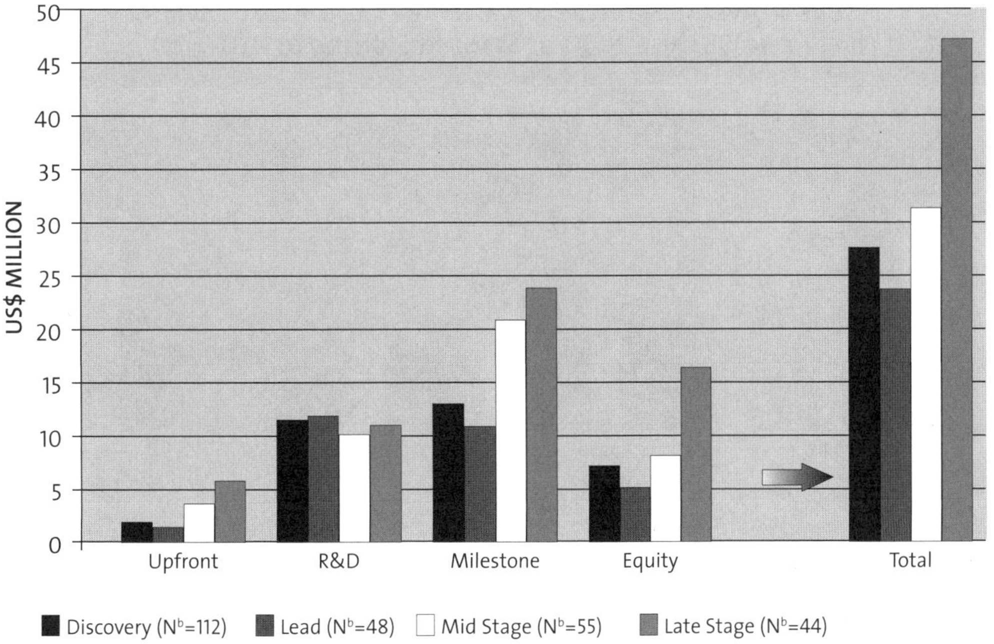

a Average nonzero payments, by type
b Number of alliances

Source: Recombinant Capital www.recap.com

FIGURE 3: MEDIAN TOP-20 PRELAUNCH PAYMENTS[a]
(BETWEEN 1981 AND 2000, BY STAGE AT SIGNING OF ALLIANCE)

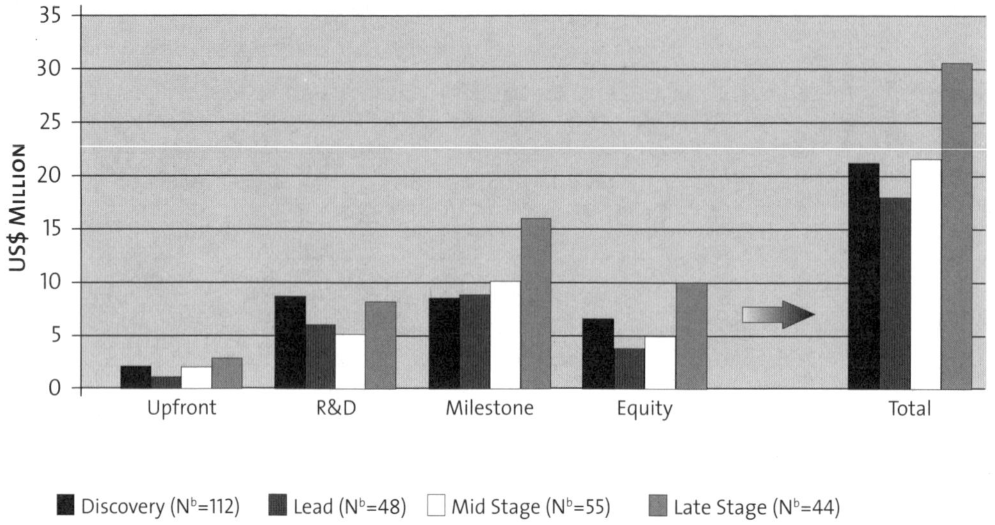

a Median nonzero payments, by type
b Number of alliances

Source: Recombinant Capital www.recap.com.

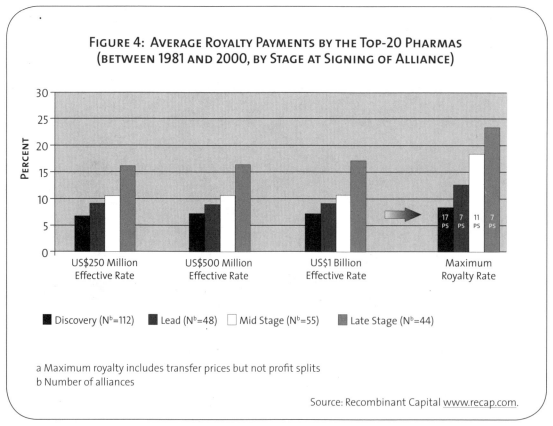

FIGURE 4: AVERAGE ROYALTY PAYMENTS BY THE TOP-20 PHARMAS
(BETWEEN 1981 AND 2000, BY STAGE AT SIGNING OF ALLIANCE)

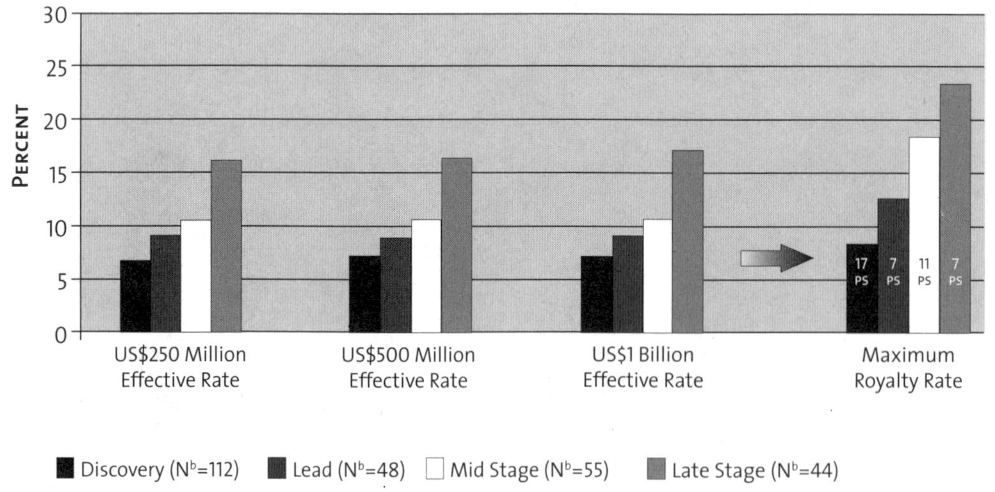

a Maximum royalty includes transfer prices but not profit splits
b Number of alliances

Source: Recombinant Capital www.recap.com.

FIGURE 5: MEDIAN ROYALTY PAYMENTS BY THE TOP-20 PHARMAS (BETWEEN 1981 AND 2000, BY STAGE AT SIGNING OF ALLIANCE)

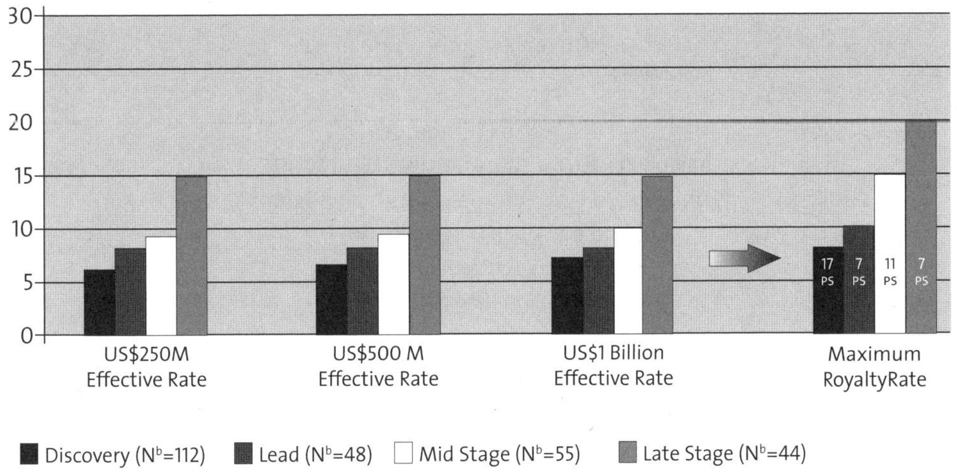

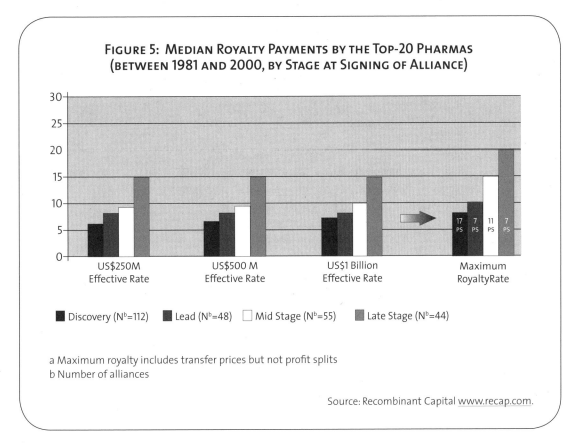

a Maximum royalty includes transfer prices but not profit splits
b Number of alliances

Source: Recombinant Capital www.recap.com.

FIGURE 6: AN ILLUSTRATIVE DISCOVERY-STAGE ALLIANCE

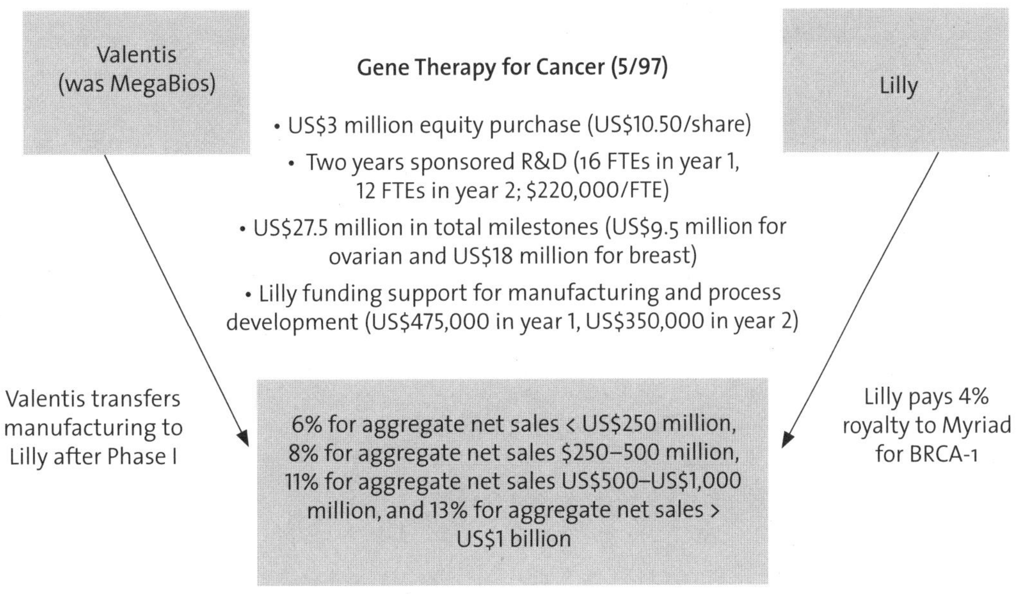

Source: Recombinant Capital www.recap.com.

FIGURE 7: AN ILLUSTRATIVE LEAD-STAGE ALLIANCE (ALL FIGURES IN U.S. DOLLARS)

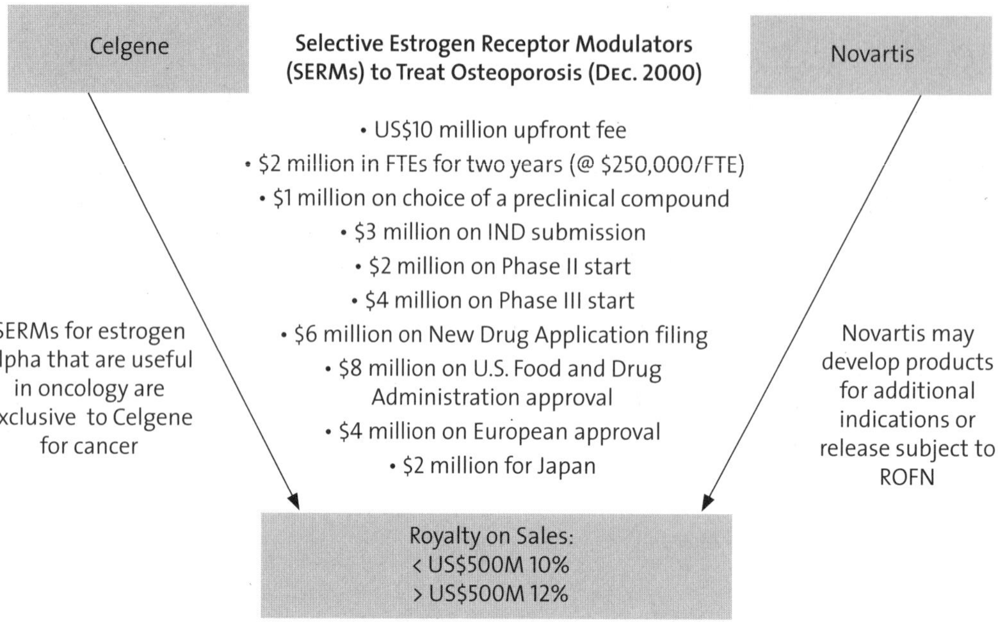

Celgene has the right to partner SERMs for estrogen alpha for cardiovascular indications based on activity in its cardiovascular assay.

Source: Recombinant Capital www.recap.com.

FIGURE 8: AN ILLUSTRATIVE MID-STAGE ALLIANCE

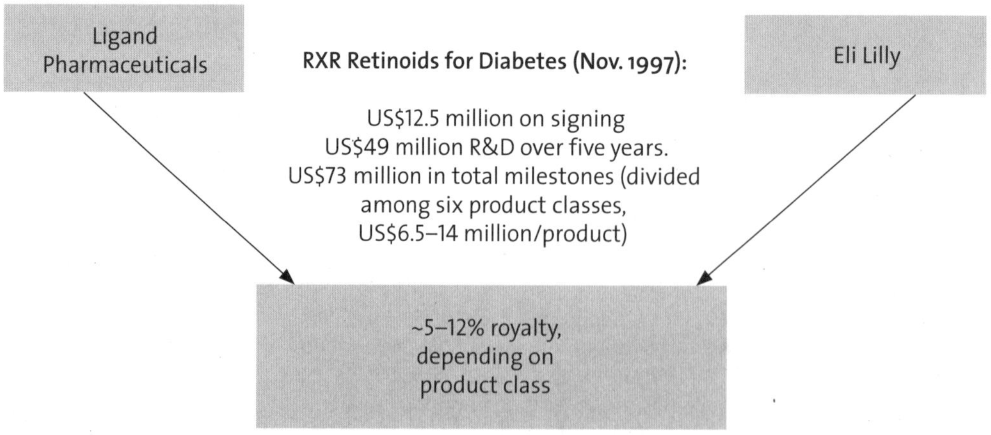

Ligand has the option to co-develop SERM oncology product, by paying 33% of development costs after Phase II, and for one-third higher royalty on cancer sales.

Source: Recombinant Capital www.recap.com.

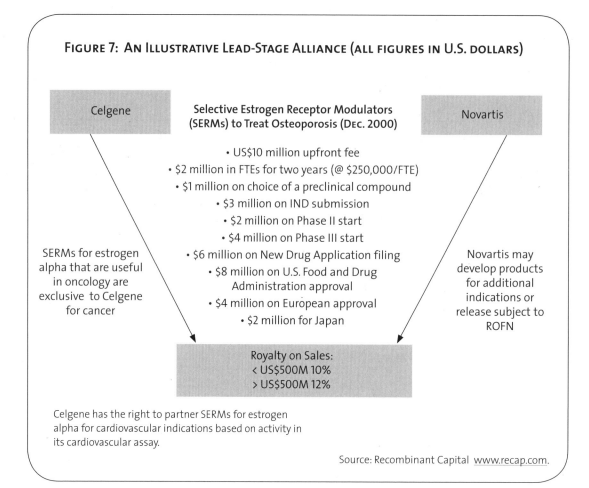

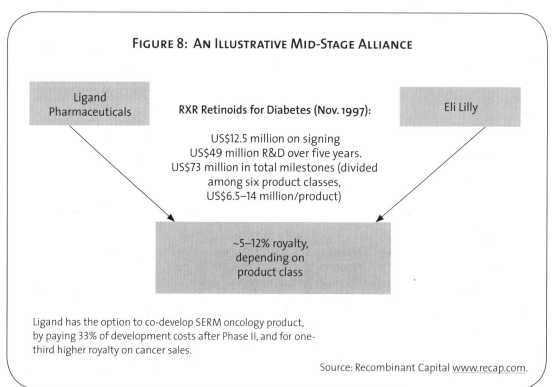

FIGURE 9: AN ILLUSTRATIVE, LATE-STAGE ALLIANCE

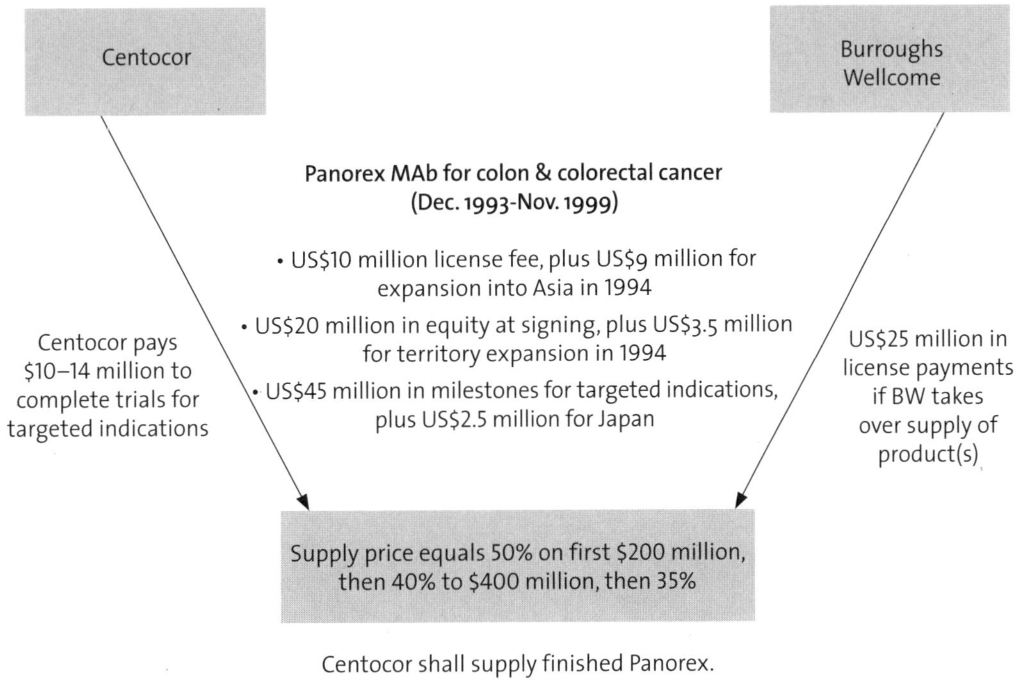

Source: Recombinant Capital www.recap.com.

Product Development and IP Strategies for Global Health Product Development Partnerships

SANDRA L. SHOTWELL, *Managing Partner, Alta Biomedical Group, LLC, U.S.A.*

ABSTRACT

The mission of global health product development partnerships (PDPs) is to develop effective, affordable health products and make them available and affordable to those in need. The not-for-profit product development partnerships (PDPs) often seek for-profit partners to access essential technology, expertise, and resources. These may be early-stage companies, leveraging philanthropic and government resources to develop a platform technology or established companies building out from existing markets or testing new technologies. Such not-for-profit/for-profit partnerships require unique product development and IP (intellectual property) strategies that both recognize the company's need for commercial benefit and deliver important health products to developing countries.

1. INTRODUCTION

"*Thus we come to the conclusion that patents are neither inherently bad nor inherently good for this purpose, but—like most tools—must be used wisely.*"[1] Lita Nelson's words are particularly appropriate for thinking about global health product development partnerships (PDPs), which today are harnessing the power of both the private sector—especially its intellectual property (IP)—and the IP system itself to help deliver public sector goods.

The mission of a PDP is to develop, manufacture, and deliver affordable and accessible health-care products that treat the diseases of the developing world. PDPs seek to serve underserved and disadvantaged markets where there is little or no competition from other pharmaceutical companies. In some instances, their products also will reach private, profitable markets in developed countries, but it is not their main goal to serve these markets.

The efforts of PDPs have significantly increased the number of products currently being developed for diseases that affect developing countries.[2] Products under development include drugs, vaccines, and diagnostics for diseases such as AIDS, tuberculosis (TB), malaria, meningitis, dengue fever, shigella, and cholera, among others.[3]

2. CHARACTERISTICS OF PDPS

Although they are not-for-profit organizations, PDPs have similarities with both for-profit companies and research institutions. For one thing, the IP (intellectual property) goals of PDPs are similar to those of other types of organizations: to respect valid third-party patents; to ensure their own freedom to operate (FTO)—in other words, to use their own IP without constraint and to use patents to leverage investment, partnerships, and political goodwill.

Shotwell SL. 2007. Product Development and IP Strategies for Global Health Product Development Partnerships. In *Intellectual Property Management in Health and Agricultural Innovation: A Handbook of Best Practices* (eds. A Krattiger, RT Mahoney, L Nelsen, et al.). MIHR: Oxford, U.K., and PIPRA: Davis, U.S.A. Available online at www.ipHandbook.org.

Research institutions usually have neither the funding nor the expertise to take products to the marketplace. Therefore, they rely on corporate partners to develop their technologies into products for public use. They use patents to attract corporate interest in their projects, seeking patent protection in countries where corporate partners will want a competitive advantage.

Like research institutions, PDPs have nonprofit missions, rely largely on philanthropic and government support, and do not plan to manufacture and market the products that reach customers.[4] PDPs prefer to partner with for-profit companies so that they can draw on their manufacturing expertise, production facilities, market channels, and sometimes their R&D expertise, as well.

For-profit companies try to gain advantages over their competitors in order to maximize their market share and profits. They reduce the risk of developing new products by assiduously protecting their intellectual property. PDPs also work to protect the intellectual property produced through their partnerships, but their goal is, like research institutions, to leverage their intellectual property for access to other intellectual property or for other uses that will contribute to their mission.

Like for-profit companies, PDPs develop products that will someday be introduced to the marketplace. They manage portfolios of products that are at various stages of development, project and establish markets, and work to overcome logistical and social barriers to product adoption. However, their IP strategies are different from those of for-profit companies, for several reasons. They have no need to protect their market share or profits. In fact, they aim to achieve the lowest, rather than the highest, possible product pricing. They welcome the presence of other organizations that are developing products for the same market. They are open to sharing knowledge, resources, and projects. Thus, there IP strategy does not include the for-profit motive of keeping competitors out of their market or increasing market share.

In spite of these differences, most PDPs are evolving product development and IP strategies that are very similar to those of for-profit companies. In pursuit of their humanitarian goals, PDPs may license their own intellectual property or access the intellectual property of their corporate partners. In fact, if a company has already developed a product that is ready for immediate use, there may be no need for a PDP to get involved at all. This situation can occur, for example, when companies are directly engaged to provide anti-AIDS drugs at greatly reduced cost.[5]

3. PDP PARTNERSHIPS WITH FOR-PROFIT COMPANIES

In order to attract the interest and investment of for-profit partners, PDPs must protect their own intellectual property. It can be expensive and time consuming to obtain patents in developing countries, and the markets tend to be small, but the existence of an enforceable patent is often a strong inducement to potential industrial partners.

PDPs follow a wide range of business models: virtual pharmaceutical-development organizations (such as TB Alliance[6]), in-house research capabilities (such as the International AIDS Vaccine Initiative[7]), the inclusion of manufacturing capabilities (such as Aeras Global TB Vaccine Foundation[8]), and nonprofit pharmaceutical companies (such as Institute for OneWorld Health, iOWH[9]). All PDP business models draw heavily on public and philanthropic support (such as the Bill and Melinda Gates Foundation [BMGF]), as well as on extensive partnering with not-for-profit, government, philanthropic, and for-profit partners.

Examples of the many partnerships PDPs develop with companies are presented in the case studies in the *Handbook Executive Guide*.[10] The product development and IP strategies vary considerably based on the technology, the stage of development, and the nature of the market. Most products developed by PDPs fall into one of two broad categories: those that incidentally have large, profitable markets in developed countries (such as those that treat AIDS or TB) and those that do not. Examples of how the

PDP strategies differ in these two situations are shown below.

3.1 *Producing healthcare products with profitable markets*

TB affects both the developed and the developing world. One PDP, the TB Alliance, seeks to develop more affordable, more effective products with shorter dosing regimens that increase the likelihood that patients will complete their courses of medication.[11] A major component of the TB Alliance's product-development strategy is the formation of partnerships with companies that own the rights to approved, IP-protected drugs that could be repurposed to treat TB. It has therefore partnered with Bayer Healthcare AG to perform clinical studies on Bayer's drug moxifloxacin; it is hoped that this drug will be effective in three or four months rather than the standard six months. The agreement states that Bayer donates the drug and covers regulatory costs; the TB Alliance will coordinate and help cover the cost of the trials, and seek to leverage support from corporate partners.

In 2006, another PDP, AERAS, exclusively licensed patent rights to a vaccine technology from Vanderbilt University so that it could develop a TB vaccine; the university retained the right to license the technology to other partners engaged in non-TB development. The exclusive license gives AERAS access to the technology and university expertise, as well as freedom to operate; if the organization is able to develop a TB vaccine (or even to make some improvements on the existing technology), it will be able to use its knowledge to attract for-profit partners.

3.2 *Producing healthcare products without profitable markets*

Different strategies are needed when developing products for markets with low (or no) profit potential. It may be difficult to find a for-profit corporate partner that is already working to develop such products. However, there are companies with relevant expertise, technology, and products, and they can be encouraged to partner with PDPs to their mutual benefit.

Malaria is found disproportionately in developing countries, though for-profit markets are growing in such places as India and among travelers and military personnel from developed nations.[12] There is currently no approved malaria vaccine. The PDP Malaria Vaccine Institute partners with universities, government labs, and both early-stage and established companies in order to advance malaria vaccine candidates. It is currently working with the for-profit company GSK Biologicals to test its vaccine in African children. The vaccine has proven to be effective for at least 18 months, reducing clinical malaria by 35% and severe malaria by 49%. *Time* magazine declared this project to be one of the most important accomplishments in the field of healthcare in 2005.

The PDP iOWH has licensed a technology based on technology developed at the University of California at Berkeley. This technology is useful for producing a precursor to artemisinin, a natural product in short supply that is used in malaria treatment. The PDP iOWH teamed up with a spinout company, Amyris Biotechnologies, in late 2004. With support from the BMGF, the three-way agreement benefited all parties: the university's technology was advanced, Amyris fine-tuned its production processes, and iOWH developed a malaria drug candidate.

4. CONCLUSIONS

The developed world has a growing commitment to meeting the healthcare needs of the developing world. Successful product development and IP strategies are just two of the many issues involved in the commitment to developing products for underserved markets. The engagement of various regulatory jurisdictions, local political and legal issues, the management of liability, the delivery of products to areas with limited infrastructure or security, and cultural acceptance of new products—all of these issues need to be addressed and managed in order for PDPs to achieve their goals. ■

SANDRA L. SHOTWELL, *Managing Partner, Alta Biomedical Group, LLC, 7505 S.E. 36th Avenue, Portland, OR, 97202, U.S.A.* shotwell@altabiomedical.com

1 See, also in this *Handbook*, chapter 1.4 by L Nelsen and A Krattiger.

2 Moran M, A Ropars, J Guzman, J Diaz and C Garrison. 2005. The New Landscape of Neglected Disease Drug Development. LSE Pharmaceutical R&D Policy Project. The Wellcome Trust: London. www.lse.ac.uk/collections/LSEHealthAndSocialCare/documents/PRPP/Thenew-landscapeofneglecteddiseasedrugdevelopment.pdf.

3 See, for example, www.mihr.org, www.tmgh.org, and www.gatesfoundation.org for information about PDPs and their projects.

4 A notable exception is the Aeras Global TB Vaccine Foundation, which setup an in-house manufacturing facility for producing rDNA vaccines.

5 www.bms.com/sr/philanthropy/data/introx.html.

6 www.tballiance.org.

7 www.iavi.org.

8 www.aeras.org.

9 www.iOWH.org.

10 Krattiger A, RT Mahoney, L Nelsen, JA Thomson, AB Bennett, K Satyanarayana, GD Graff, C Fernandez and SP Kowalski. 2007. *Intellectual Property Management in Health and Agricultural Innovation: Executive Guide to Best Practices*. MIHR: Oxford, U.K., and PIPRA: Davis, California, U.S.A.

11 "Working with the best in both the public and private sectors, we collaborate formally with leading university laboratories, large pharmaceutical companies, biotechnology companies, and government agencies. Our work is also informed by constant dialogue with other organizations working to develop TB treatments." TB Alliance. 2006. Next Steps Now. Annual Report 2005/06. TB Alliance: New York. p. 14. www.tballiance.org/downloads/publications/TBA_Annual_2005-2006.pdf.

12 www.malariavaccine.org/files/Market-Assessment-18Jan05-LB-BOS.pdf.

SECTION 13

The Public Sector and Entrepreneurship

Creating and Developing Spinouts: Experiences from Yale University and Beyond

ALFRED (BUZ) BROWN, *Director, Office of Cooperative Research, Yale University School of Medicine, U.S.A.; Currently: Managing Director, BCM Ventures, U.S.A.*

JON SODERSTROM, *Managing Director, Office of Cooperative Research, Yale University, U.S.A.*

ABSTRACT

This chapter is about university spinouts: why they are created, who founds them, and how they are developed. It also considers many of the issues that a university and its faculty have to address to successfully launch and develop new for-profit ventures. Spinouts carry risks, but they may also be the best vehicle for developing early-stage university technologies and providing a host of other benefits. The chapter offers examples from the past five years at Yale University, as well as from the private sector, that suggest ways to minimize the risks and maximize benefits.

1. INTRODUCTION

In the course of fulfilling university research and educational missions, faculty often create intellectual assets that can benefit society. These assets may include patentable inventions, copyrightable works, and ideas that form the basis for new products and services. As they emerge from university laboratories, these inventions are not mature commercial products. To fully realize their potential requires significant resources, both human and financial. These resources are not generally found within the university environment.

Therefore, commercial development of the invention requires the participation of for-profit partners who possess the requisite resources. The most common means available to universities for attracting such partners are licenses. Patents, copyrights, and other instruments of intellectual property (IP) protection safeguard investments made by the university's corporate partners. In general, universities license technologies to three classes of private sector entities: established companies with more than 500 employees (large companies), established companies with less than 500 employees (small companies), and newly formed companies (spinouts). The term *university spinout* refers to those companies that are formed around one or more faculty inventions, with involvement of the faculty inventors and the cooperation of the university licensing office, in the licensing of university assets.

This chapter is about university spinouts: why they are created, who founds them, and how they are developed. The chapter also considers many of the issues that a university and its faculty has to address to successfully launch and develop new for-profit ventures. Many of the examples are drawn from the authors' experiences at Yale University over the past five years; other examples are culled from collective experience elsewhere in the private sector.

Brown A and J Soderstrom. 2007. Creating and Developing Spinouts: Experiences from Yale University and Beyond. In *Intellectual Property Management in Health and Agricultural Innovation: A Handbook of Best Practices* (eds. A Krattiger, RT Mahoney, L Nelsen, et al.). MIHR: Oxford, U.K., and PIPRA: Davis, U.S.A. Available online at www.ipHandbook.org.

Editors' Note: We are most grateful to the Association of University Technology Managers (AUTM) for having allowed us to update and edit this paper and include it as a chapter in this *Handbook*. The original paper was published in the *AUTM Technology Transfer Practice Manual* (Part XIII: Chapter 1).

2. WHY UNIVERSITY SPINOUTS?

University spinouts provide many benefits. Among them are:

- the public may have access to new products or services
- success is maximized
- enhancement of the university's and the faculty's image
- improved faculty retention
- local, regional or national economic development
- economic returns to the university and inventor(s)

2.1 *Public benefit*

The academic mission and goals of major universities include engaging in research that is useful to society. To translate this research into beneficial commercial products requires a significant investment of human and financial resources. Commercializing inventions is generally not a central focus of academic or non-profit institutions; such endeavors are more central to the missions of companies. However, in order for a company to justify making investments in the development of inventions from universities, the university typically must first protect its IP through patents, copyrights, or trade secrets.

During the course of managing, protecting, and commercializing university discoveries, the technology transfer manager has many choices, and often there is no apparent best option. A spinout company is rarely a university's first choice for a partner in the private sector. If an existing company has the interest, capability, capacity, and financial resources—and the intent to reach broad markets—a university might prefer to work with that company. Sometimes, however, the market dictates that a spinout should be formed around a collection of technologies. One of the fundamental principles of the Office of Cooperative Research (OCR) at Yale is to make decisions that increase the probability of technology's successful commercialization.

Spinouts carry a number of risks that may exceed those found in established companies. Managers are often less experienced, and personnel may be working together for the first time. Company financing depends on funds from venture investors, who frequently react to environmental changes in ways that are not always in the best interests of the company. For example, during periods of low economic growth, venture investors may elect to invest more in existing portfolio companies and in secondary and mezzanine financings of existing companies. During economic expansions, however, investors actively seek to invest in new companies—sometimes at premiums that hurt future financing.

With certain factors in place, however, a spinout can represent the best opportunity for developing early-stage university technologies. It is crucial to identify a management team for the spinout company, including at least a chief executive officer/chief operational officer and a chief technology officer. Adequate financing must also be obtained; ideally, the business team will have experience and can convince others to invest at a premium to the initial financing of the company. Finally, a spinout's business strategy must be solid and serve a broad customer base.

Spinouts formed around university technologies have a vested interest in the success of those technologies. Company management, consultants and science advisors, board members, and staff are recruited because they believe in, and are committed to, the success of university technologies. Initial investors are especially committed to the success of the initial technologies. In contrast, when technologies are licensed to existing companies, there is often strong initial support for a new licensed technology, although the commitment is rarely as strong and as lasting as it is with spinouts. Existing companies may not identify as strongly with the recently acquired technology, and support may wane in the face of obstacles that a spinout might be able to overcome. Given the larger number of product opportunities in development at bigger and more-established companies, business priorities and personnel can change rapidly, leaving the university's assets undeveloped.

2.2 *Economic development*

New ventures formed to undertake the commercialization of inventions can promote the development of a local economy. This may not be compelling in the technology-rich environments of Boston, San Diego, and the San Francisco Bay area. However, the economy in New Haven, Connecticut, which declined significantly from 1970 through the early 1990s, clearly benefited from the development of technologies created at Yale. A regional economy can experience growth when spinout ventures decide to remain in the area. By 2007, more than 30 companies had been formed around Yale technologies, with more than half locating in New Haven. These ventures provided more than one thousand jobs for highly skilled workers in the year 2000 alone. The ventures generated many joint-research projects undertaken by these companies and the university. The companies have made New Haven both a bioscience center for the state and a magnet for the relocation of existing companies to the city and region.

2.3 *Faculty recruitment and retention*

Faculty that are being recruited by Yale increasingly inquire about opportunities to become involved with existing and spinout companies in the area. A recently recruited department chairman, with significant entrepreneurial experience at the medical school, cited the university's successful technology commercialization efforts and the robust bioscience industry as key in the decision to relocate. A vibrant local and regional technology economy can provide significant job opportunities for the spouses of new faculty hires. Regional technology-based spinouts often have state-of-the-art research tools and expert staff that can be valuable to academic researchers, and faculty members often view the opportunity to collaborate with these ventures as necessary to stay ahead of rapid developments in their fields. If spinouts remain in the region and faculty inventors remain active consultants and advisors to these companies, they can be a powerful force in keeping these inventors at the university.

2.4 *Financial incentives*

Equity, in the form of stock, options, or warrants, is frequently part of the consideration for IP licensed to spinouts; equity may also be granted as consideration for assisting in the formation of a new venture. At Yale and many other institutions, equity-only licenses are rarely used. License agreements with equity consideration usually include cash considerations as upfront license fees, minimum annual and/or milestone payments, royalties on sales, and a percentage of sublicense income. However, upfront fees are frequently reduced when equity consideration is part of the license package. Stock is viewed as a reasonable business solution to enhance the overall financial package—a solution acceptable to the company and its investors—while providing an opportunity for the university to increase its potential return.

Financial returns on equity are independent of the success of the licensed technologies; therefore, equity can be a way to capture value even if the initial licensed technology isn't successful or if the company chooses another market. A few universities view equity as a way to generate large amounts of revenue to benefit their program or the university. To date, this is not a proven strategy. Big winners in equity deals are perhaps even rarer than big winners in traditional licensing deals.

3. HOW TO CREATE A SPINOUT

3.1 *Investable CEO*

While a major part of determining whether or not a spinout represents the optimal commercial path has to do with technology and market assessments, an equally critical aspect is finding an experienced business manager to join the founding team. We often refer to this individual as an *investable CEO,* because he or she has a track record in the technology area that can create added value in the eyes of professional investors. Such an individual must be able not only to understand and communicate with the founding scientists and inventors but also be capable of strategic, tactical thinking and action. The investable CEO must have had operational, preferably profit-and-loss responsibility, in small high-growth technical companies and must be able to work successfully with university founders and scientists. Such individuals are difficult to find. At Yale we succeeded by using the knowledge of industry professionals and senior managers of comparable companies to locate potential candidates. As existing bioscience companies mature in the New Haven area, these become an important source of next-generation CEOs. Fortunately, some of the best CEOs are serial entrepreneurs; once they have had a taste of success with a spinout, they are eager for another. Furthermore, some individuals would prefer not to work at large bureaucratic organizations.

A typical spinout CEO will:

- possess a successful venture-backed, spinout track record
- understand, accept, and manage risk
- comprehend science, discovery, and developmental processes
- be capable in academic and business environments
- have realistic expectations compatible with the university and the investors
- have an entrepreneurial attitude

3.2 *IP assessment*

There are two major questions that investors will almost certainly ask of the technology: (1) Are there technologies or products that can block the development and commercialization of your technology? And (2) can your technology dominate and prevent others from entering the marketplace? While the OCR rarely commissions formal due-diligence opinions, which we consider to be the responsibility of the licensee, we do conduct literature and patent searches to investigate the relative strength of the IP. Although these searches often are initiated prior to identifying a CEO candidate, once such an individual has been identified, the office enlists him or her to assist with the assessment.

3.3 *Market-opportunity analysis*

The key decision in determining the most appropriate path for commercializing any university-controlled IP is whether to license it to an established enterprise or to a new business venture. Regardless of the commercialization path, market and opportunity assessments are conducted on most technologies. Such an assessment looks to balance the perceived technical and market risks with potential return on the investment, for both the university and the potential licensee. Conducting such an analysis includes considering the following questions:

- What are the market applications of the technology?
- Who are the potential customers, and why would they want to buy the technology?
- How are the needs currently being served for each application?
- How does the invention compare to existing technology?
- What is the character of the competition in the market?
- What is the market structure of competing technologies?
- What are the major obstacles to adopting the technology?
- What would it take to make the technology attractive to industry?
- What additional features should be designed to make the invention more attractive?
- What price would the market be willing to pay for this technology?
- What rate of adoption could be expected for the technology?
- What would the competition be in particular markets after the technology has been introduced?
- What are the regulatory requirements and success rates for technologies of this nature and at this stage of development?

All of the above questions help define a product scenario for the technology. Managers and staff need to know enough about the final product to be able to develop preliminary revenue and expense projections over the life of the IP. Obviously, assumptions must be made, and, to the extent possible, these assumptions need to be based on comparable product sales, margins, and expenses. However, when dealing with medical needs or technologies there are frequently no comparables, and sometimes an educated guess is all that is possible.

3.4 *Financial projections*

For every spinout where Yale is the founder, the licensing office puts together a set of financials that capture the basic elements of the business. Linked spreadsheets are an ideal tool for this purpose. Spreadsheets include numbers of customers, product scenarios, revenue, expenses (including personnel, administrative, equipment, and marketing), and cost of goods sold. We use a summary sheet to roll up all of the individual sheets. Identifying key variables (such as numbers of customers and pricing) and linking related elements of the plan (such as numbers of employees or the development status of a new product) can greatly facilitate scenario testing and useful projections. We have found that these projections are of great value in developing product scenarios and business and operational plans, but that they often contain more information than is required by prospective investors—at least for initial meetings.

3.5 *Business plans and investor presentations*

In our experience, business plans are most useful to the founders and company management, while investor presentations are directed to the potential funding audience. While investors will use business plans to challenge the thinking and assumptions made by the founding group, they will most generally use the investor presentation to make the initial decision on whether or not to pursue an opportunity. Accordingly, we use the business plan as a management tool to profile the business opportunity, and we use the investor presentation to raise capital. The investor presentation does, however, usually flow from the business plan, or, at least, makes use of the thinking and assumptions that went into the business plan.

We have found that the ideal investor presentation is 20 minutes long and contains no more than about a dozen overheads or computer-driven slides. The logic is that most investment groups allocate about an hour for the initial meeting, and about half of that time is usually taken up by questions. Assume another ten minutes for introductions and setup and only about 20 minutes are left for the actual presentation. Box 1 presents the elements of a successful presentation used by our group.

4. BUSINESS CREATION: TWO EXTREMES

4.1 *Hands-on approach*

For a number of important reasons, the preferred approach in recent years at Yale has been an intensive, hands-on approach to founding companies around university technologies. Yale's OCR has developed business plans for companies, secured the rights to other institutions' technologies (or parts thereof), recruited management, developed and made investor presentations, negotiated financing agreements, and even assumed the role of interim management for these companies. To be clear, two things we have not done are to invest university funds in spinouts, or to personally take equity or any other incentives from these spinout companies. To a large degree, the OCR has performed these functions because New Haven lacked a strong biomedical entrepreneurial and/or venture investment community. There was also the desire to both maximize the success of Yale technologies and to expand the economy of New Haven and the surrounding communities. Another very important lesson that we have learned from these activities is that when the office undertakes a leadership role in founding these companies—particularly when recruiting management—the companies should locate close to New Haven. This is especially important for the founding scientists and inventors who consult for the company, since it reduces travel and facilitates company–university interactions.

4.2 *Hands-off approach*

During the early years of establishing spinout companies at Yale, the hands-off approach produced variable results, and certainly few successes. There was a time when the university wouldn't even permit faculty members to hold meetings on university property to discuss the prospect of forming a company. Companies still surviving from these times are frequently considered to have persisted despite the activities of the licensing office, rather than as a result of them. By policy, many universities assume a much less proactive role in forming companies. In many cases, institutions market spinout activities (for example, license opportunities) by sending out mass mailings; in other

Box 1: Elements of a Successful Presentation

Problem/need
What is the unsolved problem or unmet need that the business/products will address? This is comparable to reverse engineering the technology—what market opportunities does the technology meet?

Technology/products
What is the technology, and how will it result in new products, or how will it be incorporated into new products? What products will result from the technology?

Long-term plans
Assuming a ten-year cycle, what will the business look like in the second half of the cycle?

Short-term plans
What will the business look like, in one-year intervals, during the initial funding period and for the remainder of the first half of the business cycle? Discuss initial product-development plans, partnering and hiring strategies, and market and revenue opportunities.

IP and market protection
What is the current status of the IP licensed or developed by the company, and how will the IP be protected in the future? Discuss freedom to operate versus the ability to exclude others from the marketplace. What are the plans for acquiring or developing proprietary IP in the future?

Competition
What is the current competition, and what will be the competition when the technology is commercialized? Distinguish the company from the competition.

Management/founders
Who are the scientific founders? Who is the management? Who are the anticipated scientific and business advisors?

Capital needs
What are the capital needs for the first two years or for the initial funding period? What are the expected funding needs after the first two years but prior to exit, initial public offering, or profitability?

Uses of funds
What are the specific accomplishments that will enhance valuation of the business during the first two years or the initial funding period?

cases, investors interact directly with university scientists to develop product scenarios and business strategies and recruit management.

5. EQUITY: FOUNDERS AND TECHNOLOGY CONSIDERATION

5.1 *Founders equity*

Our office has adopted a proactive approach with respect to spinouts. We take founders equity in the new company separate and distinct from consideration for technologies that are being licensed to the spinout. When we initiate the hands-on activities described above, we negotiate an agreement with the other founding members of the company that delineates the roles of the respective parties and the compensation (founders equity) that each party will receive. The value of the equity when the initial founders agreement is made, before the company has any IP assets or capital, is negligible. Therefore, it is best to deal in percentages of founders equity rather than absolute amounts. For example, if there is one university scientist who participates as a founder, one investable CEO, and the university, we would typically agree to split the founders equity equally and to assign a per-share value of US$0.01, par value. In our experience, not all university inventors are founders and not all founders are university inventors. This may seem inconsistent with standard licensing practices, where university inventors are generally treated equally under university patent policies. But not all inventors choose to be entrepreneurs, so our approach benefits both those who want to be founders and those who do not. Founders equity is generally issued as common stock, and although the various founders may have different vesting parameters, all have similar shareholder rights.

5.2 *Equity as technology consideration*

Our experience has been that founders equity is frequently confused with equity that may be granted as consideration for technology rights. At Yale, we have a policy against all-equity license deals, and typical terms for licenses to university spinouts are similar to those that would have been negotiated with existing companies. Therefore, our typical licenses to spinouts include license issue fees, milestone payments, royalties on revenue and sublicense fees, annual minimums, and diligence requirements. Once we have identified the investable CEO and negotiated a founders' agreement with the founders, we will begin the process of negotiating license terms with the investable CEO. Because most of the IP licensed to spinouts is early stage product leads and technologies, the upfront licensing fees are generally low—in the range of US$50,000 to US$250,000. In many cases, common stock may be substituted for the license issue fees. However, license consideration equity is often granted at a par value greater than founders' equity because the license transaction occurs sometime after the founders' agreement and company formation.

6. WHO ESTABLISHES UNIVERSITY SPINOUTS?

6.1 *University founders*

University founders represent the university in spinout activities. At Yale, the OCR performs this function. Many of the founding activities are routinely reviewed with representatives of the general counsel's office, the provost's office, and the dean of the appropriate school. The ultimate internal approval process varies from university to university. Equity received is held by the university and is liquidated according to the equity policy of the university. The following list includes activities that are routinely conducted by our office in launching university spinouts:

- provide IP development and patenting
- create product scenarios
- develop business models and strategy
- identify and develop preliminary relationships with potential development partners
- find and recruit key management
- establish a founding team
- develop revenue and expense projections
- write an executive summary
- prepare investor presentations
- initiate conflict-of-interest clearance

- manage relationships with outside counsel, IP, and/or transactional attorneys
- negotiate interinstitutional agreements and obtain technology rights from other universities
- structure and negotiate technology access term sheets and licenses
- structure and negotiate capital investment
- negotiate investment capital terms
- represent the university in technical and IP due diligence
- review and approve company documents, including shareholders agreements and stock purchase agreements
- hold board seats in spinout companies

6.2 *Inventors and faculty founders*

The structure and policies at Yale University permit faculty inventors to be founders of spinout companies. In our experience, it is rare for an inventor not to want to participate as a founder once the decision to form a spinout has been made. However, we believe our faculty members need to make that decision individually, especially in cases where there are multiple inventors, some of whom may be students, postdoctoral scientists, and untenured faculty who may not have time to participate as founders. It is also possible for faculty who are not inventors to participate as founders of a spinout. We have a number of cases where senior faculty members have expressed an interest early on in participating as heads of scientific advisory boards (SAB) and taking on many of the functions of a university founder. Participation in a spinout can be a particularly rewarding experience for faculty inventors and scientists, not only financially, but also because they can contribute more to their invention's eventual practical applications.

University faculty founders commonly:

- aggressively pursue research consistent with the university's responsibilities and mission
- participate in developing product scenarios and business strategy
- assist with identifying development partners and preliminary talks with them
- assist with the recruitment of key company management and scientific advisors
- assist with fundraising and presentations to investors
- participate in technical and IP due diligence
- participate on, or lead, a scientific advisory board

7. MANAGING THE SPINOUT COMPANY

In most cases, management decisions fall to the investable CEO. However, should the CEO have weaknesses or lack critical experience, the following capabilities/functions may be undertaken by a variety of individuals:

- develop product scenarios, business models, and strategy
- identify and develop preliminary relationships with potential development partners
- find and recruit key operations and technical team members
- help establish the founding team
- develop revenue and expense projections
- write an executive summary
- prepare investor presentations
- participate in developing an IP protection strategy
- negotiate licensing terms and agreements
- structure and negotiate capital investment
- negotiate investment terms
- represent the company in technical and IP due diligence
- review and approve company documents, including shareholders agreements and stock purchase agreements

8. SPINOUT INVESTORS

The sources of capital for university spinouts range from individual angel investors to large, multinational, professional venture funds. The practice at Yale has been to work almost exclusively with larger professional funds specializing in technology-based spinouts. These funds have the ability to lead both current and successive rounds of financing. In the last few years, we have seen initial investments in spinouts increasing in size from US$500,000 to US$5 million, with many recent spinouts raising in excess of US$10 million in the first round. This may be because many of the larger venture capital funds have more money to invest.

Correspondingly, the pre-money value of many spinouts has also increased. We carefully choose the initial group of prospective investors based on prior investments, technical strength in the field of opportunity, and their ability to make follow-on investments. Typically, we target six investment funds and hope that we will be able to obtain a lead investor and one or two co-investment firms from this initial group.

9. DEAL STRUCTURE AND EXAMPLES

Figure 1 presents an overly simplified example of the structuring of a Yale university spinout representing the period of time between the initial founders' agreement and company formation and the point of an initial public offering.

The initial distribution of equity is equal among founders: the university, university inventor, university scientist, and founding CEO. This example assumes one inventor and one scientist/noninventor from the university.

When the company is formed, each founder is issued an equal number of founding common stock at a nominal US$0.001 per share. When the scientific advisory board (SAB) is initially formed, members are issued stock options from the company stock-option pool with a nominal value, or exercise price, of US$0.01 per share. When the technology is licensed to the company, shares are issued to the university, instead of license issue fees, at US$0.50 per share. The initial capital is invested at US$1 per share. Thus, there is an increase in pre-money value in the company, because of significant events, like retaining a world-class SAB, and not because SAB members, or the university, are issued stock at these set values (Figure 2).

Given an equal distribution of initial founders equity between the founding members of the company, the initial equity distribution upon company formation will be as follows (Table 1). *Founders' equity* is the designation given to the common stock issued to founders, and it will have the same value as common stock issued to employees and advisors. The cost of acquiring this equity for the founding members is nominal

FIGURE 1: INITIAL FOUNDERS' AGREEMENT

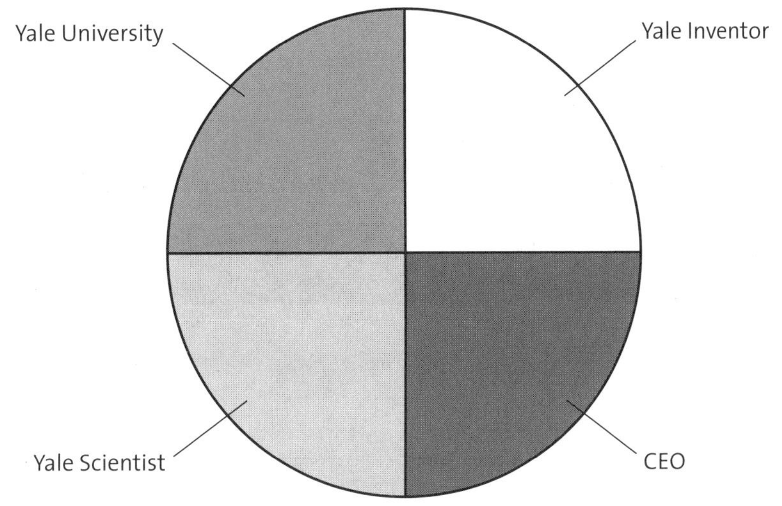

(US$0.001 per share or US$100 for each member), which can be issued at this price because the company, at this point, has minimal value.

In the example above, the company recruits a number of leading international advisors (technical, clinical, and business experts) who will serve on the SAB and on the company's board of directors. These boards are formed after company formation but before the initial financing, thus building additional value in the company prior to financing. In this example, this equity is issued in the form of stock options, as opposed to common stock, because of the immediate value that the recruitment of these key individuals brings to the company. The company then negotiates licenses for three technologies on terms outlined in Table 2.

For technologies A and B, the university receives stock instead of the initiation fee, resulting in the stock division (Table 3). For technology C, the company elects to pay the license issue fee in cash.

After setting aside an option pool for management, SAB, the board of directors, and others (at the discretion of the board), the initial investments total US$15 million, and the stock distribution is as listed in Table 4 and Figure 3.

10. RISKS OF EQUITY PARTICIPATION

While a university's active participation in creating new business ventures can significantly enhance both financial and nonfinancial benefits to the university, such participation increases the

FIGURE 2: INITIAL EQUITY CYCLE

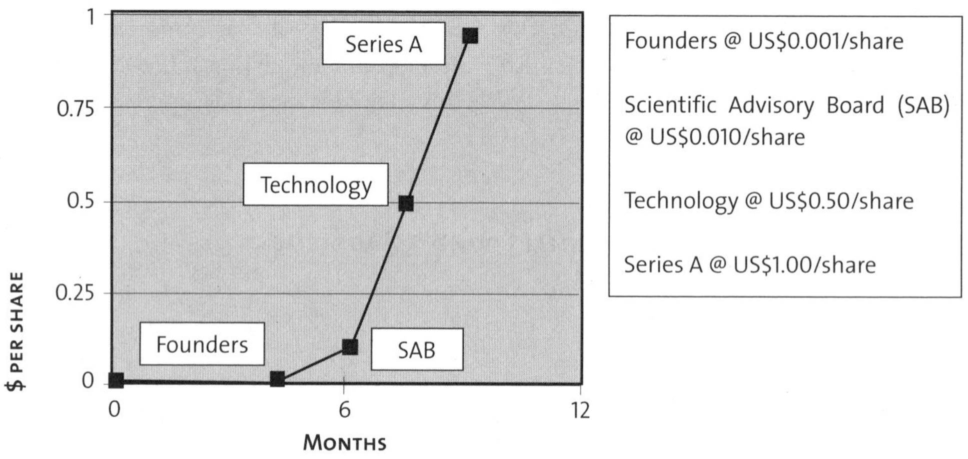

TABLE 1: COMPANY FORMATION AND INITIAL CAPITALIZATION

SHAREHOLDER	FOUNDERS' EQUITY	% CLASS	TOTAL ISSUED AND OUTSTANDING	% TOTAL
University	100,000	25%	100,000	25.0%
Inventor	100,000	25%	100,000	25.0%
Scientist	100,000	25%	100,000	25.0%
CEO	100,000	25%	100,000	25.0%
Totals	400,000	100%	400,000	100%

university's exposure to various financial, legal, and ethical risks.[1] As universities become increasingly more engaged in venture formation, they must be cognizant of the risks and prepared to aggressively manage them. The risks include:

- impacts on tax-exempt status
- creation of taxable, unrelated business income
- exposure to liability
- creation of conflicts of interest and/or conflicts of commitment
- creation of conflicts with the mission of the university

10.1 *Protecting tax-exempt status*

To protect its tax-exempt status under Section 501(c)(3) of the Internal Revenue Code, a university's activities must be charitable, educational, or scientific. The Internal Revenue Service has not defined a strict test to determine the quantity of unrelated activities that can be undertaken before jeopardizing exempt status. Loss of exemption, however, is not commonplace and considered unlikely if commercial business activities are insubstantial relative to exempt activities. Because intermediate sanctions have been developed to punish certain inappropriate activities

Table 2: License Arrangements

	Technology A	Technology B	Technology C
Initiation fee	US$100,000	US$50,000	US$10,000
Royalty	6%	3%	1.5%
Minimum royalty	US$100,000	US$50,000	None
Milestone payments			
- Investigational New Drug (IND) filing	US$250,000	US$50,000	US$50,000
- Phase 2 clinical trial	US$500,000	US$250,000	US$100,000
- Filing of New Drug Application (NDA)	US$2,000,000	US$1,000,000	US$500,000
- Drug registration/licensure	US$10,000,000	US$5,000,000	US$1,000,000

Table 3: Equity Division

Shareholder	Founders' equity	% class	Common stock	% class	Total issued and outstanding	% total
University	2,000,000	25%		0%	2,000,000	24.1%
Inventor	2,000,000	25%		0%	2,000,000	24.1%
Scientist	2,000,000	25%		0%	2,000,000	24.1%
CEO	2,000,000	25%		0%	2,000,000	24.1%
Technology A		0%	200,000	67%	200,000	2.4%
Technology B		0%	100,000	33%	100,000	1.2%
Totals	8,000,000	100%	300,000	100%	8,300,000	100%

Table 4: Stock Distribution

Shareholder	Founders Equity	% Class	Option Pool	% Class	Common Stock	% Class	Series A Preferred	% Class	Total Issued and Outstanding	% Total
University	2,000,000	25%		0%	2,000,000	19%		0%	2,000,000	7.8%
Inventor	2,000,000	25%		0%	2,000,000	19%		0%	2,000,000	7.8%
Scientist	2,000,000	25%		0%	2,000,000	19%		0%	2,000,000	7.8%
CEO	2,000,000	25%		0%	2,000,000	19%		0%	2,000,000	7.8%
Option pool		0%	2,250,000	100%	2,250,000	21%		0%	2,250,000	8.8%
Technology A		0%		0%	200,000	2%		0%	200,000	0.8%
Technology B		0%		0%	100,000	1%		0%	100,000	0.4%
Lead investor		0%		0%	0	0%	7,000,000	47%	7,000,000	27.4%
Investor 2		0%		0%	0	0%	4,000,000	27%	4,000,000	15.7%
Investor 3		0%		0%	0	0%	4,000,000	27%	4,000,000	15.7%
Totals	8,000,000	100%	2,250,000	100%	10,550,000	100%	15,000,000	100%	25,550,000	100%

by nonprofit organizations, caution is advised when a university forms new business ventures. Technology transfer managers should carefully monitor the extent of the university's control over day-to-day activities of the for-profit entity to avoid a possible finding of private inurement or exposure to other liabilities.

10.2 *Accounting for income tax*

Income generated from business activities unrelated to an exempt organization's primary purpose, conducted regularly either directly or through other partnerships, may be subject to unrelated, business income tax (UBIT). There are important statutory exceptions from UBIT. Specifically, passive investment income is not generally taxed. Such income includes most of the major sources of financial remuneration universities would expect in their spinout activities, including:

- royalties
- dividends
- interest
- receipt or sale of stock
- exercise of stock options

But even passive income, if derived from an entity that is more than 50% controlled by the tax-exempt entity, may be taxed if the controlled entity claims the payment as a deduction in computing its own taxes.

Exempt status is not at risk if unrelated activities are insubstantial in relation to the overall exempt activities. Careful records must be maintained, however, to permit the identification of taxable and exempt income, as well as related expenses. The university needs to evaluate whether a passive revenue stream that is typically exempt from UBIT, such as royalties, may be tainted by other aspects of an agreement between the university and the licensee—and thus subject to UBIT. This could be the case, for example, if services are provided by the university to the licensee.

The impact of any new venture activities on university facilities that were constructed using tax-exempt bonds should also be investigated, so that these activities do not jeopardize the bonds' exemption. Generally, no more than 5% of the proceeds of tax-exempt bonds may be used for

FIGURE 3: STOCK DISTRIBUTION DURING THE FIRST ROUND OF FINANCING

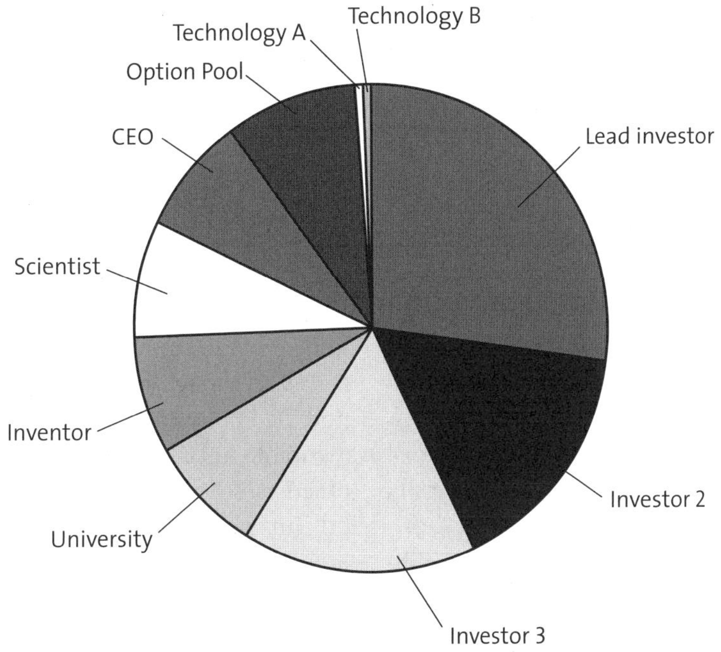

an unrelated trade or business. This test applies to the use of bond-financed facilities as well, though special exceptions may apply to the use of university research facilities for corporate-sponsored research.

10.3 *Exposure to liability*

Any time a person or organization participates in a commercial transaction with another party, the risk of injuring another party increases. The party injured by the tort may sue the wrongdoer for damages. Such injuries include nonperformance of provisions of a contract, or property damages or personal injuries caused by a faulty product. When individuals engage in business activities where they might be sued, they will most often form a corporation. Through the formation of a corporation, the shareholders are shielded by the corporate veil and granted limited liability, or insulation, from court-assessed damages that may result from the commission of a tort.

The use of the corporate form for new ventures probably maximizes the university's protection against such risks while it is actively engaged in commercialization efforts. As long as the university does not control the venture, either in terms of stock ownership or day-to-day management, the university will likely not be held liable for debt and liabilities incurred by the corporation in which it holds stock. Moreover, if it serves mainly as a passive investor, the university's tax status will not likely be jeopardized by the type or extent of business activities conducted by the corporation.

10.4 *Conflict of interest*

When a university interacts with external corporate ventures, the interests and commitments of the various parties involved—the university, individual faculty and staff, government, and industry—are complex and not necessarily aligned. These interests may conflict. A conflict of interest exists when an individual has sufficient external incentive and the opportunity to affect university activity.

Conflicts of interest may arise when an individual is involved in making a university's financial decisions regarding investments, loans, purchases or sales of goods or services, and accounting.

An individual's economic interest may be derived from:

- employment, independent contractor, or consulting relationships
- management positions, board memberships, and other fiduciary relationships with for-profit organizations
- ownership of stock or other securities and financial interests such as loans
- any other activity from which the individual receives or expects to receive remuneration

Such conflicts can arise naturally and do not necessarily imply wrongdoing on anyone's part. It is likely that the number of such conflicts will increase as universities expand their commercialization activities. When conflicts do arise, however, they must be recognized, disclosed, and either eliminated or properly managed.

10.5 *The university's public face*

Yale's Policy on Conflict of Interest and Conflict of Commitment states that Yale is committed to ensuring that its interactions with outside ventures are "*conducted properly and consistently with the principles of openness, trust, and free inquiry that are fundamental to the autonomy and well-being of a university and with the responsible management of the university's business.*"[2] Most universities have similar policies. As universities become more active in the commercial arena, occasions when the above policies might be violated will likely become more frequent.

A primary concern is that, whether violations be actual or perceived, the public could question the integrity of academic research and those conducting such research. For example, a faculty member might be involved in a new venture that brings to market a technology that is seriously flawed. Although the university may have done nothing improper in this case, it is visibly and inextricably linked to the inappropriate actions of others associated with it.

An additional conflict may arise between industry's desire to protect proprietary rights and the academic commitment to freedom of communication and publication of research results. Entwined with this issue are concerns about protecting the rights and interests of postdoctoral research associates and graduate students who may be involved in industry-supported research and whose interests may not be consistent with those of the faculty.

When such conflicts arise, they have the real potential to compromise the atmosphere of free inquiry that is vital for universities. Such conflicts must be promptly and properly addressed. Left unchecked, they may seriously damage not only the credibility of the individuals involved, but the university as well.

10.6 *Minimizing risk*

Although risks may arise, the threat, by itself, should not preclude a university's participation in venture formation. However, a university should establish procedures to identify and aggressively manage perceived risks. An active risk-management approach for new ventures makes a number of reasonable and prudent actions standard practice. These include:

- **Protecting the university's nonprofit status and avoiding intermediate sanctions.** Although not strictly required by the tax laws, a university should protect its ability to demonstrate that an investment is not an active trade or business. This is best done by limiting the equity interest in new ventures to a minority position and prohibiting active day-to-day involvement of university personnel in the venture's business activities. The university should carefully scrutinize any arrangements where private inurement or benefit might be found.
- **Accounting for tax consequences.** The university should limit its exposure to unrelated business income tax by remaining a minority shareholder in business ventures and relying primarily on the income derived from the passive, tax-exempt sources cited earlier.
- **Minimizing exposure to liability.** When creating new business ventures, the university should use the corporate form to maximize protection against the risks of product, tort, or contract liabilities.
- **Guarding against conflicts of interest/commitment.** According to most university conflict-of-interest policies, faculty are required to report annually on investments in, positions held at, and advisory or consulting relationships with any company in which the university holds license-derived stock or has a contractual relationship. This information often must be disclosed in any publication of research involving the company. These types of policies should be well-publicized and rigorously implemented.

  To help protect the university from securities law and conflict-of-interest problems resulting from the appearance of insider trading, the university should consider holding stock only until the stock is publicly traded and any trading restrictions are lifted, or until the company is acquired by a third party. University representatives on the boards of directors of spinout ventures should be prohibited from holding personal equity of any size. This prohibition should continue until the company goes public.

  Business relationships with new ventures, such as licensing or sponsored-research agreements, should be handled at arm's length. These relationships also should be permitted only after a review by an appropriate body determines that there are no perceived or real conflicts of interest.
- **Enhancing university image.** Any decision to participate in the formation of a new venture should always consider its likely impact on the university's image. The question, How would this look on the front page of the *Wall Street Journal*? should be on the minds of those university decision makers.

11. MANAGING THE PROCESS

In addition to these guiding principles, universities need to establish a management process to guide their technology transfer office's (TTO's) evaluation and management of these risks and opportunities. This review process will serve as a mechanism for dealing with issues surrounding the formation of new ventures and will help establish a formal mechanism for university officials to provide guidance on commercialization activities.

When the TTO is responsible for forming new ventures (for example, creating business concepts, recruiting management teams, and raising venture capital) the responsibility for *approving* formation and *reviewing* the status of new ventures should reside in another part of the university, such as the office of the provost. The oversight office would be best advised by a committee, which could include:

- university officers, such as vice presidents of finance and administration, and general counsel
- deputy provosts representing the major physical- and life-science research areas
- senior administrators from the relevant schools within the university

12. EQUITY MANAGEMENT

A university may receive equity in one of three ways: (1) in lieu of cash for a license to a technology, (2) for its activities in helping to found a new venture, and (3) in the case of some universities, for direct purchase of stock as a financial investor in a venture. Once a decision has been made to accept stock from a company, the university should have in place a set of policies and procedures for the management and disposition of the stock, particularly after it acquires value in public markets. Eventually, the university will want to sell some or all of its shares to generate cash, and the university should establish and publicly announce a policy for when and how it will accomplish this. Such a pronouncement avoids the potentially damaging impact on a newly publicly traded venture that may occur when the university begins to divest itself of its equity position (suggested guidelines and policies are provided in Boxes 2 and 3 at the end of this chapter).

13. CONCLUSIONS

Many technology licensing offices have begun taking a more strategic approach to commercializing IP assets. The approach has led some to focus more attention on the spinout of new ventures. Spinouts provide opportunities to receive royalty income and capital appreciation of a university's equity stake, and a university's involvement can be instrumental in deciding to locate facilities near the university. Such involvement in venture formation may, however, increase exposure to new and different risks. This should not preclude the university's participation, but the university should establish mechanisms devoted to identifying and aggressively managing them. ■

ALFRED (BUZ) BROWN, *Director, Office of Cooperative Research, Yale University School of Medicine, U.S.A.; Currently: Managing Director, BCM Ventures, Eleven Greenway Plaza, Suite 2900, Houston, Texas, 77046, U.S.A.* bbrown@bcmventures.com

JON SODERSTROM, *Managing Director, Yale University, Office of Cooperative Research, 433 Temple Street, New Haven, CT, 06511, U.S.A.* jon.soderstrom@yale.edu

---

1 This section is intended to be a brief overview of the types of risks to consider. Much of this material is adapted from an unpublished monograph titled *Trading Technology for Equity: A Guide to Participating in Spinout Companies, Joint Ventures, and Affiliates* by RM Goodman and LA Arnsbarger, attorneys with Morrison and Foerster LLP in Washington, D.C.

2 Yale University. 1995. Policy on Conflict of Interest and Conflict of Commitment. Memorandum from Provost Alison Richard to all faculty and principal investigators, August 1995.

Box 2: Suggested Guidelines for Acquiring Equity Holdings in New Ventures

1.1 If the university does decide to make cash investments in a spinout venture (outside of any venture capital funds in which the university investments office may have holdings), it is recommended that such direct financial-investment decisions be made at arm's length to avoid any perceived or real conflict of interest or commitment. Such investment decisions should be undertaken only as part of the investment office's normal investment activities, or as part of other special university initiatives. Decisions to invest in later rounds, however, should be made by personnel insulated from the management of the license-derived stock.

1.2 The equity position of the university should be a minority one, and subject to the same dilution as other shareholders, as the company raises additional capital.

1.3 Many universities, as an institution, retain the right to designate a representative, either as an observer or as a full voting member, to the board of directors of new ventures in which it holds equity.

1.3.1 If the university designates a board member, it is recommended that the representative resign from the board prior to the company's registration with the Securities and Exchange Commission for an initial public offering.

1.3.2 During the term of board participation, any fees or other forms of compensation accruing to the board member should be the property of the university and credited to the appropriate account.

1.3.3 If an individual is designated to serve on the board as a full voting member, he or she will require indemnification through the university or the venture's insurance policy to the extent permitted under state law.

1.4 Faculty and staff participation in new venture activity (whether by stock ownership, board membership, consulting agreement, or otherwise) should be governed by the university's policy on conflicts of interest and conflicts of commitment and must comply with that policy in all respects.

Box 3: Suggested General Policies for the Management of Equity in New Ventures

1.1 Stock acquired through the activities of the technology licensing office should be subject to the same policies and procedures as govern other equity holdings of the university.

1.2 If the stock is received in lieu of cash in consideration for a license, the stock will be treated as royalty income and distributed to inventors in a timely manner in accordance with the university's royalty-sharing policies. For the purposes of this distribution, the stock should be valued at the per-share value that it held when originally issued to the university. Following issuance of the stock to the inventors, it is then the sole responsibility of the inventors to manage their shares and to comply with any tax, legal, or contractual obligations associated with the distribution, ownership, or disposition of those shares.

1.3 Universities tend to follow one of two options in managing and disposing of stock held for the benefit of the university.

1.3.1 One option is to immediately transfer the shares to the university investment office to be managed in the same manner as other equity holdings in the endowment portfolio. Of course, all restrictions, such as any lock-up period where shares cannot be traded after an initial public offering, must still be observed. Because most universities maintain a legal wall between the investment office and the rest of the university, such a practice may help mitigate any perceived or real conflicts of interest. There are some potential difficulties with this approach, including the investment office's lack of knowledge and/or expertise in managing individual shares in private ventures, establishing a value for the shares at the time of transfer, and accounting for the value if the shares are not immediately liquidated.

1.3.2 An alternative approach is for the technology licensing office to hold and manage the shares until a public market exists for the shares (for example, after any restrictions on the sale of the shares has expired). When a public market exists, the shares could be transferred to the investments office in return for a transfer of funds to the appropriate income accounts equal to the value of the stock at the close of trading on the day of transfer. The investment office is then free to manage the orderly liquidation of the stock much as it would any other gift of stock to the university.

CHAPTER 13.2

Dealing with Spinout Companies

JON C. SANDELIN, *Senior Associate Emeritus, Office of Technology Licensing, Stanford University, U.S.A.*

ABSTRACT

This chapter provides a practical guide for organizations seeking to transfer their intellectual property (IP) rights to a spinout company (normally through a licensing agreement) so that the company can convert the IP into products or services that benefit the public. Based on experiences at Stanford University over the past three decades, key issues have been identified for negotiating transfer to a spinout, and guidance on best practices for reaching a successful agreement is provided. The chapter briefly reviews potential conflict-of-interest and conflict-of-commitment issues that inevitability arise when employees of public research organizations become involved in spinout companies.

1. INTRODUCTION

Public Research Organizations (PROs) often create spinout companies to commercially develop and market the PRO's inventions. The new company may be formed by PRO faculty, staff, and/or students, by entrepreneurs not affiliated with the PRO, or by a combination of these parties. In almost all cases, investors in the new company desire a relationship with the inventors of the licensed technology. The investors recognize that the know-how, "show-how," and detailed knowledge of the technology possessed by the inventors will be important to the company's success.

The technology transfer office (TTO) has an important role to play in this process, one that can take many forms. The TTO must be clear about what roles it will or will not play in the formation of new companies that utilize PRO technology and/or PRO employees. The most common model for U.S. TTOs is passive involvement. Referrals are provided to resources that can assist in the spinout process, but the TTO itself is not actively involved. Active involvement does occur when the TTO engages in some, or all, of the following activities: writing or help in writing the business plan, assisting with incorporation of the company, finding initial seed funding, recruiting a management team, and securing the first-round venture funding. Such active involvement can be very time consuming and normally requires people with special skills and experience.

Spinout companies are frequently formed because spinouts are the only alternative available for converting a technology into useful products or services. Of course, it is the *products and services* stemming from new technology that improve our health and standard of living—*not* the technology itself. Often, however, inventions are undeveloped and unproven, and established companies are unwilling to commit resources to license and develop the technologies. The inventors, on the other hand, may believe strongly in the social value of the inventions, and so will assume risk and make deep commitments to foster an invention's further development into products. The inventors often do so by getting involved in spinout companies.

Sandelin JC. 2007. Dealing with Spinout Companies. In *Intellectual Property Management in Health and Agricultural Innovation: A Handbook of Best Practices* (eds. A Krattiger, RT Mahoney, L Nelsen, et al.). MIHR: Oxford, U.K., and PIPRA: Davis, U.S.A. Available online at www.ipHandbook.org.

The AUTM (Association of University Technology Managers) surveys show that in recent years, 5% to 10% of licenses annually granted by U.S. universities are granted to spinout companies. In 2003, U.S. universities reported 374 licenses to spinout companies, or about 7.5% of the total licenses granted. Sold equity totaled US$39 million, which was about 3% of total royalty income in 2003.

Over the past 15 years, Stanford University has taken equity as part of its licensing agreements with 140 spinout companies. As of 2005, Stanford holds equity in 85 companies. Fourteen percent of the companies in which Stanford has taken equity have failed, making the equity worthless. For 18% of the companies, equity has been sold. Two companies generated more than 80% of the total amount of cashed-in equity (US$22.5 million). Spinout companies have paid earned royalty income and annual minimum payments, but no data exists for these categories. As is true for licensing in general, when licensing and supporting spinouts, the focus should *not* be on how much income can be generated, but on the value flowing from a new partnering relationship (for example, consulting opportunities for professors, sponsorship of research, hiring of graduating students, and donations and gifts of equipment) and on the public benefits from the products and services the spinout may produce. Spinout companies can be a significant source of new jobs and of local, state, and federal taxes. They can produce exports. A few spinouts (for example, Hewlett-Packard in Silicon Valley) have grown into major corporations that are regional anchors, attracting entrepreneurs and other companies.

2. EVALUATING THE ENVIRONMENT

The role the TTO plays with spinout companies will be strongly influenced by the general attitude of the PRO's senior administration and members of the governing board toward spinouts. These individuals can be encouraging, supporting, merely tolerating, or discouraging. One can see why in some cases their views may be less than positive. The involvement of PRO personnel with spinouts can create conflicts of interest, and valued faculty members who take a leave of absence to work in a spinout may not return. Moreover, leaves of absence require changes in teaching assignments and graduate-student supervising. If leaves are not taken, the commitment of faculty members to spinouts may lead faculty members to neglect teaching or research responsibilities (such conflict issues are covered in detail later in this chapter). Clearly, concerns of senior administration and board members about spinouts, involving PRO personnel, can be legitimate.

Almost all PROs in the United States at least tolerate spinouts, and the trend in recent years is toward greater acceptance of spinouts. Most faculty who are actively involved with spinouts speak positively about their experiences. If these individuals obtain significant wealth, usually through stock options, they serve as role models for others. Experience working with a spinout can also *enhance* faculty performance at the university. John Hennessy, the president of Stanford University, took a one-year leave of absence in the 1980s to be involved with a spinout named MIPS. He openly reports that the experience with MIPS was extremely valuable and useful for managing his teaching and research activities after returning to Stanford.

3. NEGOTIATING A LICENSE AGREEMENT

The TTO's first involvement with a spinout is usually to provide a license to the technology that the company plans to convert into commercial products or services. In most cases, the licensed technology will be the company's fundamental technology, and so the company will request an exclusive license. Investors want to be assured that their investments will be protected by patents or other intellectual property. In the license agreement itself, investors will normally focus on:

- the length of the exclusive period
- field-of-use limitations
- improvement inventions
- agreement assignment provisions
- financial terms

Investors almost always request a life-of-patent exclusive period. This is to be expected. In the

United States, because a large percentage of inventions is generated through research supported by the federal government, the policy is to limit the exclusive period. In the initial Bayh-Dole Act, the U.S. government specified that the exclusive period would end either at five years from first product sale or eight years from the effective date of the license agreement, whichever came first. Although this requirement was later eliminated, it is still used as a guideline by many U.S. TTOs. In the United States, government guidelines are that the term of the exclusive period should be the shorter of eight years from the effective date of the license agreement or five years from the date of the first sale of the licensed product. Experience has shown that in most cases, a period of five years from the first licensed product sale allows a fair return. However, if the company can provide convincing evidence that a longer period would be needed in the company's situation, such evidence would be evaluated and considered. If such evidence were not available at the time of licensing, but might appear at a later time, the new evidence could eventually justify extending the exclusive term.

Investors almost always prefer no limitations in the license. And if the TTO insists on a defined field of use, the investors will want a limitation as small as possible. Sometimes a compromise allows a grant of exclusive right for a specific field of use but permits access to other fields of use. Such an arrangement could be made by granting a nonexclusive right to other fields of use, or by specifying a right to add other fields at a later time, but with a requirement for a business plan, added payments, and appropriate diligence terms for licensed product development in the added fields.

Investors will also prefer to be automatically added to license-improvement patents that may emerge from continuing research in the area of the licensed technology. If the improvement has been described in the specification of the licensed patent, and the original invention and the improvement have common inventors, then the improvement could be filed as a continuation-in-part (CIP) application. In such cases CIPs would normally be part of the definition of licensed patent(s). During the exclusive period, no one else could practice the improvement patent without rights to the dominant licensed patent, so the improvement patent has no value to the PRO. To add improvement patents that are not CIPs under the license agreement, the recommended policy is to do this only with the express written consent of the potential inventors.

Experience has shown that the most common exit pathway for PRO-based spinout companies is merger and acquisition. Very few reach an initial public offering (IPO). Thus, the ability to assign the license rights to the merging or acquiring party can be very important. The options for the TTO are: (1) no assignment without the written permission of the TTO, (2) automatic assignment to a party, of all of the assets of the licensee, without an added fee, or (3) automatic assignment to a party, of all of the assets of the licensee, *with* an added fee. The typical approach is to combine (1) and (3), so an assignment that is not part of a merger or acquisition requires written approval, and an assignment that *is* part of a merger or acquisition is automatic but requires payment of a negotiated amount.

Spinouts must carefully manage their available cash; for license fees, the spinout will prefer to trade equity for cash. Although fully paid licenses for equity are sometimes written, they are rare, and usually normal financial terms apply. The cash license fee is kept low (but usually not to zero), with equity taken as a substitute. The annual fee may start low and then increase over time. The earned royalty is targeted at what would be normal for the technology; however, in some circumstances, the spinout must also license from others to have all the rights needed to create a licensed product. In such circumstances, each of the licensing parties is asked for a reduction, so that the total earned royalty rate is reasonable. And with patent cost reimbursement, the payments are sometimes delayed until a certain funding level for the spinout is reached.

How much equity should the PRO receive for the technology license? This is a challenging question. Certainly the amount of equity to the PRO should not be so great that insufficient equity remains for successfully developing the business. Equity will be needed to secure funding and to attract the best available people. Some entrepreneurs have proposed that the amount of founding equity for the technology should range from 1% to 10%. If the technology is an unproven idea, then 1% would apply. If the technology is essentially ready for market, then 10% would apply. Following this rule, most PRO technology, which is in the earliest stage of development, would fall within the 2% to 4% range. However, the specific situation may include other factors that affect how much equity is reasonable.

Another issue is whether the percentage ownership of the PRO should remain the same through subsequent funding rounds by antidilution clauses. Investors will not want the PRO to get an increasing number of shares at no cost at each funding round. This is reasonable. However, most will agree to some antidilution provision, such as nondilution through the initial venture round (usually called funding round A), or antidilution until the company reaches a certain valuation.

Investors will also be concerned about diligence terms, which require the spinout company to reach certain milestones or face the TTO's termination of the agreement. Any clause that permits the TTO to terminate the agreement is cause for investor concern, but such diligence terms for the spinout are important because they ensure that the company does not become what John Preston (former director of the TTO of the Massachusetts Institute of Technology) refers to as the "living dead." In such cases, the spinout company never grows beyond a few employees and never progresses beyond the product development phase, or only manages to sell small quantities of licensed product, mostly for evaluation purposes. The intent of the diligence terms (reaching specified funding levels, having production facilities, and reaching certain sales volumes by agreed-to dates) is to ensure that the spinout doesn't lose its viability.

Other sections of a license agreement that are typically discussed during negotiation are:

- *Definitions*, in which key words are defined
- *Infringement provisions*, in which the respective responsibilities of the parties are defined in the event that infringement by a third party of licensed patents is detected
- *Sublicensing*, in which the parameters for sublicensing (including sharing of sublicensing income) are defined
- *Warranties and indemnities*, in which the provisions for protection of the university are defined.

Definitions will normally be the first section of a license agreement. In this section, key words used in the agreement are defined. What is meant by "Licensed Products," "Licensed Patents," and "Licensed Field of Use" is extremely important. A definition should be clearly written so both parties fully understand the meaning; any possible future dispute over the meaning of a key term should be avoided utterly. It is therefore worth investing time to ensure that definitions are clear and unambiguous. Sometimes giving an example will make a definition more understandable. As is true with any of the agreement terms, if a person is presented with a definition that he or she does not fully understand (for example, it contains unfamiliar, legal wording), then the person can either rewrite it to reflect his or her understanding or ask the potential licensee to reword it so that it is understandable.

The *Infringement provisions* section describes what actions will be taken if infringement of the licensed patent(s) by a third party is detected. In the United States, infringement litigation is very expensive; if carried through to trial it can amount to many millions of dollars. Thus, the license agreement should not require the university licensor to pursue litigation for any reason, and certainly not for an infringement. The most common approach to settling accusations of infringement is for the parties to review the evidence of infringement and then decide how to proceed. The most desired outcome is a solution that does not involve litigation. The university may be able to use its influence to find such a solution—most companies wish to maintain good relationships with universities, so they will usually also seek a satisfactory solution. However, if it appears that litigation is the only possible course of action, then the licensee and the university can agree to pursue the litigation jointly (and share both costs and awards), or if one party does not wish to join, the other party can pursue the litigation. The nonjoining party will provide reasonable support as requested, but the litigating party would pay all costs and retain any awards that might result.

The Sublicensing section describes how the licensee may grant another party the right to make and sell licensed products under the third party's brand name. Sublicensing does not apply to situations where the licensee is having components for a licensed product manufactured by others or where the licensee is using a distributor or other party to sell licensed products. A sublicensing provision is only included in an exclusive license. For a nonexclusive situation, the TTO will grant further licenses to the licensed patent(s). The main issue in the sublicensing provision is how the sublicensing income will be shared. At the time the license is signed, the most common approach is to share sublicensing income equally. In practice, sublicensing is very rare, but if it does occur, it will occur well after the licensee has been selling licensed products. Typically many years will have passed since the license was signed and the 50/50 sharing will probably have been renegotiated. The sublicense, at the time of issue, would almost certainly include patents, know-how, and perhaps even training from the company issuing the sublicense. To be fair, the TTO should agree to compare the relative value of the original licensed patent(s) to what the company is adding under the sublicense to determine a fair distribution of sublicensing income.

Warranties and *indemnities* are provisions that protect the university. This is one area in which attorneys are necessary and legal terminology may be required. If any significant changes to these provisions in the template agreement are requested during negotiations, the technology transfer officer should stress that making any changes is very difficult and will need to be approved by the university's attorneys. In most cases, university attorneys will not approve significant changes. Companies will usually complain that these provisions are too one-sided in favor of the university, but without such provisions, the risks to the university would be so great that licensing would not be possible. Given that the parties are partners and not competitors, and that both have strong motivations to maintain a good relationship, disputes can often be resolved through discussion. Thus, the provisions in the warranties and indemnities section of the agreement are very rarely, if ever, invoked.

4. CONFLICT OF INTEREST AND COMMITMENT

Conflict of interest and conflict of commitment are serious concerns for the PRO. The presidents and members of the governing boards of PROs are charged with maintaining and protecting the reputations of their institutions. These individuals worry about any type of activity or situation that could reflect badly on the integrity of the PRO, because a loss of public trust would have serious negative consequences, including lost gifts, donations, and funding from potential research sponsors. So it is not surprising that considerable attention is given to identifying and managing COI (a conflict resulting from a financial interest held by a person employed by the PRO) and COC (a conflict whereby the commitments of the PRO employee to the institution are adversely affected).

Conflicts can result in:

- loss of public trust in both the PRO and/or an individual connected to the PRO
- unfulfilled commitments to research sponsors, students, and/or to general PRO responsibilities
- bias, when reporting research results or not reporting research findings at all
- exploiting the work of graduate students
- adverse and embarrassing reports in the media

Some potential outcomes due to conflict situations include:

- research directions and priorities moving toward company interests
- restrictions on the distribution of research results

- pipelining of research results and related IP to a particular company
- inappropriate access by a company or individual to PRO facilities

Most PROs recognize that conflict situations are unavoidable in the current environment. If the PRO is to contribute to the public good, the PRO must enter into relationships in which conflicts can arise. Governments worldwide are looking more and more to PROs to contribute to economic development and growth, and legislation similar to Bayh-Dole is appearing all over the world. PROs therefore are creating "early warning systems" to identify when a potential conflict situation is developing. Attention can then be directed to the situation to ensure it does not evolve into an actual conflict with negative results. A conflict situation in itself may not be bad, and in fact it may allow important benefits to flow to the individual and/or the PRO. But the conflict-management system of the PRO must review and monitor conflict situations to avoid negative outcomes.

To manage conflict situations, many PROs implement an annual survey of all faculty members. The faculty person lists all outside interests of himself or herself and his or her spouse (if any) that could create conflicts. The information is reviewed by the PRO administration, and any areas of concern are discussed with the faculty member.

Most PROs have developed COI and COC policy statements that identify specific situations requiring an ad hoc conflict review. At Stanford University, if an employee (for example, a professor) is to be involved with a spinout company that has applied for or been granted a license from the PRO, then an ad hoc conflict review would be required.[1, 2]

Box 1 sets out examples, involving conflicts of interest and commitment, that may clarify some of the issues PROs may confront.

5. CONCLUSIONS

A spinout company may be the best, or perhaps the only, alternative by which newly discovered technology is converted into products or services for public benefit. Governments everywhere have, or are creating, policies and laws to encourage spinouts based on IP rights from PROs. Successful spinouts create new jobs and contribute to economic development, and they have the potential to grow into large multinational corporations. Thus, creating an environment that nurtures and encourages the formation of spinout companies is a reasonable goal of all regional economies. The role of the TTO in such an environment can take many forms. The TTO must evaluate the environment in which it exists and determine what role it will play in the formation of the spinout company. One fundamental role is to provide the licensing agreement that will allow the spinout to seek funding from potential investors. In doing so, the TTO must balance the interests of the PRO it represents with those of the spinout, as well as with the interests of society. The TTO also must recognize potential damaging conflict situations and participate in developing and implementing policies and procedures to avoid or minimize them. ■

JON C. SANDELIN, *Senior Associate Emeritus, Office of Technology Licensing, Stanford University, 1705 El Camino Real, Palo Alto, CA, 94036, U.S.A.* *jon.sandelin@stanford.edu, sandelin@stanford.edu*

1 See Stanford University's policies on faculty conflicts of commitment and interest at www.stanford.edu/dept/DoR/rph/4-1.html.

2 Other sources of COI guidelines include: (1) the October 2001 *Report on Individual and Institutional Financial Conflict of Interest* published by the Association of American Universities (AAU), (2) the June 2003 *Recommendation of the Council on Guidelines for Managing Conflict of Interest in the Public Service* published by the Organisation for Economic Co-operation and Development (OECD), and (3) the 2004 *Approaches to Developing an Institutional Conflict of Interest Policy* published by the Council on Government Relations (COGR).

Box 1: Examples Involving Conflict of Interest and Conflict of Commitment

Example 1:

This example is from the first of a series of symposia held at Stanford University in 1982 titled Universities, Industries, and Graduate Education (reported by Lee Randolph Bean in the October 1982 *Hastings Center Report*). Stanford's then-president, Donald Kennedy, presented this example to illustrate the problems that arise as faculty members move from the role of teacher/investigator to that of entrepreneur. Although more than 20 years old, the example is as relevant today as it was then.

Dr. X and his graduate students work on a basic molecular biology project. Dr. X is a consultant and shareholder in Clotech, Inc., which has built a scaled-up facility for producing and testing a useful protein that is the primary gene product from a plasmid Dr. X first got from bacteria cells. Stanford, which has an assignment to the patent on the product, is now considering offers to invest in Clotech, and plans to offer an exclusive license to Clotech for a related process for which Stanford holds patent rights. Meanwhile, Mr. Y, a graduate student who is good at purifying the protein, has complained to the university ombudsman that Dr. X is using every means at his disposal to induce Mr. Y to accept outside employment with Clotech.

The issues Kennedy wished to bring forward for discussion at the symposia were:

Conflict of interest. Is Professor X devoting undue time and effort to Clotech because of his profitable consulting and equity arrangements, to the neglect of his teaching responsibilities? Do his outside ties create competing loyalties between Stanford and Clotech?

Secrecy. Has Dr. X kept past research results to himself, because his colleague, Dr. Z, works for a competitor company? Did Clotech ask Dr. X to delay publication of his work in order to secure an exclusive license from Stanford? [Author's comment: Should Stanford have marketed the license to the patent(s) to others to determine if another party, perhaps one better qualified, would develop licensed products? Or should Stanford seriously consider offering nonexclusive licenses to all interested parties?]

Patents. Should scientific knowledge be owned and traded for profit? Should the university share in that ownership?

Research priorities. Does Dr. X's involvement in a commercial production facility indicate a shift in his focus from basic to applied research? Will the future direction of scientific research be skewed to respond to the needs of private industry?

Graduate students. Have Mr. Y's time and talents been exploited for the gain of his advisor's company?

Public perception. Will extensive ties to the private sector erode public confidence in the detachment and trustworthiness of university research?

Scientific norms. The open and free sharing of information and a disinterested approach to research that puts the advancement of science first are norms that have traditionally governed science, according to sociologist Robert Merton. Are those norms disintegrating as the pull for commercial application of research and consequent profits intensifies?

(Continued on Next Page)

Box 1 (continued)

Example 2:
This illustration and the following one were created by the author and are based on experiences at Stanford University.

Clotech has expanded and upgraded the scale-up facility to the point that it will now permit Mr. Y to run experiments in pursuit of his Ph.D.-, qualifying research work that he cannot do with the facilities in Dr. X's lab. Mr. Y's research is fully funded under a U.S. government grant. Clotech is willing to make its facilities available for the research project of Mr. Y, as the company realizes such work will be very relevant to their product plans. Clotech has requested a right to help guide the research work of Mr. Y and also requested a document signed by the university stating that any IP created by Mr. Y resulting from the use of their facilities will be owned by Clotech. Dr. X is encouraging Mr. Y to utilize Clotech's facilities in his research, and is urging the university to accept the requests of Clotech. Clotech has indicated that it would be willing to hire Mr. Y as a paid consultant, as long as he follows the guidance of Clotech in his research, and that any IP created from the research would be owned by Clotech. Dr. X is supportive of Mr. Y being a paid consultant for Clotech under these terms.

Ms. Z in the Office of the Dean of Research has been asked to review the situation and inform Dr. X and Clotech as to what the university's policies will allow in this case. After a careful review, including discussions with Dr. X and Mr. Y, her response is as follows:

- Any IP created by Mr. Y that is related to his research program for his Ph.D. degree, as specified under the work statement in the government grant that is funding Mr. Y's research, will be owned by the university. This is regardless of where and with what facilities Mr. Y conducts such research.
- Mr. Y cannot be a paid consultant for research work that is also funded by the government.
- A designated professor in the department of Dr. X will become a co-advisor for Mr. Y and will be charged with ensuring the research work of Mr. Y is in full compliance with progress toward his Ph.D. degree.
- A collaboration agreement will be negotiated between the university and Clotech that will spell out clearly the terms of the proposed collaboration, including university ownership of IP created by Mr. Y and the right of Mr. Y to freely publish, at any time, the results of his research.
- A meeting will be held with Dr. X and the dean of research to discuss the situation and to ensure Dr. X understands that the university would not allow, under any circumstances, an outside company to direct the research of a graduate student and that ownership of any IP created by a graduate student, as part of his funded research work will be owned by the university.

Example 3:
Professor A in the university's ophthalmology department, a renowned eye surgeon, disclosed an invention four years ago to the technology licensing office. This invention holds great promise for eye surgery. A patent, assigned to the university, has issued. The patent is exclusively licensed to the spinout company EyeCare, Inc., to which Professor A is both a consultant and the chair of the Scientific Advisory Board. Professor A has been given 100,000 shares of the company stock for her services. The university received 200,000 shares of stock as partial compensation for the exclusive license. In addition, EyeCare has sponsored research in Professor A's lab for the past three years (ever since the company was formed). When EyeCare first proposed supporting the research of Professor A, the university established an oversight panel to review research proposals and results, as well as the involvement of graduate students with the company, and to advise Professor A of potential conflict situations.

(Continued on Next Page)

Box 1 (CONTINUED)

Because of this sponsorship, EyeCare has exercised its right to exclusively license three improvement patents resulting from the research. A separate conflict review was required before the exclusive license could be granted. The university licensing office submitted a report on its marketing the invention to other parties, and a statement that EyeCare is the best alternative for commercialization of the invention, in a timely manner. This conflict review very carefully evaluated how the relationship with EyeCare might impact the graduate students conducting research in Professor A's lab, as the potential for altering the work of students to benefit the company was a major concern.

The invention licensed to EyeCare has now reached the stage where clinical studies, with human subjects, will be required to obtain government approval to sell the medical device in the United States. The lab of Professor A is clearly the best source for coordinating such trails, with Professor A and her colleagues performing the procedures. However, the relationship of Professor A with EyeCare, through which she could profit handsomely if the clinical trails are successful, is a cause of great concern. The university must therefore carefully review the situation in order to determine if it will conduct the trails or not, and if it will permit conducting the trials, what level of oversight and controls will it exercise.

The university, following a review, decides to conduct the trials with the following oversight conditions:

- Professor A must sell all her shares in EyeCare and agree not to acquire any shares in the future, including options to acquire shares.
- The university will sell all its shares in EyeCare and agree not to acquire any shares in the future, including options to acquire shares.
- Professor A will participate in the clinical trails, but will not be the principal investigator for the trials.
- An oversight committee will be formed that will review the results from the trials and any publications related to the trials. The committee will include Professor B, a respected eye surgeon from another university medical center.
- Professor A will fully disclose her relationship with EyeCare in any publications or presentations related to any research connected to EyeCare.
- Professor A's relationship to EyeCare must be fully disclosed and explained on the "informed consent" agreement signed by every human subject participating in the trials.

What the Public Sector Should Know about Venture Capital

ROGER WYSE, *Managing Director and General Partner, Burrill & Company, U.S.A.*

ABSTRACT

Ready access to venture capital investments is vital to the success of start-up companies in the capital intensive high-technology sectors such as biotechnology. But there is a common misconception that an abundance of venture capital will spawn the formation of new companies. In fact, the opposite is true: new companies actually attract venture capital. This chapter provides an overview of the venture capital system, explains its importance, and identifies what qualities of a company make it attractive to venture capital investors. Some of the factors can be influenced by government action, so the chapter offers several ways that governments can encourage venture capital investment.

1. INTRODUCTION

Commercialization of biotechnology research is a long, expensive process that requires highly trained staff, sophisticated laboratory facilities, and costly regulatory approvals. A growing amount of this work is done by small companies. They are the primary source of innovation in biotechnology and are performing an ever-increasing share of total U.S. R&D. According to data from the National Science Foundation, the value of small - company R&D rose to US$40 billion, accounting for 20.7% of the value of all private sector R&D. These small start-up companies rely on venture capital investment to fund their R&D activities.

As pharmaceutical and agriculture companies merge and become larger, they increasingly focus on development and marketing, and lose their agility and ability to innovate. Thus, large companies increasingly gain access to the innovations of small companies through licensing agreements, R&D partnerships, and acquisitions.

Prior to the 1980s, most agricultural innovation in the U.S. originated at land-grant universities; there were very few small start-up companies. Innovation was offered directly to farmers and to large agriculture companies via products and license agreements. Then with the onset of the go-go genomics era in the late 1990's agriculture went through two major restructuring cycles. The first cycle was based on the premise that understanding of life processes at the molecular level could be leveraged across agriculture and pharmaceuticals. So-called life science companies were formed. Small agriculture biotechnology (agri-biotech) companies were started based on new genetic technologies; these small companies were quickly acquired by larger companies as they raced to converted into life sciences companies through the acquisition of genomics technologies and germplasm.

However, these large life science companies soon discovered the complexities inherent in managing business units with very different cost structures, market sizes, margins, and regulatory paths. Within two to three years, therefore, the large companies spun off freestanding

Wyse R. 2007. What the Public Sector Should Know about Venture Capital. In *Intellectual Property Management in Health and Agricultural Innovation: A Handbook of Best Practices* (eds. A Krattiger, RT Mahoney, L Nelsen, et al.). MIHR: Oxford, U.K., and PIPRA: Davis, U.S.A. Available online at www.ipHandbook.org.

pharmaceutical and agriculture companies. These rapid cycles of restructuring negatively affected small companies, because very few partnerships and acquisitions took place between 1998 and 2004. Fortunately, the trend now seems to be reversing and large agri-biotech companies are again acquiring innovation from small companies, particularly in an era when agriculture increasingly includes food production and biomass for fuels and materials. The ongoing challenge now is to create an environment that encourages entrepreneurship, the formation of small innovative companies and venture capital investment.

2. WHAT IS VENTURE CAPITAL?

Venture capital (VC) is high-risk capital that is invested in early-stage companies. It is not a loan; it is an equity investment, with the investor owning shares of the company. Venture capital companies invest in high-growth, early-stage private companies when the technology risk is still high and, if successful, potential financial returns are also high. The VC is managed by companies with deep expertise in the sector and with experience in forming and nurturing start-up companies. Venture capitalists are not only a critical source of funding; they are also actively involved in helping to manage and develop small companies.

Some venture companies, called *seed stage funds*, focus on very early-stage companies. These funds are generally small, ranging in size from US$10–50 million. They will usually invest US$250,000–3 million in a single company. *Growth stage funds* are larger, possessing US$75 million–1 billion. They invest in later-stage companies where investments of US$10–20 million are common.

VC companies raise money from institutional investors, corporations, pension funds, government agencies, and private individuals with high net worth. Most funds last for ten years. In the initial three- or four-year period, a fund typically invests money in a portfolio of 15 to 20 companies.

Investors get a return on their investments only when portfolio companies are either sold via a trade sale or participate in an initial public offering (IPO), usually three to five years after the initial investment. At that point, the investors are repaid their initial investment and any profits are split 80:20 between investors and the venture company. In general, venture capital companies can expect to achieve a return of 20–40% IRR (internal rate of return) over the life of a fund.

3. WHY IS VENTURE CAPITAL IMPORTANT?

The capital that drives the biotechnology industry comes from many sources, but mostly from R&D and marketing partnerships between small and large companies. In 2005, US$34 billion was invested in U.S. biotech companies from all sources (Table 1). This amount was already exceeded by the end of the first three quarters of 2006. In 2005, approximately US$4 billion in investment capital came from venture capital. Over half of the total annual investment from all sources came from R&D partnerships established between large and small companies.

Venture-backed small companies also create new jobs, generate wealth, and contribute to economic growth. Historically, 80% of new jobs in the United States are created by companies with fewer than 500 employees, many of which are venture financed. Between 1970 and 2003, venture-backed companies accounted for 10.1 million new jobs in the United States and US$1.8 trillion in revenues.

The impact of venture-backed small companies on local and national economies is most dramatic when two conditions are present: an entrepreneurial culture and a critical mass of small companies that attract venture investments. Most venture capital companies are located in the United States, and most venture backed U.S. companies are found in California (in the San Francisco Bay area and San Diego), Boston, and along the Atlantic seaboard. Only six states in the United States account for nearly 75% of all venture capital invested in all sectors (Table 2).

Venture capital is a vital element in establishing a biotechnology industry but it is very difficult to accomplish. Few geographic locations have

Table 1: Sources of Capital in the Biotech Industry

Sources of capital		Total investments (US$, millions) 2005	2006 (1st Q to 3rd Q)
Public	IPO[a]	819	567
	Follow-ons[b]	4,194	3,032
	PIPES[c]	2,376	1,817
	Debt	5,565	12,241
Private (Venture capital)		3,518	3,186
Other		1,114	303
TOTAL CAPITAL		17,586	21,146
Partnering		17,268 (50%)	12,463 (37%)
TOTAL		34,854	33,609

a IPO – initial public offering: a private company files to have a portion of its shares sold to the public on a regulated stock exchange, such as NASDAQ.

b. Follow-ons – When public companies sell additional shares on the stock exchange to raise additional cash.

c. PIPES – Private investments in public entities: the sale of public shares to private financial institutions that may take public shares off the public market as a way for companies to raise cash.

Table 2: Investment of Venture Capital by State

State	Percent of total U.S. venture capital
California	47.5%
Massachusetts	10.3%
New York	5.2%
Texas	4.7%
New Jersey	4.2%
Colorado	3.0%
Total of top six states	74.9.%

been successful. Seventy five percent of all venture capital in the world is in the United States and about 75% of that is in six states. However, the fundamentals for success are clear; the formation of new companies operating in an environment that increases the probability for success.

4. WHAT ATTRACTS VENTURE CAPITAL?

4.1 *The formation of companies with attributes for success*

4.1.1 *A strong management team*

Early-stage companies are high-risk investments: they will always run into problems and they will always be short of capital. Therefore, it is vitally important to have a management team that can solve problems quickly and use limited capital efficiently to create real value.

4.1.2 *Viable technology*

Small companies should be founded on scientific research published in peer-reviewed publications; however, many companies are started well before true proof of concept is demonstrated. Indeed, venture capitalists usually decide whether or not to invest in a company based on the quality of the science it does or plans to do. Venture capitalists will mitigate their own risk by offering funding in stages, investing more money as the company passes each technological milestone.

4.2 *IP ownership and freedom to operate*

The value of a biotechnology company is based on the amount of intellectual property (IP) it can acquire, develop, and protect—and on the potential market served and not on current revenues. Therefore, companies must acquire a strong IP position and have a good patent strategy. The company should ideally be based in a country with strong patent laws.

Patents are only valuable, however, if the company also has *freedom to operate*: that is, the ability to use the patented technology without having to rely on other technologies to which it does not own IP rights.

4.3 *A large potential market*

Companies with products or technologies that have large markets are obviously more attractive to investors than those that have smaller markets, even though the cost of development of a small-market technology is usually about the same as that of a large-market technology.

4.4 *A favorable entrepreneurial environment*

Companies within an entrepreneurial environment of "critical mass"—that is, an environment that has a sufficient number of similar companies and therefore a critically large pool of talent—are more attractive to investors than companies outside of such environments. This is true for several reasons. First, when there are a number of small companies in the same area, CEOs can share ideas and develop solutions with each other. Should one company fail, employees can easily move to other companies, and there is enough management talent in the area to fill the needs of the companies. The area also likely supports a large number of attorneys and accountants who are familiar with the issues of small companies.

Venture capitalists never fully fund an investment alone. They almost always syndicate the investment with other local companies, particularly those that have large funds. The presence of venture capitalists makes syndication easier. Venture capitalists who are not locally based will want to partner with other venture capitalists who are local, especially when investing in early-stage companies.

5. WHAT ENVIRONMENTS ATTRACT START-UP COMPANIES?

5.1 *An encouraging business culture*

The ideal business culture rewards success, sees failure as a learning experience, and strongly believes that technology and innovation are the drivers of economic growth and wealth creation.

Indeed, success breeds success. The presence of a few local heroes who have taken risks and built successful companies encourages entrepreneurs to start companies and to stay the course when problems arise, as they always do.

Finally, already-existing networks of experienced CEOs/managers can help lead new companies or provide mentoring to young CEOs.

5.1.2 *Access to intellectual capital*

Successful biotechnology clusters are fed by the intellectual capital flowing from great research universities. Such clusters are found in Boston (M.I.T. and Harvard University), the San Francisco Bay Area (Stanford Unviersity, U. C. Berkeley, and U. C. San Francisco), and the United Kingdom (the University of Oxford and the University of Cambridge).

5.3 *Access to financial capital*

Financial capital includes funding for peer-reviewed research; seed capital, usually put up by angel investors (wealthy individuals); and early-stage and growth capital, which is put up by venture investors.

5.4 *Other factors*

The area must also contain appropriate, readily available facilities, such as low-cost laboratories and offices. It should have a sufficient number of lawyers and accountants, and a low cost of living and high quality of life are added advantages.

6. VENTURE INVESTMENTS IN AGRI-BIOTECH

Health care biotechnology has a 40-year history of successful venture capital investment and experienced venture-capitalists and CEOs, and the products have well-known paths to market. However, venture capital investment in other sectors—such as agriculture and health & wellness, as well as the industrial application of biotechnology—is only just beginning.

Investing in agriculture is particularly challenging. Market sizes and values are smaller than for pharmaceuticals, developing a new trait or enabling technology is costly, and the impact of new developments on established crops can be quite small. Since most crops are commodities used for food or feed, profit margins are low, and it is difficult to get an attractive return on a venture investment. It takes ten to 12 years for an agricultural product to come to market, about the same length of time it takes to bring pharmaceuticals to market. However, the potential market value of agriculture products is less than that of pharmaceuticals.

During the last ten years, the agri-biotech industry has become greatly consolidated. The number of potential R&D deals and acquisition opportunities has been reduced, and the sector is much less attractive to potential venture capitalists. Finally, the uncertain regulatory issues surrounding genetically modified organisms mean that investors consider agriculture a risky investment.

In order to encourage venture capitalists to invest in agri-biotech, the public sector must provide more funding for *translational research*, that is, research that moves a technology or product further up the value chain and closer to market, thus reducing both the investment needed for commercialization and the risk (Figure 1). The point of the figure is that knowledge-based biotech industries in agriculture require a greater emphasis on translational research, compared to the pharma industry, to be able to attract the venture capital and corporate investment necessary to commercialize new products and technologies

7. HOW CAN GOVERNMENTS ENCOURAGE ENTREPRENEURSHIP?

Governments cannot dictate or legislate entrepreneurial activity; they can only help provide an environment in which the skilled entrepreneur has ready access to capital, technology, and support. The following actions can help promote such an environment:

- **Provide an educated workforce.** The biotechnology industry requires a pool of individuals with advanced degrees in biology, as well as people trained in mathematics, computer science, and advanced laboratory practices.
- **Provide funding for basic and translational research.** Innovation relies on the unrestricted pursuit of knowledge. Local and national governments should therefore assure support for universities. Depending on the circumstances, government grant

money may be best used to fund applied, not basic, research. Local governments should fund translational research for agri-biotech to make up for the lack of investment from large companies and venture capitalists.

- **Enforce strong patent laws.** Laboratory research, no matter how innovative, is of little social or economic value unless it is actively protected by strong patent laws.
- **Encourage proactive technology transfer.** The transfer of technology from universities to the private sector is often a weak link in the innovation path. Such transfer should be performed proactively and efficiently. Technology transfer offices must recognize that small companies are cash poor and and are working under severe time constraints. Therefore, they must be flexible in the license terms being willing to take an equity position in lieu of cash payments. Also, funding for proof of concept research will lend clarity to the real value of the technology and the remaining risk

Figure 1: Commercializing Knowledge-Based Biotech Industries in Agriculture and Pharmaceuticals

Agriculture requires much translational research.

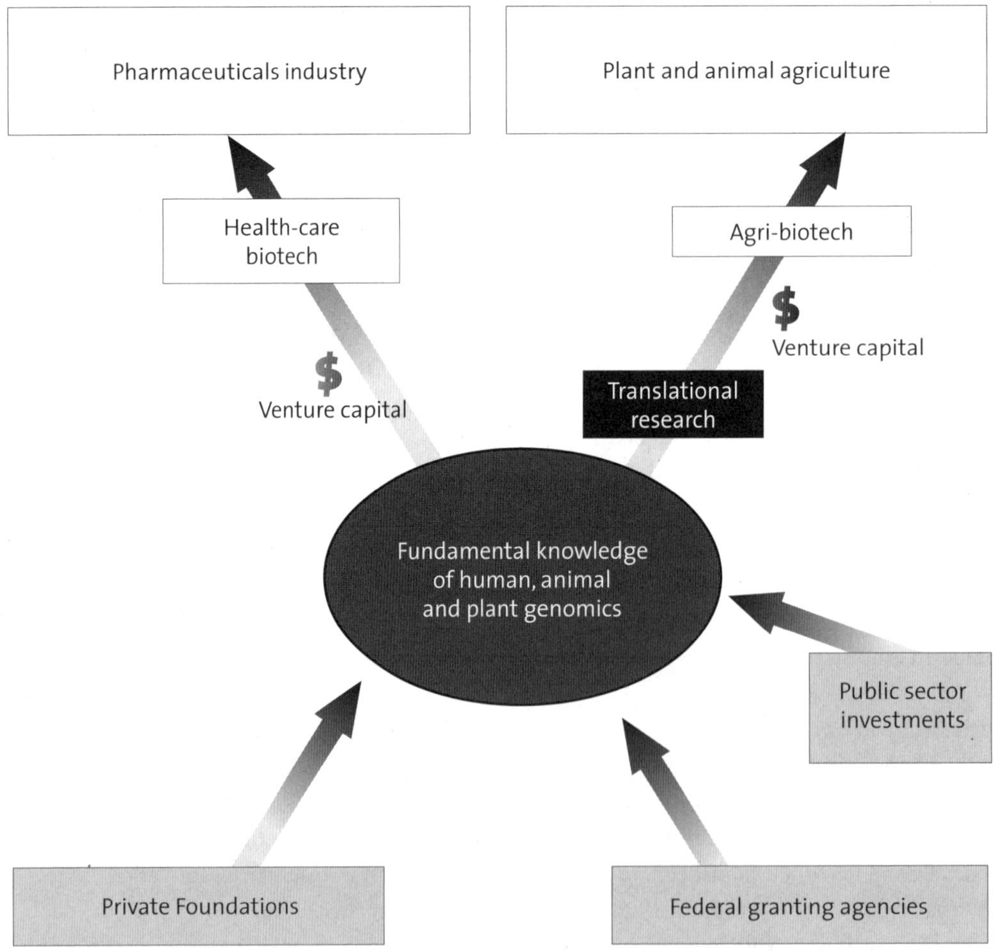

to commercialize. This information can reduce the negotiating period needed to agree on the value of the license.

- **Use the bully pulpit.** Governments must be strong advocates for biotechnology and entrepreneurs. They need to build an environment of expectation, address the naysayers, and signal that their locale is the place to grow a business in biotechnology. Press releases, exhibits, and advertisements by senior officials are just a few examples of actions that have proved successful.
- **Provide a science-based regulatory environment.** Investors and entrepreneurs are attracted by a regulatory system that is based on science, that encourages development while protecting the environment and society, and whose decision-making is transparent.
- **Provide financial incentives to investors and entrepreneurs.** Creative financial incentives that attract risk capital, such as venture capital including R&D tax rebates (which must be tradable, if they are to be of value to small companies), deferred taxes, subsidized incubators, and low- or no-interest loans. In some cases, the incentives may go directly to investors. A source of capital that matches VC investments in companies and tax offsets as enticements for investors to invest in venture funds reduce the overall risk to investors.

8. DEVELOPING A TECHNOLOGY CLUSTER

There is a common misconception that an abundance of venture capital will spawn the formation of new companies. In fact, the opposite is true: high-quality new companies will attract venture capital. It is therefore important to establish a *technology cluster*: a group of small companies working in the same area and in the same or related sectors.

In order to build a technology cluster, certain ingredients are necessary: technology licensing, business-plan development, seasoned managers who can assist in developing business strategies and mentoring management teams, a pool of angel investors, and venture capitalists with experience in seed-stage investing. All of these things will encourage entrepreneurs to start new companies and will accelerate the development of those companies.

The next step should be to encourage experienced, nonlocal venture capitalists who manage large funds to become involved with local companies. Local capital will never be sufficient to fully fund the development of a successful biotech company, and larger venture funds are managed by individuals who have a great deal of knowledge and often participate in global networks. However these large investors are located in just a few locations primarily in the coastal states of the United States. They can be engaged in several ways, but the easiest is probably to invite them to investor meetings where companies from a certain region present their business plans. Since venture capitalists are very busy people, the more companies that attend these meetings, the better. Another strategy that is likely to be more successful is investing local capital into the funds of a VC company and requiring that, in return, the company establishes a presence in the region. Once the company is established, it will be available to advise local companies. The company, however, would not be obligated to invest in local companies.

Finally, a local or national government may set aside a development fund and ask an external VC company to manage or co-mangage it. This system nurtures local venture-capital talent and brings venture capitalists with broad industry perspective to the region. This approach has several benefits and has a history of some success. It addresses the important issues of the global perspective necessary toward biotechnology and access to sufficient capital to fully fund a company through the various value-creating steps prior to an exit via IPO or acquisition. The large companies will have a network within the VC community, so they can syndicate the large follow-on investment required to complete the development of the company through an acquisition or IPO. ■

ROGER E. WYSE, *Managing Director and General Partner, Burrill & Company, One Embarcadero Center, Suite 2700, San Francisco, CA, 94111, U.S.A.* roger@b-c.com

CHAPTER 13.4

The Role of Technology Transfer Intermediaries in Commercializing Intellectual Property through Spinouts and Start-ups

TIM COOK, *Director, Isis Innovation Ltd., U.K.*

ABSTRACT

Intellectual Property (IP) can be commercialized via free distribution or licensing, or through new companies that develop and exploit it. These new companies are called spinouts, or start-ups. Establishing successful spinouts and start-ups requires a solid business plan, coordinated teams of professionals who share a common vision, a respected managing director, and technology transfer intermediaries. Intermediaries help bridge the cultural divide that often exists between the generators of intellectual property and the new companies.

1. INTRODUCTION

What are the forces that encourage or discourage the commercialization of inventions? Part of the answer to this question can be found in the culture of IP-generating institutions and particularly the cultural barriers between academia and industry. Motivated technology transfer intermediaries can help overcome these barriers to commercialization by mediating between inventors, developers, and marketers. The tactics behind such mediation efforts can be useful also for developing countries as they undertake technology transfer projects.

2. IP GENERATION AND DISCLOSURE

Individual inventors, commercial entities, academic institutions, and charitable foundations all produce commercializable IP. There are several ways for this intellectual property to be commercialized: it can be given away (either to a specific recipient or a more general audience via publication), licensed, developed, or exploited through a new company, so-called spinouts and start-ups. This chapter concentrates on the last option. It is important to remember, however, that spinouts and start-ups are not always the most appropriate IP commercialization option.

Inventors are usually creative, self-motivated, flexible individuals. However, the popular idea of the "mad scientist" who is oblivious to the surroundings and keeps going regardless of failure or discouragement is rather uncommon in real life. In fact, whether or not an inventor ever shows his or her invention to the outside world will depend on two variables: (1) whether or not he or she wants to disclose it and (2) whether the environment in which the inventor operates encourages or discourages disclosure.

Some factors, with respect to the inventor, encourage disclosure:

- passionate about the invention
- confident of the worth of the invention
- possesses self-confidence
- resource rich
- solid education
- contacts encourage him or her to disclose the invention

Cook T. 2007. The Role of Technology Transfer Intermediaries in Commercializing Intellectual Property through Spinouts and Start-ups. In *Intellectual Property Management in Health and Agricultural Innovation: A Handbook of Best Practices* (eds. A Krattiger, RT Mahoney, L Nelsen, et al.). MIHR: Oxford, U.K., and PIPRA: Davis, U.S.A. Available online at www.ipHandbook.org.

Other factors, with respect to the inventor, discourage disclosure:

- not passionate about the invention
- not confident of the worth of the invention
- lacks self-confidence
- receives no encouragement to disclose
- resource poor
- lacks time to consider disclosure
- lacks financial support for disclosure
- no reward for disclosure is likely

Positive factors can sometimes compensate for negative ones. For example, if an inventor's environment promotes creativity and is receptive to invention disclosure, it will not matter as much if an inventor has less self-confidence or is less of a risk-taker. It is a well-established fact that the creation of a more-receptive environment often increases the number of commercial ideas: this transformation occurred in the United Kingdom university system between the change of government in 1997 and the present.[1]

This list of factors does not imply that those that favor disclosure should be pursued to an extreme. The best atmosphere for disclosure requires a balance. If the environment becomes too receptive to invention disclosure, or if the invention process is overstimulated by generous government spending, a glut of noncommercializable inventions may be produced. Such inventions do little except consume resources that might have been better used elsewhere.

3. NEW COMPANIES

New companies—regardless of whether they are spinouts from universities or larger companies, or stand-alone start-ups—are new! This means they have little momentum. Their management teams are still developing. The companies themselves have no established market position, and they have the difficult job of convincing potential investors that they have a favorable future. Furthermore, they are usually understaffed and lack adequate resources. What this all means is that single-minded management direction and maximum efficiency are essential for such a company to even survive its first few years, let alone develop a strong position in its field.

In most cases, commercial success is more likely if the inventor remains enthusiastically engaged with the project. The inventor does not need to be in charge of the process; indeed, inventors are not usually the best people to implement commercial development plans. However, he or she should remain an active partner of the plan: not only can he or she prevent the repetition of unsuccessful experiments ("blind alleys"), but his or her creativity can be used to solve problems that may arise as commercialization proceeds.

The company employees need not be close friends, but they should respect each other. Choosing a respected managing director is especially important, since the director will implement the business plan. This plan must clearly and succinctly describe how the business will make money: What is the company going to sell? Where is it going to get raw materials? Who is it going to sell the finished products to, and how? Implementing the answers to these questions will require both intelligence and leadership, which are obvious essential traits for a managing director.

4. BARRIERS BETWEEN IP GENERATORS AND NEW COMPANIES

In the commercial world, research and development must follow a strict budget and schedule; if one element fails, the whole enterprise fails. However, inventors are usually less interested in the commercial ramifications of their work than the work itself. Furthermore, many inventors are academics. In academic research, changes of direction must be made almost daily: tomorrow's experiment is decided by today's results, and researchers are therefore extremely self-directed. Yet they are very willing to share their successes with their colleagues and competitors so that they can further advance their own research. Moreover, academic excellence is measured by the quantity and quality of publications; academia encourages the free exchange of ideas. Researchers in the private sector, on the other hand, will pursue experiments that are part of a larger corporate goal driven by market needs. While they may share their work with fellow researchers in the company, their efforts are usually kept secret from the general public because of the potential monetary value of the inventions the researchers generate.

Box 1 compares the forces that drive the two main types of research environment (academic and commercial). There are, of course, numerous counterexamples: some inventors in industry are publication driven and some academics are secretive.

5. BREAKING DOWN THE BARRIERS

To overcome the problems that may arise when inventors must work with businesspeople, consider a parallel situation: two countries with different cultures and languages must work together on a joint plan. Obviously, the most effective method of helping the two countries interact with each other would be to hire bilingual intermediaries who have a deep understanding of both cultures and both vocabularies. Such intermediaries must: (1) understand the value systems of both cultures; (2) be fluent in language of both cultures, so they can translate while retaining all linguistic nuances; and (3) be credible to members of both cultures (there may be a third "culture" involved: that of the financial investors).

Where do we find such intermediaries? How do we fit them into the overall process? And how do we motivate and reward them?

5.1 *Sources of competent intermediaries*

An industrialist can theoretically be taught how universities really work; an academic can theoretically be taught how industry works. Both methods have been tried (probably the latter more often than the former) with limited success. It is difficult for an individual who has spent all of his or her life in one environment to adapt to the culture of another. Experienced industrialists find it difficult to get over their belief that universities are "badly managed factories," and senior academics find it difficult to adapt to industry's need for discipline and conformity, which they see as "inflexibility." Consequently, it makes sense to recruit intermediaries from the middle ranks of academia or industry, rather than from the top.

5.2 *Where competent intermediaries fit*

Intermediaries can be based in a university, its technology transfer company, in professional service companies (banks, accounting firms, law firms), or even in civil service. They may also be investors or employees of investors who are charged with generating investment opportunities (the author of this chapter was engaged in the latter from 1990 to 1997). Ultimately, of course, intermediaries must be based where they will be most

Box 1: Research Activity Compared with Commercial Activity

Research Activity	Commercial Activity
driven by researchers	driven by market needs
today's result defines tomorrow's experiment	tomorrow's experiment is part of an overall plan
unpredictable outcomes	outcomes must be predictable
relies on individual efforts	relies on cooperative activity

effective. If the goal is to maximize the transfer of technology from a university, then it is sensible to locate the intermediary in that university, or in the university's technology transfer company.

5.3 *Motivating intermediaries*

Intermediaries can be rewarded based on their:

- **financial success.** They may be paid a performance-related salary or be given a financial share in a successful deal.
- **community-building success.** Being part of a team engaged in a worthwhile activity is its own reward.
- **civic or humanitarian contribution.** Contributing to a national or local economy is a satisfying accomplishment .

Of course, the most appropriate basis for reward will vary from situation to situation; in some cases, it will not be appropriate to give any reward at all. There may be limitations on the kinds of rewards that can be given. An intermediary who is also a staff member in a university technology transfer office (TTO) may be forbidden from having any personal interest in technology transfer agreements because of restrictions imposed by university statute or local or national law.

An intermediary, however, who is employed by a technology transfer company that is owned by a university will not have any legal restrictions on his or her personal financial interest in any technology transfer agreements. Still, any bonus that this kind of intermediary receives may negatively affect relationships with university colleagues. For example, at Oxford University, the technology transfer staff (who are not university staff members but are employed by a company owned by the university) work closely with members of the university administration on commercialization projects. If one such project were to produce a large financial gain for the technology transfer staff but not for the university employees, their relationship would be strained.

In addition, the success of one researcher might cause bad blood between the intermediary and her other clients. For example, each technology transfer project manager in Oxford manages about 40 projects at a time: that is, each manager supports at least 40 individual researchers (Figure 1). If one such project were very successful, both the technology transfer manager and the researcher who generated the technology would of course be pleased. However, the other researchers in the manager's portfolio may feel that their own projects had not been given proper attention, and their relationships with the manager might sour.

Figure 1: Oxford University Technology Transfer Staff vs. Growth of Spinout and Licensing Activity

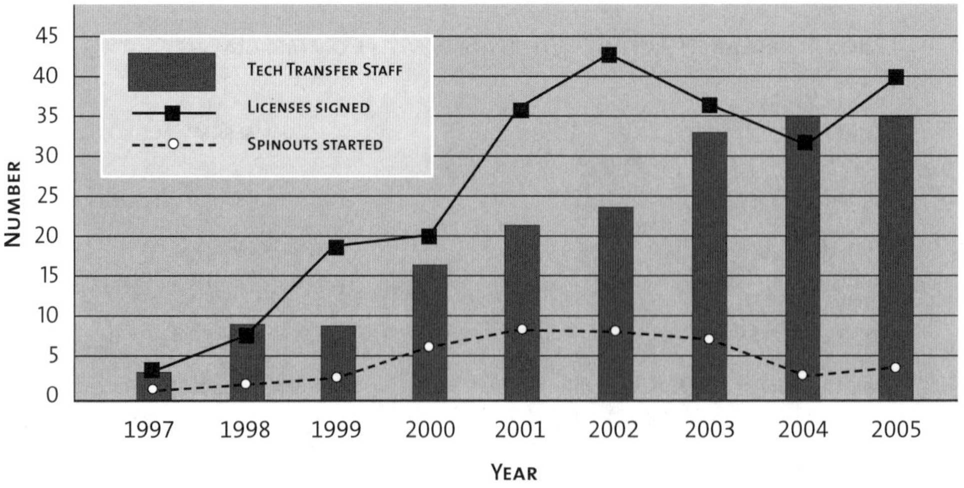

If intermediaries are employed by investors, rather than by a university or a university's technology transfer company, it is quite appropriate for them to receive compensation for their efforts and to apply those efforts where they would be expected to be most lucrative. After all, the job of this kind of intermediary is to help the company or institution realize a profit, and the intermediary is under no obligation to support all researchers from a particular university.

An intermediary who is neither employed by a university nor by investors faces a somewhat murkier situation. In general, the closer to the public sector one works, the less appropriate are technology transfer deals motivated only by financial reasons.

Probably the most powerful motivator for many intermediaries is not financial but intellectual: the pride inherent in associating with creative scientists and collaborating in the creation of new products. It is profoundly rewarding to be the person who brings an invention, whether it is a drug or a software product, from a university researcher's desk to the market. Indeed, it is rewarding to employ one's skills to bring together the academic, financial, and commercial communities and make something new happen. Of course, this sort of intangible motivation only works if the TTO pays its staff well, provides excellent working conditions, and recognizes that job satisfaction can be a powerful motivator.

6. IMPLICATIONS FOR DEVELOPING COUNTRIES

The commercialization of intellectual property (IP) is a potential contributor to economic development. In order to successfully commercialize IP, a country must have a stable economic and institutional environment, sources of investment capital, sources of commercializable IP, a commercial environment that can accept intellectual property and commercialize it, and, as this chapter has suggested, competent technology transfer intermediaries.

Technology transfer of any sort is only likely to succeed if there is sustained commitment at the most senior levels of both government and research institutions. In order for a developing country to create the right conditions, it must make certain commitments:

- general national framework conditions
- a strong commitment to education and training at both the elementary and secondary level
- a commitment to strengthen the conditions that will allow major established firms to develop: the rule of law, labor-market flexibility, infrastructure, financial market efficiency, and management skills[2]

"Business angels" (that is, individual private investors), rather than venture-capital companies, are the initial source of funding for many U.K. university spinouts. They work with fledgling companies, contributing their skills, experience, and contact network. These angel investors have an edge over more traditional venture-capital companies because they are more flexible: they can offer smaller sums of capital and can make decisions more quickly, because they do not rely on the cumbersome analytical machinery of big investment houses. Once a new spinout is established, it becomes more attractive to conventional investors, who want to see a complete management team, a clear business plan, and, ideally, a good track record.

In a developing country, business angels are less common, so new ventures must rely on international investor networks, in which researchers in a developing country team up with researchers in industrialized countries in order to raise money. Such networks may be created through academic links or through personal or industry connections. When a new company grows, it can become too large to depend on the financial resources of private investors; hopefully, by that time, it will be attractive to venture-capital companies. ■

TIM COOK, *Director, Isis Innovation Ltd., Ewert House, Ewert Place, Summertown, Oxford, OX2 7SG, U.K.*
tim.cook@isis-innovation.oxford.ac.uk

1 See Wright M, M Binks, A Vohora and A Lockett. 2003.

Annual Survey of Commercialization of University Technology. UNICO/NUBS/AURIL, Nottingham.

2 Acs ZJ. 2004. Overview of the Global Entrepreneurship Monitor. *2004 Executive Report* (Key Findings from the Global Entrepreneurship Monitor Report of 2004). London Business School: London, U.K. www.gemconsortium.org/download/1166438555062/overview%20of%20gem%202004.pdf. For the full report, visit www.gemconsortium.org/download.asp?fid=364.

New Companies to Commercialize IP: Should You Spinout or Start-up?

CATHY GARNER, *Chief Executive Officer, Manchester: Knowledge Capital, U.K.*
PHILIP TERNOUTH, *Associate Director, R&D and Knowledge Transfer, Council for Industry and Higher Education, U.K.*

ABSTRACT

Universities are eagerly seeking ways to commercialize their innovations. The recent success of spinout companies has made that commercialization option more popular, but commercialization may not be the most efficient approach for research institutions. The risks must be weighed, as well as the benefits, and this chapter offers an overview of the hidden costs of setting up a spinout. Exploring the necessary supporting conditions that can improve the potential for success, the chapter also considers start-ups and incubation centers as potentially better options.

1. INTRODUCTION

Since the late 1990s, a great deal of attention has been focused on how new companies can commercialize technology from research institutions. This route is seen as an attractive alternative to the licensing of technology to an existing company. Even within large R&D-intensive firms, "corporate incubation" has become a trend. By forming new companies, large companies have begun trying to generate value from technology that is not considered core to their existing business.

The attraction of new companies to the owners of the technology and to those concerned with regional economic development is compelling. The venture capital boom of the late 1990s created the impression that forming a new company was the route to rapid wealth for the founders because it enabled a company to go from spinout to an Initial Public Offering (IPO) in a few years. To those concerned with economic development, the formation of new, successful, high-tech companies is considered a route to local economic development: it creates high-paying, high-tech jobs, as well as a number of other jobs (three for every one high-tech job[1]) supported by high-growth, new technology spinouts. Many countries have specifically tried to support this trend by forming "business incubators" and science parks to create a supportive environment. This chapter will explore the advantages and disadvantages of bringing new technologies to market by creating new companies. The chapter also will explore the necessary supporting conditions that can improve the potential for success.

This chapter does not specifically address how to deliver direct public benefits to developing countries from technologies via spinouts.[2] Technology spinouts typically depend on venture capital, which is predicated on high rates of return through profit growth or through the growth of capital through increases in share price. A typical return expected by a venture capital investor is likely to be around 30% at exit, and such expectations leave little room to substitute social outcomes for profits and company growth. A profitable market is therefore key to obtaining the necessary venture funding in the first place. Such markets may exist in developing countries,

Garner C and P Ternouth. 2007. New Companies to Commercialize IP: Should You Spinout or Start-up? In *Intellectual Property Management in Health and Agricultural Innovation: A Handbook of Best Practices* (eds. A Krattiger, RT Mahoney, L Nelsen, et al.). MIHR: Oxford, U.K., and PIPRA: Davis, U.S.A. Available online at www.ipHandbook.org.

and it is important for individuals in developing countries to assess how spinouts might help address public health needs.

Perhaps more importantly, the creation of spinout companies has indirectly been a major economic driver, as new businesses and local jobs create public benefits. This trend of generating new companies from academic research began in the United States, partly because of the contributions of universities to national defense during World War II. That experience of early spinouts emphasized the need for a strong commitment to partnerships and linkages among industry, academia, and government-research sectors. The value of university research in this respect was first recognized by Vannevar Bush, the science policy adviser to President Franklin D. Roosevelt in the 1940s.[3] Bush saw it as a vehicle to enhance the economy by increasing the pool of knowledge that industry—supported by government—could use. Likewise, the story of Silicon Valley and its legendary spinout successes was enabled by the contributions of universities.[4]

Currently in the United States, there is a lot of spinout activity. In the financial year 2000, some 500 new companies were formed to exploit the technology based on academic discoveries made in the 121 universities that responded to the Association of University Technology Managers (AUTM) survey. Notably, for 80% of these companies, each was based in the university's home state. The more than 600 licenses to these new companies accounted for 14% of the total number of licenses reported. An additional 50% of all licenses were to small companies (those with fewer than 500 employees). Similarly in the U.K., a recent report on U.K. universities showed that licensing income fell in recent years, possibly because public authorities have been pressing for the creation of more spinouts.[5]. In U.K. universities there are signs of a more-balanced approach developing. Still needed for successful inception and growth of spinouts are increased recognition of market conditions, internal and external support, and management and intelligent early-stage finance. This is reflected in a wider range of metrics being adopted by central government for assessing knowledge transfer performance.

2. NEW COMPANIES AS THE APPROPRIATE ROUTE TO MARKET

Given the major worldwide interest in the formation of new companies to commercialize technology, surprisingly little systematic work has been published on the circumstances conducive to their success.

A number of perspectives should be taken into account when deciding whether to form a new company to commercialize a piece of technology. However, there can be little doubt that from the perspective of successfully introducing a new product to the market, the new company route is higher risk than a traditional out-license to an existing company. In general, the circumstances that favor establishing a new company to develop products and take them to market are those in which the same "offer to market" cannot be made by licensing the technology to an existing company. Conversely, where such a licensing arrangement is available, a new company is unlikely either to generate the same value for the owners of the technology or to succeed in making the product as available as it would have been through a licensing arrangement.

In most circumstances, an existing company with the necessary infrastructure already in place (such as channels to market, facilities, commercial management, sector knowledge, and an existing contacts network) is likely to be a lower risk. However, where the new technology is disruptive and/or where it is far from the market (as is the case for university research-based technologies), then creating a new company may be the only realistic alternative. In addition, the political priority for new jobs and local economic development brings additional pressures and benefits from the new-company route.

Nonetheless, universities find it hard to build such companies from the ground up, especially when new markets have to be created. Marketing expertise needs to be in place to position new product categories in crowded markets, and carrying out these tasks is costly. Moreover, figuring out how to meet the university's social mission to deliver public sector benefits may become critical in deciding whether or not to form a new company. For example, making products available in developing countries may be one social consideration that universities could take into account when considering the route to market. If this were the prime consideration, then establishing a new company would not be realistic. Markets in the developing world are unlikely to be sufficiently robust to persuade investors to commit enough funds to support establishing a new company. When a potential market for the product and the licensing arrangement is unavailable, a new company may be the only available route to market. This could be because the market or product category is completely new. In this case, a qualified licensee (one with a better package of expertise and infrastructure than could be developed with the required speed by a new company) might not exist. However, the costs for developing new markets or marketing new product categories are very high. Adequate resources have to be put in place, and the time-to-market and to significant sales and revenues might be long. These factors need detailed analysis so that initial funding needs can be calculated with a suitable break-even outlook and a realistic picture for investment returns. Such considerations are rarely systematically assessed in a university situation, because the institution's wish to meet its political goals and the inventor's wish to make money frequently override fundamental economic analysis.

3. NEW COMPANY FORMATION ROUTES—START-UPS VERSUS SPINOUTS IN A UNIVERSITY CONTEXT

For the purposes of this chapter, a start-up[6] is a company created by people outside of a research institution. A start-up is built on a license for one or more technologies, but draws its other resources (such as management) from elsewhere. In contrast, a spinout company is created when an institution invests its own resources to form and incubate the company up through the first round of venture capital investment. The creation of a spinout usually involves the transfer of existing university staff into the new company, either on a permanent or on a secondment basis. A special case of the start-up modality is that practiced by the partnerships between some universities in the U.K. and the IP Group[7] in which resources are made available to universities under package agreements giving access rights to IP. We have yet to establish the extent to which such agreements might have negative affects, for example on the university's wider missions or their research agenda.

Opting for a spinout may lead to the under-exploitation of the economy's intellectual assets and may be a drain on the experienced resources of a university. Research institutions are normally limited in terms of staff resources and capabilities that can be devoted to commercializing technology. It follows that such institutions will be able to create fewer businesses using their own resources, particularly when compared to the number and the quality that they could deliver by attracting resources into the institution. Forming a start-up company by attracting new resources to the institution is likely to be more efficient, not only in terms of the use of scarce resources, but also in terms of the available experience that can be applied to developing and managing a commercial business and company in a limited timeframe.

3.1 *Risks and rewards*

From a university's perspective, the choice may be based on balancing risk and reward. A university setting up spinouts will retain a higher percentage of equity in new companies because the university builds the value in the company before seeking external investment. Using the start-up approach, the university will have had to cede founder's equity to the incoming entrepreneur; effectively, it will have merely adopted a license for equity role. On the other hand, when building a spinout in-house, the institution is using its fixed resources (people) and trading off their time for high equity stakes. In the 1990s, the markets might have indicated that this was indeed a good risk/reward balance. However, two issues should encourage universities and research institutions to naturally prefer obtaining licenses to building spinouts. First, experience has shown that rather than the technology per se, the management of a new company is the critical element for success. Spinouts formed with inexperienced management are more likely to fail, so start-ups are preferable when managers are inexperienced. Second, the high level of risk associated with high-growth–new-technology businesses (where investors plan for nine out of ten to fail), suggests that universities would be more certain of a return from their commercialization activities if they adopted a portfolio approach. Universities should ensure that as much technology as possible is made available for licensing—whether to established firms or to new companies built by external managers. Acquiring smaller equity shares in a larger number of companies would be a safer investment strategy than using high levels of fixed resources to create one or two major spinouts. Universities have fixed and limited resources to undertake technology transfer, and so from a conventional capital appraisal, it is difficult to see how spinouts can be justified when alternatives are available, either from an economic-good or a social-good perspective.

3.2 *Economic and social return*

The intensity and challenge of managing several spinouts through to venture capital investment can be exciting and may also seem to offer greater control for the hosting institution. But given an institution with limited, fixed resources available for technology transfer, the achievement and eventual realization of value created by individual projects has to be set against the growing value of an expanding portfolio of underexploited technology that would have accumulated while resources were focused on selected projects. In fact, from the perspective of the economy and the lost opportunity for creating a social return from the use of the technology, the contrast between the economic value and the social value is likely to be far greater. The value to the economy is measured by the number of jobs or the number of quality companies created, not by the equity retained by the university. And the social return is a factor of the public benefit created (for example, making new health care products available and having used the available funds wisely and optimally).

Focusing research institutions' resources on managing their financial resources optimally is of even greater importance than the subsequent decision as to whether a limited number of spinouts is created or a potentially larger number of start-ups facilitated. The ultimate objective is to ensure that technologies have the best opportunity to come to the market. Current pressures on research institutions to become the engines of economic growth in their local regions tend to emphasize the number of new companies created rather than the successful commercialization of technologies. Too often, universities have confused objectives and a multiplicity of performance targets, all of which drive technology transfer efforts toward inefficient commercialization. Indeed, the policy of the institution needs to be clear on whether commercialization is undertaken primarily for public good or for institutional profit.

4. CONDITIONS THAT CONTRIBUTE TO SUCCESSFUL, NEW TECHNOLOGY COMPANIES

The creation of new companies from research institutions can benefit from a virtual company phase. This phase can last for a long time using the spinout approach from universities, and indeed there have often been companies, solely within universities, existing without clearly defined boundaries. The virtual phase can be useful in preparing the company for a stand-alone existence. In times of volatile venture-funding for specific technology sectors, the virtual phase may allow new technologies to be brought closer to market without the burdens of a formal legal existence. In the U.K., under certain economic-development seed funding (for example, the Scottish Enterprise Proof of Concept Fund[8]), the virtual model is a condition of funding. However, companies must take on a separate existence in due course, and they are typically legal entities (corporations) in their own right established to conduct a business. Whatever way a business is conducted and whatever legal form it takes, some key aspects (given in Box 1) are essential for the business's viability and success.

Box 1: Critical Success Factors for New Companies

Experience has shown that the following factors are critical to success or failure:

Technology.
A technology that provides a substantial but incremental improvement over an existing product category (as opposed to a platform technology) is most likely to be effectively licensed. Existing products have existing markets with existing channels and customers, and it is risky to compete with existing products. Companies will be in competition for the market, and those who are second or third, in terms of market share, will be eager to exploit innovations and take market share from the leader. Although in most cases the market leader is best positioned to turn a product/technology into maximal value, the leader might risk cannibalizing its existing market and try to keep a new product out. In such circumstances, any license to the market leader would best be supported with strong performance clauses (milestones).

With regard to platform technologies (which enable a range of different products to be produced, possibly for different markets), forming a new company will frequently be the way to get the best value and ensure that the technology is fully exploited. This may or may not address the markets directly, depending on the marginal costs and benefits arising from the technology. Platform technologies are often attractive to investors, because the range of potential markets that can be developed offers a greater security of return if the initial intended application fails. Likewise, there is an implicit chance of greater returns than with a single product technology.

Market Development.
An existing market (defined as the sales of products of a particular type to a defined group of customers) is most likely to be served by entrenched competitors with existing customer loyalties and established distribution channels. The circumstances are likely to be similar to those in which the technology is an incremental improvement, which suggests that the best option will be close to the licensing end of the spectrum. Conversely, when a market is new, the licensing route may be unavailable or will have higher marginal costs for a prospective licensee. Accordingly, forming a new company may be a better option financially, provided that potential market demand exists.

Product, System, or Component?
If the intended product is a complete system, then it will be theoretically possible to form a start-up or spinout to take it to market, because the company may be capable of providing a solution to end users. If the intended product is a component of a larger system, then the product will need to be channeled via established companies in the field who will embed it in a complete system.

Management Availability.
Developing a technology relies heavily on capable management. This is one of the potential advantages of start-ups as opposed to spinouts. By marketing the technology well (presenting it in the context of its compelling benefits in product form), the technology assets can be used to leverage these management resources from the marketplace. Conversely, attracting management to a proposition proves difficult, this may be because the other requirements for forming a new company have not been met. Choosing a licensing route effectively co-opts the management of existing companies into a new product's channel to market.

(Continued on Next Page)

Box 1 (continued)

Market Concentration.
A concentrated market has the majority of its value in a limited number of customers. A diffuse market has its value dispersed in a large number. It is easier to locate and access a limited number of large customers than to locate and sell to a large number of small ones. Exploiting an existing distribution channel via a distributorship arrangement may be the only economical way of addressing the latter, even if there is genuine new product or company potential.

Complexity of Sales Task.
If the sales task is complex and the type of product is unknown to the customer and the benefits unproven—which it may well be for a new product concept—only perhaps the originators can describe the product's features adequately and work with innovative customers to prove its utility. In such circumstances, the best option is to work with a capable marketer and adequate training mechanisms to enable the marketer to present the product correctly.

Availability of Investment.
For development that goes all the way from technology to market, investment may be unavailable for the complete project because of the high costs and risks involved. A licensing route or license to develop may be the only way that investment can be made available. If feasible, then the other factors that favor licensing are also likely present. The classic example is the drug development and marketing process, where the costs of clinical trials and regulatory processes may be over U.S.$100 million, and the attrition rate higher than 90%.

Complexity of Delivery.

If the delivery of a product or service is highly complex, the undertaking may require detailed knowledge of the technology underpinning the product and the services of a coordinated team. Such a situation, which is common, for example, in software development and in the installation of health technologies in their infancy, may argue for a more extended period of in-house development, at least in the early stages of market introduction.

New companies intending to exploit biotechnology are entering an environment that requires collaboration. There are many different processes needed in a complex value-chain, running through target identification, compound design or synthesis and screening, and drug development and market. A supporting infrastructure is needed that might include the production of animal models of disease, bioinformatics, gene sequencing, chemical synthesis, combinatorial chemistry, drug delivery, formulation and manufacturing, clinical trials management, biostatistics, and managing regulatory approvals.

The interdependencies of different skills and specializations mean that producing a start-up company to develop and market its own products is unlikely to succeed. Moreover, the global pace of scientific advance makes it hard to simply keep up-to-date with relevant discoveries. Interpreting their implications for existing projects or new opportunities is even harder. For example, the sequencing of the human genome has generated more potential disease targets than even the largest pharmaceutical company can handle. These circumstances together make collaboration essential. Through collaboration, large companies can increase their project pipeline, and small companies can obtain the resources they need to develop their products.

The ability of research institutions to collaborate and access resources in other companies is a competitive capability in its own right, and it follows that new biotech companies should plan their strategy around developing this ability. Early in their development, companies should identify potential partners. This requires an openness and a readiness to work with other companies to identify potential collaborative projects. At the same time, a high degree of professionalism is needed to protect commercial interests. This includes the protection of commercially sensitive information and materials under Non-Disclosure and Materials Transfer Agreements, and, above all, the protection of IP through the filing and prosecuting of patent applications.

5. BUSINESS INCUBATION FOR NEW COMPANIES

There is a growing trend for new, technology-based companies to be supported by incubators that are located often in close proximity to research institutes. No discussion or presentation on spinout or start-up companies would be complete, therefore, without some consideration of business incubation and incubators. Some internationally renowned research institutes, such as the Massachusetts Institute of Technology (M.I.T.) and the University of Cambridge (U.K.), are surrounded by an environment that strongly supports the development of new business. It provides a local pool of management talent, funding, professional support (such as patent agents and attorneys), and a cluster of existing companies that may act as potential collaborators. The importance of such an Innovation Ecology, has been documented in a recent publication[9] detailing the case histories of some 30 companies in and around Oxford. Where this kind of environment does not exist, a more studied and deliberate approach may be made to provide the benefits of such an environment through specifically designed incubators.

Incubators provide to a new company a number of potentially valuable services that can enable management to focus on running their core business. The best incubators also provide access to a network of contacts whose expertise can be leveraged to develop the businesses. Government and other public sector agencies often see investing in incubators as key to stimulating knowledge-based economic development. In fact, incubation can provide the facilities, resources, and expertise that may be difficult to access during the early stages of a business. Such access may have a critical part to play in ensuring that the business achieves early commercial success. But incubators should *not* be seen as a long-term source of support for businesses that, perhaps because of a lack of market opportunity, are unlikely ever to be more than marginal.

The critical business-acceleration aids that an incubator can provide include a rapid introduction to a network of individuals who may include those with relevant market and management experience. Some of these individuals may be able to guide and mentor inexperienced management either formally (perhaps as employees of, or consultants to, the incubators) or informally. Other individuals may include business people with customer contacts who might themselves assist in turning the technology around to "face the market" and in shaping the business to achieve its first revenues. Just as important are contacts with potential early-stage funders, especially those "added value" funders who can help by shaping the business, identifying and fulfilling its investment potential, and sourcing the potential members of a growing commercial team. These key functions of the virtual incubator are well described in *Networked Incubators: Hothouses of the New Economy*.[10] The best incubators also provide access to a network of professional support services (often provided pro bono), such as basic advice on patenting, incentive agreements for employees, and licensing agreements.

Incubators may also be formed to develop and accelerate business in specific market sectors. In the case of biotechnology, for example, a key contribution is made by obtaining access to an international network of contacts, which includes potential research or product development collaborators in the complex drug-development value chain. These may provide useful regulatory advice and guidance. Additionally, they may include access to very high-cost capital equipment, such as scanning electron microscopes and nuclear magnetic resonance machines.

Incubators can also assist by providing basic business and office support facilities and services, such as accommodation, payroll management, bookkeeping, and high-bandwidth Internet access. A lack of these facilities and services can steal attention from the management of a business, especially since such matters may be unfamiliar to those with a predominantly technical background.

Incubated companies will expect to pay lower-than-market rates for the services they receive from incubators, at least in the early stages of incubation. These lower rates are made possible by one or more of the following:

- Achieving economies of scale by combining the otherwise uneconomic provision of professional and business support services for a number of smaller customer companies in the incubator
- Public (for example, local or regional government) subsidies made in anticipation of economic development
- Incremental occupation and service charges that are lower at the outset and increase progressively as the company obtains commercial success
- Paying for a proportion of the occupation and support charges in the form of equity (a key strategy in the case of for-profit incubators)

A number of successful incubators operate using the model described here, but many do no more than provide accommodation. These latter incubators have been severely criticized in the United States.

6. CONCLUSIONS

There are many success stories about start-ups and their impact on the growth of local economies, such as in Silicon Valley, California, and Route 128 on the East Coast of the United States. This chapter, however, has pointed to the complexity of developing a successful start-up enterprise. Choosing the route to market strategy for new technologies requires making a set of complex decisions that many universities and research institutions are not specifically equipped for. The conventional licensing route for technologies may not only involve lower risk for the institution, but may also deliver more technologies from the institution's scientific research. Universities and research institutions with the primary mission to deliver social and economic goods rather than investment returns should carefully consider how to achieve this mission most effectively. Establishing spinouts that disproportionately consume their in-house resources might not be the best approach. The current pressure from governments to create new companies and new local jobs from university research should not be accepted without the new resources to support this activity.

Once created, new companies face many challenges to achieving sustained growth and successfully delivering value to shareholders. Technology alone is rarely sufficient to reach this goal. Good management, awareness of market forces, and a good supporting environment in the early stages are all more important. Still, while failure rates are high, for those companies that succeed, the returns to the founders, the institutions, and the local economy can be significant. ■

CATHY GARNER, *Chief Executive Officer, Manchester: Knowledge Capital, Churchgate House, 56 Oxford Street, Manchester, M60 7HJ, U.K. Cathy.Garner@ManchesterKnowledge.com*

PHILIP TERNOUTH, *Associate Director, R&D and Knowledge Transfer, Council for Industry and Higher Education Bowerham House, The Grove, Lancaster, LA1 3AL, U.K. philip.ternouth@innopolres.com*

1 Patrick O'Brien. Member of Canadian Parliament. Spoken presentation on a visit to the U.K. in March 2001.

2 See, also in this *Handbook*, chapter 13.4 by T Cook.

3 Bremer H. 2003. Technology Transfer: The American Way. International Patent Licensing Seminar, Tokyo, Japan.

4 Lewis M. 2000. *The New New Thing: A Silicon Valley Story.* Penguin Books: London. See also by the same author The Valley of Money's Delight, in the 27 March 1997 issue of *The Economist*; and Know Thyself in the 28 October 1999 issue of *The Economist*.

5 *Annual UNICO/NUBS Survey on University Commercialisation Activities.* 2002. Nottingham University Business School: Nottingham. See also www.unico.org.uk and www.hero.ac.uk/uk/business/archives/2002/spinouts_pick_up_speed2872.cfm.

6 There are some differences in the way the terms *start-up* and *spinout* are used in this *Handbook* and indeed in the technology transfer literature. In this chapter, we have used a definition of *start-up* similar to one used in the U.S. *AUTM Licensing Survey*™. Other authors have used "spinout" to include the kind of company described here.

7 www.ip2ipo.com.

8 Scottish Enterprise's Proof of Concept Funding Programme. www.Scottish-Enterprise.com.

9 Hague D and C Holmes. 2006. *Oxford Entrepreneurs.* CIHE/Said Business School: Oxford, U.K.

10 Hansen MT, HW Chesbrough, N Nohria and D Sull. 2000. Networked Incubator: Hothouses of the New Economy. *Harvard Business Review* Sept/Oct:75–83.

Formation of a Business Incubator

EDWARD M. ZABLOCKI, *Office of the Vice President for Research, University at Buffalo, State University of New York, U.S.A.*

ABSTRACT

Business incubators, as economic tools, have become increasingly common in the last decade and a half for stimulating local development. Incubators provide facilities and services (for example, business planning and legal, accounting, and marketing support) to catalyze small-business growth. In fact, incubated companies have a dramatically higher rate of survival than an average spinout does. This chapter explains what steps to take to set up an incubator, including the basic structure and the kinds of services generally offered. Successful incubator programs are discussed, and a helpful bibliography focused on case studies is provided.

1. INTRODUCTION

An invention sometimes requires the efforts of a spinout enterprise to be commercialized. Without a corporate infrastructure to execute an established commercialization process, an institution, such as a university, may be reluctant to invest in the steps needed to move technology out of the laboratory. In contrast, a spinout may be more favorably positioned to embrace new technologies because of access to capital and grant monies. Philosophically, moreover, a spinout is generally more willing to accept risk than an established concern constrained, perhaps, by shareholder interest. Forming a spinout is a critical option for moving an invention into the marketplace. To succeed, three components must be assembled: capital, organization, and facilities.

This chapter focuses on the last of these. It is intended to provide fundamental background information for use by the technology transfer practitioner and includes information on terminology, incubator formation, and successful incubator programs, as well as a helpful bibliography.

2. INCUBATORS

Smilor and Gill define an incubator as an organization that "*seeks to give form and substance—that is, structure and credibility—to start-up or emerging ventures. Consequently, a new business incubator is a facility for the maintenance of controlled conditions to assist in the cultivation of new companies.*"[1]

Commonly classified by ownership and capital sourcing, there are three types of incubators: public, private, and university. Numerous sets of subclassifications of the latter two types exist, depending on their status as for-profit or nonprofit entities. Other attributes of the business incubator that distinguish it from other commercial enterprises include the range of services, the ease by which tenants can cancel their lease, and the

---

Zablocki EM. 2007. Formation of a Business Incubator. In *Intellectual Property Management in Health and Agricultural Innovation: A Handbook of Best Practices* (eds. A Krattiger, RT Mahoney, L Nelsen, et al.). MIHR: Oxford, U.K., and PIPRA: Davis, U.S.A. Available online at www.ipHandbook.org.

Editors' Note: We are most grateful to the Association of University Technology Managers (AUTM) for having allowed us to update and edit this paper and include it as a chapter in this *Handbook*. The original paper was published in the *AUTM Technology Transfer Practice Manual* (Part IV: Chapter 3) co-authored with DE Massing.

reduced (often subsidized) rent during the incubation term.

3. INCUBATION AND ECONOMIC DEVELOPMENT

In the 1980s, small became big in economic development circles. During this period, state and regional economic development strategies shifted from seeking to attract companies from elsewhere (industrial recruitment) to focusing on assistance for the homegrown entrepreneur. This shift in economic development strategy occurred for good reason. Seminal studies by David Birch at M.I.T.[2] showed that almost all job growth in the U.S. economy was attributable to small companies. While the validity of Birch's findings has recently come into question, their impact on policy circles at the time is undeniable. Economic development officials and policy planners sought to create jobs in their states and regions by fostering the growth of small companies.

Small business incubators became a preferred vehicle for providing assistance to new companies. In the 1980s, incubators were referred to as the most potent economic development tool to be introduced in this decade. Only a handful of incubators were present at the beginning of the decade, but the National Business Incubator Association's report in 1992 on the state of the incubation industry illustrates their dramatic growth.[3] Of 147 respondents to the NBIA's survey, only four had opened by 1980, with nearly two-thirds opening between 1988 and 1991. Today, there are more than 500 incubators.

The incubator concept is simple and appealing. An incubator is a multitenant facility providing affordable space and an environment that promotes the growth of small companies. Initially, some incubators provided an inexpensive physical environment to spinouts in what had been old or vacant buildings. Later incubators concentrated on the companies themselves, helping them to grow by creating an entrepreneurial environment. A range of services was developed to assist the small company: shared support services, such as the availability of secretarial help, a receptionist, and access to copiers and professional services, including business planning and legal, accounting, and marketing support. Access to working capital was also arranged through provision of debt financing and equity financing, government grant/loan assistance, and connection to a financial network of angels, bankers, and venture capitalists. Today, however, most incubators prefer the company-centered approach, charging market rates for rent and offering services as the value-added benefit of locating in the incubator. Thus, incubators are probably best defined as programs rather than facilities.

Nonprofit entities operate almost 90% of incubators. Their purpose is to stimulate job growth in various sectors of the local economy. Some incubators, particularly those with ties to higher education, emphasize technology-based development. Communities that lack the critical infrastructure of technology-related business and research-intensive universities may direct incubators to serve developing companies in the manufacturing and service sectors. Incubators have also been used to encourage entrepreneurial activity among disadvantaged populations, including women and minorities. For example, the New Enterprises for Women Building in Greenville, Mississippi, targets assistance to low-income, minority women.

These varied economic development purposes are reflected in the 1991 NBIA survey, which found that the most important objectives of incubators were economic development (91.3%) and economic diversification (60.9%), followed by research commercialization, technology transfer, women/minority opportunities, and neighborhood revitalization, among others. The great variety of the types of companies incubated further confirms the diversity of purpose in business incubation. The most common company types are service (36%), light manufacturing (20%), technology products (15.9%), R&D (10.7%), and wholesaler/distributor (7.8%).

Small business incubators have proven to be effective economic development tools, even though they may not have fulfilled early optimistic expectations for job creation. Their greatest benefit may be enhancing company survival rates. Incubated companies have a dramatically higher rate of survival than the average spinout. Incubator managers report that somewhere between 80 and 90% of companies that have incubated with them are still in existence after five years. This figure vividly contrasts with the Small Business Administration (SBA) statistic that finds that only 50% of start-ups survive their first five years. These figures are less surprising when one considers that nine of ten companies fail because of management deficiencies, and that 90% of these deficiencies could have been foreseen. Job creation statistics are more modest. The average incubator in the 1991 study was four years old and occupied a space of about 20,000 square feet in size. Each incubation facility averaged 12 tenants with 54 employees. Graduate companies (those that relocated from the incubator) provided an average of 85.3 full-time jobs per incubator.

The establishment of new incubators peaked in 1987, and the new wave of economic development initiatives in the1990s focused on helping existing businesses survive and prosper in the face of global competition. Small business incubation is now an entrenched and accepted economic development tool used in both urban and rural areas throughout the United States. Incubators are now used to promote the growth of entrepreneurial ventures of every imaginable type.

4. PRELIMINARY WORK

4.1 *The feasibility study*

Conducting a feasibility study for a proposed incubator can achieve a number of important objectives and, if properly done, can provide a solid basis for judging the economic and political viability of the proposed project. The feasibility study represents the first in a series of early development phases that, for planning purposes, can be described as follows:

- feasibility: 3 months
- development: 9 months
- renovation: 3-12 months
- early-stage operations (up to anticipated break-even point): 18 months

Meeder[4] suggests a number of reasons why conducting a feasibility study is wise. These include:

- helps to forge a consensus among key organizations and civic leaders
- catalyzes the involvement of organizations that can provide the incubator with a range of resources including facilities, funding, equipment, and human resources
- allows for the completion of plans for both the facilities and the services to be provided
- helps secure funding from government sources at all levels
- educates public and private sector constituencies about business incubation in order to avoid confusion and unwarranted expectations
- provides an occasion to contact successful incubator programs in similar communities to learn their best practice lessons

A feasibility study should also reveal examples of critical errors made with respect to other incubator programs. Such errors might involve facility and site selection, structure of the governing board, funding arrangements, income assumptions, or the nature of the business assistance program.

Meeder suggests that a thorough feasibility study will help avoid the two classic errors of incubator formation: accepting the worst building in town and thinking that the management assistance program will somehow take care of itself. While recommending the use of a consultant, Meeder notes that selecting a consultant without direct incubator experience can result in a study that provides general analysis, but lacks concrete recommendations. Specific recommendations can make the difference in an incubator's long-term success. An adequate feasibility study will answer essential questions about how to proceed in a systematic fashion and how to secure funding during all the phases of incubator development. Indeed, a thorough study by a qualified consultant can and should provide the information necessary to determine whether the project should be pursued.

4.2 *Building support*

A core group committed to starting a business incubator must recognize that its efforts cannot be pursued in a vacuum. The dream of a few must become the dream of many. An incubator represents an important community investment, both practically and symbolically, and requires broad-based community support to be feasible. In *Forging the Incubator,* Meeder suggests that meetings with community leaders can achieve several objectives. Community meetings allow proponents of the incubator to:

- provide information on the business incubation industry
- invite reaction to the prospects for a local business incubator
- solicit referrals to people, companies, organizations, and facilities that can assist the process of feasibility and/or development
- offer the opportunity of direct participation, to seek specific leads to entrepreneur prospects, and/or gather information that had been overlooked

Engaging in this process should clarify the prospects for starting an incubator. The process should help to identify potential sites, funding sources, project champions from key organizations, and sources of assistance and support, both individual and organizational. The process may, however, also uncover serious impediments to realizing the project. Meeder suggests that project supporters make serious efforts to placate opponents; indeed, project supporters should not assume that the project will be successful in the face of persistent opposition. Real estate developers, for example, may resist the project because they believe an incubator will cut into their market. A persuasive argument, in this case, is that the incubator will only incubate companies for a limited period of time and that the incubator should serve to increase both the quantity and quality of companies seeking to rent space. Community consensus building should help locate organizations that will identify with the successes and failures of the proposed incubator. These organizations are known as *stakeholders.*

4.3 *Identifying and securing stakeholders*

A stakeholder is any group or individual who can affect or is affected by achievement of an organization's objectives. While each incubator's circumstances are unique, anticipated stakeholders would likely include local and state governments and a variety of public and private sector organizations (universities, major corporations) interested in fostering new-business development in the region. Stakeholders might also include economic development organizations that could fund the rehabilitation of a facility and/or the operation of the incubator program. The support of these stakeholders is critical to initiating an incubator program. At the same time, potential supporters of the incubator effort understandably have varied motivations and expectations. Their level of understanding of the purposes and methods of business incubation will vary greatly.

Stakeholders need to be identified and then cultivated. The first step is to secure commitment from potential stakeholders who have the strongest interest and who are most likely to provide financial support for the endeavor. Once stakeholders have committed to the project, the organizational structure needs to be formalized. A governing body, typically a board of directors, provides the organizational vehicle for maintaining, building, and strengthening commitment to the incubator program.

One of the board's tasks is getting interested parties to agree to a clear articulation of the mission and goals of the incubator. This articulation of the incubator's goals brings the stakeholders together with a common purpose. Experience has shown that incubators that fail to achieve consensus on mission and goals invite trouble from their board, since members will create their own tacit mission statement and begin to act accordingly.

Incubator managers should seek to expand the number of valid stakeholders. New stakeholders should be welcomed as long as they have something tangible to contribute. On the other hand, allowing tenants to serve on the board can create conflicts of interest, so tenant participation on the board should be evaluated on a cost-benefit basis. Additionally, incubator managers must remain sensitive to external conditions, which may strengthen or weaken the commitment of stakeholders to the incubation enterprise. Finally, by-laws are crucial. They provide an objective means of removing nonparticipatory board members and, at the other extreme, board members who are exerting undue influence.

4.4 *Identifying a market niche*

A business incubator will operate in a particular locale with its own rich history, so it must act with an eye to the regional economy and institutions. To become an accepted part of this complex social fabric, an incubator must establish its distinctiveness and unique purpose. From a business perspective, the incubator needs to identify its market niche. Successful businesses carefully attend to the work of defining the market position of their products and services relative to their competitors, as well as to modifying their market position in response to changing customer preferences.

Developing a market niche for a business incubator requires similar attention to these tasks. An incubator's competitors come from the spheres of real estate and economic development. Within the real estate market, the incubator must distinguish itself from other multiple-tenant properties. For a technology-related incubator, the distinction may be readily apparent, for example, in that incubator facilities may offer wet and dry lab space. Incubators also differ from conventional real estate agents in that they often offer short-term leases and flex-space for a company's expansion. Certainly, rent subsidization can be attractive to cash-poor start-ups. The availability of shared support services is another appealing feature of incubator facilities, although provision of such services by for-profit organizations has become a growth industry.

Economic development programs for small businesses proliferated in the 1980s. These programs have been referred to as "incubators without walls." Well-managed incubators often distinguish themselves by serving as a focal point for access to the broad spectrum of available business services. Incubator managers thus provide the point of contact for entry into various programs. Many efforts to assist small business are, by contrast, programmatic in nature and limited by the scope of their intent. A well-positioned incubator, on the other hand, will help its tenants access the range of existing programs and, in addition, provide access to informal networks for business and financial advice and assistance. For example, a retired executive may agree to help out a struggling firm or a business angel may appear, discretely looking for new investment opportunities.

The incubator program may also delimit itself and define its market by the type of company or client served. While high-tech incubators may limit their scope of service to technology-focused companies, some incubators may be even more targeted (for example, restricting their services to biotech companies). The customer for the incubator should be determined during the feasibility phase, during which new-business registrations, by industry type, are classified and certain industry sectors identified for their spinout potential.

Whatever the mix of services offered and the assessment of the market to be served, the incubator must somehow package its product to effectively position itself.

5. THE FORMATION PROCESS

The basic structure of an incubator facility is determined by owner attributes and regional demographics. The following owner/sponsor classifications can generally be applied:

- private
- local government
- university
- state government
- private nonprofit
- federal government

A typical organizational format includes executive and advisory boards, a CEO or operations manager, and support staff. Selections for board positions and other representative forums may come from the following: private enterprise, educational institutions, government, organized labor, development and investment community, and private citizens.

The role of the manager or chief executive officer of the incubator is both internal and external. This person is chiefly responsible for:

- incubator policy and planning
- marketing and recruitment
- tenant selection and lease negotiation
- facility operations management
- tenant service and administration

The manager has multiple constituent groups representing both the sponsoring (funding) segments and the user (spinout) population. Appropriately selecting advisory board members allows the manager to establish and maintain networks for the dissemination of information and policy to these disparate groups. Table 1 provides typical staffing levels for incubators.

An important function is marketing the incubator, which will be driven, in part, by the results of the market analysis conducted during the feasibility study. The market analysis should consider the following major aspects of the local economy:

- characteristics of large corporations in the area
- level of entrepreneurial activity in the community
- demand for incubator-type space
- small-business support services by industry type, if feasible.

Large corporations can supply an important market for new businesses and are also the chief sources of spinout companies in a region. The number, type, and rate of filing of new-business permits can provide important indicators of potential demand for incubator space. An inventory of available space broken down by type (office, manufacturing, and so on) is essential for determining potential demand.

Market information can also be secured by offering a workshop or seminar that highlights some of the proposed business-service components of the incubator (for example, a workshop on developing an effective business plan or one on the accounting needs of small businesses).

This information can provide the basis for a market strategy that is integrated into the overall incubator budget.

Proactively gathering market information is recommended over a reactive mode, which does not typically serve to effectively market the incubator. A reactive approach is tempting when an incubator manager is stretched thin with other responsibilities. However, a written marketing strategy allows other parties (board of directors, advisory board, related organizations) to assist. As Meeder[6] points out, the most successful sales organizations have a standard sales script or routine with which everyone involved is familiar.

The marketing effort should include typical means of communication, including brochures, newsletters, and press releases about new tenants, tenant successes, and graduations. One of the incubator's sponsoring organizations may be able

TABLE 1: TYPICAL INCUBATOR STAFFING

	INCUBATOR TYPE		
	PUBLIC	UNIVERSITY	PRIVATE
Median number of administrative staff	1.60	1.90	3.50
Median number of business consulting staff	1.40	2.10	2.10
Ratio of business consultants to firms	0.13	0.12	0.12
Managers with previous business experience	70%	67%	92%
Managers with business consulting duties	73%	67%	93%

Source: National Council for Urban Economic Development[5]

to help develop these promotional materials. In addition, the incubator story may be included in the communications of sponsoring organizations. Other organizations may also be interested in co-sponsoring seminars of interest to entrepreneurs.

Such marketing efforts are necessary but not sufficient. Studies have shown that most entrepreneurs learn about the incubator through word of mouth. To market the incubator effectively, it is incumbent on the incubator manager to continue to develop and maintain a network of contacts in real estate, banking, patent law, business and economic development, both formally, through boards of directors and advisors, and informally, through professional organizations and business contacts. Individuals in an incubator's local community are often the first to alert a nascent entrepreneur of the benefits of locating in a small-business incubator.

6. SERVICES

As the incubator concept has evolved, the range of services offered by incubators has greatly expanded. Early incubators provided access to a photocopier and a conference room, clerical support, and perhaps switchboard services. Today, incubators themselves provide, or provide access to, a broad spectrum of office, business consulting, and professional services. The most common in-house and outside services offered are given in Table 2.

In recent years, incubators have greatly expanded the variety of office services they provide. For example, the menu of office services offered by an incubator based in Pennsylvania in operation for three years includes:[8]

- clerical services
- switchboard services
- voice mailbox

TABLE 2: TYPICAL INCUBATOR STAFFING

SERVICES	IN-HOUSE (PERCENT OF TOTAL)	OUTSIDE (PERCENT OF TOTAL)
Office services	81	2
Business/strategic planning	65	32
External debt financing	59	7
Government grant/loan assistance	58	28
Training/educational programs	52	29
Financial management	51	36
Sales/marketing	51	37
External equity financing	47	27
Employment assistance	31	41
Lab equipment access	29	24
Bookkeeping	23	30
Government procurement	19	52
R&D/product development	19	43
International trade	14	52
Accounting or tax assistance	8	59
Legal/patent services	6	67

Source: NBIA[7]

- electronic mailbox
- telephone equipment
- FAX service
- postal service
- overnight courier service
- notary services
- photocopier
- VCR/TV equipment
- audio-visual equipment
- conference room
- printing services
- furniture rental
- laser printing/graphics
- auto service discounts
- sports ticket purchasing

Business consulting services may include business plan preparation, financial planning, advertising and marketing, strategic planning, technical and commercial communications, relocation planning, capital development (equity and debt services), business taxes, employee relations, R&D, and government procurement.

Professional services include legal/patent services, accounting, business development (including sales/marketing), and technical/scientific support, among others. Professional services may be provided at special discounts to incubator tenants. Some incubators arrange for new tenants to initially receive some professional services at no cost or at a deep discount. Given that entrepreneurs have no time to spare, professional service providers are often regularly available at an incubator and make themselves available for support and consultation.

In developing the spectrum of services for a new incubator, several options need to be explored. First, there is the essential question of which services will be offered. Next, incubator managers must consider which of these services will be offered in-house. This will depend on internal resources and the external availability of business services. The availability of qualified outside sources will depend on the success of forging informal alliances with a range of service providers in the public and private sectors. For those services offered in-house, the question of cost recovery will need to be addressed. Several services are typically included as a standard feature in a tenant's rental agreement. These most commonly include janitorial service, management assistance, utilities, shared office services, and financing assistance. Other services, such as clerical assistance, are charged back to the company on an at-cost or cost-plus basis. The quality, range, dependability, and accessibility of these services are the value-added features that will provide the strongest lure for attracting entrepreneurs to an incubator. The incubator should solicit feedback from tenants to ascertain whether or not the services are effectively meeting their needs and to determine whether additional services should be added.

7. STRATEGIC PLANNING

While the previous sections have addressed discrete issues related to incubator formation, the need for strategic planning—and the integration of these various elements into a coherent, multi-phased plan—should be apparent. Determinations about one aspect of the plan will affect other aspects. A rather obvious example is the effect that the facility's net available square footage will have on rental income. More subtle considerations might include expectations for the facility's long-term self-sufficiency. Managers should consider whether self-sufficiency can be achieved solely from rental income, through subsidies from sponsoring organizations, or through grants.

Strategic planning compels incubator management to confront tough issues. How will the incubator continue to operate if revenue projections from rental income are not achieved? How will major facility repairs (for example, a ruptured boiler) be paid for? Addressing these worst-case scenarios through strategic planning can provide both a clear course of action if things go as planned and, if they do not, the necessary contingency plans to navigate what may be a difficult beginning.

Strategic planning usefully determines not only *what* will be done but *when* it should be done. The initiation of a new phase of the incubator may or may not be made contingent upon the successful completion of an earlier phase.

Can the operation begin as an "incubator without walls," providing business services before the facility is ready for occupancy? At what point in the development process is the manager hired? The notion that timing is everything is certainly true in strategic planning for an incubator spinout.

8. CASE STUDIES

Detailed case studies in the literature are cited but not restated in this chapter since these studies are generally quite lengthy. Some of the incubators noted below are not in operation today, but the histories may still provide useful information. As a guide to the reader, these studies are classified in outline form to permit selection based on interest.

The first set of examples is facility-based:[9]

- university-related incubator: Renssalaer Polytechnic Institute—The Advanced Technology Development Center
- community-sponsored incubator: The Fulton-Carroll Center for Industry
- corporate/franchise incubators: Control Data Corporation Business and Technology Centers
- private incubator: The Rubicon Group

The second group is objective based:[10]

- promote economic diversification: St. Paul Small Business Incubator
- provide a base for advanced technology development: Ohio University Innovation Center
- opportunities for targeted populations: New Enterprises for Women Building (NEW Building)

In sum, principal factors for successful incubator strategies include:

- Know the community and its strategic strengths and weaknesses.
- Locate entrepreneurial opportunities.
- Design (tenant) selection criteria to match goals and objectives.
- Determine the space and service needs of tenants.
- Locate the facility in a site that can be developed within the cost parameters of target companies.
- Find opportunities to link up with existing sources of business and management services.
- Recruit an entrepreneurial personality to manage the incubator.
- Build an overall environment for entrepreneurship.

9. CONCLUSION

Incubators have been formed to serve entrepreneurs of every ilk; they have been established by a wide variety of sponsors. It is therefore not surprising that their missions, programs, and objectives have differed substantially. Nevertheless, over the past 15 years, examples of best practices have emerged. Some general factors critical to an incubator's success include:[11]

- on-site business expertise
- access to financing and capitalization
- in-kind financial support
- community support
- entrepreneurial networks
- entrepreneurial education
- perception of success
- selection process for tenants
- ties to a university
- concise program milestones with clear policies and procedures

Along a more practical vein, some of the specific practices known to affect the relative success of incubator operations include:[12]

- Incubators with less than 30,000 square feet have generally been unable to reach financial self-sufficiency.
- Incubators without an articulated policy for collecting past-due rent have experienced high levels of bad debt.
- An incubator manager's most effective use of time is to evenly balance attention to tenant services and facility upkeep. Initially, the demands of the facility will predominate. Subsequently, the manager should concentrate on achieving balance

by expanding time spent in the provision of services.

- Terms and conditions of tenant leases are critical for protecting the incubator program.
- The phone system is an essential link for companies and must be structured appropriately.
- The board of directors must be clear about its authority regarding management decisions versus policy decisions.
- The structure of service provision should include ways to increase effectiveness within the budget. Methods include the use of third-party service providers and collecting fees for services.
- Exit policies should encourage, but not mandate, tenant graduation. ■

EDWARD M. ZABLOCKI, *Office of the Vice President for Research, University at Buffalo, State University of New York, 516 Capen Hall, Buffalo, NY, 14260-1611, U.S.A.* Zablocki@research.buffalo.edu

1 Smilor RW and MD Gill. 1986. *The New Business Incubator: Linking Talent, Technology, Capital and Know-How.* Lexington Books, D.C. Heath and Company: Lexington, Mass.

2 Birch D. 1979. *The Job Generation Process.* M.I.T. Mimeo of the Program on Neighborhood and Regional Change; Boston. Birch D. 1981. *Choosing a Place to Grow: Business Location Decisions in the 1970s* Mimeo of the Program on Neighborhood and Regional Change. Birch D. 1981. Who Creates Jobs? *Public Interest* Fall. pp. 3-14.

3 NBIA. 1992. *The State of the Business Incubator Industry 1991.* National Business Incubation Association: Athens, Ohio.

4 Meeder RA. 1993. *Forging the Incubator: How to Design and Implement a Feasibility Study for Business Incubation Programs.* National Business Incubation Association: Athens, Ohio.

5 National Council for Urban Economic Development. 1985. *Creating Jobs by Creating New Business—The Role of Business Incubators.* National Council for Urban Economic Development: Washington, DC.

6 See *supra* note 4.

7 See *supra* note 2.

8 See *supra* note 5.

9 Based on selections listed in *supra* note 1.

10 Based on selections listed in *supra* note 2.

11 See *supra* note 1.

12 See *supra* note 5.

SECTION 14

Freedom to Operate and Risk Management

CHAPTER 14.1

Freedom to Operate, Public Sector Research, and Product-Development Partnerships: Strategies and Risk-Management Options

ANATOLE KRATTIGER, *Research Professor, the Biodesign Institute at Arizona State University; Chair, bioDevelopments-International Institute; and Adjunct Professor, Cornell University, U.S.A.*

ABSTRACT

Freedom to operate (FTO) is—first and foremost—a strategic management tool. It is the synthesis of scientific, legal, and business expertise coupled with strategic planning. Strictly speaking, however, FTO is a legal concept. It is a legal opinion by patent counsel on whether the making, using, selling, or importing of a specified product, in a given geographic market, at a given time, is free from the potential infringement of third-party intellectual property (IP) or tangible property rights. As such, it is one type of input among many that managers use to make strategic risk-management decisions in relation to R&D and product launch. For academic and public research institutions, bringing products to market is often not a main goal. However, as a portion of their research moves downstream into product development, FTO becomes—or should become—an integral component of their endeavors. This is particularly relevant for product-development partnerships (PDPs) in health and for various public–private partnerships (PPPs) in agriculture, as well as for the Consultative Group on International Agricultural Research (CGIAR) and national agricultural research systems (NARS), all of which are concerned about global access.

Research exemptions exist in many jurisdictions, so most university research does not generally need to be concerned with FTO unless product development takes place. But PDPs, such as the Malaria Vaccine Initiative or the TB Alliance, are in a different category since their purpose is directly related to the distribution of products in the developing world. This chapter discusses three main categories of options that are available to reduce risk and obtain a *manageable* level of FTO. In practice, a combination of two or more options will often be pursued concurrently. These are:

- Legal/IP management strategies: license-in, cross-license, oppose third-party patents, seek nonassert covenant, seek compulsory license
- R&D strategies: modify product, or invent around
- Business strategies: merge and/or acquire, wait and see, abandon project

Each option presents its own risks and opportunities. Any action—including the decision not to take action—carries risk. Delaying the licensing of third-party intellectual property, for example, could lead eventually to expensive licensing terms, the inability to obtain a license, or the possibility of being sued for patent infringement. But for some organizations, such as those developing genetically modified crops, the reverse may be the case. For the public sector, the challenge will be to balance the various types of risks that each option presents.

The chapter concludes by urging the public sector to judiciously evaluate whether and when FTO concerns should be considered, and to build in-house capacity to conduct patent searches and cursory FTO analysis (as opposed to legal opinions). This will lead to benefits like better competitive intelligence and culture change in public sector organizations engaged in product development. An FTO strategy, therefore, is a *plan* that begins with research and evolves into an *attitude* throughout a product's R&D and commercialization/distribution cycle.

1. INTRODUCTION: FTO AND RISK MANAGEMENT

Successful freedom to operate (FTO) strategies require forming partnerships, both within

Krattiger A. 2007. Freedom to Operate, Public Sector Research and Product-Development Partnerships: Strategies and Risk-Management Options. In *Intellectual Property Management in Health and Agricultural Innovation: A Handbook of Best Practices* (eds. A Krattiger, RT Mahoney, L Nelsen, et al.). MIHR: Oxford, U.K., and PIPRA: Davis, U.S.A. Available online at www.ipHandbook.org.

institutions and with third parties. Although FTO is often narrowly considered as only a legal issue, when approached from a more practical standpoint, FTO is a strategic risk-management tool; it relies on a synthesis of scientific and legal expertise, business development, and strategic planning. An FTO *opinion* is legal advice or input that managers use to make business decisions based on a full range of criteria (business goals, competitors' position, financial goals, and so forth).

FTO has two fundamental aspects. First, it is a legal concept: an FTO *opinion*, rendered by patent counsel, will advise senior management about whether the making, using, or selling of a specified product in a given geographic market would infringe a third-party's intellectual property (IP) or tangible property right. The legal opinion is based on a detailed analysis of the product or service under consideration, an analysis that primarily involves searching patents (though other forms of intellectual property, such as trademarks, will also be considered). The analysis also involves examining the claims of such patents, reviewing possible material transfer or contractual obligations, and providing a legal interpretation of the analysis.

Second, FTO indicates the nature of the business constraints imposed on the institution, such as whether regulatory approvals have been granted or import or export licenses have been obtained. Third, the word *freedom* in *freedom to operate* does not imply absolute freedom from the risk of infringing another party's intellectual property. It is a *relative* assessment based on the analysis and knowledge of IP landscapes for a given product, in a given jurisdiction, at a given point in time. This point underscores a critically important concept: there is no such thing as a risk-free decision. Whether an organization decides to perform an FTO or not, both options carry an element of risk. Not making a decision is itself a decision.

This chapter focuses on legal, research, and business strategies for resolving the legal aspects of patent infringement—in other words, on strategies for minimizing IP constraints. Companies deal with these challenges routinely. Early or cursory FTO reviews[1] are typically conducted during the conceptualization of research projects to indicate early on how to reduce IP/licensing constraints that may emerge further down the road. This makes it possible for a company to decide in advance which components, technologies, and processes are best incorporated into the product under development. Certain R&D projects may even be stopped fairly early—or may never be pursued—when the FTO situation seems too uncertain or too costly to resolve. Hence, with any FTO strategy there will be other business-related considerations, including market potential, geographic location, short- and long-term business opportunities, and the positions of competitors.

One of the big questions the public sector has struggled with is whether, when, and how to concern itself with FTO. University researchers generally do not need to be concerned with commercial FTO unless they are engaged in research that aims specifically at product development. This kind of engagement is becoming more prevalent in the public sector, not least through collaborations with product-development partnerships (PDPs), where the primary reason for funding the research is the development of products to help the poor. Such is the case for the research centers of the Consultative Group on International Agricultural Research (CGIAR) and for many national agricultural research systems (NARS). Universities, too, are shifting their research focus; some manage their innovations in novel ways. For example, Arizona Technology Enterprises LLC, the technology commercialization arm of Arizona State University (ASU), in-licenses (or assembles) IP to establish core technology platforms around ASU inventions, and then licenses the bundled IP as solutions, offering quicker market access and greater commercialization opportunities.[2]

These trends within the public sector require the building of various types of partnerships. Indeed, the very process of seeking and obtaining FTO, which requires myriad licenses and other forms of institutional arrangements, leads to partnership building. But partnerships carry risks—as does acting independently. Risk cannot be avoided completely. Instead, researchers and administrators must be aware of the different types of risks and ask themselves how they can best be balanced.

2. FTO: FROM ANALYSIS TO STRATEGY

The approach to FTO follows a logical sequence (Figure 1). It begins with an FTO analysis, which is an investigation whereby the planned or existing product is dissected into its component parts. For each of these, a search is conducted for any intellectual and tangible property rights. The results of such an analysis allow patent counsel to provide an FTO opinion that discusses the likelihood that the product or process infringes identified IP rights or tangible property rights of others. The resulting FTO status becomes the baseline for formulating an FTO strategy, which then allows management to weigh different risks and make informed business decisions.

An FTO opinion usually divides third-party intellectual property into three classes (lawyers may not use the terminology used here):

1. Patents that have a high likelihood of being infringed and therefore require a license
2. Patents that *may* be infringed, depending on how claims are interpreted
3. Patents that are clearly outside the field of the product and require no license

Unfortunately, many patents will not have a clear status that would place them squarely in category 1 or 3. Many will instead fall into the more uncertain category 2. The classification is based in part on the analysis of the meaning and scope of the patent *claims*, the detailed portion of the patent text that specifically defines what the invention is and lays out a conceptual boundary or property line around the patented invention. Legal protection is awarded only to what is captured in the claims; anything outside the claims is open to the public.

Patent claims are analogous to the "metes and bounds" described in real estate deeds. As with a deed for land, claims delineate the limits (the dimensions and borders) of the invention. However, as distinguished from the tangible property rights to a deeded piece of real estate, patents deal with intangible property rights. Finding the precise limits of IP rights is thus not a quantitative activity; it is, therefore, open to interpretation, because one cannot see or touch the actual property in a patent (it is "intellectual," or of the mind). The boundaries can only be described with words, yet the meanings of words are not precise. They are always open to interpretation, especially given their context.[3] For these reasons, it is useful to further subdivide category 2 patents into subsets defined by the possible outcome of legal action:

2(a). It could be argued with some level of certainty that, if defendant were taken to court by plaintiff, defendant would probably ***lose*** a *patent infringement lawsuit.*

2(b). It could be argued with some level of certainty that, if defendant were taken to court by plaintiff, defendant would probably ***win*** a *patent infringement lawsuit.*

Counsel can advise senior management about the number of patents that fall into each of these categories—1, 2(a), 2(b), and 3—and about the institutions that would have to be contacted to form a partnership or licensing deal. But counsel would not be able to tell which options made the most sense from an R&D, institutional, and business perspective. From a purely legal perspective, obtaining licenses for all the patents that fall into category 1 and 2 would minimize risk. Lawyers will tend, therefore, to identify licensing as the lowest risk option. To what extent this makes business, financial, and strategic sense, however, requires considering other options explained below.

3. WHEN TO SEEK FTO

For ***companies***, FTO has to be considered very early in the product-development process. Once millions of dollars have been invested in the research, development, regulatory compliance/approval, formulation, and manufacture of a product, it would be difficult to obtain beneficial licensing terms from third parties. The more resources invested, the more difficult the bargaining position, though other factors may be equally important. For example, a company that has good marketing networks already in place might find it easier to negotiate licenses.

In practice, performing a detailed FTO analysis on every product or process early in the pipeline would be impractical and prohibitively expensive. Therefore, even the early decision on whether or not to commit resources to perform an FTO analysis for a given project or product candidate must itself be based on a preliminary, or cursory, assessment. Such a preliminary assessment can help determine when to perform a more-detailed FTO analysis and at what level of sophistication and depth.

For ***public sector*** entities, the same *principles* usually apply to FTO but with important differences. For universities the organization's primary mission or focus is research, teaching, and sharing knowledge. The freedom to engage in these endeavors derives from the norms of academic freedom and, in some countries, is codified as academic research and fair-use exemptions under IP law. Downstream business development considerations are often a secondary or derivative focus.

Figure 1: FTO Strategy in Context

FTO Analysis
An FTO analysis is a focused and intense investigation, performed by meticulously dissecting a biotechnological product or process into its fundamental components and then scrutinizing each for any attached, unlicensed intellectual property (such as patents, plant variety protection, or trade secrets) and tangible property of third parties.

↓

FTO Opinion
Based on the results of the FTO analysis, patent counsel will draft an FTO opinion that indicates the likelihood that the biotechnological product or process infringes the IP rights or tangible property rights of others. The likelihood of such infringement might be either low or high, depending on the results of the FTO analysis.

↓

FTO Strategy
The FTO status establishes a baseline for formulating a strategy for product development. This involves business and legal considerations to balance potential risks with anticipated benefits. The FTO strategy considers all options and then decides on the approach that best fits the mission of the organization and its tolerance for risk.

↓

FTO Status
The FTO opinion will inform, with respect to the overall status of FTO for a given product—depending on the time and place—the level of potential risk associated with contemplated R&D and/or commercialization activities. Such risks vary; hence, FTO status is relative.

Source: SP Kowalski, personal communication.

This is why university technology transfer offices typically license inventions (patents) and, in some cases, trademarks and plant varieties, but do not develop and sell finished products. However, for PDPs and many nonprofit organizations, product distribution and access often are their main purpose, even if they may not be the party that will actually produce and distribute the products. Their missions focus on the development of products for the marketplace (whether considered nonprofit or for humanitarian purposes). The main questions, therefore, are simply when to initiate the examination of FTO and when to begin the process of assembling the necessary intellectual property.

Should the assembly[4] of intellectual property be done early or late in the product-development process? Timing the licensing of third-party intellectual property is an important strategic decision, and like any decision carries certain risks. By deciding to delay, an institution accepts the following possibilities:

- that higher licensing terms will be extracted (Once an institution invested years and millions of dollars into R&D, its bargaining power is often reduced.)
- that no license will be obtained
- that a lawsuit will be filed for patent infringement

Conversely, by seeking to in-license early on, an institution accepts other risks. In agricultural biotechnology, for example, one of the biggest obstacles for public sector institutions in obtaining IP licenses from companies is their lack of trust and confidence in the public sector's ability to produce a high-quality product and to be a responsible steward of the technology and product. Few public sector entities have experience in developing biotechnology products. Understandably, companies may therefore be reluctant to grant licenses—especially those for humanitarian purposes—to entities that have not demonstrated credible product-development plans and that lack the requisite resources for product stewardship throughout the product's life span. Public sector entities may therefore find it easier to obtain licenses on preferential terms once they have demonstrated a product's quality and their overall institutional capacities, especially their capacity in IP management, regulatory management, and high-quality productions. Demonstrated capability generates confidence and trust, which translates into a greater willingness by companies to provide licenses and to enter into partnerships. This is one reason for the creation of AATF: the stewardship of agricultural applications.[5]

In sum, there is no textbook strategy. Each case must be reviewed and evaluated, and the best strategy—or strategies—will depend on many factors, including:

- the mission of the organization
- the range of existing partnerships
- the ease with which the organization interfaces with companies
- the type of product under consideration
- the degree of overlap between public and private sector interests related to the specific product.

4. COMPLEMENTARY STRATEGIES TO OBTAIN FTO

Companies determine their overall FTO strategies, generally speaking, through a combination of decisions by boards, senior executives, business managers, marketing executives, R&D managers, and legal counsel. Although this chapter has so far stated that most IP issues related to FTO are about deal making, in-licensing, and partnership building, such deals are the *results* of choosing from among a combination of ten main options (Table 1).

To be sure, not all of the options apply equally well to public sector research institutions. Bringing products to market is not their major concern, but to the extent that their research is used downstream, such as in collaboration with the private sector, FTO is becoming more integral to their endeavors.

4.1 *Legal/IP Management Strategies*

4.1.1 *License-in*

All FTO issues can be resolved by acquiring (individually or through consortia) a commercial

Table 1: The Ten Strategic FTO Options

Option	Pros	Cons	Key challenge for the public sector
1. Legal/IP Management Strategies			
License-in	Is relatively straightforward	May not foster in-house R&D initiatives and may be costly	Determining the right time to initiate licensing discussions/ negotiations
Cross-license	Involves give and take	In certain cases, antitrust issues may arise	Requires alignment of institutional strategy
Oppose third-party patents	Can be cost effective	Can be expensive and result might be undesirable (stronger and/or broader patent)	Policies of public sector rarely allow for such measures; cost may be prohibitive
Seek nonassertion covenant	Is cheap and effective	Rarely allows for the in-licensing of valuable know-how	Might require lobbying by lead scientist and head of institution
Seek compulsory license	Allowed under TRIPS under certain circumstances	Will not allow for the in-licensing of know-how and brings many constraints and complexities with it	Many conditions need to be fulfilled for compulsory licensing to be feasible
2. R&D Strategies			
Modify product	Can be fairly simple if planned early in R&D stage	May not be possible due to lack of readily available alternatives; incurs opportunity costs	Requires early FTO review and business-driven R&D strategy
Invent around	Could lead to cross-licensing position	Could lead to delays in product launch and might be costly; incurs opportunity costs	IP/licensing department would need to drive, or at least influence, researchers and the direction of research
3. Business Strategies			
Wait and see	Gives time for strategic positioning	Could lead to litigation and jeopardize investment already made	Generally undesirable
Abandon project	Is simple and effective	May be costly (need to write off R&D investments already made, incurs opportunity costs)	Difficult to determine when, how, and by whom such a decision is made (unless the financial donor has a clear IP policy)
Merge and/or acquire	Is highly effective	May distract from main business focus	Not generally feasible
→ In Practice A combination of several options implemented concurrently			Requires strategic mindset

Source: A Krattiger

license from the certified owners/assignees for each IP right that the product under study is likely to infringe. Negotiating a license is the most common option and perhaps the most logical. It may be broad—a grant to make, have made, use, have used, import, export, offer to see, sell, or have sold all products and product parts and all related products and processes—or it may be more restrictive.

Licenses are agreed to every day, and in many circumstances entering into licensing agreements is almost a mechanical matter.[6] However, we hear of special cases when licenses have been difficult to obtain, when licenses were refused, or even when license disputes have ended up in court. Considering the number of licenses executed each year, these special cases are rare, but they seem to receive an inordinate amount of attention. The main question is not *whether* to license, but *when* to initiate licensing discussions/negotiations (or when and how to pursue other options discussed here). But, to reemphasize, licensing is just one of many options.

4.1.2 *Cross-license*

Cross-licensing occurs when two IP holders license intellectual property to each other: "A" licenses a set of patents to "B," and in exchange B licenses a set of patents to A. This approach is often adopted when one entity holds a patent on an invention and another has an improvement on it. For example, assume that A holds the rights to a promoter that is only effective in cereal species. B, however, has modified the gene so that it is now also useful for dicotyledonous species (which are non-cereal species). A can continue to practice its invention on cereals but could not use it in beans (since they are dicotyledonous species). Yet B cannot use its improvement in beans because it would require a license from A. Cross-licensing inventions in this case allows both A and B to both apply their inventions in beans.

Some companies have entire teams of researchers conducting research to place the company in a stronger cross-licensing position with certain competitors. Due to costs, public sector institutions are probably not in a position to do this; nonetheless, cross-licensing should not be dismissed outright.

4.1.3 *Oppose third-party patents*

It is generally presumed that, after issuance, a patent is valid. But patents can be challenged. Essentially, there are three components to patent validity under U.S. law: novelty, utility, and nonobviousness. A successful challenge on any of these grounds will annul a patent claim, and sometimes the entire patent. A patent claim can also be declared invalid if it can be shown that the written description requirement was inadequate. When considering litigation, two certainties must be kept in mind: the cost of litigation is high, and the outcome is uncertain. Furthermore, preparation for a patent-invalidity challenge will involve research and analysis that is comparable to, if not greater than, that involved in an FTO analysis. Cost must be carefully considered when thinking about this option. Other possible drawbacks are that the assignee/inventor comes back with additional claims (as happened with the Enola bean case at first).[7]

4.1.4 *Seek nonassertion covenant*

Many companies are, in principle, willing to license their valuable intellectual property for developing country and humanitarian uses. But quite naturally, they are reluctant to take on risks for activities that do not generate cash flow or profits. One way for them to manage some of the risks is through nonassert covenants, or nonassert agreements, through which an IP rights holder essentially assures the IP rights user that it will not enforce the IP right. These are fairly simple agreements to execute and may be in the form of public statements or bilateral or multilateral agreements.[8]

In this new era of "humanitarian" licensing, the international community is struggling to develop and distribute new products and to extend the benefits of those the developed world already enjoys. Dealing with all of the FTO issues, however, can be daunting. Just obtaining licenses can be complex, time consuming, or impossible. Companies may be reluctant to license due to liability issues. This is especially so with agricultural biotechnology applications (partly brought about by the Cartagena Protocol's ongoing international negotiations on liability and redress) and with vaccine technology. Fortunately, many of these complexities can be circumvented with a simple nonassert covenant.

4.1.5 *Seek compulsory license*

Most countries have provisions for the issuing of compulsory licenses to national producers in national emergencies, provided that certain conditions are met according to the Agreement on Trade-Related Aspects of Intellectual Property Rights (TRIPS). The country must have the manufacturing capacity to produce the patented invention and must also have attempted to negotiate a license in good faith (although the World Trade Organization's Council recently instituted a waiver to the original TRIPS Agreement that allows developing countries without manufacturing capabilities to import patented drugs from sources other than the originator company). Compulsory licensing has to be initiated by governments for public non-commercial uses and may take one or more years to complete: it is a complex process and requires significant government resources and experience.[9]

Production under compulsory licenses presents several operational challenges. Patent holders are unlikely to license and transfer their know-how under compulsory licenses, so companies in developing countries will need to develop know-how internally. Exports, moreover, may only be made to certain countries under specific conditions, which limits economies of scale and potentially increases production costs significantly. But compulsory licensing can also be a beneficial tool (for example, as a negotiation strategy). Furthermore, international IP standards mandated by TRIPS already allow member nations considerable discretion to enact laws and provisions that not only meet treaty obligations, but also support national innovation policies, development priorities, and cultural values. This includes voluntary pricing and licensing arrangements. Other options primarily relate to national policies and laws beyond the purview of this chapter (for example, permitting and regulating the government use of patented inventions, taking actions through patent courts to protect public interests, and the judicious framing of competition law and policy). Importantly, when compulsory licenses are issued, the licensor has no obligation to transfer know-how/trade secrets or any safety, efficacy, or clinical data. In other words, the compulsory license may be limited to the information disclosed in a patent specification, which frequently represents only an invention's early best mode. It will not include subsequently developed and/or ancillary technical know-how or related show-how.

Given the range of necessary licenses and the time required to issue a compulsory license, this option might not permit a developing country to quickly develop a product. That especially applies to licensing vaccines, for which know-how is a major component of the intellectual property. Moreover, even raising the possibility of compulsory licensing would significantly deter future investments. A "false alarm scenario," in which a national emergency is proclaimed to justify compulsory licensing when the conditions may not fully warrant such a proclamation, might be a harmful approach, since such compulsory licensing could act as disincentive for future investments. Granted, the threat of a compulsory license can prompt an early licensing agreement, but seeking a commercial license early is probably more effective in most circumstances.

4.2 *R&D Strategies*

4.2.1 *Modify product*

An alternative to licensing is to change the product specifications. In agriculture, for example, instead of using a certain (patented) promoter that would require a license, the vector design would include a different type of promoter unencumbered with intellectual property.

Such a strategy will succeed only if (1) there are alternatives in the public domain that would work at least as well as the encumbered promoter and (2) an FTO analysis is performed relatively early during the R&D stage (preemptive FTO analysis). Otherwise, many years of work would be lost, and a license might suddenly seem quite appealing, if not necessary, in order to gain FTO.

A license may also come with regulatory know-how/trade secrets, data, and trademarks. Of course, it is critical that this approach include analyses of any viable alternatives so that their likelihoods of FTO can also be assessed. One does not want to exchange a sick pony for an even sicker burro!

4.2.2 *Invent around*

Choosing the invent-around option would require a research team to search for alternative ways to develop the product in question. Taking again the example of a promoter, the team would seek to isolate a new, unknown promoter and concurrently seek patent protection. This option would delay product development but could lead to significant benefits in terms of new inventions, new intellectual property for cross-licensing, and perhaps even better products. The main downside is that costs would be high, so in many cases the option might not be feasible for public sector organizations. The costs of licensing versus the costs of an all-out development of a new product should be weighed using a risk/benefit analysis. Given the frequent open-ended cost structure of research and development, licensing might be more feasible. In industry, inventing around is often a strategy pursued in parallel with licensing negotiations.

4.3 *Business Strategies*

4.3.1 *Wait-and-see*

The simplest option is to commercialize the product under question and wait to see if the IP holder contacts you for a license. If and when that happens, it would still be possible, perhaps, to come to a licensing arrangement (discussed in Section 4.1.1). Alternatively, the option of opposing a third-party patent (discussed in Section 4.1.3) could be pursued as a form of defense. In addition, a cross-license (discussed in Section 4.1.2) might be offered in return. However, in the United States, the potential downside is that if it can be proven that the infringer willfully infringed the particular IP rights of the other party, then a court may assess damages as high as three times the IP owner's lost revenue. In exceptional cases, the court may also award reasonable attorney fees to the prevailing party (that is, the owner of the IP rights).

4.3.2 *Abandon project*

If all else fails, a project may simply have to be abandoned, freeing investments for safer and less-risky ventures. Naturally, the best time to decide to abandon a project is before initiating any research and development. For this reason, companies typically hold regular project/product planning meetings that include scientists, business-development managers, and legal counsel.

Public sector institutions often find it difficult to abandon projects since promises to donors have often been made for several years. Scientists in the public sector also often have a lot of autonomy compared to their corporate counterparts. That is why a donor's IP policy is so important for determining when, how, and by whom such a decision is made (unless the financial donor has a clear IP policy).[10] The requirement of the Bill and Melinda Gates Foundation (as well as other donors) for a global access strategy is particularly welcome and important in this context.

4.3.3 *Merge and/or acquire*

Any company, regardless of its size, may acquire, through mergers and acquisitions, a number of smaller companies, just to expand its IP portfolio. Although not a feasible option for academic institutions, in the private sector this practice is an important step in obtaining FTO.

Nonprofit PDPs and other nongovernmental organizations (NGOs), moreover, might gain by considering mergers, perhaps not so much as a strategy to obtain FTO, but as a way to increase the potential for innovation. For example, in the 1990s, when the world around the centers of the CGIAR became more complex, with many more actors and spheres of influence, rather than regroup and focus, the CGIAR expanded (with a constant or reduced budget in nominal terms) and has since become an increasingly diffuse entity. This is particularly problematic because the work of this group is conceivably more important than ever from strategic and humanitarian points of view. Paradoxically, over the same period, the private sector undertook mergers and acquisitions, reducing the number of key players from more than 20 to a mere five or so. This happened during a time when development agencies, NGOs, and a plethora of other service organizations increased and multiplied.[11]

5. CONCLUSIONS

For public sector institutions, planning for FTO early in the research phase is neither necessarily appropriate nor feasible. Indeed, since much of the research conducted in academic institutions is not directly intended for commercial use, there is and indeed should be little concern over FTO. But public sector institutions, particularly the NARS and CGIAR in agriculture, and the PDPs in health, are increasingly dealing with the complex interface of proprietary science and the public domain. Moreover, donors such as the Bill and Melinda Gates Foundation are requiring them to develop global access strategies that spell out how intellectual property will be managed to make the products from the grants available to the poor. This will increasingly require FTO considerations as products are moving downstream.[12] Significantly, however, while the steps involved in an FTO are straightforward, their execution is complex and time consuming, and the implications of an FTO are difficult to translate into a product-development strategy.[13] As mentioned in the introduction, FTO opinions provide only snapshots of the intellectual property related to a product at a given point in time. For example, the patent landscape changes daily as the specifications of the product become modified and improved, as the legal landscape evolves (for instance, rules are issued for what type of invention is patentable), and as patent applications are filed and patents issue, expire, or are invalidated.

A sound strategy for obtaining FTO for a given product or process should consider all options and an assessment of the risks of each in relation to the institutional context, the product type, and market dynamics. In practice, several options are pursued concurrently. Strategies will need to be regularly revised and tactics adapted in response to changing circumstances. In practice, some options may be more feasible during the R&D stages (such as inventing around), whereas others may become the only option if all else fails (such as litigation or abandonment of a product).

All of the options outlined in this chapter require, in some way, the formation of partnerships, both internal and external. First, managing potential IP infringement requires cooperation and partnerships between and among R&D personnel and professionals in business development, finance, strategy, law, and even governance. Moreover, translating this coordinated, focused, and informed risk management into a solid, reliable, and thorough FTO strategy should be a shared goal for all involved. Indeed, everything in the end is driven by relationships, both internal and with third parties outside the organization seeking FTO. If a decision is made to passively manage such risks, unexpected problems could arise and opportunities could be missed.

Above all, as with any strategic issue, the key is not so much to *have* an FTO strategy—but to *execute* it. Strategy is not so much a plan but an attitude. Take a positive attitude to facing problems, view them as opportunities, chart the best course action, and then implement it. ■

ACKNOWLEDGMENTS

I wish to thank Stanley P. Kowalski (Franklin Pierce Law Center) and Greg D. Graff (PIPRA and University of California, Berkley) for helpful discussions and constructive suggestions during the preparation of this chapter.

ANATOLE KRATTIGER, *Research Professor, the Biodesign Institute at Arizona State University; Chair, bioDevelopments-International Institute; and Adjunct Professor, Cornell University, PO Box 26, Interlaken, NY, 14847, U.S.A.* afk3@cornell.edu

---

1 Fenton et al call such searches "level one" searches that are more cursory than "level two" investigations where legal FTO opinions are typically given in a confidential document under attorney-client privilege (see, also in this *Handbook*, chapter 14.4 by GM Fenton, C Chi-Ham and S Boettiger).

2 See, also in this *Handbook*, chapter 17.8 by PJ Slate and M Crow.

3 In addition to the interpretation of claims, an analysis of patents must also include embodiments of inventions.

4 *Assembly* essentially means to have gathered the relevant pieces of intellectual property.

5 See, also in this *Handbook*, chapter 17.17 by RY Boadi and M Bokanga.

6 See, also in this *Handbook*, Section 11 on Technology and Product Licensing.

7 Rattray GN. 2005. The Enola Bean Patent Controversy: Biopiracy, Novelty and Fish-and-Chips. *Duke L. & Tech Rev* 0008 (3 June).

8 See, also in this *Handbook*, chapter 7.6 by A Krattiger.

9 See, also in this *Handbook*, chapter 3.10 by CM Correa.

10 See, also in this *Handbook*, chapter 5.2 by Z Ballantyne and D Nelki.

11 For this, and other reasons, it has been proposed to restructure the CGIAR and FAO, for a consortium of developing countries to purchase Monsanto, and for the creation of an agricultural investment service (see Krattiger AF. 2003. Technology transfer to developing countries and technology diffusion: The future role of institutions in capacity building, regulations, IPRs and funding. In *Handbook of Plant Biotechnology* eds. P Christou and H Klee. John Wiley & Sons, London. pp. 987–1010. http://www.wiley.com/WileyCDA/WileyTitle/productCd-047185199X,descCd-tableOfContents.html).

12 One of the first major FTOs conducted for a public sector consortium was for GoldenRice. It was commissioned by the Rockefeller Foundation on behalf of the International Rice Research Institute (IRRI) and the Humanitarian Board for GoldenRice (see Kryder D, SP Kowalski and AF Krattiger. 2000. *The Intellectual and Technical Property Components of pro-Vitamin A Rice* (*Golden*Rice™): A Preliminary Freedom-to-Operate Review. *ISAAA Briefs* No 20. ISAAA: Ithaca, NY. www.isaaa.org/kc/bin/isaaa_briefs/index.htm. See also www.goldenrice.org/Content2-How/how9_IP.html.

13 See, also in this *Handbook*, chapter 14.2 by SP Kowalski, and *supra* note 1.

CHAPTER 14.2

Freedom to Operate: The Preparations

STANLEY P. KOWALSKI, *The Franklin Pierce Law Center, U.S.A.*

ABSTRACT

Freedom to Operate (FTO) is the ability to proceed with the research, development and/or commercial production of a new product or process with a minimal risk of infringing the unlicensed intellectual property (IP) rights or tangible property (TP) rights of third parties. The procedure for assessing whether the product or process possesses FTO is called the FTO analysis, performed by meticulously dissecting the product or process into its fundamental components and then scrutinizing each for any attached IP or TP rights. The early preparations for an FTO analysis are crucial, because they will influence all that follows and hence determine the quality of the work product. Thorough preparation will lay a solid foundation, supporting a credible and reliable FTO analysis. This chapter explains these preparations through an example.

1. INTRODUCTION

1.1 *Freedom to operate defined*

Access to agricultural biotechnology (agri-biotech) and pharmaceutical (pharma) products, including vaccines, and processes can help developing countries improve public health and nutrition, contributing to the well-being of those most in need. Such products and processes are categorically technically complex. A cursory glance at a "materials and methods" section of any paper published in a scientific or medical journal reveals the plethora of components and processes that are routinely employed in the research, development, and eventual commercialization of an agri-biotech or pharma product. This technical complexity mirrors the corresponding intellectual property (IP) rights and tangible property (TP) rights complexity; that is, each component, process and/or combination thereof that went into the product might have either IP rights (for example, patents) or TP rights (for example material transfer agreements [MTAs]) of other parties attached. Hence, an agri-biotech or pharma product/process might not be "clean" in a legal sense, meaning that moving ahead with research, development, and commercialization could constitute infringement of another's IP or TP rights. However, the risk of infringement liability can be systematically managed and dramatically reduced. This is what freedom to operate (FTO) is all about.

Broadly defined, FTO means the ability to proceed with the research, development and/or commercial production, marketing or use of a new product or process with a minimal risk of infringing the unlicensed IP rights or TP rights of third parties.[1] The procedure for assessing whether or not the product or process possesses FTO is called the FTO analysis. An FTO analysis is performed by meticulously dissecting the product or process into its fundamental components and then scrutinizing each for any attached IP or TP rights. It is critical to make clear, however, that an FTO

Kowalski SP. 2007. Freedom to Operate: The Preparations. In *Intellectual Property Management in Health and Agricultural Innovation: A Handbook of Best Practices* (eds. A Krattiger, RT Mahoney, L Nelsen, et al.). MIHR: Oxford, U.K., and PIPRA: Davis, U.S.A. Available online at www.ipHandbook.org.

analysis neither explicitly nor implicitly denotes an absolute *freedom* to operate, but is instead a risk management tool, the purpose of which is to assess the *likelihood* for infringement-litigation liability associated with the new product or process: an FTO is therefore an informed, reasoned, and calculated *best estimate* of infringement liability, in a given jurisdiction, at a given period of time (that is, a snapshot assessment of the contours, canyons and crevasses of the IP/TP rights landscape for the specific product or process).[2]

Thus, an FTO analysis will inform an institution or company that the research, development, and commercialization of the new product or process may proceed with a minimal risk of infringing the unlicensed IP rights and/or TP rights of others.[3] However, as the IP/TP rights and legal landscape changes, shifts, and evolves, the dynamics and results of the FTO analysis may also change. (For example, patents may issue, expire, or be invalidated; licenses may be granted or terminated; patents may be assigned and then reassigned.) Also, patent rights are strictly territorial,[4] meaning that a product/invention might possess FTO in one jurisdiction (a nation where a relevant patent has not issued) but, on the other hand, would not possess FTO in another jurisdiction (a nation where a patent has issued). Therefore, the results of an FTO analysis must be periodically reassessed and updated where and when appropriate.[5]

1.2 *FTO analysis preparations: overview*

The FTO analysis must be organized, logical, methodical, meticulous, and carefully documented. An important initial step in a thorough FTO analysis (that patent counsel may then subsequently use to draft an FTO opinion) is the completion of the following preparations:

- assembling the FTO team
- analyzing, understanding, and dissecting the technology
- assessing plant pedigrees
- recognizing pharmaceutical technical considerations
- interviewing the researchers
- locating notebooks, lab records, and computer files
- finding MTAs, bag-tags, bags of seed, and any unknown property trail
- formulating the series of FTO questions
- selecting scientific databases
- selecting patent databases
- identifying special resources for pharmaceutical patent information
- understanding U.S. Patent and Trademark Office (PTO) information (file wrappers and disclosures)
- remaining aware of the 18-month "period of silence"
- maintaining due diligence throughout the FTO analysis

In this chapter, each of the aforementioned preparative steps is explained within the context of preparing for and conducting a successful FTO analysis. Applicable technologies might be either agri-biotech or pharma. Although the materials, methods, and tools used may be dissimilar from agri-biotech to pharma, the fundamental FTO principles and procedures remain unwavering for each of these. Hence, by following this FTO analysis blueprint, a series of sound FTO questions can be formulated, so as to lay a solid foundation from which a reliable FTO analysis will be able to develop. Patent counsel can then draw upon this analysis to formulate either one or a series of FTO opinions.

1.3 *Illustrative example*

Throughout this chapter, in order to help clarify and exemplify the topics covered, an illustrative hypothetical will be employed. It is a *purely fictionalized* situation, presented solely for the purpose of focusing the discussion and facilitating understanding.

1.3.1 *Background*

Recently a new viral disease has emerged in east Africa. The causative agent is a virus, simian in origin, having been asymptomatically endemic in an isolated population of pygmy desert baboons for millennia. The scourges of war, famine, and drought have impelled many people to seek sustenance from bush meat, which they eventually find by scouring the wilderness for days on end.

It is believed that the pathogenic virus made the leap from baboon to human when famished refugees consumed uncooked baboon meat infected with the virus, which likely rapidly entered the bloodstream via the portals of ulcerated oral lesions caused by advanced scurvy. Upon entering the new host, the virus migrated to skeletal muscles, where, in contrast to the primary baboon host, the virus causes progressive muscular degeneration with symptoms resembling myasthenia gravis. It is colloquially referred to as the "falldown disease" (FDD). The most serious concern with this emergent disease is that it appears to be readily transmissible from human to human via bodily secretions. Hence, it may have the capacity to spread throughout crowded refugee facilities, creating even more suffering and death.

The sudden appearance of this deadly virus has prompted a series of research and development efforts across the globe. These include developing techniques to raise the virus (it can only be cultured in monkey cells), sequencing and characterizing the viral genome, cloning the battery of genes that encode the viral proteins, and developing candidate vaccines.

An east African nation is home to the Institute of Dry Land Crop Research (IDLCR). This nation has recently acceded to the World Trade Organization (WTO) and is serious about becoming compliant with the Agreement on Trade-Related Aspects of Intellectual Property Rights (TRIPS) so that it can increase economic growth, for example by attracting greater foreign direct investment, particularly in the areas of emerging technologies, such as, biotechnology. As a result, a greater number of foreign interests are filing patent applications for their biotechnological applications and technologies in this nation, usually as part of the national-phase filing pursuant to the Patent Cooperation Treaty (PCT).

In response to the looming crisis of the emergent viral disease FDD, the IDLCR, in conjunction with this nation's leading medical research center, has launched a program to produce a large quantity of viral antigen in recombinant grain sorghum, transformed with the most immunogenic of the viral antigens. This will then be used to produce large amounts of vaccine to immunize thousands of displaced refugees. Such a research and development program will inevitably entail numerous proprietary components and techniques, likely having the IP and TP rights of third parties attached. Therefore, FTO issues will be a very real and constant concern.

2. ASSEMBLING THE FTO TEAM

2.1 *Skilled leadership of the FTO team*

From the very start of an FTO analysis, it is absolutely essential to establish credible, capable, competent leadership so that the FTO analysis is properly conceived, organized, and conducted. Because an FTO analysis is a multidisciplinary endeavor, the team leader must ensure that it remains focused, on-course, and precise. Under ideal circumstances, that is, qualified patent counsel is available and affordable, such counsel should lead the way. However, in many situations this might not be possible. Also, depending on the stage of the FTO analysis, patent counsel leadership might not be required. For example, early stages of a preliminary FTO analysis can be performed in lieu of counsel, possibly in order to assess or survey the IP rights landscape. Counsel may be sought later when and if it is warranted, possibly at later stages of the FTO analysis when questions of legal significance arise (for example, patent claims analysis). At such a stage, one possible route would be to seek pro bono counsel via services provided by public interest associations (for example, Public Interest Intellectual Property Advisors [PIIPA]).

In order to be most effective, the FTO team leader ideally should have expertise in agri-biotech and/or pharma, depending on the exact product and/or process undergoing FTO analysis. Furthermore (if patent counsel will not initially lead the FTO team) the FTO team leader must be the available professional with the greatest expertise in IP-related issues (for example, a technology-transfer professional officer, an intellectual property practitioner such as a patent agent or a scientist who has participated in various IP rights and technology-transfer courses, workshops, and/or seminars). The FTO team leader must understand the dynamics of the step-by-step process of FTO analysis, not only within the legal paradigm, but also from a sophisticated technical and scientific perspective. Because an FTO analysis is conducted at the interface of science and law, the FTO team leader must be professionally amphibious (that is, capable and comfortable in two different professional environments).[6]

2.2 *The FTO team is multidisciplinary*

The FTO team leader selects who will be part of the FTO team. FTO team members should include: scientists who had supervised the project, technology transfer personnel, and technicians/support staff. The last are absolutely essential, as they frequently know what *really* happened during product research, development, and commercialization. The FTO team might also include business personnel (depending on the stage of commercialization) and possibly administrative staff. The latter might have information pertaining to relevant communications, documents, and agreements. It is also very important to note that the FTO team may, or may not, be the same as the client. For example, the actual client might be a research institute, and the FTO team would be composed of employees.

2.3 *Work product doctrine and patent counsel*

One important reason that it is judicious to have patent counsel lead the FTO team, particularly at later steps in the analysis, pertains to maintaining the confidentiality of documentation. In the event that a claim of patent infringement arises, the FTO analyses and opinions, prepared under the guidance of patent counsel, may be protected from discovery (the compulsory disclosure of documents to an opposing party), pursuant to the attorney work-product immunity doctrine. However, it is unclear how far this immunity reaches, and so one must exercise caution. In general (in the United States), "*[pursuant to the U.S. Federal Rules of Civil Procedure, Rule 26(B)(3)] written material and mental impressions prepared or formed by an attorney in the course of performing legal duties on behalf of a client are protected from discovery as the attorney's 'work product' in the absence of undue prejudice or hardship to the party seeking discovery.*" In spite of this, "*there has been disagreement among courts construing this language as to its proper interpretation and its integration with other doctrines impacting on discovery jurisprudence … With respect to the standard of protection from discovery which an attorney's opinion work product should be given, a few courts have held that Rule 26(b)(3) mandates absolute protection, while a growing number of the more recent decisions have held that the standard of protection is less than absolute, with the strict protection generally afforded an attorney's opinion work product allowing for exceptions in certain circumstances.*"[7] Such complex issues relating to work-product immunity, and the extent to which it might reach, further illustrate the advisability of having qualified patent counsel as the FTO team leader.

After the FTO team is assembled, the leader coordinates, leads, and guides the team throughout the entire FTO analysis.

2.4 *The importance of scientific understanding*

In the case of FDD vaccine development, the IDLCR FTO team leader must carefully select a cadre of scientists who will, collectively, comprehend the spectrum of biological, genetic, agronomic, and biotechnological components and techniques that will go into the research, development, and commercialization of the vaccine. These individuals will form the basis of the FTO team. In addition, other professionals might be selected, such as technology transfer officers, administrators, and business managers. This team will then be poised to begin the arduous task of FTO analysis.

3. ANALYZING, UNDERSTANDING AND DISSECTING THE TECHNOLOGY

3.1 *Product deconstruction*

As the initial step in the FTO analysis, the FTO team must thoroughly know the precise nature of the technology itself, whether it is a product, process, or combination thereof (referred to hereinafter as the product/invention). In order to accomplish this, the FTO team must work closely with all of the research and development staff, so as to understand the nature of the technology to such an extent that it can be "disassembled" into its fundamental components, that is, *deconstructed.*[8]

Therefore, in the deconstruction phase of the early preparations for the FTO analysis, the FTO team and any other scientists, collaborators, or staff, work together to resolve the product/invention into the fundamental processes used to make it, the components that went into its construction, and any possible combinations of processes and/or components potentially pertinent.

3.2 *Research tools*

At this stage it is important to identify any research tools that were used during research and development of the product/invention.[9,10,11] Research tools, integral for the efficient development of commercial applications both in agri-biotech and pharma, are defined by the National Institutes of Health (NIH) as the "*full range of resources that scientists use in the laboratory including [a fragment of a gene, a gene], cell lines, monoclonal antibodies, reagents, animal models, growth factors, combinatorial chemistry and DNA libraries, clones and cloning tools (such as PCR), methods, laboratory equipment and machines, databases and computer software.*"[12,13] Identifying them is a critical step in the early FTO analysis preparations, because, although seemingly ubiquitous and readily, even "freely," available in many laboratories, there nevertheless appears to be no research tool usage (experimental use) exemption in the United States. To assume otherwise would be to unwisely overlook and thereby disregard important steps in the product/invention undergoing FTO analysis.[14]

3.3 *Components of the vaccine*

In the case of FDD vaccine development, the production and deployment of a vaccine from transgenic sorghum would entail numerous components and technologies, including, but not limited to:

- monkey cell culture (for viral propagation)
- antibodies against the viral proteins
- the viral genome
- individual viral genes
- research tools used to clone the viral genes (for example, the polymerase chain reaction [PCR], and related techniques)
- plant transformation techniques (for example, *agrobacterium* and/or bio-projectile methodologies)
- plant genetic transformation constructs (for example, vectors, promoters, transit peptide sequences)
- plant cell culture techniques and cell lines
- sorghum germplasm used for genetic transformation
- procedures for harvesting and purifying expressed antigen
- formulation, production and delivery of the actual vaccine

Each of the above would most likely represent a deconstructed piece of the contemplated final vaccine, and each would therefore constitute an FTO question (see section 8) that the FTO team would subject to thorough scrutiny in the FTO analysis.

4. IP AND TP RIGHTS

At this stage, it will be instructive to briefly and clearly define some of the forms of IP and TP rights that are commonly encountered in an FTO analysis.

4.1 *Patents*

Patents, as referred to in this paper, are utility patents: a grant by a government to an inventor, for the right to exclude others from making, using, or selling his or her invention, for a specified term of years. This is done in exchange for the inventor fully disclosing the invention in the patent document (typically the specification). Hence, a patent can be viewed as a contract between the inventor and the government, wherein the inventor provides full disclosure of the invention in exchange for absolute exclusivity to the IP rights for a specified term. Patents are applicable to both agri-biotech and pharma.[15]

4.2 *MTAs*

MTAs are legal instruments that typically accompany the transfer of TP. They usually (possibly *ideally*) document *what* is transferred, *who* transfers to *whom*, as well as the provisions, uses, scope of rights, confidentiality, and term of the agreement.[16] MTAs are legally defined as bailments.[17] So, the question naturally arises, what is a bailment? A bailment is the delivery of an item of TP from one party to another, for a specific purpose, pursuant to the terms of a contract. However, in a bailment it is critical to remember that although there is a change in the actual physical *possession* of the property, there is no transfer of *ownership*: title remains with the owner (bailer)—even though possession has shifted to the recipient (bailee).[18] In addition to being a bailment, an MTA also entails contractual obligations, and hence, as a binding contract, the terms and provisions of an MTA must be taken very seriously by both parties involved in the transfer/transaction, so as to avoid the possibility of breach of contract liability.[19]

The terms and provisions of MTAs can vary considerably, particularly when comparing MTAs executed by the nonprofit sector (for example, universities) with those executed by the for-profit sector (for example, corporations).[20] Confidentiality, publication rights, and reach-through rights may vary significantly, and one must exercise caution so as not to agree to an MTA with potentially onerous terms.[21] If the material used in the development of the product or process was obtained in violation of an MTA between two other parties, then the "obtainer" of the material may be liable for unauthorized use. For example, Andy transfers (technically speaking bails) a plasmid to Roberta (with specified contractual obligations attached), which is then "obtained" by Carl, via trick, theft, or other nefarious means, and Carl then uses it to either develop, or incorporate into, his product/invention. Carl might very likely have a liability problem—possibly misappropriation of Andy's tangible property.[22] MTAs are applicable to both agri-biotech and pharma.

4.3 *Bag-tags*

Bag-tags, a type of agri-biotech TP rights protection, are enforceable contracts[23] that restrict the licensee (grower) in the use and/or reuse of seed.[24] The bag-tag license is analogous to shrink-wrap, box-top, and tear-me-open software license transactions, such that an implicit contract is formed when the seal is broken, which then obligates the grower to the terms of the license as articulated on said seal.[25, 26]

4.4 *Plant/germplasm protection*

4.4.1 *Plant IP rights statutes*

Germplasm IP rights protection (agri-biotech) exists in various forms, with each form addressing different types and levels of what is protected. In the United States, the Plant Patent Act (PPA), the Plant Variety Protection Act (PVPA), and Utility Patents for Plants (UPP) are the statutory forms of germplasm IP rights available.[27] The PPA provides IP rights protection for asexually (vegetative) propagated plants, (for example, plants that are propagated from cuttings or by budding or grafting); tuber-propagated plants (potato varieties) are not covered by the PPA. The PVPA provides IP rights protection for sexually propagated plant varieties, F_1 hybrids, and also tuber-propagated plants (potato varieties); plant varieties must meet the new, distinct, uniform, and genetic-stability requirements. With UPP, the level of IP rights protection is much broader than that afforded by either the PPA or the PVPA. The PPA and PVPA only confer IP rights protection for certain plant varieties, but UPP can claim plants, plant varieties, plant parts, seeds, and tissue cultures.[28, 29]

4.4.2 *Plant IP rights treaties*

In addition to the PPA, PVPA, and UPP, there are two treaties that address germplasm IP rights protection: the International Treaty on Plant Genetic Resources for Food and Agriculture (PGRFA)[30] and the Convention of the International Union for the Protection of New Plant Varieties (UPOV).[31] In PGRFA, important provisions include an agreement not to claim IP rights for any of the germplasm resources "*in the form received*" from the multilateral system. There is also a benefit-sharing scheme triggered by the commercialization of new plant varieties.[32, 33] A treaty seeking to impart international conformity in plant variety protection, UPOV, fundamentally consistent with the PVPA, specifies that the fundamental criteria for IP rights protection are distinctiveness, uniformity and stability.[34]

4.5 *Technology-use licenses*

Technology-use licenses may need to be sought for the use of certain research tools (see section 6.2), which frequently are indispensable in order to facilitate the research and development phase of an agri-biotech or pharma product, process, or application.[35] Although there is currently considerable debate as to whether the patenting and licensing of research tools should be subject to either experimental use exceptions or compulsory licensing schemes,[36] the basic presumption should remain that there is no experimental use exemption for research tools, regardless of whether the work is performed in a profit or nonprofit entity.[37]

5. ASSESSING PLANT PEDIGREES

5.1 *The complexity of plant-related IP rights*

When analyzing an agri-biotech product/invention, it is necessary to determine the pedigree of the germplasm forming its very foundation. In other words, the trail of germplasm, with as much detail as possible, must be traced and documented. If detailed breeding records are available, this task will be much easier. Hence, the FTO team must ask these questions: What type of germplasm is the product/invention embedded in? Where did this germplasm come from? What is the detailed pedigree of the germplasm?

Furthermore, as already discussed hereinabove, plant germplasm may be protected by various overlapping forms of IP rights:

- trade secrets (primarily for proprietary inbred lines, for example, in hybrid maize breeding)[38]
- utility patents[39]
- Plant Variety Protection Act (PVPA)[40]
- Plant Patent Act (PPA)[41]
- UPOV (as consistent with the PVPA)[42]
- PGRFA (for germplasm accessed from the multilateral system)[43]

Hence, the FTO team must remain aware of the possibility of a complex IP/TP rights situation with regard to germplasm. It must therefore proactively corral as much information as possible.

5.2 *Germplasm issues*

Concerning *in planta* expression of viral antigen in transformed sorghum, varieties contemplated for genetic transformation with the viral gene(s) will likely present complex germplasm considerations during the FTO analysis. For example, overlapping forms of IP and TP rights protection might apply: an ideal sorghum line could simultaneously have third-party patent and plant variety protection rights attached. Since the nation where the IDLCR is located is seeking to comply with the TRIPS Agreement, it will likely have a UPOV-harmonized PVPA enacted as statutory law, and certainly also a patent statute. Hence, germplasm issues, occasionally (and foolishly) subordinated to patents in an FTO analysis, will be of critical importance.

6. RECOGNIZING PHARMACEUTICAL TECHNICAL CONSIDERATIONS

6.1 *Pharma components*

As with agri-biotech, when examining a pharma product/invention the FTO team will need to consider pharma-product/process-specific components.[44]

The compound itself must be considered:

- crystalline form
- amorphous form
- enantiomers
- metabolites
- prodrugs

The types of pharmaceutical compositions must also be considered:

- delivery systems
- vehicles
- adjuvants

The methods, steps, and components involved in the product synthesis are also critical (see also section 6.2):

- steps and the reagents and techniques that compose each step
- intermediates (For example, for a five-step synthesis, there are at least four intermediates to clear and four sets of the reagents that are used to convert the intermediates.)[45]
- reagents (For example, "*Before launching an all-out patent search, it is often productive to search your old organic chemistry/biochemistry textbooks and Aldrich/Sigma catalogs, and ask two questions: (a) what chemical utilities and processes are clearly within the public domain, or (b) can be purchased from vendors that can sell them to you for unrestricted use?*")[46]
- purification techniques and protocols
- handling techniques and procedures

Methods of use, that is, downstream considerations, also are important to keep in mind:

- modes of treatment
- dosimetry
- limiting side effects

6.2 *Research tools*

And finally, but no less important, research tools must be considered. Biotechnology research tools are used in the development of drug products, therapeutic devices, or diagnostic methods. These research tools are not themselves physically incorporated into the final product/device/diagnostic. Hence, they represent the full range of resources used in drug discovery and development.[47] (See also section 3.2.)

6.3 *Vaccines*

In the case of vaccines, there are additional FTO analytical considerations specific for vaccine research, development, manufacture and deployment, including:

- expression systems
- fusion partners
- immunostimulators
- adjuvant systems
- excipients
- delivery devices[48]

As with the pharma product/invention, the FTO team must carefully analyze each of these and, using the results of this analysis, formulate an appropriate series of FTO questions (see section 8). In the case of the FDD vaccine FTO analysis, there will be:

- upstream considerations (for example, the viral genes, monkey cell culture, cloning)
- midstream considerations (for example, sorghum germplasm and plant transformation, *in planta* antigen expression)
- downstream considerations (for example, vaccine formulation, production, optimization [adjuvant selection] and delivery)

As already discussed, each of these will likely have third-party IP and/or TP rights appurtenant.

7. INTERVIEWING RESEARCHERS AND LOOKING FOR RECORDS

7.1 *Interviews and laboratory history*

To ensure success when performing the FTO analysis, a continuing rapport between the FTO team and scientific and technical staff is essential. This will help keep everyone involved on the same track, maintain momentum, and keep the FTO analysis up and running. Such informal dialogues with research personnel can reveal critical snippets of information, such as the trail of acquisitions. (For example, who got what from whom, and was it with or without proper authorization as to embedded IP and/or TP rights?) Consider this hypothetical scenario: Andy obtains a product component from Roberta, who had previously obtained it from Carl. However, there was no proper authorization (for example, no MTA) for such a transfer in the first instance, which is definitely something that the FTO team needs to know.

Such anecdotal narratives can never be found in a paper trail; these are solely preserved in the "oral history" of the laboratory. Thus, the FTO team must, at times, function as investigative cultural anthropologists, sorting through the history, habits (possibly bad habits), and "traditions" of a laboratory and research group. Additionally, this sort of dialogue will also help researchers to recall more fully what they had done, allowing them to fill gaps in the written records. What is in the laboratory notebooks may only be part of the story.

7.2 *The paper trail*

Still, the FTO team must tenaciously pursue every paper trail, searching the laboratory offices, greenhouse, and even the field house, in order to track down notebooks, laboratory records, associated paperwork, computer files, MTAs, bag-tags, bags of seed, and any evidence suggesting an unknown tangible property trail, misappropriated property, or unauthorized access to a third party's confidential information. A comprehensive review of the research and development group's written and oral records and related information will thereby enable the FTO team to acquire a sophisticated understanding of what the product/invention is and what IP and TP rights might be involved.

After the FTO team has identified and understood each of the fundamental units of the deconstructed product and/or process, they then can use this information to frame a series of "FTO questions."

7.3 *Template for FTO questions*

For the FDD vaccine, the product deconstruction table (Table 1) concisely summarizes the components and process that go into its research, development, and commercialization, as well as the potentially appurtenant third-party IP and TP rights. This is the template, the roadmap, from which the FTO questions (see section 8) can be formulated, addressed and analyzed.

8. FORMULATING THE FTO QUESTIONS

Following the technical deconstruction of the product/invention, a series of FTO questions are formulated.[49] These questions are structured to systematically analyze the dissected processes, components, and any combinations thereof, for potentially embedded IP rights (for example, patents and trade secrets) and TP rights (for example, MTAs and bag-tags[50, 51]). Each FTO question, therefore, asks whether a method to make, a material used to make, or any combination of methods and materials, has, or may have, third-party IP or TP rights attached. Thus, a *single* material or method, used in the development of either an agri-biotech or pharma product/invention, may have multiple proprietary issues, that is, both an IP right (for example, a patent right) and a TP right (for example, an MTA) of potential relevance. The gravity of formulating a correct series of FTO questions, then, underscores the necessity for caution and meticulousness at this early stage in the FTO analysis, because all the work that follows is built upon this foundation.

9. SCIENTIFIC DATABASES

Note: scientific database searches and patent database searches are mutually reinforcing, that is, the two support, verify, guide, and inform each other throughout the process of the FTO analysis. For example, inventors might be authors; institutions might be assignees; scientific discoveries might be the actual invention (disclosed in a scientific publication).

Scientific database searching, along with patent database searching, are integral to the FTO analysis. This is where the FTO team assembles the piles of raw information and data, both written and anecdotal, that will subsequently be parsed, analyzed, and organized in order to address the FTO questions that the FTO team has formulated. Furthermore, the FTO team needs to know what types of scientific informational resources are available, both freely and also on a premium, value-added, pay-per-view basis. Furthermore, the FTO team needs to understand what constitutes the value added for the pay-per-view databases, so that they will be used according to specific needs at certain times in the FTO analyses in the most cost-effective manner.

There are many examples of scientific databases. For example, freely available ones include:

- Agricola[52]
- Google™[53]

Whereas premium value-added, pay-per-view databases include:

- Biosis[54]
- Current Contents[55]
- Cab Abstracts[56]

Table 1: Product Deconstruction, FDD Transgenic Vaccine

Technological component, process, or tool	Proprietary protection, likely appurtenant	Relevant documents
Monkey cell culture (for viral propagation)	IP Rights, TP Rights	patents, MTAs
Antibodies against the viral proteins	IP Rights, TP Rights	patents, MTAs
The viral genome	IP Rights	patents
Individual viral genes	IP Rights	patents
Research tools used to clone the viral genes (for example, the polymerase chain reaction [PCR], and related techniques)	IP Rights	patents, technology-use licenses
Plant transformation techniques (for example, *agrobacterium* and/or bio-projectile methodologies)	IP Rights	patents, technology-use licenses
Plant genetic transformation constructs (for example, vectors, promoters, transit peptide sequences)	IP Rights, TP Rights	patents, MTAs
Plant cell culture techniques and cell lines	IP Rights, TP Rights	patents, MTAs
Sorghum germplasm used for genetic transformation	IP Rights, TP Rights, possibly trade secrets (for example, if variety was developed using parental lines protected as trade secrets)	patents, plant variety certificates, possibly MTAs (for example, if germplasm is covered by the PGRFA), bag-tag licenses
Procedures for harvesting and purifying *in planta* expressed antigen	IP Rights	patents
Formulation, production, and delivery of the actual vaccine	IP Rights, TP Rights, trade secrets (for example, confidential third-party know-how and/or show-how protected as trade secrets)	patents, MTAs, technology-use licenses

10. PATENT DATABASES

10.1 *Free and premium databases*

As with scientific databases, the FTO team needs to know what resources are available vis-à-vis patent databases, both freely available and premium value-added, pay-per-view. The FTO team should also know the type of value added for the pay-per-view databases. These databases can then be accessed according to specific needs at key stages in the FTO analyses.[57]

For example, freely available patent databases include:

- PTO[58]
- esp@cenet®[59]

And premium pay-per-view (with value-added features) patent databases include:

- Delphion[60]

10.2 *Pay-for-view, value-added features*

For purposes of illustration, some of the value-added features of Delphion that distinguish it from either the PTO or esp@cenet are discussed here. While free patent research sites can provide patent records, they do not offer the analytical and productivity tools needed to make sense of the data in those records. What follows are some of the key features of Delphion that can make this fee-based service the right choice at the right time in the FTO analysis.[61]

Rather than presenting just a patent record, the primary display record on Delphion is an integrated view that provides a cross-collection of information without the need to perform extra queries. Included in the integrated view are:

- family information showing the countries in which an invention is protected
- the Derwent World Patents Index (DWPI) title and abstract written in English using clear, concise, industry-specific terms
- accessible references to both patent and nonpatent prior art
- extensive hyperlinking to a variety of related information—including definitions for the fields contained in the integrated view

Delphion offers pay-per-use searching of the value-added DWPI database, which covers 13 million unique inventions and has a unique hierarchical system of coding allowing extra precision and accuracy in searching. DWPI data can be used in most of the Delphion analytical and productivity tools. The Delphion Snapshot analytical tool creates quick, easy-to-read bar charts allowing summarization of key bibliographic data—and then further refinement of those summaries. Delphion Work Files allow the saving of result sets or groups of patents that are to be reviewed for future reference. One can easily share these Work Files with colleagues, thus allowing worldwide collaboration. And one can also use analytical tools, like Snapshot, to perform further analyses of these groups of hand-selected records. Delphion allows a user to save frequently used queries, thus eliminating the need to reconstruct them each time. This saves time and decreases the chance for errors to occur in queries. Saved searches can be set to run automatically, advising one of the search results. Data Extract exports more than 50 key bibliographic fields in formats designed for use in other popular applications. The Family Legal Status reports the current legal status of the family members of the invention being examined, which means that there is no need to individually search for each member of the family in order to ascertain the overall view of the protection in each jurisdiction. Delphion, as part of the Thomson Scientific family of IP solutions, offers all the advantages of working with a worldwide company, including a robust infrastructure and support network, interoperability with other Thomson Scientific solutions, and a global perspective on IP research and management.

11. PHARMACEUTICAL PATENT INFORMATION

A pharma product/invention, has, in addition to the standard patent search tools and resources listed hereinabove, its own patent resource materials. These include the Orange Book, the Merck Index, and the actual physical "shoes" at the PTO.

11.1 *The Orange Book*

The Orange Book, "*is an FDA-published document available in paper and electronic form that lists all FDA-approved drugs with any patents pertaining thereto.*"[62] The Orange Book contains approved drug products with therapeutic equivalence,[63, 64] as well as the expiration dates of patents on therapeutic small molecules and on approved indications and compositions.[65] The Orange Book is available as a printed, bound edition, complete with an orange cover, or online.[66]

11.2 *The Merck Index*

The Merck Index lists patents and publications on older drugs and reagents.[67] It is available as a printed edition or online.[68]

11.3 *PTO shoes*

When working with a pharma product/invention, a hand search of the "shoes" in the PTO may be prudent.[69] This is an actual physical paper search, within the shoes: the boxes containing patent prior art.[70] This is sometimes necessary due to the differences in nomenclature used by various patent drafters, differences that might not be readily identified and sorted out in electronic searching. Hence, under certain circumstances, the physical shoe search is an added measure of due diligence.

12. PTO INFORMATION

In addition to searching scientific and patent databases, and checking the Orange Book, Merck Index, and the PTO shoes, there are several other resources of which the FTO team needs be aware. These include patent applications and the patent file wrapper.

12.1 *The patent file wrapper*

A very specialized informational resource is the patent file wrapper. The file wrapper is a physical folder, held by the PTO. It contains documents pursuant to the patent application and prosecution, including the original patent (or trademark) applications, as well as any amendments, affidavits, and written arguments submitted by the applicant, and the actions taken by the examiner concerning the application.[71] The file wrapper becomes publicly available only after the patent issues. The file wrapper can be either physically accessed,[72] or accessed via a searchable, writable, PDF format, which requires an up-to-date version of Adobe® Reader® and sufficient RAM (random-access memory) on the searcher's computer.[73]

12.2 *Patent counsel analyzes the patent file wrapper*

Since the file wrapper is such a specialized informational resource, it will typically be accessed and analyzed during an FTO analysis specifically to address very technical issues (for example, claims interpretational queries, usually done only near the terminal phase of the FTO analysis). Furthermore, the file wrapper should be searched and analyzed only by qualified patent counsel, who ideally, at this late stage in the FTO analysis, is the leader of the FTO team. This is because counsel, by reviewing any patent claim amendments or disclaimers, will be using the contents of the file wrapper to carefully construe the precise meaning and scope of the claim language.[74, 75] It is important to recall that an FTO analysis proceeds from broad and general to narrow and precise. Correspondingly, the analysis of patents proceeds from the patent itself (the abstract, claims and specification) to the claim language construction, to the file wrapper contents.[76] Hence, the greater the precision and specificity of the analysis, the greater the advisability for patent counsel participation: the ability to understand the legal basis of claim meaning and scope become critical at the later stages of the FTO analysis.

12.3 *Patent applications*

Patent Applications are filed with the PTO, the PCT, and also in the various National Phase Applications. Although patent applications do not technically confer statutory IP protection, they nevertheless are a good indicator of what might be subject to protection pending patent issuance.

13. THE "PERIOD OF SILENCE"

It is critical to understand that patent applications will not be available prior to publication, and so their contents remain unknown for a period of 18 months after the earliest effective filing date.[77] Therefore, whereas such inventions are held in trade-secret status during this period, they nevertheless are still pending as *potential future patents.* However, under U.S. law, if the patent application is only to be filed in the United States, then the 18-month rule may not apply. (That is, the applicant may opt out of the 18-month requirement, and in that case the invention, as disclosed in the patent application, remains a trade secret until patent issuance.[78]) The 18-month period of silence, therefore, has implications in the FTO analysis, in that there may be pending IP rights, still below the surface, but nonetheless relevant to the FTO analysis. A diligent analysis of the published scientific literature, including conference papers, abstracts, and presentations, might suggest what pertinent IP rights are lurking in patent applications still hidden during the 18-month period.

14. DUE DILIGENCE

During the preparation, set-up, data accumulation, and FTO question-formulation stages of an FTO analysis, due diligence is required. *Due diligence*, broadly defined, is "*Such a measure of prudence, activity, or assiduity, as is properly to be expected from, and ordinarily exercised by, a reasonable and prudent [person] under the particular circumstances; [Due diligence is] not measured by any absolute standard, but [depends] on the relative facts of the special case.*"[79] From a practical standpoint, due diligence necessitates a methodical approach, such that all forms of IP and TP rights are garnered, organized, and assembled into a coherent document, for example, a Microsoft Excel spreadsheet.[80] The question often arises, as to how much diligence is enough. The answer? When one finds oneself treading the same ground, then the requirements of due diligence are satisfied.

15. CONCLUSIONS

The preparations for an FTO analysis will determine the quality of the final work product. Organization, thoroughness, meticulous documentation, and solid leadership by a capable FTO team leader will all combine to contribute to a successful outcome. A comprehensive checklist of what must be established during the early stages of the FTO analysis serves as a helpful tool.[81] For example, the list should include:

- possible pertinent patents, including their prosecution and/or litigation status
- patent applications
- third-party trade secrets, including whether they might have been misappropriated
- all third-party TP rights
- all research tools used to make the agri-biotech product or pharmaceutical innovation
- any agreements (for example, trade secret licenses, MTAs, bag-tag [shrink-wrap], or technology-use licenses, noting conditions and restrictions appurtenant)

And finally, it is imperative that all records are properly maintained. Consistent records of all searches and search terms must be documented and organized. This should include:

- spreadsheets of all FTO search results
- records of search terms used
- databases searched
- interviews with researchers, with notes
- notes and annotations by patent counsel

Having spent the early phases of the FTO analysis with the disciplined rigor laid out in this chapter, the later steps in the FTO analysis should proceed with a minimum of problems. Diligence will pay off in the end with a solid and reliable FTO analysis that can be routinely updated and revised and that can also provide patent counsel with the requisite information for drafting FTO opinion letters. ■

STANLEY P. KOWALSKI, *The Franklin Pierce Law Center, 2 White Street, Concord, NH, 03301. U.S.A. spk3@cornell.edu and skowalski@piercelaw.edu*

---

1 Anonymous. 2004. Freedom to Operate. *WIPO Magazine*, September–October, Issue 5.

2 Woessner WD. 2002. Product Clearance and Noninfringement Opinions in a Post-Festo World. *PLI/Pat* 715: 135–66.

3 Kowalski SP and RD Kryder. 2002. Golden Rice: A Case Study in Intellectual Property Management and International Capacity Building. *Risk: Health Safety & Env't* 13: 47–67. www.piercelaw.edu/Risk/Vol13/spring/Kowalski.pdf.

4 Burrone E. New Product Launch: Evaluating Your Freedom to Operate. The World Intellectual Property Organization. www.wipo.int/sme/en/documents/freedom_to_operate.html.

5 Kryder RD, SP Kowalski and AF Krattiger. 2000. The Intellectual and Technical Property Components of Pro-Vitamin A Rice (Golden Rice): A Preliminary Freedom-to-Operate Review. Brief No. 20, ISAAA: Ithaca, NY. www.isaaa.org/kc/bin/isaaa_briefs/index.htm.

6 Valoir T. 2005. Look Before You Leap: Investing in a Freedom-to-Operate Patent Review. *Executive Legal Advisor*, January 20–21. www.bakernet.com/BakerNet/Resources/Publications/Recent+Publications/FTO+Paper.htm.

7 Wagner JF. 2005. Protection from Discovery of Attorney's Opinion Work Product under Rule 26(B)(3), Federal Rules of Civil Procedure. *American Law Reports Federal* 84: 779–801.

8 Kowalski SP, RV Ebora, RD Kryder and RH Potter. 2002. Transgenic Crops, Biotechnology and Ownership Rights: What Scientists Need to Know. *Plant J.* 31: 407–21. www.blackwell-synergy.com/links/doi/10.1046/j.1365-313X.2002.01367.x.

9 Aljalian NN. 2005. The Role of Patent Scope in Biopharmaceutical Patents. *B.U. J. Sci. & Tech. L.* 11:1–76.

10 Mireles MS. 2004. An Examination of Patents, Licensing, Research Tools, and the Tragedy of the Anticommons in Biotechnology Innovation. *U. Mich. J. L. Reform* 38: 141–235.

11 Xiao J. 2003. Carving Out a Biotechnology Research Tool Exception to the Safe Harbor Provision of 35 U.S.C. §271 (e)(1). *Tex. Intell. Prop.* L. J. 12: 23–67.

12 See *supra* note 9.

13 See *supra* note 10.

14 See *supra* note 10.

15 McCarthy JT, RE Schechter and DJ Franklyn. 2004. *McCarthy's Desk Encyclopedia of Intellectual Property, 3rd edition*. The Bureau of National Affairs: Washington, DC.

16 Simpson PM. 1998. Use of Bailment in Transferring Technology from a University. *Journal of the Association of University Technology Managers* 10: 85–100.

17 Goldstein B. 2002. Overview of Technology Development. *Introduction to the Principle and Practice of Clinical Research* (ed. John Gallin) pp. 307–27. Academic Press.

18 Gardner BA, ed. 1999. *Black's Law Dictionary*. West Group, St. Paul, MN.

19 Schwaller M. 2004. Uniform Biological Material Transfer Agreements: An Argument for Uniform Use. *Hous. Bus. & Tax. L.J.* 4: 190–230.

20 Rodriguez V. 2005. Material Transfer Agreements: Open Science vs. Proprietary Claims. *Nature Biotechnology* 23: 489–91.

21 Houser D. 1993. Exemptions under Patents and Certificates Covering Plants and Comments on Material Transfer Agreements. In *Intellectual Property Rights; Protection of Plant Materials* (eds. SP Baenziger, RA Kleese, and RF Barnes) pp. 107–109. Crop Science Society of America: Madison, WI.

22 See *supra* note 20.

23 Janis MD and JP Kesan. 2002. Plant Variety Protection: Sound and Fury? *Hous. L. Rev.* 39: 727–78.

24 Binenbaum E, C Nottenburg, PG Pardey, BD Wright, and P Zambrano. 2003. South-North Trade, Intellectual Property Jurisdictions, and Freedom to Operate in Agricultural Research on Staple Crops. *Econ. Devel. Cult. Change* 51: 309–35.

25 See *supra* note 18.

26 Wright BD. 2002. Biotechnology and Intellectual Property. In *California Council on Science and Technology, Benefits and Risks of Food Biotechnology*. Chapter 8. July, pp. 151–65.

27 See *supra* note 15.

28 Daniels TP. 2003. Keep the License Agreements Coming: The Effects of J.E.M. AG Supply, Incorporated V. Pioneer Hi-Bred International, Incorporated on Universities' Use of Intellectual Property Laws to Protect Their Plant Genetic Research. *B.Y.U. Educ. & L.J.* 2: 771–813.

29 Koo B, C Nottenburg and PG Pardey, 2004. Plants and Intellectual Property: An International Appraisal. *Science* 306: 1295–1297.

30 Fowler C. 2004. Accessing Genetic Resources: International Law Establishes Multilateral System. *Genet. Resour. Crop Ev.* 51: 609–20.

31 Benda S. 2003. The Sui Generis System for Plants in Canada: Quirks and Quarks of Seeds, Suckers, Splicing, and Brown Bagging for the Novice. *Canadian Intellectual Property Review* 20: 323–80.

32 See *supra* note 30.

33 Brush SB. 2005. Protecting Traditional Agricultural Knowledge. *Wash. U. J.L. & Pol'y* 17: 59–109.

34 See *supra* note 31.

35 Ducor P. 1999. Research Tool Patents and the Experimental Use Exemption: A No-Win Situation? *Nature Biotechnology* 17: 1027–28.

36 Hoffman DC. 2004. A Modest Proposal: Toward Improved Access to Biotechnology Research Tools by Implementing a Broad Experimental Use Exception. *Cornell L. Rev.* 89: 993–1043.

37 See *supra* note 10.

38 *Pioneer v. Holden*. 35 F.3d 1226, 31 U.S.P.Q. 2d 1385 (8th Cir. 1993).

39 Agris, CH. 1999. Intellectual Property for Plants. *Nature Biotechnology* 17: 197–98.

40 Ibid.

41 See *supra* note 39.

42 See *supra* note 30.

43 See *supra* note 31.

44 Woessner WD. 2003. Patent Infringement and Freedom-to-Operate Opinions. *PLI/Pat* 761: 99–132.

45 See *supra* note 2.

46 See *supra* note 2.

47 See *supra* note 11.

48 The World Health Organization Immunizations, Vaccines, and Biologicals. 2004. In *Intellectual Property Rights and Vaccines in Developing Countries*. Meeting report, April 19–20. www.technet21.org/whodocs.html.

49 See *supra* note 6.

50 See *supra* note 19.

51 See *supra* note 24.

52 www.nal.usda.gov.

53 www.google.com.

54 www.biosis.com.

55 www.isinet.com.

56 www.cabi.org.

57 See, also in this *Handbook*, chapter 14.3 by H Thangaraj, RH Potter and A Krattiger.

58 www.uspto.gov.

59 www.espacenet.com/access.index.en.htm.

60 www.delphion.com.

61 Information graciously provided by Kathy Little (Thomson Scientific).

62 Derzko NM. 2005. The Impact of Reforms of the Hatch-Waxman Scheme on Orange Book Strategic Behavior and Pharmaceutical Innovation. *IDEA* 45:165–265.

63 Anderson DZ. 2003. New Orange Book Listing Requirements: Will FDA Raise Bar on Generic Drug Entry? *Lawyers J.* 5: 6–7.

64 Wharton JS. 2003. "Orange Book" Listing of Patents under the Hatch-Waxman Act. *St. Louis U. L.J.* 47:1027–63.

65 See *supra* note 2.

66 www.fda.gov/cder/ob/default.htm.

67 See *supra* note 2.

68 O'Neil MJ, A Smith, PE Heckelman and S Budavari (eds.). 2006. *Merck Index: An Encyclopedia of Chemicals, Drugs, & Biologicals*. Merck & Co: Rahway, NJ. library.dialog.com/bluesheets/html/bl0304.html and www.merckbooks.com/mindex/.

69 See *supra* note 2.

70 See *supra* note 15.

71 See *supra* note 15.

72 See *supra* note 15.

73 See *supra* note 6.

74 Berliner BM. 2004. Preparing Patent Infringement and Freedom-to-Operate Opinions. *Computer & Internet Law* 21: 1–8.

75 See *supra* note 6.

76 Deveau T, and M Hall. 2002. Noninfringement and Freedom-to-Operate Opinions. *PLI/Pat* 715: 115–33.

77 See *supra* note 15.

78 Watase R. 2002. The American Inventors Protection Act of 1999: An Analysis of the New Eighteen-Month Publication Provision. *Cardozo Arts & Ent. L.J.* 20: 649–84.

79 Sullivan RC. 2001. Intellectual Property Due Diligence. AUTM 2001 Annual Meeting, New Orleans, Louisiana, March 2. www.darbylaw.com/Files/tbl_s31Publications/FileUpload137/1629/Autmrcs.ppt.

80 Martinez de Andino JM, RL Tate and T Maddry. 2004. Conducting an Intellectual Property Due Diligence Investigation. *J. Proprietary Rts.* 8: 1–6.

81 Ibid.

CHAPTER 14.3

How and Where to Search for IP Information on the World Wide Web: The "Tricks of the Trade" and an Annotated Listing of Web Resources

HARRY THANGARAJ, *Director, Research, MIHR, U.K.*
ROBERT H. POTTER, *Senior Associate, Agriculture & Biotechnology Strategies, Inc., Canada*
ANATOLE KRATTIGER, *Research Professor, the Biodesign Institute at Arizona State University, Chair, bioDevelopments-International Institute; and Adjunct Professor, Cornell University, U.S.A.*

ABSTRACT

Emphasizing patents and patent searching, this chapter will put readers on the initial path to understanding and protecting intellectual property (IP). By exploring patent information on the Web site of the European Patent Office and other Web sites listed in this chapter, the reader can begin to learn by doing and quickly gain experience that should improve his or her searching skills. Other resources dealing with IP in general are described. This collection is by no means exhaustive, given the vast amount of information on IP that is present on the Web, but the sites listed here should be valuable in accessing unbiased, useful information about the IP landscape, especially for key areas of technological interest. The value of IP searches for a typical technology transfer office is also discussed.

1. INTRODUCTION

One of the major advantages of the information age is the ability for almost anyone to access information and resources that would otherwise be available only to specialists. The Internet—and its offspring, the World Wide Web—have become so pervasive that there is now little information that cannot be obtained from your desk for free or at a relatively low cost. Information about patents and other intellectual property (IP) is now almost instantly available; however, it takes a certain level of knowledge and experience to get there.

But the ubiquitous and egalitarian nature of the Internet raises some problems—the biggest problem is that it is overloaded with essentially unchecked and often highly partisan information. For this reason, a novice searcher needs some background on how to obtain relevant information and how to properly assess the reliability of a source. Simply typing *patent* or *intellectual property* into a search engine is likely to get hits, from many highly biased sources, on the desirability (or otherwise) of a patent system and the wealth-creating or wealth-destroying nature of IP regimes. Providing you with some good, general, and, we hope, unbiased places to start is one of this chapter's main goals.

A great deal of valuable information on IP rights can be found in the databases of patents and patent filings, which are now becoming more accessible; however, the databases do not provide comparable levels of interpretation and can be somewhat idiosyncratic. With databases, as with many other things, one gets what one pays for, and fee-based subscription services are always going to have more value. That does not mean that the free services are without value: a great deal can be achieved using these free sources alone.

One proviso has to be included before continuing: any searches you can perform yourself are not likely to be as complete or as well-prepared as

Thangaraj H, RH Potter and A Krattiger. 2007. How and Where to Search for IP Information on the World Wide Web: The "Tricks of the Trade" and an Annotated Listing of Web Resources. In *Intellectual Property Management in Health and Agricultural Innovation: A Handbook of Best Practices* (eds. A Krattiger, RT Mahoney, L Nelsen, et al.). MIHR: Oxford, U.K., and PIPRA: Davis, U.S.A. Available online at www.ipHandbook.org.

those prepared by a professional patent agent or patent attorney. For any kind of IP related opinion, about which there may be legal or financial repercussions, retaining the services of an attorney is a necessity. However, for preliminary searches, for finding background information, for keeping up with the most current technological developments, and even for personal interest, knowing where to look to find patent information is very useful.

2. PATENT SEARCHES

Many people assume that IP is all about patents and that searching patent databases is a good way to identify when a product has some protected IP components. Although this is not necessarily true, patent searching is of great importance to technology transfer offices (TTOs) and IP management offices in public sector research, academic institutions, and research councils. Indeed, patents are a central tool in technology transfer and commercialization strategies in both the public and private sectors. The reasons are described below.

2.1 *Freedom to operate*

Freedom to operate (FTO) is becoming increasingly important for both the research and commercialization phases of the development of important products and technological processes. While most countries have generous *research exemptions* incorporated into their national legislation for the use of patented technologies in research, the scope and nature of the research exemption will vary from one jurisdiction to another. In the United States, the exemption is narrow and restrictive; in other jurisdictions, such as European countries, academic establishments tend to benefit from this exemption over industry, sometimes regardless of whether there are any commercial objectives. Nevertheless, this exemption may be subject to periodic review by events in judicial law, such as litigation proceedings. It is often difficult to determine clearly when early research will result in commercial activities, and so it is necessary to exercise judgment about the best time to start evaluating FTO.

When a research activity does produce commercialization initiatives, the scope of the research exemption likely will be significantly narrowed. Due diligence and thorough searching of background patents, therefore, will establish the scope of FTO. Patenting inventions generated in public sector institutions can also establish FTO. In this instance, extensive searching of patent databases is necessary for a researcher to establish whether he or she has a patentable invention.[1]

2.2 *Transfer of technology for the public good*

Some public sector institutions take patenting strategies very seriously in order to protect technologies that can be developed and transferred for the public good. Such strategies are likely to gain increasing acceptance in many other institutions. Not surprisingly, the generation of revenue from these patents is often a secondary consideration.

Developing core technology is frequently thwarted when it is simply released into the public domain. This is because the development of the technology most often requires a commercial partner who needs incentives to invest in the costly and risky development phase. Incentives include exclusivity facilitated through patent protection. For an example of such strategic considerations adopted in the public sector, see the IP draft guidelines document by the Indian Council for Agricultural Research (ICAR).[2]

That ICAR document considers patenting strategies for securing FTO, enabling food security, and *"cater[ing] to the agricultural and technological need of Indian farmers/citizens by maximising [referent] capacity for innovation and ensuring rapid transfer of technologies."* (Note that the ICAR document is not yet final and is accessible for feedback purposes at this stage.) In order to achieve the goals of food security, institutions need to balance the need for patenting with the need to release the innovation into the public domain. The strategy the institution adopts depends on which strategy will fulfill the institution's basic mission. The optimum strategy can be determined only on a case-by-case basis, depending on, among other factors, the nature of the individual technologies to be transferred. Of course, good patent-searching tools and expertise are essential for achieving the stated aims.

2.3 *Mining technical information*

Patent descriptions contain a lot of scientific information, which makes them useful alternatives to published papers. With strong encouragement from journal editors to shorten primary papers, the materials-and-methods sections of papers are often little more than reference lists, which necessitate a paper chase if a researcher wants to discover the actual process. Patent application descriptions, however, often contain excellent methodological detail in the *enabling disclosure* section, which is the quid pro quo of the monopoly patent right. In other words, the invention is disclosed in a manner that enables the reproduction of results. Often patents may be the sole source of technical information about new technology that involves either products or processes. One example is a recently published application assigned to Moraga Biotechnology Corporation, which discloses an invention related to totipotent nonembryonic stem cells (Application number: WO 06028723A1). Not only are the methods for isolating cells presented in a level of detail that covers media compositions, cell culture techniques, and surgical procedures, but no peer-reviewed journal had published anything similar at the time of patenting.

2.4 *Avoiding wasted research efforts*

Simply searching journals will not uncover all the available technological areas of research and product development, particularly those areas that have been more recently developed. Patent searching can quickly uncover newer areas of research and can help avoid the duplication of efforts in a given area of technology. However, researchers must remember that there is a time lag of up to 18 months, sometimes more, in many jurisdictions between the filing and publication of an application.

3. PATENT SEARCHING STRATEGY

Searching for patents on a particular topic or product is not always straightforward. Patents are national rights, so it is necessary to give thought to where—in geographic terms—the product or invention needs protection. Important inventions are commonly patented in more than one country and—since these inventions involve the biggest potential markets—searching the patent offices of the United States, Europe, and Japan frequently covers nearly all of the potential patents and patent filings. A more recent system of international preliminary patent applications—the Patent Cooperation Treaty (PCT) system—is also a useful resource since inventors filing through the system have the possibility of protection in any signatory state of the PCT. Inventors have up to 30 months after filing the preliminary application to decide in which countries to file full applications.

There are no universal rules for good patent searches, and the following guidance is based on the personal experiences of one of the authors of this chapter. Starting from a position of limited or no knowledge of the technology in question, the first step is to carry out a standard bibliographic search on scientific publication databases. The online database Pubmed,[3] specialized technical journals, and other scientific search sites (including sites such as Google's™ scholar[4]) are good places to start. It is also worth trawling for information using general Internet search engines such as Google. Communicating with scientists involved in the technology is another essential requirement for the technology transfer officer/searcher. Developing effective communication between these groups may require some time and effort.

Once armed with essential information, patent databases are queried with technical search words or inventor, applicant, or company names. There are no standardized forms for company names, so one must try various options to use that search field, particularly because companies may be listed as different entities in different countries. Also, many databases do not update these fields. So it is valuable to search using older names, such as Ciba-Geigy, Novartis, and Syngenta, as many name changes occur when companies merge or are taken over. Once relevant patents are found, it is possible to obtain their "equivalents" or "family members" through databases such as the European Patent Office (EPO). Patent families are explained later in this chapter.

It is very important to have a structured search strategy. Although the strategy can vary according to invention type, a structured strategy involves breaking into its essential elements an invention or a field of technology and emphasizing those elements that are expected to be novel and inventive. This approach creates a series of useful search words. Often it is impossible to determine useful terms before the search, and further elements can be identified as the search progresses. Using the citations within individual patents will further aid the search along the complex patent trail. Lastly, IPC (International Patent Classification) codes are useful for simplifying the retrieval of documents (more on this topic to follow).

Although many national patent offices make possible online searches of granted patents and applications issued in their jurisdictions (for example, the U.S. Patent and Trademark Office[5]), some Web sites (such as that of the Singapore IP office[6]) support searches of multiple sites at one time. Some have even collected multiple patent publication information into a single database (like the European Patent Office site[7]).

Many sites contain links to other sites with useful IP information, and one of the most comprehensive lists is maintained by the British Library,[8] but, as with all such lists, it is most likely incomplete.

Below is an annotated list of selected sites and some information about their usefulness.

4. AVAILABLE RESOURCES

Depending on available resources, individual TTOs may wish to have a dedicated searcher or to outsource the searching function, although this may be an expensive option in the long term. Requisite in-house resources include at least one well-trained staff member committed to performing patent searches. Ideally, this person should have a scientific background with the aptitude to absorb and understand technological concepts from a variety of disciplines. This is a talent-driven competence, so insistence on strict minimum qualifications can sometimes be unwise.

Investment in computer hardware is essential. Also, a fast Internet connection is ideal because many of the electronic documents that researchers will download are large. A good relationship with a local patent attorney also is advisable, especially in situations where the accuracy of the information under analysis is crucial and additional input is necessary. Investing in at least one commercial database (such as Delphion and/or Derwent) is highly recommended if financial resources permit and if the frequency and volume of searches requires that searches be completed quickly.

4.1 *Free patent sites and patent/patent searching resources*

4.1.1 *The Web portal of the European Patent Office*

The Web portal of the European Patent Office (EPO) contains bibliographic (front page) information from patent publications worldwide.[9] As of late 2006, the site contained information from more than 60 million documents from 72 countries. No full-text versions of the documents are available in the html page views, but many of the documents can be viewed and downloaded as pdf files.

Below is a detailed description of the EPO site using the esp@cenet engine and an illustration of how to search patent databases.

Esp@cenet can be accessed and searched using the following steps:

1. Go to www.espacenet.com in the browser window. This will bring you to the home page of esp@cenet. Figure 1 shows what you will see on your computer screen.
2. Scroll down the page to find a list of different servers located at the EPO, European Commission, and the national offices of members of the European Patent Convention and other European ("invited") states.
3. Click on the link that takes you to the EPO server (ep.espacenet.com).
4. Choose the search option you want from the menu on the left side of the screen.

A. Quick search

A quick search may be useful for simple searches using keywords, including the name of an inventor or a company. Figure 2 shows what you will see on you computer screen after selecting "Quick Search."

The default search part of the "Worldwide" database is shown above. Other choices are "EP" and "WIPO," which are in the drop-down menu. To perform a search, select either "Words in the title or abstract" or "Persons or organisations." Next, type in the words or names as appropriate in the search box. This box allows Boolean search operators such as "AND" and "OR."

B. Number search

A number search allows rapid access to publications and applications. Click on the "Number Search" option on the left side of the screen. Type in the application, access, publication, or priority numbers. If typing in the document "type" code (A1, A3, and so on—sometimes known as the "kind" code), which is appended to the end of the publication number, ensure that the code is separated from the publication number by a single space.

As an example, searching with "EP1226178A1" may not produce a result. The "A1" publication type code should either be deleted or entered as "EP1226178 A1." The "EP" code or any national (alphabet) code at the beginning is optional. If difficulties arise, try leaving out the national code in the beginning as well, bearing in mind the possibility of duplicated numbers across different patent offices.

Checking the "Including family" box while performing the search will return all members of the same patent family relating to multiple filings of the same invention and sharing the same *earliest priority date* (the filing date of the earliest application in the family). This helps

FIGURE 1: ESP@CENET PORTAL SCREEN: ACCESS PAGE

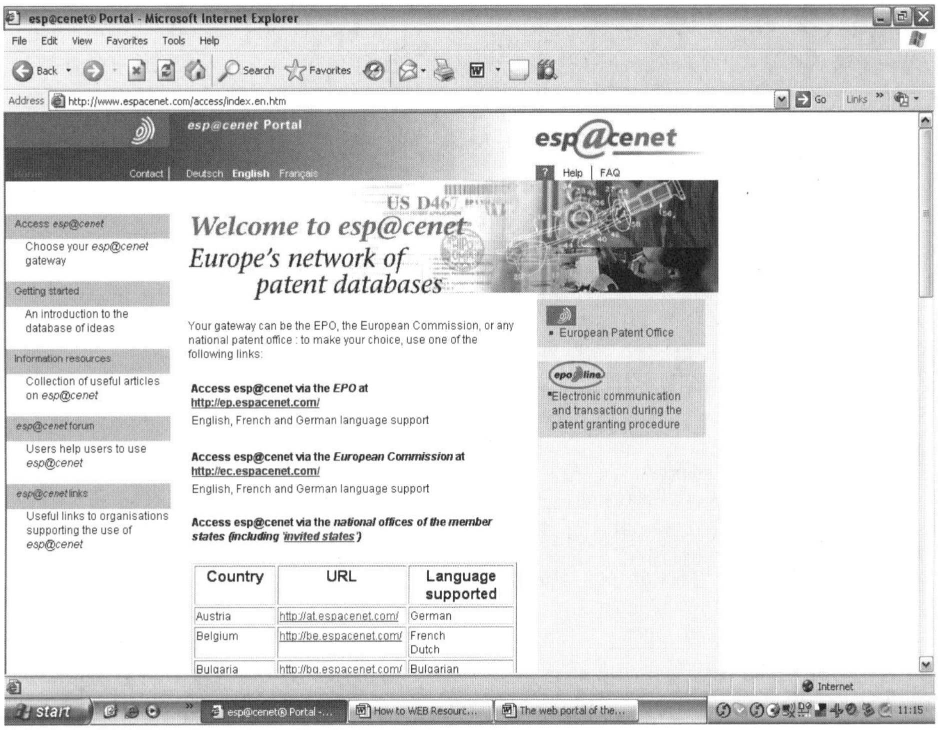

determine the geographic scope of protection worldwide, although it should be noted that patents filed in developing countries without easily accessible electronic information are often not listed.

Figure 3 shows the results window of the search with the patent publication number "EP1226178" together with the checked "Including family" box.

The results show multiple applications or publications in Great Britain, Japan, and Canada with their publication dates. These applications are related to each other by a common invention and a shared earliest priority date. In this case, all applications have an identical title, although this is not always the case for patents within a single family.

It is possible to examine the individual patent documents retrieved in more detail, including those documents retrieved through other search windows such as "Quick search," by clicking on the titles. Clicking on the title "Recombinant Fusion Molecules," application number EP1226178 A1, takes you to the page shown in Figure 4.

The page shows a variety of information, including the inventors and the abstract. In this case, the actual abstract displayed is that of an equivalent PCT application belonging to the same patent family. Clicking on the appropriate tabs shows the "Description" of the invention, "Claims," the "Legal status" of the patent, and the "Original document" in pdf format.

Click on the "Original document" tab shown in the esp@cenet document view window. You will see the screen shown in Figure 5.

This screen displays the full document in pdf format. Navigate it using the scroll bar. In order to save the document to a disk or other specified location, click on the "Save Full Document" link shown in red.

Note that if you attempt to save or print the displayed pdf document using the Adobe®

Figure 2: "Quick Search" Screen

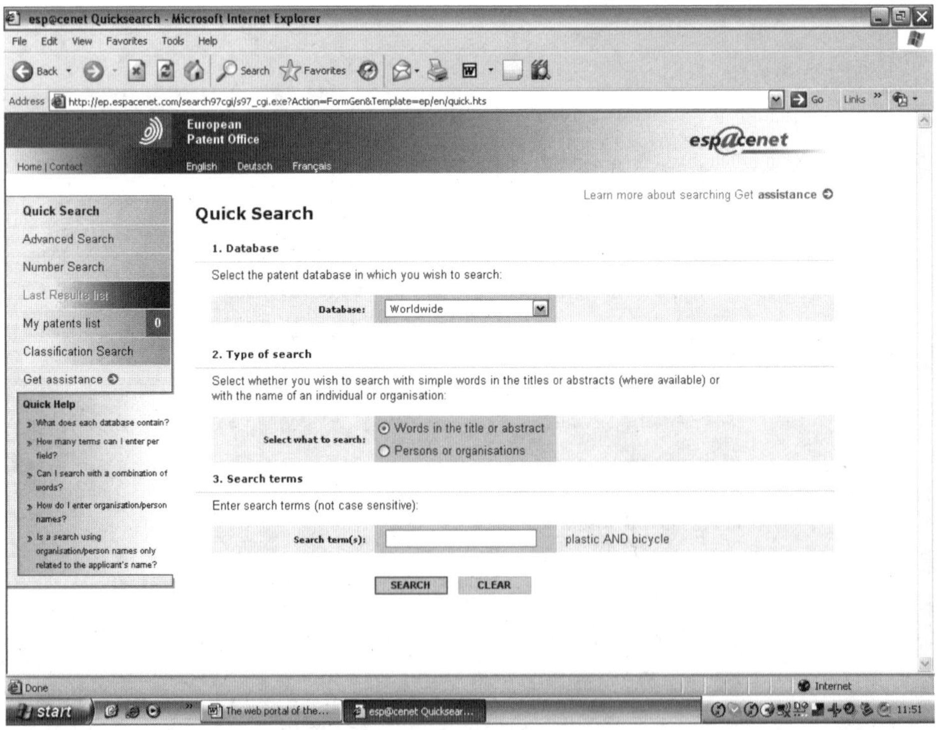

Figure 3: Family List View: Results of a Search Using the Patent Publication Number

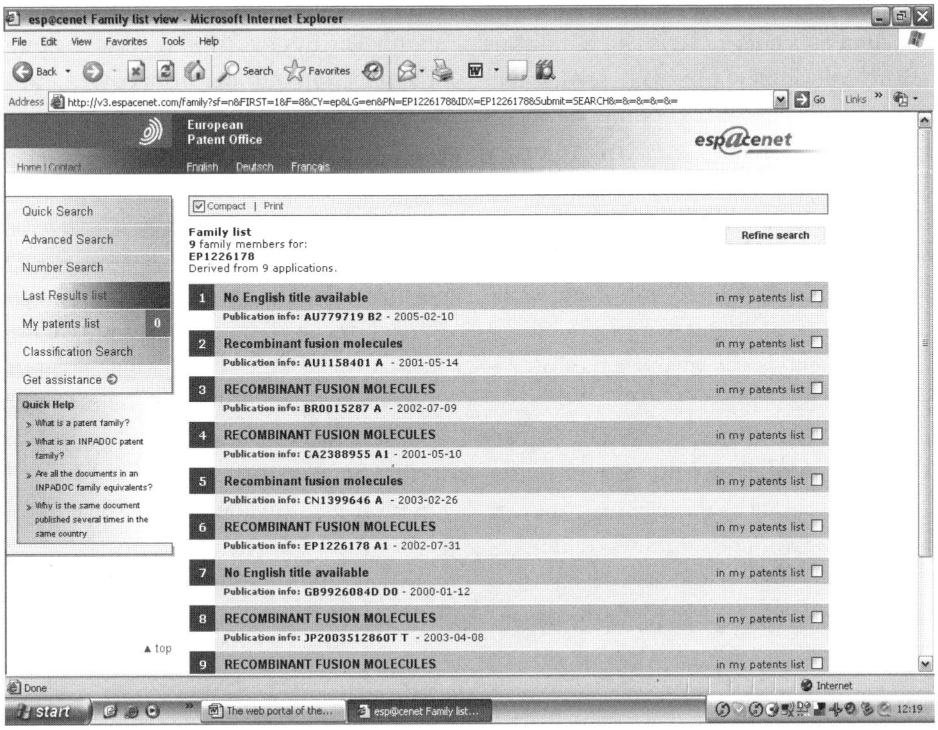

Figure 4: Document View: Results of Selecting "Recombinant Fusion Molecules" for a Specific Publication Number

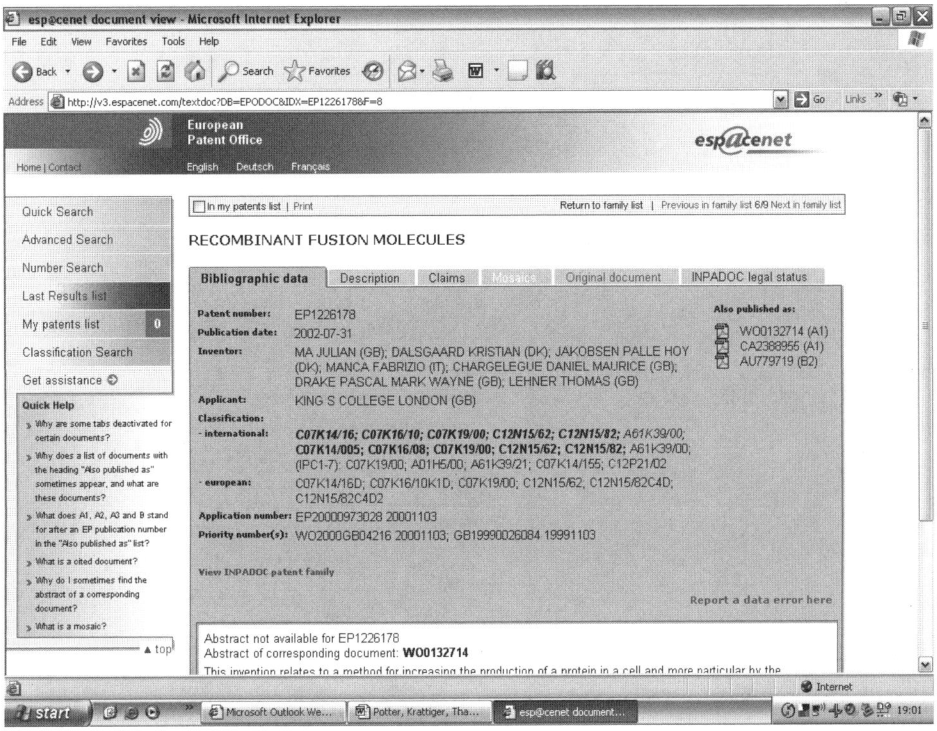

Acrobat® menu icons displayed (the "save" and "print" icons of a disk or printer, respectively), you can only do so one page at a time.

If a document is not available as a full download, the site will display a message saying "No full document available" instead of the "Save Full Document" link. In many cases, recent documents are not available for complete download, even if the entire document is available to view in pdf format on the screen.

Clicking on the "Mosaics" tab enables the searcher to view six of the drawings in a single window. In the example shown above these drawings are not available.

Clicking on the "View INPADOC patent family" (Figure 4) shows the related patent documents of the family or those patents that are linked by a common priority number or date.

Clicking on the "INPADOC legal status" gives useful information on the legal status of the patent.

If certain EP patents are not available as full downloads from the server, check their availability through the EPO publication server.[10] Simply key in or copy and paste the publication or application number from patent searches into the appropriate box.

C. Advanced searches

Advanced search techniques, which narrow searches by combining various search terms, are possible using the "Advanced Search" option. The search box can contain a maximum of four search terms using Boolean language. Figure 6 shows a keyword search for "Antibodies AND Plants" against "title or abstract." The search is narrowed further to patents applied for by the "Scripps Research Institute," where the inventor is "Mich Hein." Additionally, the patent search is limited to those patents published in "1997" and "2001." This criterion is specified by entering those years

Figure 5: Original Document View: Results of Selecting the "Original document" Tab

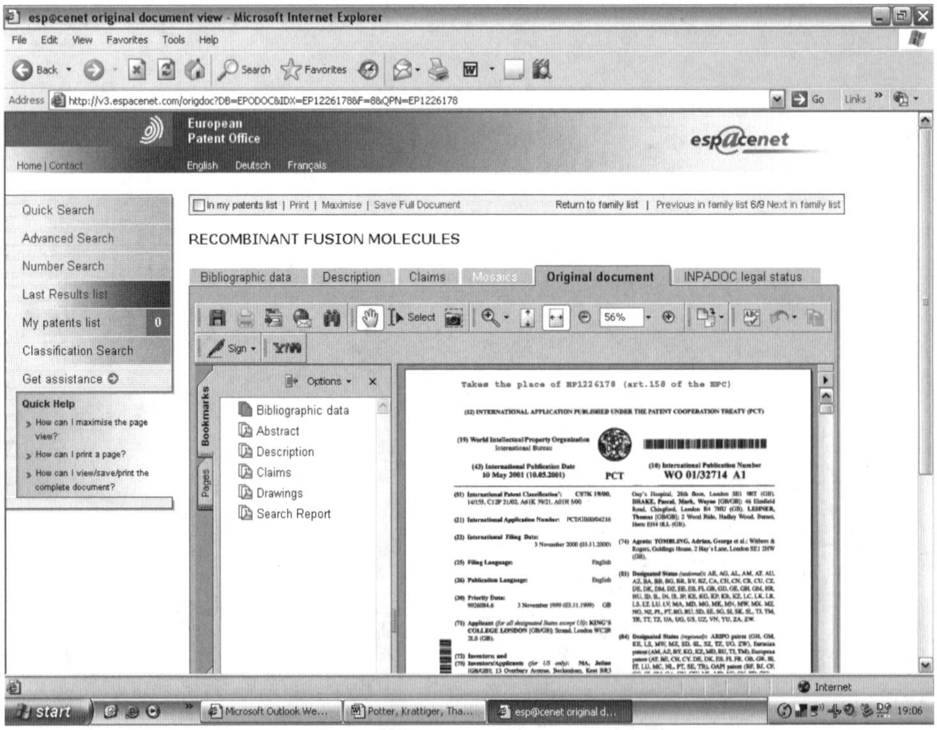

into the "Publication date" box, for which the Boolean operator is "OR" by default—alternatively, one can add the "OR" operator explicitly as "1997 OR 2001."

The results of this search are shown in Figure 7.

Here is an explanation of each of the search fields on the advanced search form (not shown in Figure 7):

- **Keywords in title and/or ritle or abstract.** A text search for the entered keywords. Up to four keywords can be entered into each box using Boolean operators such as "AND" and "OR."
- **Publication number.** The number assigned to a patent or published application. In some cases, a granted patent will retain the publication number assigned to the published application, but not all country systems do this. Granted and published patents are often distinguished by a "Kind Code" (attached to the end of the publication number), such as "A" for an application and "B" for a granted patent, followed by a numeral (for example A1 or B1) referencing to the EPO system.
- **Application number.** The number assigned to an application when filing. Rather confusingly, this number is separate from the publication number. The numbers—also shown on the front page of publications—include the country of filing and the year, but can sometimes have a different format. To further illustrate this, the application number for the example publication EP1226178 A1 shown above is EP2000000973028.
- **Priority number.** The application number of the priority application. Patents are often filed based on previously filed applications or applications filed in other countries. By including the application number of the

Figure 6: Advanced Search View: Entering Your Search Parameters

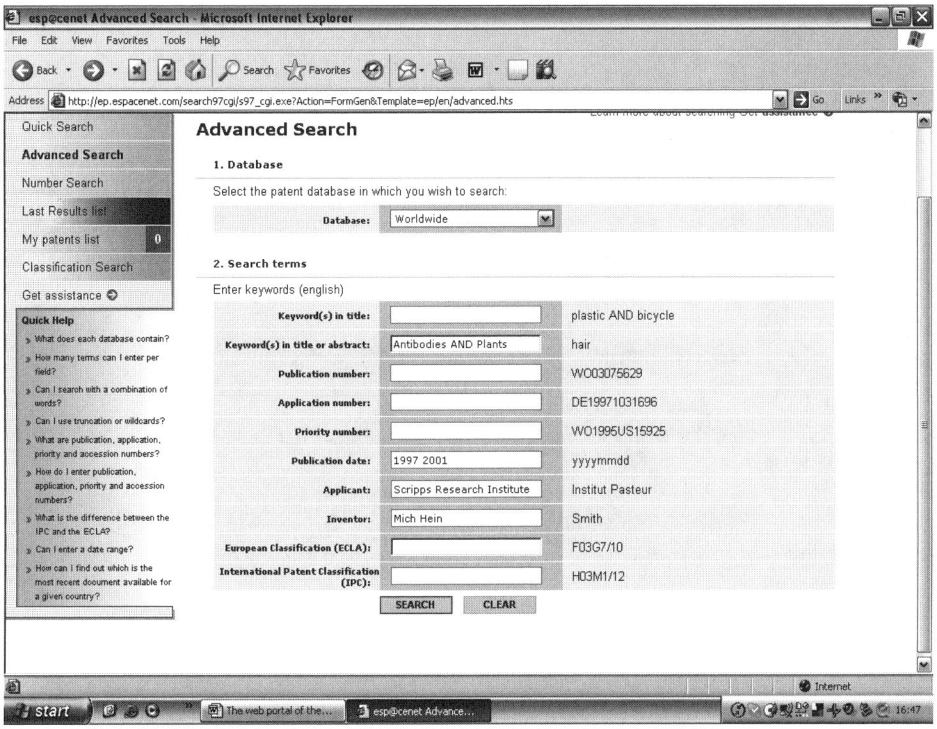

first (or priority) filing, the applicant can take advantage of the earliest filing date. A patent family is a set of publications linked by a common priority number. This is a useful way of determining patent coverage in several different countries, especially since direct patent searches are only available in the United States and Europe. To further illustrate this, the priority number for the example publication EP1226178 A1 shown above is GB1999000026084.

- **Publication date.** The actual publication date. Note that, although a range of years cannot be searched currently, up to four different years can be searched using the "OR" Boolean operator (as in the example above). It is also possible to search for precise dates using the yyyymmdd format.
- **Applicant.** Usually a company, university or other institutional entity, but can instead be one or more individuals. Words can be combined with Boolean operators to form the title of an applicant.
- **Inventor.** Name of the person or persons who discovered the invention. In the United States, the inventor must be noted on a patent or patent application. Although this is not strictly required in many other jurisdictions, inventors are usually included in practice. Often listing the authors of early papers in a field is effective in searching for related patents.
- **European Classification.** A designation used to classify technical content of patent documents based on the EPO's European Classification, which is an extension of the International Patent Classification (IPC).
- **International Patent Classification.** The technical content of patent documents clas-

FIGURE 7: RESULTS VIEW: RESULTS OF SEARCH USING KEYWORDS AND OTHER PARAMETERS OF PUBLICATION

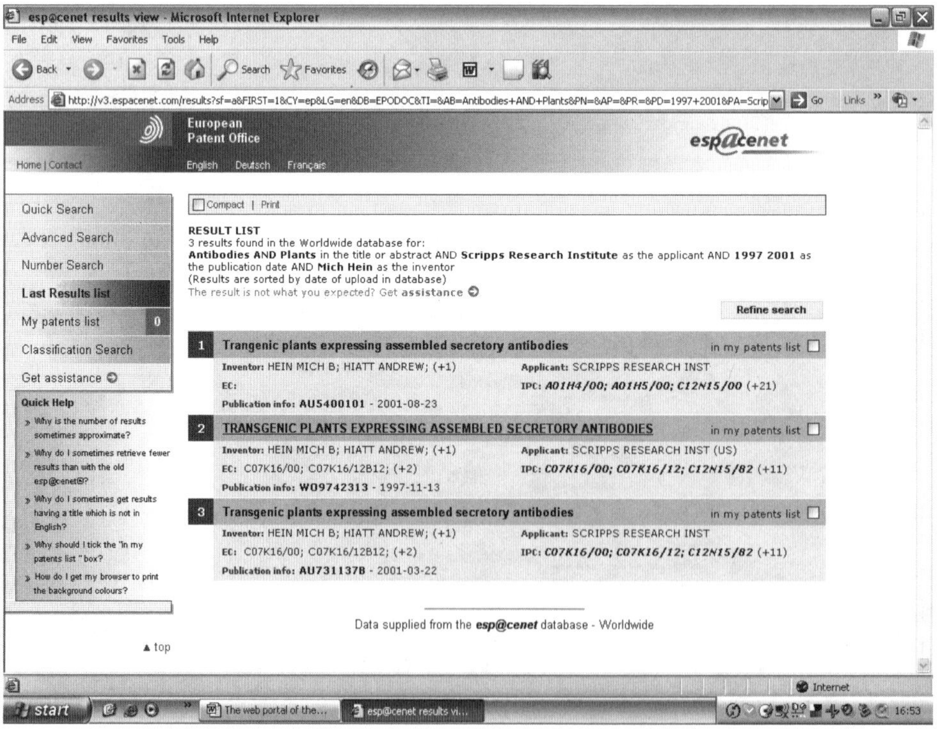

sified according to the IPC. This is assigned by the publishing office and is thus independent of the applicants, which makes it a good searching tool for patents in a specific area.[11]

D. Classification search

Finally, we consider the classification search. Using codes based on the EPO classification system, this type of search is useful when searching for all patents in a particular technical area. It is a powerful search tool, indispensable to professional searchers. Full consideration of the use of the classification system is beyond the scope of this chapter. However, the following steps are intended to give the reader some guidance on how to begin using the codes on esp@cenet.

Click on the "Classification Search" button.

A window, similar to the one in the Figure 8, should appear. Key in the words "genetic engineering" in the box labeled "Find classification(s) for keywords" and click "Go."

This takes you to the following page (Figure 9).

A number of specific codes appear, along with descriptions. Assuming we are interested in further exploring "C12N15," click on the code or the title next to it, and you will be led to another list of codes with hierarchical subclassifications and descriptions. Scroll down to examine the list of subclassifications (or finer categories), and explore each of these codes until you have identified the codes pertinent to your areas of interest. As you explore, you can copy any code or codes to your main search form by ticking the box against the code and then choosing "Copy." The codes are then copied into the search form to narrow your searches to the relevant invention areas. Note, however, that this requires a lot of experience, and searching with multiple codes simultaneously may restrict the results.

You can also use codes with the "Advanced Search" form.

FIGURE 8: CLASSIFICATION VIEW

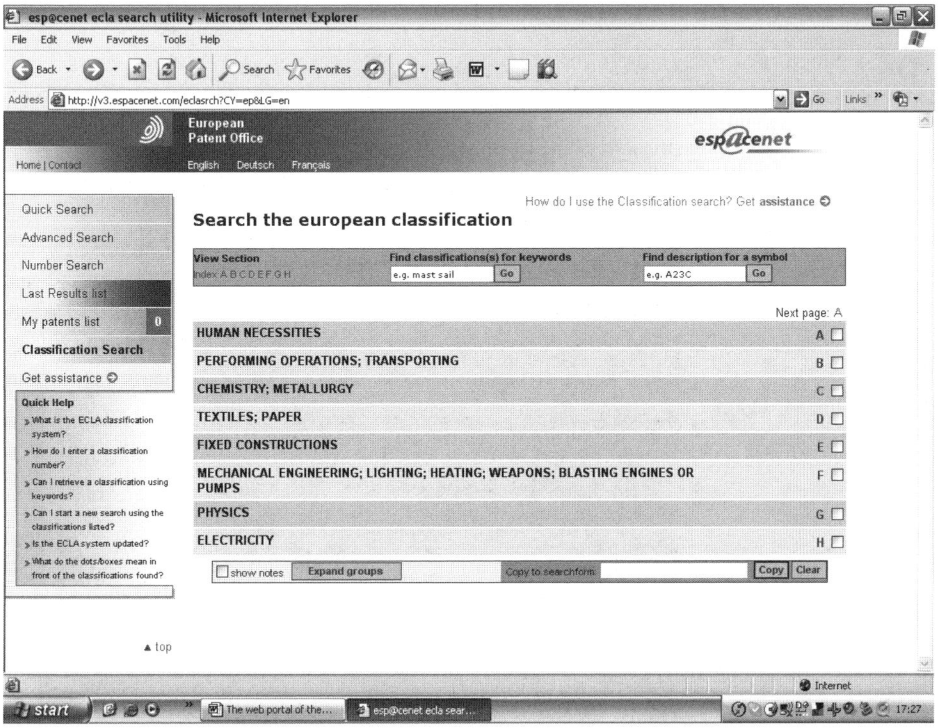

Note that all search screens (Quick, Advanced, Number, Classification) contain "Quick Help" links that guide the reader through many common queries. Clicking on the "Get assistance" link on search pages will take you to the esp@cenet Assistant, an interactive training module that is useful for beginners. There are several other help pages that the reader should find while navigating the site.

We have only commented on basic search techniques in relation to esp@cenet, but one can search the sites of a number of national patent offices, using generally similar techniques. At least two sites provide lists of links to national offices and to their patent collections.[12]

Selected patent search sites and scientific and instructional resources are briefly described below.

4.1.2 *U.S. Patent and Trademark Office*

The Web site of the U.S. Patent Office (PTO) contains the full text of granted patents from 1976 to the present, as well as full-page images since 1790.[13] Published applications date from 15 March 2001. Searches include both quick and advanced features. Page images are only available a single page at a time, and one must download a Tiff-viewer to see them. At the same site, you can also search trademarks registered in the United States. The U.S. Patent Office site contains links to helpful information including an excellent book on searching, titled *Patent Searching Made Easy*, which is available in paperback.[14]

4.1.3 *UK Patent Office*

The patent site of the United Kingdom Patent Office[15] is somewhat similar to the U.S. Patent Office site but offers less utility because an increasing number of patents are now filed as EP (U.K.) patents (derived from filings for multiple jurisdictions at the EPO). The U.K. site, therefore, ends up as more of a register of inventions registered in the United Kingdom after the granting of the

Figure 9: Keyword Search Results

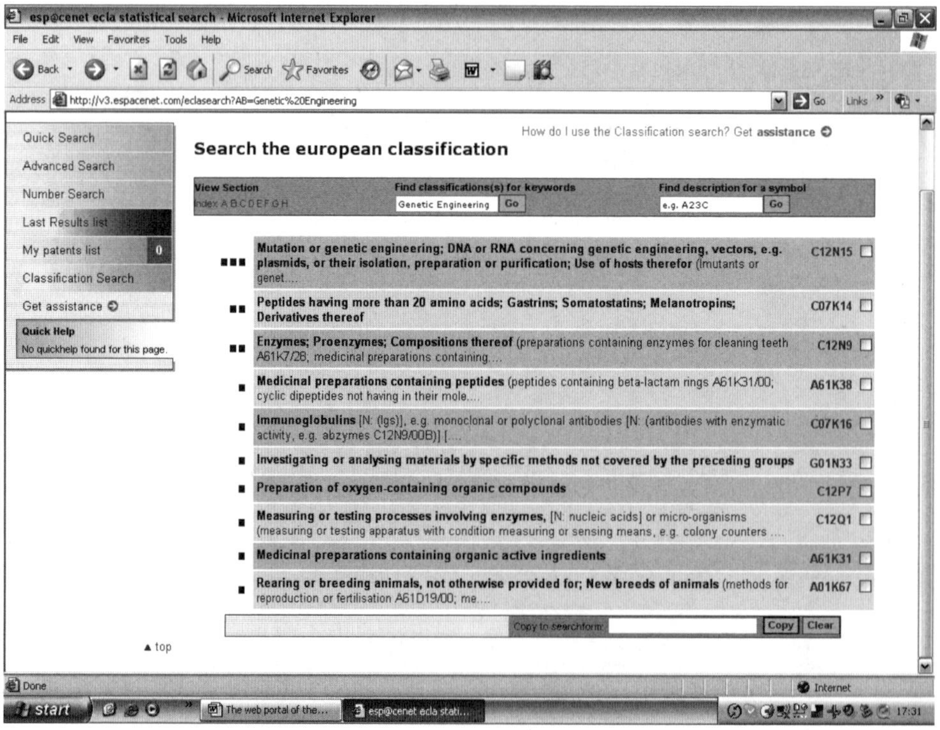

EP filing. One of its more useful features is the ability to access patent status information, such as changes in assignees (common with the current rounds of company takeovers) and the payment of maintenance fees.

4.1.4 *WIPO PCT applications*

Maintained by the World Intellectual Property Organization (WIPO), the WIPO PCT site allows you to search PCT (Patent Cooperation Treaty) applications for international filings.[16] Both simple and "structured" (advanced) search options are available for more than a million international patent applications.

Published on a weekly basis, *The PCT Gazette* gives up-to-date information on applications and publishes special notices relating to the treaty itself and its regulations. Weekly issues are also available.[17]

4.1.5 *CAMBIA Patent Lens*

CAMBIA Patent Lens is an independent resource for patent information from the EPO, PCT, and U.S. filings.[18] The files are selected on the basis of the international classification codes as related to agriculture/plant technologies. CAMBIA's database has approximately 5.5 million documents. A site search on the home page will also reveal links to various useful, comprehensive articles in key areas of interest to agricultural biotechnology, specifically in relation to patent landscapes. These papers are updated every few years to cover newer patent applications and grants. Areas covered include promoters for the expression of heterologous genes in plants, Agrobacterium gene transfer methods, and selected antibiotic and herbicide-resistance genes. The Agrobacterium document alone amounts to 350 pages of preanalyzed patent and scientific information.

For those who specialize in agricultural or plant-related research, this is a valuable resource. First, it is helpful to have plant technology-related patents from multiple databases prefiltered into a single resource. This saves a researcher from having to search multiple unfiltered databases (EPO, PTO, and PCT), each of which have different interfaces and idiosyncrasies. Users of multiple databases need experience with all such interfaces, whereas with Patent Lens, one needs familiarity with just one interface. The Patent Lens search interface has both simple and advanced search options, including the ability to filter results according to granted patents or published applications.

Second, the full text is downloadable. Until recently, even the EPO did not have this capability.

Third, Patent Lens gives extensive coverage to understanding the IP world, including how to read and interpret a patent and its claims, with a particular focus on agricultural biotechnology. For example, inside the technology landscapes, there is expert commentary on patent protection for a given technology, structured in a manner that makes the patent maze more navigable and transparent. Each landscape explains the science behind the technology, the legal and expiry status, the claim scope of key patents, the key assignees of patents, and the geographical coverage of key patent families. Help pages for searching techniques are extensive. Considering that the CAMBIA initiative is an open-access project for the public good, such an achievement is remarkable.

Despite these advantages, a professional searcher would use commercial search tools mentioned elsewhere in addition to Patent Lens. This is primarily because the collections of professional tools are more likely to be up-to-date and will also have extended geographic coverage. However, for researchers, small businesses, and TTOs, using Patent Lens would have certain advantages, not least of which is the ability to quickly search the prior art (in plant-related patents), thereby avoiding wasteful reduplication of research efforts and investments. Besides, it is unlikely that TTOs in some developing nations would have access to commercial databases.

In summary, the CAMBIA database cannot be used in isolation. EPO, PTO, other patent offices, and commercial databases are more likely to be current. In instances where there are combinations of technologies, for example cloning and expression of a particular mammalian gene in a plant, one would want to look beyond CAMBIA. It is, however, an excellent and comprehensive starting point.

4.1.6 *Intellectual Property Office of Singapore*

The Web site of Singapore's Intellectual Property Office is a good site for starting searches.[19] It has links to other sites in "search results."

4.1.7 *Other resources*

There are numerous other resources related to patent and scientific information. Some of the more useful ones, though not related to national offices or databases, are:

- **Intellectual Property and Biotechnology Handbook.**[20] Although several years old and, in places, somewhat out-of-date, this handbook remains a useful teaching tool. Individual modules of the handbook are downloadable in pdf format, and Module Four is devoted to searching patent databases.
- **Manual for Biotechnology and IP.**[21] This excellent manual, though somewhat outdated, on IP in biotechnology was prepared by Patent and Trademark Attorneys Spruson and Ferguson. The manual includes searching strategies.
- **Managing IP Web site.**[22] This is an excellent and comprehensive guide to IP in the field of biotechnology. The site has lots of useful and current information. Although not all resources are instantly accessible without registration, many are. This site contains current information on various changes in IP laws, interpretation of these laws, and litigation outcomes related to IP disputes in a selection of jurisdictions. The articles are written by well-respected IP professionals worldwide. The "International Briefings" link, which takes you to country-specific articles, is particularly useful.
- **patent blog sites.** Recent trends have led to a rapid proliferation of so-called patent blogs. These are sites that allow anyone interested in patents to share information related to patents, including information from other blog sites. While many of these have limited or no moderation and so contain information that may be highly opinionated or of no value, it is possible to find a lot of useful information, such as announcements of patent seminars, new developments in the patent world, and patent searching tips.

Blog sites are useful if you have queries that other users can help with. Some blog sites may deal with highly specialized areas such as searching or licensing. This arena keeps changing, with new blogs being introduced and older ones disappearing, but the net trend is an overall increase in these sites. A few of the blog sites are listed here—not as recommendations, but as starting points—and the reader is encouraged to search for more resources, like these, on the Internet:

- **Patent Information Users Group (PIUG)**[23]
- **Phoista®**[24]
- **PatentlyO®**[25]
- **Promote The Progress®**[26]
- **Scirus**[27]
- **Health InterNetwork Access to Research Initiative (HINARI)**[28]

4.2 *Commercial databases and search engines/services*

Commercial patent search engines offer tremendous advantages over free databases, but they are only available for a fee. Depending on the specific requirements, the costs are easily offset by the added functionalities and options that are available only with these databases. Some of the most prominent and flexible commercial services are:

- **Delphion.**[29] Delphion is a patent search database owned and maintained by Thomson Scientific. There are two versions of Delphion:
  1. A basic, free version that requires registration. This version enables the user to perform "quick searching" among granted U.S. patents, as well as "number searching" for worldwide patent collections. Through this version, pdf files of patents are available for download on a pay-per-use basis.
  2. A fee-based subscription version with various levels of access to downloads and other features depending on the particular subscription options purchased. In our opinion, Delphion is

one of the best—if not the best—search engine (subscription version). It is the most widely used search tool of patent attorneys and professional searchers worldwide. Its comprehensiveness and accessibility far outstrip individual free services/databases.[30]

- **Derwent World Patent Index (DWPI).**31 DWPI (like Delphion) is a proprietary database owned by Thomson Scientific. It is different from other databases in that it consists of summary information about patent applications that are rewritten, annotated, and formatted by Thomson's staff in a manner that makes the site more user-friendly for beginner patent searchers. It also includes some powerful value-added features for both novice and advanced searchers. DWPI is useful if one wants to quickly determine the technical content of patents. It requires a paid subscription and is sometimes bundled as a companion product with Delphion since they are designed for integration. A version of DWPI is available for use within the buildings of the British Library in London (free remote access is not possible).[32]
- **Micropatent.**[33] Micropatent is another popular source for patent and trademark information. The coverage is comprehensive and full-document delivery is available.
- **Questel Orbit.**[34] The service hosts commercial patent search tools. It offers fee-based search services.
- **Nerac.**[35] Using this fee-based service, analysts and searchers can extract market information, prior art, and patents based on specific requests.

4.3 *Patent statistical and business analytical tools*

Several software tools and services exist that businesses use to extract information from large patent datasets, examine relationships between patents and patent sets, or create visual representations of search results and summarize the information in reports. It is possible to use these tools to, for example, find out key assignees and inventors in a particular field, view industry and technological trends, and compare patent filings over time. But many of these tools come at a price. These may be of interest to universities or spinouts with advanced commercial activities, but probably not to the average TTO. A handful of tools: MAPIT,[36] BIZINT,[37] PatGraph,[38] MapOut Pro,[39] and PatentLab II.[40]

4.4 *Other useful web resources for nonpatent IP searches and information*

While patents are arguably the most important form of IP for research and product development in health and agricultural technologies, patent information may often be worth little if it is not examined in the context of other IP rights. For example, genetic information may not only be the subject of patent protection. It can also enjoy other forms of protection. A few key links related to a selection of IP rights are mentioned below to give the reader a starting point for exploring these IP rights in these fields.

4.4.1 *Trademark Searches*

A comprehensive collection of links to trademark offices in many jurisdictions, databases, search engines, and official gazettes is available on a British Library Web page.[41]

4.4.2 *Copyright Issues*

Issues of copyright are complex, involving various national laws and partial harmonization through regional and international treaties. Some good places to start examining the issues are the various conventions and treaties in the WIPO Web site.[42]

4.4.3 *International treaties*

The WIPO portal is a great way to begin exploring international treaties on intellectual property rights (including the Agreement on Trade-Related Aspects of Intellectual Property Rights [TRIPS]) available through the WIPO portal.[43]

The WTO-TRIPS agreement is increasingly affecting cross-border trade and national and international IPR regulation and enforcement. Useful resources include the WTO Web page on TRIPS.[44]

In addition, the *Resource Book on TRIPS and Development* is available through the UNCTAD-ICTSD Capacity Building Project on IP rights. This authoritative, practical guidebook to all aspects of TRIPS is downloadable in sections.[45] Although the guide is written mainly for policy-makers and negotiators, its comprehensive coverage of subject matter is useful for all those interested in TRIPS.

5. CONCLUSIONS

Patent searching is an art that requires a solid foundation in and understanding of the sciences related to specific searches. In addition to compiling patents related to certain inventions (for FTO purposes, prior art, or scientific endeavors), it is important to pay attention to other aspects, such as regular updating. One free service for this is particularly noteworthy. FreshPatents allows users to track the publication of new patents related to certain well-defined fields of scientific endeavor.[46] After registering and defining the fields of interest, weekly e-mails are sent that include a list of the patent applications by the U.S. Patent Office. Other services are also available, such as RSS feeds. ■

HARRY THANGARAJ, *Director, Research, MIHR (Centre for Management of IP in Health R&D), Oxford Centre for Innovation, Mill Street, Oxford, OX2 0JX, U.K. harry@mihr.org*

ROBERT H. POTTER, *Senior Associate, Agriculture & Biotechnology Strategies, Inc., 106 St. John Street, PO Box 475, Merrickville, Ontario, K0G1N0, Canada. rpotter@agbios.com*

ANATOLE KRATTIGER, *PO Box 26, Interlaken, NY, 14847, U.S.A. afk3@cornell.edu*

---

1 See, also in this *Handbook*, chapter 14.4 by GM Fenton, C Chi-Ham and S Boettiger; chapter 14.2 by SP Kowalski titled Freedom to Operate: The Preparations; and chapter 14.1 by A Krattiger.

2 www.icar.org.in/miscel/iprguidelines-draft.pdf.

3 www.ncbi.nlm.nih.gov/entrez/query.fcgi?DB=pubmed.

4 scholar.google.com.

5 www.uspto.gov.

6 www.surfip.gov.sg.

7 www.espacenet.com.

8 www.bl.uk/collections/patents/othlinks.html.

9 www.european-patent-office.org.

10 publications.european-patent-office.org/PublicationServer/main.jsp.

11 Details of this complex classification system can be found at www.wipo.int/classifications/fulltext/new_ipc/ipc7/eindex.htm.

12 inventors.ca/ipsearch.htm; www.bl.uk/collections/patents/polinks.html.

13 www.uspto.gov.

14 Hitchcock D. 2005. *Patent Searching Made Easy*. Lulu Press: Morrisville, North Carolina.

15 www.patent.gov.uk.

16 www.wipo.int/pct/en/.

17 www.wipo.int/pct/en/gazette/.

18 www.patentlens.net/daisy/patentlens/patentlens.html.

19 www.ipos.gov.sg.

20 www.dfat.gov.au/publications/biotech/overview.html.

21 www.sprusons.com.au/files/Biotechnology_IP_Manual.pdf.

22 www.managingip.com.

23 piug.derwent.co.uk/archive/index.html.

24 On this blog site, at the time the chapter was written, there was a video seminar by a U.S. patent attorney about what universities should know about the patent review process and how an inventor can submit a successful application. www.okpatents.com/phosita.

25 www.patentlyobviousblog.com.

26 www.promotetheprogress.com.

27 www.scirus.com/srsapp.

28 The HINARI initiative was set up by the World Health Organization (WHO) along with several major publishers of academic journals, including *Nature* and *Science*. About 3,300 complete journals are available for free to health research institutions in countries with a "GNP per capita" of less than US$1000. A second group of countries with a GNP per capita of US$2000–$3000 get low-cost access. However, it should be noted that some countries, like India, are excluded from both categories, despite fulfilling the eligibility criterion for free access, because the publishers have, in their agreement with the WHO, reserved the right to exclude markets where the publishers have significant sales. By clicking on "Eligibility" on the left-hand side of the home page, you can ascertain whether your country/institution is eligible for free or low-cost access. Registration is required for using the site. www.who.int/hinari/en/.

29 www.delphion.com.

30 For further information on the features of Delphion, including detailed instructions for use, visit www.delphion.com/help/index#derwent.

31 scientific.thomson.com/products/dwpi/.

32 For further information on the features of Derwent, including detailed instructions for use, visit www.delphion.com/help/index#derwent.

33 www.micropatent.com.

34 www.questel.orbit.com.

35 www.nerac.com.

36 www.mnis.com.

37 www.bizcharts.com/patents/index.html.

38 www.micropat.com/0/new_patgraph9809.html.

39 www.mapout.se/MapOut.html.

40 www.wisdomain.com/download.htm.

41 www.bl.uk/collections/patents/tmlinks.html.

42 www.wipo.int/copyright/en/treaties.htm.

43 www.wipo.int/portal/index.html.en.

44 www.wto.org/english/tratop_e/trips_e/trips_e.htm.

45 Download the guidebook at www.iprsonline.org/unctadictsd/ResourceBookIndex.htm.

46 www.freshpatents.com/.

Freedom to Operate: The Law Firm's Approach and Role

GILLIAN M. FENTON, *Chief Intellectual Property Counsel, Emergent BioSolutions, Inc., U.S.A.*
CECILIA CHI-HAM, *Director, Biotechnology Resources, PIPRA, U.S.A.*
SARA BOETTIGER, *Senior Advisor, PIPRA and Chief Economist, M-CAM, Inc., U.S.A.*

ABSTRACT
In the fields of health and agriculture, it has become increasingly important to understand the role of patent infringement in research, development, and commercial production. If a patented technology is used without permission, the patent holder may have the right to sue the researcher for patent infringement. Many companies routinely analyze the freedom to operate (FTO) of a research project or product, assessing whether it is likely to infringe existing patents or other types of IP rights. Private companies more routinely engage in FTO analysis than public sector research institutions because the infringement risks they face must be directly considered in the calculus of profitability. Public and not-for-profit private institutions also are becoming increasingly aware of the need for better FTO information, but FTO analysis is expensive, and its benefits must be weighed against its costs. This chapter provides an overview of the process, including considerations of when to invest in FTO analysis, and particularly focuses on the law firm's role and perspective.

1. CONCEPTS AND DEFINITIONS
In the fields of health and agriculture, it has become increasingly important to understand the role of patent infringement in research, development, and commercial production. Patenting has become so prevalent in some countries that agriculture and health researchers often use patented technologies daily in the course of their work. If a patented technology is used without permission, the patent holder may have the right to sue the researcher or their employer for patent infringement. Many companies routinely analyze the freedom to operate (FTO) of a research project or product, assessing whether making, using, or selling it is likely to infringe existing patents or other types of IP (intellectual property) rights. The resulting information contributes to a larger risk assessment that may involve a range of options: identifying in-licensing targets, considering the substitution of technologies, deciding to ignore the potential infringement, investing in work-around technologies, or perhaps deciding to abandon the project all together.

Private companies are more likely to engage in FTO analysis because the risks they face must be directly considered in their calculus of profitability. Public and not-for-profit private institutions are becoming increasingly aware of the need for better FTO information. FTO analysis, however, is expensive, and its benefits must be weighed against its costs. Researchers in public institutions, not-for-profit institutions, and in developing countries must consider different factors when weighing the benefits and costs of FTO analysis. In particular, many technologies patented in developed countries are not patented in developing countries. Therefore, institutions making, using, or selling the technologies are not at risk of infringing in those developing countries. However, if a product is *imported* to

Fenton GM, C Chi-Ham and S Boettiger. 2007. Freedom to Operate: The Law Firms Approach and Role. In *Intellectual Property Management in Health and Agricultural Innovation: A Handbook of Best Practices* (eds. A Krattiger, RT Mahoney, L Nelsen, et al.). MIHR: Oxford, U.K., and PIPRA: Davis, U.S.A. Available online at www.ipHandbook.org.

a country where patents on the technologies are in force, then the importer may be infringing in that country.

This chapter and that by Kowalski[1] together provide an overview of the FTO analysis process, including considerations of when (and whether) to invest in this type of analysis. Kowalski discusses FTO analysis from the researcher's perspective, whereas this chapter is particularly focused on the law firm's perspective. In this chapter, we draw from a case study of the E8 promoter. One of many enabling technologies used in the genetic transformation of plants, the E8 promoter provides a concrete example of FTO analysis.

While patents are the most common type of IP right encountered, a thorough FTO analysis will assess *all* types of existing property rights in order to determine the likelihood that the research project or the product being commercialized infringes. As Kowalski[2] and Krattiger,[3] we are also concerned with both intellectual and *tangible property* rights. In biotechnology, tangible property comprises the biological material of the invention: one can physically possess such material. Common examples of tangible property in health and agriculture include cell lines, transgenic mice, germplasm, and plasmids. The transfer of tangible property often occurs under a contract that governs the terms under which the property changes possession but not ownership (commonly called *material transfer agreements,* or MTAs[4]). Unlike IP rights, ownership rights over tangible property do not expire. Tangible property rights provide a further source of protection for certain elements of an invention. Sometimes elements of an invention can be the subject of both types of rights. The use of a gene, for example, may require a license to a patent as well as a material transfer agreement governing possession of the DNA itself.

IP is a category of intangible assets, and includes things such as creative works, inventions, or commercial secrets. Under United States law, IP rights are defined as *exclusionary* rather than affirmative rights. That is, the owner of IP generally has the right to exclude or prevent others from using the intellectual property. The owner can grant permission for use in the form of a license or similar contractual agreement. IP rights are granted by government entities (for example, the U.S. Government or other countries) or by multinational authorities pursuant to international treaties (for example, the European Patent Office [EPO] acting under the European Patent Convention). A grant of IP rights thus confers exclusivity only within the territory controlled by the grantor and only for a limited number of years.

The practice of IP rights in the absence of the owner's permission is defined as infringement. U.S. law provides a number of remedies for infringement, chiefly the award of damages (a monetary award of the amount necessary to fully compensate the IP owner for the harm resulting from infringement) and/or the grant of an injunction (a court order to cease infringing activity or to refrain from commencing such activity). In some cases, additional remedies may apply, such as the award of attorneys' fees and/or the enhancement of damages (doubling or tripling of the award); these additional remedies may be awarded when the act of infringement has been willful.

Because IP rights are exclusionary, the government grant of an IP right, such as a patent, in no way confers an affirmative right to practice the intellectual property. This stands in fundamental contrast to the grant of, for example, a regulatory license by the U.S. Food and Drug Administration (FDA), which does confer the right to sell a new drug or medical device in the U.S. market. Thus, when pursuing a business goal, such as the development and commercialization of a new technology, one must be cognizant of the IP rights of others, because those others may have the right to block or impede progress toward the desired business goal.

FTO is defined as the absence of third-party IP rights that impede progress toward a desired business goal. FTO is also sometimes referred to as *clearance*. As will be discussed below, FTO cannot be conclusively established, but rather should be viewed as an ongoing investigative activity for as long as the corresponding business goal is pursued.

We distinguish the concept of *exclusivity*, as distinct from FTO, defining it as the benefit conferred by a collection of IP rights amassed by a single owner, that the owner can use to prevent others, such as business competitors, from using a technology. It is possible, for instance, for an institution to have created a high degree of exclusivity for a technology through patenting but still not have FTO because the making, using, selling, or exporting of the technology infringes another's patents.

A collection of IP rights in similar subject matter or a single technology is often referred to as an *IP portfolio*. On a practical level, such IP rights protect the present or future potential market of the owner. The portfolio should be designed so that it corresponds to, and therefore supports, a business goal.

The concepts of exclusivity and FTO must be considered together when assessing the relative risk or desirability of pursuing a particular business goal. When initially formulating the business goal or assessing a new discovery, there may be little to no exclusivity or FTO (or at least knowledge about the status of either parameter) to consider at that time. It is customary to build an IP portfolio in parallel with the process of technology development; however, during the course of development it may be unwise to defer an FTO investigation for too long. As noted above, a particular technology can accrue a high degree of exclusivity in the form of a well-rounded IP portfolio but still suffer from a lack of FTO. The risk associated with further development or commercialization of this technology may lead to remedial steps, such as thoroughly investigating FTO and entering into license agreements to improve FTO status.

Conversely, some technologies, such as those in the public domain, can be commercialized with a relatively low risk of being found to infringe the IP rights of others. However, it is important to understand that public domain technologies are exposed to the full force of market competition through use by others—a product developer cannot shelter them by the exercise of exclusionary IP rights.

Accordingly, in the course of developing a new technology, it is important to consider building the exclusivity of an IP portfolio, while assessing and preserving FTO. That is why this chapter focuses on the process of investigating and monitoring FTO while concurrently building an IP portfolio. Technology that corresponds to business goals and that possesses maximal FTO *and* maximal exclusivity is the most likely to attract and retain investment capital.

It is worth noting that public sector institutions differ fundamentally from private companies in many of the elements discussed here. Consider, for instance, a university's portfolio of relatively early-stage technologies in which the licensee, not the portfolio manager, is commercializing the technology. It is the licensee who assesses risks in relation to a particular business goal and who seeks maximal exclusivity and maximal FTO. For the technology manager, FTO is important partly because blocking patents may make a university technology unmarketable or otherwise limit its future implementation.

In universities, moreover, faculty inventors often respond to a different set of incentives than technology transfer staff. Compared to a private company where incentives are more likely to be aligned around the successful commercialization of products, the bifurcated structure in universities between the production of intellectual property and its management can make it very difficult to coherently assess risk or build an IP portfolio with particular business goals in mind.

Public sector institutions may also pursue goals that are substantially different from those supported by IP management strategies in the private sector. In that sense, the calculus of their risk assessments may differ. For example, an institutional goal may be to preserve broad access to invented technologies or to ensure that new technologies are adopted as broadly as possible. While a public sector institution's use of intellectual property—and therefore its consideration of FTO and exclusivity—to achieve these goals may differ from private commercial companies, a sound understanding of the basic process and characteristics of FTO remains a common critical skill for successful technology management.

2. TYPES OF IP RIGHTS

AS THEY AFFECT FTO

As summarized below, a number of distinct categories of IP rights can be used to build a portfolio. This chapter emphasizes the types of rights typically encountered in the life sciences, such as biotechnology, pharmaceuticals, and medical devices. Naturally, similar issues and opportunities are presented in many fields of technology.

2.1 *Patents and trade secrets*

These first two main types of IP rights are based on the concepts of *inventions* and *know-how*. *Inventions* are the practical, useful aspects of discoveries and are typically embodied in the development of new technology. An invention can be protected by a utility patent if it meets the statutory criteria specified by the relevant government entity. In the United States, patents are granted by the U.S. Patent and Trademark Office (PTO), which is part of the federal government. The criteria include novelty, utility, and nonobviousness. Patents are granted in response to filed applications that provide an adequate written description of the invention, teach how to make and use the invention, and, in particular, point out and distinctly claim the essential elements of the invention in one or more written *claims*. A *patent* is a government grant of the exclusionary right to prevent others from making, using, selling, offering to sell, or importing the invention as claimed. The patent is granted for a limited time: under current United States law, the patent grant expires 20 years from the filing date of the first application disclosing the claimed invention. A *patent portfolio* includes all patent rights, including both issued patents and pending patent applications, that correspond to the invention and its various aspects and uses.

The broader category of *know-how* includes technology and information that may be related to inventions or to their use, marketing, distribution, or sales but is not patentable. Such information, if its proprietary status is maintained, may qualify for trade secret protection. *Trade secrets* are IP rights in unpatented technology and information that confer a competitive advantage to the owner, and are generally unknown. Trade secret status depends on the vigilant preservation of the secret by limiting knowledge of it to those key employees or other workers who have a *need to know* and by using suitable nondisclosure agreements and policies. Examples of trade secrets include ingredients, manufacturing methods, business methods, and customer lists. In the United States, whether information qualifies as a trade secret is determined in accordance with state law. Generally, the applicable law confers on the owner the right to prevent others from copying or pirating the secret. As with patents, the remedies available in the event of the misappropriation of a trade secret include damages and injunctions. However, no remedy is available where the secret is independently discovered by another who acts in good faith and does not engage in unfair business practices. Also, trade secret protection ceases upon publication or other public disclosure of the secret by any party. Thus, while trade secrets may be an important component of the IP portfolio for a particular technology, they may not function as business assets in the same way or to the same degree as patent rights. For example, trade secrets cannot be showcased, as patent rights often are, to attract investment capital.

2.2 *Regulatory rights and licenses*

There are other categories of exclusionary rights besides patents and trade secrets. In the United States, one important additional category includes rights granted by the U.S. Food and Drug Administration (FDA) in accordance with the Federal Food, Drug, and Cosmetics Act. For example, *orphan drug* status provides a seven-year period of exclusivity for a new drug developed to treat a disease or disorder afflicting less than 200,000 individuals in the United States. Once entitlement to orphan drug status is established to the satisfaction of the FDA, the agency generally will refrain from granting any additional regulatory approvals to competing drugs developed for the same disease or condition until the exclusivity period expires. It is not necessary that the drug granted orphan drug status be patentable. Similarly, to encourage the development of new drugs for pediatric use, the FDA may grant a six-month period of *pediatric exclusivity* to the first developer to establish safety and efficacy in pediatric-patient populations. Finally, to encourage the development of generic drugs upon expiration of patent protection for an innovative drug, the FDA may grant a six-month period of exclusivity to the generic drug developer who is the first to file an *abbreviated new drug application* (ANDA). These regulatory rights and licenses provide important business assets during the commercial lifetime of the technology, rather than at its inception or during the development phase.

2.3 *Copyright*

Copyright is defined as the protection afforded to original works of authorship that are fixed in a tangible (perceivable) medium of expression that can be copied or otherwise reproduced. Copyright exists in literary works, musical works, pictorial works, audiovisual works, software code, and so on. It is important to bear in mind that copyright protects the expression—not the underlying concept or idea. Copyrighted assets that may be relevant to life-science industries include bioinformatics or other software, documents, content posted on Internet Web pages, and advertising and promotional materials. As with patents, copyright is the government grant of the right to exclude or prevent others from making and/or distributing copies of the works and also of the right to prevent others from preparing derivative works. There are limits and exceptions to the scope of this exclusionary right: the owner cannot prevent *fair use*, which encompasses reproduction for such purposes as news reporting, criticism, teaching, and research. Also, under current U.S. law, the copyright lasts only for the life of the author plus 70 years, or in the case of a work made for hire, for the later of 95 years from first publication or 120 years from creation. Remedies for copyright infringement include money damages, injunctions preventing copying or distribution, and court orders impounding or destroying unauthorized copies or the means to create or distribute copies.

2.4 *Corporate identity*

In modern commerce, the principal types of IP rights that protect a technology owner's corporate identity, or its effort to develop goodwill and brand identity, are trademarks, service marks, and top-level domain names on the Internet. A *trademark* is any word, phrase, brief slogan, design, symbol, or logo that identifies the owner as the source of particular commercial goods. In health and agriculture, trademarks can be used to brand products such as plant varieties or drugs. Similarly, a *service mark* identifies the owner as the source of commercial services. As such, trademarks and service marks become important assets during the commercial product lifetime, rather than during the research and development phases. The same is generally true for *top-level domain names (TLDs)*, which may be identical to, or incorporate, the trademark. Under U.S. common law, trademark rights arise via actual commercial use of the mark. Preferably, however, the trademark is registered with the U.S. PTO either upon actual use in interstate commerce, or upon a showing of a bona fide intent to commence such use within a specified time limit. Federal registration provides nationwide rights of enforcement and constructive notice of the mark to infringers. The duration of a trademark right is coextensive with actual use of the mark in commerce. Registration rights are granted for ten-year terms, which may be renewed indefinitely on a showing that the mark remains in actual commercial use. Conversely, a mark can be cancelled from the register if it is shown not to have been continuously used in commerce during the first five years after registration, or at any time if it is shown to have become generically descriptive. Unauthorized reproduction or counterfeiting of the mark, or of a colorable (confusing) imitation thereof, is an act of trademark infringement, as is the unauthorized importation of trademarked goods. Remedies include the grant of a permanent injunction against copying, recovery of the infringer's profits, money damages, and costs. Infringing goods can be impounded and/or destroyed. If the infringing mark is a counterfeit, treble damages and attorneys' fees are available. In the case of a TLD, the remedy may be limited to the transfer of the registration to the rightful owner.

2.5 *Plant breeders' rights*

Plant breeders' rights (PBRs) protect plant varieties that are deemed new, uniform, stable, and distinct against unauthorized sale for replanting. PBRs do not generally prohibit the use of germplasm as breeding stock for creating new varieties. However, an exception to this was included in the 1991 version of the International Union for the Protection of New Varieties of Plants, commonly known by its French acronym UPOV. It prohibits the breeding of a variety *essentially derived* from a protected parent.[5] In the United States, *plant variety protection certificates* (PVPCs) confer protection against the use of sexually propagated seed germplasm. PVPCs are administered by the U.S. Department of Agriculture (USDA) under the legal authority of the Plant Variety Protection Act of 1970.

The foregoing is not an exhaustive list of the types of IP rights that may be relevant to a particular technology or product. For example, design patents may protect an attractive or distinctive original design of a useful article, such as a medical or diagnostic device. In the field of agricultural biotechnology, although plants are generally protected by utility patent rights, either *plant patents*—which in the United States grant protection from unauthorized use of most clonally propagated plants—or PVPCs may be obtained in addition to or in lieu of utility patent rights.

3. SUBJECT MATTER OF THE FTO

The first step in conducting an FTO investigation is to define what is to be searched. How precisely the subject matter can be defined will depend largely on the developmental stage of the product or other technology, as well as the nature of the technology itself. For example, a product candidate ready to enter preclinical development requires a more substantial search than a newly discovered gene or biological pathway. In addition, research tools and platform technologies may present unique restrictions on the scope of an FTO search. For example, the search may be limited to an anticipated field of use, or a full search of all uses may be required. Manufacturing technology and methods of use likewise may permit more or less precise descriptions of the subject matter to be searched. Manufacturing typically involves a number of different technologies, such as gene-expression vectors and host cells, as well as a number of different process steps. Each of these technologies may require an individual search, or the search may center on specific combinations of technologies and/or processes. Methods of use may be broadly or narrowly defined; related fields and collateral uses (for example, off-label uses of a therapeutic agent) may also require searching. In addition, the country or countries to search in must be identified. These should include any countries in which the technologies are likely to be made, used, or sold, as well as any countries intended as destinations for export. In general, the subject matter to be searched should be defined as precisely as circumstances permit. When a search is revisited or updated, care should be taken to refine the definition of the subject matter to be searched.

4. WHEN TO CONDUCT AN FTO SEARCH

Prudence must be the watchword guiding the decision of when to conduct an initial or updated FTO search. The decision depends, as a practical matter, on the nature of the risks involved and the level of risk tolerance acceptable to the client. The following is a brief survey of typical considerations that may guide the decision to engage in an FTO investigation as well as how such an investigation should be defined.

4.1 *Business goals*

One particularly useful rule of thumb in determining whether to conduct an FTO search is to review and rank the relative importance of an entity's business goals. This should be done by the decision maker in consultation with counsel. For each business goal, counsel must ask the decision maker whether they could *walk away* from that goal, that is, cease all activities in pursuit of that goal. This assessment is dictated by the availability of permanent injunction as a remedy for infringement of a number of different types of IP rights, such as patents, trademarks, and copyrights. Several subsidiary considerations further guide this analysis.

First, it has become clear that, under United States law, there is effectively no *research exemption*: the decision in *Madey v. Duke* indicates that exploratory or basic research may constitute patent infringement. So far, commercial companies have not sued universities for the infringement of patents used by their faculty in research.[6] Indeed, a commercial company's decision to turn a blind eye toward infringement in the public sector makes some economic sense. Were a patent owner to sue and win a patent litigation case against a university, the patentee would be titled to injunctive relief and damages, that, for the typical use of patented technologies in basic research, would likely be negligible and not worthy of multimillion dollar patent litigation. However, universities who wish to promote the further development and eventually the commercialization of their faculty's research may want to pay increasing attention to FTO issues so that they can understand how their technologies are situated with regard to other patents in the field and how they can reduce potential future impediments to commercialization.

There is, however, a *safe harbor* exemption for research and development relating to the submission of applications for regulatory approval by the FDA, including both clinical and preclinical studies. The scope and limits of this safe harbor have not been conclusively established, necessitating a case-by-case analysis. Also, many developed countries have similar laws governing whether basic research and research related to the approval of new drugs is exempted from patent infringement. The scope and precision of laws on this point may differ significantly from country to country, and a detailed discussion is beyond the scope of this chapter.

Second, and in view of the above, one must consider the geographic scope of the market to be served by the business goal under consideration. Since IP rights are granted by governments and are territorial in nature, an FTO investigation should apply the laws of the country or countries in which activities are undertaken in pursuit of the business goal. For example, all research, development, and manufacture may take place in the United States, but the commercial market may include Europe as well as the United States. In other situations, the inverse may be true. The corresponding FTO investigations should identify and assess third-party patent rights in both the United States and Europe. In the case of a worldwide market, cost and a pragmatic assessment of risk may dictate that the FTO assessment be restricted to major markets.

Third, it is important to consider how much has been invested in the business goal to date. A significant investment, or an investment representing a significant portion of total business assets, heightens the need for an FTO search. This principle is illustrated below in the context of a biotechnology or a pharmaceutical for human healthcare. Another approach, suitable to assessing FTO for a research tool or platform technology, is to determine whether use of the technology is limited to a specific (and minor) project. If the technology will be relied upon broadly, or will underpin an important long-term business goal, an FTO search should be considered early.

A related consideration is whether the early establishment and monitoring of FTO will increase the attractiveness of the business goal to potential investors. Venture capital investors and large institutional investors tend to be quite sophisticated and keenly interested in the IP risks pertaining to a technology or business plan of interest. More recently, a well-formulated IP strategy is a requirement for funding agencies that, in addition to supporting research, are dedicated to ensure the prompt dissemination of a project's outcome.

4.2 *Risk of IP infringement litigation*

Another rule of thumb is equally important. Counsel and the decision maker should assess together whether the client can tolerate the risk of litigation. Risk tolerance varies with government oversight and regulations, management style, and the nature of business activities, but is also closely tied to financial resources, including the availability and scope of relevant insurance. When assessing the risk and consequences of infringement litigation, one must bear in mind that, at least in the United States and Europe, the cost of defense is significant. Also, at least in the United States and the United Kingdom, damages awards for patent infringement tend to vary from large to quite large. Legal costs and damages, taken together, can figure in the tens of millions to the hundreds of millions of U.S. dollars.

As mentioned previously, another significant risk of infringement litigation is that a court will issue a permanent injunction, for example, ordering the client to cease its infringing activities or, under certain circumstances, ordering the seizure, impoundment, and/or destruction of infringing goods. Thus, the risk assessment must take into account the value of lost business opportunities. There may be other risks consequential to the initial infringement liability, such as shareholders lawsuits and investigation and/or enforcement actions by regulatory authorities (for example, the U.S. Securities and Exchange Commission [SEC][7]).

It must be noted, however, that infringement litigation is also costly to the plaintiff and may not be pursued when the unauthorized use of the technology does not threaten the patent holder's business goals. The use of patented technologies in the course of academic research in the United States, for instance, has been shown to constitute infringement, but infringement lawsuits against academic researchers are likely to provide little benefit to the patent holder either through injunction or through the recovering of damages. Examining the economic and legal rationales for infringement litigation may be particularly important for assessing the risk of infringement litigation by researchers in public and not-for-profit institutions and in developing countries.

4.3 *Level of investment*

A third useful framework for deciding when to conduct an FTO search is to determine what business decisions should trigger the search. It will be fairly straightforward to identify the types of decisions that would significantly increase resource commitments to a specific business goal. Such discontinuities in business strategy or financial investment should signal the need for an initial or updated FTO search. Indeed, many companies have made projects pass a series of increasingly rigorous FTO studies during the course of development. A sampling of the changes in investment that may merit new or updated FTO studies in the development of a novel biologic or pharmaceutical drug are illustrated in Box 1. Analogous investment changes that may warrant an FTO analysis also exist in other fields, such as agricultural and industrial biotechnology.

5. SCOPE OF THE TYPICAL FTO INVESTIGATION

A typical FTO search canvasses all reasonably available sources that are likely to reveal relevant third-party IP rights. For the most part, these are computerized databases and search engines capable of surveying publicly accessible patent, technical, and commercial literature. Issued patents, published patent applications, and scientific/technical publications, as well as databases of meeting presentations and grant awards, can be searched using keywords, investigators' names, assignee/owner names, and subject-matter classifications. Biological sequence databases, including both nucleic acid and protein sequences, can be searched using a query sequence. Patent assignment branch records should be searched to reveal the names of real parties in interest, as well as transfers of ownership. Patent annuity and maintenance-fee records should be searched to verify that patents identified as relevant are in fact still in force. On the commercial front, the SEC filings of identified assignee/owner businesses that are publicly traded can be searched on the electronic data gathering, analysis, and retrieval system (EDGAR)[8]. The filings of interest include companies' quarterly (8-K) and annual (10-K) reports of progress toward their business goals, which include self-assessments of risk. A search of the records of known competitors may reveal common threats to FTO status, such as third-party IP rights in broad classes of molecules (for example, fusion proteins) or manufacturing technologies. When appropriate, press releases, industry-specific news reports, and stock analysts' reports also should be investigated.

5.1 *"Level one" FTO investigation*

As mentioned above, not every FTO investigation merits the same scope or depth of search.

Box 1: Changes in Investment Meriting an FTO Analysis of a New Drug or Biologic

Selection of a druggable target

Exploratory research into a specific biological pathway may reveal one or more genes or proteins that appear to be a suitable site for intervening in a disease process or other metabolic process. A *druggable* target is a molecule identified as pivotal to a biological process, with a structural feature such as a cleft for which a pharmacophore can be identified or designed. In many cases, IP rights encompassing the use of the target or compositions of matter corresponding to all of the target or specific parts of the target may exist. Universities and research institutions frequently own such IP rights.

Screening/research tool technology

A number of companies have developed business models based on providing tools and services to the research community, and these may be aggressively protected by IP rights. Affymetrics, for example, markets and sells nucleic acid microarray chips. The Harvard oncomouse, commercially available from DuPont, is another example.

Identification of a lead compound

The selection of a lead compound typically represents the transition from research to development. It is axiomatic that the structure of a lead compound, one incorporating a successful pharmacophore, cannot reliably be predicted based on knowledge of the target. Thus, the lead compound and the structural class to which it belongs represent both new opportunities for developing an IP portfolio and new risks in light of which FTO should be established before committing resources to a development-phase project. Both specific and general features of the lead compound should be investigated. For example, IP rights may be found to cover humanized antibodies or different types of fusion proteins. The same considerations apply to any back-up compound.

Preclinical development

The commencement of preclinical development means both a significant rise in the level of financial commitment and the beginning of the *safe harbor* from patent infringement. Here, activities focus on the development of data to be included in an investigational new drug (IND) submission to the FDA. Despite the safe harbor, this step represents a formal commitment to develop a new drug or biologic for eventual commercial use. Thus, from the investment standpoint, it is a critically important stage at which to conduct a thorough FTO search or update and refine a prior search. Also, at this stage, many ancillary aspects of commercialization may be established beyond the structure of the drug candidate, such as its formulation or dosage, its primary commercial indication for use, and basic manufacturing techniques.

Selection of manufacturing technology

In many instances, the manufacturing technology needed to support commercial scale production of a new drug will differ from that practiced at the research or even developmental stage. Because of the magnitude of resource commitment required for manufacturing, many companies have patented successful manufacturing techniques broadly. One example would be the patenting of a particular type of chromatography resin to purify a particular class of molecules (for example, humanized antibodies). Another example would be the type of host cell or a formulation found to enhance shelf life or solubility.

(Continued on Next Page)

Box 1 (continued)

Selection of a clinical indication(s)	A main or primary clinical indication for the new drug or biologic may have been selected based on an understanding of the target and its mechanism of action. As development progresses, however, additional indications may become apparent, as may additional channels of commercialization (for example, neurologists may find the drug attractive for one disease, while gastroenterologists may perceive its value for another distinct disease). Each distinct clinical indication may attract its own competitors, dictating the need for corresponding FTO studies.
IND Submission	An IND application is the document the FDA uses to decide whether to allow human trials of a new drug or biologic agent. Readiness to submit an IND and, even more so, holding an approved IND represent a critical achievement in the business life-cycle. The interest of investors and potential corporate partners or acquirers is piqued, and the value of a business is significantly enhanced. It is particularly important at this juncture to establish the feasibility of the business goals corresponding to the drug development project. Indeed, a number of pharmaceutical companies treat the FTO investigation conducted at this juncture as the *go/no-go* decision on commercialization.
Pivotal clinical trial	A pivotal clinical trial is one that can generate statistically sound data that the FDA can use to decide whether to approve a drug for commercial sale. Depending on the clinical indication, such a trial may take from one to five years, and may involve from tens to thousands of patients. Initiating and conducting such a trial often represents the single largest investment made during the course of commercialization. In addition, starting such a trial signals a commitment to particular drug compositions, formulations, methods of manufacture, and methods of administration and use. This commitment alerts third-party IP rights holders to the value of their IP, raising the cost of establishing FTO by entering into license agreements or avoiding adverse IP rights by *designing around* them.
NDA/BLA submission	The new drug application (NDA) or biologics license application (BLA) is the dossier submitted to the FDA for its decision on commercial approval of a new drug or biologic agent. FDA approval, which typically takes from two to four years, signals the end of the *safe harbor* from patent infringement. Thus, the period of NDA/BLA pendency represents the last stage at which any remaining FTO issues may be resolved without exposure to infringement litigation.
Commercial launch	This is the commencement of actual commercial activity, the stage at which a company is fully vulnerable to charges of IP rights infringement. Prudence dictates that FTO must be established prior to this stage and that periodic monitoring be conducted to ensure preservation of FTO throughout the product's lifetime.

Exploratory-stage research, or consideration of a new business goal, may require no more than an overview and risk identification. The question to be answered is whether there are any so-called *blocking patents* that would preclude pursuing the new goal. This is called a "Level One" FTO study to distinguish it from more in-depth analyses. The Level One study assesses only public information, typically in the following two categories:

- **Patent database searches.** Keyword, surname, business name, and sequence searches of patent databases are conducted to reveal relevant patents and published applications (which are potential future risks).
- **Patent ownership and status searches.** Surname and business/entity name searches of assignment branch records are conducted to reveal ownership interests, transfers of ownership, and other recorded rights affecting ownership. If deemed prudent, secretary of state records may also be searched to reveal any transfers or liens that may not have been recorded at the federal level. Searches of relevant annuity/maintenance-fee databases are conducted to reveal whether any of the identified patent risks have lapsed for nonpayment.

5.2 *"Level Two" FTO investigation*

There are many ways to design and implement more in-depth FTO searches. The nature of each search is dictated both by the precise definition of the subject matter to be searched and by the decision maker's desired degree of risk characterization. Both considerations rest, in turn, on the significance of the business goal and the amount of resources required to achieve it.

A typical "Level Two" FTO investigation is considerably more sophisticated than a Level One, yet still only requires access to public information. Assessing nonpublic information requires either cooperation among the relevant parties (for example, IP due diligence in support of a business alliance) or court order (such as during the discovery phase of infringement litigation). Both are beyond the scope of this chapter.

Patent database searches are conducted as described in the Level One investigation, but the analysis conducted on this raw data goes beyond mere identifying potential *blocking* patents. Instead, the patent rights are evaluated substantively to construct a *patent landscape* in which the patent claims are grouped by subject matter. For example, one group may encompass expression vectors and be subgrouped according to the type of vector. Another may encompass host cells, including specific types of host cells and their culture methods. Yet another group may encompass the structural class to which the drug of interest belongs. For example, all patent rights on fusion proteins may be grouped together, with sub-groups defined according to the protein class of interest (for example, receptor-Ig fusion proteins). The groupings can be configured to most effectively educate the business decision maker about how to proceed.

Another very informative way to analyze the search results is to construct a timeline of patents on similar or overlapping subject matter. Ordering the patents and published applications according to their *priority dates* (also known as *effective filing dates*) reveals important relationships. For example, it reveals which patents are *prior art* against newer patents. Since patents may only be granted if the claims are both novel and nonobvious over the prior art, this analysis reveals the relative dominance of earlier, broader patents over later, narrower patents. There are many circumstances in which broadly and narrowly defined claims covering the same subject matter can coexist and be owned by different parties. Analyzing the priority timeline will reveal whether some patents should be licensed or designed around by developing alternative technology. This analysis will also reveal which parties possess more leverage to seek higher license fees. Including published applications in the timeline enables the astute decision maker to make educated guesses about the scope of claims likely to issue from applications filed later. Finally, it provides insight into possible interferences. Unique to U.S. patent law, an *interference* is an administrative proceeding before the Board of Patent Appeals and Interferences, in which two or more parties claiming the same subject matter in separate patent applications engage in a contest to determine who was the first to invent. The procedural rules are strict, and the winner is awarded the patent. Figure 1 in the case study below illustrates a typical patent-priority timeline.

Scientific and patent literature, including patents and patent applications, illustrate the existing prior art at the time that related patent applications were filed. The priority dates of each patent and patent application relative to the publications dates of the main scientific literature are shown.

The analysis of priority claims in published patent rights also reveal *family* relationships among different patents and published applications. *Patent families* include both vertical and horizontal relationships. A vertical or lineage relationship arises when a later patent application claims the benefit of an earlier, related application that names the same inventor (or at least one common inventor, in the case of joint inventors). If the *specification* (text portion of the application) is identical to the earlier application, but the claims cover different subject matter, the later application is called a *continuation* or a *divisional*. If the specification has been edited to disclose more or less information, and corresponding changes have been made in the claims, the later application is called a *continuation in part*. Horizontal relationships arise in *foreign filings*, counterparts of the original application filed in other countries or common patent territories. Such foreign filings are made under bilateral or regional treaties in which two or more governments agree to reciprocally recognize the priority of applications filed in each others' territories. The main vehicle for generating horizontal families of counterpart applications is the Patent Cooperation Treaty (PCT). The PCT provides a preliminary *clearinghouse* in which the claims are searched, and optionally examined, by a single examining authority. Both

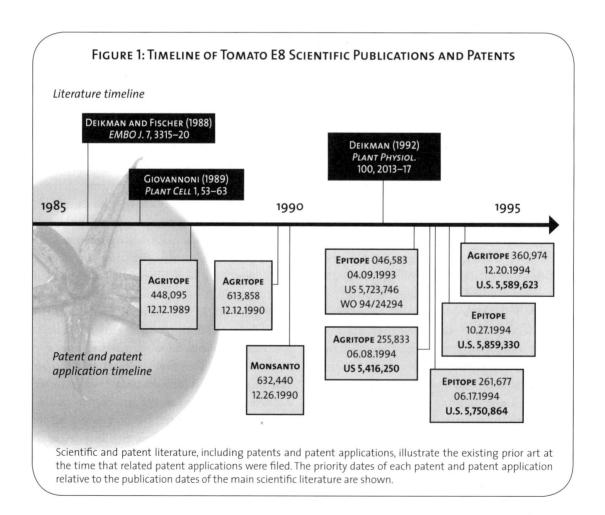

FIGURE 1: TIMELINE OF TOMATO E8 SCIENTIFIC PUBLICATIONS AND PATENTS

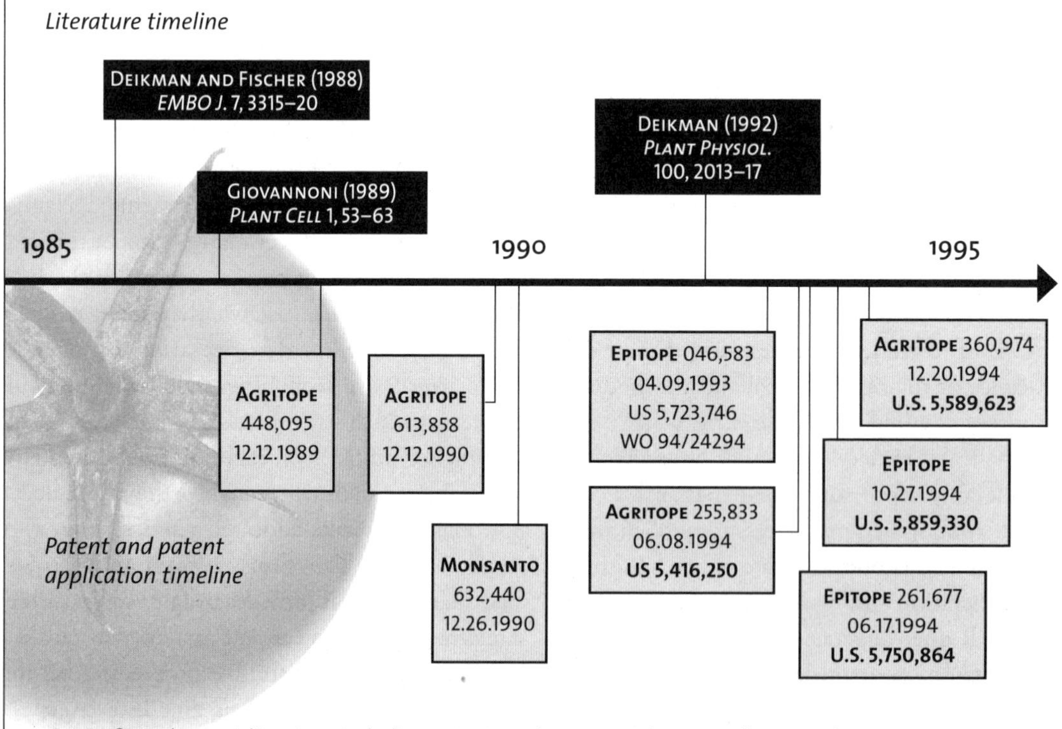

Scientific and patent literature, including patents and patent applications, illustrate the existing prior art at the time that related patent applications were filed. The priority dates of each patent and patent application relative to the publication dates of the main scientific literature are shown.

the PCT examination report and the PCT search report are publicly available. Figure 2 in the case study below provides an illustration of patent family relationships.

Very often, the foregoing analyses reveal a subset of identified patent rights that require close analysis, including advice to the decision maker about the scope of the patent claims. This type of analysis is known as *claim construction.* It requires counsel to obtain and evaluate the patent file histories. The *file history* (or prosecution history) is the written record of negotiations between the patent applicant and the examiner. The patent specification (text portion) typically does not change during prosecution; however, the claim language does change, sometimes quite dramatically. For example, the examiner may require that the claims be divided into subsets, which are then prosecuted separately in divisional applications. Other changes in claim language arise from the need to conform to patentability requirements, such as enablement, written description, clarity, novelty, and nonobviousness. Even where the claims have not been amended, the patent applicant may have made remarks that define the scope of the claim or that disclaim a broad interpretation. Such remarks are referred to as *file wrapper estoppel* or *prosecution history estoppel* because the patentee is not allowed to assert a broader claim scope when enforcing the resulting patent.

In the United States, prosecution history analysis is restricted to the histories of issued patents and published applications, since the files of provisional and unpublished applications are

FIGURE 2. THE EPITOPE/AGRITOPE PATENT FAMILY

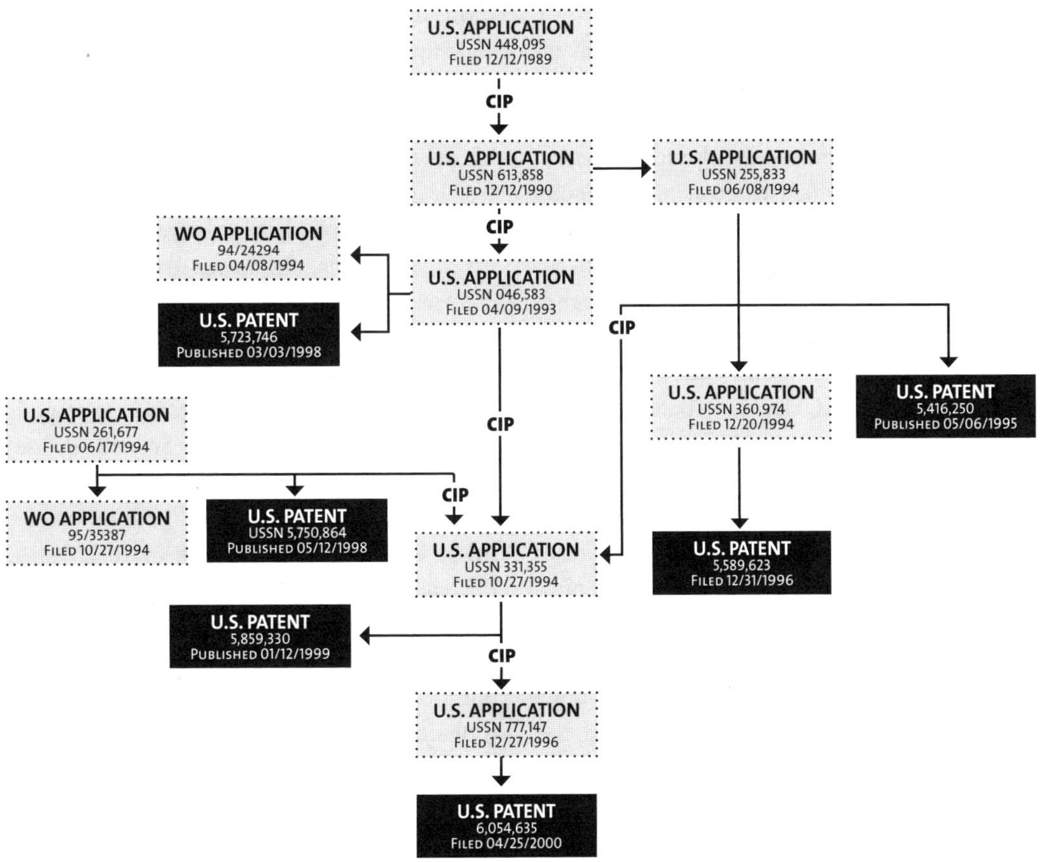

confidential by law. In most cases, the file histories of foreign counterpart applications are available to the public. Thus, one can obtain insight into the potential scope of patentable claims by obtaining and analyzing the file histories of one or more counterpart applications in a patent family. European prosecution histories are available electronically as .pdf files. Australian histories can also be obtained and are often useful because the pace of examination in Australia is frequently more rapid than it is in other PCT member states. Each foreign counterpart application is examined in accordance with the granting country's patent law, so one must expect to encounter more or less nuanced differences in the scope and format of patentable claims.

In addition to analyzing prosecution histories, it is necessary to check the appropriate patent litigation databases to determine if any patents of interest have been held invalid or unenforceable. The PTO Web site should also be checked for information on whether an interference has been declared involving a patent of interest. The interference proceedings are not public information, but the final decisions of the Board of Patent Appeals and Interferences are publicly posted. Similarly, the records of foreign patent offices should be checked to determine whether any newly-granted patents have been the subject of patent oppositions. An opposition is an administrative proceeding in which any member of the public adversely affected by the patent grant may file arguments urging that the patent should not have been granted, that is, it fails to comply with the grantor's laws on patentability. Europe and Australia are among the countries that permit the filing of oppositions within a specified time period following the patent grant. The record of opposition proceedings in each country is publicly available.

Finally, a prudent and thorough FTO investigation includes searches of business and news records as well as of patent records. General and industry-specific news reports may reveal the names of business or nonprofit IP holders not revealed through the patent database searches. They may also provide useful overviews on the state of the art or the competitive marketplace. If available, stock analysts' reports on an industry sector or an individual business are particularly helpful. Such reports often provide independent, expert assessments of business risk, including IP risks. As mentioned earlier, the annual report or SEC filings of an IP rights holder provides useful self-assessments of risk and competition. Perusing the patentee's Web site and relevant press releases will often reveal whether the patent rights in question correlate to a stated business goal. Such information provides the decision maker with valuable insight into both the business model of the patentee and the importance—and therefore value—that the patentee places on the patent rights of interest. For example, a university or nonprofit organization may have a stated policy of licensing its IP rights in order to pursue its mission of advancing public knowledge or providing public benefit. Similarly, research tool companies have adopted business models that rely on broad licensing of their IP rights. In contrast, innovator drug companies and biotech companies may be motivated to preserve their exclusionary rights, such that licenses may not be available, or offered only on unfavorable terms.

5.3 *Limitations*

It is imperative for the decision maker to bear in mind, when considering the results of any FTO investigation or clearance search provided by counsel, that such searches are, by their very nature, limited. First, the search is limited in time. New patents may have issued since FTO was last analyzed, and it is for this reason that periodic updates must be considered. Second, the search is limited to publicly accessible information. It is impossible to identify all of the new inventions made in the field of interest or to characterize the trade-secret rights claimed by competitors or other business entities. Similarly, no FTO search can identify or analyze unpublished patent applications. This category includes United States provisional patent applications, as well as utility applications that are less than 18 months old (as measured from the priority date). Under current United States law, older utility applications also may not be published if the applicant has requested nonpublication and disclaimed the right to file foreign counterpart applications. Also, as mentioned above, the file histories of unpublished U.S. applications are not available to the public. For these reasons, an FTO investigation may need to be updated regularly. If desired, an automated, computer-based monitor may be instituted to alert counsel of new patent information as soon as it becomes publicly available.

Business information also may not be available. A company's business goals and the status of its research and development projects, for example, may not be publicly disclosed. Similarly, information about the competitive risks perceived by the company may not be publicly available. While lawsuits are a matter of public record once filed, invitations to license a patent, threats of litigation, licensing negotiations, and settlement discussions are usually not in the public record. Corporate documents, such as contracts affecting the ownership of intellectual property (for example, assignments, security interests, joint development agreements, service contracts) also are usually not public records. Similarly, contracts affecting the use of IP rights (for example, licenses, settlements, options, material transfer agreements, confidentiality agreements, employment agreements, consulting agreements, noncompetition agreements, service contracts) are usually not public records. Business information influencing the results or interpretation of an FTO study may not be revealed until a due diligence investigation is carried out as part of a license negotiation, or until the discovery phase of a patent infringement suit commences. In certain circumstances, however, the existence of a corporate document that affects ownership or use of IP rights *material* to a publicly traded company's business may be revealed in the company's SEC filings. A document is considered *material* if it affects the value of the company's stock.

Another key area that usually cannot be explored when using only publicly available information is whether there are any adverse claims to inventorship of third-party patent rights. Increasingly, inventorship disputes are being considered in litigation and other adversary proceedings as a way to obtain a license from a newly added, sympathetic co-patentee. As with nonpublic business information, the existence of possible inventorship claims is often not revealed until a licensing due diligence investigation is carried out with the consent of the patentee or the discovery phase of litigation commences.

6. THE PRODUCT OF AN FTO INVESTIGATION

The product of an FTO investigation conducted by a law firm or an in-house attorney and communicated to the decision maker is uniformly recognized under U.S. state law as being *attorney-client privileged information*, and depending on the circumstances may also fall under the *work product privilege*. The results retain their privileged status as long as the client (who holds the privilege) chooses not to reveal the information to others, or it is not inadvertently disclosed. The attorney-client privilege applies to advice regarding IP rights in most European countries as well as in the United States. It is important to keep in mind, however, that in some countries patent professionals are not attorneys; thus the degree of protection afforded to the results of an FTO investigation may vary and should be established in advance. Switzerland, for example, does not recognize privilege in the communications between a patent practitioner and a client, although it does so between an attorney and a client.

In many circumstances, for example, where a Level One FTO investigation is all that is needed, the results of the investigation may simply be the oral advice of counsel to the decision maker. Depending on the purpose of the FTO investigation, or where a more-detailed Level Two investigation has been carried out, counsel may provide a written report to the decision maker. Typically, the report includes brief statements of the scope of the search, as well as a listing of the search strategies used (for example, keywords, sequences, assignee names). The report also includes a listing of the identified third-party IP risks. A written report of the identified risks is usually brief and carefully worded because of the potential for such commentary to function as admissions against interest of the client, if the attorney-client or work product privilege is lost or waived.

The most important feature of any document reporting the results of an IP FTO investigation is that it is a *living document*—it should be updated as new information comes to light through a monitor or through a regular schedule agreed to between counsel and the decision maker. The decision maker should understand that decisions may have to be modified or reconsidered in light of updated information about changes in the nature of the IP risks being monitored, their status, or newly emerging intellectual property. This process should continue for as long as the business goal is pursued.

7. IP RISK MANAGEMENT STRATEGIES

The process of securing or improving FTO does not stop once the results of an investigation are available. Rather, the results of a clearance study provide the tools and intelligence necessary to determine the most desirable course of action for the client (whether a business, a university, or other nonprofit entity) to take in light of the discovery of a so-called *blocking patent*. Counsel should work closely with the decision maker in developing IP risk management strategies. Box 2 presents a representative but by no means exhaustive survey of the principal strategies that may be considered. Any one or a combination of the risk management strategies shown in Box 2 may be employed, as deemed prudent and appropriate by the decision maker working in consultation with counsel. These and other options are further discussed by Krattiger.[9]

8. CASE STUDY: FTO ANALYSIS AND THE LEGAL LIMITATIONS OF A PUBLIC-DOMAIN TECHNOLOGY

The purpose of this case study is to illustrate basic strategies for performing an FTO search of a technology that has both research and commercial objectives. This particular example includes the decision maker's considerations when engaging in an FTO analysis, the process of gathering FTO information in-house, the evaluation by legal counsel, and the outcome of the analysis. Attorney-client confidentiality privileges have been waived for the sake of sharing the experiences of this investigation. The end results of the analysis show that while the target technology, per se, is in the public domain, FTO restrictions are present when it is combined with other technologies.

Legal counsel is often sought when developing commercial products. But the use of FTO searches is not limited to business plans; they may also be crucial to projects with research and social objectives. Platform technologies used in the early phases of product development are of special concern because failing to negotiate access could drastically affect subsequent research and development plans or the licensing value of the technology. Unlike established agricultural biotechnology companies with in-house IP counsel, public sector scientists around the world may not have easy access to legal experts and consequently are often unaware of the IP restrictions on commonly used research tools. Fortunately, to facilitate the research and development of improved crops with commercial and humanitarian objectives, the Public Intellectual Property Resource for Agriculture (PIPRA)[10] is working to design agricultural biotechnologies that are technically strong and subject to minimal IP restrictions.

Plant transformation vectors—the molecular shuttle vehicles that introduce desired genes and traits into bioengineered crops—are a key platform technology in agricultural biotechnology. Plant transformation vectors combine numerous components, such as genetic regulatory elements (promoters), selectable markers, systems to remove those markers, and more. By virtue of the fundamental role that these technologies play in bioengineered crops, they are often protected by intellectual property. Moreover, the FTO pathway quickly becomes entangled and complex because these technologies are usually not used individually but combined with different traits and in numerous host plants. To steer clear of potential blocking patents, it is important to incorporate technologies and methods that are in the public domain (free of IP restrictions) or that can be used with permission. This is why PIPRA, in collaboration with scientific and legal experts, is researching the FTO of various vector components,

Box 2: Options for strategic use of the results of an FTO investigation

Abandon or modify business goal or business practice

If a blocking patent has been discovered and cannot be licensed or avoided, the decision maker must consider whether it is acceptable to abandon pursuit of the affected business goal or the affected business practice (such as the use of a particular research tool or methodology). Alternatively, it may be commercially reasonable to modify the business goal or practice and thus obviate the blocking effect. This process is called "designing around the blocking patent." The effect and cost of the modification must be taken into account to consider the reasonableness of this approach. For example, the decision maker must consider whether FDA approval would be required to change a formulation or manufacturing processes

Take a license, if one is available on commercially reasonable terms

It is important to evaluate the likelihood that the owner of a blocking patent will accommodate the client's business goal by granting an affordable license. Intelligence on this point can be gleaned from reviewing the mission statement of the business or non-profit patentee, as well as from reviewing SEC records or press releases to determine whether the patent in question has been licensed to others. The financial effect the license will have on commercializing the product or technology must also be considered. Royalty payments and manufacturing expenses together account for the cost of goods sold (COGS), so a patent license in effect forces cost cutting in other areas. The pressure on manufacturing costs is even greater for products subject to royalty stacking, when multiple royalties under multiple licenses are needed to commercialize a single product.

Ask the owner of a blocking patent to relinquish their IP rights

Patent owners may consider relinquishing their IP rights in territories or fields of use when they do not foresee sufficiently large commercial markets. In addition, a patentee may find that the benefits of good public relations weigh in favor of relinquishing IP rights for particular humanitarian uses of a technology. In these cases, negotiating a royalty-free license or a covenant not to sue may be possible. However, product liability and stewardship issues remain concerns for many patentees. In fact, potential licensees may find that a patentee is seeking to avoid a liability risk for any defective products incorporating the patented technology that enter the stream of commerce.

Obtain a formal written opinion of counsel

If the consequences of abandoning or modifying the business goal are unacceptable, or if the decision maker suspects or has established that a license may not be available from the patentee on commercially reasonable terms, or if it seems likely that the patentee may take some offensive legal action, counsel may be asked to provide a reasoned written opinion on the non-infringement or invalidity of one or more claims of the blocking patent. It is important for the decision maker to realize that such an opinion does not shield the client against infringement litigation. However, it may provide useful insight or leverage in licensing or settlement negotiations, as well as precluding a court holding of willful infringement (which would permit doubling or trebling of damages).

(Continued on Next Page)

Box 2 (continued)

Leverage the client's own IP portfolio.	Another means of improving the odds of obtaining a license on commercially reasonable terms is to inventory the client's own IP portfolio supporting the business goal (or even other business goals) to determine whether any existing claims (pending or issued) could provide a cross-blocking effect. Can any of the client's claims impede the FTO of the blocking patentee? If none are issued or pending, a client's new patent application may provide a good basis for drafting and prosecuting new claims in pursuit of a cross-blocking effect. Alternatively, if there are issued or pending claims in the client's patent estate that overlap with the patentee's blocking claims, it may be possible for the client to trigger a U.S. patent interference with the patentee. Each of these strategies can provide important leverage in negotiations with the patentee

including promoters used to regulate the expression of desired traits in specific plant tissues.

8.1 *Defining the subject matter of the FTO or clearance search*

The target technology for this case study is a fruit-specific promoter from the tomato E8 gene. Technically, the E8 promoter is often chosen because gene expression under its control is triggered by developmental cues such as fruit ripening. Expression of the gene of interest is confined to the ripe fruit and is not detected in other organs such as leaf, root, or stem. In addition, the promoter can stimulate gene expression in response to a chemical stimulus (ethylene) also in organ-specific fashion. As such, this transcription-regulation element has been used to improve nutritional and juice qualities, extend the vine life of tomato fruit, and express edible human vaccines in tomato fruit.

As previously described, the first step in an FTO investigation is to clearly define the target technology. In this case, PIPRA proposes to use the fruit-specific promoter exactly as described in the initial publications by Deikman and Fischer[11] and Giovannoni et al.[12] The promoters in these publications are virtually identical and consist of about 2,100 nucleotides upstream of the E8 gene. Further promoter characterization disclosing the location and sequence of functional elements within the promoter and upstream nucleotide sequence was reported in Deikman et al.[13] These publications draw the technical boundaries surrounding the target promoter technology and, as we will discuss later, provide important prior art.

8.2 *Does the business plan warrant an FTO analysis?*

The plan of a project sets the direction that an associated FTO investigation will take. In this case, PIPRA foresees that, once this particular promoter is integrated into plant transformation vectors, it will be used for both research and commercial purposes, both within the United States and abroad. Since it will be part of a platform technology that may be broadly adopted, it warrants an in-depth analysis to determine FTO.

Because the technology is being evaluated at an early stage and could be used in a wide range of projects, PIPRA cannot know all of the specific genes of interest that the E8-promoter might be used to drive. Therefore, PIPRA chose to limit the analysis to FTO on the promoter *per se* and not on its use in combination with specific genes of interest, with an understanding that future analyses will be needed to determine FTO for specific combinations of the E8 promoter and heterologous genes. As described before, compounding technologies create a more complex IP landscape because of the potential for overlapping patent claims. This initial FTO analysis thus indicates only the technology's general availability. Still, evaluating limitations at an early stage provides researchers and business developers with important information about the technology's FTO position and illustrates the legal limitation of a technology presumed to be in the public domain.

8.3 *Case study: FTO information and legal opinion*

In this case, PIPRA provided the background FTO information to legal counsel, who subsequently conducted an FTO analysis. The background FTO information packet consisted of a detailed description of the proposed construct, proposed management strategy of the plant transformation vectors, scientific literature on the technology, and IP search results. Legal counsel assessed the relevant patents, grouping them according to subject matter and assignee, constructed a priority timeline integrating relevant literature and intellectual property (Figure 1), and then delivered an oral and written FTO opinion. The following is a detailed account of the process.

8.3.1 *Client's FTO background information*

The scientific literature in the background file included a list of publications describing the discovery, characterization, and applications of the E8 promoter. Literature records were identified and extracted using keyword and author searches using online databases. Assembling a timeline of publications and contacting the original inventor/author of a technology are advantageous when investigating whether patent protection was sought at the time of invention and publication. This is particularly important because it is possible that corresponding U.S. patent applications could remain unpublished and later emerge as issued patents. PIPRA contacted the principal investigator (PI) of the group that originally identified and characterized the E8 promoter, Robert L. Fischer, at the University of California, Berkeley, and discovered that the inventors did not apply for patent protection prior to their seminal publication. The absence of patent applications by Deikman and Fischer[14] was confirmed by subsequent investigations. Because at this point of the investigation it was presumed that the technology was in the public domain, documenting published literature or *prior art* was particularly crucial.

The patent landscape included patents and patent applications that were closely related to the technology. Keywords and authors of key publications were used to search for patents or patent applications. The patent search engines used were Delphion,[15] M-CAM,[16] and the EPO.[17] A separate search was conducted to identify patents or patent applications that referenced the scientific publications describing the technology. In addition, patented DNA and protein sequence databanks were searched using the E8 promoter's DNA sequence as a query. Because the target technology was identified and characterized in the late 1980s and early 90s, special attention was given to publications with a priority date around that time. After evaluating patents and patent applications, a list was distilled of patents with claims to regions from the tomato-derived E8 promoter. Furthermore, a schematic representation illustrating the claimed DNA sequence between the target technology and patent claims was incorporated (Figure 1). For legal counsel's convenience, a table of patents and patent applications was provided that included record numbers, family members, assignees, publication, priority, and application dates, as well as relevant notes. The patent landscape documentation also indicated whether the patent's nonpatent prior art section (field 56 on the patent coversheet) cited Fischer's publications (evidence that this was considered prior art).

Another independent search was conducted to identify specifically those patents that claim the use of the E8 promoter to drive genes of interest. This search was conducted in the same manner as described above, using keywords for the E8 promoter to search within claims. The pertinent patents and patent applications were extracted and analyzed. Again, a table with the patent records was compiled and a written report with the described information was submitted to legal counsel. After verbal communications and revisions, additional information (such as patent family trees) was provided for analysis (Figure 2).

8.3.2 *Legal counsel's FTO opinion*

Using this background information, legal counsel constructed a priority timeline including the key scientific literature and the most closely related patent records, which were assigned to Agritope, Epitope, and Monsanto (Figure 1). As shown, the Deikman and Fischer[18] and Giovannoni et al[19] publications initially describe the E8 technology. This precluded the novelty of any subsequent patent claims on the E8 promoter per se (for example, applications filed by Agritope and Epitope). While the detailed written FTO opinion of legal counsel is not included in this report, counsel concluded that the tomato E8 promoter constructs per se (searched as described above, without considering association with any heterologous gene) can be reasonably considered to be in the public domain.

Since the analyses did not examine FTO with the E8 promoter in conjunction with other genes or other vector elements, appropriate FTO limitations and future considerations were highlighted. Interestingly, because the initial E8 publications did not disclose the use of the technology with a variety of heterologous genes, subsequently issued patent claims were able to limit use of the technology by covering novel combinations of already known elements. Thus, while the technology itself is in the public domain, its use with particular genes of interest is not. This information indicates to PIPRA that FTO should be reevaluated in more-advanced stages of the vector construction when other technology components are known. Though not exhaustive, some of the patents claiming chimeric constructs comprising the E8 promoter and heterologous genes are shown in Figure 3. The patents can be grouped into three broad categories related to agronomic characteristics, biopharmaceuticals, and gene expression control. Notice the potential for claim overlap within these broad categories, for instance, gene expression control patent claims may span uses in agriculture and pharma.

Legal counsel conveyed the results of the analysis to PIPRA via oral communications and, subsequently, in a written report. It is important to note that FTO analysis materials are protected by attorney-client privilege and thus should only be shared on a confidential basis with personnel that have a need to know (for example, business decision makers). In the case of the E8 promoter FTO investigation, the client (PIPRA) decided after consulting with legal counsel to disclose the results of the investigation for public informational and educational purposes. ■

GILLIAN M FENTON, *Chief Intellectual Property Counsel, Emergent BioSolutions, Inc., 300 Professional Drive, Gaithersburg, MD, 20879, U.S.A.* fentong@ebsi.com

CECILIA CHI-HAM, *Director, Biotechnology Resources, PIPRA, University of California, Department of Plant Sciences, Plant Reproductive Biology Building, Extension Center Circle, Davis, CA, 95616, U.S.A.* clchiham@ucdavis.edu

SARA BOETTIGER, *Senior Advisor, PIPRA and Chief Economist, M-CAM, Inc., One Shields Avenue, PRB - Mail Stop 5, University of California, Davis, CA 95616-8631, United States.* sboettiger@ucdavis.edu

---

1 See, also in this *Handbook*, chapter 14.2 by Stanley P Kowalski.

2 Ibid.

3 See, also in this *Handbook*, chapter 14.1 by Anatole Krattiger.

4 See, also in this *Handbook*, chapter 7.3 by Alan Bennett.

5 The scope of this exception has been the subject of analysis. An informative publication on the definition of "essentially derived" can be found at www.amseed.com/pdfs/EDVInfoToBreeders_0605.pdf.

6 The *Madey v. Duke Univ.* case was based on a dispute between employer and employee, rather than a commercial firm's decision to sue a university.

7 www.sec.gov.

8 www.sec.gov/edgar.shtml.

9 See *supra* note 3.

10 www.pipra.org.

11 Deikman J and RL Fischer. 1988. Interaction of a DNA Binding Factor with the 5'-Flanking Region of an Ethylene-Responsive Fruit Ripening Gene from Tomato. *Embo J.* 7(11):3315-20.

12 Giovannoni JJ, D DellaPenna, AB Bennett and RL Fischer. 1989. Expression of a Chimeric Polygalacturonase Gene in Transgenic Rin (ripening inhibitor) Tomato Fruit Results in Polyuronide Degradation but not Fruit Softening. *Plant Cell* 1(1):53-63.

13 Deikman J, R Kline and RL Fischer. 1992. Organization of Ripening and Ethylene Regulatory Regions in a Fruit-Specific Promoter from Tomato (Lycopersicon esculentum). *Plant Physiol.* 100:2013-17.

14 See *supra* note 11.

15 www.delphion.com.

16 www.m-cam.com.

17 www.european-patent-office.org.

18 See *supra* note 11.

19 See *supra* note 12.

CHAPTER 14.5

Managing Liability Associated with Genetically Modified Crops

RICHARD Y. BOADI, *Legal Counsel, African Agricultural Technology Foundation, Kenya*

ABSTRACT

Recent years have seen intense global debate about whether or not agricultural biotechnology—particularly genetically modified organisms (GMOs) and genetically modified crops (GM crops)—should be covered by a specially designed liability regime. This chapter examines common and statutory law theories of liability, various attempts at the national and international levels to design liability regimes for GMOs, and liability risk-mitigation measures.

1. INTRODUCTION

Liability is the "quality or state of being legally obligated or accountable."[1] The word refers to the obligation of a person or institution to provide compensation for damage it is deemed to be responsible for. Historically, liability has been determined using common and statutory national laws; however, when questions of liability overreach national borders—as they often do in such fields as agricultural biotechnology—historical legal methods are not always applicable. Indeed, there has been intense global debate about the creation of a liability regime for genetically modified organisms (GMOs) and genetically modified crops (GM crops or transgenic crops).

This chapter examines the existing common law and statutory theories of liability; the various attempts to design liability regimes for GMOs at national and international levels; the potential liability risks shared by stakeholders, including small-scale farmers; and risk-mitigation measures.

2. COMMON LAW AND STATUTORY THEORIES OF LIABILITY

Common law forms a major part of the law of those countries of the world that were once British territories or colonies. It is the body of law derived from centuries of judicial rulings, rather than from statutes or constitutions.[2] The common law provides a means of compensating for wrongful acts (known as *torts*), whether they are intentional or are caused by negligence; it is also a way to regulate contracts.

The common law theories of liability include the following: *negligence*, which refers to the breach of a legal duty by one party that proximately causes damage to another party; *trespass*, which refers to an unlawful act committed against the person or property of another, including wrongful entry on another's property; *nuisance*, which refers to an unreasonable interference in another person's or other persons' use and enjoyment of their land (*private nuisance* and *public nuisance*, respectively); and the principle of *strict liability*, which is not fault-based and may apply despite the exercise of utmost care on the part of the offender.

Boadi RY. 2007. Managing Liability Associated with Genetically Modified Crops. In *Intellectual Property Management in Health and Agricultural Innovation: A Handbook of Best Practices* (eds. A Krattiger, RT Mahoney, L Nelsen, et al.). MIHR: Oxford, U.K., and PIPRA: Davis, U.S.A. Available online at www.ipHandbook.org.

The main statutory[3] theories of liability include strict liability and infringement of intellectual property (IP). Strict liability was first defined in the case *Rylands v Fletcher*,[4] in which the defendant had a reservoir built on his land that caused flooding of the plaintiff's mine. This case articulated the principle that liability would arise in cases where damage is not necessarily caused as a result of the defendant's actual negligence or intent to harm but based on the breach of an absolute duty as, for instance, when his or her nonnatural use of land causes the accumulation of dangerous things, which then *escape* and cause damage. In modern statutory law, a use is considered to be nonnatural if it is a special use that creates an abnormal risk of damage to another person's property.[5] The occupier of the land is liable for damage caused by an escape and has several defenses (for example, common benefit, act of a stranger, statutory authority, consent of the plaintiff, default of the plaintiff, or act of God). *Infringement* of IP refers to use by an unauthorized party of any of the exclusive rights enjoyed by the owner over his or her own IP.

3. LEGAL LIABILITIES AND GM CROPS

3.1 *The international debate*

There has been considerable international debate about the liabilities associated with GMOs and specifically the liabilities with GM crops.[6] One school of thought believes that GMOs pose no unique risks and argues that GMOs can be covered by liability regimes commonly used for other agricultural technologies; the other school of thought maintains that agricultural biotechnology is fundamentally different from other forms of agricultural breeding technology and argues that special legal liability regimes are required to ensure that those who experience loss arising from GMOs can obtain adequate relief. Countries such as Canada, the United States,[7] the United Kingdom, and New Zealand[8] adhere to the first school of thought and apply general agricultural liability laws to GM products. The European Union, which holds the opposing view, has proposed that GM products be subject to a special legal liability regime. Certain E.U. countries, such as Austria and Germany, have passed national laws that impose strict liability for particular types of loss (such as death, injury, and damage to property) caused by GMOs. Under Austrian law, in the event of an accident involving GMOs (such as contamination of the food chain), the releasing entity will be liable for any harm to health, property, or the environment, and must return any affected property to its "original" state. For example, Austrian companies that manufacture GMOs must obtain sufficient liability insurance. German law imposes liability for injury to property or human health caused by GMOs.[9] German regulations place liability at the "manager level" of the company, or installation, an assumption that is likely to make farmers who grow GM crops (as installation managers) liable for any accidents that may occur. German law also makes liability insurance mandatory for GM operators.

3.2 *The African Model Law approach*

In the midst of this international debate, the Organization for African Unity (OAU), now known as the African Union (AU) and the Ethiopian Environmental Protection Authority developed the African Model Law on Safety in Biotechnology in 2001 that was intended to be a basis for formulating national laws concerning biotechnology.[10] This model law proposed instating a strict liability regime for GMOs.[11] To date, however, the liability regimes being proposed in the draft biosafety laws of African countries seem to disregard the extreme position of the African Model Law.[12]

3.3 *Liability and redress under the Cartagena Protocol on Biosafety*

The issue of liability and redress for damage resulting from the transboundary movements of GMOs was addressed by the Biosafety Protocol of the Convention on Biodiversity (which referred to GMOs as living modified organisms [LMOs]). The negotiators were, however, unable to reach a consensus regarding the details of a liability regime. Therefore, in the final text of the protocol (Article 27), the Conference of Parties was urged to develop an international liability regime within four years.[13] A group known as the Ad-Hoc Open-Ended Work Group of Legal and Technical Experts on Liability and Redress in the Context of the Cartagena Protocol on Biosafety has since been created in order to achieve this goal. The group has met twice, both times in Montreal, Canada, first on 25–27 March 2005[14] and again on 20–24 February 2006. In the second meeting, the group developed a list of criteria for assessing the effectiveness of any rules and procedures referred to in Article 27 of the protocol and developed different options for operational text on scope, damage, and causation.[15] The group has yet to agree on a liability regime.

Kershen and Smyth have argued that "*Developers of new agricultural biotechnology crops and animals—be they public or private; be they industrialized or developing countries—would be hindered by the inclusion of speculative risks in a liability and redress regime, especially public researchers in developing countries.*"[16] Kershen and Smyth contend further that an Article 27 liability and redress regime would reduce the amount and availability of agricultural biotechnology and thus impede public research on behalf of the poor in developing countries.[17] They also assert, and the author of this chapter agrees, that future liability costs could adversely affect agricultural research in public research institutes in developing countries, since such facilities may not have the requisite financial resources to absorb the costs of any future liability. Furthermore, future liability costs could increase operational costs and thus raise product costs.

4. MANAGING EXISTING POTENTIAL LIABILITIES

The production and use of GMOs can create many potential liabilities. For instance, the producer or user of GM crops or animals may be liable for damage caused by GM crops or animals to the person or property of another person or to the environment.[18] Pollen flow from transgenic crops to nontransgenic crops may cause crop damage. For instance, transgenic pollen flow may ruin the "organic" status of crops or the purity of the genetic material of other seeds.[19] Questions may arise as to whether transgenic crops or their food products are toxic, allergenic, or pose a long-term health threat.

Claims for compensation in actions for personal or property damage could be based on a theory of negligence, trespass, nuisance, or strict liability, although there has not yet been a definitive judicial decision on these. A class action suit brought by farmers and other parties against Aventis Cropscience, U.S.A., alleged that their corn had been contaminated by transgenic corn approved for animal feed and ethanol production but not for human food. The court determined that plaintiffs who could prove the alleged contamination would have a claim based on the theories of negligence, private nuisance, and public nuisance.[20] This case, which was settled with the proposed payment of over US$100 million to members of the defined class,[21] underscored the potential for liability arising from the development, production, and use of agricultural biotechnology products.

4.1 *Negligence*

A person whose crops or property is damaged because a neighbouring farmer failed to take adequate precautions to contain his transgenic crops may have a claim against both the neighbouring farmer and the biotechnology company that created the transgenic crop.[22] To sustain a claim based on negligence, the claimant (plaintiff) would need to prove four elements: the defendant's *duty of care*—a legal obligation imposed on an individual requiring that they exercise a reasonable standard of care while performing any acts that could forseeably harm others—to the plaintiff, breach of that duty by unreasonable conduct of the defendant, a causal link between the alleged unreasonable conduct and damage, and damages (a harm or injury valued in monetary terms). When a farmer growing GM crops knows that neighboring farmers (such as organic and GM-free farms) may be adversely affected by GMO contamination, he or she arguably owes a duty of care to such farmers and must keep his or her GMOs from spreading beyond the bounds of his or her property. However, because there is no scientific proof regarding the extent to which pollen or seed may be dispersed, it is impossible to determine who is affected by the unintended spread of GMOs from the defendant's land. A GM farmer's breach of duty of care, and the damages that he or she must pay as a result, will be judged according to the standards of a reasonable person and may take into account such factors as the magnitude of the risk posed by the GMOs, the degree of probability that such contamination would naturally occur, and the expense, difficulty, and inconvenience to the GM farmer that would result if he or she were required to rectify the situation. Biotechnology companies and farmers may be obligated to take additional reasonable precautions to contain certain transgenic crops if, for example, the agronomic evidence shows that a particular transgenic crop causes weediness, pollen flow, or volunteer plants to a greater degree than do nontransgenic crops.[23]

Obviously, biotechnology companies and farmers must develop techniques that minimize pollen flow and the establishment of volunteer plants in order to protect themselves from liability. For example, transgenic crops could be engineered to have biological barriers against pollen flow or preventing volunteer survival through male sterility (preventing fertilization), seed sterility (preventing volunteer crops), or control of flowering time (preventing cross-pollination with other, nontransgenic crops).[24] Indeed, if such biological barriers can reasonably be incorporated into a transgenic crop, a biotechnology company that failed to incorporate these biological barriers and was subsequently accused of causing damage to property or person might be liable for a product's liability claim for design defect.[25] Furthermore, farmers of transgenic crops can adopt agronomic practices to prevent pollen flow or the establishment of volunteer plants: farmers can plant fields at *isolation distances*; plant *barrier crops*, *border rows*, or *refugia* (non-GM areas of the same crop); or establish agronomic zones dedicated to non-GM crops.[26] Biotechnology companies would likely have a duty to educate farmers, with whom the companies have entered into contracts, about these agronomic management practices and possibly have the obligation to police farmers growing the companies' crops. Farmers, for their part, would have duty of care to abide by the agronomic management practices recommended by the biotechnology companies.

However, given the nature of agriculture in most of the developing world, where subsistence farming and small landholdings are the norm, it would be impractical to expect developing-world farmers to adopt most of the agronomic practices mentioned above. Biotechnology companies that donate their technologies for humanitarian use would benefit from a technology transfer scheme that permits such companies to provide technologies, like genes and transformation systems, to developing-world farmers, while protecting them from liability risk in case the transgenic crops are misused.

4.2 *Trespass*

Persons who believe they have suffered damage from transgenic pollen flow may bring a common law cause of action based on the theory of trespass.[27] In this case, *trespass* indicates the physical invasion by transgenic crops of the possessory interests of the property (land) of the person claiming damages. Technically, proof that transgenic pollen has spread to neighboring fields could be sufficient evidence to establish trespass. However, it is a biological fact that pollen flows between varieties of the same crop and between related plant species. Therefore, if pollen flow constituted trespass upon a neighbor's crops, all farmers would be liable for trespass for almost every crop they grow. Jurisdictions such as the United States have differentiated between pollen flow that constitutes trespass and pollen flow that is accepted as a biological fact of farming;[28] to sustain a successful action in trespass, there must be proof that the alleged physical invasion caused damage (such as contaminated seed). Naturally, the extent to which a claimant could rely on this theory of liability would depend on the local laws regulating seed and crop standards.

4.3 *Private nuisance*

Unlike the common law claims of trespass, strict liability, and negligence, all of which focus on the conduct or activity that causes harm to the person or property of another, the claim of private nuisance focuses on a person's interests being protected (that is, the right of an individual to use and enjoy, free from interference by others, one's private land). Fundamental to the private nuisance claim is the notion that neighbors must be accommodating of one another so as to allow peaceful coexistence. A private nuisance claim must prove that an invasion (1) is either intentional and unreasonable or unintentional and otherwise actionable as a legal claim for trespass, strict liability, or negligence; and (2) causes significant harm (the definition of which is based on the gravity of the alleged harm and its level of normality in a particular locality). In the case of GMOs, a claimant must prove that nearby fields of transgenic crops have unreasonably interfered with the use and enjoyment of his or her own land. The courts are unlikely to endorse a private nuisance claim that, for example, insists on zero tolerance of pollen flow or of volunteer plants, or which claims "*significant* [emotional] *harm*" from personal opposition to transgenic crops.[29]

4.4 *Strict liability*

Persons who believe their land or crops have been damaged by a neighbor's transgenic crops may bring a tort claim in strict liability if the activity of growing transgenic crops is "*abnormally dangerous*" when the following factors are taken into account:

- the degree of risk of some harm to the person, land, or chattels of others resulting from the growing of the crop
- the likelihood that the harm that results from growing the crop will be great
- the grower's inability to eliminate the risk by exercising reasonable care
- the extent to which the grower's activities are unusual or unapproved
- the inappropriateness of the grower's activities to the location in which they are conducted
- the extent to which the value of the grower's activities are outweighed by their potential dangers

In the United States, where transgenic crops are grown on a wide scale and where agricultural biotechnology is not considered legally different in kind from other agricultural breeding technologies, liability claims based on any of the above theories are difficult to establish.[30] It will be interesting to see how the policy-makers and courts of the developing world will deal with the transgenic crops beginning to arrive on their shores.

4.5 *Liability for infringement of intellectual property rights*

IP rights are a category of intangible rights regarding creations of the human intellect.[31] The holder of an IP right may exercise exclusive control over its use for a limited period of time; any unauthorized use of the IP right during the statutory period of protection would constitute an infringement. It is possible, therefore, that farmers whose crops are accidentally affected by the presence of GMOs (as a result of pollen flow or seed comingling) might be held liable for IP rights infringement. Recently, Monsanto successfully brought suit in Canada against a conventional farmer who replanted seeds that had been contaminated with genetic material from Monsanto's genetically modified crops. The GMOs in question, Roundup resistant plants, contain a patented transgenic gene that confers herbicide resistance. The court held that the harvesting and sale of crops derived from seeds that were known, or suspected, to be Roundup tolerant infringed on Monsanto's exclusive IP rights.[32]

5. OTHER LIABILITY-MANAGEMENT TOOLS AND APPROACHES

5.1 *Compliance with IP, license, and regulatory requirements*

The developers of GM products must adopt appropriate scientific and technical safeguards for all products and advise stakeholders, including smallholder farmers, as to the appropriate use of technologies and products. Farmers of GM crops, for their part, need to comply with relevant license conditions, standards, guidelines, and directions regarding deployment or use of GM products.

Proper compliance with these guidelines can help protect all parties from liability risks.

5.2 *Indemnification*

Indemnification is a promise, usually contractual, to protect a party from financial loss. Indemnification may work by either direct compensation to the injured or by reimbursement for any loss incurred. One way to manage liability is to include indemnification provisions in agreements relating to the transfer, development, and deployment of technologies. Such a provision specifies that the *indemnifying* party will compensate the *indemnified* party for any loss or damage that may be sustained by it as a result of the actions of the former. Under this approach, the first party (the indemnifying party) agrees to hold the second party (the indemnified party) harmless and to defend the second party and its officials against claims resulting from the first party's actions and/or omissions.

In order to limit the risk of liability to what it can adequately control, AATF might reasonably agree to indemnify a technology donor for claims resulting from AATF's use of the licensed technology, provided that the indemnity granted under these conditions excludes claims resulting from the technology donor's own acts and/or omissions.[33]

5.3 *Warranty disclaimers*

Another approach to managing liability is the use of *warranty disclaimers*. A warranty, either express or implied, is a guarantee that a particular product or technology will serve a specified purpose. A warranty disclaimer enables one party, usually a technology developer or transferor, to expressly disclaim guarantees. Conceivably, technology developers or transferors could be held to one of two implied warranties: merchantability or fitness for a particular purpose.

An *implied warranty of merchantability* is a warranty implied by law, such that if a *merchant* (someone who makes an occupation of selling things) sells an item, he or she is guaranteeing that the item is reasonably fit for the general purpose for which it are sold. GM-technology developers qualify as merchants and their technologies deemed to be reasonably efficacious for the general purpose for which they may be transferred to a user. Thus, the failure of GM technology could subject the developer/transferor to liability for breach of the technology's implied warranty of merchantability. An implied warranty of *fitness for a particular purpose*, sometimes referred to simply as a *warranty of fitness*, is a warranty implied by law, such that if a seller knows of, or has reason to know of, a particular purpose for which an item is being purchased, the seller guarantees that the item is fit for that particular purpose. For instance, if a GM technology is developed for, or transferred to, a user for the purpose of addressing a particular agricultural constraint, the technology developer would be deemed to provide a guarantee that the technology would indeed address the constraint. To manage potential liability claims resulting from the GM technology failing to fulfill the general purpose for which it was developed or sold or effect the specific constraint the technology was meant to address, the technology developer/transferor would need, at the time it develops or transfers the technology, to expressly disclaim implied warranties of merchantability and fitness for a particular purpose.

4.4 *Letters of nonassertion*

A *letter of nonassertion* assures the user that the technology owner will not enforce its IP rights.

4.5 *Technology/product stewardship*

Technology- and/or product-stewardship procedures include: comprehensive risk analyses for projects and/or phases of projects; appropriate risk-mitigation strategies (including appropriate insurance coverage, outlining specific uses for technology, management and oversight protocols, procedures to protect confidential information, etc.); and compliance with all applicable laws.

Adherence to appropriate technology/product- stewardship best practice guidelines can help protect technology developers and users from potential liability as their actions would likely be deemed reasonable under the applicable circumstances.

6. CONCLUSION

The international legal debate continues about whether or not GMOs should have special legal liability. Actors in agricultural development have a responsibility to develop and deploy safe and environmentally friendly products through the adoption of appropriate technology-and/or product-stewardship measures. The legal, health, and environmental risks of using GMOs should be reduced as far as possible. Failure to manage risk appropriately may be extremely costly in terms of lost time and money.

The African Agricultural Technology Foundation (AATF) is an institution that gives smallholder farmers in Sub-Saharan Africa access to technologies, including agricultural biotechnology. It is imperative that the AATF examine the potential liability issues associated with GM crops, identify the key liability risks for specific members of the agricultural communities, and suggest measures that may be implemented to minimize such risks. ■

ACKNOWLEDGMENTS

The author is grateful to Mpoko Bokanga (AATF) and Francis Nang'ayo (AATF) for critical input into preparing this chapter. Any errors or misrepresentations remain the author's.

RICHARD Y. BOADI, *Legal Counsel, AATF, c/o ILRI, PO Box 30709, Nairobi 00100, Kenya. r.boadi@aatf-africa.org*

---

1 Gardner BA, ed. 1999. *Black's Law Dictionary*. West Group: St. Paul, MN.

2 The common law of England was one of the three main historical sources of English law, with the other two being legislation and the doctrines of equity. Common law constitutes the basis of the legal systems of Australia (both federal and individual states), Brunei, federal law in Canada and the provinces' laws (except Quebec civil law), Hong Kong, India, Malaysia, Malta, New Zealand, Pakistan, Singapore, South Africa, Sri Lanka, most of the United Kingdom (that is, England and Wales, Northern Ireland, and the Republic of Ireland), federal law in the United States and the states' laws (except Louisiana), and many other generally English-speaking countries and British Commonwealth countries. Essentially, every country that has been colonized at some time by Britain uses common law except those that had been colonized by other nations (Quebec follows French law to some extent and South Africa follows Roman Dutch law), in which the prior civil law system was retained to respect the civil rights of the local colonists. India's system of common law is a mixture of English law and the local Hindu law.

3 *Statutory law* is written law set down by a legislature or other governing authority, such as the executive branch of government (unlike common law that is based on judicial rulings) in response to a perceived need to clarify the functioning of government, improve civil order, answer a public need, or codify existing law or for an individual or company to obtain special treatment.

4 L.R. 3 H.L. 330 (1868).

5 See §§ 519–24. American Law Institute, 3 Restatement of the Law (Second) Torts 34–35 (1977).

6 The issue has been debated at several meetings, including a Workshop on Liability and Redress (Cartagena Protocol, 2–4 December 2002, Rome, Italy); a meeting of the Technical Group of Experts on Liability and Redress (18–20 October 2004, Montreal, Canada); the first meeting of the Ad Hoc Group on Liability And Redress (25–27 May 2005, Montreal, Canada); and the second meeting of the Working Group on Liability and Redress (20–24 February 2006, Montreal, Canada). See also [No Authors Listed]. 1987. Designer Genes That Don't Fit: A Tort Regime for Commercial Releases of Genetic Engineering Products. *Harvard Law Review* 100:1086-105.

7 See Kershen DL. 2002. Legal Liability Issues in Agricultural Biotechnology. *National Agricultural Law Center Research Article* No. 3. www.nationalaglawcenter.org.

8 See *supra* note 7, at page 3.

9 See Gentechnikgesetz (GenTG), Bundesgesetzblatt I. 1993. p. 2066 (published on 16 December 1993). Recent amendments to conform to recent E.U. laws can be found at www.bmelv.de/cln_045/nn_750598/SharedDocs/Gesetzestexte/G/GesetzNeuordnungGentechnikrechts.html.

10 www.africabio.com/policies/MODEL%20LAW%20ON%20BIOSAFETY_ff.htm.

11 Article 14 says, in part, that "*A person who imports, arranges transit, makes contained use of, releases, or places on the market a genetically modified organism or a product of a genetically modified organism shall be strictly liable for any harm caused by such a genetically modified organism or a product of a genetically modified organism. The harm shall be fully compensated.*"

12 See, for instance, the draft biosafety laws of Cameroon, Ghana, Kenya, and Uganda.

13 Article 27 of the Biosafety Protocol reads as follows: "*The Conference of the Parties serving as the meeting of the Parties to this Protocol shall, at its first meeting, adopt a process with respect to the appropriate elaboration of international rules and procedures in the field of liability and redress for damage resulting from transboundary movements of living modified*

organisms, analysing and taking due account of the ongoing processes in international law on these matters, and shall endeavour to complete this process within four years."

14 The full report of the first meeting of the Ad-Hoc Open-ended Work Group of Legal and Technical Experts on Liability and Redress in the Context of the Cartagena Protocol on Biosafety is available at the CBD Web site. www.biodiv.org/doc/meeting.aspx?mtg=BSWGLR-01.

15 The full report of the second meeting of the Ad-Hoc Open-ended Work Group of Legal and Technical Experts on Liability and Redress in the Context of the Cartagena Protocol on Biosafety is available at www.biodiv.org/doc/meetings/bs/mop-03/official/mop-03-10-en.pdf.

16 Kershen DL and SJ Smyth. 2006. Agricultural Biotechnology: Legal Liability from Comparative and International Law Perspectives. *ExpressO Preprint Series*. Working Paper 1279, at pp. 102–103. law.bepress.com/expresso/eps/1279.

17 Ibid., at page 104.

18 The issue of liability for environmental damage, including loss of diversity, is typically addressed with reference to the statutes of the relevant jurisdiction and has not been explored in this chapter.

19 Note, however, that Kershen and Smyth (see *supra* note 16) assert that "*Organic agriculture is a set of production standards and not product standards*" and thus, "*...an organic farmer who follows an approved production plan produces organic products on an organic farm regardless of adventitious presence of transgenic material...*"; see also *Hoffman v. Monsanto Canada, Inc.*, S.K.Q.B. 225 (2005) where the plaintiffs abandoned claims (among others) alleging loss of organic label.

20 In re StarLink Corn Products Liability Litigation, 211 F. Sup.2d 828 (N.D. Ill. 2002).

21 In re StarLink Corn Products Liability Litigation, Proposed Settlement and Fairness Hearing Document, MDL Docket No. 1403 (N.D. Ill. 2003). The defined class included farmers and other persons (individuals, partnerships, and corporations) with a financial interest in non-StarLink corn harvested between 1998 and the time of the settlement who either (1) operated a farm from which non-StarLink corn was harvested in 1998, 1999, 2000, 2001, or 2002, whether or not the crops or corn stores suffered actual Cry9C contamination; or (2) operated a farm from which non-StarLink corn was harvested at any time since 1998 and suffered actual Cry9C contamination.

22 See *supra* note 7, at page 10.

23 See *supra* note 7, at page 10.

24 See *supra* note 7, at page 10.

25 See *supra* note 7, at page 11.

26 See *supra* note 7, at page 11.

27 Grossman MR. 2002. Biotechnology, Property Rights and the Environment. *Am. J. Comp. L.* 50:215–48.

28 See *supra* note 7, at page 6.

29 See *supra* note 7, at page 12.

30 See *supra* note 7, at page 8.

31 Gardner BA, ed. 1999. *Black's Law Dictionary,.Seventh Edition* The West Group:St. Paul Minn.

32 *Monsanto Canada Inc. & Monsanto Co. v Percy Schmeiser & Schmeiser Enterprises Ltd* [2004] 1 S.C.R. 902, 2004 SCC 34

33 AATF would usually indemnify and hold a technology donor and its employees, directors, officers, and agents harmless against any and all claims, losses, liabilities, or expenses on account of any infringement or alleged infringement of IP or injury or death of persons or damage to property caused by, arising, or alleged to arise out of AATF's acts or omissions under, or in connection with, the applicable agreement except to the extent that such claims, losses, liabilities, or expenses are the result of the technology donor's acts or omissions.

SECTION 15

Monitoring, Enforcement, and Resolving Disputes

CHAPTER 15.1

Administration of Technology Licenses

HANS H. FEINDT, *Chief, Monitoring and Enforcement Branch, Office of Technology Transfer, National Institutes of Health, U.S.A*

ABSTRACT

The National Institutes of Health Office of Technology Transfer (NIH OTT) administers technology licenses for the NIH, generating substantial royalties (in the millions of dollars). Although this revenue flow is important, the NIH OTTs principal mission is the timely introduction of new products and technologies into the marketplace to ensure that the fruits of NIH research and development are made commercially available to serve the greater public good. The NIH OTT utilizes six types of technology licenses:

- commercial evaluation licenses (also known as options)
- patent commercialization licenses (either exclusive or nonexclusive)
- nonexclusive patent licenses (for internal use)
- biological materials licenses
- software licenses

The NIH OTT insists that licenses are drafted with well-defined financial terms and clearly delineated reporting obligations, so that both parties to the license (NIH as licensor and, for example, a biotech firm as licensee) understand their respective obligations. The NIH OTT seeks to build cooperative relationships with its licensees in order to facilitate problem solving discussions, resolve outstanding issues, and identify possible opportunities for advancing commercialization of products and/or services. As a best practices licensor, the NIH OTT carefully manages license administration by monitoring commercial development performance benchmarks, reviewing sales reports, and enforcing other license obligations. The office will also, if necessary, impose sanctions in license enforcement and implement procedures for dealing with infringement of its patents. The policies, protocols, and procedures of the NIH OTT have broad applicability to both developed and developing countries; scientists, administrators, technology managers, intellectual property professionals, and even attorneys can learn from the NIH OTT, a good example of an office operating effectively, efficiently, and profitably by employing best practices.

1. INTRODUCTION

The National Institutes of Health Office of Technology Transfer (NIH OTT) strives to fulfill its mission of transferring technology to improve public health not only by licensing to commercial enterprises but also by working with and licensing to institutions serving disadvantaged populations in the United States and abroad. The administration of technology licenses is an important part of this process. License administration focuses on the licensee's obligations to the licensor, such as periodic royalty payments and reports. In fiscal year 2005, the NIH collected over US$98 million in royalties from 750 licenses (out of a total portfolio of over 1400 licenses). Royalties from commercial products made up nearly US$77 million of this amount.

Describing the different types of licenses used by NIH to carry out its technology transfer program, this chapter explains the procedures for ensuring that licensees meet their obligations. It provides an overview of the tools used to administer large numbers of technology licenses and offers advice on how to monitor commercial-development

Feindt HH. 2007. Administration of Technology Licenses. In *Intellectual Property Management in Health and Agricultural Innovation: A Handbook of Best Practices* (eds. A Krattiger, RT Mahoney, L Nelsen, et al.). MIHR: Oxford, U.K., and PIPRA: Davis, U.S.A. Available online at www.ipHandbook.org.

performance benchmarks, review sales reports, and enforce other license obligations. This chapter also discusses the use of amendments in license administration, sanctions in license enforcement, and suggests procedures to follow when nonlicensed companies infringe on patented technology. The policies and practices of the NIH OTT aim to further develop scientific discoveries that may lead to commercial products that improve public health. This overview of license administration at NIH seeks to provide guidance for others who are considering establishing and operating their own programs for administering technology licenses.

2. TYPES OF TECHNOLOGY LICENSES

Technology licenses include commercial evaluation licenses (also known as options), exclusive and nonexclusive patent commercialization licenses, nonexclusive patent licenses for internal use, biological materials licenses for commercial sale, biological materials licenses for internal use, software licenses for commercial sale, and software licenses for internal use. Financial terms and reporting obligations vary with the type of license. Table 1 shows which obligations are typically included for each type of license. Regardless of the type, licenses should be written with well-defined financial terms and reporting obligations that both parties understand. This section briefly describes each type of NIH license; a more detailed discussion about the various types of technology licenses can be found elsewhere in this *Handbook*.[1]

Commercial evaluation licenses (also known as options) are useful for companies to explore the value or appropriateness of a new technology for a limited time without committing the financial and other resources required by a standard exclusive or nonexclusive patent license. Appropriately, these agreements have smaller financial terms and are for a short duration. If the licensee finds the technology meets their needs, then the parties will generally negotiate a new exclusive or nonexclusive patent commercialization license.

Patent commercialization licenses provide licensees with rights to patented technology or inventions described in patent applications that have been filed. An *exclusive patent commercialization license* provides a single licensee the right to practice and exclude others from practicing the technology for a period of time limited by the term of the patent. In most fields of commercial endeavor, an exclusive license provides a significant competitive advantage to the licensee and, therefore, the potential for a large financial return. Consequently, the royalty obligations and financial terms in such licenses are generally quite substantial. With exclusive licenses, the licensor also has a higher level of expectation that the licensee will diligently meet the performance milestones agreed to in the license.

Nonexclusive patent commercialization licenses give patent rights for technology to multiple licensees. These may be for a limited time or for the term of the patent. Such licenses are often given when the patent technology has the potential to significantly benefit the broader public. By providing such technology to multiple licensees entry into the marketplace will be accelerated. Royalty obligations imposed on nonexclusive patent commercialization licensees vary widely, depending on the nature of the technology.

Nonexclusive patent licenses for internal use provide a licensee with access to a patented technology that may be useful as a tool or process but is not itself a marketable product.

In the biotechnology field, ***biological materials licenses*** provide licensees with access to nonpatented materials or biological constructs that were prepared at great effort and expense and that may be available only from the laboratories that made them. Nonexclusive biological materials licenses for internal use provide a licensee with access to unpatented technology that is unique or difficult to replicate without significant expense. This saves the licensee time in its commercial development efforts. Biological materials licenses for commercial sale promote the wider use of unique materials or biological constructs in the research and commercial development community.

Similar to biological materials licenses, ***software licenses*** provide licensees with access to nonpatented software that may only be available from the laboratories that developed them. As shown

TABLE 1: TYPICAL LICENSE OBLIGATIONS

	Types of Technology Licenses					
Financial Terms and Other Obligations Found in Technology Licenses	Evaluation	Exclusive Patent for Commercial Use	Nonexclusive Patent for Commercial Use	Nonexclusive Patent for Internal Use	Biological Materials for Commercial Sale	Biological Materials for Internal Use
License execution fees	+	+	+	+	+	+
Annual (minimum annual) royalties	±	+	+	±	+	±
Past patent-prosecution fees	–	+	±	–	–	–
Ongoing patent-prosecution and patent-maintenance fees	–	+	±	–	–	–
Annual, periodic, or final reports on commercial development or research progress	+	+	+	±	±	±
Report of performance benchmark achievement	–	+	+	–	+	–
Performance benchmark royalties	–	+	+	–	–	–
Report of first commercial sale	–	+	+	–	+	–
Annual, periodic, or final reports on sales and earned royalties due	–	+	+	–	+	–
Earned royalties on product sales	–	+	+	–	+	–
Report of sublicensing activity	–	+	–	–	–	–
Report of sublicensing considerations and royalties due	–	+	–	–	–	–
Sublicensing royalties	–	+	–	–	–	–
License renewal or term extension fees	–	–	±	+	+	+

Key: + = Generally in license.
± = May or may not be in license.
– = Generally not in license.

in Table 1, the financial terms and obligations found in such licenses vary depending on the type of license.

NIH has used most of these license types to expand the transfer of technologies—specifically those for neglected diseases or that meet public health needs—to public and private institutions in developing countries.

3. TASKS OF THE LICENSOR

To administer, monitor, and enforce technology licenses requires the licensor to follow-up on the execution of a license agreement. The licensee has agreed to fulfill various financial terms and reporting obligations in exchange for the right to practice a licensed technology for a limited period of time. Regular reminders may be needed to ensure that they fulfill these obligations throughout the term of the license.

The licensor should monitor compliance with royalty payment and reporting obligations during the license term, and reports submitted by licensees should be carefully reviewed on an ongoing basis. Routine correspondence with licensees about these matters is usually handled through invoices, form letters, and e-mail. However, license administrators will sometimes need to invest considerable investigative time and practice skillful communication to understand the activities of the licensee and determine which actions should be undertaken to remedy any noncompliance. A cooperative approach that engages the appropriate licensee contact in problem solving is generally best. Such discussions will resolve most issues and also provide feedback that may be useful for future technology license negotiations. Utilizing these contacts also may allow the licensor to direct the licensee to financial, technical, and other resources that will help the licensee move its commercialization efforts forward.

Most tasks performed in the administration, monitoring, and enforcement of technology licenses typically flow out of the financial terms and reporting obligations described in Table 1. The more-routine license administration tasks include:

- arranging for shipment of licensed materials to the licensee
- invoicing licensees for royalty payment obligations specified in the license
- recording royalty payments
- verifying that the amount paid is correct
- distributing royalty receipts
- requesting overdue royalty payments through reminder notices
- requesting overdue reports through reminder notices
- notifying licensees of license expiration

Other license administration tasks related to monitoring and enforcement include:

- checking the accuracy of sales and earned royalty reports
- collecting overdue or underpaid royalties and imposing additional royalties for late payment
- reviewing progress reports against performance benchmarks
- tracking and recording achieved-performance benchmarks so that associated royalty payments are invoiced at the proper time
- contacting licensees about license noncompliance issues
- amending licenses to extend them, modify benchmark schedules or other license terms, or correct errors in the original license
- preparing and reviewing patent expense reports that support the billing for patent expense reimbursement

4. TOOLS FOR LICENSE ADMINISTRATION, MONITORING, AND ENFORCEMENT

4.1 *Licensee contacts*

One of the most important tools for effectively administering, monitoring, and enforcing licenses is the list of licensee contacts. If contact information for royalty and reporting obligations is not available when the license is executed, it should be obtained immediately after. The list could include contacts in business development, legal affairs, licensing, finance, and research. The names of senior-level executives should also be included. Ideally, full names, titles, mailing addresses, phone numbers, and e-mail addresses should be recorded for each contact. The contact list should be periodically reviewed and updated. These contacts are extremely important for beginning discussions about royalty payments and other noncompliance issues that may develop. Without a contact list, valuable time can be wasted trying to identify the appropriate contact.

4.2 *Filing system*

A well-organized system for filing and retrieving documents, reports, correspondence, and other information related to a specific license is as important as licensee contacts. Depending on how things are organized, several different files may be needed to address and keep track of different aspects of license administration. For example, a file used only for archiving the original, executed license agreement may be set up. Another "working" file may be set up for daily use in filing, reviewing, and retrieving a reference copy of the license and any correspondence associated with the license. If a computer network and systems are available, the filing system may be set up electronically by scanning and converting all correspondence and license agreements into image files (for example, Adobe Acrobat® pdf files) that can be easily stored, searched, and retrieved. It is essential, of course, for any such system to be maintained.

4.3 *Tracking system for license terms and due dates*

To effectively administer license agreements, collect royalties that are due, and monitor and enforce license obligations, the licensor must have a reliable system to record and track the financial terms, performance milestones/benchmarks, reporting obligations, amounts due, due dates, invoice or overdue notice deadlines, payments receipts, and royalty payment distributions for individual licenses.

The greater the number of licenses, the more important it is to use a computerized database for license administration. At the NIH Office of Technology Transfer, the database has been essential for monitoring, recording and updating contact lists, tracking due dates for financial terms, recording the amount of royalty payments received, tracking the due dates of performance benchmarks, recording the receipt of reporting obligations, recording completion dates for performance milestones/benchmarks, and so forth.

Ideally, the database should be designed to meet the needs of the entire technology transfer office. The NIH database consists of an integrated system of interactive modules that handle data about people (contacts), companies, inventions, invention marketing, patent prosecution, patent annuity payments, license applications, license royalty payment obligations, royalty receipts, license reporting obligations, and so forth. Queries can be made about the data, and a variety of report types can be generated. The database sends reminder e-mails to individuals in the office and allows routine form letters and reports to be prepared, edited, and printed. The database also allows comments to be recorded and the attachment of externally generated electronic files (such as scanned copies of licenses and correspondence or e-mails) to specific records in the database. These features help to maintain a historical record of each invention and license.

4.4 *Technology transfer office Web site*

The NIH Office of Technology Transfer recently reorganized and updated its Web site[2] in order to answer licensees' questions about license obligations and provide potential license applicants with information. A menu bar on the Web site provides links to licensing and royalties information; examples of Forms and Model Agreements; FAQs (frequently asked questions) about royalty payments, reporting obligations, and other license matters; and contact information. By providing links to technologies currently available for licensing, the Web site helps market those technologies. Finally, neglected disease technologies available for licensing can be shared via the web.[3]

4.5 *Royalty payment obligations*

When a license is fully executed, several royalty payments will often be due. These may include: (1) a license execution royalty payment, (2) a prorated minimum annual royalty payment, and, for patent licenses, (3) a royalty payment for past patent prosecution costs. Typically, these payments are mailed to the licensee with individual invoices that state the license number, the type of royalty payment due, the amount due, the due date, and instructions for where the payment should be mailed. The database is used to record when payments are received and to alert license administrators when payments become due or are overdue.

When royalty payments become 30 days overdue, a first overdue notice is mailed to the licensee. If there is no response within two weeks, it is often useful to contact the licensee to verify that the contact information is correct and determine why payment has not been made. If payment is not received within 60 days after the due date, a final notice is mailed out. This notice informs the licensee that failure to pay may result in license termination. If payment is not received within 90 days of the due date, a license administrator contacts the licensee to determine why payment was not made and to discuss possible sanctions that may be imposed if payment is not received within a short period of time (see below).

4.6 *Sales and earned royalties reporting*

Licenses for the development and/or sale of commercial products usually require periodic sales reports and the earned royalty due. These reports may be annual, semiannual, or quarterly, depending on the product type and anticipated sales volume. Net sales figures quantitatively measure a license's performance and are the basis for calculating the earned royalties due. Licenses prescribe in some detail the deductions allowed from the gross sales for calculating the net sales figure. However, ambiguities or misunderstandings often arise. Recognizing such issues early, when smaller amounts of money are involved, usually makes resolving them easier for both parties. If sales and earned royalty reports are not provided with the earned royalty payments submitted by the licensee, the licensee should be reminded of its obligation to provide them, and a short-term deadline should be established for submitting the reports.

The accuracy of reported sales figures can be verified in several ways. Comparison to prior period sales figures will show whether product sales are growing or declining and at what rate. Company press releases, annual reports, filings with governmental securities agencies (such as the U.S. Securities and Exchange Commission SEC), stock analysts' reports, marketing reports, news stories, and so forth, are other resources that can be studied to verify reported sales figures. Many of these sources are available on the Internet. When the reported sales figures seem inconsistent with data from other sources, the licensee should be asked to explain the discrepancies. If the license includes provisions for auditing the company's sales to verify the figures reported for the licensed product, this may be the time to conduct an audit.

4.7 *Commercial development or research progress reports*

Most technology licenses require periodic reports describing the progress of research, commercialization, or product development. These reports serve several purposes:

- they verify that the licensee is using the licensed technology or product
- they demonstrate, for commercialization licenses, that an effort is being made to bring the licensed technology or product to market
- they provide verification that a license benchmark or milestone was achieved and when

Moreover, when benchmarks or milestones have associated royalty payments, the reports alert the license administrator to invoice the licensee for a royalty payment. If a licensee fails to provide these reports, the licensee should be contacted and reminded of their obligations. A short-term deadline should be set for the licensee to submit the report.

Progress reports should be carefully reviewed and compared to the commercial development plan and the benchmarks or milestones described in the license. Are initial expectations being met? If not, why not? Are the problems technical? Are they due to insufficient financial resources? Regulatory issues? Has the company lost focus in its desire to commercialize the product? Are there other issues not mentioned in the report? Getting answers to these questions usually requires contacting the licensee for additional information. Once these answers are obtained, a decision can be made about what actions to pursue with the licensee. (See the sections "Amendments to license agreements" and "Sanctions for noncompliance" for examples.)

4.8 *Patent prosecution and maintenance cost reimbursement*

Patent claims should match the commercial goals of licensees. Since IP protection normally precedes licensing, those responsible for licensing inventions need to monitor patent prosecution to ensure that the goals pursued by patent agents and attorneys align with those of the licensees.

Patent licenses often include the reimbursement of past patent prosecution costs incurred by the licensor for a licensed technology as a financial obligation. The licensee may also agree to pay ongoing (future) patent prosecution and maintenance costs. Periodically, these costs need to be carefully tracked, documented, and billed to licensees. Patents are usually not assigned in technology licenses, so control of patent prosecution most often resides with the licensor and not the licensee. Like all legal fees, patent prosecution costs can quickly get out of control without careful monitoring. Seeking timely reimbursements of patent costs incurred by the licensor is an important part of license administration.

Occasionally, an applicant for a technology license may want to manage patent prosecution and be billed directly for the costs incurred. In this case, special oversight is needed to ensure that the licensor's interests are protected.

5. AMENDMENTS TO LICENSE AGREEMENTS

The outcome of an effort to commercially develop a new technology is often difficult to predict because of technological, regulatory, financial, patent, and business issues. Licensees usually set timelines for meeting performance benchmarks or milestones with a best-case scenario in mind. Not surprisingly, delays are common. When a company is demonstrating diligence but has encountered unexpected delays that have a reasonable chance of being overcome, the appropriate action may be simply to amend the license to update the benchmark or milestone schedule. Such amendments reflect mutually agreeable changes in the expectations of licensor and licensee. But when the company's issues appear insurmountable, it may be better to terminate the license. Other considerations may lead to different approaches to such situations, but a successful conclusion will be based on establishing and maintaining good communications between the license administrator and the licensee.

License term extensions are normally simple modifications of a license that indicate the satisfaction of both sides in the existing agreement and a desire to continue the agreement. Sometimes, term extension amendments also include changes to other terms or obligations. For example, minimum annual royalties may be raised or lowered to reflect the current institutional costs of administering the agreement and the costs associated with the amendment process, or to better capture the value of the invention for the extended time period.

Financial hardship, changes in the cost structure of doing work, opportunity costs, or priority changes can make licensees want to change the financial terms of technology license agreements. Like most tangible assets, licensed IP assets depreciate with time (due to the shrinking of the exclusivity period, changing marketplace interests, and the degree to which the technology provides a competitive advantage over the industry's standard technology). While it is not a good idea to set rules for changing financial terms, an effort to weight influencing factors can be useful. The licensor might weigh such factors as the probability of getting paid, the probability of relicensing the technology (if the license is terminated), the present value of a payment reduction, and the costs involved. Consistently administering this amendment process will also prevent opportunistic changes in licenses that are not linked to appropriate needs.

In addition to amendments, other changes can be made to existing agreements to increase the chance that a technology will be successfully developed. Some areas that may need to be addressed include:

- changing the field(s) of use
- permitting the licensee to seek a patent term extension
- eliminating or adding certain technologies from or to a license
- allowing the licensee to seek sublicensing agreements
- allowing the licensee to take on patent-prosecution responsibilities

Many of these issues may be more appropriately handled by licensing personnel than by license administrators. However, the latter should understand the ongoing development of the technology so that they know when deviation from the original agreement is warranted.

6. SANCTIONS FOR NONCOMPLIANCE

When a technology transfer office has a large portfolio of inventions and technologies available for licensing, companies often will return to license additional technologies. This gives the licensor an opportunity to obtain some leverage for collecting late or underpaid royalties due on existing licenses with that applicant. The licensor may put on hold the execution or negotiation of new agreements until the licensee has fully paid any outstanding royalty obligations under existing licenses. All that is needed to use this sanction well is effective communication between license administrators and licensing personnel.

The threat of terminating a license due to a licensee's defaulting on the material obligations of a license is an important tool for enforcing compliance. However, license termination procedures are usually not undertaken until the licensee has been given (1) several written notices describing the obligation(s) in default and (2) an opportunity to respond. If no satisfactory response is forthcoming, a written 90-day notice of license termination is given as the final step. If the licensee's response is still unacceptable after 90 days have passed, a final letter of termination is sent to the licensee.

Although other intermediate sanctions may be desirable, they are frequently unavailable. The licensor's only choice then is to threaten license termination in order to recapture the technology for relicensing. However, when a licensee's breach causes a license to be terminated, license administrators should not forgive any outstanding financial obligations that predate the effective date of the license termination. Unpaid license financial obligations—such as minimum annual royalties, reimbursable patent costs, execution fees, and others—should be identified when a license is terminated, and serious efforts should be made to collect the monies owed. When a license expires, the licensor should conduct a similar review to capture any lost or missed milestone payments, patent-prosecution costs, minimum annual royalties, or other royalties.

One of the hallmarks of a successful technology transfer program is maximizing the collection of license financial obligations. Technology transfer programs that operate as part of a government agency may have that government's power to enforce debt collection, while nongovernmental technology transfer programs may have to rely on the courts for enforcement.

7. LICENSE EXPIRATION

At license expiration and during the ongoing monitoring of active licenses, license administrators can provide helpful feedback about the terms and structure of license agreements to those who negotiate them. Likewise, the performance of licensees can be assessed during the term of a license and when it expires. Delays in development, ambiguous license terms, and failures to address license issues that may require an amendment during the term of a license are good examples of what can be identified from monitoring and expiration reviews. Capturing this knowledge and sharing nonconfidential information about best practices with other organizations can help build a knowledge base that continuously improves the technology licensing process.

8. PATENT INFRINGEMENT

One enforcement task that does not flow out of existing license financial terms and reporting obligations is the pursuit of suspected patent infringers. When a company has not licensed a patented technology but is infringing a patent owned by the licensor, legal action should be undertaken. The first step is to notify the infringing company by letter that they are infringing and should immediately cease to do so. The company usually receives an offer at that time to license the technology in order to avoid legal action against the company by the patent holder. Follow-up may require negotiating a license agreement or, if the license is refused, additional legal action by the licensor.

9. CONCLUSION

Administering technology licenses gives a TTO an opportunity to monitor and participate in an invention's development and commercialization. A successful effort requires good organization, good tools, diligent attention to detail, and the persistence to engage licensees in dialogue when license obligations are not being met. While many technology transfer organizations focus most of their time and effort on negotiating license terms, the overall success of a TTO also requires allocating resources and time to license administration, monitoring, and enforcement. Thorough, consistent follow-up with licensees will ensure that the licensor and inventors financially benefit. The licensee may also benefit from the discipline of an attentive partner and access to the knowledge and experience of the licensing office. Above all, effective license administration ensures that economic development and the public good are well served by the timely introduction of new products and technologies in the marketplace. ■

ACKNOWLEDGMENTS

The author would like to acknowledge David B. Schmickel as a contributor to the chapter on "License Follow-up" that appeared in the first edition of this *Handbook*. This new chapter updates that earlier one. The author would also like to acknowledge the helpful comments and suggestions for this chapter of the following NIH Office of Technology Transfer personnel: Bonny Harbinger, Luis A. Salicrup, and Steven M. Ferguson.

HANS H. FEINDT, *Chief, Monitoring and Enforcement Branch, Office of Technology Transfer, National Institutes of Health, 6011 Executive Blvd, Suite 325, Rockville, MD, 20852, U.S.A.* feindth@mail.nih.gov

1 See Section 11 in this *Handbook* which contains a number of related chapters.

2 www.ott.nih.gov.

3 www.ott.nih.gov/licensing_royalties/NegDis_ovrvw.html.

Policing Intellectual Property

H. WALTER HAEUSSLER, *Director, Technology Transfer and Intellectual Property, Texas Tech University and Texas Tech University Health Sciences Center, U.S.A.*

RICHARD S. CAHOON, *Executive Director, Cornell Center for Technology, Enterprise & Commercialization and Senior Vice President, Cornell Research Foundation, U.S.A.*

ABSTRACT

A university's intellectual property (IP) cannot be simply shelved and forgotten. IP, with patents as a particularly cogent example, must be managed, monitored, maintained, and policed in an ongoing "cultivation" of the IP rights. For patents, it is important to be able to identify potential infringement early, by means of coordinated surveillance by the technology transfer office. If, and when, possible patent infringement is detected, it will then be necessary to evaluate the type of infringement, that is, direct or contributory, and also to assess whether the activity legally appears to be infringing, reading on each and every element of a patent claim. Strategic and business considerations must be considered as the university decides what course of action might be appropriate in response to an alleged infringement of a patent. Specifically, in the context of litigation, the university must consider whom to sue (if there are multiple infringers), when to sue (if too late, could risk loss of IP rights), and where to bring suit (for a favorable venue). An even more critical consideration is whether to even litigate at all. It may be wiser to seek one of various forms of alternate dispute resolution, for example, negotiation, mediation, or arbitration. It is important to never forget that litigation is expensive, risky, and unpredictable. Hence, it should be viewed as not the first option, but as the final one, and it should be approached as a cold business decision and not to give teeth to emotions or carry out revenge. Throughout the process of managing and policing its IP rights, a university should have access to legal counsel. Finally, proactive, good license hygiene is the best way to proceed, and the most effective way to avoid expensive litigation. By demonstrating credibility, conviction, and focus, the university will show potential infringers that it is serious about policing its IP, and that they therefore won't be able to escape the university's diligent surveillance. Licensing, and not infringement, will then become the only sensible route to accessing the patent rights.

1. INFRINGEMENT OF INTELLECTUAL PROPERTY

Infringement is any manufacture, use, sale, offer to sell, or importation of intellectual property (IP) that has not been authorized by the legal owner of the IP. Forms of IP that are subject to infringement include patents, copyrights, and trademarks; these provide the owner of the IP rights with certain legal remedies for redressing infringement. Infringement of IP should be considered neither mysterious nor overly complex and technical. Basically, infringement is analogous to trespassing on another person's physical property or real estate: it is an invasion and misappropriation of another's exclusive property right. Correspondingly, one can obtain permission to occupy, or to use, real estate by renting it or to use IP by licensing it; the two actions are entirely parallel.

Identifying and taking action to remedy infringement is an essential part of IP ownership.

Haeussler HW and R Cahoon. 2007. Policing Intellectual Property. In *Intellectual Property Management in Health and Agricultural Innovation: A Handbook of Best Practices* (eds. A Krattiger, RT Mahoney, L Nelsen, et al.). MIHR: Oxford, U.K., and PIPRA: Davis, U.S.A. Available online at www.ipHandbook.org.

Editors' Note: We are most grateful to the Association of University Technology Managers (AUTM) for having allowed us to update and edit this paper and include it as a chapter in this *Handbook*. The original paper was published in the *AUTM Technology Transfer Practice Manual* (Second Edition Part XV: Chapter 3).

Asserting IP rights is essential for preserving these rights and for maximizing their economic value. A university's maintenance and assertion of its IP ownership rights, including a willingness to bring legal action if necessary, is essential for the licensability of its IP. The perception on the part of industry and, in particular, potential infringers, that the university will take action to remedy infringement is critical for the focus, determination, and credibility of the university's technology licensing effort and key to the value of its licensable technology. This chapter examines these issues in the context of U.S. patent law.

2. HOW TO IDENTIFY INFRINGEMENT

Infringement is a legal event, that the patent owner (patentee) bears the burden of proving. Proof of infringement proceeds by a two-step analysis. First, the alleged infringed invention must be defined, by the court's construing of the actual patent claims. Second, the patentee, through the strength (or preponderance) of the evidence must show that infringement actually occurred. Hence, the patentee can neither guess, think, nor presume infringement, but rather must *prove* infringement. For example, if the patent in question is a process, the fact that a product sold by the "infringer" is identical to the university's product does not prove the alleged infringer is liable; the alleged infringer could be using an entirely different process to make the product.

Typically, literal infringement occurs when the infringer's product or process reads on each and every element of a patent claim. The fewer the elements or steps in a patent claim, the more likely apparent infringement will turn out to be actual infringement.

With this in mind, it is important to consider claim structure and scope when a university initially files a patent application. The university will be in much better position to protect its IP rights if the attorney who prepared and prosecuted the application understood that the university has no need for narrow-claim, defensive patents. A university does not manufacture and therefore has no products to protect. Unless the claims of a university patent are sufficiently broad to have economic value (this cannot easily be avoided if one practices the technology), the patent will have little, and perhaps even negative, value. For example, negative value may arise if the inventor exclaims, "*Look at this infringer,*" and the university responds, "*Yes, the company is practicing your invention, but our claims were drafted too narrowly, and our patent is therefore not infringed.*" At that point all may painfully realize that during the actual prosecution of the patent application, it would have been better to appeal to the patent office for, and then lose on, broader claims. The inventor would thereby have realized that a patent with real economic potential was unattainable from the start, rather than only becoming disappointed later, accusing the licensing office of not doing its job with adequate diligence, when the patent is only then determined to be worthless.

The key message here is this: patent prosecution must be conducted with an eye toward winning future infringement litigation should it arise. The whole point of patent prosecution should not be to have a given claim or any claim allowed so a patent will issue, but rather to have a claim approved that is consistent with the university's mission, has economic potential in the marketplace, and will be enforceable.

Assuming the university owns a patent with strong claims, how can the university determine, especially when not actively engaged in the marketplace, whether that patent is being infringed?

2.1 *Establishing surveillance for possible infringement*

Inventors should be contacted on a regular basis and asked if they know of anyone who is or might be infringing their patent. If nothing else, the effort could lead to licensing possibilities and reveal who is interested in using the patented invention.

Technology transfer staff members should review key media related to the technology on a regular basis to watch for potential infringers. Again, this is doubly advantageous because it can also generate licensing possibilities. The focus of the marketplace reviewers in the technology transfer office must not be on marketing alone, but also on infringement and licensing opportunities.

Therefore, to the greatest extent possible, one must know the marketplace. It is critical to make an effort to talk to existing licensees, alumni, others who are knowledgeable in the relevant areas, and to potential licensees of related technology in order to learn what they and their peers are doing and/or thinking of doing. In other words, it is essential to build and maintain professional networks.

2.2 *Evaluating infringement*

Unless the technology transfer manager is an IP legal professional who can assess the possibility of infringement, it will be necessary to initially seek the opinion of counsel in order to be certain of potential infringing activity. As previously stated, the burden of proving infringement is on the patentee. Therefore, the university cannot expect the apparent infringer to willingly help prove there is actual infringement. While certain industries typically respect university patents and are forthcoming, others have a "catch me if you can" attitude. If the university has a process patent where the process does not leave a footprint on the product, proving infringement may be extremely difficult.

Literal infringement requires that each and every element or recitation in a particular claim must be infringed. If there are five steps in the claim and the apparent infringer practices only four of those steps or combines a different fifth step with the university's first four steps, there may be no infringement. Being close to infringement does not usually count towards an infringement determination.

There is, however, the ***Doctrine of Equivalents***, which is a more flexible rule of claim interpretation. The doctrine provides that, even though a claim is not literally infringed, a case for infringement can still be made if the infringer has used a variant of the patented invention that is substantially the same as what is actually claimed as the invention. If a technology transfer manager thinks that the apparent infringer is too close to be allowed to escape infringement, the manager should get an expert opinion to help the university decide whether the Doctrine of Equivalents may be applicable.

It is important to note, however, that the Doctrine of Equivalents cannot be employed where the patentee narrowed the claims in response to a substantive rejection by the patent office during patent prosecution. This creates a bar to the use of the doctrine (File Wrapper Estoppel), because it would be unfair to initially argue during prosecution that the claims were narrow enough to avoid prior art and hence be patentable, but then later, during infringement proceedings, attempt to expand the scope of the claims beyond their literal language by invoking the Doctrine of Equivalents, that is, to attempt to reclaim in litigation what was surrendered during prosecution of the patent application. Once the scope of the claims is narrowed, it is narrowed for good.

2.3 *Record keeping and evidence gathering*

In general, the better the records kept by the inventors, the better the patentee's (or applicant's) ability to win in an infringement action. However, in litigation, the patentee's records, while a source of validation of assertions in the patent, are accessible to the opponent and may be searched for contradictory statements or adverse data that was not given to or considered by the patent examiner. A possible defense raised, if such adverse data is found by the alleged infringer, may be considered *fraud against the patent office*. This is a form of inequitable conduct perpetrated by the patentee during prosecution of the patent application, by which the patentee deceives the patent office by either withholding material or submitting false information, thereby rendering the patent unenforceable. The patentee should search its own records so that it is not later surprised by any data that might be subsequently used against it. The best way to avoid this problem is to pay close attention to the duty of disclosure to the Patent Office during the prosecution of the patent application, that is, better to take a proactive and preventive approach early on than to be sorry later.

When gathering evidence of infringement, if the university has other licensees, they will usually help the university to acquire information and analyze samples. If necessary, the university may have to buy an infringing product and analyze it. The university will need to document exactly where the infringer is selling the offending product, for example, whether directly, or through agents or distributors. When the infringer is manufacturing or using the infringing product, the evidence must be hard, including documents, materials (with analysis of the materials), and eyewitness testimony (for example, a signed affidavit as to what a person would testify to if called as a witness). Hearsay will not prove the university's case. "*My brother-in-law told me that he had seen* ..." won't work. Actually proving infringement, and exactly when and where it occurred or continues to occur, is necessary but frequently quite difficult. Issues of venue, that is, where legal action can be brought, may cause the university to want to prove infringing acts in a certain geographic area; this makes the job potentially more difficult.

3. SOME LEGAL (AND PRACTICAL) CONSIDERATIONS

3.1 *Patent or contract suit*

If the person using the technology or inventions has not signed a license agreement, usually a contract of some sort, then the university's only practical litigation recourse is usually a suit for patent infringement. If there is another legal relationship such as a license agreement where the licensee has ceased to pay royalties, or a material transfer agreement where it appears that the infringer is improperly using material received from the material transfer agreement, there are alternatives to consider, such as breech of contract actions. It is possible that the location of litigation (or the issues) may be in the university's favor, or the price of litigation may be cheaper if the university brings suit on an existing contract rather than a suit for patent infringement. It is therefore important to examine, in depth, all the business relationships existing between the infringer and the university, which may include consulting contracts between the inventors and the infringer.

3.2 *Whom can the university sue?*

If there is more than one possible infringer, then it is important to weigh the pros and cons of suing each infringer. Sometimes the choice is clear; at other times consideration must be given to select the target of litigation. A patent owner need not sue all infringers at the same time. A single suit against a single member of a group of infringers is the usual tactic.

Patent litigation is expensive and, as in a poker game, it is difficult to win against a player who has an order of magnitude more money than the rest of the players. The player who has more money can unfairly distort the game. The same is true in patent litigation, and it is usually inadvisable to litigate against the party that has the largest financial resources or the largest financial interests in the outcome of the litigation. On the other hand, the party having the largest financial interest may indeed be the one to sue, because in a practical sense, if the litigation is successful, then the issue will have been essentially resolved, with the largest part of the market secured and other infringers likely to fall into line and comply with licensing terms.

Other considerations include the convenience of the forum, ease in collecting damages, and existence of issues that are particular to a given infringer that might enhance the university's chances of winning. For example, clear statements that an infringer's actions were knowing and deliberate may indicate selection of that particular infringer to sue. The alleged infringer has made himself a target for litigation.

One method of managing the venue of the lawsuit is to sue a party in the distribution chain in a location of the university's choice: for example, a party who through purchase is an infringer. Frequently the original infringer becomes involved in such a lawsuit because of an obligation to indemnify the purchaser. Therefore, the university can potentially access the most important infringer in a favorable venue, which otherwise might have been difficult or even impossible.

The patent law provides recourse and remedy not only against direct infringement, but also against contributory infringement and inducement to infringe. A party can infringe by actively and knowingly assisting in another's direct infringement. The most common type of ***contributory infringement*** is where a company sells a component to the infringer in a situation where the company knows or should have known that the only practical use for that component was to make infringing devices or create an infringing use. As for ***inducement to infringe***, the patent statute states, "*Whoever actively induces infringement of the patent shall be liable as an infringer.*" Hence, inducement to infringe is where the party actively and knowingly aids and abets another in direct infringement. Whether a company intends to induce infringement is a factual determination.

3.3 *Where can the university sue?*

In the United States, since patents are enforced in the federal courts, theoretically, a university can sue for infringement anywhere in the United States, but there are jurisdictional requirements, venue, and service requirements that usually limit the number of actual forums available. When considering where to file a suit, proximity to the court may be a major issue. The university must also consider where its trial counsel and inventors or other witnesses are located, whether there is a need to compel certain witnesses to attend, and in what jurisdiction the university can likely prevail. Of course, specifically inconveniencing the party one intends to sue should not be overlooked as a useful strategy.

Certain courts are busier than others, and therefore, if the university looks for a speedy trial, it may want to pick a forum that has a small backlog or one that has developed an attitude, capacity, and reputation for rapidly processing cases.

Furthermore, the attitude of a particular judge or a group of judges in a particular court may influence the choice of forum. If the university can determine that the judge has an identifiable track record for deciding certain underlying issues, then it may, or may not, choose that court, based upon the record of the judge's rulings, philosophy, and apparent priorities.

If a jury trial is selected, then the location of the forum can have a substantial impact on the nature and attitude of the jurors. A state university that has a long history of agricultural extension no doubt has an advantage if the jury consists of local farmers. On the other hand, if the university sues an infringing company, seeking venue in a small town where the company is the largest single employer, then it can expect that the jury might be biased against the university and favor the accused infringer.

U.S. federal law and a section on the venue of particular U.S. federal courts states that, "*Action for patent infringement may be brought in the judicial district where the defendant resides, or where the defendant has committed acts of infringement and has a regular and established place of business.*" The federal courts are split as to what is a regular and established place of business. Some courts have held that there has to be a formal office and others have held that a sales representative operating out of his or her home may satisfy the requirements.

3.4 *When can the university sue?*

A university cannot initiate patent litigation until after there is an actual act of infringement. At the other extreme, the university must bring the suit before the suit is barred by the potential equitable defenses pursuant to the statute of limitations, the doctrine of laches, or equitable estoppel.

From a strictly legal technical point, there is no such thing as a statute of limitations in the patent law. That is, there is nothing in the patent statute that absolutely bars the bringing of an infringement suit. However, the statute does bar recovery of damages for infringing activity that occurred more than six years prior to the filing of the infringement action.

Laches can be defined simply as the patentee waiting too long to take action for no good reason. The federal circuit has held that laches bars relief on a patentee's claim only with respect to damages that occurred prior to the suit. It is important to note that there are two elements to laches. First, there must be an inexcusable delay for an unreasonable length of time in initiating litigation. Second, the defendant must show that the litigation was prejudiced by the delay. The longer the delay, the less is needed to show specific prejudice. Usually there has to be a considerable delay before the doctrine of laches has any relevance. There is a presumption of laches after six years, but the patentee can overcome this with suitable evidence rebutting the two elements that establish laches.

Another defense against a patent infringement action is *equitable estoppel*, which simply means that there is a particular reason that the university, as the patentee/plaintiff, should be barred from suing the particular defendant. Equitable estoppel usually results when the patentee intentionally communicates with the infringer such that the infringer relies upon and is then mislead and materially harmed by the deeds, actions or words of the patentee. For example, the officers of the patentee through affirmative conduct induced the infringer to believe that the patentee had abandoned its claim against the alleged infringer, and therefore, the infringer kept manufacturing. Clearly, there should be an equitable estoppel. It is important to note that the silence of the patentee alone will not constitute an equitable estoppel, although that silence over a long period of time may create laches.

4. THE LAWYERS

4.1 *How soon should counsel become involved?*

There are no right or wrong answers for how soon to involve counsel. Before doing so the university should determine that there is in fact an infringement. If the answer is yes, the university should then determine whether the usual licensing routes been explored and a negative response received? If the answers to these questions are also yes, then the university should recognize that the case is not an ordinary one and that there are valid business reasons to consider infringement action. At that point, the university should have preliminary discussions with its counsel prior to making any decisions.

4.2 *Who will serve as counsel?*

There are several very important issues that must be contemplated in selecting counsel. If the university (the client) does not control the proceedings, and therefore, does not control the cost, the result typically is extraordinary financial bleeding. If the university finds that it is working with counsel who tends to say, "*Just leave it in our hands; we know best,*" the university can expect the fees to be high. It is important to pick counsel who has a perspective as to the way proceedings are conducted and the way costs are controlled that is compatible with the philosophy of the technology transfer office and the university. For example, does the university intend to be represented at every deposition held by the other side? What level of discovery is the university going to seek? Is the selected counsel comfortable working solo or with one other people in the firm, or does the intended counsel suggest that there be a team of four people, plus a backup team of two people (as a precaution)? These attitudinal differences vastly affect the kind of litigation that is going to be conducted and the cost of that litigation.

The amount of money spent has some bearing on the outcome of the litigation, but the attitude should be, "*I want to spend the least amount of money necessary to win,*" not, "*Let's do everything imaginable so that nobody can ever accuse us of losing because we failed to do (and spend) enough.*"

The university may have trial lawyers on staff. Those trial lawyers can be invaluable for interfacing between the university and outside trial counsel, and also for helping the university manage the issues, even though in-house trial lawyers may not have any experience with patent litigation.

A decision to hire outside counsel leads to the question of whether one attorney, one firm of attorneys, or multiple attorneys should be involved. One can argue that lawyer(s) rendering opinions as to whether infringement exists and, if so, a strong likelihood of prevailing in litigation, should be independent of the lawyer(s) who ultimately litigate. For example, if the lawyer rendering the opinion recognizes that he or she will not financially benefit from a statement that there should be litigation, then the university is more likely to get an unbiased answer. The same is true on the issue of infringement. If the lawyer understands that he or she will not have the benefit of the litigation if he or she gives the opinion that there is infringement, then the university may get a more objective opinion. This is not necessarily the case, for example, if the university has a solid, trusting relationship with counsel, and counsel recognizes that sooner or later, given a legitimate case, he or she indeed will have involvement in litigation, then the university can comfortably use the same lawyer(s) for both opinion work and litigation. After all, the more a university works with an attorney or firm, the more likely the technology transfer manager and other institutional legal counsel will generate useful opinions and advice.

There are no right answers to selecting counsel. The bottom line is to pick a trial lawyer who accepts the fact that he or she will be required to justify how and why the money is spent and to give the university clear choices so that it can control costs. Keep in mind that the actions of the opposing side have a large impact on costs. Once litigation is commenced, while the university may diligently work to control costs, the actions of the other side can make that job difficult. Frequently the best estimates of cost before litigation starts are discovered to be completely inaccurate after the litigation is under way and the issues are revealed. Therefore, it is important to select counsel who is willing to revisit the issues of control of the proceedings, including control of costs and strategy, so that the university can continue to make intelligent choices.

5. IS THE UNIVERSITY READY TO LITIGATE?

Patent litigation is expensive, involves substantial risk, and endangers the university's IP rights. A frequent defense to an accusation of infringement is patent invalidity. Therefore, the university can lose the litigation on a judgment that the individual is not an infringer and can also lose on a judgment that its patent is invalid. However, an issued patent is presumed valid by statute, and the accused infringer carries the burden of proving (by clear and convincing evidence) that the patent is indeed invalid. Still, in the event of a declaration of patent invalidity, the university has no further opportunity to license the technology and any existing licensees will stop paying royalties. On the other hand, if the university has a group of licensees and there is a party substantially infringing without licensing, ultimately all of the university's licensees will recognize this and possibly also stop paying royalties unless the university takes action. As a result, the university may be in the position where it will bleed to death slowly or have an instant death if it loses the litigation. In any event, the only way to preserve the long-term economic viability of the proprietary technology is to bring suit.

5.1 *Warning letters*

After identifying a likely act of infringement, the technology transfer manager may enter into a dialogue with the infringer in an effort to end the infringement; this is frequently resolved by entering into a license negotiation. At some point there will be a written communication stating that the university believes the party may be an infringer and that if it neither ceases nor licenses, the university will consider taking legal action. The manager should understand that if the university clearly and precisely accuses a party of infringement and threatens the party with litigation, then the situation may rise to the level of an actual case/controversy, triggering the accused party's right to seek a declaratory judgment. This involves asking the court to declare that there is no infringing activity and/or that the university's patent is invalid. Therefore, the right to seek legal relief becomes not only the university's, but also that of the party accused of infringement; in other words, the table has turned. Therefore, caution is important. As long as the university's letters fall short of making an actual accusation of infringement and of threatening litigation, then the decision to go to court remains solely with the university. If a manager is not comfortable, experienced, and skilled in drafting such letters, then a warning letter should be reviewed (and possibly even written) by counsel before it is mailed. Clearly, the wrong warning can lead to unintended consequences and come back to hurt the university in several ways.

5.2 *Beware of oversights in record keeping*

A university is not ready to litigate until it has investigated its own records and spoken with the people on the university's side who are associated with the potential litigation (and who might be witnesses) to discover whether there is any knowledge or written correspondence or records that would have an embarrassing or otherwise negative impact on the outcome of the litigation. The university should not let the infringing party discover these damaging oversights; it should know about them ahead of time because this may greatly impact the decision of the university's trial counsel of whether to proceed with litigation.

5.3 *Exhaust all alternative means of settling the controversy*

Before the university litigates, it should consider involving a third party for informal dispute resolution or possibly proceeding with formal arbitration or alternative dispute-resolution mechanisms in order to find a solution short of court.

5.4 *Valuing the alternatives*

The alternatives to litigation include changing the licensing terms or creating licensing scenarios that take into account issues raised by the infringer as reasons for not taking a license. It is a wise strategy to consider offering license terms that make opposition to paying royalties economically irrational (when compared to the costs of litigation) for the infringer. When valuing the alternatives and seeking alternative resolution, the university must consider other licensees and the existence of *favored nation* clauses in other license agreements. The university may have to extend the same terms to all its other licensees, and therefore, alternative dispute resolution may have a financial impact beyond the particular infringing activity, with a potentially broader impact and implications for the value of the technology.

The university must reach an approximation of the true cost of litigation, which is more than the cost of outside attorneys. Litigation requires an enormous amount of staff time, not only of the technology transfer office, but of the university's counsel office as well. Also, litigation can involve much of the inventor's time and anguish, since the inventor's skill and integrity may be challenged in the litigation. Ultimately, of course, there is the dollar cost. Importantly, the university must recognize that past infringement, the cost of the litigation, and the impact on the future value of the technology are issues that have to be separately assessed when considering alternatives.

5.5 *Making a difficult business decision: Walk away or litigate?*

Because patent litigation is expensive and puts the university's IP at risk of being declared invalid, the vast majority of patent disputes are settled before they ever come to court. For both sides, it is usually better to resolve the dispute than to litigate. But ultimately, the technology transfer manager may be required to make a very hard business decision on behalf of the university. A manager should never litigate out of anger or pride. The university should only litigate if it makes absolute business sense, that is, if it is economically better to litigate than not to litigate, and only after the university has examined all of the issues, including the risk of losing versus the value of winning, and finds, on balance, that it makes sense to litigate.

5.6 *The effect of the Markman decision*

The way patent litigation is conducted was significantly impacted by the Supreme Court decision in *Markman v. Western Instruments, Inc.*[1] Claim interpretation was taken away from the jury and handed to the court. The result of Markman and related later cases was that claim interpretation could occur at any time in the litigation, and not just before, during, or after the trial.

In *Vitronics, Inc. v. Conceptronic, Inc.*[2] the federal circuit held that it is the rare case where patent claims should be interpreted based on anything other than the patent, the specification, and the file history (the public record). Therefore, the hope was that claims could be construed early so that the parties would know the meaning and scope of the claims before starting discovery. Since discovery is often over one-half the cost of expensive patent litigation, if it can be narrowed to more-specific issues, cost should be less, enhancing the chance for early settlement.

The results of the *Markman* and *Vitronics* decisions have been mixed. District courts have shown little uniformity with respect to the timing of claim interpretations. Some courts make their interpretations very early in the process; some as part of a conference just before the trial starts (and after discovery is complete); and some courts do so during or at the end of the trial. Most courts now have formalized a "Markman procedure." Some have built claim construction hearings into their local rules. Finally, whether extrinsic evidence can be used in a claim construction hearing is far from being settled.

Clearly, claim construction is a critical element in litigation that now has assumed an independent place in the litigation process. If one can obtain early claim construction, doing so should be a significant benefit with respect to the cost of the litigation; if there is a serious issue regarding the scope of the claims, claim construction may prompt settlement or dismissal.

5.7 *The university's role if the licensee litigates*

Some universities may give their licensee the first right to litigate. This is quite common in cases where an exclusive license has been granted. But even in that case, the university should pay very close attention to what is happening and may want to participate in key strategy sessions held by the licensee and its counsel and/or have the university's own counsel participate and/or review all documents. Where the university's personnel are deposed or where discovery is held on the university's documents, the university's counsel should be involved. But, if the university granted the licensee the right to litigate, thus saving the university the cost of litigation, why should the university incur significant expense to look over the licensee's shoulder?

There are a number of valid reasons why the university should remain active in the litigation. On many points the licensee's and university's interests may not exactly correspond, and, in certain situations, a choice may be made that reflects badly on the university, though it would benefit the licensee. This is very important, as there should always be concern for the university's good reputation and the reputation of the researcher/inventor. Both can be at risk in litigation. It is critical to keep in mind that the actions, words, skill, or integrity of the researcher/inventor may be put at issue, which could become traumatic for the researcher/inventor in the unpredictable process of litigation. Another reason to maintain involvement is the potential for a loss of property. As pointed out previously, once a patent is declared invalid, it is forever invalid, so the university could lose its valuable IP rights. The licensee may not have as much at stake; it may only lose by gaining a competitor.

Just because the university lets the licensee assume the burden of litigation, the patentee should still be vigilant as to the licensee's determination, skill, and strategy for litigation, as well as its attitude toward the university and the university's researchers. The patentee should also remain aware of a licensee's financial status. Letting the licensee carry the burden of litigation may significantly ease the university's financial burden and the level of technology transfer staff involvement. However, because it is the university's patent, and because the university's staff may be vital witnesses, the university will almost always have a critical, although reduced, role.

5.8 *The licensee's promise to hold harmless*

In most instances where the licensee is litigating, there is a license obligation to hold the university harmless in the litigation. Even so, the university must look closely at the state and condition of the licensee at the time of the litigation. If things go badly and the university is at risk, can the licensee perform adequately on its promise to protect? Does it have sufficient assets to pay an adverse judgment? Is it going bankrupt? Is there collectable insurance available? There may be a rude awakening, if the university is not attentive to the meaningfulness of a hold-harmless promise, both at the time of entering into the license agreement and at the time litigation by the licensee is contemplated.

6. GOOD LICENSE AND LICENSEE "HYGIENE" TO PREVENT LITIGATION

A technology transfer manager should review the university's license agreements on a regular basis to make sure that its licensees are current in their payments and all other obligations. The technology transfer manager should be talking to the university's licensees about the marketplace and should listen if licensees are complaining that there is a party performing unauthorized acts. The manager should talk to the inventors or other people who are knowledgeable in the technology field, so that if there are infringers, the university can contact those parties early and they will not be led to believe they are free to act. The single most likely cause for litigation between a university and industry occurs when an industry member has the perception that the university won't litigate, or that the university is inadequately represented and doesn't know what it is doing. Clearly, communicating with conviction and credibility that the university indeed will sue, and emphasizing that the university has, or will, retain competent counsel and pay the price necessary, will go a long way toward bringing the infringer to the table to discuss the issues.

A final word of advice for the university: write the good things, and say the bad things. Although the attorney-client privilege is real, it is frequently penetrated. Consider anything in writing accessible to the other side in litigation and available for use against the university. ■

H. WALTER HAEUSSLER, *Director, Technology Transfer and Intellectual Property, Texas Tech University and, Texas Tech University Health Sciences Center, Lubbock, TX, 79409-2007, U.S.A. waltertx1@yahoo.com*

RICHARD S. CAHOON, *Executive Director, Cornell Center for Technology, Enterprise & Commercialization and Senior Vice President, Cornell Research Foundation, U.S.A. rsc5@cornell.edu*

1 116 S. Ct. 1384 (1996).

2 90 F.3d 1576 (Fed. Cir. 1996).

CHAPTER 15.3

Alternative Dispute-Resolution Procedures: International View

EUN-JOO MIN, *Senior Legal Officer, Arbitration and Mediation Center of the World Intellectual Property Organization (WIPO), Switzerland*

ABSTRACT

As multinational technology-development partnerships have become more common, so have disputes between the parties. Litigation, however, is not the only option for resolving such disputes. In fact, for partnerships between entities in developing and developed countries, litigation may be a complicated, time-consuming, expensive, and doubtful process. Arbitration and mediation may offer the promise of more effectively resolving disputes, and this chapter explains how these methods work, their advantages and disadvantages, and suggests which questions should be asked (especially for a developing country institution) to begin to establish a dispute prevention and resolution strategy. The chapter offers both strategic and practical insights about how to use these mechanisms to resolve disputes and preserve partnerships.

1. INTRODUCTION

Institutions in developing countries are increasingly entering the IP market, and multiparty, multinational IP relationships are becoming more common, and even essential to socio-economic development. Through transactions involving these relationships scientific, technical, entrepreneurial, creative, and traditional knowledge is exchanged. Nonetheless, a protected right also tends to increase the likelihood of disputes related to that right.[1] While parties seek to reduce the frequency of disputes by rigorously managing their IP rights and obligations, disputes will inevitably arise. When they do, they can negatively affect both sides. Parties involved in IP transactions, therefore, should be aware of dispute-resolution methods and have a specific dispute-prevention and resolution strategy. Dispute-resolution procedures too often are unwittingly selected when a relationship begins, often years before a dispute actually arises. The dispute-resolution clauses will therefore have been inserted into contracts by people no longer involved in the issues. Moreover, clauses frequently are inserted with a limited awareness of their specific implications in a dispute-resolution scenario.

Litigation, the formal, public process for resolving disputes before national courts, is the most conventional method of dispute resolution. Particularly for transnational disputes, litigation may be risky, frequently protracted, and may at times require seemingly unlimited legal costs and management time. Moreover, a dispute taking place in multiple jurisdictions may result in different outcomes depending on which court decides the case.

This chapter explores alternative dispute resolution (ADR) procedures for resolving IP disputes, focusing on the interests of developing countries. ADR encompasses a range of options for resolving disputes outside of formal court procedures. These options differ in terms of formality, party control, and finality. Each option, moreover, offers benefits uniquely appropriate to different circumstances. This chapter concentrates on two

Min EJ. 2007. Alternative Dispute-Resolution Procedures: International View. In *Intellectual Property Management in Health and Agricultural Innovation: A Handbook of Best Practices* (eds. A Krattiger, RT Mahoney, L Nelsen, et al.). MIHR: Oxford, U.K., and PIPRA: Davis, U.S.A. Available online at www.ipHandbook.org.

representative ADR procedures, arbitration and mediation.

2. DISPUTE SCENARIOS

The following dispute scenarios discuss some specific circumstances that apply to health or agricultural IP disputes. The scenarios may have particular relevance for institutions in developing countries. Parties to the types of disputes in these scenarios will most likely first consider resorting to litigation in national courts. They will, however, often find court action stymied because of the challenges involved: cost, length of procedure, legal uncertainty, decision makers' lack of expertise, confidentiality/publicity, the difficulty of seeking action in foreign jurisdictions, and the negative impact on existing business relationships. Given these difficulties, parties should consider whether there are practical alternatives to expensive and protracted court proceedings.

2.1 *Research collaboration: ownership dispute*

Researchers in a medical research center in a developing country (Center X) build a research partnership with a leading university in a developed country (University Y). They collaborate on pursuing leads for pharmaceutically active compounds. The partners exchange data and discuss research directions. University Y has a well-established policy of patenting campus research, and an invention disclosure is filed with the technology transfer office (TTO). This becomes a patent application in the name of University Y, citing three of its researchers as inventors. There is no notice to, nor recognition of, the researchers in Center X. The researchers at Center X denounce the behavior of University Y and request that their names be included as inventors. When University Y refuses this request, the researchers contemplate legal action, but are stymied by prohibitive legal costs.

2.2. *Patenting of research outputs from genetic material*

A research institute obtains patent protection for a cell line developed from genetic material obtained from one of the institute's patients. The patient is from an indigenous group that lived an isolated existence until very recently. The indigenous group seeks redress, claiming ownership of interest in the patent and breach of fiduciary obligations by the research institute. The research institute asserts that it proceeded to commercialize the research result based on the patient's prior consent to treatment. The controversy, with claims of biopiracy, rapidly escalates into a global public debate.

2.3 *Claims based on traditional rights*

An ethnobotanist collects traditional medical herbs and associated knowledge about their therapeutic use from an indigenous community. The community is led to believe that this is the personal research of the ethnobotanist; the researcher acquires some of the knowledge after he falls ill on site and is treated by a traditional medicine man. The customary law of the indigenous community constrains both the dissemination and use of this knowledge within the community. The researcher subsequently publishes the knowledge, and details about the plants he collected, in a noncommercial academic publication. This publication is widely distributed and used by several private companies in their medical research. The disclosure of the information leads to patents, not directly on the traditional knowledge, but on further innovations, which are guided by and dependent upon the traditional knowledge. These patents acknowledge the prior publication, but give no direct reference to the traditional community itself. The traditional community attempts to seek relief but quickly finds that the legal remedies at their disposal are unclear and inappropriate for dealing with the cultural and spiritual harm incurred.

2.4 *Agricultural products and patents*

Farmers in a developing country have cultivated for centuries a certain type of grain that gains popularity in global markets. A biotechnological corporation obtains patents on the grain by introducing genetic modifications. Farmers in the developing country denounce their loss of international market share resulting from the actions of the biotechnological corporation.

The farmers are concerned, however, that any inherent right they may claim will be overshadowed in court by the economic, technical, and legal prowess of the corporation.

2.5 *David v. Goliath?*

An inventor in a developing country holds patents in a number of countries on components used in consumer goods. The inventor enters into a license agreement regarding these patents with a multinational manufacturer. A dispute arises regarding royalty payments under the license agreement. The inventor wants to enforce his rights, but does not dare to engage in protracted and expensive multijurisdictional litigation. Furthermore, the inventor hopes to maintain his profitable relationship with the manufacturer.

3. THE ARBITRATION OPTION[2]

Seeking resolution to the above disputes through litigation promises much pain and little certainty for parties in developing countries. An alternative approach to litigation, however, could offer better results. Arbitration, for example, involves submitting a dispute, by agreement of the parties, to one or more arbitrators who make a binding decision.

3.1 *Arbitration procedure*

To send a dispute to arbitration, the parties must sign an agreement to submit their existing or future disputes to arbitration. Such an agreement is the foundation of an arbitration arrangement.[3] It demonstrates the parties' genuine willingness to settle the dispute through arbitration and limits the parties' right to take the dispute to court.

Arbitration may be conducted in different ways, and it is up to the parties and the arbitrator(s) to decide how the procedure should unfold, subject to any applicable rules and public policy requirements. Parties may agree on the number of arbitrators, type of arbitration (ad hoc or institutional), place of arbitration, language of arbitral proceedings, and the applicable substantive law.

Figure 1 describes the principal steps in a typical arbitration, referencing the Arbitration Rules of the World Intellectual Property Organization (WIPO)[4] (see also section 6.2 below).[5]

3.2 *Role of the arbitral tribunal*

An arbitral tribunal operates differently from a judge in national court. Judges have powers defined by national laws. The powers of an arbitral tribunal are limited to those the parties have conferred to it. An arbitral tribunal may only determine the disputes stipulated by the parties involved, and may only do so using powers conferred by the parties through the arbitral clause and adopted rules.

Since the arbitral tribunal is the dominant authority in settling the dispute, the appointment of the tribunal is probably the single most determinative step in an arbitration. Parties should, therefore, be able to exert as much influence as possible on the establishment of the tribunal. Parties can normally agree on the appointment procedure, the number of arbitrators to be appointed, any required qualifications of the arbitrators (including nationality), and persons to be appointed as arbitrators. In reviewing these factors, parties will have to weigh considerations of cost and efficiency against the weight and complexity of the dispute. The legal, cultural, and economic backgrounds of the parties will be reflected in the tribunal appointment process.

3.3 *Legal framework of arbitration*

While arbitration is a private mechanism, it is not altogether free from regulation by national laws. In international arbitration, different systems of law, most notably the law governing the *substance of a dispute* and the law governing the *arbitration procedure*, will typically interact. In general, parties are free to choose, by agreement, which laws will apply.

Parties may agree on which national law should govern the substance of the dispute. Parties may also agree that the dispute be determined on the basis of what is *just and good* (*ex aequo et bono*). In certain fields of consequence to developing countries, such as agriculture, biotechnology and traditional knowledge, the legal regime is actively evolving, and the basis and extent of rights and obligations can be controversial. In these cases the possibility of dispensing with law, and deciding the dispute in equity, may be an attractive option.

FIGURE 1: PRINCIPAL STEPS IN A TYPICAL WIPO ARBITRATION

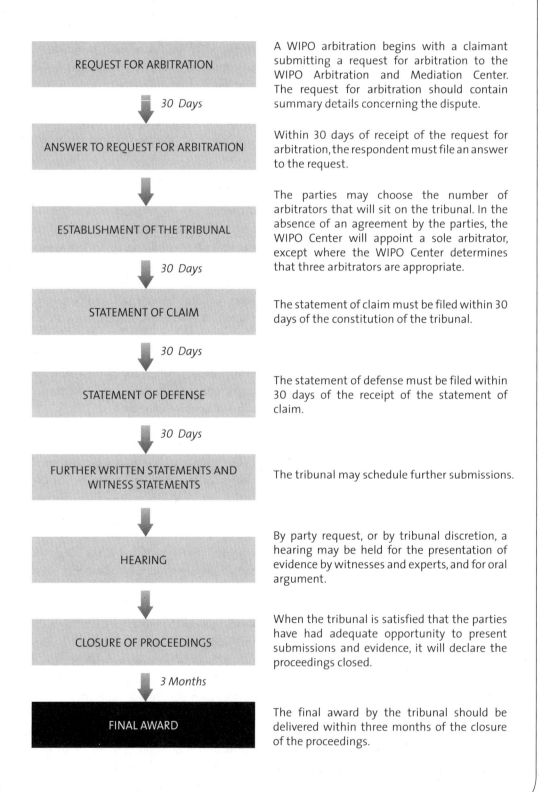

REQUEST FOR ARBITRATION

A WIPO arbitration begins with a claimant submitting a request for arbitration to the WIPO Arbitration and Mediation Center. The request for arbitration should contain summary details concerning the dispute.

↓ *30 Days*

ANSWER TO REQUEST FOR ARBITRATION

Within 30 days of receipt of the request for arbitration, the respondent must file an answer to the request.

↓

ESTABLISHMENT OF THE TRIBUNAL

The parties may choose the number of arbitrators that will sit on the tribunal. In the absence of an agreement by the parties, the WIPO Center will appoint a sole arbitrator, except where the WIPO Center determines that three arbitrators are appropriate.

↓ *30 Days*

STATEMENT OF CLAIM

The statement of claim must be filed within 30 days of the constitution of the tribunal.

↓ *30 Days*

STATEMENT OF DEFENSE

The statement of defense must be filed within 30 days of the receipt of the statement of claim.

↓ *30 Days*

FURTHER WRITTEN STATEMENTS AND WITNESS STATEMENTS

The tribunal may schedule further submissions.

↓

HEARING

By party request, or by tribunal discretion, a hearing may be held for the presentation of evidence by witnesses and experts, and for oral argument.

↓

CLOSURE OF PROCEEDINGS

When the tribunal is satisfied that the parties have had adequate opportunity to present submissions and evidence, it will declare the proceedings closed.

↓ *3 Months*

FINAL AWARD

The final award by the tribunal should be delivered within three months of the closure of the proceedings.

The law applicable to the arbitration procedure (*lex arbitri* or arbitral law) is the law that governs the procedural framework, such as whether a dispute is arbitrable, the availability of interim measures of protection, the conduct of the arbitration, and the enforcement of the award. The arbitral law need not be the same as the law applicable to the substance of the dispute. A tribunal may, for example, be subject to the arbitral law of Switzerland, but may be required, by party agreement, to apply Indian law to the substance of the dispute.

4. THE MEDIATION OPTION

Arbitration is not the only option to litigation. The parties can also opt for mediation, a non-binding, confidential procedure in which a neutral intermediary assists the parties in reaching a mutually satisfactory settlement of their dispute.[6]

4.1 *Mediation procedure*

The starting point of a mediation, like an arbitration, is the agreement of the parties to submit their existing or future disputes to mediation. Once a dispute arises and there is an agreement (either *ex ante* or *ex post*) to mediate, a party will initiate the process by informing the other party of the commencement of mediation. The mediation procedure is then largely determined by the parties, together with the mediator. Figure 2 describes the principal steps in a typical mediation.

4.2 *Role of the mediator*

Unlike a judge or an arbitrator, whose mandate is to issue a binding decision or award, a mediator does not have any power to impose a settlement on the parties. The role of a mediator is to serve as a catalyst for party negotiations. A mediator works to improve communication between the parties, helps parties clarify their understanding of their mutual interests and concerns, sheds light on the strengths and weaknesses of each party's legal position, explores consequences of not settling, and helps generate options for a mutually agreeable resolution of their dispute.

5. CHARACTERISTICS OF ARBITRATION AND MEDIATION

5.1 *Resolving multijurisdictional disputes*

With the creation and exploitation of international IP rights, disputes are increasingly multijurisdictional. Resolving transnational disputes through litigation requires the expense and complexity of pursuing parallel proceedings in a number of countries and confronting multiple rounds of appeals in each jurisdiction. Furthermore, despite broad harmonization of substantive IP laws, national prejudices and differences in approaches still remain. Therefore, in a multijurisdictional dispute, a win in one jurisdiction will not necessarily translate into a win in other jurisdictions. The risk of inconsistent results is significant.

Through arbitration or mediation, the parties can agree to resolve, in a single procedure, disputes involving intellectual property in a number of countries. For a deep-pocketed party that has an interest in broadly manifesting its strong IP enforcement policy, litigation may be a more appealing option. The threat of drawn-out court procedures in multiple jurisdictions may be an effective strategy to induce the other party with limited resources to accept a quick settlement. On the other hand, for a party seeking a timely, cost-efficient resolution of the immediate dispute, resolution through a single arbitration or mediation procedure may be more advantageous.

5.2 *A neutral dispute-resolution forum*

Litigation between parties of different nationalities means that the home party enjoys an advantage, since the other party bears the burden of a foreign and unfamiliar jurisdiction. In arbitration or mediation, parties may resolve a transnational dispute on neutral territory, so neither party is subjected to foreign court procedures, laws, customs, languages, and prejudices. In arbitration or mediation, parties may appoint an arbitrator or mediator of a neutral nationality and choose a neutral language and venue of procedure. In arbitration, parties may agree on neutral substantive and procedural law.

Figure 2: Principal Steps in a Typical Mediation

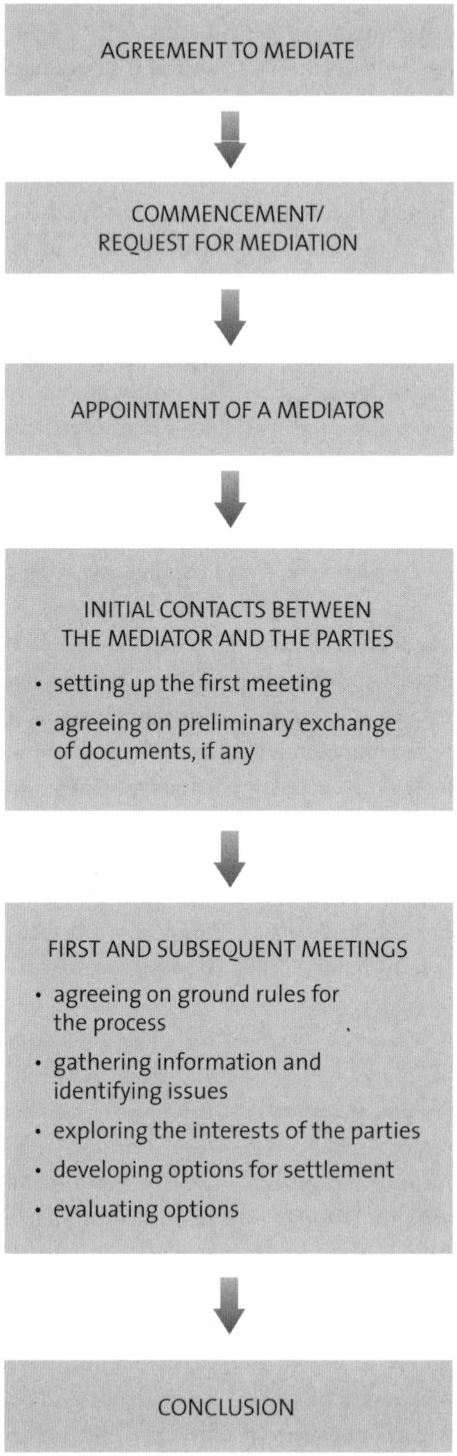
AGREEMENT TO MEDIATE
COMMENCEMENT/
REQUEST FOR MEDIATION
APPOINTMENT OF A MEDIATOR
INITIAL CONTACTS BETWEEN
THE MEDIATOR AND THE PARTIES
• setting up the first meeting
• agreeing on preliminary exchange of documents, if any
FIRST AND SUBSEQUENT MEETINGS
• agreeing on ground rules for the process
• gathering information and identifying issues
• exploring the interests of the parties
• developing options for settlement
• evaluating options
CONCLUSION

Recourse to arbitration or mediation in a convenient, neutral forum may be especially attractive when public entities are party to a dispute.[7] If a dispute is between a state entity and a private party, the private party will be disinclined to go to the court of the state entity, and the state party will not want to submit to the jurisdiction of the courts of another state. In such a case, a neutral procedure such as arbitration or mediation may be the only option acceptable to both parties. This feature may be particularly relevant in IP transactions involving entities in developing countries, where public institutions often largely own IP rights.

5.3 *Autonomy*

Arbitration and mediation are based on consent of the parties. It follows that arbitration or mediation proceedings require party autonomy and that parties largely retain control over the dispute-resolution process.

In principle, parties are free to agree on the procedure to be followed in the arbitral proceedings. Depending on their needs, parties can select streamlined or more extensive procedures and choose the applicable procedural and substantive law, place and language of the arbitral proceedings, and the arbitrator(s). Thus, the parties can adapt an arbitration procedure to fit the dispute.

Mediation offers parties control over not only the procedure to follow, but also the outcome of the process. Parties may fashion the mediation process to their specific needs. Commencement of the mediation is based on the parties' agreement to resolve the dispute through mediation, and continuation of the process depends on the parties' continued acceptance of the terms of the mediation. Unlike arbitration, a party that has submitted the dispute to mediation may withdraw at any time from the mediation. The outcome of a mediation also depends on the will of the parties. While the mediator will assist in the procedure, it is ultimately up to the parties to determine whether they will settle the dispute in accordance with their interests or seek resolution in a different forum, such as litigation or arbitration.

5.4 *Choosing relevant expertise*

Judges often have varying degrees of experience and qualification, and national courts are frequently ill equipped to deal with technically complex issues presented in IP disputes.

In arbitration, parties normally participate in selecting arbitrators and are, in principle, free to appoint arbitrators of their choice. Arbitrators may be chosen for their skill and expertise in a specific legal, technical, or business field. Arbitrators with relevant expertise will ensure proper understanding of facts and law and, therefore, contribute to a timely, cost efficient resolution of the dispute. When the dispute involves parties of different cultural and economic backgrounds, an arbitrator's knowledge of cultural or social sensitivities may also be helpful.

As in arbitration, parties select their mediators. A mediator's role, however, is fundamentally different from that of a judge or an arbitrator. The mediator's role is not to render a decision but to facilitate the process through which parties endeavor to settle their dispute. The mediator may inject a degree of detachment and objectivity into the dispute. The role of the mediator as an intermediary may be especially crucial when the share of information and bargaining power between the parties is unequal. An effective mediator will address these concerns.[8] A mediator also will help parties rebuild trust to increase the chances for settlement.[9]

The success of an arbitration or a mediation depends largely on the quality of the arbitrator(s) and mediator(s), and the challenge is often to find candidates that have both arbitration or mediation skills and experience with the specialized knowledge of the disputed subject matter.

5.5 *Confidentiality*

Parties to arbitration or mediation can keep the proceedings and any results confidential. In doing so, parties can focus on the merits of their dispute and avoid distraction from external factors, such as unwanted negative press coverage. Confidentiality may be especially important where the terms of the parties' relationship are undisclosed to the public, as in most licensing agreements, and where commercial reputation and trade secrets are at stake. Particularly in mediation, the private nature of the procedure allows parties to engage in frank, exploratory settlement negotiations and not be intimidated by formal legal procedures.[10]

On the other hand, if one of the parties wishes to establish a public precedent to dissuade other parties from engaging in similar conduct, the confidential nature of arbitration and mediation may make these options less desirable. In certain cases, it may be more effective to take the case to the public and seek the support of public organizations or nongovernmental organizations. A degree of publicity may at times assist in negotiating a settlement.[11] For disputes involving issues of broad public concern, which is often the case in health and agriculture, it may be inappropriate to keep the existence of the dispute, and its outcome, confidential. When appropriate, parties may agree to employ mediation or arbitration to resolve the dispute and consent explicitly to make the process and result public.

5.6 *Preserving relationships*

As multiparty, complex IP relationships become more common, partnerships between actors in government, academia, and industry in developing and developed countries occur regularly and, frequently, expand beyond a single short-term transaction. The multiparty nature of such relationships exacerbates the complexity of dispute resolution. When disputes arise out of these relationships, a party's desire to resolve the immediate dispute should not eclipse safeguarding the relationship.

The adversarial nature of litigation often fosters hostility and resentment between the parties, rendering the dispute intractable and potentially destroying a working relationship. On the other hand, the consensual nature of mediation, and to a certain extent arbitration, accommodates a long-term approach. Parties can resolve the dispute at hand and still maintain a working relationship. In this way, antagonism between parties can be mitigated and mutual understanding fostered. This feature of mediation and arbitration may be particularly relevant for entities in developing countries that rely on alliances with foreign enterprises. Developing countries are still dependent on foreign sources for technology, and so there is a marked need to maintain these relationships. Also, a large proportion of innovation occurs in university or government laboratories, after which rights are exploited in collaboration with foreign companies. Foreign IP rights holders will demand a particular level of protection; entities in developing countries, especially those in the public sector, may need to accommodate these demands with national development goals or other vested interests.[12]

5.7 *Arbitration's finality*

The protracted nature of litigation, which pushes parties into multiple rounds of appeals, is a common problem when litigating transnational disputes. In addition, it is difficult to enforce any court judgment outside the court's jurisdiction. The end result of arbitration is, on the contrary, a final, binding award. Normally appeals are not allowed, and awards are directly enforceable by national courts under the Convention on the Recognition and Enforcement of Foreign Arbitration Awards (New York Convention).[13] This convention, currently ratified by 139 countries, greatly facilitates the enforcement of awards across borders by providing for recognition of awards on a par with domestic court judgments, without review on the merits. The convention only permits awards to be set aside in very limited circumstances.

5.8 *Mediation's nonbinding, interest-based procedure*

In litigation or arbitration, the outcome of a case is determined by the facts of the dispute and the applicable law. Mediation, on the other hand, involves more than the exercise of rights and obligations set within legal parameters. It is often a coordinated exercise of legal rights, with consideration given to other economic and social variables.[14] With mediation, the dispute resolution options are broadened, allowing the parties, with the help of the mediator, to craft innovative, common-sense solutions that amicably settle the dispute. Parties may find a solution to their dispute by considering their business or social interests. They may also reach package deals that include nonmonetary benefits, such as technology transfer agreements, training programs, or infrastructure development.

In certain circumstances, mediation may be the only option available for resolving the dispute. Parties in a dispute may each have a claim that is valid and enforceable and, yet, impossible to fulfill.[15] The dispute may involve a subject matter where there is no established legal framework, or where there are certain interests that may not be adequately addressed by traditional legal means.[16] In such cases, the only strategy to break the impasse may be a cooperative solution, such as mediation.

The nonbinding nature of mediation means that a decision cannot be imposed on the parties and that all involved must voluntarily agree to accept the settlement. Any settlement may be recorded in a contract; if either party does not perform the contract, actions for breach of contract may be brought. Of course, if the outcome of a mediation represents the interests of the parties, the outcome is more likely to endure as a long-term solution to the conflict.

5.9 *Mediation—minimal risk*

Even when the parties have agreed to submit a dispute to mediation, if a party feels that it is not making any progress, that the procedure is becoming too costly, or that the other party is not acting in good faith, the party may withdraw from the mediation process at any time and seek to resolve the dispute through litigation or arbitration. Accordingly, mediation involves low risk. Should mediation not produce a settlement, the procedure might still assist the parties by defining the facts and issues of the dispute, thus preparing parties for subsequent arbitration or court proceedings.

5.10 *Comparing options at a glance*

Table 1 provides an overview of the different strengths and weaknesses of litigation, arbitration, and mediation.

6. PRACTICAL CONSIDERATIONS

Since arbitration and mediation are private proceedings, the support of lawyers and experts skilled in the process is essential. Institutions in developing countries will want to exercise care in retaining appropriate counsel when exploring arbitration and mediation options.

6.1 *Controlling costs*

The validity of a claim may be irrelevant if the concerned parties are unable to afford the appropriate dispute-resolution procedure. Institutions will need to confront any financial constraints that might complicate the choice of a dispute-resolution strategy.

Arbitration and mediation are essentially private processes, and a number of advantages, including party autonomy, confidentiality, neutrality, and expertise, stem from the private nature of the proceedings. This private nature, however, also means that parties are obliged to bear the costs. The parties involved in a dispute do not pay judges in national courts, but they do pay arbitrators and mediators.

In an arbitration, parties must cover legal fees, plus the additional fees and expenses of arbitrators. If an institution administers the arbitration, administrative fees must also be paid. Thus, arbitration may not necessarily be less costly than litigation. However, parties can consciously try to limit costs by expediting the procedure and by selecting cost-efficient venues for meetings and hearings. Parties can also endeavor to appoint an arbitrator that is sensitive to the financial constraints of parties, and choose an arbitral institution that charges reasonable administrative fees. Furthermore, while arbitration may be costly, the finality and enforceability of arbitral awards may make arbitration less costly than litigation, which often involves multiple appeals and requires a judgment to be enforced in a foreign jurisdiction.

In mediation, costs are more easily contained. Mediation costs include the legal fees of each party, the mediator's fees, and administrative fees (if an administering institution is present). Parties can monitor the costs and progress of the mediation to determine whether to continue it. While the cost of mediation is generally shared equally between the parties, parties may agree to change this allocation of costs depending on the economic power of each party.

Table 1: Litigation, Arbitration, and Mediation Compared

Common features of many IP disputes	Litigation	Arbitration	Mediation
International	solution limited to court's jurisdiction multiple proceedings under different laws, with risk of conflicting results possibility of actual or perceived advantage to party that litigates in its own country	global solution a single proceeding under the law determined by parties arbitral procedure and nationality of arbitrator can be neutral to law, language, and institutional culture of parties	global solution single proceeding mediation procedure and nationality of mediator can be neutral to law, language, and institutional culture of parties
Technical	decision maker might not have relevant expertise	parties can select arbitrator(s) with relevant expertise	parties can select mediator(s) with relevant expertise
Urgency	procedures often drawn out injunctive relief available in certain jurisdictions	arbitrator(s) and parties can shorten the procedure arbitration may provide provisional measures and does not preclude seeking court-ordered injunction	mediator(s) and parties can shorten the procedure while provisional measures are not available in mediation, parties not precluded from seeking court-ordered injunction
Legal framework	court generally applies only its national laws	applicable law may be determined by parties; absent party agreement, arbitrator(s) will select the law(s) that it determines appropriate to the dispute multiple national laws may concurrently apply tribunal may decide in equity (rather than specific law)	procedure less governed by law and more by the social and economic interests of parties

(Continued on Next Page)

TABLE 1 (CONTINUED)

COMMON FEATURES OF MANY IP DISPUTES	LITIGATION	ARBITRATION	MEDIATION
Finality	appeal possible	limited appeal option	any settlement agreement is binding between parties as a matter of contract law
Confidential/ trade secrets and risk to reputation	public proceedings	proceedings, disclosures, and awards confidential	proceedings, disclosures, and outcomes confidential
Continuing relationship	parties may or may not be in a continuing relationship dispute may be resolved without adverse party's active participation adversarial nature of litigation may further antagonize parties	parties often in a continuing relationship	parties often in a continuing relationship mediation shields the relationship by fostering an amicable resolution of dispute

Whether in arbitration or mediation, parties should bear in mind that the procedure is largely under their control and costs will vary depending on the choices made throughout the procedure.

6.2 *Ad hoc or institutional procedure?*

Arbitration and mediation may take place ad hoc or under the aegis of an institution. In an ad hoc procedure, the parties, with the arbitrator or mediator, administer the proceedings themselves. This requires sufficient cooperation among the parties and the arbitrator or mediator, as well as considerable experience in arbitration/mediation procedures. In an institutional arbitration or mediation, the institution provides a procedural and administrative framework for initiating and conducting the procedure, and oversees the integrity and independence of the process. Especially where parties are inexperienced in dispute resolution, they should consider opting for an institutional procedure. Administrative fees vary greatly by institution and will be a factor in selecting one. However, the cost of using a moderately priced institution will guarantee considerable benefits, including administrative and technical assistance, availability of a tested set of procedural rules, and access to qualified arbitrators and mediators.

Governments and public institutions can help make arbitration or mediation procedures accessible and available by identifying and supporting neutral institutions that can provide cost-efficient, timely dispute-resolution services, and by catering to the needs of local enterprises, government agencies, and foreign entities. The Arbitration and Mediation Center of the World Intellectual Property Organization[17] (the WIPO Center) is worth keeping in mind. Established in 1994 to promote the timely, cost-effective resolution of IP disputes through alternative dispute resolution, the WIPO Center has created, with the active involvement of many ADR and IP practitioners, the WIPO mediation, arbitration, and expedited arbitration rules and clauses. Together with its extensive network of IP and ADR experts, the WIPO Center ensures that WIPO procedures are at the cutting edge of IP dispute-resolution techniques and that these procedures meet the needs of parties of different economic and social backgrounds.

6.3 *Drafting clauses*

Arbitration and mediation are premised on party agreement; it is uncommon that these procedures are adopted after a dispute arises, when animosity between parties generally overshadows their interest in resolving the dispute. Therefore, arbitration and mediation clauses often refer to potential disputes under a particular contract, including those conflicts that might emerge regarding patents, know-how and software licenses, franchises, trademark coexistence agreements, distribution contracts, joint ventures, R&D contracts, technology-sensitive employment contracts, and mergers and acquisitions with important IP aspects. These clauses generally determine a number of the procedure's essential elements, such as its specific type, language, number of arbitrators or mediators, and the applicable law. Arbitration and mediation institutions generally make available model clauses. Adopting these clauses will help to avoid any uncertainty that might unnecessarily burden the arbitration or mediation proceeding. Parties may introduce certain cost-saving models in appropriate circumstances.[18]

Dispute-resolution clauses can provide for a multitiered process, namely, by mandating mediation followed, in the absence of settlement, by arbitration. Even mediation may be preceded by direct party negotiation, which may be particularly relevant in disputes in public settings. When opting for a multitiered process, it is useful to stipulate time periods for each procedure in order to prevent protracted discussions and delays between the procedures.[19]

Public sentiment may not always support the development of and participation in ADR procedures. Public ADR pledges may be useful to handle this. Furthermore, legislative authorities may consider adopting procedural laws referring to or integrating ADR methods.

7. CONCLUSION

Entities in developing countries face a number of challenges, when a dispute arises, with entities in developed countries. The entities in developed countries will often have greater financial power and technical expertise with which to pursue a favorable dispute resolution. Since technology transfer is tied closely with economic development, disputes may trigger public reaction. Moreover, language and cultural barriers can be obstacles to effective communication, and questions may arise about how rights asserted by developing countries may be accommodated by the existing IP regime.

Having a dispute-resolution policy can help to address these concerns. It can also provide strategic benefits and minimize the risk of disputes escalating. The dispute-resolution strategies should therefore be crafted with regard to the specific circumstances of the dispute and the background of the parties. Ideally, a procedure that assists in mitigating economic inequalities between parties should be identified and implemented. Technical, commercial, legal, and social interests may need to be considered. In certain cases the result will be compromise; in other cases, robust enforcement will be sought.

Litigation, arbitration, and mediation operate within very different paradigms. To adopt the most appropriate dispute-resolution strategy for a potential or existing dispute, parties should understand the differences between the procedures and determine which is most appropriate to the circumstances of the conflict. Remember, litigation is not the only option. Arbitration or mediation may offer a sustainable solution that will satisfy all the parties involved. ■

ACKNOWLEDGMENTS
While much of this chapter provides information which is also subject of the WIPO Center's provision of resources, the views expressed herein are not necessarily those of WIPO or any of its Member States.

EUN-JOO MIN, *Senior Legal Officer, Arbitration and Mediation Center, World Intellectual Property Organization, 34 Chemin des Colombettes, 1211 Geneva 20, Switzerland. eunjoo.min@wipo.int*

---

1 O'Connor SN. 2005. Intellectual Property Rights and Stem Cell Research: Who Owns the Medical Breakthroughs? *New England Law Review* 39: 665.

2 Information in this section is largely extracted from the *Guide to WIPO Arbitration*. WIPO Publication no. 919. www.wipo.int/freepublications/en/arbitration/919/wipo_pub_919.pdf.

3 Redfern A and M Hunter. 2003. *Law and Practice of International Commercial Arbitration*. Sweet & Maxwell: London. pp. 135.

4 See, also in this *Handbook*, chapter 3.6 by A Taubman.

5 Parties who place a premium on time and cost effectiveness can opt for the procedural framework established by the WIPO Expedited Arbitration Rules, which condenses the principal stages of an arbitration under the WIPO Arbitration Rules.

6 Information in this section is largely extracted from the *Guide to WIPO Mediation*. WIPO Publication no. 449(E). www.wipo.int/freepublications/en/arbitration/449/wipo_pub_449.pdf.

7 See *supra* note 3, p. 42.

8 Williams BA. 2000. Consensual Approaches To Resolving Public Policy Disputes. *Journal of Dispute Resolution* 8 (2): 144.

9 Ibid..

10 Crowne CH. 2001. The Alternative Dispute Resolution Act of 1998: Implementing a New Paradigm of Justice. *New York University Law Review* 76 (6): 1768.

11 Boettiger S and A Bennett. 2006. The Bayh-Dole Act: Implications for Developing Countries. *IDEA: The Intellectual Property Law Review* 46(2), cites the example of the inventor of Golden Rice™ recounting that "*publicity sometimes can be helpful: Only a few days after the cover story about golden rice had appeared in* Time, *I had a phone call from Monsanto offering free licenses for the company's IP rights involved.*"

12 Reichman JH and D Lange. 1998. Bargaining around the TRIPS Agreement: the Case for Ongoing Public-Private Initiatives to Facilitate Worldwide Intellectual Property Transactions. *Duke Journal of Comparative and International Law* 9(11): 50-54.

13 See the full text of the New York Convention, with the list of its contracting states, at www.wipo.int/amc/en/arbitration/ny-convention/.

14 See *supra* note 12.

15 See *supra* note 1, p. 669. Herein is cited the example of patentable inventions being assigned to two or three different funding sources, each assignment being legally binding, yet impossible to fulfill.

16 See *supra* note 8, p. 139.

17 See *supra* note 4.

18 See the WIPO Center's recommended clauses at www.wipo.int/amc/en/mediation/contract-clauses/.

19 See, for example, the WIPO-recommended clause for submission to mediation followed, in the absence of a settlement, by arbitration at www.wipo.int/amc/en/mediation/contract-clauses/clauses.html.

CHAPTER 15.4

Parallel Trade: A User's Guide

DUNCAN MATTHEWS, *Senior Lecturer in IP Law, Queen Mary University of London, U.K.*
VIVIANA MUNOZ-TELLEZ, *Programme Officer, Innovation and Access to Knowledge Programme, South Centre, Switzerland*

ABSTRACT

This chapter provides guidance about parallel trade to developing country policy-makers and other stakeholders in intellectual property. What is parallel trade? And how can it be utilized to promote access to medicines and support poor farmers in developing countries? Engaging in parallel trade is an option provided by the Agreement on Trade-Related Aspects of Intellectual Property Rights (TRIPS) under the World Trade Organization. Furthermore, the 2001 Doha Declaration on TRIPS and Public Health confirmed that developing countries could use parallel imports to support public health. As a result, developing countries can ensure access to lower-priced patented and/or branded products, such as medicines and basic agricultural inputs, by incorporating legislation to allow for parallel imports. When implementing measures to facilitate parallel trade, developing countries can establish and maintain an effective system by adequately regulating the quality, safety, and health of parallel imports. At the same time, developing countries need to prevent low-priced patented products available in their countries from entering high-priced developed country markets.

1. WHAT IS PARALLEL TRADE AND WHY DOES IT HAPPEN?

Parallel trade occurs when products produced under the protection of a patent, trademark, or copyright in one market are subsequently exported to a second market and sold there without the authorization of the local owner of the intellectual property (IP) right. Often, the local owner of the IP right will also be a local dealer who, through a license or other exclusive agreement, has been authorized by the patent, copyright, or trademark holder to market the protected product. Naturally, when the licensed dealer has an exclusive agreement, he or she expects to be the only party supplying the product in the local market.

Parallel trade *does not* refer to unofficial, illegal, or informal-sector activities that may take place inside a country or among countries. Moreover, parallel trade *is not* trade in pirated or counterfeit products. The latter are unauthorized versions of products that infringe an IP right. Parallel imports (also called *gray-market* imports) are genuine, often branded, products that do not violate an IP right. Importing the products from one country to another, however, may not be authorized by the right holder.

The main difference between parallel importation and "official" importation is that the parallel imports probably were produced originally for sale in a particular market and then were passed through an unauthorized dealer before reaching the consumer. Parallel imports may differ in superficial ways from those made available by the local dealer—they may be packaged differently or lack the original manufacturer's warranty—but otherwise they will be identical to the official import being marketed locally.[1]

When parallel importation occurs, the practical effect is that a patented and/or branded

Matthews D and V Munoz-Tellez. 2007. Parallel Trade: A User's Guide. In *Intellectual Property Management in Health and Agricultural Innovation: A Handbook of Best Practices* (eds. A Krattiger, RT Mahoney, L Nelsen, et al.). MIHR: Oxford, U.K., and PIPRA: Davis, U.S.A. Available online at www.ipHandbook.org.

product becomes available locally from multiple sources. Parallel importing allows dealers to bypass official or authorized local suppliers or licensees and obtain products directly from overseas suppliers. The enhanced market competition between sources of the same products tends to drive prices down.

Indeed, the incentive for parallel importation is the fact that there are price differences between identical products in different markets. Parallel importing usually occurs when the price differences are high, because then the potential gains (price savings, product availability, profit) for most stakeholders are large enough to compensate for the transaction costs, including shipping costs and complying with customs regulations. The price differences can be due to a variety of factors. In the case of the pharmaceutical market, where important price differentials exist between countries, price differences can result from government-enforced price controls, pricing manipulated by the owner of an IP right holder, fluctuations in currency values, a combination of these conditions, and other factors.

2. THE EFFECTS OF PARALLEL TRADE ON STAKEHOLDERS

2.1 *Government-supported parallel trade*

The regulation of parallel trade involves balancing the interests of producers and consumers. An important public policy mechanism for developing countries, parallel importation can be used to protect the interests of consumers, particularly with regard to pharmaceutical and agrichemical products. Countries can introduce legal provisions to permit parallel importing in order to ensure adequate access to imports. Parallel importing also allows the government to shop around in different markets for the lowest price on an IP protected product.

The prospect of parallel imports of products protected by IP rights is particularly important in the public health sector, where prices for medicines in developing countries may be higher than most people can afford. By utilizing parallel imports, developing countries can access alternative sources of medicines at lower prices, guaranteeing greater access and availability of medicines. Hospitals, pharmacies, and health insurance companies can acquire pharmaceutical products at lower prices from other markets through parallel trade, which can potentially lower prices in the local market.

Parallel imports can also be used to access basic inputs to agricultural production (such as pesticides and fertilizers) at lower prices than those charged locally by the owner of an IP right. These reduced costs could contribute to improving poor farmers' incomes and livelihoods. Developing countries can also use parallel importing to curb anticompetitive practices: it allows them to ensure adequate price competition in the local market and a competitive supply of products from a variety of sources. Section 3.0 of this chapter provides more information about how developing countries can make effective use of parallel importing.

2.2 *Benefits to consumers*

Potentially, consumers have much to gain from parallel imports. By increasing the options for alternative supplies of products, parallel imports can allow consumers to gain access to the products they need from another market at lower prices than are being charged in their own market. In developing countries, it is often the case that essential products such as medicines are unavailable or inaccessible to a large portion of the population because they are unable to afford them at the prices charged by the IP right holder, and the government is unable to subsidize their purchase.

2.3 *Retailers, wholesalers, and traders*

Parallel imports can be attractive to traders when price differences are significant enough to ensure profits. Similarly, parallel importing gives local retailers and wholesalers the ability to obtain patented and/or branded products directly from multiple overseas sources. Doing so may offer better prices than obtaining the products from the local authorized dealer. By bypassing the local licensed dealer, retailers may be better able to meet the needs of their consumers.

2.4 *The view of right holders and local licensed owners*

IP right holders, including authorized importers, licensees, and other agents, generally support restricting parallel trade because they directly benefit from having an exclusive right to import protected products. In the absence of parallel importing, local licensed dealers do not face competition, in terms of price, for the same products. In markets where no alternative sources are available, the product can be sold at the highest price the local market can tolerate. Moreover, restrictions on parallel importing allow right holders to take advantage, on a regional or international scale, of market segmentation and differential pricing strategies. Where parallel imports are not permitted, right holders may charge different prices in different markets. Right holders can also control distribution, pricing, and other aspects of the local market for products produced under IP rights.

Right holders often argue that parallel importation should be restricted because driving down prices might reduce incentive to invest in research and development in the pharmaceutical and agrichemical sectors. Parallel importation may also reduce the incentive for right holders to donate products at low cost or free of charge to developing countries, since there would be a risk that those products would be diverted back into developed country markets and sold at higher prices than were intended. Parallel importation may also hinder the ability of governments in different countries to maintain price controls on pharmaceutical products within their territory. Furthermore, rights holders or licensed local owners may pay marketing costs that the suppliers of parallel traded goods benefit from for free. In the long term, there is the possibility that this will reduce the willingness of rights holders or licensed local owners to supply particular markets.

In developing countries where some type of parallel importation is permitted, local licensed dealers may seek to overcome the competition of parallel traders by offering after-sale service, warranties, and so forth that parallel traders, generally with small profit margins, may be unable to offer. When price differences between markets tend to be large, as in the case of medicines, IP right holders can apply differential pricing policies, charging lower prices for medicines in lower-income markets than in higher-income markets. Price differentiation to ensure lower prices for patented medicines in developing countries may reduce the incentive there for parallel imports. If parallel imports are properly regulated in both exporting and importing countries, however, differential pricing agreements still can function without displacing IP right holders and local licensed dealers.

2.5 *Reimportation and other problems*

Developed countries with parallel trade in products protected by IP rights frequently identify a potential problem: IP right holders, particularly in the pharmaceutical industry, could be discouraged from pricing their products differently in different markets to benefit developing countries. Prices for medicines protected by patents or trademarks in developing countries tend to be high. Some argue that if developing countries allow parallel importation, patented medicines that the industry could potentially sell for a low price in a low-income country may find their way back to high-income markets and sold at higher prices. Reimporting medicines protected by patents or trademarks would mainly benefit intermediaries and reduce the incentive for industry to sell medicines protected by patents or trademarks at lower prices in developing countries. Furthermore, developed countries are concerned that parallel trade could channel counterfeit and/or pirated products into the market.

As noted above, however, parallel trade does not concern substandard products. Moreover, countries can and should address these concerns by adequately regulating and monitoring parallel imports and exports. To reduce the risk of reimportation and to maintain effective pro-poor (or humanitarian) differential pricing arrangements for medicines in developing countries, developed countries can adopt measures to prevent parallel imports into higher-priced markets.[2] For example, developed countries can (and do) enact national legal provisions to ban parallel imports from developing countries.

3. THE LEGAL FRAMEWORK FOR PARALLEL TRADE

The legal question with regard to parallel trade is: To what extent should countries allow or limit the ability of IP right holders within particular national/regional territories to control the movement of products across different markets on the basis of local ownership of IP rights? Countries are entitled to regulate parallel trade involving intellectual property in their own best interests. Indeed, parallel imports have been admitted in many developed and developing countries on a regional or international scale.[3]

The Agreement on Trade-Related Aspects of Intellectual Property Rights (the TRIPS Agreement) gives World Trade Organization (WTO) members the freedom to design their own regimes for the *exhaustion* of IP rights (exhaustion occurs when a right holder's control over a product ceases). Because the exhaustion of rights cannot be challenged as a violation of the TRIPS Agreement under the WTO dispute-settlement mechanism, the TRIPS Agreement allows parallel importation. According to Article 6:

For the purposes of dispute settlement under this Agreement, subject to the provisions of Articles 3 and 4 nothing in this Agreement shall be used to address the issue of the exhaustion of intellectual property rights.

Moreover, the Doha Declaration on TRIPS and Public Health[4] reaffirmed this freedom, giving developing countries greater certainty about their ability to use parallel importation to protect their interests, particularly for safeguarding public health. According to Article 5(d) of the Doha Declaration:

The effect of the provisions in the TRIPS Agreement that are relevant to the exhaustion of intellectual property rights is to leave each member free to establish its own regime for such exhaustion without challenge, subject to the MFN and national treatment provisions of Articles 3 and 4.

A country's decision about their exhaustion of rights doctrine will either restrict or allow parallel importation policies in their territories.

The doctrine describes three types of exhaustion of rights:

- **national exhaustion** (first sale doctrine). Also known as first sale doctrine, national exhaustion holds that the exclusive rights of IP right holders over protected products cease after the first sale of the product within national borders.
  Implication: Right holders can block parallel imports from entering the local market, even though their rights are exhausted in that market. Example: United States.
- **regional exhaustion**. The exclusive rights of IP right holders over protected products cease after the first sale in the regional market.
  Implication: Parallel trade is allowed within the group of countries, but right holders can ban parallel imports from countries outside the region. Example: European Union.
- **international exhaustion**. Right holders' exclusive rights over protected products cease after the first sale in any market.
  Implication: Right holders cannot exclude parallel imports from entering the local market because their rights with respect to that market are exhausted. Example: Kenya.

Accordingly, developing countries can incorporate into their national laws the principle of international exhaustion of rights, thus allowing for parallel imports on an international scale.[5] Put differently, developing countries can decide whether or not to allow parallel importation for all or particular IP rights. Allowing for parallel imports of patented or trademark protected products, that is, the application of the international exhaustion principle to the rights of patent holders, is an option made available by TRIPS to developing countries. Though relevant to all fields, the potential benefits of parallel importing are particularly important for patents and public health. As noted above, importing patented medicines from a market where they are sold at lower prices may give those who need them in the importing country greater access. Concerns about the possible negative effects of parallel imports, moreover, can be dealt with through adequate monitoring and regulation, rather than through trade restrictions.

4. MODEL PROVISIONS FOR ENABLING PARALLEL IMPORTATION OF PATENTED PRODUCTS

This section provides TRIPS-compliant model provisions that would enable parallel importation of patented products into a country when incorporated into a national patent law. The model provisions adopt the principle of international exhaustion (see Box 1).

Model provision 1 is the narrowest interpretation of the international exhaustion principle, allowing only for parallel importation of patented products that have been placed on the market by the patent holder. Model provision 2 extends the exception by allowing for parallel importation of patented products that have been placed on the market by any authorized agent (that is, a local licensed dealer) of the patent holder. Finally, model provision 3 provides the broadest exception to the exclusive rights of a patent holder allowing parallel imports originating from any country. Under this provision patent holders' rights may also be exhausted based on the sale or marketing of the product authorized by a government under a compulsory license. Hence, patented products that have been produced and placed on the market by a compulsory licensee may be parallel imported.[7]

While each of the three provisions have been adopted, it is questionable whether provision 3 is TRIPS compliant.[8]

5. CONCLUSIONS AND RECOMMENDATION

Policy makers in developing countries should seek to utilize fully the options available under the TRIPS Agreement for promoting access to medicines and supporting poor farmers. Since these options include applying the principle of international exhaustion, policy-makers in developing countries should seek to take full advantage of the possibilities afforded by parallel trade. They can ensure that a patent holder does not have the right to prevent imports of a product covered by a patent when the patent holder has put that product on the market in another country. To utilize this flexibility to the fullest, countries should consider adopting a version of the model provisions for enabling parallel importation. ■

DUNCAN MATTHEWS, *Senior Lecturer in Intellectual Property Law, Queen Mary Intellectual Property Research Institute, Centre for Commercial Law Studies, Queen Mary University of London, John Vane Science Centre, Charterhouse Square, London EC1M 6BQ, U.K.* d.n.matthews@qmul.ac.uk

VIVIANA MUNOZ-TELLEZ, *Programme Officer, Innovation and Access to Knowledge Programme, South Centre, CP 228, 1211 Geneva 19, Switzerland.* munoz@southcentre.org

Box 1: Model provisions[6]

1. A patent holder shall not have the right to prevent acts of importation of a product covered by a patent that has been put on the market in any country by the patent holder or with his or her consent.
2. A patent holder shall not have the right to prevent acts of importation of a product covered by a patent that has been put on the market in any country by the patent holder, with his or her consent or in any other legitimate manner.
3. A patent holder shall not have the right to prevent acts of importation of a product covered by a patent that has been put on the market in any country by the patent holder or by an authorized party.

1 Maskus KE. 2001. Parallel Imports in Pharmaceuticals: Implications for Competition and Prices in Developing Countries. *Final Report to the World Intellectual Property Organization*. WHO: Geneva, p. 2. www.wipo.int/about-ip/en/studies/pdf/ssa_maskus_pi.pdf.

2 UNCTAD-ICTSD. 2005. *Resource Book on TRIPS and Development*. Cambridge University Press: Cambridge. p. 446. www.iprsonline.org/unctadictsd/ResourceBookIndex.htm.

3 Correa C. 2000. *Integrating Public Health Concerns into Patent Legislation in Developing Countries*. South Centre: Geneva, p. 74. www.southcentre.org/publications/publichealth/publichealth.pdf.

4 www.worldtradelaw.net/doha/tripshealth.pdf.

5 Sisule M and C Oh. 2005. The Use of Flexibilities in TRIPS by Developing Countries: Can They Promote Access to Medicines? Study Paper 2a, Commission on Intellectual Property Rights, Innovation and Public Health (CIPIH). World Health Organization: Geneva. p. 31. www.southcentre.org/publications/flexibilities/FlexibilitiesStudy.pdf.

6 See *supra* note 3.

7 Third World Network. 2003. Parallel Import. In *Manual on Good Practices in Public Health-Sensitive Policy Measures and Patent Laws* (chapter 4). Third World Network: Penang, Malaysia.

8 Abbott F. 2002. WTO TRIPS Agreement and Its Implications for Access to Medicines in Developing Countries. Study Paper 2a, UK Commission on Intellectual Property Rights (CIPIR): London. www.iprcommission.org/graphic/documents/study_papers.htm.

SECTION 16

Bioprospecting, Traditional Knowledge, and Benefit Sharing

Biotechnology Patents and Indigenous Peoples

DENNIS S. KARJALA, *Jack E. Brown Professor of Law, Sandra Day O'Connor College of Law, Arizona State University, U.S.A.*

ABSTRACT

How do biotech patent systems affect indigenous peoples, particularly in relation to health products? This question raises two distinct issues. First, the question of biopiracy—to what extent do patent systems necessarily exploit traditional indigenous knowledge to produce valuable medicinal products? Second, the question of patenting gene-sequence and gene-product information taken from living organisms, especially human beings—how can we justify patenting naturally occurring substances? And how should we negotiate the myriad ethical issues that arise from doing so? This chapter argues that the core of the biopiracy problem is not the availability of patents based on traditional indigenous information but rather the unfair acquisition of knowledge and the inequitable sharing of profits derived from developing such information into a valuable product. Solving this problem requires ensuring that traditional information is fairly acquired and that fair compensation is paid to the group from which the information derives. In regards to patenting gene-sequence and gene-product information, this chapter concludes that such issues equally affect indigenous and nonindigenous populations and that the best way to address them is by making policy changes.

1. INTRODUCTION

Much has been written on the general subject of how modern systems of intellectual property do, can, and should affect the lives and welfare of indigenous peoples.[1] When the focus is on biotechnology, however, copyright does not play much of a role in protecting functional inventions,[2] and while trade secret is important, no biotechnology issues specific to the interests of indigenous peoples are apparent.[3] This paper therefore tries to bring to light some of the issues involving patent rights in biotechnology that have become the legitimate concerns of indigenous peoples.

Two issues, in particular, dominate the literature about biotech patents in the context of globalization and indigenous peoples' rights. The first is the use of traditional indigenous knowledge as a starting point for producing a valuable product, such as a medicine. The second is the patentability of gene-sequence and gene-product information taken from living organisms, especially human beings. While the two are perhaps related (when, for example, the genetic information is taken from an indigenous group), it may be helpful to attempt at least a conceptual separation between the two issues in order to clarify the analysis. The first issue raises questions of so-called *biopiracy* of indigenous information by developed countries. As such, the issue directly implicates the rights of indigenous peoples, even though, as discussed below, most problems can be resolved when a few basic principles of patent law are brought to the fore. The second issue, especially when information concerning the human genome is involved, necessitates important ethical inquiries and poses fundamental questions for patent law and patent policy. Most of these problems, however, are

Karjala DS. 2007. Biotechnology Patents and Indigenous Peoples. In *Intellectual Property Management in Health and Agricultural Innovation: A Handbook of Best Practices* (eds. A Krattiger, RT Mahoney, L Nelsen, et al.). MIHR: Oxford, U.K., and PIPRA: Davis, U.S.A. Available online at www.ipHandbook.org.

not specific to biotech patents as they impact indigenous peoples, and indeed many of them impact everybody, whether they live in a developing or a developed country. Parts 2 and 3 of this chapter develop these arguments.

Having set aside patents as an important cause of biopiracy and having shown that gene and gene-product patents do not pose indigenous-peoples-specific problems, Part 4 attempts to outline the real problems that the world patent system poses for developing countries. Part 4 concludes that, while it is difficult to make the case that adopting a modern patent system directly benefits developing countries, the worldwide patent system also has little direct adverse effect. The problem is not so much that the existence of patents prevents the diffusion of biotechnological advances in developing countries but that there is a danger of *leakage* through the parallel importation of patented products from developing countries back to developed countries with strong patent systems. Too much leakage can impair incentives for innovation even within the developed world, and that is not good for anybody.

This last conclusion rests upon a basic assumption that underlies the entire paper. It remains a matter of serious debate whether and to what degree patent law in general serves as an incentive to innovate or commercialize innovations. Is patent law too strong or too weak? Is the period of patent protection too long or too short? We do not know very much about how the incentives of our IP systems, especially patent and copyright, work in practice.[4] This paper does not aim to undertake a fundamental analysis of the patent system generally. It therefore assumes that the patent system in developed countries, somehow or another, generally achieves its basic goal of stimulating innovation by providing a period of exclusive rights to those whose intellectual creations qualify for patents.[5]

2. BIOPIRACY AND PATENTS

2.1 *The basic problem*

The biopiracy problem is exemplified by the taking of indigenous peoples' information about the medicinal effects of a plant or other natural substance and the developing of that substance into a patented and popular drug by a large pharmaceutical company.[6] The fundamental question is whether or to what degree it is fair for outsiders to use, and especially to profit from, knowledge of this type. Paterson and Karjala have considered this problem from the point of view of indigenous rights outside of the traditional patent and copyright regimes, concluding that a statute based on traditional principles of contract and unfair competition law could address and likely resolve this problem without raising the fundamental difficulties that would result from using traditional IP rights under patent or copyright to achieve the desired goal.[7] This paper addresses the problem from the other side: What, if anything, about patent law creates or exacerbates the problem of biopiracy?[8]

2.2 *Physical vs. informational resources*

In considering the problem of biopiracy, it is vital to distinguish between the use of a physical resource and the use of an informational resource. Physical resources are depletable, and what one person uses is no longer available for another. Informational resources are nondepletable (infinitely multipliable) in that one person's use of information does not prevent another from making the same or a different use of it.[9] In one of the strongest condemnations of *biocolonialism* that I have seen, Professor Whitt states, "*By allowing access to and exportation of data, biocolonialism concentrates knowledge about a people and their environment in the hands of an imperial power.*"[10] This is simply wrong. Publicly available knowledge cannot be "concentrated" in the hands of anyone. Perhaps Professor Whitt intended to say that the *use* of some indigenous knowledge is concentrated under the patent system in outsiders who obtain foreign patents based on some of the exported data. But even that would not be correct if the implication is that the source peoples can no longer use their traditional knowledge in their traditional ways.

On the other hand, it is also incorrect to say, in general, that a patent owner is not harmed by the sale of unauthorized copies of the patented product, on the ground that the patent owner remains free to sell any amount of the product he chooses. There is absence of harm only if the purchase of the pirated product is not a substitute for purchase of the patented product. While this is often the case because some purchasers of pirated products would wholly forego use of the product rather than pay the higher price for an authorized version, there are likely to be at least a few people who would pay the higher price if less expensive versions were unavailable. Moreover, if pirated drugs sold at a low price in poorer countries do not reach patients unable to afford the authorized version, and these drugs find their way back to developed countries, they may displace further sales and thereby reduce the patentee's profits.

IP is thus fundamentally different from tangible property, which is why the legal rules relating to IP must also be different. This point is obvious, indeed almost trite, to IP scholars, but it seems to be often overlooked in the literature on biopiracy. Nondepletability of informational resources implies that, once the information is publicly available, it is economically inefficient to afford exclusive rights in it.[11] We grudgingly accept the limited-term exclusive rights of patent and copyright, notwithstanding the ex post economic inefficiency, because we believe that they serve as an incentive to the creation of desirable works. In other words, we accept the immediate economic inefficiency for the duration of the rights in the belief that in the long run we will have more and more desirable works overall. Calls for exclusive rights in information outside the patent and copyright regimes, especially for rights in information that is already publicly known, cannot be justified by a similar creation incentive. Some other justification is necessary.

I will note only in passing that the *other justification* will be difficult to find in so-called "natural rights" theory. Natural rights theory ("*I made it so it's mine.*") carries no limitation on the duration of protection, nor does it distinguish between the rights afforded by patent and copyright for works that are equally intellectually creative. Some of the most creative works of human history, like Newton's theory of gravity or Einstein's theories of relativity, get no protection anywhere under either the patent or the copyright regime, which is difficult to explain if natural rights to one's creative ideas and discoveries are the basis for exclusive rights. In the case of indigenous populations who assert natural-rights based exclusive rights in information they have developed or discovered, mutuality demands a similar recognition of rights in information developed elsewhere. Such recognition, however, would surely cost any given group much more than it gains.

2.2.1 *Depletion of physical resources*

To the extent that criticism of biopiracy focuses on the depletion of a physical resource, the problem may be controlled under the environmental regulation of the source country.[12] In other words, this is not an IP rights question but a tangible property question. There is no significant debate today about whether taking such resources without authority (theft) or by fraud should be unlawful. But a patent elsewhere on the active ingredient of a plant simply has nothing to do with the problem of environmental depletion with regard to the plant. If the patentee can manufacture the active ingredient synthetically, that activity does not contribute to further depletion. If the patentee needs the plant itself but can grow it away from its original source, again there is no contribution to depletion in the source country. And if the plant grows only in the source country, the existence of a patent abroad or even in the source country itself gives no right to take the physical plant in order to manufacture the patented product. Although a patent on the active ingredient, if recognized in the source country, would give the patentee the legal right to prevent others from taking the physical plant for the purpose of extracting the active ingredient, exercise of that right would likely mean *less* depletion of the physical resource, because it would no longer be in anyone's economic interest to take more of it than whatever is required by traditional uses. The patent thus may add a little something to the source country's power to regulate depletion, but it cannot exacerbate the depletion if the source country chooses to prohibit the patentee's taking of the plant.

2.2.2 *Depletion of informational resources*

Where the complaint is that the source country's people are not rewarded for supplying the information leading to the invention, several points should be borne in mind. First, if the information is obtained legally and results in a patented invention, that patent cannot cover any prior use that the source country's people made of the original resource.[13] Indeed, if the end product is a naturally occurring substance, that country may be in a position to refuse a patent altogether. Even U.S. patent law denied patents on naturally occurring substances until relatively recently, regardless of whether they had been isolated and purified.[14] Trade-Related Aspects of Intellectual Property Rights (TRIPS) requires member states to have patent laws that protect inventions that are *"new, involve an inventive step, and are capable of industrial application,"*[15] but TRIPS nowhere defines what *new* means. Any member state is therefore free to deny patents covering naturally occurring substances or traditionally used methods of treatment on the ground that they are not *new*. According to TRIPS Article 27(3)(a), a member state can also deny method patents covering the use of naturally occurring substances, purified or not, for therapeutic or diagnostic purposes. Moreover, following traditional U.S. law, a member state could find that isolating and purifying such substances lacks *invention* and therefore does not involve an *inventive step*.

Second, where the end product is a substantial modification of the original source[16] and constitutes a true invention that has, let us assume, greater therapeutic value than the original source, a patent in the source country will indeed have the effect of allowing the patentee to charge, for the period of the patent, a monopoly price in that country for use of the new drug (assuming there is no effective substitute that could hold down the price). If people in the source country cannot afford the new drug, their position is no different from that with respect to any other new drug, whether or not patented, or indeed any other product, that they cannot afford. They have not lost anything that they previously had. They can continue to use the original source as they always did, and they now have, in addition, the possibility of more effective therapy (if they can afford it), as will indigenous (and other) peoples elsewhere who never before had even the original treatment.[17] The wider availability of both the original treatment and the newly developed drug after biopiracy perhaps deserves more emphasis. In her article referenced above, Professor Whitt states:

> *Across the planet, at an accelerating pace, collectively owned traditional medicines and seeds are being privatized and commodified. Altered sufficiently to render them patentable, they are transformed into the 'inventions' of individual scientists and corporations and placed on sale in the genetic marketplace.*

But it is difficult to see just how the people who *collectively owned* the forerunners of the now improved medicines and seeds have been harmed. Moreover, the improved products are now available to a much wider range of users, including indigenous peoples from other parts of the globe. The patent may, indeed, mean that the price everywhere is higher than it would be were the product available without patent protection. It remains a fair question, however, whether the improved product would exist at all but for the patent incentive. We must bear in mind that no one is forced to buy the new product. Everyone is free to continue using whatever he or she has used in the past. Those who do choose to buy patented seed, for example, presumably believe that the higher seed cost is more than compensated by the beneficial improvements brought about by the newer product. It is true that patent law does not do much to alleviate the most important problems facing the people of developing countries, such as poverty, contaminated water, and lack of education. In developing countries, 840 million people currently suffer from malnutrition and 1.3 billion are afflicted with poverty.[18] But, to the extent that patent law serves as an incentive to the development of new products, especially medicines and improved agricultural varieties, it increases the options of everyone, including indigenous peoples, marginally to improve their lives. If the goal is to alleviate the wretched conditions under which many people in developing countries live, it cannot be right to say that information held by some of them that could be useful in addressing parts of the problem should remain confined to the small group discovering it, provided at least that the information is acquired in ways that are both legal and moral. It is also important to note that most indigenous groups will have no resources at all, genetic or otherwise, on which profitable products can be built. All such people potentially benefit if patent law serves as an incentive to create products that meet important human needs.

Third, denying patents in these cases will not necessarily stop the supposed *misuse* of the original information. It may well be *commodified* by an outsider anyway, in the hope of sufficient return from first mover or secrecy advantages. If, therefore, we are to accept the economic inefficiency of recognizing exclusive rights in information held by indigenous societies, some justification that outweighs the inefficiency should be offered. As mentioned above,[19] creation incentives are not involved, which distinguishes information collected from indigenous peoples from information that can be protected by patents and copyright. Claims of unfairness in these scenarios should articulate precisely what is unfair about developing, perhaps at great expense, something new and useful out of existing knowledge (which is what the patent incentive is all about). If the unfairness in a particular case is acquisition of information by fraud or other surreptitious or dishonest means, existing legal principles may supply a remedy, or at least an approach for statutory regulation. If the unfairness is lack of equal bargaining power because of ignorance of western legal customs, again a limited statutory approach setting default assumptions on agreement to pay a royalty or some other compensation may be in order. Cases in the United States show that using information to create a patented product without adequate disclosure to the source of the information is not limited to developing countries or indigenous populations.[20] Breach of a confidential relationship, fraud, invasion of privacy, and even more general notions of unfair competition may, in a given case, justify accepting the economic inefficiency of protecting traditional information.

It is possible that the availability of patents based on information derived from indigenous peoples creates a perverse incentive for western scientists and their employers to attempt to gain information through nefarious means, such as fraud or breach of confidence. One could surely find examples of creative inventors who have been cheated out of the financial return that would have been theirs under patent law by the illegal or unsavory actions of others. By providing exclusive rights, patent law does produce the occasional bonanza for the patentee, and logically the hope of such a bonanza would lead to at least some activity aimed at getting an unfair share of the prize. But this is again simply a general feature of patent law and property rights in general. The existence of property rights is indeed a prerequisite to theft. Biotech patents would seem an unlikely candidate for supplying a *special* incentive in this regard, given that most inventions require a huge investment to convert the initial information into a commercial product and test it for health and safety. Indeed, the numerous *enclosure* laws that a number of developing countries have adopted to maintain control over their genetic heritages may be driving researchers away from bioprospecting, due to the difficulty of identifying source material that will lead to a valuable product and to the complexity of achieving the necessary consents.[21] In other words, the causal link between a biotech patent and any assumed fraud in obtaining the information on which it is based from indigenous sources is weaker than for many other products. Moreover, the vast majority of patents, biotech and otherwise, are the result of unobjectionable behavior (that is, for example, there exists no fraud or breach of confidence). We therefore return to the need to identify the behavior that is wrongful when information derived from indigenous sources is turned into a patented product and to look for an appropriate sanction for that behavior.

Some commentators assert more generally that indigenous peoples often object to the use of their traditional knowledge on ethical grounds, arguing that IP should be treated as a pure public good.[22] Indeed, as Sabrina Safrin has argued, the numerous *enclosure* laws that a number of developing countries have adopted in an effort to maintain control over their genetic heritages may be driving researchers *away* from bioprospecting, due to the difficulty of identifying source material that could lead to a valuable product and to the complexity of achieving the necessary consents. No one can say that this view is *wrong*, as it comes down in the end to a question of fundamental values. Still, the question remains whether the members of any group following this belief should retain exclusive rights, with respect to people outside the group, to use information they have discovered. If the information is freely available simply by visiting the group and observing their lifestyle, and if a visitor does this without fraud or duplicity, saying that the visitor cannot use the information as the basis for creating a new, and perhaps patentable, product is equivalent to recognizing exclusive, perhaps group, rights in the information. Maybe such recognition can be justified on the ground that the group's culture should be respected by outsiders. But if this is the claim, we should be able to articulate it in terms of western notions like breach of confidence or privacy rights. Something besides *"We discovered it so it's ours"* is necessary unless one takes the extreme step of embracing a full-fledged natural rights basis for IP or one simply has a preference for economic inefficiency over economic efficiency.

A related view is that patents impoverish indigenous cultures by ultimately providing products that displace traditional sources and methods, leading to a loss of biodiversity and, eventually, an irretrievable loss of crucial elements of traditional knowledge and culture. Few would deny that such losses occur and that these losses represent ones suffered not only by the indigenous group but by all who, but for the displacement, might later have learned from such knowledge how to improve the physical or spiritual quality of their lives. If preventing the loss of indigenous culture is the goal, however, it is quite myopic to focus attention on patents derived from traditional information. Most indigenous groups do not end up being the source of information that leads to profitable patents. Moreover, even for those groups that do supply information leading to a patent, that specific information is only a small part of their entire cultural heritage, much of which is under threat from other sources, like music, films, and clothing. Indeed, to the extent that patents inhibit technology transfer to indigenous cultures (due to higher prices or lack of local implementation know-how), those patents should actually impede slightly the deleterious effects of the onslaught of western culture. Eliminating patents for advances in biotechnology will not eliminate biotech innovation or the adverse effects of patented and unpatented advances in other fields of technology. Needless to say, eliminating biotech patents will have no effect on cultural losses resulting from the adoption of western style music, cinema, clothing, and fast food. In short, the harmful influences of western life style for indigenous cultures are serious and real. Unfortunately, they will not be ameliorated by what would inevitably be minor adjustments to patent law in western countries or in locales of traditional cultures.

The core of the biopiracy claim thus appears to be not the availability of patents based on traditional indigenous information but rather the unfair acquisition of the knowledge and the absence of fair sharing of the profits that ultimately derive from developing it into a valuable product. The problem to be addressed becomes one of ensuring that traditional information is acquired in a fair and equitable way and that fair compensation is paid to the group from which the information derives. Some developing countries have proposed amending TRIPS to mandate disclosure of the source of genetic resources used in an invention, of evidence that the country of origin had consented, and of evidence of fair sharing of the benefits as conditions to the issuance of a patent. My colleagues George Schatzki and Ralph Spritzer have suggested to me the possibility of refusing to enforce any patent based on information that has been unfairly acquired, or of placing on enforcement the condition that a fair sharing exists (as determined by court ruling) between the patent holder and the people who served as the information source. This would not be a major extension of the doctrines of patent and copyright misuse, under which the intellectual property rights owner is denied enforcement until the abuse is cured.[23] It is important to keep in mind that without the patent there would be no profit for *any* compensation to be paid. [24]

One policy implication of this analysis for developing countries is straightforward: to the extent one is concerned about biopiracy, it is a mistake to focus on patent law as a crucial, or even an important, part of the problem. Addressing the real problems associated with biopiracy is much more difficult. To the extent a given country or group considers its traditional knowledge sacred and not available for economic exploitation, rules and statutes can always be created that make illegal any attempt to learn or exploit such information. That will surely discourage what would otherwise be legal activities leading, perhaps, to products that could improve the lives of many, both within the source country and without. But that is the expected cost of attempting to respect the local view concerning traditional knowledge. The problem is that, in the long run, such an approach is unlikely to work. It takes just one person who has knowledge of information to transmit it outside the group, and once the information is out it is impossible to make secret again.

To the extent that a given group's biological knowledge or makeup is considered an economic resource, it is important to encourage exploitation of that resource by those who are willing to pay for it. Policy-makers must define, or find ways of allowing markets to define, what is fair and equitable compensation for indigenous peoples' contribution of information to what ultimately becomes a profitable product and who is entitled to such compensation. Then policy-makers must seek ways of rendering potentially valuable information inaccessible without prior agreement concerning compensation. And they must do this in ways that do not raise the costs of bioprospecting so much that they discourage people and companies that *could* potentially make valuable use of the information from seeking it. None of this is easy. The proper direction in which to look for legal approaches, however, is in areas like contract and unfair competition law.

3. TECHNICAL ISSUES INVOLVED IN GENE-RELATED PATENTS

Patents on genes, especially human genes, and gene products (such as proteins and enzymes) raise some important technical issues in the interpretation of current patent law.[25] In addition, there is always the basic policy question for patents of whether the gain from affording patent protection (new products and processes that, but for the patent incentive, would not have been invented or disclosed) justifies the harm that flows from a government-enforced monopoly for the patent period (such as higher prices for products that would have been invented anyway and inhibitions on further research). Finally, some biotechnology patents raise ethical issues of a very different type than patent law has faced in earlier periods.

3.1 *Naturally occurring substances*

Analysis of biotech patent issues under U.S. law always begins with *Diamond v. Chakrabarty*, in which the Supreme Court held that the law did not preclude patents on living organisms (447 U.S. 303 (1980)). The court stated that the patentability line was "*not between living and inanimate things, but between products of nature, whether living or not, and human-made inventions.*" The case is justifiably controversial for such a broad interpretation of section 101 of the Patent Act, which allows a patent for one who "*invents or discovers any new and useful process, machine, manufacture, or composition of matter.*" Living organisms do not fit easily into any of these categories.[26] For present purposes, however, the most important aspect of *Chakrabarty* was its express retention of the long-standing prohibition on the patenting of naturally occurring substances. Upholding and distinguishing an earlier case[27] that the *Chakrabarty* court characterized as denying a patent for merely discovering "*some of the handiwork of nature,*"[28] *Chakrabarty* emphasized that the bioengineered microorganism at issue was not "*a hitherto unknown natural phenomenon*" but rather a *"product of human ingenuity"* that differed markedly from anything found in nature.[29]

Genes and gene products, as they exist or are created in the cells of living organisms, are naturally occurring substances. They may be difficult to find, but we know they are there and that they can be found if enough effort is put into the project. One would have thought that the prohibition on patenting naturally occurring substances would have ruled out at an early stage patents for genes and gene products.[30] Yet, notwithstanding the highest court's reaffirmation of the prohibition on patenting naturally occurring substances, lower U.S. courts and the Patent and Trademark Office (PTO) have deviated substantially, further expanding patent coverage in the process. In the case of genes, the discussion got sidetracked at an early stage into the issue of whether a raw gene sequence, without disclosure of the gene's function or utility, could satisfy the *utility* requirement of the Patent Act.[31] In response to arguments that inventions are patentable, but mere discoveries (such as a particular gene) are not, the PTO held that:

> *[W]hen the inventor ... discloses how to use the purified gene isolated from its natural state, the application satisfies the "utility" requirement. That is, where the application discloses a specific, substantial, and credible utility for the claimed isolated and purified gene, the isolated and purified gene composition may be patentable.*[32]

Thus, while a gene in its natural state inside the cells of a living organism is not patentable, anyone who succeeds in isolating and purifying a gene (even by a perfectly routine methodology) and discloses an appropriate utility for it can obtain a patent on the gene.

Many commentators have decried treating an isolated and purified form of a naturally occurring substance as patent subject matter just because the purified form does not exist in nature.[33] Professors Linda Demaine and Aaron Fellmeth have recently supplied a thorough and convincing analysis criticizing this contention and demonstrating that it deviates substantially from precedent.[34] They argue that section 101 of the Patent Act mandates *invention* rather than mere *discovery*,[35] based on the express statutory requirement that the object of the patent be *new and* something that arises from application of human intellectual thought. They point out that the *isolated and purified* interpretation abrogates the requirement for *invention* and allows patents for essentially any alteration of a naturally occurring substance if increased commercial or therapeutic value results. As they point out, under this rationale to patentability, the first person to purify water or blood cells could have patented them.

Demaine and Fellmeth recommend a test of whether the naturally occurring substance has been transformed in such a way as to create a new product that is substantially different in biological function from the naturally occurring phenomenon. For biological substances, passing such a test would require in practice a change in molecular structure, because biological function is largely, if not wholly, determined by molecular structure. By requiring a substantial change in function, this test obviates the otherwise thorny problem of deciding whether a slight structural change (for example, adding or removing an extraneous atom or two) is sufficiently creative to deserve a patent. If the gene or its product still function as they do in nature, the new version will simply not be sufficiently creative under their test to be patentable. For naturally occurring substances unmodified by human-initiated structural change, another possibility would simply be to state expressly that only process patents, covering new and nonobvious *uses* of the now isolated and purified substance that occurs in nature, will be available.[36] Either approach would leave the substance itself, purified or not, free for research and for yet additional uses not envisioned by the owner of the use patent. (According to the PTO, a product patent covers *all* uses of the product, whether or not they are disclosed in the patent.) Finding a new use for such substances may well involve substantial investment and require the incentive of patent protection. While process patents are generally considered weaker than product patents, if a purified gene or gene product is used in a specific therapeutic method, there may be no readily available substitute, so the method-patent owner would maintain exclusive rights to that use.

A more substantial objection to method patents for new and nonobvious uses of genes and gene products derives from the TRIPS rule that permits excluding from patentability "*diagnostic, therapeutic, and surgical methods for the treatment of humans or animals*" (TRIPS Article 27(3)(a)). Much of Europe and many other countries have availed themselves of this exclusionary possibility. While the U.S. does not preclude patents on therapeutic processes, it does exempt medical practitioners from liability for infringement arising in the course of performing a medical activity (35 U.S.C. § 287(c)(1)). Among other exclusions, however, the immunity does not apply to infringements arising from practicing a process "*in violation of a biotechnology patent*" (35 U.S.C. § 287(a)(2)(A)). This would seem to leave unimpaired, in the U.S. at any rate, a patented method for using a naturally occurring substance derived through biotechnology. In any event, whether and to what extent therapeutic methods should be protected under patent law involves fundamental policy issues. If patent law today, under the TRIPS permissive exclusion, supplies insufficient protection to therapeutic methods, that aspect of it should be amended. It is not a satisfactory solution to make an end-run around the current spate of exclusions for therapeutic methods by protecting naturally occurring substances as products.[37]

In any event, while U.S. law has deviated from its long-standing prior position that naturally occurring substances are unpatentable and that merely extracting them in purified form does not make them patentable, arguments are available everywhere else in the world that such substances are not patentable because they are not new. TRIPS requires patents only for inventions that are new, and member states are free to decide whether or not a naturally occurring substance, like a gene or gene product, is new in the sense required by their patent statutes. Moreover, merely finding raw genes is not particularly difficult or inventive. Consequently, denial of patents on raw genes could also be predicated on absence of an inventive step.[38]

3.2 *Patent conditions for biotech inventions*

Many biotech inventions, as in *Chakrabarty*, will creatively alter a naturally occurring substance. In such cases, an objection to patenting based on the absence of something new, in the sense of *not previously existing*, is unavailable. Neither, at least in many cases, is an objection based on the absence of sufficient human creativity in the final product. Consequently, if a product, like the microorganism in *Chakrabarty*, otherwise meets the requirements for a patent, such as the technical standards for novelty and the substantive standards for nonobviousness, there are no grounds in the Patent Act itself for denying a patent.[39] TRIPS, of course, allows for the exclusion of plants and animals (other than microorganisms) from patentability,[40] and many countries may choose to do likewise on ethical grounds. But the absence of patent protection for genomic innovations does not ensure that no products based on modified genes or gene products will appear. Moreover, recognition of patents in this area does not mean that there can be no regulation or even outright prohibition by specific legislation. We should bear in mind that a huge potential exists for genetically modified organisms to contribute to the elimination of hunger and disease in developing countries, particularly if access to the technology is available. If patents on such products, at least in developed countries, serve as an incentive for their creation—meaning that without patents we would all have the benefit of less innovation—outright denial of patent rights would appear to effect a net social loss.

4. BALANCING THE COSTS AND BENEFITS OF GENE-RELATED PATENTS THROUGH POLICY

4.1 *Naturally occurring substances*

Whether or not patents on gene sequences or naturally occurring gene products conflict with the earlier prohibition on the patenting of naturally occurring substances, until the Supreme Court addresses the matter we must accept that the courts and the PTO have expanded the notion of patent subject matter to include them, provided that they have been isolated and purified. Still, does this expansion of traditional patent law make sense as a matter of policy?

Professor Epstein has articulated the basic policy issue that must be examined in deciding whether to recognize gene-related patents: Do the incentives for the creation of these inventions justify the restrictions on output that follow from exclusive rights?[41] Few, if any, have argued on economic grounds that gene-related patents should be wholly proscribed. But many able commentators have argued cogently that patents on raw gene sequences could inhibit, rather than promote, the progress of science and the development of products that are actually useful. Gene sequences alone, even in their isolated and purified forms, rarely have any direct use.[42] Useful products are normally the result of implanting the gene into the genome of an organism, such as a bacterium, that will then manufacture the protein or enzyme encoded by the gene. Then that protein or enzyme must be extracted from the cellular environment in which it was produced by the *vector* organism (in this case, the bacterium) and ultimately tested for safety and efficacy in its hypothesized use. These latter *downstream* activities that go from the gene itself to a useful product usually require a huge effort, quite often more than the *upstream* effort required to determine the gene in the first place. Thus, patents on basic upstream tools can inhibit, rather than promote, valuable downstream research.[43] Indeed, Professors Demaine and Fellmeth point out that when an upstream patent lacks ingenuity (which is the case for naturally occurring gene sequences), the patent incentive may not even be necessary to induce innovation but may still strongly preclude downstream research.[44]

It has also been argued that patents on raw genes may result in too much investment in the search for genes and insufficient investment in developing new products and carrying them to market.[45] Such patents can also inhibit information flow, which in turn duplicates research.[46] Finally, Professors Heller and Eisenberg have argued that gene-sequence patents can lead to a *tragedy of the anticommons*, in which many overlapping claims to gene fragments or *stacked* rights established by reach-through license agreements[47] between upstream patentees and downstream researchers must be coordinated to develop a useful product. Too many such claims may make negotiations among all affected parties difficult or impossible.[48] Moreover, a biotech anticommons is more likely to endure than in other areas of IP because of higher transaction costs, heterogeneous interests among owners, and cognitive biases of researchers.[49]

These policy arguments, therefore, suggest that it was a mistake for U.S. law to deviate from its traditional refusal to protect naturally occurring substances, even though purified, in the case of gene sequences. Like the argument against such patenting based on the absence of *invention* or *newness*, however, nothing in it suggests differential treatment of indigenous peoples from anyone else. If patenting genes or gene products is wrong on either statutory or policy grounds, we should correct the law, not because it imposes a particular burden on indigenous peoples, but because it imposes an unreasonable burden on everyone.[50]

4.2 *Modified genes and their products*

In the cases of human-created DNA sequences that do not occur naturally, and products derived from such sequences, we can no longer say, in general, that there is no invention or that the invention is not new. Such inventions, like the oil-spill-eating bacterium at issue in *Chakrabarty*, have much potential for ameliorating some of humankind's worst afflictions. Whether and to what extent patents supply the necessary incentive to undertake the research leading to such inventions is, as with all inventions, a difficult and unresolved question. However, I see no reason to distinguish these genomic inventions from any other on this score.

5. ETHICAL ISSUES ARISING FROM GENE-RELATED PATENTS

Patents confer upon their owners the right to exclude all others from making, selling, or using the patented invention. Thus, patents covering genes of living organisms, particularly patents covering pieces of the human genome, raise ethical questions concerning:

- whether such private control over genes or their products involves monopolization of the *common heritage of mankind*
- whether they denigrate human life by reducing life to a commodity
- whether they interfere with individual or collective privacy
- whether they promote distributive justice when they are concentrated in a few economically developed countries

Patents on crop varieties have also been said to threaten biodiversity.[51] These are serious issues that will continue to be examined for some time. I only touch upon them here, because it seems to me that indigenous and nonindigenous populations are equally affected or, at least, where there are differences in how costs or benefits deriving from gene-related patents are distributed, analysis shows that it is not the patent that is responsible for the problem.

5.1 *Monopolizing the common heritage of mankind*

We should first note that any objection to gene-related patents as monopolizing the common heritage of mankind must in fact refer only to patents on human genes, as it is those genes that have been passed down to us over the generations. If all living things were deemed part of the common heritage of mankind, there could be no property rights at all, let alone patent rights, in domestic animals, or indeed even plants. This objection to human-gene-related patents would seem to be subsumed in the *naturally occurring substance* controversy. If we upheld the traditional ban on patents covering naturally occurring substances, whether or not isolated and purified, human genes and their protein products would not be patentable.[52]

On the other hand, it is at least possible that a full-fledged cost/benefit analysis might show gains, from recognizing patents in genes and their products, that outweigh the losses. Patents may actually serve as an incentive to discover these products and their desirable uses to such an extent that the disadvantages of temporarily higher pricing and reduced information-flow should be accepted. If we assume for the moment that this is in fact the case, we must deal with the claim that human-gene-related patents should be denied, notwithstanding their economic advantages, because they would amount to undesirable monopolies on the common heritage of mankind.

This claim is most potent if a patent on a human gene or its protein product were construed to cover the naturally occurring processes that take place within human cells, where the gene itself resides and causes the manufacture of its protein product. Literally, the cell, and thus the human being to whom the cell belongs, is *making* the gene every time the cell divides, and the cell *uses* the gene in the process of *making* the gene product. Thus, it would appear that a patent covering the gene or its product would be infringed by these natural activities.[53] Although the patent only issues upon the applicant's claim that the product has been isolated and purified from its natural form, once issued the product (or composition-of-matter) patent covers any use of the chemical composition. A patent on a new drug, for example, will cover any form of chemical packaging into which the drug is incorporated or mixed. If it did not, the patent would be worthless. Thus, the logic of composition-of-matter patents on naturally occurring genes and their products leads to an absurd result when applied to living organisms and represents a basic flaw in the theory.[54]

The problem arises, however, not because genes are part of the common heritage of mankind but because gene and gene-product patents, by their nature, cover things that are not inventions. One can imagine, for example, someone or some group whose cells contain a unique mutation in a particular gene that gives the gene some special value. It is not part of the common heritage of mankind because, by hypothesis, at most a limited group carries the gene.[55] Moreover, by limiting focus on human genes, the common heritage approach would leave naturally occurring genes in other plants and animals free for the patentable taking. It would therefore seem that opposing gene patents on the ground that genes comprise the common heritage of mankind is less fruitful analytically than simply staying within the bounds of traditional patent law and seeking denial of patents on the ground that patents on naturally occurring genes and gene products give a theoretical monopoly over the life processes of the organisms from which they derive. Such a monopoly, even though apparently more theoretical than practical at the moment, is simply unacceptable, regardless of the economic cost/benefit analysis.

In any event, and of most relevance for the present topic, nothing in the common heritage argument distinguishes indigenous from non-indigenous peoples. If it is bad for indigenous peoples that anyone should get a patent in a piece of the common heritage of mankind, it is equally bad for everyone else.[56]

5.2 *Reduction of life to a commodity*

Many maintain that patents on pieces of the human genome are morally wrong because they reduce life to a commodity.[57] While this argument has a certain rhetorical ring, its high level of generality renders analytical application difficult. A patent on a gene that is useful for diagnosing potential disease, for example, may mean that anyone who wishes to undergo the genetic test will have to pay more than if the gene were in the public domain. It is not clear to me, however, how this commodifies human life any more than a patent on any other medical diagnosis device or procedure. Slavery commodifies human life. Patents on the whole genome might well be said to commodify human life. While at bottom it may come down to questions of fundamental ethical or religious values,[58] to me no single gene or gene product can be meaningfully deemed *human life*. While the entire human genome may validly be thought of in many contexts as a *blueprint for human life*, no patents are going to issue anywhere on the entire human genome. A product is *commodified* when it becomes the subject of market transactions—it is widely available, like aspirin, against payment of the purchase price. It is easy to imagine markets in unpatented products based on human genes, and such products, like aspirin, will be commodities. They are no less commodities if they were never subject to a patent, or if the patent has expired, than they are while they are under patent. Moreover, the unavailability of patents will not stop scientific activity on human genes or all market activity in gene products.[59] Conversely, the availability of patents is not synonymous with commodification.[60]

Finally, this again raises the question of how making and selling a product based on a human gene differentially affects indigenous and nonindigenous peoples. It may be more likely that an indigenous group that has managed to remain relatively isolated from the onslaught of modern society will have in its collective genome a genetic characteristic of particular interest to those who would seek to develop genes into patentable products.[61] But it is difficult to see how studying the genetic characteristic of interest reduces to a commodity the lives of the people from whom the information is derived. More often, the complaint is that these people should be able to benefit from any profits that are eventually derived from the results of such studies, which is simply the human genome variant of the more general biopiracy problem discussed above with respect to nonhuman resources. Indeed, if it is true that the benefits of developments in modern medicine are slow to reach many indigenous societies, it is difficult to see how commodification in developed countries affects them at all.

5.3 *Privacy and human dignity*

Many have decried the recognition of gene-related patents as being fundamentally in conflict with norms of privacy and human dignity.[62] The underlying notion seems to stem from the intimate relation between an individual's genes and his or her phenotype, as expressed in physical, intellectual, and emotional characteristics.[63] Because genes are also part of our collective make up, it has been suggested that gene patenting may violate some sort of collective privacy right as well.[64]

At the individual level, there is no doubt that knowledge of someone's genome, in particular the presence of specific genes known to have a causal relationship to a particular disease, can be put to unfair discriminatory use in areas like employment or insurance.[65] To the extent that such a gene is known to be differentially preponderant in a specific group, the danger of group stigmatism is also very real. Without downplaying the importance of either of these problems, it is difficult to see how gene-related patents exacerbate the problems. Genomic research has been going on for some time and is not likely to stop, regardless of the availability of patents. Indeed, it is the identification of the gene and its function that sets the stage for any subsequent discrimination that may occur, individual or collective. One of the major policy arguments against patenting such naturally occurring substances is that patents are *not* necessary as an incentive for this kind of research.[66] There is good reason to hope that much of this research, even when it identifies a particular set of genes with a given generally undesirable phenotypical response, such as a disease, will ultimately lead to valuable therapeutic interventions, or at least methods of prevention. Withdrawing the patent incentive will almost surely be detrimental for these developments.

Interference with privacy norms and affronts to human dignity through the misuse of the results of genomic research would also seem to be at least as problematic for people in developed countries as it is for indigenous peoples. The most likely worst case scenario for indigenous peoples might be the finding of a gene specific to a particular group that plays a causal role in some undesirable phenotypical attribute (as viewed from outside the group). Such a discovery could unfairly stigmatize the group in the eyes of outsiders. Patents, however, would seem unrelated to such a discovery. When outsiders have sought patents based on the genetic make up of an indigenous group, it is usually because the group is perceived as having a genetic *advantage* over the rest of humankind.[67] By the nature of the patent incentive, it is unlikely that the possibility of a patent would encourage anyone to look for a gene causing what is perceived in developed countries as a disadvantage that is unknown in those countries.

5.4 *Crop monocultures and monopolization of crop genomes*

Even outside the human genome some commentators have raised ethical questions concerning the appropriateness of gene patents. Patents on crop varieties, for example, may result in monocultures and the use of expensive inputs, such as fertilizers, that cause environmental harm.[68] It has been claimed that broad plant variety patents have conferred on a few corporations virtual monopolies on the genomes of important crops.[69]

Here again we find some potentially serious problems. If all the world's wheat is a single variety, for example, and if that variety turns out to be susceptible to a rapidly spreading blight of some sort, a significant portion of the world's food supply could be wiped out, with catastrophic consequences. Still, we must consider the role patents might play in creating or exacerbating these problems. If the use of expensive inputs is the problem, it would seem that not everyone would use the variety (in particular, those who cannot afford to pay). It should be borne in mind that a patent on a crop variety obligates no one to buy the seed. All farmers are free to continue using their traditional varieties in their traditional ways. Patents can serve as an incentive for finding or commercializing environmentally friendly crops and other inventions, and the existence of a patent can reduce resort by the distributor to economically inefficient and perhaps environmentally dangerous self-help approaches.[70] Moreover, if environmental harm is the problem (and a susceptible monoculture is one such example), environmental regulation is most likely necessary to remedy it.[71] Because of the human tendency toward free riding, no one can be expected to adopt an environmentally friendly approach to food production without the assurance that his competitors are operating at the same (economic) disadvantage. Moreover, if a given but advantageous variety is unpatented, it is likely to be adopted even more widely than if it is patented, increasing the danger of dependence on a monoculture.

6. POLICY IMPLICATIONS

This section demonstrates that the major policy problem for patent law in biologic materials is not peculiar to indigenous peoples or developing countries. Rather, it is the treatment under current U.S. and European law of naturally occurring chemicals (DNA sequences and genes, and their natural products) as patent subject matter when extracted in isolated and purified form. Nothing in the language of the extant patent statutes or in the international IP or trade agreements compels this treatment. Allowing patents for naturally occurring substances goes against a long patent tradition even within the United States, and so far no one has made a convincing policy case that such a radical change from traditional patent principles should be made. Policy-makers in developing and developed countries should therefore resist pressure to adopt such a change, not because such patents have an untoward effect on privacy and human dignity but because denying patents on naturally occurring substances is simply good patent policy.

6.1 *Patents and developing countries*

Any country that wishes to have the free-trade advantages supposedly supplied by the World Trade Organization (WTO) must comply with the IP requirements of TRIPS. Among other things, TRIPS mandates that its member states adopt patent laws in keeping with those of the developed nations of the United States and the European Union. Many commentators have argued that developing countries have little to gain from recognizing foreign patents, as required by TRIPS, except to avoid trade retaliation.[72] A lively debate continues over whether patent laws promote or inhibit technology transfer to developing countries. That, in turn, raises the question of whether the costs of establishing a patent system, largely for the benefit of developed countries, are outweighed by the benefits. In addition, some commentators have raised ethical and human rights issues outside the specific realm of biotechnology. These include issues of distributive justice[73] and access to pharmaceuticals.[74] Other commentators have asserted that developing countries may view IP as a community (public domain) asset that no individual should own.[75] Patenting, in particular, has been said to clash with indigenous knowledge and value systems.[76]

6.1.1 *Technology transfer*

There is little doubt that TRIPS impedes the ability of developing countries to determine their own IP standards and policies in the hope of achieving a better fit to their own economic and social conditions.[77] In particular, TRIPS does not allow the choice of simply not recognizing patents for inventions by nationals of other member states.[78]

The advantages to developing countries of having a patent law have also been seriously questioned. It has been claimed, for example, that recognizing patents stimulates technology transfer, allowing the patenting country to gain not only the knowledge supplied in patent applications themselves but also the necessary know-how to start going into many of these fields of technology themselves. Others have disputed these claims, however, arguing that foreign patents deter developing countries from appropriating new technologies and products.[79] The needs of developing countries are often quite basic, for example, and some lack the ability to assimilate the latest technologies. A foreign patent owner may have little incentive to transfer technological know-how related to a patented invention if profits are available from imports. Most obviously, the information contained in a patent application is always available in the developed countries in which the invention is patented. Therefore, if a developing country is indeed capable of making use of such information in local industry, it would have access to the information without having its own patent law, and its citizens could make use of the information sooner, or at least without having to license it.[80]

6.1.2 *Access to inventions*

It is routine to observe that patented goods that reach the market will have a higher price than if they were not patented.[81] To the extent that this is true, it reduces access to the patented goods if there is any elasticity in demand, because people at the margin, by definition, could afford a lower price but not the higher one. It has been argued, moreover, that a patent owner might choose neither to enter a market nor to authorize local production, thereby reducing access in that country.[82] Probably the most convincing argument against patent laws in developing countries is Professor Oddi's observation that few inventions are *patent-induced* with respect to a given developing country.[83] That is, most inventions likely would have been invented, anyway, regardless of whether any given developing country has a patent law that might protect it. To the extent that an invention is not patent-induced in this sense, patent protection in a developing country necessarily adds to that country's costs, because institutions in that country have access to the information in the patent in the countries where the invention is patented, so recognizing such a patent brings nothing more to the table.[84]

6.1.3 *Balancing the costs and benefits of patent law*

The above analysis implies that patents in developing countries can add significantly to those countries' costs with respect to new inventions,[85] and this cost is likely not offset by an increase in local technological development or in access to inventions that are, indeed, patent-induced. Still, consideration of the most dramatic case, which is access to vital pharmaceuticals, shows that the problem is more complex than this basic theoretical analysis would suggest.

In an effort to investigate the effect of patent laws on access to effective treatment in developing countries, Attaran and Gillespie-White looked at the availability of antiretroviral drugs for AIDS treatment in Africa.[86] Somewhat surprisingly, and contrary to conventional wisdom, they found no correlation between access to antiretroviral treatment and patent status across Africa.[87] Access to these drugs was found to be uniformly poor across Africa, independent of whether and where the drugs were patented.[88] Thus, at least in the poorest countries, access to potentially life-saving drugs seems not to be inhibited by patents but by the lack of funding to obtain access to these drugs at any price reflecting the cost of their production and administration.[89]

This suggests that the problem of access to inventions, and technology generally, in developing countries will not be solved by the denial of patents in those countries. It certainly will not be solved by denying patents in the developed world, if such denial eliminates the incentive for their discovery—the innovations would then be available to no one. The issue brings us back to the fundamental nature of IP and, in particular, its infinite multipliability without reduction of supply.[90] We can ask, for example, why the owner of IP should care whether the product embodying such IP is copied and distributed in that market if a given market offers no expected return from the exploitation of IP, such as a patent.

Consider an extreme case for the sake of illustration. Suppose country X has zero dollars to pay for a patented, potentially life-saving drug. The patentee could not have been thinking of X as part of his expected return while developing the drug, and indeed the patentee gets no return from X after the drug is on the market, *whether or not the drug is copied and distributed in X*. The copying and distributing of the drug in X does nothing to the patentee's exclusive right to market the drug in other countries where it is patented and where people can afford to pay something for it. This activity thus has utterly no effect on the patentee, provided that all of the drug that is copied and distributed in X actually stays in X and is used solely for the benefit of X's citizens. The problem for the patentee, then, is not the copying and distribution in X but rather the potential for grey-market leakage into markets where the drug is profitable for the patentee, because such leakage could potentially bring down the price of the drug in those markets.[91] There is no economic reason, therefore, why the patentee (on these extreme facts) would not be willing to sell the drug in X at cost, provided the patentee could ensure that none of it would leak back into his or her more lucrative markets.[92] In other words, the presence or absence of a patent law in X is essentially irrelevant to the patentee, whose only concern is with competition in his or her other markets from drugs originally distributed in X.

In any realistic situation, of course, there will always be at least a few people who can afford to pay the patentee's price, so selling the drug at cost would actually reduce the patentee's return.

For the poorest countries of the world, however, the number of such people will be very small. For other countries, where more resources are available for health care, discriminatory pricing (charging more where the demand is inelastic and less where it is elastic) will likely result in wider access to drugs in developing countries and a profit to the patentee.[93] But even these schemes will be avoided by patentee drug manufacturers if products sold at a low price in one country find their way back to their more lucrative markets elsewhere.[94] Moreover, under any price discrimination scheme aimed at maximizing the patentee's profits, the price will likely be higher than it would be in the absence of the patent's exclusive rights, which to that extent continues to reduce access below that of a completely free market.

Another variation of the problem of balancing public access with the need for incentives occurs in university research, because research universities both actively seek the financial returns that are available from patented research and engage in public service. It was recently reported that a number of research universities had formed the Public-Sector Intellectual Property Resource for Agriculture in an effort to standardize their licensing practices to allow them to engage in humanitarian endeavors. Some of these universities are owners of valuable biotech patents that they have licensed away and now find themselves needing to use in efforts to create new crops that could feed impoverished people. The patent rights thereby stand in the way of their humanitarian mission. One idea is to include a *humanitarian use* clause in future licenses to make sure that universities retain the right to engage in such activities.[95]

TRIPS does allow for some amelioration of the exclusive rights of a patent through compulsory licensing.[96] The Doha Declaration on the TRIPS Agreement and Public Health expressly gives member states the freedom to determine the grounds on which compulsory licenses can be granted.[97] For countries that lack the facilities and technological expertise to manufacture complex pharmaceuticals locally, the TRIPS Council adopted a decision, which was implemented by the WTO in 2004,[98] waiving the obligations of an exporting member under Article 31(f) with respect to a compulsory license to produce and export pharmaceuticals to *eligible importing members*, subject to conditions like producing no more than necessary to meet the needs of the eligible importing country.

We may conclude that access to patented inventions, especially pharmaceuticals, is not as readily available as it might be were these inventions unpatented everywhere in the world. TRIPS is part of the problem, and the perceived danger of parallel importing is another.[99] It is important for these problems to be resolved in a way that maximizes worldwide access to all types of innovation, but especially to life-saving pharmaceuticals. Solutions should avoid undercutting incentives for more innovation in developed countries. To many it seems just plain wrong not to provide universal access to life-saving innovations in pharmaceuticals.[100] We are forced, however, to make a tradeoff between universal access to existing technology and future access to new technology. If the attempt to supply universal access to a given innovation reduces or eliminates future innovation, the ultimate result is no, or at least reduced, access to innovation for anybody.

7. CONCLUSIONS

Understanding the effect of patent rights in biotechnological inventions on the interests of indigenous peoples requires a more nuanced analysis than has generally appeared in the literature. The problem of so-called biopiracy, for example, is not one of the availability of patents based on traditional indigenous information but rather the failure to share fairly the profits that ultimately derive from developing the information into a valuable product. Patents on naturally occurring genes and gene products raise serious problems under traditional patent law on both technical and policy grounds, and they raise important ethical questions as well. These problems and questions, however, are not unique to indigenous peoples. Rather, they should, and must, be addressed by all peoples in the world, developing and developed. The basic problem with respect to indigenous peoples is patent law generally, beyond mere biotech patents, and whether its forced adoption by TRIPS will result in a net benefit to developing countries. Serious questions have been raised concerning whether local adoption of a patent law will improve technology transfer or increase access to desirable inventions in those countries. The issue boils down to the extent that the absence of patent protection in developing countries erodes the incentive for innovation in developed countries, either through the absence of a profitable market in countries lacking a patent law or through grey-market arbitrage that allows patented products to flow back into the markets that do serve as incentives to innovate. ■

ACKNOWLEDGMENTS

The author gratefully acknowledges numerous helpful comments on an earlier draft from his colleagues Aaron Fellmeth, Robert Clinton and Douglas Sylvester. An earlier draft of this paper was presented at a conference titled Intellectual Property and Biotechnology in the Age of Globalization: Challenges, Opportunities and Risks, September 19–20, 2003, at the University of British Columbia, Vancouver, B.C., Canada. A modified version of this paper was published in Volume 7, Issue 2, of the *Minnesota Journal of Law, Science & Technology* in 2006.

DENNIS S. KARJALA, *Jack E. Brown Professor of Law, Sandra Day O'Connor College of Law, Arizona State University, Tempe, AZ, 85287-7906, U.S.A. dennis.karjala@asu.edu*

---

1 See for example, Paterson RK and DS Karjala. 2003. Looking Beyond Intellectual Property in Resolving Protection of the Intangible Cultural Heritage of Indigenous Peoples. *Cardozo J. Internat'l & Comp. L.* 11:633.

2 See Karjala DS. 2003. Distinguishing Patent and Copyright Subject Matter. *Conn. L. Rev.* 35:439 (arguing that functional subject matter belongs under the patent, and not the copyright, regime).

3 A modification of trade secret law aimed at protecting group privacy interests in sacred symbols and rituals might be effective. See Paterson and Karjala, *supra* note 1, at 665–66. Fellmeth has suggested that some might argue for a collective trade secret in the indigenous use of herbs or other natural materials, including biological materials. Such knowledge might qualify for protection under ordinary modern trade secret law because it may have independent economic value resulting from its not being generally known and the group may take reasonable measures to maintain secrecy. See Uniform Trade Secrets Act § 1(4), 14 U.L.A. 537–51 (1980 & Supp. 1986), defining *trade secret*. Article 39(2) of the Agreement on Trade-Related Aspects of Intellectual Property Rights (TRIPS) requires such protection for "natural and legal persons." Agreement on Trade-Related Aspects of Intellectual Property Rights, Apr. 15, 1994, Marrakesh Agreement Establishing the World Trade Organization, Annex 1C, 1869 U.N.T.S. 299, 33 I.L.M. 1197 (1994) [hereinafter TRIPS]. It is not much of a step further to recognize protection for cultural or ethnic groups collectively if the other conditions of trade secret protection are satisfied. To the extent that such knowledge is ineligible for trade secret protection, the problem is essentially that of biopiracy discussed in Part 2 of the chapter.

4 Lemley MA. 2000. Reconceiving Patents in the Age of Venture Capital. *J. Small & Emerging Bus. L.* 137, 139.

5 For a recent critique of this assumption, see Moore AD. 2003. Intellectual Property, Innovation, and Social Progress: The Case Against Incentive Based Arguments, *Hamline L. Rev.* 601.

6 See *supra* note 1, at 662–67.

7 See also Chen, J. 2006. There's No Such Thing as Biopiracy ... And It's a Good Thing Too. *McGeorge L. Rev.* 37, at p.1 (2006)(arguing that the biopiracy narrative is largely a myth and that the term should be stricken from the ethnobiological discussion).

8 As Mark Lemley has stated: "*The economic rationale underlying much privatization of land, the tragedy of the commons, simply does not apply to information goods. It is possible to imagine physical bandwidth or server capacity being overconsumed, although the danger of that currently seems remote. But it is not possible to imagine overconsumption of a nonrivalrous thing like data. ...From an economic perspective, the more people who can use information, the better.* Lemley MA. 2003. Place and Cyberspace. *Cal. L. Rev.* 91:521, 536 (citations omitted).

9 Whitt LA. 1998. Indigenous Peoples, Intellectual Property & the New Imperial Science. *Okla. City U. L. Rev.* 23:211, 220.

10 Seeratan NN. 2001. *Comment,* The Negative Impact of Intellectual Property Patent Rights on Developing Countries: An Examination of the Indian Pharmaceutical Industry. Scholar: *St. Mary's Law Review on Minority Issues* 3:339, 353–54.

11 See *supra* note 8. On the Economic Inefficiency of Protecting Works That Have Already Been Created; see also Karjala DS. 1997. The Term of Copyright, in *Growing Pains: Adapting Copyright for Libraries, Education, and Society* 33, 42–44 (ed., LN Gasaway). For a basic analysis of the underlying theories of property as they relate to traditional property (rivalrous in consumption) and intellectual property (nonrivalrous in consumption), see Epstein RA. 2005. Liberty Versus Property? Cracks in the Foundations of Copyright Law. *San Diego L. Rev.* 42:1 (arguing that for both types of property utilitarian tradeoffs are necessary).

12 Professor Whitt describes how the Brazilian Guajajara treated glaucoma with a local plant now depleted by exports to the tune of US$25 million per year, with corporations holding patents earning even more. See *supra* note 9, at 213–14. To the extent depletion of the plant is the problem, this dispute would seem to be between the Guajajara and the Brazilian government, not between the Guajajara and the foreign patentees. Brazil has the legal right and power to regulate or even prohibit the exporting

of the plant in question, especially if it is in danger of depletion.

13 As is discussed in detail below, the whole notion of composition-of-matter patents on naturally occurring substances is shaky under U.S. patent law itself, resting on a rationale that it is the isolated and purified form of the substance that is patented, not the substance as it exists in nature. In any event, the source country's long use of the plant for particular medicinal or other purposes would not be novel. Any claim that covered such a use (in the original source country) should be invalid for want of novelty.

14 For an argument that naturally occurring substances were long deemed by courts to be unpatentable and that Congress showed no intent in the 1952 Patent Act revisions to change that, see Demaine LJ and Fellmeth AX. 2002. Reinventing the Double Helix: A Novel and Nonobvious Reconceptualization of the Biotechnology Patent. *Stan. L. Rev.* 55:303, 366–84.

15 Agreement on Trade-Related Aspects of Intellectual Property Rights, Including Trade in Counterfeit Goods (TRIPS) Article 27(1).

16 This is likely to be the case, at least in the United States, for any biotech patent based on indigenous information. Simply claiming a procedure observed in use by an indigenous non-U.S. group is likely to result in an invalid patent, because U.S. patent law requires that the patent applicant be the inventor. 35 U.S.C. § 102(f) (listing as an exception to patent entitlement that the applicant *"did not himself invent the subject matter sought to be patented"*).

17 See also Heald PJ. 2003. The Rhetoric of Biopiracy, *Cardozo J. Int'l & Comp. L.* 519, 527: *"[N]o international patent can diminish [the indigenous group's] ability to cultivate, maintain, and use their existing resources."*

18 Kowalski T. 2002. International Patent Rights and Biotechnology: Should the United States Promote Technology Transfer to Developing Countries? *Loy. L.A. Int'l & Comp. L. Rev.* 25:41, 42. (citing Clive J. 2000. Global Status of Commercialized Transgenic Crops, Section 1. www.isaaa.org/kc/Publications/pdfs/isaaabriefs Briefs%2021.pdf)

19 See *supra* text accompanying note 11.

20 The *Moore* case, in which spleen cells extracted during therapy were used without the patient's knowledge to develop a new line of cells that became the object of a valuable patent, is surely the most notorious. *Moore v. Regents of the University of California*, 793 P.2d 479 (Cal. 1990). However, there are reports of other cases in which people discovered that cell or tissue donations were used in ways beyond their expectations and the original purpose of their donations. See Kolata G. Sharing of Profits Is Debated As the Value of Tissue Rises, *New York Times*, May 15, 2000, at A1.

21 Safrin S. 2004. Hyperownership in a Time of Biotechnological Promise: The International Conflict to Control the Building Blocks of Life. *Am. J. Internat'l L.* 98:641, 657.

22 See, for example, Gutterman AS. 1993. The North-South Debate Regarding the Protection of Intellectual Property Rights. *Wake Forest L. Rev.* 28:89, 122; Sturges ML. 1997. Who Should Hold Property Rights to the Human Genome? An Application of the Common Heritage of Humankind. *Am. U. Int'l L. Rev.* 13:219, 244. (asserting that developing countries view intellectual property as a publicly owned community asset that no single person should own); Whitt, *supra* note 9, at 252–53 (discussing a type of knowledge that the Maori call "tapu" and regard as sacred, believing that its misuse would cause the knowledge to lose its power).

23 See *Morton Salt Co. v. G.S. Suppiger,* 314 U.S. 488, 490–92 (1942); *Lasercomb America, Inc. v. Reynolds,* 911 F.2d 970, 977 (4th Cir. 1990). Another approach to limiting biopiracy directly under the patent law would be to eliminate the geographical limitations on disqualifying prior art; Bagley MA. 2003. Patently Unconstitutional: The Geographical Limitation on Prior Art in a Small World. *Minn. L. Rev.* 87:679, 724–27.

24 Kieff notes the need to find ways, possibly through contract, to solve the problem of allocating the wealth generated by a patent based on access to biodiversity, but he points out that without a patent system the wealth itself would be sacrificed. Kieff FS. 2002. Patents for Environmentalists. *Wash. U. J. L. & Policy* 9:307, 318.

25 I address technical questions of patent law primarily by reference to U.S. patent law, which is the only patent law with which I am even modestly familiar. I assume, but am willing to stand corrected, that my comments will apply in at least some general way to the patent laws of most countries.

26 This broad interpretation of section 101 also conflicts with the special statutes aimed specifically at protecting plants. These included the Plant Patent Act of 1930, 35 U.S.C. §§161–64 (protecting a new and distinct variety of plant that is asexually reproduced) and the Plant Variety Protection Act of 1970, 7 U.S.C.A. 57 (giving patent-like protection to sexually reproduced plants constituting a "new variety"). As a result of the *Chakrabarty* decision, plants are also patentable under the general Patent Act, a conclusion that the Court recently confirmed in *J.E.M. Ag Supply, Inc. v. Pioneer Hi-Bred Int'l, Inc.*, 534 U.S. 124 (2001). These interpretations of section 101 render both specific plant protection statutes largely extraneous. Had Congress thought that the Patent Act covered all living organisms invented by man, it is unlikely it would have seen any need for special plant protection statutes.

27 *Funk Brothers Seed Co. v. Kalo Inoculant Co.,* 333 U.S. 127 (1948).

28 447 U.S. at 309–10.

29 See Demaine and Fellmeth, *supra* note 14, at 316–17.

30 While the proteins or enzymes that constitute gene products do occur naturally in living organisms in the form for which patents may be sought, genes themselves rarely do. A typical gene as found in the DNA of a living organism contains both exons and introns, which are regions that, respectively,

are and are not "expressed" in protein production through the process of RNA transcription. See Karjala DS. 1992. A Legal Research Agenda for the Human Genome Initiative. *Jurimetrics J.* 32:121, 129-33. If a gene researcher seeks to patent a DNA sequence composed only of the natural gene's exons, he would be technically correct in saying that such a sequence of DNA does not occur naturally and he has therefore created something "new." Excluding such DNA sequences from patentability, therefore, requires more than appeal to the traditional exception for "naturally occurring substances." The basis for exclusion must lie in the fact that this DNA sequence stands in a complementary one-to-one correspondence with the messenger RNA that serves as the template for protein production. Karjala DS at 130–32. The issue is whether exon-only DNA is sufficiently different from natural substances—the messenger RNA—to justify a patent. The substantial transformation test offered by Demaine and Fellmeth addresses this question and would deny a patent unless the new sequence shows a substantially different biological function from its natural forebear in the organism. See Demaine and Fellmeth, *supra* note 14, at 444–45.

31 Section 101 of the Patent Act requires that the invention be "useful."

32 PTO Final Examiner Guidelines on Utility Requirement, 66 Fed. Reg. 1092, Dec. 29, 2000, at 1093 (Response to Comment 1).

33 For example, see Drahos P. 1999. Biotechnology Patents, Markets and Morality. *E.I.P.R.* 441, 443, which argues that treating an isolated and purified form as an invention exalts form over substance. Epstein contends that granting patents to the discovery of cDNA tags would be like giving Madame Curie a patent for radium because she first isolated it from pitchblende, in Epstein RA. 1996. Property Rights in cDNA Sequences: A New Resident for the Public Domain. *Roundtable* 575, 579. Meyers argues for distinguishing between a discovery and an invention, in Meyers, AS. 1996. Intellectual Property at the Public-Private Divide: A Response. *Roundtable* 581; and Looney makes the case that a gene unaltered by human intervention does not necessarily lose its status as an object of nature simply by taking it outside the body identifying its function, in Looney B. 1994. Should Genes Be Patented? The Gene Patenting Controversy: Legal, Ethical, and Policy Foundations of an International Agreement. *Law & Pol'y Int'l Bus.* 26:231, 264.

34 See *supra* note 14, at 366–84.

35 Section 101 provides for a patent to whoever "invents or discovers" patentable subject matter. In addition, the Constitution actually uses the word "discoveries" for the object of the exclusive rights Congress may afford to inventors: Congress shall have the power *"to promote the Progress of Science and useful Arts, by securing for limited Times to Authors and Inventors the exclusive Right to their respective Writings and Discoveries."* U.S. Constitution, Article I, § 8, cl. 8. The PTO has latched onto the "discover" aspect of section 101 as a basis for gene-sequence patenting. With painstaking care, Demaine and Fellmeth argue that the word "discovery" was more narrowly understood at the time the Constitution and the first Patent Act were adopted and in those contexts required some creative act by the inventor ("invention") and not merely that he had "found" something. See *supra* note 15, at 366–84. Demaine and Fellmeth argue further that the word "discovery" in the current Patent Act still requires "invention" and that Congress could not have intended to abrogate the requirement for human intellectual creativity if a patent is to be obtained. See also King J and D Stabinsky. 1999. Patents on Cells, Genes, and Organisms Undermine the Exchange of Scientific Ideas. *Chronicle of Higher Ed.*, at B6, B7; (*"'Products of nature' such as animals, plants, elements, and minerals could not be patented [before Chakrabarty], because they are found or discovered, not invented"*); compare to Sturges, *supra* note 22, at 242 (asserting not entirely correctly, see *supra* note 30, that gene researchers do not create anything new but only indicate where a gene might lie along a naturally occurring sequence).

36 This suggestion was made to the PTO but they rejected it. Their response was simply that *"Patent law provides no basis for treating DNA differently from other chemical compounds that are compositions of matter."* PTO Utility Guidelines, *supra* note 32, at 1095 (Response to Comment 10). This, of course, is completely erroneous, insofar as naturally occurring sequences of DNA are concerned. Technically, a naturally occurring DNA sequence is usually not patented in the form it is found in nature.

37 I am indebted to my former student Fariba Sirjani for making me aware of section 287(c) and the alternative approaches to limiting therapeutic-method patents elsewhere.

38 Erramouspe M. 1996. Comment on Staking Patent Claims on the Human Blueprint: Rewards and Rent-Dissipating Races. *UCLA L. Rev.* 43:961, 997.

39 Rai AK. Patenting Human Organisms: An Ethical and Legal Analysis, Draft paper prepared for President's Council on Bioethics, June 21, 2002. www.law.upenn.edu/fac/akrai/rai.patents.cob.doc.

40 TRIPS Article 27(3)(b). Fellmeth has pointed out to me in a private communication that Article 27(3)(b) of TRIPS may soon be ineffective as a result of bilateral free trade agreements between the United States and many other countries, especially in the western hemisphere. These agreements require protection generally equivalent to that available in the United States after *Chakrabarty*.

41 Epstein RA. 2003. Steady the Course: Property Rights in cDNA Sequences. *U Chicago Law & Economics,* Olin Working Paper No. 152, p. 577. Here Professor Epstein argues against patentability for the discovery of cDNA sequences, equating it to giving the first person to capture a fox an exclusive right to all foxes (an analogy that admittedly conflates physical and informational resources).

42 Obviously, gene sequences inside the organisms from which they derive often have very important uses. The issue here is whether there is another use, therapeutic or otherwise, to which the purified form of the gene can be put.

43 Barton J. 1995. Patent Scope in Biotechnology. *International Review of Industrial Property* 26:605, 614. Barton argues that *"highly basic patents that preempt a large area of research are unlikely to be beneficial."* Dickson describes Human Genome Organization (HUGO) officials' opposition to patents on cDNA sequences as "routine discoveries" that could inhibit incentives to establish gene function or develop applications, in Dickson D. 1995. HUGO and HGS clash over "utility" of gene sequences in US patent law, *Nature* 374:751. Epstein decries cDNA patents as opposed to patents for the fashioning of some new bacterium or virus with commercial applications. See *supra* note 41, at 578. See also Horn ME. 2003. *Note to* DNA Patenting and Access to Healthcare: Achieving the Balance among Competing Interests. *Clev. St. L. Rev.* 50:253, 263–64, 274–76.

44 See *supra* note 14, at 417–18.

45 Carroll AE. 1995. Comment on Not Always the Best Medicine: Biotechnology and the Global Impact Of U.S. Patent Law. *Am. U. L. Rev.* 44:2433, 2482; See Drahos, *supra* note 33, at 443.

46 See Carroll, *supra* note 45, at 2483–84; Chapman AR. 2000. Approaching Intellectual Property as a Human Right: Obligations Related to Article 15(1)(c), U.N. ESCOR, *Comm. on Econ., Soc. & Cultural Rts.*. U.N. Doc. E/C.12/2000/12, at ¶¶ 6, 57; but see Looney, *supra* note 33, at 244–45 (concluding that the impact of gene patenting on the dissemination of information is unclear).

47 See Marshall E. 1997. Need a Reagent? Just Sign Here.... *Science* 278:212 (describing the complex bureaucratic web resulting from general implementation of *materials transfer agreements* requiring the surrender of property rights in subsequent discoveries in exchange for materials intended for research use).

48 Heller MA and RS Eisenberg. 1998. Can Patents Deter Innovation? The Anticommons in Biomedical Research. *Science* 280:698, 699-700; see also *supra* note 14, at 419–21 (noting that *"multiple patentable sequences [ESTs, codons, SNPs, etc.] can originate in the same gene, resulting in upstream patentees owning rights to different parts of the same gene"*); See Horn, *supra* note 43, at 265–67.

49 See Heller and Eisenberg, *supra* note 48, at 700–701; see also Burk DL and MA Lemley. 2002. Is Patent Law Technology-Specific? *Berk. Tech. L.J.* 17:1155, 1195–96 (arguing that the Federal Circuit's application of a stringent disclosure requirement and a lax nonobviousness requirement to biotech inventions exacerbates the anticommons problem by resulting in a multitude of narrow upstream patents that can strangle downstream product development).

50 But compare *to* Kieff FS. 2001. Facilitating Scientific Research: Intellectual Property Rights and the Norms of Science: A Response to Rai and Eisenberg. *Nw. U. L. Rev.* 95:691, 704. (concluding, contrary to the premise in the text, that patent availability for basic biotechnological inventions increases the funds available for research and commercialization and will more likely promote traditional scientific norms, such as independence and objectivity, than would be observed in a world without such patents); Adelman DE. 2005. A Fallacy of the Commons in Biotech Policy. *Berkeley Tech. L.J.* 20:985, 988 (states *"there are few signs that biotech patenting has impeded biomedical innovation"*). One commentator has argued that biomedical patents, by raising the cost of research tools, actually promotes fundamental scientific advances by giving scientists additional incentive to innovate at the level of basic scientific theory. Lee P. 2004. *Note to* Patents, Paradigm Shift, and Progress in Biomedical Science. *Yale L.J.* 659, 694–95.

51 Center for International Environmental Law. *The 1999 WTO Review of Life Patenting Under TRIPS.* ciel.org/Publications/WTOReviewofLPunderTRIPS.pdf (hereinafter cited as 1999 CIEL Report).

52 Demaine and Fellmeth's substantial transformation test would allow a product patent on genes, including human genes, biochemicals, and tissues, that are so substantially transformed from their natural state that they perform a different biological function than they do naturally. Thus, anything taken out of the "common heritage" would have to be so changed from its natural state that a patent could not be used to control its natural use. See *supra* note 14, at 444–45. Effectively, the substantial transformation test they recommend for patentability should mean that no composition-of-matter patents would issue on naturally occurring genes or their products, because in order to perform a different biological function the substances almost certainly would have to have a different structure. Their test is thus one of the degree of *inventiveness* an applicant must show in order to get a patent on a composition of matter that he has modified from its natural form. This test does seem to leave the theoretical issue of whether a composition-of-matter patent could issue on a naturally occurring substance that has been isolated and purified and found to perform not only its natural function but also a completely different biological function. In this purely theoretical case, there remains a danger of control over its natural use. Product patents give rights to make, use, or sell the product, covering even uses not disclosed in the patent application. Limiting protection to a method patent covering only the use of the isolated and purified substance in a specific therapy would avoid even this theoretical objection. Expressly restricting naturally occurring substances to method patents would not in any way preclude application of the substantial transformation test to substances that *are* structurally transformed. Indeed, that test is then vital in determining whether the applicant has truly "invented" something new or has simply made minor modifications of nature's handiwork.

53 For example, Demaine and Fellmeth (see *supra* note 14, at 434: *"From a purely positivistic perspective, a patent on a DNA molecule or protein entitles the patentee to forbid cell building, transcripting, and reproducing by any individual whose genome contains that DNA molecule or uses that protein, as such activities constitute using and making unauthorized copies of the DNA molecule or protein."*

54 See *supra* note 14, at 435. While no court will be led to find infringement based on the natural operations of living organisms that have been taking place for eons, Demaine and Fellmeth point to other examples that may be closer to reality: A patient whose cells have been patented, for example, would be prohibited from donating or selling blood or sperm without a license from the patentee.

55 In a private communication, Aaron Fellmeth has offered some variations on the "universal heritage" argument. Some might argue, for example, that a gene is still part of the common heritage of mankind even though only a limited group carries it. The underlying principle would be that a gene is nature's, or God's, handiwork and cannot therefore be legally owned or monopolized by anyone other than the whole of humankind. One can get to this same result much more mundanely, but analytically more cleanly, by reactivating the traditional rule against the patenting of naturally occurring substances. And insofar as the argument is based on not monopolizing something created by God or nature, it still leaves open the question of whether and when patents should be available for structurally modified products of nature. For *that* determination we need something like the substantial transformation test of Demaine and Fellmeth. Another argument might be that genes are not just physical products but constitute information about nature and that such information should not be monopolized. This, however, is at bottom an attack on all of intellectual property law, because monopolization of information is precisely what patent and copyright laws do. Every invention carries with it information about the operation of nature, because technology works by natural laws. Consequently, the "information about nature" argument is not easily limited to genes and gene products.

56 It might be noted that the Biodiversity Convention requires that members facilitate access to genetic resources, subject to fair sharing of the benefits after genetic resources have been obtained by prior informed consent. Convention on Biological Diversity Arts. 15(2), 15(4), 15(5), & 15(7). The convention thus rejects any form of the "common heritage" doctrine that would prohibit all forms of commercialization. Downes DR. 1993. New Diplomacy for the Biodiversity Trade: Biodiversity, Biotechnology, and Intellectual Property in the Convention on Biological Diversity, *Touro J. Transnat'l L.* 4:1, 9. Similarly, Article 4 of the *Universal Declaration on the Human Genome and Human Rights,* adopted by the United Nations General Assembly, G.A. Res. 152, U.N. GAOR, 53d Sess. U.N. Doc. A/53/625/Add.2 (1998)[hereinafter referred to as *Universal Declaration*], declares that the human genome in its natural state shall not give rise to financial gains. This too seems to allow commercialization of the human genome outside its "natural state," which would presumably include its "isolated and purified" form. This goes well beyond what would be permitted by traditional patent law under the exception for naturally occurring substances.

57 See Chapman, *supra* note 46, at 3; see Downes, *supra* note 56, at 4; see Sturges, *supra* note 22, at 242, 244–45; see 1999 CIEL Report, *supra* note 51, at 4.

58 The argument might be that every part of the human body is sacred and therefore may not be commodified. If this is the argument, however, it rejects even commodification of an unpatented human-gene-related product. It is markets, not patents, that make something a commodity. This approach risks losing many products that have a potential for reducing human suffering and disease, which is a heavy price to pay in support of what is essentially a metaphysical principle.

59 Poste G. 1995. The Case for Genomic Patenting. *Nature* 378:534, 536; see Rai, *supra* note 39, 55.

60 Rai, *supra* note 39, at 55.

61 Compare Gross N and J Carey. Who Owns the Tree of Life? *Business Week,* Nov. 4, 1996, p. 194 (describing the Papua New Guinea Hagahai's apparent immunity to a virus that usually causes leukemia); See King and Stabinsky, *supra* note 35, at B6 (describing patent applications for cells and genes of New Guinea tribes because of an apparent immunity against certain viruses); see Frow J. 1995. Elvis' Fame: The Commodity Form and the Form of the Person. *Cardozo Stud. L. & Lit.* 7:131, 150 (describing applications for patents on the cells of individuals from Papua New Guinea and the Solomon Islands, each of them carriers without apparent harm of the HTLV-I virus). In addition, remote, isolated populations often make it is easier to trace disease heredity, which means that studying the genes from these groups can speed up gene discovery and drug development. See Gross and Carey, *supra* note at 61; See Safrin, *supra* note 21, at 660–61 (DNA from homogeneous and isolated populations can facilitate discovery of disease-causing genes).

62 See *supra* note 14, at 437–38 (discussing the worldwide concern about these issues).

63 See Looney, *supra* note 33, at 238.

64 *Id.* at 238–39.

65 See Karjala, *supra* note 30.

66 See *supra* text accompanying note 44.

67 See *supra* note 61 (describing attempts to patent cells and genes of indigenous groups based on an apparent immunity to diseases that afflict developed countries).

68 See CIEL Report, *supra* note 51, at 4.

69 See Chapman, *supra* note 46, at ¶ 64.

70 See Kieff, *supra* note 24, at 318–19 (arguing that a patent can obviate the perceived need of the innovator of a new and valuable seed to use potentially dangerous technologies to protect against competitive sale of seed by initial purchasers).

71 *Id.* at 318 (arguing that where new technologies are harmful to environmental goals, the existence of a patent at least does not exacerbate the harm, because a patent's right to exclude does not provide an affirmative right to *use* the technology by the patentee, so such use can be regulated or prohibited).

72 See Carroll, *supra* note 45, at 2471 (citing ET Penrose, *The Economics of the International Patent System* 116–17 (1951)).

73 See Looney, *supra* note 33, at 240.

74 Lazzarini Z. 2003. Making Access to Pharmaceuticals a Reality: Legal Options Under TRIPS and the Case of Brazil. *Yale Hum. Rts. & Dev. L.J.* 6:103, 115–119 (arguing that access to pharmaceuticals should be thought of as a human right).

75 See Sturges, *supra* note 22, at 244.

76 See Whitt, *supra* note 9, at 240.

77 See Chapman, *supra* note 46, at ¶ 16. One commentator has said that forcing countries to adopt patent laws and accept conditions of technology transfer laid down by the holder of the patent is "technological colonialism." See Carroll, *supra* note 45, at 2466-67.

78 Anawalt HC. 2003. International Intellectual Property, Progress, and the Rule of Law. *Santa Clara Computer & High Tech. L.J.* 19:383, 404 (*"The linkage of WTO membership to mandatory intellectual property rights and procedure should be ended"*).

79 See Gutterman, *supra* note 22, at 122, 137; *compare* Downes, *supra* note 56, at 22–23 and Lazarini, *supra* note 74, at 111 (both concluding that the empirical evidence on the inhibiting or beneficial effects of intellectual property rights on technology transfer is scanty); see Seeratan, *supra* note 10, at 383 (noting that industrialized countries did not adopt strong intellectual property laws until they themselves had reaped the benefits of nonprotectionist policies). Even within the United States there is much anecdotal information that recent advances in medicine do not reach many of those who need it or their physicians, often even years after the information is publicly available. For example, Begley S. Too Many Patients Never Reap Benefits Of Great Research. *Wall Street Journal,* Sept. 26, 2003, at B1.

80 On these issues see Oddi AS. 1987. The International Patent System and Third World Development: Reality or Myth? *Duke L.J.* 831–52.

81 See Carroll, *supra* note 45, at 2468; see Chapman, *supra* note 46, at ¶61; see Seeratan, *supra* note 10, at 375 (asserting that the TRIPS requirement for both product and process patents will substantially increase the cost of pharmaceuticals).

82 See Gutterman, *supra* note 22, at 122-23. One might question why a patent owner would adopt this strategy, however. It would seem that if he or she is unwilling to import into a given country, one would be better off economically by licensing local production. One possible explanation is fear of grey market "leakage" that is difficult to control by contract. But even this explanation is unsatisfying, because under TRIPS, if the country has the local ability to manufacture the invention, it may grant a compulsory license. TRIPS Article 31. Of course, any such compulsory license is supposed to be primarily for local consumption. *Id.* Article 31(f). However, if grey market leakage is a problem under a negotiated license, where the patentee has direct contact with the licensee, it would seem to be an even bigger problem under a compulsory license.

83 See Oddi, *supra* note 80, at 844; see also Seeratan, *supra* note 10, at 386 (*"None of the pharmaceutical companies really depend on achieving profits in developing countries, which generally only account for a minimal percentage of drug sales worldwide"*); *compare to* Anawalt, *supra* note 78, at 397 (*"Adequate incentives for innovation do not depend on mandatory international intellectual property rules"*).

84 See Oddi, *supra* note 80, at 846.

85 Additional costs of a patent system come in the form of training patent officials, lawyers, and judges. See Carroll, *supra* note 45, at 2468.

86 Attaran A and L Gillespie-White. 2001. Do Patents for Antiretroviral Drugs Constrain Access to AIDS Treatment in Africa? *J. Am. Med. Ass'n* 286:1886.

87 *Id.,* at 1890. They also discovered that the option to patent antiretroviral drugs often went unexercised, surely the result of the meager expected financial return from very poor countries. This supports the conclusion of Professor Oddi that increased incentive for innovation from the possibility of obtaining patents in poor countries is negligible, that is, none of these drugs is "patent-induced" with respect to the patent law of any given African country. See Oddi, *supra* note 83 and accompanying text.

88 See *supra* note 86, at 1891. Attaran and Gillespie-White blame lack of international funding, even to purchase drugs at cost, rather than patents, for the low level of antiretroviral treatment in Africa.

89 See *supra* note 86; see Lazzarini, *supra* note 74, at 135. Aaron Fellmeth has reminded me in a private communication that an effective monopoly might result not only from a patent but also from trade secret law or pursuant to exclusive pharmaceutical marketing approvals.

90 See Lemley, *supra* note 8 and (text at) note 11.

91 See Scherer FM and J Watal. 2002. Post-TRIPS Options for Access to Patented Medicines in Developing Nations, *J. Internat'l Econ. L.* 913, 928 (*"When prices are higher in one nation than in others, there is a tendency for arbitrage to occur through what is known as 'parallel trade'."*); see also *supra* note 82.

92 More generally, enforceable and accurate price

discrimination should push output to the full competitive output level, but for this to occur arbitrage between high- and low-value users must be prevented. See Kieff, *supra* note 24, at 311 and note 23.

93 See Scherer and Watal, *supra* note 91, at 9:25–28; see Lazzarini, *supra* note 74, at 125.

94 They will also be avoided to the extent the developed countries adopt notions of "reference pricing," requiring, for example, that their own domestic prices to be no higher than those charged elsewhere. See Scherer and Watal, *supra* note 91, at 929.

95 Blumenstyk G. 2003. Coalition Seeks to Make Agricultural-Biotechnology Tools More Widely Available. *Chr. Higher Ed.*, July 11. chronicle.com/daily/2003/07/2003071105n.htm.

96 TRIPS Article 31; See Lazzarini, *supra* note 74, at 125.

97 World Trade Organization, Ministerial Conference, Doha Declaration on the TRIPS Agreement and Public Health, No. 01-5770, Nov. 14, 2001, ¶ 5(b). In most cases compulsory licenses can be granted only after good faith negotiations with the patentee have failed to result in a voluntary license "on reasonable commercial terms and conditions." TRIPS Article 31(b). However, nothing in TRIPS supplies any standard of reasonableness, so the failure of the patentee to agree to a member state's good faith offer to pay what it believes it can afford, given its other obligations and the country's needs, should suffice to permit going ahead with the compulsory license. Moreover, even the obligation to negotiate is waived in cases deemed to be a "national emergency."

98 World Trade Organization General Council, Implementation of Paragraph 6 of the Doha Declaration on the TRIPS Agreement and Public Health, Decision of 30 August 2003, WT/L/540, 43 I.L.M. 509 (2004).

99 Some drug manufacturers have begun experimenting with "out-licensing," under which the patentee licenses generic manufacturers who agree to supply medicines to poorer countries. Friedman MA, H den Besten and A Attaran. 2003. Out-licensing: a practical approach for improvement of access to medicines in poor countries. *The Lancet* 361:341. Requiring pills to have different colors and shapes could be helpful in inhibiting parallel importing back into the more lucrative markets. *Id.* at 343; see also Hensley S. Pharmacia Nears Generics Deal On AIDS Drug for Poor Nations, *Wall Street J.*, Jan. 24, 2003.

100 See Seeratan, *supra* note 10, at 403–4 (*"Many human rights activists assert that the TRIPs provisions on the patenting of pharmaceuticals violates basic human rights by compromising the ability of poor countries to access essential medicines"*). The *Universal Declaration on the Human Genome and Human Rights demands* that *"Benefits from advances in biology, genetics, and medicine, concerning the human genome, shall be made available to all, with due regard for the dignity and human rights of each individual."* See *Universal Declaration*, *supra* note 56, Article 12(a). Another commentator argues that distributive justice requires providing all countries with access to the benefits of gene research. See Looney, *supra* note 33, at 240 (*"Gene patenting is ethically suspect if it concentrates genome benefits in those few countries fortunate enough to have the resources to obtain gene patents, when all humans should enjoy such benefits"*). In these situations, however, it is not clear why gene patents or even medicine generally are singled out. Starvation is a huge problem in the world, which has a production capability more than sufficient to supply everyone alive with at least a minimal food supply. Unequal distribution of resources, both natural and human-made, almost inevitably raises questions of distributive justice. To the extent that patent law serves as an incentive for innovation, a patent does not create the injustice. It only brings more clearly into focus that there is widely different access to valuable resources between rich and poor countries. Without the patent, by assumption, *nobody* would have access to the innovation. With the patent, some relatively wealthy people do. But the poor are no worse off than they were before the innovation became available.

Access and Benefit Sharing: Understanding the Rules for Collection and Use of Biological Materials

CARL-GUSTAF THORNSTRÖM, *Senior Research Advisor, Agriculture, Sida/SAREC; Docent-Associate Professor, Guest Researcher and Advisor on Genetic Policy, Swedish Biodiversity Center, Swedish University of Agricultural Science, Sweden*

ABSTRACT

The rules that govern the collection and use of biological matter have changed dramatically in the last 15 years. Arising out of the Convention on Biological Diversity (CBD), the Access and Benefit-Sharing (ABS) project applies to research carried out for either purely scientific or commercial reasons, for which organisms or parts thereof and/or related traditional knowledge are obtained from countries that are party to the CBD and their local and indigenous communities. Other agreements have added new ABS legislation to govern the acquisition and use of biological material and related information. Everyone—including tourists, nature conservationists, scientists, photographers, and journalists—is subject to these new regulations. But scientists and researchers who seek to access and use proprietary genetic resources, biological matter, and related information (such as traditional knowledge and farming know-how) are especially affected by the ABS project. It is essential for scientists and researchers to understand the fundamental principles of ABS. This includes knowing the relevant rules, regulations, laws, customs, and conditions for benefit sharing in the country where one intends to conduct research and/or collect samples. One must carefully plan ahead for any such activities by contacting key organizations and filing the proper documentation. Lack of planning may lead to unfortunate and undesired outcomes, including fines, imprisonment, deportation, and denied future access. Planning is critical.

1. INTRODUCTION

According to the Convention on Biological Diversity (CBD), biological resources belong to the states in whose territory the resources are found. So, with regard to ownership, biological resources are no different from mineral resources, oil, or timber. However, in recent years there have been times when this principle of ownership has not been respected. Resources were exported, developed, and commercialized without the consent of the countries that provided them and without enabling those countries to partake in the benefits that resulted from these activities. In order to prevent this *biopiracy* and create a climate of mutual trust, the community of states undertook to regulate the handling of genetic resources in a binding international agreement referred to as the CBD.

CBD implementation is not only a moral obligation, but also a legal one that binds member states. The goal of the CBD is to conserve biological diversity and to promote its sustainable use in conjunction with the fair and equitable sharing of benefits. Responsibility for implementing the agreement is given to the state in which the biological material originates. However, all states have a responsibility to cooperate in implementing and enforcing the agreement. For industrialized countries, this means supporting biodiversity-rich, but often economically poor countries in their efforts to conserve and manage biodiversity. The keys to these collaborative efforts are technology transfer and cooperative research. The CBD contains rules that clarify the rights and responsibilities of all of the parties involved in these efforts.

Thornström CG. 2007. Access and Benefit Sharing: Understanding the Rules for Collection and Use of Biological Materials. In *Intellectual Property Management in Health and Agricultural Innovation: A Handbook of Best Practices* (eds. A Krattiger, RT Mahoney, L Nelsen, et al.). MIHR: Oxford, U.K., and PIPRA: Davis, U.S.A. Available online at www.ipHandbook.org.

To advance its mission, the Conference of the Parties (COP) of the CBD decided in 2004 to create the Access and Benefit-Sharing (ABS) project, an international program overseeing access to genetic resources and the sharing of benefits arising out of their utilization. Negotiations over ABS began in 2005. It is anticipated that it will take up to ten years for it to be completely established.

Correspondingly, over the last decade a number of new legally binding agreements regarding biological material/related information have been signed and ratified by United Nations member countries. Examples are the CBD, the Agreement on Trade Related Aspects of Intellectual Property Rights (TRIPS), treaties of the International Union for the Protection of New Varieties of Plants (UPOV, particularly the 1991 treaty), the International Treaty on Plant Genetic Resources for Food and Agriculture (the Treaty), the Intergovernmental Committee on Intellectual Property and Genetic Resources, Traditional Knowledge and Folklore (ICGTK) that meets under the World Intellectual Property Organization (WIPO), the Cartagena Protocol on Biosafety under the CBD, and the nonbinding Global Crop Diversity Trust, among others.

All these agreements add new legal dynamics to ABS legislation that addresses the acquisition and use of biological material and related information (such as ethnobiology and traditional knowledge). Indeed, there is a new world order emerging in relation to biological matter, a fact that changes the nature of public and private sector research and development efforts.

Everyone, including tourists, nature conservationists, scientists, photographers, and journalists, are subject to these new regulations. Particularly targeted are scientists and researchers who make significant use of proprietary genetic resources, biological matter, traditional knowledge, and farming know-how. Such knowledge may, in national legislation, be considered intellectual property (IP) or trade secrets, and, as such, neither in the public domain nor available for unauthorized appropriation.

Violation of the new access laws (for example, by scientists conducting unauthorized collection activities) can result in fines, imprisonment, and denial of future visits to the collection area. A violation may result in increased transaction time for obtaining formal access permits. A violation may also result in a prohibition on other scientists working in a country.

Unfortunately, it can take a lot of time to get the requisite permissions for collecting biological specimens. In Brazil, approximately 400 applications to use biological materials are received annually. The processing rate for these applications is 25–50 per year. This is due to strict ABS legislation. A similar situation prevails in Colombia, which has received some 50 access applications over the last five years. Of the 50 applications, 22 were denied due to improper access behavior, and one application (for biological research on dolphins) was approved. The remaining applications are still being processed.

2. THE NEW GENETIC-POLICY LANDSCAPE

Below is a brief summary of each agreement in the new genetic-policy landscape, with regard to use of biological matter.

- The CBD, adopted in 1992 at Rio de Janeiro, provides national sovereignty over genetic resources and access conditions for other sovereign parties.
- TRIPS, adopted in Marrakesh in 1994, provides a minimum IP protection standard for biological matter such as plant varieties, microorganisms, and microbiological processes.
- ICGTK was set up in 2001 by WIPO to discuss IP issues relating to access to genetic resources and the protection of traditional knowledge, including disclosure requirements in patent applications.
- UPOV provides legal protection for plant varieties fulfilling the NDUS criteria (new, distinct, uniform, and stable), while including a breeder's exemption and farmer's privilege.
- The International Treaty on Plant Genetic Resources for Food and Agriculture, adopted in Rome in 2001, provides a multilateral system of access and benefit sharing under a

revised material transfer agreement (MTA) in relation to some 35 defined crops.

- The Global Crop Diversity Trust, set up in 2002, is an attempt by the Food and Agriculture Organization (FAO) of the United Nations and the World Bank to establish a trust fund for global *ex situ* collections of germplasm of relevance for food and agriculture.
- The Cartagena protocol, adopted in Montreal in 2000, provides rules for the transfer of genetically modified living organisms across borders.
- In 2002, the CBD adopted the Bonn Guidelines on Access to Genetic Resources and Fair and Equitable Sharing of Benefits Arising out of their Utilization. A voluntary supplement to the CBD, the Bonn guidelines offer basic information about the rules on access and concrete procedures (or protocols) to follow. The objectives of the Bonn guidelines in relation to academic research are:
  - to promote awareness of the implementation of relevant provisions of the CBD
  - to provide parties to the CBD and stakeholders with a transparent framework to facilitate access to genetic resources and ensure fair and equitable sharing of benefits
  - to provide information about the practices and approaches to be adopted by users and providers in the context of access and benefit sharing
  - to promote capacity building and the transfer of appropriate technology to providing parties

3. IP RIGHTS

IP rights are temporary, exclusive ownership rights to the *application* of an idea. Such rights may be granted in the form of patents, trademarks, industrial designs, copyrights, geographical indications, or trade secrets. Given the breakthroughs in biotechnology and information technologies in the last few decades, intellectual property has expanded considerably into the area of biological matter. For example, in the area of agricultural research the following biological matter falls under various IP regimes:

- plant seeds or other propagative plant parts collected after 1994
- plant and animal cell lines
- plasmids
- other recombinant vectors
- gene promoters
- gene markers
- transformed bacteria
- isolated plant DNA
- plant cDNAs
- isolated animal DNA
- bacteria (other than the transformed bacteria)
- isolated/purified proteins (other than those obtained by purchase of laboratory reagents)
- equipment for specialized laboratory purposes
- information regarding laboratory methods
- genomic sequence database(s)
- other nucleotide sequence database(s) such as PCR primer databases, cDNA sequences

Traditional and farming knowledge is also protected under the CBD and the Treaty, subject to national legislation. In general, researchers in the public sector, using proprietary biological materials and related information owned by private sector companies, may have to sign agreements stipulating further use and confidentiality conditions. Furthermore, public research products using proprietary materials and methods may be required to sign license and royalty agreements with those who hold the relevant IP rights.

It should always be remembered that IP protection is territorial; it may be recognized in some countries and not in others. This territoriality of intellectual property has implications for scientists' freedom to operate: what they may be able to do in one country may not be possible in another country without an appropriate license.

4. THE EMERGING NEW WORLD ORDER REGARDING BIOLOGICAL MATTER

The new national sovereignty over biological and genomic matter mandates new rules for the access and use of biological matter and related information. Examples of recent legislation in Latin America include the Andean Pact Decision 391/96: Common Regime on Access to Genetic Resources.[1] Peru, in accordance with its National Strategy on Biological Diversity (Decreto Supremo No. 102-2001-PCM), recently added legislation relating to traditional knowledge (Law 27.811, August 2002), and a special national authority (INRENA) has been established to deal with ABS issues. In Africa, the Organization for African Unity (OAU, now the African Union) Model Law for the Protection of the Rights of Local Communities Farmers and Breeders and the Regulation of Access to Biological Resources (adopted in Addis Abeba, December 2001) has been used by some nations as a model for regulating access to biological material. In 2001, India adopted a bill to protect plant varieties and farmer's rights (Bill No. 123 of 1999) and, in 2000, a biodiversity bill (Bill No. 93 of 2000).

These examples illustrate the different kinds of regulations now facing foreign parties, whether scientists, commercial prospectors, or nature conservationists, who seek access to biological material and information. The examples suggest a need for a coherent understanding of researchers' obligations under TRIPS and the International Treaty on Plant Genetic Resources for Food and Agriculture.

5. OBTAINING RESEARCH PERMITS WITH ABS PROVISIONS

The following issues should be addressed before collection leading to R&D begins:

- Under which conditions may I, as a scientist, *enter* another sovereign state in my scientific capacity?
- Under which conditions may I, as a scientist, *collect* biological material and related information?
- Under which conditions may I, as a scientist, *carry out or export* biological material and related information from that sovereign state?
- Under which conditions may I, as a scientist, *make further use* of collected biological material and related information?

Before collecting for purposes of research, contact your counterpart in the country to find out which rules apply. It is useful to also contact that country's embassy/consulate/legation in your own home country. Information on the following topics would be useful:

- requirements for foreign parties to access biological material and information
- conditions of benefit sharing
- conditions regarding applying IP rights
- national focal point for handling ABS issues
- ABS conditions (are written instructions available to foreign parties?)

6. PREPARING YOUR RESEARCH PERMIT APPLICATION

After having checked with your counterpart or the relevant embassy, fill in any research permits provided by relevant authorities in the country you plan to visit. If ABS issues are not specified, then do the following:

- Present briefly the scientific objectives, refer to your national counterpart, and include specifics about what biological matter and related information is planned for collection.
- Indicate how you will collect the material and with whom, and state if duplicates will be deposited in the country where collection is carried out.
- Indicate that, if necessary according to the country's laws, you will apply for an export permit.
- Indicate how further use of the collected material will be made upon your return, such as:
  - showing material and sharing information at seminars and lectures

- sharing collected material and information with other scientists, botanical gardens, and/or private companies
- using the collected materials and/or information in the R&D of products that may eventually be commercialized

- Indicate, in case of possible commercialization, what steps you have taken to comply with relevant national ABS provisions in the country concerned.

7. IF YOU GET INTO TROUBLE

Should you encounter difficulties, or just have questions related to ABS, consult the clearinghouse or the legal department of your research institution, university, or college for specific advice and information about the policies and guidelines your home institution has implemented to comply with the CBD and other agreements.

If the answers you get are inadequate, then consult your country's ABS focal point or ask research funding agencies about colleagues who have contacts in the country concerned. Contact the embassy of the country concerned in case their national authorities do not answer; try direct contact by telephone. Remember that it is usually far easier to be cautious and proceed correctly than it is to fix a problem after it has happened.

8. ISSUES OF UNCERTAINTY

Unfortunately, there is still uncertainty concerning the potential restrictions of accessing, using, and transferring biological material and related information. These include, but are not necessarily limited to:

- **international seas and arctic areas, which are not covered by national laws.** ABS issues regarding these areas are not fully regulated in international conventions
- **protection of traditional/indigenous knowledge, which is still being established.** Such protection is possible under CBD Article 8 (j), subject to national legislation. At present there are some 20 national legislations in place using the sui generis provisions. However, these have not yet been tested by the TRIPS Council and are still under discussion in the Intergovernmental Committee.
- **global consensus on Access and Benefit Sharing for all genetic resources, which is still being developed.** This initiative, following the CBD Bonn guidelines on ABS, is mainly discussed in the Intergovernmental Committee. Today access and exchange of the Treaty through the Treaty will be multilateral, according to a standardized Prior Informed Consent/ Mutually Agreed Terms and a standardized MTA agreed by the governing body of the Treaty. Access to and exchange of all other genetic resources and material (excluding human material) is presently subject to bilateral provisions set in national legislation. Some 35 countries have legislation in place, including India, Brazil, and the Andean Community. The ABS project under CBD and the Intergovernmental Committee is an attempt to try to standardize ABS for non-Treaty material.[2]
- **legal protection of plant varieties inside/ outside UPOV.** Landraces and farmer varieties/primitive cultivars are protected, subject to national legislation under CBD Article 8 (j) and the Treaty Article 9. The NDUS criteria of UPOV do not normally cover landraces and farmer varieties/primitive cultivars, but these are still the result of intellectual innovation, mainly by local farmers. In TRIPS Article 27.3 (b) provisions are given to introduce sui generis protection of such plant material. India's plant variety protection and farmer's rights bill provide such protection.
- **Certificate of Origin / Disclosure of Origin (CO) in IP applications.** Discussions are ongoing in CBD and in the Intergovernmental Committee regarding a compulsory requirement in IP/patent law that applicants must provide a Certificate of Origin that verifies bona fide access (CBD's Prior Informed Consent/Mutually Agreed Terms) of genetic resources used. Controversy exists with regard to CO

"when possible" vs. "always required" for granting intellectual property.

- **nonlist material in the Treaty, nonparties, and repatriation of genetic resources.** Questions remain, for example, regarding material that is currently designated under the agreement between the Consultative Group on International Agricultural Research (CGIAR) and FAO of 1994, but that is not on the Treaty crop list (such as groundnuts and soybean). The roles and rights of parties who have not signed/ratified the Treaty still remain open questions, as are provisions in the MTA accompanying repatriation to parties/nonparties of the Treaty.
- **requests for germplasm samples.** The CGIAR genebank collections will form the base of multilateral crop material under the Treaty. The majority of requests for germplasm presently come from developing countries, which increasingly (referring to the Cartagena protocol) require that centers of the CGIAR shall fully guarantee that delivered germplasm does not contain genetically modified crops. Checking every such delivery for a CGIAR center represents significant costs.

9. CONCLUSIONS

The implementation of the ABS system is ongoing, both at the national and international levels. Thus, the relevant authorities may therefore not be clearly designated, and the established procedures may not be transparent and smooth. If the scientists can choose where to carry out research and collection activities, he or she should examine the relevant experience of other researchers and institutes. The national law of the providing countries regulates the ABS procedure. This includes the definition of the competent government agency and of the other stakeholders that must be involved. If relevant national legislation does not yet exist, access permits may be issued on a case-by-case basis, based on general principles of law and similar proceedings and rules.

The ABS procedure may also be combined with other licenses and permits, including for research, collection, and export, as well as Convention on International Trade in Endangered Species of Wild Fauna and Flora (CITES) permits and so forth. However, ABS will not yet apply in most cases and countries. Standardized MTAs and benefit-sharing agreements for similar resources and similar uses may already exist (taxonomy, collection, research, commercialization).

The Bonn guidelines recommend public participation at the local level with regard to all government decisions concerning issues involving resources and permits that affect the public. This may lead to the need for stakeholders at different levels to grant their *prior informed consent*, which may ultimately cause the ABS procedure to become more complex and time consuming. Based on its current complexity, ABS legislation can be divided into four broad categories:

1. **No ABS situation.** The research does not involve any access situation or genetic resources. Thus no ABS contract is necessary. However, other research permits may be required.
2. **Simple ABS situation.** The research involves the collection and transfer (including export) of samples for an inventory. A (standardized) MTA is sufficient.
3. **ABS situation.** The export of samples is required for further analysis and study in a laboratory abroad. No further exploitation is planned. A simple ABS contract is sufficient.
4. **Complex ABS situation.** The proposed research involves various steps, including possible research for commercial purposes or the use of traditional knowledge. A full ABS contract is required.

Whatever the ABS situation turns out to be, in the final analysis the most critical aspect will be to understand the ABS regime; to thoroughly research the laws, rules, regulations, and customs of the country where you intend to conduct research and/or collect; and to *plan ahead* for all foreseeable contingencies. This will make a rewarding trip far more likely, and your subsequent research activities will have broad benefits that are consistent with the spirit and goals of the ABS project. ■

ACKNOWLEDGEMENTS
This chapter is based on a heavily edited synthesis of several documents prepared for a range of purposes. We are grateful to our colleagues and the publishers for having allowed us to edit and use extracts from these copyrighted materials.[3]

CARL-GUSTAF THORNSTRÖM, *Swedish Biodiversity Center, Swedish University of Agricultural Sciences, PO BOX 7007, 750 07 Uppsala, Sweden.* *Carl-Gustaf.Thornstrom@cbm.slu.se*

1 www.comunidadandina.org/INGLES/normativa/D391e.htm.

2 See, for example, www.wipo.int/tk/en/genetic/proposals/index.html.

3 Thornström CG and L Björk. 2006. Accessing Others Proprietary Biological Matter and Related Information–Towards a Handbook in Access and Benefit Sharing and Related Intellectual Property: Part Three: Entering into Agreements. Unpublished.

Biber-Klemm S and S Martinez. 2006. *Access and Benefit-Sharing Good Practice for Academic Research on Genetic Resources*. First Edition. Swiss Academy of Sciences: Bern. www.scnat.ch.

Sennerby Forsse L. 2003. *Inventory Regarding ABS Legislation in Selected Countries*. Swedish Research Council/FORMAS: Stockholm.

Thornström CG and L Björk. 2006. Accessing Others Proprietary Biological Matter and Related Information–Towards a Handbook in Access & Benefit Sharing and Related Intellectual Property. Part One: Chapeau.

CHAPTER 16.3

Access and Benefit Sharing: Illustrated Procedures for the Collection and Importation of Biological Materials

CARL-GUSTAF THORNSTRÖM, *Senior Research Advisor, Agriculture, Sida/SAREC; Docent-Associate Professor, Guest Researcher and Advisor on Genetic Policy, Swedish Biodiversity Center, Swedish University of Agricultural Sciences, Sweden*

LARS BJÖRK, *Associate Professor of Ethnobotany and Pharmacogonosy, Systematic Botany, EBC, Uppsala University; Member of the Swedish Scientific Board of Biodiversity, Sweden*

ABSTRACT

The Convention on Biological Diversity (CBD) contains rules that clarify the rights and responsibilities of parties accessing biological resources from member nations. One aspect of the convention addresses the system that governs access to genetic resources and how the benefits arising from their use are shared. This legislation is commonly called the Access and Benefit-Sharing (ABS) program. Anyone pursuing collection activities, whether of tangible materials or intangible information, may be subject to these new regulations. Especially targeted are scientists and researchers who make significant use of proprietary genetic resources, biological matter, and related information, such as traditional knowledge and farming know-how. Therefore, it is important for all potential collectors to be familiar with the fundamental principles of ABS law as well as the procedures that must be followed in order to be fully compliant with the rules and regulations of the countries where collecting occurs. Well in advance of any collection activities, researchers should review the ABS situation, determine who could best answer questions about ABS, find authorized partners in the country of interest, locate relevant information on the specific ABS regime, and, most importantly, execute the documents, letters and agreements necessary to proceed with collection activities.

1. INTRODUCTION

According to the Convention on Biological Diversity (CBD), the rights to biological resources belong to the state in whose territory the resource is found. In order to prevent biopiracy and create a climate of mutual trust, the global community undertook to regulate the handling of genetic resources in the CBD, a binding international agreement.

The goals of the CBD are to conserve biological diversity, promote its sustainable use, and ensure the fair and equitable sharing of benefits arising from its use. Responsibility for implementing the agreement is given to the state in whose territory the biological material is found. The CBD, however, contains rules that clarify the rights and responsibilities of all of the contracting parties. One relatively recent addition to the convention addresses the system governing access to genetic resources and the sharing of the benefits arising from their use: Access and Benefit Sharing (ABS). With this new legislation, a new world system for the use of biological matter now exists that has changed the nature of public and private sector R&D efforts.

Anyone pursuing collection activities, whether of tangible materials or intangible information, may be subject to the new regulations. Especially targeted are scientists and researchers who make significant use of proprietary genetic resources, biological matter, and related information, such as traditional knowledge and farming know-how. Such knowledge may, in national legislation, be considered intellectual property (IP) or trade secrets, and as such not in the public domain or available for unauthorized appropriation.

Thornström CG and L Björk. 2007. Access and Benefit Sharing: Illustrated Procedures for the Collection and Importation of Biological Materials. In *Intellectual Property Management in Health and Agricultural Innovation: A Handbook of Best Practices* (eds. A Krattiger, RT Mahoney, L Nelsen, et al.). MIHR: Oxford, U.K., and PIPRA: Davis, U.S.A. Available online at www.ipHandbook.org.

Violation of the new access and benefit-sharing law, for example, by scientists conducting unauthorized collection activities, can result in fines, imprisonment, and the denial of future visits. Violation may also increase the transaction time needed to obtain a formal access permit. Therefore, it is essential not only to know the relevant policies, principles and laws, but also to have a practical understanding of the various potential ABS scenarios and the agreements, documents, applications, and other required procedural steps necessary for full compliance.

2. FOUR ABS SCENARIOS

We provide here the basics of ABS law, following four categories suggested in a recent publication by the Swiss Academy of Science,[1] providing some examples of agreements currently or soon to be in effect. Please note that on the Internet you may find thousands of examples of letters of intent, research permits, prior informed consent/mutually agreed terms (PIC/MAT) agreements, material transfer agreements (MTAs), and confidentiality agreements. To find out which type of agreement fits your project best, please consult the legal department at your university or college.

2.1 *No ABS situation*

For some projects, research does not involve any access to genetic resources for which ABS contracts are necessary. However, other research permits may be required. A research permit request may ask for more details than necessary in this situation. Possible situations might include:

- **research performed on human biological resources; human biological resources and genetic material are not covered by CBD.** You would need, instead, a research permit and approval by an ethical committee. Therefore, make appropriate contact with local academic partners and/or the national center for medical research.
- **research performed locally on national biological resources, without any involvement of indigenous people.** If you are employed by a national academic institution, normally a research and work permit is necessary. Contact a local academic colleague.
- **using Sweden as an illustrative example, research in Sweden, on Swedish material, or on material introduced before 1992.** If animals are included, an ethical committee permit is needed. Research that includes collection of red listed species necessitates a permit from Swedish regional authorities (called *Länsstyrelse* in Swedish).

2.2 *Simple ABS situation*

In the simplest scenario in which access and benefit sharing are relevant to research involves the collection and transfer (including export) of samples for an inventory. A (standardized) MTA is normally sufficient. In some countries this could be done with a standard research permit application that includes the MTA (see section 3.2 dealing with the research permit).

Other situations will require different actions:

- No standard research permit is available; the researcher/collectors will need to find a national colleague and formulate both a PIC and a MTA.
- When working with genetic resources deposited at the institutes of the Consultative Group on International Agricultural Research (CGIAR), standardized agreements are often available. This is, however, only the case for certain species used as crop plants.
- If the collection necessitates cooperation with indigenous people, a separate contract must be signed and the situation is more complex.
- If humans or animals are included in the research, a permit from the national ethical committee will need to be obtained.

2.3 *ABS situation*

A third scenario involves a situation in which the export of samples is required for further analysis and study in a laboratory abroad. No further exploitation is planned. In this scenario, PIC, MAT, and MTA are all necessary. For the most part, completion of the documents mentioned in the simple ABS situations described in Section 2.2 is sufficient; however, each document will be more extensive. In the more-elaborate research permit applications, these additional documents are included. Confidentiality agreements also might be requested.

2.4 *Complex ABS situation*

The most complex scenario involving access and benefit sharing is a situation in which proposed research involves several steps, including research for commercial purposes and possible use of traditional knowledge. Initially, confidentiality agreements and letters of intent could be signed, followed by PIC, MAT, and MTA. In the MAT, issues concerning benefits have to be elucidated and agreed upon. Terms like interest, profit, and return, as well as payment times, have to be discussed and jointly interpreted by all stakeholders.

3. ILLUSTRATIVE EXAMPLES AND TEMPLATE AGREEMENTS

In order to better assist the reader of this chapter in understanding the various ABS scenarios and the documents, letters, and agreements that might be applicable, we present examples of:

1. Letter of intent
2. Research permit
3. PIC
4. MAT
5. Template MTA
6. Confidentiality agreement

In addition, we also provide examples of ABS legal principles in various countries, along with useful online links where information can be obtained. It should be noted that the examples of template contracts or agreements presented below are for illustrative purposes only and in no way refer to specific existing agreements. The examples are meant to provide input for the development of real documents, which will need to be adapted to each specific circumstance.

3.1 *Letter of intent*

The letter of intent is a document in which the partners in the project describe their intentions. It is not legally binding as are the PIC and the MTA. It is mainly useful as a vehicle by which the parties can convey to one another their expectations and anticipated degree of involvement. A letter of intent could be used later as a basis for a PIC. A project could be financed through a planning grant, based on a letter of intent, and signed by all cooperating partners and stakeholders. The planning grant should finance the negotiations resulting in a PIC, MAT, and MTA. Planning grants have to be prepared three to four months before the deadline for submitting project proposals.

The examples given here (Boxes 1A–1D; all Boxes are at the end of this chapter) are hypothetical but use features from the real world. It describes a study in a developing country where local scientists and indigenous people working together study herbs used for malaria and vector control. Stakeholders in the project could be the National Government (represented by the University of Vientiane and the Ministry of Environment and the Ministry of Health), local authorities in the province, national park officials where the study is performed, and indigenous people (represented by village representatives and individual healers).

3.2 *Research permit*

To be able to research in several foreign countries outside the E.U., you will need a research permit. In some countries, this is easy to obtain using a standard procedure with standard fees. In other cases, it can only be obtained in cooperation with a national partner or through prior informed consent contracts. For an example, see Box 2.

3.3 *Prior informed consent*

PIC is a description of the project signed by all stakeholders and other concerned parties. It can be difficult to determine who exactly is affected by the project; another problem is financing the information and negotiations. It can also be problematic to have to devote all the work and generate expectations for a project that does not have any guaranteed financing. The prior informed consent is normally written to fit a commercial bioprospecting project. But how does it work with a basic noncommercial program?

Near-term, medium-term, and long-term benefits should be considered, including up-front payments, milestone payments, and royalties. The benefit-sharing time frame should be definitively stipulated. Furthermore, the balance between near-term, medium-term, and long-term benefits should be considered on a case-by-case basis.

The prior informed consent includes:

- conditions for export of biological material and related information
- conditions for use of the material and related knowledge
- conditions for how and what to make public
- patents and country of origin
- how and where to solve disputes

Prior informed consent means that everyone concerned has to be informed about the project and its terms before the project starts. If indigenous people are concerned, they must be informed so they understand the project. It may be necessary to translate the project into native languages or make a clear presentation with pictures. If the indigenous people do not give their consent, the project cannot start.

A letter of intent could be used to introduce the project and start negotiations, even before the project is financed. When the project is financed, the letter of intent could be integrated into the PIC. Remember that most academic organizations are not familiar with using letters of intent. Also, it is also important to identify and understand the respective roles of the legal entities involved:

- The scientist who is collecting should sign, in addition to the director of the institute, unless he has delegated the right to sign. The government, in the country where the collection is performed, could be represented by the ministry responsible for natural resources or another delegated unit. The government is legally considered the owner of the rights to the genetic material.
- If the project is performed in cooperation with a local university or institute, a local legal representative from the university should sign. In some countries, cooperation with a local university is a prerequisite for a project to be accepted. Depending on local laws, a cooperating scientist is sometimes expected to sign.
- If the project is performed within a national park, park authorities have to sign. This could also make the collection easier.
- If indigenous people are involved, their local representatives have to sign. This can be a complicated task, as several local communities may be involved and sometimes it is not clear who is a legal representative. A local community can also refuse to sign, and that will prevent the project from being performed in their legally defined area.
- If local individuals contribute to the project, they also are considered concerned parties. This may be the most complicated part to determine, as it is not easy to judge who will contribute prior to the project start. In Sweden, the scientist, if not otherwise stated in his or her contract, has the right to his or her inventions and intellectual property, which should then also be regulated in the PIC.

The above concerns and others are addressed and discussed in the following examples (Boxes 4A–4E) and their analysis.

The PIC should always set a time schedule. The duration could be a couple of months, for a specific collection, to up to five years or so if the project includes a Ph.D. program. The PIC must also define what happens with material and results after the time schedule has ended (see Box 4C).

Geographical area or areas shall also be defined realistically. This could be the whole nation or a local area. It is better to include any areas that could be of interest, rather than make it necessary to start new PIC negotiations, since these take a lot of time. An area could also be defined as a certain biotope in different geographical areas (see the examples in Box 4D).

Scientists often specialize in collecting genetic resources within a certain family or selected genera. However, often material also is collected for colleagues interested in other species. If this is the case, it should be mentioned in the PIC, which should also state if the material will be given to a third party, how it will be used, and by whom. The genetic resources can be living or dead specimens, and also parts of specimens, such as genes, enzymes, or specified chemicals or extracts. Whole material from families or material from several genera can be included. The PIC can also include new derivatives made from the collected material. Questions to ask include: Why is the collection being made? How shall results be used? Is material to be taken out of the country? What information can be published? What species/samples can be transferred to a third party? What research methods may be involved? Have these provisions been set out in your project proposal for financing? Does publication of material obtained from indigenous people necessitate their consent? Is the project classified as commercial or as noncommercial? Box 4E offers several relevant examples.

3.4 *Mutually agreed terms*

In accordance with Article 15, Paragraph 7, of the Convention on Biological Diversity, each contracting party shall "*take legislative, administrative or policy measures, as appropriate [...] with the aim of sharing in a fair and equitable way the results of research and development and the benefits arising from the commercial and other utilization of genetic resources with the Contracting Party providing such resources. Such sharing shall be upon mutually agreed terms.*" It is therefore important to assist parties and stakeholders in the development of mutually agreed terms to ensure the fair and equitable sharing of benefits.

3.4.1 *Basic requirements*

Everyone signing PICs and MTAs should understand the content, consequences, and meaning of certain terms. Mutually agreed terms take into account the different capacities and needs of those involved, including governments, indigenous and local communities, holders of ex situ collections, and the intended user organizations. This approach will contribute to fair negotiations and equitable shared benefits. Mutually agreed terms facilitate:

- legal certainty and clarity
- minimization of transaction costs
- inclusion of provisions on user and provider obligations
- development of different contractual agreements (for example, template agreements)
- different uses: taxonomy, collection, research, commercialization
- negotiated efficiently, within a reasonable period of time
- codification of written agreements

The following principles or basic requirements could be considered for the development of mutually agreed terms:

- legal certainty and clarity
- minimization of transaction costs by:
  - establishing and promoting awareness of the government's and relevant stakeholders' requirements for prior informed consent and contractual arrangements
  - ensuring awareness of existing mechanisms for applying for access, entering into arrangements, and ensuring the sharing of benefits
  - developing framework agreements, under which repeated agreement under expedited procedures can be made
  - developing standardized MTAs and benefit-sharing arrangements for similar resources and similar uses (the online version of the *Handbook* includes the BIO-EARN MTA with suggested elements of such an agreement)
- inclusion of provisions on user and provider obligations
- development of different contractual arrangements, for different resources and for different uses, and development of template agreements
- different uses may include taxonomy, collection, research, and commercialization, among other things
- mutually agreed terms should be negotiated efficiently and within a reasonable period of time
- mutually agreed terms should be set out in a written agreement

The following elements could be considered as guiding parameters in contractual agreements and as basic requirements for mutually agreed terms:

- regulating the use of resources in order to take into account ethical concerns of the particular parties and stakeholders, in particular of the indigenous and local communities concerned
- making provision to ensure the continued customary use of genetic resources and related knowledge
- provision for the use of IP rights, including joint research and the obligation to obtain rights on inventions and to provide licenses by common consent
- the possibility of joint ownership of IP rights according to the degree of contribution

3.4.2 *Typical terms*

A list of typical mutually agreed terms would include the following:

- type and quantity of genetic resources and the geographical/ecological area of activity
- any limitations on the possible use of the material
- recognition of the sovereign rights of the country of origin
- capacity building in various areas to be identified in the agreement
- a clause addressing whether the terms of the agreement, in certain circumstances, could be renegotiated
- whether the genetic resources can be transferred to third parties and conditions to be imposed in such cases.
- whether the knowledge, innovations, and practices of indigenous and local communities have been respected, preserved, and maintained, and whether the customary use of biological resources in accordance with traditional practices has been protected and encouraged
- treatment of confidential information
- provisions regarding the sharing of benefits arising from the commercial and other utilization of genetic resources and their derivatives and products

Mutually agreed terms for access to and specific uses of genetic resources (or derivatives), in accordance with Article 15, Paragraph 4 of the Convention on Biological Diversity, may also include conditions for transfer of such genetic resources to third parties, subject to national legislation of countries of origin.

3.4.3 *The Bonn guidelines on MAT*

The development of mutually agreed terms should be based on the principles of legal certainty and minimization of cost. These principles were included in the Bonn Guidelines to respond to the concerns of scientific researchers and users of genetic resources that national procedures for obtaining access could be too complex and burdensome. The guidelines enumerate a detailed description of the type of provisions that could form part of a contractual arrangement. Some of the proposed provisions are quite innovative and include the specification of uses, the regulation of those uses in light of ethical concerns, the continuation of customary uses over genetic resources, the possibility of joint ownership of IP rights according to contributions, and the existence of confidentiality clauses and sharing of benefits from commercial and other utilization of genetic resources, including derivatives thereof. The principle subjects to be agreed upon as listed in the Bonn Guidelines are:

- type and quantity of resources
- limitations on possible use
- recognition of sovereign rights of country of origin
- capacity building
- whether terms of agreement can be renegotiated
- whether genetic resources or derivatives can be transferred to third parties
- whether traditional knowledge is respected
- treatment of confidential information
- types of benefits
- timing of benefits
- distribution of benefits
- mechanisms for benefit sharing

3.4.4 *Convention on Biological Diversity: MAT guidelines*

Box 5 provides the relevant sections on mutually agreed terms from Decision VIII/4 of the CBD.

3.5 *Material transfer agreements*

If you need to transfer biological material from a foreign country, you must sign an MTA with the authorities of the foreign country. This could cover extracts for isolation of chemical compounds, as well as dried or otherwise preserved biological material. The material could be used in the national herbarium or for breeding purposes. The MTA should include:

- a definition of the material to be transferred
- reasons for the transfer
- restrictions or stipulations on how it can be used
- an explanation of the costs of the transfer and who will pay the costs
- start and termination dates
- settlements of disputes provisions

See Box 6 (at the end of chapter) for a sample MTA that puts into place the above considerations.

3.6 *Confidentiality agreements*

Before information of possible commercial value is given to another party, normally a confidentiality agreement is signed. The confidentiality agreement states what must be kept secret and stipulates a time frame for confidentiality. The agreement also includes paragraphs on how to proceed if confidentiality is broken. The Bonn Guidelines suggest that a confidentiality agreement be included in the PIC. Before signing any confidentiality agreement, a researcher should contact the legal affairs office at his or her university.[2] See Box 7 for a sample confidentiality agreement.

5. FINDING ABS INFORMATION

Now that you have an understanding of the basic steps to take to ensure ABS-compliance, how do you find out about the ABS rules of a given country? Which kind of legislation exists in the country you want to work in? Which conventions on biological material have they signed? In order to find answers to these questions, many countries have specific Web sites.[3]

5.1 *Europe*

The E.U. and member states have signed CBD. The E.U. is now implementing ABS, but there is no common law. The E.U. Parliament and the Council directive have suggested introducing the country of origin in patent law (Directive 98/44/EG). The EC ABS portal covers: Austria, Belgium, Cyprus, Czech Republic, Denmark, Estonia Finland, France, Germany, Greece, Hungary, Ireland, Italy, Latvia, Lithuania, Luxemburg, Malta, Netherlands, Poland, Portugal, Slovakia, Slovenia, Spain, Sweden, and the United Kingdom. Some specific country information follows:

- Austria: With regard to benefit-sharing arrangements, access to natural genetic resources is free in Austria, as long as the animal and plant species are not protected by nature-protection laws, such as endangered species, national parks, hunting, and, of course, private-property laws. If somebody gets financial support from the State for scientific research and profits from the results, she or he has to pay back only the subsidy.
- Greenland: In late 2006, Greenland Home Rule Parliament adopted an Act on Commercial and Research-Related Use of Biological Resources.
- Iceland: Iceland has introduced access legislation related to microbe prospecting in volcanic areas.
- Norway: Norway recently adopted access legislation, regarding boreal coral reefs, among other things.
- Sweden: There is no specific legislation on ABS. Sweden follows E.U. legislation with few of its own initiatives. There is no authority that can certify country of origin. Material deposited in the Nordic gene bank or in Swedish botanical gardens after 1992 is available under international law.[4]

5.2 *Asia*

The ASEAN framework agreement on access to biological and genetic resources has been signed by Brunei, Cambodia, Indonesia, Laos, Malaysia, Myanmar, Philippines, Singapore, Thailand, and Vietnam. However, there is still great uncertainty in several countries about how to formulate contracts. Thailand, Malaysia, and Philippines are uncomplicated, while the situation in Laos and other countries is relatively complex.[5]

The Philippines was the first country to implement legislation to regulate access to biological resources. Executive Order No. 247, signed by the president of the Philippines in May 1995, was the product of discussions between government agencies, nongovernmental organizations, indigenous peoples, and academic consultants. The legislation established a framework to regulate biodiversity prospecting having four basic elements:

1. An interagency committee to consider and enforce research agreements and coordinate further policy development
2. A procedure to get prior informed consent for access to traditional knowledge
3. A two-tiered system of mandatory research agreements, incorporating benefit-sharing terms, between collectors and the government: an academic research agreement, valid for five years, and a commercial research agreement, valid for three years
4. Minimum requirements to conform to environmental protection laws and regulations.

A material supply (or transfer) agreement is required for material leaving an institution. It should set out any relevant original terms of acquisition and state any additional terms of use, transfer, and benefit sharing. The Indigenous People's Rights Act, 1997, includes a Code of Conduct for Academic Collector of Biological and Genetic Resources for collectors working in the Philippines.[6]

Other countries that passed such legislation include Bangladesh,[7] Pakistan,[8] and India.[9] Some countries in Asia plan to regulate access to genetic resources to ensure PIC and benefit-sharing mechanisms. These include Fiji, Nepal, Pakistan, Papua New Guinea, Samoa, the Seychelles, the Solomon Islands, South Sri Lanka, and Vanuatu.

5.3 *Africa*

In Africa, the OAU model law on the protection of the rights of local communities, farmers, and breeders and the regulation of access to biological resources (OAU, Addis Ababa, December 2001) has been used by some nations as a model for regulating access to biological material.[10] So far, mainly Cameroon's legislation follows the African Union principles. Case studies from Cameroon include contrasting benefit sharing in the pharmaceutical and phytomedical industries in relation to *Ancistrocladus korupensis* and sustainable harvesting of *Prunus africana* on Mount Cameroon.[11] Nigeria[12] and South Africa[13] also recently passed legislations.

5.4 *Latin America*

Costa Rica has been one of the first countries globally to take a lead in biodiversity-related legislation. Information about regional groups, national governments, or state governments already regulating access to genetic resources to ensure prior informed consent and benefit-sharing can be found in the Ley de Biodiversidad No 7788, which has been in force since 1998.[14] The rules on access to biodiversity (Presidential Decree No. 31-514) have been in force since 2003.[15] The decree covers the following topics: access to genetic resources, equitable sharing of benefits arising out of the utilization of genetic resources, equitable sharing of benefits from the utilization of traditional knowledge, innovations and practices, intellectual property rights related to genetic resources and/or protection of traditional knowledge, innovations, and practices related to genetic resources.

More broadly in Latin America, the Andean Pact decision 391/96 on the Common Regime on Genetic Resources is leading the tone of the discussions. Peru, under its National Strategy on Biological Diversity (Decreto Supremo No. 102-2001-PCM), recently added a regime on traditional knowledge (Law 27.811, August 2002). In Peru, a special national authority (INRENA) has been established to deal with access and benefit-sharing issues.

The countries of the Andean region (Bolivia, Colombia, Ecuador, Peru, and Venezuela) decided to take a regional approach to regulating access to their genetic resources. The Andean Pact Decision 391 Agreement (1996) established a common rule on access to genetic resources for member countries, leaving implementation up to national regulation. The thinking behind this approach was that it made little sense for one country to regulate access strictly, when a neighboring country, with similar flora and fauna, had little or no regulation in place.[16]

Other countries where laws have been passed include Argentina,[17] Bolivia,[18] Brazil,[19] Mexico,[20] Panama,[21] and Peru.[22]

5.5 *Australia and the United States*

Despite the fact that the United States has never signed the Convention, most organizations and universities follow the CBD and the Bonn Guidelines.[23] Several U.S. projects are financed with universities, together with NCI or NIH, and coordinated by the Fogarty International Center.[24] Australia already regulates access to genetic resources to ensure prior informed consent and benefit sharing (the states of Western Australia and Queensland).[25]

6. CONCLUSIONS

Depending on the ABS situation (that is, no ABS situation, simple ABS situation, ABS situation, or complex ABS situation), a series of procedural steps will need to be taken pursuant to relevant national legislation. Accordingly, researchers must have a clear understanding of what documents need to be executed. These documents might include:

- Letter of intent
- Research permit
- Prior informed consent/PIC
- Mutually agreed terms/MAT
- Material transfer agreement (MTA)
- Confidentiality agreement

For each of these, it will be important to know who the authorized counterparts are in the country where collection activities are anticipated. In addition, it will be necessary to know where to find accurate and current information about the precise ABS legislation that prevails.

Although this might initially seem daunting, full compliance is necessary. Careful planning and proactive management will pay off in the long term, by minimizing the possibility of misunderstandings and possible legal problems, including detainment or expulsion.

Perhaps most importantly, these ABS regimes are in place to facilitate the building of equitable, sustainable, and solid networks for sharing biological resources for R&D programs. We all hope that the regimes ensure that any benefits that accrue will extend to all involved. ■

ACKNOWLEDGEMENTS

This chapter is based on a heavily edited synthesis of several documents prepared for a range of purposes. We are grateful to our colleagues and the publishers for having allowed us to edit and use extracts from these copyrighted materials.[26]

CARL-GUSTAF THORNSTRÖM, *Swedish Biodiversity Center, Swedish University of Agricultural Sciences, PO BOX 7007, 750 07 Uppsala, Sweden. Carl-Gustaf.Thornstrom@cbm.slu.se*

LARS BJÖRK, *Associate Professor, Ethnobotany and Pharmacognosy, Evolutionary Biology Centre, Uppsala University, Kabovaegen 18D, SE 752 36 Uppsala, Sweden. bjork.lars@tele2.se*

---

1 Biber-Klemm S and S Martinez. 2006. *Access and Benefit Sharing—Good Practice for Academic Research on Genetic Resources*. 1st edition. Swiss Academy of Sciences: Bern. www.scnat.ch.

2 The Bonn Guidelines use the term *secrecy agreement*. Although there are different terms, including nondisclosure agreement, the editors of the *Handbook* prefer the term *confidentiality agreement*, which is more neutral and is used more widely.

3 For example, for Europe: www/abs.eea.eu.int. Worldwide: www.biodiv.org/programmes/socio-eco/benefit/. National focal points can be found at www.biodiv.org/doc/lists/nfp-abs.pdf. Case studies can be found at www.biodiv.org/doc/case-studies/abs/. If you are unable to open a link here, use www.biodiv.org/doc/info-centre.shtml and click on case studies.

4 See www.ngb.se/Material/seedrequest.php and www.ngb.se/Material/MTA.

5 The ASEAN Framework Agreement on Access to Biological and Genetic Resources (2000). www.grain.org/brl/?docid=785&lawid=1261. See also www.wipo.int/globalissues/databases/contracts/index.html and www.biodiv.org/programmes/socio-eco/benefit.

6 Further information: The Philippines Executive Order No 247 (May 1995) can be found at www.grain.org/brl/?docid=915&lawid=1482; www.chmbio.org.ph/eo247.html; www.chmbio.org.ph/dao20-96.html.

7 www.grain.org/brl/?docid=538&lawid=1274.

8 www.grain.org/brl/?docid=683&lawid=1456.

9 During 2001 India adopted a bill for the protection of plant varieties and the farmer's rights (Bill. No.123 of 1999). A Biodiversity Bill (Bill. No.93 of 2000) has also been passed by the Parliament. www.grain.org/brl/?docid=322&lawid=1378 (in force beginning in 2002).

10 See www.grain.org/brl/?docid=798&lawid=2132. An explanatory booklet of the African Union Model Law, written by Professor J. A. Ekpere, can be found at www.grain.org/brl/?docid=798&lawid=2132.

11 www.biodiv.org/doc/case-studies/abs/cs-abs-aristo.pdf.

12 www.biodiv.org/doc/case-studies/abs/cs-abs-ng-a.pdf.

13 www.grain.org/brl/?docid=621&lawid=1222.

14 Both Spanish and English versions are at www.grain.org/brl/?docid=879&lawid=1859.

15 www.grain.org/brl/?docid=209&lawid=1872.

16 The Common Access Regime for Genetic Resources, Andean Pact Decision 391 (July 1996), can be found in Spanish at www.sice.oas.org/trade/JUNAC/decisiones/DEC391S.asp and in English at www.sice.oas.org/trade/JUNAC/decisiones/DEC391e.asp. These laws and others can also be found on the Web site of GRAIN (Genetic Resources Action International): www.grain.org/brl/index-en.cfm .

17 www.grain.org/brl/?docid=327&lawid=1677.

18 The *Reglamento de la Decision 391 Regimen Comum de Acceso a los Recursos Genticos* (in force beginning in 1997). www.lclark.edu/org/ielp/boliviaspanish.html; in English at www.lclark.edu/org/ielp/boliviaeng.html.

19 The *Acre State Law (Accesso a recursos genéticos lei estadual)* has been in force since 1997 and can be found at www.lclark.edu/org/ielp/acre.html. The *Amapá State Law on Access to Genetic Resources*, in force since 1997, can be found in Portuguese at www.lclark.edu/org/ielp/amapaportuguese.html and in English at www.lclark.edu/org/ielp/amapaenglish.html. This legislation can also be found on WIPO's site: www.wipo.int/tk/en/laws/. The *Medida Provisória No. 2.186-16* of 23 August 2001 regulates Access to Genetic Heritage, Protection of and Access to Associated Traditional Knowledge, Sharing of Benefits, and Access to and Transfer of Technology for their Conservation and Use and can be found at www.wipo.int/tk/en/documents/word/brazil-provisional-measure-por.doc in English. For Portuguese, visit www.planalto.gov.br/ccivil_03/MPV/2186-16.htm.

20 See Torres Nachon C and G Cantoe. Towards a Law on the Access and Use of Genetic Resources in Mexico. Center for Environmental Law and Economic Integration of the South. http://www.biodiv.org/doc/case-studies/abs/cs-abs-biodassur.pdf.

21 *Law No. 20* (26 June 2000) on a Special Regime of Collective Rights of Intellectual Property of Indigenous People for Protection of their Cultural Identity and their Traditional Knowledge. www.grain.org/brl/?docid=461&lawid=2002. See also www.wipo.int/tk/en/documents/word/panama-executive-decree-12.doc.

22 The *Propuesta de Regimen de Proteccion de los Conocimientos Colectivos de los Pueblos y Comunidades Indigenas Vinculados a los Recursos Biologicos* (in force beginning in 2002), can be found in Spanish and English at www.grain.org/brl/?docid=175&lawid=2041.

23 www.state.gov/g/oes/rls/or/25962.htm. For Americans working in other nations, contact: Access and Benefit-Sharing Officer, U.S. Department of State's Office of Ecology and Terrestrial Conservation, (202) 647-1804 or FAX (202) 736-7351. For research-permit applications, visit science.nature.nps.gov/research/.

24 More information about these can be found in *Chemical Prospecting: An Overview of the International Cooperative Biodiversity Groups Program, Francesca T. Grifo Fogarty International Center*. National Institutes of Health: Bethesda, MD www.fic.nih.gov/programs/research-grants/icbg/index.htm.

25 For information about performing research in national parks, see www.deh.gov.au/epbc/permits/parks/research.

26 Thornström CG and L Björk. 2006. Accessing Others' Proprietary Biological Matter and Related Information–Towards a Handbook in Access and Benefit Sharing and Related Intellectual Property. Part three: Entering Into Agreements. Unpublished.

Sennerby Forsse L. 2003. *Inventory Regarding ABS Legislation in Selected Countries*. Swedish Research Council/FORMAS: Stockholm.

Thornström CG and Björk L. 2006. Accessing Others' Proprietary Biological Matter and Related Information–Towards a Handbook in Access and Benefit Sharing and Related Intellectual Property. Part One: Chapeau. Unpublished.

Also the publication in *supra* note 1.

27 *Editors' Note:* This and other forms in this chapter have been lightly edited for English, consistency and clarity.

28 www.epaguyana.org/downloads/BiodiversityGuideLinesForResearch.pdf.

29 www.biodiv.org/decisions/default.aspx?m=COP-08&id=11016&lg=0.

Box 1A: Simple Letter Agreement for the Transfer of Materials

Ministry of Environment, People's Democratic Republic, represented by Mr./Mrs. __________; and the Department of Systematic Botany, Uppsala University, represented by Dr. Barbro Sundberg, hereby declare their intentions to develop a cooperative project in systematic botany and ethnobotany within the Nam-Nam National Biodiversity Conservation Area.

Dr. Barbro Sundberg and her Ph.D. student, Hugo Brun, Uppsala University, and Ph.D. Mak Naeng, of the National University of the PDR, are given permission to collect plant material, in the form of herbarium vouchers, within the Nam-Nam NBCA in order to study the floristic biodiversity of the area and the documentation of the PDR genetic resources. All specimens are collected in triplicate and processed, and will be detained in the NU herbarium, in the Uppsala University (UPS), with one specimen going to the Stockholm Natural History Museum (S).

All samples are marked with catalog number and the text: "*The rights to this material belong to the PDR. Any distribution or DNA sampling of this material necessitates a specific permit from the Ministry of Environment of the PDR.*" All publications deriving from the study of this material should acknowledge the Ministry of Environment of the PDR, and botanical publications should be published with consent of the curator of the NU Herbarium.

All expenses for the above-mentioned project are planned to be financed by the Swedish International Development Cooperation Agency. The project is planned to take place from July 1, 2007, to June 30, 2010. This letter of intent covers that time period only.

The Capital, February 10, 2007

The Ministries of PDR	Barbro Sundberg	Mak Naeng
NU Herbarium officer	Hugo Brun	

Box 1B: Letters of Intent—Derivatives

The Ministry of Health, the People's Democratic Republic, represented by Mr./Mrs. _________; and the Department of Systematic Zoology, Uppsala University, represented by Dr. Åke Mattsson, hereby declare their intentions to conduct a cooperative project in biology within the Nam-Nam NBCA that concerns traditional techniques for malaria control and development of vector control. Dr. Åke Mattsson, Professor Dr. Thomas Lundberg, and Ph.D. students Nils Svensson, Uppsala University, and Mai Moeng, National University of the Capital, are given a permit to collect plant material and insect samples within the NBCA for documentation.

All specimens are collected in triplicate and processed to be detained in the NU herbarium, in the Uppsala University (UPS), and at the Stockholm Museum of Natural History (S).

All samples are marked with catalogue number and the text: "This material belongs to the PDR. Any distribution of this material to a third party requires a specific permit from the Ministry of Environment of PDR." Prepared plant extracts are transferred to Uppsala University for analysis. All extracts are marked with catalog number and the text: "This sample belongs to the PDR. Any transfer of material to a third party necessitates a permit from the Ministry of Health, PDR."

All expenses for the project are to be financed by the Swedish International Development Agency. The project is planned to take place from April 1, 2007 to August 31, 2009, and this letter of intent covers that time period only.

The Capital, March 3, 2007

The Ministry	Åke Mattsson	Thomas Lundberg
Mai Moeng	Nils Svensson	

Box 1C: Letters of Intent—Bioprospecting and Traditional Knowledge

The representative of the Council of Village Heads of Nam Rew and Nam Chaa Valleys, representing the people of the villages in the Nam Rew and Nam Chaa Valley, Nakay-Mai District, Nua Province, PDR and the Department of Systematic Botany, Uppsala University, Sweden, represented by Martin Stigberg, hereby declare their intention to cooperate on a project concerning plants used for traditional medicine and mosquito control. The project aim is to improve mosquito and health control for the people in the villages.
All field equipment used for this control will be donated to the villages after the project time expires.
All rights to findings, in the form of possible patents and marketable products, and profits from possible commercialization will be divided according to the following schema:

- 5% given to local informants and/or their families
- 25% put into a village development fund controlled by the Council of Village Heads of the villages in the Nam Rew and Nam Chaa Valleys
- 25% is to be used by the Ministry of Health for active disease control in the PDR
- 25% to be used by the Ministry of Environment for preservation of biological biodiversity
- 20% of gains are put into a research fund with Dr. Martin Stigberg (Lecturer in Ethnobotany, Uppsala University), Prof. Maria Karlsson (Professor in Medical Entomology, Uppsala University) and Dr. Sue-Trong (Dean of the Faculty of Sciences, National University) are board members. The fund should be used for the education of promising Ph.D. students from the PDR within the field of biology.

The project is planned to be financed by the Swedish International Development Cooperation Agency. Project time July 1, 2007, to June 30, 2011.

Nam-Nam NBCA, February 28, 2007

Chief ....
Representative of the Council of
Village Heads of Nam Rew and
Nam Chaa Valley

Martin Stigberg
Department of Systematic Botany,
Uppsala University, Sweden

Box 1D: Letters of Intent—Education/Training Situation

The Faculty of Sciences, National University of the PDR (represented by Dr. Sue-Trong) and Department of Systematic Botany, Uppsala University, Sweden (represented by Dr Lisa Svensson), hereby declare their intention to cooperate on an ethnobotany project within the Nam-Nam NBCA, the People's Democratic Republic.

From the Faculty of Science, Mr. Mak Naeng MSc and Mr. Mai Moeng MSc will take part as Ph.D. students, with Dr. Lisa Svensson and Prof. Birgitta Eriksson from Uppsala University as supervisors. From Uppsala University, Hugo Brun is financed as a Ph.D. student.

Financing for Mr. Naeng and Mr. Moeng is from the bilateral program of the Swedish International Development Cooperation Agency (Sida) and NUOL. Financing for Mr. Brun is from a grant from the Sida/SAREC to Dr. Lisa Svensson.

Vientiane, February 25, 2007

Dr. Sue-Trong
Dean of the Faculty of Sciences,
National University of PDR

Dr Lisa Svensson
Department of Systematic Botany
Uppsala University, Sweden

Box 2: Research Permit Examples

FORM B-001[27]

ENVIRONMENTAL PROTECTION AGENCY/EPA APPLICATION FOR SCIENTIFIC AND/OR COMMERCIAL RESEARCH ON BIODIVERSITY IN THE COOPERATIVE REPUBLIC OF GUYANA

The EPA welcomes applications from persons interested in conducting biodiversity research in Guyana.

NOTES TO THE APPLICANT

a. A non-refundable fee of US$75 is required for the processing of each application. The fee, along with the method of payment, can be found online.
b. All questions must be answered. Separate sheet(s) may be used for answers to any or all questions.
c. All applications must be typewritten. Failure to do so will result in a delay in processing the application.
d. Two (2) copies of the completed Application Form must be submitted, not later than three (3) months prior to the commencement of the research, to the Environmental Protection Agency for review.
e. All current sponsors, employers, collaborating institutions, and affiliations with commercial entities, relating to any or all of the researchers, and for the proposed research, must be specified (see 11 below).
f. Any change in the details of the application (for example, in the membership of the research team, or current sponsors/institutions), which occurs after approval has been given, should be reported to the EPA in writing.
g. The kinds and quantities of information, samples, and specimens proposed to be collected as part of the research are expected to be justified by the aims and objectives of the research, and quantities of materials to be removed are to be reasonable in relation to the abundance of any particular species (see 5 to 12 below).
h. It is recommended that applications be submitted before funding arrangements for the research are finalized with funding agencies, or, at the latest, prior to the departure of the research team for Guyana.
i. If you are intending to conduct research as an individual, you must submit a letter of recommendation from a recognized Institution/Body/Society. In the case of student applicants, the name and signature of the supervisor is required.
j. The Researcher must ensure that all necessary precautions be taken with regard to the health (vaccinations) of the research team.
k. The researcher/research team must work in accordance with the approved Guidelines for Biodiversity Research.

Please provide the information specified in the items below:

1. Name of authorized signatory to this application
2. Agency/institution on whose behalf the application is being made, if any
3. Postal address, telephone, fax and e-mail
4. Descriptive title of the proposed project
5. Summary of the proposed project (please attach a copy of the project proposal)
6. Objectives; proposed site(s) of the research (give as precise geographical delineation as possible); description of the proposed research, including methodology(ies):
7. What kinds of material/information are to be collected/produced/imported? (Please check appropriate boxes)
    - [ ] Specimen/sample collection (specify nature and numbers)
    - [ ] Recordings (audio and video)
    - [ ] Photographs
    - [ ] Written notes

(Continued on Next Page)

Box 2 (continued)

[ ] Computer entries
[ ] Reports
[ ] Articles and scientific papers
[ ] Other outputs (specify) ______________________________

8. Anticipated intermediate and final destinations of all information/reports and specimens and materials:
9. Is your project intended for commercial or exclusively academic purposes? Please specify your exact intentions. Commercial purposes here include but are not limited to:
   (i) The use of samples or specimens, photographic and audiovisual materials and illustrations, for commercial purposes
   (ii) Chemical, pharmacological, and biotechnological study
   (iii) The use of materials or specimens for propagation or breeding purposes

Academic purposes here refer to only taxonomic, conservation, ecological, and biogeographical investigations

10. Time schedule (arrival in/and departure from Guyana, including dates in hinterland)
11. Composition of research team (attach very brief CVs). Also attach a statement on current sponsors.
12. Expected environmental impact of the research (brief statement)
13. Expected source of funding (see Notes to the applicant [d]). Please attach the budget proposal that will be or has been submitted to the funding agency, including foreign and (estimated) local costs.
14. Proposed linkage(s) with local institution(s), if any. (State whether each institution has been formally approached and indicate (very briefly) its response.)
15. Training component for local counterparts
16. Do you intend to conduct research on lands legally owned or occupied by indigenous or local communities? If so, where?
17. Give a brief description of how Guyana will benefit from your research, including what compensation you anticipate immediately and in the long term for Guyana (cash, barter, services, specimens, sharing future production possibilities from research, royalties, equipment, or materials).

Signature of applicant
Signature of supervisor, if applicable
Office held in the Agency/Institution
Date
Environmental Protection Agency
IAST Building, U.G. Campus, Turkeyen
Greater Georgetown, GUYANA

Source: EPA, Republic of Guyana.[28]

Box 4A: Prior Informed Consent—Biodiversity and Ethnobotany: Garcinia sensu lato (Clusiaceae) in Cuba

EXCHANGE CONTRACT FOR ACCESSING BIOLOGICAL RESOURCES BETWEEN THE NATIONAL BOTANICAL GARDEN OF CUBA (JBN) AND THE INSTITUTE OF EVOLUTION, GENOMICS AND SYSTEMATICS, UPPSALA UNIVERSITY, SWEDEN

ON BEHALF OF THE FIRST PART: The National Botanical Garden under ownership of Havana University, Ministry of Education, JBN in advance, with legal address in Carretera El Rocio Km 3, Calabazar, Boyeros, 19230—Havana, Cuba, represented in this document by Dr. Angela T. Leiva Sánchez, as head director of the institution.

ON BEHALF OF THE OTHER PART: The Department of Evolution, Genomics and Systematics, Uppsala University, Uppsala, Sweden, IEGSU in advance, with legal address in Norbyvägen 18D, SE-752 36 Uppsala, Sweden, represented in this document by Dr. Britta Ekholm as head of the Ethnobotany group of the Department of Systematic Botany at the Institute EGS, Uppsala University, Uppsala, Sweden.

Both Parts Manifest:
- that they have mutual interest to establish a bilateral collaboration for accessing biological resources, with the specifications, obligations, and conditions that figure in the present document
- that both parts have the means and resources needed to get the exchange of experiences in the best conditions with the requested quality
- that they commit themselves to observing the strict fulfillment and respect of the Convention on Biological Diversity which both parts have signed
- that they acknowledge the mutual benefits that such a collaboration will represent for the contracting institutions and both countries

BOTH PARTS: Acknowledging the person and legal entity which they sign on this document, agree to subscribe to the present contract following the next specifications, obligations and conditions:

FIRST: The objective of the present bilateral contract is to access the Cuban alive biological resources for scientific purposes, for taxonomical studies, ethnobotanical studies, the investigation of chemical compounds and molecular studies on Cuban tropical plants of the genus Garcinia L. (Clusiaceae), in cooperation between JBN and EGS; the biological alive plant resources being accessed will be sent from Cuba to Sweden, as a sample big enough to achieve the above mentioned studies, from the wild harvest or donations of the Botanical Gardens in the National Network of Cuba.

SECOND: The alive plant biological resources of Cuba from wild harvesting or donations of the Botanical Gardens in the National Network of Cuba will always have a herbarium sample that will be kept as part of the herbarium collections HAJB of the National Botanical Garden and UPS under ownership of the Uppsala University Museum of Evolution, and they will not be utilized for commercial purposes or exchange; if new species are described from this material, the holotypes must be deposited at the HAJB herbarium.

THIRD: JBN will manage and pay the expenses for the official permits needed to access the natural areas, the biodiversity, exportation, and plant care.

FOURTH: EGS will pay the expenses in Cuba of the Cuban partner for the supervisor, the driver that will take part in the expeditions and the plant care revision, the customs fee, and the transportation for the plant biological material.

(Continued on Next Page)

Box 4A (continued)

FIFTH: The live Cuban biological resources sent from JBN to EGS, collected from germination and cultivation will not be used for commercial purposes under any circumstances, if either the material's origin is wild collected or is a donation from the Botanical Gardens in the Cuban National Network.

SIXTH: The results derived from the chemical and molecular studies will be for mutual benefit and will be shared by JBN and IEGSU, in the way of scientific publications or otherwise, as agreed by the parts.

SEVENTH: The transportation from JBN to IEGSU of the living plant biological resources will be done by EGS researchers directly from the International Airport José Martí, Havana, to Stockholm.

EIGHTH: Possible modifications or additions to the present contract should be made through a formal agreement between the parties as included as an appendix to the present Agreement.

NINTH: The present Contract of collaboration between JBN and EGS will be valid for two years from the signature date, extendable by equal periods, provided that no party terminates the agreement early.

TENTH: Any difference or difficulty caused in relation with this Contract interpretation or execution, while in effect, will be resolved by means of friendly negotiations between the parties. In case an agreement cannot be reached, the conflict will be solved in the Arbitration Court of the Chamber of Commerce of the Cuban Republic.

ELEVENTH: The applicable law is the portion of Cuban Law that agrees with the Convention on Biological Diversity, which both parties have signed.

Two exact copies of the Contract will be signed, and both copies will be legally valid and will carry the approval of the Cuban Authority of the Centre for Inspection and Environmental Control. Each party will keep a copy in its possession.

The present document is signed on 3 March of the year 2007.

National Botanical Garden of Cuba	Institute of Evolution, Genomics and Systematics
Fdo. Dr. Enrico Chavez Head Director	Fdo. Dr. Britta Ekholm Head of the Ethnobotany group in the Department of Systematic Botany

Vto. Bno. Ing. Tomás Rivera Amarán
Director del C.I.C.A.
Science, Technology and Environment Ministry

Box 4B: Prior Informed Consent— Description and Inventory of the Flora of Malgonia

PROJECT 1: DESCRIPTION AND INVENTORY OF THE FLORA OF MALGONIA

Parties: Institute of Systematic Botany, Uppsala University, Sweden, The prefect, Dr. Sven Berg, and the performing scientist, MSc Anna Skool, Institute of Applied Botany, University of Malgon, Malgonia, The director Dr. Marin Marais, the government of Malgonia represented by the director of the Malgonian Environment Protection Agency (MEPA).

PROJECT 2. Inventory and use of the flora in Nam Noi valley, Lao People's Democratic Republic (hereafter Laos or Lao PDR).

Parties:
- Institute of Systematic Botany, Uppsala University, Sweden, The prefect, Dr. Sven Berg and the performing scientists, Eva Lund, Dr. Mikael Engström, and MSc Birgitta Karlsson
- National University of Laos, faculty of Science, the dean, Dr. Boukaone Nourinam, National University of Laos, faculty of Medicine, Dr. Bourisak Nam, NUOL represents the Government.

Box 4C: Time-Frame Examples

Project 1. Description and inventory of the Flora of Malgonia
The project is planned and financed for a three-year period with an additional six months for publishing and reporting.
Time schedule: January 1, 2007 until June 30, 2010

Project 2. Inventory and use of the flora in Nam Noi Valley, Laos
The project is financed over a three year period, but two Ph.D.s in a sandwich program are expected. The time schedule of the PIC could only be signed for three years but a renewal is prepared.
Time schedule: January 1, 2007 until December 31, 2009, with possible renewal from January 1, 2009 until December 31, 2012.

Box 4D: Examples of Geographical Definitions

Project 1. Description and inventory of the Flora of Malgonia.
The project area is defined as "mountain areas throughout the country." With respect to the Northeastern part of the country, a special permit is necessary and hereby given.

Project 2. Inventory and use of the flora in Nam Noi Valley, Laos
The project area is defined as the Nam Noi Valley and the nearby Nam Pheo Valley between Nam Theun, Laos and the Vietnamese border. The project is permitted to expand to the Nam Theun basin, which is planned to be flooded.

Box 4E: Description of Material

Example 1: Description and inventory of the Flora of Malgonia
The PIC concerns flowering plants without specific limitations. Species within the family Acrididiae will be collected for Dr. Grazia Hopper at the Museum of Natural History in Amsterdam. Reference specimens will also be given to the Museum of Natural History in Malgon. Collected specimens will be marked: *Property of Malgonia*.

Example 2. Inventory and use of the flora in Nam Noi Valley, Laos
The relevant genetic resources are species and their derivatives used for malaria and vector control. A specific permit is given to Professor Gunnar Sellström for collection of insects within the genus *Anopheles*. Collected specimens are treated the same as collected plant species. Collected specimens should be marked: *Property of Lao PDR*.

Example 3. Description and inventory of the flora of Malgonia
The project is noncommercial and intended to improve knowledge of the Malgonian flora. The project will result in the description of species and the collection of herbarium specimens. Triplicates of the specimens will be deposited at the herbaria in Malgon, Uppsala, and Stockholm. A separate MTA is signed for material deposited in Stockholm and Uppsala. The results should be published in well-known, scientific journals and will also be at the disposal of the committee of the Malgonian flora. Collected specimens will be marked: *Property of Malgonia*. Transfer of material to a third party requires a permit, granted by the Malgonian Environment Ministry. Material transferred to Dr. Grazia Hopper, Amsterdam University, is described in a separate MTA. The project is used to introduce PCR techniques and training of Staff at NUM.

Example 4: Inventory and use of the flora in Nam Noi Valley, Laos
While collecting an inventory of the flora is noncommercial, the ethnobotanical study of the use of plant may contain commercial aspects. The long-term objective of the project is improved knowledge of the genetic resources in Laos, with the hope of establishing local production. The project will have two basic dimensions:

1. Inventory of species within the Zingiberaceae family, botanical and chemical. The chemical evaluation will concentrate on the essential oils from seeds and roots. Steam distillation techniques will be introduced on sight. Identification and structure determination of chemical compounds in the essential oils will be done through GC-MS in Lund, Sweden.

2. Ethnobotanical study of plant material used to cure and control malaria and mosquitoes. Open-ended or semistructured interviews of members of different ethnic groups in the area will be followed by statistical analyses, identification of species, evaluation of processing influence on chemical composition, literature studies on species and isolated compounds, and screening for biological activity.

Active compounds are identified using GC-MS and HPLC-MS. The project will serve technology transfer and technique training for Ph.D. students at NUOL. Blood sampling and analysis is performed by scientists of the medical faculty at NUOL. Ethical permits for blood sampling are obtained by the faculty. Expected outcome: improved malaria control among the population in the studied valleys and identification of possible products for malaria control. At least two PhD students will use the project to complete their exams.

Box 5: CDB on MAT

1. [Minimum conditions for the fair and equitable sharing of the benefits arising out of the use of genetic resources, derivatives or products shall be stipulated in relevant national [access] legislations [or] [and] under the international regime] and [shall] [may] be taken into consideration in mutually agreed terms [shall] [may] be based on prior informed consent between the provider and user of given resources.]

2. [Mutually agreed terms conditions may stipulate benefit-sharing arrangements regarding derivatives and products of genetic resources.]

3. The conditions for the sharing of the benefits arising out of the use of traditional knowledge, innovations or practices and associated [with] genetic resources [derivatives and products] [will] [may] be stipulated in mutually agreed terms [between users and the competent national authority of the provider country with active involvement of concerned indigenous and local communities] [between the indigenous or local communities and the users, and where appropriate with the involvement of the provider country].

4. [Mutually agreed terms may contain provisions on whether intellectual property rights may be sought and if so under what conditions.]

5. Mutually agreed terms may stipulate monetary and/or non-monetary conditions for the use of genetic resources, [their derivatives and/or products] and associated traditional knowledge, innovations and practices.

6. [The international regime should establish basic benefit-sharing [obligations] [conditions], including the distribution of benefits through the financial mechanism, to be applicable in the absence of specific provisions in access arrangements.]

7. [Where the country of origin of the genetic resources or derivatives accessed cannot be identified, the monetary benefits there from shall accrue to the financial mechanism and the non-monetary benefits shall be made available to those Parties that need them.]

Source: CBD.[29]

Box 6: Sample MTA

This overall Material Transfer Agreement (MTA) will govern the exchange of selected biological material between the University of Nangijala and the University of Uppsala, jointly referred to below as the Parties. This MTA is based on a collaborative research contract between the parties and may be amended where any national laws or regulations require it, or upon the mutual agreement of the contracting Parties. It is understood that all exchange of biological material will be done strictly in accordance with the principles set out in the Convention on Biological Diversity.

The Parties therefore agree to the following terms and conditions:

1: Definitions

Biological material means any material of a plant or animal, or microorganisms or other genetic resources or derivatives thereof.

Provider means provider of biological material and may be the country providing a genetic resource collected from *in situ* or *ex situ* sources, including populations of both wild and domesticated species, according to the principles of the Convention of Biological Diversity. Provider may also be an institution providing part of a plant or animal, or microorganisms or other genetic resources or derivatives thereof.

2: Designation of Implementing Agency

Depending on the situation in the countries of the respective Parties, several options for the designation of implementing agency are possible:

2.1 University X hereby designates an authorized representative from Faculty Y as the competent University X representative for the purposes of this MTA. Such a representative should be at the level of Director/Dean/Chairperson, or be an appropriate representative. For clarification, it is agreed that the Faculty Y shall be responsible for ensuring that all national laws and procedures in force in Country U, relating to the exchange of biological material, are respected. Faculty Y shall make reasonable efforts to inform individual researchers/investigators of the national laws and procedures relevant to this Agreement.

2.2 If the University has not designated an authorized representative, the University is represented by the Head of the Administration, and the Administration shall be responsible for ensuring that all national laws and procedures in force in Country U relating to the exchange of biological material are respected. The Administration shall make reasonable efforts to inform individual researchers/investigators of the national laws and procedures relevant to this Agreement.

3: Purpose

The primary purpose of this Agreement is to provide a framework for the exchange of selected biological material for the purposes of research and education.

4: Ownership

4.1 Biological material exchanged in accordance with this Agreement, including any material contained or incorporated in modifications, wherever located, shall at all times be the property of the provider and shall not be used by, or transferred to, third parties without the knowledge, consent, and written authorization of the provider in accordance with the principles in the Convention on Biological Diversity. The ownership of any new intellectual property derived from material transferred under this Agreement shall be governed by the terms described in Article 7 of this Agreement. For the purpose of this MTA, the provider is defined as the Department or the University that has provided the biological material as defined in each Implementing Letter of Agreement and also defined in Article 5 of this Agreement.

(Continued on Next Page)

Box 6 (continued)

[Note: In the absence of specific legislation vesting ownership of biological material held by research institutions, it is prudent to have it vested in the Faculty or University. The Faculty in both Parties' countries need to rigorously follow the developments in the emerging access legislation and respond to any legal developments accordingly.]

4.2 The Parties agree to refer to each other any requests for the use of material from third parties not defined under this MTA.

5: Implementing Letter of Agreement
For all material to be exchanged or transferred under this Agreement, the Parties shall execute an Implementing Letter of Agreement (ILA), describing the nature of the material to be collected or transferred under this Agreement. Each ILA shall be concluded before any authorization for the transfer of material is granted. ILAs must contain the signatures of the relevant principal researchers that are providing and receiving the defined material in each ILA. The ILA must explicitly reference the rights and responsibilities of the Parties as defined by this MTA.

[The purpose of this section is to avoid a situation in which an MTA would need to be concluded for every single exchange of material. The section will accurately define the nature of the material that is transferred under each MTA.]

6: Conditions relating to the use of biological material

6.1 The Parties agree that the material collected and transferred under this agreement is to be used for teaching and academic research purposes.

6.2 It is agreed that any other application or use of the material provided, including any modification thereof, for commercial purposes shall be allowed at the sole discretion of the provider. If either of the Parties wishes to use the material, or derivatives thereof, for purposes other than that described in Article 6.1 of this MTA, the authorization for such use shall be at the sole discretion of the providing institution as described in this MTA, and such authorization shall not be reasonably withheld.

6.3 Each of the Parties agrees to comply with the terms of this Agreement. This includes any scientists or any person(s) of either Party who may come to possess the material in the ordinary course of his/her business as an employee of the Parties. Such person(s) shall not make available the material or any part thereof, or related information to any person(s) or third parties other than those personnel under the Parties' immediate and direct control.

7: Intellectual Property Rights
Any inventions that are derived in whole or in part from the biological material transferred under this MTA shall be assigned in accordance with the relevant laws governing intellectual property. Each assignment shall (1) identify the provider of the material and (2) identify the country of origin of the material used in any commercialized product(s). The assignees of inventions of any commercialized product(s) shall negotiate a good faith, mutually acceptable agreement with the provider of the material, according to the principles set out in the Convention on Biological Diversity.

8: Publication
Copyrighted publication generated from research exchanged under this agreement or extracted from biological material collected in the pursuance of this agreement shall not include any restrictions whatsoever regarding use of such publication by the Parties.

(Continued on Next Page)

Box 6 (continued)

9: Duration of the Agreement
This MTA shall be valid until the end of 2001, according to the BIO-EARN project contract. The agreement may be renewed for a new BIO-EARN Programme period (2002–2005) upon mutual agreement of the contracting Parties.

10: Termination
10.1 Unless otherwise agreed, this MTA will terminate at the expiration of the present cooperation program.

- The Parties shall remain bound to each other by the least restrictive terms applicable to the material obtained in the pursuance of the purposes of this Agreement, and any modifications thereof, in accordance with Article 7 of this Agreement.
- The Parties will discontinue their use of the material and may destroy or return any remaining material to the country of origin.
- If for any reason, either of the Parties wishes to terminate this Agreement before the completion of the research, each of the Parties agrees that it will to the other Party give written notice six months prior so as to enable the completion of ongoing research. Such written notice shall be provided to each representative of the Parties' signatory to this Agreement.

10.2 Nothing in this Agreement shall be interpreted as having the effect of preventing or delaying the publication of research findings resulting from the use of the material or modification thereof.

11: Settlement of disputes
11.1 In the event that a dispute arises regarding the interpretation or application of the provisions of the Agreement, the Parties shall initially resolve their disputes in an amicable manner through consultations.

11.2 If the Parties fail to resolve their disputes amicably within a period of six months, they shall resort to arbitration.

11.3 Each Party shall nominate two arbitrators, and a fifth arbitrator shall be nominated by the United Nations Legal Affairs Office. The latter shall be the Chair of the Arbitral Tribunal. The decision of the arbitrators shall be final. Decision shall be passed by consensus. If consensus cannot be achieved, the decision shall be made by vote.

12: Miscellaneous
The Parties acknowledge that the biological material provided in pursuance of this Agreement may have characteristics that are unknown or difficult to determine and which may be potentially hazardous. Neither Party makes any warranties, express or implied, as to the safety, quality, viability, or purity of the material, or its merchantability or fitness for any particular purpose.

University/ Research Institute in the Nation concerned

Name of University______________________________

Full Address______________________________

Authorized Officer______________________________

Title______________________________

Signature______________________________

Date______________________________

(Continued on Next Page)

Box 6 (continued)

University in Sweden____
Name of University____
Full Address____

Authorized Officer____
Title____
Signature____
Date____

Box 7: Sample Confidential Disclosure Agreement

This Confidential Disclosure Agreement ("Agreement") is made and entered into as of the _____ day of ___________, 20__ (the "Effective Date") by and between ____________________ ____ (hereinafter referred to as "LENDER") having its principle address at ____ ________, and ____ (hereinafter referred to as "BORROWER") having its principle office at ____________________ ____.

The LENDER and the BORROWER are each hereinafter sometimes referred to individually as a "Party" and collectively as the "Parties," for the purpose of protecting the patent, trade secret, and other proprietary rights of the LENDER and the BORROWER in the following subject matter, which may be mutually beneficial to the Parties to disclose for evaluation:
Subject Matter Description:

The Parties agree as follows:

Neither Party will directly or indirectly divulge to unauthorized persons any information received from the other Party that relates to the subject matter of this Agreement, except as otherwise required by law. As a condition to receiving such information, each Party to the Agreement hereby acknowledges that all information provided by either Party to the other in connection with the subject matter of this Agreement is confidential and proprietary with regard to the Party providing such information. Information to be subject to this Agreement shall be disclosed in writing or, if it is verbally or electronically disclosed as confidential at the time of disclosure, its confidentiality shall be confirmed in writing within twenty (20) days of disclosure by the Party making the disclosure.

Each Party, as recipient of such proprietary information from the other Party, will disclose such information only to its employees, directors, agents, consultants, bankers, and advisors ("Representatives") for the purpose of evaluation, and any Representatives to whom such information is disclosed shall be informed of the proprietary nature of the disclosure and of this Agreement and shall agree to hold such information in confidence and be bound by this Agreement in the same manner that each Party is bound. Each party shall be responsible for any breach of this Agreement by its Party Representatives.

Neither Party will use such information received from the other Party for any purpose except evaluation, testing, research, and related activities and will not disclose such information to anyone except its Representatives, unless prior written consent is obtained from the Party providing such information or as required by law.

This Agreement shall be binding on both Parties for a term of ______ ( ) years from the effective Date of this Agreement, except under the following conditions:

1. If a Party can show that such information was in its possession at the time of the disclosure; or
2. If the information disclosed by one Party to this Agreement is or becomes publicly known during the term of this Agreement other than through a breach of that Party's obligations under this Agreement; or
3. If the Party later receives such information from a third Party as a matter of right; or if such information is developed by one Party independently or any disclosures made under this Agreement, as evidenced by that Party's written records.

This Agreement shall be governed by the laws of ______________________________.

(Continued on Next Page)

Box 7 (continued)

To evidence their Agreement to the foregoing, the Parties have, through duly authorized representatives, executed this Agreement.

LENDER ______________________________
By: ______________________________
Name: ______________________________
Title: ______________________________

BORROWER ______________________________
By: ______________________________
Name: ______________________________
Title: ______________________________

Deal Making in Bioprospecting

CHARLES COSTANZA, *Consultant, Atlanta, Georgia, U.S.A.*
LEIF CHRISTOFFERSEN, *Associate, E. O. Wilson Biodiversity Foundation, U.S.A.*
CAROLYN ANDERSON, *President, Capia IP, U.S.A.*
JAY M. SHORT, *Founder, President, and Chairman, E. O. Wilson Biodiversity Foundation, U.S.A.*

ABSTRACT
There is an upward trend in demand for intellectual property protection in agriculture. While international agreements exist to protect agricultural biodiversity, the specific rights, benefits, and responsibilities of parties entering into commercial agreements that involve the use of genetic resources still must be clarified. This chapter provides practical guidance for creating agreements around the use of biodiversity resources, as well as guidance that may provide valuable insights for creating similar agreements on the use of unique agricultural resources.

1. INTRODUCTION
Intellectual property (IP) rights protection is increasingly available for many aspects of agriculture, particularly through utility patents and plant variety protection (PVP), known also as plant breeders' rights. Globally, however, the kinds of intellectual property rights that can be exercised over living things vary greatly. This is especially true for the living things that make up the *biodiversity* of the planet—the millions of naturally existing species and their attendant gene pools—as well as for *agricultural biodiversity*—that subset of biodiversity involving cultivated crops used for food, materials, fertilizers, energy, and so on. It is useful to recall that the United Nations (UN) Convention on Biological Diversity defines biodiversity as "*the variability among living organisms from all sources, including,* inter alia, *terrestrial, marine, and other aquatic ecosystems, and the ecological complexes of which they are part: this includes diversity within species, between species and of ecosystems.*" With respect to IP rights, naturally occurring living organisms *cannot* be protected; nonhuman living things that have been modified by man *can* be protected. *Bioprospecting* is the exploration or screening of natural biodiversity or agricultural biodiversity in order to identify potential commercial applications from those genetic resources. Bioprospecting should not be confused with *biopiracy*, which is the unauthorized and uncompensated taking of biological or genetic resources.[1]

This chapter seeks to aid parties in creating biodiversity access agreements (BAA) for the use of unique genetic resources that require additional development to commercialize. There is considerable—although not widespread—experience to date in creating BAAs involving microbial genetic resources. This general discussion of biodiversity access agreements will not encompass all of the factors necessary to create every kind of commercial agreement, but it may prove useful for the following:

- **a reference model.** For creating a relationship for the use of a resource for which there are international guidelines, but for which, in most cases, clear procedures for structuring specific agreements do not exist. This lack of guidance has forced the public

Costanza C, L Christoffersen, C Anderson and JM Short. 2007. Deal Making in Bioprospecting. In *Intellectual Property Management in Health and Agricultural Innovation: A Handbook of Best Practices* (eds. A Krattiger, RT Mahoney, L Nelsen, et al.). MIHR: Oxford, U.K., and PIPRA: Davis, U.S.A. Available online at www.ipHandbook.org.

and private sectors to cooperate to achieve a mutually beneficial and sustainable relationship based on the commercial use of a unique genetic resource.
- **resource valuation.** For valuing resources that may hold significant commercial potential and may also require significant investment for developing a marketable product (capital, technology, and management).
- **stakeholder identification and value contribution.** For valuing resources in which many stakeholders have overlapping interests. (Proper valuation of these resources requires the consideration of traditional knowledge, farmers' rights, and other historic rights. The present condition and composition of a resource, such as an isolated natural compound or unique variety of plant, may be the result of multigenerational trials and errors. These and other factors need to be considered when determining the appropriate value of the resource so that benefits from commercial development can be fairly distributed.)
- **benefit sharing.** For the sharing of benefits between parties to an agreement.

2. BIODIVERSITY AND IP

2.1 *The international agreements*

Biodiversity is addressed by the UN Convention on Biological Diversity (CBD). The objectives of the CBD are:
- conservation of biodiversity
- sustainable use of the components of biodiversity
- fair and equitable sharing of the benefits from the use of genetic resources

By recognizing a national government's sovereignty over all genetic resources within its borders (Article 15) and facilitating access to these resources based on *"mutually agreeable terms"* subject to the *"prior informed consent"* of the country of origin, the CBD provides firm conceptual grounding which can be adapted to guide commercial agreements.

Agricultural biodiversity in particular is governed also by the International Treaty on Plant Genetic Resources for Food and Agriculture (the Treaty). This agreement encourages open access to plant genetic resources and requires sharing the benefits of these resources through the exchange of information, access to technology transfer, capacity building, and the sharing of financial and other benefits of commercialization.[2]

The Trade-Related Aspects of Intellectual Property Rights (TRIPS) agreement of the World Trade Organization (WTO) provides minimal guidance on the issue of agricultural biodiversity, exempting both plants and animals that are not classified as modified microorganisms. Article 7 of TRIPS states that the protection and enforcement of IP rights should contribute to:
- the promotion of technological innovation and the transfer and dissemination of technology
- the mutual advantage of producers and users of technological knowledge that is conducive to social and economic welfare
- a balance of rights and obligations

The TRIPS agreement requires that signatories either provide patent protection of plant varieties or devise an effective sui generis (a specifically dedicated and unique) system for plant variety protection.

Currently, there is an effort to standardize countries' sui generis plant variety protection systems through the International Union for the Protection of New Varieties of Plants (UPOV) Convention, the purpose of which is to "*ensure that the members of [UPOV] acknowledge the achievements of breeders of new varieties of plants, by granting them an intellectual property right, on the basis of a set of clearly defined principles.*"[3,4,5]

The CBD, the Treaty, TRIPS Agreement, and UPOV Convention provide general guidance for parties engaged in developing their own agreements for access to genetic resources. It is important to realize, however, that the existing (international) agreements are based on broad standards of conduct. The agreements provide overarching principals but not instructions on how to meet the requirements of every unique situation. The

Bonn Guidelines, adopted by the COP in 2001, serve as a first step in bridging the gap between international agreements and the requirements of parties negotiating access to biodiversity resources. In 2005 the Biotechnology Industry Organization (BIO) developed and published its own guidelines for members engaged in the discovery of natural products such as enzymes, chemicals, and small molecules.[6]

From the perspective of two parties attempting to come to an agreement on providing or obtaining access to a unique genetic resource, which may or may not become a successful commercial product, the international agreements leave many questions unanswered. Parties must use common sense to strike a balance between protecting rights and providing fair compensation, on the one hand, and working within limits imposed by markets and legal frameworks on the other.[7] In the case of commercializing biodiversity, the parties must agree upon ownership of the resource and the subsequent product, the amount of investment required to bring the product to market, and the distribution of benefits resulting from the sale of the product.

One commentator[8] has noted a difference in negotiating access to agricultural genetic resources and nonagricultural (particularly microbial) genetic resources: whereas microbial biodiversity governed under the CBD has been seen as bilateral bargaining, the Treaty puts a premium on open access, seeking to keep access costs low and bolster global food security by encouraging breeding and research. The model provided in this chapter does emphasize sharing in a manner consistent with the Bonn Guidelines of the CBD and many of the financial and nonfinancial benefits outlined in the Treaty.

2.2 *Beyond international agreements*

Given the limited guidance on terms for biodiversity agreements, the private and public sectors have had to collaborate to create biodiversity access agreements (BAA) on a case-by-case basis. Over time, some companies have developed frameworks based on internationally accepted principles for creating BAAs. For example, Diversa, a publicly traded U.S. biotechnology firm (NASDAQ: DVSA), has entered into many BAAs with partners including Alaska, Antarctica, Australia, Bermuda, Costa Rica, Ghana, Hawaii, Iceland, Indonesia, Kenya, Mexico, Puerto Rico, Russia, the San Diego Zoo, South Africa, and Yellowstone National Park. The company, which is involved in the discovery and evolution of novel genes and genetic pathways from unique environmental sources, sees access to microbial biodiversity as critical to ensuring a greater diversity of genetic material; this access increases the chances of discovering a novel and unique gene for a new product or application. During a time when few or no models, guidelines, or requirements existed, Dr. Jay M. Short, then chief executive officer and chief technology officer of Diversa, and his team of intellectual property, commercial, and scientific specialists developed and refined a set of principles for selecting areas of the world in which to work, selecting partners, and creating agreements with governments, academic institutions, and private companies to help ensure long-term relationships based on the sustainable use of biodiversity.

Through its decade of experience with BAAs, the Diversa biodiversity team determined that there are three main factors that lead to a successful biodiversity collaboration:

1. Efficient and reasonable benefit-sharing negotiations
2. Efficient and reasonable permit systems (requiring three months or fewer to secure a permit and oblige the permit holder to reasonable reporting criteria). It should be understood that all national, regional, or local regulation that affects an agreement should be sufficient to provide reasonable regulatory oversight without creating an unnecessary burden on the parties
3. Capacity building

Based on the experience of Diversa, the following characteristics have been useful for evaluating the best locations to establish biodoversity collaborations:

- **legal framework and political will.** As is the case with access to agricultural biodiversity, many countries have not yet fully addressed the legislative and regulatory issues required for BAAs. Other countries may have significant legislation on biodiversity that is so comprehensive and complicated that it becomes too cumbersome for BAAs. In other cases, problems may lie with IP protection. Countries that have not previously concluded BAAs often lack the basic administrative procedures, such as approvals for the export of DNA samples, required to fulfill such agreements. In these cases, the government's political will to help orient and train their officials about bioprospecting is critical to the success of any international bioprospecting initiative.

- **equal treatment for all companies.** Although no national laws regulating access to biodiversity may exist in a particular country, it should view all potential commercial collaborators equally (these frequently include academics who are conducting research funded by a private commercial research interest), such that all commercially oriented researchers collecting samples should be required to enter into a government-sanctioned BAA that follows the guidelines and supports the objectives of the CBD.
- **strong scientific and conservation partners.** Appropriate scientific capabilities speed the process of narrowing the search for target organisms. As these collaboration partners receive training, they are able to provide more value-added services.
- **unique and protected habitats.** A greater diversity of habitats translates to a greater diversity of genetic material, and, consequently, increases the chances of discovering novel and unique genetic material for a new product or application. Protected habitats are important because they indicate that there are sufficient genetic resources to support a long-term biodiversity (or bioprospecting) collaboration.

Once a collaboration partner has been identified, the terms of the BAA must be decided. Highlighted below are key issues that influence the success of BAAs. This list has evolved significantly both through the implementation of BAAs (based on assessments and guidance from companies and biodiversity collaborators[9]) and through monitoring and adapting to changes within international conventions. The main issues include:

- **legal rights to genetic resources.** Countries that are able to efficiently assign and clearly define a company's legal rights with respect to the use of environmental samples and associated genetic material make attractive potential collaboration partners. Assigning and defining these rights reduces the risk of future claims being made against any commercial discoveries.
- **prior informed consent.** Recognizing that land owners and managers have a stake in bioprospecting activities, companies should require that biodiversity collaborators secure informed consent from landowners and managers prior to collecting samples.
- **rights to patent and commercialize.** The rights to patent and commercialize are critical to the creation of benefits that can be shared among the parties to a BAA. The way benefits are to be distributed will be outlined in the agreement. Diversa, in its BAAs, maintains the rights to patent and commercialize its inventions, including genes and gene products derived from samples.
- **competition between biodiversity collaborators.** Many companies have proprietary technologies that are necessary to commercialize their biodiversity-derived products. Companies do not want their biodiversity collaborators to use the proprietary technology transferred as part of the BAA to compete against them (the companies). Accordingly, strict and conservative interpretations of confidentiality are critical ingredients for developing a productive relationship.
- **transfers to third parties.** For some companies, their greatest competitive advantage is proprietary technology, and it is critical that it not be shared with third parties. Technology transfer to a collaborator is for the benefit of the collaborator in the context

of its own capacity building. Companies should respect and protect the confidentiality of their biodiversity collaborators' proprietary knowledge and information. Further, terms should be included in their agreements that prevent companies from transferring samples to third parties without the written permission of the biodiversity collaborators.

- **exclusivity requirements**. The terms of the BAA should not restrict biodiversity collaborators from cooperating with other companies. The more biodiversity collaboration agreements that exist, the more viable is the biodiversity collaborator and the more resources it has to preserve biodiversity in its country because of the added benefits and experience it receives from other industrial or commercial collaboration. However, many companies may resist collaborator involvement with competitors with regard to specific projects, due to their own confidentiality requirements or their need to secure a competitive advantage through access to a unique source of genetic material.

Countries also must evaluate the potentially collaborating corporations, nongovernmental organizations (NGOs), or academic institutions to judge their suitability as partners. Criteria for evaluation include

- **low-impact sample collection**. Biodiversity collaborators should understand that while biodiversity can be the raw material for commercial products and the potential source of untold scientific discovery, biodiversity is also a precious, limited resource. Therefore all sample-collecting regimes should be adapted to minimize the impact on the environment in order to preserve biodiversity (for example, sample sizes and collection frequency should be kept to a minimum).
- **adherence to international conventions and best practices**. Partners must demonstrate an understanding of and adherence to the principles of the CBD and the TRIPS Agreement. Partners with experience in BAAs may also have their own criteria based on international convention and practical experience.
- **track record**. Countries and collaborators should understand their commercial partner's experience with BAAs. BAAs have been and continue to be closely watched by the international community, and many companies have an established track record. If they do not, countries and collaborators should scrutinize, and if possible, compare to other agreements the proposed terms of benefit-sharing arrangements, protocols for sample collection, and conditions related to transfers to third parties. If partners have been criticized for past BAAs, countries should determine how they have changed their policies or their approach. What assurances are they willing to provide to ensure that those mistakes are not repeated?

3. BIODIVERSITY ACCESS AGREEMENTS

Once the parties have determined that they want to create a BAA, the challenge is to formulate a relationship that will provide access to a necessary stream of processed raw material (for example, novel genetic material) while ensuring the sustainability of that resource and compensating the party granting access by sharing benefits. BAAs contain basic elements that are common to all standard contracts, but they also contain very specific information that changes from agreement to agreement. This section discusses the necessary elements for a BAA.

3.1 *Parties to an agreement*

The most basic element of the BAA is to determine the appropriate parties to the agreement. It is critical to identify who has the proper authority to grant access to the particular biodiversity resource. In addition, it is important to identify all parties affected by access to the biodiversity resource, such as those people who live and work in proximity to it. Specifically, the parties need to identify the following:

- individuals or groups who legally control access to the resource in question (Ownership rights and authority can be documented

through permits, and that documentation should be included as an appendix to a BAA.)
- authorities who are authorized to grant access (the so-called competent authorities)
- individuals or groups who have been the "stewards" of the resource
- individuals who have been tenants of the land on which the resource is located
- individuals or groups who are currently using the resource
- individuals or groups who want access to the resource for commercial development
- universities, NGOs, researchers, conservationists, and so on, who will use access for nontraditional purposes

The National Focal Point for Access and Benefit Sharing (ABS) is frequently a good starting point for clarifying issues of authority, jurisdiction, stewardship, and tenancy.[10] As a practical matter, the company should request that the prospective biodiversity collaborator[11] provide evidence that it has authority to enter into a BAA, collect samples from designated areas, and share in the benefits that may arise from such collaborative work.[12]

3.2 *Duration of the agreement*

The period of time that the BAA is in effect should be indicated in the initial agreement. It is important for this time horizon to be referenced in later sections regarding the future ownership and disposal of genetic or other material obtained under the agreement, as well as the future benefits that may be derived from the commercialization of a biodiversity-based product. It is advisable for the parties to:
- determine how long the access agreement will be in place
- indicate how parties may terminate the agreement
- determine whether the agreement can be renewed or negotiated and what the terms are for a possible renewal or renegotiation

3.3 *Jurisdiction*

Parties must agree on the legal framework within which the agreement will function. Doing so requires that the companies determine:
- which country's laws will take precedence in the contract
- to what degree international conventions will be incorporated into the contract
- what method of dispute resolution will be required in the event of disagreements (arbitration versus litigation)

3.4 *Contribution of each party*

The parties must agree not only on what they propose to contribute to the deal but also on how to value the contribution. For the creation of BAAs, firms will see biodiversity as raw material for a biodiversity-derived product, the realization of which will require their processing, manufacturing, and marketing to make the collaboration commercially viable. Countries contributing the biodiversity resource must consider the many values of the genetic resource when creating the BAA. A variety of benefit-sharing mechanisms, both financial and nonfinancial, can be used to compensate parties for their contributions to the venture. Valuation of the biological or genetic resource and equitable benefit sharing are ultimately the responsibility of the parties to the BAA and must be detailed in the BAA.

As companies, research institutes, academic institutions, and government agencies cooperate on exploring biodiversity for commercial applications and products, they enter into agreements that govern access and also define a regime for sharing benefits. This requires the valuation of a genetic resource as an input into the development of the product. Significant effort in the form of, for example, processing, manufacturing, or marketing required to transform the microbial biodiversity into a marketable product must also be considered. The market will determine the value of a biodiversity-derived product. Companies will know the commercialization costs and their target profit margin. For the company to see the project as economically viable, biodiversity access royalties, collection fees, and other benefits to collaboration partners would have to be covered by market value of the product *less* commercialization costs *less* target profit. The uniqueness of the biodiversity (that is, the fact that it has not previously been commoditized) will influence the value placed on it by a company, with a higher degree of "uniqueness"[13] being more highly valued.

In practice, as this is a relatively new market in terms of the formation of such collaborations and formal agreements, it may be difficult to convince companies to recognize the full value of the biodiversity resource and the contribution of the biodiversity collaborator to the satisfaction of the international environmental community. Companies and biodiversity collaborators must find a middle ground where the negotiated benefits to the collaborator are not economically prohibitive to product development but do provide incentives to the collaborator to participate in the BAA. As the market matures, biodiversity collaborators should be able to increase the value of their contribution as they increase their capacities through training and the transfer of technology that they receive from companies. Moreover, as companies become more accustomed to these collaborations, the companies are likely to be more open to increasing benefits to their collaborators. Many BAAs have been abandoned due to ambitious demands for benefit-sharing terms that are economically unfeasible. Parties to the BAA, therefore, must carefully and collaboratively determine the value of their contributions to the overall development and marketing of the product as a percentage of the entire contribution.

Finally, financial benefits are finite and may not be realized immediately. They also may require significant, long-term investment to be realized. Fortunately, there are a number of nonfinancial benefits potentially available that could encourage participation in a deal, as described below in the section on benefit sharing.

3.5 *Rights and responsibilities of each party*

In addition to each party's contribution, the BAA should provide specific information about the expectations of action and conduct that the parties have for themselves and one another.

3.5.1 *Rights*

The BAA will generate many questions about IP rights. Typically, the collaborator will provide access to the resource, and depending upon its scientific capacity, collection samples and isolated strains. These samples or isolates are then further developed by the company. Between the stages of granting access and the commercial sale of a product resulting from a BAA, there are intermediate stages, many of which create IP rights issues.

- **use of samples**. Parties should determine how samples collected under the BAA can be used by the parties. For instance, can the samples be distributed to third parties (such as research partners of either party)? If so, does doing so require written notification from the other party, and what is the required time for a response?
- **IP rights for inventions, samples, and derivatives**. Any IP rights resulting from the BAA must be fully explained and addressed within the BAA. Diversa, for example, maintains its right to own its inventions based on unique genetic material obtained under a BAA. It is important to note that this does not limit a biodiversity collaborator's right to benefits from the invention. This is negotiated under the benefit-sharing section of the agreement. Diversa also maintains the ownership rights of the derivatives that it makes from samples. The samples themselves remain the property of the biodiversity collaborator.
- **publication of knowledge**. Parties must determine who will have the rights to publish novel information resulting from the BAA.

3.5.2 *Responsibilities*

The parties must also determine their respective responsibilities. Examples of operational responsibility include sample collection and processing, regular reporting, communications, and administrative filings. Below is an excerpt from a BAA which outlines the responsibilities of the parties:

Collaborator will be responsible for the collection, processing and shipment to [the Company] of environmental samples from diverse habitats and/or DNA samples isolated from such environmental samples using the [the Company's] technology. Collaborator shall further be responsible for planning and execution of collection trips with and without the participation of [Company] personnel. Collaborator will provide laboratory space for the collaboration activities. Environmental samples shall include, but not be limited to, soils, sediments, mire, earth, microbial mats and filaments, plants, ecto and endo symbiont microbial communities, endophytes, fungi, animal and/or insect excrement, marine and terrestrial invertebrates, air and water. Collaborator will provide to [the Company] a minimum of [number] environmental samples per year.[14]

3.6 *Benefit sharing*

Once the parties have agreed upon the value of their contributions to the deal, they must discuss the sharing of benefits that encourage the sustainable use of the genetic resource. There are many options for sharing benefits, both financial and nonfinancial.[15] Table 1 provides an extensive list of financial and nonfinancial benefit-sharing possibilities, and divides them into short-, medium-, and long-term categories. An appropriate, deal-specific mixture of financial and nonfinancial benefits will enable a company to provide incentives for biodiversity collaboration while working within international guidelines and remaining responsible to shareholders.

3.6.1 *Sharing financial benefits*

The short-term financial benefits listed in Table 1 deal with up-front access payments, sample collection fees, contribution to collaborator research budgets, and use-based contributions to funds set up to preserve biodiversity. In the medium term, financial benefits include milestone payments for the achievement of certain goals during collaboration and research funding. Longer-term benefits include a share in the profit from sales and increased opportunities to earn money for performing value-adding tasks in the production process.

Several observations can be made about the negotiation process for determining these benefits. For markets with relatively small potential payouts, biodiversity collaborators may favor receiving sure payments for performance up front versus some portion of unknown future royalties. Conversely, when there are many potential applications coupled with potentially large revenues, biodiversity collaborators may be interested in a larger share of royalties at the expense of up-front payments, hoping for a percentage of a larger payout. In this case, biodiversity collaborators would have to weigh the importance of receiving money sooner versus the potentially larger payout of up to 15 to 20 years or more later.[16]

In many cases, the market potential of the collaboration will be obvious at the outset; in other cases it will not. Where the potential is not obvious, graduated royalties could be used, which change the percentage of proceeds from product sales according to such variables as the sales volume or end-product market segment.

3.6.2 *Sharing nonfinancial benefits*

There are many nonfinancial benefits at the parties' disposal. Many have noted that for access and benefit-sharing agreements for both microbial biodiversity and plant genetic resources, nonfinancial benefits may be more valuable to developing countries than financial benefits.[17, 18] Nonfinancial benefits can be shared in the short-, medium-, and long-term as well. Over the life of the collaboration, these benefits will accrue to the biodiversity collaborator on all levels (national, regional, institutional, and individual). Professional development for individuals and capacity building and technology transfer at the country, regional, and institutional levels will enable the collaborator to perform more value-added work. As a result, the biodiversity collaborator can generate additional revenues and access more upside potential by contributing more to the development of products resulting from the BAA.

Short-term, nonfinancial benefits may include biodiversity collaborator access to facilities and proprietary databases that may otherwise be inaccessible. In the medium term, technical know-how, training in specific technologies, new equipment, and more reliable stocks of laboratory supplies can enhance the biodiversity collaborator's scientific capacity. In addition, including biodiversity collaborators in planning and decision making increases their administrative capacity for additional projects. Longer-term benefits, aside from the cascading effects of the above, may include ownership of IP rights and access to technologies and products that result from the

Table 1: Short-, Medium-, and Long-Term Benefits: Nonfinancial and Financial

Time frame	Benefit Type	Monetary	Nonmonetary
Short-term	access to corporate facilities and databases		X
	advance payments	X	
	bioprospecting fees (up-front fees)	X	
	payments per sample (sample fees)	X	
	share in research budget or equipment	X	
	fees to trust funds for conservation and sustainable use of biodiversity	X	
	research support for a project that is considered important or critical for the biodiversity collaborator	X	X
	publications that stem from the research activities of the biodiversity collaboration that is written by all parties to the agreement		X
	joint development and pursuit of grant opportunities to support and expand the biodiversity collaboration		X
Medium-term	acknowledgment in publications		X
	joint research and scientific capacity building		X
	administrative capacity building		X
	participation in planning and decision making		X
	protection of local existing applications of IP rights		X
	technology transfer (equipment, material donation, sharing of know-how)		X
	training in bioprospecting, collection, and preparation of samples; biodiversity monitoring, socioeconomic monitoring, and/or nursery and agronomic techniques (increased conservation capacity)		X

(Continued on Next Page)

Table 1 (continued)

Time frame	Benefit Type	Monetary	Nonmonetary
Medium-term	research support for a project that is considered important or critical for the biodiversity collaborator	X	X
	publications that stem from the research activities of the biodiversity collaboration that are written by all parties to the agreement		X
	joint development and pursuit of grant opportunities to support and expand the biodiversity collaboration		X
	commitment to resupply in source country		X
	research funding	X	
	milestone payments	X	
Long-term	co-ownership or sole ownership of IP rights		X
	development of alternative income generating schemes	X	
	free access to technology and products resulting from agreements		X
	research support for a project that is considered important or critical for the biodiversity collaborator	X	X
	publications that stem from the research activities of the biodiversity collaboration that are written or approved by all parties to the agreement		X
	joint development and pursuit of grant opportunities to support and expand the biodiversity collaboration		X
	percentage royalties on net sales	X	
	gross sales, license issue fees, and other revenues	X	
	participation in value added	X	

Source: Adapted from Liebig and from Tides Center/Biodiversity Action Network.[19]

collaboration. Across all three time frames, the parties could consider pursuing grant opportunities to expand their research activities, as well as working together to produce publications. The biodiversity collaborator might consider asking the company to provide research support for a project that is important to the biodiversity collaborator and is more easily implemented by incorporating the company's technology.

Box 1 contains an excerpt from a benefit-sharing section of a BAA and provides instances of both financial and nonfinancial benefits. While the actual percentages and dollar volumes have been removed (as they provide no useful insight without the details of the entire deal), this example illustrates a very specific royalty payment scenario in which sources of income have been separated and shared differentially. The agreement envisions revenue from both direct sales of the product by the company and from licensing to third parties. Proceeds from direct sales are shared on a graduated basis. The biodiversity collaborator receives a percentage of net direct sales up to a certain dollar limit. Should net direct sales exceed that amount the biodiversity collaborator will receive additional income. As an example, assume the net direct sales of US$150 million. If the agreement held that the biodiversity collaborator receives 0.5% of the first US$75 million in net direct sales, and 1.0% of net direct sales exceeding US$75 million, the biodiversity collaborator would receive US$1.125 million. For revenues derived from licensing, the agreement provides a similar graduated benefit-sharing mechanism.

The agreement presented in Box 1 has a royalty stacking provision. *Royalty stacking* occurs when there are multiple patents that affect the final product. It is often the case that a number of different patented items have been licensed for the development of a new product. The company developing the product may have to pay for the use of each of these patents, adding to the cost of commercialization. When multiple patents are held by third parties, the royalty structure may make a deal financially unattractive.[20] When one company holds multiple patents involved in the process, determining final royalty allocation is simplified. For the purposes of this discussion, each patent owner's rights to the product should be understood and considered in the business decision to proceed with the BAA. (For a more-detailed discussion on royalty stacking, see the World Intellectual Property Organization's Web site.[21])

In addition to royalties, which are based on the overall success of product sales and licensing efforts on the company's part, the biodiversity collaborator also receives milestone payments. These payments are performance-based payments rewarding the biodiversity collaborator for competently executing its responsibilities. The milestone payment is pro-rated to the level of collaborator performance. In the example in Box 1, the maximum amount is established as a percentage of the annual funding that the biodiversity collaborator receives from the company and can be based on a range reflecting the degree of success or progress achieved by the biodiversity collaborator. Alternatively, the milestone can be based on the completion of stages toward product development. One of the drawbacks associated with this latter approach is that it is frequently predicated on the company's success and leaves the biodiversity collaborator with little ability to influence the amount of payment received. Hence the former option is sometimes considered the preferred approach.

The excerpt in Box 1 also provides two examples of nonmonetary benefits. These nonmonetary benefits address technology transfer and on-site training (both at the company's and the biodiversity collaborator's laboratories). In this case, the company is training the collaborator in both advanced scientific methods and in the use of its proprietary technology. In addition, the company encouraged the collaborator to send employees to the company for training. This not only improves the scientific capacity of the employees, but also gives the employees access to professional resources that may not be available in their own laboratories. The training that takes place in the biodiversity collaborator's laboratory is critical. Often collaborator laboratory infrastructure requires updating, and lab protocols need to be changed, with the guidance of the company, to support different equipment

Box 1: Typical Benefit-Sharing Section in a BAA

1. **Royalties**

For each calendar year during the term of this Agreement, The Company shall pay to Collaborator a royalty based on Product(s) sold by The Company, its Affiliates and/or licensees as follows:

On The Company direct sales:

(i) A% of the first X U.S. dollars (US$ X) in Net Sales of Product(s) sold by The Company;

(ii) B% of Net Sales of Product(s) sold by The Company in excess of X U.S. dollars (US$ X);

On revenue The Company receives from licensees:

(iii) C% of the first X U.S. dollars (US$ Y) in Product Sales Net Revenues that The Company receives, recognizes as revenues, or is otherwise entitled to receive (without duplication) in such calendar year;

(iv) D% of Product Sales Net Revenues in excess of X U.S. dollars (US$ Y) that The Company receives, recognizes as revenues, or is otherwise entitled to receive (without duplication) in such calendar year; or

(v) In the event that The Company's compensation from its licensees does not include royalty payments on sales of Product(s) by such licensee, then The Company shall further pay to Collaborator a royalty of E% of all license fees actually received by The Company in consideration of such a license, including, but not limited to, license issue fees, annual maintenance fees and sublicense revenue.

No royalties are due on products made available to third parties for testing only.

All royalties are subject to a royalty stacking provision and a pro rata share of products made using the company's proprietary technology.

2. **Milestones**

Further, The Company shall provide to Collaborator, on an annual basis, a list of goals that shall be directly related to Collaborator's work under this Agreement. Such goals may include, but not be limited to, items such as the following:

(i) 100% complete environmental/isolate sample data sheets submitted for all environmental samples received by The Company within five (5) business days of receipt of the sample each calendar year;

(ii) Providing DNA for each sample when requested (for soil samples ensuring that both DNA and soil are sent for each sample);

(iii) 100% compliance with The Company protocols for DNA isolation;

(iv) 100% compliance with shipping protocols;

(v) Fulfilling specific sample requests according to sampling capabilities of Collaborator;

(vi) Achieved maximum coverage of biotopes or habitats; and

(vii) Responds to requests in a timely and professional manner.

In the event that Collaborator achieves all of such goals, then The Company shall pay to Collaborator a milestone payment in an amount of Z percent (Z%) of Collaborator's annual funding hereunder. In the event that only a portion of such goals are achieved, then The Company will determine what portion of the milestone shall be paid based upon percentage of the milestones completed and the relative value of the completed milestones.

3. The Company shall also provide Collaborator with training in technology for the molecular phylogenetic analysis of different habitats, including the following techniques ("Technology"): a) techniques for nucleic acid extraction from environmental samples; b) techniques for generating gene libraries; c) techniques for PCR cloning of genes directly from environmental samples; and d) information technology for DNA analysis.

4. Additionally, Collaborator may designate employees, at its sole discretion and expense, to visit The Company's facilities for purposes of training in the technology for an equivalent of one person for one month's time (for example, two people for two weeks, four people for one week, etc.).

Source: Excerpted and generalized from a redacted Diversa BAA that was submitted by the University of Hawaii to the Office of Information Practices in the State of Hawaii.

and supplies. It is also not uncommon for the biodiversity collaborator to improve protocols for the company and provide training and education in the opposite direction. This further enhances the biodiversity collaborator's probability for increasing its share of the benefits. While a superb example of a highly desirable and valuable nonmonetary benefit, it is not often available due to confidentiality requirements within companies.

4. PRACTICAL PITFALLS OF BIODIVERSITY ACCESS AGREEMENTS

The above guidance is meant to provide a practical framework highlighting the major issues for consideration when constructing a BAA. It has been distilled from more than a decade of experiences of companies and biodiversity collaborators. However, no discussion of BAAs could be complete without a cautionary note on the business and political circumstances under which the BAA will be created and implemented. These factors are as important as any listed above, and failure to adequately deal with them could prove fatal for the BAA. They can also add substantially to the costs of creating a BAA as they require significant time, effort, and resources to resolve. A brief discussion of these issues is presented below.[22]

4.1 *Valuation versus negotiation*

Given that there is no established market for biodiversity resources or databases with details of other BAAs, valuation of the biodiversity resource will ultimately come down to discussions between the biodiversity collaborator and company. As with all negotiations, parties are well advised to understand the motivations and interests of their negotiation partners. Biodiversity collaborators and companies will need to have the overarching goal of making cooperation work, and will have to be flexible enough to incentivize their partners (and to respond to any incentives partners offer) fairly, in the context of the agreement.

From the collaborator's point of view, the best knowledge to have when negotiating for monetary benefits would be the level of profit that the company expects. In practice, this figure would be very difficult for the collaborator to obtain. Companies will be reluctant to share projections for many reasons, not the least of which is their desire to maximize profit. Even the best projections of future profit are just that, projections, and subject to varying degrees of risk, only a portion of which can be mitigated. Moreover, a corporate proclamation of an attractive potential profit will provide incentive for other companies to compete, possibly reducing the value of their future profit. Regardless of the reasoning, collaborators are unlikely to get an accurate picture of the expected profits from the deal.

Companies, too, would do well to study the terms of any previous BAAs available, especially those concluded with the intended biodiversity collaborator. Information about which nonmonetary benefits a collaborator would value would enhance the company's position and relieve some of the pressure to negotiate away projected profit.

Ultimately, the parties will either identify the right mixture of monetary and nonmonetary benefits to be distributed in the BAA, or lose patience with or confidence in their partners and walk away from the negotiating table without an agreement.

4.2 *Politics and perception*

Although the mechanics and structure of negotiating BAAs have become somewhat clearer over the past decade, not much has been clarified when it comes to the difficulties in politics and perception that companies face when attempting to create BAAs with biodiversity collaborators. Although biodiversity permit systems may be in place, the proposal of a BAA almost always creates controversy. Once a company states that it would like to create a BAA and establish a new standard for securing genetic resources from around the world, the most common response is for the governing authority to move extremely slowly, fearing that it will be accused of authorizing an inequitable agreement that undervalues their biodiversity and does not support their country's development. This problem can be further complicated by watchdog groups that consider the private sector to be inherently corrupt. No matter what benefits the company offers, such groups will criticize the deal as inequitable to the biodiversity collaborator. Ironically, this reaction reflects negatively on the very companies that are taking the lead in supporting the CBD. Unfortunately, those companies wishing to construct BAAs based on the principles of international conventions are seen in the same light as those companies that continue their research without any benefit-sharing arrangement and without permits. All of this has created an atmosphere in which life science corporations have been given every incentive to avoid engaging in bioprospecting and divulging or sharing any information about such endeavors. This actually makes it more important for biodiversity collaborators to seek out companies that are willing to take the step towards building a new approach to discovering products from nature, an approach that respects the economic interests and property rights of the nation providing the biodiversity (genetic resources).

Another complicating factor is that parties to the CBD have been slow to implement legal frameworks that facilitate legal access to their biodiversity and provide guidance on accepted or preferred benefit-sharing arrangements. Furthermore, the measures taken to date have been diverse in terms of their scope and their clarity. Compared to those countries that have created a simple, efficient approach, countries that have chosen a more cumbersome, comprehensive approach have generally had little participation from bioprospectors. Nonetheless, many countries remain without any legal frameworks to govern bioprospecting, allowing some companies to engage in bioprospecting without securing legal access to collect environmental samples and without providing associated equitable benefits.

The case of politics and perception is similar to that of benefit sharing in that both parties must demonstrate a willingness to make the BAA and successive agreements work. This requires each party to set aside short-term self-interest.

4.3 *The shortcomings of business as usual*

In addition to the practical challenges of negotiation and politics, there are several issues with current research sampling practices that will continue to grow in importance as more BAAs are concluded and the market for products developed as a result of the BAAs develops further. Often, samples collected for research purposes will be "contaminated" with types of biodiversity other than the target type. This unintended transfer of genetic material may constitute giving away potentially valuable (with respect to its potential for commercialization) biodiversity. Another issue that will become increasingly contentious is that limiting access to biodiversity may have a detrimental effect on scientific research. While these issues may not surface in a BAA between a company and its biodiversity collaborator in the near term, they will certainly have to be addressed in the longer term for the sake of scientific advancement and the conservation of global biodiversity.

4.4 *Addressing the pitfalls*

Many of the problems identified above could be mitigated or eliminated by improving the information available to parties to the BAA as well as to the larger pool of stakeholders interested in the outcomes of these agreements. Parties to the BAA want to know that they are being fairly treated. Collaboration partners want to understand the fair value of access and local value-added processing. Companies need to understand the amount and composition of compensation required to create the BAA. Companies can face higher commercialization costs in the absence of this information. The relative lack of standard information on BAAs can engender feelings of mistrust not only among the parties to the BAAs but also in stakeholders outside the agreement. Standards for creating BAAs, based on the experiences of many biodiversity collaborators and companies, would give the parties to the agreement a reliable and acceptable framework to aid decision making, negotiations, and communications about the agreements. These standards could even extend beyond the terms of the BAA to include model legislation and regulations to provide consistency to the legal and administrative environments in which BAAs will be created. Standards for BAAs could address the longer-term issues as well by explicitly discussing the rights and responsibilities of researchers and providing guidance on accessing IP-protected biodiversity for noncommercial purposes.

Participation by NGOs may be one way to address these issues. The main benefit of NGO involvement would be credibility. NGOs operating independently as neutral third parties can build trust among partners on both sides of the BAA. This neutrality could satisfy stakeholders outside the BAA, concerned with broader issues of biodiversity conservation and continued access to biodiversity for scientific research. NGOs would be able to leverage the expertise and experience of governments, research organizations, other NGOs, and companies globally to provide standards that are broadly applicable.

An example of an NGO making progress in this direction is the E. O. Wilson Biodiversity Foundation. Through the creation of its BioTrust, envisioned and initiated by one of the authors, Jay Short, the foundation seeks to ensure fair terms between countries and companies for access to biodiversity while preserving the biodiversity resources. BioTrust is a set of strategic relationships predicated on the notion that all countries (especially developing counties) contain wealth in the form of biodiversity and that they should be compensated for its exploitation. Saddled with the burden of long-term stewardship, most countries are currently without a financial incentive to continue.

By acting as an honest broker using a master agreement that binds the interested parties to a quid pro quo relationship, BioTrust ensures the fairness sought by the parties to the access agreement and the continuation of biodiversity conservation. Under this model, companies, as well as academic and research institutions, can sample and analyze genes, small molecules, and proteins, but a portion of revenues produced from any resulting products flows back to the country of origin for purposes of conservation. BioTrust participants agree to participate in capacity building through technology access and/or education for source nations.[23]

5. CONCLUSIONS

The experience of companies and countries in creating BAAs to share access to microbial biodiversity offers lessons that can be adapted for use with agrobiodiversity. These lessons will help interested parties bridge the gap between broad international guidance on the commercial use of biodiversity and the practicalities of deal making. Just as important as any technical aspect of deal making is the commitment of both parties to a sustainable and rational use of biodiversity in a way that encourages commercial development and protects the unique resource. Both parties need to conduct the due diligence on each other to foster the trust required for cooperation.

Companies should devise a set of operating principles based on the CBD and provide partners with real incentives for cooperation, which should include both equitable monetary and nonmonetary benefits. Countries must develop, clarify, or streamline administrative and permit procedures to encourage the sustainable, commercial use of biodiversity. They must also have the resolve to operate in a principled manner, consistent with international consensus (CBD). Both parties should be willing to engage in open debate with domestic and international critics to demonstrate the value of making progress in this field, despite having limited knowledge about the market potential of biodiversity-derived products.

There are a number of practical challenges to concluding BAAs. Many of these challenges could be addressed by improving information available to all stakeholders. NGOs could play a critical role in facilitating fair access to biodiversity for commercialization while preserving scientific access to biodiversity for research purposes. ■

CHARLES COSTANZA, *Consultant, 1304 North Avenue, NE, Atlanta, GA, 30307, U.S.A. chuckcostanza@yahoo.com*

LEIF CHRISTOFFERSEN, *Associate, E. O. Wilson Biodiversity Foundation, 10190 Telesis Court, San Diego, CA, 92121, U.S.A. leif@eowilson.org*

CAROLYN ANDERSON, *President, Capia IP, 10190 Telesis Court, San Diego, CA, 92121, U.S.A. Carolyn@Capiaip.com*

JAY M. SHORT, *Founder, President, and Chairman, E. O. Wilson Biodiversity Foundation, 10190 Telesis Court, San Diego, CA, 92121, U.S.A. jshort@eowilson.org*

---

1 Christoffersen LP and S Fish. 1999. Standing Up to Biopiracy: Fostering Sustainable Development through Bioprospecting. *Resource Africa* Issue 7, June 25.

2 Liebig K, D Alker, K Chih, D Horn, H Illi and J Wolf. 2002. Governing Biodiversity: Access to Genetic Resources and Approaches to Obtaining Benefits from Their Use: The Case of the Philippines. Reports and Working Papers, May, German Development Institute.

3 Smolders, W. 2005. Disclosure of Origin and Access Benefit Sharing: The Special Case of Seeds for Food and Agriculture. QUNO Occasional Paper 17, Quaker United Nations Office, Geneva. p. 3.

4 UPOV. 2005. What It Is, What It Does. International Union for the Protection of New Varieties of Plants, UPOV Publication No. 437(E), September 15.

5 Commission on Intellectual Property Rights. 2002. Integrating Intellectual Property Rights and Development Policy. Final Report of the Commission on Intellectual Property Rights, DFID: London. www.iprcommission.org.

6 BIO. 2005. See www.bio.org/ip/international 200507 guide.asp.

7 Mathur E, C Costanza, L Christoffersen C Erickson, M Sullivan, M Bene and JM Short. 2004. An Overview of Bioprospecting and the Diversa Model. *IP Strategy Today* 11:1–20. www.bioDevelopments.org/ip.

8 See *supra* note 2.

9 For the sake of clarity, the term *biodiversity collaborator* is used to describe the country granting access, as well as its institutions, universities, and researchers. The term *company* is used to describe any corporation, NGO, university, or research organization that could commercialize biodiversity or genetic resources.

10 Contact information for the National Focal Points for Access and Benefit Sharing can be found on the CBD Web site at www.biodiv.org/world/map.aspx. Once at that Web site enter the name of the country for the list of the National Focal Points for each country. See, also in this *Handbook*, chapter 16.2 by C.G. Thornström.

11 See *supra* note 9.

12 There are 188 parties to the Convention out of a possible 195. The nation states that have yet to ratify the CBD and become members are 1) Andorra, 2) Brunei Darussalam, 3) Holy See (Vatican), 4) Iraq, 5) Somalia, 6) Timor-Leste, and 7) the United States.

13 Uniqueness in this chapter means a resource has not been commoditized in a particular place or situation.

14 This text was excerpted and generalized from a redacted Diversa BAA that was submitted by the University of Hawaii to the Office of Information Practices in the State of Hawaii.

15 Tides Center/Biodiversity Action Network. 1999. Access to Genetic Resources: An Evaluation of the Development and Implementation of Recent Regulations and Access Agreements. *Environmental Policy Studies Working Paper No. 4*. Columbia University School of International and Public Affairs, p. iii. www.biodiv.org/doc/case-studies/abs/cs-abs-agr-rpt.pdf.

16 See *Supra* note 7.

17 Smolders W. 2005. Plant Genetic Resources for Food and Agriculture: Facilitated Access or Utility Patents on Plant Varieties? *IP Strategy Today* 13:1–17. www.bioDevelopments.org/ip.

18 Thayer AM. 2003. Diversa Promises Products, Profits. *Chemical and Engineering News* 81, (14).

19 See *supra* notes 2 and 15.

20 Clark V. 2004. Pitfalls of Drafting Royalty Provisions in Patent Licenses. *Bioscience Law Review*.

21 www.wipo.int/sme/en/documents/pharma_licensing.html - P819_48478.

22 The authors will address these issues and strategies for dealing with them in a forthcoming paper.

23 For more information on the E. O. Wilson Biodiversity Foundation, visit itsWeb site at www.eowilson.org.

CHAPTER 16.5

Bioprospecting Arrangements: Cooperation between the North and the South

DJAJA DJENDOEL SOEJARTO, *College of Pharmacy, University of Illinois at Chicago, U.S.A.*
C. GYLLENHAAL, *University of Illinois at Chicago, Chicago, U.S.A.*
[illegible] A. TARZIAN SORENSEN, *Global Health Initiatives, Johns Hopkins University, U.S.A.*
H.H.S. FONG, *University of Illinois at Chicago, Chicago, U.S.A.*
L.T. XUAN, *National Center for Science and Technology, Hanoi, Vietnam*
L.T. BINH, *National Center for Science and Technology, Hanoi, Vietnam*
N.T. HIEP, *National Center for Science and Technology, Hanoi, Vietnam*
N.V. HUNG, *National Center for Science and Technology, Hanoi, Vietnam*
B.M. VU, *Supporting Center for Community Economic Development, Hanoi, Vietnam*
T.Q. BICH, *Cuc Phuong National Park, Ninh Binh, Vietnam*
B.H. SOUTHAVONG, *Traditional Medicine Research Center, Vientiane, Lao People's Democratic Republic*
K. SYDARA, *Traditional Medicine Research Center, Vientiane, Lao People's Democratic Republic*
J.M. PEZZUTO, *University of Hawaii at Hilo, U.S.A.*
M.C. RILEY, *College of Pharmacy, University of Illinois at Chicago, U.S.A.*

ABSTRACT

The ICBG (International Cooperative Biodiversity Groups) program, through which institutions located in biotechnology-rich countries in the North collaborate with institutions located in the biodiversity-rich countries in the South (with the support of an industrial partner) to discover and develop natural-product drugs, is an experiment in the design of bioprospecting efforts. This chapter describes the general aims and organization of the ICBGs and describes in great detail the agreements that governed the University of Illinois at Chicago-Vietnam-Laos ICBG. The chapter includes material concerning IP (intellectual property) rights issues, informed consent, various forms of benefit sharing (including the sharing of short- and long-term, namely, royalty, benefits), capacity building, and community reciprocity. It offers a model for other such agreements.

1. INTRODUCTION

The term *bioprospecting* or *biodiversity prospecting* has been defined as "*the exploration of biodiversity for commercially valuable genetic and biochemical resources*,"[1] or "*the search for wild species, genes, and their products with actual or potential use to humans*,"[2] or the search for commercially valuable biochemical and genetic resources in plants, animals, and microorganisms.

One model of a biodiversity prospecting effort is a program called ICBG (International Cooperative Biodiversity Groups). Based in the United States, ICBG falls under the auspices of the Fogarty International Center (FIC) of the United States National Institutes of Health (NIH). It also collaborates with the National Science Foundation (NSF) and the U.S. Department of Agriculture (USDA).[3] A five-year cycle program, it went into operation in 1993 in response to a request for applications issued by FIC in 1992.[4] The ICBG second cycle began on 1 October, 1998, as a result of new and recompeting proposals in response to a request for applications issued by FIC in 1997.[5] On 17 October 2002, a request for applications for a 2003–2008 ICBG cycle recompetition was again issued.[6]

2. THE ICBG PROGRAM

International Cooperative Biodiversity Groups, or ICBGs, address the interdependent issues of drug discovery, biodiversity conservation, and sustainable economic growth. They are founded in the belief that efforts to examine the medicinal

Soejarto DD, C Gyllenhaal C, JA Tarzian Sorensen, HHS Fong, LT Xuan, LT Binh, NT Hiep, NV Hung, TQ Bich, BM Vu, BV Southavong, K Sydara, JM Pezzuto and MC Riley. 2007. Bioprospecting Arrangements: Cooperation Between the North and the South. In *Intellectual Property Management in Health and Agricultural Innovation: A Handbook of Best Practices* (eds. A Krattiger, RT Mahoney, L Nelsen, et al.). MIHR: Oxford, U.K., and PIPRA: Davis, U.S.A. Available online at www.ipHandbook.org.

potential of the earth's plants, animals, and microorganisms are urgently needed, and that continuing habitat destruction and ever-diminishing biodiversity will make it increasingly difficult to do so in the future. If bioprospecting directly benefits local communities and source country organizations, ICBGs believe that they will have strong incentives to preserve and support sustainable use of the environment.[7]

As a result of the 1992 and 1997 ICBG award competitions, eight ICBGs were established.[8] Each ICBG has as its administrative base a U.S.-based institution that is paired with other organizations (governmental and nongovernmental, including industrial/pharmaceutical) that are located both inside and outside the United States; one of these organizations is a host institution in one or more developing, biodiversity-rich countries, usually in the South. The personnel, organizational structure, specific aims, and methods of operation of each of these ICBGs have been fully described elsewhere.[9]

3. THE NORTH/SOUTH ICBG BIOPROSPECTING ARRANGEMENT

ICBG proposals must address access and benefit-sharing (ABS).[10] ABS is based on contractual agreements that take into account:

1. **The benefits that may be derived from bioprospecting.** These may include royalties from the sales of drugs developed from bioprospecting, advance payments (access fees or payments for samples when a commercial partner is involved), capacity building (equipment, training, infrastructure), and focus on the priority areas in the country(ies) of the host institution(s), such as priority diseases or collections and identification in geographic areas or biological groups that are high priorities for conservation needs.
2. **The recipients of the benefits.** These may include individuals and communities, government institutions (including national parks, forest services, national herbaria), and nongovernmental institutions (including universities, conservation and development service organizations, and private companies). Whether or not useful ethnomedical knowledge comes from the bioprospecting efforts, communities must receive both short- and long-term benefits for collaborating in the research process.
3. **The negotiation process.** Negotiators should consider the following elements:
   - *Informed consent*, from informal disclosure of the potential uses of their knowledge offered by individuals or communities, to formal documentation in the form of project descriptions and related materials
   - *Consensus building* among communities and government and nongovernmental organizations
   - Independent legal advice for all consortium members
4. **The structure of the agreement between the recipients.** ICBG models include the *one-contract model*, the contract wheel, the dual-contract model, and the wheel-triangle model.[11] All of these agreements include research and benefit-sharing terms, intellectual property (IP) rights, material transfer, confidentiality, and other terms. Often, specific agreements may address components of the above, including material transfer agreements (MTAs), know-how licenses, and so on.

4. THE UNIVERSITY OF ILLINOIS AT CHICAGO-VIETNAM-LAOS ICBG

4.1 *Background*

The members of the UIC ICBG Consortium were the University of Illinois at Chicago (UIC); the Vietnamese National Center for Science and Technology (NCST), based in Hanoi, Vietnam; Cuc Phuong National Park (CPNP), in Ninh Binh, Vietnam; the Traditional Medicine Research Center (TMRC), based in Vientiane, Laos, (formerly named the Research Institute for Medicinal Plants [RIMP]); and Glaxo Wellcome Research and Development (GW), based in Greenford, U.K. (today known as GlaxoSmithKline [GSK]).

The grant award (made on 29 September 1998) represented a cooperative agreement between the U.S. Government and UIC. The letter of award (Terms and Conditions of Award) indicated that the U.S. government agreed to fund the work (via the FIC) of the UIC-based ICBG, so long as certain criteria were met: the principles of ABS were fulfilled, the progress of the project was satisfactory, and funds were available.[12]

The general background of the ICBG (the events that led to the writing of the proposal, the selection of partner institutions, and the submission of a Letter of Intent to submit a proposal to the FIC), as well it's the structure of the ICBG (personnel, organization, research plan, and policies toward IP rights and informed consent) have been described in an earlier paper.[13]

4.2 *The aims of the consortium*

The specific aims of the UIC-Vietnam-Laos ICBG were:

- The discovery of biopharmaceuticals in the plants of Vietnam and Laos and the development of drugs to treat cancer, AIDS, malaria, tuberculosis, pain, and diseases that affect the central nervous system (particularly Alzheimer's disease)
- Creating a biodiversity inventory and conserving biodiversity, with a specific focus on plants of Cuc Phuong National Park and medicinal plants of Laos
- Aiding economic development in cooperating communities
- Capacity building among the collaborating institutions in the host countries

4.3 *Negotiations among consortium members*

After the Letter of Intent was submitted to the FIC on 3 October 1997, discussions were held between the principal investigator and the director of UIC's IPO (Intellectual Property Office). An important element of discussions was the principle stated in the so-called Manila Accord (at the 1990 Regional Workshop for the Chemistry of Natural Products in Southeast Asia), which states that at least 51% of the income generated from the commercialization of a drug derived from a plant collected in a particular country should go to the institution located in the plant's country of origin. The eventual outcome of these discussions was a Memorandum of Agreement (MOA) that bound the five members of the UIC ICBG. The ICBG proposal and the draft MOA (which had been accepted by member institutions but not signed) were sent to the FIC on 20 January 1998.

The new ICBG, Studies on Biodiversity of Vietnam and Laos: The UIC-based ICBG Program, was created on 1 October 1998. Its bioprospecting program was not fully functional until nine months later, when the MOA was signed by all parties on 28 June 1999. In the negotiation process, the principal investigator of this ICBG advised NCST, CPNP, and TMRC to consult attorneys regarding the draft MOA.[20]

4.4 *The Memorandum of Agreement*

The MOA consists of 15 pages of text plus 5 Addenda (which total 5 pages). Addenda I and II are included at the end of this chapter (Figures 1 and 2) and are further discussed below. It should be noted that the natural product program at GSK was phased out in 2000 as a result of the merging of GW and Smith Kline Beecham, so GW/GSK withdrew from the consortium in November of 2001.

4.4.1 *The MOA structure*

The University of Illinois at Chicago, which is bound in a contractual agreement with the U.S. government, is the administrative seat of the consortium. The transfer of funds (grants, not IP rights or benefit-sharing agreements) from UIC to the other member institutions (except Glaxo) was outlined in separate subcontract agreements. Glaxo was not a recipient of ICBG funds and did not provide any funding to the consortium; it did, however, agree to contribute to capacity building of scientists and institutions in Vietnam and Laos.

4.4.2 *Clauses of the MOA*

Part I of the MOA defines the consortium members' Scope of Cooperation. Part II defines the General Areas of Cooperation of the consortium members, including the exchange of faculty members or scientific personnel, joint research activities, joint participation in seminars and scientific meetings, the exchange of academic and research materials and other information, and the participation in special short-term academic programs. Part III describes the details of the joint research activities and consists of five sections (III-A/Precedents, III-B/Purpose, III-C/Objectives, III-D/Responsibilities, and III-E/ Finance and Services).

- *III-A/Precedents* contains clauses that describe the considerations that led to the cooperation, such as the previous track record of collaboration between UIC and the member organizations, the proposal writing, the funding award, the key personnel and organizational structure/component roles, and a reference to the terms and conditions of the ICBG award.
- *III-B/Purpose* defines the purpose of the cooperation: to discover and develop new medicines, to conserve and sustainably use the flora of the Cuc Phuong National Park in Vietnam and the medicinal flora of Laos, and to increase development in both cooperating communities and in the ICBG host institutions.
- *III-C/Objectives* spells out the specific aims of the consortium, including its approaches to plant selection, disease targets, the inventory of the seed plants of CPNP, biomass production of biologically active and promising species, capacity building, conservation education, economic improvement of local communities, in the CPNP area in Vietnam, and medicinal-plant inventory and databasing (and community reciprocity) in Laos, as well as human-resource development and infrastructure strengthening of the ICBG host institutions in Vietnam and Laos.
- *III-D/Responsibilities* spells out the responsibilities of each member organization and their joint responsibilities.
  - III-D-1 defines the responsibilities of UIC (23 clauses).
  - III-D-2 defines the responsibilities of NCST, IBT, ICH and IEBR (14 clauses).
  - III-D-3 defines the responsibilities of CPNP (12 clauses).
  - III-D-4 defines the responsibilities of RIMP/TMRC (11 clauses).
  - III-D-5 defines the responsibilities of Glaxo/GW (ten clauses).
  - III-D-6 defines the joint responsibilities of the member institutions and the industrial partner (eight clauses). It includes the time period the MOA is in force, conditions for withdrawal of any of the member organizations, amount of samples at initial collection for screening and recollection for isolation and structure determination, conditions for exchange of personnel as part of capacity building, the requirements for technical reports, how the materials and data may be used in the event the agreement is terminated, the limitations on the collaborative use of genetic materials, requirements for acknowledging the grant in publications, and the requirement that international arbitration must be sought in the event of disputes.
- III-E specifies the source of funding as the FIC/NIH (ICBG Grant 1UO1-TW01015-01).

Part IV defines the period of validity of the MOA; the conditions for termination, extension, and amendment of the MOA; and the number of copies of the MOA that must be signed by members of the consortium.

The signature page states that the five addenda to the text of the MOA will become binding upon the signing of the legal representatives whose names are affixed therein. These include the chancellor and two representatives of the board of trustees (for UIC), the director of the Institute of Biotechnology and an ICBG-NCST liaison (for NCST, representing IBT, ICH, and IEBR), director and vice director of Cuc Phuong National Park, director and deputy director of TMRC, and director for scientific research of GW.

- **Addendum I** (Figure 1) describes a long-term benefit-sharing scheme that will go into effect in the event that discovery of a biopharmaceutical is made by UIC (in cooperation with ICH) and that Glaxo develops and commercializes the drug. In this scheme, the royalty stream is distributed among the organization members of the Vietnam-Laos ICBG (excluding Glaxo, which waived its share of any royalties) and the communities in the ICBG host countries.
- **Addendum II** (Figure 2) presents a long-term benefit-sharing scheme to go into effect in the event that Glaxo discovers, develops, and commercializes the drug. As in Addendum I, in this second scheme, the royalties are distributed among the member organizations (excluding Glaxo) and the communities in the ICBG host countries.
- **Addendum III** grants rights to GW in the event of the licensing of discoveries made at UIC-ICH under the framework of the ICBG and GW's rights of first refusal.
- **Addendum IV** defines the milestone payments that GW will make in the event a drug is discovered at UIC. The amount of payment is determined by the following variables: the site of the screen (UIC versus GW), the selection of compound for clinical trial, entry to Phase II and Phase III clinical trials, and approval of NDA (New Drug Application).
- **Addendum V** defines milestone and royalty payments for any drug developed and commercialized by GW. The payments are determined by the patent rights on, and the chemical structure of, the GW development compound, as well as by the target activity (in other words, whether or not the target is one of those in which ICBG is interested). Milestone and royalty payments will be made on new drugs that are derivatives of natural compounds discovered in collected plants, as well as on the natural compounds themselves.

4.5. *IP rights issues*

In the event of a relevant UIC discovery, the IPO of UIC-PCRPS will determine the ownership of any resulting IP with the assistance of all members of the Group. The named inventors may consist of individuals from any or all of the consortium members. The question of ownership shall be determined in accordance with the applicable laws of the country in which the invention or discovery is made. With the assistance of all members of the consortium, the UIC IPO will obtain patent protection for the invention or discovery and/or seek such other IP protection, as UIC deems appropriate. UIC IPO will be responsible for the management and licensing of the invention or discovery in accordance with the terms of the agreement.

In the event that an invention or discovery is made at GW based on plants that were collected or acquired within the ICBG framework, GW will determine the ownership of any resulting intellectual property with the assistance of all members of the consortium. The named inventors may consist of individuals from any or all of the consortium members. The question of ownership shall be determined in accordance with the applicable law of the country in which any invention or discovery is made. GW will obtain patent protection for such invention or discovery and/or seek such other intellectual property protection, as GW deems appropriate with the assistance of all members of the Group. GW will be responsible for the management and licensing of such protected inventions. The parties further agree that they will make available all relevant information to GW (including the country of origin of the sample and its taxonomic identity, where appropriate) so that GW will be able to register IP rights.

GW will have the rights to file for patent protection for a discovery it makes that is based on plant samples or extracts received by GW under the framework of the ICBG, but it will consult with the consortium in determining co-inventorship of the discovery. GW also agrees to notify the consortium in the event a decision is made to proceed with the development of a compound or compounds derived from plants supplied by the ICBG.[21]

4.6 *Informed consent*

There are two provisions regarding informed consent in the Vietnam-Laos ICBG agreement: (1) informed consent in the case of collection and use of plant/genetic materials and (2) informed consent of individuals and their communities regarding the traditional medicinal use or uses of a plant.

Thus, in Vietnam, "*informed consent (collecting permits) of the Government of Vietnam, the owner of the samples (genetic materials) and derivatives thereof, will be secured before the implementation of the work proposed as described in the ICBG proposal,*" and ICBG through IBT, IEBR, and CPNP "*will liaison with the Government of Vietnam in matters related to permit for the collection and export of plant samples or their extracts for use in the ICBG project.*" In Laos, TMRC/RIMP will collect plant samples from various sites in Laos "*through prior informed consent of the Government of Lao PDR, the owner of the samples (genetic materials) and derivatives thereof.*" Prior informed consent (collecting permits) will be secured before the implementation of the work. The governments of Vietnam and Laos are acknowledged as the owners of genetic materials and their derivatives in their respective countries.

In Vietnam, ICBG investigators "*will seek the informed consent of individuals and/or communities for the recording and use of data on the medicinal and other uses of the plants in the Cuc Phuong National Park, for the intended study as described in the ICBG proposal.*" In Laos, ICBG investigators "*will seek the prior informed consent of individuals and/or the communities for the recording and use of data on the medicinal and other uses of plants of Laos, for the intended study as described in the ICBG proposal.*"

4.7 *Royalty distribution*

The full scheme of royalty distribution in Addenda I and II of the MOA (Figures 1 and 2) has been presented in an earlier paper.[14] At the time of ABS negotiations, UIC channeled the net royalty stream (after deduction of out-of-pocket costs) received from an industrial partner or licensee into two equal portions. The first 50% (referred to as the "common fund") is to be distributed to the collaborating institutions, the inventors, and the UIC administration, while the other 50% is to flow back to communities in the country of origin of the genetic material of the commercialized product, through a trust fund.

The distribution of the first 50% share may happen in two different ways. In the first scenario, UIC investigators discover a drug, and a pharmaceutical company develops and commercializes the compound. In the second scenario, a drug is discovered, characterized, developed, and commercialized by a pharmaceutical company that is an ICBG industrial partner (in other words, UIC inventors do not hold IP rights).

In the first instance (UIC inventors hold IP rights) the common fund is to be distributed as follows: (1) the inventors will receive a 40% share of the 50% portion (equal to 20% of total net royalty), as an incentive for future inventions; (2) the collaborating institutions (PCRPS and counterpart institutions) will receive a 20% share of the 50% portion (equal to 10% of total net royalty) for their research contributions; and (3) the UIC administration will receive a 40% share of the 50% portion (equal to 20% of total net royalty) for their administration and legal contributions.

In the second scenario (UIC inventors do not hold IPR), the common fund is to be distributed as follows: (1) the collaborating institutions will receive a 40% share of the 50% portion (equal to 20% of the total net royalty); (2) UIC-PCRPS will receive a 20% share of the 50% common fund (equal to 10% of the total net royalty) for its research contribution; and 3) the UIC administration will receive a 40% share of the 50% common fund (equal to 20% of the total net royalty).

The full details of the UIC-based Vietnam-Laos ICBG benefit-sharing scheme are spelled out in a 2002 paper.[15] In November 2002, further discussions and analyses of the above royalty distribution schemes at UIC led to the application of the policy to joint drug-discovery efforts:

Sixty percent of the split of the net royalty would go to the collaborating institutions, while 40% would go to UIC. Despite the change of this benefit-sharing policy, the original benefit-sharing schemes set down and agreed to by the UIC ICBG consortium and embodied in the ICBG MOA remain in force to this date.[16, 22]

Funds provided by GSK at the time of its withdrawal are being used to establish two trust funds: the Nature Conservation Foundation (NCF), Vietnam, and the Laos Biodiversity Fund (LBF). The objectives of the NCF and LBF include conservation of resources, capacity building, biodiversity research, and community reciprocity.[17] These funds will serve as the conduit for the 50% of the royalties that are due to flow back to the communities in question.

4.8 *Community reciprocity*

Community reciprocity measures are implemented in the Vietnam-Laos ICBG.[18] Both the UIC and the host-country institutions have responsibility for implementing community reciprocity.

5. CONCLUSION

The success of an ICBG depends on the goodwill and understanding of the collaborating parties toward the achievement of a common goal, namely, the conservation of biodiversity, the discovery and development of pharmaceutically beneficial products, and the equitable sharing of the benefits that may result. In setting up the arrangement, multiple, complex requirements must be satisfied, the most important of which is the contractual agreement. Eight ICBG bioprospecting groups have so far been created, each with various models of contractual arrangement. The common features of these models, however, are their satisfactory arrangements for IP rights issues, informed consent, and benefit sharing.

The UIC-based Vietnam-Laos ICBG is one example of such a North–South collaborative arrangement. Parties to this ICBG have successfully achieved goodwill and understanding. Despite the short time it has been in operation, the accomplishments of this ICBG to date indicate that the ICBG model works.[19, 23]

Bioprospecting endeavors such as these are also unique in the way in which they involve local communities. In order to effectively carry out this sort of activity, collaboration at the local level—with poor farmers, rural villagers, many of whom have only limited education or opportunities in life—is crucial. The ICBG allows rural villagers participation in conservation, economic and development initiatives in a way that is not often seen in "macro," nation-wide efforts to promote conservation, development or new economies. (Often, villagers are told what to do or are displaced by these new initiatives.) And the ICBG also allows villagers input on a number of issues—health care delivery, education, local economics, conservation, and development—which is a natural by-product of forming the ICBG project and determining what benefits "make the most sense" to the local communities with which the ICBG works.

Often times in the implementation of international, national or even provincial development, conservation or economic initiatives, the peasant-farmer is left out of the dialogue entirely, or is told to change/is displaced from life-long patterns of living and working. Under these circumstances, the peasant-farmer does not have a voice, and new schemes for economy, conservation, and development are imposed upon villages from the outside rather than collaboratively developed with villagers, in accordance and consideration of the local needs of villagers in different regions of the country. Projects such as the ICBG can provide a model for how to successfully implement national policy initiatives at the "micro" level—that is, figuring out the best ways to improve health care access and delivery systems, or to implement new economic, development, and conservation initiatives that are in keeping with local village practices and rhythms of life, especially when it turns out that local villagers have their own, traditional practices that may directly or indirectly contribute to conservation, economic, and development efforts. While the governments of Vietnam and Lao PDR do attempt to take into consideration the

Figure 1: Royalties Sharing in the Event that UIC Discovers and Characterizes a Compound, and Glaxo Develops It

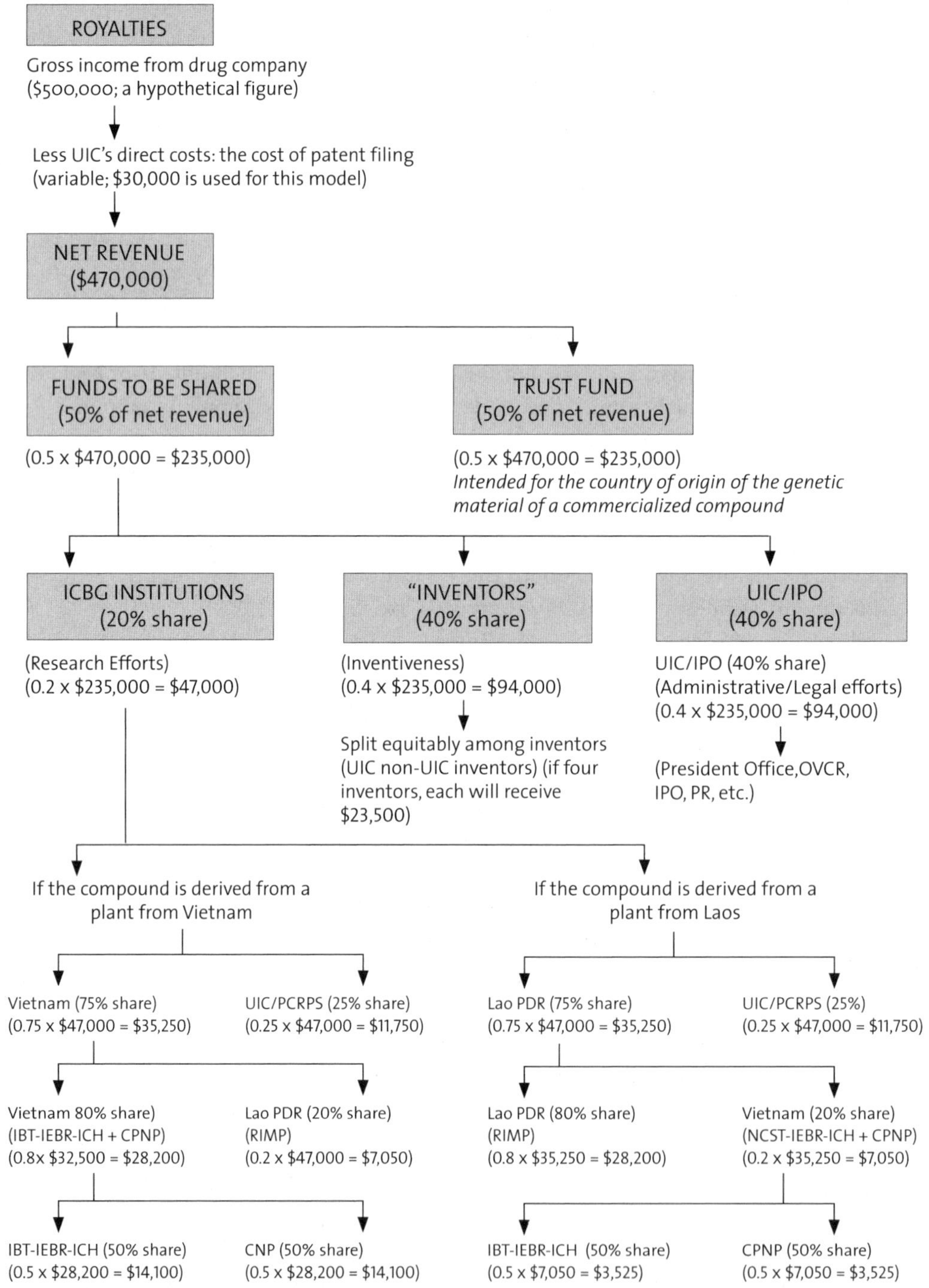

Note: In this scenario: a) the total amount of funds from the net royalty income that remains in the United States will be: $30,000 (direct) + $94,000 UIC share + $11,750 PCRPS share + $70,500 inventors' share = $206,250 or 41.25% of gross royalties. [If all inventors are UIC scientists, the UIC share will be: $30,000 direct costs + $94,000 UIC share + $11,750 PCRPS share + $94,000 Inventors' share = $229,750 or 45.95% of gross royalties.] and b) the total amount that will go back to the source country (Vietnam and Laos) will be: trust fund ($235,000) + ICBG institution share ($35,250) + Inventors' share (in the above scheme with one non-UIC inventor, $23,500), for a total of $293,750 or 58.75% of gross royalties. [If all inventors are UIC scientists, the share of the source country (Vietnam and Laos) will be: $235,000 trust fund + $35,250 ICBG institution share = $270,250 or 54.05% of gross royalties.

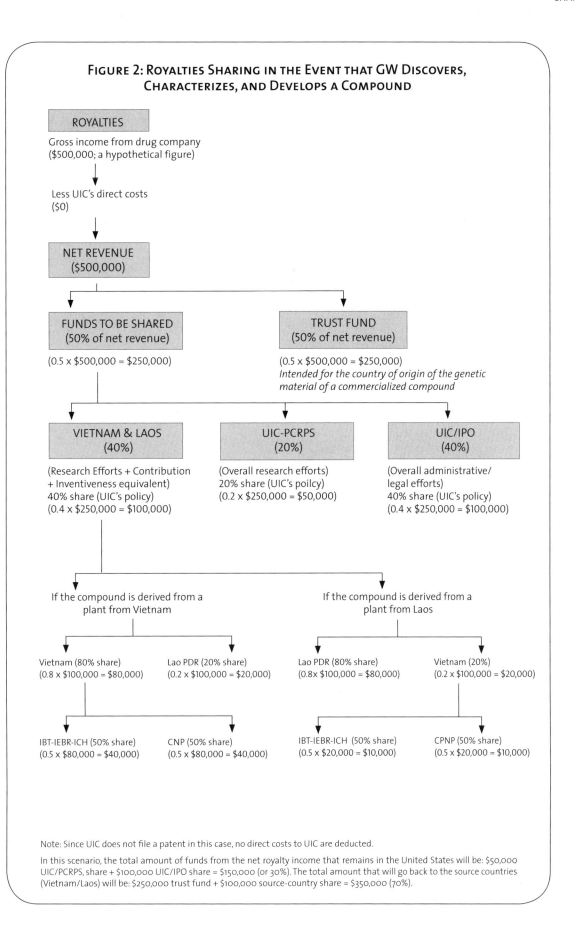

Figure 2: Royalties Sharing in the Event that GW Discovers, Characterizes, and Develops a Compound

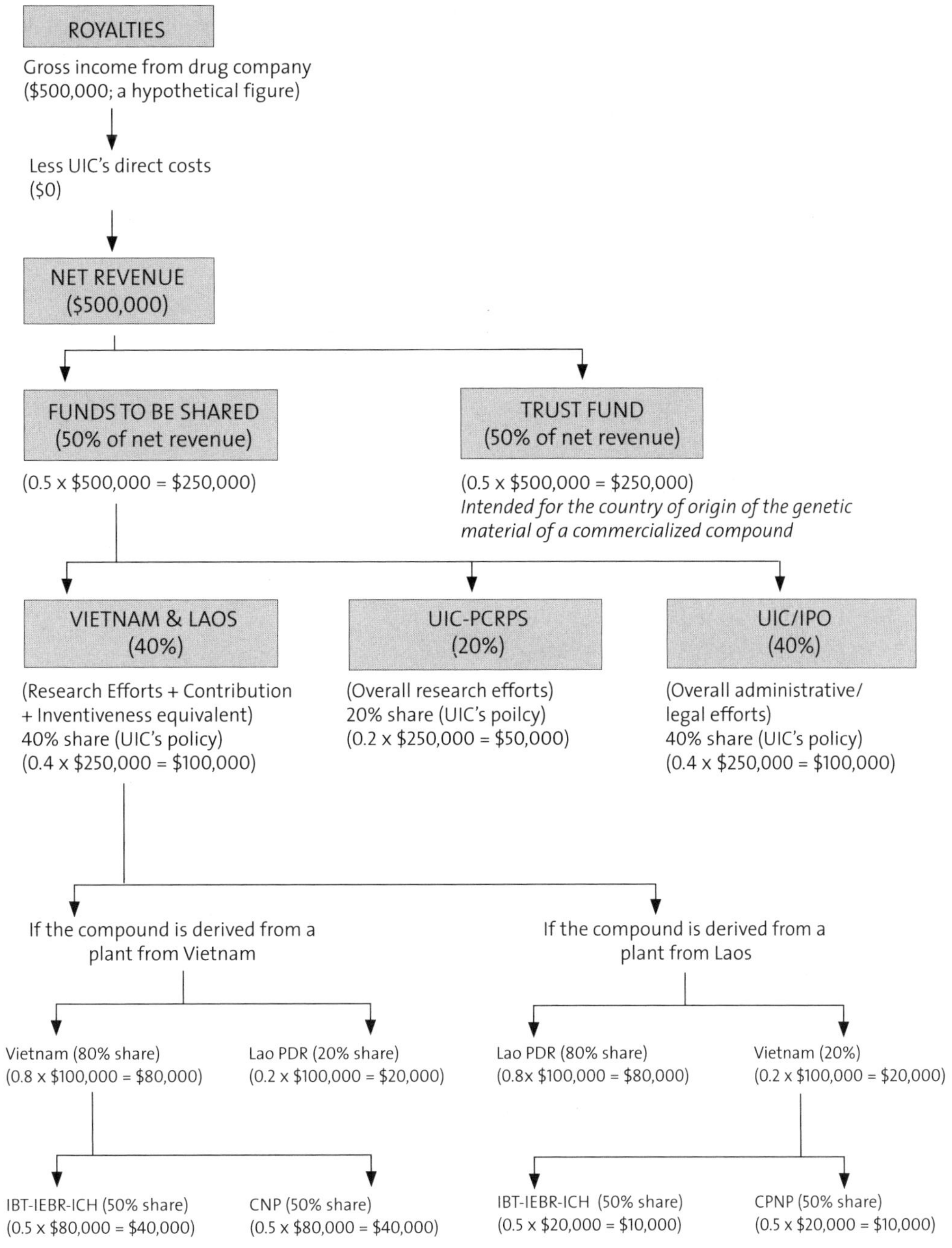

Note: Since UIC does not file a patent in this case, no direct costs to UIC are deducted.

In this scenario, the total amount of funds from the net royalty income that remains in the United States will be: $50,000 UIC/PCRPS, share + $100,000 UIC/IPO share = $150,000 (or 30%). The total amount that will go back to the source countries (Vietnam/Laos) will be: $250,000 trust fund + $100,000 source-country share = $350,000 (70%).

needs of local villagers when implementing new policies designed to improve the quality of life in rural areas, projects such as the ICBG can act as a model for obtaining additional data on the actual living conditions of rural villagers, and how to work with and for local communities, because of the close association between the ICBG and local village authorities and councils.

Moreover, the rural villagers begin to see themselves as stakeholders interested in the outcomes of conservation, economic, development, and health care delivery efforts because of their direct participation on the process of locally implementing national policies. Instead of feeling alienated by the process of reform, rural villagers realize their direct contribution to the process itself when they are actively engaged and participating in local projects—and when their contributions to the process are valued.

The ICBG might not be the only model for implementing change at the local level, and in a way that is welcomed and guided by villagers (since it is in cooperation with improving the quality of life at the village level); but it is a model currently in use and from which lessons and "best practices" may be gleaned and then replicated elsewhere worldwide. In this way, the ICBG contributes to the larger knowledge base of solutions for effective cooperative endeavors between North and South. ■

ACKNOWLEDGEMENTS

The UIC-based Vietnam-Laos ICBG is funded by U.S. government grant 1-UO1-TW-01015-01 (1998-2003) through funds from the National Institutes of Health (NIH), the National Science Foundation (NSF), and the Foreign Agriculture Service of the U.S. Department of Agriculture (FAS/USDA) and is administered by the Fogarty International Center of the NIH. The authors express deep appreciation to all members[24] of the UIC-Vietnam-Laos ICBG for their goodwill, understanding, hard work, and dedicated effort during the structuring of the MOA presently in force and the implementation of the project activities. Thanks to Dr. Joshua Rosenthal, ICBG Program Director, Fogarty International Center, for kindly reviewing the manuscript.

The authors wish to thank BIOTEC (National Center for Genetic Engineering and Biotechnology), 113 Thailand Science Park, Phaholyothin Road, Klong 1, Klong Luang, Pathumthan 12120, Thailand, for the invitation and partial sponsorship in the presentation of this paper in the BioThailand Conference in Bangkok on July 17, 2003, and for permission to reproduce this paper, originally published in the *BioThailand 2003 Proceedings*, and edited for this Handbook by MIHR/PIPRA.

Edited and reprinted with permission from the National Center for Genetic Engineering and Biotechnology (BIOTEC), 113 Thailand Science Park, Phaholyothin Road, Klong 1, Klong Luang, Pathumthan 12120, Thailand.

Corresponding author:
DJAJA DJENDOEL SOEJARTO, *PCRPS, College of Pharmacy, University of Illinois at Chicago, 833 S. Wood St., Chicago, IL, 60612, U.S.A.* dds@uic.edu, doelsoejarto@gmail.com

---

1 Reid WA, SA Laird, R Gamez, A Sittenfeld, AH Janzen, MA Gollin and C Juma. 1993. A New Lease on Life. In Reid WA, SA Laird, CA Meyer, R Gamez, A Sittenfeld, AH Janzen, MA Gollin, and C Juma, *Biodiversity Prospecting: Using Genetic Resources for Sustainable Development.* World Resources Institute: Washington DC.

2 Sittenfeld A and R Gamez. 1993. Biodiversity Prospecting by INBio. In Reid WA, SA Laird, CA Meyer, R Gamez, A Sittenfeld, AH Janzen, MA Gollin, and C Juma, *Biodiversity Prospecting: Using Genetic Resources for Sustainable Development.* World Resources Institute: Washington DC.

3 Anonymous. 2003. Fogarty International Center Web site. www.fic.nih.gov/programs/icbg.html.

4 Grifo FT. 1996. Chemical Prospecting: An Overview of the International Cooperative Biodiversity Groups Program. In *Biodiversity, Biotechnology, and Sustainable Development in Health and Agriculture: Emerging Connections* (ed. J Feinsilver). Pan American Health Organization: Washington, DC. pp 12-26.

Rosenthal JP. 1997. Integrating Drug Discovery, Biodiversity Conservation, and Economic Development: Early Lessons in the International Cooperative Biodiversity Groups. In Grifo F and J Rosenthal, *Biodiversity and Human Health.* Island Press: Washington, DC. pp 281-301.

Rosenthal JP, DA Beck, A Bhat, J Biswas, L Brady, K Bridbord, S Collins, G Cragg, J Edwards, A Fairfield, M Gottlieb, LA Gschwind, Y Hallock, R Hawks, R Hegyeli, G Johnson, GT Keusch, EE Lyons, R Miller, J Rodman, J Roskoski, and D Siegel-Causey. 1999. Combining High Risk Science with Ambitious Social and Economic Goals. *Pharmaceutical Biology* 37(Supplement):6-21.

5 See Rosenthal et al. in *supra* note 4.

6 www.grants.nih.gov/grants/guide/rfa-files/RFA-TW-03-004.html and www.fic.nih.gov/programs/research_grants/icbg/index.htm.

7 Rosenthal JP. 1997. Equitable Sharing of Biodiversity Benefits: Agreements on Genetic Resources. Presentation at the International Conference on Incentive Measures for the Conservation and the Sustainable Use of Biological Diversity in Cairns, Australia, 25-28 March

1996. *Investing in Biological Diversity: Proceedings of the Cairns Conference*. OECD: Paris.

8 See Rosenthal, et al. in *supra* note 4.

9 Berlin B, EA Berlin, JC Fernandez Ugalde, L Garcia Barrios, D Puett, R Nash, M Gonzalez-Espinosa. 1999. The Maya ICBG: Drug Discovery, Medical Ethnobiology, and Alternative Forms of Economic Development in the Highlands Maya Region of Chiapas, Mexico. *Pharmaceutical Biology* 37 (Suppl.): 127-141.

Kingston DGI, M Abdel-Kader, B-N Zhou, S-W Yang, JM Berger, H Van der Werff, JS Miller, R Evans, R Mittermeir, L Famolare, M Guerin-McManus, S Malone, R Nelson, E Moniz, JE Wisse, DM Vyas, JJK Wright, and S Aboikonie. 1999. The Suriname International Cooperative Biodiversity Group Program: Lessons from the first five years. *Pharmaceutical Biology* 37 (Suppl.): 22-34.

Kursar TA, TL Capson, PD Coley, DG Corley, MB Gupta, LA Harrison, E Ortega-Barria, and DM Windson. 1999. Ecologically Guided Bioprospecting in Panama. *Pharmaceutical Biology* 37 (Suppl.): 114-126.

Lewis WH, G Lamas, A Vaisberg, DG Corley, C and Sarasara. 1999. Peruvian Medicinal Plant Sources of New Pharmaceuticals (International Cooperative Biodiversity Group-Peru). *Pharmaceutical Biology* 37 (Suppl.): 69-83.

Schuster BG, JE Jackson, CN Obijiofor, CO Okunji, W Milhous, E Losos, JF Ayafor, and MM Iwu. 1999. Drug Development and Conservation of Biodiversity in West and Central Africa: A Model for Collaboration with Indigenous People. *Pharmaceutical Biology* 37 (Suppl.): 84-99.

Sittenfeld A, G Tamayo, V Nielsen, A Jimenez, P Hurtado, M Chinchilla, O Guerrero, MA Mora, M Rojas, R Blanco, E Alvarado, JM Gutierrez, and DH Janzen. 1999. Costa Rican International Cooperative Biodiversity Group: Using Insects and other Arthropods in Biodiversity Prospecting. *Pharmaceutical Biology* 37 (Suppl.): 55-68.

Soejarto DD, C Gyllenhaal, JC Regalado, JM Pezzuto, HHS Fong, GT Tan, NT Hiep, LT Xuan, DQ Binh, NV Hung, TQ Bich, NN Thin, PK Loc, BM Vu, BH Southavong, K Sydara, S Bouamanivong, MJ O'Neill, J Lewis, X-M Xie, and G Dietzman. 1999. Studies on Biodiversity of Vietnam and Laos: The UIC-based ICBG Program. *Pharmaceutical Biology* 37 (Suppl.): 100-113.

Timmermann BN, G Wächter, S Valcic, B Hutchinson, C Casler, J Henzel, S Ram, F Currim, R Manak, S Franzblau, W Maiese, D Galinis, E Suarez, R Fortunato, E Saavedra, R Bye, R Mata, and G Montenegro. 1999. The Latin American ICBG: The First Five Years. *Pharmaceutical Biology* 37 (Suppl.): 35-54.

10 Anonymous. 1997. International Cooperative Biodiversity Groups (ICBG): RFA TW-98-001. *NIH Guide* 26. Anonymous. 2002. International Cooperative Biodiversity Groups (ICBG)- RFA: TW-03-004. grants1.nih.gov/grants/guide/rfa-files/RFA-TW-03-004.html.

11 See *supra* note 7.

12 See *supra* note 10.

13 See Soejarto et al. in *supra* note 9.

14 Soejarto DD, LT Xuan, BM Vu, LX Dac, TQ Bich, BH Southavong, K Sydara, S Bouamanivong, HJ Zhang, HHS Fong, G Tan, J Pezzuto, SG Franzblau, C Gyllenhaal, MC Riley, NT Hiep, PK Loc, and NV Hung. 2002. Implementing IPR and Benefit-Sharing Arrangements: Experiences in the University of Illinois at Chicago-Vietnam-Laos ICBG Program. In Anonymous. *Intellectual Property Rights and Traditional Knowledge on Genetic Resources in Pharmaceutical and Cosmetic Business*. JBA/NITE, International Symposium 2002, Tokyo, Japan, 14 November 2002. pp. 47-83.

15 Soejarto DD, JA Tarzian-Sorensen, C Gyllenhaal, GA Cordell, NR Farnsworth, HHS Fong, AD Kinghorn, and JM Pezzuto. The Evolution of University of Illinois' Policy of Benefit-Sharing in Research on Natural Products. *Proceedings of the 7th International Congress on Ethnobiology*, Athens, Georgia, 23-27 October 2000. University of Georgia Press: Athens, Georgia. pp. 21-30.

16 See *supra* note 14.

17 See *supra* note 14.

18 See *supra* note 14.

19 See *supra* note 14 and Soejarto DD, C Gyllenhaal, JC Regalado, JM Pezzuto, HHS Fong, GT Tan, NT Hiep, LT Xuan, NV Hung, TQ Bich, PK Loc, BM Vu, BH Southavong, K Sydara, S Bouamanivong, MJ O'Neill, and G Dietzman. 2002. An International Collaborative Program to Discover New Drugs from Tropical Biodiversity of Vietnam and Laos. *Natural Product Sciences* 8:1-15.

20 The 1998-2003 UIC ICBG is referred to as UIC ICBG Phase I. On 1 October 2003, a new funding as a result of 2003-2008 re-competition was awarded; the 2003-2008 UIC ICBG is referred to as PHASE II. In Phase II, UIC, NCST [this institution name was changed in 2004 to VAST (Vietnamese Academy of Science and Technology)], CPNP, and TMRC continued to be members of the consortium, with Purdue University added as a new member, and Bristol-Myers Squibb (B-MS) Pharmaceutical Co. became the industrial partner.

21 The terms and conditions of Phase II UIC ICBG are, in large part, similar to the 1998-2003 Phase I, with the exception that, in Phase II, each member of the consortium has the right to file an IP protection.

22 In Phase II, the percentages of the royalty stream also flow to Purdue Universty; B-MS waived the rights to any royalties.

23 The accomplishments of the UIC ICBG (Phase I and Phase II) were examined in a 2006 paper: Soejarto DD, HJ Zhang, HHS Fong, GT Tan, CY Ma, C Gyllenhaal, MC Riley, MR Kadushin, SG Franzblau, TQ Bich, NM Cuong, NT Hiep, PK Loc, LT Xuan, NV Hai, NV Hung, NQ Chien, LT Binh, BM Vu, HM Ly, B Southavong, K Sydara, S Bouamanivong, JM Pezzuto, WC Rose, GR Dietzman, BE Miller, and TV Thuy. 2006. Studies on biodiversity of Vietnam and Laos 1998-2005: Examining the impact. *Journal of Natural Products* 69: 473-481.

24 Funding of Phase II UIC ICBG through NIH grant 2-UO1-

CHAPTER 16.6

Issues and Options for Traditional Knowledge Holders in Protecting Their Intellectual Property

STEPHEN A. HANSEN, *American Association for the Advancement of Science (AAAS), Science and Human Rights Program, U.S.A.*
JUSTIN W. VAN FLEET, *Principal Consultant and Director, The Advance Associates, LLC, U.S.A.*

ABSTRACT

Traditional knowledge (TK) is the information that people in a given community, based on experience and adapted to local culture and environment, have developed over time and that continues to develop. This knowledge is used to sustain the community and its culture, as well as the biological resources necessary for the continued survival of the community. Since 1948, international human-rights standards have recognized the importance of protecting intellectual property. Yet, to date, intellectual property (IP) rights are not adequately extended to the holders of TK. The requirements for IP rights protections under current IP regimes remain largely inconsistent with the nature of TK. As a result, it is neglected and considered part of the public domain with no protections or benefits for the knowledge holders, or expropriated for the financial gains of others, often referred to as biopiracy. This chapter presents basic IP concepts in the context of TK with specific attention to identifying, classifying, and protecting elements of TK. The advantages and disadvantages of the various IP protection options are discussed, and a number of case studies are presented to facilitate a better understanding of each option or issue.

1. INTRODUCTION

Traditional knowledge (TK) is information that people in a given community, based on experience and adaptation to a local culture and environment, have developed over time and continue to develop. The knowledge is used to sustain the community and its culture and to maintain the genetic resources necessary for the community's continued survival. Key examples of TK are these uses of biological resources:

- *plao-noi* in Thailand for the treatment of ulcers
- the *hoodia* cactus by Kung Bushmen in Africa to stave off hunger
- turmeric in India for wound-healing
- *ayahuasca* in the Amazon basin for sacred religious and healing purposes
- *j'oublie* in Cameroon and Gabon as a sweetener

TK includes mental inventories of local biological resources, animal breeds, and local plant, crop, and tree species. It may include such information as which trees and plants grow well together and which are "indicator plants" (plants that show soil salinity or are known to flower at the beginning of the rains, for example). TK includes practices and technologies, such as seed treatment and storage methods and tools used for planting and harvesting. It also encompasses belief systems that play a fundamental role in peoples' livelihoods, maintain their health, and protect and replenish the environment. TK is dynamic in nature and may include experimentation in the integration of new plant or tree species into existing farming systems or a traditional healer's tests of new plant medicines.

Hansen SA and JW Van Fleet. 2007. Issues and Options for Traditional Knowledge Holders in Protecting Their Intellectual Property. In *Intellectual Property Management in Health and Agricultural Innovation: A Handbook of Best Practices* (eds. A Krattiger, RT Mahoney, L Nelsen, et al.). MIHR: Oxford, U.K., and PIPRA: Davis, U.S.A. Available online at www.ipHandbook.org.

The term *traditional* used in describing this knowledge does not imply that it is old or untechnical in nature, but that it is tradition based. It is traditional because it is created in a manner that reflects the traditions of the originating communities, therefore not relating to the nature of the knowledge itself, but to the *way* in which that knowledge is created, preserved, and disseminated.[1]

TK is collective in nature and is often considered the property of the entire community, not belonging to any single individual within the community. TK is transmitted through specific cultural and traditional information-exchange mechanisms—for example, orally through elders or specialists (breeders, healers, and so on)—and often to only a select few people within a community.

The knowledge and uses of specific plants for medicinal purposes (often referred to as traditional medicine) is an important component of TK. Once, traditional medicines were a major source of materials and information for the development of new drugs. In the 20th century, however, new sources for pharmaceuticals led to a decline in the importance of ethnobotany in drug-discovery programs. However, new discoveries of potentially potent anticancer agents in plants (such as turmeric and taxol), as well as a rapidly growing herbal remedies market, have revived industry interest in traditional medicinal knowledge and practices. As interest in traditional medicine is rekindled, indigenous knowledge of the cultivation and application of genetic resources is being exploited at an alarming rate.

IP (intellectual property) rights should guarantee both an individual's and a group's right to protect and benefit from its own cultural discoveries, creations, and products. But Western IP regimes have focused on protecting and promoting the economic exploitation of inventions with the rationale that doing so promotes innovation and research. Western IP law, which is rapidly assuming global acceptance, often unintentionally facilitates and reinforces a process of economic exploitation and cultural erosion. It is based on notions of individual property ownership, a concept that is often alien to indigenous communities and can be detrimental to them. An important purpose of recognizing private proprietary rights is to enable individuals to benefit from the products of their intellect by rewarding creativity and encouraging further innovation and invention. But in many indigenous worldviews, any such property rights, if they are recognized at all, should be extended to the entire community. They are a means of maintaining and developing group identity, as well as group survival, rather than promoting or encouraging individual economic gain.

2. IP PROTECTION OPTIONS FOR TK HOLDERS[2]

2.1 *Patents*

Patents provide a legal monopoly over the use, production, and sale of an invention, discovery, or innovation for a specific period of time (usually about 20 years). A monopoly is the right to exclusive control over the use, development, and financial benefits derived from a patented item. In order for an invention or innovation to be patentable, it generally must meet three criteria: novelty, nonobviousness, and industrial application (or utility). Indeed, it must meet all of these criteria, and if one can be disproved, the patent cannot be approved.

Novelty refers to the "newness" of an invention, in other words, there is no prior art. Prior art is the knowledge base that existed before the invention was discovered or before the invention was disclosed by filing a patent application.

Nonobviousness refers to the presence of an inventive step, that is, the invention or innovation must not have been obvious at the time of its creation to anyone having "*ordinary skill in the art.*"[3]

Industrial application, or utility, refers to the very reason for patent protection, that is, to promote the progress of the useful arts. For a product or process to be useful it must, at least, work, although it does not have to work perfectly or even better than any competing products or processes, nor does there have to be a market for the invention (nor even a potential market).

For several reasons, patents might not represent the most advantageous form of IP rights protection for TK. First, applying for a patent requires full disclosure of (making public) the invention or innovation. Shortly after the patent is approved, the information is placed in the public domain by making the patent application publication available to the public. In the United States, a patent is made public 18 months after it is approved. If the TK is considered a *trade secret*, a patent may not be the most appropriate IP solution. Second, the invention or innovation must be novel according to patent-office standards. The patent applicant must prove that the invention or innovation is not part of the current prior-art base as defined by each country's legal definition of *novelty*. In many countries, TK may be considered, de facto, part of the prior-art base. This task can either be simple or somewhat difficult, but nonetheless, it must be demonstrated.

2.2 *Petty-patent models*

Petty patents allow for protections similar to those of patents, but for knowledge consisting of a less-detailed inventive step.[4] The knowledge must still meet the novelty and industrial-application criteria. The term of protection for a petty patent is typically between four and six years, which is shorter than the term for the standard patent.

The petty patent exists only in a few countries and is not mentioned in the Agreement on Trade-Related Aspects of Intellectual Property Rights (TRIPS) as a minimum standard for IP protection. However, some countries are pushing for the inclusion of petty patents in the TRIPS Agreement. Petty patents may be more suitable for TK, as TK is not typically documented in the same manner as Western science. Despite the fact that petty patents are not globally recognized as a minimal standard for IP protection, some countries have enforced the mechanism as a way of protecting TK. For example, a type of petty patent is mentioned in Kenyan legislation in order to protect indigenous claims to traditional herbal medicine.[5] Although the current application of petty patents is relatively small, their implementation at a broader level could serve TK as a viable IP protection option.

2.3 *Plant variety protection/ plant breeders' rights*

Many countries protect plant varieties with the plant variety protection certificate. This mechanism is used to protect the rights of breeders of sexually reproducing (by seed) varieties of plants. Breeders' rights protect the commercial interests of the breeder so that economic incentives exist for continued breeding of new plant varieties, ultimately serving farmers or those who grow the varieties. Importantly, unlike utility patents, plant variety certificates do not require the authorization of the breeder for use of the variety by others for further breeding purposes.

The criteria for a plant variety protection certificate are fairly uniform across countries that offer them. The variety must meet all of these criteria:

- distinct from existing, commonly known varieties
- sufficiently uniform
- stable
- novel[6]

The International Convention for the Protection of New Plant Varieties (UPOV) is not a legal mechanism per se. Rather, UPOV is an international treaty and an organization that sets certain standards. A country can only become a member of UPOV if its plant variety protection schemes meet these minimum standards. Importantly, under the TRIPS agreement, countries are bound to enact sui generis protection for plants, and the UPOV requirements are generally considered to meet such standards.

Proposals for legislation in Nicaragua have included provisions that require ten unique characteristics in order to distinguish a variety as *distinct*; to exclude protection for "discovered" plants; and, not to extend plant breeders' rights to plants used for food or sown directly by farmers. Zambia has cited the Convention on Biological Diversity (CBD)[7] in developing its plant variety protection mechanism and states that any final legislation must recognize and reward indigenous innovation. India's Plant Variety Protection Act (2001) declares that the rights of the farmer supercede those of the breeder. The Plant Varieties Protection Act of Bangladesh (1998) states that a variety must have *"immediate, direct and substantial benefit to the people of Bangladesh,"*[8] and protects both community and farmers' rights.[9] These examples demonstrate that options other than UPOV can be established that effectively address the needs of TK holders.

2.4 *TK registries*

Public registries place information in the public domain and serve as a form of prior art or *defensive disclosure*. They can be public or private. A defensive disclosure, by describing information in a printed publication or other publicly accessible medium, helps to establish prior art capable of preventing patents.

2.4.1 *Public registries*

TK registries are official collections of documentation that describe TK (see Box 1). Registries can be established and maintained either locally (within a community) or outside a community (external), even for an entire country (see Box 2). With a locally maintained registry, the community may collectively decide what is to be included in the registry and what knowledge is to be shared and/or disclosed to people outside the community.

2.4.2 *Private registries*

Private registries do not place knowledge in the public domain. But private registries can be effective as:

- protection mechanisms for TK in instances where a sui generis system is in place
- preservation mechanisms when cultural and historic preservation is a goal
- tools for access and benefit-sharing agreements

Since the information in a private registry is documented but is not in the public domain, it may not constitute prior art capable of preventing a patent based on the knowledge by an outsider. The knowledge in a private registry cannot prevent the approval of a patent under most IP systems unless the knowledge constitutes prior art through a sui generis mechanism and disclosed to patent authorities. However, it may be possible to challenge and revoke a patent with knowledge documented in a private registry if patent law recognizes prior art not disclosed to the public as being admissible under a sui generis system. Reexamination requests of patents can be both costly and time consuming. Also, the knowledge may need to be disclosed to the public if no sui generis protection mechanism exists that would prohibit its public disclosure during reexamination.

Because the recognition and effectiveness of private registries varies from country to county, private registries are most effective as a mechanism for preservation of knowledge and as a tool for access and benefit-sharing agreements. A private registry can serve as a catalog for knowledge that can be licensed to outside parties for research and product development. As a mechanism for cultural preservation, the private registry serves as a cultural library that documents and maintains TK belonging to a community and helps prevent loss of the TK (see also Box 2).

A typical form of registry is a computer database. The Internet is an ideal location for public databases containing TK, as they can serve as a vehicle for defensive disclosure and are accessible to patent offices worldwide as a source of prior art. The World Intellectual Property Organization (WIPO) is in the process of compiling a list of TK-related databases for international patent offices, and several large public databases collect TK as a means of defensive disclosure against the misappropriation of IP.

The benefit of both public and private registries lies in their ability to prevent or revoke inappropriate claims of IP rights. In order to be effective in this manner, it is essential that national patent offices are made aware of the public registry for use in prior-art searches. The public registry has the additional benefits of negating the application of IP rights on TK prior to patent approval and promoting free use of the knowledge in the public domain for everyone's benefit.

Box 1: An Example of TK Documentation

To illustrate how a claim may be documented, an entry from the Honeybee Network's Innovation Database is provided here. That database is a large online database of grassroots innovations detailing contemporary and traditional innovative practices.

Claim[a]	To illustrate how a claim may be documented, an entry from the Honeybee Network's Innovation Database is provided here. That database is a large online database of grassroots innovations detailing contemporary and traditional innovative practices.
Inventor[a]	Hirabhai Kodarbhai Raval
Address of innovator	Sabarkantha Gujarat
Details of innovation	Hirabhai Kodarbhai Raval has a special way of treating his animals for stiffness of the body. He prepares a mixture of 250 g variyali (*Foeniculum vulgare*), 50 g turmeric powder, and 500 g Dalda ghee. This, when given to the animal to drink, loosens the stiffness in the body of the animal and relieves joint pains. Half this dosage is prescribed for very young animals.
Reference from	Honey Bee, 9(4): 15, 1998

Note that this database entry contains the following information:

Claim being made: Curing joint pains (In this format for documentation, the claim also serves as the name or descriptive title for the claim.)

Name of the inventor or claimant: In this example, the inventor is an individual, but this could be the name and/or location of a community as well.

Details of the invention: It is a mixture consisting of the following ingredients and amounts: 250 grams of variyali (*Foeniculum vulgare*), 50 grams of turmeric powder, and 500 grams of Dalda ghee.

How applied: It is given to the animal to drink.

Dosage: As mixed and half dosage for very young animals

Results: Loosens the stiffness in the body of the animal and relieves joint pains

a Term added by the authors.

2.5 *Trade secrets*

Trade secrets protect undisclosed knowledge through access agreements, which may involve paying royalties to knowledge holders for access to and the use of their knowledge. Three elements are required for knowledge to be classified as a trade secret. The knowledge:

- must have commercial value
- must not be in the public domain
- is subject to reasonable efforts to maintain secrecy

TK that is maintained within a community could be considered a trade secret. But once the knowledge is made public, this option no longer exists. A trade secret is only enforceable as long as it remains a secret. Trade secrets have no legal protection except in cases of "*breach of confidence and other acts contrary to honest commercial practices.*"[12] This means that one must be able to prove some form of malicious intent on the part of a contracting party as the cause for a trade secret's diffusion to the public in order to be compensated for the loss of secrecy.

It is important to remember that knowledge considered a trade secret can be used by anyone if the knowledge is leaked into the public domain, is independently discovered by another individual, or is reverse engineered. It is difficult to protect trade secrets against misappropriation due to lack of legal entitlement to the bearer of the secret. When applied to knowledge belonging to a community, the community must make a reasonable effort to maintain the secrecy. If there is not a reasonable effort to maintain secrecy with respect to the TK, then trade secret protection is not applicable to it.

2.6 *Trademarks*

The U.S. Patent and Trademark Office (PTO) defines *trademark* as "*a word, phrase, symbol or design, or a combination of words, phrases, symbols or designs, that identifies and distinguishes the source of the goods of one party from those of others.*"[13] In other words, trademarks are a way of protecting the use of words, phrases, symbols, designs, or any combination of these associated with a product. Once a trademark is established, it can be used to

Box 2: A Public Registry in India

One example of a public registry is the people's biodiversity registers (PBRs) in India. Recognized in the Indian Biological Diversity Bill of 2000, the PBRs consist of records of people's knowledge of biodiversity, its use, trade, and efforts for its conservation and sustainable utilization. The PBRs are developed at the village level by a local school and college teachers, students, and nongovernmental (NGO) researchers, and villagers. Biodiversity registers are then compiled in the form of computerized databases at the levels of talukas, districts, states, and the entire country, in order to provide information to the public, government, and industry. These PBRs have been recognized by the Indian Biological Diversity Bill as a form of prior art in the evaluation of patent applications, as well as serving to ensure equitable access and benefit sharing.

External registries are maintained outside the community, often on the national or international level, by governments, NGOs, museums, or libraries. These registries can be collections of TK specific to one particular community or to several communities. Local communities may have control over what is entered into the registry, but may not be responsible for the registry's maintenance. Distinguishing between local or external registries is at the discretion of the TK stakeholders.

A disadvantage of the public registry is the disclosure of knowledge to others outside the community. When placing knowledge in the public domain, the knowledge may lose its commercial value, limit options for IP protection for the community, and may be used by the public without permission.

identify and differentiate similar products. Think how often names, images, and photos are always used in marketing products.

Trademarks are based on two principles: distinctiveness and avoiding confusion. Being distinct means that the trademark does not resemble any other existing word, phrase, symbol, design, and so on, associated with a similar product. Avoiding confusion as to the source of a product is important for consumers purchasing these products. Trademarks distinguish products in order not to mislead consumers into thinking that a product is something that it is not or that it comes from another source.

How can trademarks be applied to TK? Suppose a company sells a product composed of *maca*, a plant native to the Andean region. An indigenous community in the Andes, the original knowledge holders of *maca*'s uses, may also want to sell *maca* or profit from their own natural resources and knowledge. They could register a trademark like the example below:

Authentic Andean Maca

A traditional cure from the high peaks of the Andes

The indigenous group can register the above trademark and sell *maca* using this symbol to distinguish the brand.

2.7 *Geographical indicators*

A geographical indicator identifies a good as originating in a territory or region, or locality in that territory, where a given quality, reputation, or other characteristic of the good is attributable to its geographical origin.[14] Like trademarks, geographical indicators are typically words or terms, but when associated with a product, positively attribute a known quality to the product that is associated with a specific geographical location.

A geographical indicator cannot be used to describe a product unless it originates in the region associated with the name. For example, Swiss watches are associated with a tradition of high quality, so the term *Swiss watch* is a geographical indicator that assumes a watch came from Switzerland. Roquefort cheese (from France) is another product associated with high quality and constitutes a geographical indicator. Roquefort cheese can only be used to describe cheese produced in Roquefort-sur-Soulzon, France, and aged in the traditional caves (a practice also associated with the geographical indicator).

Other examples of geographical indicators include Bordeaux wine (France), Parma ham (Italy), Stilton cheese (United Kingdom), Darjeeling tea (India), Cognac (France), and Queso Murcia (Spain).

Geographical indicators serve four main purposes. They:

- identify where the product is from (its source)
- indicate the unique qualities of a product
- promote the product with a distinguishing name (for business purposes)
- prevent infringement and unfair competition by establishing a legal basis for using a location name to avoid confusion with similar products [15]

A specific form of geographical indicator is called an *appellation of origin*. Appellations of origin specify the quality of a product based on its geographical environment and are protected under the Lisbon Agreement of 1958. Twenty countries are party to the Lisbon Agreement. In 1998, of the 766 protected appellations of origin, 95% belonged to European countries.[16] Countries such as India and Bulgaria have recently been highly active in seeking *appellation of origin* protection for many of their products.

Preemptive protection of geographical indicators will ensure that they are commonly known and documented. This can be done by placing the geographical indicator in the public domain via a database or other publicly accessible medium.

The second option is to apply for a certification mark that is an official registration (as opposed to an unofficial disclosure of the indicator in the public domain). The certification mark is a type of trademark. Currently, international registry protection is available only for wines, and all other products are subject to national registry laws.[17]

If a country is party to the TRIPS Agreement, it is the country's international legal obligation to formulate legislation protecting geographical indicators. Article 22 of the TRIPS agreement states that members must provide legal means to prevent:

> *the use of any means in the designation or presentation of a good that indicates or suggests that the good in question originates in a geographical area other than the true place of origin in a manner which misleads the public as to the geographical origin of the good.*[18]

Additionally, the TRIPS Agreement requires the protection of what is defined as unfair competition in the Paris Convention.[19] "*All acts of such a nature as to create confusion by any means whatever with the establishment, the goods, or the industrial or commercial activities, of a competitor*" shall be prohibited under this article.[20]

What does all this mean in the everyday life of a TK holder? Let's examine an example that adequately explains the importance of a geographical indicator. The *maca* plant is native to the high peaks of the Andes Mountains where it thrives in the high altitudes. Suppose a Western company was to modify the plant so that it could grow in lower elevations. Then, that company was to grow large quantities of the plant in the United States and market the plant product as "Andean *maca*." This is a clear violation of the provisions that protect against the improper use of geographical indicator. Andean *maca* is associated with a distinguished quality, and by using the name, the product, which is not produced in the Andes, misleads consumers into believing both that:

- the product was actually cultivated in the Andes
- the product is of the quality as that produced in the Andes

Only *maca* grown in the Andes, then, is permitted to be marketed as "Andean *Maca*" if:

- Andean-grown *maca* is commonly known to be of superior quality to other *maca*, and this fact is documented in the public domain
- a certification mark has been officially registered with a federal government for "Andean *maca*"

3. PRIOR ART AND DEFENSIVE DISCLOSURE

When determining whether a claim is novel, either through the filing of a patent application or during the patent application review process, the prior-art base (the public domain) is examined. If the invention or claim is found described in the prior-art base or has been offered for use or sale for more than one year, it is not entitled to a patent. In U.S. patent law, *prior art* is defined as a publication printed either in the U.S. or a foreign country describing the invention or discovery and dated more than one year before a patent's filing date or, simply, dated before the act of invention or conception. A publication may include any document accessible to persons working in a certain profession or field and therefore skilled in the relevant art. These could include magazines, trade or scientific journals, newsletters, newspapers, and Web sites, to name but a few.

The European patent system does not limit evidence of prior art solely to printed publications, but includes everything made available to the public by the means of a written or oral description, by use or by any other way, anytime before the patent application filing date.[21] The difference between the U.S. and European definition of *prior art* has serious implications for the recognition of TK as prior art, as much TK is not documented nor published, but is shared orally, or publicly known through demonstrated and public use.

Prior art is taken into account for the non-obvious requirement in applying for a patent. In many cases, the prior art may prove to be very similar, but not exactly like the claim or invention itself, but the differences would be obvious to someone with ordinary skill in the area and who knew, or had relatively easy access to, the prior-art base.

3.1 *Defensive disclosure*

Defensive disclosure refers to information or documentation intentionally made available to the public as prior art in order to render any subsequent claims of invention or discovery ineligible for a patent. A defensive disclosure provides evidence of the invention, knowledge, or use of the invention by others before it was claimed by another inventor or offers evidence of public use or sale more than one year before the filing date of the patent.[22]

Defensive disclosures can be made anonymously without attributing the knowledge to a particular person or community. Anonymous disclosures might have a benefit for those who want to disclose information but at the same time not want to attract unwelcome attention to a community.

There are basically two types of mechanisms for defensively disclosing information. One consists of the traditional methods of publication: scientific, academic, technical, and business journals, and so on. The other mechanism is electronic publication through the Internet. In recent years, many Internet sites have been developed solely for the purpose of defensive disclosure. There are many Internet-based Web sites and databases that contain information on TK.

A community registry could serve as a viable means of defensive disclosure. This would involve placing the registry on the Internet for all to access (this would also include patent examiners during prior art searches), or if a country has a sui generis system in place, limiting outside access to only the patent office.

3.2 *Prior informed consent*

The CBD declares the obligation to obtain prior informed consent for accessing genetic resources. The Bonn Guidelines (2002)[23] further link genetic resources with TK in the obligation to acquire informed consent. Prior informed consent is the approval in advance for the use of one's genetic resources and any associated TK. *Prior* indicates that the approval must come before access is allowed or others use the knowledge. *Informed* means that information is provided on how the resource and/or knowledge will be used. *Consent* means permission to use the resource or knowledge. Sufficient information should be provided to a community, either by the IP office or other party, regarding the aims, risks, or implications of using the knowledge, including its potential commercial value.

Does a community possessing TK legally have the right to prior informed consent if someone accesses its genetic resources and related TK and wishes to use them? The answer: maybe. If the country where the community is located has ratified and implemented the CBD, access to TK should be subject to prior informed consent of the knowledge holders under Article 8(j).

Perhaps an example is the best way to understand how prior informed consent works. Suppose a scientist is traveling in South America and begins to work with a community in the Amazon region. The scientist is particularly amazed when he or she observes the methods used by a local community to process and apply a local plant to heal wounds. The scientist, now aware of the genetic resource and local knowledge of its use, can do one of two things: he or she can do nothing with the knowledge or can use the knowledge. If the scientist does nothing, there is obviously no need to obtain prior informed consent. If the scientist wishes to use the resource or knowledge (publish the knowledge in a journal article, apply for a patent, etc.), he or she must obtain prior informed consent of the appropriate national authorities if that Amazonian country has implemented the CBD.

4. SUI GENERIS PROTECTION SYSTEMS

Sui generis literally means "of its own kind" and consists of a set of nationally recognized laws and ways of extending plant variety protection (PVP) other than through patents. TRIPS itself does not define what a sui generis system is or should be. And although TRIPS does not mention UPOV, it is generally agreed that the UPOV standards meet the requirements for a sui generis system for plants. However, countries do not have to join UPOV to implement a sui generis system to comply with TRIPS.[24]

A sui generis system might consist of some standard forms of IP protections combined with other forms, or none at all, for genetic resources. For example, a country could provide patent protections for inventions, plant variety certificates (PCV) for plant varieties or just certain varieties, and/or exclude plants from any form of IP protection at all (although this could conflict with TRIPS compliance).

Potentially, a sui generis system could be defined and implemented differently from one country to another. In addition, a sui generis system might be defined to create legal rights that recognize any associated TK relating to genetic resources and promote access and benefit sharing. The government may choose to extend protections to genetic resources and/or knowledge to a community in the form of patents, trade secrets, copyrights, farmers' and breeders' rights, or another creative form not currently established in the IP regime.

In addition, a sui generis system may adopt measures of protection specific to TK in order to nullify inappropriate patents. For example, the Andean Community's Decision 486 states:

patents granted on inventions obtained or developed from genetic resources or traditional knowledge, of which any member state is the country of origin, without presentation of a copy of the proper access contract or license from the community shall be nullified.[25]

A sui generis system may legally acknowledge and protect knowledge related to the use of genetic resources even when it is not officially documented, but instead exists in the form of oral information, and traditional and historic use. Even though protections might be extended here, the government's IP office needs to know about the knowledge or practice in order to enforce protection. Therefore, if a country has some form of a sui generis system in place, it is important for local communities to establish a working relationship with the IP office. In addition, these offices may privately maintain inventories or registries of locally held knowledge, and can assist in its protection. For example, this office can deny a patent application if the knowledge it is based on is already held in the registry.

Under a sui generis system, and as called for by the CBD, any person interested in gaining access to a community's biological resources or knowledge for scientific, commercial or industrial purposes would need to obtain the prior informed consent of the indigenous peoples who possess the knowledge in question unless the knowledge is already in the public domain. This would allow the community to decide on access to and use of its genetic resources and knowledge, with the option to share or not to share them. If consent is granted, the person or persons wishing access to lands held by indigenous communities or a conservation area, its biological resources, and associated knowledge would need to present evidence of this consent to either the IP office or to the proper authority.

5. ACCESS AND BENEFIT SHARING

Access refers to granting permission to enter an area for the purpose of sampling, collecting, and removing genetic or other resources. Benefit sharing refers to all forms of compensation for the use of genetic resources, whether monetary or nonmonetary. This might also include participation in scientific research and development of genetic resources, as well as the sharing the findings of any potential benefits resulting from this work.

Articles 1 and 8(j) of the CBD encourage the equitable sharing of benefits arising from TK for conservation and sustainable use of biological diversity. In benefit-sharing arrangements, all parties share the benefits arising from the use of genetic materials and TK of their uses. For the local community, this involves the sharing of TK and resources with contracting parties and others who wish to use it for research and/or developing new products based on this knowledge. The contracting parties in turn would share any advancements, benefits (including financial), or products that made use of the resources developed from local resources with the local community.

Article 15 of the CBD states that access to genetic resources and any transfer of technology be provided and/or facilitated under fair and mutually agreed-upon terms. This may include types of financial arrangements described later in the CBD (Articles 20 and 21).

Benefits include a wide range of options and often beneficiaries receive more than one type of benefit. They may include:

- **Start-up/upfront benefits.** Payments paid as a lump sum (if a financial arrangement) or delivered (if a cooperative or capacity building project). (These benefits would include equipment such as computer hardware, software, or extraction and screening facilities.)
- **Process benefits.** Derived during the process of research and development. (In addition to financial payments, process benefits may include capacity, expertise, or know-how building, and training through joint research.)
- **Product benefits.** Paid after commercialization of the final product. (These may include royalty payments that may be negotiated according to the contribution of the genetic resource or the amount of or role of local knowledge that was used in creating the final product.)
- **Moral and relation benefits.** Unlike the financial benefits described above, not transferred according to a formalized arrangement, but based on the interaction of the participants.[26]

As an example, let us consider a case in Ecuador. In that country, the Inter-American Development Bank (IDB) and several NGOs have launched a project titled "The Transformation of TK into Trade Secrets." The goal of the project is to catalogue TK and then maintain the database at regional centers, access to which will be safeguarded. Each participating community will have its own file in the database and will not be able to access files of any other community. The collected knowledge will be reviewed, and knowledge that is not common to multiple communities may be negotiated as trade secrets through material transfer agreements (MTA). The benefits from any MTAs are to be split between the Government of Ecuador and the communities that deposited the knowledge in the database. Payments to communities will then be used to finance public projects previously identified by each community.[27]

Contractual agreements[28] are at the heart of any benefit-sharing mechanism. They are legally binding documents between parties. In relation to TK, they are generally used to outline and enforce access and benefit-sharing agreements, as well as trade secrets. Contracts relative to TK may explain or clarify the following points:

- parties to the agreement
- duration of the agreement
- knowledge included in the agreement
- uses of the knowledge
- restrictions placed on the knowledge's use
- restrictions placed on confidentiality
- specifics for benefit sharing

Some types of contracts that might be employed for access and benefit sharing in compliance with the CBD include:

- confidentiality (also known as non-disclosure agreements)
- exclusive licenses
- nonexclusive licensing agreements
- material transfer agreements[29]

The type of contractual arrangement will vary according to the knowledge and/or genetic resources in question, as well as the interests and cultural components related to the knowledge. If considering a contractual agreement, make sure that the selected type of contract corresponds to both the short-term and long-term interests of the community (see also Box 3).

6. LOCATING AND IDENTIFYING TK

In order to protect or preserve TK utilizing the Western framework of IP rights, it is necessary to first locate and identify this knowledge according to the epistemological constructions recognized under this system. TK can be identified in:

- daily activities including, among other things:

    - farming
    - gardening
    - animal breeding and care
    - food and nutrition
    - healthcare and reproductive health
    - water-resource use
    - spiritual and religious activities
    - folklore, songs, poetry, and theater
- community records (Although TK is mostly transmitted by word of mouth, some other forms of record keeping may exist, for example, maps, boundary markers [trees, poles, stones, and so on], drawings, paintings or carvings, and many other forms.)
- people working with the community, such as NGO researchers, academics, scientists, and development specialists who may have been collecting TK
- secondary sources such as journal articles and books, unpublished documents, databases, videos, photos, museums, and exhibits.[31]

An element of TK for which IP protections could potentially apply is called a *knowledge claim.* A TK claim contains three essential components: a genetic resource, a preparation or process, and an end result or product derived from a preparation or process. The genetic resource is typically a plant. The process encompasses the various ways of using the plant for an end result. Processes may include methods of growing, harvesting, extracting, preparing, or applying the plant. The end result is the benefit from using the biological resource and the process. Let's look at an example (Figure 1).

The three categories (Plant, Process, Product) can be combined in a variety of ways producing several claims. For example, from the simple figure below, it is possible to deduce six claims of process methods involving the plant:

- growing *maca* to cause an increase in livestock reproduction
- preparing *maca* to cause an increase in livestock reproduction
- administering *maca* to cause an increase in livestock reproduction
- growing *maca* to improve human fertility
- preparing *maca* to improve human fertility
- administering *maca* to improve human fertility

7. IDENTIFYING WHO HOLDS THE KNOWLEDGE

After identifying a TK claim, the next step is to determine whom the knowledge holders and stakeholders are for the claim. The knowledge holders are the people who hold and/or use the knowledge, and stakeholders are the people in the community with a direct interest in the knowledge. When making a decision in relation

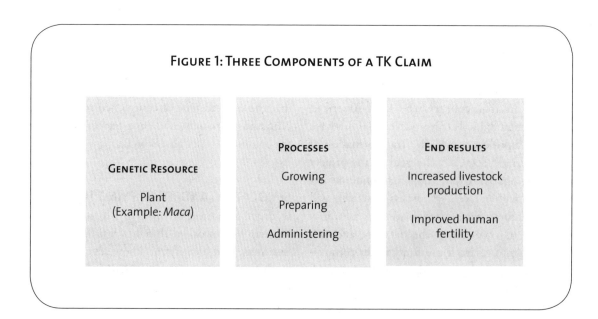

FIGURE 1: THREE COMPONENTS OF A TK CLAIM

GENETIC RESOURCE	PROCESSES	END RESULTS
Plant (Example: *Maca*)	Growing Preparing Administering	Increased livestock production Improved human fertility

to a specific knowledge claim, one must consult all of the stakeholders of that claim (which is often the entire community and/or other communities as well) before making a final decision about how any IP rights should be applied.

TK can either originate within a community or enter a community from the outside. If the knowledge is not originally from within the community in question, then it may not be subject to any IP rights and may already be part of the public domain. If the knowledge is from within the community, then the next step is to determine who holds the knowledge. The holder(s) of the knowledge can be an individual, multiple individuals, or the community as a whole.

The next step is to determine who uses or has access to the knowledge. Knowledge claims can either be held or practiced by no one, an individual, multiple individuals, a community, or people outside the community.

Any potential IPR options will depend on how many people are aware of the knowledge and who these people are. Based on these variables, a knowledge claim can fit into on of three groups:

1. Known and used by an individual
2. Known and used by several individuals or a community
3. Diffused broadly and in the public domain.

Figure 2 can assist in determining who holds the knowledge and who the stakeholders are in order for help in deciding which options to pursue for an identified knowledge claim.[32] The dashed box in the figure represents knowledge that may fall within IP rights protections and that is not part of the public domain. If the knowledge crosses outside the box, the knowledge may already be in the public domain (with or without prior informed consent[33] and with no options for IP rights protection [see Section 3.2]).

8. IDENTIFYING IP OPTIONS

8.1 *Determining cultural aspects*

The scientific aspect of TK is only one aspect of a larger culture of knowledge. For this reason, culture cannot be ignored when applying IP rights to TK. Cultural aspects that are important to TK are described below under six general categories. Each category should be considered independently, and in combination, when evaluating the place of a specific claim in its cultural context and in the IP rights regime.

1. **Spiritual.** knowledge that not only has a useful or functional purpose but also some form of spiritual, religious, or sacred importance
2. **Subsistence.** knowledge necessary for the basic survival of the community, including knowledge used for food production or any knowledge vital for life and survival
3. **Economic.** knowledge with strong ties to the economic survival or benefit of the TK stakeholders
4. **Traditional secret.** knowledge that is held as a secret among the community (Disclosing knowledge within this category to the general public would be culturally inappropriate.)
5. **Medicinal.** knowledge used to cure or prevent medical ailments within a community
6. **Historic.** knowledge that is of historic importance to the community

8.2 *Determining community goals*

When evaluating a knowledge claim and determining potential options for protection, the goals and interests of the community are important to consider. Five categories may be used for determining community goals for a claim:

1. **Profit.** commercializing and receiving financial gains or other economic benefits from TK
2. **Dissemination for public good.** sharing TK in order to benefit others (This goal is particularly applicable to TK with medicinal or agricultural uses.)
3. **Avoiding exploitation.** preventing the harming or usurpation of culture and environment (Control over knowledge, the way it is used, and its concurrent effects on the culture and environment are important to the TK stakeholders.)

Figure 2: Identifying Who Holds the Knowledge

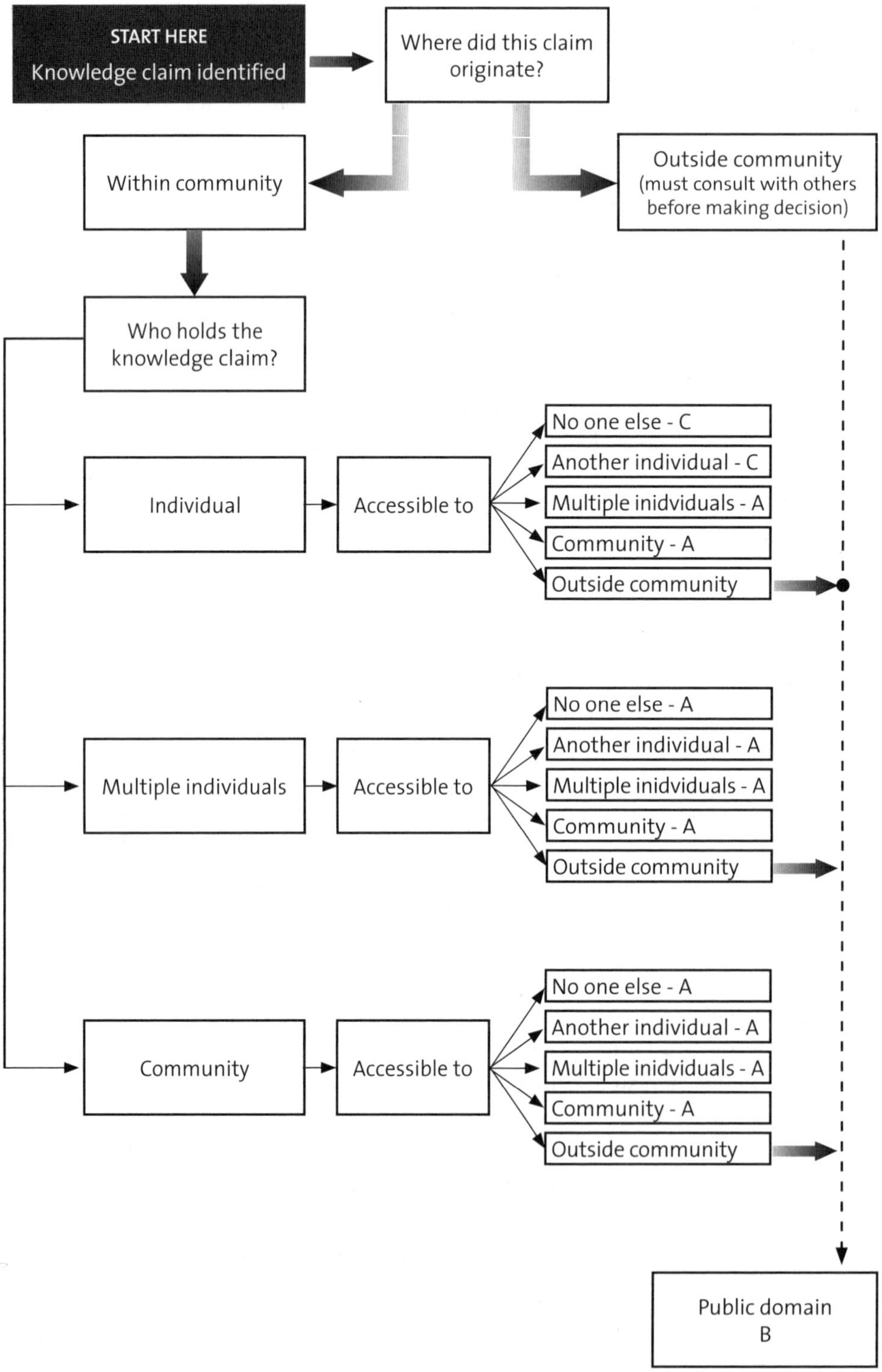

4. **Avoiding inappropriate IP claims.** avoiding IP claims on community knowledge or resources by outsiders (The protection of moral and material interests is of primary importance.)

8.3. *Preserving TK above other interests or desires*

Once TK has been identified and the cultural and goal-oriented dimensions of the knowledge explored, stakeholders should cross-reference these cultural values and goals with relevant IP options available in a given country.

9. CONCLUSIONS

This chapter explains possible IP mechanisms that might be applied to protect TK and biological resources. Our experience shows that it has served more as an educational resource to alert TK holders to the possible risks of others seeking IP rights protection than as a resource for seeking IP rights protections themselves. Yet, it is true that over the past several years a growing number of TK holders have started to explore the potential use of IP protections. Still, for many reasons, TK remains elusive to current IP laws.

Local and indigenous peoples' management and protection of IP rights associated with their biological resources and TK remain a challenge. In order to address this challenge, it will be necessary to properly recognize and protect TK and also to employ global mechanisms for equitable benefit sharing. In the more-immediate term, existing mechanisms of IP rights protection will need to be effectively utilized in order to confer adequate protection and benefit sharing. However, in the longer term, changes to both the domestic and global IP regimes might be required. Yet, regardless of the exact type of IP rights protection employed, the end result must always be aimed toward a balance, that is, to better protect and provide equitable benefit to the originators of that TK while serving the broader public interest. In other words, access, development, and distribution must be balanced against equitable benefit sharing, sustainable development, and conservation. ■

ACKNOWLEDGMENTS

The authors would like to thank the following people and organizations for their support and assistance throughout the creation of this chapter: the secretariat of the World IP Organization's Intergovernmental Committee on Genetic Resources, Traditional Knowledge and Folklore; Merida Roets, president of *ScientificRoets* (South Africa) and former AAAS Science Radio Fellow; participants from the AAAS Roundtable on Traditional Knowledge at the Fifth Session of the WIPO IGC on Genetic Resources, Traditional Knowledge and Folklore, 14 December 2002; Rosemary Coombe, Tier One Canada Research Chair in Law, Communication and Cultural Studies at York University in Toronto; Michael Gollin, project legal consultant and attorney at Venable, LLP; Matthew Zimmerman, computer specialist at the AAAS Science and Human Rights Program. This publication was made possible in part from grants provided by the Center for the Public Domain and the Richard and Rhoda Goldman Fund.

STEPHEN A. HANSEN, *American Association for the Advancement of Science (AAAS), Science and Human Rights Program, 1200 New York Avenue NW, Washington DC, 20005, U.S.A.* shrp@aaas.org

JUSTIN W. VAN FLEET, *Principal Consultant and Director, The Advance Associates, LLC, 2112 New Hampshire Avenue NW, Suite 308, Washington DC, 20009, U.S.A.* justin@theadvanceassociates.com

---

1 WIPO. 2002. Elements of a Sui Generis System for the Protection of Traditional Knowledge. Intergovernmental Committee on IP and Genetic Resources, Traditional Knowledge and Folklore, Third Session. WIPO/GRTKF/IC/3/8.

2 The field of IP rights is rapidly changing and laws vary from country to country. This chapter attempts to provide an accurate summary of general IP concepts and options. All options are subject to national laws and legislation. Therefore, before pursuing any option, it is important to check with local legislation. Additionally, any IP option mentioned in this chapter should not be pursued without consulting appropriate legal advisors. This chapter should not be used to advise a community on a specific action to take regarding a specific case, but instead as a tool for forming a general IP strategy to protect and sustain a community's knowledge and biological diversity.

3 U. S. Code, 35 U.S.C. § 103.

4 Kadidal S. 1997. Subject-Matter Imperialism? Biodiversity, Foreign Prior Art and the Neem Patent Controversy. *IDEA: The Journal of Law and Technology*. p. 371–403. www.idea.piercelaw.edu/articles/37/37_2/9.Kadidal.pdf.

5 U.N. 2000. Systems and National Experiences for Protecting Traditional Knowledge, Innovations and Practices. United Nations Conference on Trade and

Development, Commission on Trade in Goods and Services, and Commodities Expert Meeting on Systems and National Experiences for Protecting Traditional Knowledge, Innovations and Practices. TD/B/COM.1/EM.13/2.

6 UPOV. 2000. Brief Outline of the Role and Functions of the Union, October. www.upov.int/eng/brief.htm.

7 CBD. 1992. Convention on Biological Diversity, Conference of the Parties (COP). Rio de Janeiro.

8 Plant Varieties Act of Bangladesh, 1998, Article 7-3.

9 GRAIN, (Genetic Resources Action International). 1999. Beyond UPOV: Examples of Developing Countries Preparing Non-UPOV Sui Generis Plant Variety Protection Schemes for Compliance with TRIPS. July www.grain.org/publications/nonupov-en.cfm.

10 www.sristi.org/honeybee.html.

11 Utkarsh G. 2002. Documentation of Traditional Knowledge: People's Biodiversity Registers (PBRs). Foundation for Revitalization of Local Health Traditions (FRLHT). India. www.ictsd.org/dlogue/2002-04-19/Utkarsh.pdf. See, also in this *Handbook,* chapter 16.2 by CG Thornstrom.

12 World Trade Organization. 2002. Trading into the Future: The Introduction to the WTO, IP Protection and Enforcement. www.wto.org/english/thewto_e/whatis_e/tif_e/agrm6_e.htm.

13 U.S. Patent and Trademark Office. What Are Patents, Trademarks, Service Marks, and Copyrights? www.uspto.gov/web/offices/pac/doc/general/whatis.htm.

14 1994. Agreement on Trade-Related Aspects of IP Rights (TRIPS). Art. 22-1.

15 U.S. Patent and Trademark Office. What Are Geographical Indications? www.uspto.gov/web/offices/dcom/olia/globalip/geographicalindication.htm.

16 International Trademark Association. 2000. Lisbon Agreement for the Protection of Appellations of Origin: Violation of the TRIPS Agreement. INTA Issue Brief.

17 2002. Commission on IP Rights, London. Integrating IP Rights and Development Policy: Report of the Commission on IP Rights. TRIPS, 1994. Article 22.

18 Ibid.

19 Paris Convention for the Protection of Industrial Property, as revised at Stockholm on 14 July 1967 (Stockholm Act).

20 Ibid., Art. 10 bis, 3(1).

21 European Patent Convention. Article 54(2).

22 Pryor G. 1991. The Case for Defensive Disclosure. Software Patent Institute (SPI). www.spi.org/defdis.htm.

23 CBO. 2002. Bonn Guidelines on Access to Genetic Resources and Fair and Equitable Sharing of the Benefits Arising Out of Their Utilization. Convention on Biological Diversity, Conference of the Parties (COP), Decision VI/24.

24 TRIPS, Plant Variety Protection and UPOV. The South Centre.

25 Florez M. 2000. Andean Community Adopts New IPR Law. *Ag BioTech InfoNet*, Oct. 5.

26 1998. Synthesis of Case Studies on Benefit Sharing. Conference of the Parties to the Convention on Biological Diversity, Fourth Meeting, Bratislava, 4–15 May.

27 Vogel J. 1997. The Successful Use of Economic Instruments to Foster Sustainable Use of Biodiversity: Six Case Studies from Latin America and the Caribbean, Case Study 6: Bioprospecting. *Biopolicy Journal* Vol. 2, Paper 5 (PY97005). www.bdt.org/bioline/py.

28 For specific language and sample contracts, see Gollin MA. 2002. *Elements of Commercial Biodiversity Prospecting Agreements in Biodiversity and Traditional Knowledge: Equitable Partnerships in Practice* (ed. SA Laird). Earthscan: London.

29 Brascoupé S and H Mann. 2001. A Community Guide to Protecting Indigenous Knowledge. *Indian and Northern Affairs Canada,* June.

30 Jones P. 2002. Brazilian Tribe Feels Betrayed by Plant Search, *Seattle Times,* Sept. 16.

31 Adapted from *Recording and Using Indigenous Knowledge: A Manual. International Institute of Rural Reconstruction*. 1996. IIRR: Cavite, Philippines. www.panasia.org.sg/iirr/ikmanual.

32 Adapted from Gupta A. 2002. How to Make IPR Regime Responsive to the Needs of Small, Scattered and Disadvantaged Innovators and Traditional Knowledge Holders: Honey Bee Experience. Conference on the International Patent System, WIPO, Geneva, 26 March 2002.

33 See, also in this *Handbook*, chapter 16.2 by CG Thornström. See, also in this *Handbook*, chapter 16.3 by CG Thornström and L Björk; and chapter 9.4 by A Krattiger.

CHAPTER 16.7

Reconciling Traditional Knowledge with Modern Agriculture: A Guide for Building Bridges

KLAUS AMMANN, *Guest Professor, Delft University of Technology, Department of Biotechnology, The Netherlands*

ABSTRACT

In the years since the Convention on Biological Diversity was adopted, issues of traditional knowledge have come to affect the legitimacy of the multilateral trading system, in general, and its IP (intellectual property) aspects, in particular. In order to engage indigenous knowledge in furthering socio-economic development, policy-makers will need to reconsider the prevailing notion of a fundamental dichotomy between indigenous and scientific knowledge and begin to challenge both types of knowledge. This chapter concentrates on traditional knowledge—and how it relates to the ecology of agriculture, in all of its variants—and compares it to recent advances in scientific knowledge and the resulting applications of biotechnology in global agriculture.

The chapter argues that this dichotomy between traditional and scientific ways of knowing is not only artificial but problematic, in that it hinders exchange and communication between the two. The dichotomy between traditional knowledge and scientific knowledge is most apparent in, and lies at the root of, perceived differences between the approaches of today's organic farming and technology-intensive farming systems. While indeed there are important differences, traditional knowledge and scientific knowledge share important similarities. Knowledge, in both cases, is based on human observation and experience and is tested, replicated, and transmitted within its respective community through social institutions and mechanisms put in place for that purpose. Moreover, deeper examination of the genetic integrity of plants used within organic and biotechnology-based agricultural systems shows that the respective crop varieties being used under each system are more similar than they are different. Increasingly, organic farming is building on scientific knowledge, and agricultural biotechnology is seeking to draw on traditional knowledge.

This chapter challenges policy-makers and scientists to examine and, ultimately, to move beyond those conceptual worldviews, or constructs, that maintain the current divide between traditional knowledge/organic agriculture and scientific knowledge/agricultural biotechnology.

By building the bridge between traditional knowledge and science and becoming free to draw upon the best existing ideas and practices from both, a larger palate is available to draw from. But, more importantly, by integrating the innovation systems of both traditional and scientific communities, a much larger range of new ideas and practices could be generated. The chapter calls such dynamic integration the "participatory approach" to agricultural innovation, building upon the "unifying power of sustainable development" and leading to balanced choices in agricultural production chains and rural land use.

Such an integration would require adaptations of Western social institutions and mechanisms of intellectual property in order to interface in a more nuanced fashion with quasi-public-domain knowledge that is external to the published records of Western science and IP systems. At the same time, indigenous communities will need to learn to adapt their social institutions and mechanisms that govern what is, in a sense, sovereign or communal property to coexist with and at times be translated into formal IP rights and practical uses that are external to their traditional systems.

1. INTRODUCTION: GLOBAL TRENDS IN BIODIVERSITY PROTECTION

Since the adoption of the Convention on Biological Diversity in 1992[1] the legal status of plant genetic resources and traditional knowledge

Ammann K. 2007. Reconciling Traditional Knowledge with Modern Agriculture: A Guide for Building Bridges. In *Intellectual Property Management in Health and Agricultural Innovation: A Handbook of Best Practices* (eds. A Krattiger, RT Mahoney, L Nelsen, et al.). MIHR: Oxford, U.K., and PIPRA: Davis, U.S.A. Available online at www.ipHandbook.org.

has received increasing attention in international fora, non-governmental organizations (NGOs), and academic research. Several factors have stimulated this ongoing debate: the steady loss of biodiversity in plant genetic resources;[2] the contrast between protected plant varieties and genetically engineered products, on the one hand, and traditional crops and landraces in the public domain, on the other hand; the advent of the Agreement on Trade-Related Aspects of Intellectual Property Rights (TRIPS) under WTO; and the International Treaty on Plant Genetic Resources for Agriculture.[3] The Doha Agenda Ministerial Declaration[4] explicitly endorsed the issue of traditional knowledge as a subject for further negotiation. What was, some years ago, a concern limited to the ecological aspects of preserving biodiversity has moved to center stage. Today, policy-makers recognize that traditional knowledge affects the legitimacy of the multilateral trading system, in general, and its intellectual property aspects, in particular, as well as its interface with modern agricultural and environmental policies.

One of the difficulties in advancing toward any resolution or consensus in this debate is the relationship between varying negotiation processes in different fora. Another related problem involves the contradictory relationships between regulatory agencies at different levels (international, regional, and local) in dealing with traditional knowledge.[5] While it will be of prime importance to move toward a reconciliation between the CBD and the TRIPS agreement,[6] any progress must take into account the full complexity of issues related to biodiversity.[7, 8] Such reconciliation will not come easily.

To productively engage indigenous knowledge in efforts for economic development, policy-makers will need to reconsider the notion of a dichotomy of indigenous and scientific knowledge and begin to challenge both types of knowledge. Doing so will mean developing both greater autonomy for participating in the production of new knowledge and envisioning new approaches to regulating science. The Cartagena Biosafety Protocol, in particular, is today seen by many in the scientific community as having gone too far, imposing inordinately high levels of regulation, focusing excessively on transgenic plants (as opposed to other potential biosafety risks), and taking into account only the risk side of the equation of human welfare. Agricultural innovation has always been knowledge based, relying foremost on farmers' experience. With the development of modern science and its applications to agriculture, the situation has changed considerably. Without a doubt, agriculture owes many of its recent advances to the rapid growth of scientific knowledge, in both ecology and molecular biology. Yet, this advancement has been accompanied by a lack of awareness of traditional agricultural knowledge and even an active disregard for it.

To move toward a possible resolution, terms of the debate, it is of prime importance to reconcile the terms of the CBD and the TRIPS Agreement. In critiquing what some would call a utopian attempt to strengthen the position of indigenous peoples relative to other populations, it is necessary to examine the basic question of *how power structures knowledge*. Otherwise attempts to address the interests of indigenous people will inevitably fail. This will also necessitate challenging and changing government policies, questioning science, and strengthening independent decision-making processes among indigenous peoples. Simply to document traditional knowledge will not be enough. To bring indigenous knowledge to bear on agricultural and economic development, we must go beyond the dichotomy of indigenous versus scientific knowledge and work toward a better integration of the two.

It is also essential to adapt the regulation and application of IP systems to include humanitarian (that is, nonmarket) aspects of knowledge use in order to reconcile science-based agriculture with the needs and practices of traditional agriculture. Industry leaders and academicians in the field of biotechnology have recognized this, voluntarily developing and introducing new approaches to IP management that begin to affirm the inextricably public aspects of knowledge generation and to acknowledge that the extremely low cash flow of smallholders in the developing world will not generate significant royalties.[9, 10]

It will be necessary to overcome the compartmentalized views held within the halls of Western science and begin to integrate traditional knowledge into the scientific learning process. The Rio Convention is a remarkable framework document toward these ends. It succeeds in creating an opening for this kind of shift by focusing, not merely on conservation, but also on the sustainable use of genetic resources and the fair sharing of benefits that may arise from them. In particular, the provisions concerning access and benefit sharing (ABS) and the protection of traditional knowledge emerged as a viable way forward, creating room for the development of innovative solutions.

In addition, the dichotomy between Western science and traditional knowledge has caused a growing divide in the views held by the leaders of the international agricultural research community. The concept of biodiversity has too often in the public arena evolved into an unreflected mantra of environmentalists. While many today can agree that agriculture needs to become more sustainable—and that sustainability, in a broad sense, does have an important relationship with measures of biodiversity—what is needed is a precise analysis of the role of biodiversity within the actual context of all the complex elements of global agriculture, including the compelling need for ever-higher productivity.

This chapter concentrates on traditional knowledge—and how it relates to the ecology of agriculture in all of its variants—and compares it to recent advances in scientific knowledge and the resulting applications of biotechnology in global agriculture. The notion of a deep contrast between agriculture that is based on traditional knowledge and agriculture based on scientific knowledge is challenged. While on the surface there are major cultural and philosophical differences in the conceptual underpinnings of traditional and scientific knowledge, there are also striking similarities. In order to overcome major misunderstandings and to create new and sometimes surprising understandings, this chapter advocates a discursive system of debate that takes into account different kinds of knowledge and proceeds under a recognition of the "symmetry of ignorance."[11]

2. DEFINITION OF TRADITIONAL KNOWLEDGE

Comparing indigenous cultures and Western culture, the contrasts in mode and structure seem obvious, leading to the assumption that the thinking of human beings from such diverse situations must somehow be intrinsically different. The religious rites and rituals of indigenous peoples can be perceived to be without parallel in contemporary postindustrial Western society. Worse yet, the tendency of some Western intellectuals is to romanticize indigenous cultures, celebrating the untapped richness—yet thereby making the perceived contrast even greater and obscuring or ignoring the commonalities in human thinking across all cultures.

According to Berkes, et al.,[12] traditional knowledge is a way of knowing similar to that of Western science in that it is based on an accumulation of observations, but it is different from science in several other fundamental ways. The anthropologist Levi-Strauss[13] argued that traditional knowledge and Western science are two parallel modes of acquiring knowledge about the universe, yet he observes that "*the physical world is approached from opposite ends in the two cases: one is supremely concrete, the other supremely abstract.*"

Similarly, the philosopher Feyerabend[14] distinguished between two different traditions of human thought: abstract traditions (to which science belongs) and historical traditions (which include most systems of knowledge by people outside Western science), the latter being those through which knowledge becomes encoded in rituals and in the cultural practices of everyday life.

Traditional knowledge may be holistic in outlook and adaptive by nature, gathered over generations by observers whose lives depended directly on the quality of information and its use. It often accumulates incrementally, its reliability is assessed through trial and error, and it is transmitted to future generations orally or by shared practical experiences.[15]

Case studies reveal that there exists a diversity of local, or traditional, practices for ecosystem management.[16] These include multiple-species management, resource rotation, succession management, landscape-patchiness management, and other ways of responding to and managing ecological pulses and surprises. Social mechanisms behind these traditional practices include a number of adaptations for the generation, accumulation, and transmission of knowledge, the use of local institutions to provide leaders/ stewards and rules for social regulation, mechanisms for cultural internalization of traditional practices, and the development of appropriate world views and cultural values. The use of the term *traditional ecological knowledge* has become established, among others, through the work of an international conservation union (IUCN) working group[17, 18] and *traditional ecological knowledge and wisdom* (TEKW) has become established as a major term in all fields of ecology, including agriculture.[19, 20, 21] (Figure 1)

3. RESOLVING THE CONTRASTS BETWEEN TRADITIONAL AND SCIENTIFIC KNOWLEDGE

Agrawal[22] and Agrawal[23] both claim that by distinguishing indigenous knowledge from scientific knowledge, theorists are caught in a dilemma. Focus on indigenous knowledge has gained indigenous peoples an audible voice in development circles. Yet, this distinction creates and perpetuates the dichotomy between indigenous and scientific ways of knowing. This dichotomy is especially problematic because it often hinders exchange and communication between the two. Further, both Agrawal and Agrawal argue that the basic distinction between indigenous and scientific knowledge is artificial.

This artificial barrier, I will contend, is one of the primary reasons why there appears to be such a distinct contrast between traditional organic or subsistence farming and technologically intensive agricultural methods, including biotechnology. Most scientists depict traditional knowledge as somehow unable to learn from experience, fuzzy in its concepts, and closed to conceptual inputs from the outside, whereas science is open to new thought, precise in its empirically tested progress, and responsive to the real needs of farmers. Critics of science, however, mistrust it for being too abstract, analytical, and divorced from the needs of real people.

The reality in both cases is different from the perception. Closer consideration reveals that the differences are indeed much smaller. Traditional knowledge that has accumulated since ancient times and been transmitted by oral tradition has often turned out to be strikingly precise when tested against empirical observation. Indeed, given the test of time, traditional knowledge is verified or falsified by experiment and observation. And, in Western science, oral tradition is certainly present: scientific communities with different views and lexicons continue to exist regionally despite the homogenizing influences of the scientific literature and the Internet (for instance in botanical nomenclature). Feyerabend notes critically, that scientists are often closed to matters outside science.[24] However, as Karl Popper[25, 26] rightly claims, a line must be drawn when a theory cannot be falsified: in such a case a theory should not be called scientific. Traditional knowledge is of course open to similar scrutiny.

Indeed, there are a number of authors who emphasize the commonalities between scientific and traditional knowledge without making the mistake of turning the terms into synonyms. Horton,[27, 28] for instance, cannot understand why some persons, familiar with theoretical thinking in their own Western tradition, have failed to recognize its African equivalents. He contends that they simply have been blinded by differences in idiom and that exhaustive exploration of features common to Western and traditional African thought should come before any enumeration of differences. The same can be argued for the comparison between Western, science-based agriculture and all kinds of traditional agricultural practices.

The following sections seek to advance such a comparison between two apparently very different approaches to agriculture. In this case, the comparison is between organic agriculture and biotechnology-based agriculture, leaving out, for reasons of simplicity, the wider range of other agricultural approaches. Based on the lines of reasoning developed above, effort is made not to be distracted by the "idiomatic" contrasts or

FIGURE 1: COMPONENTS OF TRADITIONAL ECOLOGICAL KNOWLEDGE AND WISDOM OF THE NATIVE PEOPLES OF NORTHWESTERN NORTH AMERICA (FROM TURNER AND BURKES, 2006.)

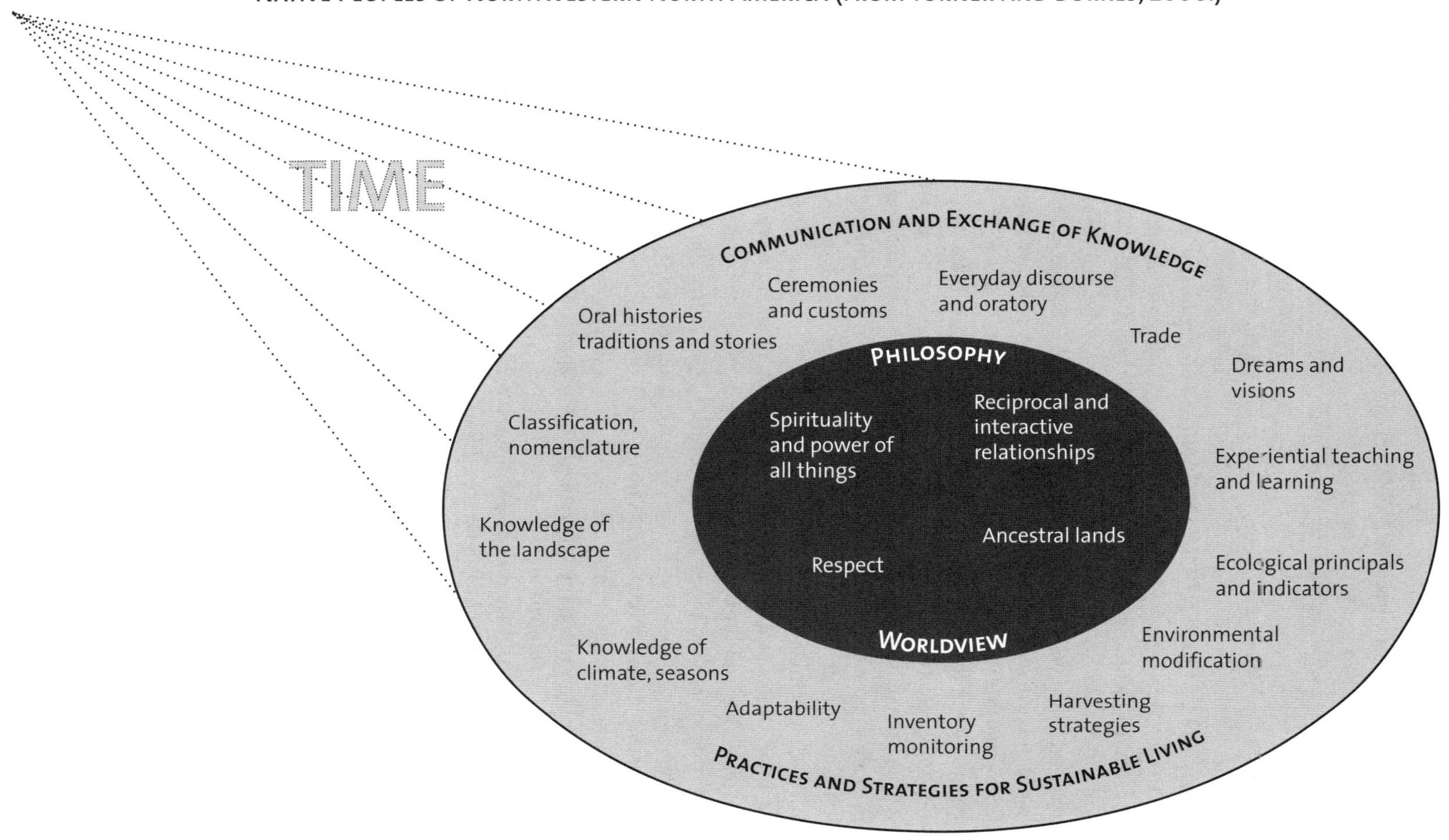

distinctions drawn between the two, but to explore the commonalities. In fact, both strategies considered here comprise elements of traditional knowledge and empirical precision. Differences drawn between the two are based on emphasizing methodology, a view that will be tested and challenged.

4. DEFINITION OF PRESENT-DAY ORGANIC FARMING

Organic farming (including some aspects of agroecological approaches to farming as referred to by Altieri and Nicholls[29]) started as a heterogeneous set of alternative-management methods in agriculture. This explains the multiple origins of organic farming and the fact that certifications of organic-farming practices have been introduced separately in various times and places. Organic farming is now growing rapidly and becoming a viable industry in its own right. Harmonizing standards and regulations are being developed and imposed more or less strictly on organic farms, both by states, like California,[30] and by national government agencies, like the U.S. Department of Agriculture.

Today, the International Federation of Organic Agriculture Movements (IFOAM) is serving to unite the various organic movements of the world, with members in 108 countries and support from the UN Food and Agricultural Organization (FAO). IFOAM advances basic views on organic farming, such as the following four principles:[31]

1. **Principle of health.** *Organic Agriculture should sustain and enhance the health of soil, plant, animal, human and planet as one and indivisible.*
2. **Principle of ecology.** *Organic Agriculture should be based on living ecological systems and cycles, work with them, emulate them, and help sustain them.*
3. **Principle of fairness.** *Organic Agriculture should build on relationships that ensure fairness with regard to the common environment and life opportunities.*
4. **Principle of care.** *Organic Agriculture should be managed in a precautionary and responsible manner to protect the health and well-being of current and future generations and the environment.*

Specific rules for organic agriculture are still the subject of international debate, given efforts to improve them, to find the right mix between regulatory strictness and diversity of applications. Some important documents in circulation intentionally go beyond the basic agreed-upon principles of organic farming [32,33,34,35] in order to stimulate discussion and to propose targets.

The main Swiss rules for organic agriculture are as follows:[36]

- *Natural cycles and processes are respected.*
- *The use of chemical-synthetic substances is avoided.*
- *The use of GMOs is not allowed, nor their derivatives, exception: products for veterinary medicine.*
- *The products shall not be treated with radiation, and no products having undergone irradiation shall be used.*

Since 2005 an official definition document on organic agriculture[37] has been in a process of transparent deliberation and elaboration. The latest language, which has not yet received definite approval, describes it as follows:

Organic agriculture, as defined by IFOAM, includes all agricultural systems that promote environmentally, socially and economically sound production of food and fibers. Recycling nutrients and strengthening natural processes helps to maintain soil fertility and ensure successful production. By respecting the natural capacity of plants, animals and the landscape, it aims to optimize quality in all aspects of agriculture and the environment. Organic Agriculture dramatically reduces external inputs by refraining from the use of synthetic fertilizers and pesticides, Genetically Modified Organisms and pharmaceuticals. Pests and diseases are controlled with naturally occurring means and substances according to both traditional as well as modern scientific knowledge, increasing both agricultural yields and disease resistance. Organic agriculture adheres to globally accepted principles, which are implemented within local socio-economic, climatic and cultural settings. As a logical consequence, IFOAM stresses and supports the development of self-supporting systems on local and regional levels.[38]

It is notable that debate over the very definition of organic agriculture persists. The problem is that top-down regulation of organic agriculture means coming to terms with standards met also in traditional agriculture, such as defining levels of toxicity for biopesticides, which is often not easy.[39]

Altieri summarizes agroecology, following Reijntjes, Haverkort, and Waters-Bayer,[40] with the following principles:[41,42,43,44]

- *Enhance recycling of biomass and optimizing nutrient availability and balancing nutrient flow*
- *Securing favorable soil conditions for plant growth, particularly by managing organic matter and enhancing soil biotic activity*
- *Minimizing losses due to flows of solar radiation, air and water by way of microclimate management, water harvesting and soil management through increased soil cover*
- *Species and genetic diversification of the agroecosystem in time and space*
- *Enhance beneficial biological interactions and synergisms among agrobiodiversity components, thus resulting in the promotion of key ecological processes and services*

Details of modern breeding methods are still controversial in organic agriculture communities. While genetic engineering itself is widely rejected, IFOAM agrees to the use of tissue culture and genetic assays, including genetic-marker-assisted breeding.[45] Note that Altieri and colleagues do not explicitly exclude transgenic plants in principle, while they clearly do not agree with the practices of multinational corporations advancing this technology. Some organic rules do not take any position on mutagenesis (traits introduced by genetic changes resulting from exposure to radiation or chemicals). This may not be unusual, since many successful crop traits have come from this method in the past.

Another breeding-related controversy is that of new hybrid crops: whereas many organizations in organic agriculture accept hybrid maize, since this is a biological phenomenon that cannot be easily reversed or avoided, most are opposed to the introduction of more hybrids in other crops.

In summary, organic farming has strong roots in traditional-agricultural knowledge. Today, it is drawing more and more on scientific research. Finding the right balance between these two sources of knowledge will continue to precipitate discussion within organic agriculture communities. Furthermore, the spectrum of different variants within organic and agroecological farming continues to expand and widen, ranging from integrated-pest-management techniques, used in conventional farming, to mainstream organic forming, to agroecological farming, and even to extreme forms of biodynamic farming.

In a number of developing countries, there are clear intentions to develop transgenic plants for use in subsistence farming, as indicated by statistics published by Cohen[46] and the FAO.[47]

5. DEFINITION OF BIOTECHNOLOGY-BASED AGRICULTURE

5.1 *Transgenic crops and genomic integrity at the molecular level*

Van Bueren, et al.,[48] explore the nature of genetic engineering at the molecular level, in an effort to explain why organic farming cannot accept plant varieties manipulated by biotechnology. Following Verhoog, et al.,[49] they posit "naturalness" as not only the avoidance of synthetic chemical inputs and the application of agroecological principles in cultivation, but also the maintenance of the "intrinsic integrity" of the organisms being cultivated, including the integrity of their genomes. Their definition of the integrity of plant genomes is as follows:

*The general appreciation for working in consonance with natural systems in organic farming extends itself to the regard with which members of the movement view individual species and organisms. Species, and the organisms belonging to them, are regarded as having an intrinsic integrity. This integrity exists aside from the practical value of the species to humanity, and it can be enhanced or degraded by management and breed*ing measures. This kind *of integrity can only be assessed from a biocentric perspective ... Organic agriculture assigns an ethical value to this integrity, and encourages propagation, breeding, and production systems that protect or enhance it.*

And further:

... biocentric perspective, organic agriculture acknowledges the intrinsic value and therefore the different levels of integrity of plants as described above. The consequence of acknowledging the intrinsic value of plants and respecting their integrity in organic agriculture implies that the breeder takes the integrity of plants into account in his choices of breeding and propagation techniques. It implies that one not merely evaluates the result and consequences of an intervention, but in the first place questions whether the intervention itself affects the integrity of plants. From the above described itself affects the integrity of plants.

Then, based on the nature of plants and their characteristics, a number of criteria, characteristics, and principles for organic plant breeding and propagation are excluded for violating the integrity of plants: for example, all breeding methods using chemicals or radiation—such as colchicine or gamma-radiation-induced mutants—all methods not allowing a full life cycle of the plant, and all methods manipulating the genome of the organisms. Unfortunately, the authors do not inquire very deeply into questions of the extent to which the structures and assembly of common crop species DNA has in fact been changed or manipulated by centuries of traditional selection and breeding.

For example, all varieties of wheat used today—by organic as well as conventional farmers—are a product of processes by which the genome has been subjected to numerous fundamental changes, and those changes have been successfully integrated inside the organism known today as wheat. These modifications include the addition of chromosome fragments, the integration of entire foreign genomes, and radiation-induced mutations (in the case of *Triticum durum*). Indeed, chromosome inversions and translocations are well documented in most major crops.

Thus, the reality of all systems of agriculture is such that most of the principles of genomic integrity, as advocated by Van Bueren and colleagues,[50,51,52] have long since been violated in almost all existing crops, and the naturalness or genomic integrity cannot be regained, unless theoretically one goes back to the ancestral genomes (which, in the case of each of the major crops, have not survived the intervening centuries of classical breeding). So, in reality, the principle of the "intrinsic integrity" of agricultural plant genomes is, at best, a fiction.

Other advocates of preserving the intrinsic integrity of organisms advise against crossing the natural hybridization barriers between species. Yet, species barriers have been overcome by traditional-breeding methods for decades, as well as by methods of biotechnology. Here the most salient example is somatic hybridization, which involves the nonsexual fusion of two somatic cells. The advantage of this method is that, by the fusion of cells with different numbers of chromosomes (for instance, from different species of *Solanum*) fertile products of the crossing can be obtained immediately. As a result, the polyploid plants that are obtained contain all of the chromosomes of both "parents," instead of the usual half set of chromosomes obtained through sexual reproduction. In order to achieve such somatic hybridization, required are cells, the walls of which have been digested away by enzymes, that are then enclosed only by their cell membranes (so called protoplast cells). With the loss of their cell walls, protoplasts also lose their typical shape and become spherical, like egg cells. The mixture of cells is then exposed to electric pulses to induce fusion. In order to get the "right" fusion product (since the fusion of two cells from the same parent plant can also occur) distinct selectable markers are necessary from each of the original parent plants. Only cells that survive this double selection are genuine products of fusion. The easiest way of implementing two such selectable markers is by genetic engineering, such as incorporating antibiotic resistance genes into the original parent plants. Such processes of protoplast fusion have been investigated and applied to potatoes, for instance. Under European Union (E.U.) regulations concerning the deliberate release of GMOs into the environment, somatic hybrids are not considered GMOs and do not require authorization. In fact, the most recent draft of E.U. organic regulations, in which the introduction of GMOs in organic cultivation is forbidden, follows the definition given earlier.[53, 54]

The concept of the naturalness or intrinsic integrity of plant genomes is also challenged by observations of Arber (a 1978 Nobel laureate) of the insertion of genes across natural species barriers in the case of naturally transgenic grasses.[55] Arber compared designed genetic alterations (including genetic engineering) with spontaneous genetic variations, those variations on which natural selection then operates to drive evolution:[56]

Site-directed mutagenesis usually affects only a few nucleotides. Still another genetic variation sometimes produced by genetic engineering is the reshuffling of genomic sequences, e.g. if a given open reading frame is brought under a different signal for expression control or if a gene is knocked out. All such changes have little chance to change in fundamental ways, the properties of the organism. In addition, it should be remembered that the methods of molecular genetics themselves enable the researchers anytime to verify whether the effective genomic alterations correspond to their intentions, and to explore the phenotypic changes due to the alterations. This forms part of the experimental procedures of any research seriously carried out.

Interestingly, naturally occurring molecular evolution, i.e. the spontaneous generation of genetic variants has been seen to follow exactly the same three strategies as those used in genetic engineering. These three strategies are:

(a) small local changes in the nucleotide sequences,

(b) internal reshuffling of genomic DNA segments, and

(c) acquisition of usually rather small segments of DNA from another type of organism by horizontal gene transfer.

However, there is a principal difference between the procedures of genetic engineering and those serving in nature for biological evolution. While the genetic engineer pre-reflects his alteration and verifies its results, nature places its genetic variations more randomly and largely independent of an identified goal. Under natural conditions, it is the pressure of natural selection which eventually determines, together with the available diversity of genetic variants, the direction taken by evolution. It is interesting to note that natural selection also plays its decisive role in genetic engineering, since indeed not all pre-reflected sequence alterations withstand the power of natural selection. Many investigators have experienced the effect of this natural force which does not allow functional disharmony in a mutated organism.

Genetic modifications of plant genomes may in fact be common. Recently, another natural transgenic plant was discovered by Ghatnekar, Jaarola, and Bengtsson,[57] involving the introgression of a functional nuclear gene from Poa to *Festuca ovina*. Yet other work reinforces the comparison, at the genomic level, between natural evolutionary processes and modern modifications of plant genetics through biotechnology.[58,59,60]

Still, despite such similarities, there is one major difference: natural genetic variation and selection acts on a completely different timescale from transgenic agriculture. Naturally occurring mutants that survive in the wild can take from hundreds to millions of years to survive selection pressures and finally take over against their preexisting competitors. With transgenic crops the timescale is totally different. They run through a research, development, and regulatory process that lasts, on average, 15 to 20 years after which the successful ones are completely deregulated. These can then be propagated nationally and cover millions of hectares within an extremely short time span on the evolutionary clock.

This basic insight of molecular biologists has been confirmed in analysis of modern breeding processes. The best example here is a comparison at the genomic level between transgenic and nontransgenic wheat by Shewry et al.:[61]

Whereas conventional plant breeding involves the selection of novel combinations of many thousands of genes, transgenesis allows the production of lines which differ from the parental lines in the expression of only single or small numbers of genes. Consequently it should in principle be easier to predict the effects of transgenes than to unravel the multiple differences which exist between new, conventionally-produced cultivars and their parents. Nevertheless, there is considerable concern expressed by consumers and regulatory authorities that the insertion of transgenes may result in unpredictable effects on the expression of endogenous genes which could lead to the accumulation of allergens or toxins. This is because the sites of transgene insertion are not known and transgenic plants produced using biolistics systems may contain multiple and rearranged transgene copies (up to 15 in wheat) inserted at several loci which vary in location between lines.[62,63] *Similarly, this apparently random insertion has led to the suggestion that the expression of transgenes may be less stable than that of endogenous genes between individual plants, between generations and between growth environments. Although there is evidence that the expression of transgenes introduced by biolistic transformation is prone to silencing in a small proportion of wheat*[64,65]*… recent reviews*[66,67,68,69] *… demonstrate the utility of biolistics transformation as a basis for stable genetic manipulation.*

Such studies confirming the stability of transgenic integrations[70,71] have been extended to other methods of transformation, such as the direct insertion of DNA fragments,[72] with some questions remaining about the long-term stability of agrobacterium-mediated transformations.[73] But, some of the most interesting observations in this line of inquiry about genome integrity have been documented by Baudo, et al.,[74] showing that the measured genomic disturbances from traditional breeding can be greater than the genomic disturbances from genetic transformation:

Detailed global gene expression profiles have been obtained for a series of transgenic and conventionally bred wheat lines expressing additional genes encoding HMW (high molecular weight) subunits of glutenin, a group of endosperm-specific seed storage proteins known to determine dough strength and therefore bread-making quality. Differences in endosperm and leaf transcriptome profiles between untransformed and derived transgenic lines were consistently extremely small, when analyzing plants containing either transgenes only, or also marker genes. Differences observed in gene expression in the endosperm between conventionally bred material were much larger in comparison to differences between transgenic and untransformed lines exhibiting the same complements of gluten subunits. These results suggest that the presence of the transgenes did not significantly alter gene expression and that, at this level of investigation, transgenic plants could be considered substantially equivalent to untransformed parental lines.

An ironic consequence of such results is that organic farming—by definition seeking to maintain the integrity of the plant genome by minimizing artificial DNA disturbances—should in such cases favor the genetically engineered variety. A more general conclusion may be that transgenic crops should not have been subject to regulations based purely on the fact that they resulted from the methodology of genetic engineering. Rather, it would have been more consistent to have a close look in each case at the product itself.

5.2 *The Green Revolution and agricultural biotechnology*

The social impacts and implications of modern agricultural biotechnology have their origins in the *Green Revolution*, a term coined by William Gaud at a 1968 meeting of the U.S. Agency for International Development (USAID) referring to the extremely successful agricultural movement through which new crop varieties, improved irrigation, adopted fertilizers and pesticides, and installed mechanization resulted in crop yields increasing dramatically, particularly in Asia.

One of the key innovations that drove the Green Revolution was the genetic improvement of plant varieties, especially the introduction of dwarf and semi-dwarf traits, in which stem height was reduced but the size of panicles, and thus seed production was not reduced. However, the yield gains of the Green Revolution also depended upon the application of high doses of chemical fertilizers and copious irrigation. Abundant yields attracted a variety of pests, and, therefore, chemical pesticides needed to be applied in greater volume. In addition, new crop varieties were also selected for photo-insensitivity, so that they could be adapted for multiple cropping sequences, patterns, and latitudes.

Evenson and Gollin[75] provide a thorough assessment of the Green Revolution, showing how over the period 1960 to 2000 the international agricultural research centers, in collaboration with national agricultural-research programs, contributed to the development of modern varieties in many crops. These varieties contributed to large increases in crop production. Productivity gains, however, were uneven across crops and regions. Consumers generally benefited from the resulting decline in food prices, but farmers benefited only where cost reductions exceeded those price reductions.

Two names are intimately linked to the Green Revolution: Norman Borlaug (who was awarded the Nobel Peace Prize in 1970)[76,77,78] and Monkombu Sambasivan Swaminathan (who was awarded the World Food Prize in 1987).[79,80] Yet, very early on, Swaminathan warned of unwelcome developments related to the Green Revolution:

The initiation of exploitive agriculture without a proper understanding of the various consequences of every one of the changes introduced into traditional agriculture, and without first building up a proper scientific and training base to sustain it, may only lead us, in the long run, into an era of agricultural disaster rather than one of agricultural prosperity. [81]

As the successes of the Green Revolution were becoming manifest together with its detrimental effects—including the upsurge of insect pests, growing insect resistance against widely used pesticides, and negative effects on the soil fertility—Swaminathan felt obliged to call for an Evergreen Revolution, beginning as early as 1968, yet continuing all the way through 1990.[82, 83] Unfortunately, farmers' access to free electricity to draw groundwater for irrigation, the negligence of legumes in crop rotations, and the indiscriminate application of chemical fertilizers and pesticides culminated in the degradation of soil and water. The damage to the ecological foundations essential for sustainable advances in productivity led to the onset of fatigue in agricultural systems.

Lessons drawn from the Green Revolution are that steps taken toward productivity enhancement should concurrently address the conservation and improvement of soil, water, and biodiversity, as well as providing for the atmosphere and renewable energy sources. Keeping these goals in focus, the goals of the Evergreen Revolution for achieving higher productivity in perpetuity were developed. What this calls for is a system of agriculture that involves sustainable management of natural resources, while progressively enhancing soil quality, biodiversity, and productivity.

Only much later has biotechnology proven to be able to contribute to the goals of the Evergreen Revolution, since it helps to enhance some of the ecological factors.[84,85,86,87] Biotechnology has proven to reduce pesticide use, positively influence nontarget insect populations, and induce no-tillage management practices that are beneficial to soil fertility.[88, 89]

An example of new biodiversity strategies fostered by a company known for the production of pesticides has been published by Dollaker and Rhodes.[90, 91] They propose to integrate crop productivity and biodiversity within pilot projects, jointly addressing the challenges of achieving crop productivity and biodiversity conservation objectives. Three pilot initiatives, developed by Bayer CropScience in Brazil, Guatemala, and the U.K. in collaboration with a variety of local stakeholders, illustrate how conservation objectives can be embedded in land-management practices that enhance agricultural productivity and profitability, thereby addressing both food security and biodiversity-conservation challenges.

A new variant of industrial farming, developing in the United States, is called precision farming. It is a management system based primarily on a combination of information technologies, including networked computing, satellite monitoring, and automated guidance systems for farm machinery. Precision farming can save time and energy and, by reducing unnecessary applications of chemicals and irrigation, can lead to a more ecological farming with higher yields.[92,93,94] Methods of precision farming do not contradict the main principles of organic farming and, thus, could be seriously considered as helpful auxiliary methods.

6. SUSTAINABILITY AND BIODIVERSITY

All agricultural systems must include the ability to provide an economic return to the farmer; unprofitable agricultural systems will not survive unless they are subsidized. In the cases of the United States and Europe, such policies are problematic in the long run for many reasons. Today's farming systems must provide opportunities to produce more food on smaller acreages.

Related to this imperative are issues concerned with maintaining and enhancing output, such as soil fertility and reducing losses to weeds and pests. It is less easy to argue that a natural or diverse ecosystem is a critical input to sustainable agriculture.

While ecologists frequently stress the interrelationships between species, it is difficult to see how the existence of species such as the swallowtail butterfly or a rare orchid could contribute to a farming system's sustainability.[95] The degree of redundancy in ecological communities is largely unknown and remains a rich field of investigation for ecologists. Agricultural systems can benefit from a higher biodiversity (not necessarily within the production surface) by presenting in the near vicinity of the production fields, biological networks hosting highly diverse arthropod populations, making the whole region more resistant to rapid pest invasions.[96, 97] This is not to say that agriculture could continue in the absence of all nonfarmed species. Rather, there is a suggestion that only a subset of all existing species is essential for food and fiber production.[98, 99]

6.1 *About sustainability in farming systems*

Definitions of *sustainability* are manifold. Some, such as that of the FAO[100] concentrate on ecological factors alone, while others concentrate only on management factors. The question that concerns us is whether organic farming or biotech farming is more sustainable. The answer is not clear, since the comparison often does not involve the same basic elements.

In one example that challenges the common view, Edward-Jones and Howells[101] come to the conclusion that organic-farming systems are not sustainable in the strictest sense. Considerable amounts of energy are put into organic-farming systems. The majority of the compounds utilized in crop protection are derived from nonrenewable sources and incur significant processing and transport costs prior to application. Nevertheless, the long-term balance of inputs clearly favors organic-farming systems.[102,103,104,105] Whereas nutrient (nitrogen, phosphorus, and potassium) inputs into the organic systems seem to be 34 to 51 percent lower than with conventional systems, mean crop yield was only 20 percent lower over a period of 21 years, indicating on balance an efficient production. In the organic systems, the energy to produce a dry matter unit of crop harvest was 20 to 56 percent lower than in conventional agriculture and correspondingly 36 to 53 percent lower per unit of land area.

On the other hand, many of the "biopesticides" used to control pests are not without toxicological hazards to humans and the environment. As an example, there are a number of research groups working on the difficult question of how to avoid, or at least reduce, the input of copper sulphate as a biopesticide. It is clear from some studies, that copper deposited in high concentrations has a negative impact on soil microbes. Pedersen, et al.,[106] found that total microarthropod abundance was highest at intermediate copper concentrations and linearly related to grass biomass. For single-species populations, no clear picture of abundance in relation to soil copper was seen, but two collembolan species, *Folsomia quadrioculata* and *Folsomia fimetaria*, were among the most sensitive. The resulting Shannon-Wiener index of biodiversity decreased linearly with increasing soil copper concentrations. Those results imply that a short-term strategy would be to avoid high concentrations of copper in the soil, but in the long run it will be better to avoid copper sulfate as a biopesticide altogether.

Sustainability can also be measured on a larger scale with methods developed in Europe to measure landscape quality.[107] Results need to be verified, but show positive influence of organic farming in Norway. What we can learn from this is that sustainability on all kinds of farming strategies depends on the local circumstances and may not submit to overall categorization. It certainly depends on the weight given to specific factors of sustainability. In the author's view, population size and feeding the growing number of people should have a very high priority on any such scale. Again, the claim is made that traditional knowledge can contribute in important ways to developing sustainable practices in agriculture and silviculture.[108]

6.2 *Biodiversity and farming systems*

It is important to distinguish between overall biodiversity in a given farming-landscape system, including the production area and biodiversity within the production system itself, the farm fields. The latter is often illusionary. Weeds within harvested fields are to be avoided, either by old-fashioned tilling or by various environmentally acceptable herbicides. The reason is simple: for example, in wheat production systems some of the weeds cherished by conservationists such as *Agrostemma ghitago* are highly toxic because of their saponin and githagenin contents and can spoil the harvested grain even in low quantities.[109]

Many of the crops growing in farming systems around the world have ancestral parents that lived originally in natural monocultures.[110] There are many examples of natural monocultures, such as the classic stands of kelp, *Macrocystis pyrifera*, which was, in fact, analyzed by Darwin.[111] Ecologists now recognize that simple, monodominant vegetation exists throughout nature in a wide variety of circumstances. Indeed, Fedoroff and Cohen[112] reporting on Janzen[113, 114] use the term *natural monocultures* as analogous with the term *crops*. Monodominant stands may be extensive. In one example, Harlan recorded that for the blue grama grass (*Bouteloua gracilis*) "*stands are often continuous and cover many thousands of square kilometers*" of the high plains of the central United States. It is of the utmost importance to agricultural sustainability to determine how these extensive, monodominant, natural grassland communities persist when we might expect their collapse.

More examples are given of wild species in Wood and Lenne,[115] including *Picea abies*, *Spartina townsendii*, *Sorghum verticilliflorum*, *Phragmites communis*, and *Pteridium aquilinum*. Early cultivars are also cited extensively,[116] wild rice (*Oryza coarctata*), for instance, reported in Bengal as simple oligodiverse pioneer stands on temporarily flooded riverbanks.[117] Similarly, Harlan [118] described and illustrated harvests from dense stands of wild rice in Africa (*Oryza barthii*, the progenitor of African cultivated rice, *Oryza glaberrima*). *Oryza barthii* was also harvested wild on a massive scale and served as a local staple across Africa, ranging from the southern Sudan to the Atlantic. Evans[119] reported that the grain yields of such wild-rice stands in Africa and Asia could exceed 0.6 tons per hectare—an indication of the stand density in monocultures of wild rice.

Botanists and plant collectors have, according to Wood and Lenne,[120] repeatedly and emphatically noted the existence of dense stands of wild relatives of wheat. For example, in the Near East, Harlan[121] noted that "*massive stands of wild wheats cover many square kilometers.*" Hillmann[122] reported that wild einkorn (*Triticum monococcum* subsp. *boeoticum*) in particular tends to form dense stands, and when harvested its yields per square meter often match those of cultivated wheats under traditional management. Harlan and Zohary[123] noted that wild einkorn "*occurs in massive stands as high as 2000 meters [elevation] in south-eastern Turkey and Iran.*" Wild emmer (*Triticum turgidum* subsp. *dicoccoides*) "*grows in massive stands in the northeast*" of Israel, as an annual component of the steppe-like herbaceous vegetation and in the deciduous oak park forest belt of the Near East.[124] According to Wood and Lenne[125] they are the strongest examples embracing wild progenitors of wheat. And Anderson[126] recorded wild wheat growing in Turkey and Syria in natural, rather pure stands with a density of $300/m^2$.

There are grounds for seriously rethinking the view of many agrobiologists that appear to uncritically accept that there was a loss of genetic diversity following the introduction of high-yielding Green Revolution wheat and rice varieties in the 1960s and 1970s. The same is feared to follow the rapid adoption of superior GM crops today. There are several reasons for caution in these interpretations.

There is evidence for genetic simplifications having occurred in ancient times. According to the analysis of Fedoroff,[127] thousands of years ago maize underwent a streamlining of its genome. Similar phenomena often occur in weeds like the chenopod *Atriplex prostrata* and are considered to have contributed to their exceptional migration ability since the last Glacial Maximum some 18,000 years ago.[128]

We can also paradoxically encounter an enhancement in genetic diversity in modern soybean breeding. For example, Sneller[129] looked at the genetic structure of the elite soybean population in North America, using a coefficient of parentage (CP) analysis. Whereas common sense would tell us that soybean genetic diversity has diminished considerably in the wake of genetic engineering, there is hard data proving that the trend is not so simple, in fact, to the contrary, genetic diversity can also be enhanced through the introduction of herbicide-tolerant traits. The introduction of herbicide-tolerant cultivars with the Roundup Ready® trait was shown to have had little effect on soybean genetic diversity because of the widespread use of the trait in many localized breeding programs. Only 1% of the variation in CP among lines was related to differences between conventional and herbicide-tolerant lines, while 19% of the variation among northern lines and 14% of the variation among southern lines was related to differences among the lines from different companies and breeding programs.

In more-simple numbers of soybean traits: the new management conveniences associated with the herbicide-tolerant soybeans allowed for a more-liberal use of varieties, most of them transgenic.[130] These include nearly 400 nematode-resistant varieties of soybean from 48 seed companies and five universities. All but seven of the varieties listed contain nematode resistance derived from a certain breeding line PI 88788. Of the varieties listed, 286 are resistant to the herbicide Roundup®, six are tolerant to sulfonylurea herbicides, and the remainders are conventional, nonresistant varieties.

Similarly, when Bowman, May, and Creech[131] examined genetic uniformity among cotton varieties in the United States, they found that genetic uniformity had not changed significantly with the introduction of transgenic cotton cultivars. In fact, when they compared the years before and after the introduction of transgenic cultivars, they observed that both the percentage of the crop planted with a small number of cultivars and the percentage planted with the most popular cultivar had declined. Thus genetic *uniformity* actually decreased by 28% over the period of introduction of transgenic cultivars. In light of the data, the theoretical concepts of Gepts and Papa,[132] that GM crops are likely to be responsible for a biodiversity decline within crops is not very convincing. It remains to be said that the continued use of locally adapted traits gained in traditional breeding should play an important role.[133, 134]

Several reviews[135,136,137] contend that the negative impact of modern biotech agriculture on biodiversity has been overestimated, and perhaps even overstated, by the organic-farming community for the purpose of marketing its alternatives on the grounds of their environmental characteristics. We begin to see that, contrary to the preponderance of negative views, there are beneficial effects stemming from no-tillage, the reduction of pesticide amounts applied to fields, and enhanced biodiversity.

But there are also many studies that show that organic farming has definite advantages over conventional agriculture, particularly regarding biodiversity. One extensive review[138] cites many field studies showing a wealth of evidence that now points to agricultural intensification as the principal cause of the widespread declines in European farmland bird populations,[139,140,141] as well as of the reduction in abundance and diversity of plant and invertebrate taxa over the past decades (well documented by Donald,[142] Preston, et al.,[143] and Wilson, et al,[144] and others).

Only a few studies have sought to integrate the changes in soil conditions, biodiversity, and socio-economic welfare linked to the conversion from nonorganic to organic production (Cobb, et al.).[145] Conclusions may not be representative for all organic conversions, but the findings are of relevance at a time of debate over changing patterns of subsidies and other incentives in agricultural policy. The study showed that there were demonstrable differences in overall environmental conditions in the comparison of organic and nonorganic farming, showing evidence of increased regional species diversity, and an eventual improvement in the profitability of the organic-farming regime. The study also showed that variations in farm-management practices strongly influence the notion of on-farm and off-farm environmental consequences.

The same positive effects of organic farming are shown in a 21-year study in Switzerland (the so called DOK study).[146] Part of the data has been published in *Science*.[147] The organic farming benefits related to biodiversity are well documented, especially with soil microbial diversity: root length colonized by mycorrhizae in organic-farming systems was 40 percent higher than in conventional systems.[148] Biomass and abundance of earthworms were higher by a factor of 1.3 to 3.2 in the organic plots as compared with conventional.[149] At the same time yield is, compared to traditional farming, dropping 20 percent. This fact triggered a debate in *Science* concerning whether such a drop in yield is tolerable with regard to the protection of biodiversity, since today we should realize the imperative to produce more food on a shrinking amount of arable land.[150,151,152] Potato yields in the organic systems were 58 to 66 percent of those in the conventional plots, mainly due to low potassium supply and the incidence of *Phytophtora infestans*. Winter wheat yields in the third crop-rotation period reached an average of 4.1 metric tons per hectare in the organic systems. This corresponds to 90 percent of the grain harvest of the conventional systems. In an overall comparison, provided the lower energy input is also taken into account, one can conclude that, theoretically, in some favorable conditions organic farming can be the more-efficient production strategy. A rather negative point is the safety of organic food: infections with the infamous *Echerichia coli* O157-H7, with its sometimes deadly consequences, seem to be a problem with respect to organic food. A number of papers demonstrate the legitimacy of these concerns.[153,154,155,156,157,158,159]

Only a very few studies exist (such as Roush)[160] that concentrate on a circumscribed agricultural practice comparing organic and biotech farming. This early paper compares directly Bt sprays used in organic farming and Bt transgenic crops, and the case is clear: Bt transgenic crops have advantages. Also, it has to be said that detailed studies of the impact of organic farming on various environmental factors are still scarce.

7. CONSEQUENCES AND CONCLUSIONS

Following the lines of reasoning presented here to their logical ends would, foremost, advocate a refrain from fostering the notion of a divide between agriculture using transgenic crops and organic-management systems. It is difficult to consistently maintain any divide along the lines of breeding technologies or the use of agrochemicals. The current perception of large differences in practices are mostly the result of differences in world view, often built, as has been argued here, on unfounded theories and even quasi-religious beliefs.

A successful integration of present-day management systems needs a new communication strategy. Such a strategy should embrace a dialogue with the public utilizing the "Three *E* Strategy" (entertainment, emotion, and education), which, according to Osseweijer[161,162] could initiate a decision-making process along the lines of the "Systems Approach," a discursive decision-making process for socially contentious issues.[163]

But a dialogue, in itself, will not create agricultural-management systems that build on local conditions, help poverty alleviation, respect elements of traditional knowledge, and combine it in a successful relationship with science. Building those bridges, in reality, need more than public acceptance. And more than decision-making processes, the effort will require making real decisions and following through on them.

Such an effort also needs the initiation of a mechanism like the *participatory projects* proposed by Slingerland et al.,[164] a working team from Wageningen that started a participatory farming project in Ouagadougou in West Africa with sorghum. Addressing iron deficiency caused by malnutrition in West Africa, this became an interdisciplinary program targeting the food-chain. In Africa current interventions are dietary diversification, supplementation, fortification, and biofortification. But such interventions alone have only moderate chances of success due to low purchasing power of households, lack of elementary logistics, lack of central processing of food, and the high heterogeneity in production and consumption conditions. Slingerland[165] proposed, based on excellent theoretical views, a staple food-chain approach, integrating parts of current interventions as an alternative. The research was carried out in several villages in Benin and Burkina Faso to take ecological, cultural, and socio-economic diversity into account. The interdisciplinary approach aimed at elaborating interventions in soil-fertility management, improvement, and choice of sorghum and other crop varieties and food processing, to increase iron and decrease the phytic acid-iron molar ratio in sorghum-based foods. The phytic acid-iron molar ratio was used as a proxy for iron-bioavailability in food. Synergy and trade-offs resulting from the integrated approach showed their added value. Phosphorous fertilization and soil organic amendments applied to increase yield were found to also increase the phytic-acid content of the grain and thus decrease its nutritional value, countered by new food processing reducing the phytic-acid levels again.

Ultimately, only a participatory approach building on the "unifying power of sustainable development" will lead to balanced choices between "People, Planet, and Profit" in agricultural production chains and rural land use, in building the bridge between traditional knowledge and science. The Golden Rice project[166] and the SuperSorghum project[167] both need to take account of these ideas in order to make those projects real successes. They include transgenic plants and, thus, need special efforts in participatory management in order to bring them to fruition.

Synergies will be of considerable importance, as soon as we begin to refrain from unproductive controversies over breeding and management methodologies. In the face of the urgent situation in many countries in the developing world, there is no time for contention and the overload of regulations. These prevent or at least slow the introduction of socially beneficial nutritional innovations, in the very countries where they are needed most. ■

ACKNOWLEDGMENTS

Thanks go to Maja Slingerland from the Wageningen University in the Netherlands, to Gregory Graff (PIPRA and the University of California, Berkeley), and to Anatole Krattiger (Arizona State University and Cornell University) for their many helpful remarks on the manuscript.

KLAUS AMMANN, *Guest Professor, Delft University of Technology, Department of Biotechnology, Julianalaan 67, 2628 BC Delft The Netherlands. klaus.ammann@ips.unibe.ch*

---

1 CBD. 1992. Convention on Biological Diversity. United Nations, New York. www.biodiv.org/convention/default.shtml.

2 CBD-SBSTTA. 1999. Consequences of the Use of the New Technology for the Control of Plant Gene Expression for the Conservation and Sustainable Use of Biological Diversity. www.biodiv.org/programmes/areas/agro/gurts.asp.

3 Giannakas K. 2003. Infringement of Intellectual Property Rights: Developing Countries, Agricultural Biotechnology, and the TRIPS Agreement. Biotechnology and Genetic Resource Policies (ed. p. Philips.). IFPRI: Washington.pp. 25–28.

4 WTO Doha Ministerial Declaration. 2001. Ministerial Declaration, WTO. WT/MIN(01)/DEC/1, 20 November 2001.

5 Brand U and C Gorg. 2003. The State and the Regulation of Biodiversity: International Biopolitics and the Case of Mexico. *Geoforum* 34(2):221–33.

6 Curci Staffler J, 2003. Towards a Reconciliation between the Convention on Biological Diversity and TRIPS Agreement. An Interface among Intellectual Property Rights on Biotechnology, Traditional Knowledge and Benefit Sharing, in Institut Universitaire de Hautes Etudes Internationales. University of Geneva: Geneva. p. 108.

7 Girsberger M. 1999. Biodiversity and the Concept of Farmers' Rights in International Law; Factual Background and Legal Analysis. *Studies in Global Economic Law.* Vol. 1 (ed. C Thomas). Peter Lang Publishing: Bern p. 415.

8 Biber Klemm S and T Cottier. 2006. *Rights to Plant Genetic: Resources and Traditional Knowledge: Basic Issues and Perspectives.* DEZA: Bern; World Trade Institute: Bern; and CABI: Wallingford. p. 464.

9 Atkinson R, RA Beachy, RN Conway et al. 2003. Intellectual Property Rights: Public Sector Collaboration for Agricultural IP Management. *Science* 301(5630):174–75.

10 Beachy RN. 2003. IP policies and Serving the Public. *Science* 299(5606):473.

11 Ammann K and B Papazova Ammann. 2004. Factors Influencing Public Policy Development in Agricultural

Biotechnology In *Risk Assessment of Transgenic Crops* (ed. S Shantaram). Wiley and Sons: Hoboken, NJ. p. 1552.

12 Berkes F, J Colding and C Folke. 2000. Rediscovery of Traditional Ecological Knowledge as Adaptive Management. *Ecological Applications* 10(5):1251–62.

13 Levi-Strauss C. 1962. La pensée sauvage. Plon: Paris. p. 269.

14 Feyerabend P. 1987. *Farewell to reason*. Verso: London.

15 Ohmagari K and F Berkes. 1997. Transmission of Indigenous Knowledge and Bush Skills among the Western James Bay Cree Women of Subarctic Canada. *Human Ecology* 25(2):197–222.

16 Berkes F, MK Berkes and H Fast. 2007. Collaborative Integrated Management in Canada's North: The Role of Local and Traditional Knowledge and Community-Based Monitoring. *Coastal Management* 35(1):143–62.

17 Johannes RE. 1989. *Traditional Ecological Knowledge: A Collection of Essays*. W.C.U. IUCN: Gland.

18 Williams NM and G Baines (eds.). 1993. Traditional Technological Knowledge: Wisdom for Sustainable Development. In *Centre for Resource and Environmental Studies*. A.N. University: Canberra.

19 Turner NJ and F Berkes. 2006. Coming to Understanding: Developing Conservation through Incremental Learning in the Pacific Northwest. *Human Ecology* 34(4):495–513.

20 Turner NJ, IJ Davidson-Hunt and M O'Flaherty. 2003. Living on the Edge: Ecological and Cultural Edges as Sources of Diversity for Social-Ecological Resilience. *Human Ecology* 31(3):439–61.

21 Turner NJ, MB Ignace and R Ignace. 2000. Traditional Ecological Knowledge and Wisdom of Aboriginal Peoples in British Columbia. *Ecological Applications* 10(5):1275–87.

22 Agrawal A. 1995. Dismantling the Divide between Indigenous and Scientific Knowledge. *Development and Change* 26(3):413–39.

23 Agrawal DP. 1997. Traditional Knowledge Systems and Western Science. *Current Science* 73(9): 731–33.

24 See *supra* note 14.

25 Popper K. 1972. *Objective Knowledge: An Evolutionary Approach*. Clarendon Press, Oxford University Press: London. p. 390.

26 Popper K. 1994. *Objektive Erkenntnis, ein evolutionärer Entwurf*: Campe Paperback.

27 Horton R. 1967. African Traditional Thought and Western Science 2: Closed and Open Predicaments. *Africa*, 37(2):155–87.

28 Horton R. 1967. African Traditional Thought and Western Science I: From Tradition to Science. *Africa* 37(1):50–71.

29 Altieri M. and CI Nicholls. 2005. Biodiversity and Pest Management in Agroecosystems, Second Edition, first Indian Reprint. Army Printing Press (ed. Lucknov) International Book Distributing Co.: Indiana. p. 236.

30 Guthman J. 1998. Regulating Meaning; Appropriating Nature: The Codification of California Organic Agriculture. *Antipode* 30(2):135.

31 www.ifoam.org.

32 IFOAM. 2004 D2 Draft Biodiversity and Landscape Standards IFOAM, Editor. IFOAM (International Federation of Organic Agriculture Movements): Bonn. p. 7.

33 IFOAM. 2004. D1 Plant Breeding Draft Standards. IFOAM (International Federation of Organic Agricultural Movements): Bonn. p. 2.

34 IFOAM. 2007. Principles of Organic Farming IFOAM, Editor. IFOAM, (International Federation of Organic Agriculture Movements): Bonn.

35 IFOAM. 2004. D3 Resource Use Draft Standards. IFOAM, Editor. IFOAM (International Federation of Organic Agriculture Movements): Bonn. p. 7.

36 Verordnung vom 22. September 1997. über die biologische Landwirtschaft und die Kennzeichnung biologisch produzierter Erzeugnisse und Lebensmittel (Bio-Verordnung). www.admin.ch/ch/d/sr/c910_18.html.

37 IFOAM. 2005. Definition of Organic Agriculture. IFOAM (International Federation of Organic Agricultural Movements) proposals: Bonn. p. 12.

38 IFOAM. 2005. Directory 2005 (most recent example).

39 See supra note 30.

40 Reijntjes C, B Haverkort and A Waters-Bayer. 1992. *Farming for the Future*. London: MacMillan Press Ltd.

41 Altieri M. 2000. Agroecology: Principles and Strategies for Designing Sustainable Farming Systems. University of California, Berkeley. www.cnr.berkeley.edu/~agroeco3/principles_and_strategies.html.

42 Altieri M. 1981. Mixed Farming Systems. *Environment* 23(10):5.

43 Altieri M. 1992. Classical Biological-Control and Social Equity—Reply. *Bulletin of Entomological Research* 82(3):298.

44 Altieri M, C Ines and PD Nicholls. 1994. *2004 Biodiversity and Pest Management in Agroecosystems*. International Book Distributing Co; Indiana.

45 See *supra* note 33.

46 Cohen JI. 2005. Poorer Nations Turn to Publicly Developed GM Crops. *Nature Biotechnology* 23(1):27–33.

47 Dhlamini Z, C Spillane, J Moss, et al. 2005. Status of Research and Application of Crop Technologies in Developing Countries, Preliminary Assessment. In FAO Reports. FAO, Editor. FAO: Rome. p. 62.

48 Van Bueren ETL, PC Struik, M Tiemens-Hulscher and E Jacobsen. 2003. Concepts of Intrinsic Value and Integrity of Plants in Organic Plant Breeding and Propagation. *Crop Science* 43(6):1922–29.

49 Verhoog H, M Matze, EL Van Bueren and T Baars. 2003.

The Role of the Concept of the Natural (Naturalness) in Organic Farming. *Journal of Agricultural & Environmental Ethics* 16(1):29–49.

50 Van Bueren ETL and PC Struik. 2004. The Consequences of the Concept of Naturalness for Organic Plant Breeding and Propagation. Njas-Wageningen. *Journal of Life Sciences* 52(1):85–95.

51 Van Bueren ETL and PC Struik. 2005. Integrity and Rights of Plants: Ethical Notions in Organic Plant Breeding and Propagation. *Journal of Agricultural & Environmental Ethics* 18(5):479–93.

52 Van Bueren ETL, PC Struik and E Jacobsen. 2002. Ecological Concepts in Organic Farming and Their Consequences for an Organic Crop Ideotype. *Netherlands Journal of Agricultural Science* 50(1):1–26.

53 Shewry PR, S Powers, JM Field, et al. 2006. Comparative Field Performance over Three Years and Two Sites of Transgenic Wheat Lines Expressing HMW Subunit Transgenes. *Theoretical and Applied Genetics* 113(1):128–36.

54 Barcelo P, S Rasco-Gaunt, C Thorpe and P Lazzeri. 2001. Transformation and Gene Expression. In *Advances In Botanical Research Incorporating Advances In Plant Pathology* (ed. PR Shewry, PA Lazzeri and KJ Edwards.) pp. 59–126.

55 Arber W. 2004. Biological Evolution: Lessons to Be Learned from Microbial Population Biology and Genetics. *Research in Microbiology*,155(5):297–300.

56 Arber W. 2002. Roots, Strategies and Prospects of Functional Genomics. *Current Science* 83(7):826–28.

57 Ghatnekar L, M Jaarola and BO Bengtsson. 2006. The Introgression of a Functional Nuclear Gene from Poa to Festuca Ovina. Proceedings: *Biological Sciences* 273(1585):395–99.

58 Arber W. 2000. Genetic Variation: Molecular Mechanisms and Impact on Microbial Evolution. *Fems Microbiology Reviews* 24(1):1–7.

59 Arber W. 2003. Elements for a Theory of Molecular Evolution. *Gene* 317(1-2):3–11.

60 Arber W. 2004. Biological Evolution: Lessons to Be Learned from Microbial Population Biology and Genetics. *Research in Microbiology* 155(5):297–300.

61 See *supra* note 53.

62 See *supra* note 54.

63 Rooke L, SH Steele, P Barcelo, et al. 2003. Transgene Inheritance, Segregation and Expression in Bread Wheat. *Euphytica* 129(3):301–9.

64 Anand A, HN Trick, BS Gill and S Muthukrishnan 2003. Stable Transgene Expression and Random Gene Silencing in Wheat. *Plant Biotechnology Journal*, 1(4):241–51.

65 Howarth JR, JN Jacquet, A Doherty, HD Jones and ME Cannell. 2005. Molecular Genetic Analysis of Silencing in Two Lines of Triticum Aestivum Transformed with the Reporter Gene Construct pAHC25. *Annals of Applied Biology* 146(3):311–20.

66 Sahrawat AK, D Becker, S Lutticke and H Lorz. 2003. Genetic Improvement of Wheat via Alien Gene Transfer: An Assessment. *Plant Science* 165(5):1147–68.

67 Kohli A, RM Twyman, R Abranches, E Wegel, E Stoger and P Christou. 2003. Transgene Integration, Organization and Interaction in Plants. *Plant Molecular Biology* 52(2):247–58.

68 Altpeter, F., N. Baisakh, R. Beachy, et al. 2005. Particle Bombardment and the Genetic Enhancement of Crops: Myths and Realities. *Molecular Breeding* 15(3):305–27.

69 Jones HD. 2005. Wheat Transformation: Current Technology and Applications to Grain Development and Composition. *Journal of Cereal Science* 41(2):137–47.

70 See *supra* note 54.

71 Baker JM, ND Hawkins, JL Ward, et al. 2006. A Metabolomic Study of Substantial Equivalence of Field-Grown Genetically Modified Wheat. *Plant Biotechnology Journal* 4(4):381–92.

72 Paszkowski J, RD Shillito, M Saul, et al. 1984 Direct Gene-Transfer to Plants. *Embo Journal* 3(12):2717–22.

73 Maghuly, F., A. da Câmara Machado, S. Leopold, et al. 2007. Long-Term Stability of Marker Gene Expression in Prunus Subhirtella: A Model Fruit Tree Species. Journal of Biotechnology 127(2): 310-321.

74 Baudo MM, R Lyons, S Powers, et al. 2006. Transgenesis Has Less Impact on the Transcriptome of Wheat Grain than Conventional Breeding. *Plant Biotechnology Journal* 4(4):369–80.

75 Evenson RE and D Gollin. 2003. Assessing the Impact of the Green Revolution, 1960 to 2000. *Science* 300(5620):758–62.

76 Borlaug NE, I Narvaez, O Aresvik and RD Anderson. 1969. Green Revolution Yields a Golden Harvest. *Columbia Journal of World Business*. 4(5):9-19.

77 Reynolds, MP and NE Borlaug, 2006. Applying Innovations and New Technologies for international Collaborative Wheat improvement. *Journal of Agricultural Science* 144: p. 95-110.

78 Reynolds MP and NE Borlaug. 2006. Impacts of Breeding On International Collaborative Wheat Improvement. *Journal of Agricultural Science* 144: 3–17.

79 Swaminathan MS. 1972. Agriculture Cannot Wait. *Current Science* 41(16):583.

80 Swaminathan MS. 2006. An Evergreen Revolution. *Crop Sci* (%R 10.2135/cropsci2006.9999) 46(5):2293–303.

81 Swaminathan MS. 1968. The Age of Algeny, Genetic Destruction of Yield Barriers and Agricultural Transformation. Presidential Address, Agricultural Science Section. in 55th Indian Science Congr. Jan. 1968. Varanasi, India.: Proc. Indian Science Congr.

82 Swaminathan MS. 2006. An Evergreen Revolution. *Crop Sci* 46(5):2293–303.

83 Kesavan PC and MS Swaminathan. 2006. From Green Revolution to Evergreen Revolution: Pathways and

Terminologies. *Current Science* 91(2):145–46.

84 Fawcett R, B Christensen and D Tierney. 1994. The Impact of Conservation Tillage on Pesticide Runoff into Surface Water. *J. Soil Water Conserv.* 49: 126–35.

85 Ammann K. 2005. Effects of Biotechnology on Biodiversity: Herbicide-Tolerant and Insect-Resistant GM Crops. *Trends in Biotechnology* 23(8):388–94.

86 Sanvido O, M Stark, J Romeis and F Bigler. 2006. Ecological Impacts of Genetically Modified Crops, Experiences from Ten Years of Experimental Field Research and Commercial Cultivation. In *ART-Schriftenreihe1* (ed. T Reckenholz) Agroscope Reckenholz-Tänikon Research Station ART, Reckenholzstrasse 191, CH-8046 Zurich, Phone +41 (0)44 377 71 11, Fax +41 (0)44 377 72 01, info@art.admin.ch, www.art.admin.ch. Zürich: Reckenholz. p. 108.

87 Cerdeira AL and SO Duke. 2006. The Current Status and Environmental Impacts of Glyphosate-Resistant Crops: A Review. *J. Environ. Qual.* 35: 1633–58.

88 Schier A. 2006. Field Study on the Occurrence of Ground Beetles and Spiders In Genetically Modified, Herbicide Tolerant Corn in Conventional and Conservation Tillage Systems. *Journal of Plant Diseases and Protection* 113(3):101–13.

89 Fawcett R and D Towery. 2002. Conservation Tillage and Plant Biotechnology: How New Technologies Can Improve the Environment by Reducing the Need to Plow. Purdue University. www.ctic.purdue.edu.

90 Dollaker A. 2006. Conserving Biodiversity Alongside Agricultural Profitability through Integrated R&D Approaches and Responsible Use of Crop Protection Products. *Pflanzenschutz-Nachrichten Bayer* 59(1): 117–34.

91 Dollaker A and C Rhodes. 2007. Integrating Crop Productivity and Biodiversity Conservation Pilot Initiatives Developed by Bayer CropScience, in Weed Science in Time of Transition. *Crop Science* 26(3):408–16.

92 Leithold P and K Traphan. 2006. On Farm Research (OFR)—A Novel Experimental Design for Precision Farming. *Journal of Plant Diseases and Protection*. p. 157–64.

93 Thenkabail PS. 2003. Biophysical and Yield Information for Precision Farming from Near-Real-Time and Historical Landsat™ Images. *International Journal of Remote Sensing* 24(14):2879–904.

94 Godwin RJ, GAWood, JC Taylor, et al. 2003. Precision Farming of Cereal Crops: A Review of a Six Year Experiment to Develop Management Guidelines. *Biosystems Engineering* 84(4):p. 375–391.

95 Walker BH and JL Langridge. 2002. Measuring Functional Diversity in Plant Communities with Mixed Life Forms: A Problem of Hard and Soft Attributes. *Ecosystems* 5(6):529–38.

96 Nentwig W. 1999. Weedy Plant Species and Their Beneficial Arthropods: Potential for Manipulation in Field Crops. In *Enhancing Biological Control*. University of California Press, Berkeley.

97 Wood D and J Lenne. 2006. The Value of Agrobiodiversity in Marginal Agriculture: A Reply to Bardsley. *Land Use Policy* 23(4):645–46.

98 Edwards-Jones G and O Howells. 2001. The Origin and Hazard of Inputs to Crop Protection in Organic Farming Systems: Are They Sustainable? *Agricultural Systems* 67(1):31–47.

99 Walker BH. 1992. Biodiversity and Ecological Redundancy. *Conservation Biology* 6(1):18–23.

100 Narain P. 2001. Agri-Environmental Indicators, Concepts and Frameworks FAO's Handbook on The Collection of Data and Compilation of Agri-Environmental Indicators. In Working Paper No. 23,. C.o.t.E.C. Eurostat, Editor. Statistical Commission and Economic Commission of Europe, Paper submitted by FAO. Ottawa: Canada. p. 9.

101 See *supra* note 98.

102 Mader P, A Fliessbach, D Dubois, et al. 2002. Soil Fertility and Biodiversity in Organic Farming. *Science* 296(5573):1694–97.

103 Mader P, A Fliessbach, D Dubois, et al. 2002. The Ins and Outs of Organic Farming, Response to Goklany I. *Science* 298(5600):1889–90.

104 Mader P, A Fliessbach, D Dubois, et al. 2002. Organic Farming and Energy Efficiency. *Science* 298(5600):1891.

105 Goklany I. Mader P., and D. Zoebl. 2002. Organic Farming and Energy Efficiency. *Science* 298(5600):1890–91.

106 Pedersen MB, JA Axelsen, B Strandberg, et al. 1999. The Impact of a Copper Gradient on a Microarthropod Field Community. *Ecotoxicology* 8(6):467–83.

107 Clemetsen M and J van Laar. 2000. The Contribution of Organic Agriculture to Landscape Quality in the Sogn Og Fjordane Region of Western Norway. *Agriculture Ecosystems & Environment* 77(1-2):125–41.

108 Duffield C, JS Gardner, F Berkes and RB Singh. 1998. Local Knowledge In The Assessment of Resource Sustainability: Case Studies in Himachal Pradesh, India, and British Columbia, Canada. *Mountain Research and Development* 18(1):35–49.

109 Firbank LG. 1988. Agrostemma-Githago L. *Journal of Ecology* 76(4):1232–46.

110 Wood D and J Lenne. 2001. Nature's Fields: A Neglected Model for Increasing Food Production. *Outlook on Agriculture* 30(3):161–70.

111 Darwin C. 1845. *Journal of Researches into the Natural History and Geology of the Countries Visited During the Voyage of H.M.S. Beagle Round the World*. John Murray: London.

112 Fedoroff NV and JE Cohen. 1999. Plants and Population: Is There Time? *Proceedings of the National Academy of Sciences of the United States of America*, 96(11):5903–07.

113 Janzen D. 1999. Gardenification of Tropical Conserved

Wildlands: Multitasking, Multicropping, and Multiusers. *Proceedings of the National Academy of Sciences of the United States of America* 96(11):5987–94.

114 Janzen D. 1998. Gardenification of Wildland Nature and the Human Footprint. *Science* 279(5355):1312–13.

115 Wood D and J Lenne. 1999. Agrobiodiversity and Natural Biodiversity: Some Parallels, in Agrobiodiversity, Characterization, Utilization and Management (eds. D Wood and J Lenne) CABI: Oxon, UK and New York. p. 425–45.

116 See *supra* note 110.

117 Prain D. 1903. Flora of the Sundribuns. *Records of the Botanical Survey of India*. p. 357.

118 Harlan JR. 1989. Wild-Grass Harvesting in the Sahara and Sub-Sahara of Africa. In *Foraging and Farming: the Evolution of Plant Exploitation.* (ed. DR Harris and GC Hillman) Unwin Hyman: London. Pp. 79–98, and Figures. 5.2–5.3.

119 Evans LT. 1998. Feeding the Ten Billion: Plants and Population Growth. Cambridge University Press. Cambridge. p. 34.

120 See *supra* note 110.

121 Harlan JR. 1992. *Crops and Man*, Second Edition. American Society of Agronomy: Madison, Wisconsin. p. 295.

122 Hillmann G. 1996. Late Pleistocene Changes in Wild Food Plants Available to Huntergatherers of the Northern Fertile Crescent: Possible Preludes to Cereal Cultivation. In *The Origin and Spread of Agriculture and Pastoralism in Eurasia* (ed. DR Harris) University College Press: London. Pp. 159–203, 189.

123 Harlan J and D Zohary. 1966. Distribution of Wild Wheats and Barley. *Science* 153: 1074–80.

124 Nevo E. 1998. Genetic Diversity in Wild Cereals: Regional and Local Studies and Their Bearing on Conservation Ex Situ and In Situ. *Genetic Resources and Crop Evolution* 45(4):p. 355–70.

125 See *supra* note 110.

126 Anderson PC. 1998. History of Harvesting and Threshing Techniques for Cereals in the Prehistoric Near East. In *The Origins of Agriculture and Crop Domestication* (ed. AB Damania, et al.). ICARDA: Aleppo. Pp. 145–59.

127 Fedoroff NV. 200.3 Prehistoric GM Corn. *Science* 302(5648):1158–59.

128 Mulder C. 1999. Biogeographic Re-Appraisal of the Chenopodiaceae of Mediterranean Drylands: A Quantitative Outline of Their General Ecological Significance in the Holocene. *Palaeoecology of Africa* 26: 161–88.

129 Sneller CH. 2003. Impact of Transgenic Genotypes and Subdivision on Diversity within Elite North American Soybean Germplasm. *Crop Science* 43(1):409–14.

130 Tylka GL. 2002. Soybean Cyst Nematoderesistant Varieties for Iowa. Iowa State University, Ext. Publ. Pm-1649, 1649: p. 1–26.

131 Bowman DT and OL May and JB Creech. 2003. Genetic Uniformity of the US Upland Cotton Crop since the Introduction of Transgenic Cottons. *Crop Science* 43(2):515–18.

132 Gepts P and R Papa. 2003. Possible Effects of Trans(Gene) Flow from Crops to the Genetic Diversity from Landraces and Wild Relatives. *Environmental Biosafety Research* 2: 89-113.

133 Swaminathan MS. 1968. Changing Concepts and Canvass of Plant Breeding. *Indian Journal of Genetics and Plant Breeding* A 28: 7.

134 Swaminathan MS. 1998. Genetic Resources and Traditional Knowledge: From Chennai to Bratislava. *Current Science* 74(6):495–97.

135 Miller HI. 2007. Biotech's Defining Moments. *Trends in Biotechnology* 25(2):56-59.

136 See supra note 85.

137 See supra note 86.

138 Hole DG and AJ Perkins, JD Wilson et al. 2005. Does Organic Farming Benefit Biodiversity? *Biological Conservation* 122(1):113–30.

139 Donald PF, RE Green and MF Heath. 2001. Agricultural Intensification and the Collapse of Europe's Farmland Bird Populations. *Proceedings of the Royal Society of London Series B-Biological Sciences,* 268(1462):p. 25–29.

140 Robinson RA and WJ Sutherland. 2002. Post-War Changes in Arable Farming and Biodiversity in Great Britain. *Journal of Applied Ecology* 39(1):157–76.

141 Krebs J, J Wilson, R Bradbury and G Siriwardena. 1999. The Second Silent Spring? *Nature* 400: 611–12.

142 Donald PI and WOS. 1998. Changes in the Abundance of Invertebrates and Plants on British Farmland. *British Wildlife* 9: 279–89.

143 Preston CD, MG Telfer, HR Arnold, et al. 2002. *The Changing Flora of the UK*. DEFRA.

144 Wilson JD A.J Morris, BE Arroyo, et al. 1999. A Review of the Abundance and Diversity of Invertebrate and Plant Foods of Granivorous Birds in Northern Europe in Relation to Agricultural Change. *Agriculture Ecosystems & Environment* 75: 13-30.

145 Cobb D, R Feber, A Hopkins et al. 1999. Integrating the Environmental and Economic Consequences of Converting to Organic Agriculture: Evidence from a Case Study. *Land Use Policy* 16(4):207–21.

146 Fliessbach A, P Mader, D Dubois, et al. 2000. Organic Farming Enhances Soil Fertility and Biodiversity, in Fibl Dossier 1. Fibl, Editor. Research Institute of Organic Agriculture, Federal Research Station for Agroecology and Agriculture: Frick, Switzerland. p. 16.

147 See *supra* note 104.

148 Fliessbach A and P Mader. 2000. Microbial biomass and size-density fractions differ between soils of organic and conventional agricultural systems. *Soil*

Biology & Biochemistry, 32(6):p. 757-768.

149 Pfiffner L. and P Mader. 1997. Effects of biodynamic, organic and conventional production systems on earthworm populations. *Biological Agriculture & Horticulture*, 15(1-4): 3-10.

150 See *supra* note 104.

151 Ibid.

152 See *supra* note 105.

153 Lienert J, M Haller, A Berner, et al. 2003. How Farmers in Switzerland Perceive Fertilizers from Recycled Anthropogenic Nutrients (Urine). *Water Science and Technology* 48(1):47-56

154 Mukherjee A, D Speh, E Dyck and F Diez-Gonzalez. 2004. Preharvest Evaluation of Coliforms, Escherichia coli, Salmonella, and Escherichia coli O157 : H7 in Organic and Conventional Produce Grown by Minnesota Farmers. *Journal of Food Protection* 67(5):894–900.

155 Mukherjee A, D Speh, AT Jones, et al. 2006. Longitudinal Microbiological Survey of Fresh Produce Grown by Farmers in the Upper Midwest. *Journal of Food Protection*, 69(8):1928–36.

156 Blaise D, CD Ravindran and JV Singh. 2006. Trend and Stability Analysis to Interpret Results of Long-Term Effects of Application of Fertilizers and Manure to Cotton Grown on Rainfed Vertisols *Journal of Agronomy and Crop Science* 192(5):319–30.

157 Islam M, J Morgan, MP Doyle, and XP Jiang. 2004. Fate of Escherichia coli O157 : H7 in Manure Compost-Amended Soil and on Carrots and Onions Grown in an Environmentally Controlled Growth Chamber. *Journal of Food Protection*, 67(3):p. 574–78.

158 Islam M, MP Doyle, SC Phatak, P Millner, and XP Jiang. 2004. Persistence of Enterohemorrhagic Escherichia coli O157 : H7 in Soil and on Leaf Lettuce and Parsley Grown in Fields Treated with Contaminated Manure Composts or Irrigation Water. *Journal of Food Protection* 67(7):1365–70.

159 Islam M, MP Doyle, SC Phatak, P Millner and XP Jiang. 2005. Survival of Escherichia coli O157 : H7 in Soil and on Carrots and Onions Grown in Fields Treated with Contaminated Manure Composts or Irrigation Water. *Food Microbiology*, 22(1):p. 63-70.

160 Roush RT. 1994. Managing Pests and Their Resistance to Bacillus-Thuringiensis—Can Transgenic Crops Be Better Than Sprays. *Biocontrol Science and Technology* 4(4):501–16.

161 Osseweijer p. 2006. A New Model for Science Communication that Takes Ethical Considerations into Account—The Three-E Model: Entertainment, Emotion and Education. *Science and Engineering Ethics*, 12(4):591–93.

162 Osseweijer p. 2006. Imagine Projects with a Strong Emotional Appeal. *Nature* 444(7118):422.

163 See *supra* note 11.

164 Slingerland MA, K Traore, APP Kayode and CES Mitchikpe. 2006. Fighting Fe Deficiency Malnutrition in West Africa: An Interdisciplinary Programme on a Food Chain Approach. Njas-Wageningen. *Journal of Life Sciences* 53(3-4):253–79.

165 Slingerland MA, JAE Klijn, RHG Jongman, et al. 2003. The Unifying Power of Sustainable Development. Towards Balanced Choices Between People, Planet and Profit in Agricultural Production Chains and Rural Land Use: The Role of Science. In WUR-report Sustainable Development. Wageningen University: Wageningen. Pp. 1–94.

166 www.goldenrice.org/.

167 www.supersorghum.org.

SECTION 17A

Putting Intellectual Property to Work: Experiences from Around the World

COUNTRY STUDIES

CHAPTER 17.1

Current Issues of IP Management in Health and Agriculture in Brazil

CLAUDIA INÊS CHAMAS, *Researcher, Oswaldo Cruz Institute, FIOCRUZ, Brazil*
SERGIO M. PAULINO DE CARVALHO, *Researcher, National Institute of Industrial Property (INPI), Brazil*
SERGIO SALLES-FILHO, *Professor, Geopi, State University of Campinas, Brazil*

ABSTRACT

This chapter presents Brazil's intellectual property (IP) system and identifies relevant experiences of IP management in the fields of health and agriculture. Brazil takes advantage of the flexibilities offered by relevant international agreements, such as the Agreement on Trade-Related Aspects of Intellectual Property Rights (TRIPS), and attempts to implement an equitable system. During the 1990s, Brazil revised its industrial property and copyright laws, and other related laws, and enacted new legislation that includes provisions for plant variety protection and for access to biological resources.

1. INTRODUCTION

Brazil is considered to be an innovative developing country,[1] with a robust scientific research structure in both health and agriculture. The Brazilian trend toward innovation will become even more relevant in the years ahead as a result of the recent Policy for Industry, Technology, and Foreign Trade of 2004, which prioritizes these economic sectors. In addition, the country has engaged in continuous revision of its IP policies to keep up with advances in science and technology, approved an Innovation Law in 2004, and continues to strengthen its presence in international research and innovation.

IP is a social institution, changing in form and function through, for example, the Paris Convention in 1883, the Bern Convention in 1886, the UPOV Convention in 1961, the Convention on Biological Diversity of 1992, and the 1994 Agreement on Trade-Related Aspects of Intellectual Property Rights (TRIPS). These international agreements are the instruments of such changes.[2, 3, 4, 5] An important characteristic of a system of IP protection is its impact on various industries and countries. The degree of impact depends on, among other factors, infrastructure and the level of training of individuals working in technology and science. Thus, the National System of Innovation places the IP system in context, providing necessary substance.[6] Heterogeneity of national laws also impacts IP protection as a function of the differences in terms of the way laws are applied in each country, because, in spite of the homogenization process that has accompanied TRIPS, flexibility in the formulation and implementation of national laws is possible.[7]

The reform of the legislation related to IP, which took place in Brazil in the second half of the 1990s as a consequence of TRIPS, brings with it opportunities as well as obstacles. These relate to the type of protection (including, for industrial property: patents, trademarks, geographical indications; for copyrights, in general; for computer programs; and for sui generis protection of plant varieties and biological diversity), or to the

Chamas CI, SM Paulino de Carvalho and S Salles-Filho. 2007. Current Issues of IP Management in Health and Agriculture in Brazil. In *Intellectual Property Management in Health and Agricultural Innovation: A Handbook of Best Practices* (eds. A Krattiger, RT Mahoney, L Nelsen, et al.). MIHR: Oxford, U.K., and PIPRA: Davis, U.S.A. Available online at www.ipHandbook.org.

national scientific and technological capability to generate new and useful knowledge.[8, 9]

An important aspect of TRIPS is its linking of IP protection to international commerce. Traditionally, agreements in the field of IP, especially the Paris Convention, linked IP to the technological and economic development of the countries participating in those agreements. This change in emphasis gave rise to some relevant issues. One issue is the enlargement of asymmetries between countries, in terms of the kinds of economic development occurring. These asymmetries can be of obvious concern to developing countries, particularly those that are lacking the infrastructure, scientific, technological and industrial capability for assimilating the technologies more strongly protected pursuant to TRIPS standards.[10, 11]

There is a new structure of international trade regulation that restricts the use of incentive policies for stimulating local production. This is similar to industrialization in developing countries, especially where import replacement is based upon direct subsidies and the closing of national markets. In addition, policies supporting industrialization, competition, and scientific and technological growth embed innovation, converging towards policies of science, technology and innovation. In the context of innovation and industrial policy, IP is important, augmenting the positive impacts and reducing the potential embarrassment that might be caused by restrictions to technological development deriving from the TRIPS agreement.[12]

Specific policies can and should be developed by nation states, particularly starting from the national scientific and technological asset base. Brewster and colleagues[13] believe that the promotion of access to innovations in the fields of health and agriculture to groups of lower income in developing countries should be the basis of those IP policies.

Brazil presents two outstanding examples of IP policy applied in those specific sectors in the controversy over the drug cocktail for the AIDS program of the Brazilian government: (1) the role of EMBRAPA (Institute of Agricultural Research of the Ministry of Agriculture) in the Brazilian seeds market; and (2) the role of FIOCRUZ (an institute of the Ministry of Health that works in research, education, technological development, and production in the field of the human health). In the first case, supported by an IP policy in the area of plant varieties, EMBRAPA was able to assemble partners, both public and private, who worked on the development of new plant varieties, allowing the country to keep the majority of national plant varieties after the promulgation of the Plant Variety Protection Law in 1997, pursuant to TRIPS requirements. FIOCRUZ, through Far Manguinhos, its drugs production unit, provided the Ministry of Health with a cost structure for the drugs that constitute the drug cocktail used in the AIDS program and identified the necessary technology for production of the drug cocktail.[14]

In both FIOCRUZ and EMBRAPA, a new standard of research organization is being implemented: the search for partnerships and the sharing of proprietary results. The search for complementing competences, which would be impossible to find in a single research institution or national economic agent, is a main factor. The rationale underlying the role of public research may be centered in the relevant markets, without losing focus on the mandate and rationale for the generation of technical and scientific knowledge.[15]

2. RECENT DEVELOPMENTS

2.1 *Legal aspects*

In Brazil, TRIPS is viewed as representing an initiative on the part of developed countries to increase the protection of IP. Further, TRIPS is seen as having sought to expand international commerce and the technological content of these exports, as well as to consolidate the new concepts of global production, where the control of technology obtains a differentiated qualitative dimension as compared to the environment in which the Paris Convention was ratified. (Brazil was one of the originators of that convention and has adhered to all of its revisions[16]). Two benefits of TRIPS, however, seem unequivocal: first, the maintenance of compulsory licensing with the possibility of implementing parallel import mechanisms, and second, the use of sanctions panels within the World Trade Organization (WTO), which minimizes the negative effects of unilateralism. Importantly, both of these elements can be exploited to the greatest advantage of developing countries if the countries have a certain level of technical and scientific capacity.

Prior to the present Industrial Property Law of 1996 (Law No. 9279), Brazil had already reformed its legislation concerning the protection of industrial property, instituting the Industrial Property Code in 1971 (Law No. 5772). The code prohibited the patenting of chemical products, food- and chemical-pharmaceutical products or processes, and did not recognize transgenic microorganisms as patentable. Due to Article 27 of the TRIPS agreement, the new Industrial Property Law recognized these fields as patentable matter.

Further relying on TRIPS, Brazil introduced a new legislation for authors' rights (the Authorship Rights Law of 1998 (Law No. 9610), a Computer Programs Law of 1998 (Law No. 9609), and the Plant Variety Protection Law of 1997 (Law No. 9456). The latter aims to encourage private investment in plant breeding. The law is widely perceived in Brazil as a radical change with regard to the protection of IP.

2.2 *Institutional aspects*

The following federal agencies are responsible for the administration of IP systems in Brazil:

- **for industrial property and computer programs.** Instituto Nacional da Propriedade Industrial (National Institute of Industrial Property [INPI]), an economically self-sufficient and independent government agency subordinate to the Ministério do Desenvolvimento, Indústria e Comércio Exterior (Ministry of Development, Industry and Foreign Trade [MDIC]). The INPI handles the processes for the granting of patents for inventions and utility models, the protection of trademarks, the protection of industrial designs, the protection of geographic indications, and the registration of computer programs. Furthermore, the country's legal dispositions established the requirement of prior approval by the Agência Nacional de Vigilancia Sanitária (National Health Surveillance Agency [Anvisa]), subordinate to the Ministério da Saúde (Ministry of Health[MS]), to subsidize the analysis process of patents on drugs, in accordance with the prerequisites established by Law No. 9279/96.
- **for plant variety protection.** The Serviço Nacional de Proteção de Cultivares (National Plant Varieties Protection Service [SNPC]), created by Law No. 9456 (of 1997) and subordinate to the Ministério da Agricultura, Pecuária e Abastecimento (Ministry of Agriculture, Livestock, and Food Supply [MAPA]), is accountable for its administration.
- **for authors' rights.** This is a field of protection that does not demand registration in order to guarantee rights. Computer programs, which are included in this category of IP protection, are registered at the INPI, as mentioned above. All other work protected by authors' rights, may be registered at various institutions, however, registration is not required. Works can be registered at the National Library (literary works), the Councils of Engineering and Architecture (plans, maps, and designs), and the School of Music of the Federal University of Rio de Janeiro (music, musical arrangements), and at other institutions. The policies for authors' rights are established by the authors' rights board within the Ministry of Culture. Additionally, the Interministerial Committee Against Piracy, subordinated to the Ministry of Justice, coordinates and implements enforcement policies, focusing on those works that are protected under the various fields of protection (that is, plant variety protection, industrial property, and so on) with greatest emphasis being placed on authors' rights.
- **genetic resources.** With the publication of Provisional Measure No. 2186-16 (of

2001), legislation relating to genetic assets was altered with respect to the conservation of biological diversity, the integrity of genetic assets, and associated traditional knowledge. As with Provisional Measure No. 2186-16 and Decree No. 3945/2001, access to and dispatch of the country's genetic assets are determined by the Council for the Management of Genetic Assets, whereby the benefits are liable for distribution, and the exchange and dissemination of components of genetic assets as well as associated traditional knowledge of indigenous and other local communities are preserved, provided doing so benefits them and is based on common practice.

One action that has had, and should continue to have, repercussions in the field of health and agriculture research is the promulgation of the Innovation Law of 2004 (Law No. 10973). An increase in the number of partnerships between companies, universities, and scientific and technological institutes is expected. The greater likelihood of attracting university researchers to establish companies dedicated to innovation is also expected. The law serves as a stimulus to the creation of technology-based companies that would be capable of marketing the results of research undertaken in universities and research institutes. Participation of these researchers in the management or administration of private companies is now allowed, so the new law provides the freedom for these professionals to realize their entrepreneurial potential. In addition, the law allows the sharing of space and infrastructure between public research and private companies. The law promotes the elimination of various bureaucratic hindrances, such as the requirement of a bidding process for the licensing of patents when these belong to a public agency.

The Innovation Law demands the establishment of technological innovation offices at universities and research centers. This innovative and potentially powerful incentive is expected to encourage the protection and commercialization of academic inventions, fostering economic dynamism and new job opportunities.

3. ISSUES CONCERNING HEALTH AND AGRICULTURE

3.1 *IP in agriculture*

The use of biotechnology as a tool for the improvement of traditional plant varieties has been an important issue. Expectations concerning the implementation of the Plant Variety Protection Law were very diffuse at first. Some authors argued that the law would promote the privatization process derived from the recognition of proprietary rights, thus displacing the public research sector, cooperatives, and producers' associations.[17] Others argued that the impact tended to be differentiated, in terms of the dynamism of the cultures and of the technical and scientific conditions. The technical and scientific training of the public sector and synergy among associations and producers' associations, would help it to maintain its production release capacity of new plant varieties.[18] Either way, only time will tell how the impact of the law will play out.

Currently, the main assignees of protected plant varieties are the national public research institutes (39%), foreign private companies (38%), and producer associations or related foundations (20%) (see Table 1). Local companies and universities each hold marginal positions, with a participation of less than 2% of the total protected plant varieties. Seven of the protected plant varieties are among the 10 most important in terms of the amount harvested during the 2001–2002 harvest season.

EMBRAPA is the economic player of greatest relevance in the production of protected soy seeds. Individually, it holds 23% of the registered protected plant varieties of all cultivated species. If its partnerships are included, EMBRAPA's participation increases to 36%. By itself, EMBRAPA holds the registry of 27% of the protected plant varieties employed in the production of seeds, and, including its partnerships, EMBRAPA's participation amounts to 41%.

For the harvest of 2001–2002, in terms of bearing registration of protected plant varieties, Monsanto Co., through the firm Monsoy, has a position superior to that of EMBRAPA, when the latter is considered on its own. Monsoy is the bearer of 55 protected plant varieties (30% of the total), 13 of which are genetically modified. This participation, however, falls to 23% when considering only the protected plant varieties used as seeds. Thus, Monsoy assumes second place in terms of the protected plant varieties used in the production of seeds and third place in terms of the quantity of seeds produced using protected plant varieties.

Another relevant economic player is the Central Cooperative for Agricultural Research (Coodetec), linked to the Cooperative Organization of Paraná (OCEPAR). For the harvest of 2001–2002, Coodetec participated with 10% of registered protection for soy plant varieties, having three intended for derivation and three genetically modified. The company's participation was slightly more than 13% when considering the use of protected plant varieties. Coodetec's participation in the amount of seeds of protected plant varieties was 12%.

Table 1: Protected Plant Varieties of Soybeans in Brazil, by Bearer and According to the Number of Plant Varieties and Use as Seeds, 2000–2001 Harvest

Main bearers	Description					
	Protected plant varieties		Plant varieties used as seed		Approved production	
	Unit[a]	%	Unit[a]	%	1,000 metric tons[b]	%
EMBRAPA, with partners[3]	67	37	43	41	217	51
Monsoy	55	30	24	23	89	21
Coodetec	19	10	14	13	94	22
Pioneer Hi-Bred International, Inc.	8	4	6	6	11	3
Fundação mato grosso (fmt)	10	5	5	5	1	0
Other bearers	25	14	13	12	15	3
Total of protected plant varieties	**184**	**100**	**105**	**100/52[c]**	**427**	**100/56[d]**
Nonprotected varieties (as percentage of total)	0	0	96	48	338	44
Total	**184**	**100**	**201**	**100**	**765**	**100**

Source: Carvalho[19]

a Number of protected plant varieties and varieties in use as seeds
b Volume of basic seed obtained from plant varieties in use as seeds
c FMT, CPTA, Epamig, Agrop. Boa Fé, Copamil, APSEMEG, Emater-GO, Agrosem, Ag. Rural-GO, CPTA, Empaer-MS
d Percentage of protected plant varieties as part of total plant varieties harvested of 2000–2001

Participation of the players may be understood by reviewing the trajectories of EMBRAPA, Coodetec, and Monsoy with regard to soy production. Both the public research institutions and the rural producer organizations tend to have a relevant role in the generation and adoption of new technology processes, particularly where the capacity for the appropriation of the generated innovation tends to be small. With the exception of seeds for hybrids, where biological characteristics increase the capacity for appropriation, private companies demonstrate little interest in the improvement of autogamous species, the seeds of which are capable of being reused by the rural producer.

The three economic players mentioned maintain trajectories with supplementary involvement that allow a highly competitive environment. There is a coevolution process of these players paralleling the institutional changes, particularly those changes that have affected statutes for the protection of plant varieties.

However, the introduction of new Brazilian players and economic units fuels the debate on the range of protection of innovations in the agricultural field, and, especially, the role of the national company. When prohibiting gene sequence patenting in 1996, the Brazilian legislation of industrial property aimed at ensuring the preservation of the national industry, as it was thought that it would not otherwise be able to compete with mostly transnational companies of larger size and more invested in technology.

The initial investment effort in scientific and technological training in the identification and genome sequencing in Brazil (*Xylella fastidiosa* and *Xanthomonas citri* among others) brought about conditions for the establishment of companies as a result of this research, for example, the venture capital fund of Votorantim Ventures, linked to the huge homonymous Brazilian industrial group, Scylla Bioinformática and Alellyx Applied Genomics.[20]

Scylla Bioinformática was formed by a group of researchers from the State University of Campinas (Unicamp)[21] and offers computing solutions and software development for companies and research centers that use or develop biotechnology. Alellyx Applied Genomics is a research and development company in applied genomics. The company's initial investment was around US$2 million. It is currently focused on research with soy, orange, eucalyptus, and sugarcane. Complementarily, the company performs contracts for the use of the genes by customers, invests in the development of an IP culture, and monitors global databases. Alellyx uses public domain information as well as information that is internally generated. IP is considered fundamental to the company's growth, particularly with respect to patent protection for genes. The strategy of the company has been to apply for patents in the United States on genes with potential value.

Evidently, restrictions on gene patenting in Brazil are somewhat of a bottleneck, because the Brazilian legislation on industrial property does not protect the genes themselves, but only the genetically modified organisms. Besides, the Brazilian Plant Variety Law forbids double protection, making the legislation on plant variety protection the only form of protection for plants.

In one sense, the current institutional picture tends to affect those activities in a regressive way, because the system of IP protection does not create incentives for those companies.

3.2 *IP in health*

3.2.1 *Antiretroviral access*

Since the end of the 1980s, the Brazilian Ministry of Health has supported policies for the provision of antiretroviral drugs as well as drugs for opportunistic infections. In 1991, Zidovudine was already provided with government support to serum-positive patients, although the supply suffered from eventual discontinuities. Decree No. 9313 (of 1996) ensured to all HIV-infected patients free access to all the medication necessary for their treatment. The distribution of drugs for triplex therapy with protease inhibitors began in December 1996.

Currently, 17 antiretrovirals (ARVs) are available from the Ministry of Health, eight of which are produced locally. Some are not protected by patents, entering the market before

Law No. 9279 was enacted. The ARVs that have patent protection are considerably more expensive. There is a natural tendency for newer drugs to overtake older ones (in the marketplace), because many patients develop resistance to drugs and begin to seek out new (drug) treatments. Access to drugs has become increasingly expensive.

The strategy for maintaining the antiretroviral access policy has various dimensions:

- systematic follow-up of patents in force
- monitoring what is in the public domain
- negotiations with suppliers
- local production and importation of generic medicines
- intensification of local R&D activities in an effort to minimize the technological gap
- adjustments in the legal procedures to facilitate access measures

Five companies in Brazil have industrial and technological capabilities for the production of generic ARVs. The national access policy also includes intense participation by various public laboratories.

Government expenditure for its access policy was around US$34 million in 1996 and has grown steadily to US$332 million in 2000. In 2004, government expenditure with the acquisition of ARVs jumped to US$238 million (80% from imports, 20% from local production). The increase in expenditure is mainly due to the increase in the number of patients under treatment, the increase in the proportion of patients needing more complex therapies, and the updating of therapy recommendations. The threat of compulsory licensing, a government recourse, forced the dropping of the price of three drugs in 2001: indinavir, produced among others by Merck and Co., Inc., (by 64.8%); efavirenz, also from Merck (by 59%); and nelfinavir, from Roche, (by 40%).

Aside from the direct benefits of the Brazilian program to individuals in Brazil infected by the HIV virus, as evidenced by the reduction in the AIDS mortality rate and the rate of opportunistic infections, the program has indirectly benefited other countries by providing a model in their efforts to combat AIDS. These countries include Angola, Nigeria, Venezuela, Guyana, and Mozambique, all of which are in now cooperating with the Brazilian government to develop production capability for antiretrovirals.

3.2.2 *Intangible assets in health biotechnology*

Concerning health research evolution indicators in Brazil, the most indicative at this stage is the number of publications. A recent article, published by the National Science Foundation (NSF),[22] indicates the increase of scientific publications in Latin America. The number of Latin American articles tripled during the period from 1998 to 2001, with most articles being written by Brazilian, Argentine, Chilean, and Mexican authors. Considering only the Brazilian contribution, the number of articles quadrupled during this same period.

In the last two decades, Brazil rose from 27th to 18th place in the world ranking for science and technology publishing. There were 1,887 articles published in periodicals indexed by the Institute for Scientific Information (ISI) in 1981, which corresponds to 0.44% of the world output. By 2001, this number had risen to 10,555 articles, or 1.44% of the world total. The number of articles in the medical and biomedical research areas has also increased.

In Brazil during the period from 1997 to 2001, the medical research community produced 7,365 articles (0.9% of the worldwide total) and ranked 23rd in the world. Medical research was 3rd in an internal ranking, representing 16.9% of the total articles indexed for the country on the basis of the ISI figure. The biomedical community had an even greater output than did medical research, with 8,366 articles for this period (0.9% of the worldwide total). With this output biomedical research was in the 21st place in the world ranking and second place in the internal ranking. Biomedical research contributed 19.0% of all the country's articles indexed on the basis of the ISI Deluxe.[23, 24]Despite a large part of Brazilian scientific production taking the form of published articles, it *is* possible to protect knowledge by means of IP rights.

Other indicators are somewhat less positive. Brazilian participation in triadic patents[25] remains very low at 0.2%. This low participation reinforces the necessity of developing specific incentive programs for technological research. In Brazil, the assessment of projects undertaken by agencies still judges researchers chiefly by their results in terms of publications. Progressively, the matter of IP is beginning to be incorporated into the analysis criteria of researcher productivity, but this is not an established routine in the academic community yet.

Data from the Directory for Research Groups of the National Council for Scientific and Technological Development (CNPq) indicates that groups that undertake health research produce a considerable amount of work with predominantly bibliographic-academic characteristics. Among each 10 published works only one represents research of a technical nature that results in some kind of protection for the purpose of eventually obtaining IP rights. Not all institutions have adequate support for providing protection to IP or for the identification of patentable subject matter.

The low participation by companies, in the areas of science, technology, and innovation (ST&I), and the lack of ability to transfer knowledge generated in universities to industry and various service sectors, partly explain the predominance of bibliographic-type work production. The ST&I activities are relatively concentrated in the university setting and in some research institutions that are dedicated to specific purposes. The development of these activities inside private companies of the productive sector is small despite efforts aimed at their expansion.

One of the more important effects of modern biotechnology is that it has greatly contributed to the closing of the gap between science and the market.[26] Because of this, academic medical and biomedical research may be viewed as appropriable technology, subject to formal IP protection. A lack of appropriation of academic research in Brazil, however, indicates both that the culture for IP is still undeveloped in academic institutes and that the sponsors of medical and biomedical research have a biased perception, still bound by the obsolescent dichotomy between basic research (freely disseminated) and applied research (appropriated for IP protection). This is reflected by the scant participation of Brazilian patents in the area of unquestionable scientific and technological competence (assuming that the inventions in these areas have a strong academic component).

Our research group is presently evaluating protection by means of patents in biotechnology in Brazil. Preliminary research was undertaken in some of the fields of the International Patent Classification related to the protection of biotechnological inventions. Despite being in the early stages of the research, our analysis of the database of the National Institute of Industrial Property (INPI) revealed patent applications and/or patents in all the verified fields. The research involved overall numbers, regardless of the origin of the application priority and numbers relating to the application priorities of Brazilian origin. Table 2 summarizes the results collected for a period from 1992 to 2005.

The correlation between publications and applications for patents is not linear. However, as the above data show, in the field of biotechnology in health, the volume of Brazilian publications grew intensely. The numbers of patents or patent applications shown (Table 2) having Brazilian priority are relatively modest. In all the fields of patents, the ratio of Brazilian priority to overall priority is low. Despite the very early stage of our research, it is possible to discern that biotechnological inventors seeking patent protection are predominantly foreign. It can be noted that there is a bias for protection in fields C12M, C12P and G01N33/50 with regard to patent applications being first filed in Brazil (which can be interpreted as technology developed in Brazil). Thus, the data seems to indicate that Brazilian technological production is focused in enzymology, microbiology, fermentation, or chemical analysis of biological material. Applications in the field of genetic engineering represent a mere 8.8%. These figures should be investigated more closely, as should the reason for these results. Deeper analysis may explain the disparity between scientific domain (publications) and technological domain (patents).

The recent approval of the Innovation Law and the structuring of technological innovation offices in universities and research centers indicates that patenting intensity of biotechnology should soon increase.

4. CONCLUSIONS AND FUTURE DIRECTIONS

One of the most important elements of the regulatory process is the area of IP rights. Especially since the 1980s, the results of research in biotechnology have been liable to protection through various mechanisms of IP. There is a trend toward a progressive increase in the scope of what can be considered patentable. The patent proves to be the most relevant and controversial asset; with other assets also being considered as such: trademarks, plant varieties, traditional knowledge, geographical indications, trade secrets, and so on. Common practice shows an intensive and complementary use of several of these assets; the possible combinations depend on the sector of activity (human health, animal health, agribusiness, and so on).

In the recent reorganizations of IP systems, countries and blocks seek to adopt more or less consistent positions in accordance with industrial and technological development. Both the 1980s and 1990s were marked by strong pro-patent movement tendencies; however, this approach was heavily criticized by many groups. The passing of the Bayh-Dole Act prompted the opening of more than 200 IP offices in U.S. universities.[27] Patenting with academic ownership became aggressive, altering standards of generating restrictions for the access to research results. University patents started to become the subject

Table 2: Health Biotechnology Patents in Brazil

Section, class, subclass, main group, or subgroup*	Patent or patent applications overall	Patent or patent applications, Brazil priority	Brazil priority (percent of total)
C12M	228	58	25.4
C12N	4,020	353	8.8
C12P	1,521	318	20.9
C12Q	940	82	8.7
C07K	2,523	171	6.8
G01N33/50 (including subdivisions)	171	27	15.8
A61K39	1,290	128	9.9
A61K48	260	7	<0.1
A01H	710	63	<0.1
Others	N/A	42	3.5
Total	**11,663**	**1,249**	**100.0**

* Fields of International Patent Classification

of negotiations between the academic and corporate fields. Universities began to be summoned to court, being frequently questioned concerning the exaggerated broadness of the scope of various patents, which hindered access to certain markets (very high royalty rates, questionable conditions of exclusivity, and so forth). In this context, benefits such as the research exemption faced extinction. The patent race U.S. universities entered into was also taken up by European institutions and, on a smaller scale, by Brazilian institutions. In Brazil, during the mid-1990s, a series of legal mechanisms motivated the IP protection of academic inventions. More recently, the Innovation Law was enacted.

In accordance with evidence advanced by several authors, patents have a crucial role in the biomedical industry.[28, 29, 30, 31] The introduction of a new drug demands great expenditure for research, development, and preclinical and clinical tests. There exists a relative ease of imitation without requiring the same amount of investment made by the innovating company, especially if the imitator possesses a technological capability similar or even close to that of the innovator. Patents, therefore, serve as the equivalent of a mediation contract between public and private interests. Thus, having made a technique public through publication of a patent document, the bearer of the patent is granted the right to exclude third parties from exploiting the invention.

The biomedical sciences also see the fractioning of existent rights, chiefly patent rights. Heller and Eisenberg[32] point to an intriguing phenomenon concerning the present commercialization of patents in the biomedical field. The grant of broad-scope patents and the grant of many patents with overlapping claims, whereby the determination of the exact limits of each one is difficult, has lead to what the authors term the "tragedy of the anticommons."

The metaphor corresponds to a situation in which many persons fight for the rights of exclusion in an environment of meager resources. The negotiations to ensure the rights of different bearers may stall, imposing obstacles to further development of the invention. The development of new drugs dependent on the multiple patents referring to DNA fragments and other intermediaries and research tools becomes vulnerable due to this "patent thicket." The eventual payment of the various license rates raises costs, making many products far too expensive.

The group of patents to be negotiated to make a product viable may belong to one or several bearers. If the bearers of the rights to be negotiated are distinct companies or institutions, there arises a further difficulty: that of dealing with a heterogeneous environment, each party having its own purpose, culture, and administrative experience. It should not be forgotten that the area of biomedical research is a heterogeneous environment composed of multinational corporations, small- and medium-sized technology-based companies, universities and research institutes. A further obstacle exists in the form of each invention as such. After licensing a biotechnological invention, the investor still has much work to do, with development needed—and uncertainty concerning success ever present—until the final product is marketed.

In Brazil, IP rights are consistent with a specific level of technological and industrial development. The country takes advantage of the (now, almost minimal) degree of freedom offered by the international agreements for the conformance/harmonization of IP rights (the TRIPS Agreement, for example) to innovate more equitably at the national level. Since the 1990s, Brazil has promoted a broad and deep revision of various legal instruments (Industrial Property Law, Copyright Law, and so on) and has inaugurated certain approaches (for example, through the Plant Variety Law and the Regulation for the Access to Biological Resources).

IP protection in biomedical fields differs from protection in the agricultural field due to the distinctive nature and dynamics of each. In health biotechnology, patents perform a fundamental role. The agents organize themselves to achieve protection (especially simultaneous protection, through patents and trademarks) and try to maximally extend the term of protection. On the other hand, the rationale of the developing countries is confounded by the dilemma of prices and the access to technologies. The issue of access has been broadly described in literature and in practice. The Brazilian Antiretroviral Access Policy reflects these dilemmas and difficulties. Thus, is it possible to reconcile IP protection and also provide the population with access to advanced technology at prices compatible with the local economies?

The impact of the incentive brought about by the IP is idiosyncratic, differing in terms of sections, of industries (and inside of a same section and a same industry), of companies (differing in their use of the strategies in different markets and segments), and of countries. Thus, the ability to appropriate innovation will equally present variations. The protection offered by the different protection fields (in the case in analysis, industrial property and plant improvers' rights) is different and related to the scientific and technological qualification and to the market and industrial structure in Brazil. Equally important is the way that institutional structures for the formulation and execution of public policies differentiates in the economic sector impact as linked to the protection fields.

In this way, specific characteristics of creation and incorporation of inventions/innovations tend to develop different intervention backgrounds. In the case of inventions/innovations in plant varieties, willingly or not, sponsored or not, there is no way for a foreign organization to introduce plant varieties that are not adapted to the area and the productive pattern where the plant variety will be used. This is a fundamental distinction between the areas of health and agriculture. In the case of the health, the companies do not find themselves under the contingency of setting up R&D structures in the countries where the drugs will be used.

In the case of the seeds industry, companies are structured either alone or in partnership with public and/or private research institutions. To be granted protection, plant varieties must pass tests that evaluate performance in the actual conditions of the country. Furthermore, the way legislation was negotiated, for the international treaties (TRIPS and UPOV), differs from negotiations for industrial property, hence creating more favorable conditions for a national project in the particular sector. For that, one should recognize the crucial contribution of institutional training by EMBRAPA, which organized partnerships for the development and licensing of new proprietary varieties, allowing for the main agents (public research, multinational corporations, and rural producers organizations) to establish complementary, yet synergistic, paths.

The drug market presents a rather different situation. It is worthwhile to stress the point concerning the need for the pharmaceutical industry to maintain R&D structures, either alone or in partnership. To enter the Brazilian market, multinational corporations do not need such structures locally. Besides, before the 1996 Industrial Property Law, national industries manufactured similar products, in other words, copies, modified or not, of the innovative products launched in both foreign and internal markets. As from 1997, when the new legislation came into effect, the traditional national producers' catalogue of drugs tended toward obsolescence as copying became illegal except for drugs already available (that is, nonpatented).

The government policies universalizing drug distribution to serumpositives in Brazil, on the other hand, was unable to foster the development of the national industry (national capital private companies) even with a massive government purchase program. The rationale underlying the negotiations on the industrial property legislation resulting in the current legislation, was highly regressive, with respect to industry and the national interest. Giving up the flexibilities offered by the TRIPS Agreement, especially the possibility of obtaining up to 10 years for the recognition of new drugs (even adopting the pipeline) the country's local production of active principles by the national industry was vastly hindered.

In spite of the contradictions of the adopted policies, they were able to answer the challenges imposed by the industrial property legislation. The country managed to overcome much embarrassment, transforming industrial development opportunities. Those opportunities, however, will not be sustainable long without a clear articulation between industrial property and the innovation policy, focusing on the enlargement of the competence and training of the national private companies in the maintenance of the present standard of excellence of the state laboratories and, mainly, in the creation of incentives, inductive or mandatory, to the international pharmaceutical companies, so that they focus R&D efforts toward the national scientific and technological structure. The protection instruments to the IP will play a central role in that process.

On the other hand, there are business opportunities consequential to the national scientific and technological training, as well as the venture investment in innovation undertaken by national companies, that are not protected by Brazilian laws. This creates a contradictory picture, in which fear of occupation of economic space in the Brazilian market by transnational corporations inhibits the activities of the national companies. That phenomenon is clear in the case of Alellyx Applied Genomics. Perhaps, the best way to ensure the access of developing countries to technology is less in the legislation and more in the defense against competition and in market regulation. The case of Brazilian agriculture seems to point in that direction.[33] On the other hand, the impact of IP in the field of health is central. Any discussion on the subject of protection should take into account the deep technological dependence of Brazil in the field. IP policy should be linked to scientific, technological development, and innovation and, also, be an integral part of the agricultural, health, industrial, and foreign trade policies.

Countries that present rich biodiversity, such as Brazil, still need to acquire the ability to act more actively in the dynamic environment of protection and exploitation of IP, whether to protect local inventions or to gain the knowledge to acquire technology developed by third parties. The demand for highly qualified professionals in this field of work is most urgent, as is the strengthening of the National Institute of Industrial Property. More energetic and integrated actions on the part of Brazil's public administration would contribute to a more mature policy in the area of industrial property and to the development of a configuration for a more competent system for innovation and IP management. ■

ACKNOWLEDGMENTS
We would like to acknowledge the support of FIOCRUZ, INPI, Unicamp, CNPq, and Faperj.

CLAUDIA INÊS CHAMAS, *Researcher, Oswaldo Cruz Institute, FIOCRUZ, Ministry of Health, Av. Brasil, 4365, Mourisco, 4365, sala 122, Rio de Janeiro RJ, 21040-900, Brazil. chamas@ioc.fiocruz.br*

SERGIO M. PAULINO DE CARVALHO, *Researcher, National Institute of Industrial Property (INPI), Praça Mauá 7, Rio de Janeiro RJ, 20083-900, Brazil. sergiom@inpi.gov.br*

SERGIO SALLES-FILHO, *Professor, Geopi, State University of Campinas, Instituto de Geociências/Unicamp, Caixa Postal: 6152, Campinas SP, 13083-970, Brazil. sallesfi@ige.unicamp.br*

---

1 Morel CM, T Acharya, D Broun, A Dangi, C Elias, NK Ganguly, CA Gardner, RK Gupta, J Haycock, AD Heher, PT Hotez, HE Kettler, GT Keusch, AF Krattiger, FT Kreutz, S Lall, K Lee, R Mahoney, A Martinez-Palomo, RA Mashelkar, SA Matlin, M Mzimba, J Oehler, FG Ridley, P Senanayake, P Singera and M Yun. 2005. Health Innovation Networks to Help Developing Countries Address Neglected Diseases. *Science* 309:401–404.

2 CIPR. 2002. Integrating Intellectual Property Rights and Development Policy. DFID: London. www.iprcommission.org/papers/pdfs/final_report/Ch7final.pdf

3 Dutfield G. 2001. *IP Rights and Development*. UNCTAD/ICTSD: Genevra. (preliminary version).

4 Aded AO. 2001. The Political Economy of the TRIPS Agreement: Origins and History of Negotiations. Dialogue at the Aberdare Country Club in Kenya, 30–31 July 2001, under the sponsorship of the International Centre for Trade and Sustainable Development (ICTSD) and the African Centre for Technology Studies (ACTS) in collaboration with the Quaker United Nations Office (Quno).

5 Juma C. 1999. IP Rights and Globalization: Implications for Developing Countries. Science, Technology and Innovation. Discussion Paper no. 4. Center for International Development, Harvard University, Cambridge, Mass., U.S.A.

6 Buainain AM, SP de Carvalho, SR Paulino and S Yamamura. 2005. Propriedade Intelectual e Inovação Tecnológica: Algumas questões para o debate atual. In *O Futuro da Indústria: Cadeias Produtivas* (D Henrique de Oliveira, ed.). MDIC/STI: Brasília. pp. 11-38.

7 Carvalho SMP, AM Buainain and CI Chamas. 2005. Políticas de propriedade intelectual no Brasil: Análise comparativa entre saúde e agricultura. In: Seminario de gestión tecnológica ALTEC:Salvador.

8 Carvalho SMP. 2002. Tendências focalizadas em propriedade intelectual, transferência de tecnologia

e informação tecnológica no Brasil. In: XXII Simpósio de gestão da inovação tecnológica. Anais/núcleo de política e gestão tecnológica da Universidade de São Paulo (PGT/USP).

9 Carvalho SMP and LDR Pessanha. 2001. Propriedade intelectual, estratégias empresariais e mecanismos de apropriação do esforço de inovação no mercado Brasileiro de sementes. *revista de economia contemporânea, Rio de Janeiro*, 5(1):151–182.

10 See *supra* note 4.

11 See *supra* note 3.

12 Carvalho SMP. 2003. *Propriedade intelectual na agricultura*. Doctoral thesis. Departamento de Política Científica e Tecnológica (DPCT), Instituto de Geociências (IG), Universidade Estadual de Campinas (Unicamp). Campinas.

13 See, also in this *Handbook*, chapter 2.2 by A Brewster, SA Hansen and AR Chapman.

14 See *supra* note 8.

15 Salles-Filho SLM, R Albuquerque, T Szmrecsanyi, MB Bonacelli, S Paulino, M Bruno, D Mello, R Corrazza, S Carvalho, S Corder and C Ferreira. 2000. *Ciência, tecnologia e inovação: a reorganização da pesquisa pública no Brasil*. Campinas/Brasília: Komedi/Capes.

16 Barbosa ALF. 1981. *Patentes: Crítica à racionalidade em busca da racionalidade*. Rio de Janeiro: Mimeo.

17 See, for example, Velho PE. 1992. O direito do melhorista e o setor públic,\o de pesquisa. *Cadernos de ciência e tecnologia*, v. 9, (1/3).

18 See, for example, *supra* note 9.

19 See *supra* note 12.

20 Salles-Filho SLM, SMP Carvalho, C Ferreira, and E Pedro . 2006. *Sistema de propriedade intelectual e as pequenas e médias empresas no Brasil*. Campinas, IG/DPCT/ GEOPI.

21 www.unicamp.br.

22 NSF. 2004. Latin America Shows Rapid Rise in S&E Articles. *Infobrief* 04-336. National Science Foundation: Washington, DC.

23 Guimarães JA. 2004. Medical and Biomedical Research in Brazil: A Comparison of Brazilian and International Scientific Performance. *Ciênc. saúde coletiva* 9(2):303–327.

24 Guimarães R. 2003. Bases para uma Política Nacional de Ciência, Tecnologia e Inovação em Saúde. *Cadernos de Estudos Avançados* 1(2).

25 Patents issued in the United States, Europe, and Japan.

26 Mello MTL. 1995. *Propriedade intelectual e concorrência: Uma análise setorial*. Doctoral thesis, Unicamp.

27 Mowery DC and A Ziedonis. 2000. *Numbers, Quality, and Entry: How Has the Bayh-Dole Affected U.S. University Patenting and Licensing?* Harvard Business School Press: Boston.

28 Mansfield E. 1986. Patents and Innovation: An Empirical Study. *Management Science* 32(2).

29 Levin R, A Klevorick, R Nelson and S Winter. 1987. Appropriating the Returns from Industrial R&D, *Brooking Papers on Economic Activity*, 3.

30 Scherer FM, S Herzstein Jr, A Dreyfoos, W Whitney, O Bachmann, C Pesek, C Scott, T Kelly and J Galvin. 1959. *Patents and the Corporation: A Report on Industrial Technology Under Changing Public Policy*. 2nd Edition. Harvard University, Graduate School of Business Administration: Cambridge, Mass.

31 Cohen, WM, RR Nelson and JP Walsh. 2000. Protecting Their Intellectual Assets: Appropriability Conditions and Why U.S. Manufacturing Firms Patent (or Not). NBER Working Paper 7552.

32 Heller M and RS Eisenberg. 1998. Can Patents Deter Innovation? The Anticommons in Biomedical Research. *Science* 280(5364):698–701.

33 See *supra* notes 7 and 15.

CHAPTER 17.2

A Model for the Collaborative Development of Agricultural Biotechnology Products in Chile

CARLOS FERNANDEZ, *IP and Regulatory Affairs, Biotechnology Development, Fundación Chile; Currently: Director, Startegic Studies, Foundation for Agricultural Innovation (FIA), Chile*
MICHAEL R. MOYNIHAN, *Director, Biotechnology Development, Fundación Chile, Chile*

ABSTRACT

This chapter presents an operational model used by Fundación Chile to develop commercial biotechnology products. The first section highlights the challenges faced by a developing economy of which the main crops are so-called *orphan crops*. Fundación Chile's experience has shown that establishing public–private collaborations and a solid international network are critical to overcoming obstacles and increasing the probability of success. Indeed, accessing various technology components and managing intellectual property and regulatory issues are serious challenges for a small, export-oriented economy like Chile, and Fundación Chile´s response has been to implement a model that includes the participation of companies and local research organizations with specific expertise at different points along the value chain. International agencies complement the activities and contributions of these local organizations. The chapter's second section gives some specific examples of new products being developed with the new tools of biotechnology.

1. INTRODUCTION

In ten years the area planted with genetically engineered varieties in Chile has grown to more than 81 million hectares.[1] Just four crops—soybean, maize, cotton, and canola/rape—account for almost 100% of this area. Agricultural biotechnology can potentially add significant value to a wide range of crops, but the development of genetically engineered varieties requires a wide range of skills, access to many technologies, and many years of research and development. Because of the lower economic returns for developing products grown in limited areas, such crops have difficulty competing for investors. In fact, the major agribiotechnology companies focus on global *vision crops* that involve large planted areas. Crops covering limited areas can nevertheless be important for specific regions. These crops can be developed by focusing local R&D and leveraging resources through public–private collaborations, which can help to overcome major challenges, such as critical mass in R&D, freedom to operate, and regulatory issues. A similar approach is useful for commercially developing other types of regionally important biotechnology applications.

2. TECHNOLOGY AND IP ISSUES

Developing a commercially viable transgenic plant product requires inputs that include:

- high-quality germplasm
- gene cassettes for the engineering of a specific trait, including appropriate coding sequences and regulatory regions
- a transformation system for the species and genotypes of interest

Materials and technologies in each of the categories may be covered by one or more types of IP (intellectual property) rights, including patents, plant breeders' rights, and copyrights, as well as

Fernandez C and MR Moynihan. 2007. A Model for the Collaborative Development of Agricultural Biotechnology Products in Chile. In *Intellectual Property Management in Health and Agricultural Innovation: A Handbook of Best Practices* (eds. A Krattiger, RT Mahoney, L Nelsen, et al.). MIHR: Oxford, U.K., and PIPRA: Davis, U.S.A. Available online at www.ipHandbook.org.

contractual agreements, such as material transfer agreements. IP rights are granted by individual countries and so, can vary from country to country, which often complicates the situation for export-oriented industries (for example, Chile's fruit industry).

Consolidation of the agri-biotechnology industry now means that a few large multinational companies control a large part of the intellectual property related to the genetic engineering of crops.[2] These companies are often reluctant to provide technology for specialty crops or so called *orphan crops* because of liability concerns arising from others' use of the technology.

Public sector laboratories have made, and continue to make, important contributions to agriculture, but they have emphasized the development of novel specific components, without consideration of the IP rights for other components needed to further develop or commercialize complex products such as transgenic crops. As a result, although these public institutions frequently can offer rights to components (for example, a DNA sequence coding for a specific gene of interest or a promoter that drives expression in a particular tissue), the institutions are rarely able to license a complete transgenic plant, or even an entire cassette, for transformation. It is essential to consider IP issues in the R&D program from the outset, because restrictions on freedom to operate can be a barrier to attracting the investment necessary to develop and commercialize products.

Such difficulties have been described for specific cases, such as pro-Vitamin-A containing *golden rice*.[3] Organizations that are attempting to address these issues on a more general level include the International Service for the Acquisition of Agri-biotech Applications (ISAAA),[4] CAMBIA,[5] and the Donald Danforth Plant Science Center.[6] Recently, a group of several leading universities and research institutes in the United States formed the Public Intellectual Property Resource for Agriculture (PIPRA),[7] which has expanded to include a number of nonprofit institutions in other countries, including Fundación Chile.[8] Although a major motivation for such initiatives has been to ensure the availability of biotechnology for humanitarian purposes in developing countries, these organizations are also facilitating the commercial development of minor crops through public–private partnerships.

3. REGULATORY ISSUES

Regulatory issues are currently a major factor when commercializing transgenic plants and the products derived from them. To avoid problems that can prevent or delay commercialization, potential regulatory issues must be considered at the inception of R&D planning and throughout the R&D process. Even during the research phase, it is critical to understand and comply with regulations regarding the handling and movement of genetically modified organisms.

Regulatory issues related to R&D in the genetic engineering of plants can be complex, involving biosafety, environmental impacts, food safety and so on. Transferring materials, especially among international collaborators, can involve phytosanitary regulations and international agreements such as the Convention on Biological Diversity.

Considerations that may affect choices logical of R&D strategies include the source of genes or gene products (allergenic organisms, food crops, nonfood plants, animals), properties of gene products or related proteins (toxicity, allergenicity, antinutritional effects, resistance to digestion), choice of selectable markers, and the design of vectors and transformation procedures, as well as the selection of specific transformation events to minimize the presence of DNA and gene products from other species.

As Chile's agricultural industry is largely export oriented, the policies and regulations of both domestic and major export markets must be taken into account. There are big differences, moreover, between the United States and Europe, and these present significant challenges. Regulations change continuously and must be monitored continually.

4. A COLLABORATIVE MODEL

Solving the difficulties requires the participation of many different types of professionals. Indeed, it is difficult for a small biotechnology program with a narrow focus to maintain in-house all the types of expertise required. Fundación Chile's approach has been to develop international networks of parties, with complementary capacities and resources, for the initial development of products. These products are commercialized through new companies with specific commercial foci. The collaborations involve existing companies with strategic positions at different places along the value chain (for example, nurseries with access to germplasm and experience in introducing new varieties to market). The general scheme is illustrated in Figure 1.

4.1 *The R&D consortium*

In this model, the initial task is to form a research and development consortium with a specific focus. Each of the partners in the initial R&D consortium has a largely complementary primary role critical for success:

- **R&D organizations:** research capabilities for the adaptation of technologies to local conditions and the development of products addressing local priorities
- **technology partner:** identification, assessment, and global access to additional appropriate research capabilities and technologies
- **local technology transfer organization:** initial R&D funding, assistance in obtaining grants and other funding, incubation of new technology company
- **strategic private sector partner:** understanding of market demands, ability to

FIGURE 1: FUNDACIÓN CHILE'S COLLABORATIVE MODEL

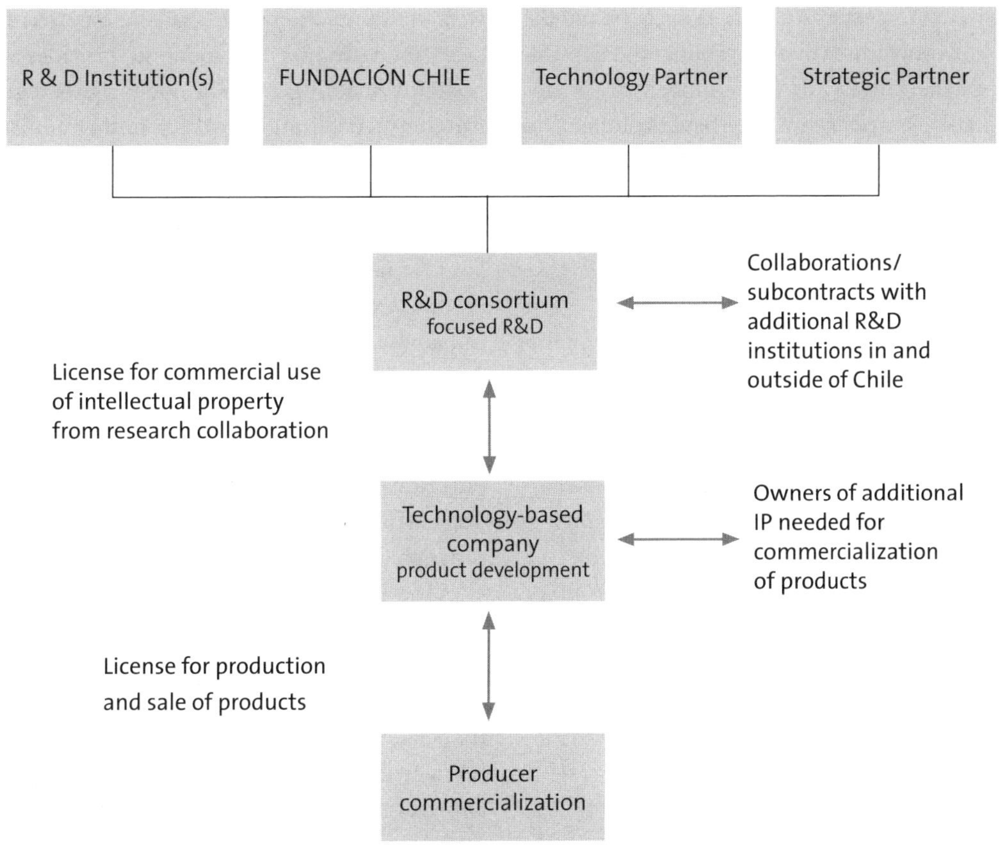

introduce or use the novel products in the target sector, initial R&D funding

Depending on the specific situation, each of the participants may contribute in additional ways. For example, researchers in the R&D organizations are likely to know about specific technologies of interest and may already have relevant relationships with other R&D centers. The private sector partner may already have rights to some intellectual property useful for developing the new products. The technology consultants may be from an entity that will also contribute to R&D funding.

In the biotechnology programs of Fundación Chile, the R&D consortia have made it possible to leverage investment through public support and the use of existing public research institutions. National agricultural research institutes and universities provide infrastructure (laboratories, green houses, equipment) and human resources to carry out the work.

The consortium is the repository of new intellectual property generated during the project. However, in most cases it is expected that the R&D consortium will not produce final products. In the Fundación Chile model, this is undertaken by a new technology-based company, to which the consortium will license rights to intellectual property in exchange for a royalty or other compensation.

4.2 *The R&D network*

The goal of the consortium is to provide the critical inputs necessary for successful R&D in the specific area. In most cases, achieving significant results in a reasonable time frame requires taking advantage of relevant results from other laboratories. Moreover, licensing and option agreements, research contracts, and collaborative research agreements between the R&D consortium, or its members, and other research institutions and companies are critical to establishing an adequate research network. Whenever possible, Fundación Chile has incorporated provisions for training local personnel as part of such agreements.

4.3 *The channel for commercialization*

Later commercial development usually will require different capabilities and considerable additional resources in the early R&D phase. In general, in Chile there is likely to be a significant gap between the results of projects conducted in public research institutions and industry's ability to use them. The model includes creating one or more new technology-based companies focused on commercially developing specific products. The companies will license the results of the R&D consortium and will be responsible for their commercialization. Achieving the latter will require different partners with different interests and resources. Once development has advanced to a stage at which existing companies can produce or use the product, licensing to a company with an established reputation in the area and with its own existing infrastructure may be most appropriate. In cases where an established company with plant breeders, nurseries and so forth does not exist, a new company may be created to produce and sell the product directly.

Establishing, early on, a commercial entity with rights to the outputs of the R&D consortium has advantages. Doing so provides a vehicle for licensing any additional rights required for commercialization and for raising additional investments.

5. THE GENETIC ENGINEERING OF GRAPES

A program to genetically engineer grapes was initiated in 2000 by the Chilean Institute for Agricultural Research (INIA),[9] Fundación Chile, InterLink Associates, Inc. (Princeton, U.S.A.), and Agrícola Brown Ltda. (Los Andes, Chile) with support from the FONDEF program of CONICYT.[10] The relationships of the entities involved in the program are summarized in Figure 2.

The program is one of several initiatives in plant biotechnology for which Biogenetic S.A. (Santiago, Chile)—a joint venture formed in 1998 between Fundación Chile and InterLink Associates, Inc.—has contributed to the development and implementation of strategies for applying biotechnology to problems of strategic importance for Chilean agriculture. InterLink provides expertise in technology scouting and

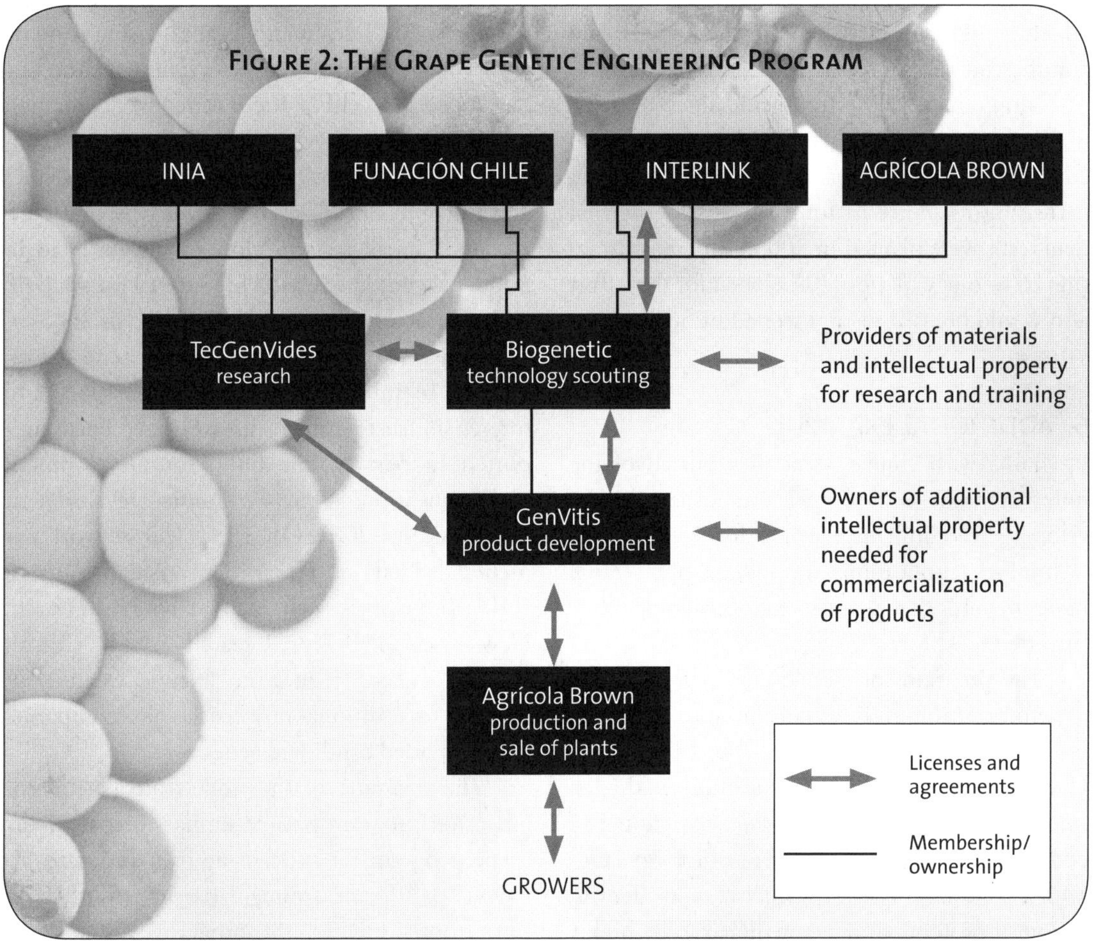

assessment. In addition, it assists in negotiating agreements with a number of different providers of technology components (such as tissue-culture methods and gene candidates) for engineering specific traits in the United States and Europe.

Making use of INIA's existing human resources and infrastructure, the collaboration improved INIA's capacities through the acquisition of additional equipment, construction of new culture rooms and greenhouses, and training of INIA personnel in specific grape-tissue-culture methods at a laboratory in the United States.

The leading producer of grape planting stock in Chile, Agrícola Brown has pioneered the introduction of proprietary varieties of table grapes and also produces and exports grapes. Agrícola Brown's knowledge and experience help to ensure that the grape R&D program addresses the right targets and that any products introduced would be thoroughly evaluated.

The participants in the project formed TecGenVides (Sociedad Tecnológia Genetica en Vides Ltda.) as the entity that would own the results of the initial R&D project. TecGenVides would license intellectual property and materials generated in the project to GenVitis S.A., a new subsidiary of Biogenetic, for further commercial development.

GenVitis would pay a royalty based on its revenues related to the licensed property. In this case, it has been agreed that the major part of the royalty received by TecGenVides will be distributed to the research institution (INIA), with a minor part shared by the other members, who also have the opportunity to benefit from value captured at later stages.

It was agreed at the start of the project that the new technology company, GenVitis, would license the production and sale of transgenic plants in Chile to the strategic partner, Agrícola Brown, which would also participate in the downstream commercial development of products.

More than 1,000 transgenic lines of table grapes have been produced, with most of them containing combinations of candidate genes for increasing tolerance to fungal diseases. The first field trials were planted in 2005. The transformation technology platform developed for this effort also would be used to engineer additional traits.

6. ADDITIONAL EXAMPLES

Programs with similar structures but involving different partners have been established for developing recombinant vaccines for salmon, biotechnology applications for radiata pine (*pinus radiata*), and the genetic engineering of stone fruit trees.

The program for developing novel vaccines to protect salmon is a collaboration between Fundación Ciencia para la Vida and Fundación Chile. The genome of the salmon pathogen *Piscirickettsia salmonis* was sequenced though a contract with a U.S. Department of Energy laboratory. Annotation of the sequence, and identification of protein domains predicted to be highly immunogenic, was carried out by a network of Chilean and foreign researchers. AquaGestión, a company affiliated with Fundación Chile, performed the initial testing of vaccine candidates. Rather than developing production capabilities for a single product, the production and marketing of the vaccine was licensed to Syngenta A.G. (Basle, Switzerland), which was not a participant in the R&D project. A multiple recombinant protein vaccine for *P. salmonis* was expected to be introduced soon thereafter.

The radiata pine biotechnology program includes improvement through clonal selection and genetic engineering. In this case, Fundación Chile established a forestry biotechnology laboratory on the campus of Universidad Austral in space rented from Cefor S.A. (Valdina, Chile), a company affiliated with the university. Some of the investigators were employed directly by Fundación Chile.

The clonal forestry program includes commercialization in Chile by a new company, GenFor S.A. (Talcahuano, Chile). Using somatic embryogenesis and cryopreservation technology developed by CellFor Inc. (Vancouver, Canada), the material was developed by CellFor in collaboration with Bioforest S.A. (Concepción, Chile) and Rayonier Inc. (Jacksonville, U.S.A. and New Zealand). Field tests of clones were initiated in 2000. The initial selection of material for scale-up and commercialization is being made in 2005.

Projects for engineering radiata pine for resistance to insects, for wood composition, and for resistance to fungal diseases have been supported in part by the Fund for Development and Innovation of the Economic Development Corporation (CORFO). The R&D network has included GenFor, Cefor, Universidad Austral, INIA, InterLink, New Zealand Forest Research, New Zealand HortResearch, and Carson Associates Ltd. (Rotorura, New Zealand). A number of additional universities and companies have provided candidate genes.

The structure of the stone-fruit genetic engineering program is very similar to that of the grape program, but the stone-fruit program involves a different strategic partner. With support from CORFO, the program was initiated in 2002 by Fundación Chile, Biogenetic, INIA, and the Andes Nursery Association (ANA; Paine, Chile). ANA is a company focused on developing new fruit varieties that are owned by six nurseries. In addition to an extensive testing program in stone fruit, ANA has initiated a breeding program in peaches and nectarines, in collaboration with the Universidad de Chile, that is focused on improving the fruit's storage life and post-storage quality.

As in the case of the grape program, the products built upon the results of the research consortium will be commercially developed by a new subsidiary of Biogenetic, CaroGen. ANA has a right to license traits developed by CaroGen for commercialization in Chile. The research network includes Okanagan Biotechnology Inc. (Summerland, Canada), which has research collaborations with the Pacific Agri-Food Research Centre of Agriculture and Agri-Food Canada and the U.S. Department of Agriculture Appalachian Fruit Research Station.

Tissue culture and transformation work in the stone-fruit program is being carried out in the same laboratory at which the grape genetic engineering program is based (INIA, La Platina, Chile). This colocation has allowed some synergy among the programs. ■

CARLOS FERNANDEZ, *IP and Regulatory Affairs, Biotechnology Development, Fundación Chile, Av. Parque Antonio Rabat Sur 6165, Santiago, Chile (Present address: Director, Strategic Studies, Foundation for Agriculture Innovation [FIA], Loreley 1582, La Reina, Santiago, Chile). carlos.fernandez@fia.gob.cl*

MICHAEL R. MOYNIHAN, *Director, Biotechnology Development, Fundación Chile, Av. Parque Antonio Rabat Sur 6165, Santiago, Chile. mmoynihan@fundacionchile.cl*

---

1 James C. 2005. Global Status of Commercialized Biotech/GM Crops: 2004. *ISAAA Briefs No. 32*. ISAAA: Ithaca, New York. www.isaaa.org/kc/Publications/pdfs/isaaabriefs/Briefs%2032.pdf.

2 Graff GD, SE Cullen, KJ Bradford, D Zilberman and AB Bennett. 2003. The Public-Private Structure of Intellectual Property Ownership in Agricultural Biotechnology. *Nature Biotechnology* 21: 989–95.

3 Kryder RD, SP Kowalski and AF Krattiger. 2000. The Intellectual Property and Technical Property Components of Pro-Vitamin A Rice (Golden Rice): A Preliminary Freedom-to-Operate Review. *ISAAA Briefs No. 20*. ISAAA: Ithaca, New York.

4 www.isaaa.org.

5 www.cambia.org.

6 www.danforthcenter.org.

7 www.pipra.org.

8 ww.fundacionchile.cl.

9 www.inia.cl.

10 FONDEF (the Fund for the Promotion of Scientific and Technological Development of Chile) was founded in 1991 as a direct government initiative to improve the level of R&D. CONICYT is the National Commission for Scientific and Technological Research. www.conicyt.cl.

IP Rights in China: Spurring Invention and Driving Innovation in Health and Agriculture

ZHANG LIANG CHEN, *President, China Agricultural University, Beijing, China*
WANGSHENG GAO, *Director, Regional Agricultural Development Center, China Agricultural University, Beijing, China*
JI XU, *Professor and Advisor of International Relations, China Agricultural University, Beijing, China*

ABSTRACT

During its relatively brief history of IP (intellectual property) rights protection, China has achieved early success, thanks to the strengthening of governmental IP rights legislation, the establishment of an IP rights management system, the promotion of public knowledge about IP rights, and increasing opportunities for international exchange and cooperation. IP rights protection in the fields of health and agriculture has increased investment in these sectors, encouraged innovation in health and agricultural science, increased farmers' incomes, and improved the quality of life for Chinese citizens. Dramatic increases in patent applications in China suggest that widespread implementation and greater enforcement of IP rights are stimulating inventive activity, encouraging technology transfer, and driving greater and greater innovation.

1. A BRIEF HISTORY OF IP RIGHTS PROTECTION IN CHINA

The China Patent Administration (CPA) was founded in 1980. China joined the World Intellectual Property Organization (WIPO) in March 1980. The first Chinese patent law was passed in March 1984 and became effective on 1 April 1985. China joined the Patent Cooperation Treaty (PCT) in 1994, indicating that China's IP rights legislation was consistent with international standards. China became a member of the World Trade Organization (WTO) in 2002 and pledged to follow the Agreement on Trade-Related Aspects of Intellectual Property Rights (TRIPS) while promoting the development of its own IP rights protection system. The CPA was renamed the State Intellectual Property Office (SIPO)[1] in 1998.

China's patent system has developed quickly in the past 20 years. IP rights regulations, management systems, and publicly available information have gradually improved. In 2006, China ranked fifth in the world for the number of patent applications filed.

Chinese IP rights protection covers the following five categories of intellectual property: (1) patents and technological secrets; (2) trademarks and business secrets; (3) software; (4) copyrights; and (5) know-how about technologies, information, instructions, and so on involved in cooperation activities that need to be kept confidential.

2. AN OVERVIEW OF PATENT DEVELOPMENT IN CHINA

In 2006, 573,178 patent applications were filed for three kinds of patents (invention, utility-model, and design). This figure was 4.6 times the number of patent applications filed in 1998. Numbers of patent applications increased by an average of 19.4% each year from 1998 to 2006. There was an average annual increase of 23.9%

Chen ZL, W Gao and J Xu. 2007. IP Rights in China: Spuring Invention and Driving Innovation in Health and Agriculture. In *Intellectual Property Management in Health and Agricultural Innovation: A Handbook of Best Practices* (eds. A Krattiger, RT Mahoney, L Nelsen, et al.). MIHR: Oxford, U.K., and PIPRA: Davis, U.S.A. Available online at www.ipHandbook.org.

for inventions, 14.0% for utility models, and 21.4% for designs.

Between 1985 and 2006, the total number of patent applications was 3,334,374, including 1,089,521 inventions (32.6%), 289,868 utility models (38.7%), and 954,985 designs (28.7%).

The total number of patents granted by the SIPO from 1998 to 2005 was 1,469,502, including 238,717 inventions (16.2%), 730,573 utility models (49.7%), and 500,212 designs (34.1%).

In 2006, 82% of patent applications came from domestic applicants; 18% came from foreign applicants. The number of foreign applications (all of them for inventions) was four times higher in 2006 than it was in 1985. (Table 1)

In the period between 1985 and 2006, 296,507 Chinese patents were awarded. Of these, 37.9% represented domestic applicants and 62.1% represented foreign applicants.

The ten regions with the greatest number of patent applicants are all located in eastern China (Table 2), in areas with strong science and technology bases and stronger economies than average.

3. IP RIGHTS IN THE HEALTH SECTOR

There are four ways to protect intellectual property in the Chinese health industry: (1) through "administrative" protection, which is used to protect new and traditional medicine; (2) with patents; (3) as trade secrets; and (4) through laws and regulations, such as trademark protection.

Patents for medicine, veterinary science, and health are represented by the code "A61," according to the international patent classification. Table 3 below shows that the total number of A61 patent applications was 24,875 in 2005, four times the number of patent applications for 1994 (6,227). There is a strong annual growth trend. The total number of patents granted in 2005 was 10,179, or 3.5 times the number granted in 1994.

Ninety-seven percent of domestic applications for A61 patents in the "medical" subsector were for traditional Chinese medicines. Foreign applicants filed 92% of the applications for nontraditional pharmaceuticals; there were few domestic applications for nontraditional pharmaceuticals. Chinese applicants filed nearly half

Table 1: The Top-Ten Countries in which Foreign Applicants for Chinese Patents Were Based (2006)

Country	Number of patent applications filed
Japan	36,221
U.S.	20,395
Republic of Korea	9,300
Germany	7,502
Netherlands	3,988
France	3,190
Switzerland	2,106
Italy	1,632
U.K.	1,613
Sweden	1,101

Source: *SIPO Annual Report 2006.*[2]

Table 2: The Top Ten Chinese Regions in which Patent Applicants Were Based (2006)

Province/Municipality	Number of patent applications
Guangdong	72,220
Zhejiang	43,221
Jiangsu	34,811
Shanghai	32,741
Shandong	28,835
Beijing	22,572
Taiwan	20,599
Liaoning	15,672
Tianjin	11,657
Hubei	11,534

Source: *SIPO Annual Report 2006.*[2]

Table 3: Patents Granted in Medicine, Veterinary Science, and Health (1994–2005)

Year	Number of patent applications	Number of patents granted
1994	6,227	2,891
1995	6,177	2,517
1996	6,203	2,084
1997	7,589	2,250
1998	5,720	2,554
1999	8,757	4,865
2000	9,296	5,285
2001	12,509	4,781
2002	13,196	5,418
2003	16,583	6,838
2004	17,448	9,094
2005	24,875	10,179
Total	134,580	58,756

Source: *China Statistics Yearbook 2005.*[4]

(48%) of the patent applications for modern medicines. However, the number of domestic applications for *creative* patents fell well short of the number of foreign applications; this is an area for future improvement.

Overall, the Chinese medical and health sector seems to lack qualified personnel and an IP rights concept. The government needs to promote research and development, capacity building, technical innovation, and the promotion and modernization of industry, in order to increase China's competitiveness in the medical and health sector.

4. THE CURRENT STATE OF AGRICULTURAL IP PROTECTION

The Patent Law of the People's Republic of China, passed in 1984, stipulated regulations for IP protection of plant varieties. China entered the International Union for the Protection of New Varieties of Plants (UPOV) in April 1999 as its 39th member. The State Regulation for Protection of Place of Origin and Products was issued in 1995 and the Seed Law was passed in 2000.

To date, China has granted protection for a total of 62 categories and species of crops and 78 species of trees. In the agricultural sector, there are more than 150 kinds of products protected by trademarks, and more than 600 varieties have plant variety protection certificates.

New regulations that protect plant varieties have encouraged investment in agricultural research and development. A survey conducted by the Ministry of Agriculture (MOA) of more than 500 patent applications and patent grants revealed that companies contributed 83% of the money invested in the research and development of new plant varieties; the government contributed only 17%.

These new regulations have promoted agricultural innovation. In the last 40 years, China has successfully cultivated more than 40 new varieties of different crops and more than 6,000 new varieties. One outcome of this innovation is a 30-40% of increase in grain production in recent years.

The regulations mean that plant breeders have begun to receive economic benefits for their work, which in turn has encouraged them to put still more effort into research and innovation, thus benefiting farmers. As a result, farmers' incomes have increased. In addition, the MOA survey mentioned earlier found that nearly 43 million hectares (ha) had been planted with new plant varieties, increasing yields by 56.3 million tons and increasing farmers' profits by US$2,886 million. Another investigation found that the new, protected varieties of paddy rice protected by IP rights could produce an average profit of US$562 per ha in east China's Jiangsu Province; while ordinary varieties of rice produce an average profit of only US$420 per ha, which is US$142, or 13%, less. The investigation also indicated that the new varieties of paddy rice in southwest China's Sichuan Province produced a 37% higher yield than ordinary varieties.

As Table 4 illustrates, the number of agricultural patent applications has steadily increased. There were 6,802 applications filed in 2005, 4.4 times the number of applications filed in 1994. In 2005, the total number of patents granted was 3,157, which was 4.5 times the number granted in 1984.

China is one of the most prolific filers of applications for IP protection of new plant varieties. According to statistics provided by MOA, the number of applications for variety rights protection increased from 115 applications in 1999 to nearly 1,000 in 2006. There were 3,879 variety rights applications filed in the period from 1999 to the end of 2006, and 899 patents were eventually granted. During the same period, foreign applicants filed 144 patents and five patents were granted (see Table 5). Most applications for variety rights are filed for field crops (90.5%); paddy rice accounts for 31.5% and corn accounts for 39.5% (Table 6).

5. CASE STUDIES

5.1 *Genetically modified cotton*

China has a long history of producing cotton and has been a major cotton-producing country for some time. After China joined the WTO, Monsanto quickly established two subcom-

TABLE 4: CHINESE PATENTS IN THE AGRICULTURE, FORESTRY, LIVESTOCK, AND FISHERIES INDUSTRIES (1994–2005)

YEAR	NUMBER OF APPLICATIONS	NUMBER OF PATENTS GRANTED
1994	1,538	693
1995	1,845	1,045
1996	2,107	904
1997	2,685	942
1998	2,581	1,266
1999	3,534	2,163
2000	3,420	2,235
2001	4,027	2,068
2002	4,782	1,989
2003	4,835	2,530
2004	5,856	2,758
2005	6,802	3,157
TOTAL	**44,012**	**21,750**

Source: *China Statistics Yearbook* 2005.[5]

TABLE 5: NUMBER OF TOTAL PLANT VARIETY PROTECTION APPLICATIONS FILED FOR NEW PLANT VARIETIES (1999–2006)

YEAR	PATENT APPLICATIONS
1999	115
2000	112
2001	227
2002	290
2003	567
2004	735
2005	950
2006	883
TOTAL	3,879

Source: Ministry of Agriculture.[6]

panies in China and introduced its transgenic pest-resistant (GMPR) cotton. Ninety-six percent of the cotton planted in Hebei Province from 1999 to 2001 was American GMPR cotton. In 1999, 400,000 ha of Chinese soil was planted with American GMPR cotton. In 1999, 65% of the pest-resistant cotton planted was American GMPR cotton; 80% was American GMPR cotton in 2000. Monsanto has since obtained a total of nine biosafety certificates from the MOA: four for corn, one for soybeans, one for oilseeds, and three for cotton.

The Chinese government realized that it was important to protect the pest-resistant cotton varieties developed by Chinese scientists. Less American GMPR cotton is now planted, and there is healthy competition between Chinese and American scientists for the GMPR cotton business. To date, China has protected 55 new varieties of GMPR cotton, which makes up 10% of the total amount of all cultivated cotton. More than 6.7 million ha of Chinese GMPR have been planted, yielding profits of close to US$2 billion.

5.2 *Hybrid rice*

Hybrid rice has contributed remarkably to Chinese food security. To date, hybrid rice has been planted on more than 300 million ha of Chinese soil. The current annual yield has been increasing since 1976, and it now feeds 60 million people per year.

Table 6: Patent Applications and Granted Patents for Plant Varieties (1999–2006)

Crops	Number of patent applications	Percentage of total patent applications (%)	Number of patents granted	Percentage of total patents granted (%)
Field crops[a]	3,510	90.0	831	92.5
-Paddy rice	1,222	31.5	261	29.0
-Corn	1,531	39.5	344	38.3
-Soybeans	126	3.2	34	3.8
-Wheat	357	9.2	89	9.9
Vegetables	164	4.2	34	3.8
Flowers	101	2.6	13	1.4
Fruit	101	2.5	21	2.3
Grasses	3	0.8	0	0.0
Total	3,879	100.0	899	100.0

a This list of individual crops is not complete but represents the major crops. Hence, the totals of field crops is higher than the combined total of paddy rice, corn, soybeans, and wheat.

Source: Ministry of Agriculture.[7]

After approval by the Ministry of Agriculture and the State Import & Export Commission, U.S. Western Petroleum's Ring Round Co. paid for the rights of transferring the Hybrid-Rice Technology via the China Seed Corporation in March 1980. It was the first time in China's history that it made such a paid-technology transfer to the outside.

Since the passage of the Regulation for the Protection of New Variety of Plants of the People's Republic of China, a total of 3,879 patent applications have been received for plant varieties; 899 patents have been granted, 280 of them for paddy rice.

The Food and Agriculture Organization of the United Nations has listed Chinese hybrid rice as the most important technology for combating food insecurity in developing countries, especially low-income and food-deficit countries. Vietnam sowed 600,000 ha of hybrid rice in 2003 and achieved a high average yield of 6.3 tons per ha. The country plans to increase the area planted with hybrid rice to one million ha in 2010. India sowed 280,000 ha of hybrid rice in 2003 and 700,000 ha in 2005; the hybrid rice produced a 15–20% higher yield than ordinary rice would have produced. With China's assistance, the Philippines has greatly expanded its hybrid-rice production areas. In the Philippines, 200,000 ha of hybrid rice were planted in 2004 and one million ha will be planted in 2007. In the United States, 20,000 ha of hybrid rice were planted in 2001 and 87,000 ha in 2006. An estimated 30% of all paddy rice planted in the United States in 2007 will be hybrid rice.

The protection of variety rights has encouraged research institutions and private companies to make continuous innovations with regard to hybrid rice. The Hunan Hybrid Rice Research Center developed 36 varieties of hybrid rice in five years (2001–2005), which was 1.5 times the amount developed in the previous ten years (1990–2000).

The protection of new varieties of plants, not only creates direct economic benefits for China, but also helps coordinate the efforts of those working in different areas of the hybrid-rice sector: seed breeding, research, and extension. Sixty-seven million ha of the Pei'ai 64S, the most popular photoperiod- and temperature-sensitive strain, have been planted in China, producing US$10.3 billion, up to year 2004.

5.3 *Pharmaceuticals*

According to the Derwent Innovation Index, the United States is ranked first in the world for the production of new pharmaceuticals, with 1,676 patent applications. China is ranked second, with 1,083 patent applications. China is followed by Japan, with 88 applications. Of the ten companies in the world with the greatest number of patent applications, eight of them are American and two of them are Chinese.

The Shanghai Shengyuan Gene Development Co. Ltd. in China is mainly involved in the research and development of human cDNA. It has a strong technical team and is well equipped. It has identified more than 500 gene elements and has submitted 851 patent applications for genes, more than any other company in the world.

5.4 *The case of Jiangsu Provincial Academy of Agricultural Sciences*

The Jiangsu Provincial Academy of Agricultural Sciences applied for its first patent in 2000 in order to protect a new variety of double-line hybrid paddy rice named Liangyou-Beijiu. By the end of 2004, the Academy had applied for 32 patents and received 23 grants.

The academy could get a benefit of more than US$2.5 million by transferring a series of new variety rights of new wheat seeds cultivated by the academy to a total area of 4.5 million ha. This would provide great social benefits represented by more than US$1.2 billion in value.

6. CONCLUSIONS

Developing countries must protect their IP rights in order to promote domestic innovation, increase resource utilization, improve farmers' income, and promote international cooperation and competition. The following four steps are essential for protecting IP rights: (1) the passing of government legislation; (2) the establishment of a national IP rights-management system; (3)

publicity and promotion of the IP rights concept; and (4) international cooperation.

In general, IP rights protection in developing countries is inferior to that in developed countries. This is because the international IP system may not be fully understood, the legal system may be incomplete, and the human capacity for IP work may be weak. To overcome these obstacles, it is important for developing countries to draw on the experiences of developed countries.

China still lags behind many other countries in IP matters. According to the *WIPO IPRS Report* of 2006, an average of 148 patents were filed for each million people in 2004. Japan filed 2,884 patents per million people; Korea filed 2,189; the United States filed 645; and China filed only 51, putting it in 27th. The global average for patent applications per US$1 billion GDP was 19 applications in 2004. For the Republic of Korea, the number was 116.2 applications; for Japan, 107.3 applications; and for China, only 9.4 applications, putting it in 17th place.

Over the last decade, led by a cadre of world-class scientists and researchers, China's investment in biotechnological R&D has dramatically increased. This has generated remarkable developments and successes, benefiting the people of China in many ways. However, in order to sustain and continue to drive this enormous leap in progress, greater human and institutional capacity in IP law and management will be necessary. Such capacity will serve to further foster and encourage even more inventive activities, innovative initiatives, and the development of the next generation of advances in health and agriculture, for the benefit of all in China. ■

ZHANG LIANG CHEN, *President, President's Office, China Agricultural University, No.17 Qinghua East Road, Haidian District, Beijing, 100083, China. chen@cau.edu.cn*

WANGSHENG GAO, *Director, Regional Agricultural Development Center, China Agricultural University, No.2 Yuanmingyuan Xi Lu, Haidian District, Beijing 100094, China. wshgao@cau.edu.cn*

JI XU, *Professor and Advisor of International Relations, China Agricultural University, No. 17 Qinghua East Road, Haidian District, Beijing 100083, China, jxu@cau.edu.cn*

---

1 See www.sipo.gov.cn.

2 China Statistics Yearbook. 2006. SIPO Statistics: Beijing.

3 Ibid.

4 China Statistics Yearbook. 2005. SIPO Statistics: Beijing.

5 Ibid.

6 Ministry of Agriculture. Office for the Protection of New Varieties of Plants.

7 Ibid.

Experiences from the European Union: Managing Intellectual Property Under the Sixth Framework Programme

ALICIA BLAYA, *IPR-Helpdesk Project, Universidad de Alicante, Spain*

ABSTRACT

Health and agriculture are at the very core of the European Union's policies for socio-economic development. One of its most active efforts is the Framework Programmes for Research and Technological Development. With a specific focus on international cooperation, this is the European Union's main financial instrument to promote and strengthen research and technological cooperation within the European Union (E.U.). Through the E.U. Framework Programmes, actors from different countries and sectors (industry, research centers, small- and medium-sized enterprises, universities, and so on) work together to improve science and create a better standard of living.

Given the massive movement of scientists and experiences exchanged through these Programmes, it seems that the E.U. is on the right track. However, these Programmes can only be used to their fullest potential when participants understand and appropriately handle the intellectual property rules governing them.

1. INTRODUCTION

In the increasingly large group of countries that compose the European Union, there are not only large differences in the climate and natural resources, but also large contrasts in terms of cultural traditions and economic development. Together, these create the specific needs and challenges of E.U. citizens. As an example, in the summer of 2005, a good part of Spain and Portugal saw woods and mountains burn and not a drop of rain to interrupt a sustained period of drought and add to reservoirs, many of which were below 25% capacity. That same summer, Central and Eastern Europe experienced one of the worst floods in recent years.

The summer of 2006 was not better in terms of forest fires and climate conditions. Countries like France and Belgium experienced unusually high temperatures. In recent years, in southern Europe, global climate change has made obtaining (and adequately storing) drinkable water a key concern and a central focus of its research policies. The countries of the E.U. face many of the same environmental challenges as other countries of the world—plagues, ecological accidents and attacks, and natural disasters. This illustrates the problems E.U. member states encounter and the need to take a coordinated approach to managing natural resources and planning their use and exploitation.

E.U. countries have their own policies and initiatives for the optimal and responsible use of their natural resources. Many technological efforts focus on rural areas and businesses that could develop E.U. agriculture, fisheries, and food industries. Using new technologies in rural areas is one of the most common ways to help farmers and small enterprises compete with large corporations.

Apart from the Framework Programmes (hereafter FPs), which are the subject of this

Blaya A. 2007. Experiences from the European Union: Managing Intellectual Property Under the Sixth Framework Programme. In *Intellectual Property Management in Health and Agricultural Innovation: A Handbook of Best Practices* (eds. A Krattiger, RT Mahoney, L Nelsen, et al.). MIHR: Oxford, U.K., and PIPRA: Davis, U.S.A. Available online at www.ipHandbook.org.

chapter, there are other Community actions that benefit the E.U. and partner countries (like those actions promoted under the European Regional Development Fund [ERDF], aimed at regional development, or those projects funded under the MEDA Programme, the objective of which is to improve the socio-economic conditions of countries in the Mediterranean region).

2. THE FRAMEWORK PROGRAMMES AND TRANSNATIONAL COOPERATION

Created by the treaty that established the European Community (the European Community Treaty), the E.U. Framework Programmes for Research and Technological Development are a financial tool to support research and innovation. The multiannual Programmes commenced in 1984. Currently, the Sixth Framework Programme (FP6) is being implemented. FP6 started in 2002 and will run until the end of 2006. (FP7 will start in 2007 and end in 2013.)

While the general objective of the FPs is to boost research and innovation in the E.U., FP6 aims particularly at contributing to the creation of the European Research Area (ERA), which would be a single market for R&D. FP6 seeks to play a significant role in achieving the ambitious challenge of Lisbon 2000: for the European economy to become, by 2010, the world's most competitive and dynamic knowledge-based economy. To meet this objective, R&D in Europe needs to be overhauled. Europe has prominent scientists and researchers, but establishing stable, durable cooperation schemes and turning research into tangible and exploitable results must be an ongoing priority.

To foster European excellence in R&D and innovation, FP6 is based on scientific and technological cooperation at a transnational level. To achieve this cooperation, FP6 has a total budget of €17,883 million.[1] Of this amount, €12,438 million is devoted to the so-called "FP6 Thematic Priorities." The priorities represent seven areas in which research is considered a key need. They are, along with amounts budgeted to accomplish the goals:

1. Life sciences, genomics, and biotechnology for health (€2,514 million)
2. Information society technologies (IST) (€3,984 million)
3. Nanotechnologies and nanosciences, knowledge-based multifunctional materials, and new production processes and devices (€1,429 million)
4. Aeronautics and space (€1,182 million)
5. Food quality and safety (€753 million)
6. Sustainable development, global change, and ecosystems (€2,329 million)
7. Citizens and governance in a knowledge-based society (€247 million)

The FP6 budget acknowledges that small- and medium-sized enterprises (SMEs) are principal engines of the E.U. economy (accounting for approximately 99% of all businesses, giving jobs to almost 95 million people, and accounting for 66% of private employment).[2] In order to help SMEs innovate and develop, they are assigned at least 15% of the general amount budgeted for thematic priorities. In addition, SMEs have €473 million of the total FP6 budget for funding SME-specific actions.

Besides the thematic priorities, other activity areas (such as SME-specific actions, researchers' mobility and training, and international cooperation) share the remaining €5,445 million of the FP6 budget. Nuclear energy and training in this field has a special programme: FP6/EURATOM, with a budget of €1,230 million.

2.1 *Health and agriculture within the FP6 thematic priorities*

Of the total budget for the first thematic priority (life sciences, genomics, and biotechnology for health), €1,209 million is set aside for research on advanced genomics and its applications for health (first subpriority), and €1,305 million is assigned to combating major diseases (second subpriority). One of the main interests of E.U. society is the advancement of cancer research and treatment, and so from the budget of the first thematic priority, up to € 475 million goes exclusively to cancer-related research. Agriculture is covered by the fifth priority, food quality and safety. For the sixth priority (sustainable development, global change, and ecosystems), €890 million is planned for research on sustainable energy systems (first subpriority). €670 million is devoted to sustainable surface transport (second subpriority) and €769 million is for research related to global change and ecosystems (third subpriority).

2.2 *Participation and funds*

Fundamental participants in projects funded under FP6 are legal entities (universities, research centers, enterprises, and sometimes individuals) from E.U. member states. Entities from the E.U.-associated candidate countries (Bulgaria, Romania, Turkey and Croatia[3]), and entities from other countries associated with the FP6 by means of particular agreements (Iceland, Israel, Liechtenstein, Norway, and Switzerland) participate in projects funded under FP6 on the same footing as entities from E.U. member states: They have the same funding options and, in addition, there is the possibility for a consortium made up exclusively of entities from those countries.

However, one of the features that make the FPs attractive to any research entity is the possibility of participation by entities from countries that are not associated with the FP6. Although there are different modalities for participation and funding, entities from these non-E.U. member countries can also participate via thematic priorities and through the International Cooperation (INCO) activity.

2.2.1 *Measures supporting the International Cooperation activity*

The E.U. is a world leader in development aid, and, under FP6, entities from non-E.U. member states can participate even if they are not specially linked with the Programme. The INCO activity, however, best reflects the Programme's international dimension.

INCO is an FP6 activity specifically aimed at cooperation with third countries, and in particular with INCO target countries: developing countries, Mediterranean partner countries, Russia and the other New Independent States (former members of the Soviet Union), and the western Balkan countries.[4] For this specific activity, FP6 reserves €346 million.

Up to €312 million is allocated to support the participation of entities from non-E.U. countries in thematic priorities and other activities, which provide a total of €658 million for the participation of non-E.U. member entities. In addition, resources from the general budget of €1,732 million for Marie Curie actions are available to fund research training and mobility in Europe for researchers coming from non-E.U. member countries.

2.2.2 *How it works*

FP6 funds research and related activities. Actions for funding are open to potential participants (usually, groups of entities, or consortia, coming from different countries) through *calls for proposals,* which establish the main requirements of an activity (for example, the minimum number of participants, origin, objectives of the activity, and deadlines for submitting the proposal). These calls are published on the Internet in the *Official Journal of the European Union* and on the CORDIS Web site[5] (a key service for anyone interested in E.U. R&D and innovation), amongst others. Consortia are generally made up of a minimum number of participants from different E.U. member states or associated states. Once the minimum number is reached, more participants from the same or other countries, even from non-E.U. countries, are welcomed, always taking into account the optimum magnitude of each project.

Generally, once a person or group is considering opting for a research project funded under any FP6 priority or subpriority, the person or group has to find enough partners to form a *project consortium.* Many entities know others in the field with which they would like to partner in research. If this is not the case, CORDIS and other sites provide a partners' search tool.

Deciding on the type of project is a next step. FP6 has a wide range of project types, including integrated projects (IP), networks of excellence (NoE), specific targeted research projects (STREP), specific targeted innovation projects (STIP), cooperative research projects (CRAFT), collective research projects (the last two, represent SME-specific actions), specific support actions

(SSA), which can be carried out by a single entity, and Marie Curie actions (fellowships).

Each project type has its own "personality" and focuses on specific aims. Proposers will need to choose the type that best fits their needs in terms of size (some projects, like integrated projects, are designed for large consortia; others are better managed by a small ones, like the specific targeted research projects), time (some projects can last longer than others; for example, SME-specific actions are relatively short, lasting about two years), and objectives (some projects, such as integrated projects, are focused on developing a specific product or technique through in-depth research; other projects, such as networks of excellence, aim to achieve long-lasting integration of research forces).

Taking all of the above into account, interested parties submit their proposals by a deadline established in the relevant call. These proposals are then evaluated by independent experts. Depending on the proposal's scientific interest, input in R&D, level of innovation, and potential for fulfilment of the aims of the call in question, the proposal may be selected for funding.

Addressing intellectual property (IP) rights issues is crucial for the success of any research project. A competitive proposal has to consider IP aspects carefully in order to convince evaluators that it deserves to be funded. Generally, applicants will be asked about their plans for using and disseminating the expected research results. The applicants need to know what they have, what the state of the art is in the field in question, whether or not there are patents that cover something (for example, a molecule) they may need during the course of their research, what IP they need to work with, what would make them ask for a license, how to share their IP resources for work purposes, what results may be expected, and how these results can be managed and exploited. Of course, the level of detail and scientific certainty of these plans would not usually be very high, but they should be as complete as could be reasonably expected at that stage.

In order to have a well-managed project (and to make the most of the results to be obtained), participants need to be familiar with the FP6 rules for participation and EC model contracts.[6] Furthermore, apart from the FP6-specific rules, participants should take into account other elements, such as other research concurrent with their FP6 project, some national laws (for example, regarding employees' creations or joint ownership), and competition rules, since they may affect the FP6 project.

It is worth mentioning that the IP related rules under FP7, even if maintaining features of FP6, will be likely to change somewhat to the benefit of the project participants, partly by giving them more autonomy. Entities interested in having their research activities funded under FP7 can start now to get familiar with the new rules. (Relevant documents on FP7 can be found, i.e., on the IPR-Helpdesk Web site.[7])

2.2.3 *Do not forget*

Taking part in an E.U.-funded project involves sharing, collaborating, exchanging know-how, and effort. Besides the rules, participants have to be aware of this basic requirement from the very beginning (even before the proposal is selected) to pave the way for their cooperation.

3. IP RIGHTS ISSUES IN AN FP6 PROJECT

Dealing with IP rights-related issues is essential for any research project, and this is even more true for a transnational project than for a project with a narrower focus. The diverse nature of the participating entities (enterprises, public/private research centers, universities, and so on) and their origin (different countries with different laws and cultures) are responsible for the richness of these projects but can be also an obstacle if consortia and resources are not managed adequately.

The relevance of IP related questions is reflected in the attention those questions receive under FP6. The E.U. Framework Programmes provide participants with a set of rules and guidelines that are very detailed in comparison with other funding programs. The rules are laid out in the contract that participants enter into with the European Community (EC)—the EC contract. The contract mirrors the rules for participation in the Framework Programme. Participants will find in the contract the basic norms that are to govern their research project and also several obligations and rights to be exercised at the conclusion of the project (the exploitation-of-results phase).

The EC contract is a pre-established contract that cannot anticipate all the specificities of a single project and consortium. For this reason, participants sign a complementary contract (the consortium agreement) to which the European Community is not a party. Due to the importance of this agreement for implementing the project, it is compulsory under FP6, unless the relevant call specifies otherwise. (Indeed, signing this agreement is particularly obligatory in SME-specific actions, integrated projects, and networks of excellence, while it is usually optional, but highly recommended, in other actions.)

The IP rules concentrate on managing IP resources during the project, with a forward focus on the use of the results obtained from the project. These rules deal with four main aspects:

1. Ownership of the results obtained during the project
2. Protection of results (by means of IP rights)
3. Access rights (licensing)
4. Use and dissemination of results

There are ancillary issues (such as confidentiality, IP related costs, and so forth) that are also important for good IP management and are also considered in the rules.

3.1 *Basic terms*

To understand the IP related rules and their practice, it is necessary to explain some FP6 terminology:

- ***pre-existing know-how***. Even though the definition of *pre-existing know-how* given in the FP6 rules may seem complex, it is actually quite simple: any information and IP resources that participants have *before* entering the FP6 project or that they obtain in parallel to it (that is, any information participants acquire independently of their participation in the FP6 project). The definition applies to any information, not just technical *know-how*.
- ***knowledge***. In the context of FP6, *knowledge* means any results of the project and the related IP rights.
- ***access rights***. The frequently used term *access rights* refers to licenses or user rights to knowledge or pre-existing know-how.
- ***use***. The meaning of *use* is also very specific and distinct from its common meaning. In the terminology of FP6, *use* means: the commercial/industrial exploitation of results obtained or their application in further research activities, either by their owner or by an authorized third party.
- ***dissemination***. The concept of dissemination refers to another activity that FP6 project participants need to carry out: *disclosure* of the results of a project by any appropriate means. The rules specify, "*appropriate means other than publication resulting from the formalities for protecting knowledge.*" This wording helps to clarify that, for example, publication of the patent application by a patent office is not considered dissemination. Scientific publications, general information on Web sites, conferences, and the like are good examples of dissemination.

3.2 *Who owns the project results?*

One of the questions that arises within research collaboration activities is who owns the results. FP6 ownership provisions strive to be logical and lucid, which makes it easier for people who are unfamiliar with legal issues to understand them. The provisions also mirror the general principles of modern IP laws, which provide a fair degree of legal certainty.

The basic rule is that the results obtained in a project are owned by the participant who has carried out the work leading to those results. Importantly, the participant is the entity that enters into the EC contract—for example, a university—not the department or research group actually working on the project.

Where several participants work together toward the results of a project, and their respective portions of the work cannot be ascertained, the participants are considered joint owners and must agree on the allocation and terms of exercising ownership.

In SME-specific actions, the cooperative (CRAFT) and collective research projects, only the SMEs and the enterprise groupings, respectively, get (joint) ownership of the results (even if the results have been generated by other participants). This is because these actions are designed to benefit SMEs.

3.2.1 *Practical issues of joint ownership*

Joint ownership established by the EC contract is a guarantee for the working parties; they can agree to continue under a proper co-ownership regime (therefore establishing the rules to be followed) or agree on other options. The EC contract tries to avoid situations of conflict between weaker and stronger participants by guaranteeing that, where work is carried out in common, all parties must give their opinion before any decision is made.

Joint ownership, however, may arise from either common work or voluntary decision. Its regulation will generally be left at first to the agreement of the parties concerned. Any loophole in the regime will be closed by the applicable law, which changes from one country to another. Accordingly, and to avoid difficulties as much as possible, if the parties decide to continue with a co-ownership regime, they should seek the assistance of a professional in order to draft an adequate agreement that deals in detail with the most important aspects of the ownership regime.

3.2.2 *Taking personnel rights into account*

It goes without saying that the EC contract does not replace participants' national laws, rules, statutes, and so on. Of all these rules, perhaps the most relevant ones are those dealing with employees' and other personnel rights. Policies differ from country to country, so each participant has the responsibility to check its position toward its personnel. The participant and relevant personnel should sign appropriate agreements—and, if necessary, transfer ownership—in order to avoid future claims about the ownership of the results.

For the purpose of this rule, "personnel" may be:

- staff employed by the participants (employees)
- doctoral students
- personnel made available by a third party (invited professors or lecturers)
- subcontractors, and so on

Special care should be taken with those who are not regular employees. In many countries, the situation of employees regarding IP ownership is controlled under labor or IP laws. However, the situation is usually less clear when the work is carried out by scholars or when it is a commissioned work.

3.2.3 *Transfer of results*

Transfers of ownership (including transfer because of takeovers and mergers) are allowed but with some conditions (participants implement their projects thanks to E.U. funds).

The participant transferring ownership has to pass on to the assignee its obligations under the contract (including those related to compulsory licensing, use, and dissemination). Therefore, the assignee gets a "pack of rights and obligations" with regard to the EC and the participants in the project. The transferring party has to give prior notice about the transfer and the assignee to the European Commission (hereafter the Commission) and to the other participants. The Commission may particularly object when the assignee is an entity not established in a E.U. member state or associated state, if such a transfer is not in accordance with the interests of the E.U. economy or is inconsistent with ethical principles. The other participants may object if their licensing options could be affected.

3.3 *How to protect the results obtained*

Adequately protecting results with commercial or industrial application is one of the participants' obligations. After all, a new product, process, or technique can only be properly commercialized when it is adequately protected.

3.3.1 *Options for protecting results*

The participant who owns the results of a project is obliged to ensure their protection. However, the Commission may take over these duties should the owner fail. According to the FP6 rules, the owner should adequately and effectively protect results, while having due regard for its own legitimate interests. This allows for flexibility and gives participants room for decision.

A decision-making process to consider the most appropriate way to protect the results of an FP6 project follows the same path that a university, laboratory, enterprise, or research center does to protect an invention or a piece of work. The decision to seek protection would take into account such factors as the nature of the results obtained (which would lead to the consideration of certain types of IP rights and the dismissal of others; see Figure 1), the level of novelty and inventiveness of the results, the likely market and possibilities for commercial expansion, financial resources, and so on.

The above should lead to the application of the most appropriate IP rights. It should also point to countries for which it would be advisable to seek protection for the results (remember that IP rights are territorial rights). For the best outcome, the participants should get the advice of an expert in the field.

Finally, there is flexibility in the EC contract concerning the kinds of protection and exploitation that are appropriate. If the circumstances of the case warrant it, participants may, for example, decide to opt for trade secret protection rather than applying for a patent. Participants may choose other options in different situations, for example, follow a standardization process or distribute their software under open source licenses.

3.3.2 *Protection and publishing*

Protecting and publishing are two activities that should be carefully balanced under FP6. Academic participants in particular should be aware of the following:

- **Protection prevails over dissemination.** When results come up, before disclosing them to the general public or specialized public, participants need to appraise the commercial/industrial potential of the

FIGURE 1: BASIC TYPES OF IP

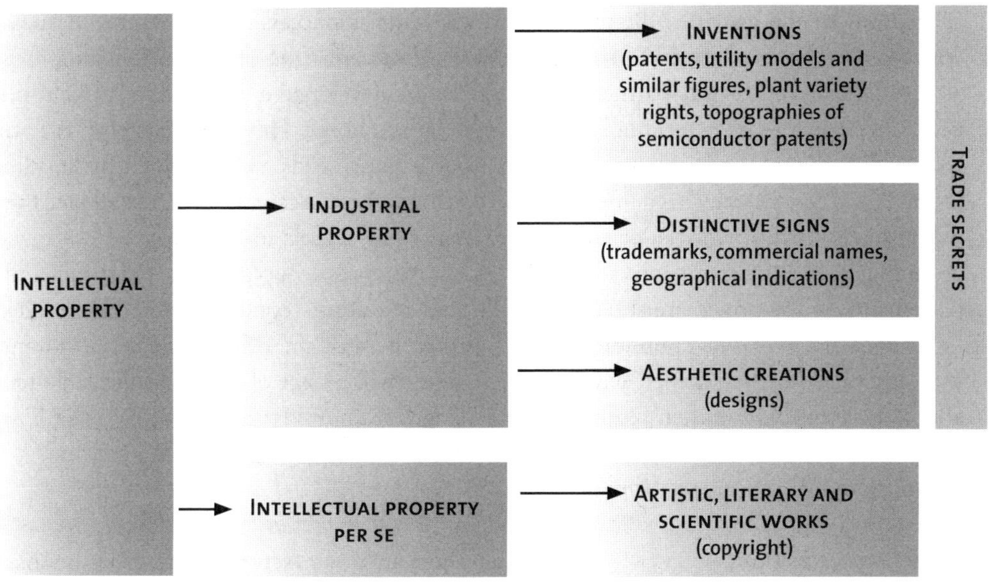

Note: Certain differences in classification exist because of differing national laws.

results. If they can be commercially/industrially applied, dissemination will need to be postponed until protection is ensured. For example, if the option of applying for a patent is being studied, a prior publication may preclude the novelty needed to obtain the patent. Therefore, publication should be postponed until the patent application is submitted to the patent office. Even though this principle may be difficult to follow for those working in academia, universities and research centers, they should not be deterred from participating in FP6. In the European Union (in contrast with the United States), there is no grace period allowing for publication without prejudicing novelty. Publishing in Europe has been considered the traditional activity of academia, but in the last two decades patenting in universities has become more commonplace. For these innovative universities, the waiting approach is already practiced, because protecting first; publishing after is the general principle they follow to turn their research results into profits.

- **Publications are conditioned**. The FP6 rules establish that publication is to be carried out by the owner of the results (or with the owner's consent). In SME-specific actions, the technological partners (RTD performers, in the FP6 terminology) can also publish the results they have generated (even if, as has already been mentioned, ownership vests in the SMEs or enterprise groupings). The Commission and other participants in the project must be notified in advance of any planned publication, and they can object if the planned publication affects the protection of their results.

3.4 *Sharing resources among participants*

3.4.1 *Granting access rights*

Whether generated by their own team or by other participants in the project, the result obtained benefits all participants; participants may need to be licensed or be granted user rights, or *access rights,* by one another. It is compulsory for participants to grant licenses to each other if either of the following conditions exist: It is necessary to carry out the project, or it is necessary for using one's own results.

In the first case, a participant needs information or IP resources from other participants in order to carry out its work in the project, and they shall be required to grant the requester access to the resource in question by means of a license or user right.

Example: The research project aims to develop a new product for the massive cleaning of contaminated water. One of the project participants is in charge of testing a pilot process in its laboratories but needs biomaterials (bacteria) from one of the research centers taking part in the project. In this situation, the latter shall grant access to the bacteria.

The access is granted at no cost if the requester needs results obtained in the project by another participant. Accessing pre-existing know-how is also free (unless partners agreed on a fee before the EC contract was signed).

In the second case, a participant needs information or IP resources from other participants so that it can use the results it has obtained in the project, and the latter shall be required to grant the requester access to the resource in question.

Example: One of the participants in a project has developed a robotic arm to help disabled people at home. However, to exploit the arm, the participant needs a chip owned by another participant. In this case, the latter shall give the other participant access to the chip.

Access is to be granted under fair and non-discriminatory conditions if the pre-existing know-how of the other participant is requested. Access will be free of charge (unless an alternative is agreed upon before the EC contract is signed) to a participant's results.

3.4.2 *Other issues*

There are other factors which affect the sharing of resources and information:

- Compulsory licensing is activated by written request, and regarding pre-existing know-how, the required participant has to be free to grant access to it. This condition may seem quite obvious, but the FP6 rules make a point of requiring this. It is common for research entities to enter into agreements (for example, MTAs or common licenses) with other entities (whether from research or industry) involving day-to-day research. It may happen that participants in an FP6 project have already concluded agreements on their pre-existing know-how that prevent them from granting the other project participants further access to it. In such cases, the participant concerned should inform the other participants of its limitations as soon as possible, in order to avoid false expectations or conflict.

- Participants may condition the grant of licenses on the conclusion of certain further agreements (for example, on confidentiality) that guarantee the proper use of the licensed resources.
- It is possible (and desirable) to grant more favorable or additional licenses. Licensing third parties (that is, licensing the results obtained outside the project partners' group) is also permitted and encouraged.
- As a general rule, sublicensing is not allowed unless expressly agreed upon by the participants concerned. Whatever commitments may be reached, participants' potential rights have to be preserved and rules of competition observed.

3.4.3 *Terms for request*

The Programme's rules include various other provisions related to the sharing of intellectual property among participants:

- Access rights for carrying out project work may be requested until the end of the project (even if the participant concerned leaves before the project is completed).
- Access rights for use can be requested up to two years after the end of a project or end of participation of the contractor (whichever is sooner) if the contractor leaves before the project is completed, unless the partners had previously agreed to extend the period.
- Duration of access rights has to be agreed upon by the parties involved and stated in the licensing agreement.

3.4.4 *Exclusion of pre-existing know-how*

Even though sharing and cooperating is the basis of FP projects, policy-makers are aware that participants' legitimate interests may sometimes be compromised by giving access to specific resources. FP6 offers participants the possibility of excluding certain pre-existing know-how from their obligation to grant access rights to the other participants.

This possibility only exists under two circumstances—before the EC contract is signed and before a new contractor joins the project—and the exception always has to be responsibly exercised. It requires *good faith negotiation* among all participants (some or all may oppose it if the project or their interests are significantly affected), and it can only apply to *specific or concrete pieces* of resources (massive or implicit exclusions are not allowed). Remember that the rule was designed to promote sharing, not excluding.

What if the cause of the exclusion is that an entity fears losing valuable information? In principle, this should not be a reason for excluding access to IP resources, because participants shall preserve the *confidentiality* of the sensitive information they share. It is advisable to sign confidentiality agreements from the moment valuable information is exchanged (if possible, before the project even starts). Once the project is under way, the EC contract requires participants to preserve the confidentiality of the information identified as such (diligence is required). The participant shall guarantee confidentiality for any third party to which sensitive project information is communicated.

3.4.5 *Licensing third parties*

The FP6 rules expressly admit the possibility of granting third parties licenses to project results. However, E.U.-oriented benefits also imply that the Commission can object when the planned license is not in accordance with the interests of the E.U. economy or is unethical. This measure is rarely taken (or needed) but in any case participants have the obligation to inform the Commission in advance when a grant is planned and they think the above-mentioned risks may be present.

How can participants be sure that nothing contrary to the wellbeing of the E.U. economy or unethical is going on? Participants may have an idea about practices that are unethical (as this is a matter frequently in the news). Knowing (even roughly) when the interest of the E.U. economy would be affected would seem to be another story. Aware of this difficulty, the Commission published a note that provides examples of possible scenarios that might be risky. (A typical example of a situation that might affect the economic interests of the E.U. could be that of a planned exclusive license to a company established in a third country.)

In any case and to be on the safe side, it is advisable to inform the Commission whenever a minimum doubt arises. Informing the Commission does not necessarily mean that it will object. Experts will always evaluate the case in the light of its specific circumstances.

3.5 *After the results are in*

The E.U. funding should lead to the use and dissemination of the project's results. The Commission's supervisory role is obvious with regard to the participants' obligation to state their goals and intentions in the *plan for using and disseminating the knowledge.*

The first draft of this plan is to be included in the project proposal. This shows how important it is to have clear ideas on IP management and exploitation at the very beginning. Once the project is under way, a periodic report is required. The report must communicate the participants' intentions regarding the protection, use, and dissemination of the results generated under the project.

A final report (at the end of the project) creates post-contractual obligations for the participants and may be subject to a technological audit (up to five years after the end of the project). The final report must be approved by the Commission.

3.5.1 *Use of the results*

Participants shall use the results they own in accordance with their interests. This can be done through the exploitation of the results or by carrying out further research activities. Both types of activities can be carried out directly by the owner or by a third party that is authorized by the owner. This usually means licensing the results to other participants or third parties. Other options may exist, such as assignments or the creation of a new entity (for example, a spinout).

3.5.2 *Dissemination of the results*

The E.U. funding aims to provide for the dissemination of the results to a wider audience. This means disclosing the results obtained, an obligation when protection and use are not affected. Participants should disseminate the results within two years after the project ends. Should they fail to accomplish this, the Commission may take over these duties.

Results can reach the public through many different channels: Web sites, conferences or seminars, articles for specialized journals, and so on. When studying dissemination (whether by the participants themselves or by the Commission), it is necessary to consider the IP rights involved, promptness, confidentiality, and the participants' legitimate interests.

3.5.3 *Helpful sites*

There are many Web sites and services that help consortia to use and disseminate the results of their research by giving publicity or facilitating contacts (Web addresses for these sites can be found in the endnotes[8]). Among the most useful sites is CORDIS, which offers its Technology Marketplace. This feature records research results with commercial potential into a database arranged thematically using the fields: biology/medicine, energy, environment, IT-telecommunications, and industrial technologies. Other CORDIS services are the *RTD Results Supplement* (a supplement to the *CORDIS Focus* magazine) and *CORDIS Wire.*

Apart from these services, many technology platforms exist at the Community and national levels. The European Technology Platform for Sustainable Chemistry is an example of the former. The Gate2Growth Initiative is also a useful resource; a pan-European business platform for business matching, knowledge sharing among technology investors or knowledge transfer offices, amongst other services. The Commission has published a catalogue to help innovators find local technology transfer institutions.[9]

3.6 *Financing post-research phases*

The projects work under a co-financing principle (something covered by the participants themselves, and main part of costs covered by E.U. funding). To be eligible, costs must fulfill the general requirements stated in the FP6 rules. Among these costs are included costs that are "*actual, economic, incurred within the duration of the project, and necessary*" for the project. If IP related costs comply with these general requirements, they can be funded. Eligible costs may be related to IP protection (patent searches, IP rights filing), the dissemination of results (seminars, publications, and so on), and activities promoting exploitation (for example, feasibility studies, take-up activities).

3.7 *Other IP related obligations*

Having a particular research initiative funded by the E.U. goes, to some extent, beyond the interests of the participating entities. Ancillary provisions try to ensure wide access to the results obtained. These obligations may last longer than the project itself and are always covered by confidentiality guarantees.

These complementary rules include communicating results data to the Commission for evaluation purposes or to standardization bodies (whenever participants have results that may constitute technical standards), giving information to the Commission about results that might be relevant with regard to public policy in member states or associates states, and providing the necessary publicity to the funded project.

4. CONCLUSION

Fostering E.U. research and development requires managing the IP resources of different projects. The entire process, from pure research to the exploitation of research results, has to be well planned.

The E.U. Framework Programmes are an ambitious tool for helping to implement this process. Mirroring modern IP laws, FP6 (and FP7) rules seek to facilitate IP management and increase legal certainty. They also try to balance public and private interests, but the success of these research actions cannot be left to the rules. The goals of the E.U. Framework Programmes can be met only if the participants involved are aware of these rules and do their best to implement them. An open sharing of information and experience will develop the essential trust, good relationships, proper planning, and solid cooperation needed to achieve the Programmes' goals. Indeed, success very much depends on the participants' commitment and effort. ■

ACKNOWLEDGMENTS

IPR-Helpdesk is a project of the European Commission and DG Enterprise and Industry. The project is financed under the Sixth E.U. Framework Programme for Research and Technological Development.

The opinions and views expressed in this chapter are those of the author.

ALICIA BLAYA, *IPR-Helpdesk Project, Universidad de Alicante, Edificio Germán Bernácer, Apartado de correos 99, E-03080 Alicante, Spain.* alicia.blaya@ua.es

---

1 Source for FP6 budget and relative breakdowns: Decision No 786/2004/EC of the European Parliament and of the Council of 21 April 2004. See also CORDIS, at cordis.europa.eu/fp6/budget.htm.

2 See online, Observatory of European SMEs, 2003/7.

3 Bulgaria and Romania should normally join the E.U. in January 2007.

4 For the full list of countries, go to cordis.europa.eu/inco/fp6/intro_en.html#eli.

5 cordis.europa.eu/.

6 Available on the IPR-Helpdesk Web site, www.ipr-helpdesk.org.

7 Ibid.

8 For additional information and further reading:

CORDIS innovation services: cordis.europa.eu/guidance/services1.htm and CORDIS Web site: cordis.europa.eu/ and CORDIS FP6 SME TechWeb: sme.cordis.lu/home/index.cfm.

DG Research (European Commission): ec.europa.eu/research/.

European Regional Development Fund (ERDF): ec.europa.eu/regional_policy/funds/prord/prord_en.htm.

Gate2Growth Initiative: www.gate2growth.com/.

INCO Portal (CORDIS): cordis.europa.eu/inco/home_en.html.

IPR-Helpdesk: www.ipr-helpdesk.org.

MEDA Programme: ec.europa.eu/comm/external_relations/euromed/meda.htm.

Technology Marketplace (CORDIS): cordis.europa.eu/marketplace/.

World Intellectual Property Organization (WIPO): www.wipo.int.

9 ec.europa.eu/enterprise/enterprise_policy/competitiveness/index_en.htm.

CHAPTER 17.5

Current IP Management Issues for Health and Agriculture in India

KANIKARAM SATYANARAYANA, *Chief, IP Rights Unit, Indian Council of Medical Research, India*

ABSTRACT

This chapter describes the current status of IP (intellectual property) management in the areas of health and agriculture in India with a focus on post-2005, at which time India became fully complaint with the Agreement on TRIPS (Trade-Related Aspects of Intellectual Property Rights). The major policy trends existing in India include (1) public sector expenditure for R&D is on the rise and is currently about US$5.0 billion (one US$ equals about 4 Rs); (2) pharma industry R&D expenditures were on the rise and had reached Rs 15.0 billion, or close to 4.0% of their turnover; (3) several major policy initiatives had been undertaken by the government, including the National Health Policy (2002), National Policy on Indian Systems of Medicine and Homeopathy (2002), and National Biotechnology Policy (2005). Other major initiatives to promote IP generation include the creation of a Central Drug Administration, a new national body for the registration of medical devices, a National Registry for Clinical Trials, and a law similar to the Bayh-Dole Act that provides for the sharing of IP with inventors. The Departments of Science and Technology and Biotechnology, the Council of Scientific & Industrial Research, the Indian Council of Medical Research, the Indian Council of Agricultural Research, and so forth, have initiated large R&D programs in the health sector for the generation of new diagnostics, vaccines, and drugs largely focused on current health problems of India. A few indigenous products are being tested for safety and efficacy before use in the public health system. A new thrust and focus are being given for public–private partnerships involving both national and international partners. In agriculture, besides a substantial allocation of funds for R&D, two new initiatives—the National Agricultural Innovation Project (NAIP) and the Indo-U.S. Agricultural Knowledge Initiative (AKI) were started in 2005. The NAIP is a World Bank-supported project worth approximately Rs 11.7 billion that is expected to strengthen basic and strategic research in agriculture in India. The AKI is expected to address a large number of issues including education, research services, and commercial linkages in agriculture.

1. INTRODUCTION

Compared to many developing countries, India has a strong science and technology base. When India gained its independence in 1947, many science and technology institutions already existed there. Moreover, during the past 50 years, India has made rapid strides in science through a series of policy initiatives promoting high-quality research. This chapter focuses on developments in the last few years, especially since 2005, when India became fully complaint with the Agreement on TRIPS (Trade-Related Aspects of Intellectual Property Rights).

The past five years have seen an important change in science and technology primarily due to the anticipated impact of the TRIPS agreement on IP regimes in India. Globalization and liberalization, which have primarily affected the economy and business, also have triggered innovative R&D, as Indian companies have realized that unless they learn to become globally competitive, they may not survive. The pharmaceutical

Satyanarayana K. 2007. Current IP Management Issues for Health and Agriculture in India. In *Intellectual Property Management in Health and Agricultural Innovation: A Handbook of Best Practices* (eds. A Krattiger, RT Mahoney, L Nelsen, et al.). MIHR: Oxford, U.K., and PIPRA: Davis, U.S.A. Available online at www.ipHandbook.org.

and biotechnology industries have been among the first to understand the implications of the new patent regime and the need to carry out innovative R&D. Some companies have increased their research budgets to as much as 10% of their total budgets. Even public sector institutions have realized that there is a need to reconsider their IP policies. Agencies like the Indian Council of Medical Research (ICMR) have adopted IP polices to promote innovative R&D, encourage partnerships with industry, create incentives for patent filing and systems of royalty sharing with inventors, and so on. In the more recent past there have been attempts to create, through partnerships of the ICMR with agencies such as the National Institutes of Health (NIH) and MIHR (the Centre for the Management of Intellectual Property in Health Research and Development), a strong force of technology transfer professionals. The formation in 2005 of the Society for Technology Managers (STEM), accomplished with the help and cooperation of AUTM (the Association of University Technology Managers) is a watershed in IP management in India.

2. EXPENDITURE ON R&D

2.1 *Government sector*

Publicly funded biomedical research and development (R&D) in federal laboratories in India is carried out by the ICMR, the Council of Scientific and Industrial Research (CSIR), the Department of Biotechnology (DBT), and a few institutes of the Department of Atomic Energy (DAE). The ICMR has 21 research institutes and six regional medical research centers. There are at least six laboratories of the CSIR, four DBT institutions, two DAE centers, six autonomous institutes that carry out significant medical research, and approximately 25 medical colleges, a few of which belong to the private sector. The colleges are supported by nongovernmental scientific research organizations (perhaps about a hundred) that are registered with the Government. Still, a significant chunk of biomedical research is carried out with only government support. The public-sector R&D effort primarily focuses on mapping disease burdens, profiling infectious diseases, carrying out preventive and/or therapeutic interventions, testing the efficacy of available new therapeutic interventions (such as drugs and diagnostics), finding new drugs and diagnostics for more cost-effective interventions, and carrying out basic research to improve understanding of biological systems.

No reliable data exists with regard to total expenditures for health and biomedical R&D. Estimates by one researcher, based on expenditures on R&D by major agencies, suggest that total expenditures (excluding expenditures by the pharma sector) about US$5.0 billion (one US$ equals approximately 4 rupees [Rs]), or approximately 2.5% of the estimated direct government expenditure on health.

2.2 *Research and development in the pharma industry*

In the private sector, the pharma industry spends the majority of its R&D funds on biomedical research. This started a few years ago, primarily because India's impending globalization and TRIPS-compliance spurred the pharma industry to carry out more innovative research to create new molecules in order to remain globally competitive. Recently, technologically competent small and mid-sized firms collaborating with multinational corporations and Indian generic companies have emerged. Some Indian pharma companies have already reoriented their R&D strategy from business-driven research (generic manufacture) to research-driven business (developing new molecules, novel drug-delivery systems, and so on). From 1999 to 2003, the number of U.S. patents granted for drugs and pharmaceuticals to India grew significantly. In 2003, India filed the most drug master files (DMF) applications (126) with the U.S. Food and Drug Administration (FDA), which was more applications than had been submitted by China, Italy, Spain, and Israel combined. India has the largest number of FDA-approved manufacturing facilities (more than 60) outside of the United States. R&D investment inside India is on the rise. For the year 2003–2004, the top-ten pharma companies spent more than Rs

9.7 billion on R&D (greater than 6.0% of their turnover). The pharma sector spent more than Rs 13.0 billion (almost 4.0 % of its turnover), which was the highest R&D investment of any Indian industry sector.

In addition, publicly funded institutions like the DST, CSIR, and DBT, support research in pharmaceutical R&D (including biopharmaceuticals) through various schemes, such as the New Millennium Indian Technology Leadership Initiative (NMITLI) and the Pharmaceutical Research & Development Support Fund (PRDSF).

3. POLICY INITIATIVES

The Government of India recently implemented a series of policy initiatives that have energized and focused R&D on generating new knowledge that could lead to new products and processes of public-health importance.

3.1 *National Health Policy in* 2002

Recognizing changing demographics, altered disease patterns, the health needs of its diverse populations, and the intensification of technology interventions in delivering health care, the Government of India announced the National Health Policy (NHP-2002). This initiative seeks to: (1) expand and improve primary healthcare facilities; (2) meet the health needs of disadvantaged sections of the population (women, children, elderly, and tribals) through special programs; and (3) mount programs to eradicate polio, yaws, leprosy, kala-azar, and filiarasis and to control diseases like HIV/AIDS, TB, and malaria within specified time periods. To achieve these objectives, the Government has committed to raising public spending on health to 2–3% of GDP (gross domestic product). Although the NHP-2002 does not explicitly state it, there has been a focus on generating new drugs, diagnostics, and vaccines for diseases of public-health importance like TB, HIV/AIDS, and so on. This attention is evident from the new initiatives to generate new diagnostics and vaccines through various public–private partnerships with national and international partners.

3.2 *Biotechnology Policy*

The new Biotechnology Policy of the Government of India (2005) draws a clear roadmap for developing biotechnology R&D in India. Some of the major initiatives proposed include encouraging R&D in academia, entering into partnerships with industry and support to industry per se, granting tax breaks and other incentives to biotechnology companies, and setting up biotechnology parks, special economic zones, and so forth. To help industry quickly bring products to market, the regulatory framework is being streamlined through a new set of simplified guidelines for the approval of all recombinant DNA products. Also, a single biotechnology regulatory authority (BRA) for clearing biotechnology products is being created. A series of bioclusters will be developed around existing biotechnology centers and some identified institutes of excellence. A strong focus on human resources development is evident from new programs, such as a new M.D.-Ph.D. program, an Asian-level United Nations Educational, Scientific and Cultural Organization (UNESCO) Center for teaching and training in biotechnology, and training fellowships abroad for cutting-edge areas like stem-cell technology and nanobiotechnology. The National Jai Vigyan Science and Technology Mission also provides support for developing new products and processes. Finally, the new Small Business Innovation Research Initiative (SBIRI) aims to provide early-stage funding to scientists in private industries for high-risk, innovative, or commercializable product proposals.

3.3 *Policy on traditional medicine*

Traditional systems of medicine have always figured prominently in India's healthcare delivery system because the practitioners of Indian Systems of Medicine and Homeopathy (ISM&H), comprising Ayurveda, Unani, Siddha, and Homeopathy, have a significant presence in India's rural areas. Recognizing the importance of Indian systems of medicine (ISM) in healthcare, in 1995 the Indian Government established a full-fledged Department of ISM&H, not only to promote curative aspects of ISM but also to energize R&D in this area. In 2002, the Government announced its National Policy on ISM&H to address inadequacies in existing mechanisms, initiating new strategies to (1) improve the quality of teaching in ISM courses, (2) ensure the availability of quality raw materials for therapeutics, (3) formulate and implement standards (for example, good manufacturing practices), (4) encourage research in ISM&H to generate new drugs, and (5) address IP protection for traditional remedies.

Some steps have already been taken. These include setting up the National Medicinal Plants Board to provide quality material for herbal drugs, as well as establishing drug testing laboratories to ensure quality-assurance standards for bringing out pharmacopoeia in ISM, and so forth. Traditional systems of medicine are important because they offer therapeutic alternatives for some lifestyle, degenerative, and age-related ailments, such as rheumatism, for which other satisfactory therapies are lacking. Industry has been encouraged to carry out innovative research to bring traditional-medicine formulations to contemporary dosage standards through concentration of the liquids, modifications in the physical forms, developing appropriate delivery formats, increasing shelf life, ensuring stability in storage, enhancing sensorial acceptance, undertaking limited clinical trials for validating drug safety resulting from new forms and procedures for preparations, standardizing formulations based on active markers and fingerprint profiles, and, most importantly, adapting, modifying, and designing processing equipment to handle the botanical materials at appropriate processing conditions. A dynamic, continuing process, the initiative is spearheaded by about a dozen large, leading Indian pharma companies and a few publicly funded R&D and academic institutions. As an example of the work that's being done, a recent innovation by CSIR provides quantitative scientific representations of various Ayurvedic concepts using three-dimensional high-throughput liquid chromatography (HPLC) techniques. This invention has been patented in the United States and other countries. The Golden Triangle program (see below) is an example of how traditional systems of medicine and modern research and medical systems can work together to bring new drugs to the market.

3.4 *New IP rights regime*

In 1970, India enacted the Indian Patents Act, which came into force in 1972. Some significant features of the Patent Act include restricting to process patents these products in the areas of food, drugs, and agrochemicals; limiting patent life to seven years; and providing more liberal compulsory licensing provisions. The act's primary objective was to promote the development of the domestic pharma industry. Without product patents, it was hoped that the Indian public could get affordable drugs, and indeed the act encouraged industry to manufacture and distribute generics. The policy helped build a strong domestic industry that tapped India's scientific strength, especially in chemistry, to churn out generic equivalents that not only catered to local needs but also built a formidable bulk-drug export market. The act triggered significant growth and revenues through the export of bulk drugs. In the bargain, it also created a demand for testing and evaluation technologies and quality-control systems in the pharma industry. More importantly, this growth created opportunities for the industry to invest in reverse engineering R&D, which created a world-class generic industry. The Patent Act fully served its purpose of providing affordable medicines to the poor.

As a founder-member of the World Trade Organization and signatory to the TRIPS Agreement, India was expected to make its patent laws fully TRIPS compliant by January 2005. Accordingly, the Patents Act was amended three times to become TRIPS compliant. It now provides for product patents in all fields of technology and for other provisions stipulated under the Agreement. Due to these changes, multinational pharma companies have started to consider investing in India for R&D and manufacturing facilities. The companies see the ready availability of qualified people, infrastructure, and the advanced regulatory environment. These policy changes have been received very positively by the pharma industry, as is evident from the increased

Abbreviated New Drug Application (ANDA) filings over the past four years by the top-ten Indian companies. In the short period of six years beginning in 1995, patent filing in the drug industry nearly quadrupled. Today, some pharma multinational companies have started to expand their presence: their share in the Indian market is expected to double in the next five years.

3.5 *Regulatory environment—creation of the Central Drug Administration*

R&D in the pharma sector can be promoted only if an appropriate and reliable regulatory system is in place. India's disorganized drug-control administration has been seriously criticized, especially regarding spurious drugs. Manufacturing licenses for drug formulations are being issued by the drug controllers of various states and Union territories, with no coordination between them or the drug controller general of India (the regulator of the federal Ministry of Health & Family Welfare). Often, the drug controllers of various states were violating their authority by issuing licenses for drugs that were banned by the federal government. As a result, thousands of irrational and harmful combinations are on the market.

A committee under the chairmanship of Dr. R. A. Mashelkar, director general, CSIR, was formed to examine this problem and suggest how to revamp India's drug-regulatory system. The committee's recommendations focused on how to bring central monitoring and intervention activity to bear on the actions of state drug-control agencies in order to uniformly implement the Drugs & Cosmetics Act. The committee proposed elevating the Central Drugs Standard Control Organisation to the level of a Central Drug Administration (CDA), a federal body reporting directly to the federal health ministry that would, among other things, have complete control over the licensing of manufacturing units in the country—a power that was earlier vested with drug departments at the state level. The committee called for the creation of a specific medical-devices division to properly manage the approval, certification, and quality assurance of medical devices in India. The committee underscored the need to globally harmonize regulatory and scientific requirements. It also addressed regulating the activities of healthcare providers and the Indian systems of medicine and food supplements. The report recommends that drug regulatory administration be system based, with every activity justified within a clear policy framework and supervised for uniform implementation and the timely, transparent disposal of license applications, renewals, and so on. Finally, the committee recommended upgrading the present Central Drugs Standard Control Organisation (CDSCO) to the level of a Central Drug Administration (CDA), a federal body reporting directly to the federal Health Ministry, somewhat like the U.S. FDA.

3.6 *Medical devices registry*

More than 80% of the estimated amount spent on medical devices and other critical-care equipment purchased in India (about US$1.5 billion) is now made up of products that are imported. Several academic and research organizations, as well as private entrepreneurs, have started taking an active interest in the development and production of medical devices. Important devices, such as heart valves, orbital implants, coronary stents, oxygenators, cardiac catheters, eye lasers, external cardiac pacemakers, and critical-care ventilators, have emerged from high-technology research spinouts of the research laboratories of CSIR, the Defense Research and Development Organization (DRDO), DST, and others. Many products are also at advanced stages of development/clinical evaluation. Successes in this field—especially when sustained—are impressive.

Biomedical devices are technology based and have a shorter market life span. Unlike drugs, biomedical devices do not work via chemical action within or on the body by pharmacological/chemical/immunological means or by being metabolized within the body. Regulations of biomedical devices with regard to safety, health protection and performance, characteristics, and authorization differ from country to country.

In the United States, the FDA has a separate department to evaluate and regulate medical and radiological devices; in Europe, the safety and efficacy of a product and its quality assurance are the responsibility of the manufacturers themselves. Avoiding the pitfalls and drawbacks of the U.S. FDA system, the European regulatory model has evolved into one of the most effective, efficient systems in the world today. Although expensive for the manufacturer, the onus of quality assurance is on the producers; any infringement in quality control leads to judicial penalties (as with the U.S. FDA). Indian officials recognize that the country acutely needs a regulatory body to control biomedical devices and ensure that the public is protected from poor-quality products—both indigenous and imported. In addition, a regulatory body could help the various segments of the developmental chain: the R&D groups, the manufacturers, the clinicians, and finally the patients.

To address this issue, the ICMR, New Delhi, and the Society of Biomedical Technology (SBMT) of the DRDO jointly worked to find a suitable structure for regulating medical devices. The proposal was made available on a Web site for comments, and suggestions were invited from a representative group of physicians, surgeons, and other experts using medical devices. A draft report was discussed with a group of experts. In the end, the proposal for an Indian medical devices regulatory authority (IMDRA) was submitted to the Government of India. The IMDRA will be responsible for implementing the country's regulations for medical devices. The proposal remains with the Government for implementation.

3.7 *Clinical-trial registry*

Attempts to register clinical trials being conducted in India have been minimal, because not many trials are carried out, and even the few that are carried out are not reported. But given the availability of large numbers of patients, qualified professionals, and hospitals with infrastructure that can perform clinical trials in accordance with global standards of good clinical practices, India is expected to become a global clinical trial hub. In fact, several contract research organizations have already set up their offices in India. There is, however, serious concern about a lack of transparency for trial data, especially in light of the conduct of unethical trials that the media has uncovered. The flouting of ethical guidelines has been on the rise, and so the Government of India is seriously considering bringing all clinical trials under strict regulatory control through a trial registry. Mandatory trial registration is bound to positively affect the quality of clinical trials conducted in both private and government sectors. So far, the government has entrusted the ICMR with the responsibility of setting up a clinical-trials registry. Efforts are already underway to establish a registry at ICMR, New Delhi.

3.8 *Awards*

There are more than 15 different awards established by Indian Government agencies like the CSIR, DSIR, National Research Development Corporation (NRDC), and DBT. Significantly, the nature of these awards has been changing. The thrust and focus of earlier awards was on import substitution and/or the indigenization of a technology. Recent instituted awards, however, emphasize the generation of innovative technology. For example, the CSIR Diamond Jubilee Technology Award (Rs 1.0 million) is given for the "most outstanding technological innovation that has brought prestige to the nation." Moreover, the CSIR Diamond Jubilee Invention Awards for School Children encourage a culture of innovation at a young age.

3.9 *Innovation and IP ownership*

A major portion of innovative R&D carried out in Indian universities is not IP protected. This is partly because India's university system lacks technology transfer offices that could help university researchers protect and exploit new innovations. In addition, as a matter of policy, most government agencies own all of the IP generated through research funded to the universities extramurally. Therefore, little incentive exists in terms of inventors sharing the royalties of new IP. This situation is widely considered to be the crippling factor that explains both why little innovative work takes place in the university sector and why enthusiasm is lacking to commercialize the few innovations that are IP protected. To address this issue, the Government of India is seriously considering enacting legislation modeled after the Bayh-Dole Act in the United States, which would allow university inventors to own IP generated from federally funded projects.

3.10 *Entrepreneurship development-new policy on contract research*

The policy of contract research, a system through which public sector R&D institutes collaborate with industry, has been in existence in India for more than 20 years in major scientific organizations like the CSIR and the ICMR. This scheme has recently been liberalized to allow scientists and institutes to work with industry on projects of mutual interest. The time scientists could spend on R&D projects with industry (person days/year) and the amount of honoraria they could earn from such projects per year has been increased. Scientists are encouraged to take up R&D projects from industrial partners from India and abroad that would create products and processes for industrial application. In addition, some institutes have also made provisions for scientists to be entrepreneurs while holding their regular position within the organization (for example, the CSIR and the Indian Institute of Science, Bangalore). The impact of these initiatives has been positive. Some scientist-entrepreneurs from the institutes are already pursuing spinout companies.

4. STRATEGIES AND OUTCOMES

4.1 *Policy initiatives*

Infrastructure–Creation of new institutes of excellence: National Institutes of Sciences. If India hopes to become a global leader in science and technology, it must raise science education standards. A good science education is available in only a few institutes of excellence, such as the Indian Institutes of Technology, the Indian Institute of Science, Bangalore, and a few federally supported universities. Many population centers in India have no institute of excellence nearby, which discourages bright students from taking up science. Accordingly, four new centers of excellence in science education are being set up in different parts of the country. These institutes would be established at Allahabad near Allahabad University (in northern India), at Chennai near Anna University (in southern India), at Pune near Pune University (in western India), and at Bhuvaneswar near Utkal University (in eastern India). The centers primarily would offer an integrated, five-year basic and applied program in sciences leading to a master's degree. Linked with national research labs, science agencies, and industry right from their inception, these institutes will be "incubated" within the existing premier universities. But, although they will be connected to the universities, the institutes will enjoy complete academic, administrative, and financial autonomy. The corresponding university will initially award educational and research degrees to an institute's scholars and students, but the institute will have complete, total freedom to set out its academic programs, frame suitable course structures, and establish its own methods of teaching and evaluation. This organic link with the universities will be crucial in the initial phases. Administrative and financial details have been worked out and the proposal is in the approval process.

4.2 *National Biotechnology Development Strategy*

Ever since the full-fledged Department of Biotechnology (DBT) was set up in 1986 under the Ministry of Science & Technology, the DBT has played a pivotal role in R&D, education, technology management, and support to nascent industry. Both health and agri-biotechnology have received considerable support, and now there is a vibrant industry, a growing number of competent biotechnologists, and a regulatory framework that helps put products on the market.

The DBT has drawn up a ten-year strategy to put India on the global-biotechnology map. This National Biotechnology Development Strategy was unveiled by the Minister for Science and Technology, Kapil Sibal, on March 31, 2005. Highlights of the strategy include 100% of biotechnology units funded by foreign direct investments (FDI), priority sector lending tags, tax credits for money spent on international patent filings, and the creation of ten biotechnology parks with special economic zone status, among others.

The DBT and the Ministry of Environment and Forests have released a set of guidelines for the approval of all recombinant DNA products. This is expected to give a huge boost to the biotechnology industry, because about 90% of the organisms used by biotechnology companies will be outside the purview of the Genetic Engineering Approval Committee (GEAC). For recombinant pharma products derived from living modified organisms (LMOs) but for which the end product is not an LMO, applications can be submitted for approval directly to the drug controller general of India (DCGI).

The Government is setting up a single biotechnology regulatory authority for clearing biotechnology products, and a high-level committee is figuring out how to create such an authority and rationalize the legislative and regulatory regime. Presently, several agencies under federal Ministries—Agriculture, Health and Family Welfare, Environment and Forests, Science and Technology, and Biotechnology—are involved in clearing biotechnology products.

Another important step being taken by the DBT to build the biotechnology industry is that of fostering bioclusters. Developed around existing biotechnology centers, a series of bioclusters will be formed by strengthening research in medical colleges (both translational biology and clinical research in the cities of Bangalore, Hyderabad, and other centers that have potential). The National Center for Biological Studies (NCBS), Bangalore, and the Christian Medical College, Vellore, are working to strengthen and transform CMC Vellore into a molecular medicine, translational, and clinical-research center. Likewise, a translational research institute is planned in Gurgoan with links to the National Brain Research Centre (NBRC) there. Biotech parks are being planned also on the Delhi-Gurgoan belt with the purpose of attracting industry. In Hyderabad, the DBT is creating a stem-cell R&D cluster (in addition to the one in Bangalore) and an agri-biotechnology corridor is being developed in Punjab.

Support to industry would be extended through (1) a quick, responsive regulatory framework; (2) support for late-stage development; 3) training in clinical validation; (4) a third-party associate for technology transfer projects with international companies/scientists; (5) encouraging industry participation in international science meetings; 6) creating an industry research support cell; and 7) direct industry funding for SMEs. The institutional sector will be strengthened by:

- expanding existing support for science education and training
- supporting the creation of new innovation centers and centers of excellence
- increasing contact and engagement between cross-disciplinary professionals through special grants and interdisciplinary centers of excellence
- a niche-area overseas training program
- large infrastructure grants
- five-year grants for translational research

The DBT has also funded the creation of "good manufacturing practice" facilities at several institutions and is working with Reliance and two other companies to support clinical research on DBT's products. On the international front, various collaborative programs are being considered that emphasize building strategic partnerships. A major program for animal vaccines and immunostimulants in aquaculture has been firmed up with the government of Norway, and another agreement was signed with Australia. Strategic partnership agreements have also been signed with Denmark in the area of agriculture and food biotechnology, with the UK in relation to cutting-edge biology, and with Finland in diagnostics.

Other new initiatives include:

- consolidating support services for regulations relevant to trade
- partnering with the Ministry of Health on GM food testing
- introducing biotechnology methods into the judiciary through a DNA academy funded at the Center for DNA fingerprinting
- improving the capacity for clinical trials in the country

- setting up new life-sciences institutions (like the translational health-science institute in Faridabad and the UNESCO center for training and education in Delhi)
- creating an animal biotechnology institute
- creating two policy centers (a center for health technology policy and a center of agriculture and allied areas).

Furthermore, the DBT is launching the Small Business Innovation Research Initiative (SBIRI), which provides early-stage funding to scientists in private industries for high-risk, innovative, or commercializable product proposals.

Some initiatives in human resources include identifying an Asian-level UNESCO Center for teaching and training in biotechnology. There is also a proposal to award 25 special overseas fellowships to students doing research in stem-cell technology and nanobiotechnology, as well as a plan to support 20 undergraduate colleges across the country (one per state) focusing on high-quality teaching in the life sciences (this is in addition to summer project support for students and skill-enhancing training for teachers). Further efforts to develop quality human resources include:

- a masters program begun in 2007 in health and clinical sciences, as well as a Ph.D. program in health sciences
- initiating similar educational programs for the environment, agriculture, marine, and other sectors
- providing summer project support in diverse life-science fields
- upgrading teachers' skills by developing one high-quality life-science college in every city
- launching an institutional innovation grants scheme
- substantially increasing the number of Ph.D. and postdoctoral fellowships
- creating a national pool of jobs.

4.3 *DBT's technology-mission programs*

Recognizing India's native intellectual capacity, the Ministry of Science and Technology has identified 21 technology missions for integrated technical development that would benefit rural people. The mission covers the areas of plant genetic-resource conservation, the development of new-generation vaccines, biotechnological approaches to herbal product development, genomic research, the development of light transport aircraft, and ocean-thermal-energy conservation. The basic aim of these technology-mission programs is "*Science in the service of common man.*" Of the 21 National Jai Vigyan Missions initiated by various departments, four were launched by the DBT to generate new vaccines, develop herbal products, improve coffee, and establish mirror sites of genomic databases in India.

4.4 *Developing new-generation vaccines and diagnostics*

The main objective of DBT's mission has been to develop candidate vaccines for cholera, rabies, Japanese encephalitis, tuberculosis, malaria, and HIV infections using novel strategies. Such strategies include recombinant proteins; DNA vaccines; recombinant/peptide vaccines for cholera, malaria, tuberculosis, Japanese encephalitis, rabies (for animals and humans); and preventive/therapeutic DNA candidate vaccines for HIV infection.

Current work in support of this mission includes:

- **cholera vaccine.** An indigenous recombinant oral vaccine based on the VA 1.3 strain of Vibrio cholerae was jointly developed and tested by the National Institute of Cholera and Enteric Diseases, Kolkata (NICED); the Institute of Microbial Technology, Chandigarh (IMTECH); SAS, Kolkata; SGPGIMS, Lucknow; and PGIMER, Chandigarh. Phase I clinical trial results indicate that the vaccine is safe, and an extended Phase I/Phase IIa study is currently underway in about 1,000 volunteers. Simultaneously, site preparation work in Kolkata for Phase III clinical trials has been initiated by determining the baseline antibody levels in a cohort. The IMTECH Chandigarh is also scaling up the production of the VA 1.3 strain of V. cholerae.

- **DNA rabies vaccine.** Rabies continues to be a serious public-health problem in many countries, especially poorer ones. An indigenous, unique, low-cost antirabies vaccine has been jointly developed by the Indian Institute of Science (IISc) and Indian Immunologicals Ltd (IIL). The world's first combination rabies vaccine, it contains DNA vaccine and a low dose of cell-culture vaccine. Costing much less than the existing vaccine (Rs 300-400), this new vaccine will be affordable for India's people. In addition, it may be stable at room temperature, which would make refrigeration unnecessary. Human trials are being initiated.
- **Japanese encephalitis (JE).** This candidate DNA vaccine for JE virus was developed by the National Institute of Immunology, New Delhi. The tissue-cultured vaccine could provide about 70% protection in animals following intracereberal challenge. This new vaccine will be able to replace the existing Japanese encephalitis vaccine.

4.5 *Public–private partnerships*

The New Millennium Indian Technology Leadership Initiative (NMITLI) is an innovative public–private partnership started by CSIR in 2000 to make India a global leader in the field of science and technology. The strategy of the NMITLI is to catalyze innovation centered in scientific and technological developments in order to allow Indian industry to attain a global leadership position in selected niche areas. The Initiative seeks to identify and synchronize the strengths of publicly funded R&D institutions, academia, and industry. NMITLI supports two types of projects: those initiated by the program and those initiated by industry. In both types of projects, the best public institutions and industry are identified and a joint project formulated. To date, more than 40 projects in various fields (biotechnology, bioinformatics, agriculture and plant biotechnology, drugs and pharmaceuticals, and so on), with more 400 groups in R&D labs, academia, and industry, have been supported.

Some areas that have received support include:

- new targets, drug-delivery systems, bioenhancers, and therapeutics for latent Mycobacterium tuberculosis
- novel herbal therapeutics for degenerative disorders
- osteoarthritis and rheumatoid arthritis
- diabetes mellitus type II (NIDDM)
- hepatic disorders and hepato-protective agents
- development of an oral, herbal formulation for the treatment of psoriasis
- a new process for manufacturing Tamiflu® (a drug for avian influenza)
- the oral delivery of insulin
- the development of Lysostaphin (a novel biotechnology therapeutic molecule)

One major achievement is the development of the new antimycobacterial molecule Sudoterb (LL 4858) by Lupin laboratories in collaboration with other R&D partners. Sudoterb is the first anti-TB drug in the past 40 years. Tests in laboratory animals have shown that, when given in combination with conventional drugs like rifampicin and pyrazinamide, Sudoterb was able to reduce the duration of TB treatment, from the current six to eight months, to three months. The new molecule is undergoing a Phase I clinical trial. Another significant new drug developed through the NMITLI program is LL 4218 (Desoside-P), a single plant-based herbal drug for psoriasis that is undergoing Phase II clinical trials. Currently there is no drug treatment for psoriasis, which affects millions of people the world over. Trials have shown that this Ayurvedic drug was able to reduce psoriasis symptoms by about 70% in 16 weeks. Just Rs 700 million was spent by Lupin Labs to develop this drug. Lupin Labs collaborated on this project with an R&D laboratory (Central Drug Research Institute, Lucknow) and an academic institute (National Institute of Pharmaceutical Education and Research, Chandigarh).

Recognizing the need to support indigenous R&D in the drug and pharma sector, DST initiated the Drugs and Pharmaceuticals Research Program in 1994–95, providing funds of Rs 1,500 million. The program aims to promote R&D collaborations between industry and institutions for all areas of drug R&D that would help indigenous industry pursue innovative R&D and develop new molecules. Support is available for R&D projects proposed by industry, academic institutions, and laboratories. Funding is also provided to establish state-of-the-art facilities for drug R&D in India. In addition, soft loans at a simple interest rate of 3% per annum are being offered to industry with in-house R&D laboratories and nonprofit industrial research organizations. A drug-development promotion board has been set up to run this program.

Funded by the DSIR, New Delhi, the Technology Development and Innovation Program aims to promote the development and demonstration of indigenous technologies, the development of capital goods, and the absorption of imported technologies by Indian industry. The DSIR provides partial financial support to research, development, design, and engineering projects related to new or improved product and process technologies (including those for specialized capital goods) for both domestic and export markets. The program also supports projects that absorb and upgrade imported technology. The partial financial support by DSIR is primarily meant to cover costs for prototype development and pilot plant work, the testing and evaluation of products flowing from such R&D, user trials, and so on. Industry funds a major portion of the cost for these projects.

4.6 *Golden Triangle*

The Golden Triangle partnership was conceived in 2003, when it was decided to set up and provide special budgetary support for an integrated technology mission focused on the development of Ayurveda and traditional medical knowledge that synthesizes modern medicine, traditional medicine, and modern science. The CSIR and ICMR are working with the Department of Ayurveda, Siddha, and Homeopathy to bring out safe, efficacious, and standardized classical products for identified disease conditions. New Ayurvedic and herbal products for diseases of national/global importance are also being pursued. Innovative technologies are being used to develop single and poly-herbal-mineral products, which have the potential for IP protection and commercial exploitation by national/multinational pharma companies.

Some areas identified include Rasayana (rejuvenators/immunomodulators) for healthy aging, joint disorders, memory disorders, bronchial allergy, fertility/infertility, cardiac disorders (cardio-protective and antiatherosclerotic), sleep disorders, and diabetes. Identifying the strengths and weaknesses of existing modern medical products, the strategy seeks to develop new products to address gaps; formulate an appropriate R&D strategy for standardization, quality control, IP, and other related issues; take up toxicity/efficacy studies in government laboratories, medical colleges, and universities; prepare detailed dossiers of effective formulations; and negotiate with an identified industry partner to begin commercialization after clinical trials are carried out using standard protocols.

This ambitious multiagency program proposes to spend more than Rs 350 million in the next three years. Several areas have already been identified and research is underway.

4.7 *Promoting innovation in traditional knowledge*

The Traditional Knowledge Digital Library (TKDL) is a CSIR initiative aimed at providing easy access to traditional Indian systems like yoga, Ayurveda, and Unani. The initiative also is intended to prevent IP piracy and promote innovation through the use of traditional knowledge. TKDL will publish an encyclopedia with more than 30 million pages in electronic format. The encyclopedia will contain information on traditional medicine, along with exhaustive references, photographs of plants, and scanned images from original texts of traditional systems. Traditional text in the original Persian, Hindi/Sanskrit, or other Indian languages is being translated into English, French, German, Japanese, and Spanish. Ten million pages have already been converted into electronic format, which is a big step towards the TKDL's goal of minimizing the biopiracy of India's indigenous wealth.

The TKDL is expected to be an authentic source for patent examiners in major global patent offices (like the U.S. Patent Office) to conduct prior art searches. Currently, examiners are often unable to determine the novelty of inventions based on traditional knowledge/plant-based drugs because they have no ready access to authentic sources. Although well documented in various regional languages, the Indian traditional knowledge sources are readily available to patent examiners from other countries; this has resulted in the granting of patents like the patent on haldi granted by the U.S. Patent Office.

The TKDL encyclopedia should help examiners cross-check the validity and originality of patent applications. It should assist examiners in determining whether an invention is already known and recorded in ancient literature. The availability of the TKDL may also help avoid litigation regarding granted patents, thus saving time and money in litigation. This is especially important for India, which has spent almost US$6 million fighting legal battles against just two patents on turmeric and neem. Significantly, as of 2000, the number of patents on plant-based products granted by the U.S. Patent Office was about 5000, of which an estimated 80% were possibly plants of Indian origin.

To conform to international standards, the TKDL follows the International Patent Classification (IPC) system, having considerably expanded the IPC group AK61K35/78 on medicinal plants to incorporate detailed information about traditional knowledge with a new section titled Traditional Knowledge Resource Classification. The IPC Union of WIPO (World Intellectual Property Organization) is closely associated with this project through a multinational task force. More than 36,000 ancient Ayurvedic formulations have been translated into current scientific/medical terminology, classified as per the modified IPC subclass, and put in digital format. The TKDL has made it possible for all traditional knowledge to be brought under IPC, which should significantly help protect the traditional knowledge of India from being unfairly exploited by others.

4.8 *ICMR as Department of Health Research*

To encourage medical and health research and, more importantly, to ensure better coordination and promotion of India's national health programs, the government is considering upgrading the ICMR to the Department of Health Research (DHR). This would put the ICMR on par with other departments in science and technology.

Creation the DHR will help better coordinate such sister scientific departments as the departments of science and technology; biotechnology; scientific and industrial research; agricultural research and education; and space, atomic energy, and ocean development, all of which are headed by secretaries to the Government of India. The ICMR's collaborative health projects with these departments will be further strengthened, and they will be better placed to foster such complementary interagency partnerships. The secretary of the DHR will also be in a better position to articulate the policies of the Ministry of Health and Family Welfare and to further the Government's programs and policies in this area. During national emergencies, when critical science and technology inputs are required from other agencies, the DHR would function more effectively. Technologies and products developed by other science and technology agencies will transition more easily into the healthcare sector. In addition, a coordinated effort with other agencies in cutting-edge science (stem-cell research, functional genomics, molecular medicine, proteomics, and so on) will be vital for identifying and supporting the best scientists with timely and adequate budgetary support. This effort should send better drugs and devices to market more quickly. The DHR could help translate research results into policy through a vibrant health-research system. Unlike ICMR, the DHR could address labor and infrastructure requirements for medical and health research in India because it would be seamlessly linked with other agencies (and thereby avoiding potential duplication of efforts).

4.9 *Small Business Innovation Research Initiative*

The DBT has introduced a new scheme to boost public–private partnership efforts. It supports both high-risk, pre–proof-of-concept research and late-stage development for small and medium companies led by innovators with backgrounds in science. The Small Business Innovation Research Initiative (SBIRI) has a unique process for generating ideas. Bringing together technology users and producers, it seeks to promote products that could be created only with the help of the private sector. National consultations are to be held every three to six months to generate ideas in different sectors of biotechnology (medical, agriculture, food, industry, and environment).

The SBIRI aims to:

- strengthen private industrial units whose product development is based on in-house innovative R&D
- encourage other smaller businesses to increase their R&D capabilities and capacity
- create opportunities for starting new technology-based or knowledge-based businesses by science entrepreneurs
- use private industries to stimulate innovation and thereby fulfill Government objectives in fostering R&D
- increase private-sector commercialization derived from Government-funded R&D

The scheme covers all areas in biotechnology that are related to healthcare, agriculture, industrial processes, and the environment. This unique scheme, which directly funds industry, is a big boost for small companies. It took off very well: the DBT received 70 proposals in just the first month. In the year 2005–2006, about 12 companies were financed. The DBT is planning to expand the scale of this program to Rs 1000 million per year.

4.10 *National Innovation Foundation*

Created by the Department of Science and Technology, the National Innovation Foundation (NIF) seeks to recognize, respect, and reward grassroots technological innovators and traditional-knowledge experts. Established as an autonomous society in 2000, its mission is to make India an innovative, global leader in sustainable technologies. It was patterned upon the Honey Bee Network established in 1989, which sought to connect creative people across language cultures, acknowledge the contribution of innovators, expand policy and institutional space for local knowledge experts, and ensure the fair sharing of benefits. The honeybee model was chosen for the NIF because it reflects how innovations are collected without making the innovators poorer and how innovators themselves create connections. It provides a platform to foster innovators who have solved a technological problem through their own intellect with little government or industry help. Located at Ahmedabad, Gujarat, the NIF has a corpus fund of about Rs 20 crores, the interest on which is used to fund the activities of NIF.

Similarly, the Gujarat Grassroots Innovations Augmentation Network (GIAN) picks up innovations from the Honey Bee Network database, performs market research, builds links with design, research, and development institutions to improve the technological efficiency of the innovation, helps test the product, and develops business plans and a market-launch strategy. Conceived in 1997 with support from the Government of Gujarat, IIM Ahmedabad and SRISTI, the GIAN helps with filing patents and licensing the innovators' technologies. GIAN now has separate offices in the north (Jaipur), west (Ahmedabad), and northeast (Guwahati) India. Although protecting IP rights still remains difficult, 29 technologies have been licensed since GIAN was launched.

More than 12,000 contemporary innovations and outstanding traditional-knowledge examples/practices have been documented by the network, but none of the innovations documented have led to viable businesses, because the innovators had neither the resources nor the expertise to commercialize their inventions. To address this issue, the NIF was set up in 2000 to help promote these inventions and to build an entire value chain around them. So far, about 37,000 innovations and traditional knowledge examples have been identified from more than 350 country districts. Currently, the NIF database has more than 50,000 innovations from more than 400 districts. The challenge is to incubate these technologies so that they generate commercial and noncommercial opportunities to improve productivity, generate employment, overcome poverty, and conserve the environment.

4.11 *Society of Technology Management*

To steer tech transfer towards a brighter future and promote better tech transfer management, the Society of Technology Management (STEM) was launched at the international workshop on Intellectual Property Rights, Technology Transfer, Licensing, and Commercialization convened by Cornell-in-India and Sathguru Management Consultant on April 17, 2005. The society was conceptualized by a group of visionary professionals to promote best practices among technology management professionals in south Asia.

The objectives of STEM include:

- offering guidance and assistance to inventors and corporations IP matters
- providing learning opportunities to dealing with the real-world aspects of IP law
- increasing the general awareness of IP laws and their increasing importance
- promoting best practices in technology management and engaging in capacity-building among technology management professionals in India and neighboring countries
- catalyzing the professional development of technology managers for the commercial benefits of innovations

STEM hopes to achieve its objectives through a well-formulated strategy that allows genuinely interested Indian researchers and technology experts to network with global technology managers. Annual meetings and seminars will be organized to benefit tech transfer professionals nationwide, and STEM will promote the economic growth of its constituent members and the organizations those professionals represent. STEM has the support of all the major research funding bodies, academic institutions, and private-research enterprises in India. To build links with similar organizations, STEM participated in the Asian tech transfer meeting in Singapore in 2005 and the Association of University Technology Managers (AUTM) meeting in 2006.

The International Federation of Technology Transfer Organizations, the Southern African Research and Innovation Management Association, AUTM, and the Association of European Science and Technology Transfer Professionals have extended their wide support to STEM.

5. INTERNATIONAL COOPERATION FOR CAPACITY BUILDING

India continues to greatly benefit from technical, financial, material, managerial, and human-resource inputs and assistance from international agencies, developed countries, and, more recently, international not-for-profit organizations for capacity building in the healthcare sector. Initially, such assistance was mainly for human-resources development through training, infrastructure development, and financial and material assistance. But as India has advanced in the healthcare sector, the programs have shifted toward capacity building in the community for health delivery and networking, policy frameworks, and so on. These ongoing initiatives encompass a large number of programs and projects (for example, there are more than 30 ongoing programs with more than 700 activities being implemented in collaboration with the World Health Organization [WHO]).

5.1 *International collaboration in promoting technology management*

With the support of the NIH in the United States, the Technology Forecasting and Assessment Council (TIFAC) of the Department of Science & Technology has just initiated a joint program to train young technology managers at the NIH tech transfer system for five weeks. The first batch of two interns was at the NIH in the summer of 2006.

The ICMR, in collaboration with MIHR, organized a very successful joint symposium on TRIPS and Public Health followed by a one-day workshop at the ICMR headquarters, New Delhi. More than 20 young, mid-level scientists and technology transfer professionals participated and shared experiences with Richard Mahoney and Lita Nelson on technology transfer issues. The Government of India has decided to enter into a formal agreement with MIHR to utilize the expertise of U.S. technology managers to train a new cadre of health technology managers.

In agriculture, the major government departments in India engaged in agricultural technology are the Department of Agriculture Research and Education (DARE) and the Department of Biotechnology. DARE coordinates and promotes agricultural research and education. It provides the necessary government links for the Indian Council of Agricultural Research (ICAR), the country's premier research organization with more than 6,000 members and a countrywide network of 47 institutes (four with university status), five national bureaus, 31 national research centers, 12 project directorates, 89 all-India coordinated-research projects, and 38 agriculture universities.

DARE is the nodal agency for international cooperation in the area of agricultural research and education. The department liaises with foreign governments, the United Nations, CGIAR, and other multilateral agencies concerned with agricultural research. DARE coordinates the admission of foreign students in various Indian agricultural universities and ICAR Institutes. Some of its specific activities include:

- international cooperation and assistance in the field of agricultural research and education, including relations with foreign and international agricultural research and educational institutions (It participates in international conferences, associations, and other bodies dealing with agricultural research and education, and follows up on decisions at such international conferences.)
- fundamental, applied, and operational research in higher education, including coordination of such research in agriculture (agroforestry, animal husbandry, dairying and fisheries, agricultural engineering) and horticulture (agricultural statistics, economics, and marketing)
- coordination and determination of food and agricultural standards in higher education, research, and scientific and technical institutions (This includes animal husbandry, dairying, and fisheries.)
- development of human resources in agricultural research/extensions and education
- access for financing to the Indian Council of Agricultural Research, and community research programs other than those relating to tea, coffee, and rubber
- sugarcane research

5.2 *New policy initiatives*

In addition to DARE, recent new policy initiatives include:

- increased allocation for agricultural research
- research program on microorganisms
- one Krishi Vigyan Kendra (KVK) in each district of India
- National Museum on Agricultural Sciences
- National Agricultural Innovation Project (NAIP)
- new intellectual property rights management, that is, new IPR management is being developed to enable the smooth transfer of agricultural technology for benefit sharing with all stake holders.
- Indo-U.S. Agricultural Knowledge Initiative (AKI)
- a range of activities related to human resources and institutional capacity building

With a specific focus on agricultural technology, the following is the proposed work plan under the agreed priority areas:

- **Education, learning resources, curriculum development and training: Building human and institutional capacity and strengthening public–private partnerships.** Private-sector-sponsored chairs in India or the United States will be created for R&D on strategic/niche areas. This will help establish close collaboration between the public and private sectors, which in turn will lead to the commercialization of technologies at a faster pace. In addition, each year, industry scientists and faculty from premier U.S. and Indian business/management schools and agri-business institutions will be invited to a workshop (in India or the United States) to devise synergistic strategies for exploring the emerging trends and needs in the agriculture sectors of both countries. It will seek to orient education, training, and research to contribute to economic growth. The workshop will inventory, upgrade, and build on existing agri-business programs to match students or professionals with practical internship experiences.

- **Food processing, use of by-products and bio-fuels.** The AKI Board agreed that developing agricultural marketing and processing industries is now a priority for India's increasingly need-based, demand-driven, market-oriented agricultural sector. The following initiatives seek to meet this need:

  Joint research programs. Technology to rapidly detect and control biotoxins, chemical contaminants, and heavy metals in agricultural produce and by-products: Food quality and safety are essential for both domestic and export markets. Developing or acquiring rapid test equipment and protocols to ensure food quality at various points in the value chain would be developed through training and joint-applied research programs.

  Biotechnology. The Initiative recognizes that both partner countries share the common goal of translating lab results into beneficial products delivered to farmers. Subject to funding from the U.S. and Indian governments, and bearing in mind possible private sector engagement, focus will be on transgenic crops, genomic, molecular breeding, diagnostics, and vaccines and training.

  Water management. The improvement of water quality and water-use efficiency will be vital to the continued growth and productivity of the agricultural sector in both India and the United States. The Board agreed to cooperate on capacity building and joint research activities to develop improved technologies and management practices in a framework that incorporates the needs of multiple stakeholders from lab to farm.

6. CONCLUSIONS

The transition of India from a protected economy to be an open, global-economic power has prompted India to take a series of steps to face the new challenges of globalization. All the public sector science and technology agencies have realized the importance of IP and its creative management and have initiated steps toward generation of knowledge that could be IP protected. This is especially important as the health products of diseases of poor countries need to be indigenously developed in view of the lack of interest by large multinational companies that have little interest in the development of such products. Public–private partnerships with both Indian and foreign collaborators is being explored with some measure of success. In addition, active steps are being taken to strengthen IP protection systems and policies and also to create a trained cadre of technology transfer professionals in the areas of health and agriculture. An important means of skill building in the area of IP include international collaboration and networking with agencies abroad. Early experience has shown that it is only through indigenous development that new health products could be developed, introduced, and marketed. Strengthening R&D and establishing policies for the creation and management of IP and public-partnerships are important steps for making available products of public-health importance in all poor countries. ■

KANIKARAM SATYANARAYANA, *Chief, IP Rights Unit, Indian Council of Medical Research, Ramalingaswami Bhawan, Ansari Nagar, New Delhi 110029, India.* *kanikaram_s@yahoo.com*

CHAPTER 17.6

Current Issues of IP Management for Health and Agriculture in Japan

JUNKO CHAPMAN, *Research Associate, MIHR, U.K.*
KAZUO N. WATANABE, *Professor, University of Tsukuba, Japan*

ABSTRACT

This chapter describes current and historical trends and issues related to intellectual property (IP) management in Japan. It gives a history of Japan's national IP system in order to provide an understanding of the nature of the system and why and how it was established. The chapter also describes current government efforts to provide insights into the system's future. With regard to current IP issues, two topical issues are discussed: industry-university collaboration on R&D and employees' inventions. Japan's efforts to resolve these issues may be helpful for other countries that are grappling with similar issues.

The chapter also details health and agricultural IP issues in Japan. It discusses and compares with the practices of other countries the patentability of medical methods and exemptions for the experimental use of patented products. Furthermore, the chapter offers an overview of Japan's national policy on agricultural R&D and bioresource centers (the functioning of which greatly involves the transfer of materials with IP rights). RIKEN (The Institute of Physical and Chemical Research) is offered as a case study to clarify the policies and issues discussed.

Finally, for the benefit of other countries that are coming to terms with IP management issues, the chapter offers some lessons learned by Japan that have helped shape its national IP policy, strategy, and institutional IP management.

1. INTRODUCTION

Japan's recognition of the importance of IP—and the importance of good IP management to economic and scientific development—at one time lagged behind that of other developed countries. This was partly due to Japan's national isolation policy, in effect between 1603 and 1867, a time during which other advanced countries were beginning to establish their patent systems. Once international trade resumed in Japan, it established its own patent system, incorporating standards set by other countries and adapting them to domestic circumstances. Since the 1980s, Japan's national IP policy has changed significantly. Former Prime Minister Junichiro Koizumi's policy of "*Chitekizaisan-Rikkoku* (Nation Built on IP)" in 2002 reflected the country's new pro-patent policy. Since 2002, IP policy and a legal framework for IP rights protection have been reasonably well established for all categories of industrial invention.

In pursuing this recent national IP policy and strategy, however, issues have been raised by various stakeholders, involving industry–academia collaborative partnerships and the status of employees' inventions. To address the former, the Japanese government has made great efforts over the last decade to promote university–industry partnerships to effectively commercialize research results. In regard to employees' inventions, provisions in Japan's patent law were enacted rather early in its patent-legislation history. After several revisions, the current provisions came into effect in April 2005. Still, even after these revisions, several lawsuits by former employees claiming better remuneration from their

Chapman J and KN Watanabe. 2007. Current Issues of IP Management for Health and Agriculture in Japan. In *Intellectual Property Management in Health and Agricultural Innovation: A Handbook of Best Practices* (eds. A Krattiger, RT Mahoney, L Nelsen, et al.). MIHR: Oxford, U.K., and PIPRA: Davis, U.S.A. Available online at www.ipHandbook.org.

employers for their inventions have raised significant debate.

Japan's status as a highly industrialized, developed country has been achieved partly through an IP rights protection system that, since 1975, has been harmonized with major international legal instruments,[1] including the World Intellectual Property Organization (WIPO). Japan participates in the following treaties associated with IP laws: the Paris Convention for the Protection of Industrial Property (1899 [years in parenthesis are those when Japan ratified/acceded to the convention or institution]); the Bern Convention for the Protection of Literary and Artistic Works (1899); the Universal Copyright Convention of 1952 under the United Nations Educational, Scientific, and Cultural Organization (UNESCO, 1956); the Patent Cooperation Treaty (PCT, 1978); and the Union for the Protection of New Varieties of Plants Convention (UPOV,1982). Japan has a branch office of AIPPI (Association Internationale pour la Protection de la Propriété Industrielle) (1956), called AIPPI-JAPAN.[2] The country is a member of the Convention on Biological Diversity (CBD, 1993), which emphasizes the importance of genetic resources, traditional knowledge, and access and benefit-sharing—including IP rights protection.

On the other hand, Japan has not signed the International Treaty on Plant Genetic Resources for Food and Agriculture (PGRFA).[3] These abstentions are principally due to concerns about the protection of IP rights that may not synchronize with WIPO and the Agreement on Trade-Related Aspects of Intellectual Property Rights (TRIPS) under the World Trade Organization (WTO). In the near future, when IP matters are better understood in domestic debates and corresponding laws are made, Japan may agree to actively participate in these major international treaties.

In addition, IP laws in Japan have peculiarities with regard to health and agriculture: 1) some aspects of medical technology, such as surgical operation methods, cannot be protected due to public equity concerns in IP laws (this is not the case in the United States); and 2) as in the majority of developing nations, traditional knowledge in agriculture is recognized as a public good.

2. JAPAN'S IP POLICY AND STRATEGY

Japan's IP policy and strategy developed from a relatively primitive level through the formation, addition, and revision of patent laws since the Meiji era (1868–1912), when Japan abandoned its policy of national isolation after the Edo era (1603–1867). For more than 200 years (1616–1854), the government had banned foreign contact, except for very limited contact with only a few countries.[4] Japan refused to import or utilize advanced technologies developed in the United States and Europe. After reopening the country to trade in 1858, however, Japan began to work to catch up with industrially advanced countries by introducing invention-promotion systems and a national patent system.

During the last five years, in addition to developing patent laws, the government has promoted its national IP policy and strategy by developing general national frameworks and establishing a special function in the Cabinet. All of this was initiated by former Prime Minister Koizumi.

2.1 *History of Japanese patent law*

In 1624, England adopted a patent ordinance that is the basis for today's British patent system. The adoption of this first patent ordinance was followed by the adoption of patent legislation in the United States in 1790, and in France in 1791.[5] During this period (the Edo era), Japan pursued a policy of national isolation, and the manufacturing of new products based on technologies developed in European countries and in the United States was prohibited. In the 1870s, the Meiji government sought to establish the Japan's first patent law.[6]

In 1871, the first patent law — known as the Exclusive Right Law—was passed and enacted. The government, however, was not prepared to implement such a law: there was no government office to accept patent applications and no officials to handle them. Furthermore, the public were generally against proprietary inventions, and so the new patent system was not widely accepted. Ultimately useless, the law was abolished one year after it was passed. Without a patent law, imitations and misappropriations of inventions were widespread, and inventors frequently lost profits from royalties. In 1885, a new patent law was passed that followed the U.S. and French patent laws. Having learned from the failure of the Exclusive Right Law, the government established a patent bureau in the Agriculture and Commerce Ministry and staffed it with a director, three judges, an examiner, and an assistant examiner. By 1899, the bureau had expanded to five judges, 15 examiners, and 20 assistant examiners; and the number of patent applications was doubled in 1887, reaching 1,515 in 1899.[7] The patent ordinance, however, was still imperfect and far from its modern version.

Since 1887, Japan's patent system and law have been revised many times, mainly because of pressure from domestic proponents and developed countries. The modernization of the patent law began in 1921 through a revision that aimed to accommodate the increased demand for Japanese products as substitutes for foreign products during World War I (international trade had been suspended and high-quality foreign technologies and materials could not reach Japan during those years [1914–1918]). After World War II (1945–1949), Japan's principal economic objective was "*quantitative recovery, ignoring efficiency.*" This changed only after the 1950s, when economic control and subsidies were gradually abolished, the market mechanism was largely restored, private international trade began, political independence was regained under the San Francisco Peace Treaty (1951), and U.S. economic assistance to Japan ended.[8]

Japanese industry began to pursue efficiency and competitiveness, which required cost reductions and higher-quality products. Moreover, "*it was a time when the number of patent applications resulting from active industrial investment in research and development was increasing, causing a variety of problems to emerge, such as late examination, etc.*"[9] Despite these circumstances, the patent law remained unchanged until 1959. The revision in 1959 was intended to cope with the needs of a newly liberalized economy and developments within international patent systems. More revisions followed in 1970, when technological development had become increasingly rapid and industrial property issues were extremely significant for Japan.

Japan's rapid economic growth stalled in the early 1970s, demonstrating that Japan had caught up with developed countries and had matured economically and industrially. At such a point in a modern economy's development, the economy can no longer grow through imitation but must innovate to spur growth. Japan's revision of patent law in 1975 aimed not only at the creation of new technologies but also at international harmonization. The revision included a substance-patent system and a multiclaim-application process.

As international harmonization proceeded in the 1980s and 1990s, various kinds of new institutions for pro-patent policies were introduced. The most influential factor was pressure from advanced countries represented by the United States, which feared the incremental rise of Japan's export market and strongly promoted a domestic pro-patent policy during that period. Local voices called for the strengthening of Japan's patent system to further development and prevent an increasing risk of the country's original technologies and products being copied abroad, especially by developing countries, such as China, that were trying to catch up with developed countries.[10] Japan's pro-patent policy has expanded the scope of patent protection, extended the patent period for pharmaceutical products, and strengthened deterrence against infringement.

In 1990, the Japan Patent Office (JPO) was the first patent office in the world to start a paperless system to accept and handle patent applications.

2.2 *Recent IP policy and strategy*

Having recognized its need for more creative and advanced technological innovations, Japan has emphasized a pro-patent policy since the early 1990s. In line with this position, former Prime Minister Junichiro Koizumi's policy statement in February 2002, proclaiming that he would make Japan a country built on IP, followed the passage in 1998 of a "law on promoting technology transfer from universities to industry," so-called "TLO Law," and the Japanese version of the Bayh-

Dole Act (Article 30 of the 1999 Law of Special Measures for Industrial Revitalization).[11]

During its period of high economic growth, Japan had been good at exporting technologies based on imported technologies. After reaching the global technological frontier, however, Japan's advantage came under attack, especially by neighboring countries, such as China, that had plentiful, cheap labor and increasing technical and economical power. Japan suffered from an economic recession in the 1990s and created a plan to break the impasse of the recession. The *Chitekizaisan-Rikkoku* plan would add value to the technologies, products, and culture created in Japan for export overseas by further strengthening the nation's IP regime and management. This entailed specific, concrete provisions for planning and policy implementation.

Having been regarded as fundamental for national development, the former patent system had been established largely to stimulate domestic industries. Under the *Chitekizaisan-Rikkoku* plan, Japan began to make more substantial efforts to develop and implement an IP strategy, focusing on IP rights generated not only from the private sector but also the public/university sector.

In March 2002, one month after the government's policy statement, the prime minister's cabinet inaugurated the Strategic Council on Intellectual Property, which discussed the details of the plan. The Council created an Intellectual Property Policy Outline in July 2002.[12] It referred to an "*intellectual creation cycle*": the cycle of the creation, protection/establishment, and exploitation/utilization of IP/IP rights (Figure 1). Aligned with other global IP systems, the cycle established a mechanism to create high-quality IP protected by patents. Protected IP is exploited throughout society, and the resulting profits are used to recoup the cost of original R&D and to invest in the creation of new IP. The cycle is considered fundamental to the government's intellectual property policy outline and to Japan's recent IP strategy.

Furthermore, the December 2002 Basic Law on Intellectual Property[14] was promulgated in pursuit of implementing the IP strategy and

FIGURE 1: INTELLECTUAL CREATION CYCLE

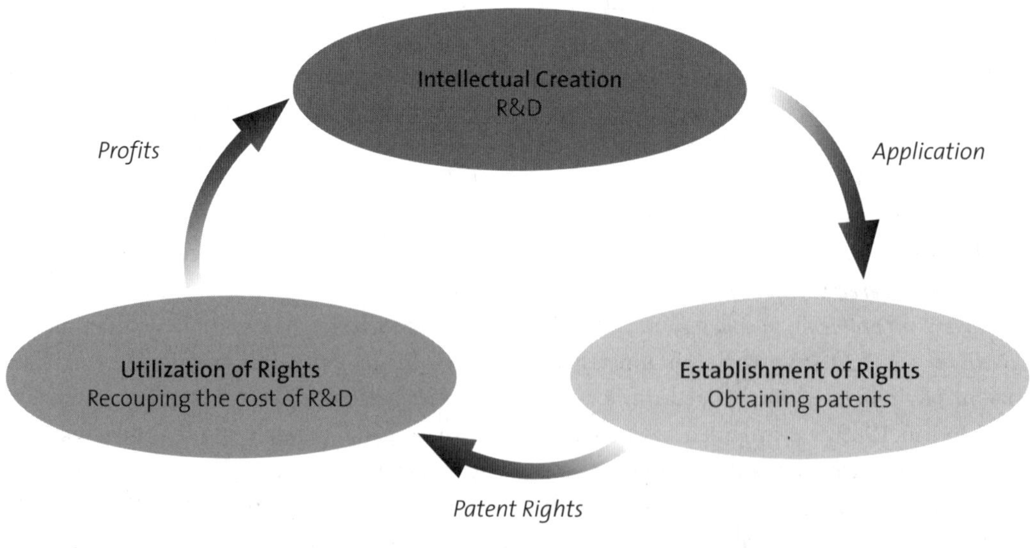

Source: METI[13]

stipulated the establishment of the Intellectual Property Policy Headquarters (established March 2003 in the cabinet). In July 2003, at the fifth meeting of the Intellectual Property Policy Headquarters, a promotion program (called the "Program for Promoting the Creation, Protection, and Exploitation of Intellectual Properties"[15]) was initiated. This program set out specific goals and time frames for implementing the new IP strategy. The program has been implemented and reported upon annually since then as the "Intellectual Property Strategic Program."[16] The reports are composed of five sections: Creation (of IP), Protection (of IP), Exploitation (of IP), Expansion of Content Business, and Developing Human Resources and Improving Public Awareness.

3. INDUSTRY–UNIVERSITY COLLABORATION OF R&D

Japanese central and local governments have promoted partnerships among industry, academia, and government—particularly between industry and academia. Industry provides information on public or market needs; academia provides the seeds for commercializing technology (that is, inventions); and the government plays the role of agent or mediator between industry and academia.

Measured in terms of publications and in acquiring publicly available competitive grants, national universities have been the leading academic institutions in basic research. Out of more than 500 universities registered by the Ministry of Education, Culture, Sports, Science, and Technology (MEXT), the top 20 universities acquiring extramural funding are national universities involved in all fields of research. In the medical, pharmaceutical, and physical sciences, certain private universities have an advantage over others due to specialization, but national universities generally lead. National universities have also been more engaged in collaborations with industry for some time. In 2004, 92.2% of national universities had established an office for cooperation with industry, such as a technology transfer office (TTO) or a technology licensing office (TLO); this compares to only 42.8% of private universities and 59.6% of national research institutes.[17] However, the effectiveness of such collaboration has been hindered due to unclear R&D policies with industry, poor IP controls, lack of incentives for researchers at universities, legal constraints stemming from the nature of national universities, and general administrative slowness.

At leading private universities, implementing industry–university collaborations has been much easier due to the relative ease of contractual negotiations, administrative procedures, and the lack of restrictions on the dissemination and use of funds. Still, only a limited number of private universities have been able to accommodate very active collaborations.

3.1 *Reforming national universities*

In 2004, all national universities were separated from the direct supervision of MEXT and became independently managed administrative institutions. Currently 89 national universities and four educational research institutions have reformed. The numbers will be further reduced by mergers and acquisitions. The key aspects of increased independence are: (1) all decision making can be made by each university's administration and council instead of requiring approval from MEXT; (2) a medium-term plan for each six years is used as an achievement evaluation point; (3) funding is granted by MEXT based on the medium-term plan; (4) profit acquiring and commercial activities are permitted; (5) academic faculty members have more flexibility in creating business ventures; (6) TTO and IP controls are enforced at each institution, with the resulting expectation that university–industry collaboration will be boosted; and (7) faculty members are provided incentives to innovate. Despite all this, the overall system still needs to be revised, and governance needs to be improved to enhance the implementation of R&D and technology transfer from academic institutions.

With the reform of the national universities, the government now increasingly promotes academic institutions to enhance industry–academic institution collaborations and to establish TTOs.

The development of small business ventures by faculty members has also been encouraged in order to commercialize their research. According to the Intellectual Property Strategic Program 2005, the number of new venture companies derived from universities was 199 in fiscal year 2003, and 129 in fiscal year 2004, for a total of 1,112 by the end of fiscal year 2004. Universities provide support grants for such business attempts, but often overall strategic plans are missing on the university side. Insufficient consideration is given to IP rights, which are a strong driving force toward venture-business success. Each university has taken its own approach to alleviating these weaknesses.

3.2 *Technology/IP rights transfer between universities and the private sector*

Under Japan's former national academic institution system, it had been difficult to exploit IP rights because: (1) IP rights, particularly patents, were owned by the Japanese government; and (2) many academic institutions lacked the systematic capacity to form university–industry liaisons. The old national university system deterred the promotion of invention and proper legal handling. Additionally, it seems that universities did not give scholars much incentive to innovate and invent. Faculty members also often would abandon patent applications due to high costs and the university's propensity for rejecting patent applications. Instead, faculty members often allowed ownership rights to be transferred to the private sector in return for gift donations for their research. This in turn hindered the development of research and business opportunities from universities. A survey of the top-ten major national universities in terms of extramural research-grant acquisition, revealed that legal and administrative systems often lagged far behind the private sector's ability to facilitate collaboration or complex contractual matters.

Due to changes in the law, however, the past ten years have seen robust growth in the establishment of TTOs at universities. University IP offices take care of governance issues, and TTOs support the technology transfer process. In general, university TTOs have four functions: (1) IP rights protection, (2) marketing of university-derived technologies, (3) licensing, and (4) promotion of commercial ventures by faculty members. TTOs have been legally supported by the government since 1998. The TTOs that have been approved by MEXT and the Ministry of Economic, Trade, and Industry (METI), "approved" TLOs are entitled to special treatment under the TLO Law and the "Japanese Bayh-Dole" Law. The treatment could include direct funding by ministries and free or discounted fees for the maintenance of patent rights and examination requests. Between 1998 and April 2006, 41 such TLOs were established. In addition, there were four "accredited" TLOs as of April 2006. These TLOs are assigned nationally owned patents and then out-license them, while the approved TLOs register patents for university faculty—and exploit their inventions. Guidelines and reports for these TTOs have been published.[18] (A detailed list of approved and accredited TLOs is available upon request from the authors and from the METI JPO Web site.[19])

Japanese universities are recognizing the importance of their own IP for commercializing research and establishing technology-based companies. There are an increasing number of university-derived companies (generally referred to as spinouts), particularly in the area of biotechnology, compared with five years ago.[20]

3.3 *Human resource needs*

Generally, the key IP issue at academic institutions is processing ability. Establishing a contract on applied R&D takes time and requires specialists on legal matters. Universities are short of practical lawyers and officers, and it is common for most of the officers to be transferred to a different section of the university within two or three years, which prevents these individuals from gaining sufficient skills and knowledge.[21] This hinders efforts to implement and disseminate research results and applications promptly and smoothly. Japanese universities are in great need of institutional reform related to the administration of contractual matters and industry–university collaborations.

While the number of patent attorneys who specialize in various disciplines of modern technologies in Japan has dramatically increased, an overall understanding of the IP management by patent attorneys is crucial. Patent attorneys may have specific know-how related to recent changes in the patent law, but joint activities with lawyers are often required to identify or challenge infringements of IP rights. With regard to the commercial aspects of IP management, much needed are multiskilled specialists who are competent in both the legal and technical aspects of technology transfer in marketing, licensing, and integration of IP rights.

The Intellectual Property Strategic Programs emphasize using university infrastructure to develop IP specialists. Such individuals are not only needed to manage IP in universities but also in the wider business market. Multidisciplinary graduate school programs are increasingly being offered at many universities, but professionals with such know-how are still few, so TTOs often offer seminars/workshops on IP education and practical operations for their faculty members and senior graduate students. Through these efforts, IP courses are becoming popular at many universities.

4. EMPLOYEES' INVENTIONS

Information surveys, such as those published by the Mitsubishi Research Institute,[22] point to Japan's lack of strong incentives for researchers and engineers as a potential pitfall. The problem is caused by the weak support for employees' inventions created through the work service. Article 35 of Japan's patent law defines employees' inventions, but the law is often criticized for not promoting employees based on their record of inventions and formal IP, especially at public institutions. Compared to the United States, where public institutions file for many patents, relatively few patents are filed by Japan's public institutions. This is true especially for the national universities. Instead, Japanese academia recognizes and rewards publishing, which is used as almost the sole criterion for promotion.[23] MEXT and its subsidiary organization, JSPS (Japan Society for the Promotion of Science), have noted the low number of patent filings at academic institutions and have used grants to encourage promotions based on the patenting of inventions.[24] Over the last few years, patent filing and registration have drastically increased under the Research for the Future Program promoted by MEXT and JSPS.[25]

The debate between employers and employees about their proportional ownership of inventions at universities dates back to the 1970s.[26] After some argument, MEXT reported in 2002 that inventions created by faculty at universities should be owned by universities. This principle has since been the basis of university IP management strategies. Meanwhile, according to a survey conducted in 1997 by the Japan Institute of Invention and Innovation (JIII),[27] more private companies have been providing relevant regulations and rules and have been increasing remuneration and employee incentives to generate inventions. The survey revealed that:

1. An increasing number of companies have regulations and rules established.
2. Remuneration is made at different milestones, such as patent application, patent registration, and exploitation/working. The proportion of companies adopting such remuneration rules has increased for all the milestones.
3. The amount of remuneration is fixed for some companies; others value it in proportion to the profit acquired from the invention.
4. In both cases, the average amount of remuneration generally has increased.

The recognition and awareness of employees' inventions and their remuneration have been rising for the last decade; nevertheless, various issues remain.

4.1 *Laws on employees' inventions*

The Patent Law of 1909 gave the patent right to an employee invention to his or her employer, but ownership reverted to the employee under the 1921 Patent Law. The 1921 law aimed to protect employees by ensuring that they received reasonable remuneration when the right of ownership was passed to the employer (in accordance with contracts made in advance).[28] The Patent Law of 1959, Article 35, revisited these provisions governing employees' inventions. The law declared that if an employee's patented invention was classified as an *"employee's invention"* (as defined in the patent law[29]), the employer had the right to a nonexclusive license. The same law stipulated that the employee is entitled to reasonable remuneration if he or she assigns the patent right, or an exclusive right to such invention, to the employer in accordance with contracts, regulations, and other stipulations. The law also provided that the remuneration amounts would be decided by referring to the profits that the employer would make from the invention and to the amount of the employer's contribution to the invention.

4.2 *New policy and strategy on employees' inventions*

As mentioned earlier, the Intellectual Property Strategic Program 2003 was adopted in July 2003. The Creation part of the program states the following provision to employees' inventions:

Abolishing or Amending the Provision Regarding Employees' Inventions under the Patent Law.

For the purpose of securing R&D incentive for inventors, reducing patent management cost and risk in individual companies, and strengthening the industrial competitiveness of Japanese industry, the GOJ (Government of Japan) will consider necessary issues on an employee's invention, while taking into account the changes in the social environment, and submit a bill to abolish or amend the provision in Article 35 of the Patent Law to the ordinary session of the Diet in 2004.

Consequently, in December 2003, a METI committee of professionals from universities and from the public and private sectors, with expertise in law and in science and technology, created a report titled "What employees' inventions should be."[30] The report suggested amending the provision regarding employees' inventions instead of abolishing it. The National Forum for Intellectual Property Strategy appeared at the same time. With a range of expertise including lawyers/patent attorneys, research scientists, business executives, and journalists, members of the forum asserted that the provision should be abolished. The details of the various views are discussed below. Based on the METI committee's report, the amendment of Article 35 of the patent law went into effect in April 2005.[31]

4.3 *Amendment of Article 35*

An employee's invention is defined in the law as an invention "*which by reason of its nature falls within the scope of the business of the employer, etc. and an act or acts resulting in the invention were part of the present or past duties of the employee, etc. performed on behalf of the employer, etc. (Article 35.1).*" In other words, an employee's invention results from R&D conducted by an employee as part of his or her work within the scope of the employer's business. There are two other types of inventions mentioned in the law: those created by an employee, but outside of his or her work service, and those created by an employee outside of the employer's scope of business. These differences in the three types are explained in the provisions. Although the employee's invention is created by the employee's own efforts and abilities, the employer contributed to the creation by providing salary, facilities, equipment, and expenses. Considering such contributions, the law provides that the employer shall have a nonexclusive license on the patent right in order to gain appropriate remuneration (Article 35.1).

Article 35.2 stipulates that, provided the invention is the employee's invention, the contractual provision, service regulation, or other stipulation made in advance shall be valid, and the employer shall be given the right to the patent or to the exclusive license. This provision is said to protect employees from being exploited if inventions fall outside of the scope of the employee's invention. The employee shall have the right to reasonable remuneration when he or she has transferred the right to the employer in accordance with the contract, service regulations, or other stipulations (Article 35.3).

Although there has been no amendment for Article 35.1 to 35.3 since the 1959 Patent Law, the subsequent two sub-clauses, Article 35.4 and 35.5, were amended. As mentioned above,

Article 35.4 of the 1959 Patent Law stipulated that thc amount of thc remuneration shall be decided by referring to the profits that the employer will make from the invention and to the employer's contributions to making the invention. The new Patent Law of 2005 stipulates that when the contractual provision, service regulation, or other stipulation between the employer and employee determines the criteria for remuneration, the criteria should be reasonable. Reasonableness shall be determined by considering the decision process of the criteria, such as the conditions of discussion between the employer and employee, hearing of the employees' views on the calculation, and the disclosure status of the criteria.

If judged as unreasonable in accordance with Article 35.4, the amount of remuneration shall be decided in light of the profit, expenses, and other contributions of the employer regarding the invention, the treatment of the employee, and other circumstances (Article 35.5).

4.4 *Current issues regarding employee's inventions*

In the last few years, the increasing number of employees who have resigned from their companies have been suing their former employers due to dissatisfaction with the remuneration paid for inventions the employees created during their employment. The surge in the number of lawsuits reflects an increasing awareness of IP among employees and has aroused the public's interest in IP and employees' inventions. The most famous case, known for the exceptional amount claimed by the employee, is the lawsuit between Dr. Shuji Nakamura and his former employer, Nichia Corp., Ltd., a chemical maker, concerning his invention of a blue light-emitting diode (LED). Originally claiming 20 billion JPY, the court decided Dr. Nakamura was entitled to receive about 600 million JPY (plus interest payments of about 240 million JPY) from his former employer. The case had been reviewed by the Tokyo District Court (2004) and the Tokyo High Court (2005) before it was settled in 2005. It is noteworthy that there was an enormous difference between the percentages that the two courts identified as Dr. Nakamura's contribution: 50% in the district court and 5% in the high court. Dr. Nakamura certainly lost a large amount, but generally the case is considered to be a victory for the employee.

Over the last few years, other former employees have gained more than their former employers had expected to pay. The Japan Intellectual Property Association (JIPA)[32] cautions against extreme legal moves to support remunerations for employees' inventions because overestimated valuation of inventions may destroy some employer companies. The purpose of Article 35 is primarily to appropriately balance the interests of employers and employees. Both the employer and employee require significant—and often different—incentives to ensure that appropriate, relevant investments are made to enable and stimulate innovation.

History suggests that the provisions for employees' inventions under the patent law have been ineffective. Some groups, such as the National Forum of Intellectual Property Strategy, and some private companies fearing huge employee remuneration costs have argued that Article 35 should be abolished or, at least, amended.[33] The critics contend that the individual contractual provision, service regulation, and other stipulations made in advance between the employer and employee (or individual agreements) should be considered reasonable unless they were made under conditions of fraud, duress, or other unreasonable processes.[34] Individual agreements, not Article 35 *per se*, should be applied to settle disputes between the employer and employee regarding the employee's invention.

The same critics argue that the following issues regarding employee's invention under the current Patent Law (Article 35) are also important:

- Criteria for calculating the amount of remuneration have varied from court to court and from case to case. Without any rigid criteria, the decision is vulnerable to the subjective calculations of the judge (as seen in the Nakamura case)
- Criteria for judgment of an "unreasonable" payment in accordance with Article 35.4 are obscure

- Ultimately, it is dubious whether or not a court has the ability and capacity to judge the reasonableness and appropriateness of the remuneration amount

5. HEALTH-RELATED ISSUES

5.1 *Patent protection on methods for medical activities or practices*

5.1.1 *Patentability and unpatentability*

In Japan, medical methods are out of the scope of patentability; however, pharmaceutical products and medical equipment products are patentable. This is inconsistent with U.S. and E.U. practices. In the United States, methods relating to medical activities and practices are generally patentable. Under 35 U.S.C. 287 (c)(1),[35] however, a medical practitioner can use patented medical methods without risking infringement. In the European Union, under the European Patent Convention (EPC), Article 52 (Patentable inventions)[36] stipulates that methods to treat the human as well as animal body by surgery or therapy, as well as diagnostic methods practiced on the body, shall not be regarded as inventions that can be applied industrially. In other words, the methods of operation, treatment, and diagnosis of the human body are not protected by patent rights. However, as an exception to that rule, the first two of the three stages in diagnostic methods: data collection, their comparison, and decision making of medical treatment, have been interpreted as patentable according to the EPC.[37]

In Japan, first and most fundamentally, medical methods fall out of the scope of patentability according to patent law Article 29 (1).[38] In other words, they are regarded as inventions that are *not* industrially applicable because of their humanitarian implications in the medical field. It was feared that patients' wellbeing might be jeopardized by patent protection, which could have effectively deterred medical practitioners from utilizing certain methods if they did not have a license from the patent owner. Secondly, medical activities including R&D are generally regarded as being not for profit, and it is widely held that incentives should be based on academic appraisal and rewards rather than economic gain. Additionally, innovation in the medical field was largely conducted by universities and public institutions that were sufficiently funded by the public sector, which eliminates the need to rely on the modern, private model of patenting and receiving royalty earnings from licensing.[39] Consequently, the decision was made that medical methods should be excluded from patent protection.

However, many players in both academia and industry regard this decision as outdated because of various changes that have taken place in Japan over the last decade.

5.1.2 *Trends in perspective*

The most prominent issue relating to the nonpatentability of medical methods is the lack of incentives for pursuing costly, risky innovation in the medical field. In addition to the major roles of universities and public research institutions in medical innovation, bioventures (biotechnology ventures) and spinouts have increased their role over the last decade because of the increased recognition of IP rights and the establishment of TTOs in universities and public research institutions. Needless to say, such privately run companies cannot expect public funds to cover the costs of this increasing investment, much of which is directed at the universities and public institutions. Instead, it is increasingly expected that investment costs will be covered by patenting and licensing. However, companies have no way to generate returns on investments into medical method inventions. Moreover, their inventions can be easily copied and utilized freely by others. Not surprisingly, therefore, potential bioventure companies are not eager to enter the field.[40] In the absence of actively nurturing this sector, many believe that Japan's competitiveness in this field will weaken because investments in medical innovation will always be deterred. In the long run, patients may lack access to new, highly effective diagnosis or treatment methods that could be developed locally. There may also be negative economic consequences.

Some critics argue that excluding methods and processes from patent protection does not comply with the TRIPS Agreement, which stipulates that patents shall be available for all inventions, whether products or processes, in all fields of technology, provided they are new, involve an inventive step, and are capable of industrial application (Article 27).[41]

Thus, it is increasingly felt that not just medical products, but also methods, should be considered inventions with industrial application that should be given patent protection.

Based on the above analyses, the government of Japan is reconsidering patent protection for medical methods. In response to recent changes in circumstances and views, the government established a task force on "*the protection of patents of medical-related acts.*" The task force committee was established under the Intellectual Property Policy Headquarters and began consultations in October 2003.

The main purpose of the meetings was to discuss whether or not medical methods should be covered by patent protection. The committee published a summary report of their discussions in November 2004,[42] which involved hearings from not only committee members but also other professionals from various fields, such as medical science, the medical industry, medical economists, and the legal field. The report also included public comment. After 11 meetings, the summary report made the following recommendations:

- From a humanitarian standpoint, the methods relating to medical activities by medical practitioners should be excluded from patent protection.
- Operational methods of medical equipment should be covered under the scope of patent protection, with the exception of those related to medical activities by medical practitioners.
- With regard to methods for generating new potent and efficacious medicines for production and sale, the possibility of expanding patent protection should be pursued by allowing product patents rather than process patents to begin with. Process patents could be discussed and pursued later on. The limited protection reflects the potentially obscure distinction between medical activities by medical practitioners and others.

In April 2005, based on the committee's recommendations, the government amended the practical examination criteria of medical inventions for patents and utility models.[43] The amendment makes explicit provisions for patenting methods and processes related to the use of medical equipment, but methods and processes related to medical activities by medical practitioners are not patentable.[44]

5.1.3 *Issues for the near future*

Although the examination criteria have been amended, some issues and arguments still require resolution. The report recognizes that medical methods for patients who need access to state-of-the-art medical practices should be excluded from patent protection. However, no such law has yet been passed, and legal guidance similar to the U.S. provision in 35 U.S.C. 287 (c)(1) is urgently needed.

Despite the report's conclusion, expanding the scope of process patents in the medical field to cover whole methods is still widely debated. Some argue that amending the examination criteria is insufficient and that Japan's competitiveness in the medical field will not be enhanced without protecting medical process inventions.

5.2 *Limitations of the patent right*

The limitation of the patent right or the exemption from patent infringement for the experimental use of a patented invention affects all fields in science and technology. Given its impact on public health outcomes, however, this limitation is important especially for biotechnological and medical experimentation.

5.2.1 *Background*

Article 69 (1) of the Japanese patent law provides that "*the effects of the patent right shall not extend to the working of the patent right for the purposes of experiment or research* (Limits of Patent Right)." The original purpose for establishing the patent law was "*to encourage inventions by promoting their protection and utilization so as to contribute to the development of industry* (Article 1)," and extending the patent right to experimentation and research is considered contrary to this purpose. Such limitations to the patent right were originally inserted into the patent law of 1909, which was reaffirmed in Article 69 (1) of the patent law of 1959. Article 68 of the patent law also provides that a patentee shall have an exclusive right to "*commercially*" work the patented invention. The word *commercially* leads some to conclude that experiments and research conducted in universities and public research institutions will be excluded from patent protection because they are largely considered as nonprofit.

The patent law, however, does not clearly distinguish between profit and nonprofit purposes in terms of the effects and limits of the patent right. The above interpretation has depended solely upon legal theory, and very few judicial precedents have emerged regarding the interpretation of "*experiment or research*" provided for in Article 69 (1). Therefore, failing to obtain a proper license for utilizing a patented invention in experiments and research in universities and public research institutions can potentially be considered as infringement. Moreover, patent owners have a clear right to require universities and research institutions to obtain licenses for each invention used in their experiment or research. These procedural requirements and the related royalty payments deter researchers. If patent protection extends to experimentation and research linked to technological advancement, it could eventually thwart the evolution of national industry.

5.2.2 *The current situation and precedents*

The accelerated progress of biotechnology, the increased collaboration between academia and industry, and the enhanced awareness of IP strategy among various players over the last decade have heightened concerns over obscurity in the patent law. In the Intellectual Property Strategic Program 2003, the government decided to review and clarify the extent to which experiments or research are exempted from patent infringement. This review would investigate current situations and precedents not only in Japan but elsewhere, and the results would be widely disseminated to both the public and private sectors in order to reduce the possibility of conflict. Composed of experts and leaders from various areas, including executives of private companies, patent attorneys, faculties of universities, and representatives of TTOs, a working group on patent strategy established under the METI in 2003 discussed the issue in a report on issues relating to effective use of patented invention.[45] Completed in November 2004, the report focused principally on three aspects: the experiment or research, generally, clinical trials for approval of generic medicines, and experimentation and research in universities and public research institutions.

According to the report, very few judicial precedents in Japan interpret experiment or research, so guidance has been sought in legal theory instead of judicial rulings. The most widely accepted theory was described by Keiko Someno in 1988.[46] It limits experiment or research to the purpose of *"progress in technology,"* such as the examination of an invention's patentability, the examination of an invention's function, and experiments to improve or develop the invention.

The results of the investigation of other countries are summarized in Box 1 (see end of chapter). While the wording and scope vary from country to country, on the whole the laws provide an exemption from patent infringement for experimental use. In some countries, however, the interpretation of the provision is incoherent due to a lack of case history—and even the theories are variable in such countries. Still, in most of the countries, clinical trials to obtain regulatory approval are exempted, while there is no or very little case history regarding experimental use in universities.

The report concluded that Someno's theory is appropriate for Japan and in line with the situation and precedents of other countries. The report recommended its use to clarify the scope of the experiment and research exempted from the patent infringement. According to the theory (and given the fact that Japanese patent law does not distinguish between for-profit private companies and nonprofit universities and public research institutions when it comes to experiment or research using a patented invention for the effects of patent right), experimentation and research conducted in universities and public research institutions are potentially infringement unless licenses are obtained from the patent owner. If the subject of the experiment or research is a patented invention itself and the purpose is technological progress, however, utilization is exempted from the license requirement. Likewise, Article 69 (1) is not likely to apply to the utilization of research tools unless the subject of the research is the patented invention itself and its purpose is for technological progress.

There have only been a few occasions when universities and public research institutions utilizing a patented invention for their experiment or research have been sued by private companies owning the patent right in Japan. However, the report notes an increased concern about such lawsuits, particularly because universities are more likely to create profits from experimentation and research using patented products through increased collaboration with private industry than in the past. Besides, the report emphasizes the importance of disseminating information and generating a consensus on this issue in both the public and private sectors in order to minimize the number of such conflicts.

6. AGRICULTURAL BIOTECHNOLOGY

6.1 *National policy on R&D*

Japan has pursued R&D in agricultural biotechnology in the public and private sectors since the 1980s, with the government and relevant public-funding supports determining priorities. While basic R&D has contributed to global plant biotechnology communities, Japan has not taken the leadership in the business development of agri-biotechnology.[48] Furthermore, even though academic publications are recognized within global R&D networks, Japan's national policy lacks a strategic vision in the area of technology commercialization.

Despite the huge investment made by the public and private sectors between 1980 and 1999, no fruitful commercialization has taken place in Japan,[49] except for small cases relating to transgenic flowers. Many factors have been suggested for this: the weakness of decision making by the public sector's senior administration—and the private sector's correlating impatience; an overall shortage of adequate human resources; the lack of a strategic approach to commercialization; disorganized IP strategies; poor accountability, particularly in public-funded research; poor public communication approaches and consequent negative sentiment; and unfavorable regulations for R&D, despite government policies to support overall biotechnology.[50] Compared with other biotechnology areas, no major venture capitalists or investment banks are actively funding Japanese plant biotechnology R&D.[51] On the other hand, investors need patience. In general, agri-biotechnology R&D is a slow process, which is reflected in the slow growth of related industry.

On the upside, policy related to general support for biotechnology as a national priority has been reformed by the Council for Science and Technology Policy (CSTP)[52] under the cabinet office. Under supervision from METI, government funding agencies, such as the Research Institute of Innovative Technology for the Earth (RITE),[53] the New Energy and Industrial Technology Development Organization (NEDO)[54] and the Ministry of Agriculture, Forestry, and Fisheries (MAFF), have refocused research on crop-genome and crop-biotech applications, while MEXT and JSPS continue to fund basic research. This may drive policy toward the developmental outcomes of the Kyoto Protocol on environmental biotechnology applications (including transgenic applications). In the long term, these developments could revive overall agricultural biotechnology, including genetically modified (GM) crops. Also, as is the case in the United States and Europe,[55] the private sector in plant biotechnology could restructure by redefining and limiting its business context and partners.

6.2 *Agri-biotechnology industry and IP rights*

The Japanese biotechnology industry is very large in terms of assets and investments and is growing rapidly. Biotechnology research in Japan covers a wide range of areas from the elucidation of biological mechanisms to the development of new functional materials. Due to the broad spectrum of biotechnology, however, it is becoming increasingly difficult for a private company to monopolize, or even to know about, all the patents in a single product. Without intending to do so, a company can use another's patented technology inappropriately. The possibility of such patent infringement reaching the courts is increasing, and a complicating factor is the variety of national and international laws. In the field of agri-biotechnology, for instance, for new plant varieties it is unclear how laws/treaties on patent and those on plant variety protection should coexist or be applied.[56]

The number of ventures and spinouts in the area of biotechnology has increased, particularly since the reform of national universities into independently managed administrative institutions. Nevertheless, investors see agri-biotech companies as a high risk; their long-term efforts and contribution have been stagnant.[57] Major venture capitalists or investment banks are less likely to fund Japanese plant biotechnology R&D in comparison to other areas. Japanese companies have lost opportunities as a result, and key patents on plant biotechnology have been swept away by U.S. and European private companies, which strongly and adversely affected Japan's agricultural biotechnology industry. Numerous obstacles have contributed to this situation: (1) the complication of patenting inspection; (2) the tendency to grant wider coverage of patentable subjects, such as DNA sequences; (3) the changes in laws regarding patentable "process"; and (4) slow follow-up on litigation in agri-biotech IP rights.[58]

6.3 *Bioresources centers/genebanks*

Genetic resources have been well recognized as a key resource for R&D in Japan. To ensure synergy among germplasm banks, a consortium has been established that includes individual academic agencies. Similar to GRIN (Germplasm Resources Information Network)[59] in the United States, this information system is being further elaborated. There is common understanding of the uses of the germplasm acquisition agreement (GAA) and materials transfer agreement (MTA) from public bioresources centers/genebanks to different stakeholders in Japan. Details within MTA documents vary because each academic agency has to determine its own policies and rules under the common government framework.

The private sector also establishes its own MTA documents. These are based on different cases of use, such as basic research collaboration, R&D toward commercial orientation, collaboration with other private companies, and so forth. Although largely confidential, surveys made by the Japan Bioindustry Association (JBA) clearly reveal a system designed to accommodate various scenarios, particularly in relation to microorganisms. Plant genetic resources, however, are different, and Japanese seed companies still need to comprehend and tackle access and benefit-sharing issues under international debate—including the CBD and Treaty.

Case examples of access and benefit sharing (ABS) with southeast Asian countries emphasizing industrial applications include Indonesia with some pharmaceutical companies, Pathein University in Myanmar with the National Institute of Technology and Evaluation (NITE)[60] bioresources center, and the Forest Research Institute of Malaysia (FRIM) with Nimura Genetic Solutions (NGS),[61] a biotech venture-ABS company.

With the efforts of such intersectoral liaisons as JBA, some progress has been made in promoting and developing models for ABS-based R&D. However, Japanese academic institutions will be better able to address this matter by paying more attention to contemporary international discussions, such as those of the PGRFA, that are working towards an agreement on a standard MTA document.[62]

7. CASE STUDY: RIKEN

7.1 *Outlines of RIKEN*

RIKEN[63] is one of Japan's most distinguished public research institutes in the natural sciences. Its history began in 1913, when Jokichi Takamine, a Japanese scientist who discovered Taka-diastase and adrenaline, pointed out the need for a national science-research institute. Through the efforts of Takamine and others, including Eiichi Shibusawa, a businessman who greatly contributed to Japan's industrialization in the early 20th century, a bill to establish RIKEN was passed by the 37th Imperial Diet in 1915. A "*Proposition relating to the establishment of RIKEN*" was submitted to the government in 1916, followed by a "*Bill for governmental subsidy of a semipublic organization to conduct research in the physical and chemical sciences.*" RIKEN was eventually founded in 1917 as a private research foundation.

In 1927, Rikagaku Kogyo was incorporated exclusively to make marketable products from RIKEN's inventions. In other words, Rikagaku Kogyo had a similar function to a TTO.[64] Subsequently, other new companies were created to manufacture the products. By 1939, there were 63 companies and 121 plants. The group was called RIKEN Industrial Group, otherwise known as "*RIKEN* Konzern." It included some successful companies, such as RICOH, that survived and flourished even after the dissolution of the Konzern. RIKEN registered 0.7% of all patents registered in Japan during the period from 1918 to 1944 and actively transferred its technologies to the RIKEN Konzern companies, many of which were commercialized. Simultaneously, the proportion of royalties from patents as a percentage of RIKEN's entire revenue dramatically increased from 0% in 1927 to 48.4% in 1939, reaching a high of 60.4% in 1940.[65]

Dissolved by the General Headquarters of the Allied Powers after Japan's defeat in World War II, RIKEN was later reorganized and incorporated as a private corporation called Kaken Kagaku Ltd. (Scientific Research Institute Ltd.) in 1948. The corporation covered its research expenses with royalties earned by out-licensing its inventions. However, royalties gradually became insufficient to cover research costs, so government funding became necessary.

RIKEN was reinvented and inaugurated in 1958 as a special public institution operated by the RIKEN Law, for comprehensive research in science and technology under the jurisdiction of the Science and Technology Agency (STA, later integrated as the MEXT). In October 2003, special public institution reforms by the government reorganized RIKEN into an independently managed administrative institution. Since the reorganization, RIKEN and other public research institutions and national universities (see Section 3) have had more independence and autonomy to make decisions about research activities and finances. On the other hand, this greater responsibility requires more transparency and accountability in relation to fiscal and administrative management.

RIKEN's total budget in fiscal year (FY) 2005 was 86,769 million JPY. Medical science and bioscience account for large shares of the budget. Funding is provided by the government (about 80%) and by RIKEN itself (about 20%).

RIKEN has full-time and part-time employees. Full-time employees are either permanent or contract-based employees (usually one-year and renewable). The number of full-time employees is approximately 3,000, more than 70% of which were contract-based in FY 2005. Part-time workers also number about 3,000. Both full-time and part-time employees include foreign researchers. The total number of foreign researchers has increased from 352 in FY 1993 and 519 in FY 1998 to 576 in FY 2002. Chinese researchers account for a quarter of the foreign researchers at RIKEN. Many other foreign researchers come from Korea, the United States, France, and Russia. The portion of researchers from European countries has expanded gradually, but China consistently is most strongly represented. RIKEN's personnel reflect a diversity of positions and backgrounds—a significant asset in today's globalized world.

RIKEN is headquartered in Wako, Saitama, and there are eight other RIKEN research sites across Japan's mainland. Each one specializes in a specific research field. In addition to the domestic branches, RIKEN has three overseas branch institutes: one in the United Kingdom and two in the U.S. Research facilities have been established at these locations in collaboration with the host laboratories. In April 2006, RIKEN launched an office at Biopolis, a biomedical research hub in Singapore with both public and private sector researchers. In partnership with regional research institutions and Singapore's Agency for Science,

Technology, and Research (A*STAR), this new office is a hub for research collaboration in Asia.

RIKEN has always collaborated with domestic universities and built close ties by accepting their research students. In addition to graduate-student partnerships with 23 Japanese universities as of 2005, RIKEN has established similar partnerships with several universities in other Asian countries. RIKEN jointly conducts various official research projects with over 50 overseas research institutes—unofficial collaboration and exchanges of material and information greatly swell this number.

7.2 *RIKEN's IP policy and strategy*

Under the RIKEN law, the institute's objectives are to conduct comprehensive research in science and technology and to disseminate research results. RIKEN carries out research in many fields, including physics, chemistry, medical science, biology, and engineering, that ranges from basic research to practical application. In its previous role as a special public institution, RIKEN emphasized basic research over practical research. In the last few years, however, the institution has focused more on practical applications. Especially since becoming an independently managed administrative institution in 2003, RIKEN has emphasized earning its own funds through commercialization, instead of relying on government funds. As part of this effort, RIKEN established the Center for Intellectual Property Strategies (CIPS) in April 2005.[66] CIPS was charged with handling IP policy, strategy, and management. CIPS addresses these issues comprehensively and has been able to deal successfully with the increasing numbers and varieties of researchers, laboratories, centers, and institutes within RIKEN.

7.2.1 *IP status*

Figure 2 shows the number of patents newly filed each year and retained by RIKEN domestically and overseas. The number of newly filed domestic and overseas patents has gradually increased, while the number of domestically owned patents has generally decreased. Overseas ownership has gradually increased. These trends have two important implications:

For one, RIKEN's efforts to file IP rights (principally patent rights) for as many inventions as possible, whether domestic or overseas, have increased the number of patent filings. Also, RIKEN has become increasingly selective in retaining its patent rights because to do so is costly. Every year, owners are required to pay on patents, not only filing and registry fees, but also maintenance fees. RIKEN's status as an independently managed administrative institution has made it adopt a more cautious approach to retaining patent rights. It has decided which patents to abandon by reviewing and assessing the value of each invention in terms of its potential profit and licensing prospects. This is another reason why the number of domestic patent rights has declined. For overseas patents, the selection was less pressured because it is more difficult to identify the value of each invention for the international market. Consequently, the number of overseas patent rights retained has increased—in FY 2003 it outnumbered the domestic.

Figure 3 shows the number of licensed patents owned by RIKEN and the royalties earned through licensing each year. RIKEN's exploitation/licensing rate[68] is currently about 12%. This is below RIKEN's own expectations—as are the royalty amounts earned—so it is assumed that many of the inventions generated at RIKEN are not practical for commercialization. RIKEN has made the following efforts to raise the rate:

7.2.2 *Objectives for IP policy*

RIKEN's fundamental IP policies are driven by three main objectives: (1) to promote greater protection of IP rights on inventions, particularly patent rights; (2) to partner with industry; and (3) to generate profits through licensing.

1. **Promotion of IP rights.** The promotion of activities relating to filing patent rights is aimed at contributing to the public domain by disclosing RIKEN's inventions through patent applications and at generating profits through licensing patented inventions to industry.

(A) ***Patent liaison staff.*** To promote IP protection, RIKEN deploys about 10 staff members called "patent liaison staff."

FIGURE 2: RIKEN's PATENT RIGHTS (RETAINMENT AND NEWLY FILED)

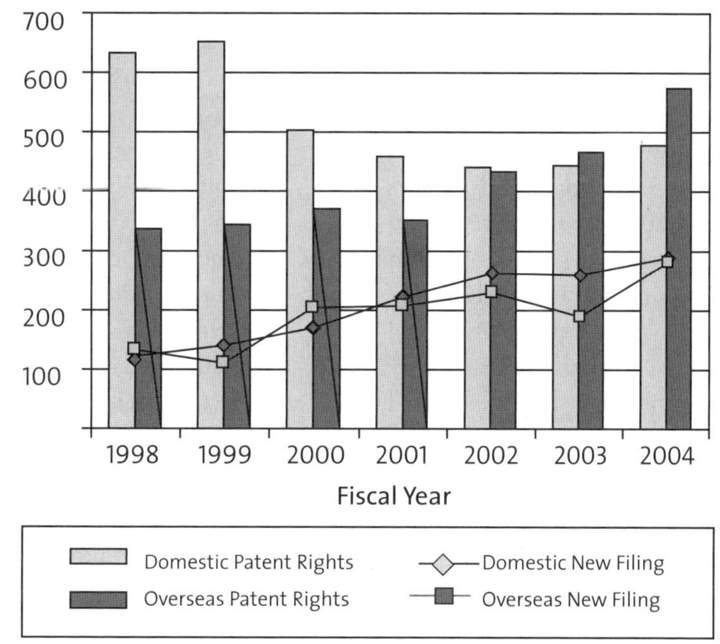

Source: RIKEN CIPS[67]

FIGURE 3. THE NUMBER OF LICENSED PATENTS OWNED BY RIKEN AND THE ROYALTIES EARNED

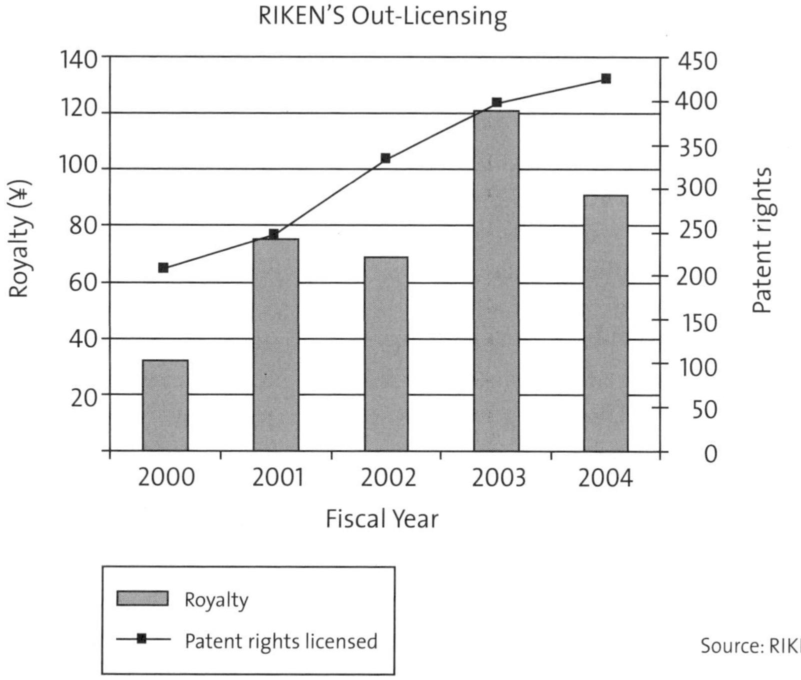

Source: RIKEN CIPS[69]

Their responsibilities range from identifying inventions to protecting them through consultation with RIKEN's inventors. The staff is made up of qualified patent attorneys; incumbent staff employed and temporarily transferred by private companies or attorneys' offices with relevant experience; and retirees of private companies. There is no staff member with tenure deployed for patent liaison. It is felt that none of RIKEN's tenured staff have adequate knowledge and experience in IP and technology management because staff are rotated to other divisions every three or four years under the organization's personnel policy.

(B) ***Employee invention regulations.*** Compared to other public institutes in Japan, RIKEN set up regulations for employee inventions comparatively early. The regulations were amended in April 2004. Previously, employee inventors had to decide whether to retain ownership of an applied or registered patent (or other form of IP right) jointly with RIKEN—and shared equally—or to waive the whole right to their invention and assign it to RIKEN. If they decided to own half, they were required to bear half of the expenses for applying, registering, and retaining the IP (RIKEN paid the other half). Meanwhile, the employee could benefit from a variable percentage of the royalties that would be paid based on the amount of received royalties. Furthermore, a fixed amount of remuneration was paid to the inventors for both the application and registration of the patent. RIKEN was seeking to promote patent rights, and to encourage researchers to make more inventions, with the potential for economic returns by providing remuneration to the inventors.

The new regulations of April 2004, however, eliminated the inventor's option to own half the IP rights. The whole right would from then on be owned solely by RIKEN. The rationale for the change was that sole ownership by RIKEN would enable the institution to manage the entire technology transfer process, enabling it to determine licensing issues itself and to decide upon licensing details. The licensees are also likely to welcome RIKEN's sole ownership because the process is easier. Moreover, the number of one-year employment contracts within RIKEN has greatly increased over the last few years. Most researchers and inventors are newly employed and could resign one or a few years later. This fluidity makes it difficult to jointly own IP because the institute has to chase down inventors who have left RIKEN in order to obtain consent for exploiting or waiving rights. Besides, a number of inventors had not, in fact, chosen the option of joint ownership in the previous system. This was largely because of the risk and ambiguity involved in exploitation, as well as the high costs of applying for, registering, and retaining patent rights.

Another amendment relates to provisions for remuneration. The remunerations for application and registration were combined and paid together one year after the application, while the provision related to remuneration for licensing remained as it was. This amendment was a result of the increased fluidity of personnel: the registration process takes a few years—during or after which time the inventor may have left RIKEN—making the payment procedure ineffective.

(C) ***Raising Awareness.*** RIKEN has made efforts to promote IP by raising awareness. Seminars and consultations about various IP rights issues are regularly held in not only the headquarters but also the branch institutes and centers. Because the frequent turnover of employees hinders the diffusion of knowledge about

RIKEN's IP policy and strategy, RIKEN requires newcomers to attend specific explanatory lectures that are held several times a year. This is in addition to the regular IP rights seminars. As a consequence of those efforts, the number of IP rights applications by RIKEN has been increasing.

2. **Partnership with industry.** RIKEN belongs to the academic sector. It makes a public contribution by providing the seeds of innovation to industry. Since becoming an independently managed administrative institution, generating profits through licensing has become increasingly significant for RIKEN. Its IP strategy focusing on partnerships with industry is a tool that allows RIKEN to generate social and economic returns simultaneously.

   Such partnerships involve not only technology transfer but also research collaboration. CIPS is highly involved in coordinating, funding, providing research space, and hosting industrial researchers for the collaboration. One of the programs RIKEN/CIPS formally organizes is the Fusional Cooperative Research Program. Started in 2004, the program transfers researchers employed by private companies to RIKEN to conduct collaborative research for several years. Under contract with RIKEN, the researchers can become team leaders of their research in RIKEN. RIKEN has published on its Web site[70] a database of its researchers who have registered for this program. The database includes their research activities and interests. A private company interested in a RIKEN researcher and his/her research applies for the program with a collaborative research proposal. The collaborative research under the program enables the rapid commercialization of the technology by the parallel creation of "seeds" and "needs" from the very beginning of the research planning stage. RIKEN contributes research expertise and facilities, and the private company contributes commercialization expertise and shares management tools to increase efficiencies. Expenses are borne by both RIKEN and the private company. The contracted term is generally five years. As of April 2006, ten teams have been created and are pursuing collaborative research under the program.

3. **Promotion of exploitation.** RIKEN has adopted some strategies to promote the exploitation of inventions that its researchers generate. These strategies include disseminating information about patents owned by RIKEN, coordinating and facilitating technology transfer, and the "RIKEN Venture" system.

(A) ***Disseminating information about patents owned by RIKEN.*** RIKEN has actively tried to promote the exploitation of inventions by disseminating information about its patents, which is expected to increase private companies' abilities to find and exploit them. Information is disseminated via the *Journal of RIKEN Patents* published by CIPS. A patent database is published online at the R-BIGIN (RIKEN-Business Information for Global IP Network) Web site, and RIKEN also exhibits its technologies at external fairs relating to technology transfer.

(B) ***Coordination of technology transfer.*** RIKEN deploys several coordinators in CIPS to increase the transfer and exploitation of its technology. Similar to the patent liaison staff, the coordinators include current private sector employees, who have been temporarily transferred to RIKEN, and experienced retirees from the private sector. Their responsibilities are to search for licensees, negotiate terms, and conclude licensing contracts. In addition to the coordinators, RIKEN outsources contracts to some large enterprises to coordinate technology transfer with private companies. These enterprises have varied, detailed information about potential licensees, and this external coordination facilitates technology transfer from RIKEN to industry.

(C) ***The RIKEN Venture system.***
Set up in 1998, RIKEN Venture system supports and encourages employees to establish and operate private companies based on inventions generated at RIKEN. In addition to enabling RIKEN employees to retain a post at the private company, RIKEN provides preferential treatment to the company:
- RIKEN licenses its patent rights relating to the invention exclusively to the company
- RIKEN allows the company to utilize its research space and facilities for collaborative research with RIKEN
- RIKEN provides the company with office space and equipment for management at preferential rates

These advantages make it easier for inventors to exploit and distribute their own inventions to the public, which creates yet another incentive for researchers to make or adapt practical, profitable inventions. Additionally, innovations that existing companies find difficult to exploit can be given another chance by their inventors. The program offers support to each company for five years, which can be extended for an additional five years. As of July 2005, the program has supported 16 companies: seven are in the field of biomedicine.

7.3 *RIKEN BioResource Center (BRC)*

In 2001, RIKEN founded the BioResource Center (BRC)[71] at the Tsukuba Research Institute. After a gene bank service was established at RIKEN in 1987, the BRC was founded to expand the scope of the collected resources. The Japan Collection of Microorganisms, which had initially been established at RIKEN headquarters, was integrated within the BRC in 2004. Integration enabled the BRC to offer a distribution service for a wide range of resources including animals, plants, cells, genes, and microorganisms. The RIKEN BRC has been supported by Japan's national bioresources project.

The principal contribution of the BRC to life sciences research is to collect, preserve, breed, and distribute biological resources to and from researchers in Japan and overseas. Other BRC activities include the development of bioresources and new technologies to increase their value. The BRC has made a great effort to foster transfers of bioresources for both collection and distribution since its foundation. All transfers are carried out based on the conclusion of MTAs, for which RIKEN has created its own forms and procedures. Although some details vary among the types of resources, the grounds for transfer are generally as follows:

1. Collection (resources are deposited or assigned by originators).
   - An MTA must be concluded between RIKEN BRC and the originator for the deposit/assignment. The MTA form for deposit or assignment is provided by the BRC.
   - The originator is entitled to choose to deposit the resources and retain the IP rights to the resource or to assign the resource with the IP rights to RIKEN.
   - Whether it is a deposit or an assignment, resources are collected by the BRC without any remuneration to the originator. (RIKEN bears the expenses of shipment for collection.)
   - In addition to the requirements set by RIKEN BRC, a third party's minimum requirements for using resources, such as acknowledgement in publication of research results, can be added to the MTA by the originator.
   - By the deposit/assignment, the originator can, for no charge, be credited and provided with other resources collected by the BRC, according to the number of resources that he or she provides.
2. Distribution (resources are transferred from the BRC to a third party [recipient/user] for their research use).
   - An MTA between RIKEN BRC and a user must be concluded and signed for the distribution to occur. The MTA form is provided by the BRC.

- The user bears the expenses of shipping, handling, part of production, and other costs related to preparing or distributing the resources. Allocations of costs are differentiated between public and private partners, with private partners assuming the greater burden.
- The user is required to specify a research theme for which the resources are used. If resources are used for another theme, prior notification to the BRC is required.
- When research results that used the resources are published, the user is required to make it clear that the resources were provided by RIKEN BRC.
- The user cannot transfer or make the resources available to other parties for any purposes.

The BRC is becoming recognized as one of the major bioresource centers in the world. Furthermore, the BRC/Experimental Animal Division is one of the founding members of the Federation of International Mouse Resources (FIMRe),[72] along with such outstanding mouse resource centers as the Jackson Laboratory (U.S.) and European Mouse Mutant Archive (EMMA). The FIMRe is a collaborating consortium group of mouse repository and resource centers worldwide whose collective goal is to archive and provide to the research community strains of mice, as cryopreserved embryos and gametes, embryonic stem (ES) cell lines, and live breeding stock. The mouse-strain resources deposited or assigned to the RIKEN BRC—and related pieces of information—are registered and published on the database of the FIMRe, known as the International Mouse Strain Resource (IMSR),[73] Registration promotes and facilitates global access by researchers to BRC resources. Additionally, the RIKEN BRC receives complementary support for the specific management of IP protected microorganism collections from NITE, which is under supervision of METI. This interagency collaboration facilitates the coordination of R&D.

8. CONCLUSION

Japan's patent system was established at the end of its national isolation policy. The system is reasonably effective. Emphasizing the importance of national and institutional IP management in its policy and strategy for national development over the last decade, the government has revised aspects of the patent law and reformed related systems, including those related to national universities and public institutions. Some of the revisions and reforms have been geared towards international harmonization and the adoption of precedence established in other countries. Others have been intended to establish *sui generis* laws and systems to suit the country's unique interests. Despite this progress, some issues and arguments have yet to be conclusively addressed.

Regarding collaborations between industry and academia, for example, the reform of national universities and public institutions in the early 2000s has catalyzed partnerships, largely because of the expanded freedom and responsibilities given to universities by the government. Over the last decade, universities have established TTOs in order to create, transfer, and exploit IP rights derived from their research projects, an increasing number of which are carried out in partnership with industry. Nevertheless, human resource shortages plague the system. Personnel with expertise in both legal and technical aspects are especially in demand.

Since the early stages of Japan's industrial development, the Patent Law has made provisions for employees' inventions. Over the last decade, an increasing number of institutions and companies have recognized the significance of rules and regulations for employees' inventions and taken steps to establish them. This has been supported by the new government's policy and strategy: a Nation Built on IP. In 2005, the provisions (Article 35 of the patent law) were amended in favor of inventors so that the criteria for remuneration for inventions would be reasonable for them. Since a few years prior to the amendment, the number of lawsuits in which a former employee sued his or her former employer because of dissatisfaction with their remuneration has increased. Some lawsuits have been settled, but various questions remain unresolved.

Japan's IP system and management is in some ways unique in relation to the health and agriculture sectors. IP rights for health care have been recognized as publicly shared knowledge and skills with equitable properties rather than personalized trade secrets or proprietary knowledge and skills, although some incentives have been furnished to enable the sharing and development of individual invention and know-how. Agriculture has traditionally been in the public domain, while specific technology has been protected as individual trade secrets. In the past, crop varieties were recognized as common heritage. Due to plant variety protection law and the recent paradigm shift in international and domestic arenas affecting IP laws, however, the use and status of the varieties has been in question, with business incentives rather than the public good driving the changes.

Overall, the stakeholders in health and agriculture will recognize IP increasingly in Japan. Diverse ways of adapting IP protection are being considered, and a *sui generis* approach may be adopted to tackle many subjects. Public awareness is likely to be promoted through public engagement in IP management, particularly in health and agriculture. ■

JUNKO CHAPMAN, *Research Associate, MIHR (Centre for the Management of Intellectual Property in Health Research and Development), Oxford Centre for Innovation, Mill Street, Oxford, OX2 0JX, U.K. junko.chapman@mihr.org*

KAZUO N. WATANABE, *Professor, Gene Research Center and Division of Bioindustrial Sciences, Graduate School of Life and Environmental Sciences, University of Tsukuba, 1-1-1 Tennoudai, Tsukuba, Ibaraki, 305-8572, Japan. nabechan@gene.tsukuba.ac.jp*

1 Watanabe KN and A Komamine. 2004. Issues on Intellectual Property Rights Associated with Agro-Biotechnology in Japan. In *Intellectual Property Rights in Agricultural Biotechnology*, 2nd edition. (eds. FH Erbisch and KM Maredia). Michigan State University: East Lansing, Michigan, and C. A. B. International: Wallingford. pp.187–200.

2 www.aippi.or.jp/.

3 www.fao.org/AG/cgrfa/itpgr.htm.

4 Ohno K. Economic Development of Japan. Lecture Notes. National Graduate Institute for Policy Studies (GRIPS): Tokyo. www.grips.ac.jp/teacher/oono/hp/lec_J.htm.

5 *Kougyoushoyuukenseido-no rekishi* (History of Industrial Property System). JPO: Tokyo. (In Japanese) www.jpo.go.jp/seido/rekishi/rekisi.htm.

6 APIC. 1999. *History of Japanese Industrial Property System*. APIC: Tokyo. www.apic.jiii.or.jp/p_f/text/text/1-07.pdf.

7 Ibid.

8 See *supra* note 4.

9 See *supra* note 6.

10 Goto A. 2003. Kyoushinka-no process-toshiteno nihon-no tokkyoseido-to gijutsu-kakushin. Chitekizaisanseido-*to Innovation (Intellectual Property System and Innovation)* University of Tokyo Press: Tokyo. (In Japanese). pp. 311–35.

11 Okazaki, Y. 2000. Research and Study on the Intellectual Property System in the 21st century, *IIP Bulletin 2000*. Institute of Intellectual Property (IIP):Tokyo. www.iip.or.jp/e/summary/pdf/detail99e/e12.pdf. pp. 114–23.

12 www.kantei.go.jp/foreign/policy/titeki/kettei/020703taikou_e.html.

13 Intellectual Creation Cycle (Figure). Japan Patent Office: Working Toward the Establishment of a Nation Based on Intellectual Property. METI: Tokyo. www.meti.go.jp/english/aboutmeti/data/aOrganizatione/keizai/tokkyo/01.htm and METI JPO (Web site).

14 www.kantei.go.jp/foreign/policy/titeki/hourei/021204kihon_e.html.

15 www.kantei.go.jp/foreign/policy/titeki/kettei/030708f_e.html.

16 Strategic Program for the Creation, Protection, and Exploitation of Intellectual Property (Intellectual Property Strategic Program 2003). www.kantei.go.jp/foreign/policy/titeki/kettei/030708f_e.html; Intellectual Property Strategic Program 2004. www.ipr.go.jp/e_material/ip_st_program2004.pdf#search='Intellectual%20Property%20Strategic%20Program%202004'; Intellectual Property Strategic Program 2005. www.ipr.go.jp/e_material/ip_st_program2005.pdf#search='Intellectual%20Property%20Strategic%20Program%202005'; Intellectual Property Strategic Program 2006.www.kantei.go.jp/jp/singi/titeki2/keikaku2006_e.pdf.

17 Kondo M. 2006. University-Industry Partnerships in Japan. Presented at the symposium 21st Century Innovation System for Japan and the United States. National Institute of Science and Technology Policy (NISTEP): Tokyo, January 10–11, 2006. www.nistep.go.jp/IC/ic060110/pdf/5-2.pdf#search='kondo%20Universit

y-Industry%20Partnerships%20in%20Japan'.

18 METI. 2004. *Guide to Technology Licensing Organizations (TLOs) in Japan.* FY 2004. Industry-University Cooperation Division. METI: Japan.

19 www.jpo.go.jp/kanren/tlo.htm (In Japanese).

20 See *supra* note 1.

21 Kneller R. 2003. San-gaku-renkeiseido-no nichibei-hikaku. Chitekizaisanseido-*to Innovation (Intellectual Property System and Innovation)* University of Tokyo Press: Tokyo. (In Japanese). pp. 51–99

22 Mitsubishi Research Institute. 2004. Shokumu-hatsumei-seido. ***Keywords*** 058. (In Japanese) sociosys.mri.co.jp/keywords/058.html.

23 Normile D. 2004. Japan Ponders Starting a Global Journal. *Science* 303: 1599.

24 www.jsps.go.jp/english/index.html.

25 JSPS. 2004. Jigyou gaiyou: Tokkyo-shutsugan settei-touroku-no joukyou. Japan Society for the Promotion of Science: Tokyo. (In Japanese) www.jsps.go.jp/j-rftf/gaiyo/gaiyo_tokyo_i.html.

26 *Chitekizaisan-senryaku-kenkyukai.* 2005. Hyakumannin-no shokumu-hatsumei. Ohmu: Tokyo. (In Japanese).

27 JIII. 1997. Shokumu-hatsumei Hoshou-kingaku-no Shousa-kekka. Japan Institute of Invention and Innovation: Tokyo. (In Japanese) www.jiii.or.jp/syokumu.htm.

28 Iwai T. 2003. Modalities for the Employees' Inventions System. *IIP Bulletin 2003*, Institute of Intellectual Property (IIP): Tokyo. pp. 18–26. www.iip.or.jp/e/summary/pdf/detail2002/e14_03.pdf#search='modalities%20for%20the%20employees'%20inventions%20system.

29 "... an invention which by reason of its nature falls within the scope of the business of the employer, etc. and an act or acts resulting in the invention were part of the present or past duties of the employee, etc. performed on behalf of the employer, etc. (hereinafter referred to as an *employee's invention*) ..." (Patent Law, Article 35[1]).

30 www.jpo.go.jp/cgi/link.cgi?url=/shiryou/toushin/toushintou/patent_houkoku.htm. (In Japanese).

31 See *supra* note 26.

32 www.jipa.or.jp/content/english/.

33 Sumikura K. 2003. Bio-tokkyo nyumon kouza (Lecture on Bi- patent). Yodosha: Tokyo. (In Japanese).

34 ACCJ. 2003. Amendment to Article 35 of Japan's Patent Law. The American Chamber of Commerce in Japan. Viewpoint: Tokyo.

35 www.uspto.gov/web/offices/pac/mpep/documents/appxl_35_U_S_C_287.htm.

36 www.european-patent-office.org/legal/epc/e/ar52.html#A52.

37 Intellectual Property Policy Headquarters. 2004. *Iryoukanrenkoui-no tokkyohogo-no arikata-ni-tsuite (Torimatome)* (Summary of the discussion on patent protection of medical activities). The Cabinet Office: Tokyo (In Japanese) www.kantei.go.jp/jp/singi/titeki2/tyousakai/iryou/torimatome.pdf.

38 "Any person who has made an invention which is industrially applicable may obtain a patent therefore ..."

39 Izukawa T. 2001. Research and Study on Patent Protection in Medical Field. *IIP Bulletin*, Institute of Intellectual Property (IIP): Tokyo. p. 45–55. www.iip.or.jp/e/summary/pdf/detail2000/report4.pdf.

40 *Ibid.*

41 *Ibid.*

42 See *supra* note 37.

43 www.jpo.go.jp/shiryou/index_g.htm (In Japanese).

44 Bio-Life Science Committee. 2005. Iryoukanrenkoui-no tokkyohogo-no kakudai-ni tsuite. *JPAA. Patent.* 58 (7): 69–75. (In Japanese) www.jpaa.or.jp/publication/patent/patent-lib/200507/jpaapatent200507_069-075.pdf .

45 METI JPO. 2004. Tokkyohatsumei-no enkatsu-na shiyou-ni kakawaru shomondai-nitsuite. JPO: Tokyo. (In Japanese) www.jpo.go.jp/shiryou/index.htm.

46 Someno K, 1988, AIPPI, Vol.33, No.3.

47 See *supra* note 45.

48 See *supra* note 1.

49 Watanabe KN, Y Sassa, E Suda, CH Chen, M Inaba, and A Kikuchi. 2005. Global PEST issues of GM crops with special references to Japanese cases. *Plant Biotechnology* 22: 515–22.

50 Watanabe KN. 2003. Pitfalls on implementing the Cartagena Protocol on Biosafety in Japan. *Nature* 421: 689.

51 JBA. 2003 and 2004. Report on Bio-ventures and Bio-clusters. Japan Bioindustry Association: Tokyo. (In Japanese) www.jba.or.jp/bv/rep_bc-bv.pdf and www.jba.or.jp/bv/2004-bv-summary.pdf.

52 www8.cao.go.jp/cstp/english/index.html.

53 www.rite.or.jp/English/E-home-frame.html.

54 www.nedo.go.jp/english/index.html.

55 Lee DP and MD Dibner. 2005. The Rise of Venture Capitol and Biotechnology in the U.S. and Europe. *Nature Biotech* 23: 672–76.

56 See *supra* note 1.

57 See *supra* note 1.

58 See *supra* note 1.

59 www.ars-grin.gov/.

60 www.nbrc.nite.go.jp/e/deposit-e.html.

61 www.ngs-lab.com/en/test/ngs_eg/egsplash.html.

62 www.iisd.ca/biodiv/itpgrgb1/.

63 Information in this section is based on information on RIKEN's Web site and on interviews held in July 2005 with RIKEN's Center for Intellectual Property Strategies

(CIPS), except where stated otherwise. www.riken.jp/engn/index.html.

64 Baba R. 2001. *RIKEN-ni miru venture-no genkei*. Daijoubu-ka nihon-no tokkyo-senryaku, President: Tokyo. (In Japanese). pp. 215–22.

65 See *supra* note 17.

66 r-bigin.riken.jp/bigin/ (In Japanese).

67 See *supra* note 17.

68 The "exploitation/licensing rate" is calculated as the number of exploitations/licenses divided by the total number of registered patents and patent applications under examination.

69 See supra note 17.

70 www.riken.jp/engn/index.html.

71 www.brc.riken.go.jp/inf/en/index.shtml.

72 www.fimre.org/.

73 www.informatics.jax.org/imsr/index.jsp.

Box 1: Situation and Precedents Relating to Limits of Patent Rights Regarding Medical Methods

Country/Region	Law on limits of patent rights Situation and precedents on: (A) Experiment or research (B) Clinical trials of generic medicines (C) Experimental use in universities
Japan	**Article 69** (1) The effects of the patent right shall not extend to the working of the patent right for the purposes of experiment or research. (A) The theory that has been the most widely accepted is the one that limited the experiment or research applicable to Article 69 (1) to those for the purpose of "progress in technology." (B) Many theoreticians assert that private companies cannot use others' patent rights in clinical trials (for obtaining regulatory approval for manufacturing generic medicines), but past legal judgments have been variable and reflected both sides of the argument. The Supreme Court's judgment in 1999, however, set a legal precedent that confirmed that trials for the purpose of obtaining regulatory approval should be exempt. (C) Historically, in determining cases of exemption, the courts have not distinguished between university and industry (private companies). That is, there is no exemption for universities because of their academic and educational nature. However, based on the principle that experiment/research aimed at technology advancement is exempted from infringement, university-based experiment/research is congruently exempted.
United States	**35 U.S.C. 271 (e)(1) ("Bolar Provision")** It shall not be an act of infringement to make, use, offer to sell, or sell within the United States or import into the United States a patented invention (other than a new animal drug or veterinary biological product [as those terms are used in the Federal Food, Drug, and Cosmetic Act and the Act of March 4, 1913] which is primarily manufactured using recombinant DNA, recombinant RNA, hybridoma technology, or other processes involving site specific genetic manipulation techniques) solely for uses reasonably related to the development and submission of information under a Federal law which regulates the manufacture, use, or sale of drugs or veterinary biological products.

(Continued on Next Page)

Box 1 (continued)

United States (continued)

(A) Experiment or research using patented products for commercial purposes is considered to be an infringement. The Federal Circuit Court of Appeals has reconfirmed in several cases that the scope of exemption from infringement in relation to experimental use should be very narrow.

(B) The case of *Eli Lilly & Co. vs. Medtronic, Inc.* in 1990 confirmed that the Bolar Provision (inserted into U.S. patent law in 1984) covers clinical trials using not only medicines but also medical tools, but the application of the Bolar Provision is limited to the development and submission of information to the FDA (Food and Drug Administration).

(C) With regard to experimental use in universities, there have been very few cases. One is the case of *Madey vs. Duke University* in 2002, which confirmed that the scope of exemption should be very narrow. The exemption was not applied in this case.

European Union

EPC (European Patent Convention) Article 64

Rights conferred by a European patent: (3) Any infringement of a European patent *shall be dealt with by national law.*
CPC (Community Patent Convention), Article 27, Limitation of the effects of the Community patent. The rights conferred by a Community patent shall not extend to:

(a) acts done privately and for noncommercial purposes;
(b) acts done for experimental purposes relating to the subject matter of the patented invention; etc.

United Kingdom

Patent Act 1977
Article 60

Section 5. An act which, apart from this subsection, would constitute an infringement of a patent for an invention shall not do so if:

(a) it is done privately and for purposes which are not commercial;
(b) it is done for experimental purposes relating to the subject matter of the invention, etc.

(A) Experiment or research using patented products for commercial purposes are distinguished between those trials in which products are merely being tested for quality, which are exempted, and others in which they are being demonstrated

(Continued on Next Page)

Box 1 (continued)

United Kingdom
(continued)

to a third party or used for quality enhancement in other products provided to a third party, which are considered within the scope of infringement.

(B) The trials and manufacturing of patented products for the purpose of obtaining regulatory approval is out of the scope of exemption and regarded as patent infringement.

(C) There have been no cases establishing precedent relating to experimental use in universities.

Germany

(1) Patent Law
11. (Amended in 1981)

The effects of a patent shall not extend to:

1. acts done privately and for noncommercial purposes;
2. acts done for experimental purposes relating to the subject matter of the patented invention; etc.

(A) Experiment or research using patented products either to obtain information regarding the subject of the patented products (for noncommercial use) or to enable scientific investigation is exempted from patent infringement. From the viewpoint of public benefit, patent rights do not extend to cases interfering with technological progress.

(B) In the Clinical Tests II case in 1997, it was confirmed that trials to clarify areas of uncertainty or trials aiming at acquisition of new knowledge relating to the subject of the patented products fall under the scope of exemption.

(C) There have been no cases establishing precedent relating to experimental use in universities.

France

Intellectual Property Law
Art. L. 613-5. (Amended in 1978)

The rights afforded by the patent shall not extend to:

(a) acts done privately and for noncommercial purposes;
(b) acts done for experimental purposes relating to the subject matter of the patented invention; etc.

(A) The case of *Babolat vs. Redeye* (1992) set a precedent that experimental use for the purpose of evaluating the commercial effect of a patented product on consumers would be considered an infringement.

(Continued on Next Page)

Box 1 (continued)

France (continued)

(B) The cases of the *Wellcome Foundation Ltd. vs. Parexel International, Flamel Technologies & Créapharm* (2001) and *Science Union & Servier vs. Expanpharm* (2002) confirmed that trials for the purposes of obtaining regulatory approval for substitutes of marketed medicines (that is, generics) and of obtaining regulatory approval fall under the scope of exemption.

(C) There have been no cases establishing precedent relating to experimental use in universities.

Republic of Korea

Patent Law 96

(1) The effects of the patent right shall not extend to the following:
  (i) working of the patented invention for the purpose of research or experiment; etc.

(A) There have been no cases establishing precedent relating to experimental use.

(B) There have been no cases establishing precedent relating to clinical trials. Theoreticians regard this as exempt from infringement.

(C) There have been no cases establishing precedent relating to experimental use in universities.

China

Patent Law, Article 63

None of the following shall be deemed an infringement of the patent right:
  (4) Where any person uses the patent concerned solely for the purposes of scientific research and experimentation.

(A) Exemption is not always applicable in cases relating to experimental use in general R&D activities. Exemption from infringement applies when experimentation relates to technical appraisal of patent rights and regarding the patented technology per se.

(B) It is generally considered that clinical trials for the purpose of obtaining regulatory approval are not an infringement if undertaken within two years of the patent expiration date.

(Continued on Next Page)

Box 1 (continued)

China (continued)	(C) There have been no cases establishing precedent relating to experimental use in universities.
Singapore	**Patent Act** 66.-(2) An act which, apart from this subsection, would constitute an infringement of a patent for an invention shall not do so if: (a) it is done privately and for purposes which are not commercial; (b) it is done for experimental purposes relating to the subject matter of the invention; (h) it consists of the doing of any thing set out in subsection (1) in relation to the subject matter of the patent to support any application for marketing approval for a pharmaceutical product, provided that any thing produced to support the application is not: (i) made, used, or sold in Singapore; or (ii) exported outside Singapore, other than for purposes related to meeting the requirements for marketing approval for that pharmaceutical product; etc. (A) There have been no cases establishing precedence or coherent theory regarding the exemption of experimental use as Singapore's patent system and law is rather young (since 1994). (B) The Amendment to the Patent Act in 2004 exempted clinical trials for the purpose of obtaining regulatory approval. (C) There have been no cases establishing precedent relating to experimental use in universities.
India	**Patent Act** 47. The grant of a patent under this Act shall be subject to the condition that: (3) any machine, apparatus or other article in respect of which the patent is granted or any article made by the use of the process in respect of which the patent is granted, may be made or used, and any process in respect of which the patent is granted may be used, by any person, for the purpose merely of experiment or research including the imparting of instructions to pupils; and etc.

(Continued on Next Page)

Box 1 (continued)

India
(continued)

> 107A. For the purposes of this Act:
> (a) any act of making, constructing, using, selling, or importing a patented invention solely for uses reasonably relating to the development and submission of information required under any law for the time being in force, in India, or in a country other than India, that regulates the manufacture, construction, use, sale, or import of any product;
> ... shall not be considered as an infringement of patent rights.

(A) There have been no cases establishing precedent or coherent theory regarding the exemption of experimental use.

(B) The Amendment to the Patent Act in 2002 (Sec. 107A) exempted clinical trials for the purpose of obtaining regulatory approval.

(C) There have been no cases establishing precedent relating to experimental use in universities.

TRIPS

> **TRIPS Article 30**
>
> Exceptions to Rights Conferred
> Members may provide limited exceptions to the exclusive rights conferred by a patent, provided that such exceptions do not unreasonably conflict with a normal exploitation of the patent and do not unreasonably prejudice the legitimate interests of the patent owner, taking account of the legitimate interests of third parties.

Source: *Tokkyohatsumei-no enkatsu-na shiyou-ni kakawaru shomondai-ni tsuite* (Report on issues relating to effective use of patented invention)[47]

Technology Transfer in South African Public Research Institutions

ROSEMARY WOLSON, *Intellectual Property Manager, R&D Outcomes, Council for Scientific and Industrial Research (CSIR), South Africa*

ABSTRACT

This chapter provides an analytical overview of technology transfer in South Africa. Technology transfer offices (TTOs) are relatively new in the country, and not all South African universities have explicit IP policies. The chapter discusses and analyzes the current performance of TTOs. Among other things, the results show that the income accruing to universities from technology transfer activities is not substantial, that there is a time lag before a TTO can generate sufficient income to become self-supporting, and that the performance of TTOs at different institutions varies widely. A history of public policy efforts to strengthen technology transfer in South Africa is provided, and the government's 2006 publication of the Framework for Intellectual Property Rights from Publicly Financed Research receives considerable analysis. Other measures being undertaken to support technology transfer are also discussed, as are the problems that such efforts still face.

1. CURRENT STATUS OF TECHNOLOGY TRANSFER ACTIVITY IN SOUTH AFRICAN RESEARCH INSTITUTIONS

1.1 *Background*

Institutional technology transfer offices (TTOs) are a relatively new development in South African universities and research organizations and are not yet found in all research institutions. While some efforts were made to promote technology transfer activities as early as the 1980s, it was not until the late 1990s that a handful of institutions set up TTOs. There are currently six universities and science councils with well-established technology transfer activities.[1] The main catalyst for setting up these TTOs appears to have been an awareness of international trends—the first offices were established before any meaningful attempts by government to better utilize research outputs. Some TTOs function as dedicated offices within their organizations. They are sometimes responsible for other functions, such as sponsored research, development, contract management, or industry liaison, and activities are sometimes dispersed among some of these offices. Other institutions have set up associated companies that are wholly or partly owned by the organization concerned to perform their technology transfer activities. In one case, a company was set up to manage jointly the IP from a science council and a university, but the partnership has since dissolved. The number of TTOs continues to grow. Several institutions have newly established offices, and those without TTOs are in the process of setting up offices. Institutions without TTOs either contract external service providers for assistance on a case-by-case basis or do not actively engage in technology transfer as an institution, although individual researchers or departments might do so.

Wolson R. 2007. Technology Transfer in South African Public Research Institutions. In *Intellectual Property Management in Health and Agricultural Innovation: A Handbook of Best Practices* (eds. A Krattiger, RT Mahoney, L Nelsen, et al.). MIHR: Oxford, U.K., and PIPRA: Davis, U.S.A. Available online at www.ipHandbook.org.

1.2 *Ownership of intellectual property*

1.2.1 *Within the institution*

Pending the introduction of legislation governing the ownership of IP developed by staff and students in the course of university activities, not all South African universities have explicit IP policies. Where policies are in place, these are not uniform across institutions. In some cases, IP is owned by individuals (unless specifically assigned, for example as a condition for the award of certain funding); in other cases, the university owns IP, depending on internal policies, conditions of employment, and student rules. Ownership rights of student IP vary widely, even for universities with clear policies that allow for institutional ownership of staff IP. When rights are assigned to the university, proceeds generated from the exploitation of IP are generally shared between the institution (possibly divided among multiple entities within the institution, such as research grouping, department, faculty, and to the central administration) and the individual inventor/s concerned, according to a formula set out in the IP policy.

1.2.2 *In respect of third parties*

While most institutions prefer to retain ownership of their IP and facilitate exploitation through licensing, and while most make every effort to negotiate this whenever possible, research sponsors frequently insist upon the assignment of IP as a key condition of a research funding agreement. This applies both to certain public sector and private sector funders, and may or may not include an obligation on the part of the assignee to share with the institution any future benefits derived from the exploitation of the IP. Ownership policies for IP that arises from government-funded research vary widely, ranging from unfettered ownership by the research institution, to shared ownership between the research institution and the funding agency, to full ownership by the funding agency, with benefit-sharing mechanisms applicable in some cases. The trend is for government entities to take a greater interest in IP matters than in the past, which often leads to more complicated funding contracts and longer negotiation periods to finalize them and release the research funding.

Industry research sponsors typically insist on owning technology that arises from research they fund, on the grounds that they have financed it. This does not, however, take into account the fact that universities also contribute to supporting these projects financially, because universities do not generally apply principles of full cost recovery when pricing these contracts. Research universities are therefore grappling with how to cost and price research contracts more effectively without alienating industry funders.

Companies wishing to access technology developed at a research institution that they have not funded are more likely to be open to a licensing arrangement, depending on the technology and the license terms.

1.3 *Performance of South African TTOs*

No comprehensive benchmarking of the performance of South African TTOs has yet been performed.[2] Table 1 provides rough data and estimates for four universities offering technology transfer services. These data have been compiled from anecdotal evidence and collegial information sharing among technology transfer professionals. While the data is incomplete (lacking some of the most important benchmarks, such as invention disclosures and patenting activity) and is not necessarily fully comparable in all cases across the surveyed institutions, it provides initial evidence to demonstrate that South African activity corresponds with experience elsewhere. Among other things, the Table indicates that the income accruing to universities from technology transfer activities is not substantial, that there is a time lag before a TTO can generate sufficient income to become self-supporting, and that the performance of TTOs at different institutions can vary widely. This is in line with what might be expected for a technology transfer system in its early days.

1.4 *The Southern African Research and Innovation Management Association*

Established in 2002, the Southern African Research & Innovation Management Association (SARIMA) is a stakeholder organization that provides a platform for individuals from government, academia, and industry, with an interest in research and innovation management, to interact on common issues. SARIMA's objectives include the professional development of those persons involved in managing research and in the creation of intellectual capital; promotion of best practices in the management and administration of research and in the use of intellectual capital to create value for education, public benefit, and economic development; advocacy of appropriate national and institutional policy to support research and generate intellectual capital; and advancement of science, technology, and innovation.[3] SARIMA has links with several local, African, and international organizations with related objectives.

2. KEY POLICY INSTRUMENTS

2.1 *Summary of main policies relevant to technology transfer*

With a new democratic regime in place since 1994, policy developments in South Africa have been numerous. Much attention has been given to supporting innovation, in acknowledgement of its critical role in promoting development, enhancing competitiveness, and improving quality of life. The 1996 White Paper on Science and Technology established the concept of a National System of Innovation (NSI).[4] The paper created the framework for a set of key enabling policies and strategies to inform the strategic development of science and technology in South Africa. In an effort to sustain the White Paper's vision for an effective, well-managed NSI and to improve the impact of the policy, the National R&D Strategy was released in 2002. This recommended specific strategic interventions to address identified weaknesses, including the commitment of substantial additional resources from government to support research and innovation.[5] Under the umbrella of the R&D strategy, various other initiatives have emerged, including the National Biotechnology Strategy[6] and the Nanotechnology

Table 1: Summary of TTO activity for Four South African Universities

	University A	University B	University C	University D	Note
Staff 2003	1,246	1,924	1,014	530	
Students 2003	19,978	24,769	16,660	27,729	
Licenses					4.0 licenses per US$100 million adjusted research expenditure
2001	2	0	3	3	
2002	4	0	3	1	
2003	3	0	3	1	
Spinouts					3.1 spinouts per US$100 million adjusted research expenditure
2001	1	0	4	3	
2002	0	2	2	4	
2003	1	0	1	0	
License income 2001–2003	R209,000	?	R1,656,948	R32,173	0.1% of research income
Patent budget 2002–2004	R450,000	R355,000	R500,000	R800,000	0.3% of research income
TTO staff FTEs					
Professional	1	4	3	4	
Support	1	1.5	2	1	

Strategy.[7] These aim to build on and enhance existing strengths in these key sectors, while developing human resources and generating research outputs to help South Africa to become more globally competitive and address some of its socio-economic problems. Of particular relevance to technology transfer practitioners was a proposal contained in the National R&D Strategy to introduce measures to encourage better protection and exploitation of IP arising from publicly funded research projects. This has recently been expanded upon with the release in 2006 of the Framework for Intellectual Property Rights from Publicly Financed Research.[8]

This framework is intended to bridge the "innovation chasm," which describes the gap in South Africa between knowledge generators (in particular, universities and research institutions) and the market. Although research organizations are performing some high-quality basic and strategic research, and while industry has some relatively sophisticated manufacturing operations, South African technology-led companies typically access their technology from abroad—local innovation has had relatively little impact on economic growth. The framework calls for a consistent approach to protecting IP developed with public financing, based on good practice globally while remaining responsive to the local context. Institutions will be required to put in place IP policies consistent with this legislation within a limited timeframe after the legislation takes effect. This will ensure a level of harmonization across institutions. One of the more significant provisions is that these policies would obligate employees and students to disclose all IP that they develop.

The framework draws heavily on the U.S. Bayh-Dole Act and proposes the adoption of several similar provisions. These include:

- conferring on institutions the responsibility to seek protection for their IP in exchange for the right to own and exploit it
- a reporting duty to a designated government agency about IP management activity
- an obligation to share revenues earned from the exploitation of IP with the individual inventors or creators of the IP concerned
- a right for government to a "free license" to IP should this be in the national interest
- a preference for licensing to local companies and small business

Additional provisions are proposed to address unique local conditions. In this vein, a further preference for licensing to Broad-Based Black Economic Empowerment (BEE) companies is recommended.[9]

A short public consultation process was carried out to give stakeholders the opportunity to comment on the framework. Legislation based on the framework, and taking into account responses received as part of the public consultation process, was being drafted at the time of writing.

2.2 *Innovation Fund*

The Innovation Fund is one of the main agencies responsible for implementing the R&D Strategy. It aims to promote competitiveness by investing in "technologically innovative R&D projects, the effect of which will be new knowledge and widespread national benefits in the form of novel products, processes or services."[10]

In its early days, the Innovation Fund was essentially a funding agency that supported research projects carried out by consortia (typically a combination of universities, science councils, and/or firms).[11] More recently, though, it has assumed a more proactive role in promoting technology transfer and assisting eligible South African institutions and researchers in their technology transfer activities.

The Intellectual Property Management Office (IPMO) and the Innovation Fund Commercialization Office (IFCO) are units within the Innovation Fund that support IP management and technology commercialization, respectively. They also assist in building capacity for the exploitation of IP, having co-hosted a series of training courses for technology managers with MIHR (the Centre for the Management of IP in Health Research and Development) and other organizations. An internship program in partnership with a multinational business consulting and advisory service firm has also been put in place. The Innovation Fund holds subscriptions to patent and marketing databases that can be accessed by universities and public research organizations at no cost or at subsidized rates. The Patent Support Fund allows universities and science councils to reclaim up to 50% of their patent expenditures annually. As an incentive to increase patenting activity, the Patent Incentive Scheme makes cash awards to inventors who have assigned their rights in an issued patent to a South African university or public research organization. The Innovation Fund has also provided financial support for various ad hoc initiatives, such as the establishment of university technology transfer offices and a university chair in intellectual property. It is proposed that the Innovation Fund be the designated reporting agency responsible for overseeing the implementation of the IP framework.

Other support measures for commercializing R&D include several directed-funding programs for research, development, and innovation, accessed on a competitive basis, funds from these programs are accessed on a competitive basis. Business incubators and government venture-capital funds are examples of other forms of support available.

3. TAKING STOCK

3.1 *A summary of progress to date*

Technology transfer in South Africa shows encouraging signs of progress:

- A handful of TTOs have been operating for several years and are now regarded as established entities within their organizations.
- Several new TTOs have recently been set up or are in the process of being launched.
- A track record of licensing deals and spin-out companies is gradually being built up.
- A core exists of professional, experienced technology transfer practitioners who are enthusiastic about sharing their skills with newcomers to the profession.
- A vibrant stakeholder organization provides a platform for networking and professional development in the field.
- Links have been forged that strengthen research collaborations and technology transfer partnerships with organizations elsewhere on the African continent and internationally.
- All of this is underpinned by support from government.

3.2 *Constraints*

Despite these advances, however, it must be acknowledged that technology transfer performance can, and indeed must, be improved. It is therefore instructive to identify the constraints and discuss how to overcome them.

3.2.1 *Few invention disclosures*

South African TTOs generally receive a weak flow of invention disclosures. There are several reasons for this. Some overburdened academics juggling heavy teaching loads, research responsibilities, and administrative duties are reluctant to take on the additional obligations that follow an invention disclosure. Other researchers are unaware or skeptical of the role of the TTO. Research funding levels are also fairly low, which limits overall research output (and thus the subset with commercialization potential). Furthermore, the typical funding mix of South African universities leaves them with a relatively small proportion of unencumbered IP. Few South African universities substantially contribute to research from their own internal budgets. Government funding makes up a relatively small proportion of total research expenditure, and so the greatest share of research funding comes from external sources, including local and international companies, philanthropic organizations, development agencies, and nongovernmental organizations. The research projects carried out with such funding are governed by research agreements that, among other things, lay out terms for the use and ownership of project IP. Commercial entities frequently insist on the assignment of any project IP, and even not-for-profit funding entities are increasingly demanding more stringent IP provisions (although generally for different reasons, such as ensuring their own freedom-to-operate for utilizing or disseminating the results of the research they fund).

The rate of invention disclosure could likely be improved to some extent by proactive actions on the part of the TTO (for example, more effective marketing of its services to potential clients within the institution, more frequent IP audits of research groups, or the introduction of internal procedures for compulsory disclosure prior to publication). But ultimately, more examples of successfully commercialized technologies are needed to persuade skeptical researchers that disclosing inventions is worthwhile.

3.2.2 *High costs associated with patenting*

Patenting costs are a problem. A new TTO typically struggles to secure a reasonable budget allocation for patent filing and prosecution. The TTO is sometimes viewed as competing with researchers, many of whom would prefer this funding to go directly to research. Patent protection is rarely worthwhile if pursued only in South Africa because the local market is not very large. The volatility of the currency makes it difficult to budget properly for international patent filing. Moreover, because of the pressure academics face to publish their research, patenting often takes place earlier than would be optimal, with the result that the technology is insufficiently developed to interest a licensee by the time it must be filed internationally. Universities cannot rely on licensees to assume foreign patent costs; at best, they can hope to be reimbursed at a later date, if and when the technology is finally licensed. TTOs are therefore severely constrained in terms of the number of patenting opportunities they can pursue.

This has been partially addressed by the Innovation Fund's Patent Support Fund, which allows universities and public research organizations to reclaim up to 50% of their expenditure on patent-related costs retrospectively.

3.2.3 *Limited capacity*

Local training opportunities are limited. There are only a few experienced technology transfer practitioners to act as mentors and share good practice. At the same time, the number of new entrants and available positions in the profession are too few to sustain specialized extended training programs. As a result, capacity-building initiatives consist of short courses that try to draw a wide audience by covering a broad range of general subject matter. Opportunities for continuing education on more advanced topics are rare and are often included as part of courses with a large proportion of beginners' content.

Longer-term capacity-building programs are being investigated, and some organizations have set up internship programs, but the system is probably still too immature to assess future needs accurately. The costs of an ambitious dedicated program will only be justified if there is a large enough pool of candidates. It is difficult to determine how quickly the system will be able to absorb new entrants as well as to estimate the number of technology transfer professionals needed to establish and sustain an effective system. Much of this will depend on when institutions without TTOs begin requiring technology transfer services (whether through an institutional TTO or via external service providers). Ongoing monitoring and refinements are likely to be required. Meanwhile, training opportunities overseas are also being explored.

3.2.4 *Unclear expectations and objectives for TTOs*

The rationale for university technology transfer is frequently misunderstood, which makes it difficult to obtain support from the broader university community. Income-generating objectives often assume greater importance than they should, and revenues accruing to an institution from technology transfer activities remain one of the main measures of success, despite the fact that most institutions explicitly acknowledge that income generation is not a major driver of their technology transfer activities. Among other things, this leads some academics to criticize the TTO on the ideological grounds that universities should not be undertaking commercial activity. Others resist the idea that the university has any right to IP that they feel entitled to own personally. Executive management often has unrealistic expectations about the financial returns that are likely to be generated by the TTO. When these fail to materialize quickly, they withdraw support or redirect the focus of the TTO. Clear objectives must therefore be set (preferably in conjunction with stakeholders) and communicated to all frequently and effectively.

3.2.5 *Difficulties with IP management in the life sciences*

The IP landscape has become increasingly complex, particularly with respect to biotechnological inventions. Available expertise, however, is limited. Only a handful of local patent attorneys have life sciences training, and those with advanced degrees are even rarer. Freedom-to-operate constraints are often encountered. Access to proprietary biological material, reagents, or tools for research purposes (for example, under an MTA) could facilitate the development of a new invention, but negotiating the rights for commercial use may prove too time-consuming or complicated to pursue, or the terms offered might be prohibitive.

3.2.6 *Limited licensing opportunities*

Licensing opportunities for existing companies are lacking. Domestic firms often lack the markets or distribution channels for viable exploitation. Without a track record or personal contacts to facilitate meaningful links, marketing to overseas companies can be difficult. At the same time, spinout opportunities for new businesses are few and far between. Financing is not easily raised from risk-averse financial institutions and venture capitalists, who are particularly wary of biotechnology because they do not understand it. Angel investors are few and far between.

4. CONCLUSION

Clearly, the impact of the IP Framework will be one of the most critical factors shaping the future prospects of South African technology transfer. Still, the ultimate success of this initiative is likely to depend on the implementation of details that are not provided in the Framework. These will have to be sufficiently flexible to accommodate the varying levels of resources, expertise, and capacity in research, research management, and technology transfer in different organizations.

Expectations will have to be managed carefully. A growing body of evidence shows that 1) substantial investments in technology transfer are needed to generate downstream benefits, 2) there is typically a significant time lag before net benefits are realized, and 3) the distribution of returns is very skewed (for example, analysis of AUTM surveys).[12] But in South Africa it remains a fairly common perception that the main motivation for undertaking technology transfer activities at a university is to generate income. This is fortunately not a universal perception, but technology transfer practitioners, government, and agencies such as the Innovation Fund will have to dispel such misperceptions via effective communication strategies.

One of the greatest benefits that the envisaged legislation might provide would be to align the IP policies of public funding agencies, which would reduce the transactions costs of navigating the complex and varied structures that are currently in place and that often require protracted negotiations. It is not apparent, however, that the legislation will achieve this.

Similarly, by providing clear guidelines for the use and ownership of IP developed at public research institutions with industry funding, negotiations around sponsored research agreements could be simplified and expedited. The Framework proposes a default position of ownership by the public organization, which can be altered if certain criteria are met. This establishes a useful starting point, as long as the process for exceptions to the default position is not made too cumbersome. Private-sector funding represents a higher proportion of overall research funding in South Africa than in many other countries (estimated at 28% overall according to CENIS[13]), and universities will want to avoid creating disincentives for their industry research collaborators and sponsors. At the same, such research support comes at a price because it seldom fully recovers costs and overhead charges. The IP Framework will strengthen the bargaining position of institutions in this respect by making it easier to price research contracts appropriately.

The Framework for Intellectual Property Rights has successfully drawn attention to the need for more effective exploitation of publicly funded research, stimulating a robust debate among stakeholders around the country. The real test of its impact, of course, will come with implementation. A positive outcome may be expected if a cooperative, enabling approach is taken that draws on the experience of organizations active in the field for some time. An approach that is too prescriptive and lacks sufficient flexibility to take into account unique circumstances will likely yield much less valuable results. ■

ROSEMARY WOLSON, *Intellectual Property Manager, R&D Outcomes, Council for Scientific and Industrial Research (CSIR), PO Box 395, Pretoria 0001, South Africa. rwolson@csir.co.za*

---

1 These have been in operation for at least five years and were included in an Innovation Fund assessment exercise conducted in 2003.

2 A handful of information-gathering exercises attempting an assessment have yielded data of limited value, because clear definitions of requested values were not always given and were thus not interpreted in the same way by all respondents, which made the information supplied incomparable.

3 SARIMA Charter.

4 Department of Arts, Culture, Science and Technology. 1996. White Paper on Science and Technology. www.dst.gov.za/legislation_policies/white_papers/Science_Technology_White_Paper.pdf.

5 Department of Science and Technology. 2002. South Africa's National Research and Development Strategy. www.dst.gov.za/legislation_policies/strategic_reps/sa_nat_rd_strat.pdf.

6 Department of Science and Technology. 2001. A National Biotechnology Strategy for South Africa. www.dst.gov.za/programmes/biodiversity/biotechstrategy.pdf.

7 Department of Science and Technology. 2006. The National Nanotechnology Strategy. www.dst.gov.za/publications/reports/Nanotech.pdf.

8 Department of Science and Technology. 2006. Framework for Intellectual Property Rights from Publicly Financed Research Department of Science and Technology: Brummeria, South Africa.

9 Broad-Based Black Economic Empowerment (BEE), a key component of the government's growth strategy, is considered a tool to broaden the country's economic base and accelerate growth, job creation, and poverty eradication, by addressing inequalities resulting from the past systematic exclusion of the majority of South Africans from meaningful participation in the economy. BEE is implemented and measured by means of a 'balanced scorecard' approach, which takes into account the elements of equity ownership, management, employment equity, skills development, preferential procurement, enterprise development and other residual elements.

10 www.innovationfund.ac.za/callforproposals.asp.

11 The Innovation Fund has been operating since 1997.

12 AUTM 2005. AUTM Licensing Survey™: FY 2004. Association of University Technology Managers: Northbrook, Illinois. www.autm.net. See also Scherer FM and D Harhoff. 2000. Technology Policy for a World of Skew-Distributed Outcomes, *Research Policy* 29; Heher AD. 2003. Return on Investment in Innovation: Implications for Institutions and National Agencies. Paper prepared for the First Globelics Conference on Innovation Systems and Development Strategies for the Third Millennium, Rio de Janiero; Advisory Council on Science and Technology. 1999. Public Investments in University Research: Reaping the Benefits. Report of the Expert Panel on the Commercialization of University Research Presented to the Prime Minister's Advisory Council on Science and Technology. acst-ccst.gc.ca/comm/rpaper_html/report_title_e.html Crown Copyright. 2003. Lambert Review of Business-University Collaboration.

SECTION **17B**

Putting Intellectual Property to Work: Experiences from Around the World

PUBLIC SECTOR INSTITUTIONS AND UNIVERSITIES

CHAPTER 17.8

The New American University and the Role of "Technology Translation": The Approach of Arizona State University

PETER J. SLATE, *Chief Executive Officer, Arizona Technology Enterprises, LLC, U.S.A.*
With contribution by **MICHAEL CROW**, *President, Arizona State University, U.S.A.*

ABSTRACT

This chapter provides a conceptual overview of Arizona State University's mission, and explains how the university's "technology translation" efforts support that mission. The chapter offers a rationale for why effective technology translation and commercialization are economically and socially relevant. A case study illustrates how a program established by Arizona State University's technology commercialization group has led to significant returns for the university and the local community. The authors conclude that public and private institutions in both developed and developing countries can implement the concepts and strategies for technology commercialization described in the chapter.

1. BACKGROUND AND INTRODUCTION

Arizona State University (ASU) is becoming recognized for having adopted one of the most forward-thinking university models in the United States, a new model of excellence and access, where connection to community is an expectation. Since one of the co-authors of this chapter, Michael Crow, became president of ASU in July 2002, the university's stature as a leading transdisciplinary research institution has grown significantly. Along with investments in transdisciplinary research infrastructure and new faculty, ASU has completely overhauled its technology commercialization capabilities and implemented programs that have improved the economic and social vitality of the state of Arizona in the southwestern part of the United States.

In the 2002 inaugural address to ASU faculty and administrators, Crow[1] unveiled a vision and strategy for a dynamic, inclusive university that assumes a share of responsibility for the economic and cultural development of the society it serves. The university would commit itself to outcome-focused excellence, both in the use-inspired research agenda it pursues and in the diversity of its student body. The university would become—to put it simply—a New American University.

As a New American University, ASU has been structured on fundamental design imperatives (Box 1). The spirit of these design imperatives is embodied throughout ASU's programs and strategic plans.

2. THE ROLE OF THE NEW AMERICAN UNIVERSITY

In order for ASU's research to be transformative, the university must have the staff, institutional and resource capacity to identify cutting-edge innovations and find creative ways to convert them into products that improve the quality of life. Within the framework of the New American University, the term technology transfer is abandoned in favor of technology translation. The

Slate PJ and M Crow. 2007. The New American University and the Role of "Technology Translation": The Approach of Arizona State University. In *Intellectual Property Management in Health and Agricultural Innovation: A Handbook of Best Practices* (eds. A Krattiger, RT Mahoney, L Nelsen, et al.). MIHR: Oxford, U.K., and PIPRA: Davis, California, U.S.A. Available online at www.ipHandbook.org.

Box 1: Design Imperatives for the New American University

1. **Leveraging Place**: Addressing the challenges of the region
2. **Societal Transformation**: Transcending physical location to affect society locally and globally
3. **Knowledge Entrepreneur**: Embodying a culture of academic enterprise, breaking from traditional and organizational constraints
4. **Use-Inspired Research**: Seeking research opportunities that meet community needs and enhance quality of life
5. **Focus on the Individual**: Looking beyond the academic background of incoming students to seek greater diversity of the student body
6. **Intellectual Fusion**: Adopting a research agenda that is solution-focused rather than discipline-focused
7. **Social Embeddedness**: Building an interactive and mutually supportive partnership with the community
8. **Global Engagement**: Establish programs and practices with global application through the development of innovative approaches to universal societal problems.

Source: ASU[9]

latter more appropriately captures the university's role, which is not simply innovating and transferring but, more importantly, framing innovations within the context of social and economic relevance.

Technology translation is predicated on building strong partnerships with the community and commercial entities so that the technology needs of the business and investment community are well understood. These partnerships are built around the university's core-technology competencies so that opportunities for technology development can be identified more effectively. Indeed, through technology translation, ASU provides a partnering experience more in line with the expectations of a commercial enterprise. In order to pursue this more market focused approach to building links with industry partners, ASU established a private enterprise so it could bring technologies to market more efficiently. In November 2003, ASU created Arizona Technology Enterprises, LLC (AzTE).[2] Figure 1 provides an overview of AzTE's technology-translation process and structures, which are discussed in the following sections. The translation process begins with the design of process elements that position AzTE between the market and the university. It is in this space where the work of translation can occur.

2.1 *Arizona Technology Enterprises*

AzTE is a private nonprofit, wholly owned subsidiary of the ASU Foundation.[3] The ASU Foundation was established to manage ASU's endowment and to make strategic investments for the benefit of the university. AzTE is responsible for evaluating, protecting, and translating ASU's technology portfolio. AzTE handles all of ASU's licensing, spinout company formation, consortia development, and joint venturing activities with commercial partners. Fundamentally, AzTE was founded on the notion that strong partnerships can only be established by being flexible, removing obstacles to doing business, and focusing on speed to market as a key driver in a university's dealings with its partners. AzTE's autonomy as a private organization, with most decisions being made internally, enables it to operate with the speed and efficiency of a market-based commercial enterprise.

The individuals who make up AzTE's business-development team have industrial backgrounds and strong product-development expertise. This expertise gives the company significant insight into the commercial drivers and hurdles of technology adoption in the private sector. The skills and network of AzTE's core team are supplemented by a board of directors, which is composed of venture capitalists, industry executives, technologists, and ASU leadership, as well as members of other ASU entrepreneurial programs (such as ASU Technopolis,[4] an education and networking program offered to the local business community). AzTE's strong network enhances its ability it to build relationships with industrial and financial partners.

AzTE provides to its spinout enterprises and commercial partners myriad services, including technology assessment, strategic business development, creative deal structuring, and capital formation. AzTE also offers advice on business strategy and is often instrumental in acquiring capital and management for ASU's spinout companies. Moreover, through the extensive network of ASU, the AzTE team, and its board of directors, AzTE acts as a source of business-development contacts for its partners.

In order to further develop promising technology platforms that may not have sufficient funding to achieve market viability, AzTE established the Catalyst Fund. Capital from the Catalyst Fund is invested by AzTE to conduct proof-of-concept experiments, develop prototypes, and provide seed funding to emerging ASU ventures. The Catalyst Fund has also been used to co-invest with industrial partners to develop ASU technology platforms. The company has found that small amounts of strategically allocated capital can exponentially improve the chances of a technology reaching the market.

2.2 *AzTE's market-focused model*

In addition to helping faculty incubate technologies in the existing ASU research portfolio, AzTE spends a significant amount of time meeting with industry-leading– and venture-capital companies to better understand their technology needs. By maintaining an ongoing dialogue with the business community, AzTE can continually connect these partners with sponsored-research and

Figure 1: AzTE Technology Translation Process

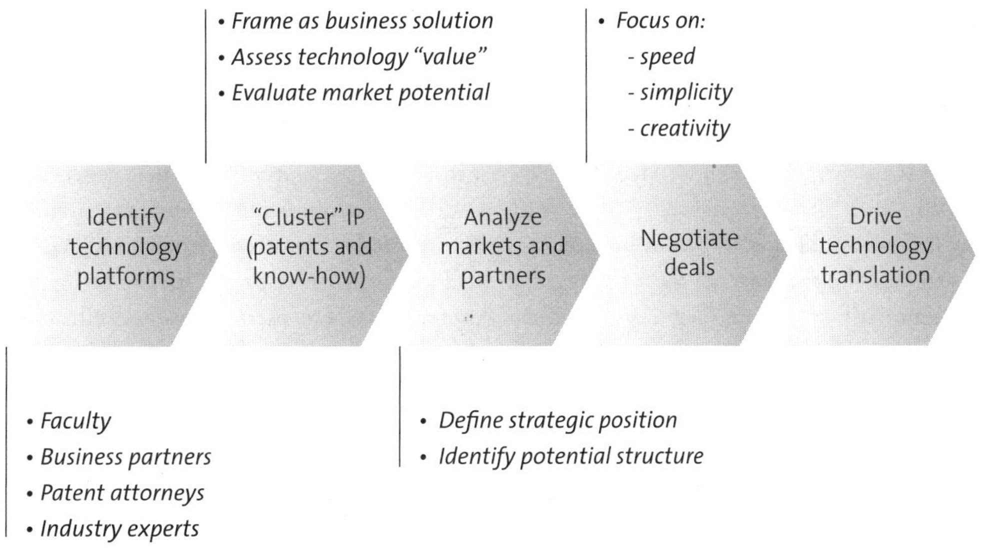

translation opportunities at ASU. This outside-in approach has significantly benefited the university. The approach provides a better understanding of societal needs and helps the university decide how to fill those needs. Moreover, these interactions have significantly contributed to the selection of ASU by many leading institutions as a partner of choice for technology acquisition.

2.3 *Knowledge entrepreneurship*

AzTE has developed programs that offer students many opportunities to gain unique, practical experience in technology-based transactions. In addition to hiring graduate students in business administration to work in AzTE's offices, AzTE has established the Technology Venture Clinic (TVC). The TVC is a multidisciplinary clinic that utilizes students from ASU's Sandra Day O'Connor College of Law, the W. P. Carey School of Business, Fulton School of Engineering, and the College of Liberal Arts and Sciences. TVC students evaluate ASU technologies, perform market research, identify commercialization opportunities, and assist with transaction negotiations. In exchange for their service, TVC students receive credit toward graduation. Privately funded by a leading corporate law firm in Phoenix, Arizona, Rogers & Theobald, LLP, the activities of the TVC offer unique experiences for its students and provide highly skilled assistance to the university's technology commercialization efforts.

In addition to the TVC, AzTE developed the Lisa Foundation Law Fellowship. Sponsored by a private foundation, the fellowship is offered each year to two top ASU law students with an interest in intellectual property (IP) law. With the guidance of an IP law firm, Steven G. Lisa, Ltd. (Chicago, Illinois), Lisa fellows learn how to draft and assess patent claims, search for prior art, and bolster claims of existing ASU filings. Like the TVC, the Lisa Foundation Law Fellowship gives a unique experience to students while providing an invaluable service to the university.

2.4 *External technology acquisition*

There are few institutions (either public or private) with an internally generated technology portfolio that, standing alone, can solve the world's most pressing health care and technology challenges. In order to develop an entity that can sustainably commercialize technology, be continually transformative, and create long-term value for the university and the community, AzTE strives to identify technologies developed by other institutions that can bolster the quality and value of ASU's technology portfolio. Bundling ASU IP with external portfolios is part of an ongoing dialogue between ASU and its commercial partners, and it has lead to joint development projects between ASU and other institutions, such as the Sun Health Research Institute (a leader in Alzheimer's research) and the Mayo Clinic. AzTE has begun to manage technology portfolios from other institutions that can be strategically bundled with ASU technologies to create new licensing and spinout opportunities. For example, one of AzTE's recent spinout companies was based on a sensor portfolio developed at Northern Arizona University.[5]

AzTE acquires access to external portfolios using a variety of structures including:

- management-service agreements to provide commercialization service in exchange for fees and/or on a contingency basis
- joint-commercialization agreements, whereby AzTE takes the lead on commercializing joint inventions
- acquisition or optioning of specific technologies of interest from another institution
- taking donations of technology portfolios from a public or private entity

Bundling technologies from other public and private sources that are synergistic with ASU's portfolio is an important part of AzTE's continued success. That is why AzTE is continually looking for opportunities to bring portfolios together where their combined effect is worth more exponentially than the sum of their individual effects.

2.5 *Speed, simplicity and certainty*

Technology translation and commercialization is sometimes called a contact sport. Transactions can take up to 18 months to consummate, and the proportion of patented innovations that

TABLE 1: PROGRESSION OF AN AZTE TRANSACTION

TIME	ACTIVITIES
MONTH 1	• hold introductory meetings • provide potential partner with nonconfidential information on technology and value proposition • respond to due diligence questions • evaluate partner's interest in moving forward and ability to maximize technology value
MONTH 2	• sign confidentiality agreement • provide confidential information on technology • assess value market opportunity and transaction economics • engage in detailed discussions between potential partners and inventors
MONTHS 3 & 4	• develop term sheet with business terms • negotiate agreements • consummate transaction

actually make it to market is relatively small. It is therefore essential for any organization engaging in technology commercialization to adopt a disciplined approach to deal making. AzTE strives to move from first contact to consummating a deal in four months. Table 1 illustrates the progression of an AzTE transaction.

The AzTE transaction team has developed three key guiding principles that govern all of its business negotiations, regardless of deal size or structure: "Speed, Simplicity, and Certainty."

- **Speed.** AzTE's autonomy and culture allow it to move quickly to consummate transactions. This is essential in today's dynamic technology marketplace. Speed in deal making is crucial for establishing strong partnerships. If a party is unable to move swiftly through the due diligence- and documentation processes, it may lack commitment to the project, or there may be insufficient buy-in at higher levels within the organization. This can affect a project's success.
- **Simplicity**. Early-stage technology transactions and joint-development projects are inherently complex. Given the numerous risks involved (for example, a technology not achieving its commercial endpoint or a partner's change in priorities), the odds of most early-stage technology transactions achieving success are low. Because of this, it is important that the structure of a transaction be kept as simple and flexible as possible. Many transactions fail because parties are unable to agree on terms that, in the end, do not fundamentally matter to a project's success.
- **Certainty.** The promise of value can be elusive if the counterparty to the transaction is difficult. Successful technology development transactions are based on successful relationships. Indeed, effective deal making requires the discipline to prefer a lower offer from a party with whom one might succeed, to a higher offer from a party that is less likely to see the project through.

2.6 *Faculty engagement in the technology assessment process*

For many faculty researchers, a significant portion of the time they will spend with the AzTE team involves the process of evaluating their inventions. As a result, AzTE has developed a technology evaluation process that, in addition to evaluating the commercial relevance of a disclosure, is designed to provide an opportunity for faculty to get to know the AzTE team and gain insight into how evaluation decisions are made. ASU researchers work alongside the AzTE team to evaluate the technology. The team shares with the researchers all of the technology and market due diligence performed. If a technology does not meet the university's investment criteria after being thoroughly evaluated, the technology is generally returned to the inventor along with all due diligence materials compiled during the evaluation process. Including inventors in the process has helped to minimize disputes over whether an investment decision was fairly determined. Additionally, close interaction between the AzTE team and researchers has taught inventors to better appreciate market needs and expectations, which has increased the quality of invention disclosures filed by ASU faculty and researchers.

3. BENEFITS OF TECHNOLOGY TRANSLATION

3.1 *Private sector benefits*

Between the research institutions that create innovations and the customers who eventually use them sit the technology adopters. These are the industrial companies, development companies, and other enterprises that adopt early-stage ideas and convert them into useable products and services that address market needs. A number of trends are providing significant opportunities for universities with effective technology commercialization programs to build strong partnerships with these technology adopters. A few of these trends are discussed below.

Only about 15% of the market capitalization of companies that make up Standard & Poor's 500 share index (a division of the McGraw-Hill Companies, Inc.) can be tracked to balance sheet net asset value.[6] This means that approximately 85% of these companies' market values can be attributed to intangible assets. The growing appreciation of the importance of intellectual assets has prompted leading companies to manage their patents with a level of scrutiny that was once reserved only for "brick and mortar" assets. Many companies are hiring senior level intellectual asset managers. Such a manager would continually evaluate whether the company's IP strategy is aligned with its business strategy, and whether the acquisition of additional technology portfolios is necessary for success. Some of the factors influencing technology-focused companies to look beyond their internal R&D efforts to find the next big thing include:

- **Market Competition.** In order to become more competitive in the global marketplace, today's companies are more likely to in-license core technology platforms so that they can get to market quicker and access greater opportunity.
- **Technology Convergence.** Cutting-edge technology platforms are complex and require multidisciplinary expertise. For example, the next generation of flexible display technology will require in-depth expertise in engineering, material sciences, microelectronics, and nanotechnology. Such a diversity of disciplines is prohibitively expensive for many companies to develop internally.
- **Innovators' Dilemma.** Many larger companies have difficulty innovating in a way that significantly changes their business. As a result, many internal R&D programs focus on incremental improvements to existing product lines. To remedy this problem, companies look outside of their internal programs to identify disruptive, "game-changing" technologies.
- **Lack of R&D Productivity.** Better tools and access to information have enabled companies to more efficiently assess the return on their internal R&D programs. Internal development projects that are not productive can be terminated in favor of acquiring technology elsewhere.

3.2 *Public sector benefits*

Companies in developed nations struggle with the economics of selling products in developing regions. Because universities are not as pressured by the competitive, profit-focused aims of the private sector, they can deploy significant resources to tackle some of the most vital challenges in these societies. Moreover, well-run technology translation programs can implement strategies to enhance the adoption of licensed technologies in developing countries. The following are some examples of strategies that ASU and other research institutes have pursued:

- reserving *carve-out rights* in licensing agreements to continue to allow the university to use and provide, for charitable purposes, private access to the technology
- favoring commercial partners that are willing to commit to providing technology access in developing regions over those who will not
- encouraging partners to set up regional joint ventures with companies capable of bringing technologies to market in developing regions
- providing to partners financial flexibility in the form of reduced royalties and other discounts to help make product development and marketing in developing countries more attractive
- providing field-of-use licenses and regional/geographic use licenses to ensure that the best commercialization partners are selected for geographic regions

Public and private research-granting organizations recognize the importance of technology translation for ensuring that funded research programs result in products that improve the quality of life throughout the world. Many granting agencies require that grant applicants provide in their applications a technology adoption and commercialization plan along with the research plan. As part of this trend, AzTE participates in ASU's application and acquisition of grants from public and private sources. In 2004, AzTE participated in developing the Intellectual Property Sharing Plan for a US$43 million grant, which ASU received from the U.S. Army, to establish the Flexible Display Consortium, a university/industry consortium developed and led by ASU to create the next generation of flexible display technologies. Box 2 provides a summary of IP management terms that public research institutions can adopt when structuring a public/private consortium.

3.3 *Local economic development benefits*

University technology translation and market-based commercialization can significantly affect the local economy. Consider the following example of a recent ASU transaction that is helping to grow the economy in Phoenix, Arizona.

Agilent Technologies, Inc. is a premier measurement-instrument and technology company with revenue in excess of US$5 billion per year. In November 2005, Agilent Technologies purchased Molecular Imaging Corp. (based in Tempe, Arizona), an ASU spinout company that has become a leader in atomic-force microscopy (a technology widely used to measure properties of materials at the nanometer scale).

In 1993, an ASU professor, Dr. Stuart Lindsay, developed his groundbreaking measurement technology. With the assistance of ASU's technology commercialization office, Dr. Lindsay and his team founded Molecular Imaging. Through a sponsored-researcher relationship with ASU, the company continued to leverage the university's research capability and infrastructure to develop its products. To build the company, Dr. Lindsay attracted entrepreneurial talent and capital to Arizona from across the United States. In fact, many employees were offered research positions at ASU. Discussions with Agilent during and after negotiations revealed that it valued the strong partnership between Molecular Imaging and ASU. Partly because of this, Agilent declared its commitment to keeping the Agilent business unit in Tempe and to growing the business locally. Agilent's investment in Arizona will yield significant benefits, including new-technology partnering opportunities, partnership opportunities for local businesses, and more technology-related jobs. Soon after the acquisition closed, AzTE began working with some of the founders of Molecular

Imaging on the next promising entrepreneurial spinout venture. AzTE is also in discussions with Agilent regarding additional technology licensing opportunities. Despite the obvious benefits of the deal to Agilent and Molecular Imaging shareholders, this transaction serves as a billboard for the power of technology translation and its impact on local economic development.

4. CONCLUSION

The importance of effective technology translation is profound. Since the enactment of the Bayh-Dole Act in 1980,[7] products derived from the research community have accounted for more than $40 billion[8] in market value alone, even without considering the positive impact on the economy. In the three years of AzTE's existence, the company has started 13 other companies, entered into over 80 commercialization transactions, and generated more than US$8 million in revenue. During the last 24 months, three of the 13 companies were sold to acquirers located in Arizona that plan to continue to grow these companies locally.

From a research institution's perspective, an effective technology translation program not only generates significant revenue for research, but also develops an entrepreneurial culture among university researchers and private researchers. For the international community, technology translation can be an important catalyst for economic development and a significant source of partnerships with the business community. Although President Crow's model for the New American University may not be adoptable completely for all institutions, its principles of social engagement and creative technology partnering can be adapted for use by other public and private institutions and can yield significant returns for those institutions in developing regions throughout the world, while benefiting people in those regions. ■

PETER J. SLATE, *Chief Executive Officer, Arizona Technology Enterprises, LLC, 699 South Mill Avenue, Suite 601, Tempe, AZ, 85281, U.S.A.* pslate@azte.com

MICHAEL CROW, *President, Arizona State University, 300 E. University Drive, Fulton Center, Suite 4104, Tempe, AZ, 85287-7705, U.S.A.* Michael.Crow@asu.edu

---

1 Crow M. 2002. A New American University: The New Gold Standard (Inaugural Address). ASU: Tempe, Arizona, U.S.A. http://www.asu.edu/inauguration/address/.

2 www.azte.com.

3 www.asufoundation.org.

4 www.asutechnopolis.org.

5 www.nau.edu.

6 www.maxiam.com/index.cfm?fuse=aboutIAM.

7 35 U.S.C. 200-212

8 *The Economist*. 2002. Invention's Golden Goose. December 14.9 See www.asu.edu and www.asu.edu/president/newamericanuniversity/.

9 ASU. 2005. *Arizona State University: A New American University*. Arizona State University, Tempe, AZ. www.asu.edu/president/newamericanuniversity.

Box 2: Summary of Terms- IP Management Plan for Public/Private Consortium

1. Selected Definitions

"Background Technology" means all Member Technology and UNIVERSITY Technology that may reasonably be expected to be required to conduct a Center Project.

"Center Projects" means projects identified in the annual plan created and amended from time-to-time, as referenced in the Cooperative Agreement, that details projects, milestones, principal investigators, and resources committed for Center activities.

"Center Technology" means all Technology that has been conceived: (1) by one or more Center Members or UNIVERSITY on a Center Project using the center facilities, or personnel of the Center or UNIVERSITY or personnel of a Member that are dedicated to the Center or (2) by one or more Center Members or UNIVERSITY using government funds allocated to the Center for Center Projects.

"Improvement(s)" means any Technology that constitutes an improvement, modification, or derivative of an item of Center Technology, but which is not itself Center Technology.

"Member Technology" means all Technology conceived, owned, or controlled by a Member that is not Center Technology.

"Technology" means all intellectual property rights, discoveries, innovations, know-how, works of authorship, and inventions, and derivative works, whether patentable or not, including computer software and code, as intellectual creations to which rights of ownership accrue, including, but not limited to, patents (including U.S. or other international or foreign patents or patent applications, whether provisional, non-provisional, or continuing, or any addition, division, continuation-in-part, substitution, renewal, reissue or extension thereof), trade secrets (as defined in the Uniform Trade Secrets Act), maskworks, and copyrights and copyrightable material.

"University Technology" means all Technology conceived, owned, or controlled by University that is not Center Technology.

---

2. Ownership

(a) Ownership of Center Technology. Inventorship of Technology is determined in accordance with U.S. patent laws. Each Member whose personnel are inventors of a particular item of Center Technology jointly owns that item in undivided shares. UNIVERSITY is deemed to be in inventor on any case where Technology was developed with Significant Use (that is, a use that materially contributes to the generation, creation, or development of Center Technology) of center facilities unless use of Center Facilities was separately paid for at full cost.

(Continued on Next Page)

Box 2 (Continued)

(b) Ownership of Member Technology and UNIVERSITY Technology. All Member Technology and UNIVERSITY Technology shall continue to be owned by such Member or UNIVERSITY and, except for specified circumstances, there is no obligation to license such Technology to others.

(c) Special Rule for Subcontracts. All Members with ownership rights in Center Technology solely by virtue of performing a subcontract for experimental, developmental, or research work, grant the licenses below regardless of the terms in any such subcontract.

(d) Ownership of Improvements. Members or UNIVERSITY who independently conceive of an Improvement on Center Technology that has been publicly disclosed shall own such Improvements, except to the extent the Improvement constitutes Background Technology of a Member or UNIVERSITY disclosed solely for the purpose of granting non-commercial uses on Center Projects.

Improvements by Members or UNIVERSITY based on Center Technology that has not yet been publicly disclosed shall be owned by the Inventing Members or UNIVERSITY, subject to the grant of license described below.

---

3. Licensing

(a) License for Research and Educational Use of Center Technology. UNIVERSITY and all non-Inventing Members (other than Channel Members) are granted a royalty-free, nontransferable, nonexclusive right to make, use, and have made on their behalf items of Center Technology solely for internal research and development purposes as required by such Member to perform research and development under a Center Project. Provided appropriate steps are taken to protect the Technology, UNIVERSITY shall have the same rights with respect to not-for-profit teaching and other educational purposes.

(b) Licensing for Commercial Uses.

(i) Non-Inventing Members. Non-Inventing Members have the right to negotiate with any Inventing Member for commercial use of an item of Center Technology on terms as they shall mutually agree. Commercial use licenses of non-Inventing Members extend to Affiliates of the Members. Subject to certain legal limitations that products be manufactured substantially in the U.S., Members may negotiate with the Inventing Members for an exclusive or co-exclusive right to any Center Technology provided all other Members agree to terms of such license.

(Continued on Next Page)

Box 2 (Continued)

(ii) Non-Member Third-Parties. Inventing Members may negotiate with non-Member third parties on such terms as they shall mutually agree for commercial use of Center Technology 18 months after the Center Director circulates a disclosure of such Center Technology to all Members.

(iii) Royalties. All remuneration received by any Inventing Member for licensing an item of Center Technology, less an administrative fee, is shared equally among all Inventing Members of such Center Technology.

(c) Licensing of Background Technology. Members are not obligated to license Background Technology, except that with respect to certain Background Technology identified by a Member to be included in Center Projects, Members and UNIVERSITY are granted a non-exclusive use for non-commercial activities on Center Projects identified in the Annual Program Plan. Members are not prohibited from negotiating licenses to such Background Technology on such terms as they shall agree.

(d) Licensing for Improvements to Center Technology. With respect to Improvements of UNIVERSITY or Members on Center Technology that have not yet been publicly disclosed: (a) All Members and UNIVERSITY are granted a royalty free, nontransferable, non-exclusive license solely for non-commercial purposes to conduct activities on Center Projects; and (b) all Members and UNIVERSITY have the right to negotiate in good faith for a non-exclusive license to use Improvements for commercial purposes.

With respect to Improvements of UNIVERSITY or Members on Center Technology that has been publicly disclosed, neither UNIVERSITY nor the Member(s) are required to license the Improvement except to the extent of any non-commercial license required under Section 3 above if the Improvement constitutes Background Technology.

---

4. Disclosure of Center Technology

Members must promptly disclose to the Center Director: (a) all Center Technology on a Center Invention and Discovery Disclosure Form, (b) patent filings, and (c) details of licenses entered into for Center Technology.

---

5. Management and Prosecution of Center Technology

Inventing Members of Center Technology appoint a Member to manage and facilitate the filing, maintenance, and prosecution of patents and copyrights (the "Designated Prosecution Member"). If the Inventing Members cannot agree on a Designated Prosecution Member, the determination is made by the Center Technology Committee. Costs related to filing, prosecution, and maintenance

(Continued on Next Page)

Box 2 (Continued)

of patents and copyrights are shared equally by Inventing Members. Each Member is responsible for the prosecution for patent application for its own Background Technology.

---

6. Follow-on Center Members

With respect to Center Technology (and Background Technology or Improvements subject to the licenses described above) developed prior to a new Member becoming a Member, the new Member: (a) is granted licenses solely for internal research and development purposes under a Center Project, and (b) may negotiate for licenses with respect to commercial use for such Center Technology.

---

7. Infringement

Members have a duty to notify the Center Director of suspected infringement of Center Technology. With the consultation of Inventing Members and the Center Director, the Designated Prosecution Member determines the proper course of legal action. The expenses and any settlement shall be shared equally, less an administrative fee. In certain cases, Inventing Members need not participate in legal actions. Inventing Members cooperate to defend validity challenges by third parties.

IP Management at Chinese Universities

HUA GUO, *Patent Specialist, Jones Day, China*

ABSTRACT

For the People's Republic of China, intellectual property (IP) is a new legal and social concept. Formal legislation was first introduced in the 1980s and was later strengthened. Due to recent publicity, however, social awareness of IP rights in China has grown. Following a series of ministerial and commission rules concerning technology transfer, universities now usually own the IP resulting from government-funded research. Not surprisingly, the number of patent applications filed by Chinese universities has increased rapidly, exceeding 13,000 in 2004. But such numbers may reflect a trend for researchers and institutions to use patents as a way of enhancing their reputations, rather than for actually transferring or commercializing technology. Most universities still lack institutional IP policies and independent offices responsible for IP management. Rates of technology transfer and commercialization, while difficult to observe, remain low. Still, some world-class universities, such as Tsinghua University and Beijing University, have become adept at IP management. These are both an exception to and an example for other universities in China, having successfully adapted IP management policies and practices to the country's legal and economic circumstances.

1. A BRIEF LEGISLATIVE HISTORY OF IP LAW IN CHINA

The formulation of laws and regulations to govern IP rights in China began in the early 1980s. The protection of trademarks and copyright has, to some extent, existed in China for a long time, but a formal Chinese trademark law wasn't promulgated until 1982. The enactment of the trademark law was a milestone for the establishment of a modern IP rights regime in China. A Chinese patent law followed 1984, and a Chinese copyright law was adopted in 1990.

These three laws have been amended several times to improve the protections they provide. The first amendment of the patent law, in 1992, expanded the scope of patent protection to chemical products and extended the term of utility patents to 20 years and design patents to ten years. In 2001, the patent law was amended again to offer new judicial and administrative protections, improved application procedures, and simplified enforcement procedures. The trademark law has been amended twice since its adoption; the copyright law has been amended once. The latest amendments of these three laws have offered stronger protection of IP rights in line with the requirements of the World Trade Organization (WTO) and the Agreement on Trade-Related Aspects of Intellectual Property Rights (TRIPS).

In an effort to bring its IP protection into accord with international systems, China has actively participated in most of the major international IP organizations and treaties since 1980. It is now a member of the World Intellectual Property Organization (WIPO), the Paris Convention for the Protection of Industrial Property, the Madrid Agreement Concerning the International Deposit of Industrial Designs, the Berne Convention for

Guo H. 2007. IP Management at Chinese Universities. In *Intellectual Property Management in Health and Agricultural Innovation: A Handbook of Best Practices* (eds. A Krattiger, RT Mahoney, L Nelsen, et al.). MIHR: Oxford, U.K., and PIPRA: Davis, U.S.A. Available online at www.ipHandbook.org.

the Protection of Literary and Artistic Works, the Patent Cooperation Treaty (PCT), and WTO, including TRIPS.

In all of Chinese legislative history, no laws have received more attention than those concerning IP. The Chinese government has tried to establish a legal system that meets the current level of IP protection in the world system. Of all the laws in China, these IP laws are the closest to corresponding laws in developed countries. In other words, China has tried, in 20 years' time, to reach the level of IP protection that it took developed countries more than 100 years to reach. One result of such rapid progress is that the resulting laws actually go beyond the common recognition and practice of society. In the past 20 years, China has been transitioning from a planned economy to a market economy. For the Chinese, IP is something of a new phenomenon, and all issues involving IP in China should be understood in the light of this. Under the centrally planned economic system, typically only a few patent applications were filed, and they had little meaning. With the new legislation and the publicity associated with it in recent years, social awareness of IP rights in the country has gradually increased.

2. REGULATING ENTITLEMENT OF IP RIGHTS

In Chinese patent and copyright law, only general definitions are given of an employee's invention or work.

2.1 *Patent law*

2.1.1 *Article 6*

An invention made by a person during the execution of tasks for an employer, or involving the use of materials and technical means belonging to, or provided by, the employer is considered to be a service invention, or *work for hire*. For a service invention, the right to apply for a patent in China belongs to the employing entity. After the patent application is approved, the employing entity shall be the patentee.

For an invention that is not a service invention, the right to apply for a patent belongs to the inventor. After the patent application is approved, the inventor shall be the patentee.

With respect to an invention made by a person using the material and technical means of the employer, where the employer and the inventor have entered into a contract that provides for the right to apply for and own patents, the terms of the negotiated provision shall apply.

2.1.2 *Article 8*

For an invention made by an entity or an individual working under commission or contract for another entity or individual, the right to apply for a patent belongs, unless otherwise agreed upon, to the entity or individual that made the invention. After the patent application is approved, the entity or individual who made the application shall be the patentee.

2.2 *Copyright law*

2.2.1 *Article 16*

Work created by an individual in the fulfillment of tasks assigned to him or her by a legal entity or organization shall be deemed to be a work created in the course of employment. The author shall hold the copyright to such work, provided that the employing legal entity or organization shall have a priority right to exploit the work within the scope of its professional activities. For the two years after the completion of the work, the author shall not, without the consent of the employing legal entity or organization, authorize a third party to exploit the work in the same way as the employing legal entity or organization does.

The author of a work created in the course of employment shall enjoy the right of authorship, while the employing legal entity or organization shall enjoy other rights included in the copyright and may reward the author, as in the following cases:

- drawings of engineering designs, product designs, and maps, computer software, and other works are created in the course of employment mainly with the material and technical resources of the legal entity or organization and under its supervision

- works created in the course of employment, in accordance with laws, administrative regulations, or contracts, enjoyed by the legal entity or organization.

2.3 *Ownership of IP created under government funding*

Before 1994, there was no uniform government policy regarding IP created with government funding, and the government took title to all IP rights resulting from work that it funded. In China, almost all universities and research institutes undertaking government projects were legally considered state-owned entities. The government was thus entitled de jure to IP rights from them. However, rights were held de facto by the universities or research institutes. Because there was no government policy regarding the entitlement and transaction of IP made under government funding, universities had no impetus to engage in IP management. In addition, few universities or institutes understood the importance of IP. Accordingly, IP management in universities and research institutes was virtually nonexistent.

In 1994, the former National Commission of Science and Technology issued a regulation titled *Measures for Intellectual Property Rights Made under the Governmental Funding of the National High Technology Program*. It provided specific rules for the ownership of intellectual property rights to inventions developed with government funding and contains several important provisions:

- When the government signs a contract with the university or institute, the ownership of the IP rights should be provided for.
- Unless otherwise stipulated in the contract, the university or research institute is entitled to all IP rights pertaining to inventions funded by the government.
- The university or institute should disclose the results to the funding government agency within 30 days after completing the project, and decide whether to file a patent application. In addition, the university or institute has the option to keep the results as a trade secret.
- The university or institute must submit a report with a plan for utilizing the invention to the funding government agency within six months after completing the project.
- The university or institute is entitled to the copyright on the work, including software funded by the government.
- The university or institute can use, assign, and exclusively license IP or trade secrets funded by the government.

Although this is only a ministerial rule, it is the first uniform government policy in China regarding the ownership of IP on inventions funded by the government.

In 2002, the Ministry of Science and Technology and the Ministry of Finance jointly issued Measures for Intellectual Property Made under Government Funding, which are often called the "Chinese Bayh-Dole Act." Based on the previous regulation, it goes even further:

- The university or institute is entitled to IP made under government funding.
- The funding government agency may decide, for compelling reasons (such as the security of the state, other vital interests of the state, or vital interest of the public), that title to the IP should be vested in the government.
- The university or institute can use the results or IP by itself or can assign or exclusively license them to a third party.
- The government retains a nonexclusive, royalty-free license to practice inventions made under government funding.
- The university or institute is entitled to receive revenue from commercializing the IP, but the university or institute must share with the inventor(s) a portion of any revenue received.
- Under certain circumstances, the government can require the university or institute to grant a license to a third party.
- Universities or institutes must give preference to the inventor when commercializing an invention.

- When a university or institute applies for government research funding, the application should contain an analysis of the feasibility of obtaining a patent.
- IP costs are to be borne by the university or institute.

In 2002, the Commission on Science, Technology, and Industry for National Defense issued a regulation titled Measures for Intellectual Property Rights Made under Governmental Funding of Defense Technology Projects. It states that:

- Unless separately provided for in a contract, the contractor is entitled to IP contained in inventions developed as defense technology projects and made under government funding.
- The funding government agency may decide, for compelling reasons (such as the security of the state, other vital interests of the state, or vital interest of the public) that title to the IP should be vested in the government.
- Under certain circumstances, the government can require the contractor to grant a license to a third party.

According to the above measures, IP resulting from research funded by the government is in practice usually owned by universities. However, there is still no law in China specifically covering the ownership of IP rights created under governmental funding.

3. IP MANAGEMENT AT CHINESE UNIVERSITIES

3.1 *Growth of patent applications by Chinese universities*

After ministerial and commission rules were issued and IP rights enforcement was strengthened, the number of patent applications filed by universities increased rapidly. Figure 1 shows statistics on patent applications by Chinese universities.

FIGURE 1: PATENT FILINGS BY CHINESE UNIVERSITIES

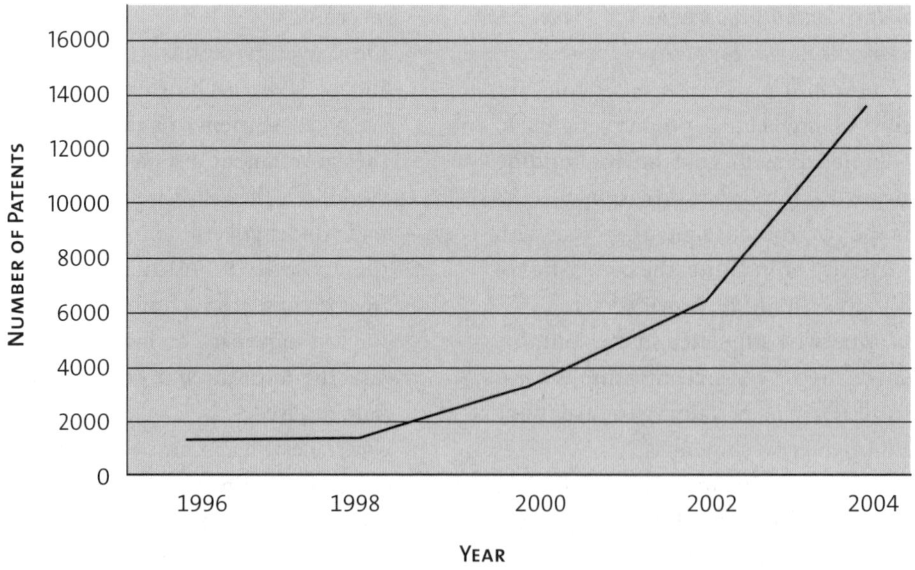

Source: China Education Online

This increase is mostly due to Chinese universities' growing awareness and recognition of the value of IP and has been accompanied by a growing acceptance of the idea of IP in Chinese society. Still, the rapid increase of university patent applications is also partly due to government policy. Recently, with regard to a university's reputation or an individual faculty member's chances at promotion, the number of patent applications has become almost equal in importance to outside reviews and number of publications. For some universities and their faculty, patent applications have become a substitute for publications (patent applications are considerably easier to obtain than publications). Additionally, in some universities students are required to submit a publication or patent application to graduate. Recently, the Ministry of Education has ranked universities based on the number of patent applications filed. Universities have begun to pay significant attention to patents because they closely correlate with institutional reputation. But because patents now garner institutional prestige, universities are filing patent applications for inventions that are not patentable or have little commercial value. In fact, despite this surge in the number of patent applications, some real problems with IP management in Chinese universities remain.

3.2 *Lack of institutional IP policies and understanding*

Despite the numbers, the loss of potential IP by universities is a serious problem. Most universities do not have a clear IP policy. While Chinese patent and copyright laws articulate, in principle, what constitutes an employee's invention and a work for hire, most Chinese universities lack clear interpretations and policies to implement. There is often no definition in place for what constitutes an employee invention or a work for hire, and no common procedure for disclosing inventions or for filing patent applications. Clauses related to IP appear seldom in employment contracts between universities and faculty, and many universities lose IP due to the mobility of faculty members, students, and visiting scholars among universities and between universities and industry. Even at universities that do have an established employee IP policy, it is often unclear to whom that policy applies. With the increasing recognition of the value of IP, more and more issues are arising about who owns patents or software.

While growing, faculty and student awareness of IP issues still falls short of what is needed, and even basic concepts are not always understood. While many Chinese universities offer some form of IP education to faculty members and students, it is seldom systematic or regular and, in most cases, has little effect.

Most faculty members do not have a concept of a publication bar. Patent applications are often filed at the same time or even after results are published or disclosed publicly. Many patent applications are therefore rejected because of the publication bar[1] or because of a lack of novelty.

3.3. *Lack of institutional offices of IP management or technology transfer*

Most universities lack an independent office responsible for IP management or technology transfer. Such functions usually fall to the office of research and technology. The primary responsibilities of this office are to apply for government research funding, supervise projects, report the results of projects to the funding agency, and facilitate publication of articles based on the projects by faculty. Typically, there is not even a single full-time staff member responsible for IP management.

Under the past, centrally planned economic system, the most important duty of the office of research and technology was to report on a project's results to the corresponding committee of experts at the government agency that funded the project. Even now, such reporting is still common in most Chinese universities and research institutes. All project results are appraised for awards through a process of review and discussion by the agency's committee. Prizes are awarded to those projects whose results were determined to be sufficiently advanced. These prizes have a significant impact on the reputation of universities and the promotion of individual faculty members, as do publications. Publications and award appraisals emphasize final outputs. Indeed, in administering research projects, most attention is generally focused on the final stages and outputs of the research process. In most universities, there is almost no administrative oversight of the early stages of research projects.

Since the university office of science and technology, which is in charge of IP management, focuses its work on the outputs of the research process, not much thought is given to the patentability of a technology when applying for project funding. And during the course of research, little attention is paid to prior art as it is articulated in patent law. For most projects, assessing patentability and prior art is only done after the work is completed.

Since the office of research and technology concentrates most of its efforts on project administration, members of that office are called upon to manage IP only as a part-time job. As a result, IP management is a secondary consideration compared with the daily work of the office. The lack of professionals who specialize in IP at Chinese universities partly reflects the rarity of professionals with expertise in IP in China.

3.4 *Growth in patent applications does not mean growth in technology transfer*

For most Chinese universities, IP management essentially means making patent applications. Most Chinese universities do not have anyone specifically responsible for technology transfer. Without a technology transfer office (TTO) or anyone in charge of technology commercialization efforts, little effort is made to promote the actual transfer or commercialization of the resulting patents. In many cases, patents and commercially valuable research results are simply left on the shelf. Little is done to publicize them, making it hard for industry to learn about new technologies. Even when an entrepreneur might be informed and interested in licensing a patent, it is often unclear who in the university has the authority to negotiate.

Some universities authorize external IP agencies to manage their IP, but usually such agencies merely concentrate on filing patent applications. The available external agencies likewise lack professionals with expertise in IP transactions.

Another reason for the low level of technology transfer is that patents are often applied for without any investigation into the market demand for the invention, which means that much of the university's IP lacks commercial value.

The very low commercialization rate at Chinese universities produces insufficient revenues to cover patent costs, and this, in turn, affects the university's ability to obtain and maintain patents. Patent costs are covered exclusively by the universities and usually come out of research funding lines. Many patents are not being maintained because universities lack funds to pay maintenance fees. As patent costs increase in China, and because it is much more expensive to file in foreign countries, universities may increasingly hesitate to file patent applications because of budget concerns. Unless it is backed up by viable technology transfer, patenting alone will be unsustainable.

3.5 *Lack of policies on revenue sharing, conflict of interest, and sponsored research*

Most Chinese universities lack clear policies about how revenue from IP will be shared with the inventor. And for those that do have such policies, the proportion of revenues shared often does not accurately reflect the inventor's effort or contribution. With no definition provided by university policy about what *hired to invent* or *work for hire* means, or to whom these terms should be applied, and with no restrictions imposed by the employment contract, many faculty members prefer to increase their personal advantage by collaborating directly with industry. Recently, faculty members have engaged more and more in part-time employment or contracting with industry. Resulting conflicts of interest between faculty obligations to the university and to industry are very common. Still, most Chinese universities lack any policy regarding faculty conflict of interest.

In collaborative research agreements between industry and universities that are funded by the industry partner, universities often give up the rights to ownership of IP made under such sponsorship. IP clauses in industry sponsored research agreements can sometimes be interpreted as inequitable to the university, especially when the contract is with a large or influential company. Typically, the company will not give ownership of resulting IP to the university, nor will it share revenue with the university. In some cases, the company may even seek to include a nonexclusive license to background IP that it did not fund. Some companies seek guarantee clauses in research agreements governing collective projects that place on the university all responsibility for the infringement of other's IP rights. Additionally, some companies will not agree to give the university rights to use the resulting IP for teaching or research purposes. There is no university association in China to advocate for the interests of universities. Not surprisingly, outside of a few famous universities, most Chinese universities are in an inferior position when negotiating sponsored research agreements with large companies. They have to accept any adverse contract terms in order to obtain the research funding.

3.6 *Regional imbalances*

Finally, there is significant imbalance among Chinese universities in terms of IP management. Economic development in China is proceeding at very different rates in different regions. In some of the most developed regions, like Shanghai and Beijing, there are world-class universities that do quite a good job managing their IP.

4. CASE STUDY: IP MANAGEMENT AT TSINGHUA UNIVERSITY

Tsinghua University has an IP committee that consists of a university vice president and managers drawn from the university's functional departments. The committee oversees an office simply called the Intellectual Property Office, which is in charge of the university's IP policy and management. The specific responsibilities of the office include:

- drafting university policies regarding IP
- monitoring policy implementation
- establishing systems and procedures for IP management
- educating faculty
- examining IP clauses in contracts between the university and industry

More recently, the Intellectual Property Office has also begun providing services in patent searching and infringement consulting. The intellectual property office appoints at least one member of each department to manage IP as part of his or her daily work.

Tsinghua University created its IP policy about ten years ago. The policy applies to all university employees, including faculty and nonfaculty researchers, provisionally hired employees, students, post-docs, and visiting scholars. All employees to whom it applies sign a pledge that they will comply with the policy. IP is defined under the policy to include patents, trade secrets, know-how, trademarks, copyrights, and any related rights. It clearly defines what constitutes employee work. It also states that when a project is completed, the investigator should disclose all results to the administrative department first, and the administrative department should then decide whether to apply for a patent. Publication and any public appraisals that would trigger the publication bar are forbidden before filing a patent application. If results appear to have commercial value but are not suitable for a patent, it is to be kept as a trade secret, and measures to maintain confidentiality are to be taken. An industry-sponsored research agreement must have a clause on ownership of resulting IP, allocation of patent costs, sharing of revenue made from the IP, and so on, and the contract must be examined by the intellectual property office before it becomes effective. When a faculty member or other employee goes to another domestic or foreign university or institute and does research, any IP resulting from that research should be assigned, or at least jointly assigned, to Tsinghua University, unless there is an agreement between that researcher and the other university or institute. Under the university policy, at least 25% of revenue generated by a piece of IP is to be shared with the inventor(s) as cash or equity.

Tsinghua University has spared no effort to educate its faculty members and students about IP and the university's IP policy. The policy is printed as a brochure. All members and students get one on their first day of joining the university. The university also propagates information about IP on its Web site. The intellectual property office periodically reports news, IP-related laws, and updates on its work. IP is also covered in a course called Fundamentals of Law, which is taught by Tsinghua Law School and required for all students. All appointed faculty members in charge of IP management for each department receive training periodically.

The intellectual property office has also set up procedures and rules for examining collaborative research agreements and sponsored research agreements between the university and other institutions or companies. Taking into consideration past contract disputes, the office has designed a standard contract for research agreements. There is also a special fund to pay patent costs, including application fees, examination fees, agency fees, and maintenance fees for the first three years after a patent is issued.

Together, the above measures have resulted in Tsinghua University owning the most patents of all Chinese universities. From 1985 to 2000, Tsinghua University filed 1,587 patent applications. Since 2001, the average annual growth rate of the university's patent filings has been 26%. In 2004, the university filed 43 foreign applications (including Patent Cooperation Treaty filings). The numbers of patents issued to the university were 121 in 1999, 187 in 2001, 501 in 2003, and 537 in 2004. Other universities with a similar level of IP management include Peking University (University of Beijing) and the Chinese University of Technology.

5. UNIVERSITY TECHNOLOGY TRANSFER AND ECONOMIC DEVELOPMENT

Given that China has only a limited number of high-tech companies, there is limited industry demand for the technology generated by universities. Most Chinese companies have neither sufficient R&D capabilities nor sufficient commitment to the long duration and great expenses of developing new products from patents. As stated above, most university inventors do not consider market demand but, instead, file patent applications to bolster the university's reputation or to assure the inventor's promotion. Taken together, these factors negatively affect the rate of technology transfer from Chinese universities.

Because the R&D capabilities of most Chinese companies are so low, when they do get involved, they tend to simply *acquire* patented technologies from universities. Ownership of IP rights is usually included up front as an assignment or licensing clause in a research agreement. There usually is no additional negotiation or contract for licensing between the university and an industry sponsor. Therefore, the exact data on rates of technology transfer cannot be found. But based on interviews with the faculty of Tsinghua University, it appears that the transfer rate of the university's IP is not high. The contract value of industry-sponsored research agreements and collaborative research agreements might proxy for the level of technology commercialization to some extent. For Tsinghua University, the contract value of industry-sponsored and collaborative research agreements was US$31.5 million in 1999 and US$45 million in 2004.

In 1999, the Ministry of Education issued a plan to develop Chinese higher education for the 21st century. One highlight of the plan was to accelerate the transfer of university technologies by encouraging universities to set up high-tech companies. Noticing the tendency for high-tech companies to advance the local economy, many local and regional governments within China supported this plan by providing their local universities with low interest or interest-free loans, housing, land, and tax concessions. The most developed city in China, Shanghai, provides interest-free loans of about US$15 million each year to local universities. Many local governments also adopt policies to encourage local university faculty members to start up companies and to encourage universities to take part in establishing technology parks. There are now about 40 technology parks associated with universities throughout China. A large number of the companies in these technology parks are startups founded by university faculty members based on their own technologies.

In one prominent example, a US$50,000 investment by Peking University in 1986 started Founder Group.[2] The technological basis of Founder Group, protected under a patent named Laser typesetting system, was invented under government funding at Peking University. After successfully developing products from this patent, Founder Group revolutionized printing technology in China. Founder Group now dominates about 85% of the domestic print market, and their products are exported to over 30 countries. Flush with capital, Founder Group has built up a strong R&D department. Founder Group now acts as an IP incubator for Peking University. Founder Group is committed to developing its own IP, transferring Peking University's IP, and sponsoring research at the university based on market demand. Founder Group now owns, or jointly owns with Peking University, 128 Chinese patents in the fields of print technology and information technology, as well as copyrights on software in the fields of digital information management, multimedia, and the Internet. With a total staff of over 20,000, Founder Group now owns five companies listed on the securities exchanges of Shanghai, Shenzhen, Malaysia, and Hong Kong, as well as more than 20 companies wholly funded by Founder Group or through joint-ventures. It achieved revenues of almost U.S. $3 billion in 2004.

In another prominent example, Tsinghua Tong Fang Co., Ltd., was floated in an IPO (initial public offering) in 1997 on the Shanghai Stock Exchange with Tsinghua University as the main shareholder. The company acts as an incubator of Tsinghua University's IP in two ways. One way is by attracting capital for the commercialization of university inventions; the other is by sponsoring research at the university related to the company's understanding of market demand. The company now owns more than 300 Chinese patents in information technology, energy resources and the environment, and applied radiation technologies, as well as 44 copyrights on software. In just the first half of 2005, the company achieved sales of about US$450 million.

6. SUGGESTIONS FOR IMPROVING IP MANAGEMENT AT CHINESE UNIVERSITIES

Given the difference between the present status of IP management in most of China's universities and the much more successful cases mentioned above, the following proposals and suggestions might be usefully implemented by those universities with less-successful IP management policies:[3]

1. Constitute an institutional IP policy that provides for at least the following key points:
   - a definition of *employee invention* and *work for hire*
   - identification of parties for whom the institution's IP policy is applicable
   - procedures ranging from disclosure of inventions to filing of applications
   - measures to avoid publication bar
   - terms for the sharing technology transfer revenues with university inventor(s)
2. Establish an independent IP management office staffed with full-time professionals familiar with IP.
3. Educate the university's faculty and students on the IP policy.
4. Establish companies to incubate technologies and accelerate technology transfer. Given circumstances in China, this is often more effective than just a licensing strategy. ■

ACKNOWLEDGMENTS

This chapter was written while the author was at Franklin Pierce Law Center, Concord, New Hampshire, U.S.A.

HUA GUO, *Patent Specialist, Jones Day, 30th Floor, Shanghai Kerry Centre, 1515 Nanjing Road West, Shanghai, 200040, People's Republic of China.* kguo@jonesday.com

---

1 A "publication bar" or "bar date" means the date beyond which patent rights are lost due to a prior "enabling" publication. In the United States, if the inventor has a potentially patentable invention and publishes the enabling information describing that invention (say, on 1 January), the inventor has a one-year period to filing a patent application (in this case until 31 December of the same year). In China and in most countries outside the United States, patent rights are lost upon publication.

2 www.founder.com.

3 *For further reading on the topic of IP management in China see the following:*

Zhao Chunsahan. 1999. IP Management at Tsinghua University. *Technology and Law* (May).

Inglis-San K, LL Laureate, C Au-Yeung and TM Frow. 1996.

Intellectual Property Protection in China: The Law. Asia Law and Practice: Hong Kong.

Wang Pinhua and Cheng Xiaoxia. 2005. Research Project Management in Chinese Universities. *Investment and Technology in China* (May).

Yan Rusong. 2005. Present Status of Intellectual Property Management in Chinese Universities. *Intellectual Property Protection*, vol. 26.

Zhao Shuru. 2005. Intellectual Property Protection and Management in Chinese Universities. *Investment and Technology in China* (May).

Zhu Xianguo and Tang Daisheng. 2005. Exploration of Intellectual Property Management Modes in Chinese Universities. *University Management* (May).

Wang Ximei and Li Zihe. 2006. Technology Transfer Analysis in Chinese Universities. Sciencepaper Online. www.paper.edu.cn/ztlw/download.jsp?file=Jishu05-07.

Mei Yuanhong and Zheng Yongping. 2006. Limiting Factors to Technology Transfer in Chinese Universities and Countermeasures. Sciencepaper Online. www.paper.edu.cn/ztlw/download.jsp?file=Jishu05-05.

For further reading on the topic of statistics on university patenting, see the following:
China Education Online. 2006. *Scientific and Technological Activities*. www.edu.cn/ke_ji_huo_dong_1832/index.shtml (last accessed October 30, 2006).

For further reading on Chinese law as it regards intellectual property rights, see the following:
Copyright Law of the People's Republic of China §16 (2001); Patent Law of the People's Republic of China §6, §8 (2001); Trademark Law of the People's Republic of China (2001).

Application and Examples of Best Practices in IP Management: The Donald Danforth Plant Science Center

KAREL R. SCHUBERT, *Vice President for Scientific Partnerships, Member and Principal Investigator, Donald Danforth Plant Science Center., U.S.A. Currently at Schubert Consulting, U.S.A.*

ABSTRACT

An independent nonprofit research institution, the Donald Danforth Plant Science Center has an international mission to address global challenges in human health, nutrition, agricultural sustainability, and the environment. The Danforth Center contributes to fulfilling this mission through collaborative research, training, and capacity building. As part of this objective, the Office of Technology Management and Scientific Partnerships at the Danforth Center, lead by the author of this chapter, has emerged as a leader in developing and implementing terms for humanitarian access to technology and has been actively involved in licensing enabling technologies for humanitarian projects. These activities include active participation and support for the creation of PIPRA, among other nonprofit organizations. The current chapter discusses the Danforth Center's philosophy with respect to the protection and sharing of IP (intellectual property) rights, the reservation of rights for humanitarian projects, and best practices to enhance and maximize value creation through technology licensing. The chapter provides examples of the Danforth Center's best practices and model documents for the establishment of interinstitutional and international collaborations and scientific partnerships. Included with the chapter are specific examples of the Danforth Center's humanitarian-use language, interinstitutional agreements, nonasserts, enabling technology licenses, memorandums of understanding (MOUs), and other framework documents.

1. INTRODUCTION

Founded in 1998, the Donald Danforth Plant Science Center (Danforth Center) is a not-for-profit research institute with a global vision to improve the human condition through plant science. This vision is exemplified in the Danforth Center's logo "Discover, Enlighten, Share and Nourish." Research at the Danforth Center includes efforts to enhance the nutritional content of plants, improve human health and well being, increase agricultural production for a sustainable food supply, preserve and renew our environment, and build scientific capacity and thereby contribute to economic growth in the developing nations of the world. The Danforth Center is built on the principles of collaboration and sharing. The center attains its goals through collaborations and scientific partnerships and continuously offers opportunities for scientific exchange and training, capacity building, technology transfer, and translational research.

2. THE DANFORTH CENTER AND IP RIGHTS

To begin, I would like to describe the Danforth Center's general philosophy regarding intellectual property (IP) and then, more specifically, address our philosophy on reservation of rights for humanitarian use. My individual philosophy is to protect and maintain the Danforth Center's IP rights to maximize value and potential for application while equally respecting the IP rights of others. We, at the center, expect in return no less

Schubert KR. 2007. Application and Examples of Best Practices in IP Management: The Donald Danforth Plant Science Center. In *Intellectual Property Management in Health and Agricultural Innovation: A Handbook of Best Practices* (eds. A Krattiger, RT Mahoney, L Nelsen, et al.). MIHR: Oxford, U.K., and PIPRA: Davis, U.S.A. Available online at www.ipHandbook.org.

than the same values and respect of our intellectual property. Inherent in this philosophy is the innate understanding that the center shall not violate or infringe the IP rights or misuse the materials and rights entrusted to the Danforth Center, even if the actions would involve no illegalities.

The Danforth Center's policies and objectives regarding intellectual property are consistent with those of the Public Intellectual Property Resource for Agriculture (PIPRA), which are to promote the management of intellectual property related to agriculture and to achieve freedom to utilize agricultural innovations for research, commercial use, economic development, specialty crops, and humanitarian purposes. In line with these objectives, the Danforth Center encourages the development of research innovations for use in agriculture while also retaining rights needed to fulfill the mission of research and product development for the broader public benefit. The center seeks to facilitate access to enabling technologies for research and commercial use and/or humanitarian purposes by our scientific collaborators and the international scientific community and work to identify strategies that effectively achieve these objectives.

3. THE DANFORTH CENTER: VISION AND PHILOSOPHY

3.1 *The vision*

The Office of Scientific Partnerships is a preferred and valued partner for plant-science research and collaboration, recognized and respected internationally for its research integrity and innovative policies and practices for the protection, management, and stewardship of intellectual property rights. The office strives to be:

- a world-class provider and developer of novel cutting-edge solutions seeking to meet global challenges in agriculture, the environment, and human and animal health and nutrition
- an engine and catalyst for economic growth and the creation of wealth and value from the intellectual and human capital and a return on the research investment of the Danforth Center
- a recognized leader at facilitating national and international research collaborations and public–private partnerships that bring the world closer together.

In all agreements between the Danforth Center and public and/or private institutions, the Danforth Center strives to reserve and protect the IP rights conceived and reduced to practice directly by Danforth Center staff or jointly with researchers from partnering institutions.

3.2 *The Danforth Center and developing countries*

An integral part of the center's philosophy relates to the desire to be able to share in the benefits of research and discovery endeavors with developing nations. This includes providing assurances that all parties benefit from the intellectual property developed through the center's collaborations and scientific partnerships. To ensure that the Danforth Center retains and maintains the rights to use technology developed by Danforth Center researchers or through collaborations, the center includes a section in sponsored-research and license agreements that provides for the reservation of rights to use technology developed for the benefit of poor and underserved peoples of the developing nations. Under these provisions, the Danforth Center and our cooperators retain the rights to develop, have developed, produce, have produced, distribute and/or have distributed (in other words commercialize) the products of our basic and applied research and our joint collaborative research and to share this freely with partnering organizations in developing countries.

In each agreement, the detailed terms may be modified to reflect the interests and needs of the parties and to achieve a mutually beneficial relationship. The terms of licensing reflect our interest in maximizing the opportunities to capture and create value from our intellectual pursuits and ensure that the benefits of our scientific research will benefit the broader international community, especially addressing grand challenges in health, nutrition, and the environment in developing countries.

This philosophy is exemplified by the center's policy not to grant broad worldwide exclusive licenses to its technology that could limit the center's ability to create the maximum benefit from any intellectual property conceived by the researchers and through collaborative research projects, as well as from any technology developed through these activities. Instead, the center grants only nonexclusive or limited exclusive licenses and, further, restricts the license rights granted to specific and/or limited fields-of-use, specific crops, and specific territories. Thus, the center retains the maximum opportunity to exploit the technology.

3.3 *Facilitating access to new technologies*

Traditionally, technology transfer and IP rights in agriculture have centered around intensive agriculture in the developed countries and reflect the commercial forces and drivers that dictate a focus on commercially relevant agricultural priorities and targets for commercial crops in the developed nations. From these research activities, new technologies are developed including enabling and platform technologies, which may have relevance in addressing needs in the developing world. Access to this "developed country technology" has been the target of many foundations and organizations focused on humanitarian efforts and programs. These programs aim to facilitate transfer of the technology to developing countries, including making IP rights and materials available to these countries. PIPRA and other groups are playing a key role in facilitating access to such technologies, while still protecting the IP assets of the inventor's institution for use in commercial agriculture, both for major crops and for minor or specialty crops.

The goals of much of the Danforth Center's research and the research of the center's scientific partners seek to address specific agronomic and nutritional targets of the highest priority and importance in developing countries that offer the greatest potential benefit to resource-poor subsistence farmers in these countries. These targets include increasing the yield of staple crops, decreasing the need for chemical pesticides in agriculture, increasing crop resistance to pests and pathogens, increasing tolerance to abiotic stresses such as drought, salinity and cold, increasing food quality and food safety, and enhancing nutritional content of staple and subsistence crops. From the results of our research and our research partnerships, intellectual property may be created that also has commercial value in the developed nations.

4. GLOBAL PERSPECTIVE

Value is enhanced by retaining the rights and options to apply and to make the technology available in as many ways, in as many applications, in as many markets, and in as many territories as possible. To accomplish this, the center does not generally grant options for an exclusive and/or worldwide license to Danforth Center or to joint intellectual property. Our policy and practices encourage granting options to license and licenses on a nonexclusive basis to use, make, and sell products incorporating the technology and to further segment and limit these licenses to specific applications of the technology, to specific fields-of-use, and to specific territories. In the latter case, these may actually provide limited and defined exclusivity to the licensee. The use of nonexclusive licenses or limited licenses enables the broadest application of the technology and does not prematurely limit the benefits of the technology either for humanitarian or commercial use.

Unfortunately, access to innovative and enabling technologies is too frequently restricted by the granting of exclusive and often worldwide options and licenses to the private sector. Such restrictive terms mean that promising technologies may be inaccessible to address developing country needs. In some cases the technology may be shelved to prevent access by competitors, while in other cases access may be hindered by fiduciary, liability, and stewardship considerations. The need for indemnification and technology stewardship frequently forms a major barrier restricting open access to enabling and platform technologies. As part of the activities of the Danforth Center, the staff is attempting to find innovative strategies to reduce these barriers and to facilitate access to technology for both humanitarian and commercial purposes.

Box 1 represent examples of general language for a reservation of IP rights for humanitarian purposes (with a specific focus on developing countries) incorporated by the Danforth Center into its research and license agreements (specific language was taken from sponsored research agreements and/or license agreements).

Agreements must also provide for the indemnification of technology providers and for technology stewardship. Here is an excellent example:

Agreement Relating to COMPANY Patent Rights and to PROJECT/TRAIT Between COMPANY and the Donald Danforth Plant Science Center:

5.1 Danforth Center agrees to indemnify and hold COMPANY and its employees, directors, officers and agents harmless against any and all claims, losses, liabilities or expenses (including court costs and reasonable fees for attorneys and other professionals) on account of any injury or death of persons or damage to property to third parties or to COMPANY caused by, arising or alleged to arise out of Danforth Center's or DEVELOPING COUNTRY COLLABORATING INSTITUTION's activities under or in connection with this Agreement. Such right of indemnification under this Agreement shall be in addition to, rather than to the exclusion of, the rights of COMPANY at law or in equity. The

Box 1: Danforth Center's Reservation of IP Rights for Humanitarian Purposes

Terms from the Article on Intellectual Property:

DANFORTH CENTER shall retain the right to use Danforth Center IP and Joint IP for both academic and commercial research purposes, which shall include the right to use such technology for the benefit of countries eligible for International Development Association funds as reported in the most recent World Bank Annual Report ("Developing Countries"). Such use of any Danforth Center IP and/or Joint IP for such humanitarian purposes shall require sixty (60) days written notice to SPONSOR of DANFORTH CENTER's intent to so use such Danforth Center IP and/or Joint IP.

Terms from the Article on Grant of Rights:

DANFORTH CENTER and SPONSOR shall diligently and in good faith negotiate the terms of any such license(s), provided, (a) any such license shall contain the terms set forth in Appendix [__], attached hereto, and (b) the parties shall in good faith negotiate provisions for preserving the availability of Danforth Center IP and/or Joint IP for meeting the needs of Developing Countries. Such option shall extend, on a patent application by patent application basis, for one (1) year after the filing of a utility patent application to protect Danforth Center IP and/or Joint IP, or for one (1) year from the termination of this Agreement, whichever is sooner (the "Option Period"), and may be exercised at any time during such period by SPONSOR in its sole discretion.

Terms from the Article on Options and Licenses:

Humanitarian Use Clause and Research Exemption. Notwithstanding anything herein to the contrary, the Parties agree that each of the Parties shall have and retain the right under Project Information and Project Patents to use Project Information for research purposes. In the case of Danforth Center this right shall be limited to the right to use such technology in research by or under the control of Danforth Center for the benefit of countries eligible for International Development Association funds as reported in the most recent World Bank Annual Report ("Developing Countries"), and the right to work with other not-for-profit Third Parties in connection with such research, and to publish the results of such research subject to the confidentiality, nonuse and nondisclosure provisions of this Agreement, provided that, Danforth Center shall grant no rights under the results of such research to any Third Party. Each Party shall provide the other Party at least sixty (60) days prior written notice of its intention so to use any Project Information.

provisions of this paragraph will survive the term or termination of this Agreement for any reason.

5. TECHNOLOGY TRANSFER

To understand the Danforth Center philosophy regarding technology transfer, it is critical to keep in mind that the driving objective is to facilitate and enable access to technology and materials. Therefore, within this context, several examples of different agreements that facilitate such access and enable the center's ability to share its technologies with collaborators and others are of specific interest. Pertinent examples include approaches for facilitating/enabling technology access, such as the *enabling technology license* and the *letter of nonassert*. In addition, although generic material transfer agreements (MTAs) are commonly available, an example is included here, as some of the specific terms are useful when there are limitations on the transfer of enabling technologies or grant-back rights to the technology provider.

Before considering any detail regarding specific strategies and practices to facilitate access to enabling technologies, the following should be noted: According to U.S. Patent Law, without the explicit right or grant of license to do so, the transfer by an entity within the United States of patented materials (that is, product or process inventions protected by a U. S. patent) or components thereof that could be used to reconstruct the patented technology to another party, even if this party is in a country in which the materials are not patented, might constitute an act of infringement by the provider, but not necessarily the recipient, of the patent rights of the patent holder. However, this possibility depends on whether, or not, pertinent patent rights have been exhausted via legitimate sale of the patented item(s). Thus, this issue needs to be carefully considered with respect to any transfer of tangible property pursuant to an MTA (that is, the omnipresent possibility of third-party IP rights embedded in the transferred materials, for example patent rights). In some cases, MTA's may have grant-back obligations based on requirements of the provider or third party requirements.

Examples of these more-restrictive MTA requirements are provided below:

- Research Materials represent a significant investment on the part of Danforth and/or Providers and are considered proprietary to Danforth and/or Providers. Recipient therefore agrees to retain control over this Research Material and further agrees not to transfer the Research Materials to third parties without advance written approval of Danforth. Under no circumstances should materials be transferred outside the United States or to an agent acting on behalf of a foreign country, except as permitted by U.S. export control laws. Recipient agrees to give Danforth reasonable advance written notice of any proposed transfer of Research Materials outside the United States or to an agent acting on behalf of a foreign country. Danforth reserves the right to distribute the Research Material to others and to use it for its own purposes. Nothing in this Agreement will prevent Recipient from engaging in any activity with regard to material that is obtained from a source other than Danforth.
- The Research Materials will be used for internal research purposes only and specifically for the Research Project as described above and in detail in the Description of Research Project, appended hereto and incorporated herein.
- Recipient will provide Danforth with a written semi-annual report ("Research Report") of the progress and results of the Research Project and the Recipient's experience in using Research Materials. The Research Report shall be due six (6) months from the *Effective* Date of this Agreement and every six (6) months thereafter with a final report due upon termination of this Agreement. Each Research Report should be provided to the attention of Dr. Karel R. Schubert, at the address included herein. Danforth may compile information contained in such Research Report for distribution among the members of the Consortium with appropriate attribution and acknowledgement.

- Nothing in this Agreement is intended to prevent publication of results of Recipient's research. Recipient will provide to Danforth, at least sixty (60) days in advance of submission or disclosure, an electronic copy for review of any abstract, presentation or manuscript describing the progress or results of the Research Project or Recipient's use of the Research Materials ("Publication") to be submitted for publication or otherwise publicly disclosed. Danforth agrees to a timely review of such proposed Publication by Recipient disclosing any confidential information of Danforth and/or Providers, as defined herein, and/or any Improvement (as defined in Section 9 hereof) for which Danforth and/or Provider may wish to seek intellectual property protection. Recipient agrees to remove, at Danforth's sole request, any confidential information and to delay publication for up to an additional thirty (30) days to permit filing for intellectual property protection on any Improvement. Public disclosures of research results will acknowledge Danforth's and/or Provider's contribution of Research Materials, in the accepted style, as appropriate under the circumstances. While Danforth does not transfer ownership of the Research Materials to Recipient, should Recipient's use of Research Materials result in patentable inventions, Recipient agrees to promptly provide Danforth with an enabling disclosure at least thirty (30) days prior to submission for public disclosure for Recipient and Danforth to determine the need to seek statutory protection.
- Recipient may make modifications or enhancements ("Improvements") to Research Materials during the course of the Research Project. Recipient understands and agrees to promptly notify Danforth of any such Improvements of Research Materials (whether or not patentable) that Recipient makes to Research Materials within no more than ninety (90) days of making such Improvement and to keep Danforth timely informed of any applications to obtain intellectual property protection to the extent claiming such Improvements. Such notification may be through (i) submission of the required semi-annual Research Reports to Danforth; (ii) through submission of Publications to Danforth for review; or (iii) through written notification to Danforth.
- In consideration of the contribution of Research Materials, Recipient grants to Danforth a royalty-free license, with the right to grant sublicenses to make and use such Improvement, and products and processes developed from or incorporating such Improvement for internal research purposes.
- Recipient grants to Danforth an option, for one (1) year following Danforth's receipt of written notification of an Improvement, to obtain a royalty bearing nonexclusive commercial license, with the right to grant sublicenses to make, use, import, offer for sale, or sell products, and processes incorporating such Improvement. The terms of the license will be negotiated with diligence and in good faith among and between the Parties at the time Danforth, at its sole discretion, elects to exercise its option. The Danforth is under no obligation to negotiate or enter into any definitive agreement with Recipient with respect to licensing.

6. PARTNERSHIPS AND IP RIGHTS

This sections explains the center's philosophy regarding the creation of scientific partnerships, collaborations, and alliances and provides some of the key elements of these agreements as they relate to IP rights and humanitarian use. Most research collaborations start with the signing of a general memorandum of understanding and agreement (MOA) between the parties (Box 2). These agreements are generally nonbinding and reflect the intent of the parties to enter into more definitive agreements. The key elements of these agreements include the statement of purpose and the intent of the parties to enter into more definitive agreements. Examples of two such generic MOA's are provided as supporting materials. Also included is

Box 2: Extracts of a General Memorandum of Understanding between an Institution and the Danforth Center

Considering:

1. Recent contacts established between the two organizations;
2. The expressed desire by the authorities of both organizations to establish long-term, fruitful collaboration in fields of common interest including improvement of cassava and sweet potato.
3. The anticipated benefits of such collaboration promoting agricultural research-for-development in Sub-Saharan Africa, among others, through advances in applied plant biotechnology and the exchange of scientists and students, hence broadening the relative expertise of each organization;
4. The prospects and mutual benefits from the potential expansion of our respective expertise as well as our financial resources base.

The Parties agree as follows:

Article 1: INSTITUTION and Danforth Center (the "Parties") agree to explore opportunities for funding of collaborative projects, and once funding opportunities are identified, to jointly develop proposals for scientific research, development, and technology transfer in areas of interest for sustainable agriculture and development in Sub-Saharan Africa.

Article 2: Each one of the parties involved can initiate the search for request for proposals, and the development of such a proposal.

Article 3: This memorandum of understanding does not prevent any party from initiating and finalizing separate bilateral (or multilateral) agreements with other institutions. Nonetheless, both parties may continue to inform each other, as appropriate, about separate collaborative agreements in areas of mutual interest.

Article 4: Each one of the collaborative projects developed under this memorandum of understanding (the "MOU") will be the subject of a specific addendum to this MOU, where the resources and responsibilities of each party or partner will be clearly defined.

Article 5: Each organization will designate a member of the institution's staff to be responsible for the management and completion of each specific project. The development of any proposal will be a joint effort where full participation between scientists from both institutions is expected.

Article 6: As projects are jointly developed, both organizations will endeavor to successfully complete the collaborative research project.

Article 7: Pending availability of funds, and to the extent possible, both organizations will promote exchange of relevant technologies and interaction and cooperation amongst personnel from each organization.

Article 8: To the extent possible, each organization will grant visiting scientists, students and trainees from the other partner institution all facilities, privileges and responsibilities that it normally grants to its own personnel, students and trainees.

CONTINUED ON NEXT PAGE

a general letter of intent used for the creation of a multi-institutional alliance or partnership.

At the Danforth Center, the next stage in the development of a scientific partnership and research collaboration between different institutions is the creation of an interinstitutional agreement (IIA) to serve as a broader umbrella agreement. The IIA provides background information on the interests of the participating organizations and general information on the purpose of the collaboration. The IIA generally does not include details about specific projects individuals are involved in, as these details are covered in subsequent, more-definitive agreements. The IIA does provide details on the general principles of confidentiality, ownership and rights of the parties, IP management practices and IP protection, financial considerations including sharing of patent costs, publications and authorship, the use of marks and publicity, handling of disputes, and the sharing of value derived from jointly created intellectual property along with other general terms.

As IP rights and ownership are essential considerations of any such agreement, the center's philosophy, as expressed in all such agreements, is that the parties, whether public or private, involved in the collaboration and pursuant to the creation of joint IP rights shall jointly own such intellectual property (with the relevant limitations of joint ownership) and shall share equally in any value created through the use and/or licensing of such technology, unless the parties mutually agree (either beforehand or

Box 2 (continued)

Article 9: The Danforth Center is not a degree-granting institution. When the exchange involves students who are interested in enrolling in coursework as part of a degree program within a local institution with the intent of obtaining an academic degree, the candidates must comply with the normal conditions of admission and candidacy within said degree-granting institution for the stated degree.

Article 10: All rights to data, including laboratory and field notebooks, and material contained in such notebooks, and research results (including formal or informal reports) and products ensuing from partnership projects between both parties, shall belong jointly to INSTITUTION and the Danforth Center.

Article 11: Both parties consider that excluding others from accessing research products and results from their joint research-for-development is contradictory to their mandate and mission. Therefore, INSTITUTION and the Danforth Center agree not to secure patents or plant breeders' rights from their partnership research unless such protection is deemed necessary to keep these materials or technologies available and freely accessible to its beneficiaries.

Article 12: Each of the parties reserve the rights to develop and commercialize the products and results from their joint research for use by small farmers.

Article 13: This memorandum of understanding will be effective upon signature by designated representatives from both organizations.

Article 14: This memorandum of understanding will be effective for a period of three years and will be renewed for the same period upon mutual consent of both parties.

Article 15: This memorandum of understanding can be modified, discontinued, or cancelled, by written notification of either party, at least six months prior to the effective date of suspension. If specific projects are ongoing at that time, the terms for their termination will be negotiated.

subsequent to the invention) to a different formula for value sharing based on, for example, differences in the intellectual and/or financial contributions of each party and/or the party's employees. It is inherent in these agreements that the terms for any value sharing between the institutions and their inventors will be determined by the respective institutions and will be revenues and royalties that will be split and distributed according to defined principles and formulas of the inventors' parent organization. The parties also agree to define the strategy and lead organization for the management of the intellectual property, including filing, prosecution, and maintenance of patents, marketing and licensing the technology, and how costs for protecting intellectual property will be shared. These key general practices and considerations are addressed upfront in the umbrella agreement and then specific details and/or modifications may be incorporated into the subsequent, definitive project-specific agreements.

An example of the generic IIA used by the Danforth Center is included in its entirety in the supplemental materials.[1] Excerpts from this generic IIA are represented in Box 3 at the end of the chapter, as they relate to some of these key elements. These excerpted, sample articles provide an overview of how such a document forms the basis for the general umbrella agreement and forms the framework for specific agreements. As such these sections can thereby be incorporated into the specific agreements. Once a technology is developed, a nonconfidential disclosure may be developed to aid in marketing joint technology. An example of a nonconfidential disclosure is included in the supplemental materials.

7. CONCLUSIONS

The Danforth Center regards its role in international development as a critical component of its overall mission, which categorically involves promoting the transfer of technological innovations arising out of the R&D efforts at the Danforth Center to developing countries around the globe. Protecting and managing intellectual property, regardless of whether it is owned by the Danforth Center, its partners/collaborators, or other third-parties, is interwoven into this process of technology transfer. Thus, IP rights, managed effectively, efficiently, and strategically, represent a mechanism for facilitating this process. Within this context, individuals at the Danforth Center have strived to organize and then implement an integrated, comprehensive and adaptable system for best practices in managing IP rights. The examples of agreements presented in this chapter are a manifestation of this system. They provide practical examples that other institutions might wish to emulate. ■

ACKNOWLEDGEMENTS

The author acknowledges his colleague, James A. Kearns, III, Partner, Bryan Cave LLP, for his assistance in the development of the Danforth Center's model agreements and humanitarian use language. His personnel insights and legal perspectives as reflected in these agreements have been invaluable.

KAREL R. SCHUBERT, *Vice President for Scientific Partnerships, Member and Principal Investigator, Donald Danforth Plant Science Center. Currently at Schubert Consulting, 817 Berry Hill Drive, St. Louis, Missouri 63132, U.S.A., krswv@earthlink.net*

---

1 The following agreements from the Danforth Center are available on www.ipHandbook.org:
- MOU Examples
- Alliance Letter of Intent
- Enabling Technology License
- Letter of Nonassert (LONA)
- Interinstitutional Agreement (IIA)
- Nonconfidential Disclosure

Box 3: Excerpts from an Interinstitutional Agreement

Article 1. Purpose and Scope of Agreement

1.1. The purpose of this Agreement is to provide a contractual framework to govern collaborative research projects and other forms of collaboration undertaken by the Danforth Center and Collaborating Institution and is intended to apply in the absence of separate agreements between the Institutions governing specific cases.

1.2. In the event of any conflict or inconsistency between the provisions of this Agreement and the provisions of a separate agreement between the Institutions governing a specific matter, the provisions of such separate agreement shall control with respect to such matter.

Article 2. Definitions

"Developing Countries" means the countries eligible for International Development Association funds as reported in the most-recent World Bank Annual Report as of the date applicable to such determination, or the substantively equivalent designation by the World Bank if such report is no longer published.

"Joint Intellectual Property" means any Intellectual Property made or obtained jointly by Researchers of both the Danforth Center and Collaborating Institution or jointly owned by both Institutions by agreement or under applicable law.

Article 3. Material Transfer

3.1. From time to time a Researcher at either Institution may wish to request from the other Institution the transfer of certain Research Information or Research Material for research purposes. Both Institutions agree to use their reasonable efforts to cause each such request to be made by and to their respective Technology Management Offices and in accordance with such procedures and forms of written agreements as each Institution may establish from time to time for transferring material to, or receiving material from, another institution. In the event of any conflict or inconsistency between the terms of this Article and any separate written agreement between the Institutions that pertains specifically to a particular transfer of Research Information or Research Material, the terms of such separate agreement shall control with respect to that particular transfer of Research Information or Research Material.

3.2. In the event that, notwithstanding the foregoing, Research Information or Research Material is in fact transferred at any time from one Institution (the "Provider") to the other Institution (the "Recipient") without a written agreement between the Institutions relating specifically to such transfer, the terms of this Article shall govern each such transfer.

3.3. The Recipient shall have a nonexclusive, royalty-free license to use the Research Information or Research Material only in connection with academic and noncommercial research conducted by the Recipient. Research Material shall not be used in humans. The Recipient shall comply with all applicable laws, rules and regulations applicable to the use and handling of the Research Information and Research Material.

3.4. The Recipient shall not transfer the Research Information or Research Material to a third party except for academic and noncommercial research and the Recipient agrees to promptly notify the Provider of each such transfer.

CONTINUED ON NEXT PAGE

Box 3 (continued)

Article 4. Confidential Information

4.1. In the event that Research Information or Research Material is also Confidential Information as defined above, then the provisions of this Article shall apply to such Confidential Information, notwithstanding the provisions of Article 3 with respect to Research Information and Research Material.

4.2. Each Institution agrees to use its reasonable efforts to obtain, or to assist the other Institution in obtaining, from each Researcher, employee and contractor of such Institution who receives Confidential Information from the other Institution, an agreement that such Researcher, employee or contractor: (a) will maintain such Confidential Information in the confidence normally accorded to internal confidential materials of the Researcher's own Institution, but in any event using not less than reasonable care; (b) will not use the Confidential Information for any purpose other than academic and noncommercial research at such Researcher's own Institution; (c) will not disclose the Confidential Information to others, other than to other Researchers at such Researcher's own Institution, making them aware of the confidentiality obligations under this Agreement; and (d) will not make any copies of the Confidential Information composed of Research Materials without the other Institution's prior written permission.

4.3. It is also agreed that each Institution will return or destroy the Confidential Information received from the other Institution within 60 days after the disclosing Institution so requests. Notwithstanding the foregoing, each Institution shall be entitled to keep one copy of the other Institution's Confidential Information which must thereafter be restricted to use for legal purposes as a record of the Confidential Information returned under this Agreement.

Article 5. Ownership of Intellectual Property

5.1. Rights to all Danforth Intellectual Property shall vest according to the policies of the Danforth Center relating to such Danforth Intellectual Property.

5.2. Rights to all Collaborating Institution Intellectual Property shall vest according to the policies of Collaborating Institution relating to such University Intellectual Property.

5.3. All Joint Intellectual Property shall vest according to applicable principles of United States law and the policies of the respective Institutions as applicable to the legal interests and rights of each such Institution in and to such Joint Intellectual Property.

Article 6. Patents and Other Protection of Intellectual Property

6.1. The Danforth Center shall be responsible for all decisions and costs relating to the preparation, filing, prosecution, and maintenance of U.S. and foreign patents and patent applications and other forms of protection with respect to Danforth Intellectual Property and for the selection and compensation of legal counsel and other representatives with respect thereto.

6.2. Collaborating Institution shall be responsible for all decisions and costs relating to the preparation, filing, prosecution, and maintenance of U.S. and foreign patents and patent applications and other forms of protection with respect to University Intellectual Property and for the selection and compensation of legal counsel and other representatives with respect thereto.

CONTINUED ON NEXT PAGE

Box 3 (continued)

Article 7. Identification of Prospects for Commercial Development

7.1. The Institutions agree to form a "Joint Marketing Team" for the purpose of collaborating in the identification and pursuit of prospects for the commercial development of Danforth Intellectual Property, University Intellectual Property and Joint Intellectual Property. The Joint Marketing Team shall have an equal number of persons appointed by the Technology Management Office of each Institution. Each Institution shall have the right to change any or all of its representatives on the Joint Marketing Team at any time and from time to time, upon written notice to the other Institution. A quorum of the Joint Marketing Team shall consist of not less than a majority of the members of the Joint Marketing Team, provided that at least an equal number of members appointed by each Institution are present.

7.2. The Joint Marketing Team shall have the following responsibilities:

(a) to stay informed on the research being conducted by Researchers at each Institution in plant biology and its application to sustainable productivity in agriculture, forestry and allied fields;

(b) to stay informed on current developments in, and prospects for, the commercial application of technologies resulting from research in the plant sciences;

(c) to evaluate the prospects for commercial development of the research in the plant sciences being conducted by the Researchers at each Institution;

(d) to identify opportunities for the commercial development of Danforth Intellectual Property, University Intellectual Property and Joint Intellectual Property, and to promptly bring such opportunities to the attention of the Technology Management Offices of the respective Institutions;

(e) to identify future research projects that could be carried out by Researchers at one or both Institutions for which there is a favorable prospect of commercial development; and

(f) to identify and pursue funding support for the conduct of such research projects by Researchers at one or both Institutions.

Article 8. License Grants and Revenue Sharing

8.1. The Danforth Center shall be responsible for all decisions and costs relating to the grant of licenses with respect to Danforth Intellectual Property and shall, as between the Institutions, be entitled to retain all revenues derived therefrom.

8.2. Collaborating Institution shall be responsible for all decisions and costs relating to the grant of licenses with respect to University Intellectual Property and shall, as between the Institutions, be entitled to retain all revenues derived therefrom.

Article 9. Reservation of Use for Research and for Developing Countries

9.1. Each Institution shall have the right to use Joint Intellectual Property for both academic and research purposes, including research conducted with corporate, governmental, or other external sponsorship.

CONTINUED ON NEXT PAGE

Box 3 (CONTINUED)

9.2. Each Institution shall have the right to use Joint Intellectual Property for the benefit of Developing Countries, and shall have a nonexclusive, royalty-free, irrevocable license, under such right, title and interest as the other Institution may have in and to the Joint Intellectual Property, to make, use, sell, offer for sale, import, or practice any Joint Intellectual Property within the Developing Countries, with the right to grant further sublicenses thereunder within the Developing Countries. The license granted hereunder includes any patent applications and issued patents claiming priority from, or the benefit of, the Joint Intellectual Property, and any reissues, extensions, substitutions, continuations, divisions, or continuations-in-part derived therefrom, or any foreign patents and patent applications corresponding thereto. In specific cases the Institutions may agree in writing to limit the license granted hereunder to specified fields of use or to one or more specified Developing Countries.

9.3. Each Institution agrees that any licenses granted by it to third parties with respect to any Joint Intellectual Property shall make provision for preserving the availability of such Joint Intellectual Property for meeting the needs of Developing Countries.

Article 10. Publications and Publicity

10.1. Recognizing each Institution's desire to publish previously unpublished Research Information, and each Institution's desire to develop the results of research for the earliest introduction to the public, each Institution agrees to submit to the other Institution copies of proposed publications or presentations as follows:

(a) each manuscript or details of each proposed public oral presentation first disclosing Joint Intellectual Property or disclosing Confidential Information of the other Institution shall be submitted to the Technology Management Office of the other Institution at least 30 days prior to submission for publication or the date of the proposed public oral presentation; and

(b) each abstract first disclosing Joint Intellectual Property or disclosing Confidential Information of the other Institution shall be submitted to the Technology Management Office of the other Institution at least 14 days prior to its submission for publication.

10.2. If within such 30-day or 14-day period, respectively, the reviewing Institution makes a good faith determination that such proposed publication, presentation or abstract contains patentable Joint Intellectual Property which needs protection or Confidential Information which requires removal or revision and notifies the submitting Institution accordingly, then the Institutions shall have an additional 60 days to agree upon revisions to the publication, presentation or abstract in order to protect Confidential Information and in order to file patent applications directed to patentable Joint Intellectual Property contained in the proposed publication, presentation or abstract. Upon the reviewing Institution's receipt of written acknowledgement from the submitting Institution of the removal or revision of Confidential Information, or the filing of a relevant patent application, or the expiration of such 60-day period, as the case may be, the publication, presentation or abstract shall be released. The Institutions may agree to reasonable extensions, of the periods provided herein in order for reviewing Institution to complete the necessary review of publications, presentations and abstracts and for the filing of patent applications.

10.3. The determination of the persons who are to be identified as the authors of each publication or presentation that discloses Joint Intellectual Property will be made on a case-by-case basis.

CONTINUED ON NEXT PAGE

Box 3 (continued)

10.4. Neither Institution shall use the name, trademarks, service marks, logos or other indicia of identity of the other Institution or of any Researcher of the other Institution, or any adaptation thereof, in any advertising or promotional literature or publicity without the prior written approval of the other Institution.

Article 11. Disclaimer of Warranties and Limitations of Liability

11.1. EXCEPT AS EXPRESSLY SET FORTH IN THIS AGREEMENT OR IN ANY SEPARATE AGREEMENT, NEITHER INSTITUTION MAKES ANY REPRESENTATIONS OR WARRANTIES OF ANY KIND WHATSOEVER, EITHER EXPRESS OR IMPLIED, WRITTEN OR ORAL, WITH RESPECT TO RESEARCH INFORMATION, RESEARCH MATERIAL, CONFIDENTIAL INFORMATION, DANFORTH INTELLECTUAL PROPERTY, UNIVERSITY INTELLECTUAL PROPERTY OR JOINT INTELLECTUAL PROPERTY, INCLUDING WITHOUT LIMITATION ANY IMPLIED WARRANTY RELATING TO MERCHANTABILITY, REGULATORY STATUS OR EFFICACY CLAIMS, WARRANTY OF FITNESS FOR A PARTICULAR PURPOSE, OR WARRANTY OF TITLE OR NONINFRINGEMENT.

11.2. No warranty is given by either Institution in relation to the collaborative research work by the Researchers of the two Institutions or the uses to which it may be put by the other Institution or its fitness or suitability for any particular purpose or under any special conditions notwithstanding that such purpose or conditions may have been made known to such Institution.

11.3. IN NO EVENT SHALL EITHER INSTITUTION BE LIABLE TO THE OTHER INSTITUTION UNDER THIS AGREEMENT OR UNDER ANY SEPARATE AGREEMENT FOR ANY DIRECT, INDIRECT, INCIDENTAL, SPECIAL OR CONSEQUENTIAL DAMAGES, WHETHER BASED UPON PRINCIPLES OF CONTRACT, WARRANTY, NEGLIGENCE, STRICT LIABILITY OR OTHER TORT, BREACH OF ANY STATUTORY DUTY, PRINCIPLES OF INDEMNITY OR CONTRIBUTION, OR ANY OTHER THEORY OF LIABILITY, IN CONNECTION WITH THIS AGREEMENT OR SUCH SEPARATE AGREEMENT, EVEN IF SUCH INSTITUTION HAS BEEN ADVISED OF THE POSSIBILITY OF SUCH DAMAGES.

CHAPTER 17.11

IP Management in the National Health Service in England

TONY BATES, *Managing Director, Tony Bates Associates Ltd., U.K*

ABSTRACT

This chapter summarizes how intellectual property (IP) arising from within the National Health Service in England is managed within the context of a national framework for managing IP from public sector research in the United Kingdom. Describing how the policy framework was developed and how National Health Service organizations were set up to manage IP, this chapter also charts progress in the administration of health R&D and the management of IP and summarizes how IP management complements R&D in the National Health Service.

1. INTRODUCTION TO THE NATIONAL HEALTH SERVICE

The National Health Service (NHS) is England's national healthcare provider. It is managed by a government department, the Department of Health, under the secretary of state for health. Similar arrangements exist for the provision of health services in Scotland, Wales, and Northern Ireland. The healthcare system is almost entirely administered and operated within the public sector—free at the point of use—with the minimal involvement of a small private sector. The National Health Service in England is one of the world's largest employers, with over 100,000 people currently employed.

Healthcare provision is divided into services available in the home and community (such as general medicine, maternity services, home health care, and prescriptions), and services available through hospitals. Healthcare provision is managed by a number of NHS trusts, self-governing organizations funded by the Department of Health. In 2006/2007 allocations were made to 176 acute trusts (consisting mainly of traditional hospitals), 31 ambulance trusts, 82 mental health trusts, and 303 primary care trusts. There are also nine care trusts, which are new organizations that provide combined health and social care. The primary care trusts are the conduit for the bulk of the national budget for all health provision. They allocate funds to other trusts according to needs and priorities. The trusts do not make a profit, although they are required to break even, and must deliver high-quality services, using the resources provided, based on a series of targets. Organizations are managed across nine regional areas: north, northwest, Yorkshire & Humberside, east Midlands, west Midlands, east, London, southeast, and southwest.

2. R&D WITHIN THE NATIONAL HEALTH SERVICE

Most hospitals engage in research, and clinicians of all disciplines participate in many thousands of projects of differing sizes and complexity. Most research takes place in the teaching hospitals, which train mainly doctors. There are around 20

Bates T. 2007. IP Management in the National Health Service in England. In *Intellectual Property Management in Health and Agricultural Innovation: A Handbook of Best Practices* (eds. A Krattiger, RT Mahoney, L Nelsen, et al.). MIHR: Oxford, U.K., and PIPRA: Davis, U.S.A. Available online at www.ipHandbook.org.

major medical schools in England, each affiliated with several NHS trusts.

Until 1992, this research was unmanaged. Universities and other research organizations, including commercial organizations, used the NHS infrastructure essentially as a free good to meet their requirements. But when the NHS R&D program began in 1992, the NHS became the only national health service to have its own R&D program. Its initial objectives were to:

- identify research questions that needed to be answered
- undertake research to answer the questions
- implement the answers to improve healthcare.

The resulting program based on these objectives has contributed significantly to the development of evidence-based healthcare internationally.

This is only part of the program. In 1994, an R&D budget (a levy on the total healthcare budget) was established by collecting together all the declared R&D expenditures of every hospital (R&D support funding). A budget for the NHS R&D program was also added. R&D support funding was divided into two budget headings: a budget to support noncommercial research done by others in receipt of their own external funding, and a budget to support research. The first budget meets all NHS costs for externally funded programs in universities and other agreed-upon research partners (research councils, medical charities, the Department of Health, and other government departments). The second budget only supports programs of the required quality. The NHS R&D program has now been extended from its original objectives to straddle such programs as Health Technology Assessment, Genetics Knowledge Parks, Research Networks, the Cochrane Collaboration and Systemic Reviews, and the expansion of a clinical research facility to support clinical trials. In 2004/2005, the total NHS R&D budget was UK£604 million, made up of UK£487 million for R&D support funding and UK£117 million for the NHS R&D program. Details of the whole program and how it is developing (including plans to bring together the NHS budget and the Medical Research Council budget within a new National Institute for Health Research) can be found on the Department of Health Web site.[1]

The research governance framework for carrying out research in the NHS[2] requires all those undertaking research to understand the importance of IP in their research and to take steps to identify and protect valuable IP.

3. IP IN THE NATIONAL HEALTH SERVICE: THE EARLY STAGES

3.1 *The nature of NHS IP*

NHS IP is generated in two ways:

- through R&D programs carried out by NHS researchers
- through the delivery and management of healthcare by NHS employees

The NHS carries out little fundamental noncommercial medical research. This is normally led by universities whose funding is provided principally by research councils and charities, but often with NHS staff as collaborators (and funded by R&D support funding). There is a small (but growing) band of NHS employees employed principally to do research. Much R&D expenditure in the NHS supports others, for example, university researchers, and IP arising from this work is often generated jointly with the research partner. Even though about one in three academic papers in bioscience has an NHS author, the R&D programs in which NHS researchers participate are not the major source of NHS IP. The NHS employs around 100,000 people who can generate IP in their day-to-day jobs, and their potential to come up with ideas for new products, processes, and treatments is significant. This potential was the principle driver for developing the program to manage IP in the NHS.

3.2 *The development of a policy framework*

The development of a policy framework to manage IP in the NHS began in 1998 with the appointment of the author of this chapter as NHS intellectual property advisor. R&D issues were addressed first. A health service circular (the method of publishing policy at that time) entitled A Policy Framework for the Management of Intellectual Property within the NHS arising from Research and Development[3] was published in 1998. It set out, for the first time, the principles of IP management. The circular was supported by two additional publications:

- Handling Innovation and other Intellectual Property: A Guide for NHS Researchers[4]
- The Management of Intellectual Property and Related Matters: An Introductory Handbook for R&D Managers and Advisers in NHS Trusts and Independent Providers of NHS Services[5]

The guide for researchers was designed to inform researchers about what constitutes IP, how to recognize it, and what to do. The handbook was for R&D managers (generally each research-active trust has an R&D manager), but not for IP practitioners.

The policy framework had the following basic principles:

- IP generated in research belongs to the organization employing the researcher
- if the IP has commercial value, it should be protected and commercialized by a suitable organization on behalf of the owner
- income generated by commercialization should be shared between the inventor, the owner organization, and the commercializing organization

These principles were foreign to the NHS because they raised the possibility that one NHS organization could retain income generated by the commercialization of IP and not share it with fellow organizations. Although these principles were agreed to and the framework was published, it was recognized that it would be more difficult to get others to sign up for these principles outside an R&D context.

In 1998, it was too early to extend the policy to IP arising from all sources, particularly from patient care. Moreover, the set-up and use of companies by NHS organizations to aid commercialization was specifically not allowed at this time, although this restriction was seen as an impediment to commercialization; because NHS owners of IP could not have a stake in spinout companies, collaborative work with universities would be inhibited.

At the time that the NHS was publishing its policy framework for IP arising from R&D, the national climate for innovation was changing and the government was determined that the public sector should develop a knowledge-based economy that would recognize IP, treat it as a national asset, and translate it into the benefits of jobs and prosperity. The treasury and the Department of Trade and Industry published a series of documents that changed the IP landscape across the entire public sector.

4. PUBLIC SECTOR IP IN THE UNITED KINGDOM

In 1985, U.K. universities had been given freedom to own and commercialize IP arising from their research funded by research councils. If universities wanted this freedom, they had to set up approved management systems. Almost all of them now have approved systems, and almost all are based on the principle of ownership by the university, an obligation to commercialize (or to give back to the researcher), and benefit sharing with the researcher.

Research is carried out in the United Kingdom not only by universities but also by a number of public sector research establishments (PSREs). In 1999, PSREs and NHS trusts spent UK£2.2 billion out of a total of UK£6.75 billion of research funding. Apart from the Institutes of the Medical Research Council, there was little history of government laboratories managing IP outputs. To try to change this position, the treasury set up a task force, which in 1999 published the Baker Report.[6] It made a series of far-reaching proposals that were accepted by the government,[7] which required all PSREs, many employing civil servants, to have systems in place to identify and commercialize IP of value. Moreover, the transfer of research outputs to benefit the wider national economy had to be part of a PSRE's mission.

The government confirmed that IP should be owned by the PSRE, which was usually the most appropriate organization (rather than its sponsoring department) to transfer the benefit to the national good. It was also generally recognized that income derived from this activity should be retained by the PSRE and not reclaimed by the sponsor. It confirmed that researchers, many of whom were civil servants, should be allowed to share in the income generated, and that when commercialization is achieved through the setting up of a spinout company, the researchers could have an equity stake. An initial fund of UK£10 million, against which PSREs could bid, was made available to allow PSREs to set up commercialization offices and increase their capacity to manage IP outputs.

The Baker report and the government response to it meant that all research outputs from public sector research would be managed under a common policy (mirroring that of universities). This policy fundamentally changed the position of researchers and their organizations. In addition, guidelines for the treatment of IP in government research contracts were published by the U.K. Patent Office.[8] These were all major changes.

5. DEVELOPMENT OF THE NHS FRAMEWORK

The work behind the development of the Baker report paralleled and in part informed NHS developments. The ability of an NHS trust to retain income generated by the commercialization of IP was used as a precedent by the treasury task force, and the support for spinout activity in PSREs led to renewed efforts to allow NHS trusts this freedom. The NHS trusts had by now been categorized as PSREs and were eligible to participate in funding schemes, particularly the fund that had been made available to set up IP management offices.

Despite the publication of the Baker Report, it was not until 2002 that the Department of Health was able to publish a policy document covering the full range of NHS outputs. This took nearly three years of intense effort. By 2002 it was no longer possible to prescribe policy; it had to be guidance against which progress (and compliance) could be measured. Aimed at providing a common framework for all NHS organizations, the document was called *The NHS as an innovative Organization: A Framework and Guidance on the Management of Intellectual Property in the NHS.*[9]

The title refers to the NHS as an innovative organization, which reflects the fact that by 2002 innovation was recognized by the Department of Health at the highest level as an important part of the work of the NHS. This was largely the result of the new government agenda for supporting innovation within a knowledge-based economy.

The content of the framework and guidance covers three main areas:

1. The management framework
2. Employment and ownership issues for organizations and employees
3. Partnership in IP management with universities and other research funders

Each area had complex issues to be resolved; we explain below how the most important of these were overcome. The content itself was developed by closely working with Department of Health commercial lawyers, and although the document is more legalistic than might have been imagined initially, it was vital to ensure that the secretary of state for health was protected from any legal challenge in this frequently contentious area.

5.1 *Extension to the existing policy framework for R&D*

NHS IP can arise from NHS R&D and from the delivery of patient care by all those employed by the NHS. The 1998 circular sets forth the responsibility of NHS organizations receiving R&D funding to identify IP arising from research, but NHS organizations were not responsible for systematically capturing IP associated with the delivery of patient care. This situation remains the same. However, trusts are expected to have access to a management structure (such as that described in section 5.7), so that when any IP is found employees have a place to go for expert advice. The position in the Framework and Guidance is that outputs from patient care should be managed in the same way as those from R&D. The Framework and Guidance recognizes that not all outputs will have a commercial endpoint and that some (indeed the majority) should be treated as opportunities to change practices by freely disseminating them across the NHS.

The statutory purpose of commercializing IP in the NHS (captured in the 1977 and 1990 NHS acts) is to make more income available for the health service. When an invention is exploited successfully and new products of commercial value are produced and sold, income will be generated (and shared with inventors). However, when the IP relates to a change of practice (usually covered by copyright), income generation is unusual. Nonetheless, because costs could be saved, it was agreed eventually that cost savings could be treated as income generation and so satisfy the statutory requirement.

5.2 *Income retention by the trust following commercialization*

Sharing income with inventors is a fundamental part of IP management within the public sector in the United Kingdom, and the principle was readily accepted and supported by health ministers and NHS leaders. However, it was more difficult to get the wording of the Framework and Guidance accepted by those charged with managing NHS finances. This is because it ran counter to a fundamental tenet of the NHS: any surplus income should be shared with other NHS organizations. In the end, it was agreed that surplus income arising from the commercialization of IP after paying all costs, for example, inventors, could be retained by the trust and used at the discretion of the trust to improve healthcare. It could be used to improve a service but not, for example, to build car parks. A retention limit of 0.2% of the trust turnover was set before it was necessary to bring a successful commercialization to the attention of those providing funding to the trust. In practice this meant that, unless there was a blockbuster invention, the principle had been accepted.

5.3 *Ownership of NHS IP*

Under U.K. law, IP (patents, copyright, trademarks, design rights, and know-how) generated by an employee in the course of employment or normal duties belongs to the employer unless the employer and employee have agreed otherwise. The latter was rare because in 2002 few employees had contracts that addressed IP.

However, for patented inventions the law gives additional conditions that must be met in order for the employer to own the rights. Not only must the invention be made in the course of normal duties, but it must also have been reasonably expected that an invention would result from such duties. For example, this would be reasonably expected for an employee engaged in R&D, but it could be doubtful for a surgeon performing an operation who suddenly realized how it could be done better.

The view contained in the Framework and Guidance is that should a surgeon (or any other employee) invent something during normal duties that requires a patent and needs development and testing before it can be used on patients, then the patent should be assigned to the employer to manage (as for all other IP) should the inventor want to use NHS resources to develop the invention. There is no requirement to assign the patent unless NHS resources are used, but since almost all such inventions would require development, and since the NHS would provide the most convenient test bed, the need to argue ownership through potentially costly legal procedures would be minimized. If an employee chose not to assign the invention to the employer, it would need to be developed, perhaps in the garden shed, without using NHS resources. Such considerations become redundant if all NHS employees have appropriate conditions in their employment contracts.

5.4 *Employment conditions*

If a trust has employment conditions that set out the responsibilities of the employer and the employee on all aspects of IP, then questions of ownership generally disappear and the focus can be on using the IP.

The Framework and Guidance include model employment contracts and a model entry to a staff handbook or similar document. It also considers staff appointed jointly with universities or other organizations and staff who combine NHS duties with private practice.

During his tenure as NHS intellectual property advisor, the author encountered examples of physicians who had made an invention, used NHS resources to develop it without informing the employer, and then claimed ownership when challenged because it was developed in private practice. In one case a new device had been patented and licensed to a U.S. medical device company without the knowledge of the employer. It was brought back into the ownership of the trust because it clearly arose from a research program.

A number of factors, foremost among them the high turnover of human resources staff and the lack of IP experience in the NHS, made it difficult and time consuming to clear this part of the Framework and Guidance through the Department of Health. Clearance was eventually given to the content when it was realized that only guidance was being given, so trusts could choose not to follow the guidance if they wished. In reality trusts are pragmatic and follow the guidance because to do otherwise would involve them in a great deal of legal work.

5.5 *Partnership with universities and other NHS research partners*

The NHS undertakes research jointly with universities and other research partners, such as charities and research councils. IP arises from this joint work, and ownership might not be clear. Before 1998, the NHS had no structure to recognize or manage IP, and almost nobody in the NHS was in a position to do anything about it. Almost by default, ownership was claimed by the research partner. There were many examples where inventions were realized through joint work but where no benefit came to the NHS.

Universities agreed in 2002 to a statement of partnership, which specified that when IP is generated by joint R&D between NHS trusts and universities (for example, by individuals holding a joint appointment), or where both the NHS and the university are partners in the research, then the organizations together should decide:

- which organization owns the IP
- which organization is to manage the IP and how costs are to be met
- how any benefit is to be shared after paying all costs (for example, inventors)

These arrangements are for research performed jointly, even if the inventor is solely employed by one organization. Frequently, the other organization contributes to developing the IP, and so by agreement it can be a beneficiary. The statement of partnership expected that a collaborating university and NHS trust would have similar revenue-sharing agreements with their inventors so that inventors from different organizations would be rewarded in a similar way when their invention generated income.

There is no rule that determines how benefits are to be shared, but current recommended practice starts with equal shares for both parties. If the parties agree otherwise, it is adapted. In practice, university and NHS bodies are moving ever closer in their ways of working—the 50:50 sharing model is becoming the norm, which is far removed from the previous 100:0 model!

5.6 *Spinout companies: The Health and Social Care Act 2001*

Publication of the government response to the Baker Report opened the way for a bill to be placed before parliament in 2001 that allowed NHS organizations (NHS trusts, primary care trusts, and so on) to set up, participate, and invest in companies to generate income. The scope was intended to be wider than just IP; in fact, the legislation does not even mention IP or spinout companies.[10] The advantage of a wider provision became clear when some of the earliest uses of the legislation were for companies that had nothing at all to do with IP.

If they are badly set up, the use of spinout companies carries inherent risk. NHS organizations are generally not free to set up companies without a business plan authorized by the Department of Health. The business plan must comply with Directions (which are legally binding) contained within the Framework and Guidance. This essentially protects the secretary of state for health against unnecessary risk, and it has meant that the Department of Health (like other government departments) has had to develop expertise in a new area of activity.

The Framework and Guidance includes detailed guidance to trusts and employees on how companies should be established, the role of the trust, and the position of employees as directors or shareholders. The content follows national guidance provided in the government response to the Baker Report.

5.7 *Management of IP in the NHS*

5.7.1 *The concept*

The Baker Report recommended that PSREs should establish management systems to deal with their IP. But how were the outputs from the acute trusts, the ambulance trusts, the mental health trusts, the care trusts, and the primary care trusts—a total of 601 organizations in 2006—to be dealt with? Although research outputs might be expected to concentrate around teaching hospitals and their partner universities, no such assumption could be made for innovations in patient care from doctors, nurses, scientists, technicians, and so on. It was clearly not cost effective or appropriate in terms of likely business to locate a management organization in each of the trusts.

Extending the scope of university technology transfer offices was rejected because their interest would be primarily in research-based innovations; patient-care-led innovations were likely to be lost. Universities were already being stretched by the government innovation agenda.

The agreed management solution was for nine regionally based NHS innovations hubs. These map on to the regional government structures (regional development agencies) in England. Each hub covers on average 60–70 organizations. Section six describes their operation in more detail.

5.7.2 *The hub as an organization*

A hub is either an unincorporated association of NHS bodies or a company limited by guarantee. It has a management board that decides structures and hires its own employees. In an unincorporated association, the employees are NHS employees; in a company limited by guarantee they are employed outside the NHS. Currently, the hubs are split approximately equally between the two models. Generally, the hubs have "branch offices" that reflect the region's different geographies. The London hub, for example, has one central office with outposts located close to the five principal teaching hospitals. The southwest hub, which covers one of the largest geographical regions, relies more on electronic communication than direct contact.

5.8 *License agreements*

The Framework and Guidance includes terms to be used in license agreements with commercial partners. It also includes, for developing countries, a specific appendix taken from MIHR (Centre for the Management of Intellectual Property in Health R&D) documentation that was produced for The Rockefeller Foundation in November 2001.

In the terms for license agreements, the Framework and Guidance recognizes that most commercializable items of NHS IP will have an international market and that licenses will cover manufacture and sale in more than one country. The Framework and Guidance states that license agreements should seek to include terms that are likely to give patients in developing countries access to products at reasonable cost.

As stated earlier there was some dispute as to whether all NHS trusts should benefit from an invention made by another trust, particularly whether products arising from the invention should be royalty free to the NHS. The Framework and Guidance says that those negotiating the license agreement (the NHS innovations hub or another body) should seek to include preferred terms for sales to other NHS organizations. Essentially, however, the main way for a trust to benefit is through developing its own inventions.

5.9 *Independent providers of health services within the NHS*

Some health professionals (such as general practitioners, dentists, and pharmacists) are not NHS employees but work under contract with a primary care trust. Some of these professionals generate

IP through NHS research and others through their services. The framework and guidance recognizes that the NHS is unlikely to own IP outside R&D, but it offers these professionals the services of the NHS hubs under the same terms and conditions as NHS employees, if they assign the IP to the primary care trust.

6. THE WORK OF THE NHS INNOVATIONS HUBS

IP management is a complex task and has not been a core business for the NHS. Ideas for new technologies (new or improved devices, for example) need to be protected, often by filing a patent application. Converting ideas into new products—and rejecting unsuitable ideas—require specialist skills, and the NHS innovations hubs have been set up to provide these services.

The first hub in the northwest began its operation in 2001, which was followed by the other eight. The last hub was only recently established in the southwest.

A driving force for their creation was the UK£10 million PSRE fund set up by the Department of Trade and Industry against which PSREs could bid. In the first round of funding, UK£6 million was provided to create capacity for IP management in the public sector, and UK£4 million for setting up seed funds. NHS trusts could apply, and bids for funding to create capacity were made from all regions through a lead trust. The fund was oversubscribed, but many NHS bids were successful in the first round of funding, receiving about one half of the total available funding. There have been two further rounds, and all hubs have now received funding from this source. This adds to core funding provided by the Department of Health; initially this was UK£2 million per year but has since increased.

The hubs are developing their operation in close partnership with the nine regional development agencies, government organizations set up to stimulate and support local business. Several of the regional development agencies provide additional funding for the hubs in the expectation that they will be the source of new products, processes, and businesses in their region.

The services that a hub provides include:

- identifying IP through clinics and similar activities
- providing training for NHS employees in the importance and understanding of IP
- evaluating IP and initiating additional R&D to produce evidence of clinical application
- protecting IP
- commissioning the production of prototypes
- advising on and exploiting IP through licensing or setting up of companies
- collaborating with universities and other third parties in the exploitation of IP generated jointly with trusts

Each hub establishes its own networks and determines its mode of operation. A national network, the IP Forum, meets monthly or every two months. Most hubs charge a membership fee to their member organizations, and a large majority of the trusts have chosen to join their hub. Hub networks usually partner with networks of R&D managers. Geography plays its part, but members of a hub typically have much in common. They extend their scope through establishing a "product champion" in a member trust who acts as the first contact point for the hub. Currently, hubs employ five to 20 people, depending on the hub's state of development. Often the enthusiasm displayed by the trusts has to be constrained by the available resources of the hub.

A hub has the considerable task of usually working with between 60 and 70 trusts. Getting all NHS employees without previous training and experience to understand IP is an arduous task. Web-based training and other methods are being used to publicize the work of the hubs and to encourage employees to think about innovation. The wage packet and trust newsletters are also effective communication tools. Regional competitions, in which employees are encouraged to submit their innovations to their hub for adjudication, have proved an excellent stimulus. The opportunity for publicity is very high, and the excitement generated in a small trust when it wins one of these competitions is remarkable.

Regional competition winners go forward to a national competition, and the publicity and enthusiasm generated by the competition, capped off with health ministers presenting prizes at a national event, bring IP and innovation in the NHS to the fore.

Here are some highlights from 2004/2005:

- the number of hub pipeline opportunities increased from 497 in 2003/2004 to 1250, of which 257 were selected for further development
- 40 licenses were brokered
- many hundreds of entries were made to regional innovation competitions
- three new spinout companies were approved
- income generated approximately doubled from its 2003/2004 level to UK£1.5 million

Of the opportunities selected for further development medical devices accounted for 49%; biotechnology and pharmaceuticals 8%; diagnostics, 8%; IT and training, 28%; and other areas, 7%. Around 30% of the potential innovations had a university link. The following examples show the breadth of these opportunities:

- a handheld device that measures accurately the size of the pupil of a patient's eye at the scene of a road traffic accident
- an electrical device to overcome the effect of "dropped foot syndrome," which is already being manufactured locally for the hub and is the basis of a spinout company
- a simple and low-cost device to eliminate incidents of patients receiving the wrong type of blood
- a device to allow the transfusion to a patient of all blood in a bag
- a company set up by a major hospital to measure glycemic index, important among other things in the management of diabetes, using the expertise of a world-renowned laboratory
- a virtual reality treatment for lazy eye

A comment on time frame is important. When the PSRE Fund was established by the Department of Trade and Industry, the Fund recognized that it would take at least ten years before its success (or failure) could reasonably be measured. The first hub was established in 2001 and the last in 2005. Many of the products arising from the NHS program require extensive testing through trials and other research programs before they can be used on patients, and so success is never easily nor instantly obtained. Even the device to ensure that all blood is completely emptied from a transfusion device would take time to develop and manufacture before it can provide the expected yearly savings of UK£20 million. Research and prototype testing of the dropped foot device began many years ago and it is only now being manufactured.

The growth in income generated in 2004/2005 was satisfying, but much of this income arose from innovations developed some years ago, before the hubs were established. The impact of the hubs and the performance indicators used to measure it are themselves an important piece of ongoing work. The impact reaches far beyond income and numbers of patents.

Each hub has its own Web site and all of them are accessible through the NHS innovations Web site at www.innovations.nhs.uk. The site holds several of the important documents referred to here.

7. APPLICATION OF THE NHS MODEL TO DEVELOPING COUNTRIES

The way that the NHS developed its framework and set up the hubs could be useful for developing countries. Perhaps the most useful aspects are:

- The NHS model will help developing countries if they can agree on a common way to treat IP. IP is difficult to deal with and differences in approaches across countries and organizations increases the degree of difficulty. The United States and United Kingdom models have similar operating principles and are recommended as tried and tested.
- Scientists and other generators of IP in developing countries cannot all be IP experts, nor do they need to be. Generators of IP do

need to recognize that a particular output might be important, and be able to identify individuals who can help promote the innovations. The training for researchers should not focus on how to draw up license agreements but on how to record results, how to avoid disclosure, and how to recognize valuable outputs. The principles contained in the NHS Guide for Researchers (applicable to researchers outside the health fields) seem appropriate.

- Researchers need someone in their organization, to act as a "product champion" who can be their eyes and ears, similar to an R&D manager who often takes on this role in an NHS trust. This advocate does not need to understand the intricacies of drawing up license agreements, but does need to understand the principles contained in the agreements and ensure that the researcher's practices are aligned with those principles. Having a product champion is particularly important when new collaborations are being set up and a collaboration agreement is being established. The handbook produced by the NHS[11] could be adapted for use by developing countries. Product champions could form learning networks as they do in the NHS.
- The IP office needs to be of sufficient scale that it offers the experience and expertise to deal with complex issues. Once a wrong agreement is in place, it cannot be corrected (though it can be amended but this often takes significant negotiation efforts), which is particularly important for developing countries with the variety of new technologies contained, for instance, in agriculture and plants. An IP office similar to that of an NHS hub, dealing with a number of organizations, has much to recommend it. Such an office could attract or have access to the necessary level of expertise, much of it from outside the country, to draw up the agreements that protect the interests of researchers, the developing country, and a collaborator or investor in a technology.

8. CONCLUSION

The NHS hubs are meeting a need and show strong indications of success. Widely valued, they are rapidly becoming the "one-stop shop" for innovation in the U.K.'s national healthcare system. To further support innovation, the Department of Health is setting up a national innovation center that will have a dedicated budget to put on the fast track to the marketplace particularly promising projects. The future looks good provided people are patient! ■

TONY BATES, *Managing Director, Tony Bates Associates Ltd., The Old Vicarage, Vicarage Lane, Olveston, Bristol, BS35 4BT, U.K.* tony@tbatesassociates.co.uk

---

1 Best Research for Best Health: A New National Research Strategy. www.dh.gov.uk/PolicyAndGuidance/ResearchAndDevelopment/ResearchAndDevelopmentStrategy/RDStrategyArticle/fs/en?CONTENT_ID=4127109&chk=RKJISx.

2 Research Governance Framework for Health and Social Care. Department of Health, U.K. www.dh.gov.uk/PolicyAndGuidance/ResearchAndDevelopment/ResearchAndDevelopmentAZ/ResearchGovernance/fs/en.

3 HSC. 1998. Policy Framework for the Management of Intellectual Property within the NHS Arising from Research and Development. www.innovations.nhs.uk/pdfs/106HSC.pdf.

4 Handling Innovation and other Intellectual Property: A Guide for NHS Researchers. www.innovations.nhs.uk/pdfs/integuid.pdf.

5 The Management of Intellectual Property and Related Matters: An Introductory Handbook for R&D Managers and Advisers in NHS Trusts and Independent Providers of NHS Services. www.innovations.nhs.uk/pdfs/intehand.pdf.

6 Creating Knowledge, Creating Wealth: Realising the Economic Potential of Public Sector Research Establishments. Baker Report. HM Treasury, U.K. www.hm-treasury.gov.uk/documents/enterprise_and_productivity/research_and_enterprise/ent_sme_baker.cfm.

7 The Government's Response to the Baker Report: Creating Knowledge, Creating Wealth: Realising the Economic Potential of Public Sector Research Establishments. www.hm-treasury.gov.uk/media/23D/D3/57.pdf.

8 Intellectual Property in Government Research Contracts. Guidelines for Public Sector Purchasers of Research and Research Providers. www.patent.gov.uk/about/notices/2001/ipresearch.pdf.

9 NHS. 2002. The NHS as an Innovative Organisation A Framework and Guidance on the Management of Intellectual Property in the NHS. National Health Service, Department of Health: London. www.innovations.nhs.uk.

10 Health and Social Care Act 2001. 2001. chap. 15. www.legislation.hmso.gov.uk/acts/acts2001/20010015.htm.

11 See *supra* note 9.

CHAPTER 17.12

Partnerships for Innovation and Global Health: NIH International Technology Transfer Activities

LUIS A. SALICRUP, *Senior Advisor for International Technology Transfer Activities, Office of Technology Transfer, National Institutes of Health, U.S.A.*

MARK L. ROHRBAUGH, *Director, Office of Technology Transfer, National Institutes of Health, U.S.A.*

ABSTRACT

Technological innovation is increasingly recognized as an important tool for improving global health. The Office of Technology Transfer of the U.S. National Institutes of Health (NIH OTT) has increased its licensing of technologies for the prevention and treatment of neglected diseases to partner institutions in developing regions of the world. Other efforts have focused on providing assistance to indigenous institutions in building their technology transfer capacity. In addition to helping to achieve the primary objectives of meeting global public health needs and strengthening local R&D capacities, NIH OTT expects such efforts to have a positive impact on national policies on intellectual property rights, and, ultimately, to increase multinational investments in developing countries, which will likely result in an even greater effort to develop accessible therapies for those in need.

1. NIH MISSION, INTERNATIONAL TECHNOLOGY TRANSFER, AND GLOBAL HEALTH

The mission of the U.S. National Institutes of Health (NIH), Department of Health and Human Services (HHS), is to support biomedical research that will reduce illness worldwide and extend healthy life. NIH's Office of Technology Transfer (OTT) works with institutes and centers at NIH and the Food and Drug Administration (FDA) to manage the patenting and licensing of inventions made by their intramural scientists. As part of this effort, NIH seeks to understand challenges that hinder the public availability of these inventions.

One might naturally ask why NIH, a domestic agency, should involve itself in international technology transfer. Enhancing technology transfer to developing countries, however, is an important humanitarian endeavor consistent with NIH's mission to improve health and save lives. Such transfers allow these countries to introduce technologies appropriate to their own regional needs, building more independence and enabling local and regional public health solutions.[1, 2, 3] Because many of these markets are not a priority for most companies in developed countries, technology transfer efforts can be extended outside the United States, consistent with humanitarian and economic goals.

By necessity, the NIH mission of NIH extends beyond U.S. borders. The U.S. works to improve health worldwide not only for humanitarian reasons but also because diseases do not observe national boundaries. Moreover, improved public health allows nations to better maintain economic growth and political stability.

One specific NIH goal for technology transfer is to "strengthen the capacity of developing countries to identify technologies and pursue their

Salicrup LA and ML Rohrbaugh. 2007. Partnerships for Innovation and Global Health: NIH International Technology Transfer Activities. In *Intellectual Property Management in Health and Agricultural Innovation: A Handbook of Best Practices* (eds. A Krattiger, RT Mahoney, L Nelsen, et al.). MIHR: Oxford, U.K., and PIPRA: Davis, U.S.A. Available online at www.ipHandbook.org.

development into products, through education and technical assistance."[4] By extending R&D activities outside U.S. borders, we transfer technological know-how to developing countries. This learn-by-doing approach enhances technological capabilities[5] and facilitates the development of technologically capable partners, which, in turn, better leverages the value of technologies and extends scientific knowledge and practice. Overall, such technology transfer activities are likely to add value and provide social returns on existing inventions,[6] either by addressing U.S. market needs or by improving the health of people worldwide and preventing the spread of disease across U.S. borders.

2. PARTNERSHIPS IN TECHNOLOGY TRANSFER

The most immediate incentive for OTT to engage in international activities is to help reduce the burden of disease globally. Developing countries stand to benefit from licensed NIH inventions, because when developed locally the technologies are more readily available to local markets. Such technology transfers may play a particularly important role in turning early-stage technologies into biomedical products in developing countries. Additional benefits accrue locally from the development of technologies for the developing world by indigenous institutions. These include enhanced local capacity in research and development, increased market competitiveness, the growth of an experienced work force, improvement of scientific excellence, and the consequential growth of the biotechnology infrastructure, all of which ultimately strengthen and stabilize developing countries' economies.[7] Figure 1 illustrates the potential impact of technology innovation on global heath.

The impact of globalization is not limited to international trade and economics.

Globalization also exacerbates existing public health challenges that in turn impact the national interests of industrialized nations. These challenges, though not limited to the developing world, can be addressed in part by the transfer

FIGURE 1: POTENTIAL IMPACT OF TECHNOLOGICAL INNOVATION ON GLOBAL HEALTH

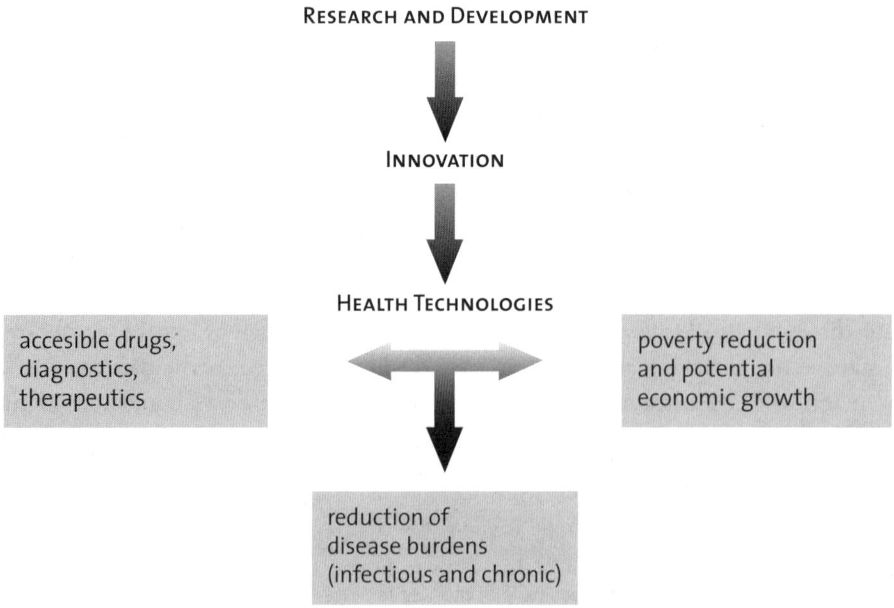

of technologies to developing countries. Indeed, the international community now widely recognizes that some diseases that once were contained within regional borders now threaten the United States in two ways:

- Emerging and reemerging infectious disease epidemics: With increased movement of goods, animals, and people, diseases spread rapidly across borders, posing direct threats to U.S. citizens. It suffices to mention epidemics of diseases such as HIV/AIDS, influenza, tuberculosis, cholera, and SARS, which threaten not only the regions where they originated but also the entire globe.[8]
- Risks from civil unrest: The spread of disease often fuels a cycle of poverty, suffering, and civil disorder. (Gaining access to drugs and medical technologies are genuine public welfare concerns in many developing countries.[9,10] Providing access to these countries will reduce the burden of disease and help improve the quality of life, thus diminishing the threat of unrest in volatile areas of the globe.)

While NIH focuses on making new methods of treating and preventing disease available to world markets, the agency also emphasizes the importance of making existing vaccines for pandemic diseases available to the countries in need For example, an effective vaccine for measles has been in use in industrialized nations for the past 40 years, but most of the developing world has only recently gained limited access to the vaccine.[11] In addition, the financial and logistical challenges of international efforts to provide antiretroviral drugs to treat HIV/AIDS in developing countries are well known.

Other diseases in developed countries remain serious public health burdens in developing countries. Malaria was virtually eradicated through the use of insecticides and antimalaria drugs in North America and Europe, while Africa, Asia, and Latin America saw the development of increasingly resistant mosquito vectors and malarial parasites. As malaria became a relatively low health risk in developed nations, the development of a malaria vaccine became a lower priority. This situation led the Bill and Melinda Gates Foundation to launch and support the Malaria Vaccine Initiative (MVI), an effort to address this serious shortcoming and accelerate vaccine development.[12] The foundation's efforts supplement ongoing research supported by NIH and other Non-Governmental Organizations (NGOs).

Another approach to these public health challenges is for institutions, both national and international, to encourage and facilitate the relatively more technologically advanced developing countries to enhance their product commercialization capacity to meet local needs. Several research studies indicate that this is the best approach to combating long-term neglected diseases in poor countries in Sub-Saharan Africa, parts of Asia, Latin America and the Caribbean, and Eastern Europe.[13,14,15] Indeed, recent work by leading private foundations, such as the Gates and Rockefeller foundations, emphasizes developing countries' "*need for self-reliance and national production [of health technologies] to ensure that country-specific disease needs can be met.*"[16,17,18,19] Ultimately, such investment will provide less-developed countries with sustainable benefits.[20]

The World Intellectual Property Organization's (WIPO) Cooperation for Development Program is committed to tailoring the implementation of its IP strategies to the diverse infrastructures and needs of developing countries.[21] Similarly, the Organisation for Economic Co-operation and Development (OECD) concludes that "*the transfer of technology to developing countries is a key element so that countries can develop their own R&D infrastructure and capabilities to meet their own needs.*"[22] Developing countries that have reached a sufficient level of technological capacity are now encouraged to enhance their capabilities more dynamically by nurturing domestic assets and creatively blending domestic and foreign knowledge.[23]

NIH Office of Technology Transfer recognizes the significance of assisting U.S. and foreign institutions in the development of technologies as a means to make medicines more accessible to everyone. By working with local institutions, international organizations, and private foundations, OTT has identified technology transfer needs and opportunities related to HIV/AIDS, pertussis, malaria, dengue, childhood diarrhea (rotavirus), meningitis, typhoid fever, cancer, and diabetes. Based on the extensive patent portfolio in neglected diseases (Table 1), OTT has already transferred technologies to public and private institutions in India, Mexico, Brazil, China, Korea, Egypt, and South Africa. The office expects to execute licenses in the near future with other institutions in Africa.

This experience demonstrates that governmental or not-for-profit research institutions should seriously consider transferring early-stage biomedical technologies to institutions in the developing world rather than focusing exclusively on pharmaceutical and biotechnology companies in the western world. Of course, this should not be done haphazardly. NIH OTT learned a key lesson while expanding its licensing activities with developing countries—licensee institutions should have at least some research and development capability, as well as clear national and regional public health objectives. When these two conditions are met, access to key technologies and models of successful product development are more likely to produce new products to improve public health. By encouraging technology transfer throughout the world, NIH contributes to its long-term global mission of reducing the burden of diseases that are particularly devastating for people living in developing countries.[24]

3. INTERNATIONAL TECHNOLOGY TRANSFER RESULTS AT NIH: LESSONS LEARNED

With the goal of global public health in mind, there are many different strategies and tools that can be utilized in the management of IP. For instance, commercialization licenses can involve the transfer of rights to utilize IP, not only in relation to patents, but also for unique biological materials such as cell lines and microorganisms to be used in production or as candidate vaccines, and any associated gene expression constructs. Patent rights can only be enforced in countries where patents have been obtained for compositions of matter (materials) or methods of producing or using a given technology. Thus, in order to enforce a patent in a particular country, the patented composition or method must be used or sold in that country (or in some countries, an unpatented product produced by a patented method can infringe the method patent when that product is imported into the country where the method patent is held). For example, if a live

Table 1: Examples of NIH Intellectual Property in Neglected Disease Areas

Disease/therapeutic Area	Distinct technologies	Issued patents	Patents pending
Dengue	27	20	40
Rotavirus	19	2	28
Human Papilloma virus (HPV)	28	23	46
Lyme disease	7	1	6
Tuberculosis	16	1	14
Malaria	36	64	39

Source: Salicrup and colleagues.[33]

attenuated virus developed for use as a vaccine has been patented only in the United States and European countries, a commercialization patent license could be given to one company for the United States and Europe and possibly another company (as a biological materials commercial license) for the rest of the world, where no patent is in force. Since many institutions, particularly government or academic laboratories, have not obtained patent protection in many, or any, developing countries, the biological materials commercialization license is an important commercialization incentive tool.[25]

In addition to commercial licenses under either type of license, an institution can grant rights on a geographic basis, either exclusive, coexclusive or nonexclusive, in another country or to multiple countries within a geographic region, continent, or throughout the world. A strategy for a particular technology may be to permit multiple institutions around the world, each with a different geographic market segment, to develop the technology in parallel. This strategy is used to increase the opportunity for introduction of a product in multiple regions nearly simultaneously with the aim of meeting public health needs with less delay. Each regional producer may want to tailor the product slightly differently to meet the public health and regulatory demands of the region it represents. Finally, with this type of strategy, there will be back-up institutions to meet worldwide needs if one of the regional producers is delayed significantly or fails to produce the product.

By law and policy, NIH favors nonexclusive licensing to promote market competition, unless an exclusive or coexclusive license is a necessary incentive for one or two parties, respectively, to bring a product to market. Thus, when an exclusive license is not needed to encourage commercialization in a given country or region, nonexclusive licensing, regionally or worldwide, will allow multiple parties to compete in the market to develop a product. Like the regional strategy with multiple codevelopers, nonexclusive licensing within a given market has similar advantages.

When framing a marketing strategy for international product development, all of these mechanisms can be utilized in complex ways to provide the appropriate incentives for each country or region. Otherwise, the licensing terms for institutions serving the public health needs of developing countries are comparable to NIH OTT licenses to institutions in developed countries. Royalty fees are negotiated on a case-by-case basis, depending on such factors as the marketing plan, market size, potential use for the public interest, and the need to license additional technologies. In developing markets, some of these factors (for example, market size and public health interests) may play a greater role in determining the license terms than licenses for markets in OECD countries. This paradigm allows OTT to fulfill its statutory requirement to favor U.S. small businesses and to use exclusive licensing strategies as a commercialization incentive only as needed and supported by the market players.[26]

In recent years, NIH has increased its filing of patents for globally important vaccines and therapeutics in countries like China, India, Brazil, and Mexico so that the exclusive or coexclusive patent license mechanism is available for use as an incentive, as needed, to develop such products. This is particularly important for technologies where no unique biological materials are needed for commercialization and biological materials licensing is thus not an option. Additionally, NIH makes efforts to transfer know-how and critical documentation for manufacturing and marketing approval (when available) to help institutions in developing countries expedite their commercialization plans.

Through an ongoing analysis of its own portfolio and the needs and capabilities of developing countries, OTT has found that a niche exists for international technology transfer that is consistent with U.S. technological, public health, and economic interests. Such transfers, moreover, can provide solutions to the most socio-economically harmful diseases. OTT has already transferred early-stage technologies to public and private institutions in India, Brazil, China, Korea, Egypt, South Africa, and Mexico (see Table 2), and negotiations are in progress with institutions in Brazil, China, Argentina, India, Egypt, and Nigeria. For example, OTT licensed a vaccine conjugation technology to PATH, a nonprofit global health organization, to develop a conjugated meningococcal vaccine in collaboration with the World Health Organization (WHO). PATH and WHO selected the Serum Institute in India to manufacture the vaccine for eventual distribution in Sub-Saharan Africa, the Middle East, Latin America and the Caribbean, and Eastern Europe. Another license agreement involves the transfer of NIH materials for the development of a conjugated vaccine against typhoid fever to the International Vaccine Institute (IVI) in Seoul, Korea, which plans to sublicense manufacturing to public and private entities in Indonesia and India for ultimate distribution of the product in Asia.

Table 2: Examples of NIH OTT Interinstitutional and Multiprong License Strategies

Technology	License type	Licensee(s)	Manufacturer	Technology distribution region
Conjugated Meningitis Vaccine	Nonexclusive patent	PATH/WHO, public and private institutions in South Africa and Nigeria (applied)	Serum Institute in India, public and private entities in Mexico and South Africa	Sub-Saharan Africa, Middle East, Asia, Latin America and the Caribbean
Human-Bovine Rotavirus Vaccine	Nonexclusive, coexclusive, or exclusive patent	Public and private institutions in Brazil, India, China, U.S.	Multiple companies; public entities in Brazil, China, India, U.S.	Latin America and the Caribbean, Asia, Africa, Middle East
Typhoid Fever Conjugated Vaccine	Nonexclusive biological materials	IVI	Biopharma in Indonesia, Serum Institute in India	Southeast Asia
Dengue Tetravalent Vaccine	Internal evaluation for Brazil (applied), nonexclusive for India and certain Latin American countries	Public institutions in Brazil, private institutions in India	Public institutions in Brazil, two companies in India, one company in U.S.	Latin America and the Caribbean, Asia
Varicella Vaccine	Commercial evaluation license	Public and private institutions in Egypt	Public entity in Egypt	Africa and the Middle East

Source: Adapted from Salicrup and colleagues.[34]

In some cases, a multiprong licensing strategy can be developed for the same technology that utilizes different license types to multiple institutions in different countries based on institutions' needs and market dynamics. For example, OTT is licensing technology related to the development of a human-bovine rotavirus reassortant vaccine to several public and private institutions in Brazil, China, India, and the United States.[27] Depending on the country and geographic region, the license is exclusive, coexclusive, or nonexclusive. The degree of exclusivity was determined by the needs of the prospective licensees and the market dynamics in each country. Surprisingly, not all nonprofit institutions were willing and able to accept a nonexclusive licensing arrangement. By granting exclusive rights only when needed to spur commercialization in world market segments, the strategy allows the market to drive the degree of exclusivity. This strategy also increases the likelihood that the technology will be developed in parallel from multiple sites for eventual worldwide distribution from multiple companies and institutions. In the case of an effective human-bovine rotavirus vaccine, such a goal is critical to significantly reducing childhood deaths from this infection, throughout the developing world, without unnecessary delays.[28, 29]

NIH OTT has found that international technology transfer requires a holistic and flexible approach—a donor-recipient paradigm that eschews unequal partnerships and the consequent challenges with trust, commitment, and reliability. Local scientists provide scientific support for the licensing strategy, and business managers directly participate in negotiations with NIH OTT as it pursues agreements with as much flexibility as possible to meet local needs. Hopefully, this strategy of enhancing technology transfer to emerging markets will build international capacity and capabilities. It should also provide regional, multilateral, and philanthropic organizations with more options to work with licensee companies to distribute products at a lower cost in developing countries and emerging markets.

4. CAPACITY BUILDING AS A TOOL FOR SUSTAINABLE ECONOMICAL AND SOCIAL DEVELOPMENT

NIH OTT also recognizes the relevance of assisting in the development of a cadre of scientists and technology managers experienced in IP management and other matters related to technology transfer. Overcoming this obstacle is necessarily a long-term project but also, eventually, a self-sustainable one.[30] As a first step, OTT is working in partnership with other stakeholders throughout the world to assess the technology transfer and training needs of institutions in developing countries. Moreover, OTT has initiated an international technology transfer capacity building program to train scientists and managers from developing countries. The first phase will include training of staff from institutions in China, Brazil, Argentina, India, South Africa, Philippines, Chile, Mexico, and Hungary. Future expansion of the program is envisioned for relevant personnel from additional institutions in Africa, Latin America, Asia, and Eastern and Central Europe.

NIH OTT, in collaboration with technology transfer offices at NIH Institutes and Centers, regularly invites individuals with particular expertise and experience with various aspects of technology transfer to give seminars at NIH. These experts include biotechnology and pharmaceutical business people, lawyers, technology transfer managers, governmental technology transfer experts, representatives of charitable foundations and NGOs dedicated to supporting product access in the developing world, representatives from nonprofit and for-profit institutions involved in commercialization efforts, and public health officials from throughout the world. Topics have included licensing strategies and terms, patents, public/private partnerships, MTAs, policy issues, and international agreements. As part of their internship at OTT, international trainees attend these lectures as they are able. OTT is currently discussing how to enhance the participation in these training and presentation sessions of both technology managers from institutions in developing countries and scientists and administrators from "resource limited" institutions in the United States. Additionally, as part of the Curriculum Planning Workgroup of the Technology Managers for Global Health, a special interest group within the Association of University Technology Managers, NIH OTT participated in the design and development of an educational booklet geared to serve as a resource tool for technology managers of institutions in developing countries.[31]

OTT is working with the Patent Facilitation Centre at the Indian Ministry of Science & Technology, the Bi-National S&T Endowment Fund (generally called the Indo-U.S. science and technology fund), the South African Council for Scientific & Industrial Research (CSIR), and the Developing Countries Vaccine Manufacturers Network (DCVMN) to develop and implement short courses, seminars, and workshops on issues pertaining to IP management that are geared to training technology managers from several universities and from research and development centers in India, South Africa, Tanzania, Egypt, Brazil, Mexico, Argentina, Chile, China, Vietnam, and Thailand.

Information and access to knowledge has been recognized as a crucial step in enhancing capacity in developing countries. NIH OTT and the technology transfer offices from several universities in the United States recently developed and implemented a database of neglected-disease technologies available for licensing from these institutions. This database is already available at the OTT Web site with discussions underway with other potential hosts.[32] The database should be an important resource and capacity building tool for technology managers of universities and research centers in developing countries for identifying more readily such technologies and for coordinating work with the licensor institutions. The expectation is that other universities and non-profit institutions with technology licensing opportunities in the area of neglected diseases will eventually join this initiative to provide information at a single Web site while retaining licensing from the institution owning the technology.

5. INNOVATION, R&D COLLABORATIONS: NEXT STEPS

As NIH OTT's relationship with institutions in developing countries matures and the relationships between the office and those institutions expand, the next steps may include an evaluation study to explore the needs and opportunities related to technology transfer and training for people from institutions in developing countries. This evaluation would explore areas that affect technology transfer outcomes, such as IP policies, regulations, clinical trials capacity, IP management capabilities, and policies influencing public/private sector partnerships (PPPs). Thus, OTT has the potential to contribute to the scientific, technological, and health needs of developing countries by improving its own ability to bring to market technologies that will benefit local and regional public health.

NIH OTT is committed to contributing expertise and sharing ideas, strategies, and practices mutually with other organizations, in both developing and developed nations, to advance the goals of international technology transfer. Such coordination can only enhance the individual efforts of each of the institutions involved. In addition, OTT will continue to learn from partners throughout the world about creative alternative solutions to the challenges of transferring biomedical technologies to benefit global health.

6. CONCLUSIONS

Building on a strong track record, NIH OTT is expanding its efforts at licensing technologies to institutions in developing countries, and it continues to work with other stakeholders to help build technology transfer infrastructures. These activities are helping NIH to fulfill an important goal of its global public health mission: to reduce the devastating disease burden on people living in developing countries. Bringing biomedical inventions to populations in less-developed regions of the world can be achieved through various technology licensing models that fit the specific competencies of the research and development infrastructure of the particular countries. Moreover, it is expected that OTT's activities in global technology transfer will promote well-recognized, good licensing practices that meet regional and national health priorities and standards. As a result, these activities should enhance public availability of new technologies, attract new biotechnology R&D resources, obtain returns on early-stage public investment, and stimulate economic and social development. ■

ACKNOWLEDGMENTS

The authors wish to thank Uri Reichman, Susan Ano, Peter Soukas, Chekesha Clingman, and Steve Ferguson from NIH OTT as well as scientists such as Al Kapikian (NIH), Isaias Raw (Brazil), Suresh Jadhav and R. Saha (India), and others for their commitment to global public health.

LUIS A. SALICRUP, *Senior Advisor for International Technology Transfer Activities, Office of Technology Transfer, National Institutes of Health, 6011 Executive Boulevard, Suite 325, Rockville, MD, 20852, U.S.A. salicrul@mail.nih.gov*

MARK L. ROHRBAUGH, *Director, Office of Technology Transfer, National Institutes of Health, 6011 Executive Boulevard, Suite 325, Rockville, MD, 20852, U.S.A. mark.rohrbaugh@nih.hhs.gov*

---

1 Maskus K. 2004. Encouraging International Technology Transfer. In *Intellectual Property Rights and Sustainable Development*. May 2004. UNCTAD-ICTSD: Geneva. www.iprsonline.org/unctadictsd/docs/CS_Maskus.pdf.

2 Varmus H, R Klausner, E Zerhouni, T Acharya, AS Daar and PA Singer. 2003. Grand Challenges in Global Health. *Science* 302(5644): 398–99.

3 Saha R, K Satyanarayana and CA Gardner. 2004. Building a "Cottage Industry" for Health (and Wealth): The New Framework for IP Management in India. *IP Strategy Today* No. 10:23–58. www.bioDevelopments.org/ip.

4 GPRA. 2003. Performance and Accountability Results. Government Performance and Results Act. Federal Trade Commission: Washington, DC. www.ftc.gov/opp/gpra/.

5 Marshall A. 2004. Open Secrets. *Nature Biotechnology* (Supplement) 22:1.

6 Gardner C and Garner C, 2004 Technology Licensing to Nontraditional Partners: Non-Profit Health Product Development Organizations. For *Better Global Health*. MIHR: Oxford, U.K. www.mihr.org/?q=taxonomy_menu/1/12.

7 DNDi. 2003. Drugs for Neglected Diseases Initiative Working Group and MSF-Medicins San Frontiers. Intensified Control of Neglected Diseases: WHO Report of an International Workshop. Berlin, 10-12 December, 2003. www.who.int/lep.

8 www.globalhealth.org. At the same time, chronic diseases such as cardiovascular diseases and diabetes, which historically have been diseases primarily of the developed world, are also increasing among people in developing countries.

9 WHO 2004. Intensified Control of Neglected Diseases: Report of an International Workshop. Berlin, 10–12 December 2003. World Health Organization: Geneva, Switzerland.

10 Hirschberg R, J La Montagne and AS Fauci 2004. Biomedical Research: An Integral Component of National Security. *The New England Journal of Medicine* 350(21):2119–2121.

11 WHO-UNICEF 2003. Global Leaders Intensify Commitment to Prevent Leading Childhood Killer—Measles. Cape Town, 17 October 2003 Meeting.

12 MVI 2004. Malaria Vaccine Initiative 2004 Backgrounder: Accelerating Vaccine Development to Save Lives. Malaria Vaccine Initiative. Bethesda, Maryland. www.malariavaccine.org/ab-ov2-compellingneed.htm.

13 Boulet P, C Garrison and E 'tHoen 2003. Drug Patents Under the Spotlight: Sharing Practical Knowledge About Pharmaceutical Patents, Medicins Sans Frontiers, May 2003 Report, SRO-Kundig, Geneva.

14 See *supra* note 9.

15 OECD. 2002. Report of the Conference on Biotechnology for Infectious Diseases: Addressing the Global Needs. Organization for Economic Co-operation and Development: Paris, France. www.oecd.org/document/60/0,2340,en_2649_34537_2500476_1_1_1_1,00.html Also see Rapporteur's Report at www.oecd.org/dataoecd/53/20/2500434.pdf.

16 See *supra* note 3.

17 Thorsteinsdóttir H, U Quach, DK Martin, AS Daar, and PA Singer 2004. Introduction: Promoting Global Health Through Biotechnology. *Nature Biotechnology*. (Supplement) 22:3–7.

18 Maurer SM, A Rai and A Sali 2004. Finding Cures for Tropical Diseases: Is Open Source an Answer? *PLoS Medicin* 1(3): 180–83.

19 Varmus H, R Klausner, E Zerhouni, T Acharya AS Dear and PA Singer. 2003. Grand Challenges in Global Health. *Science* 301:398–99. The Panel analyzing these "Grand Challenges" suggested seven overarching goals and challenges. All of these were related to developing new and better technologies, such as effective vaccine technologies, efficient vaccine- and drug- delivery systems, diagnostic tools, therapeutics, bioavailable nutrition systems (via genetic modification of plants), and so forth. The views of the panel were reiterated by Dr. Elias Zerhouni and a former director of the National Cancer Institute (Zerhouni E. 2003. NIH Roadmap.

Science 302[5642]:63–72). Furthermore, one of the key messages from world leaders at the World Summit for Sustainable Development (WSSD), held in 2002 in Johannesburg, South Africa, was the need to build capacity of the science and technology (S&T) enterprise in the developing world for its own sustainability.

20 Salicrup LA, RF Harris, C Gardner and ML Rorhbaugh. 2004. Developing Health R&D Systems: Partnerships for Capacity Building in International Technology Transfer. Global Forum for Health Research. (Oral Presentation).

21 WIPO. 2004. WIPO Program & Budget: Cooperation with Developing Countries, World Intellectual Property Organization, pp. 87–100. http://www.wipo.int/cfd/en/.

22 OECD 2002. Conference on Biotechnology for Infectious Diseases: Addressing the Global Needs. Rapporteurs' Report. Lisbon, 7–9 October, 2002.

23 See *supra* note 5.

24 Salicrup LA, RF Harris, ML Rohrbaugh. 2005. Partnerships in Technology Transfer: An Innovative Program to Move Biomedical and Health Technologies from the Laboratory to Worldwide Application. *IP Strategy Today*, No. 12:1–15. www.biodevelopments.org/ip/.

25 See for example, NIH OTT model patent licenses and commercialization licenses at www.ott.nih.gov.

26 Non-profit institutions receiving US Government funding also have similar requirements under the Bayh-Dole Act to favor licensing to U.S. small businesses. 35 USC §202(c) (7) (D) and implementing regulations at 37 CFR §401.14(k) (4).

27 Federal Register 2004a. vol. 69, pp. 57335–36 (Sept. 24) and Federal Register 2004b. vol. 69, pp. 34381–82 (June 21).

28 Parashar UD, EG Hummelmann, JS Bresee, MA Miller and RI Glass. 2003. Global Illness and Deaths Caused by Rotavirus Disease in Children. *Emerging Infectious Diseases* 9(5):565–71.

29 WHO 2003. State of the Art of New Vaccines: Research and Development. Online at www.who.int/vaccine_research/documents/new_vaccines/en/.

30 See *supra* note 19.

31 www.tmgh.org.

32 http://www.ott.nih.gov/licensing_royalties/NegDis_ovrvw.html.

33 See *supra* note 24.

34 See *supra* note 24.

The Making of a Licensing Legend: Stanford University's Office of Technology Licensing

NIGEL PAGE, *Intellectual Asset Management (IAM) Magazine, U.K.*

ABSTRACT

The history of technology transfer at Stanford goes back to an initial pilot program launched by Niels Reimers in 1970, a program that put the university in an excellent position to take advantage of the Bayh-Dole Act. Enacted in 1980, the act gave U.S. universities ownership of any patents developed using federal funds. Today, Stanford University and successful technology transfer are almost synonymous. But success is more than just a matter of timing. Stanford's Office of Technology Licensing (OTL) takes a flexible, broad outlook on the development of its intellectual property that has made Stanford a favorite business partner. This chapter reveals the secrets behind the success of Stanford's OTL.

1. INTRODUCTION

Stanford University's Office of Technology Licensing has a string of blockbuster success stories to its name—from DNA gene splicing to Cisco, Yahoo!, and Sun Microsystems. Since the office was founded in 1970, it has received US$594 million in cumulative gross royalties. No wonder the university is considered a world leader in technology transfer.

Technology transfer is big business in the United States. The concept of taking intellectual property from laboratory to market originated in that country, and the practice is now so institutionalized that the Association of University Technology Managers (AUTM) can regularly attract a cross-section of the world's leading companies, lawyers, and venture capitalists to its annual conference. A number of universities can claim to represent the gold standard in this field, among them M.I.T., Columbia, Stanford, and the University of Wisconsin. But arguably none makes a stronger claim for shaping the global technology transfer market than Stanford, the California powerhouse, which *Fortune* magazine dubbed "the intellectual incubator of the digital age."[1] Credited with kick-starting the Silicon Valley high-tech industry, and subsequently spawning a hugely influential brood of physical- and life-science businesses across the United States and the world, Stanford's technology transfer efforts have clearly transformed our world.

2. BUILDING ON DNA

The brainchild of Niels Reimers, Stanford's Office of Technology Licensing (OTL) was born more than 30 years ago, in 1970. It was Reimers who famously recognized the

Page N. 2007. The Making of a Licensing Legend: Stanford University's Office of Technology Licensing. In *Intellectual Property Management in Health and Agricultural Innovation: A Handbook of Best Practices* (eds. A Krattiger, RT Mahoney, L Nelsen, et al.). MIHR: Oxford, U.K., and PIPRA: Davis, U.S.A. Available online at www.ipHandbook.org.

Editors' Note: We are most grateful to *Intellectual Asset Management Magazine (IAM Magazine)* and especially to Joff Wild, editor, for having allowed us to update and lightly edit this paper and include it as a chapter in this *Handbook*. The original version of the paper appeared as an article in *Licensing in the Boardroom 2005*, a supplement to *IAM Magazine*, published by Globe White Page Ltd, London (www.iam-magazine.com).

huge potential in gene-splicing research being undertaken by professors Cohen and Boyer (of Stanford and the University of California, respectively). It was Reimers who persuaded them to let Stanford try for a patent (which Stanford did and ultimately secured). And it was Reimers who went on to launch a licensing program that, by the time the so-called Cohen/Boyer DNA patent expired in December 1997, had generated more than US$250 million in royalties (split with the University of California), with Stanford licensing a total of 468 companies on behalf of both universities. Having become an international consultant, Reimers saw merit in Stanford setting up an OTL that would be a marketer—not just a patent office. The office would actively pursue discoveries, market them to potentially interested companies, and collect the royalties on them. Fundamental to its structure would be a preparedness to give its licensing associates the authority and responsibility they needed to do their job effectively, free—so far as that was possible—from the red tape that entangled so many other operations. Reimers' initial pilot program, launched in 1968, produced, in one year, more than ten times the amount received by Stanford in its previous 15 years of licensing through an outside corporation. The idea was clearly a winner. Not surprisingly, M.I.T. would later go on to seek out Reimers' services and effectively transform its own technology licensing office into a global force in its own right, with gross revenues of US$33.52 million in 2002.[2]

Stanford, however, is still out in front. According to the industry-standard 2002 AUTM Licensing Survey, Stanford received US$50.2 million in adjusted gross license income for FY 2002. Even in a tough economic climate, this amount was the second-highest in the OTL's history, including an unexpected US$5.8 million in one-time royalties, with 42 of the OTL's 442 income-generating technologies each producing more than US$100,000 per year (see Box 1 for an overview of the economic impact of Stanford's OTL). Since 2002, things have gotten even better: in 2005, the OTL received on behalf of the university US$384 million.[3]

3. THE RIGHT PLACE AT THE RIGHT TIME

So what is the secret of Stanford's success? The university's symbiotic relationship with Silicon Valley has played a vital role, giving life to many of the OTL's most marketable technologies and providing the all-important local infrastructure of ideas, can-do thinking, and capital. But this climate of entrepreneurship did not grow up overnight. Back in the 1920s, Fred Terman was

Box 1: Economic Impact of Stanford University's OTL

For FY 2001 (latest figures available), the largest companies founded or co-founded by those with a current or former affiliation with Stanford University (as alumni or faculty/staff) were responsible for generating 42% (US$106 billion) of the total revenue of the Silicon Valley 150 (an annual list of the largest Silicon Valley firms).

From FY 1975 to 2005, Stanford's top six cases have been:[a]

- recombinant DNA cloning technology (total royalties US$255 million)
- chimeric receptors (total royalties US$124.7 million)
- fluorescent conjugates for analysis of molecules (total royalties US$46.4 million)
- functional antigen-binding proteins (total royalties US$30.2 million)
- fiber optic amplifier (total royalties US$32.6 million)
- FM sound synthesis (total royalties US$22.9 million)

a Sally Hines, Stanford University, Office of Technology Licensing, (personal communication).

an electrical engineering professor at Stanford. Trained at M.I.T., Terman played a key role in demolishing the ivory tower mentality, unleashing links with business that would ultimately enable Stanford's OTL to market technologies with such phenomenal success. Needing local jobs for his engineering graduates, Terman recognized the importance of attracting companies to the area, and so he introduced the core founders of Varian Associates (the radar and microwave technology business). He encouraged William Shockley, co-inventor of the transistor, to come to Palo Alto (before joining Stanford's faculty in 1963). And Shockley brought two of his own students together, William Hewlett and David Packard, who went on famously to launch HP (Hewlett Packard) in a Palo Alto garage. Indeed, it is easy to see why Terman is referred to as the father of Silicon Valley.

Without Terman and Reimers, it is questionable whether Stanford's OTL (and indeed the whole U.S. technology transfer industry) would be even close to where it is today. Of course, a fortuitous geographical position, coupled with a thirst for entrepreneurial activity, is a quintessential prerequisite for success in the field of intellectual property. But without a vehicle to encourage, enable, and market inventions, the bridge from laboratory to market would be rickety indeed. That Stanford was thinking along the right lines back in the 1960s made it ideally positioned to take advantage of the pivotal Bayh-Dole Act passed by Congress in 1980. It gave U.S. universities ownership of any patents developed using federal funds.

4. GETTING IT RIGHT

External circumstances notwithstanding, a key feature of Stanford's success has clearly been the preparedness of its leaders to think long and hard about the best possible means of implementing and running the university's licensing operations. Katharine Ku, Director of the OTL since 1991 and a major international name on the technology transfer circuit, is initially hesitant when asked about Stanford's success:

People often ask me what is our best practice? In some ways, it's hard to know, since on paper our processes and attitudes are similar to those in place at other universities." After reflection, she continues: "*It is people that make the difference. Our team is scientifically trained, but we don't always look for Ph.D.s Our work is, by its nature, very generalist. We have to know a little about a lot of different areas. And this is the opposite of a Ph.D.'s training. And we don't look for lawyers—in fact, on the licensing side, we discriminate against them. Legal training is by its nature risk-averse—whereas to succeed, we have to be risk-takers.*

Ku's department is compact. Although it is one of the most active offices in the technology transfer field (managing more than 1,900 technology dockets), the core team includes fewer than 30 staff members, with no more than seven or eight licensing staff. These licensing associates evaluate technologies that have been disclosed to the OTL, before tailoring licensing strategies to fit the ones that, in their view, have commercial potential. Each associate is given what might appear to an outsider to be a surprising degree of autonomy: he or she assumes full responsibility for a portfolio of dockets, from cradle to grave. The associates each have an area of technical expertise in life sciences, physical sciences, or both. One of Ku's team, senior associate Hans Wiesendanger explained how the process begins: "*First of all, the invention must be disclosed. To encourage disclosures, every research contract stipulates mandatory disclosure (whether from government contracts or industry sponsorships), but that said, academics tend to do what they want. We can try to manage them, but we can't control them.*" (For case studies of the private sector working with Stanford's OTL, see Boxes 2 and 3.)

5. TAKING ON TECHNOLOGY

Once an invention reaches the OTL, it is assigned to a licensing associate who assumes responsibility for it, initially evaluating the technology to identify its technical advantages. "*First, we talk to the inventors,*" explained Wiesendanger. "*They will often, but not always, have a good perspective. We also talk to outside people—colleagues, companies we've worked with in the past and so on. Then we decide on the strategy—whether to go for an exclusive or a nonexclusive license and whether to license by territory. Then we assemble a list of potential licensees that we might be interested in contacting.*" The licensing associate's responsibilities are, at this point, still far from over: "*They remain in charge of the project throughout the life-cycle of the license. They check that the royalties are being paid, which may mean arranging for an audit or a renegotiation of the agreement in line with any changed circumstances.*"

Wiesendanger's explanation gives weight to what Katharine Ku identified as her department's "X Factor." Finding associates who are willing and able to take on this level of responsibility is no small challenge. As mentioned above, Stanford rarely uses lawyers to draw up agreements. As Wiesendanger explained: "*Some of our licensing deals are quite standard—we have boilerplates that can be modified as required and that are clearly very different depending on whether they apply to software or biological material. The licensing associate negotiates these agreements, with the full*

Box 2: Alumnus Case Study 1: Dr. Mark Zdeblick

"*I've been lucky to experience Stanford's technology transfer operation from both sides of the fence,*" laughed Mark Zdeblick, founder of Redwood Microsystems, entrepreneur-in-residence with VC firm Spring Ridge Ventures, and CTO of, inter alia, Proteus Biomedical. "*I've worked there as a grad student in a research team developing a blockbuster technology* [atomic resolution microscopy]. *I've set up my own company* [Redwood] *with Stanford licensing the* [micro-valve chip] technology *I'd developed there to the business. And with Proteus, we've approached Stanford to license their technology to the company. Typically professors/inventors hold most of the power, exerting considerable influence over the choice of licensee. But with Stanford's OTL,*" he said, "*they have enough understanding to be able to influence the professors. When people have been prepared to trust them to do the right thing, they have done very well.*" The fact that the OTL can strike a balance (most of the time) between the professor's desire to tie strings to the license deal (obliging the company to pump research funds back into his or her department), and the logic behind commercializing the technology effectively, is a key variable. Commenting on Stanford's successful management of the "brain drain" experienced elsewhere, Zdeblick commented, "*Stanford often allows its professors the opportunity to take a leave of absence for two years to help spinout such technology. That level of commitment is often necessary to get backing from the private equity community. Most professors return after the two years, in which case they are in many ways much more valuable to the university. Of course, sometimes they don't return.*" When Stanford was licensing on his behalf, Zdeblick was impressed with the amount of marketing they took on: "*They made a lot of calls on my behalf, seeking out interest among potential licensees, as well as undertaking a lot of the groundwork to establish the utility of the underlying patents. That's more common now, but it was much rarer 20 years ago.*" Another view of Zdeblick is Stanford's ability to get results out of the more run-of the-mill technologies that come through the OTL's doors: "*It is easy with grand-slam technologies, where you can pull together nonexclusive licenses with everyone. The tricky thing is to get the whole portfolio working well and, as a rule, Stanford seems more willing than most other universities to take a bet and grant an exclusive license for an obscure technology.*"

authority to do so. It is only where something new crops up that he or she will consult a lawyer—there is certainly no obligation to get every deal approved by an external lawyer."

6. PATENT OPPORTUNITIES

This practice would hardly seem to be music to the ears of California's finest IP law firms. That said, there is still plenty of work for external law firms (Stanford OTL has annual patent expenses of around US$5 million)—although, as Carol Francis, a name partner with Bozicevic, Field & Francis, LLP (a local law firm with a track record advising on OTL-linked patent prosecution matters) explained, the patenting activity generated by OTL maintains its focus on commercial viability:

Stanford stands out for its ability to make quick assessments on when, and if, to go ahead and file a patent application, or to continue to prosecute an application already filed. Their experience means that

Box 3: Alumnus Case Study 2: Dr. Dari Shalon

Now running Shalon Ventures (an early-stage life-science VC) with his brother, Dari Shalon's experiences with Stanford OTL served him well. A former graduate student at the university, he went on to license his own invention from the OTL to launch Synteni, sold three years later to Incyte Genomics for US$100 million. According to Shalon, "*The technology that ended up being licensed to Synteni was developed by me and Professor Patrick Brown* [an arraying technique that became the basis of DNA microarray technology]." Although the OTL marketed the invention widely, no company expressed any serious interest, leading, in 1995, to Shalon starting his own company to develop the technology. "*I had done an MBA at M.I.T.,*" he explained, "*and then chose Stanford as an interesting entrepreneurial university. My research project was deliberately selected to have commercial application.*" Shalon remembered wandering into the OTL as a grad student in ripped t-shirt and jeans asking if he could file a disclosure: "*I had a number of unsuccessful efforts where the technology didn't work, but the OTL guys encouraged me to go back to the lab and keep trying. Finally I got it to the point of commercial feasibility and went ahead and filed.*" At that stage, he recalled, he tried to get serious: "*I turned myself into a businessman, with business cards and a suit, thinking I would step straight into the commercial sphere. What I'd failed to understand was Stanford's own fiduciary obligations to its trustees. They had to market the technology to firms that I knew would be competitors further down the line. I held my breath for six months, but to my surprise and relief, no other company had the vision to take it on.*" Things went from good to better—Shalon snagged Merck as his first customer, and shortly after pulled in US$5 million in venture financing from Kleiner Perkins Caufield & Byers. "*Had I not held an exclusive license on the technology, there's no way I would have been able to raise the capital I so desperately needed.*" Throughout this process, he was impressed with the OTL's flexibility and willingness to take a bet on him as exclusive licensee. "*At the crucial point when Incyte showed interest in us,*" he said, "*and our license was key to the sale going through, Stanford was more than happy to transfer the license to the purchaser. And subsequently, when we got involved in litigation with a major competitor relating to our licensed intellectual property, Stanford stood by us. It made a huge difference to know there was a solid partner right behind us.*"

they are adept at identifying an invention disclosure's commercial potential early on; it also means that *they're prepared to take a flexible approach to filing, often in negotiation with the ultimate licensee/s. Stanford OTL accomplishes this while at the same time respecting the academic inventors' need to publish or make presentations at meetings. While Stanford OTL may file an application to preserve patent rights that might otherwise be impacted by an imminent public disclosure, they are at the same time particularly mindful that once an application is filed, it tends to take on a life of its own, with all the expense that that entails. This analysis at Stanford OTL benefits from the experience and leadership of its Director Kathy Ku, as well as the insights and connections of the inventors themselves. Stanford OTL's insistence that the inventors be involved—and the level of involvement they receive in response—is, I think, one of the keys to their success.*[4]

(See Box 4 for an overview of how inventions move from ideas to commercial products at Stanford.)

7. NOTHING VENTURED ...

Silicon Valley has no shortage of lawyers—or venture capitalists (VCs). Not surprisingly, both camps frequently visit the corridors of Stanford, taking a keen interest in the activities of the OTL. That said, Ku pointed out that the OTL itself is not there to make contact with VCs: "*Most usually, our researchers will identify their preferred VCs in Silicon Valley and then come to see us together. That's the best approach—for technology transfer to work, where start-ups are concerned, the entrepreneur needs to feel comfortable with her chosen VC. It's up to them to get the chemistry right, which is not always something we can help them with.*" Rob Chaplinsky, a general partner with Sand Hill Road early-stage VC firm Mohr, Davidow Ventures, has had considerable experience working with the OTL, and he characterized the relationship in these terms: "*Because Stanford is bang in the heart of Silicon Valley, we have access to their researchers and professors long before the OTL. By the time we go to see the OTL, it's a matter of looking to see how we can amicably align everyone's interests. In fact, we have a saying here that if you wait until the OTL guys have the patents and call you up, you're way too late.*" Prompted to outline Stanford's formula, Chaplinsky said: "*I get a lot of calls from other institutions asking how they can copy Stanford's program—but it's not as easy as that. Some of their formula is down to geography, they're integrated in the world's venture epicenter and their professors are embedded in the community. Then there's the culture of the university—from the Dean down, they're mostly academics and entrepreneurs. At Stanford you're almost expected to start a company before becoming a tenured professor. There is something special there which can't be replicated in a hurry.*"

8. FLEXIBLE CONTROL

Where negotiations with Stanford OTL are concerned, Chaplinsky has no doubt that terms are getting tougher. Still, he stressed Stanford's willingness to be flexible, with innovative blends of upfront license fees, royalties, and equity splits very much up for discussion: "*Nothing's ever cast in stone with their OTL. There's always a door open to go back and renegotiate.*" That said, an established modus operandi underpins the OTL's position, and, as Ku explained, a big part of its rationale is the necessity to keep getting technologies out into the market: "*Our job is to plant seeds, so—because it's so hard to know which new technologies will eventually succeed—we do as many deals as possible. In some ways we've been helped in this by changing attitudes. Researchers nowadays are more interested in the potential of their technology, so we see more invention disclosures than we used to. We have to be realistic—only about seven inventions here generate US$1 million-plus a year.*" Put bluntly, this means that only about 10% of the inventions taken on by the OTL have the potential to generate significant income. Twenty to thirty percent won't bring in a great deal and the remaining 60–70% will bring in almost nothing.

Depending on the sector and the technology, the technology transfer process can be straightforward or downright complex. Ku pointed out that, as a general rule, pharmaceutical and life-sciences companies have tended to be more in tune with the process: "*They understand the long timelines*

Box 4: From Idea to Market—IP Progression at Stanford

Invention by inventor (INV)

Conception documentation: lab notebooks, dated papers, or drafts witnessed.

Disclosure: required by all sponsorship agreements for research; must include description (papers attached), information on who are inventors, what funding was used, when conceived, when first disclosed or published, signature(s) and date, and assignment to Stanford; fill in printed form or use Internet disclosure form;
must submit to OTL

Disclosure coming to OTL

Sign in: OTL logs in, gives docket number, and assigns to specific licensing associate (LA) who now has complete responsibility and authority for handling the invention from evaluation to licensing and monitoring licensee performance

Evaluation: LA discusses with INV; gets as much information as needed on details of technology, novelty, potential utility, and companies in the field

LA also gets similar information from outside sources, usually by contacting sources in the field and supplying confidential data and details after executing a confidential disclosure agreement (CDA)

Strategy: LA decides how to license: exclusive or nonexclusive, by territory or worldwide, for limited and specific uses and applications or unlimited; sublicensing permitted or not; kind of company to approach and how; key licensing terms to shoot for; suitability for a standard license that can be filled out on the Web site

Contact potential licensees: LA assembles list and makes first contact (mail, e-mail, fax, telephone, Internet); information on what invention may do, but not how; offers details after execution of CDA

Patent prosecution: LA decides whether and when to apply for a U.S. patent; selects outside patent attorney and charges him/her with filing (normal or provisional); monitors filing and prosecution, and decides filing of foreign applications; files only if deems reasonable chance of success for licensing or prospect of getting expenses paid (for example, in return for an option to a potential licensee)

Negotiations: LA negotiates with companies who respond positively; draws up a license agreement (starting with boilerplate and modifying that if/as necessary or advisable); if deemed necessary, consults with attorney for legal advice for special or unusual situations

Executed agreement: OTL logs into database, documents terms and contact information, and programs database to generate reminders and invoices, as needed

(Continued on Next Page)

Box 4 (continued)

License period:	LA monitors performance: receipt of royalties and reports. OTL sends out automatic computer-generated invoices for fees and earned royalties. If performance deficient, LA follows up with reminders or, in extreme cases, termination. LA may have to renegotiate parts of license agreement if situation has changed significantly since signing (at OTL's request or at licensee's)

involved, whereas physical-sciences companies, because they are more accustomed to a cross-licensing model, can find dealing with us quite demanding. It's really up to universities to work out how they can deal better with this side of the commercial spectrum." Other aspects of the academic/commercial relationship also have potential to complicate negotiations, as Ku said: "*Because physical science companies are major sponsors of university research, some of them expect to own the inventions that flow from that research. I'd always hoped that that battle was over with Bayh-Dole, but perhaps because universities in other parts of the world are still prepared to give up title, some companies are still laboring under a misconception on the IP ownership side when they deal with us.*"

9. REMOVING CONFLICT

Like any university technology transfer office, Stanford has an effective system in place for managing potential conflicts of interest. As Hans Wiesendanger explained, "*Anyone starting a technology transfer program for a university will be concerned about professors undertaking applied research to make money—that can be very damaging to a university's reputation. That said, there is no doubt that you can continue to be one of the world's top research centers while playing a leading role in technology transfer. To do so, however, you do have to recognize that the potential for conflicts of interest does exist. Formal procedures for dealing with conflicts if and when they arise need to be instituted. That said, it is always important to remember that any researcher will be mainly interested in just one thing: his academic standing among his peers. So, in my view, the fear of conflicts can be somewhat overblown.*"

10. IP MANAGEMENT

Patenting is a core activity, handled as necessary by outside patent attorneys. Key issues that come to the fore here are whether the invention can be licensed as *tangible research property*, whether it can be licensed as copyright, whether it is likely to be both patentable and enforceable, and whether the invention has already been publicly disclosed. There's no fixed way of handling this process; Wiesendanger explained, "*It can happen at any time—and that decision is up to the licensing associate involved. But we do have to be careful; it costs a lot and represents an ongoing commitment. Some universities patent everything, but we are under pressure not to do so. Usually we'll sign licenses before we have the patents in place, and we often start negotiations before we have even applied for them. Quite often we'll look to the ultimate licensees to cover the filing expenses in exchange for a six-month option on the technology. That can be very attractive, as that six-month period often represents a very significant competitive advantage.*"

Although Stanford supports entrepreneurs, it does not "encourage" spinouts. Nor does it start companies itself, although comparatively recently the OTL was authorized by the university to take equity as part of license fees or royalties (provided that the licensee did not conduct clinical trials at the university), as well as to license companies in which the inventors have an interest. Stanford currently holds equity in approximately 75 companies with cash-out to date of around US$22 million. Wiesendanger explained, "*Stanford is very concerned about its image and we don't want to be seen to be too involved in business. Just as Stanford has a tradition of encouraging cooperation with industry, we do not want business and university interests to affect each other in an operational way.*"

11. SPLITTING THE REVENUES

Once royalties start flowing, there's a fixed split in operation. Fifteen percent is siphoned off by the OTL to cover its own administrative expenses, although, as Wiesendanger pointed out, not all of that gets used up—the remainder is channeled into a number of funds created by Stanford, including the "birdseed fund" and the "OTL gap fund." The former provides small amounts of money (typically up to US$25,000) to fund prototype development or modest reduction-to-practice experiments for unlicensed technologies; the latter supports development efforts up to US$250,000 for unlicensed technologies with commercial potential. The remaining 85% of incoming royalties divides three ways—between the inventor, the inventor's department, and the inventor's school/faculty. In FY 2001–2002, inventors received personal income of US$11.3 million, departments received US$13.5 million, and schools received US$13.1 million. "*This split is designed to incentivize researchers,*" Wiesendanger explained, "*and some academics can do very well. But often inventors don't take their share—they ask instead for it to be signed over to their personal lab account. Research money with no strings attached is, as you can imagine, very desirable in a university.*"

The nature of the beast means that it would be commercially naïve to set targets—either for licensing deals, or for royalty income. "*On average, we expect to receive five or six new disclosures a week. We file patents on about half of them and license about one-third of them. Of course we look at how many licenses each licensing associate brings in relative to this average,*" said Wiesendanger, "*but there can be no absolute measures; fields vary hugely, and cyclically, in their appetite for new technologies.*"

12. WORK IN PROGRESS

Stanford's model is working, but there has been, and will continue to be, some turbulence. In 1995, for example, a faculty committee released a damning report on the barriers between the medical school and industry, a situation exacerbated by "*a growing mutual distrust.*" A survey of CEOs at Californian pharmaceutical companies underlined the problem when the results came back showing an almost unanimous aversion to dealing with Stanford. In particular, Stanford's attitude towards the ownership/patent status of intellectual property arising from clinical research projects was a source of friction. In response, the university focused on structuring research sponsorships that allowed funding companies to get rights to the technology. Realizing that the federal budget for research funding was in steep decline, the medical faculty had little choice but to be proactive with its industry benefactors.

With hindsight, it's clear that the acid test for Stanford's model came at midnight December 2, 1997—the moment when the (nonrenewable) Cohen/Boyer patent for recombinant DNA expired. This moment, referred to at the time by Stanford officials as "the cliff," might have defeated some operations, but at Stanford the event acted to stimulate several years of intense activity, with the university opening up the campus to industry ideas as it never had before. As Ku said, the OTL was prepared: "*We'd been moving steadily toward being more user-friendly to industry.*" That it took just six years for Stanford to top its record royalty year with Cohen/Boyer underlines the firm foundation set down by Reimers—and points the way forward to an another exciting decade in Silicon Valley. ■

NIGEL PAGE, Intellectual Asset Management (IAM) Magazine. *For information on IAM, contact: Joff Wild, Editor, New Hibernia House, Winchester Walk, London Bridge, London, SE1 9AG, U.K.* jwild@iam-magazine.com

---

1 Aley J. 1997. The Heart of Silicon ValleyWhy Stanford—the Nexus of Capital, High Technology, and Brainpower—Is the Intellectual Incubator of the Digital Age. *Fortune*

Magazine, July 7. http://money.cnn.com/magazines/fortune/fortune_archive/1997/07/07/228653/index.htm.

2 Page N. 2003. Demolishing the Ivory Towers. *IAM* July/August. entrepreneurship.mit.edu/Downloads/NigelPageArticle.pdf.

3 http://otl.stanford.edu/about/resources.html.

4 See also otl.stanford.edu/about/documents/Memo_to_Outside_Counsel.doc.

Technology Transfer at the University of California

ALAN B. BENNETT, *Associate Vice Chancellor, Office of Research, University of California, Davis; and Executive Director, PIPRA, U.S.A.*

MICHAEL CARRIERE, *Business Development and IP Manager, Office of Technology Transfer, University of California, U.S.A.*

ABSTRACT

The University of California (UC), based on its mission as a land grant university, has a long history of seeking intellectual property protection for its research discoveries and managing those technologies for the public benefit. By some measures, the UC technology transfer program is the largest public program in the world. The program has evolved over the years but has always been at the forefront of intellectual property protection. This article focuses on the history, policy, and organizational framework of the UC technology transfer program, and the information discussed herein may be instructive to administrators and others seeking to learn from the UC experiences. The program has been administered through six functional departments: Information Technology and Communications, General Counsel (legal), Licensing, Patent Prosecution, Financial Management, and Policy Analysis and Development. Perhaps the most distinctive feature of the UC technology transfer system is the development of a distributed institutional network of ten university campuses, which operate under a common policy framework and share resources. At the same time, each office functions relatively independently of the others. This structure could be emulated and implemented at different scales, from a relatively small-scale research consortium made up of a network of institutions, to a larger-scale national network of universities, to a global-scale international network of research institutions linked by common policies and objectives.

1. INTRODUCTION

The University of California (UC) is composed of ten semi-independent campuses: UC San Diego, UC Santa Barbara, UC Los Angeles, UC Riverside, UC Irvine, UC Merced, UC Santa Cruz, UC San Francisco, UC Berkeley, and UC Davis. While each campus represents a significant education and research institution in its own right, collectively, the University of California system is one of the strongest institutions of higher education in the world. This is particularly true with regard to research. The University of California is likely the largest public research enterprise in the world. With annual research expenditures in excess of US$2.9 billion, the size of its collective research programs is comparable to the total research expenditures of entire countries. One of the results of this robust research activity is the generation of a significant technology portfolio that supports the university's mission to use its research to benefit society. In 2004, University of California researchers reported nearly 1,200 new inventions, or approximately one invention for each US$2.5 million in research expenditure—a number that remains relatively consistent from year to year (the full range has been one invention for each US$2.5–4.5 million). As a consequence, the University of California has developed an extensive technology transfer program that provides a potentially

Bennett AB and M Carriere. 2007. Technology Transfer at the University of California. In *Intellectual Property Management in Health and Agricultural Innovation: A Handbook of Best Practices* (eds. A Krattiger, RT Mahoney, L Nelsen, et al.). MIHR: Oxford, U.K., and PIPRA: Davis, U.S.A. Available online at www.ipHandbook.org.

useful example for large multi-institutional networks and even entire nations.

2. HISTORY OF TECHNOLOGY TRANSFER AT THE UNIVERSITY OF CALIFORNIA

2.1 *The mission of a land grant university*

The University of California was established as a land grant university by the Morrill Act, which was signed into law by Abraham Lincoln in 1862. This Act provided each state of the United States with a grant of large acreages of public lands that the state could sell on the open market to raise funds to support at least one college at which the leading objective would be to broadly educate students in "agriculture and the mechanical arts." But it was the Hatch Act of 1887 that extended the Morrill Act and the mission of land grant universities to encompass research as well as education—specifically, research that contributed to an effective agricultural industry (Box 1).

While originally focused on agriculture, the mission of land grant universities in the United States continues to be reflected in broad mission statements that recognize the university's fundamental role in transferring research results to support applications in all industrial sectors. The principles embodied by the U.S. land grant universities have become important elements of the mission of many American universities and have played an important role in defining the context within which university technology transfer programs have developed.

2.2 *Technology transfer policy development*

Formal intellectual property protection and the management of patented technologies at the UC dates back to the 1920s. The first patent assigned to *"the Regents of the University of California"* covers technology for a *"Film Holder for Dental Work"* (U.S. Patent No. 1,657,230) awarded to Frank Simonton. Thus, there is a long history of biomedical research inventions. Other early UC patents describe methods of producing wood products (U.S. Patent No. 1,805,550 from 1931), an apparatus for cracking nuts (U.S. Patent No. 2,238,368 from 1941) and a method of preserving microorganisms (U.S. Patent No. 2,376,333 from 1945).

In 1943, the first UC patent policy was adopted, which provided mechanisms for supporting the licensing of patented inventions.[1] However, assignment of inventions to the university was determined on a case-by-case basis and UC policy was silent on royalty sharing between the university and inventors. In 1963, the university adopted a new patent policy that foreshadowed some of the requirements the Bayh-Dole Act (1980) later made mandatory, including making the assignment of rights to the university mandatory and specifying a royalty-sharing formula (50/50 sharing of any licensing revenue between the inventor[s] and the university, after deduction of a 15% administrative fee). The patent policy has changed a few times over the intervening years but has continued to include mandatory disclosure and assignment of inventions to UC and a royalty-sharing formula that provides,

Box 1: The Hatch Act Extended the Mission of Land Grant Universities to Include Research

It shall be the object and duty of the State agricultural experiment stations ... to conduct original and other researches, investigations, and experiments bearing directly on and contributing to the establishment and maintenance of a permanent and effective agricultural industry of the United States, including researches basic to the problems of agriculture in its broadest aspects, and ... as have for their purpose the ... maximum contribution by agriculture to the welfare of the consumer.

- Hatch Act of 1887, as amended in 1955

after deduction of direct expenses, 35% to the inventor(s), 15% to a campus research fund, and 50% to a general pool for the campus at which the inventor is located. This patent policy is administered by a "patent acknowledgement" (Figure 1) that is signed by all UC employees and that contains a provision which specifically allows the UC to change the policy at any time in the future, including the royalty-distribution formula. This last feature is important because the UC has been sued by an inventor who objected to the change in royalty-distribution policy.[2]

2.3 *Role of leadership*

The evolution of a policy framework to support technology transfer at the UC has been critical in developing the institutional capacity for technology transfer. However, the most important element has been the academic leadership role of the UC in recognizing the importance of technology transfer and promoting it as an activity that is central to the university's educational and research missions. The last two presidents of the UC, Richard Atkinson and Robert Dynes, clearly articulated how and why the UC should be actively engaged in technology transfer (Box 2). University technology transfer programs take nearly a decade to begin to generate sufficient licensing revenue just to break even, and without strong support from academic leadership, technology transfer programs are unlikely to be consistently supported at a level necessary to achieve successful outcomes. Because of its academic leadership, the UC technology transfer program has enjoyed several decades of solid support and, as a result, has been a net revenue generator for the university since the late 1980s.

2.4 *Evolution of a distributed institutional network for technology transfer*

An ongoing trend in the UC technology transfer program has been its gradual movement from a highly centralized network to a decentralized, or distributed, network of semi-independent, campus-based technology transfer programs. The central UC Office of Technology Transfer (OTT) was established in 1978 and for many years provided all technology transfer services from a central location in the San Francisco Bay Area. A single, central OTT providing services to such a large research enterprise allowed the investment of sufficient resources in a single program to reach critical mass and achieve early success. However, while this location is very close to the UC Berkeley campus, it is over 500 miles from the UC San Diego campus, and the lack of direct connections to researchers and the technology itself at more distant campuses proved to be problematic, especially as research programs grew dramatically in the 1980s. As a consequence, there has been an ongoing movement to establish local offices of technology transfer on each of the UC campuses. This trend began in 1990 (Table 1) and is still continuing.

3. THE UC TECHNOLOGY TRANSFER PROGRAM: ELEMENTS AND ORGANIZATION

The UC technology transfer program has been relatively successful in transferring technology to the private sector. In its best year (2002) the program generated over US$100 million in revenue, which, after expenses and distribution to inventors, provided approximately US$30 million to support education and research at the UC. While this represents good business for the university, the financial returns are modest when placed in perspective of the total UC research budget of approximately US$2.9 billion. Expenses for the program in 2004 included US$14.3 million in operating costs and US$13.9 million in unreimbursed legal expenses, reflecting the substantial investment that is required to manage a program on such a scale. A range of technology transfer performance metrics are reported annually by the UC, and there are several published reports that look at technology transfer trends in the UC in relation to other university programs.[3, 4]

The administrative structure of the UC technology transfer program has been in a constant state of flux and evolution since its inception, but the program appears to be approaching a steady state, balancing the range of activities pursued and combining centralized and distributed approaches. The UC technology transfer program has been administered through six functional departments

Figure 1: University of California Employee "Patent Acknowledgement"

UNIVERSITY OF CALIFORNIA
STATE OATH OF ALLEGIANCE, PATENT POLICY, AND PATENT ACKNOWLEDGMENT
UPAY585 (R11/97) E0420 71443-180

EMPLOYEE'S NAME (Last, First, Middle Initial)		DATE PREPARED MO DY YR
EMPLOYEE ID	DEPARTMENT	EMPLOYMENT DATE MO DY YR

STATE OATH OF ALLEGIANCE I do solemnly swear (or affirm) that I will support and defend the Constitution of the United States and the Constitution of the State of California against all enemies, foreign and domestic; that I will bear true faith and allegiance to the Constitution of the United States and the Constitution of the State of California; that I take this obligation freely, without any mental reservation or purpose of evasion; and that I will well and faithfully discharge the duties upon which I am about to enter.

Taken and subscribed before me on: ______________ MO DY YR

Signature of Officer or Employee: ______________________

(DO NOT Sign Until in The Presence of Proper Witness.)

Signature of Authorized Official: ______________________

NOTE: No fee may be charged for administering this oath.

Title: ______________________

County: State:

Oath must be administered by either (1) a person having general authority by law to administer oaths - for example Notaries Public, Civil Executive Officers (Section 1001 of Government Code), Judicial Officers, Justices of the Peace, and county officials named in Sections 24000, 24057 of Government Code: such as, district attorneys, sheriffs, county clerks, members of boards of supervisors, etc., or (2) by any University Officer or employee who has been authorized in writing by The Regents to administer such oaths

WHO MUST SIGN THE OATH: All persons (other than aliens) employed by the University, in common with all other California public employees, whether with or without compensation, must sign the oath. (Calif. Constitution, Article XX, Section 2, Calif. Government Codes, Sections 3100-3102.)

All persons re-employed by the University after a termination of service must sign a new Oath if the date of re-employment is more than one year after the date on which the previous Oath was signed (Calif.Government Code, Section 3102).

WHEN OATH MUST BE SIGNED: The Oath must be signed BEFORE the individual enters upon the duties of employment (Calif. Constitution, Article XX, Section 3: Calif. Government Code Section 3102.)

WHERE OATHS ARE FILED: The Oaths of all employees of the University shall be filed with the Campus Accounting Office.

FAILURE TO SIGN OATH: No compensation for service performed prior to his subscribing to the Oath or affirmation may be paid to a University employee. And no reimbursement for expenses incurred may be paid prior to his subscribing to the Oath or affirmation. (Calif. Government Code, section 3107.)

PENALTIES: "Every person who, while taking and subscribing to the Oath or affirmation required by this chapter, states as true any material which he knows to be false, is guilty of perjury, and is punishable by imprisonment in the state prison not less than one or more than 14 years." (Calif. Government Code, Section 3108.)

PATENT ACKNOWLEDGMENT

This acknowledgment is made by me to The Regents of the University of California, a corporation, hereinafter called "University," in part consideration of my employment, and of wages and/or salary to be paid to me during any period of my employment, by University, and/or my utilization of University research facilities and/or my receipt of gift, grant, or contract research funds through the University.

By execution of this acknowledgment, I understand that I am not waiving any rights to a percentage of royalty payments received by University, as set forth in the University of California Patent Policy, hereinafter called "Policy."

I also understand and acknowledge that the University has the right to change the Policy from time to time, including the percentage of net royalties paid to inventors, and that the policy in effect at the time an invention is disclosed shall govern the University's disposition of royalties, if any, from that invention. Further, I acknowledge that the percentage of net royalties paid to inventors is derived only from consideration in the form of money or equity received under: 1) a license or bailment agreement for licensed rights, or 2) an option or letter agreement leading to a license or bailment agreement. I also acknowledge that the percentage of net royalties paid to inventors is not derived from research funds or from any other consideration of any kind received by the University. The Policy on Accepting Equity When Licensing University Technology governs the treatment of equity received in consideration for a license.

I acknowledge my obligation to assign inventions and patents that I conceive or develop while employed by University or during the course of my utilization of any University research facilities or any connection with my use of gift, grant, or contract research funds received through the University. I further acknowledge my obligation to promptly report and fully disclose the conception and/or reduction to practice of potentially patentable inventions to the Office of Technology Transfer or authorized licensing office. Such Inventions shall be examined by University to determine rights and equities therein in accordance with the Policy. I shall promptly furnish University with complete information with respect to each.

In the event any such invention shall be deemed by University to be patentable or protectable by an analogous property right, and University desires, pursuant to determination by University as to its rights and equities therein, to seek patent or analogous protection thereon, I shall execute any documents and do all things necessary, at University's expense, to assign to University all rights, title, and interest therein and to assist University in securing patent or analogous protection thereon. The scope of this provision is limited by California Labor Code section 2870, to which notice is given below. In the event I protest the University's determination regarding any rights or interest in an invention, I acknowledge my obligation: (a) to proceed with any University requested assignment or assistance; (b) to give University notice of that protest no later than the execution date of any of the above-described documents or assignment; and (c) to reimburse University for all expenses and costs it encounters in its patent application attempts, if any such protest is subsequently sustained or agreed to.

I acknowledge that I am bound to do all things necessary to enable University to perform its obligations to grantors of funds for research or contracting agencies as said obligations have been undertaken by University.

University may relinquish to me all or a part of its right to any such invention, if, in its judgment, the criteria set forth in the Policy have been met.

I acknowledge that I am bound during any periods of employment by University or for any period during which I conceive or develop any invention during the course of my utilization of any University research facilities, or any gift, grant, or contract research funds received through the University.

In signing this agreement I understand that the law, of which notification is given below, applies to me, and that I am still required to disclose all my inventions to the University.

NOTICE: This acknowledgment does not apply to an invention which qualifies under the provision of Labor Code section 2870 of the State of California which provides that (a) Any provision in an employment agreement which provides that an employee shall assign, or offer to assign, any of his or her rights in an invention to his or her employer shall not apply to an invention that the employee developed entirely on his or her own time without using the employer's equipment, supplies, facilities, or trade secret information except for those inventions that either: (1) Relate at the time of conception or reduction to practice of the invention to the employer's business, or actual or demonstrably anticipated research or development of the employer; or (2) Result from any work performed by the employee for the employer. (b) To the extent a provision in an employment agreement purports to require an employee to assign an invention otherwise excluded from being required to be assigned under subdivision (a), the provision is against the public policy of this state and is unenforceable.

In any suit or action arising under this law, the burden of proof shall be on the individual claiming the benefits of its provisions.

RETENTION: Accounting: 5 years after separation, except in cases of disability, retirement or disciplinary action, in which cases retain until age 70.
Other Copies: 0-5 years after separation.

Employee/Guest Name (Please print): ______________________

Employee/Guest Signature: ______________________ Date: __________

Witness Signature: ______________________ Date: __________

PLEASE SIGN STATE OATH AND PATENT ACKNOWLEDGMENT -- ATTACH TO PAF, UPAY560.

that support all aspects of invention reporting, licensing, and administration. These departments are: Information Technology and Communications, the Office of General Counsel (legal), Licensing, Patent Prosecution, Financial Management, and Policy Analysis and Development. Each is described in more detail below.

3.1 *Information Technology and Communications*

The Information Technology and Communications department has focused on the development and maintenance of an intellectual property management database called the Patent Tracking System (PTS). This system is critical to all aspects of intellectual property management. A single system that integrates invention disclosure, patent prosecution, licensing, and financial information is invaluable for effective IP management—but rarely available. Early attention to developing such a system was of particular importance for the UC system, since all IP, originating from multiple campus locations, is the property of a single legal entity, the Regents of the UC. As a consequence,

Box 2: Effective Technology Transfer Programs Require Supportive Institutional Leadership

California's economic rise is closely tied to the rise of its research universities. New industries have been invented, new products have been developed, and new medical techniques have been invented to both save lives and enhance their quality.

- UC President Atkinson (1995–2003)

Our mission is education, research, and public service. Technology transfer is a vehicle that helps us do all three. It boosts research support. It creates internships and educational opportunities for our students. It stimulates the regional economy. And hopefully, it benefits society.

- UC President Dynes (2003–present)

Table 1: Establishment of Local Offices of Technology Transfer within the UC System

Campus	Local office name	Year established
UC Berkeley	Office of Technology Licensing	1990
UC Los Angeles	Office of IP Administration	1990
UC Irvine	Office of Technology Alliances	1994
UC San Diego	Technology Transfer and IP Services	1994
UC San Francisco	Office of Technology Management	1996
UC Davis	Technology Transfer Center	1999
UC Santa Cruz	Office for Management of IP	2003

a single, integrated database provided the basis for integrated reporting and improved handling of the risks associated with management of IP at multiple locations within the system. With changing information technology infrastructure, it is difficult and costly to update and keep these systems current, but it should be a high priority for any technology transfer program.

The department is also responsible for communications and reporting, which involves, for example, the publishing of an annual report and submission of survey information to the Association of University Technology Managers (AUTM). Because most of this reporting is dependent on information aggregated in the database, the department is the logical group to carry out this task. However, it has become increasingly important to also have regular strategic communications with both internal and external clientele of the technology transfer program to ensure continued support for the mission and activities of the program.

3.2 *Office of General Counsel (legal)*

Legal support for the technology transfer program is critical since it routinely enters into contracts (licenses) on behalf of the university. In the case of the UC, legal oversight for the technology transfer program is carried out by a dedicated intellectual property group within the Office of General Counsel (OGC). The OGC reports directly to the Regents and is charged with oversight of all legal issues and legal risks to the university. This structural arrangement assures that the business opportunity associated with a license agreement is not a consideration in the assessment of legal risk or exposure that the agreement carries with it. Because universities, in general, have a lower tolerance for legal risks than does industry, this arrangement is one feature that often makes negotiations with the UC difficult.

3.3 *Patent Prosecution*

This department is responsible for managing the outside counsel who draft and prosecute patent applications on behalf of the university. Primarily, the department performs a "docketing" function to ensure that external counsel meets critical filing or response dates and that fees are paid on time. The department works closely with licensing officers, inventors, and counsel during patent prosecution to ensure that UC maximizes its IP rights and that it does not inadvertently lose rights due to failure to meet bar dates in the United States or foreign patent jurisdictions.

3.4 *Policy Analysis and Development*

Because the UC is a large, risk-averse institution, it operates in a policy-rich environment. The Policy Analysis and Development department is responsible for interpreting existing policy and providing consultation to licensing officers and researchers in order to assist them in their efforts to comply with university policy, as well as with state and national law. In addition, the department plays an important role in analysis of national and state legislation and in developing new institutional policy to meet these changes as they occur. This analysis is important in developing positions for the UC with regard to new legislation that will impact the university's capability to effectively transfer technology to industry.

3.5 *Financial Management*

Depending on the scale of a technology transfer program, there can be significant infrastructure required simply to manage the program's finances. For the UC this involves monitoring the receipt of approximately US$100 million annually, payment of approximately US$20 million in attorney fees, and the distribution of net revenues to inventors and to campuses where the technology originated. This is an area where inconsistencies in financial management can lead to substantial losses in revenue, loss of IP rights, and exposure to lawsuits by licensees as well as the university's own inventors. The Financial Management department provides a dedicated financial management infrastructure for uniform and consistent financial management for the technology transfer program. It is important to recognize that the finances managed by this group are somewhat less "routine" than those managed in other university programs. The department needs to understand the legal processes surrounding IP management and also balance the differences in culture and demands arising from private industry, law firms, the university community, and individual inventors, all of whom have significant interests in the financial outcomes of the technology transfer process.

3.6 *Licensing*

The largest department within the UC technology transfer program is licensing. The UC has historically maintained sector-specific licensing groups in life sciences/pharmaceuticals, physics and engineering, and agriculture. It is particularly helpful to have technical expertise in each group as well as to have knowledge of licensing norms in the various industry sectors, which differ significantly. Licensing officers typically have one or more technical degrees (usually a Ph.D.), a law degree, and/or a business degree (M.B.A.), and are assigned primary responsibility for a case—defined as an invention disclosure—from its inception, through to licensing, and on to expiration. This practice has been referred to as "cradle to grave" management and differs markedly from the practice, typical of many institutions, of segregating invention disclosure and patenting processes from licensing negotiations and postagreement management.

Another chapter in this *Handbook*[5] provides a case study on the strawberry licensing program at UC Davis that illustrates an example of the types of licenses and licensing programs that the university has entered into as a means to transfer technology to the private sector.

4. TECHNOLOGY TRANSFER IN A DISTRIBUTED INSTITUTIONAL NETWORK

Perhaps the most distinctive features of the UC technology transfer system are its size and the development of a distributed institutional network of campuses that operate under a common policy framework and share certain resources, but function relatively independently. Valuable lessons can be learned from this system that may have applications in, or provide guidance to, other institutional networks seeking to develop capacity in technology transfer.

The first lesson is that a situation where a decentralized technology transfer program is in close geographic proximity to major research centers can lend itself to success. Decentralization and proximity are particularly important because active engagement by researchers in the technology transfer process typically requires a cultural shift that can only be made through continuous and systematic contact between technology managers and researchers.

There are, however, elements of a technology transfer program that can be effectively centralized. Candidates for centralization are, specifically, those elements of the program for which (1) uniform activities are required to minimize legal or financial risk or (2) economies of scale can be achieved by a consolidation of the activities.

Using these criteria, the UC technology program has, in general, retained centralized financial management, information technology (database) services, policy analysis and development, and legal oversight. These activities are generically referred to as our "back office" functions, which are essential for the program but do not require direct interface with our institutional clients (researchers) or our external clients (licensees).

In contrast, we have identified for local management those program elements that directly interface with researchers, research sponsors, licensees, or regional business interests. Based on this criteria, the following activities have been the focus of most of the campus-based technology transfer offices: invention disclosures and evaluation, patent prosecution, technology licensing, and business development activities.

The centralized/decentralized structure, or distributed network, described here could be emulated and implemented at different scales, from a relatively small-scale research consortium made up of a network of institutions, to a larger-scale national network of universities, to a global-scale international network of research institutions linked by common policies and objectives.

5. CONCLUSION

The UC has a long history of seeking intellectual property protection for its research discoveries and managing the technologies for the public benefit. By some measures, the UC technology transfer program is the largest public program in the world. Although it has evolved over the years, it has always been at the forefront of this endeavor. This article has focused on the history and policy and organization frameworks of the UC technology transfer program. We hope this discussion will be instructive to administrators and others seeking to learn from the UC experiences. ■

ALAN B. BENNETT, *Associate Vice Chancellor, Executive Director, PIPRA, Office of Research, University of California, Davis, 1850 Research Park Drive, Davis, CA, 95616, U.S.A.* abbennett@ucdavis.edu

MICHAEL CARRIERE, *Business Development and IP Manager, Office of Technology Transfer, University of California, Office of the President, 1111 Franklin Street, Oakland, CA, 94607-5200, U.S.A.* carriere@ucop.edu

1 Mowery DC, RR Nelson, BN Sampat and AA Ziedonis. 2004. *Ivory Tower and Industrial Innovation: University-Industry Technology Transfer Before and After the Bayh-Dole Act in the United States*, Stanford Business Books: Stanford, California.

2 *Shaw v. University of California*, 67 Cal. Rptr.2d 850. 58 Cal. App. 44 (1997).

3 Graff G, A Heiman and D Zilberman. 2002. University Research and Offices of Technology Transfer. *California Management Review* 45:88–116.

4 See *supra* note 1.

5 See, also in this *Handbook*, chapter 17.25 by AB Bennett and M Carriere.

FIGURE 2: SCHEMATIC OF THE FUNCTIONAL DIVISION OF PROGRAM ELEMENTS IN A DISTRIBUTED INSTITUTIONAL NETWORK FOR TECHNOLOGY TRANSFER

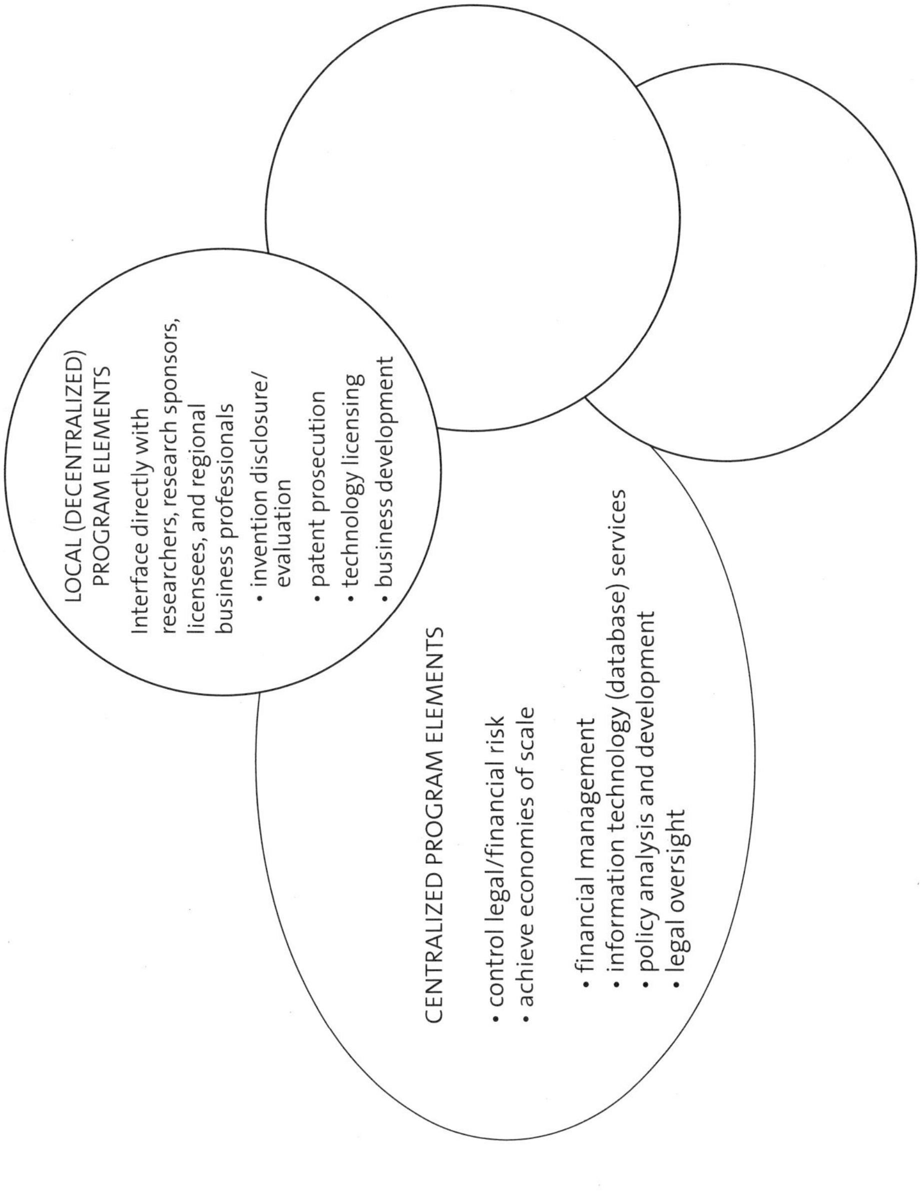

CHAPTER 17.15

Intellectual Property and Technology Transfer by the University of California Agricultural Experiment Station

GREGORY D. GRAFF, *Research Economist, PIPRA, and Visiting Research Fellow, Department of Agricultural and Resource Economics, University of California, Berkeley, U.S.A.*

ALAN B. BENNETT, *Associate Vice Chancellor, Office of Research, University of California, Davis; and Executive Director, PIPRA, U.S.A.*

ABSTRACT

One of the primary missions of the University of California Agricultural Experiment Station (AES) is to create knowledge and develop technologies that improve the productivity and environmental sustainability of agriculture in California. In addition to the public release of information and the educational activities of cooperative extension services, the University of California places the inventions of AES faculty directly into commerce through the process of patenting and technology transfer. This channel is particularly useful—and often essential—when further financial investments are necessary to develop the technology for practical applications or to manufacture, market, and distribute new products that incorporate the new technology. This report documents the patenting and formal technology transfer activities of the University of California Agricultural Experiment Station over the last 40 years.

More than 800 inventions have been reported by AES researchers between 1960 and 2001. These inventions are categorized into the five broad technology areas: biotechnology (49%), plant varieties (19%), chemicals (14%), equipment/machinery (13%), and environmental (1%). Biotechnology inventions were entirely absent until the mid-1980s, but the category has grown rapidly over the last 15 years. The growth in the number of biotechnology-related inventions has occurred not at the expense of inventions reported in the areas of plant varieties, agricultural equipment, or novel chemicals, all of which have shown a relatively stable level of activity.

Financial returns from the licensing of AES inventions was US$1.4 million in fiscal year 1982 (2.5 million in 2001 dollars) but had grown to US$12 million by fiscal year 2001. After accounting for expenses associated with patenting new inventions and distribution of a share of income to inventors, AES inventions returned over US$6 million to the university in fiscal year 2001. Since 1982, the cumulative financial return has totaled US$105.2 million in fees and royalties. About 87% of that income has been derived from the licensing of plant varieties in spite of the fact that they compose only 19% of the AES inventions, indicating the commercial importance of UC plant varieties. To date, relatively few biotechnology- or environmental-related inventions have been commercialized, but the extensive and growing UC portfolio in these areas should provide a strong base for future licensing activity.

1. INTRODUCTION

1.1 *The Agricultural Experiment Station at the University of California*

The Agricultural Experiment Station (AES) at the University of California (UC) is composed of nearly 700 researchers in 60 disciplines, carrying out over 1,000 research projects. These AES researchers are in the College of Natural Resources on the Berkeley campus, the College of Natural and Agricultural Sciences on the Riverside campus, and the College of Agricultural and Environmental Sciences and the School of Veterinary Medicine on the Davis campus. The common research goal of the AES is to create knowledge and develop technologies that

Graff GD and AB Bennett. 2007. Intellectual Property and Technology Transfer by the University of California Agricultural Experiment Station. In *Intellectual Property Management in Health and Agricultural Innovation: A Handbook of Best Practices* (eds. A Krattiger, RT Mahoney, L Nelsen, et al.). MIHR: Oxford, U.K., and PIPRA: Davis, U.S.A. Available online at www.ipHandbook.org.

improve the productivity and environmental sustainability of agriculture in California for the public benefit.

1.2 *AES research as an engine of commercialized inventions*

Much of the AES faculty research makes its impact on California and the world through the public release of new technologies or plant varieties, through cooperative extension services, and through the teaching of university students who apply their new skills and knowledge in the field.

In addition, the University of California places the inventions of AES faculty directly into commerce through the process of patenting and technology transfer. This channel is particularly useful—and often essential—when further financial investments are necessary to develop the technology for practical applications or to manufacture, market, and distribute applications that take advantage of the new technology. In this situation, the researcher is able to make an invention disclosure to the University of California's Office of Technology Transfer (OTT) at the UC Office of the President or to their individual campus's Office of Technology Licensing (OTL). Either office—the UCOP Office of Technology Transfer or the campus Office of Technology Licensing—provides a number of services to the faculty inventor. The staff evaluates the invention, and, if the invention seems to hold commercial promise, engages in efforts to protect and to market the invention. Companies that think they may be able to use one of the university inventions can take the technology for a test drive by buying an option on the technology; if a company decides that they indeed can use the technology profitably, they will sign a license agreement with the university. If the company feels that the technology is risky, is undeveloped, will require a lot of investment, or may have very uncertain returns, it may request that the option or license be sold only to itself (exclusive). Otherwise, options and licenses can be signed with more than one company (nonexclusive).

Following changes in U.S. laws in the early 1980s, the results of publicly funded research can more easily be patented and managed by universities. Other changes made biological inventions much easier to patent. A number of UC researchers have been at the forefront of making research discoveries and, under these new laws, obtaining patents with applications in agriculture. This chapter was produced in order to document the patenting and formal technology transfer activities of the California Agricultural Experiment Station over these last 20 years.

2. FINDING THE DATA ON UC'S AES INVENTIONS

The UC Office of Technology Transfer maintains the Patent Tracking System (PTS) database containing information on all inventions made by UC researchers and disclosed to the university since the early 1960s. PTS also includes complete annual financial records on every UC invention since 1982.

In order to identify those inventions made by AES faculty, rosters were obtained from the three host campuses—Berkeley, Riverside, and Davis—listing the names of all faculty members that had held AES appointments between 1980 and 2000. These names were then matched against the names of all UC inventors in the PTS database. The matches compiled showed that 283 of the AES faculty had registered at least one invention with the university (198 from Davis, 61 from Riverside, and 24 from Berkeley.) Then, using this list of active AES inventors, it was possible to exhaustively search the PTS database for all of the inventions on which the inventors were listed as contributing inventors. This yielded 808 invention disclosures, on which a total of 574 patent applications were filed in the United States, resulting in 243 U.S. utility patents and 76 U.S. plant patents issued to UC between the years of 1960 and 2001. For some of these AES inventions, foreign filings were submitted, resulting in the issue of 190 foreign utility patents and 354 various foreign plant-variety rights in a total of 83 countries.

3. IN WHICH AREAS ARE AES INVENTORS WORKING?

The 808 AES inventions are distributed among five broad technology areas (Figure 1a): 49%

are biotechnologies, including plant, animal, and human medical biology; 19% are plant varieties, primarily strawberries, avocados, peaches, grapes, and various rootstock; 14% are chemicals (primarily for pest control); 13% are equipment and machinery (for agriculture, food processing, and medicine); and 1% are environmental technologies for toxic cleanup and remediation.

Changes in emphasis over the years by the AES in these broad technology areas are illustrated in Figure 2. The number of invention disclosures in each of these five categories is shown for each year since 1960. Before the 1980s, AES inventions consisted entirely of equipment and machinery, chemicals, and plant varieties. Beginning in the 1980s, there was a large, sustained boom in biotechnologies. The rise of biotechnology, however, does not seem to have affected inventiveness in the other areas. Chemical inventions, while always sporadic, have continued, and there was a surge of new plant varieties in the late 1990s. The new, small area of environmental technologies emerged only in the 1990s. Equipment and mechanical inventions have remained remarkably steady throughout the 40-year timeframe. However, within the category of equipment and machinery, there has been a definite shift toward advanced technologies (computer and scientific equipment) for agriculture and medicine and away from farm machinery.

4. WHAT ARE THE FINANCIAL RESULTS OF AES INVENTIONS?

Four types of accounts are reported in the PTS data for each invention.

- **expenses**. All expenditures made in investigating the legal and market potential of a new invention, applying for patents, paying patent maintenance fees, and, in rare cases when necessary, enforcing UC's legal rights in patent litigation
- **reimbursements**. From firms licensing a UC invention that agree to pay for some or all of the expenses incurred in patenting the invention
- **fees/royalties**. Payments made to UC by firms for a license to use (or the option to license) a UC invention
- **disbursements**. A designated proportion of the fees/royalty revenues that is paid directly to the UC inventors as personal income

Out of the 808 AES inventions on record, only 174 have generated any fee or royalty income after 1982, when financial data began being recorded. The first 50 of these are listed in Table 1, ranked in order of revenue generated, from most to least. The most consistent "big hits" on the list are the strawberry varieties. The UC strawberry licensing program has been one of the brightest spots in the university's entire technology transfer enterprise.[1] Figure 3 plots the total licensing revenues collected for each of the 174 inventions and plots revenues from greatest to least. It is important to notice how skewed the distribution of revenues has been. The top 12 AES inventions alone account for 88% of all AES licensing revenues over 20 years of the program. It is also important to note that the inventions with lower revenues, toward the bottom of Table 1 and toward the right of Figure 3, tend to be more-recent inventions, which naturally show much less income, as they have had less time to generate royalties.

Of particular note, the tomato harvester, invented in 1960, is the first invention recorded in the dataset. Even in its third and fourth decades on the market (1982–2001), that invention brought in over US$160,000 in royalties to the university.

From 1982, when detailed annual records began to be kept, through 2001, the licensed inventions by AES researchers have earned a total of US$125 million in fees and royalties,[2] with 87% of that coming from the licensing of plant varieties, 10% from biotechnologies, 3% from chemicals, and 1% from equipment and machinery (Figure 1b). It is very interesting to note that while plant varieties make up just 19% of the inventions, they generate 87% of the revenues, while chemicals and machinery, and particularly biotechnology, fall far behind in terms of revenue generation relative to numbers of inventions (compare Figures 1a and 1b). Of the total amount, US$42 million was disbursed as inventor shares.

Figure 1a: Proportions of AES Inventions in Five Broad Technology Areas

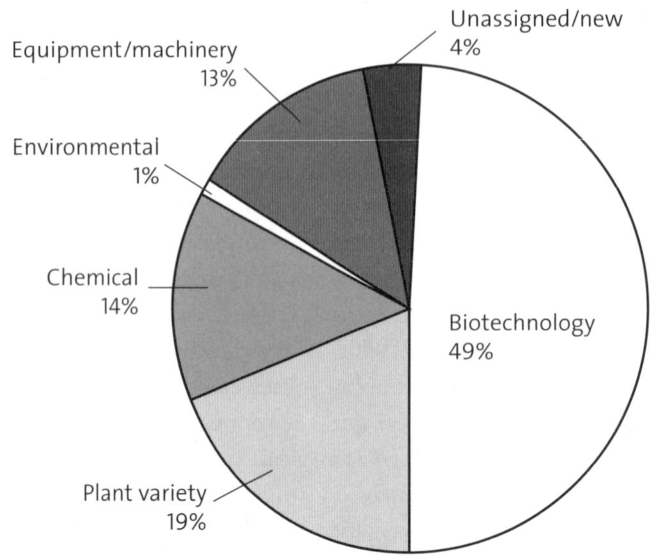

Figure 1b: Proportions of Income from AES Inventions in the Five Broad Technology Areas

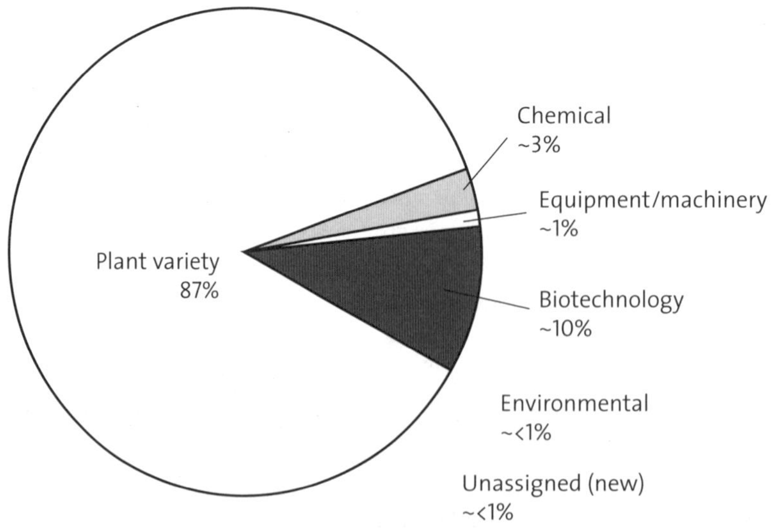

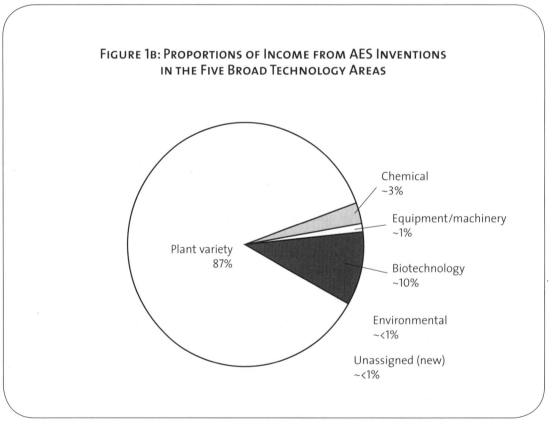

Expenses incurred in the patenting and marketing of these inventions totaled US$23 million, of which US$5.4 million was reimbursed by the licensing companies.

Over time, the annual fees and royalties generated by AES inventions has increased from approximately US$3 million (adjusted) per year in the early 1980s to almost US$12 million a year today, with particularly strong growth in the 1990s. Expenses have also grown, but at a slower rate, and reimbursements continue to offset approximately one third of expenses. The increase in expenses in the 1990s (Figure 4) was largely a result of increased foreign patent filings, particularly for plant varieties. The resulting foreign patents, however, have contributed directly to the large increase in revenues. Net income, that is, each year's total amounts received (includes fees and royalties plus reimbursements) minus each year's expenses, has continued to grow. Inventors' shares are paid out of the net income, and what is left over is returned to the university and reinvested into new research projects or used to cover university operating expenses.

5. CONCLUSIONS

The formal process of technology patenting and licensing is just one of the many ways that the University of California AES contributes to the state's agricultural economy and to the public welfare. In increasing numbers, inventions are being patented by the University of California on behalf of AES researchers and the income generated by this intellectual property is helping to support research and education at the university. A significant trend in invention disclosures is the tremendous increase in biotechnology-related inventions and the emergence of inventions in environmental technologies. At the same time, inventions reported in the areas of plant varieties, agricultural equipment, or novel chemicals have grown or remained at a stable level of activity.

FIGURE 2: ANNUAL NUMBER OF AES INVENTIONS DISCLOSED BY BROAD TECHNOLOGY AREA

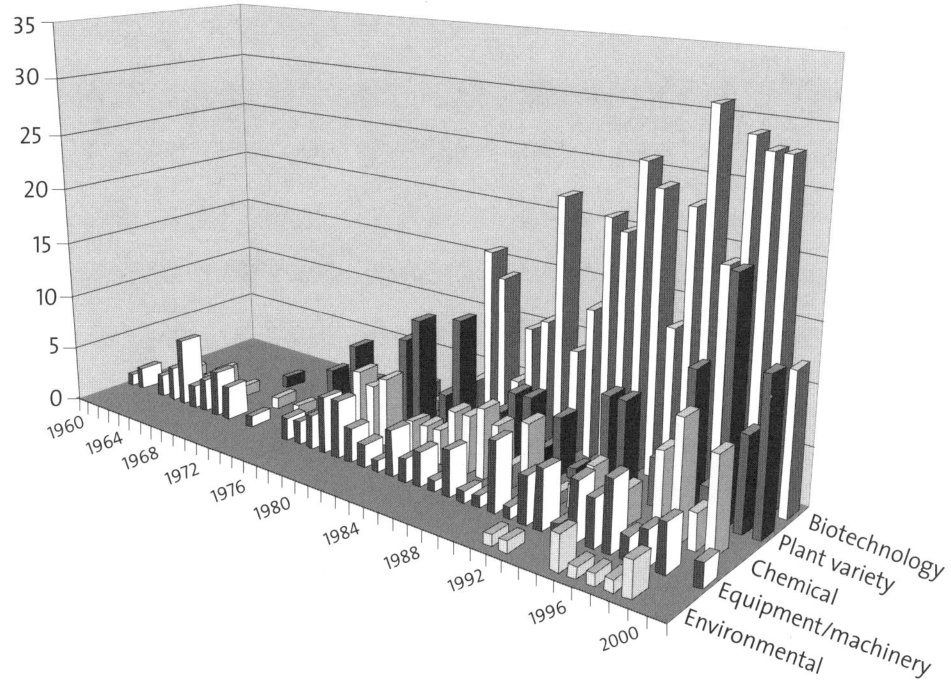

TABLE 1: THE TOP 50 INVENTIONS AT THE UNIVERSITY OF CALIFORNIA AES BY POSITIVE FINANCIAL EARNINGS (1982–2001)

FEE AND ROYALTIES RANK	INVENTION: SHORT TITLE	YEAR INVENTED	CAMPUS
1	STRAWBERRY:CAMAROSA	1992	Davis
2	STRAWBERRY: CHANDLER	1982	Davis
3	STRAWBERRY: PAJARO	1978	Davis
4	STRAWBERRY: OSO GRANDE	1987	Davis
5	STRAWBERRY: SELVA	1982	Davis
6	STRAWBERRY: DOUGLAS	1978	Davis
7	LIPOSOME STORAGE METHOD	1984	Davis
8	STRAWBERRY: SEASCAPE	1989	Davis
9	N-AMINO-S INSECTICIDE	1972	Riverside
10	GRAPE: TABLE: REDGLOBE	1979	Davis
11	REPLACE PHOSPHATE BY PHOSPHITE	1990	Riverside
12	STRAWBERRY: TUFTS	1972	Davis
13	STRAWBERRY: PARKER	1982	Davis
14	ASPARAGUS: F 109	1979	Riverside
15	CHERRY: BROOKS	1987	Davis
16	ASPARAGUS: M 120	1979	Riverside
17	ROOTSTOCK: GRAPE: 039-16	1985	Davis
18	STRAWBERRY: FERN	1982	Davis
19	FOOD SURFACE DISCOLORA REDUCER	1993	Davis
20	STRAWBERRY: DIAMANTE	1997	Davis
21	STRAWBERRY: IRVINE	1988	Davis
22	ROTARY SHAKER TOMATO HARVESTER	1978	Davis
23	STRAWBERRY: AIKO	1975	Davis
24	AVOCADO: LAMB/HASS	1993	Riverside
25	MODULATION OF ETHYLENE LEVELS	1990	Davis, non-UC
26	AVOCADO: GWEN	1982	Riverside
27	STRAWBERRY: AROMAS	1997	Davis
28	STRAWBERRY: MUIR	1987	Davis
29	ROOTSTOCK: AVOCADO: THOMAS	1986	Riverside
30	STRAWBERRY: GAVIOTA	1997	Davis
31	ANTIMICROORGANISM FINISH	1996	Davis
32	VOLATILE ELECTROLYTES	1976	Davis
33	RICE RESISTANCE TO XANTHOMONAS	1995	Davis
34	TOMATO HARVESTER	1960	Davis
35	INHIBIT FROST DAMAGE TO PLANTS	1981	Berkeley
36	GRAPE: TABLE: CHRISTMAS ROSE	1979	Davis
37	STRAWBERRY: CARLSBAD	1992	Davis
38	STRAWBERRY: HECKER	1978	Davis
39	DNA/ICE NUCLEATION BACTERIA	1982	Berkeley, non-UC
40	STRAWBERRY: BRIGHTON	1978	Davis
41	PLANT CELL FERMENTATION	1993	Davis
42	STRAWBERRY: SANTANA	1982	Davis
43	CONTROL RELEASE BIOMATERIAL	1992	Davis
44	SOLUBLE EPOXIDE HYDROLASE	1992	Davis
45	HIPPELATES EYE GNAT-CHEMICAL	1974	Riverside
46	STRAWBERRY: CUESTA	1992	Davis
47	STRAWBERRY: TORO	1975	Davis
48	LYME DISEASE: ASSAY & VACCINE	1990	Davis
49	BOVINE PARASITE DIAGNOSTIC	1993	Davis
50	GRAPE: TABLE: DAWN	1979	Davis

Figure 3: Distribution of Total Fees and Royalties Earned by the 174 Income-Generating AES Inventions (1982–2000)

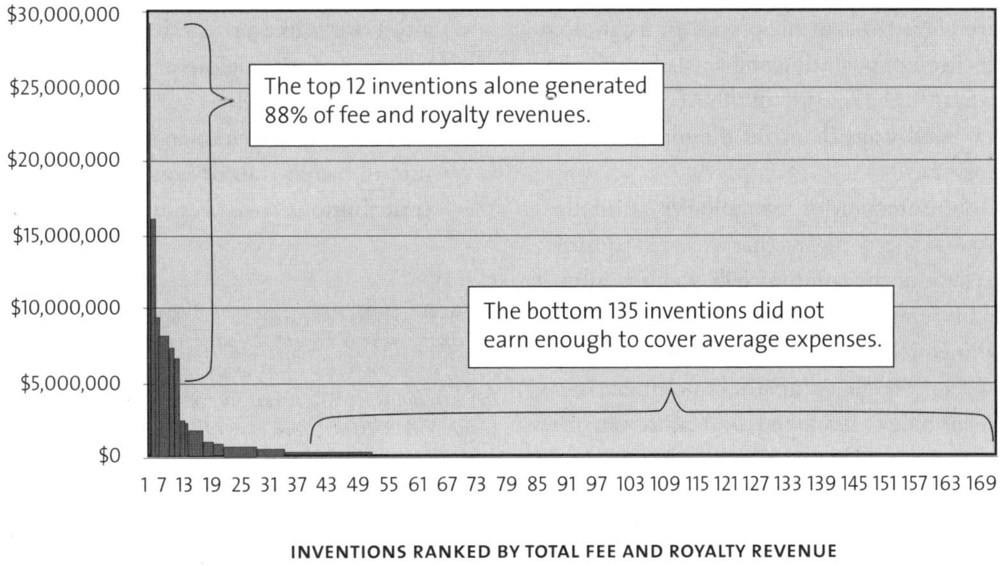

Figure 4: Annual Financial Performance of All AES Inventions, 1982–2001

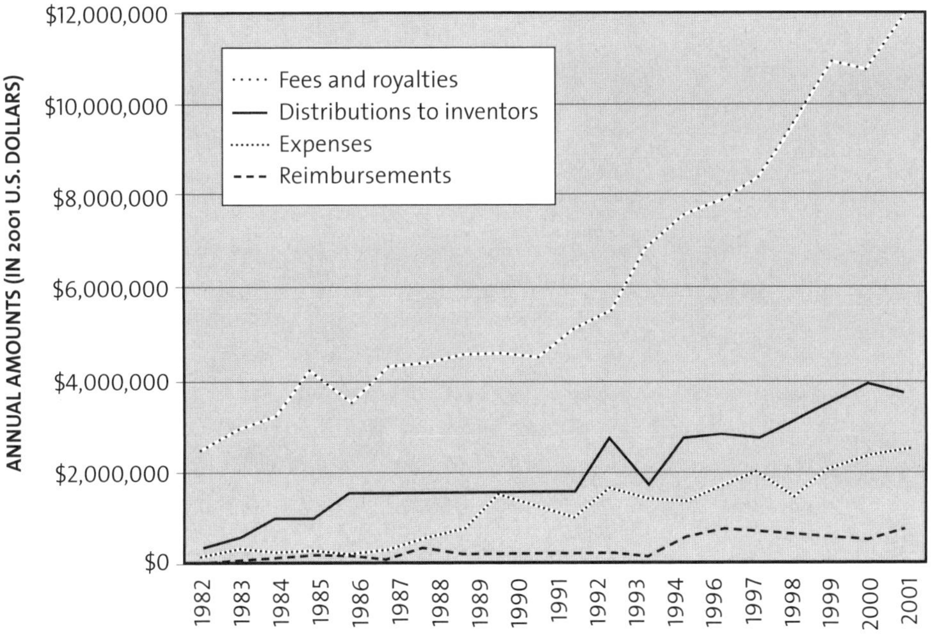

Relatively few biotechnologies or environmental technologies have been commercialized to date, but the extensive and growing portfolio in these areas should provide a strong base for expanded licensing activity in the future.

Several elements of the process of technology transfer through patenting and licensing are helping to advance the mission of the AES in new and more targeted ways than did the older mode of public release:

- The protection of technologies as intellectual property means that a clear accounting is kept of the commercially viable results of AES research.
- Protection under foreign filings means that, when foreign competitors want to use a technology developed by California, they need to compensate California to use it.
- Protection also provides the opportunity to entice companies to invest in developing earlier-stage technologies that would otherwise not likely be developed and thus not benefit the state's economy.
- The collection of licensing fees and royalties works like a highly targeted tax. The companies and growers that benefit most from AES research are thereby directly supporting the kinds of research and education at UC that commercially benefits them.
- The payment of an inventor's share of royalties works like a research prize, even if it is not quite as prestigious as the Nobel Prize. It rewards researchers for innovations that are effectively taken up in the state's agriculture in proportion to how significant their contributions have been to the economy. ■

GREGORY D. GRAFF, *Research Economist, PIPRA, and Visiting Research Fellow, Department of Agricultural and Resource Economics, University of California, Berkeley, PIPRA, Plant Sciences, Mail Stop 5, University of California, Davis, CA 95616, U.S.A. gdgraff@ucdavis.edu*

ALAN B. BENNETT, *Associate Vice Chancellor, Executive Director, PIPRA, Office of Research, University of California, Davis, 1850 Research Park Drive, Davis, CA, 95616, USA. abbennett@ucdavis.edu*

---

1 See, also in the *Handbook*, section 4.1 of chapter 17.13 by AB Bennett and M Carriere.

2 Amounts are normalized to 2001 dollars to adjust for inflation.

From University to Industry: Technology Transfer at Unicamp in Brazil

ROSANA CERON DI GIORGIO, *IP and Partnership Development Director, Inova Unicamp Technology Licensing Office, Universidade Estadual de Campinas, Brazil*

ABSTRACT

This chapter discusses how Brazil has dramatically increased technology transfer and innovation through the State University of Campinas, or Unicamp. The leader in patenting and licensing activities in Brazil and Latin America, Unicamp has vaulted to this position in the short span of two and a half years through its technology transfer office, Inova. Providing background information about Brazil's legal framework and practices, especially as it concerns the ownership of intellectual property and benefit sharing, the chapter discusses government incentives for innovation in light of Inova's impressive results. Two successful cases of technology transfer are presented as guides to realistic expectations about investments, terms of license, and royalties.

1. INTRODUCTION

In the last two years, the University of Campinas, or Unicamp, a Brazilian university publicly funded by the state of São Paulo, has become a leader in technology transfer. The critical agent in this process is Inova Unicamp,[1] the university's technology transfer office. In the last two and a half years, Inova has signed 128 technology transfer agreements and licensed 45 technologies (41 patents and four cases of know-how) to both private companies and the government. These agreements will last for more than ten years and have already generated royalties for the university. In the same period, Inova applied for 153 new patents, 22 trademarks, and 24 software registrations. Additionally, ten companies from Unicamp's business incubator have become self-sustaining. They may leave the university, after which they will pay Unicamp a percentage of their income for the next five years. Although Inova is still very young, in its first six months it achieved more results in technology transfer than had been achieved in Unicamp's entire history.

These outstanding results are unique for both Brazil and Latin America. The success of Inova has encouraged other Brazilian universities, as well as small- and medium-sized companies, to look to Unicamp as a management model.

2. PATENTING ACTIVITIES AT UNICAMP

Founded in 1967, Unicamp has, on average, 31,000 students; half of these are undergraduate students and half are graduate students. There are about 1,800 faculty members. With a total of 20 research units, Unicamp offers more than 50 undergraduate degrees and more than 100 graduate degrees. As a multidisciplinary university, Unicamp pursues a variety of technologies in many fields. Inova has assessed all of them and has aggressively pursued new patent applications

Di Giorgio RC. 2007. From University to Industry: Technology Transfer at Unicamp in Brazil. In *Intellectual Property Management in Health and Agricultural Innovation: A Handbook of Best Practices* (eds. A Krattiger, RT Mahoney, L Nelsen, et al.). MIHR: Oxford, U.K., and PIPRA: Davis, U.S.A. Available online at www.ipHandbook.org.

Editors' Note: This chapter is an edited and shortened version of an article that first appeared in *les Nouvelles*, Vol. XLI, No. 2, June 2006, pp. 90–93.

and licensing deals for those that have been most promising.

Table 1 lists the most frequent patentors in Brazil between 1999 and 2003: Unicamp is ranked as number one. It is interesting to note that among the top 20 institutions, there are five universities and two donor agencies. This runs contrary to the norm in developed countries, where industries patent more than universities and R&D centers.

The size of Unicamp's IP portfolio is also growing rapidly. As of last year, Unicamp has a substantial IP portfolio:

- 48 patents granted and 377 filed
- 17 registered trademarks and 36 filed
- 66 registered software applications/inventions and 66 filed

This portfolio is considered large for Brazil, showing that Unicamp's community has a good

Table 1: Patenting Activities in Brazil: A Ranking of Institutions (Total Patents Issued from 1999 to 2003)

Institution	Number of Issued Patents
Unicamp	191
Petróleo Brasileiro SA (PETROBRAS)	177
Arno SA	148
Multibrás Eletrodomésticos SA	110
Semeato SA Ind. e Com.	100
Companhia Vale Do Rio Doce	89
FAPESP (Fundação de Amparo à Pesquisa do Estado de São Paulo)	83
Brasil Compressores SA	81
Dana Ind Ltda	71
Universidade Federal de Minas Gerais	66
Johnson & Johnson Ind. e Com. Ltda	56
Universidade São Paulo	55
Jacto Máquinas Agrícolas	54
Minas Gerais Siderurgia (Usiminas)	48
Electrolux do Brasil SA	45
EMBRAPA	42
Conselho Nacional de Desenvolvimento Científico e Tecnológico	42
Universidade Federal do Rio de Janeiro (UFRJ)	38
UNESP - Universidade Estadual Paulista "Júlio de Mesquita Filho"	34
Dixie Toga SA	31

Source: Unpublished data from INPI (Instituto Nacional de Propriedade Industrial), Brazil.

understanding of the importance of protecting research results. Figure 1 shows the evolution of patenting activity at Unicamp. One can recognize an increase in activity after 1996, when the new Brazilian IP law was released, allowing protection for food, drugs, and chemicals, areas in which the university is very strong.

In terms of patent distribution by institute within Unicamp, patenting activities are not uniform within the university's research units. The greatest contributor to the portfolio is the Chemistry Institute, which was responsible for 48% of patents. As a result, most of the licensing agreements are made with the pharmaceutical, chemical, and medical devices industries, employing technologies originating from the Chemistry Institute. Other technologies such as medical applications (17% of licensing agreements), agribusiness (8%), and food (8%) occupy smaller places.

Inova's patent database is available online.[2] Patents are organized by market sector and can be searched by key word. This structure simplifies the localization of the available technologies by sector, which is useful for Inova's commercial team and also for external customers (industry or investors).

3. TECHNOLOGY TRANSFER ACTIVITIES AT UNICAMP

Unicamp is not only Brazil's biggest patentor but also the country's biggest licensor. According to Brazilian Law,[3] an employer is the rightful owner of all of its employee's results. Unicamp, therefore, owns 100% of its professors' and researchers' results. Although the Brazilian Innovation Law[4] allows public institutions to give up ownership to the inventor, Inova has not practiced this option. Its inventors lack commercial expertise, and it is more attractive to both the university and to the inventor for Inova to commercialize the technology and give the inventor part of the licensing fee.

Unicamp also commonly practices sponsored research. In such cases, ownership is normally split 50/50. In exceptional cases, where the industry partner or investor requires 100% ownership, Inova compensates the university by selling Unicamp's ownership to the partner.

Inova is driven by market demand. Instead of selecting Unicamp's technologies and offering them to the market, Inova finds out the market demand first and then looks for the solutions available inside of the university in response to that demand. Our focus is on the customer first

FIGURE 1: PATENTS FILED BY UNICAMP

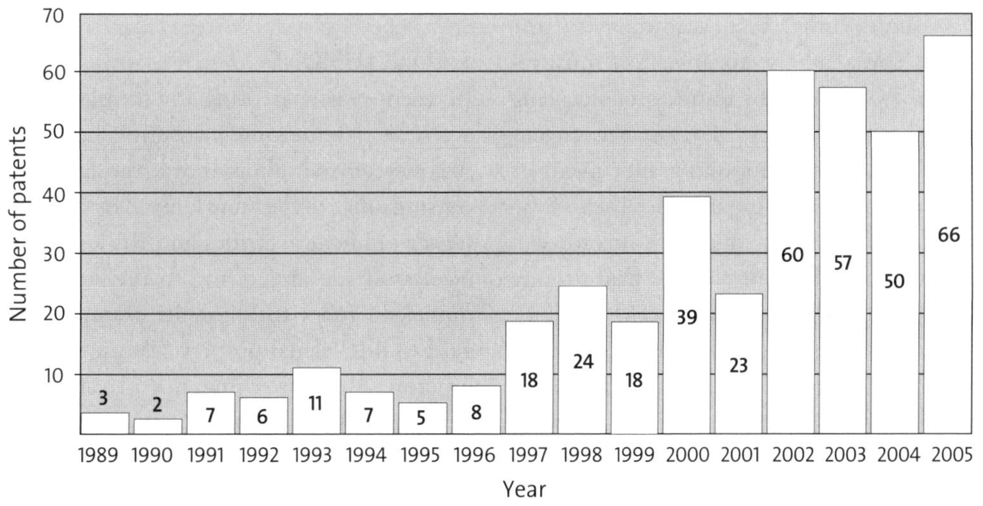

and on the technology second. Since Unicamp has a good and big technological production, normally we can provide to the market more options than they expect. Inova always tries to provide to the market a technology protected by a strong patent.

This market-demand model is partly why its results are so impressive. Another factor that contributes to Inova's success is the professional staff involved in technology transfer: the commercialization team comes from private institutions and has business skills. Most other universities in Brazil and Latin America use people from their research staff in technology transfer positions. But negotiation, market investigation, evaluation, and so on, are best done by business people who are specially trained to do it.

The 128 technology transfer agreements signed in two and a half years make Unicamp the biggest technology transfer provider in Brazil and Latin America. As Unicamp has multidisciplinary competence, the agreements were made both with private companies and with the government, as well as in many different industry sectors.

The licensing agreements last for more than ten years, and have already generated royalties for the university. These royalties range from 1.5% to 10% of the net income derived from the licensed technology. Each case has particular issues: all licensing contracts include royalty auditing in order to confirm that the sales results that the licensees present are correct.

It is noteworthy that according to the Innovation Law public sector inventors must receive from 5% to 33% of royalties or licensing income, as an incentive to develop new inventions and innovations. Unicamp grants inventors 33% of royalty and licensing income. The following cases make clear how much this income can realistically represent. Professors are paid for any consulting.

Two successful examples of technology transfer involving technology developed at Unicamp and handled through Inova are detailed in Box 1. These cases are presented as guides to suggest realistic expectations for investment, terms of license, and royalties.

4. GOVERNMENT INCENTIVES FOR INNOVATION

In recent years, the Brazilian government has provided new incentives for innovation. These include:

- tax benefits to companies that pay royalties (licensees)
- tax benefits to companies that invest in R&D, inside or outside the company (The latter includes funding R&D in universities, R&D centers, spinout companies, and independent inventors.)
- compensation for taxes paid for royalties abroad during the execution of technology transfer contracts
- no taxes on money paid to maintain patents, trademarks, and cultivar registrations abroad
- sponsorship/subsidy of 60% of the salary of a scientist hired by a company

All sectors are targeted by the law, but the Brazilian government has paid special attention to information technology, energy (electricity, oil, natural gas), semiconductors, biotechnology, and pharmaceuticals.

Importantly, the Innovation Law established that all government universities and R&D centers must have an office to take care of IP. This will increase patenting and licensing activities in public universities and R&D centers in the next few years.

5. CONCLUSIONS

In recent years, patenting and technology transfer activities have become institutionalized in Brazil. A concrete example is Inova, the technology licensing office of the State University of Campinas. Other public universities and R&D centers have been studying and trying to understand Inova's model, in order to follow its example. With regard to intellectual property, Brazil is at a crucial juncture. The government, especially in recent years, has released many incentives to innovation, which are reaching universities, R&D centers, and private companies. This certainly will increase patenting and technology transfer activities in the

Box 1: Examples of Unicamp Technology Transfer Successes

BiPhor

Licensee. Bunge (a global agricultural company)

Technology. white pigment based on aluminum phosphate nanoparticules (nanotechnology)

Target market. water-based paints (world market estimated at US$5 billion per year)

Advantages over existing technologies (TiO2). whiter, cheaper, "green" (or environmental) chemistry, improved quality and durability

Bunge's terms.

- investments of US$450,000
- exclusive license for 20 years
- target market share of 10%, worth an estimated US$500 million per year

Unicamp terms.

- 1.5% royalties (approximately US$4.5 million per year for 20 years)
- 33% of royalties (approximately US$1.5 million) to the inventors

Status.

- pilot plant running at 1,000 tons per year in sample production
- commercial plant to be running in five years at 100,000–200,000 tons per year
- sales price to be a little lower than TiO2 (product's competitor), which costs US$3,000 per ton

Aglycon Soy

Licensee. Steviafarma (medium-sized Brazilian pharmaceutical company)

Technology. concentrated phytoestrogen, extracted from soybeans using biotechnology (The unique process, developed at the university, employs a genetically modified microorganism owned by Unicamp and available at ATCC.)

Target market. hormonal therapy

Advantages over existing technologies. improved efficacy without side effects caused by conventional drugs, anticancer agent, LDL cholesterol reducer, fungicide, anti-inflammatory, and antioxidant

Steviafarma's terms.

- investment of R$100,000
- exclusive license for ten years
- target market share of R$36 million per year (Brazil only)

Unicamp's terms.

- 6% royalties over ten years (approximately R$1.2 million per year for 10 years)

Status.

- ANVISA registration granted
- production scheduled for September 2006

country and strengthen the relationship between public institutions (where the Brazilian research is mainly concentrated) and private companies, contributing strongly to innovation.

ROSANA CERON DI GIORGIO, *IP and Partnership Development Director, Inova Unicamp TLO, Universidade Estadual de Campinas, 100 Bernardo Sayao St., P.O. Box 6131, Campinas, Sao Paulo 13083-080, Brazil. rosanadg@unicamp.br*

1 Although a young organization, Inova now employs 20 people in technology transfer. They include business managers, support and marketing specialists, lawyers, patent analysts, and administrative assistants. These people and the others responsible for accounting, financial planning, management, and other crucial tasks, are strongly motivated and always have an eye for synergy. Together they are responsible for Inova's success. For further information, please visit www.inova.unicamp.br.

2 www.inova.unicamp.br.

3 Law 9.279, May 1996.

4 Law 10.973, December 2004; and Decree 5.563, October 2005.

SECTION **17c**

Putting Intellectual Property to Work: Experiences from Around the World

PRODUCT-DEVELOPMENT PARTNERSHIPS

CHAPTER 17.17

How Public–Private Partnerships Handle Intellectual Property: The PATH Experience

STEVE BROOKE, *Commercialization Advisor, Commercialization and Corporate Partnerships, PATH, U.S.A.*
CLAUDIA M. HARNER-JAY, *Commercialization Officer, Commercialization and Corporate Partnerships, PATH, U.S.A.*
HEIDI LASHER, *Principal, Lasher Consulting, U.S.A.*
ERICA JACOBY, *Senior Program Associate, Commercialization and Corporate Partnerships, PATH, U.S.A.*

ABSTRACT

PATH is an international, nonprofit organization that creates sustainable, culturally relevant solutions, enabling communities worldwide to break longstanding cycles of poor health. By collaborating with diverse public and private sector partners, PATH helps provide appropriate health technologies and vital strategies that change the way people think and act. PATH's work improves global health and well-being. Over the past 28 years, PATH has demonstrated that public–private partnerships (PPPs) can effectively address unmet public health needs, particularly when managed with a clear understanding of both public and private sector objectives. Indeed, collaboration between public sector and private sector partners is an especially valuable way to develop and advance appropriate health technologies for use in developing countries. When developing and managing PPPs, PATH recognizes that intellectual property (IP) is an especially important component in the range of variables that affect the economic, technical, and programmatic feasibility of a new health technology intervention. Our goal, therefore, is to incorporate IP considerations as a fundamental part of the PPP process. We seek to manage IP strategically to avoid or quickly overcome any IP-related roadblocks. Using three case studies, this chapter illustrates PATH's strategies for private sector collaboration, as well as PATH's approaches to managing IP.

1. INTRODUCTION

In many parts of the developing world, public health services reach less than 50% of the population. Weak infrastructure, poor living conditions, limited individual and public resources, extreme environmental conditions, population growth, new migration patterns, violent conflicts, and a host of other conditions all pose challenges to achieving "health for all." While healthcare for people in the developing world over the past quarter century has improved enormously, recently there have been significant setbacks: the AIDS epidemic and development of resistant strains of diseases, to name a couple. Continued growth in populations and decaying infrastructure due to lack of reinvestment have exacerbated the problem.

In this context, improving the effectiveness of healthcare services requires responsive, constantly evolving public health initiatives that can harness recent advances in biotechnology to solve difficult healthcare problems in developing countries. For example, new vaccines for meningitis, malaria, and rotavirus would greatly reduce the impact of these deadly diseases, which kill millions of people each year in developing countries. New, rapid diagnostic tests would detect conditions at the point of care, allowing treatment and counseling before the client has left the clinic. Heat stable and multivalent vaccines, prefilled injectors, and ice-free cooling would enhance health services and improve the effectiveness of immunization programs.

Brooke S, CM Harner-Jay, H Lasher and E Jacoby. 2007. How Public–Private Partnerships Handle Intellectual Property: The PATH Experience. In *Intellectual Property Management in Health and Agricultural Innovation: A Handbook of Best Practices* (eds. A Krattiger, RT Mahoney, L Nelsen, et al.). MIHR: Oxford, U.K., and PIPRA: Davis, U.S.A. Available online at www.ipHandbook.org.

1.1 *Why are public–private partnerships so critical for health technologies?*

Our experience suggests that one of the best ways to ensure that appropriate, affordable health technologies are developed and made available in developing countries is through public–private partnerships, or PPPs. Globally, most new health technologies come from the research and development efforts of private industry. Commercial enterprises not only have the expertise, capacity, and resources to carry a product forward to market, they also have strong market-driven incentives to do so. Unfortunately, this drive to pursue projects with the highest potential profit means that private companies usually do not put a high priority on products and services for developing countries. Markets in those countries are often unstable, and so perceived risks diminish projected return on investment. Pharmaceutical companies, for example, would rather invest in products that are targeted to large, lucrative therapeutic markets than pour research dollars into malaria or AIDS vaccines.

Without private sector collaboration many badly needed public health products/ideas simply fail to come to fruition. By itself, the public sector lacks the capacity, resources, and experience to design, develop, produce, and distribute most new technologies. The "technology challenge" for public sector health organizations, therefore, is to shift market forces enough to attract private sector involvement in developing appropriate, cost-effective healthcare technologies and to make them available to resource-poor populations. To accomplish this, the public sector must co-invest in necessary and suitable technologies, reduce risk, and invigorate private commercial investment through effective PPPs.

1.2 *What has PATH learned about PPPs?*

In the past two decades, the public sector has learned that the commercial sector can very effectively produce and distribute high-quality goods at low cost. It has also learned that before deciding to get involved in a project, the commercial sector must perceive a reasonable return on its investment and an acceptable level of risk. Acting as a "bridging agency," PATH helps to reconcile these differences by leveraging its technical innovation, knowledge of markets in developing countries, understanding of commercial imperatives, and experience of managing intellectual property (IP). PATH negotiates mutually beneficial solutions for both the public sector and private entities. Through public–private partnerships, the costs and risks of development are shared—and sometimes entirely funded by PATH with funds from donors, private foundations, and governments—at the early stages of a project, which helps private companies see the potential for a reasonable return on their investment. In return, PATH can guide technology development towards meeting the priority health needs of resource-poor populations.

Acting as a "value-added" intermediary between industry and the public sector, PATH has been involved in successfully commercializing and advancing over 50 new technologies for public health in developing countries over the past 28 years.

1.2.1 *Prioritizing availability, accessibility, and affordability*

Typically, a project will begin by clearly identifying a need or gap in the health system of a developing country that a new technology, at least in part, can address. PATH identifies potential partners, demonstrates the value of the technology, and forms collaborations with commercial companies to become codevelopers and/or sustainable suppliers of the technology to the developing world. Alternatively, the commercial company may own a technology that can be adapted for use in a developing country. In these cases, PATH may approach the company to collaborate or gain access to their technology. Within these partnerships, PATH aims to meet three objectives:

1. **Availability**: To guarantee supply for the developing world. Initially, PATH works to ensure that the company has adequate capacity to supply demonstration projects and/or clinical trials. Later, a company must be able to meet potential demand in targeted countries. Over the long term, companies must have capacity to meet

wider public sector demand in relevant developing countries.

2. **Accessibility**: To ensure that the product is available through distribution channels that actually reach target populations. Although many vulnerable populations get their services through public sector channels, they also access healthcare through private sector channels. PATH helps facilitate access to both channels by working with traditional government health services and by creating alliances with social marketing groups that are able to reach target populations more broadly.
3. **Affordability**: To create health products that the developing world can afford. PATH will often negotiate with partners to agree upon different prices for different markets (that is, tiered pricing by country, or between private sector versus public sector consumers). PATH also conducts cost-effectiveness studies to help decision-makers understand the value of the new product in relation to other potential health products.

1.2.2 *Principles for collaboration with private sector partners*

Once PATH has identified potential private sector partners, it follows a process of due diligence to examine a potential partner's operations and management and to verify material facts. Such upfront diligence significantly increases the chance of a successful partnership and assists planning. PATH needs to decide, for example, whether a company has enough resources to dedicate to a project, whether the company is stable and financially viable, whether the collaboration is appropriate given its current situation, and whether the company represents the best choice for a PATH partnership. Due diligence is an accepted—and often required—practice in the private sector, and it helps ensure the sustainability and impact of PATH's PPPs.

In addition, PATH professionals have a responsibility to preserve PATH's integrity and status as a publicly funded nonprofit, nongovernmental organization and fulfill this responsibility by evaluating partnerships with respect to nine principles for private sector collaboration.[1] From the perspective of IP management, the following two principles are most important:

1. **Clear link to mission.** PATH's collaborations with private sector companies must positively affect the availability, accessibility, and affordability of important health products for public health programs in developing countries.
2. **Recognition of private sector needs.** PATH recognizes the company's need to benefit commercially, which ensures a sustainable commitment to the collaboration. PATH's goals for availablity, accessibility, and affordability of products for developing country public health programs will likely be met if PATH's expectations of the private sector collaboration are realistic and take into account the full range of costs necessary from product development to commercialization.

2. HOW DO PATH'S PPPS HANDLE IP?

Given its mission, PATH has an inherent interest in managing IP to achieve maximum public health benefits. PATH's approach to IP management has common themes for all projects. PATH professionals review the existing and competing IP rights of all partners, negotiate with partners over the exact terms of ownership for all IP generated over the course of the project, agree on what happens if the partnership terminates before the project's completion, and specify responsibilities for protecting project IP generated by partners and PATH. After a technology is developed, IP is managed in the context of a commercialization strategy and a licensing plan.

Within each of these activities are myriad complexities that influence the specific strategies and tactics PATH adopts to negotiate IP. Perhaps the best way to understand PATH's approach to handling IP, then, is through case studies. Two of the following case studies, the first involving cervical-cancer screening diagnostics and the second involving a meningitis vaccine, are well along the product development pipeline. In these projects,

IP is managed to *advance specific products* through subsequent stages of development and commercialization leading to use in developing countries. A third case study, involving vaccine stabilization, describes technologies in an earlier stage of R&D that will become *components* of final products rather than complete products themselves. In this case, PATH is pursuing the development of a portfolio of technologies simultaneously in order to distribute risk and ensure progress toward a successful outcome. IP is managed to *advance the technology portfolio,* with the understanding that technologies developed over the course of the project will become important components of future final vaccine products.

2.1 *Cervical-cancer-screening tests: two is better than one*

Although cervical cancer is preventable, about 200,000 women die each year from it—often in their most productive years. Pap-smear screening programs help keep cervical cancer rates relatively low in wealthier countries; however, the success of these screening programs rely on regular visits to healthcare facilities, expensive pathology laboratories, and follow-up visits. Due to the cost, implementation challenges, and the complexity of properly screening and treating women in developing countries, the Pap-smear method has had only a limited impact in these areas. Not surprisingly, more than 80% of new cervical cancer cases occur among women living in developing countries.

2.1.1 *How the public and private sectors came together*

Because cervical cancer affects women in developed countries and developing countries, private industry had already invested in research to improve diagnostic screening tools for human papillomavirus (HPV), the virus is associated with over 99% of cervical-cancer cases. However, these commercial enterprises had not taken an interest in adapting their technology to make it more affordable and appropriate for developing-country health settings. This would have required a large investment in both product development and clinical studies—for a market that can afford prices that are only a fraction of those in developed countries. Hence, investing in HPV diagnostic technology for public sector markets in developing countries would never be a top priority for a commercial entity.

In 2003, PATH received funding from the Bill & Melinda Gates Foundation for its Screening Technologies to Advance Rapid Testing (START) project. This project includes support for clinical studies involving over 22,000 women in China and India, as well as support for developing low-cost, easy-to-use, culturally acceptable tests for cervical cancer screening. Since the private sector had already developed relevant technologies, and since PATH possessed useful data, a PPP was a logical choice. Two testing formats appeared promising, so PATH orchestrated partnerships with two companies to develop the test formats to detect HPV (one using DNA, and the other using a biomarker protein).

Both companies in the PPP are working to create a test that is safe, accurate, affordable, simple to use, and acceptable to women and healthcare providers. Tests will be based on a cervical swab provided by a healthcare provider or a vaginal swab obtained by the woman herself. Health workers with minimal training and equipment should be able to process either test in one day. Both tests are expected to have a higher than 90% accuracy rate in detecting cervical precancer or cancer (the Pap-smear test has a 55%–65% accuracy rate). This means that women who get tested only once in their lifetime, using one of the new methods, will still have a high probability of avoiding cervical cancer disease.

2.1.2 *PATH's management of IP*

When negotiating with partners, PATH often finds it helpful to articulate the different roles and responsibilities and the expected durations of the various phases involved in the project. For the START project agreements, there was the R&D phase, which would last approximately five years, and the commercial sales phase, which would last 10 years from the date of first sale. In the R&D phase, PATH assumed responsibility for seven primary activities:

- funding a portion of each industry partners' direct R&D costs
- providing biological samples during research
- conducting market and industry assessments
- conducting some key product development tasks, specifically with lateral flow technology
- conducting program and product cost-effectiveness studies
- developing for the new tests an evaluation framework for public health program use
- conducting multicenter, multicountry (India and China) clinical evaluations of the performance of the new test that would be suitable for the compilation of data required for product registration in those countries

In turn, PATH's industry partners agreed to:

- conduct product development activities as outlined in their agreements
- assemble and protect any needed IP
- manufacture and supply the products for clinical evaluations
- finalize the products for registration and commercial supply

Each of PATH's private sector partners in this project already controlled key IP for the technologies included in its respective diagnostic test. This eliminated the need to broker IP for reagents from multiple parties. However, the two partnerships are more complex when it comes to creating PATH's backup IP rights if either industry partner were to decide not to go forward. In one agreement, PATH obtained, under certain backup conditions, a long-term supply agreement to the partner's key reagent, as well as the ability to sublicense others to produce a final diagnostic test incorporating this reagent. In the other agreement, the industry partner agreed to appoint a third party to manufacture and supply the diagnostic test if it does not want to continue commercialization. The latter partner would never be comfortable allowing its core background IP to move out of its direct control, so rather than asking the company to grant PATH rights to background IP, PATH focused on ensuring continued supply. Both agreements set pricing targets that are significantly lower than anything currently available.

Following the successful completion of research, development, and validation, PATH's industry partners will be responsible for obtaining the necessary regulatory approvals and for manufacturing and selling the test at an affordable price in India, China, and other developing countries. By the end of 2008, two easy-to-use, inexpensive, and appropriately designed diagnostic products to detect cervical precancer and cancer should be available in developing countries.

2.1.3 *Key insights*

All projects come with their own unique challenges, particularly when multiple partnerships are involved. In the case of the START project, PATH was able to avoid some common pitfalls by carefully selecting its partners. For example, because PATH came forward with links to clinical researchers and policy-makers, and because it had a solid understanding of the specifications that any new cervical-cancer-screening test would need, PATH was able to attract two top-tier industry partners that had the expertise and capacity to move product development forward. These partners were attractive to PATH because they owned proprietary control of the key reagents needed for their specific technologies. This allowed the project to avoid the even more uncertain, complex, and lengthy negotiations necessary to bring multiple IP holders into a workable product development project.

PATH also provided access to well-characterized, highly sought-after clinical specimens from countries outside the industry partner's normal research networks. In addition, PATH offered the opportunity for major field-based clinical assessments of final products, assessments that would be sufficient for product registration in those countries. As a result, the two industry partners realized that working with PATH would provide a unique opportunity to reengineer their product (in the case of one partner) or develop a new product (in the case of the other partner) to address lower-price market segments, thus gaining valuable inroads into the challenging but attractive markets of India and China. Without the PATH program incentives, it is unlikely that either company would have undertaken these major efforts to adapt and develop their technologies for use in developing countries.

2.2 *Meningitis vaccine: a new model for vaccine development*

Meningitis, also referred to as spinal meningitis, is an infection in the fluid that surrounds the brain and spinal cord. When caused by a bacterial infection, the disease can be quite severe and may result in brain damage, hearing loss, learning disabilities, and death. Epidemic meningitis has been present on the African continent for about 100 years.

Over the last 20 years, countries located in Africa's "meningitis belt," roughly located between Senegal and Ethiopia, have depended on a disease control strategy involving surveillance and, once outbreaks are detected, reactive mass immunization campaigns using meningococcal polysaccharide vaccines. These interventions are massive, expensive, and disruptive, and they deflect scarce resources from public health efforts to control other diseases. Moreover, recent studies have shown that after an epidemic has begun, follow-up mass vaccinations are ineffective at preventing meningitis.

Unfortunately, while the public health need for a meningitis vaccine in Africa is great, no manufacturers have been willing to develop an affordable, effective group A meningococcal vaccine. In the 1990s, when more than 100,000 people died in Africa from a group A meningitis outbreak, there was also a group C meningitis outbreak in the United Kingdom, which resulted in 1,000 deaths. By 2001, three vaccine manufacturers had developed group C meningococcal vaccine for the United Kingdom. No vaccine for group A, however, had been developed.[2]

2.2.1 *How the public and private sectors came together*

The disease-specific components for a highly effective group A meningococcal conjugate vaccine existed before the PATH/World Health Organization (WHO) Meningitis Vaccine Project began. The conjugation technology also existed, which was a key production process step—it chemically links the two components, which makes the vaccine highly immunogenic and effective in young children, provides long-lasting protection, and decreases carriage and transmission rates. Yet no one was bringing these components together to develop and produce a meningococcal A vaccine. The challenge was to develop a program capable of motivating a vaccine producer to take a risk on an indigent market unable to pay high prices for the meningococcal A vaccine.

To address this challenge, in 2000 WHO commissioned an independent assessment of existing IP on conjugation technology and of the costs for project development and production for a group A or group A/C meningococcal conjugate vaccine intended for Africa.[3] The assessment showed that development was feasible and that a vaccine costing around US$0.40 per dose was possible—a price that health managers in sub-Saharan African countries were willing to pay. Soon after, the Bill & Melinda Gates Foundation awarded PATH a ten-year grant to establish, in partnership with WHO, the Meningitis Vaccine Project, which will advance the development, production scale-up, testing, licensure, and introduction of conjugate meningococcal A vaccines for Africa.

2.2.2 *PATH's management of IP*

The Meningitis Vaccine Project brought three critical partners to the table: SynCo Bio Partners B.V., which supplied meningococcal polysaccharide A (one of the two main components of the vaccine); the Serum Institute of India Limited (SIIL) to supply tetanus toxoid (the second main component of the vaccine) and to scale-up the manufacturing processes for the final vaccine; and the U.S. Food and Drug Administration's (FDA) Center for Biologics Evaluation and Research to transfer their conjugation technology. This consortium was a new model for vaccine development: a key raw material came from one source, the technology from another, and the final scale-up for production from another. Moreover, it included a north-to-south transfer of technology and capacity.

PATH first negotiated a nonexclusive license for the FDA conjugation technology from the U.S. National Institutes of Health Office of Technology Transfer (on behalf of the FDA), which PATH then sublicensed to SIIL. To protect the charitable mission of the project, PATH and SIIL agreed that if SIIL were to cease developing or producing the vaccine, SIIL would transfer to PATH the manufacturing know-how developed during their collaboration to enable another manufacturer to make the vaccine. SIIL also granted back to PATH a nonexclusive, sublicensable license to SIIL-owned technology necessary to make the vaccine. In addition, the PATH-SIIL agreement set out an explicit initial pricing of US$0.40 per dose for sales to the public sector. PATH's agreement with SIIL also includes explicit procedures and remedies should SIIL not meet public sector demand or charge the public sector more for the vaccine than the maximum agreed-upon price.

2.2.3 *Key insights*

It is somewhat unusual for vaccine manufacturers to accept a nonexclusive sublicense for a key production process such as a conjugation technology. However, the PPP and technology transfer gave SIIL incentive to accept this. First, since no manufacturer had been willing to make this vaccine, SIIL considered the risk that a competing manufacturer would step forward to use nonexclusively available FDA technology for a group A meningococcal conjugate vaccine was very small. Second, although SIIL is one of the world's leading vaccine manufacturers and had prior research experience working with conjugation technology, both SIIL and PATH knew they would be facing complex development challenges and an aggressive timetable. To help address these challenges and make the project more attractive to SIIL, PATH formed a technical team composed of the FDA inventors and other industry and government experts, who creatively and efficiently helped the Meningitis Vaccine Project surmount the inevitable technology scale-up and standardization hurdles. Third, the U.S. National Institutes of Health Office of Technology Transfer (NIH OTT) would have likely required higher up front fees, milestone payments, and higher royalty rates if PATH and/or SIIL had demanded an exclusive license to the conjugation technology. By nonexclusively in-licensing the conjugation technology under lower-cost terms and bundling it with further technology transfer support, pharmaceutical development, and clinical trials funding, PATH provided a package that would allow SIIL to keep the finished vaccine price at the targeted US$0.40 per dose, even after paying royalties to the NIH OTT. At this price, the new vaccine would cost less than current expenditures in hyperendemic areas, even before adding lost livelihood income and disability savings.

2.3 *Creativity and flexibility accelerate vaccine stabilization technologies*

The global health community is trying to make vaccines available to all the world's children, but this commitment is stressing an already fragile cold chain: the distribution network of equipment and procedures used to maintain vaccine quality from the vaccine manufacturer to the recipient. While strengthening and expanding existing cold-chain capacity is one option for reducing these stresses, improving vaccine thermostability—the inherent ability for vaccines to withstand extreme temperatures—is likely to be the more effective and sustainable approach. In recent years, stabilization technology has advanced so far that it could reduce the reliance of vaccines on the cold chain and facilitate expanded delivery options. These products could reduce the logistical burden of vaccine delivery, reduce vaccine waste, improve safety, and facilitate extended coverage.

2.3.1 *How the public and private sectors came together*

Vaccine producers typically seek to obtain sufficient product stability to meet the standards of developed countries. This means that vaccines typically require storage at frozen (–20° C) or refrigerated (2–8° C) temperatures. Some heat-sensitive vaccines (such as measles, BCG, and yellow fever vaccines) must be lyophilized (freeze-dried) in order to achieve this level of stability. Vaccine producers have been reluctant to further improve thermostability to reduce reliance on the cold chain for two main reasons. First, there is no perceived need for such products in developed countries where cold chain breaks are infrequent. This means that vaccine producers would rely solely on developing country sales to recoup their development investment. Second, the commitment of vaccine purchasers to buy stabilized vaccines for use in the developing world is uncertain—especially at higher prices.

In the absence of a market for thermostable vaccine products, PATH initially investigated the feasibility of stabilizing vaccines with funding from the U.S. Agency for International Development (USAID) under a program called HealthTech: Technologies for Health. In 2003, PATH received funding from the Bill & Melinda Gates Foundation to investigate the technical, programmatic, and market feasibility of stabilization technologies. PATH is pursuing a portfolio approach to the project, working with a range of private sector companies and universities to accelerate the development of different stabilization technologies that could be applied to a variety of vaccines. PATH has also developed its own proprietary technology to protect vaccines against freeze damage (U.S. and Patent Cooperation Treaty [PCT] patent applications are pending). As certain technologies show themselves to be more promising than others in terms of availability, accessibility, and affordability, the portfolio will be narrowed. When the technologies are mature enough to transfer, vaccine producers will need to help validate and scale up the technologies for commercial production.

2.3.2 *PATH's management of IP*

The primary focus of PATH's IP management strategy for the vaccine stabilization project has been to keep options open by holding some ownership of the new IP generated with partners. This makes it possible to move forward with the technology if the partner is unwilling and to improve the efficiency of research within the portfolio (that is, use the project IP with other partners). Since the landscape of patents in the stabilization field is fairly crowded, the strategy also involves creating partnerships with those that hold foundational IP to which others may eventually need access.

In practice, this strategy requires a great deal of creativity and flexibility. In many cases, for example, PATH and its partner jointly own project IP. Moreover, in certain circumstances, access to background IP is negotiated at the start. This is ideal because it gives PATH control without jeopardizing the partner's access. However, two specific partnerships illustrate the extremes of managing IP. On one end of the spectrum is a technology that PATH created in-house and is developing in collaboration with a partner. Since PATH owned the technology, it was able to negotiate full ownership of all improvements, even those to which the partner may contribute. On the other end of the spectrum, a private sector partner maintained very tight control over its proprietary IP. Rather than accept funding from PATH, the company tested its technology against the applications of interest to PATH, assuming the entire R&D burden in order to fully control the IP. In this case, PATH was able to obtain *an opportunity* to negotiate access to their IP in the future. Although not ideal structurally, this collaboration allowed PATH to build a relationship with a partner whose technology may be important to other technologies in the portfolio. This may allow PATH to avoid a potential roadblock to access in the future.

In addition to IP management, the project's global access strategy makes concerted efforts to align partners along the vision of how the end products might be made available in developing countries. For such purposes, PATH developed a Preferential Technology Access Program, which is written into each partner's agreement. For example, partners must agree to license their technology on nonexclusive terms to vaccine manufacturers in order to maximize access, place a royalty cap on those licensing arrangements, and restrict licensing and milestone fees. The exact terms vary with each partnership. The goal is to enable access to these technologies as they move downstream in the development pipeline.

2.3.3 *Conclusions*

When it comes to upstream research projects, we know very little about which technologies will emerge as promising, which may need to be eventually combined, and which may prove foundational for others. PATH's strategy has been to invest in a wide variety of promising approaches, promising to maximize the chances for success and integration and to negotiate some degree of access. PATH can thereby prevent those technologies that are emerging from the portfolio—and even technologies that already exist—from limiting the widespread adoption of stabilization technologies by vaccine manufacturers serving the developing world. This requires a constant reexamination of product scenarios and players. PATH uses as much flexibility and creativity as possible to move forward a market that in its absence would stall. ■

ACKNOWLEDGEMENT

Some of the writing in this chapter was borrowed from internal documents written under the HealthTech project, which was made possible by the generous support of the people in the United States through USAID under the terms of HealthTech Cooperative Agreement # GPH-A-00-01-00005-00. The ideas expressed in this chapter reflect the principles and goals of PATH and do not necessarily reflect the views of USAID or the U.S. Government.

STEVE BROOKE, *Advisor, Commercialization and Corporate Partnerships, PATH, 1455 NW Leary Way, Seattle, WA, 98107, U.S.A.* sbrooke@path.org

CLAUDIA M. HARNER-JAY, *Program Officer, Commercialization and Corporate Partnerships, PATH, 1455 NW Leary Way, Seattle, WA, 98107, U.S.A.* charner@path.org

HEIDI LASHER, *Principal, Lasher Consulting, 410 N Willson Ave., Bozeman, MT, 59715, U.S.A.* heidilasher@yahoo.com

ERICA JACOBY, *Senior Program Associate, Commercialization and Corporate Partnerships, PATH, 1455 NW Leary Way, Seattle, WA, 98107, U.S.A.* ejacoby@path.org

---

1 A full description of PATH's Guiding Principles for Private-Sector Collaboration is available online at: www.path.org/files/ER_gp_collab.pdf. Additional reading and relevant articles are:

Free MJ. 2004. Achieving Appropriate Design and Widespread Use of Health Care Technologies in the Developing World: Overcoming Obstacles that Impede the Adaptation and Diffusion of Priority Technologies for Primary Health Care. *International Journal of Gynecology & Obstetrics* 85(Suppl. 1): 3–13.

Free MJ. 1992. Health Technologies for the Developing World. *International Journal of Technology Assessment in Health Care* 8:4: 623–34.

Greenwood BM. 1999. Manson Lecture: Meningococcal Meningitis in Africa. *Transactions of the Royal Society of Tropical Medicine and Hygiene* 93: 341–53.

PATH. 2006. Global Partnerships for Malaria Control. *Directions in Global Health* 3:1. www.path.org/files/ER_directions_spr_06.pdf.

PATH. 2006. Eliminating meningitis in Africa. *Directions in Global Health* 3:2. www.path.org/files/ER_directions_summer06.pdf.

Ramsay ME, N Andrews, EB Kaczmarski and E Miller. 2001. Efficacy of Meningococcal Serogroup C Conjugate Vaccine in Teenagers and Toddlers in England. *Lancet* 357: 195–96.

Lloyd J. 2000. WHO Department of Vaccines and Biologicals. Technologies for Vaccine Delivery in the 21st Century. whqlibdoc.who.int/hq/2000/WHO_V&B_00.35.pdf.

2 Jódar L, FM LaForce, C Ceccarini, T Aguado and DM Granoff. 2003. Meningococcal Conjugate Vaccine for Africa: A Model for Development of New Vaccines for the Poorest Countries. *Lancet* 361: 1902–4.

3 See *supra* note 2.

The African Agricultural Technology Foundation Approach to IP Management

RICHARD Y. BOADI, *Legal Counsel, African Agricultural Technology Foundation, Kenya*
MPOKO BOKANGA, *Executive Director, African Agricultural Technology Foundation, Kenya*

ABSTRACT

For smallholder farmers in Africa, yields of major staple crops (maize, sorghum, millet, cassava, cowpea, bananas/plantains) have remained stagnant or even declined in the past 40 years. Numerous biotic and abiotic stresses have contributed to this dire trend. Local research efforts to overcome these stresses have been hampered by declining support for agricultural research, limited access to elite genetic material and other technologies protected by IP rights, and the absence of commercial interest in these crops from private owners of agricultural technologies. The African Agricultural Technology Foundation (AATF) is a new initiative addressing the challenge of reversing the negative trend in agriculture by negotiating access to proprietary technologies and facilitating their delivery to smallholder farmers in Sub-Saharan Africa.

This chapter addresses the IP issues and partnership arrangements associated with the access, development, and deployment of agricultural technologies in Sub-Saharan Africa by AATF. The chapter explores the model developed by AATF, which incorporates the acquisition, development, and deployment of new technologies from private sector partners, to try to address the agricultural needs of resource-poor smallholder farmers in Sub-Saharan Africa.

1. INTRODUCTION

The agricultural sector in developing countries is the key source of food, incomes, employment, and often, foreign exchange. Put another way, agriculture is crucial for sustaining livelihoods and stimulating overall economic growth. Traditionally, agricultural progress has been dependent on on-farm experimentation and the selection and adaptation of crop landraces. More recently, progress has been accelerated through the development of new varieties of crops, mainly through crossing and selecting parent crops with desirable characteristics. In Africa, smallholder farmers constitute approximately 70% of the general population and 90% of the agricultural workforce. According to Omanya and colleagues,[1] despite the availability of agricultural technologies (such as improved seeds and farm inputs), crop productivity in Africa has remained low or stagnant. This is true mostly because improved crop varieties that are resistant to biotic and abiotic constraints are not being planted. High costs and the unavailability of technologies in times of need have made drought-tolerant or disease- and pest-resistant seeds inaccessible, particularly to smallholder farmers in developing countries. The problem is compounded by the complexities associated with the protection of IP rights. Patents and plant breeders' rights attempt to strike a balance between protecting the rights of an invention and providing a benefit to the society as a whole, but such protected materials also often raise the cost of accessing new plant varieties.

The decline in agricultural productivity and the rise of IP rights has created a new challenge: How can the development community stimulate the development of innovative technologies while providing mechanisms that support the

Boadi RY and M Bokanga. 2007. The African Agricultural Technology Foundation Approach to IP Management. In *Intellectual Property Management in Health and Agricultural Innovation: A Handbook of Best Practices* (eds. A Krattiger, RT Mahoney, L Nelsen, et al.). MIHR: Oxford, U.K., and PIPRA: Davis, U.S.A. Available online at www.ipHandbook.org.

smallholder farmers' access to these technologies? A balance must be achieved in order to reach the goal of improving the economic productivity, and therefore, the lives, of these farmers. Delmer and colleagues[2] have described several initiatives designed to meet this challenge. Besides AATF, two other groups are working toward these initiatives:

- the Public Intellectual Property Resource for Agriculture (PIPRA), a U.S.-based initiative with global reach that seeks to pool publicly owned and patented technologies for use by research institutions in developing countries
- the Centre for the Application of Molecular Biology to International Agriculture (CAMBIA), an Australia-based initiative, which aims to provide technical solutions that empower local innovators to develop new agricultural innovations

AATF focuses specifically on negotiating access to proprietary technologies and facilitating delivery of the technologies to smallholder farmers in Sub-Saharan Africa. The next section describes AATF in more detail and discusses in-depth AATF's policy on IP management. Finally, the chapter describes a few specific projects under development by AATF.

2. THE AATF MODEL

2.1 *Background*

Improving agricultural productivity in Africa is key to expanding the economy and reducing poverty. Since the 1970s, significant investment and new technologies have caused agricultural productivity to rise dramatically in Asia and Latin America. But investment and innovation have been limited in Africa and agricultural productivity suffered. Sub-Saharan Africa has the highest hunger and malnutrition rates and the least productive agriculture in the world: approximately one-third of the population lacks food security (defined by the Food and Agriculture Organization [FAO] as having enough food to lead healthy and productive lives), and one-half lives on less than US$1 per day. According to World Bank figures, the 25 countries with the highest death rates are in Africa, and 24 of the 25 countries with the lowest life expectancy are in Sub-Saharan Africa. Most of the region's population depends for their livelihood on agriculture, which accounts for only 30% of the region's gross domestic product (GDP). Farmers make up about 90% of those individuals earning less than US$1 per day. Between 1980 and 1995, Sub-Saharan Africa was the only region of the world where crop production actually decreased: yields fell by 8% compared to increases of 27% in Asia and 12% in Latin America.

Developments in agricultural science and technology, however, hold out hope for significant improvements in food security and poverty reduction in Sub-Saharan Africa. African Poverty Reduction Strategy Papers, documents from the New Partnership for Africa Development (NEPAD), and multilateral policies and plans all emphasize the need for Africa to access new, better agricultural technology from the international community. Some of these technologies can be readily adapted to the region's conditions and can be provided immediately to poor farmers.

The private and public sectors hold the key to accessing these technologies—but neither alone can exploit this potential. Private sector companies have significant technological resources but currently no commercial incentive to invest in the specific technologies, varieties, and traits suitable for the unique agricultural conditions of the relatively small Sub-Saharan Africa market. On the other hand, public sector organizations have vast experience working on regionally important crops but need improved access to the proprietary technologies held by the private sector and other public sector institutions. Further, the region's public sector research institutions could benefit from assistance in adapting technologies to the needs of resource-poor farmers in Sub-Saharan Africa. But issues related to the availability, complexity, high transaction costs, licensing, testing, safety, and potential liability associated with these agricultural technologies bar access to these technologies by the region's researchers, development specialists, and resource-poor farmers. To address these issues, we need new, innovative approaches based on the support and collaboration of both the public and private sectors.

The fundamental rationale for the creation of AATF was to establish links between private sector and public sector institutions (that own technological innovations) in developed nations and African stakeholders in agricultural development, such as the National Agricultural Research and Development Organizations, farmers' associations, nongovernmental organizations, and national, private sector agribusinesses. The goal of AATF is to facilitate access to advanced scientific and technological resources and to promote their adaptation for use in specific projects intended to increase the productivity of smallholder farmers in Sub-Saharan Africa. AATF is an Africa-based, Africa-led entity, registered as a charity under the laws of England and Wales, with the specific objective of relieving poverty in Africa by facilitating public/private partnerships for the transfer and use of innovative agricultural technologies by smallholder farmers, particularly resource-poor farmers. AATF thus contributes to increased productivity, higher farm output, increased food security, and higher incomes. Headquartered in Nairobi, Kenya, AATF was officially launched there in June 2004.

2.2 *Operating principles and strategy*

AATF's strategy for achieving its objectives is to act as the principal and "responsible party" in facilitating, on a case-by-case basis, public/private partnerships. AATF works closely with other African institutions, responding on a project-by-project basis to the expressed needs of African farmers. The foundation endeavors to assemble all the necessary components for each project, balancing concerns for expense, simplicity, and effectiveness. More specifically, AATF:

- consults with African stakeholders to identify priority crops and key constraints for resource-poor farmers
- consults with potential technology providers, in both the private and public sectors, to identify technologies that can address those constraints
- negotiates with potential partners to develop a project business plan that specifies the role of each partner institution and determines how and where the technology will be used
- enters into licensing agreements to access and hold proprietary technologies within paying royalties and to ensure *freedom to operate* (FTO) for all the components of the technologies
- sublicenses partner institutions:
  - to carry out research, as needed, to adapt the technologies to smallholder farming conditions
  - to test adapted technologies for regulatory compliance
  - to produce and distribute the technologies
- monitors compliance with the requirements of sublicenses to minimize the risk of technology failure
- facilitates the work of appropriate partner institutions to ensure that links in the value chain are connected, are effective, result in technology products that reach farmers, and allow farmers' surplus harvests to reach markets.
- creates partnerships within African countries and with external stakeholders to develop necessary indigenous capacities over time

As implied above and further illustrated in Figure 1, AATF operates along the entire product value chain, from the transfer and adaptation of technology to farmers' access to output markets, with each implementation step undertaken with the relevant partner organizations. The nature of AATF's involvement varies from project to project, depending on the specific requirements and issues that need addressing.

Depending on the needs of African farmers, AATF promotes the development and transfer of all types of technology. The choice of technologies reflects African priorities, is demand driven, and is guided by the potential to improve food security and reduce poverty. AATF gives preference to technologies that are simple, cost effective, and provide sustainable value to the farmer. So far, eight broad areas

FIGURE 1: COMPLETE PRODUCT VALUE-CHAIN

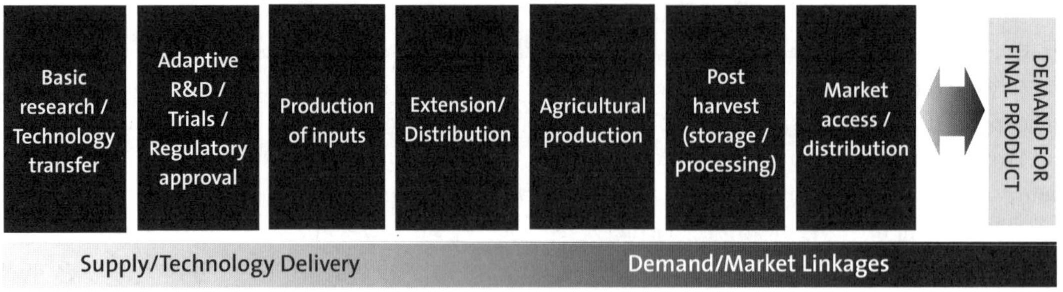

Source: AATF [2]

have been identified as priorities for intervention by AATF:

- *Striga* control in cereals (*Striga* is a parasitic weed)
- improvement of cowpea productivity and utilization
- bananas/plantain productivity
- nutritional quality enhancement in maize and rice
- drought tolerance in cereals
- reduction/elimination of mycotoxins in food grains
- cassava productivity improvement
- insect resistance in maize

AATF's policy is based on the belief that developing countries in Sub-Saharan Africa must make their own decisions about whether or not to adapt and adopt particular agricultural technologies, including genetically modified (GM) organisms. AATF expects that these decisions will be based on appropriate national or regional assessments of the costs, benefits, and social acceptability of each technology. In the case of food biotechnologies, AATF will always require countries that license technologies to have the capacity to manage their safe development and use through appropriate, operational national biosafety regulations and other necessary instruments.

2.3 *Liability and other concerns*

A major concern of AATF's project collaborators, whether they are public entities or multinational companies, is liability exposure once proprietary technologies have been licensed to AATF and subsequently sublicensed to other parties for use in Sub-Saharan Africa.[4] A related concern is the possible misuse of the technology and associated confidential information. AATF has devised the following product stewardship mechanisms to address these concerns. For each project, AATF:

- develops a business plan, which outlines the specific uses of the technology, together with management and oversight protocols that will govern and monitor such use
- conducts risk analyses to aid in formulating and implementing risk mitigation plans

Liability issues can arise due to damage caused by the use of agricultural technologies to persons, property, or the environment, for example, damage that may result from the contamination of seed and organic crop purity. Due to the close proximity in Sub-Saharan Africa of smallholder farmers to one another, pollen from one holding can move easily to neighbouring holdings, contaminating seedlings, produce, and air. Complaints of allergies and health-related problems arising from pollen flow and food consumption have led to liability suits in some countries outside Africa. While this issue is yet to be tested in Sub-Saharan Africa, farmers and biotechnology companies have the responsibility to take steps to ensure, in so far as practicable, that techniques are developed and used to prevent such damage. AATF, through its product stewardship role, helps farmers and companies carry out this responsibility.

AATF is proactive in its role of product stewardship. It ensures that smallholder farmers and research partners comply with all relevant licensing conditions, standards, guidelines, regulatory requirements, and any instructions regarding the use of GM crops. Scientific and technical safeguards are developed for all projects, and stakeholders are advised on the appropriate use of technologies and products.

AATF further protects technology donors from liability through *indemnification* provisions and warranty disclaimers in agreements and by conducting a comprehensive risk analysis for each project. Most not-for-profit organizations are typically averse to providing indemnification in the agreements they sign, but AATF is not a typical not-for-profit organization. On a case-by-case basis, AATF indemnifies technology donors. AATF also uses warranty disclaimers, allowing donors to disclaim guarantees that would otherwise arise by law. AATF's risk analysis procedures identify risks early and allow for the development of risk-mitigation strategies for each project, thus reducing exposure to possible liability claims.

3. MANAGEMENT OF IP

3.1 *Formulating an IP policy*

AATF firmly believes that effective and responsible management of IP starts with the formulation of an IP policy that:

- sets clear objectives and principles of conduct in obtaining access to and use of IP and protected technologies
- establishes guidelines as to how and when IP protection will be sought and exercised
- promotes basic principles concerning the use of IP and protected material by recipients to ensure that this use is consistent with furthering AATF's mission

AATF is formulating a policy to guide the management of IP by both AATF and its project collaborators. The policy seeks to ensure that any knowledge and products that result from AATF projects will be used for the maximum public benefit of resource-poor smallholder farmers in Sub-Saharan Africa.

Through a series of collaborative projects and technology transfers, activities coordinated by AATF are expected to contribute to the development of improved technologies used by resource poor smallholder farmers. Partners in projects coordinated by AATF are required to commit to facilitating a sharing and transfer of technology and research products for both research and commercial use that will benefit resource-poor smallholder farmers.

Further, AATF's IP policy stresses the responsible, respectful use of other's IP rights. Additionally, in the acquisition and management of IP, AATF complies with all relevant international laws and treaties as well as the national laws of the countries in which it operates.

Finally, AATF is guided by its core values of accessibility, accountability, credibility, dedication, transparency, and trustworthiness. Our approach to IP management is best illustrated by the Cowpea Improvement Project that is currently being developed.

3.2 *Implementation and management: case study of the Cowpea Improvement Project*

3.2.1 *Obtaining access to and the right to use proprietary technology (freedom to operate)*

As a *responsible party*, AATF ensures that proprietary technology is properly acquired and used by AATF and its project collaborators in order to achieve the results needed to further AATF's mission. AATF and its partners always endeavor to develop and deploy products that are *free and clear* of restrictions imposed by third-party IP rights. AATF makes genuine efforts to disclose any outstanding restrictions that might apply to such technologies and, where possible, obtains any required permissions.

The Cowpea Improvement Project currently under development best illustrates this commitment (see also Box 1 for further details). AATF has negotiated with Monsanto and obtained a

Box 1: The Cowpea Improvement Project

"Cowpea is the most important food grain legume in the dry savannas of tropical Africa. The legume is consumed by nearly 200 million Africans, provides cash income to smallholder farmers, serves as nutritional fodder for livestock, and provides an ideal way to complement protein-deficient diets."

The overall goal of the AATF cowpea project is to facilitate the development, distribution and adoption of appropriate technologies that will substantially increase cowpea productivity and utilization in Sub-Saharan Africa. In order to achieve this goal, smallholder cowpea farmers in the region need higher-yielding varieties that can perform well under adverse conditions and, in particular, that are genetically resistant to major insect pests, such as the Maruca pod borer. Farmers need to learn and apply new cropping systems that can significantly increase cowpea productivity and profitability.

AATF's role in this project includes negotiating access to the cry1Ab gene, which confers resistance to the Maruca pod borer; providing liability protection to the technology provider; ensuring high-quality seed production and availability; licensing improved seed and technology distribution in Africa, and helping to develop relevant markets. The foundation has supported three consultative meetings with stakeholders that defined project activities, roles, and responsibilities to deliver expected outputs.

Partner Institutions

- African Agricultural Technology Foundation (AATF), Kenya
- Network for the Genetic Improvement of Cowpea for Africa (NGICA)
- Monsanto Company, U.S.A.
- The Kirkhouse Trust, U.K.
- National Agricultural Research Systems (NARS) in West Africa
- International Institute for Tropical Agriculture (IITA), Nigeria
- Commonwealth Scientific and Industrial Research Organization (CSIRO), Plant Industries, Australia.

royalty-free, nonexclusive license to Monsanto technology, a *Bacillus thuringiensis* (Bt) gene (cry-1Ab) for use in the development and deployment within Africa of cowpea (*Vigna unguiculata*) varieties with resistance to the cowpea pod borer (*Maruca vitrata*), in order to provide a sustainable crop for resource-poor farmers of Africa to grow for consumption and sale.

AATF coordinated a comprehensive *technology due diligence*, whose primary objective was to inform on FTO, vis-à-vis research, to produce improved cowpea cultivation, harvesting, and storage processes, and to improve use and consumption of the final product. Achieving this objective required taking an inventory of all technologies to be used in the project, completing a search and analysis of patent databases for filed or issued patents, and preparing a report analyzing the potential strengths, weaknesses, opportunities, and threats related to the project.

The FTO assessment helped to determine the ownership rights in the gene and other component technologies that promoters needed to develop the improved cowpea variety. In the future the assessment will serve as a guide to AATF and its project collaborators to ensure that the technologies used do not infringe the IP rights of the owners. Further, the assessment will serve as a basis for seeking all required permissions from the owners of the technologies, thus removing, or at least reducing, the potential for IP infringement should a product be exported from Sub-Saharan

Africa into a territory where third-party IP rights are in place.

In line with good IP management practice, AATF will keep the FTO up-to-date by utilizing existing "Watch Lists" to track applicable patent and litigation trends.

3.2.2 *Preserving the confidentiality of IP and related project information*

AATF considers it good IP management practice to preserve the integrity of confidential information contained in third-party IP, IP resulting from AATF-coordinated projects, and general project information. Therefore, AATF includes a confidentiality clause in all employment contracts and stresses compliance with this clause as a condition of continued employment for AATF personnel. Further, AATF advocates that its project collaborators require personnel involved in any AATF-coordinated project to sign confidentiality agreements. Finally, AATF routinely enters into nondisclosure agreements with its collaborators to facilitate the free exchange of information and materials, including IP, and to preserve the integrity of confidential information at the institutional level.

3.2.3 *Defining ownership rights*

Good IP management requires that all ownership rights are defined at the start of any engagement, taking into consideration any attendant responsibilities, including liability and risk management. The rights of AATF employees are defined in an employment contract, which stipulates that any rights (intellectual or tangible property) in research products, publications, and other works created or contributed to by AATF personnel in the course of their normal and assigned professional duties will be vested in AATF.

The ownership rights of AATF and technology providers are negotiated and determined on a project-by-project basis. For instance, in the Cowpea Improvement Project, Monsanto will retain its existing IP rights, while AATF will own all right, title, and interest associated with any improvement realized through the use of Monsanto's technology under the terms of the license agreement.

The ownership rights of AATF and of project collaborators are negotiated on a project-by-project basis with the goal of equitably sharing such rights. The goal is achieved by taking into consideration the following principal factors:

- the intellectual contribution of each partner to the particular project (foreground IP)
- the contribution of IP, materials, research effort, and preparatory work of each partner brought to the project (background IP)
- the facilities provided by each partner
- the financial contribution of each partner
- other considerations determined by the partners to be relevant

Any rights (intellectual or tangible property) in research products, publications, and other works commissioned by AATF will be assigned and vested in AATF. Any rights (intellectual or tangible property) in research products, publications, and other works jointly commissioned by AATF and the project collaborators will be assigned to and vested in AATF and the project collaborators as joint right holders.

3.2.4 *Execution of agreements*

AATF believes it is essential and indeed good IP management practice to finalize all contractual terms, set them out in writing, and have an agreement duly signed by the authorized representatives of the parties before commencement of any engagement. Therefore, AATF ensures that all arrangements with third parties associated with the access to or the creation, use, or exploitation of IP protected materials are appropriately documented. Documentation for the Cowpea Improvement Project, for example, will, in the end, involve several agreements between AATF and its collaborating partners. First, AATF obtained a license from Monsanto, and thereafter sublicensed the licensed Bt gene to CSIRO and IITA in order to introduce the Bt gene into the cowpea genome. The AATF, potentially, will sublicense the resulting successful transgenic events to African agricultural research institutions, which will introgress the Bt gene in cultivated cowpea varieties. These varieties would then be licensed to commercial, nongovernment, humanitarian, or public institutions charged with disseminating the improved cowpea varieties in Africa.

3.2.5 *Identification of IP assets*

Maintaining IP asset inventories or a register of IP assets is essential for effectively managing those assets. AATF and its collaborating partners encourage the adoption of procedures and practices—such as DNA fingerprinting, the keeping of appropriate laboratory notebooks, and controls over the release of information—to properly identify, record, safeguard, and manage IP generated under projects coordinated by AATF.

3.2.6 *Publication of project research results*

AATF anticipates facilitating access to and use of improved germplasm and research products for the public benefit through publication and public disclosure. Therefore, to the extent determined appropriate and feasible by AATF and its project collaborators, research outputs and products from AATF projects will be placed in the public domain.

3.2.7 *Statutory protection of IP*

In certain cases, statutory IP protection may be necessary to ensure the continued availability of germplasm, inventions, publications, and databases to AATF and its partners. Such protection may also be needed to provide AATF with the necessary leverage to negotiate access to other proprietary rights and technologies required for product development. Therefore, in appropriate cases, AATF may seek IP protection for products (termed improvements) generated from projects for which the foundation has obtained ownership rights. For instance, as noted earlier, AATF will own all right to, title to, and interest in any improved cowpea varieties or other improvements developed using Monsanto's technology. In consultation with the project collaborators, AATF may seek to protect these improvements by obtaining IP protection through patents, plant breeders' rights, copyrights, trademarks, statutory invention registrations or their equivalent, and/or trade secrets.

In seeking IP rights,[5] AATF will be guided by its commitment to serve African resource-poor smallholder farmers—not by opportunities to obtain revenues. To the extent that IP licensing generates financial returns, they will be used by AATF and the project collaborators to achieve AATF's charitable objectives. AATF will ensure that all third-party licenses to the improvements make provisions for:

- ready access by others for humanitarian use
- avoidance of possible restrictions arising from "blocking" patents and ensuring the project collaborators' ability to pursue research without undue hindrance
- the transfer of technology, research products, and other benefits to African resource-poor smallholder farmers through public channels and, where appropriate, through the commercialization or utilization of research products.

With regard to the protection of cells, genes, molecular constructs, plants, varieties, and traits, AATF and its project collaborators will, to the extent permitted by applicable law, consider the effects that protection has on the distribution of, use of, and access to the protected product before proceeding with an application for statutory protection.

AATF and its project collaborators may allow third parties to take IP rights on research products or material derived from research products if it is determined that doing so would best serve the public good. In such cases, AATF and its collaborators will ensure that agreements granted to recipients to protect intellectual property do not in any way waive the rights of AATF and the collaborators to challenge excessive protection through administrative and/or court proceedings. AATF and its collaborators may also reserve the right to retain research products for use by AATF and its collaborators, and they may also enter into agreements to deploy research products in a targeted manner to certain partners and/or in certain markets.

3.2.8 *Publications, databases, reports, training material, public awareness material, artwork, audio-visual material*

AATF encourages the wide dissemination of publications (printed and electronic), including databases, reports, training materials, public awareness material, artwork, and audio-visual material to be used for maximum public benefit. For instance, AATF and its project collaborators have issued publicity materials, including press releases in English and three Kenyan local languages (Kiswahili, Dholuo and Luhya), to help publicize the deployment of Imidazolinone Resistant (IR) maize technology in the western part of Kenya. Named *Ua Kayongo* ("kill *Striga*" in English), it will help to control the parasitic weed *Striga.*

In creating such publicity materials, AATF and its project collaborators seek to use the copyright material of others only within "fair use" limitations, or with the consent of the copyright owner, and to properly attribute the source of the material.

AATF and the project collaborator publications (printed and electronic) will normally carry standard copyright notices that indicate AATF and/or project collaborators as the copyright owner(s) of the compilation (for the specific edition and year of publication).

AATF and the project collaborators will generally incorporate standard copyright notification statements in their publications:

- permitting, especially in the case of the National Agricultural Research Systems (NARS), the making of a reasonable number of copies of such copyrighted material for noncommercial purposes
- requiring attribution where such copyright material is reproduced in other publications
- prohibiting interference or tampering with the material without the express consent of AATF and/or the project collaborators
- addressing any other issues relevant to the best use being made of the material, such as procedures for the dissemination and recall of material subject to updating

AATF and its project collaborators may, to the extent available in national laws, enforce the copyrights in such publications (printed and electronic) and protect them from unfair competition in order to:

- respond to a breach of the above terms
- prevent misappropriation of such material for commercial purposes
- protect the integrity of such material

To the extent practicable, AATF will develop databases that assist the resource-poor and will make best efforts to keep these databases in the public domain.

3.2.9 *Trademarks*

AATF and the project collaborators may register as trademarks all distinctive marks associated with their initiatives, in order to protect the goodwill and reputation associated with the use of these marks by AATF and its collaborators.

4. CONCLUSIONS

Conventional methods for technology development and transfer have not always sufficiently supported sustainable food security and contributed to the alleviation of rural poverty in Sub-Saharan Africa. Although there have been numerous attempts in the past to promote public/private partnerships in the region, most have had little tangible or lasting effect. It has become increasingly obvious that new approaches are needed to mobilize new science for new applications in Africa. It is also increasingly obvious that developing these approaches will require the potential complementarities of public and private sector research and development efforts.

AATF represents an innovative approach based on forging collaborations between these sectors to identify and transfer proprietary technologies that would otherwise be unavailable for trying to address the problems of resource-poor smallholder farmers. AATF is surely not the only possible answer or a "silver bullet." And it may not be the only or even the best means to achieve the goal of easing access to important technologies for humanitarian purposes. But its African focus, leadership, and operational location promise a more comprehensive, realistic appreciation of the constraints to technology transfer in Africa, which will allow for the design of more feasible solutions and closer follow-up and continuity in implementation. A wide range of stakeholders in the private and public sectors and in civil society have already pledged their commitment to making the AATF concept work, and AATF seeks to retain the confidence of these stakeholders through effective leadership and responsible IP management. ■

ACKNOWLEDGMENTS
We thank Dr. Hodeba Mignouna (Technical Operations Manager, AATF), Dr. Francis Nang'ayo (Regulatory Matters Specialist, AATF), Dr. Gospel Omanya (Projects Manager and Seed Systems Specialist, AATF) and Mrs. Nancy Muchiri (Communications and Partnerships Manager) for their critical input into preparing this chapter. Any errors or misrepresentations remain the authors'.

RICHARD Y. BOADI, *Legal Counsel, AATF, c/o ILRI, P. O. Box 30709, Nairobi 00100, Kenya.* r.boadi@aatf-africa.org

MPOKO BOKANGA, *Executive Director, AATF, c/o ILRI, P. O. Box 30709, Nairobi 00100, Kenya.* m.bokanga@aatf-africa.org

1 Omanya, G, R Boadi, F Nang'ayo, H Mignouna and M Bokanga. 2005. Intellectual Property Rights and Public/Private Partnerships for Agricultural Technology Development and Dissemination. Paper presented at the Kenya National Conference on Revitalizing the Agricultural Sector, 21–24 February 2005, Nairobi, Kenya.

2 Delmer, DP, C Nottenburg, GD Graff, and AB Bennet. 2003. Intellectual Property Resources for International Development. *Plant Physiology* 133:1666–70. www.pipra.org/docs/Plant Physiology - IP.pdf.

3 AATF.2002. *AATF Business Plan*. Cambridge Economic Policy Associates, Ltd.: Cambridge, U.K.

4 See, also in this *Handbook*, chapter 14.5 by RY Boadi.

5 Should the need arise, AATF could utilize the Patent Cooperation Treaty (PCT) process, administered by the World Intellectual Property Organization (WIPO), which offers inventors and industry an advantageous route for obtaining patent protection internationally. By filing one "international" patent application under the PCT, protection can be sought simultaneously in any of the 130 PCT member countries designated in the application.

AATF could also use African regional filing mechanisms such as the African Regional Intellectual Property Organization (ARIPO) and/or the African Intellectual Property Organization (OAPI), wherein one application could result in the grant of an IP right in multiple countries. ARIPO currently has 15 member states: Botswana, the Gambia, Ghana, Kenya, Lesotho, Malawi, Mozambique, Sierra Leone, Somalia, Sudan, Swaziland, Tanzania, Uganda, Zambia, and Zimbabwe. Applicants can file their applications with either their national offices or directly with the ARIPO office. Under this system, one application is effective in all member states designated in the application. OAPI is the central registration system for 16 French-speaking African countries: Benin, Burkina Faso, Cameroon, Central African Republic, Chad, Republic of the Congo, Cote d'Ivoire, Equatorial Guinea, Gabon, Guinea, Guinea Bissau, Mali, Mauritania, Niger, Senegal, and Togo. Under the OAPI system, the IP rights of an applicant are simultaneously protected in all member states through a single deposit, which is considered as a national deposit for each member state.

For countries such as Liberia, Madagascar, Seychelles, and South Africa, which operate the "national route only" system of registration, AATF may have to apply to the respective IP offices. The same would apply in the case of Sub-Saharan Africa countries that are non-WIPO members (Angola, Burundi, Cape Verde, Comoros, Democratic Republic of the Congo, Djibouti, Eritrea, Ethiopia, Mauritius, Nigeria, Rwanda, São Tomé e Príncipe, and Somalia).

Pragmatic and Principled: DND*i*'s Approach to IP Management

JAYA BANERJI, *Communications Manager, DNDi, Switzerland*
BERNARD PECOUL, *Executive Director, DNDi, Switzerland*

ABSTRACT

The mission of the Drugs for Neglected Diseases initiative (DND*i*) is to develop safe, effective, and affordable new drugs for patients suffering from neglected diseases and to ensure equitable access to these drugs. DND*i* believes that intellectual property (IP) rights should not pose a barrier to access to these medicines. Hence, a balanced approach to IP management is critical for effective implementation of DND*i*'s mission. The organization has written an IP policy that both encapsulates and articulates DND*i*'s approach to IP based on core principles and beliefs. The policy reflects the DND*i* philosophy, vision, and mission, ensuring that its products are accessible and affordable to patients who need them most. DND*i* recognizes the reality of IP and seeks to implement its humanitarian mission using best, pragmatic practices for IP management. Indeed, DND*i* has already demonstrated that this is feasible, having successfully negotiated with both private and public sector institutions in order to actualize its principled mission.

1. INTRODUCTION

In 1999, a need was identified for an alternative method to research and develop new drugs for infectious, tropical diseases. The doctors of Médecins Sans Frontières (MSF) gave testimony to the fact that the handful of drugs (to treat such diseases) that did exist was inaccessible to patients suffering from the diseases. Most of the drugs still have to be delivered in hospital situations, which is difficult where health care is rudimentary. In addition, the medicines are unaffordable, as a consequence of the need to recoup the supposedly high costs of researching and developing the drugs and the need for pharmaceutical companies to make a profit. These drugs were, and are, not only out of reach for individual patients but also for governments of disease-endemic countries. Intellectual property (IP) rights are among the factors driving these high prices, leaving patients in the developing world to their own limited resources and, ultimately, to undesirable outcomes with disability and death as the worst consequences.

The statistics show that hundreds and thousands of disadvantaged people in developing countries are suffering their diseases in silence. These patients are unable to afford even the (largely inadequate) existing treatments, most of which have toxic side-effects, are ineffective, and need to be delivered in hospital conditions. These patients, though they urgently need new, safe, and field-adapted medicines, do not constitute lucrative markets that the current drug R&D model targets, hence the plight of these patients remains unanswered (Figure 1). Of the 1,556 new drugs that came to the market from 1975 to 2004, only 21 (1.3%), were for tropical diseases such as human African trypanosomiasis, Chagas' disease, leishmaniasis, helminthic infections, schistosomiasis, onchocerciasis, malaria and tuberculosis – diseases that account for 12% of the global

Banerji J and B Pecoul. 2007. Pragmatic and Principled: DNDi's Approach to IP Management. In *Intellectual Property Management in Health and Agricultural Innovation: A Handbook of Best Practices* (eds. A Krattiger, RT Mahoney, L Nelsen, et al.). MIHR: Oxford, U.K., and PIPRA: Davis, U.S.A. Available online at www.ipHandbook.org.

disease burden. Ten of the 21, including four of the five developed since 1999, were marketed for malaria and tuberculosis.[1] This fatal imbalance is responsible for the deaths of more than 35,000 people a day.

The Drugs for Neglected Diseases initiative (DND*i*) firmly believes that drug research can be an activity in the public domain that leads to the advancement of health, and recognizes that patented products do not always benefit those who need them most. DND*i* considers its products as public goods. It does not wish to profit from its new products and wants to share the knowledge it creates by transferring technology to other researchers and manufacturers when required. The R&D process requires access to knowledge from both private and public research organizations so that DND*i* can use the best available science to research and develop new drugs for neglected diseases. Based on its core principles and beliefs, DND*i* has crafted an IP policy that pragmatically captures the organization's philosophy, vision and mission, and, thereby, ensures that products offered by DND*i* are made accessible and affordable to patients who need them most (see Box 1 at the end of this chapter).

2. DND*i*'S VISION AND MISSION

DND*i* was set up to address the imbalance in access to critically needed medicines, by giving patients in developing countries the opportunity to be the direct beneficiaries of new products of drug R&D for diseases that do not represent a viable drug market. DND*i*'s mission is to develop safe, effective, and affordable new drugs for patients suffering from neglected diseases (Figure 1) and to ensure equitable access to these. By 2014, it aims to develop and make available six to eight such field-relevant treatments.

This, of course, is easier said than done, primarily because to most scientists, pharmaceutical companies, and institutions with whom DND*i* collaborates, the idea of placing a potentially commercial product, such as a drug, into the public domain is both novel and bizarre. DND*i*'s "no profit, no patents" stance calls for, and is committed to, a significant amount of long and sensitive

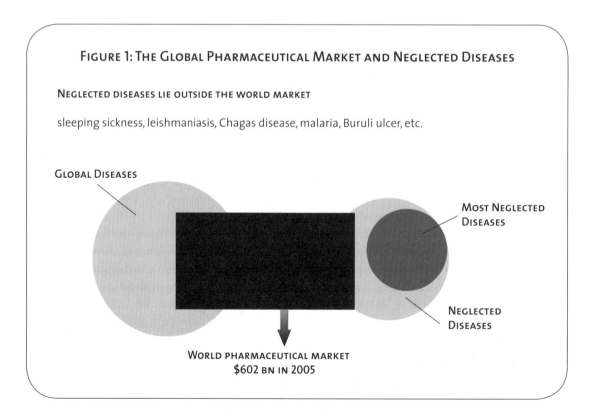

Figure 1: The Global Pharmaceutical Market and Neglected Diseases

Neglected diseases lie outside the world market

sleeping sickness, leishmaniasis, Chagas disease, malaria, Buruli ulcer, etc.

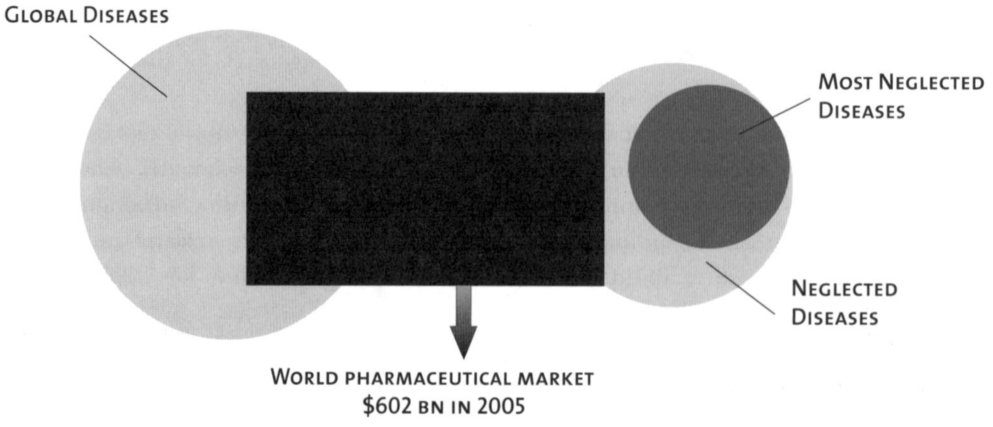

negotiation, to ensure that it becomes more acceptable and widely supported.

Furthermore, whereas the response to calls, from numerous civil society groups and nongovernmental organizations (NGOs), to dispense with patents for essential health tools has been met in some spheres with scorn and disregard, certain organisations are slowly beginning to agree that patents often hinder R&D. For example, the recent 2006 report from the World Health Organization (WHO) Commission on Intellectual Property, Innovation, and Public Health (CIPIH) explicitly stated, "*There is no evidence that the implementation of the TRIPS* [Trade Related Aspects of Intellectual Property] *Agreement*[2] *in developing countries will significantly boost R&D in pharmaceuticals on Type II and particularly Type III [neglected] diseases.*[3] ***Insufficient market incentives are the decisive factor.***"

This assertion is echoed loud and clear in the preamble of the recently adopted (27 May 2006) World Health Assembly resolution (WHA 59.24) titled Public Health, Innovation, Essential Health Research, and Intellectual Property Rights: Towards a Global Strategy and Plan of Action which notes, "*… that intellectual property rights are an important incentive for the development of new health-care products; … however …* ***this incentive alone does not meet the need for the development of new products to fight diseases where the potential paying market is small or uncertain… .***"

Nevertheless, DND*i*'s approach to IP, although highly principled, is quite pragmatic (as articulated in its IP policy), and will, therefore, contribute to encouraging further innovations and, even more importantly, to ensuring that patients have access to new products. At the heart of DND*i*'s IP policy lies the belief that the lives of neglected patients are more important than the profit motive. However, convincing industrial and academic partners that this belief should influence their investment decisions continues to be a challenge. Decisions regarding the possible acquisition of patents, ownership, and licensing terms will be made on a case-by-case basis.

3. NEGOTIATING WITH INDUSTRY

DND*i*'s first opportunity to put into action its vision of IP came in October 2003 when it established preliminary contact with the French pharmaceutical giant sanofi-aventis regarding artesunate-amodiaquine, a fixed-dose artesunate-based combination therapy (FACT) for chloroquine-resistant malaria. Artesunate combination therapies are considered an important addition to the arsenal of treatments for chloroquine-resistant malaria, especially in Africa, where more than a million children die each year from the disease. FACT was one of DND*i*'s earliest projects operating under a grant from the European Commission's International Cooperation and Development (EC-INCO DEV) programme. The terms of its contract with INCO-DEV described collaboration with an industrial partner (sanofi-aventis) for industrial validation, production, and distribution.

In keeping with these terms, DND*i* established preliminary contact with sanofi-aventis, offering them the stable artesunate/amodiaquine fixed-dose combination (AS/AQ) for completion of development and industrial scale-up and to make the registered medicine available to malaria patients. At the time, the drug company already had a combination AS/AQ on the market but it was not a fixed-dose combination (two drugs in one tablet), as was the one DND*i* had been able to develop.

Negotiations led, finally, to a contract agreeable to both parties that was signed in December 2004, and DND*i* transferred its AS/AQ dossier to sanofi-aventis. The agreement was an innovative breakthrough. Both parties wished to make this easier-to-use combination available to the poorest patients, at an affordable price and manufactured to highest international quality standards. At its own cost, sanofi-aventis took full responsibility for the drug registration, as well as for the constitution of a WHO pre-qualification file.

DND*i* was able to convince sanofi-aventis to agree to *exclusivity* of AS/AQ until first registration in a 'reference state'[4] or prequalification by WHO,[5] after which the drug would be *non-exclusive* and available for production by any generic manufacturer without paying either sanofi-aventis or DND*i* for the right to do so. The agreement also stated that sanofi-aventis would supply the drug at cost (as a *generic*) to the public sector, NGOs (for example, MSF), and international organizations, such as WHO and the United Nations Children's Fund (UNICEF). Under the terms of the agreement, sanofi-aventis could market the product under a trade name in all territories—including disease-endemic countries where the generic product would be available—in the private sector (pharmacies) at a commercial price. For the information and the data made available to the private sector by DND*i*, sanofi-aventis would pay DND*i* a fee, amounting to 3% of net sales, for seven years after launch of the product. DND*i* has decided to use this amount to further reduce the price of AS/AQ to the public sector.

This was DND*i*'s first success. Each negotiation that followed for other projects was an equally uphill task as illustrated in the following example of a research agreement with the University of California, San Francisco.

4. NEGOTIATING WITH ACADEMIA

Following the biotechnology boom during the last couple of decades, most universities see considerable financial potential in much of their medical research. University research departments now ensure that if this research is to be licensed to outside partners, then proceeds from any commercialization flow back to the university and the inventors. Protection of IP by the university is central to this, together with ensuring that the best commercial license is negotiated with the partner if a marketed product is the likely outcome of the research.

When DND*i* first approached the University of California, San Francisco (UCSF), it sought to support research that might lead to new treatments for human African sleeping sickness. Both parties found ready agreement in the use of IP for research purposes. However the goal of DND*i* was to commercialize the product of the research in a way that makes it accessible to patients, and its marketing strategy is somewhat contrary to normally accepted practice in the United States. DND*i* aims to manufacture and sell its products for the lowest price possible. It would have been difficult for UCSF to find a less attractive partner!

At the start, major issues of contention were the requests by DND*i* for:

- a royalty-free license to develop drugs arising from the research for commercialization in all disease-endemic countries
- freedom to manufacture the drugs in any country
- freedom from the requirement to patent the research outcomes for commercialization in any of the disease-endemic countries (Patents can add several million dollars to the cost of a drug.) UCSF retains the right to patent for other uses but not in a manner that will restrict DND*i*'s use of the research.

Throughout the protracted negotiations, staff at the university's business development department were supportive of DND*i*'s IP policy and commercial goals. The main obstacle was simply the difficulty faced by the legal representatives when asked to step away from the standard pro forma protocol and negotiate an agreement that flew in the face of their obligation to negotiate the best return on IP. Fortunately, at the end, a compromise was reached that favoured DND*i* very strongly, and all its requests were met. Equally gratifying was the comment from the staff of the UCSF commercialization departments that they gained tremendous personal satisfaction from the terms of the contract and from being involved in making new treatments available for the seriously neglected disease.

DND*i* has learned some lessons from this experience. In many instances, legal opinions were drafted by third parties who were not familiar with the mission of DND*i*, which slowed the pace of negotiations. Furthermore, the people with whom DND*i* held negotiation talks agreed with the organization's goals but often did not convey them effectively to outside legal representatives and other decision makers. During the final negotiations, however, DND*i* interacted with all decision makers directly—their strong support of DND*i*'s goals is reflected in the final draft of the agreement.

4. CONCLUSIONS

As clearly articulated in its IP Policy statement, DND*i* is committed to managing IP in a manner that pragmatically and effectively advances its mission of providing the most vulnerable populations in developing countries with equitable access to critically needed medicines. Perhaps this is most clearly stated in the preamble of the DND*i* IP policy statement:

The DNDi IP approach will be pragmatic, and decisions regarding the possible acquisition of patents, ownership, and licensing terms will be made on a case-by-case basis. DNDi will put the needs of neglected patients first and will negotiate to obtain the best possible conditions for them. The DNDi's decisions regarding IP will contribute to ensuring access and encouraging further innovations.

By taking this realistic, yet creative, view of IP, DND*i* seeks to advance best practices in IP management that will directly address global public interest. More importantly, by engaging in sophisticated, successful negotiations with both the public and private sectors to fulfil its dynamic vision, DND*i* has demonstrated that this mission and policy is not simply an academic exercise. These negotiations skills, based on the foundation of the DND*i* IP Policy statement, will ultimately ensure the implementation of DNDi's mission with long-term benefits accruing to those who most need, yet can least afford, essential medicines. ■

JAYA BANERJI, *Communications Manager, DNDi, 1 Place St. Gervais, 1201 Geneva, Switzerland. Currently: Medicines for Malaria Venture, 20, route de Pré-Bois, 1215 Geneva 15, Switzerland. banerjij@mmv.org*

BERNARD PECOUL, *Executive Director, DNDi, 1 Place St. Gervais, 1201 Geneva, Switzerland. bpecoul@dndi.org*

---

1 Chirac P and E Torreele. 2006. Global Framework on Essential Health R&D. *The Lancet* 367 (9522):1560–61.

2 The TRIPs Agreement of the World Trade Organization (WTO) is an agreement that addresses intellectual property concerns. It provides a set of minimum standards for intellectual property protection to which all but the poorest member countries of the WTO must conform.

3 Type II diseases are incident in both rich and poor countries, but with a substantial proportion of cases in the poor countries, for example, HIV/AIDS and tuberculosis. Type III diseases are those that are overwhelmingly or exclusively incident in the developing countries such as sleeping sickness or African river blindness. Type II diseases are often termed *neglected diseases* and Type III, *very neglected diseases*.

4 Countries that have stringent regulatory requirements are considered "reference states." Registration of a drug in a reference state facilitates its approval in other countries that do not have the regulatory infrastructure to fully assess a new registration dossier.

5 Prequalification is a rigorous process of review and approval of the quality, safety, and efficacy of drug products conducted by the World Health Organization at the request of the manufacturer. It was originally intended to give United Nations procurement agencies a guarantee of quality and now extends to other bulk purchasers, including countries and NGOs.

Box 1: DNDi's IP Policy

DND*i* hereby adopts the following intellectual property (IP) policy:

I. Preamble
The mission of DND*i* is to develop safe, effective, and affordable new treatments, for patients suffering from neglected diseases, and to ensure equitable access to these.

The DND*i* IP policy will be guided by the following principles as laid down in the business plan:

1. The need to ensure that drugs are affordable to and access is equitable for patients who need them
2. The desire to develop drugs as public goods when possible

The DND*i* IP approach will be pragmatic, and decisions regarding the possible acquisition of patents, ownership, and licensing terms will be made on a case by case basis. DND*i* will put the needs of neglected patients first, and will negotiate to obtain the best possible conditions for them. The DND*i*'s decisions regarding IP will contribute to ensuring access and encouraging further innovations. DND*i* regards drug research as a public good that should primarily lead to the advancement of health.

In addition to a pragmatic day-to-day approach on IP, the DND*i* is committed to contributing to the thinking and development of IP approaches in health R&D that are aimed at serving the public good.

II. Definitions
For the purpose of this policy, the term "intellectual property" includes, but is not limited to, intangibles that are protected by the principles of patents, copyrights, trademark, and trade secrets.

III. Intellectual Property and DNDi's Work
Basic Principles

In implementing the IP strategy, DND*i* will adhere to the following basic principles:

1. DND*i* will ensure that the results of the work carried out under its auspices are disseminated as widely as possible and its products made readily available and affordable in developing countries.

   Where the acquisition of IP is not necessary to promote its mission and goals, DND*i* will make all possible efforts to ensure that the results of its work are placed and remain in the public domain. However, it is possible that promoting DND*i*'s mission and goals will sometimes require outputs to be protected by IP (*see* Sections IV and V). Given the costs involved, patenting is likely to be the exception rather than the rule. Other nonpatent types of IP such as confidential information ("trade secrets") and copyrights will also need to be considered.

2. To make the results of its work useful and encourage the research community to engage in additional or follow-on research in the field of neglected diseases, DND*i* will seek—whenever possible, and without undermining its rationale for acquiring IP—to disseminate its research through publications, presentations, the Internet (emulating the Human Genome Project) and other appropriate channels.
3. DND*i* does not seek to finance its research and operations through IP rent revenues. Although they will constitute an exception rather than the rule, patents might be sought to strengthen DND*i*'s ability to ensure control of the development process and to negotiate with partners.

(Continued on Next Page)

Box 1 (continued)

4. IP is generated through DND*i*-sponsored research projects, it should be used to achieve DND*i*'s mission. To this end, DND*i* will pursue creative and innovative strategies to make the fruits of research projects readily available to patients affected by neglected diseases. This will require avoiding prohibitively costly approaches, restrictive IP strategies, or other issues that may inhibit or delay the rapid adoption of the invention to the benefit of developing countries.

IV. Rationale for Acquiring or Otherwise Dealing with Intellectual Property
DND*i* recognizes that in pursuing its mission it may find it necessary to acquire or otherwise manage and enforce IP. In this regard, DND*i* acknowledges that it will have to deal with IP to:

1. conclude contracts and undertake research with its research partners, contractors, collaborators and founders;
2. obtain rights to work on and develop molecules, including facilitating DND*i*'s or its partners' access to proprietary research materials;
3. ensure equitable access to, and affordability of, the end products of its research for patients.

V. Acquisition, Management, and Enforcement of Intellectual Property
Where it is considered necessary to acquire or otherwise manage IP, DND*i* will put in place measures to ensure the timely acquisition of IP by itself or its project partners, collaborators or founders for and on behalf of DND*i*. When necessary to achieve DND*i*'s objectives, enforcement may include legal actions to protect the DND*i* IP.

DND*i* will ensure that IP, however acquired, allows the initiative full freedom to operate, including retaining the right to use the inventions on which IP is obtained for DND*i*'s further research, including with other partners. To this end, DND*i* will use various mechanisms such as assignment of the IP to DND*i*, exclusive licenses and licenses of right. It will negotiate terms with partners to ensure that they will not use the acquired and/or held IP in a manner that impedes equitable and affordable access to the products of the research, or that impedes additional or follow-on research by DND*i*, its partners and other researchers, especially those undertaking research on neglected diseases.

DND*i* will not accept projects in which IP is obviously going to be an insurmountable barrier to follow-up research on behalf of DND*i* and/or equitable and affordable access. Either at the onset of a project or when problems arise, it will be important that negotiations with the public and/or private sector are backed with advocacy support.

VI. Transfer and Licensing of Intellectual Property
DND*i* seeks to enhance R&D activities for neglected disease therapeutics and may wish to in-license technologies developed by others that would help bring such products to the public. To ensure the availability and affordability of neglected disease therapeutics, it will transfer or out-license its technologies to facilitate manufacturing and distribution of its products. As a general policy:

1. DND*i* will ensure that the terms of each transfer or licensing agreement take into consideration the impact of the technology on research in medicine, and more broadly, public health; the level of support provided by DND*i*; the stage of scientific and clinical development of the technology; DND*i*'s portfolio and drug pipeline requirements; and timing and other business and economic considerations;
2. DND*i* will ensure that the terms and conditions of any licensing or transfer agreement allow the continuing availability of technology that supports further research in the field of neglected diseases;

(Continued on Next Page)

Box 1 (continued)

3. DND*i* will ensure that technologies developed under DND*i* sponsorship are brought to practical application in a timely manner and made affordable and accessible to the public;
4. DND*i* will negotiate and award licenses which may be exclusive, for specific indications, fields of use, or geographic areas, and other terms as circumstances allow;
5. DND*i* will monitor the performance of licensees and ensure that licensed technology is fully developed;
6. DND*i* will develop and use model agreements, where appropriate, to enable alternative forms of dispute resolution and therefore avoid litigation.

VII. Communities' Involvement in DNDi's Research and Benefit Sharing
When DND*i* will consider patenting an invention resulting from work with communities on traditional medicine or on community genetics, that community will be assured of receiving all eventual benefits from this work.

VIII. Amendments and Changes to the Policy
DND*i* retains the right to review, revise and/or amend this policy or any of its terms at its discretion, at any time. When warranted and in agreement with the Chair of the Board, the Executive Director will recommend the review, revision or amendment of this policy for further approval of the DND*i* Board of Directors.

IX. Administration and Implementation of the Policy
The Executive Director will ensure the full implementation of this policy and put in place, subject to Board approval, administrative, financial, technical, and other mechanisms and procedures to ensure its full implementation.

CHAPTER 17.20

From Science to Market: Transferring Standards Certification Know-How from ICIPE to Africert Ltd.

PETER MUNYI, *Chief Legal Officer, International Centre of Insect Physiology and Ecology (ICIPE), Kenya*
RUTH NYAGAH, *Chief Executive Officer, Africert Limited, Kenya*

ABSTRACT

This brief case study describes how the International Centre of Insect Physiology and Ecology (ICIPE) helped African growers maintain access to foreign markets and improve livelihoods by being able to achieve standards certification for agricultural export commodities. The process involved a characterization of the problem and a conceptualization and execution of a solution. The solution included creating a regional certification body in East Africa capable of providing globally recognized certification at costs that were locally affordable. The level of technical know-how needed by the certification body in order to be effective was significant, so the expertise of ICIPE was instrumental in creating the local certification body. Ongoing certification services provided by the certification body are highly market oriented, and because of this orientation the group was spun off as a private company, as Africert Limited.

1. INTRODUCTION: NEW CERTIFICATION REQUIREMENTS

For a long time, smallholder farmers in developing countries, including Kenya, have experienced difficulty in accessing international markets for goods produced on their farms. Whereas most of the factors involved have been attributed to archaic production and processing systems that invariably increase costs of production, other factors have recently been implicated. They involve new legal and private (consumer and market) requirements (or industry standards) for food safety, traceability, maximum residue limits (MRLs) for pesticide levels in food products, ethical and social issues in agricultural production methods, north-south market chains, and the environmental sustainability of commercial agricultural production.

Although, the global trend has been toward freer markets with fewer economic trade barriers, emerging trade standards (both legislated standards and private standards) have the potential to act as nontariff barriers to trade, between African growers and European markets, for agricultural products.

From the late 1990s, both large- and small-scale producers of export products in Africa found themselves faced with new consumer standards alongside the established ones. These standards all required separate verification (certification of conformity) from independent entities. However, these requirements invariably involved high costs related to implementation of the standards (both in terms of capacity and structures) and to their independent certification. Most farmers, particularly in the horticultural sector, found themselves faced with a possibility of being locked out of the very markets from which they were deriving their livelihoods.

2. THE SOLUTION: A LOCAL CERTIFICATION BODY IN EAST AFRICA

2.1 *The concept*

To address this problem, in 2001 the German international cooperation agency (best known

Munyi P and R Nyagah. 2007. From Science to Market: Transferring Standards Certification Know-How from ICIPE to Africert Ltd. In *Intellectual Property Management in Health and Agricultural Innovation: A Handbook of Best Practices* (eds. A Krattiger, RT Mahoney, L Nelsen, et al.). MIHR: Oxford, U.K., and PIPRA: Davis, U.S.A. Available online at www.ipHandbook.org.

through its German acronym GTZ [Deutsche Gesellschaft für Technische Zusammenarbeit GmbH]) developed the concept of facilitating the creation of a "*local certification body for products from organic agriculture in East Africa.*" The mandate to develop and implement this concept was given to the International Centre of Insect Physiology and Ecology (ICIPE)[1] with a view to ultimately establishing a regional certification body for organic products in Africa, able to offer internationally recognized certification services to small-scale producers at locally competitive costs. The terms of reference under the project included:

- identifying stakeholders in Kenya, Tanzania, and Uganda
- identifying possible business partners
- elaborating and modifying regional standards for organic agriculture
- elaborating and implementing a quality management system according to ISO 65[2] and EN 45011[3]
- establishing and publicizing the regional certification body among possible clients within the region
- monitoring and evaluating the local projects' progress.

Execution of the terms of reference entailed several key aspects, one of which was identification and training of personnel who would be able to undertake the duties of the certification body. Around the same time (from January 2004), EurepGAP[4] was seeking to extend its standards to the horticulture and floriculture industries of East Africa, particularly Kenya. This was seen as an opportunity through which the intended activities could actually be carried out. As a result, training small-scale farmers in the horticulture and floriculture industries formed a key platform activity from which it was then possible to initiate the launch of a certification body.

2.2 *The creation of Africert*

Upon successful completion of the EurepGAP training, the next step involved formation of an independent company to carry out the certification process. Africert Ltd. was thus incorporated, in November 2003, with its main objectives being to carry on, either alone or with others in Kenya and elsewhere, in providing certification services and operating certification systems and processes, as well as quality assurance services; to carry on, in any part of the world, the activities of a certification company, testing products and suppliers' quality systems and surveillance, and testing product samples, with a view to ensuring that the products tested, or certified, meet national or international standards, specifications, or technical regulations.

A key condition for the formation of Africert Ltd. was to ensure impartiality in offering its services. Thus, a strict impartiality condition was included in the so-called memorandum of the company. This statement of Africert's mission reads:

To be impartial, responsible for decisions relating to its granting, maintaining, extending, suspending and withdrawing certification, to identify the management (committee, group or person) which shall have overall responsibility for the performance of testing, inspection, evaluation and certification, the formulation of policy matters relating to its operation, the decisions on certification, the supervision of the implementation of its policies, the supervision of its finances, the delegation of authority to committees or individuals as required to undertake the objectives as listed in this Memorandum, and for the technical basis for granting certification.

On the question of ownership and governance, local ownership was emphasized. Thus, initial shares in the company were granted to ICIPE, holding its shares in trust, and an individual with the technical and managerial qualifications to guide the company toward achieving its objectives. Subsequently, as of mid-2006, ICIPE completely divested its shares in the company following identification of a qualified local institution to purchase the shares.

Africert thereafter embarked upon the process of setting up its business infrastructure as well as undertaking activities geared toward achieving accreditation under ISO 65 and EN 45011, in order to be able to certify agricultural products against various standards, beginning with the EurepGAP standards for fruits and vegetables.

Africert has added other standards to its list of certification services. It has completely spun off from ICIPE physically, occupying its own offices outside the ICIPE campus, and employing its own staff. And, ICIPE senior management no longer sits on Africert's board of directors.

2.3 *Current activities of Africert*

Africert Ltd. is currently carrying out certification and inspection services throughout eastern Africa, including Ethiopia, Rwanda, and Zambia for the following standards.

- EurepGAP fruits and vegetables.
- Utz kapeh. Utz kapeh means "good coffee." Coffee farms and cooperatives use utz kapeh certification to prove that they grow their coffee professionally and with care for their local communities and the environment. Utz Kapeh empowers growers with knowledge of good agricultural practices and the global coffee market. Certification gives growers a stronger position in the market due to buyers' specific demand for certified coffee.
- British Retail Consortium (BRC) Food Technical Standard. This standard is used to evaluate processors of fresh produce for compliance with major European Union retailers' requirements for food safety and quality.
- Starbucks C.A.F.E. (Coffee and Farmer Equity) Practices. C.A.F.E. is a verification program based on social and environmental good practice in coffee growing, processing, and marketing.
- Ethical Trade Partnership in the tea sector. The fundamental principles of the ETP standard are those of the Ethical Trading Initiative (ETI) base code, which is based on local laws and collective bargaining agreements that are relevant to workers' welfare. The code is used to support, clarify, and enrich the standard and ensure that it is appropriate to the country in which the standard is to be applied.
- MPS GAP/SQ in cut flowers. MPS GAP/SQ is a body of standards that looks into issues of social and environmental management of resources within the cut flower industry. Africert works under a subcontracting agreement with MPS-Holland.
- Organic agriculture.

3. CONCLUSIONS

The creation of a regional certification body in East Africa and the evolution of Africert Ltd. serves to illustrate two issues. First, that publicly funded research and development institutions in the developing countries have opportunities to employ their areas of expertise to improve livelihoods and incomes whether by facilitating access to markets or otherwise. Whereas in the case of Africert, the length of the project was short, the impact of results was directly felt both at the production level and in the markets.

Secondly, transfer of know-how as an aspect of technology transfer is easier to achieve than other complex technology transfer aspects that require heavy capital equipment and other infrastructure. However, this may be a function of the fact that transfer of know-how may be more appropriate in service industries than in other industries, such as engineering and biotechnology. Most importantly, technology transfer can facilitate access to markets and improves incomes. ■

PETER MUNYI, *Chief Legal Officer, International Centre of Insect Physiology and Ecology (ICIPE), PO Box 30772-00100, Nairobi, Kenya.* pmunyi@icipe.org

RUTH NYAGAH, *Chief Executive Officer, Africert Limited PO BOX 74696-00200, Nairobi, Kenya.* rnyagah@africert.co.ke

---

1 ICIPE is an international organization, based in Nairobi, with a mandate to help alleviate poverty, improve general food security and nutrition, and promote better human health for peoples of the tropics through research and development of environmentally friendly management strategies for arthropod pests and disease vectors.

2 ISO 65 is one of the many standards developed by the International Standard Organization, which maintains standards for state-of-the-art products, services, processes, materials and systems, and for good conformity assessment, managerial and organizational practice in agriculture.

3 EN 45011 is the recognized European Standard for

product certification. The objective of the standard is to promote confidence in the way product certification is carried out, giving assurance to the consumer that products meet identifiable and consistent quality levels. The standard requires inspection, testing, and surveillance to ensure that quality standards are met. When products meet standards, the products earn a certificate and carry a mark of conformity. More and more often retailers and global food-service chains are requiring that products be independently (by a third party) inspected and accredited against a recognized standard. Accreditation to EN 45011 meets this requirement. Accreditation of quality assurance schemes to the EN 45011 standard is a detailed process.

4 EurepGAP, founded in 1997, is a private organization that sets voluntary standards for the certification of agricultural products around the globe. EurepGAP started out primarily as an initiative undertaken by retailers belonging to the Euro-Retailer Produce Working Group (EUREP) along with British retailers, in conjunction with supermarkets that were the driving forces, in continental Europe. The organization observed consumers' growing concerns with product safety, and environmental and labour standards and it decided to take greater responsibility for what happened in the supply chain. The development of common certification standards were also in the interest of many producers. Those with contractual relations to several retailers complained that each year they had to undergo multiple audits of different quality criteria. Against this background EUREP started to work on harmonizing standards and procedures to serve the development of good agricultural practices (GAP) in conventional agriculture.

SECTION

17D

Putting Intellectual Property to Work: Experiences from Around the World

FOCUS ON SOLUTIONS:
ACCELERATING PRODUCT DEVELOPMENT AND DELIVERY

CHAPTER 17.21

Patent Consolidation and Equitable Access: PATH's Malaria Vaccines

SANDRA L. SHOTWELL, *Managing Partner, Alta Biomedical Group LLC, U.S.A.*

ABSTRACT

This chapter shares the results of a project that analyzed the potential for consolidating patents in the malaria vaccine field. Goals include streamlining access to critical patents, advancing the development of products, and providing equitable access to the innovations. The study assessed the current status of the relevant patents and surveyed the holders of key patents to determine the availability for licensing. Other key activities included prioritizing patents with respect to a vaccine's potential for success, identifying potential patent roadblocks by discussing the issue with patent holders, and proposing a mechanism for accessing key patents in the field of malaria vaccines. The potential role for some form of patent consolidation or technology trust, including pooling patents and technology, was explored. This chapter does not recommend developing a broad-based technology trust for existing malaria-antigen patents. Instead, several other steps are recommended to consolidate available rights and improve access for future patent families.

1. INTRODUCTION

Malaria is one of the most widespread and deadly tropical diseases. There are more than 300 million cases and more than one million deaths each year. Ninety percent of the cases occur among children in Sub-Saharan Africa. Developed countries have largely eradicated the disease through hygiene, effective drugs, and the reduction of mosquito breeding grounds via wetlands clearing, chemical treatment to control mosquito populations (early on, with DDT), and water-system management. For many reasons, including costs as well as the challenge of managing potential environmental and health effects of chemical parasite removal, these approaches have not been as effective in developing countries. Alarmingly, various factors are now spreading malaria into areas previously free of infection. New approaches to prevention and treatment are sorely needed.

No safe, effective vaccine for malaria exists. Developing a vaccine is a priority because of one especially exacerbating problem: the malaria parasite and the insects that carry it are becoming resistant to existing drug treatments and therapeutic-control measures. A malaria vaccine could greatly reduce the effects of the disease in terms of suffering and lives lost. It also could prevent the spread of malaria more cost effectively than any existing treatment. Vaccine use would reduce the need for expensive, often unaffordable medicines and remediate the problem of drug-resistant parasites. Moreover, vaccine use would reduce the need for chemical treatment to control mosquito populations, thus minimizing negative environmental effects.

Developing a malaria vaccine, however, presents big challenges. Above all, there is an economic challenge. Developing a vaccine for which there is a great medical need but no profitable market requires a clear, sustained source of funding. Fortunately, a variety of public, private,

Shotwell SL. 2007. Patent Consolidation and Equitable Access: PATH's Malaria Vaccines. In *Intellectual Property Management in Health and Agricultural Innovation: A Handbook of Best Practices* (eds. A Krattiger, RT Mahoney, L Nelsen, et al.). MIHR: Oxford, U.K., and PIPRA: Davis, U.S.A. Available online at www.ipHandbook.org.

and philanthropic efforts are targeting the problem. In particular, the Bill & Melinda Gates Foundation is providing philanthropic funding to product-development partnerships. The Malaria Vaccine Initiative (MVI) is the main recipient of funding and the catalyzing force for malaria vaccine development. MVI seeks to accelerate vaccine development through multiple approaches including partnering and the funding of promising projects. Addressing challenges simultaneously on multiple fronts, MVI has dozens of partners in ten ongoing vaccine projects worldwide.

2. THE CHALLENGE OF DEVELOPING A VACCINE

2.1 *Technical challenges*

Developing an effective malaria vaccine presents significant technical challenges:

- Malaria is caused by different parasite species in different countries and has variants within those species. The main species in terms of global health are *Plasmodium vivax*, found mainly in Asia and South America, and *P. falciparum*, found mainly in Sub-Saharan Africa.
- Malarial parasites have several different stages in their life cycle, some of which are short in duration or occur within the host's cells, making the parasites difficult to target with a vaccine.
- During each stage of the malaria parasite's life cycle, it produces a number of different antigens (substances that can evoke an immune response in humans), some of which may be useful in developing a vaccine. There may be several thousand potential target antigens, only a few dozen of which have been studied for use, either separately or in combination, as potential vaccines.

Because of these technical challenges, malaria-vaccine research has continued for decades. Only very recently has a vaccine been shown to be effective in Phase 2 clinical trials[1] in adults and then in children in Africa.[2]

2.2 *Commercialization challenges*

Given the encouraging results of the Phase 2 clinical studies, there is a strong possibility that a malaria vaccine may be ready for regulatory approval in five to ten years. The prospect of manufacturing, delivering, and paying for a vaccine, however, now raises commercial challenges:

- Different populations need very different vaccine products. For example, a vaccine for children in endemic areas is not likely to be suitable as a traveler's vaccine.
- Funding mechanisms are needed. Without clear definitions and estimates of the various markets, it is difficult for companies to justify the expense associated both with speculative vaccine development and with more straightforward manufacture and marketing costs. To help provide certainty for the various markets, MVI is working on a model that takes a variety of vaccine products and market needs into account. Current market projections make it clear that even after development costs have been handled and an approved vaccine is ready for manufacture and marketing, continued public and philanthropic funding will be required in many markets in developing countries.
- Delivery channels are needed to get vaccines to the areas where they are needed.

2.3 *IP challenges*

The possibility of commercializing an effective malaria vaccine raises significant IP challenges. Many patents, some with overlapping claims, cover malaria antigens that may be needed for vaccine development. Such a "patent thicket" is daunting because it is likely that more than one antigen will be needed for an effective vaccine. Unfortunately, accessing many patents one at a time via traditional licensing or partnering could tie up resources needed to develop and deliver the vaccines. Moreover, the negotiations required to access key patents could delay the delivery of the vaccine. Indeed, access to key patents might not even be available, which would affect investment decisions upstream in the development pipeline about vaccine candidates. Because of this, it may not be possible to pursue the most powerful vaccine candidates if companies holding valuable malaria-vaccine IP are unwilling to license to others even if they are not developing a malaria vaccine themselves. Assessing the availability of access to key patents becomes a priority.

3. PATENT AVAILABILITY

3.1 *The antigen patent landscape*

Ten malaria antigens were selected for review based on their use in the most-advanced vaccine development projects—clinical trials or late-stage preclinical studies. The antigens come from several key malaria parasites, most significantly *P. falciparum* and *P. vivax*, and from multiple phases of the parasite life cycle. Public patent databases were used to collect and organize patents and patent applications with claims covering these ten antigens. The patent landscape contained 167 patent families filed by 75 different organizations (sometimes in combination with other organizations).

Alta Biomedical worked with key MVI business and scientific staff and Falco Archer to review and prioritize the 167 patent families. A total of 39 out of 167 patent families (23%) were ranked as moderate to high priority based on the patent status (pending, issued, lapsed, or expired), length of estimated patent life, territory, and overlap between claims and vaccine-candidate attributes. The 39 patents were held by 21 organizations. Alta Biomedical met in person or by telephone with 16 of these organizations. Four of the remaining organizations were in direct contact with MVI; the fifth was not approached.

In early 2005, information from direct interviews and from MVI contacts led to grouping the 39 patent families into four categories (Figure 1). Some of the priority patents covered only one antigen; some covered multiple antigens. The distribution of patents over the ten antigens is shown in Figure 2.

FIGURE 1: DISTRIBUTION OF PRIORITY PATENTS: PRIORITY PATENT FAMILIES BY LICENSING CATEGORY

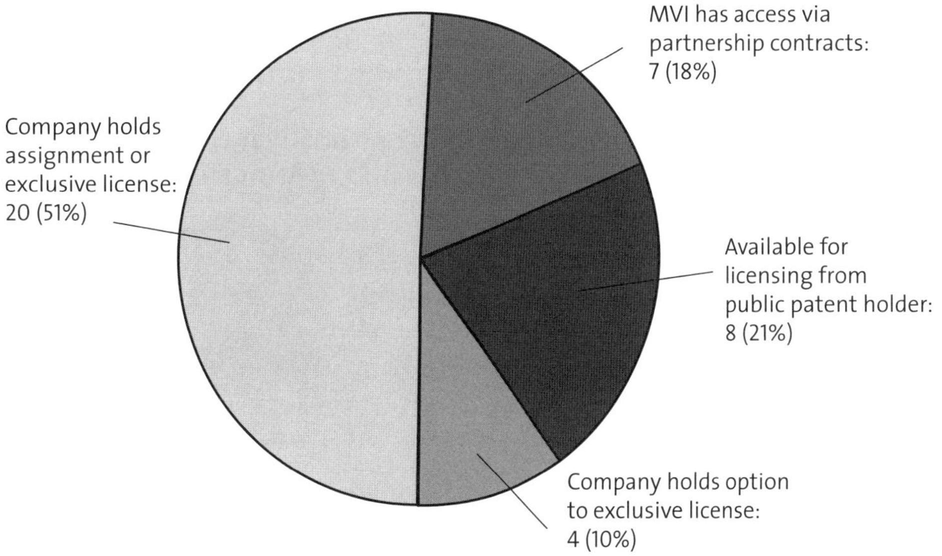

3.2 *Ensuring equitable access*

Before this study, almost half of the priority patents were removed from access by public patent holders (not private companies). Significantly, 69% (27) of the moderate- to high-priority cases originally were filed by a public entity. Five of those were filed jointly with a company. By the time of the study, only 21% (8) remained available for licensing from the public entity. Thus, almost half of the priority cases were removed from access due to actions taken by the patent holder.

To ensure that in the future public entities provide ongoing access, MVI is working with multiple groups of stakeholders to develop recommended practices. This work has involved active participation in meetings with licensing practitioners through the Licensing Executives Society (LES)[3] and the Association of University Technology Managers (AUTM)[4], including the latter's special interest group Technology Managers for Global Health (TMGH).[5] In addition, MVI and Alta Biomedical have participated in smaller group discussions on equitable-access approaches, and in global health IP meetings such as those organized by the Centre for the Management of Intellectual Property in Health Research and Development (MIHR).[6]

3.3 *Patent pooling*

To speed the delivery of vaccines to market, it would help to simplify licensing transactions for the malaria-antigen patents needed for potential vaccine products. One possible approach to simplifying licensing transactions would be to consolidate the necessary patents in a patent pool that could be accessed by any party with one license on reasonable terms. To understand this approach and assess its usefulness in the malaria-antigen area, one must consider information about past patent pools, about how patent pools are being used today, and about how patent pools are contemplated for use in health care.

In the past, patent pools sometimes have been used for anticompetitive purposes, such as collusion and price fixing. To prevent this, the U.S. Department of Justice (DOJ) and the Federal Trade Commission have set up guidelines to ensure that patent pools are "procompetitive." The guidelines include the following:[7]

- Patents in a pool should cover *complementary* technologies that can be used together as the basis for products.
- Patents should not cover *competing* technologies that could be used separately to address the same market need.
- Under the best of circumstances, an *independent standard-setting body* would establish criteria, or standards, in the field to set guidelines for what technology can be included in a patent pool.

FIGURE 2: DISTRIBUTION OF PRIORITY PATENTS: PRIORITY PATENT FAMILIES BY ANTIGEN

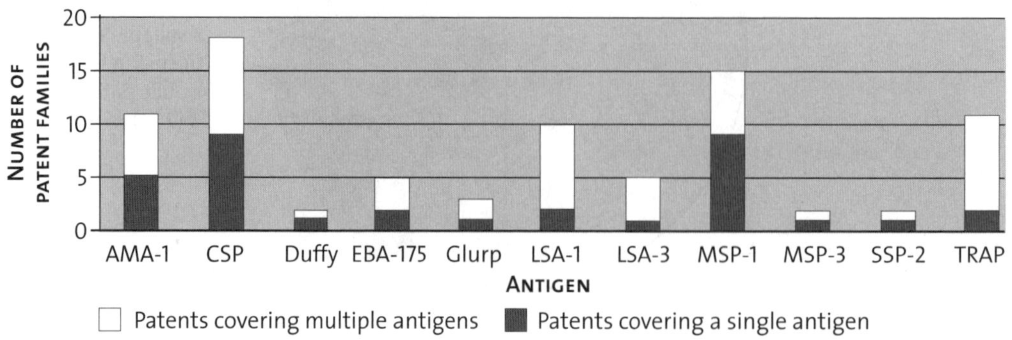

- An *independent expert* should determine which patents fit the guidelines for inclusion in the pool.
- The pooled patents should be available on a *nonexclusive* basis.
- The pooled patents should be *available separately from the individual patent holders* on a nonexclusive basis so potential licensees are not forced to license the entire pool.
- The pooled patents should be available to all parties on *nondiscriminatory terms*.

It is unclear how these guidelines would apply to the malaria-antigen patents. While the last four points can be addressed, whether patents for multiple malaria antigens can meet the requirements of complementary versus competitiveness is uncertain, and what would be considered an independent standard-setting body is unclear.

As far as complementary versus competitive technologies, individual antigens may well be viewed as both. Arguably, they could be used together or separately to develop distinct vaccines. In particular, Richard Johnson of Arnold & Porter has raised a general concern that, based upon analysis of DOJ guidelines, universities may have difficulties creating a pool that includes "*a large fraction of the potential research and development in an innovation market*."[8] This may be viewed as an antitrust concern. Given the modest number of key patents for any single antigen, the large number of target antigens, and the inclusion of more than one antigen in many vaccine product candidates, efforts that consolidate patents for only one antigen do not seem of broad value to the field.

As far as standards in the field, it is possible that an organization such as the National Institutes of Health or the World Health Organization might develop a consensus or set standards that require a vaccine to include antigens from more than one stage of the parasite life cycle, although even then there are multiple candidate antigens from each stage that could be used separately.

Also, a licensee may not need access to, for example, all of the ten most-advanced antigen candidates to develop its planned vaccine. In that situation, it seems possible that the DOJ might view the separate antigens as requiring separate pools.

Two other areas have been proposed for formal patent pools in the health care field: the Severe Acute Respiratory Syndrome (SARS) genome (proposed by holders of SARS genome patents)[9] and the Acquired Immune Deficiency Syndrome (AIDS) essential patents (proposed by Essential Inventions, Inc.).[10] Both suffer from some of the same issues: many patent holders, the lack of an independent standard-setting body, and (perhaps most critically) the inclusion of potentially competing technologies within the same pool (Table

Table 1: Proposed Patent Pools in Health Care

Technology	Number of patent holders (approximate)	Complementary technology	Competing technology	Independent standard-setting body
SARS genome	5	Yes	Yes	Needs to be identified
Malaria antigens	21	Yes	Yes	Needs to be identified
AIDS essential technologies	23	Yes	Yes	Needs to be identified

1). The proposed SARS pool may have the advantage of being early in the product-development life cycle, with patent holders and others aware that resolving patent access may be essential to stimulating investment in product development.

3.4 *Business issues with patent pools*

Several business issues could make a formal malaria-antigen patent pool challenging. For companies currently developing vaccines covered by patents, the patents are likely part of a core business strategy for which a patent pool may be an anathema. Their participation in such a pool may be unlikely.

Moreover, setting up a patent pool can be expensive, with large up front costs for developing the pool's legal framework, taking the pool through regulatory review, and performing a legal review of the patents considered for inclusion in the pool. In the electronics industry, a large-company member of the pool typically contributes much of the up-front funding. That option, however, seems unlikely in the case of malaria antigens. While a small portion of the pool's licensing income typically covers the expense of a commercially successful pool, it seems unrealistic to seek significant licensing income from a malaria vaccine for some of the world's poorest nations. Furthermore, such a goal would run counter to the mission of developing a vaccine that is broadly affordable and available.

A final concern about a potential malaria-antigen patent pool is a simple business issue—very few entities would be interested in accessing any particular antigen patent. For example, if a company was developing a vaccine using two antigens, it would not need access to patents that cover others. An antigen used in one vaccine candidate may be included in a second vaccine candidate, but in combination with a different antigen or antigens. One can easily imagine a scenario where companies would not need access to a broad set of patents, but would prefer to pick and choose. This suggests that an individual access, or clearinghouse, approach might be preferable to a patent pool.

3.5 *Patent pool alternatives*

A pragmatic course would be to obtain access to the key patents that are available through license or assignment. Access by MVI or another organization on behalf of the field could ensure that these patents do not present a potential roadblock. In addition, MVI has developed constructive partnerships with key corporate holders of malaria patent rights and can continue to develop these partnerships as needed.

This strategy could lead to a clearinghouse approach, with IP rights accessible on a pick-and-choose basis by multiple potential partners or licensees, thus avoiding the DOJ approval issues. The approach also could simplify the licensing transaction by setting up, in advance, arrangements that provide assured access at a known cost (similar to setting up a patent pool in advance). But a clearinghouse does not resolve the concern that key patents could remain outside the clearinghouse. Ideally, a clearinghouse would include all the necessary patents for each antigen. Obtaining access to all the necessary patents would require working with companies to include their patents in the clearinghouse, which is not an impossible task but one that puts the transaction burden up front on the party trying to set up the clearinghouse. It seems more reasonable to work directly with companies when it becomes clear that access will be needed to a specific company technology. The relationship may involve not just straight licensing but, among other things, co-development, manufacturing contracts, partnering, and marketing. It might make sense to wait to develop such a relationship until the needs are clearer.

4. CONCLUSIONS

The results of the MVI study suggest that developing a broad-based technology trust for existing malaria antigen patents is not a good idea for several reasons. As the findings above should make clear, with few exceptions the patents held by public and academic institutions have been assigned or exclusively licensed to private companies. The patents are not currently available for licensing from the original public-institution patent holders. While it may be possible to sublicense the patents from the current private holders, doing so is likely to be difficult and costly; engaging patent holders in contributing to a patent pool or clearinghouse also could be difficult. While the concept of a technology trust or patent pool may still be useful for patents to be filed in the future, even some of those would be under option for license by the private companies holding the existing patents. In addition, the number of high-priority cases for any malaria antigen is small, as is the number of entities likely to seek access to any given patent family. This makes the expense of a patent pool even less justifiable.

Other than a broad-based technology trust, there are several effective ways to consolidate available rights and improve access for future patent families in the malaria vaccine field, including:

- Taking assignment to or licensing the limited number of high- or moderate-priority patent families to ensure access. Holding these patents could be useful for developing products or for cross-licensing with private patent holders.
- Developing policy and public statements about why these priority patents are being held on behalf of the field, including a statement regarding the intention to allow access by others.
- Continuing to develop constructive partnerships with the corporate holders of the remaining key patents, as needed.
- Reviewing the geographic limitations of existing patents held by private companies, and considering approaches to vaccine development that do not infringe on these patents, for example, considering production by firms capable of high-quality, less-expensive production and manufacture in middle-income countries not covered by patents.
- Negotiating with patent holders for access to their know-how for development outside the patent coverage area.
- Educating public and academic patent holders about malaria-vaccine development issues in patenting and licensing as well as about balanced approaches that can meet institutional goals and accelerate the development of patents into useful vaccines. This would help to ensure that future actions by public research institutions do not create ongoing access problems.
- Working to develop consensus about when patenting makes sense, as well as the benefits of pooling for future inventions not yet patented or licensed.
- Gathering and developing model language to use in patent strategies and licenses covering malaria-vaccine technology that can ensure the development of appropriate, affordable products for markets in developing countries.
- Working with national and international leaders to encourage broad usage and a common approach for the field. Possible partners in this endeavor include MVI, The Rockefeller Foundation, MIHR, AUTM, LES, U.S. federal laboratories, and leading U.S. and international universities. ■

ACKNOWLEDGMENTS

This chapter is based on a grant report by MVI to The Rockefeller Foundation MVI. As part of the Rockefeller Foundation-sponsored project, MVI engaged Alta Biomedical Group to manage the project. This included evaluating the patent landscape previously mapped out by Falco Archer, Inc., and interacting with many colleagues at PATH, whose contributions to this chapter are gratefully acknowledged. The contributions of Lynnor Stevenson to this project and chapter are also gratefully acknowledged.

SANDRA L. SHOTWELL, *Managing Partner, Alta Biomedical Group LLC, 7505 S.E. 36th Avenue, Portland, OR, 97202, U.S.A.* shotwell@altabiomedical.com

1 Phase II trials are conducted on population groups of around 20–300 and are designed to determine dosing levels and assess clinical efficacy of a vaccine. Phase II builds on the initial safety studies of the vaccine (Phase I) and forms the basis for Phase III studies (typically randomized controlled trials on 300–3,000 or more people).

2 www.malariavaccine.org.

3 www.lesi.org.

4 www.autm.net.

5 www.tmgh.org.

6 www.mihr.org.

7 From presentations by Jorge Goldstein of Sterne Kessler Goldstein and Fox, PLLC; Richard Johnson of Arnold & Porter; Brian Stanton of the National Institutes of Health, Office of Technology Transfer; Lawrence Sung of the University of Maryland School of Law; numerous Department of Justice publications; and other Internet sources.

8 From a presentation by Richard Johnson of Arnold & Porter at the Association of University Technology Managers Annual Meeting on 4 February 2005 in Phoenix, Arizona.

9 www.who.int/bulletin/volumes/83/9/707.pdf.

10 www.essentialinventions.org.

Lessons from the Commercialization of the Cohen-Boyer Patents: The Stanford University Licensing Program

MARYANN P. FELDMAN, *Miller Distinguished Professor in Higher Education, Institute of Higher Education, University of Georgia, U.S.A.*
ALESSANDRA COLAIANNI, *Center for Genome Ethics, Law, and Policy, Duke University, U.S.A.*
CONNIE KANG LIU, *Joseph L. Rotman School of Management, University of Toronto, Canada*

ABSTRACT
The Cohen-Boyer licensing program, by any variety of metrics, was widely successful. Recombinant DNA (rDNA) products provided a new technology platform for a range of industries, resulting in over US$35 billion in sales for an estimated 2,442 new products. Over the duration of the life of the patents (they expired in December 1997), the technology was licensed to 468 companies, many of them fledgling biotech companies who used the licenses to establish their legitimacy. Over the 25 years of the licensing program, Stanford and the University of California system accrued US$255 million in licensing revenues (to the end of 2001), much of which was subsequently invested in research and research infrastructure. In many ways, Stanford's management of the Cohen-Boyer patents has become the gold standard for university technology licensing. Stanford made pragmatic decisions and was flexible, adapting its licensing strategies as circumstances changed.

1. INTRODUCTION
The licensing of the Cohen-Boyer patents by Stanford University represents one of the most successful university technology licenses. The discovery covers the technique of recombinant DNA and allows for the useful manipulation of genetic material. Examining Stanford's licensing of the intellectual property is best understood in context and as part of the university's larger strategy. Moreover, designing and setting up the licensing program involved uncharted territory at that time. The first patent issued on December 2, 1980, after 6 years under review at the U.S. Patent and Trademark Office: the original application was filed in November 1974. This date was two weeks before the effective date of the Bayh-Dole Act, which assigned intellectual property (IP) rights over faculty discoveries from federally funded research to universities and emphasized the university's responsibility for commercialization.[1] The intention was to provide a means for economic growth, technological change, and enhanced U.S. competitiveness.

The Cohen and Boyer's discovery provided tools for genetic engineering and was the subject of controversy that led to a lively public debate during the decade of the 1970s. Sally Smith Hughes documents Cohen and Boyer's scientific discovery, Stanford's decision to pursue patents, and the public controversies surrounding recombinant DNA.[2] The debate was symbolically resolved with the June 1980 U.S. Supreme Court ruling on *Diamond v Chakrabarty*, a landmark 5–4 decision, which made the patenting of life forms possible with the Court's oft-quoted clause, "*anything under the sun, that is made by man.*" This decision cleared the way for the Cohen-Boyer application, which covered a fundamental technique, with the potential to become a platform technology that essentially led to a new paradigm in biotech research.

Feldman MP, A Colaianni and C Liu. 2007. Lessons from the Commercialization of the Cohen-Boyer Patents: The Stanford University Licensing Program. In *Intellectual Property Management in Health and Agricultural Innovation: A Handbook of Best Practices* (eds. A Krattiger, RT Mahoney, L Nelsen, et al.). MIHR: Oxford, U.K., and PIPRA: Davis, U.S.A. Available online at www.ipHandbook.org.

Of course, once the patent was granted, Stanford University, as the assignee, was required to design a licensing program that would be consistent with the public-service mission of the university and provide sufficient incentives for private industry to invest the requisite resources to bring products to market while producing revenue for the university. Feldman, Colaianni and Liu[3] detail the history of Stanford's licensing program, focusing on the process and the logic that guided the commercialization regime. Given the early controversy surrounding the Cohen-Boyer patent, the eventual success required a great deal of creativity, strategy, and persistence. Certainly, the professionals involved all contributed to the success, from Donald Kennedy, then president of Stanford, Robert Rosenzweig, then vice president for public affairs, Nils Reimer, founding director of the Stanford Office of Technology Licensing (OTL) to Katherine Ku, then licensing associate and current director of the OTL.

The purpose of this chapter is to summarize lessons learned from Stanford's design and implementation of the Cohen-Boyer licensing program. Many universities attempt to emulate Stanford University's success at technology transfer; however, there is a limited appreciation for the high degree of creativity and adaptability of the Stanford Office of Technology and Licensing (OTL) in setting up its licensing program and making the myriad decisions that guided the ultimate outcome. In spite of many obstacles, Stanford University pursued the recombinant DNA patents and designed a strategy that met the public-service goals of the university by broadly licensing the technology; provided incentives for private companies to commercialize derivative products; and contributed to the creation of an innovation system that benefited Silicon Valley and reached across the American economy.

2. A LIST OF LESSONS LEARNED FROM COHEN-BOYER

2.1 *Keep wider university goals in mind*

Despite the economic success of the licensing program, profit was not the primary motive. Stanford University had four goals that guided the development of the Cohen-Boyer license:

- to be consistent with the public-service ideals of the university
- to provide the appropriate incentives in order that genetic engineering technology could be commercialized for public benefit in an adequate and timely manner
- to manage the technology in order to minimize the potential for biohazard
- to provide income for educational and research purposes

Robert Rosenzweig, vice president for public affairs at Stanford, in a 1976 open letter addressed to "*Those Interested in Recombinant DNA,*" wrote "*It is a fact that the financing of private universities is more difficult now than at any time in recent memory and that the most likely prediction for the future is that a hard struggle will be required to maintain their quality.*" As a result of these financial concerns, he concluded, "*we cannot lightly discard the possibility of significant income that is derived from activity that is legal, ethical, and not destructive of the values of the institution.*"

The balance of financial objectives against other goals is further demonstrated when Stanford decided not to pursue extending the patent life. The original 1974 patent application had claimed both the process of making recombinant DNA and any products that resulted from using that method. These applications were subsequently divided into the process patent and two divisional product applications: one claimed recombinant DNA products produced in prokaryotic cells and the other claimed the products in eukaryotic cells. Stanford filed a terminal disclaimer, which meant that all subsequent applications claiming recombinant DNA, regardless of how long the patent prosecution process took, would expire on December 2, 1997—the same date as the original 1980 patent.[4] In effect, Stanford agreed to give up royalty rights on the life of the subsequent patents (issued in 1984 and 1988) that would have extended past the original patent's expiration date. This limited Stanford's collection of royalties because of the time delay inherent in commercialization, especially of pharmaceutical products. Stanford honored its obligation to the licensees with the realization that, as Kathy Ku wrote at the time "*...it would not be good public policy or public relations if we were to ask for or even get such an extension.*"

Stanford did not require other nonprofit research institutions to take a license in order to use the technology. Niels Reimers and Kathy Ku report that the thought of licensing the technology out to other nonprofit research institutions had never entered into discussions about the licensing program. This licensing practice established a research exemption, or research-use exemption, which is consistent with the norms of open science,[5] and stands in contrast to recent developments in research-use exemption policies, such as *Duke v. Madey* and the WARF stem-cell licensing program.[6]

To summarize, engaging in commercial activity encourages higher education institutions to act like for-profit entities. Intellectual property has no value unless it is defended. Stanford set up a litigation reserve fund that provides a credible threat of enforcement of the license. Despite several attempts to withhold payments from a variety of large and small companies plus one attorney who made challenges to the patents a "hobby," Stanford was able to settle these disputes informally and without formal litigation. This stands in contrast to the recent upswing in litigation by U.S. universities, including a recent law suit filed by the University of Alabama to prevent an artist from using the universities athletic colors.

2.2 *Consult widely to build consensus*

While intellectual property typically involves limited disclosure, Stanford University engaged in a pattern of consulting widely across various stakeholders to achieve consensus and to ensure that its actions were supported. For example, Rosenzweig worked to achieve consensus with both the faculty and the National Institutes of Health (NIH) as the sponsoring agency. In a 1976 open letter, he asked the faculty to comment on whether the university should proceed with the patent process. Rosenzweig also sent a letter to Donald Fredrickson, NIH director, asking his opinion on patenting the Cohen-Boyer discovery and enclosed a copy of the memorandum sent to faculty. Fredrickson responded by sending a mass mailing to "a broad range of individuals and institutions," asking them for their comments on the patent question.[7] Fredrickson's letter laid out five possible alternatives that NIH could take regarding recombinant DNA patenting and subsequent licensing: In response, Fredrickson received approximately 50 letters.

A compromise consensus emerged from among a list that Frederickson generated that Stanford should be able to patent recombinant DNA research but with nonexclusive licensing. A nonexclusive license ran counter to economic logic, contrary to the subsequent preferences articulated in the Bayh-Dole, Act and ignored petitions from Genentech and Cetus who stood to gain from exclusive licenses. The logic was that rDNA was a platform technology and that any one company could not exploit all the possible applications. Broad nonexclusive licensing not only contributed to the economic success of the patents but also created a population of companies who drove the technology forward.

There are other instances when Stanford sought transparency that was consistent with the actions of a university. While applicants generally keep patent applications secret from the date they are filed until they are granted and therefore protected, Stanford opened the patent prosecution file to the public. This was an unusual move that was consistent with reducing subsequent questions about the technology and was also consistent with the public mission of the university.

Stanford engaged in an open process that attempted to build consensus across a wide range of stakeholders. While the university did stand to profit from the licensing program, their actions were consistent with the university's larger and more traditional societal goals.

2.3 *Don't behave opportunistically*

The most successful university technology transfer involves relationships that develop over time. Signing a licensing agreement represents a transaction that is a first step in a relationship that requires maintenance and oversight. Each licensee received an annual letter from the Stanford OTL.

That went a long way in establishing long-term relationships and encouraging dialogue.

When Stanford initiated its licensing program, no precedent existed for specific licensing terms of the IP. Keeping with its practice of consulting widely and building consensus, Stanford interviewed a variety of companies representing different markets when the license terms, particularly the royalty rates on end products, were being formulated. Through this effort, licenses were pre-sold and unrealistic terms were avoided. To make the licensing process easier, the OTL took great pains to categorize the different potential recombinant DNA products and to offer appropriate royalty rates. In the end, the OTL settled on four different product categories: basic genetic products, bulk products, end products, and process improvement products. By scaling the rates to reflect the visibility of the licensee's product and the expected revenue from each license, the OTL encouraged compliance. A graduated royalty system ensured that smaller companies weren't penalized with low sales volume.

Stanford made pragmatic decisions about pricing its intellectual property and kept the annual fees and royalty rates reasonable. While this might have reflected a strategy to deal with some of the weaknesses with the patent, the university could have been greedy and pursued higher rates. Nils Reimers recalled at least one alumnus writing, "*You've got a patent; you can dominate everything here. Why are you charging such a low royalty? You know Stanford could use the money. Charge a higher royalty.*"[8] This advice was not taken. The rates that were chosen were selected after consultation with industry about accepted practices and did not exploit the university's monopoly position.

Furthermore, Stanford created special provisions for lower licensing fees and royalty rates for small firms in 1989. At this time, 209 fledging biotech firms, most of them in the San Francisco Bay Area, signed licensing agreements.

2.4 *Be flexible and experiment*

Over the 17 years of the licensing program Stanford experimented with five versions of the standard license agreements and provided three special licensing agreements. A total of 468 companies licensed the Cohen-Boyer technology. Licensing the patents was very much a learning process that balanced the capabilities of companies, especially in the embryonic biotech industry, with the economic potential of the technology. Ku later noted, "*Stanford was trying to license an invention for which products had never been sold and which would apply to many diverse, established industries, in addition to the newly emerging biotechnology industry.*"[9] Table 1 summarizes the various licensing regimes and the number of companies that signed up under each version. Certainly the economic impact would have been less without this flexibility and adjustments.

The first version of the license provided two incentives to encourage companies to sign up. Remember that the technology was already in the public domain through publication and that the open patent files and companies were already using rDNA. It was not clear that companies would comply with the terms. The first incentive for companies to take a license in 1981 was a credit toward future royalties over the first five years, up to a total of US$300,000. The second incentive came when companies were advised that the licensing terms would change and encouraged them to sign up early. In response to this news, 82 companies signed up. The largest share of earned royalties from product sales accrued to these firms.[10]

The first license's terms were a US$10,000 up-front fee with a minimum annual advance (MAA) of US$10,000. Earned royalty rates on products were provided on a graduated basis for bulk products, end product sales, and process improvements on existing products based on production cost savings. Under the licensing agreements, Stanford received unprecedented royalties on downstream drug sales in a stipulation known as *reach-through* licensing: Stanford received end-product royalties based on a percentage of final product sales. The Cohen-Boyer IP rights extended to all products developed using the technology. If companies did not sign a license agreement, any end products they developed that used rDNA could potentially be contested.

Table 1: Cohen-Boyer Standard Licensing Agreements History (in U.S. dollars)

Version	Effective date	Sign-up fee & minimum annual advance (MAA)	Earned-royalty rates: End products	Earned-royalty rates: Bulk products	Earned-royalty rates: Basic genetic products and process improvements	Number of compan es signed	Revenue (share)
1	12/2/1980	Each $10,000; with special five times credit	Graduated rate: 1% (first $5M); 0.75% (next $5M); 0.5% (over $10M)	Graduated rate: 3% (first $5M); 2% (next $5M); 1% (over $10M)	10% for basic products sales; 10% of cost savings and economic benefits	73	$215,663,697 (84.66%)
2	1/1/1982	Each $10,000	Graduated rate: 1% (first $5M); 0.75% (next $5M); 0.5% (over $10M)	Graduated rate: 3% (first $5M); 2% (next $5M); 1% (over $10M)		15	$14,229,566 (5.59%)
3	8/1/1985	Each $10,000	Same as above, but started write-in	Same as above, but started write-in		10	$3,338,347 (1.31%)
4	11/1/1986	Each $10,000	1%	3%		21	$5,355,889 (2.1%)
5	9/1/1989	Each $10,000 if < 125 employees; Each $50,000 if > 125 employees	2%	6%		209	$12,120,719 (4.76%)
Alternative license	Mid-1991	No MAA	4%	6%	N/A	12	$2,630,195 (1.03%)

CONTINUED ON NEXT PAGE

Table 1 (continued)

R & D agreement	End of 1994	Sign-up payment waived; all future MAA as one-time payment	N/A	N/A	NA	36	$553,083 (0.22%)
Final year	1996	No sign-up fee of $10,000; MAA is prorated and payable upon execution	N/A	N/A	NA	6	$39,680 (0.02%)

The second standard licensing agreement dropped the royalty-credit incentive and an additional 15 companies signed the agreement. In August 1985, the OTL issued its third standard version of the license agreement, which allowed for negotiation by providing a space to write in agreed-upon rates. In practice, though, the earned-royalty rates were almost always at the same graduated rates that were used in the second version. This fact may be attributed to the sharing of information among potential licensees about prevailing terms and what terms might be expected. Another ten firms signed up under this licensing agreement. Another adjustment was made in November 1986, with the fourth standard licensing agreement. Instead of a graduated-royalty rate, a flat rate of 1% on end products and 3% on bulk products was used. These were the highest rates under the prior version and reflected the realization that the patents could earn higher rates. In response, perhaps motivated by the possibility of further increases in the future, 21 more firms signed licensing agreements. The fifth version of the Cohen-Boyer standard licensing agreement, adopted in September 1989, demonstrated further strategic changes. In order to encourage licensing by small start-up companies, consideration of company size was introduced. For companies with fewer than 125 employees, the sign-up fee and MAA fee remained the same, at US$10,000 each. The strategy worked—209 small biotech firms became licensees under this version, along with 12 large companies.

In addition to the standard agreements, there were three nonstandard licensing agreements that provided alternative agreements, making sure that Stanford could collect as much revenue as possible without being unfair to companies with special circumstances. The first was an alternative license for small distributors or resellers of recombinant DNA products. Fifty companies signed on under this alternative agreement, accounting for 17.5% of the total 275 licensees signed after 1991 and providing US$462,000 in licensing revenue. At the end of 1994, a research and development license agreement, with greatly reduced rates, was developed to encourage start-ups that would not realize product sales within the patent lifetime. Another 39 companies signed the research and development license agreements, and, although these licenses did not yield much licensing revenue, they were important to the legitimacy of the small companies. A third nonstandard licensing agreement was offered in the final year to tie up a few loose ends.

In total, the Cohen-Boyer licenses generated US$254 million in revenue during its 17-year term. The initial sign-up and annual fees generated US$26 million, which was 10% of the total licensing income. The licensing program certainly would have been less successful without these revisions and accommodations. A whopping 90% of the total revenue (US$228 million) was from royalty income from product sales. This mirrors the commercial success of recombinant DNA products.

2.5 *Technology transfer is all about skewed distributions*

While others have noted that the distribution of technology transfer revenues are highly skewed, with a few blockbusters accounting for most revenues, our examination of the companies that licensed the recombinant DNA technology and their products demonstrates that even within a single license, highly skewed outcomes account for the high revenues. Commercial products developed by the licensees generated over US$35 billion dollars in sales of recombinant DNA products over the life of the patent. Stanford reported 2,442 products based on recombinant DNA by the time the Cohen-Boyer patent expired in December 1997, reflecting a range of applications in a variety of industries.[11] Starting in 1991, 400 new products, on average, were being brought to the market every year. Recombinant DNA product sales reached US$500 million dollars in 1987 and then doubled from 1988 to 1990. Sales doubled again from 1991 to 1994 and yet again from 1994 to 1998.

The revenue received from each of the Cohen-Boyer licensees ranged from US$4.24 million to US$54.78 million dollars. Of the 468 licensees of Cohen-Boyer technology, ten companies alone provided 77% (US$197 million) of the total licensing income. One company, Amgen, accounted for over one-fifth of the total revenues received under the licensing program. Figure 1 provides a breakdown of the royalty share provided by different companies.

Table 2 lists these ten companies and the products developed under the license. Many of the products were developed under strategic alliances between start-up biotech firms and large pharmaceutical firms, or between biotech firms. All of the top-ten companies, except Merck (which signed the agreement in 1984) signed the first standard agreement in December 1980. The next 10 companies accounted for another 10%, while the remaining companies generated less than 13% of total royalty revenue.

3. CONCLUSIONS

In the 1970s, universities became more entrepreneurial, looking for different streams of revenue that supported the university's mission. As a result, a new system of technology transfer emerged. Certainly the Cohen-Boyer patents and Stanford University's licensing program were at the heart of the debate and central to the evolving system.

It would be a mistake to look back at Stanford's success with the Cohen-Boyer licenses and think that its success was inevitable or that the licensing process was easy. An examination of history reveals many episodes where Stanford University could have behaved opportunistically or taken a wrong turn. The mistaken notion that Stanford and the University of California system were pursuing revenue as a primary goal ignores the controversies that faced Stanford at that time and the creativity and discipline that Stanford had to employ to surmount them. Stanford's licensing program is a good example, not just in terms of its monetary success, but in terms of the lessons it affords to others who work in the area of licensing and technology transfer. While many universities have now instituted licensing programs and are aggressively pursuing intellectual property rights, our study demonstrates that this process is not at all easy or straightforward. In retrospect, Stanford's licensing venture might have failed at several turns and Stanford was forced to be innovative to accommodate the great uncertainties it faced. Had Stanford and the University of

FIGURE 1: DISTRIBUTION OF ROYALTIES

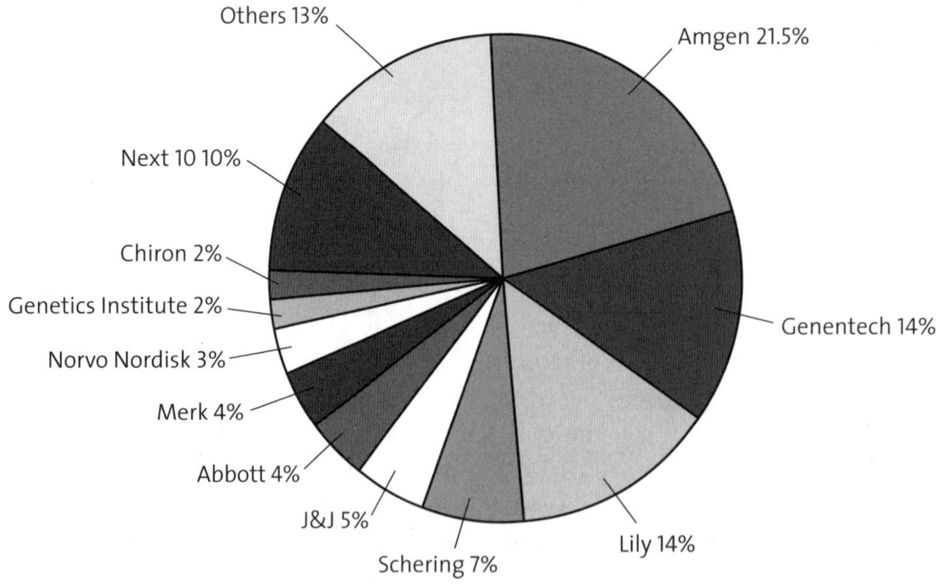

Source: Compiled by the authors.

Table 2: Blockbuster Drugs of Top Ten Licensees of Cohen-Boyer Patent

Company	Paid royalties (US dollars)	Product trade name	Year started to pay earned royalties
Amgen	$54,783,507	Epogen Procrit [a] Neupogen	FY 1989–1990
Lilly	$36,685,982	Humulin [b] Humantrope Abciximab [c] Humalog	FY 1983–1984
Genentech	$34,737,780	Humulin [d] Protropin Roferon A [e] Activase Nutropin Pulmozyme Nutropin AQ Actimmune Kogenate	FY 1985–1986
Schering	$17,960,351	Intron A [f]	FY 1986–1987
Johnson & Johnson	$13,418,280	Procrit [g]	FY 1992–1993
Merck	$10,085,657	Recombivax HB [h]	FY 1986–1987
Abbott	$9,804,444	Various in vitro HIV diagnostics	FY1987–1988
Novo-Nordisk	$8,669,119	Novolin	FY 1990–1991
Genetic Institute	$5,946,978	Recombinate	FY 1993–1994
Chiron	$5,099,071	Proleukin Betaseron [i]	FY 1987–1988

a. Partnered with Ortho and Johnson and Johnson.
b. Partnered with Genentech.
c. Partnered with Centocor.
d. Partnered with Lilly.
e. Partnered with Roche.
f. Partnered with Biogen.
g. Partnered with Amgen and Ortho.
h. Partnered with Biogen.
i. Partnered with Berliex.

California taken only financial considerations into account, it is likely that they would have opted for much higher royalty rates or a more lucrative limited-use exclusive license. Stanford made very pragmatic decisions about pricing its intellectual property. In addition, it might have had to aggressively litigate instead of playing a defensive litigation strategy. Moreover, the process was not finished once the first licensing agreement was formulated; Stanford made pragmatic decisions and proved flexible, adapting its licensing strategies as circumstances changed.

Had it not been for Stanford's enlightened licensing practices, the Cohen-Boyer technology might have been placed in the public domain where the technology could have remained undeveloped or in the laboratories of large established pharmaceutical companies. Or it might have been licensed exclusively and the rise of a biotechnology industry might have been delayed for years or decades. Small companies gained legitimacy through licensing the Cohen-Boyer patents, making it easy for the companies to attract funding and strategic alliances. Hundreds of small biotech firms were founded on the recombinant DNA technology, some of which have grown into large and successful firms. In total, 2,442 known products were developed from the recombinant DNA technology, among them drugs to mitigate the effects of heart disease, lung disease, anemia, HIV-AIDS, cancer, diabetes, and numerous other diseases and disorders. Stanford and the University of California received a quarter of a billion dollars that was used to fund internal research and provide infrastructure. It would be interesting to trace how those funds were actually used and what additional benefits may thus have been generated.

Stanford University's licensing program still provides a reference point for the future practices of university technology transfer. While the amount of licensing revenue received and the value of the commercial product generated are awe inspiring, it should be remembered that this process was neither easy nor straightforward. The Stanford OTL was very creative and adaptive in designing their licensing program. They never lost sight of their larger goals to society and to the scientific enterprise.

ACKNOWLEDGEMENTS

This work was supported in part by the Center for Public Genomics, Duke University, under grant P50-HG003391 from the National Human Genome Research Institute and the U.S. Department of Energy.

MARYANN P. FELDMAN, *Miller Distinguished Professor in Higher Education, Institute of Higher Education, University of Georgia, 2024 Meigs Hall, Athens, GA, 30602, U.S.A.* mfeldman@uga.edu

ALESSANDRA COLAIANNI, *Center for Genome Ethics, Law, and Policy, Duke University, P.O. Box 97082, North Building, Research Drive, Durham, NC 27708. U.S.A.* cac28@duke.edu

CONNIE KANG LIU, *Joseph L. Rotman School of Management, University of Toronto, 105 St. George Street, Toronto, Ontario, M5S 3E6, Canada.* connie.liu07@rotman.utoronto.ca

---

1 Previously, individual faculty members had been able to file patents, negotiating the IP (intellectual property) rights with the federal agency that had sponsored the research.

2 Hughes SS. 2001. Making Dollars out of DNA. The First Major Patent in Biotechnology and the Commercialization of Molecular Biology, 1974-1980. *Isis* 92:541-75.

3 Feldman MP, A Colaianni and C Liu. 2006. Commercializing Cohen-Boyer: 1980-1997. Unpublished.

4 "The Patent Office often requires terminal disclaimers to prevent an applicant seeking to extend patent life from filing continuation applications." See Reimers N. 1987. Tiger by the Tail. *Chemtech* 17(8):464–71.

5 The OTL *did* recognize and account for, in their subsequent licensing programs, the possibility that a research institution would develop a commercially useful transformant (a cell modified by recombinant DNA techniques) that would then be licensed or sold to a company. The OTL would then require any such company to take out a license on the patents.

6 The Wisconsin Alumni Research Foundation (WARF), in 2002, signed its first licensing agreement, for stem cells, with a commercial provider and also signed a separate license agreement permitting U.C.-San Francisco, an academic provider listed on the NIH registry, to distribute human embryonic stem cells worldwide for use in research.

7 Letter from Donald Fredrickson to Robert Rosenzweig, dated 2 March 1978. Obtained from: United States. Office of the Director, NIH. Recombinant DNA Research: Documents Relating to "NIH Guidelines for Molecules,"

June 1976 to November 1977. Department of Health, Education, and Welfare (DHEW).

8 Ibid.

9 Ku K. 1983. Licensing DNA Cloning Technology. Presented at the LES USA/Canada Central/Western Regional Meeting, Scottsdale, Arizona, February. A copy was obtained from the Stanford University OTL, 17 August 2004.

10 Amgen was grandfathered into this version of the licensing terms.

11 Compiled from OTL Archives.

Specific IP Issues with Molecular Pharming: Case Study of Plant-Derived Vaccines

ANATOLE KRATTIGER, *Research Professor, the Biodesign Institute at Arizona State University, Chair, bioDevelopments-International Institute; and Adjunct Professor, Cornell University, U.S.A.*

RICHARD T. MAHONEY, *Director, Vaccine Access, Pediatric Dengue Vaccine Initiative, International Vaccine Institute, Republic of Korea*

ABSTRACT

The public sector is making substantially increased investments in health technology innovation through public/private partnerships to bring improved health technologies to underserved people in developing countries. These product-development partnerships, however, face a common problem: how to manage intellectual property (IP). Such management involves many issues. In relation to a case study, presented in this chapter, of plant-derived hepatitis B virus vaccine, the challenges involve obtaining freedom to operate, securing new intellectual property, and deploying intellectual property to developing countries. We conclude that while challenges abound, the IP issues are fairly clear and can be addressed with straightforward IP management approaches. The cost of managing the intellectual property is expected to be minimal on the price of the finished vaccine. In the medium term, an IP protection strategy might offset costs and generate modest income. Most important for the partnerships is to develop a clear, transparent IP policy, with emphasis on the licensing principles, so that products can be made available to developing countries at affordable prices.

1. INTRODUCTION

The goal of molecular pharming is to develop valuable new drugs and vaccines for significant diseases in developed and developing countries. A number of substances have already been produced in plants and include flavors, nutraceuticals, biodegradable plastics, and metabolites. From a health perspective, plants have been engineered to produce therapeutic proteins for clinical evaluation including human serum proteins (epidermal growth factor), monoclonal antibodies, such as antigenic peptides for rabies virus, tuberculosis and HIV, antibodies to treat cancer, cardiovascular diseases, gastric lipase in the fight against cystic fibrosis, and hepatitis B antibodies, and a range of vaccines.[1] Recombinant protein drugs are one of the fastest growing segments of the pharmaceutical industry, currently generating over US$20 billion in annual revenues. They are the so-called third generation of recombinant plant products.[2]

From a global perspective, plant-derived vaccines represent an attractive mode of production to address diseases of the poor and to stimulate manufacturing in developing countries.[3] Over the last decade, the concept of plant-derived vaccines has grown more sophisticated and many research partnerships have emerged that involve advanced research centers in developing countries. Several potential characteristics of plant-derived vaccines could make them particularly attractive for controlling infectious diseases in developing countries.

- The vaccines would be orally active, thus eliminating the need for injection and the associated cost and safety concerns.
- Oral activity is associated with the ability of plant-derived vaccines to evoke mucosal immunity, which is valuable for a number

Krattiger A and R Mahoney. 2007. Specific IP Issues with Molecular Pharming: Case Study of Plant-Derived Vaccines. In *Intellectual Property Management in Health and Agricultural Innovation: A Handbook of Best Practices* (eds. A Krattiger, RT Mahoney, L Nelsen, et al.). MIHR: Oxford, U.K., and PIPRA: Davis, U.S.A. Available online at www.ipHandbook.org.

of infections that are transmitted through the mucosa.

- Plant-derived oral vaccines should be heat stable, thus largely eliminating the need for a cold chain for these vaccines.
- It might be possible to make multi-antigen vaccines either by multiple gene splicing or by mixing various plant-derived vaccines.
- A very important potential aspect of plant-derived vaccines is that developing countries could launch and carry forward their development and ultimately their production.
- Plant-derived vaccines could be produced on a very large scale and at very low cost, perhaps as little as a few cents per dose.

Indeed, a multi-disciplinary team led by Charles Arntzen[4] recently carried out detailed calculations of the comparative costs of the production of vaccines by traditional methods and by plants. The chapter here is an extension of that report. In that study (as indeed in this chapter), hepatitis B vaccine (HBV) was used as a model. The cost-of-production study computed the costs for facilities in the United States, Korea, and India capable of producing 75 million doses per year. The "effective cost" was also computed (in other words, the cost per dose to deliver in a developing country immunization program and the percent savings that could be enjoyed over the effective cost using plant-derived vaccines). The results are summarized in Table 1. It shows that the potential economic benefits of plant-derived vaccines justify the establishment of a comprehensive program to bring one or more products to the market soon.

It is not surprising therefore that government- and foundation-funded molecular pharming represents a new generation of public sector initiatives that seek to rectify a widely acknowledged imbalance: a lack of investment in R&D for health technologies for the poor. Since the private sector is, by definition, profit driven, it cannot, on its own, address this imbalance because of the need to make a competitive return on investment, which the market for the poor does not provide.

The public sector is now making substantially increased investments in health technology innovation through public/private partnerships. These product-development partnerships face a common problem: how to manage intellectual property (IP). This is no small challenge. IP management is a complex field in which learning, understanding, and using best practices is essential.

Table 1. Comparison of Production and Effective Cost for Three Countries and Two Presentations

	Korea or India	United States		Korea		India	
	Yeast-derived 10-dose vials	*Plant-derived single-dose packet*	*Plant-derived 10-dose packet*	*Plant-derived single-dose packet*	*Plant-derived 10-dose packet*	*Plant-derived single-dose packet*	*Plant-derived 10-dose packet*
Cost	$0.27	$0.15	$0.06	$0.09	$0.04	$0.075	$0.03
Effective Cost	$0.42	$0.16	$0.08	$0.10	$0.05	$0.08	$0.04
% savings for plant-derived vaccine against yeast-derived for effective cost		62%	81%	76%	88%	81%	90%

Source: Arntzen et al., 2006.[5]

IP management involves many issues, including patenting, the protection of confidential information, and the formation of cooperative R&D programs. For any area where many organizational actors converge, there are three primary challenges to IP management:

1. **Securing new intellectual property.** New research initiatives will naturally develop new intellectual property. It is essential to public sector goals that this intellectual property be identified and secured, either by filing the appropriate patent applications or by obtaining licenses from patent holders. If, for example, one group develops a method for promoting the synthesis of an antigen, and another group develops a technique for purifying the antigen from plant material, it is essential to be able to bring together both intellectual properties for developing the final product. This IP challenge can be largely overcome by undertaking an inventory of the existing intellectual property of key groups. To accomplish this work there must be access to technical experts who can identify the specific ways the intellectual property can be useful for product development.
2. **Freedom to operate (FTO).** If a molecular pharming initiative is to achieve its goals, the partnership will need to undertake a thorough Freedom-to-Operate review to provide a clear picture about which patents do, may, and do not stand in the way of developing products. These assessments are always associated with a high level of uncertainty, for a number of reasons, including the large number of patents that may exist, the numerous jurisdictions (countries) in which the patents have been or have not been filed, and the varying practices of patent offices. A blocking patent may exist and might be voided in key markets only through long and costly legal battles. The value of an FTO assessment is that it provides a good sense of the IP issues relevant for any development project, which helps minimize costly, unforeseen problems.
3. **Deploying intellectual property.** Public sector groups are often dedicated to achieving social goals, such as developing safe and effective health technologies to address disease. Further, these groups would like to see these products made widely available at affordable prices to all levels of society. To accomplish these ends, public sector groups should use humanitarian licensing practices. For example, if a group helps to develop a new monoclonal antibody against the rabies virus, it could license the technology to companies in Europe and the U.S., but the group could also reserve the right to license companies in developing countries under different terms. These countries may enjoy some advantages, such as lower costs of production. Licensing to companies in developing countries could also help to make the product available to the poor at prices near the marginal cost of production.

2. SPECIFIC INTELLECTUAL PROPERTY ISSUES WITH PLANT-DERIVED PHARMACEUTICALS

2.1 *Background*

As with most biotechnology products, the IP situation in plant-derived vaccines is complex. Managing IP and tangible property presents added challenges and expense because plant-derived vaccines build on many distinct areas of innovation, including:

- Engineering of proteins and specific antigens (including immunogens and specific genes encoding antigenic proteins). Many patents in this area are the same as those that apply to vaccine production through conventional means.
- Antigen production and accumulation in plants (including the expression of foreign genes and the optimization of genes). The technologies associated specifically with the expression of antigenic determinants in plants are the subject of several issued patents.

- Genetic transformation of plants (including vectors for use in plant transformation, transformation protocols, molecular toolkits, and various equipment). Basic plant transformation technologies have been under development for more than 20 years. The procedures commonly in use today are covered by a range of issued and pending patents. Virtually all of the groups that have been involved in plant-derived vaccine activities have utilized the agrobacterium-mediated approach to plant transformation.
- Selectable marker systems (that allow for the identification of plant cells that have successfully taken up the DNA, and comprising the gene expression systems), such as kanamycin (nptII), mannose-phosphate-6-isomerase, among others.
- Transcription regulatory elements (to ensure that the introduced genes are expressed in plants), including promoters (constituitive and/or tissue specific), and transcription terminators (terminator nucleotide sequences), which are quite often NOS or rubisco E9 terminator sequences.
- Sub-cellular targeting systems (used to "guide" the transcribed products into specific cellular organs), such as rubisco subunits and plastid signal sequences
- Related technologies (such as adjuvants, and product formulation and immunomodulatory technologies).
- Bioprocess engineering for extraction and processing.

An additional complication is that most plant-derived vaccine projects are developed through the collaborative efforts of a range of research institutions, including private companies and academic institutions. Materials often change hands periodically during the development program, possibly in conformity with material transfer agreements that stipulate certain restrictions. Research agreements must be developed for all of these collaborative efforts. The agreements must address what will happen if such inventions are developed jointly. Further, nasal administration of vaccines may require access to a number of patents, which may be difficult to obtain.

Despite the complexity, the task is manageable. Corporations typically manage their intellectual property in a strategic manner. This entails, among others, significant in- and out-licensing activities to obtain FTO as part of an integral element in their product development strategy. In contrast, public institutions are generally less experienced with FTO procedures. A better understanding of IP management will allow these institutions to take advantage of the flexibilities in IP systems. In the United States, for example, groups can undertake research without a license on patented technologies if the goal is to generate data for the regulatory requirements of the U.S. Food and Drug Administration (FDA).

While a patent thicket exists for plant-derived vaccines in industrialized countries, very few of these patents have been filed in developing countries. The absence of many patents in developing countries simplifies matters significantly with respect to humanitarian use and also facilitates commercial applications in developing countries. It does not, however, reduce the overall need for IP management in order to obtain FTO.

There are several models of humanitarian-use licensing where patent rights are effectively pooled. One example is the approach used by the developers of the biotech rice containing pro-Vitamin A, called "Golden Rice." The developers of Golden Rice encountered many of the FTO issues that face developers of plant-derived vaccines. An FTO assessment revealed that Golden Rice was related to over 70 patent applications and issued patents, most notably in the United States and Europe, and that patent applications were owned by over a dozen institutions. Few patents were applied for or issued in developing countries. However, because the material was developed in Europe, it could not be transferred for use in developing countries without proper licenses. There were a few reasons for this, not the least of which was that several material transfer agreements were limited to research use only. Thanks to the publicity surrounding Golden Rice and the seriousness of vitamin A deficiency in developing countries, these patent constraints were resolved in only a few months. The public and private organizations that held relevant patents made them available at no cost to the inventor, who, in turn, granted one single license for all the necessary intellectual property to developing country institutions. Golden Rice serves as a useful model of how to approach the owners or assignees of proprietary technologies for royalty-free access for humanitarian uses.

One important difference between nutritionally enhanced rice and plant-derived vaccines is that the vectors and gene-expression components used to produce Golden Rice were assembled without advance consideration of intellectual property and FTO. Thus, the way forward with plant-derived vaccines should proceed more smoothly than it did with Golden Rice with respect to IP issues. Preliminary analysis and continued review of the IP landscape, however, are essential elements in the development of plant-derived vaccines. While it is relatively easy to put the different pieces into place, managing the process, in tandem with scientific advancements and the development of the product, remains a major challenge.

Based on a preliminary review of a specific plant-derived vaccine against hepatitis B virus, it was concluded that (1) the IP issues are fairly clear, although additional FTO analysis will be required to address specific cases, (2) the issues can be addressed with straightforward IP management approaches, and (3) the impact on the cost of finished vaccine is expected to be minimal. If a great deal of the work is conducted in developing countries, the IP management issues will be significantly simplified, since a number of the relevant patents may not have been filed in developing countries and thus the need for licenses would be reduced significantly (unless the products are exported to countries where a patent thicket existed).

2.2 *Types of intellectual property and material property rights associated with plant-derived vaccines*

Increasingly, IP rights influence every stage of vaccine development. In this section, the specific aspects of IP management are considered as tools to (1) achieve freedom to operate, (2) capitalize on new inventions, and (3) achieve the highest possible level of accessibility and affordability in developing countries. The relevant IP includes patents, trademarks, know-how/trade secrets, plant variety protection (PVP), and tangible property (such as research materials obtained through agreements). For practical purposes, we consider IP management at three different levels:

- incoming third-party intellectual property
- newly generated intellectual property, and
- outlicensed intellectual property

2.2.1 *Third-party intellectual property*

Third-party intellectual property considerations relate to tangible and intangible property and the relevant contractual obligations.

Tangible property. The components of tangible property typically comprise plants, genes, vectors, and the conditions under which such material property was obtained. In most cases, public germplasm or varieties are available (including corn, tomatoes, and tobacco). Whereas scientists in public research institutions typically prefer to obtain such materials from colleagues, the resulting material transfer restrictions should not be underestimated. In the private sector, it would be more typical to have genes synthesized, which avoids the material transfer restrictions on the genes.

Other tangible property issues involve the machinery required for bioprocesses.

Intangible property. The intangible property aspects are often more complex. Among the reasons for this complexity is that intangible property takes many forms, including utility patents, trademarks, trade secrets/know-how, plant variety protection/plant breeders' rights and plant patents (including utility patents on plants).

- ***Utility patents.*** Much of the third-party intellectual property will be in the form of utility patents. A detailed FTO opinion will be based on the specific antigen, process, and market in which the products are to be sold. In countries where certain patents are not issued, licenses will not be required either for the production or the sale of such vaccines.

- ***Plant variety protection/plant breeders' rights, plant patents (United States only) and utility patents on plants (mainly United States).*** Depending on which crop is being used, different types of intellectual property may apply. For example, it is becoming increasingly common for companies and universities alike to seek utility patents on inbreds and hybrids of corn, and for varieties of soybeans, cotton, fruit trees, and ornamental plants. If such protected material were used, a license may need to be obtained to use the plant or export it for production in other countries. Similarly, with the advent of new PVP regulations (under the 1991 UPOV [International Union for the Protection of New Varieties of Plants] treaty), a variety with PVP could not be used to produce plant-derived vaccines within the duration of the certificate's validity, because inserting one gene or a set of genes would make it an "essentially derived" or protected line.[6]

  However, many of the IP problems described here can be avoided if appropriate strategies are pursued from the outset. This could, for example, entail the use of public germplasm instead of proprietary varieties. Such a step may not be a feasible nor cost effective since some newer varieties might be the highest yielding or provide the highest regeneration efficiency during genetic modification work.
- ***Trade secrets/know-how.*** Some of the critical steps of bioprocesses lie in the know-how or trade secrets. Know-how refers to the knowledge of how something is produced, and not the specific components that constitute a product. Know-how can be licensed through appropriate confidentiality or secrecy agreements. Requirements for licensing, however, vary widely from country to country and certain information may not be legally protected in many jurisdictions.

Cost implications. Traditionally, in-licensed intellectual property has considerable impact on the cost and pricing of vaccines. Estimates of the licensing fees vary widely—from as high as 20% of sales prices for newly introduced vaccines, to as low of 2% for haemophilius influenzae type B. However, this comparison of royalty rates does not help much when it comes to plant-derived vaccines, since the total royalties of all in-licensed IP will depend on the type of product, the number of patents, and type of market. Manufacturing costs per vaccine can be reduced by economies of scale/increased production, but, in such cases, royalty fees are unlikely to be affected since they are generally fixed percentages of the sale price of each dose.

In terms of possible royalty rates for the hepatitis B model that has been mentioned in this chapter, it is perhaps premature to speculate on royalty ranges and licensing terms, since such speculation may influence the type of deal that could be obtained. Nevertheless, it seems that reasonable royalty rates *in aggregate* would add no more than 1% to 5% to the estimated total production costs.

Finally, in addition to the costs related to in-licensed IP, IP-management-related expenditures will be incurred during the R&D phase. These include expenditures for FTO opinions, which will need to be commissioned well ahead of production. Typical FTOs cost $20,000 to $100,000, depending on the complexity of the technology.

3. DETAILED ANALYSIS FOR HEPATITIS B VIRUS VACCINE

3.1 *Research*

Since the decision of the Supreme Court of the United States on *Merck v. Integra Life Sciences* in 2005,[7] analysts contend that, with the broadened definition by the Supreme Court of the Hatch-Waxman Act[8] as it relates to data exclusivity, research in preparation of FDA approval is exempt from the requirement for research licenses. Although this broad conclusion has not been tested within specific circumstances in the lower courts, it is reasonable to assume for hepatitis B that there are no IP constraints during the research phase, until clinical trials are complete and, possibly, the submission of an investigational new drug (IND) application to the FDA.

3.2 *IP components*

3.2.1 *Patents related to the hepatitis B vaccine (HBV)*

Many of the existing patents related to HBV are unlikely to be relevant for a number of reasons. First, several surface antigens are either in the public domain or their patents are limited to parenteral[9] administration, rather than oral delivery, or the claims do not cover their production in plants. In addition, the patent issued in 1989 to Merck & Co, and the 1986 Chiron patent for the first recombinant vaccine (hepatitis B), will have expired by the time a plant-derived vaccine reaches the market. Furthermore, these patents seem to be limited to the production of virus-like particles in yeast only. A full FTO assessment will nevertheless be required to provide clearer answers and reveal other intellectual property related to the specific methods of production envisaged here.

3.2.2 *Plant transformation and antigen production in plants*

A preferred method of production for the HBV is through stable lines produced through agrobacterium-mediated transformation. The IP thicket related to agrobacterium is relatively complex and still evolving; at least one of the interference proceedings of agrobacterium-related patents filed prior to March 1995 is still ongoing, and no details on possible claims have been made public.[10] However, based on counterpart patents issued in Eurpoe, it is fair to assume that at least one license could be required from either Monsanto or Syngenta (since they, or companies they acquired, are presumed to have filed patents for agrobacterium-mediated transformation prior to March 1995).

The currently used plasmid is a derivative of the antigen pBin19 and may be covered by Monsanto patents. The promoter that drives the gene expression (CaMV 35S) and the selectable marker that allows for the selection of transformed cells (nptII) are both covered by Monsanto patents. Other patents may also cover the applications; these will be identified during an FTO. However, broad patents are not known to these authors that cover all transgenic corn, tomato, or tobacco. There might be some differences in agrobacterium-related patents depending on whether a monocotyledoneous or dicotyledoneous plant is used. These differences, however, will not materially affect the conclusions related to the key licensing requirements.

3.2.3 *Broad plant-made pharmaceutical patents*

Three of the most often cited patents related to plant-made pharmaceuticals for oral administration are the Curtiss-Cardineau patents (U.S. patents No. 5,654,184, 5,679,880 and 5,686,079), assigned to Washington University in St. Louis, but now owned by Dow. However, all claims of the three patents are limited to oral administration of "transgenic plants" or of "transgenic plant tissue." It is unclear whether the Curtiss-Cardineau patent would cover the oral administration of antigens "extracted" from plant tissue.

3.2.4 *Bioprocess facility*

Many aspects of a bioprocess facility (which is required for the extraction, purification and processing of the vaccine) are covered by the very broad U.S. patent No. 6,617,435 B2 and U.S. application No. 2004/0166026 A1 and, possibly, patents that are continuations, divisionals, foreign counterparts, reissues, reexaminations, and continuations-in-part of known patents. The former is assigned to the now-defunct Large Scale Biology (LSB) Corp. in Vacaville, California; the latter, if issued, also would be assigned to the successors of LSB Corp.[11] Since much of this bioprocess facility design would draw on the trade secrets and know-how of LSB Corp., a license from LSB Corp. or its successor would be highly desirable.

3.2.5 *Cost implications*

Production of plant-derived hepatitis B vaccines through plant transformation and antigen production in plants is expected to require a number of licenses. These should be obtainable, especially because the proof of concept has already been demonstrated and confidence built into the technology. In aggregate, licenses for HBV technology, plant transformation and broad molecular pharming patents, the total royalties should not add more than 1% to 3% to the cost of production. This estimate is based on common industry licensing practices.

Bioprocess patents are in a different category because know-how is important for the construction and operations of bioprocess facilities. Nevertheless, favorable terms for a license that would not exceed 1% to 3% of the cost of production could likely be obtained.

3.3 *New intellectual property*

3.3.1 *Utility patents*

During the development of plant-derived vaccines, certain new inventions will emerge that might be patentable. Aside from the typical inventions related to antigens, plant transformation systems, and related technologies, innovative business models and production processes might also be developed. Care should be taken in making decisions about whether or not the inventions should be patented, kept as trade secrets, or made public and consideration given especially to the best ways to make the plant-derived vaccine available at affordable prices to the neediest countries in the developing world. This goal is more likely to be achieved if a certain level of control over the vaccine is retained.

3.3.2 *Trade secrets/know-how*

Many critical aspects of the operations of bioprocessing facilities are valuable knowledge. In some jurisdictions, this knowledge can be protected under trade secret law. It is customary for any pharmaceutical production plant to keep its standard operating procedures as trade secrets, given the considerable time and resources involved in fine tuning operations. By extension, employees of such plants will need to be informed of procedures for keeping information confidential and should have related clauses in their employment contracts.

3.3.3 *Trademarks*

One expense that might be worth considering is the creation of a quality seal for all plant-derived vaccines that are made using the processes outlined in this chapter. Such trademarks could be valuable and would afford a level of quality assurance and control not otherwise available.

3.3.4 *Cost implications*

Obtaining IP protection through utility patents, and trademarks incurrs legal and government filing fees (especially if trademarks are pursued in multiple countries). (Trade secret protection, on the other hand, costs nothing.) There will also be expenses related to ongoing licensing negotiations. Nonetheless, the added cost for the protection of new intellectual property will undoubtedly be small compared to overall production costs. The expenses would likely add no more than US$10-100,000 per year to the cost of production. In time these costs can be recovered, and the IP may even lead to a modest royalty stream if licensed.

4. CONCLUSIONS

The chapter's survey of intellectual and material property issues was based on a cursory FTO review. We attempted to highlight key issues and estimated the possible costs associated with the resolution of these. As the current research emphasis evolves into a product development program with more downstream considerations, a detailed FTO will be required leading to in- and out-licensing of intellectual property. To successfully move the candidate vaccine through the various stages from research to commercialization will also require the development of a global access strategy to reach developing country markets.[12] For this, various components will need to be integrated, including regulatory aspects, manufacturing, access to markets/distribution, and trade. IP management then essentially becomes nothing but a useful tool for reinforcing the vaccine development and deployment/marketing strategy. ■

ANATOLE KRATTIGER, *Research Professor, the Biodesign Institute at Arizona State University, Chair, bioDevelopments-International Institute; and Adjunct Professor, Cornell University, PO Box 26, Interlaken, NY 14847, U.S.A. afk3@cornell.edu*

RICHARD T. MAHONEY, *Director, Vaccine Access, Pediatric Dengue Vaccine Initiative, International Vaccine Institute, San Bongcheon-7dong, Kwanak-ku, Seoul 151-818, Republic of Korea.* rmahoney@pdvi.org

---

1 Arntzen C, B Dodet, R Hammond, A Karasev, M Russell and S Plotkin. 2004. Plant-derived Vaccines and Antibodies: Potential and Limitations. *Vaccine* 23:1753-1885.

2 The *first generation* products are the agronomic traits (such as insect resistance, herbicide tolerance, and drought tolerance), and the *second generation* are nutritionally enhanced plants (including omega-3 fatty acid enrichment, vitamin A and E production, high oleic soybean oil, low saturate canola oil, and high beta carotene oilseeds).

3 ASU. 2006. *Blueprint for the Development of Plant-derived Vaccines for the Poor in Developing Countries.* Prepared by PROVACS-Production of Vaccines from Applied Crop Sciences, a Program of The Center for Infectious Diseases and Vaccinology. The Biodesign Institute at Arizona State University: Tempe. www.biodesign.asu.edu/centers/idv/projects/provacs.

4 Arntzen C, R Mahoney, A Elliott, B Holtz, A Krattiger, CK Lee and S Slater. 2006. *Plant-derived Vaccines: Cost of Production.* The Biodesign Institute at Arizona State University: Tempe. www.biodesign.asu.edu/centers/idv/projects/provacs.

5 Ibid. It is interesting to note that a sensitivity analysis that reduced the yield of antigen in the plant by a factor of three was also conducted. This is equivalent to increasing the required dose by a factor of three; all other variables such as capital and labor costs have little impact on final cost if they are varied within reasonable ranges. This sensitivity analysis shows that under worst-case conditions, the cost per dose of a product made in the US and prepared in a ten dose packet would rise to $0.09 from $0.06.

6 See, also in this *Handbook*, chapter 4.7 by M Blakney.

7 Justice Scalia drafted the Court's opinion. He wrote: *"As an initial matter, we think it apparent from the statutory text that 35 U.S.C. § 271(e)(1)'s exemption from infringement extends to all uses of patented inventions that are reasonably related to the development and submission of any information under the FDCA. Cf. Eli Lilly, 496 U.S., at 665-669, 110 S.Ct. 2683 (declining to limit § 271(e)(1)'s exemption from infringement to submissions under particular statutory provisions that regulate drugs). This necessarily includes preclinical studies of patented compounds that are appropriate for submission to the FDA in the regulatory process. There is simply no room in the statute for excluding certain information from the exemption on the basis of the phase of research in which it is developed or the particular submission in which it could be included."* Refer to 545 U.S. 193, 125 S.Ct. 2372, *Merck KGaA v. Integra Lifesciences I, Ltd., et al.* No. 03-1237. Argued 20 April 2005. Decided 13 June 2005.

8 The Hatch-Waxman Act introduced data exclusivity for medicines in 1984 and allowed for patent extensions of up to five years to compensate for the loss of patent life in meeting regulatory requirements. This came with a trade-off: data exclusivity for pharmaceutical drugs and vaccines was reduced, allowing producers of generic medicines to use the abbreviated new drug approval (ANDA) process of the U.S. Food and Drug Administration (FDA) to gain approval for generic equivalents within six months. See also Derzko NM. 2005. The Impact of Reforms of the Hatch-Waxman Scheme on Orange Book Strategic Behavior and Pharmaceutical Innovation. *IDEA* 45:165–265.

9 In other words, administration of a vaccine by a route that bypasses the gastrointestinal tract such as through the use of injections, patches, creams or sprays.

10 The filing date is important, since patents filed prior to March 1995, once issued, would be valid for 17 years from the date of issue (or 20 years from the filing date, whichever is longer) and may lead to so-called submarine patents that seem to appear from nowhere. This is because any patent filed prior to March 1995 is not published until issued. The rules changed as of March 1995: Any non-provisional patent application filed since then is published 18 months after filing.

11 It is likely that Kentucky Bioprocessing has at least non-exclusive licenses to a number of LSB Corp.'s patents.

12 Mahoney RT, A Krattiger, JD Clemens and R Curtiss III. 2007. The Introduction of New Vaccines into Developing Countries IV: Global Access Strategies. *Vaccine* (in press).

How Intellectual Property and Plant Breeding Come Together: Corn as a Case Study for Breeders and Research Managers

VERNON GRACEN, *Professor, Department of Plant Breeding and Genetics, Cornell University, U.S.A.*

ABSTRACT

Plant breeders and research managers need to understand how intellectual property (IP) restrictions on germplasm and traits affect freedom to operate for a breeding program. Access to patented germplasm and traits is restricted and can only be used under some form of material transfer agreement or similar contract. Patented materials have to be maintained under strict provisions of the contract. This adds to the cost of breeding, parent seed, and production programs. Moreover, maintaining separate versions and precise records of patented materials increases the number of seed lots that a program must maintain. For example, different versions of inbred lines of maize must be maintained for each patented trait. Otherwise, stacking two or more traits produces lines with each trait and also lines with every combination of those traits.

1. INTRODUCTION

As the manager of research and development at a major seed company for several years during the 1980s and 1990s, I saw firsthand how proprietary biotechnology transformed our business. Drawing on my experience, this chapter describes:

- the complexities of managing proprietary transgenic inbred lines, hybrids, and genes through the breeding, testing, parent seed, and hybrid production processes
- licensing and contracts relevant to the use of proprietary biotechnology in breeding programs
- tips to enable you to avoid costly errors in managing licensed biotechnology applications

Initially, you may wonder why it is essential that breeders and research managers learn how to manage proprietary biotechnology efficiently in any breeding program. The reasons are actually quite simple.

For breeders, a working understanding of the extra workload, costs, constraints, and potential benefits of using proprietary biotechnology is necessary to establish priorities for developing transgenic inbreds and hybrids. A breeder's lack of basic information about the licensing of proprietary biotechnology could be a costly waste of time, opportunity, and money. Ignoring issues associated with managing proprietary biotechnology will not make them go away. Indeed, the failure to make informed decisions about what traits to adopt and how to handle them will result in de facto decisions that may be neither desirable nor reversible.

For research managers, a working understanding of intellectual property (IP) in biotechnology is necessary to obtain freedom-to-operate (FTO) and to commercialize traits. Managers must understand the real costs of obtaining, backcrossing, increasing, and testing multiple biotech traits in order to properly allocate resources to breeding,

Gracen V. 2007. How Intellectual Property and Plant Breeding Come Together: Corn as a Case Study for Breeders and Research Managers. In *Intellectual Property Management in Health and Agricultural Innovation: A Handbook of Best Practices* (eds. A Krattiger, RT Mahoney, L Nelsen, et al.). MIHR: Oxford, U.K., and PIPRA: Davis, U.S.A. Available online at www.ipHandbook.org.

parent seed, and production programs. Finally, to make decisions about product development and release, managers must understand contractual obligations related to product quality and efficacy.

While I use corn as the example throughout this chapter, most of the principles discussed here are equally applicable to the breeding of almost any crop. So, as you read through this module, think how the experiences I share apply to your specific job.

2. OVERVIEW OF CORN BREEDING PROGRAMS

Traditional corn breeding programs in the developed world breed hybrid varieties for farmers' use. Hybrids in the United States today are mostly crosses between two inbred lines. New inbred lines are developed by selfing plants from a source population. Source populations could include open pollinated varieties, synthetics, or crosses between two or more inbred lines.

Successful commercial corn breeding programs today often start with source populations created by crossing two relatively elite inbred lines that both combine well with another line (tester) to produce hybrids exhibiting high levels of heterosis. The source population is then self-pollinated for seven to eight generations, with several hundred selfed families being selected and advanced during each selfed generation. After one to three selfed generations, the selfed families are crossed onto an inbred of a complementary heterotic group (tester) and the hybrid progeny are evaluated in replicated trials for yield and desirable agronomic traits. Lines from the selfed families that produce the best tester hybrids are advanced to further selfing generations and recrossed onto additional testers to produce new hybrids to evaluate. As the families are selfed, each generation becomes more and more homozygous, or inbred, eventually giving rise to new inbred lines. New inbred lines that produce new hybrids 5%–10% better than the best current hybrids are advanced. New hybrids are evaluated over several hundred locations over two to three years before a selected few are released as new commercial hybrids.

The above process requires eight to ten generations of selfing and three to five concurrently run years of hybrid testing. Each year of testing is called a stage, so that hybrids advance from stage one to stage five of testing. Each successive stage is marked by fewer hybrids grown at more locations. The first three stages typically are composed of two replicated plots of each hybrid, approximately 1/1000th acre in size, grown at ten to 100 locations. The last two stages are usually produced on strips of ten to 20 rows of each hybrid, planted under farm conditions. Historically, the development of new inbred lines has taken eight to ten years. Advances in data collection and analysis technology, and the use of off-season nurseries to grow additional generations per year, can cut the development time for new inbred lines to five or six years. With concurrent testing of new hybrids, the entire process can be shortened to six or eight years.

The development of transgenic corn containing proprietary insect resistant (Bt genes) and herbicide tolerant (Roundup Ready® and LibertyLink®) genes creates additional expense and workload for corn breeding programs. Each new gene or combination of genes must be incorporated into existing and newly developed elite inbred lines, requiring multiple generations of backcrossing. In addition, new versions of hybrids carrying each proprietary gene need to be generated and tested in replicated trials over many locations for several years. Since the proprietary genes are legally protected, usually by utility patents, corn breeders must obtain FTO for use of the new genes. This requires licenses and contracts that are sometimes quite complex.

3. MODIFICATIONS TO CONVENTIONAL CORN BREEDING PROGRAMS

3.1 *Conventional breeding programs*

Commercial corn breeding programs are fast paced and very competitive. Competitive breeding programs rapidly adopt new information technologies and biotechnologies. Developing new corn inbred parents and competitive new hybrids historically took ten years or longer. The basic process can require eight to ten generations to obtain new homozygous inbred lines to use as parents, and four to five years of testing combinations to select new hybrids for commercial release. If done in sequence, this would require 12 to 15 years to develop a new hybrid. If breeders initiate hybrid trials during the years when new inbred lines are being self-pollinated, they can effectively cut the time required to ten years or less. A fast-track breeding protocol using off-season breeding nurseries (to provide two, and sometimes three, generations of self-pollinating lines per year) can decrease the time required to develop new homozygous inbred lines and hybrids to seven or eight years (Figure 1). To produce an additional one or two generations per year, it is essential that breeding programs utilize new technologies to harvest trials of experimental hybrids and to select lines to advance in off-season nurseries.

3.2 *Super-fast-track conversion programs*

Starting in the 1990s, breeders developed, through plant transformation, corn lines into which proprietary genes from organisms un-

Figure 1: Fast-Track Inbred Development and Hybrid Testing Protocol for Corn

Year 1		
Winter-1		Cross inbred 1 and inbred 2
Winter-2	F_1	Self
Summer	S_1	Self and cross onto tester
Year 2		
Winter 1	S_2	Self
Summer	S_3	Self and evaluate early generation tests
Year 3		
Winter 1	S_4	Self and cross onto more testers
Summer	S_5	Self, evaluate stage 1 hybrids, cross new lines onto (such as cytoplasmic male sterility [cms], insect resistance [Bt], or Roundup Ready® [rr])
Year 4		
Winter	S_6	Self
Summer	S_7	Self and evaluate stage 2 hybrids
Year 5		
Summer		Evaluate stage 3 hybrids
Year 6		
Summer		Evaluate stage four hybrids "on farm"
Year 7		
Summer		Evaluate stage five hybrids "on farm," make hybrid release decisions

related to corn were inserted into the corn genome. Important traits, such as insect resistance (Bt) and tolerance to herbicides (Roundup Ready® and LibertyLink®), were developed and made available to the seed corn industry. These traits were rapidly accepted by the industry worldwide, dramatically changing traditional corn breeding.

The genes for insect resistance and tolerance to herbicides provided traits that were advantageous for corn farmers; however, the first sources of these genes were in corn lines that were not very competitive. In order to be commercially useful, the genes had to be incorporated into elite inbred lines that produced competitive hybrids. The process of incorporating a new gene into a corn inbred line usually requires between seven and eight backcross generations, during which a source of the new gene is crossed to an elite inbred line. After this, selected progeny are back crossed onto the elite inbred line for seven or eight generations (Figure 2). Even if you used two or three backcross generations per year by employing off-season nurseries, you would still need three years to recover a version of an elite line that was essentially identical to the original inbred line but also expressed the new gene. Unfortunately, because every year new hybrids are developed that out-perform older hybrids by 5%–10%, the half-life of many corn hybrids today is three to five years. This means that by the time you could convert the parents of a commercial hybrid to a new gene through traditional backcross procedures, the sales of the hybrid would likely be in decline.

Figure 2: Backcross Breeding Protocol

Year 1			
Winter	Cross elite inbred	Source of a new gene	F_1
Summer	Elite inbred	F_1	BC1
Year 2			
Winter	Elite inbred	BC1	BC2
Summer	Elite inbred	BC2	BC3
Year 3			
Winter	Elite inbred	BC3	BC4
Summer	Elite inbred	BC4	BC5
Year 4			
Winter	Elite inbred	BC5	BC6
Summer	Elite inbred	BC6	BC7
Year 5			
Winter	Elite inbred	BC7	BC8
Summer	BC8 Selfed as new version of elite inbred		

Thanks to new technologies involving molecular markers, however, it is possible to backcross a new gene into an elite inbred in three to four total generations, rather than seven to eight.[1] This means that a breeding company can utilize a super-fast-track conversion program to backcross proprietary genes into elite inbred parents before the hybrids produced become obsolete (Figure 3). Seed companies are therefore able to acquire new genes and transfer them very rapidly into elite inbred lines. Of course, super-fast-track conversion programs are not cheap. The use of off-season nurseries and molecular markers to obtain the rapid conversions adds considerable labor and expense to the process of commercial corn breeding. Also, breeding companies must obtain regulatory approval for the gene construct being converted. Obtaining regulatory approval in countries normally used for off-season nurseries, such as Mexico, Chile, and Argentina, is difficult and time consuming. This means that off-season nursery conversion must be done on U.S. soil, basically in Hawaii, Florida, and Puerto Rico, creating additional expense.

4. CRITICAL BREEDING DECISIONS

4.1 *Which lines and how many to convert?*

A typical corn-breeding company sells a number of specific hybrids of different maturities and geographical adaptation. The major seed corn companies usually have ten to 20 elite inbred lines in commercial use, plus several hundred new lines nearing inbred status in the developmental pipeline. The decision about which, and how many, inbreds to enter into a fast-track conversion program requires a lot of thought and often some bold decisions. Since financial resources dedicated to research and development are limited, directing funds to fast-track conversion often requires redirecting resources away from use in conventional breeding. Critical decisions about how much fast-track conversion you can afford are often difficult to make.

4.2 *Which genes and how many to convert?*

A number of transgenes that are available from biotech companies have been inserted into corn. Each of these genes has different uses in different genetic backgrounds. The usefulness

FIGURE 3: SUPER FAST-TRACK CONVERSION PROTOCOL

YEAR 1				
Winter	Elite inbred	Source of a new gene	F_1	
Summer	Elite inbred	F_1	BC1	BC1 progeny selected with PCR markers
YEAR 2				
Winter	Elite inbred	Selected Progeny at @ BC4 generation	BC5	BC5 progeny selected with PCR markers
Summer	Elite inbred	Selected progeny at @ BC8 generation	BC9	BC2
YEAR 3				
Winter	BC9 Selfed as new version of elite inbred			

of each gene must be monitored during the conversion process, since each gene may offer a trait desired by at least one segment of the population of farmers a seed company serves. Additionally, contractual restrictions often determine how and where genes can be deployed. Breeders must test lines undergoing conversion to measure the level of gene expression and to demonstrate that all plants undergoing conversion carry the gene in an active form. It is expensive to incorporate each gene into elite and newly developing inbred lines. It is even more expensive to do all the testing required by licensing agreements. In addition, each converted line must be tested in hybrid combinations that contain each gene, as compared to the same hybrids without the genes, to demonstrate that genetically modified, or GM, hybrids perform as well as non-GM counterparts. Of course, if different genes provide traits that are desirable individually, then the combination of two or more genes in the same hybrid offers an even more desirable product. Unfortunately, each gene needs to be transferred individually (Figure 3), exponentially increasing the costs of converting each line.

5. PROPRIETARY BIOTECHNOLOGY AND HYBRID DEVELOPMENT AND TESTING

5.1 *Conventional hybrid release process*

As new lines reach the second or third selfed generation, they are crossed onto one or several tester lines to generate hundreds of hybrids for evaluation. In stage one of hybrid testing, hybrids are evaluated at three to four locations in replicated, paired row plots (Figure 4). In stage two of testing, the best 10% of these hybrids are remade and tested in paired-row plots at ten to 20 locations. Subsequently, in stage three, the best 10% of stage two hybrids are advanced to paired-row plots at 50 to 100+ locations. The best of these, presuming that they have significant performance advantage over currently grown hybrids, are produced in quantities to allow testing at 100 to several hundred locations. For a period of two years, the hybrids are planted in paired-row plots and in strip plots (roughly one-tenth of an acre) and harvested using current farming practices; this comprises stages four and five of testing. After five years of small-plot and strip-plot testing at several hundred locations per year, the best-performing hybrids are approved for sale.

FIGURE 4: STAGES OF HYBRID TESTING

STAGE 1	Hundreds of new hybrids, tested in paired-row plots, 1/1000th of an acre each in replicated trials, at three to five locations
STAGE 2	The best at 10% of stage one hybrids, tested in paired-row plots in replicated trials, at ten to 50 locations
STAGE 3	The best at 10% of stage two hybrids, tested in paired-row plots in replicated trials, at 30 to 100 locations
STAGE 4	The best ten to 15 hybrids from stage three, tested again in paired-row plots, replicated at 30 to 100 locations, and also tested in one-tenth-acre strip plots on farms at 100 to 200 locations
STAGE 5	The best five to ten hybrids from stage four, tested again in paired-row plots and in strip plots

5.2 *GM hybrid test process*

GM hybrids are hybrids that contain proprietary biotech traits introduced into corn from other species through plant transformation. These GM hybrids present several challenges to the hybrid release process. First, with new gene constructs, hybrid evaluation trials must be done under an experimental use permit. This imposes restrictions on the number of hybrids and testing locations, which means that fewer hybrids can be evaluated and more years are needed to obtain data sufficient to justify commercial release.

Second, licensing agreements often impose, for each hybrid, stringent requirements for degree of expression of proprietary genes. This requires expensive, time-consuming tests to be run on all hybrids being evaluated.

Finally, the number of hybrids that must be tested increases with every new proprietary gene or combination of genes used. Even if only three new genes are used, the number of hybrids to be tested in early generations goes from several hundred to nearly one thousand. If combinations of each of the three genes are developed, you can approach two thousand hybrids to test in early generations. Even at the later stages of testing, strip tests at several hundred locations per year can increase from eight to ten new hybrids, in conventional programs, to 40 to 50, if three genes with some two-way combinations are tested. Consequently, the number of genes and hybrids must be carefully selected or the costs and logistics become prohibitive.

Fortunately, breeders can use a fast-track hybrid release process to speed the release of new GM hybrids. If there are no detrimental effects from the proprietary genes being incorporated, and the backcross conversion process is carefully monitored to get converted lines that differ from the elite line by only one to a few genes, then performance of hybrids involving the converted lines will be very similar to the performance of hybrids involving the elite, nonconverted lines. Therefore, it is possible to decrease the five-stage, five-year testing process to three years. Usually, the converted versions of hybrids are tested only at stages three, four, and five. This means that once elite inbreds are fully converted to a proprietary gene, hybrids carrying that gene could be released within three years.

6. PARENT SEED AND HYBRID PRODUCTION

6.1 *Conventional process*

Traditionally, new inbreds are advanced from research programs to parent seed programs when the inbred performs successfully in one or more hybrid combinations in stage three of research testing, usually the third year of multilocation testing across a wide geographic area. Once advanced, the parent seed department starts increasing seed of the new inbreds and producing seed of the new hybrid combinations to build up quantities needed for commercial release. Often, three generations of seed increase are needed to produce enough inbred seed of a new female parent to allow for seed sufficient for commercial release.

Normally, only one of three or four new inbreds that make it to stage three of testing actually makes it to commercial release. During testing stages four and five (strip tests on farms at many locations for two or more years), many hybrids containing new inbreds are dropped. The seed of these inbreds and new hybrids is subsequently discarded.

6.2 *GM parent seed and hybrid production process*

Each biotech trait added to an inbred produces another version of the inbred that must be increased prior to potential commercial release. So, rather than increasing one version of a new inbred, you have to increase two, three, or even more versions, many of which are never sold in any hybrids. This greatly increases the costs associated with producing hybrid and inbred seed.

7. LICENSING AND CONTRACTUAL ISSUES WITH GM TRAITS

Proprietary GM traits and converted varieties are usually protected by some form of intellectual property (IP) protection, which defines ownership of the traits, plants, or technologies. This protection may be in the form of utility patents, plant variety protection certificates, or trade secrets. Most transgenic plants embody numerous components and processes, each of which may have IP protection. You must make sure that anyone that supplies you with proprietary traits has legal access to all proprietary components and processes used in developing the genetically modified, or GM, traits. Suppliers of proprietary traits should be willing to include appropriate warranty clauses into any agreement you execute that protects you from any IP protection infringement that may arise from commercializing the traits.[2]

Several types of legal agreements are available for gaining access to proprietary traits and technologies. These may be as simple as material transfer agreements (MTAs), or as complex as commercial licensing agreements. Often, you can gain early access to proprietary genes and technology under research agreements. These allow you to obtain and incorporate proprietary genes into your germplasm, evaluate performance, and then choose only those genes that meet your commercial objectives before having to negotiate terms of commercialization. Proprietary genes and technology that you choose not to commercialize must be returned and plants containing those genes destroyed. This allows you to test a wide range of genes/technologies without having to pay royalties or fees. However, you should ensure that such research agreements contain a mechanism that allows you to commercialize those genes/technologies that you do select. Often, commercial agreements require an up-front payment to access the genes, and afterwards royalty payments based on volume and the price of products sold containing the proprietary genes. If you do not reach an agreement with the gene supplier regarding terms of commercialization before starting your research, you ought to at least agree that you will be offered terms comparable to the seed industry standard.

The contracts or licenses required to get access to proprietary genes often contain strict limitations on what you can and cannot do with the genes. It is important that all personnel who have access to the proprietary genes understand these requirements. Also, these contracts often contain specific tests or measurements that you must conduct to verify the purity and efficacy of the genes after you have crossed them into your germplasm. These tests take time and money to perform and sometimes require breeders to learn new skills.

Newly developed proprietary traits also must be approved by governmental regulatory agencies. Until approval is obtained, the traits must be grown under experimental use permits. These restrict the size and number of test plots that you can plant and require a lot of supervision and documentation. Experimental use permits also restrict your use of off-season nurseries. You cannot grow a GM trait in any country that has not approved the trait. This prevents the use of Mexico, Chile, or Argentina for off-season nurseries, which forces you to use Hawaii, Florida, or Puerto Rico. This raises costs and limits the flexible use of off-season nurseries.

8. CONCLUSION

Since this chapter was originally written, several proprietary biotech traits have been commercialized on large acreages throughout the world. As traits like the Bt gene have become commonplace in breeding programs, new source populations have been established in which both parents contain the Bt gene. This eliminates the need for fast-track or super-fast-track conversions and reduces the complexity of producing hybrids with that trait. However, as Bt and Roundup Ready® became commonplace, new transgenic traits have appeared. Thus, as companies reduce the workload and expense associated with the first generation of transgenic traits, new traits are increasing the complexity again. Also, transgenic traits for such crops as soybeans, cotton, and canola have been developed, extending the complexity to other crop breeding programs. This cycle of managing trait complexity will continue until the traits are no longer competitive, or until the patents expire. Many of the patents on first generation traits, and on the first patented inbred lines and hybrids, were issued in the last half of the 1980s, which means that both the traits and patented inbreds became public property starting in 2006. This could have a large and positive impact on plant breeding programs, since programs will be able to access and utilize these off patent materials without restrictions. Several inbreds from Pioneer Hi-Bred International Inc. (now a DuPont Company) and DeKalb Genetics (now owned by Monsanto) were applied for in 1986 and subsequent years. The patents are valid for 20 years after the application date. That means that the first inbreds patented came off patent in 2006. Each year additional inbreds will come off patent. Even though 20 years old, some of these inbreds represent significant sources of elite gene combinations representing some unique heterotic groups that could upgrade public plant breeding germplasm in the temperate world. As I understand it, seed of the patented inbreds is supposed to be maintained by the American Type Culture Collection and made available upon request from the U.S. Patent Office for the purpose of demonstrating the validity of the material patented. Presumably, seed will not be maintained after the patents expire. ■

VERNON GRACEN, *Professor, Department of Plant Breeding and Genetics, 520 Bradfield Hall, Cornell University, Ithaca, NY, 14853, U.S.A. vg45@cornell.edu*

---

1 See the section Plant Breeding 2 in citnews.unl.edu/hscroptechnology/lessonFrames.html for a review of marker assisted back crossing.

2 Kowalski SP, RV Ebora, RD Kryder and RH Potter. 2002. Transgenic crops, biotechnology and ownership rights: what scientists need to know. *The Plant Journal*. 31 (4): 407–21)

Successful Commercialization of Insect-Resistant Eggplant by a Public–Private Partnership: Reaching and Benefiting Resource-Poor Farmers

AKSHAT MEDAKKER, *Associate Consultant-Technology Management, Sathguru Management Consultants Pvt. Ltd., India*
VIJAY VIJAYARAGHAVAN, *Founder and Director, Sathguru Management Consultants Pvt. Ltd., India*

ABSTRACT

This chapter looks at the results of a unique public–private partnership instituted to provide resource-constrained farmers in the developing world with access to proprietary agri-biotechnologies. Eggplant, a widely consumed vegetable crop in the tropics, is commonly infested by the eggplant fruit and shoot borer (EFSB), which devastates both plants in the field during development and eggplant fruits after harvesting. The chapter considers the application of insect-resistance technology (based on the Cry1Ac protein from Bacillus thuringiensis) in eggplant, focusing on its sublicensing from a private company to a partnership of public institutes and agricultural universities in Bangladesh, India, and the Philippines.

1. INTRODUCTION

Eggplant (*Solanum melanogena*) is an important vegetable crop widely cultivated and consumed in the subtropical and tropical regions of Asia and Africa. It grows in a wide range of climatic conditions and is a staple of human consumption. About 510,000 hectares of arable land in India and 20,000 hectares in the Philippines are devoted to cultivating eggplant.

A long-duration crop, eggplant is grown using either hybrid varieties or open-pollinated varieties (OPVs, for which seeds can be saved and used later). Although much preventive care is taken, eggplant is commonly attacked by more than a dozen insect-pest species. Among these species, the eggplant fruit and shoot borer (*Leucinodes orbonalis*), or EFSB, is the most widespread and devastating in South and Southeast Asia, with infestation inflicting about a 70% crop loss.[1] EFSB larvae feed inside the eggplant shoot and fruits, retarding the vegetative growth of the plant and decreasing the marketability and edibility of the fruit.

Many attempts to crossbreed eggplant varieties with EFSB-resistant wild varieties have been unsuccessful. So farmers have had to rely heavily on chemical pesticides to control EFSB. According to a study conducted on pest control for eggplant in South Asia, farmers spend about US$400 per hectare on pesticides, two-thirds of which are used to control ESFB.[2] In addition, EFSB populations have gradually become resistant to certain chemicals, so farmers have resorted to using other chemicals, some of which are more hazardous to human health and to the environment, as well as illegal, to control the insect.

2. THE TECHNOLOGY

MAHYCO, a private Indian company, was the first in India to develop a hybrid eggplant containing a gene that provides resistance to EFSB. The gene it used (*cry1Ac* which produces the corresponding protein called Cry1Ac[3]) is obtained from *Bacillus*

Medakker A and V Vijayaraghavan. 2007. Successful Commercialization of Insect-Resistant Eggplant by a Public–Private Partnership: Reaching and Benefiting Resource-Poor Farmers. In *Intellectual Property Management in Health and Agricultural Innovation: A Handbook of Best Practices* (eds. A Krattiger, RT Mahoney, L Nelsen, et al.). MIHR: Oxford, U.K., and PIPRA: Davis, U.S.A. Available online at www.ipHandbook.org.

thuringiensis (Bt). Bt is a spore-forming bacterium that produces crystal proteins (called Cry proteins) that are toxic to many species of insects, including EFSB. Bt action is very specific. To become lethal, the Bt protein has to be ingested; the Bt toxin is activated in the high pH environment of the insect gut. The activated protein perforates the lining of the gut, which causes the death of the insect within a of couple days.

A main advantage of this technology is that it reduces the use of chemical pest control, thereby making the technology environmentally harmless. Through its safety tests, the U.S. Environmental Protection Agency has found no human health hazards related to Bt use. The agency has exempted Bt from its standards for food-residue tolerances and groundwater concentration, from endangered species labeling, and from special review requirements, indicating that cultivation of crops using Bt is safe for resource-constrained farmers in the developing world.

3. THE LICENSING ARRANGEMENT

MAHYCO is the first Indian company to have received the rights under license for the use of the Bt *cry1Ac* gene technology for insect-pest management from Monsanto Company. This licensed *cry*-gene technology was used by MAHYCO to develop and generate hybrid eggplant events. Under the aegis of the Agricultural Biotechnology Support Project II (ABSP II), funded by the U.S. Agency for International Development, Sathguru Management Consultants Pvt. Ltd. partnered with MAHYCO. The *cry*-gene technology was licensed then to several public institutes in South and Southeast Asia that were participating in a public–private consortium created to develop EFSB-resistant OPV eggplant that would improve the conditions of resource-constrained farmers in developing countries. The ABSP II played a pivotal role in this venture by funding all the consortium partners for their R&D roles in developing the EFSB-resistant eggplant.

The technology was sublicensed by MAHYCO on a royalty-free basis to public research institutes in India (the Indian Institute of Vegetable Research, Tamil Nadu Agricultural University, and the University of Agricultural Sciences, Dharwad), in Bangladesh (the Bangladesh Agricultural Research Institute), and in the Philippines (the University of Philippines, Los Banos). MAHYCO also sublicensed this technology to East West Seeds, a private corporation in Bangladesh, on commercial royalty-bearing terms. To safeguard the licensor's interests, specific strategies for the stewardship and monitoring of the technology by the licensees were addressed and formulated early in the sublicensing process.

4. TRANSGENIC EGGPLANT

Most eggplant farmers in India grow OPVs. The area planted with hybrid varieties is less than 30% of the total area. Growers that plant these hybrid varieties also tend to use more purchased inputs and have higher yields compared to growers who plant OPVs.[4] The main reason that the cultivation of OPVs is more widespread is that OPV seeds can be saved and replanted in future growing seasons. As a result, OPV seeds are much more available and affordable. The market price of hybrid seeds is five to ten times the market price of OPV seeds.

The first transgenic Bt hybrids developed by MAHYCO are slated to be commercially released in India by the end of the 2006–2007 season,[5] after the fulfillment of all regulatory requirements. The transgenic Bt OPVs under development by the public–private partnership are expected to be commercialized about six months later. Because of the existing price differential between conventional OPVs and hybrids, and because of the zero premium being charged for the Bt trait in the OPVs, it is still expected that most of the existing growers of hybrid eggplant will adopt the Bt hybrids rather than the Bt OPV, even though the Bt OPVs would be priced much lower than the Bt hybrids. This is primarily due to production and yield differences between the two systems. Farmers growing OPV eggplant are most likely to adopt the Bt OPV because of the cost factor. Growers of both types of eggplant can be expected to shift to the corresponding Bt versions because of the expected savings in pesticide expenses.[6]

The public–private partnership also addresses distribution issues: the participating public institutions will be able to deliver high-quality Bt eggplant seeds that are resistant to EFSB through their own public distribution systems on a cost basis (in other words, without adding profit margins).

Most resource-constrained farmers in the developing world cultivate OPVs because of the lower costs involved. By recognizing these agricultural practices, and by providing the public sector with access to Bt technology for use in OPVs, via a unique public–private partnership, MAHYCO both commercializes its Bt hybrid eggplant (sold on a for-profit basis) and through its donation addresses the need to improve crops of vital importance to poor farmers. ■

AKSHAT MEDAKKER, *Associate Consultant-Technology Management, Sathguru Management Consultants Pvt. Ltd., 15 Hindi Nagar, Punjagutta, Hyderabad 500034, India. akshatm@sathguru.com*

VIJAY VIJAYARAGHAVAN, *Founder and Director, Sathguru Management Consultants Pvt. Ltd., 15 Hindi Nagar, Punjagutta, Hyderabad 500034, India. vijay@sathguru.com*

1 Dhandapani N, UR Shelkar and M Murugan. 2003. Bio-Intensive Pest Management in Major Vegetable Crops: An Indian perspective. *Journal of Food, Agriculture & Environment* 1 (2): 330–39.

2 ABSPII. 2007. ABSPII Bt Eggplant Insect Resistance Management Strategy for the Eggplant Fruit and Shoot Borer. ABSPII, Cornell University: Ithaca, NY. www.sathguru.com/absp2/irmstrategy.pdf.

3 For a review, see Krattiger AF. 1997. Insect Resistance in Crops: A Case Study of *Bacillus thuringiensis* (Bt) and Its Transfer to Developing Countries. ISAAA Briefs No. 2. ISAAA: Ithaca, NY. www.isaaa.org/kc/Publications/pdfs/isaaabriefs/Briefs%202.pdf.

4 Kolady DE and W Lesser (Cornell University). Personal communication, October 2006.

5 Krishna V and M Qaim. 2006. Estimating the Adoption of Bt Eggplant in India: Who Benefits from Public-Private Partnership. University of Hohenheim: Stuttgart. www.envfor.nic.in/divisions/csurv/geac/so_brinjal_part-II.pdf.

6 Kolady DE and W Lesser. 2006. Who Adopts What Kind of Technologies? The Case of Bt Eggplant in India. *AgBioForum* 9 (2): 94–103.

The University of California's Strawberry Licensing Program

ALAN B. BENNETT, *Associate Vice Chancellor, Office of Research, University of California, Davis; and Executive Director, PIPRA, U.S.A.*

MICHAEL CARRIERE, *Business Development and IP Manager, Office of Technology Transfer, University of California, U.S.A.*

ABSTRACT

The strawberry improvement program located at the University of California, Davis focuses on breeding cultivars for the strawberry industry in California, yet today it supports the majority of production of fresh-market strawberries globally. Around the world, UPOV-compliant Plant Breeders' Rights (PBR) are the most common form of IP protections sought by University of California (UC) to protect its strawberry cultivars. Inside the U.S. and Canada, cultivars are licensed on a nonexclusive basis directly to nurseries. Outside of the U.S. and Canada, UC relies on business partners, referred to as "master licensees," as intermediaries. A master licensee is provided with exclusive rights within a defined territory that includes the right to issue nonexclusive sublicenses to nurseries within that territory. Overall, a three-tier royalty structure is utilized, with growers inside California paying the least, growers in the U.S. outside of California and in Canada pay slightly more, and all other growers pay even more, a percentage of which is shared with the master licensee. The ultimate future of the UC strawberry breeding program is tied to the continued development of competitive cultivars, but the team is highly skilled and, partly due to the licensing program, funding is stable.

1. INTRODUCTION

The strawberry improvement program located at University of California, Davis[1] focuses on breeding cultivars for the California strawberry industry. University of California (UC) strawberry cultivars are developed for the cool coastal Mediterranean and arid subtropical regions of California and have become the basis of a global fresh-market strawberry industry. UC cultivars represent 75%–80% of the production of the US$1.3 billion California strawberry industry and represent 50%–60% of worldwide production. The UC strawberry licensing program is active in the United States, Europe, Asia, Africa, South America, and Australia and generates an annual licensing revenue stream of US$4.5 million. This case study summarizes patent portfolio development, licensing strategy, and income trends for this successful university licensing endeavor.

2. IP PORTFOLIO DEVELOPMENT

Newly developed UC strawberry cultivars are protected in the United States under U.S. plant patents administered by the U.S. Patent and Trademark Office (PTO). A U.S. *plant patent* is available for asexually propagated plant species while *plant variety protection certificates,* administered by the U.S. Department of Agriculture (USDA), are reserved for the protection of sexually propagated species. Outside counsel is utilized by UC to secure U.S. plant patents for strawberry cultivars.

In ex-U.S. jurisdictions, U.S.-based patent counsel directs the prosecution of intellectual property in cooperation with ex-U.S. counsel. Counsel outside of the United States is often identified by the licensee in the respective terri-

Bennett AB and M Carriere. 2007. The University of California's Strawberry Licensing Program. In *Intellectual Property Management in Health and Agricultural Innovation: A Handbook of Best Practices* (eds. A Krattiger, RT Mahoney, L Nelsen, et al.). MIHR: Oxford, U.K., and PIPRA: Davis, U.S.A. Available online at www.ipHandbook.org.

tory. Worldwide, the process of obtaining IP for plant cultivars is a specialized area of IP prosecution and this reduces the pool of capable attorneys in a given territory. Additionally, plant-based IP is a new legal construct in some territories where UC seeks protection for strawberry cultivars. These factors complicate the process of identifying competent, cost-effective representation and emphasize the importance of in-country licensees in selecting legal representation. Ex-U.S. licensees are ultimately responsible for bearing the cost of IP prosecution in their territory. Since their business models depend on strong IP, they are motivated to aid in the search for capable legal representation. In some territories outside the United States, UC has identified non-attorney plant IP specialists, but in most cases it relies on the services of registered patent attorneys that also specialize in plant-based IP.

UPOV-compliant Plant Breeders' Rights (PBR) is the most common form of IP sought for UC strawberry cultivars. (The Union for the Protection of New Varieties of Plants [UPOV] has set forth standards for licensing new plant varieties.) Although UC and its licensing partners worldwide seek UPOV-compliant PBR for UC strawberry cultivars, such protection is unavailable in some territories. As a result, the UC licensing program and its master licensees are active in expanding the scope of protection for plants in some countries worldwide. A successful approach has been to build grassroots support for plant IP by coupling access to cultivars with availability of IP for those cultivars. For example, in China a strawberry industry organization successfully lobbied governmental authorities to add strawberry to the list of protectable species. This action was encouraged by UC's licensee for China. With PBR now available for UC strawberry cultivars in China, the Chinese strawberry industry gains access to UC cultivars, which leads to rural economic development in China, and UC licensing expands into the Chinese market. In Egypt, UC strawberry cultivars represent Egypt Plant Patent Nos. 1, 2, and 3, as a result of aggressively pursuing access to the nascent Egyptian plant patent system. In Brazil, UC strawberry cultivars are among the first protected strawberries under the new system of protection.

The decision to file or to engage in expanding the scope of IP for a given territory is made jointly between UC and the respective master licensee. The primary criterion is the expected value of the future licensing revenue stream. UC rarely files its strawberry cultivars "at-risk" (that is, without a licensee already identified in that jurisdiction). Master licensees are required to pay the cost of obtaining and maintaining both IP protection and commercial registration.

3. STRUCTURE OF DOMESTIC AND INTERNATIONAL LICENSING

In the United States and Canada, cultivars are licensed on a nonexclusive basis directly to plant nurseries. Nurseries are licensed the right to propagate plants and to sell the propagated daughter plants to fruit growers. Strawberry growers annually replant fruiting fields, so a royalty is collected annually. Royalties are assessed on a per-1,000 plants (purchased) basis rather than on the basis of sales.

Outside of the United States and Canada, the UC relies on business partners as an intermediary in support of the strawberry licensing program. These partners, referred to as master licensees, are provided with exclusive rights within a defined territory. The master licensee is granted the right to issue nonexclusive sublicense agreements to nurseries within the territory. In exchange for this exclusive right, the master licensee supports IP development and provides enforcement of IP rights including access to the local court system, as required. Critical responsibilities of the master licensee are market development, technical support, and the transfer of production know-how. In addition to being the local eyes and ears of UC's licensing function, the master licensee facilitates testing and evaluation of promising new cultivars. In exchange for the services provided by the master licensee, UC agrees to share a percentage of collected royalties.

A three-tier royalty structure is utilized. Growers of UC cultivars in California currently pay, in royalties, US$3.00/1000 plants. Growers in the United States outside of California and in Canada pay US$4.50/1000 plants. Outside of the United States and Canada growers pay US$10.50/1000, a percentage of which is shared with the master licensee. In addition to the royalty component described above, a research fee is collected to directly support new cultivar development. The research fee of US$1.00/1000 plants entitles the licensee to a lower royalty rate (rates stated above). The licensee receives a US$1.50 reduction in royalties for the US$1.00 research fee contribution.

The structure of the strawberry licensing program is driven in part by UC's presence as a public institution in the state of California. Nurseries and fruit growers in California are given preferential treatment, in addition to the reduced royalty rates for California. California-based nurseries (licensees) are the only nurseries in the worldwide licensing program that have access to all licensed markets. The sales territories of non-California nurseries are limited to a defined region. After the initial release of a new UC strawberry cultivar, its use is restricted to California for the first two years. This policy is designed to benefit fruit growers in the state who are concerned about competition in their own markets from UC cultivars grown abroad.

UC strawberry plants are shipped worldwide from California nurseries. To facilitate monitoring of worldwide strawberry plant shipments, an electronic, Web-based system is currently being developed with the goal of providing real-time shipping information for UC and its master licensees worldwide. Licensed nurseries will electronically declare sales before shipment. This pre-shipping electronic notification enables master licensees to accept or reject a proposed sale based on the intended use of the plant material and the licensing status of the recipient. The system is expected to reduce the occurrence of out-of-compliance shipments and provide the supplying nursery with assurance that its shipments are consistent with UC licensing policy worldwide.

4. INCOME TRENDS

For the latest fiscal year, gross annual income for the strawberry licensing program was US$4.7 million. Gross income increased from US$3.4 million in 2000 due to the combination of a rate increase in 2000 and market expansion in Europe, North Africa, South America, and Mexico. Approximately 45 percent of annual income is generated by California sales. Five percent derives from sales in the United States outside of California and in Canada. The remaining 50% of licensing income is derived from sales outside the United States and Canada. The largest non-U.S. markets, by country, are Spain, Mexico, Morocco, and Australia (from largest to smallest, within this group). In addition to royalty income, total research fee collection now totals US$650,000 annually and represents the lion's share of funding for the strawberry breeding program at UC Davis. This amount contrasts with the US$350,000 support from the California Strawberry Commission, the second largest contributor to the breeding program.

After 2007, income is expected to increase based on a 2006 rate increase and further market expansion. Over the next five years, market expansion is anticipated in Brazil, Northern Europe, China, and Turkey. Additionally, new licensing strategies are expected to boost income from established markets as master licensees will be given the opportunity to *price-to-market* in the high-value territories of the European Union and elsewhere.

The ultimate future of the program is tied to the continued development of competitive cultivars. The UC breeding team is highly skilled, and funding for the endeavor is stable. As a result, the UC breeding and licensing programs are positioned for success for at least the next 10 years.

5. CONCLUSIONS

The strawberry licensing program of the University of California provides a clear example of how intellectual property protection by a public sector institution enables the global dissemination of innovative results by providing an economic stimulus to those who adopt the technology. It also allows those who benefit most directly from the technology to help sustain financially the program that serves them.

ALAN B. BENNETT, *Executive Director, PIPRA, and Associate Vice Chancellor, Office of Research, University of California, Davis, 1850 Research Park Drive, Davis, CA, 95616, U.S.A.* *abbennett@ucdavis.edu*

MICHAEL CARRIERE, *Business Development and IP Manager, Office of Technology Transfer, University of California Office of the President, 1111 Franklin Street, Oakland, CA, 94607-5200, U.S.A.* *carriere@ucop.edu*

---

1 Further information on the strawberry licensing program is available at www.ucop.edu/ott/strawberry/welcome.html.

The IP Management of the PRSV-Resistant Papayas Developed by Cornell University and the University of Hawaii and Commercialized in Hawaii

MICHAEL GOLDMAN, *Partner, Nixon Peabody LLP, U.S.A.*

ABSTRACT

In the late 1990s, a consortium of public sector organizations commercialized the first and still-major food biotechnology product developed by public sector organizations. The author represented the Papaya Administrative Committee, an organization of papaya growers in Hawaii, in obtaining patent licenses necessary for the commercial introduction of a disease-resistant transgenic papaya. This chapter describes the approach taken in deciding what patents needed to be licensed, how the licenses were obtained, and how they were administered.

1. INTRODUCTION

In the fall of 1995, I was retained by papaya growers in Hawaii to provide legal assistance on patent and licensing issues related to a transgenic, disease-resistant papaya that had been developed for use in Hawaii. Although this technology was developed by Dennis Gonsalves while at Cornell University along with researchers in Hawaii, my client was actually the Papaya Administrative Committee (PAC) in Hilo, Hawaii. PAC had been created many years earlier under a federal marketing order by the U.S. Department of Agriculture (USDA) to assist the Hawaiian papaya industry in marketing papaya.

As a result of the devastating effect of papaya ringspot virus (PRSV) on the industry, PAC undertook to obtain the patent licenses necessary for commercial introduction of the transgenic, disease-resistant papaya. As PAC's legal advisor, I was required to identify which patent rights needed to be licensed, to negotiate and obtain licenses, and to help PAC administer the licenses that were obtained. This paper describes how I assisted PAC with these tasks and brings to light some practical considerations relating to the patenting and licensing of transgenic plant technology.

2. IDENTIFICATION OF PATENT RIGHTS THAT NEEDED TO BE LICENSED

Under 35 U.S.C. § 271, a U.S. patent gives its owner the right to prevent others from making, using, selling, or offering to sell the subject matter of the patent in the United States. The recipient of a license of such patent rights has the ability to engage in at least some of these activities without risking an injunction and/or being held liable for damages. PAC wanted to be able to go forward quickly with the transgenic papaya without fear of such risks. Therefore, my first task was to determine which patent rights needed to be licensed by PAC.

The task involved determining which patents would be infringed by the transgenic papaya technology in the absence of a license. In order to proceed, it was first necessary to identify which technology was used in making the transgenic papaya. Based on the findings, a group of patents was identified that potentially needed to

Goldman M. 2007. The IP Management of the PRSV-Resistant Papayas Developed by Cornell University and the University of Hawaii and Commercialized in Hawaii. In *Intellectual Property Management in Health and Agricultural Innovation: A Handbook of Best Practices* (eds. A Krattiger, RT Mahoney, L Nelsen, et al.). MIHR: Oxford, U.K., and PIPRA: Davis, U.S.A. Available online at www.ipHandbook.org.

be licensed. Such identification of candidate patents often requires conducting an infringement search on computer databases and in the U.S. Patent and Trademark Office (PTO). In the case of transgenic papaya, we also had some guidance from industry sources. Once a group of candidate patents was identified, I proceeded with the legal analysis to determine which of those patents would actually be infringed.

The exclusionary rights afforded by a U.S. patent are defined by the claims. Therefore, in analyzing a patent for infringement, it is first necessary to interpret the scope of the patent (in other words, the claims of the patent). This involves examining the literal language of the claims, reviewing the specification (or the body) of the patent, and studying the prosecution history of the corresponding patent application (in other words, the correspondence to and from the PTO during the patent application process). Through this analysis, the meaning of the terms in the patent claims and, accordingly, the scope of the claims as a whole is determined. With this information, it can then be decided whether the claims are infringed by the subject technology. A U.S. patent can be directly infringed in two ways:

- by literal infringement
- under the doctrine of equivalents

Literal infringement occurs if the language of the claims covers, literally, the subject technology. The absence of literal infringement does not, however, mean that infringement is avoided. Infringement can occur under the doctrine of equivalents if the differences between the subject technology and the claimed invention are insubstantial. One approach to determining whether infringement has occurred under the doctrine of equivalents is to analyze whether the subject technology and the patented invention do substantially the same thing in substantially the same way to achieve substantially the same results. The scope of the doctrine of equivalents is limited by what the prior art teaches and by what the patentee surrendered during prosecution of the patent.

In the context of a patent covering a transgenic plant, infringement can occur if a party makes, uses, or sells that plant. These actions constitute direct infringement (see Figure 1). Even if there have been no acts of direct infringement by a particular party, liability can ensue if that party induces or contributes to another's acts of direct infringement.

"Inducing infringement" occurs when one party aids and abets the direct infringing acts of another. Such liability can occur in the context of patents covering a method of making transgenic plants disease resistant (Figure 2). Researchers who are making such transgenic plants using a particular vector would be directly infringing such a patent. However, the supplier of this vector would not be directly infringing but could be liable for inducing infringement if the vector is provided with instructions to use it in order to produce disease-resistant transgenic plants.

"Contributory infringement" occurs when a party sells a nonstaple article of commerce which has no substantial noninfringing use. In the context of a patent covering a method of making a transgenic plant that is disease resistant, a party planting seeds for such transgenic plants would be a direct infringer. However, a party selling seeds for such plants, though not liable for direct infringement, would have contributory infringement liability (Figure 3).

The technology used by Dennis Gonsalves and colleagues to develop a transgenic papaya, in brief, consisted of the preparation of a vector and the introduction of it into papaya by biolistic transformation.[1] The vector, a map of which is shown in Figure 4 was an *Agrobacterium*-binary vector which included the 35S promoter, the 5' untranslated leader sequence, the PRSV coat protein encoding gene, and the β-glucoronidase (GUS) gene. Thus, we needed to consider licensing patent rights relating to various DNA components, plant transformation procedures, modes of plant disease-resistance mediation, and transgenic plants.

With the assistance of Dennis Gonsalves, I analyzed the technology utilized in developing the transgenic, disease-resistant papaya and determined which of the candidate patents needed to be licensed. It was determined that licenses were needed from Company Y for patent rights

FIGURE 1: ACTS OF PATENT INFRINGEMENT IN THE UNITED STATES INCLUDE DIRECT INFRINGEMENT

DIRECT INFRINGEMENT
U.S. Patent No. 1
Claim: Transgenic Plant

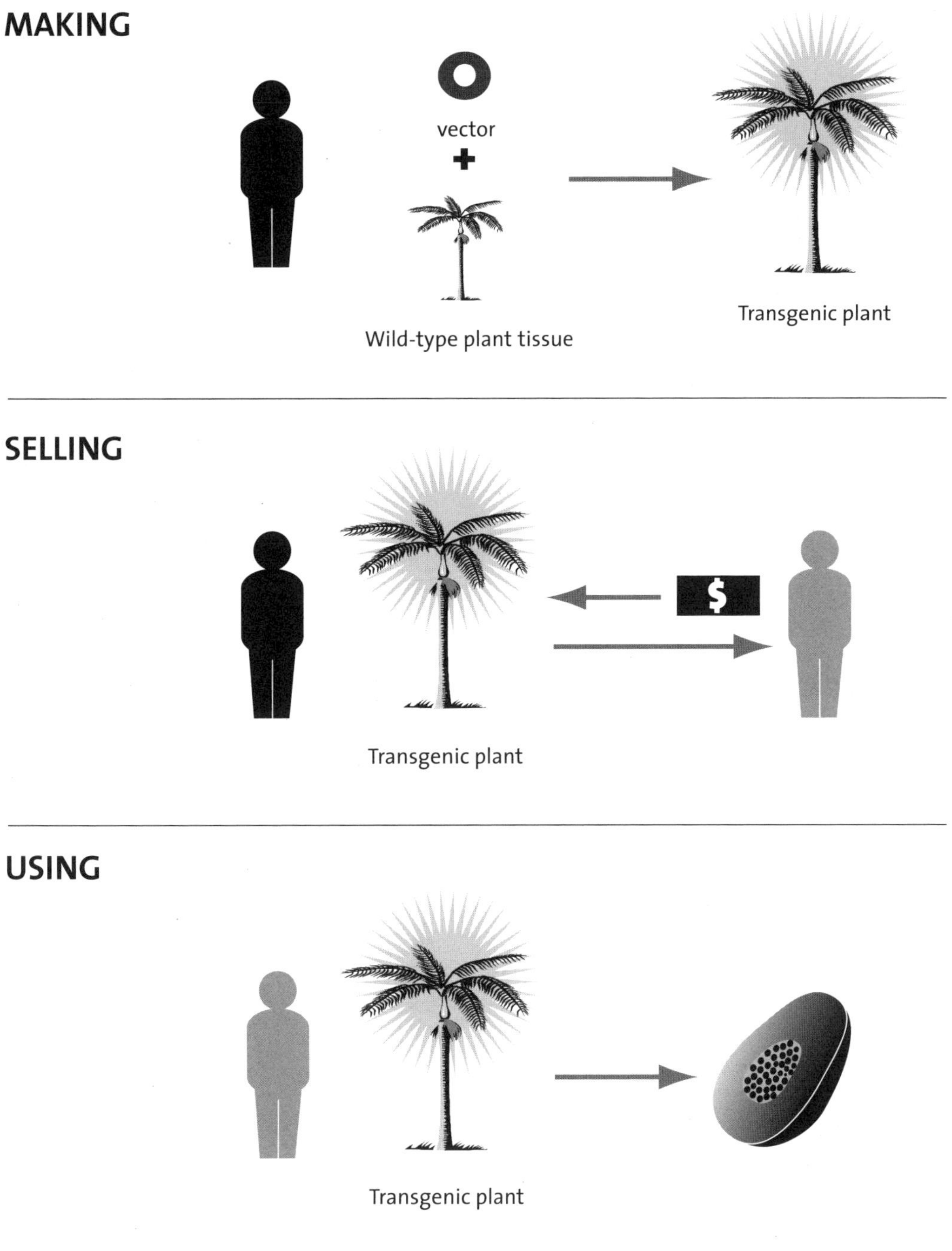

Figure 2: Acts of Contributory Patent Infringement in the United States

Contributory Infringement

U.S. Patent No. 1
Claims:
1. Process of Producing a Disease Resistant Plant Comprising
   a. Providing a Transgenic Plant and
   b. Planting the Transgenic Plant whereby the Plant is Disease Resistant

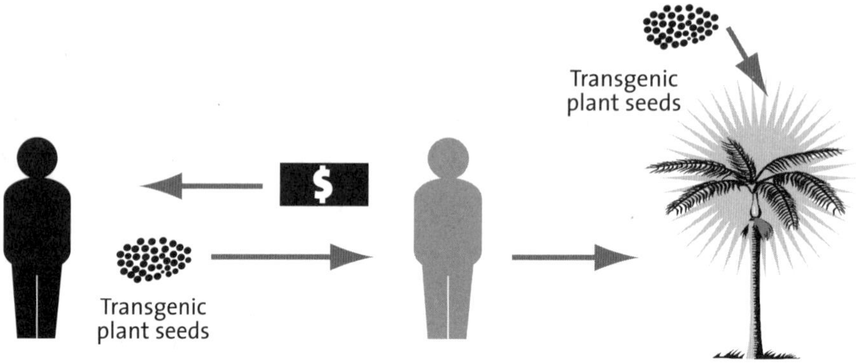

Figure 3: Acts of Inducement to Infringe in the United States

Inducement to Infringe

U.S. Patent No. 1
Claims:
1. Process of Producing a Disease Resistant Plant Comprising
   a. Providing a Transgenic Plant and
   b. Planting the Transgenic Plant whereby the Plant is Disease Resistant

Product Literature
Uses of Vector:
- Incorporating Plant Trait Gene in Plant
- Incorporating Virus Gene in Plant

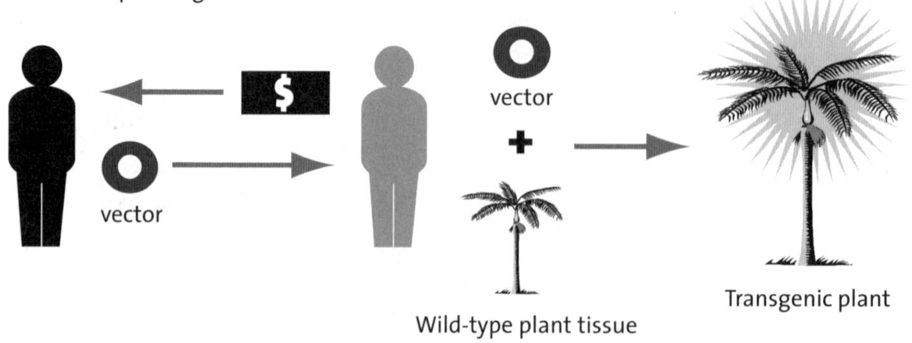

FIGURE 4: AGROBACTERIUM BINARY VECTOR PGA482GG/CPPRV-4

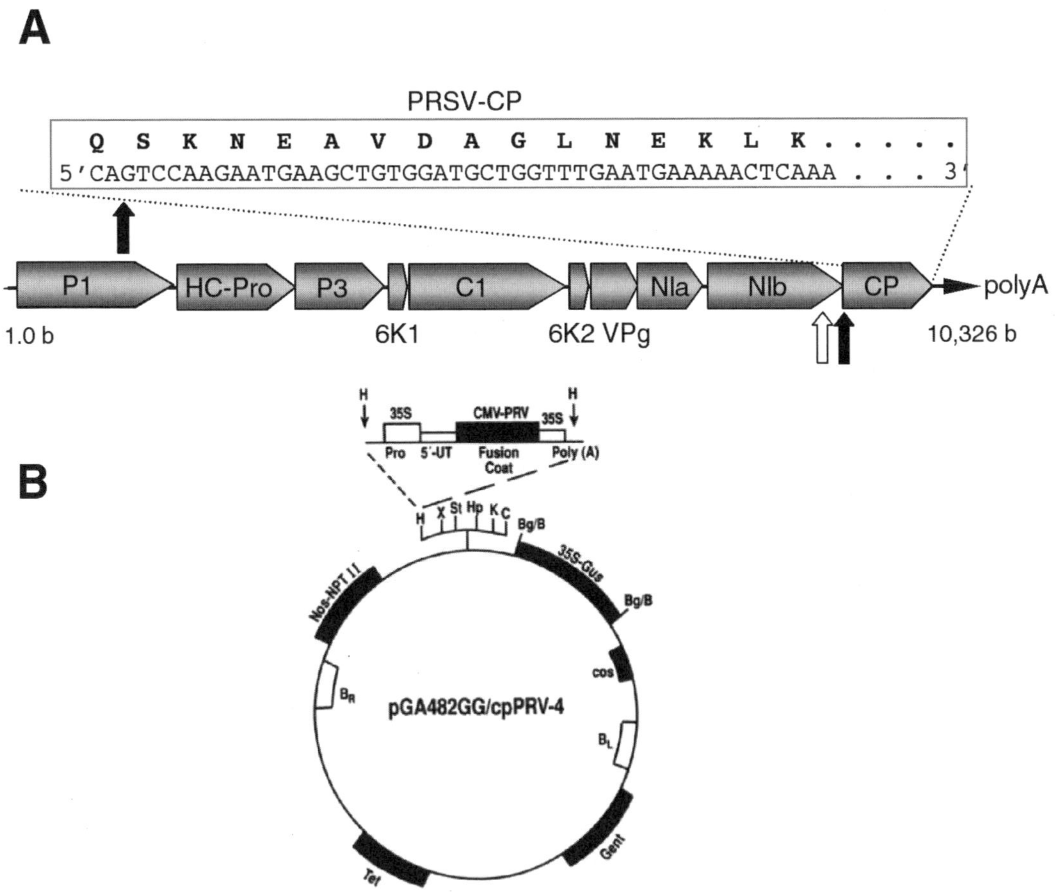

Organization and proteolytic protein products of the 10, 326 base monocistonic PRSV genome. (A) Shown in detail, the N-terminal sequence of the coat, the protein (PRSV-CP). Box arrows represent the proteolytic sites producing the mature coat protein (CP).(B) Map of the functional genes of the Agrobacterium transformation vector pGA482GG/cpPRV-4 used for PRSV-resistant papaya. The coat protein gene cassette consists of the coat protein structural gene of PRSV HA 5-1 translationally fused to the N-terminal end of the cucumber mosaic virus coat protein (CMV-CP), including the translation initiation codon, the CMV 5' untranslated sequence (5' UTR), and the Cauliflower Mosaic Virus 35S promoter. The PRSV-CP gene cassette is flanked by selectable and visible marker genes, neo (encoding NPTII) and gus, respectively. BR and BL are the left and right borders of the transformation vector T-DNA sequence.

Source: Courtesy Dennis Gonsalves (November 2006).

relating to various components of the vector and the general mode of plant disease resistance. From Massachusetts Institute of Technology (M.I.T.), PAC decided to license rights to the 5' untranslated leader sequence. Company X had rights to technology to impart resistance to PRSV by use of a gene from the virus. We wanted to license that technology. We also wanted rights to the GUS gene from Cambia Biosystems LLC (Canberra, Australia). For various reasons, we decided that licenses were not needed for other candidate patents.

3. LICENSE NEGOTIATIONS

After identifying which patent rights PAC should license, the next job was to obtain the necessary licenses. This proved to be a very difficult task because the parties had different strategic objectives.

PAC wanted to be able to distribute transgenic papaya seed without charging recipients and without having to maintain the accounting records normally needed for licenses involving a royalty on net sales. Therefore, we sought licenses involving a one-time, up-front payment. With this approach, PAC also wanted to be assured that it would receive licenses under any patents infringed by the transgenic papaya that issued after the license agreement was signed. Otherwise, PAC would be at risk of having to negotiate a new license and making further payments to a party that had already granted a license to PAC. Another issue was PAC's financial resources. Since PAC's licensing activities were financed by public funds and contributions from its members, many of whom were farmers in Hawaii, the licensing fees needed to be manageable. While PAC needed a substantial level of accommodation from licensors on financial issues, its demands on the scope of any grant under a license agreement were modest. In particular, PAC needed to obtain the right to grow transgenic papaya plants in Hawaii and to sell the resulting fruit worldwide. Finally, since PAC did not itself grow or sell papaya, it needed to be able to sublicense its rights to constituents, including growers.

All of the licensors were sympathetic to the need to introduce a transgenic, disease-resistant papaya in Hawaii. However, each had its own strategic interests, which needed to be protected. Some licensors did not, at that time, have a policy of or experience with licensing out, and they were reluctant to proceed with setting a corporate-wide strategy based on a license for a very small crop. Undoubtedly, there was concern over having any deal with PAC dictate which terms would have to be offered for future licenses on strategically important crops. Many of the individuals working on business development for the licensors were very busy and did not have much time to focus efforts on a deal for a very small crop with potentially little economic return. Some licensors had a tremendous commitment to developing a plant biotechnology business and wanted to ensure that any licensees of its rights did not jeopardize the industry as a whole. Lastly, the licensors needed to know that the financial terms of any license were fair. Given the relatively low strategic interest a transgenic papaya license had to the licensors, PAC had to engage in an extensive effort to educate them about the Hawaiian papaya industry, the impact of PRSV in Hawaii, and the benefit of the transgenic papaya to papaya growers in Hawaii. In particular, we tried to gain sympathy from the licensors by explaining that the virus had devastated the Hawaiian papaya industry and that the transgenic papaya needed to be introduced in Hawaii to ensure that farmers could maintain their livelihood. Our promotional efforts often led to questions about PAC's purpose and membership. When the licensors saw that large, well-known fruit packing companies were members of PAC, there were usually questions from the licensors about who was being aided by the licenses. However, we were able to explain that the true beneficiaries of the licenses were growers whose farms were being severely hurt by PRSV.

In some cases, sympathy for the plight of growers was not sufficient and the licensors needed to be further motivated. The USDA was helpful in several instances. Because it is an important regulatory agency in the plant biotechnology industry, the licensors wanted to remain in the USDA's good graces in order to avoid jeopardizing regulatory approvals for their own projects. Since the USDA created PAC, was already actively involved in the Hawaiian papaya industry, and wanted to see the transgenic, disease-resistant papaya introduced in Hawaii, the agency was very willing to help PAC. Without that help, a number of the licenses may never have been obtained.

Once we had communicated with the licensors, we were generally able to persuade them to prepare a draft license agreement from which license negotiations could proceed. Although we usually prefer to generate the first draft of a license agreement, doing so tends to be more costly, and we were trying to limit PAC's costs for the project. In any event, once the initial draft license agreement was received, we proceeded with license negotiations and ultimately were able to enter into license agreements with all of the targeted licensees.

Company X was anxious to put the transgenic papaya on the market as a philanthropic effort and was PAC's first licensor.

Cambia Biosystems LLC is a technology licensing company without any particular interest in exploiting the GUS gene technology in the plant biotechnology industry. They were interested in helping the Hawaiian papaya industry, as long Cambia could be assured of a fair deal, from an economic standpoint. Cambia was our next licensee.

Company Y was sympathetic to the plight of the Hawaiian papaya industry, but as a result of the company's extensive involvement in the transgenic plant industry, its strategic interests were the most difficult to harmonize. Once the company was able to resolve its objectives, it moved enthusiastically forward with license negotiations. It regarded the license to be negotiated with PAC as a prototype for future deals involving outlicensing of Company Y technology. Company Y became PAC's third licensor.

The last license was obtained from M.I.T., which had no particular strategic concerns about licensing in the plant biotechnology industry but was concerned about whether a paid-up license provided fair compensation. We ultimately were able to develop an arrangement by which M.I.T. could be assured of an economically fair deal.

As a result, PAC had obtained the licenses it needed to begin growing transgenic papaya in Hawaii. Shortly after the last license agreement was executed, PAC began distribution of transgenic papaya seed to growers. The commercial use of this product of biotechnology has had a substantial beneficial economic impact on the Hawaiian papaya industry.

4. LESSONS LEARNED

There are a number of lessons to be learned from the transgenic-papaya licensing effort. These lessons, relating to both patent and licensing issues, can benefit researchers, technology transfer professionals, business people, and lawyers. Researchers in the transgenic biotechnology area should recognize that there are patents covering many commonly used genetic components and plant transformation procedures. The manufacture, use, sale of, or offer to sell such patented materials by researchers without a research license is an act of direct patent infringement. Engaging in any of these activities would put the researchers' employers at risk of being sued and having to pay the patentees' damages, as well as attorneys' fees. If the researcher were employed at an academic institution, the prospect of incurring such expenses would be daunting. Even if a research license were obtained, it would not allow introduction of the product of research into a commercial product. Any effort to do so would be an act of patent infringement. In the case of an academic institution working with commercial entities, the licensing out of technology utilizing the patent rights of others, or the transfer of materials incorporating patented subject matter, also raises issues of patent infringement. In particular, the institution can be deemed to be inducing infringement (aiding and abetting the infringing acts of another) or engaging in contributory infringement (selling or offering to sell a material having no substantial use other than in conjunction with a patented process). To avoid these issues, researchers should use unpatented or easily licensed technology wherever possible.

On the other hand, developers of technology wishing to enhance their licensing royalties and their leverage over competitors may wish to make their technology freely available once the necessary patent applications have been filed. The widespread use of such technology can lead to its adoption as an industry standard for which substantial licensing revenue can be derived. Moreover, the use of a company's patented technology in the commercial product of a competitor can give the company significant leverage over the competitor in accessing technology owned by the competitor, in maximizing royalty payments from the competitor, and in preventing the competitor from introducing an important commercial product.

In licensing patent rights from others, it is important to examine what the various patents you are considering would actually cover. In the transgenic plant industry, there is a great deal of "street talk" about patents and what they purport to cover. Reliance on "scuttlebutt" could result in the procurement of and payment for licenses on patent rights that are not needed. On the other hand, failure to obtain all the necessary licenses raises the threat of an injunction, of liability for damages, and of the costs of litigation. A careful analysis of the patent landscape is well worth the expense. Entities licensing technology on behalf of others need to properly control how it makes the technology available. In the case of PAC, it has made transgenic papaya seed available to growers only after they attend an educational program and sign a material transfer/sublicense agreement with PAC. Likewise, researchers wishing to obtain transgenic papaya seed from PAC are required to sign a material transfer/research sublicense agreement with PAC. These measures were undertaken to ensure that growers and researchers understood the obligations pursuant to the license agreements and complied with those obligations.

5. CONCLUSION

The above events may not be of great economic significance to global agriculture. However, as one of the first efforts to develop a transgenic fruit crop, procure the necessary licenses, and introduce a product into commerce, Hawaii's transgenic papaya story is certainly an important event for the plant biotechnology industry. The successful results achieved by PAC may well serve as a model for future transgenic plant technology. ■

MICHAEL L. GOLDMAN, *Partner, Nixon Peabody LLP. Corner of Clinton Ave. and Broad Street, PO Box 31051, Clinton Square, Rochester, NY, 14603, U.S.A., mgoldman@nixonpeabody.com*

1 See Ling K, S Namba, C Gonsalves, JL Slightom and D Gonsalves. 1991. Protection Against Detrimental Effects of Potyvirus Infection in Transgenic Tobacco Plants Expressing the Papaya Ringspot Virus Coat Protein Gene. *Bio/Technology* 9:752–758. See also Fitch M, RM Manshardt, D Gonsalves, JL Slightom and JC Sanford. 1992. Virus Resistant Papaya Plants Derived from Tissues Bombarded with the Coat Protein Gene of Papaya Ringspot Virus. *Bio/Technology* 10:1466–1472.

Fundación Chile: Technology Transfer for Somatic Embryogenesis of Grapes

CARLOS FERNANDEZ, *Director, Strategic Studies, Foundation for Agriculture Innovation (FIA), Chile*

ABSTRACT

Fundación Chile is a private, non-profit organization active in developing applications of biotechnology that can improve productivity and add value to existing agricultural and natural resource products of Chile. Fundación Chile seeks to create technology-based companies that would have significant economic and social impact in Chile. This case study details Fundación Chile's initiative in grape biotechnology: globally assessing the availability and priority of different technological components and initiating efforts to access, license, and transfer those key technologies for the initiative.

1. THE INSTITUTION

Fundación Chile[1] is a private non-profit organization. Its mission is to add economic value to Chile's products and services by promoting innovation and technology transfer focused on Chile's natural resources and productive capacity. *Fundación Chile*'s primary strategy is to develop new technology-based companies in Chile that can have a significant economic and social impact. These new companies are generally joint ventures with strategic partners, although other models, such as licensing, are used.

The main activities are focused in the areas of Agribusiness, Marine Resources, Forestry and Forest Products, Environment, Information Technology, Education and Human Resources, and Tourism.

Fundación Chile is unusual in that it is a non-profit institution with active participation in the creation of innovative private companies and involvement in a wide range of activities relevant to different stages in the development of new businesses. These activities include technology services, R&D, creation and incubation of companies, seed capital, scale-up, and financial innovation.

Fundación Chile's activities are focused on increasing the volume and value derived from Chilean production of products that can be exported or can replace imports, but possibilities are also considered for production in other countries.

1.1 *Fundación Chile and biotechnology*

Since 1997 *Fundación Chile* has been active in developing applications of biotechnology that can improve productivity, add value to existing products, and promote introduction of new products in its business areas. Biotechnology activities are mainly focused in forestry, fruit, and aquaculture, with an increasing emphasis on quality and utilization. Biotechnologies used include recombinant proteins, tissue culture, molecular genetics, functional genomics, and genetic engineering.

Strategic alliances in biotechnology in the private sector include

- a licensing agreement for a salmon vaccine with Syngenta

Fernandez C. 2007. Fundación Chile: Technology Transfer for Somatic Embryogenesis of Grapes. In *Intellectual Property Management in Health and Agricultural Innovation: A Handbook of Best Practices* (eds. A Krattiger, RT Mahoney, L Nelsen, et al.). MIHR: Oxford, U.K., and PIPRA: Davis, U.S.A. Available online at www.ipHandbook.org.

- a JV in grape biotechnology with Interlink Associates LLC (Princeton, USA)
- an R&D collaboration in stone fruit biotechnology with Okanagan Biotechnology Inc. (Summerland, Canada)
- a strategic alliance in forestry biotechnology with CellFor Inc. (Canada)

Fundación Chile seeks to establish strong Intellectual Property (IP) positions through the licensing of key existing IP and the development of new IP in areas of specific strategic importance in Chile in which it participates in R&D

Fundación Chile's biotechnology activities involve an extensive network of Chilean and foreign research centers and universities, as well as participation in key international consortia. Collaborators in biotechnology R&D in Chile include

- *Fundación Ciencias para la Vida*
- the Chilean National Institute for Agricultural Research
- the University of Chile
- the University of Concepción
- the University of Santiago
- the University of Talca
- University Federico Santa Maria
- Andres Bello University
- Austral University

Alliances with foreign research centers and universities include

- the University of California
- Cornell University
- the University of Florida
- the United States Department of Agriculture
- New Zealand HortResearch
- New Zealand Forest Research

Fundación Chile is a member of PIPRA (Public Intellectual Property Resource for Agriculture) and the California Institute of Food and Agricultural Research, and it is a participant in the ALCUE-Food Specific Support Action funded by the 6th European Framework.

By establishing these networks, Fundación Chile has been able to participate in the development of new product candidates over a relatively short time frame. A recombinant protein vaccine for salmon developed in a collaboration of Fundación Chile and Fundacion Ciencias para la Vida has been licensed to Syngenta and is being introduced into the market. Elite clones of radiata pine developed through somatic embryogenesis in collaboration with CellFor are in advanced stages of testing and are being scaled up for market introduction by a Fundación Chile company, GenFor. Other biotechnology programs of Fundación Chile, including the genetic engineering of grape varieties, peaches, and pine trees are in earlier stages of development.

2. THE CASE: TECHNOLOGY TRANSFER FOR SOMATIC EMBRYOGENESIS OF GRAPES

2.1 *Importance of institutional support for a long-term R&D program*

Agricultural biotechnology R&D programs are long-term, expensive and controversial; it is essential that the institution is committed to the process. In the late 1990s *Fundación Chile* made a strategic decision to invest in development of biotechnology applications in strategic sectors of the Chilean economy: forestry, agriculture, and aquaculture. Genetic engineering was clearly a key technology with a large potential impact, as demonstrated by the rapid adoption of genetically engineered varieties of maize, soybeans, and cotton in some parts of the world. However, these crops play a relatively minor role in Chile. Little effort was being expended anywhere in the world in perennial crop species, such table grapes, which make up an important part of Chilean exports, and in which Chile is a major player.

2.2 *Identification of specific technologies and resources needed to build a foundation for the program*

Typically, three different types of technological components are needed for development of a genetically engineered plant product:

- Germplasm that provides a competitive genetic background
- Specific genes that confer new traits of interest

- Enabling tools such as genetic markers, promoters, tissue culture and regeneration systems, and transformation methods

In addition, human resources, laboratory infrastructure, and financing are needed to carry out the R&D to adapt and combine these components to produce a product.

Laboratory infrastructure existed in Chile, but improvements were needed. There were capable researchers in Chile, but a limited number. Research efforts were spread over many different objectives, and sustained support for a specific program was rare.

In the case of grapes, the foundational technologies were not available in the local R&D institutions at the start of the program, except, to a limited degree, germplasm. A global search led to the identification of sources of technologies and expertise. The availability of different components and priority for access were assessed, and efforts were initiated to access, license, and transfer key components.

2.3 *IP and freedom to operate*

The IP and freedom-to-operate issues were complex, due to the need to address the situations both in Chile and in major export markets, the long and uncertain time frames for development and commercialization of genetically engineered perennial fruit crops, and the concentration of rights to core technologies in companies with little or no interest in "orphan crops." A complete solution was not possible in the short term with the resources available. However, it was possible to establish a position in key technologies that maximized the likelihood of being competitive in a specific niche.

Based on our experience, a critical aspect was the active involvement of personnel with experience in commercial R&D programs and major agribiotech research centers in other countries, and experience in licensing agricultural biotechnologies. Practices vary from country to country and institution to institution within a country. At the time of the initiation of the program there was little experience in Chile with patenting and licensing of technology developed in public research institutions. The ability of partners with international experience to provide appropriate examples drawn from a variety of sources played an important role in bridging gaps in experience and expectations.

The description of our experiences below will, we hope, assist others in similar situations to make significant progress towards obtaining components needed to develop a biotechnology program appropriate for the development of commercial products of interest for their particular situations.

2.4 *Key technologies required for establishment of a grape genetic engineering platform*

At the time the program was initiated there were only a few published reports of transformation of *Vitis vinifera*. In order to be able to obtain R&D funding from public and private sources, and to be considered seriously as a potential licensee by technology providers, it was considered critical to demonstrate the ability to reproducibly transform the target species.

For many transformation systems, an important factor is the availability of a robust tissue culture system that makes it possible to regenerate plants efficiently. In our experience, tissue culture systems involve considerable art and are often difficult to reproduce in other laboratories. Thus, establishment of a strong position in grape tissue culture was selected as the highest initial priority. The process and progress in this area are discussed below.

The second priority was access to specific gene candidates for engineering a trait of commercial interest in the Chilean market. This was carried out in parallel in order to ensure that the tissue culture and transformation platform developed could be applied to production of prototypes with traits of interest with a minimum lag.

2.5 *Identification of leading laboratories with expertise in tissue culture systems suitable for grape transformation*

The search used different and complementary channels, including reviews of research publications, project databases, conference proceedings, patents and patent applications, news items, and personal contacts. All of them are relevant and provide useful information.

Access to many of these sources has been facilitated by the rapid improvement of the Internet, in terms of content and ease of access. Even for people without good Internet access, the availability of high-quality documents in electronic form has greatly reduced the cost of access.

Open sites such as PubMed (www.ncbi.nlm.nih.gov) and HighWire Press (highwire.stanford.edu) provide convenient access not only to bibliographic information, but to many full papers. An increasing number of full papers are available at no charge, and most others can be downloaded for a fee from sites of journal publishers or specialized clearing houses.

Online databases such as those at the the World Intellectual Property Office (www.wipo.int/ipdl), the European Patent Office (www.espacenet.com), the United States Patent and Trademark Office (www.uspto.gov), and many other national patent offices provide increasingly convenient access to issued patents and published applications.

Less widely appreciated, but valuable due to their more specialized content, are online databases of research projects. These often include information that is otherwise difficult or impossible to find. Examples include the European Union Community Research & Development Information Service (cordis.europa.eu), the Current Research Information System of the USDA (cris.csrees.usda.gov), the FAO-BioDeC database of biotechnology projects in developing countries (www.fao.org/BIOTECH), and the RedBio (*Red de Cooperación Técnica en Biotecnología Vegetal para America Latina y el Caribe*) database of biotechnology activities by member country (www.redbio.org). In Chile the web sites of the major funding agencies for R&D—CONICYT (www.conicyt.cl), CORFO (www.corfo.cl), and FIA (www.fia.cl)—include databases of projects. Many research institutions provide databases of internal research activities and funded projects, which may be useful once specific institutions of interest have been identified.

Advanced Internet search sites such as Google™ have changed the way that most people think about Internet searching. Today it is often an easy way to get started. It is important to remember that searches conducted on such sites generally do not access information stored in specialized databases such as those described above.

All of the above are useful in the identification of potential technology providers, collaborators and competitors. However, direct contacts are critical early in the process to validate the information and to establish a foundation for future relationships. It is important to establish contacts both at the level of the researcher/inventor and at the level of the institution.

2.6 *Negotiation of a research and option agreement*

Once the identification of the laboratory or institution has been made, documents are typically exchanged via electronic mail. Most large private companies and universities have standard forms that are adapted to the specific needs of a given project. Typical research agreements include the following information:

- Date and identification of the parties
- Definitions of terms
- Reports and conferences for proper follow-up of activities
- Costs, payments and other support
- Publications
- Intellectual property
- Grant of rights
- Confidentiality and publicity
- Term and termination
- Insurance and indemnification
- Governing law
- Assignment
- Agreement modification
- Notices
- Counterparts and headings

It is important to emphasize that this standard approach was designed for the United States. Intellectual property laws vary among countries, so it is important that the contents of any agreement are reviewed by a local attorney knowledgeable in intellectual property matters.

Most universities in the United States, and many other public research institutions, will require that the public institution be able to continue to use the technology for research and education purposes even if exclusive rights for commercial use are granted.

Our general approach has been to negotiate agreements that provide rights to use technologies for R&D and an option for a commercial license. We want to avoid a situation where resources are invested in research if the results cannot be commercialized. Due to the high degree of uncertainty in development and commercialization of agribiotech products, we also want to avoid paying for rights that in the end will not be used. In agreements for access to technology we have generally tried to structure compensation in ways that reduces the up-front cost in favor of sharing of benefits realized from commercialization of products. This is important for making effective use of the resources available, but more importantly, helps to align the interests of the technology provider with our interests. The agreements typically contain modest up-front payments, milestone payments based on successful transfer of the technology, additional milestone payments if a commercial license is entered into and a product is introduced into the market, and royalties based on revenue derived from commercialization of products produced using the technology.

In the case of grape tissue culture, the institution in which the technology had been developed already had agreements with a private company. Thus, we initially had to negotiate an agreement with the third party. Changes in the scope of activities of the company later led to a return of rights to the university and additional negotiations with the university. Similar events have affected other agreements related to the project. Thus, it is important to recognize that management of these agreements is a dynamic process.

2.7 *Material Transfer Agreements (MTA)*

In addition to intellectual property, the transfer of technology in agricultural biotechnology often requires or is facilitated by the transfer of materials. Terms for the use of the materials, their disposal, etc., are generally covered by a material transfer agreement (MTA).

In countries with limited innovation, lawyers have not been exposed or do not have enough experience on matters related to MTAs. If this is the case, the practical approach was to use as a reference form prepared by the technology transfer offices of universities in the United States and other countries with experience on these matters. Some of these offices have sample forms posted on their Web site.[2]

An MTA typically includes the following information:

- Date
- Identification of the provider and recipient
- Definition of the material
- Agreement to be bound by the laws of a specific, legal district
- Recipients agreement to the defined uses and conditions, such as compliance with local laws and regulations regarding the use of the material, limits on individuals with access to the material, limits on import/export of the material
- Conditions of ownership in case of derivatives
- Conditions of exclusivity or non-exclusivity, commercial or non-commercial use, and disposal of the material
- Experimental nature of the material, and no warranty expressed
- Terms if borrower intends commercialization of the material or derivatives
- Terms if borrower intends to publish results or deliver a presentation
- Reporting of observations and results and conditions of use of such information
- Material physical integrity and recordkeeping
- Conditions for termination
- Signatures and agreement to execute the agreement

The MTA should be carefully reviewed. In the past, investigators have sometimes accepted terms that have had critical effects on the value of the R&D they conducted, particularly terms regarding reporting requirements and right of the provider to use information generated by the recipient. It is also critical to consider whether the material provided incorporates material or technology owned by third parties. If so, it is advisable to request clarification of any restrictions that may be "inherited" with the material.

2.8 *Importation of materials*

Each country has its own regulations regarding the importation of biological materials. In Chile, there are forms and procedures that must be followed. Samples of tissue cultures of grapes were imported following these procedures without major obstacles, although significant time and resources were required.

2.9 *Exchange of professionals between laboratories*

Good communication between the parties is essential for a successful outcome. For transfer of some technologies, the exchange of written information and materials, supplemented by communication via phone calls and e-mail may be sufficient. However, in many cases, successful transfer is greatly facilitated by the active participation of investigators from the provider and recipient laboratories in activities in both laboratories. In the case of the grape tissue culture system, a Chilean investigator first spent time in the laboratory of the inventor to get hands-on experience with the procedures, and then returned to set up the system locally. Several months later, the inventor came to Chile and spent a full week working side by side with the local investigators, reinforcing the training and providing an opportunity to resolve issues that had arisen during the implementation. Some time later, the project leader visited the inventor's laboratory to observe procedures there, with the accumulated experiences in Chile providing a foundation for increased "receptivity." At the end of each exchange, written reports were prepared, disseminated, and discussed.

3. CONCLUSIONS

Currently the lab in Chile has been able to master grape embryogenic tissue culture and regeneration techniques and apply them to genetic engineering. Transformation of these tissue cultures has allowed the production of thousands of transformed grape lines, from which promising lines have been advanced to the field for additional testing. ■

CARLOS FERNANDEZ, *Director, Strategic Studies, Foundation for Agriculture Innovation (FIA), Loreley 1582, La Reina, Santiago, Chile. carlos.fernandez@fia.gob.cl*

---

1 See, also in this *Handbook*, Chapter 17.2 by C Fernandez and MR Moynihan.

2 See, for example, F. H. Erbisch. 2005. *Basic Workbook in Intellectual Property Management*. Michigan State University; 156 pages. Available online at http://www.iia.msu.edu/iprworkbook.htm.

APPENDIX

Sample Agreements

Editor's Note

This section shares several sample agreements for illustration:

We selected the above for illustration because they expand on or complement those that have been discussed or provided in the *Handbook* chapters, or because they explain in greater depth certain types of clauses and provisions (for example, due diligence provisions in the MRC sample license no. 3), or because they contain clauses and provisions that might help to illustrate the various licensing principles discussed in the chapters (for example, licensing terms related to field of use).

The agreements given here are also available on the online version of the *Handbook* (www.ipHandbook.org) together with sample agreements from different institutions from countries around the world. They are downloadable in Microsoft® Word or Adobe® PDF formats. Among others, the online version's agreements include confidentiality, material transfer (for germplasm, biological resources, materials for testing, research tools, and experimental animals), IP licenses for copyright, software, trademarks, trade secrets, and various forms of exclusive, co-exclusive, and nonexclusive licenses. Documents such as the Model Provisions for an Equitable Access and Neglected Disease License (developed by a working group at Yale University and convened by Universities Allied for Essential Medicines) are also included online. Other chapters in this *Handbook* contain sample agreements of nonasserts, invention disclosures, licensing checklists, and more. Please refer to the index at the back of this *Handbook* for a list of agreements.

None of these template or sample agreements should be considered as a "correct" agreement. **They are provided solely as a reference resource intended *for illustrative and educational purposes only.*** They may be used as a starting point for discussions, but any organization will benefit from developing its own template agreements, since these place the regularly required major elements within the institution's context and needs. In any case, the institution's counsel should always review draft agreements before signature (and in some cases even before sending it to the other party for review).

MIHR/PIPRA. 2007. Sample Agreements. In *Intellectual Property Management in Health and Agricultural Innovation: A Handbook of Best Practices* (eds. A Krattiger, RT Mahoney, L Nelsen, et al.). MIHR: Oxford, U.K., and PIPRA: Davis, U.S.A. Available online at www.ipHandbook.org.

Sample Agreement 1
Co-Development Agreement

Source: Mahoney RT (ed.). 2004. *Handbook of Best Practices for Management of Intellectual Property in Health Research and Development.* MIHR: Oxford, U.K. Reproduced with permission.

Public Sector Research Centre

(To be used in circumstances where PSRC and a party, ABC, are agreeing to a research program primarily to be carried out in the laboratories of ABC)

1.0 Introduction

This Co-development Agreement ("Agreement") between ABC Company ("ABC") and the Public Sector Research Centre ("PSRC") will be effective when signed by all Parties. The research and development activities that will be undertaken by each of the Parties in the course of this Agreement are detailed in the Research Plan ("RP"), which is included as Appendix A. The funding and staffing commitments of the Parties are set forth in Appendix B. Any exceptions or changes to the Agreement are set forth in Appendix C.

2.0 Definitions

As used in this Agreement, the following terms shall have the meanings provided hearein:

2.1 ***Invention*** means any invention or discovery that is or may be patentable.

2.2 ***Principal Investigator(s)*** or ***PIs*** means the person(s) designated respectively by the Parties to this Agreement that will be responsible for the scientific and technical conduct of the RP.

2.3 ***Proprietary/Confidential Information*** means confidential scientific, business, or financial information provided that such information does not include:

2.3.1 information that is publicly known or available from other sources that are not under a confidentiality obligation to the source of the information

2.3.2 information that has been made available by its owners to others without a confidentiality obligation

2.3.3 information that is already known by or available to the receiving Party without a confidentiality obligation

2.3.4 information that relates to potential hazards or cautionary warnings associated with the production, handling, or use of the subject matter of the Research Plan of this Agreement

2.4 ***Research License*** shall mean a nontransferable, nonexclusive license under any Intellectual Property (IP) license to make and use a licensed invention for purposes of research and not for purposes of commercial manufacture or distribution or in lieu of purchase.

2.5 ***Research Materials*** means all tangible materials other than Subject Data first produced in the performance of this Agreement.

2.6 ***Research Plan*** or ***RP*** means the statement in Appendix A of the respective research and development commitments of the Parties to this Agreement.

Co-Development Agreement

2.7 ***Subject Invention*** means any Invention of the Parties, conceived or first actually reduced to practice in the performance of the Research Plan of this Agreement.

2.8 ***Subject Data*** means all recorded information first produced in the performance of this Agreement by the Parties.

3.0 Cooperative Research

3.1 **Principal Investigators.** ABC research work under this Agreement will be performed by the ABC laboratory identified in the RP, and the ABC PI designated in the RP will be responsible for the scientific and technical conduct of this project on behalf of ABC. Also designated in the RP is the PSRC PI who will be responsible for the scientific and technical conduct of this project on behalf of the PSRC.

3.2 **Research Plan Change.** The RP may be modified by mutual written consent of the Principal Investigators. Substantial changes in the scope of the RP will be treated as amendments under Article 13.5.

4.0 Reports

4.1 **Interim Reports.** The Parties shall exchange formal written interim progress reports on a schedule agreed to by the PIs, but at least within twelve (12) months after this Agreement becomes effective and at least within every twelve (12) months thereafter. Such reports shall set forth the technical progress made, identifying such problems as may have been encountered and establishing goals and objectives requiring further effort, any modifications to the Research Plan pursuant to Article 3.2, and all Agreement-related patent applications filed.

4.2 **Final Reports.** The Parties shall exchange final reports of their results within four (4) months after completing the projects described in the RP or after the expiration or termination of this Agreement.

5.0 Financial and Staffing Obligations

5.1 **ABC and PSRC Contributions.** The contributions of the Parties, including payment schedules, if applicable, are set forth in Appendix B. ABC shall not be obligated to perform any of the research specified herein or to take any other action required by this Agreement, if the funding is not provided as set forth in Appendix B. ABC shall return excess funds to PSRC when it sends its final fiscal report pursuant to Article 5.2, except for staffing support pursuant to Article 10.3.

5.2 **Accounting Records.** ABC shall maintain separate and distinct current accounts, records, and other evidence supporting all its obligations under this Agreement, and shall provide the PSRC a final fiscal report pursuant to Article 4.2.

5.3 **Capital Equipment.** Equipment purchased by ABC with funds provided by the PSRC shall be the property of ABC. All capital equipment provided under this Agreement by one party for the use of another Party remains the property of the providing Party unless other disposition is mutually agreed upon in writing by the Parties. If title to this equipment remains with the providing Party, that Party is responsible for maintenance of the equipment and the costs of its transportation to and from the site where it will be used.

Co-Development Agreement

6.0 Intellectual Property Rights and Patent Applications

6.1 **Reporting.** The Parties shall promptly report to each other in writing each Subject Invention resulting from the research conducted under this Agreement that is reported to them by their respective employees. Each Party shall report all Subject Inventions to the other Party in sufficient detail to determine inventorship. Such reports shall be treated as Proprietary/Confidential Information in accordance with Article 8.4.

6.2 **PSRC Employee Inventions.** If the PSRC does not elect to retain its IP rights, PSRC shall offer to assign these IP rights to the Subject Invention to ABC pursuant to Article 6.5. If ABC declines such assignment, the PSRC may release its IP rights as it may determine.

6.3 **ABC Employee Inventions.** ABC may elect to retain IP rights to each Subject Invention made solely by ABC employees. If ABC does not elect to retain IP rights, ABC shall offer to assign these IP rights to such Subject Invention to PSRC pursuant to Article 6.5.

6.4 **Joint Inventions.** Each Subject Invention made jointly by ABC and PSRC employees shall be jointly owned by ABC and PSRC. PSRC may elect to file the joint patent or other IP application(s) thereon and shall notify ABC promptly upon making this election. If PSRC decides to file such applications, it shall do so in a timely manner and at its own expense. If PSRC does not elect to file such application(s), ABC shall have the right to file the joint application(s) in a timely manner and at its own expense. If either Party decides not to retain its IP rights to a jointly owned Subject Invention, it shall offer to assign such rights to the other Party pursuant to Article 6.5. If the other Party declines such assignment, the offering Party may release its IP rights as provided in Articles 6.2 and 6.3.

6.5 **Filing of Patent Applications.** With respect to Subject Inventions made by PSRC as described in Article 6.2, or by ABC as described in Article 6.3, a Party exercising its right to elect to retain IP rights to a Subject Invention agrees to file patent or other IP applications in a timely manner and at its own expense and after consultation with the other Party. The Party may elect not to file a patent or other IP application thereon in any particular country or countries provided it so advises the other Party ninety (90) days prior to the expiration of any applicable filing deadline, priority period or statutory bar date, and hereby agrees to assign its IP right, title, and interest, in such country or countries, to the Subject Invention to the other Party and to cooperate in the preparation and filing of a patent or other IP applications. In any countries in which title to patent or other IP rights is transferred to PSRC, PSRC agrees that ABC inventors will share in any royalty distribution that PSRC pays to its own inventors.

6.6 **Patent Expenses.** The expenses attendant to the filing of patent or other IP applications generally shall be paid by the Party filing such application. If an exclusive license to any Subject Invention is granted to PSRC, PSRC shall be responsible for all past and future out-of-pocket expenses in connection with the preparation, filing, prosecution and maintenance of any applications claiming such exclusively licensed inventions and any patents or other IP grants that may issue on such applications. PSRC may waive its exclusive license rights on any application, patent or other IP grant, at any time, and incur no subsequent compensation obligation for that application, patent, or IP grant.

6.7 **Prosecution of Intellectual Property Applications.** Within one month of receipt or filing, each Party shall provide the other Party with copies of the applications and all documents received from or filed with the relevant patent or other IP office in connection with the prosecution of such applications. Each Party shall also provide the other Party with the power to inspect and make copies of all documents retained in the patent or other IP application files by the applicable patent or other IP office. Where licensing is contemplated by PSRC, the Parties agree to consult with each other with respect to the prosecution of applications for ABC Subject Inventions described in Article 6.3 and joint Subject Inventions described in Article 6.4. If PSRC elects to file and prosecute IP applications on joint Subject Inventions pursuant to Article 6.4, ABC will be granted an associate power of attorney (or its equivalent) on such IP applications.

7.0 Licensing

7.1 **Option for Commercialization License**. With respect to ABC's IP rights to any Subject Invention not made solely by the PSRC's employees for which a patent or other IP application is filed, ABC hereby grants to the PSRC an option to elect an exclusive or nonexclusive commercialization license. The terms of the license will fairly reflect the nature of the invention, the relative contributions of the Parties to the invention and the Agreement, the risks incurred by the PSRC, and the costs of subsequent research and development needed to bring the invention to the marketplace.

7.2 **Exercise of License Option**. The option of Article 7.1 must be exercised by written notice mailed within three (3) months after PSRC receives written notice that the patent or other IP application is filed. Exercise of this option by the PSRC initiates a negotiation period that expires nine (9) months after the patent or other IP application filing date. If the last proposal by the PSRC has not been responded to in writing by ABC within this nine-month (9) period, the negotiation period shall be extended to expire one (1) month after ABC so responds, during which month the PSRC may accept in writing the final license proposal of ABC. In the absence of such acceptance, ABC will be free to license such IP rights to others. In the event that the PSRC elects the option for an exclusive license, but no such license is executed during the negotiation period, ABC agrees not to make an offer for an exclusive license on more favorable terms to a third party for a period of six (6) months without first offering PSRC those more favorable terms.

7.3 **Joint Inventions Not Exclusively Licensed**. In the event that the PSRC does not acquire an exclusive commercialization license to IP rights in all fields in joint Subject Inventions described in Article 6.4, then each Party shall have the right to use the joint Subject Invention and to license its use to others in all fields not exclusively licensed to PSRC. The Parties may agree to a joint licensing approach for such IP rights.

8.0 Proprietary Rights and Publication

8.1 **Right of Access**. ABC and PSRC agree to exchange all Subject Data produced in the course of research under this Agreement, whether developed solely by ABC or jointly with PSRC. Research Materials will be shared equally by the Parties to the Agreement unless other disposition is agreed to by the Parties. All Parties to this Agreement will be free to utilize Subject Data and Research Materials for their own purposes, consistent with their obligations under this Agreement.

8.2 **Ownership of Subject Data and Research Materials**. Subject to the sharing requirements of Paragraph 8.1 and the regulatory filing requirements of Paragraph 8.3, the producing Party will retain ownership of and title to all Subject Inventions, all Subject Data and all Research Materials produced solely by their investigators. Jointly developed Subject Inventions, Subject Data and Research Materials will be jointly owned.

Co-Development Agreement

8.3 **Dissemination of Subject Data and Research Materials.** To the extent allowed under law, PSRC and ABC agree to use reasonable efforts to keep Subject Data and Research Materials confidential until published or until corresponding patent applications are filed. Any information that would identify human subjects of research or patients will always be maintained confidentially. PSRC shall have the exclusive right to use any and all Agreement Subject Data in and for any regulatory filing by or on behalf of PSRC, except that ABC shall have the exclusive right to use Subject Data for that purpose, and authorize others to do so, if the Agreement is terminated or if PSRC abandons its commercialization efforts.

8.4 **Proprietary/Confidential Information.** Each Party agrees to limit its disclosure of Proprietary/Confidential Information to the amount necessary to carry out the Research Plan of this Agreement, and shall place a confidentiality notice on all such information. Confidential oral communications shall be reduced to writing within 30 days by the disclosing Party. Each Party receiving Proprietary/Confidential Information agrees that any information so designated shall be used by it only for the purposes described in the attached Research Plan. Any Party may object to the designation of information as Proprietary/Confidential Information by another Party and may decline to accept such information. Subject Data and Research Materials developed solely by PSRC may be designated as Proprietary/Confidential Information when they are wholly separable from the Subject Data and Research Materials developed jointly with ABC investigators and advance designation of such data and material categories is set forth in the RP. The exchange of other confidential information, for example, patient-identifying data, should be similarly limited and treated. Jointly developed Subject Data and Research Material derived from the Research Plan may be disclosed by PSRC to a third party under a confidentiality agreement for the purpose of possible sublicensing pursuant to the Licensing Agreement and subject to Article 8.7.

8.5 **Protection of Proprietary/Confidential Information.** Proprietary/Confidential Information shall not be disclosed, copied, reproduced, or otherwise made available to any other person or entity without the consent of the owning Party. Each Party agrees to use its best efforts to maintain the confidentiality of Proprietary/Confidential Information.

8.6 **Duration of Confidentiality Obligation.** The obligation to maintain the confidentiality of Proprietary/Confidential Information shall expire at the earlier of the date when the information is no longer Proprietary Information as defined in Article 2.3 or three (3) years after the expiration or termination date of this Agreement. PSRC may request an extension to this term when necessary to protect Proprietary/Confidential Information relating to products not yet commercialized.

8.7 **Publication.** The Parties are encouraged to make publicly available the results of their research. Before either Party submits a paper or abstract for publication or otherwise intends to publicly disclose information about a Subject Invention, Subject Data, or Research Materials, the other Party shall be provided thirty (30) days to review the proposed publication or disclosure to ensure that Proprietary/Confidential Information is protected. The publication or other disclosure shall be delayed for up to thirty (30) additional days upon written request by any Party as necessary to preserve patent or other IP rights.

9.0 Representations and Warranties

9.1 **Representations and Warranties of ABC.** ABC hereby represents and warrants to PSRC that the official signing this Agreement has authority to do so.

Co-Development Agreement

9.2 Representations and Warranties of PSRC.

(a) PSRC hereby represents and warrants to ABC that PSRC has the requisite power and authority to enter into this Agreement and to perform according to its terms, and that PSRC's official signing this Agreement has authority to do so. PSRC further represents that it is financially able to satisfy any funding commitments made in Appendix B.

(b) PSRC certifies that the statements herein are true, complete, and accurate to the best of its knowledge. PSRC is aware that any false, fictitious, or fraudulent statements or claims may subject it to criminal, civil, or administrative penalties.

10.0 Termination

10.1 **Termination by Mutual Consent.** ABC and PSRC may terminate this Agreement, or portions thereof, at any time by mutual written consent. In such event the Parties shall specify the disposition of all property, inventions, patent, or other IP applications, and other results of work accomplished or in progress, arising from or performed under this Agreement, all in accordance with the rights granted to the Parties under the terms of this Agreement.

10.2 **Unilateral Termination.** Either ABC or PSRC may unilaterally terminate this entire Agreement at any time by giving written notice at least thirty (30) days prior to the desired termination date, and any rights accrued in property, patents, or other IP rights shall be disposed of as provided in paragraph 10.1.

10.3 **Staffing.** If this Agreement is mutually or unilaterally terminated prior to its expiration, funds will nevertheless remain available to ABC for continuing any staffing commitment made by PSRC pursuant to Article 5.1 above and Appendix B, if applicable, for a period of six (6) months after such termination. If there are insufficient funds to cover this expense, PSRC agrees to pay the difference.

10.4 **New Commitments.** No Party shall make new commitments related to this Agreement after a mutual termination or notice of a unilateral termination and shall, to the extent feasible, cancel all outstanding commitments and contracts by the termination date.

10.5 **Termination Costs.** Concurrently with the exchange of final reports pursuant to Articles 4.2 and 5.2, ABC shall submit to PSRC for payment a statement of all costs incurred prior to the date of termination and for all reasonable termination costs including the cost of returning PSRC property or removal of abandoned property, for which PSRC shall be responsible.

11.0 Disputes

11.1 **Governing law.** This Agreement shall be governed by the law of ___________.

11.2 **Settlement.** Any dispute, controversy, or claim arising under, out of, or in connection with this agreement, including, without limitation, its formation, validity, binding effect, interpretation, performance, breach, or termination, as well as noncontractual claims, that is not disposed of by agreement of the Principal Investigators, shall be submitted jointly to the signatories of this Agreement to reach an amicable settlement. If an amicable settlement cannot be reached within 30 days for any reason, the dispute shall be referred

Co-Development Agreement

to and finally settled by arbitration in accordance with the UNCITRAL Arbitration Rules then obtaining. The appointing authority shall be the Secretary-General of the Permanent Court of Arbitration, the number of arbitrators shall be three, and the language to be used in the arbitral proceedings shall be English. The place of arbitration shall be determined by mutual agreement, but if agreement cannot be reached the proceedings shall take place in ______________.

Either party to this agreement may request any judicial authority to order any interim measures of protection for the preservation of its rights and interests to the extent permitted by law, including, without limitation, injunctions and measures for the conservation of such property and information that form part of the subject matter in dispute. Such requests shall not be deemed incompatible with, or as a waiver of, this agreement to arbitrate. In respect of any requests for interim measures of protection, and without limitation to proceeding in any other forum, the parties hereby consent to the exercise of jurisdiction by the judicial authorities of _________________.

In the event a party fails to proceed with arbitration, unsuccessfully challenges the arbitrator's award, fails to comply with the arbitrator's award, or fails to comply with any interim measure of protection issued by any competent authority, the other party shall be entitled to costs of suit, including reasonable attorney's fees, for having to compel arbitration or defend or enforce the award or interim measure.

11.3 **Continuation of Work**. Pending the resolution of any dispute or claim pursuant to this Article, the Parties agree that performance of all obligations shall be pursued diligently in accordance with the direction of the ABC signatory.

12.0 Liability

12.1 **No Warranties**. Except as specifically stated in Article 9, the parties make no express or implied warranty as to any matter whatsoever, including the conditions of the research or any invention or product, whether tangible or intangible, made, or developed under this agreement, or the ownership, merchantability, or fitness for a particular purpose of the research or any invention or product.

12.2 **Indemnification**. PSRC agrees to hold the ABC harmless and to indemnify ABC for all liabilities, demands, damages, expenses, and losses arising out of the use by PSRC for any purpose of the Subject Data, Research Materials, and/or Subject Inventions produced in whole or part by ABC employees under this Agreement, unless due to the negligence or willful misconduct of ABC, its employees, or agents. PSRC shall be liable for any claims or damages it incurs in connection with this Agreement. ABC will hold PSRC harmless for liabilities, demands, damages, expenses and losses caused by the negligence or willful misconduct of ABC, its employees or agents.

12.3 **Force Majeure**. Neither Party shall be liable for any unforeseeable event beyond its reasonable control, not caused by the fault or negligence of such Party, that causes such Party to be unable to perform its obligations under this Agreement, and that it has been unable to overcome by the exercise of due diligence. In the event of the occurrence of such a force majeure event, the Party unable to perform shall promptly notify the other Party. It shall further use its best efforts to resume performance as quickly as possible and shall suspend performance only for such period of time as is necessary as a result of the force majeure event.

Co-Development Agreement

13.0 Miscellaneous

13.1 **Entire Agreement.** This Agreement constitutes the entire agreement between the Parties concerning the subject matter of this Agreement and supersedes any prior understanding or written or oral agreement.

13.2 **Headings.** Titles and headings of the articles and subarticles of this Agreement are for convenient reference only, do not form a part of this Agreement, and shall in no way affect its interpretation.

13.3 **Waivers.** None of the provisions of this Agreement shall be considered waived by any Party unless such waiver is given in writing to the other Party. The failure of a Party to insist upon strict performance of any of the terms and conditions hereof, or failure or delay to exercise any rights provided herein or by law, shall not be deemed a waiver of any rights of any Party.

13.4 **Severability.** The illegality or invalidity of any provisions of this Agreement shall not impair, affect, or invalidate the other provisions of this Agreement.

13.5 **Amendments.** If either Party desires a modification to this Agreement, the Parties shall, upon reasonable notice of the proposed modification or extension by the Party desiring the change, confer in good faith to determine the desirability of such modification or extension. Such modification shall not be effective until a written amendment is signed by the signatories to this Agreement or by their representatives duly authorized to execute such amendment.

13.6 **Assignment.** Neither this Agreement nor any rights or obligations of any Party hereunder shall be assigned or otherwise transferred by either Party without the prior written consent of the other Party.

13.7 **Notices.** All notices pertaining to or required by this Agreement shall be in writing and shall be signed by an authorized representative and shall be delivered by hand or sent by certified mail, return receipt requested, with postage prepaid, to the addresses indicated on the signature page for each Party. Any Party may change such address by notice given to the other Party in the manner set forth above.

13.8 **Independent Contractors.** The relationship of the Parties to this Agreement is that of independent contractors and not as agents of each other or as joint venturers or partners. Each Party shall maintain sole and exclusive control over its personnel and operations. PSRC employees who will be working at ABC facilities may be asked to sign a Guest Researcher or Special Volunteer Agreement appropriately modified in view of the terms of this Agreement.

13.9 **Use of Name or Endorsements.** By entering into this Agreement, ABC does not directly or indirectly endorse any product or service provided, or to be provided, whether directly or indirectly related to either this Agreement or to any patent or other IP license or agreement that implements this Agreement by its successors, assignees, or licensees. PSRC shall not in any way state or imply that this Agreement is an endorsement of any such product or service by the ABC or any of its organizational units or employees. PSRC issued press releases that reference or rely upon the work of ABC under this Agreement shall be made available to ABC at least seven (7) days prior to publication for review and comment.

Co-Development Agreement

13.10 **Exceptions to this Agreement**. Any exceptions or modifications to this Agreement that are agreed to by the Parties prior to their execution of this Agreement are set forth in Appendix C.

13.11 **Reasonable Consent**. Whenever a Party's consent or permission is required under this Agreement, such consent or permission shall not be unreasonably withheld.

14.0 Duration of Agreement

14.1 **Duration**. It is mutually recognized that the duration of this project cannot be rigidly defined in advance, and that the contemplated time periods for various phases of the RP are only good-faith guidelines subject to adjustment by mutual agreement to fit circumstances as the RP proceeds. In no case will the term of this Agreement extend beyond the term indicated in the RP unless it is revised in accordance with Article 13.5.

14.2 **Survivability**. The provisions of articles 4.2, 5–8, 10.3–10.5, 11.1, 12.2–12.4, 13.1, 13.10, and 14.2 shall survive the termination of this Agreement.

Agreement SIGNATURE PAGE

FOR ABC: ______________________ Date ______

Mailing Address for Notices: ______________________

FOR PSRC: ______________________ Date ______

Mailing Address for Notices: ______________________

[Include additional signature and address blocks as necessary for all Parties to this Agreement.]

Co-Development Agreement

APPENDIX A
RESEARCH PLAN

Title of Agreement: ____

ABC Principal Investigator: ____

His/Her Laboratory: ____

PSRC PRINCIPAL INVESTIGATOR: _____________________________________

Term of Agreement: __________ (____) years.

The Research Plan that follows should be concise but of sufficient detail to permit reviewers of this Agreement to evaluate the scientific merit of the proposed collaboration. The RP should explain the scientific importance of the collaboration and the research goals of ABC and PSRC. The respective contributions in terms of expertise and/or research materials of ABC and PSRC should be summarized. Initial and subsequent projects contemplated under the RP, and the time periods estimated for their completion, should be described and pertinent methodological considerations summarized. Pertinent literature references may be cited and additional relevant information included. Include additional pages to identify the Principal Investigators of all other Parties to this Agreement.

APPENDIX B
FINANCIAL AND STAFFING CONTRIBUTIONS OF THE PARTIES

APPENDIX C
EXCEPTIONS OR MODIFICATIONS TO THIS AGREEMENT

Sample Agreement 2:
Public Sector Technology License

Source: Mahoney RT (ed.). 2004. *Handbook of Best Practices for Management of Intellectual Property in Health Research and Development*. MIHR: Oxford, U.K. Reproduced with permission.

THIS AGREEMENT is effective this ______ day of ______, 20__ ("Effective Date"), by and between the Public Sector Research Centre ("PSRC"), an International Organization established according to the procedures of the Vienna Convention on the Law of Treaties, and ____________________, a corporation organized under the laws of [COUNTRY] ("COMPANY").

WITNESSETH

WHEREAS PSRC has the rights to a newly developed technology for [USE OF TECHNOLOGY] ______________________ based on such core technology;

WHEREAS the core technology was developed in collaboration with [NAMES OF COLLABORATORS] ______________________, who have certain rights to the technology;

WHEREAS PSRC has the right to make, use, and sell and to sublicense others to make, use, and sell products arising from the technology;

WHEREAS COMPANY is engaged in [MAKING, USING, OR SELLING]____________ products such as [TYPE OF PRODUCTS] ________________________________ and

WHEREAS COMPANY desires to obtain a license to make, use, and sell [PRODUCTS] ______________________ based on the core technology.

NOW, THEREFORE, the parties intending to be legally bound agree as follows:

Article I: DEFINITIONS

1.1 "Technology" means the technology for [DETAILED DEFINITION OF USE OF TECHNOLOGY] ____ _______.

1.2 "Licensed Product" means a product that uses the Technology.

1.3 "Exclusive Territory" means [COUNTRIES] __________________________.

1.4 "Non-exclusive Territory" means [COUNTRIES]__________________________.

Article II: GRANT OF LICENSE

2.1 Subject to [ANY RESTRICTIONS THAT MAY EXIST FOR PSRC IN LICENSING THE TECHNOLOGY] __________________________, PSRC grants to COMPANY on the terms and conditions herein stated:

a. An exclusive license, except as hereinabove provided, to make, use, and sell the Licensed Product in the Exclusive Territory for the purpose of [LIST OF SPECIFIC USES TO WHICH THE PRODUCT CAN BE PUT] ____; and

b. A nonexclusive license to sell Licensed Product in the Nonexclusive Territory,

Public Sector Technology License

2.2 If at any time during the period of this Agreement COMPANY desires to license the Technology in countries not heretofore included in the Agreement, PSRC agrees to negotiate with COMPANY in good faith for such extension, provided that (a) the license rights in such additional countries have not previously been licensed to others; (b) COMPANY has existing distribution capability in each of such additional countries; (c) COMPANY has manufacturing capacity to serve such additional markets; (d) COMPANY is successfully marketing the Licensed Product in the Territory; and (e) COMPANY is not in default or breach of any covenant or any obligation of this Agreement.

Article III: KNOW-HOW

PSRC agrees to provide COMPANY with all know-how owned or controlled by PSRC that is reasonably required for the transmission of the Technology in accordance with the schedule and conditions specified in this Article III. Such material shall include a manual outlining the Technology; identifying all materials used; and indicating a supply source, the specifications for each material and product, and the required quality control procedures. In addition, PSRC will arrange:

(a) one (1) or more visits to the facilities of COMPANY by one (1) or more technicians knowledgeable in the Technology for a total period to be determined by PSRC; and

(b) up to three (3) visits by COMPANY's staff to suitable training sites chosen by PSRC by up to two (2) COMPANY technicians fluent in English and experienced and qualified in the technology to be transferred for a total period of up to forty (40) days.

Article IV: TECHNICIAN VISITS

4.1 COMPANY will pay the fully allocated cost of PSRC's or, at PSRC's option, a designated agent's technicians during the visits specified in Article III, paragraph (a). COMPANY shall reimburse PSRC for the out-of-pocket expenditures incurred by the technicians and for materials used during the visits specified in Article III, paragraph (a). PSRC, through a designated agent, will use its best efforts at keeping costs to a minimum level throughout the technology transfer without sacrificing quality.

4.2 COMPANY will pay the costs of its technicians and/or staff during the visits specified in Article III, paragraphs (a) and (b) and all costs for materials and equipment used pursuant to Article III, paragraph (a).

Article V: IMPROVEMENTS

Should COMPANY or PSRC develop or obtain rights to additional know-how or improvements in the manufacture or composition of any of the items comprising the Technology, such know-how and/or improvements shall be licensed to the other party royalty free with the right to sublicense said improvements to others subject to:

(a) the terms and conditions of any agreements by which PSRC or COMPANY, as the case may be, obtained rights to such licensed improvements or know-how; and

(b) any sublicense granted under this Article shall incorporate terms that require the sublicensee to respect this Agreement.

PUBLIC SECTOR TECHNOLOGY LICENSE

Article VI: PAYMENT

6.1 COMPANY shall pay PSRC a technology transfer fee of __________ U.S. dollars ($_____) to help offset PSRC's costs and expenses. This fee shall be paid within thirty (30) days of the Effective Date.

6.2 COMPANY shall pay for the costs incurred in training COMPANY staff. Periodic expense estimates will be submitted by PSRC to COMPANY, and COMPANY shall pay PSRC seventy-five percent (75%) of the estimated amounts within ten (10) days thereof in U.S. dollars to PSRC. The balance shall be invoiced by PSRC periodically and shall be paid to PSRC under the payment terms and conditions stated above. If, at the conclusion of each stage, the amounts paid by COMPANY up to that point exceed the amount due, PSRC shall repay the excess within ten (10) days.

6.3 COMPANY shall pay PSRC a royalty of ____ percent (_%) on net sales of Licensed Products. Net sales means the gross amounts of bona fide sales to others not owned or controlled by COMPANY less cash discounts and rebates actually given, in addition to duties, returns, and free replacements. No royalties shall be payable with respect to any sale of Licensed Products that takes place between COMPANY and any company owned or controlled by COMPANY, but such royalty shall accrue upon the resale of the Licensed Products to a third party. Whenever a company owned or controlled by COMPANY itself uses the Licensed Products, royalty shall accrue on such Licensed Products. Said royalty shall be calculated by using, as a sales price, the average price at which COMPANY sold Licensed Products to third parties during the calendar quarter within which the Licensed Products were used by the company owned or controlled by COMPANY.

6.4 Royalties shall be payable each calendar quarter on the last business day of the month following the calendar quarter for the royalties covering the preceding calendar quarter. Royalties are to be paid in U.S. dollars and sent by wire transfer on or before the date due to an account specified by PSRC unless otherwise agreed in writing.

6.5 On the due date of each royalty payment, COMPANY shall furnish to PSRC a full accounting showing separately the total sales of Licensed Products to the Public Sector, the total sales of Licensed Products to the Private Sector, a calculation of the royalties payable in respect thereof, production quantities, and sales prices.

6.6 COMPANY shall maintain sufficiently detailed records and books of accounts to enable PSRC to verify the payments made to PSRC by COMPANY and the reports filed therewith. COMPANY shall permit an independent accountant, appointed by PSRC and to whom COMPANY has no reasonable objection, to inspect, at reasonable times and upon reasonable notice at the principal place of business of COMPANY, the books and records of COMPANY relating to the manufacture, use, and sale of Licensed Products. The accountant shall report to PSRC only whether the amounts reported and paid to PSRC by COMPANY are accurate and, if not, what the correct figures should be. A copy of the report shall also be supplied to COMPANY. In no event shall an examination of COMPANY's books and records be made for a period prior to three (3) years from the date such audit is requested by PSRC. In the event the accountant reports that COMPANY has underpaid the amounts due to PSRC, then COMPANY shall bear the costs of such audit.

PUBLIC SECTOR TECHNOLOGY LICENSE

Article VII: OBLIGATIONS OF COMPANY

7.1 COMPANY agrees to use its best efforts to promote diligently the sale and distribution of Licensed Products throughout the Exclusive Territory in accordance with the milestones established in Exhibit A.

7.2 COMPANY will use its best efforts to promptly establish a validated manufacturing process for Licensed Products.

7.3 Governmental Approvals

(a) The cost of obtaining governmental approval and/or registration to make, use, and sell Licensed Products in the Territory shall be borne by COMPANY. COMPANY agrees to use its best efforts to obtain such governmental approval and/or registration and shall use due diligence therein;

(b) COMPANY agrees that PSRC shall have full access to any application for governmental approval or registration and the data contained therein together with the right to use such application, data, and any subsequent approval to obtain the approval of any government of any country outside of the Territory to make, use, and/or sell Licensed Products in such country, provided that PSRC shall require that any company licensed by PSRC shall respect this Agreement; and

(c) If this Agreement is terminated under Article 11.1, COMPANY agrees, to the extent permitted by local law and/or government regulation, to assign to PSRC or to PSRC's designee and without cost to PSRC or PSRC's designee all governmental approvals and/or registrations owned or controlled by COMPANY for the manufacture, use, and/or sale of Licensed Products.

7.4 COMPANY agrees that it will not knowingly sell or allow the sale of any Licensed Product that it produces which sale would be outside the terms of this Agreement except with the written approval of PSRC.

7.5 COMPANY agrees to obtain the prior written approval of PSRC for any labels, containers, packaging, package inserts, or any product associated or sold with the Licensed Products. If PSRC has not responded within thirty (30) days after receipt of a request for approval, the materials shall be deemed to be approved. Upon written request by PSRC and with reasonable notice, COMPANY further agrees to print on the packaging of the Licensed Products a legend crediting PSRC and/or the Collaborators with providing the technology on which the Licensed Products are based, provided that COMPANY has no reasonable objection to the language used.

7.6 COMPANY agrees to only use, promote, or knowingly sell Licensed Products for uses consistent with the guidelines for uses of such products as published by [REGULATORY AGENCY, WHO, OR OTHERS]. A summary of these guidelines is attached hereto as Exhibit B. PSRC may update these guidelines from time to time as agencies update their guidelines. PSRC will allow COMPANY ninety (90) days from issuance of any updated guidelines to correct any of COMPANY's promotional materials or selling practices not consistent with the updated guidelines. COMPANY agrees that, if requested by PSRC, COMPANY will communicate these guidelines to its existing and potential purchasers of the Licensed Products in a form mutually agreeable to COMPANY and PSRC.

Public Sector Technology License

Article VIII: GOOD MANUFACTURING PRACTICES AND QUALITY ASSURANCE

8.1 COMPANY agrees that all Licensed Products manufactured by COMPANY shall be produced in accordance with GMP prescribed under ISO 9002 and subsequent revisions. COMPANY shall comply with the additional manufacturing requirements outlined in Exhibit C herein and subsequent written directives from PSRC.

8.2 During the term of the license, COMPANY agrees to periodic quality control ("QC") testing by PSRC and/or a third party selected by PSRC and to whom COMPANY has no reasonable objection. COMPANY shall reimburse PSRC for the cost of said third-party testing. It is anticipated that QC testing will be performed on not fewer than the first three (3) batches produced and, thereafter, not fewer than one (1) batch per year. All QC testing under this paragraph 8.2 is in addition to testing required under paragraph 8.1 and/or the applicable laws and regulations of each country in the Territory where COMPANY manufactures and/or sells the Licensed Product.

8.3 COMPANY shall permit a duly authorized representative of PSRC, upon reasonable notice, to inspect the premises of COMPANY from time to time to ascertain that the provisions of this Article are being complied with by COMPANY.

8.4 COMPANY agrees to mark Licensed Products' packaging in accordance with the applicable laws of each country in the Territory where COMPANY manufactures and/or sells the Licensed Product.

8.5 In the event that COMPANY, at any time, fails to meet GMP and QC standards as provided for in this Article, COMPANY shall immediately, at the direction of PSRC or the government of any country in which COMPANY manufactures and/or sells a Licensed Product, recall all such Licensed Products not yet sold and cease further sales until such standards are met.

Article IX: CONFIDENTIALITY

9.1 COMPANY shall keep confidential and refrain from using, except under the terms of this Agreement, all know-how disclosed to it by PSRC, its employees, and its agents under this Agreement. The obligation of confidence shall not apply to:

(a) know-how that, at the time of disclosure, COMPANY can demonstrate was in the public domain;

(b) know-how that, after disclosure, becomes part of the public domain other than by COMPANY;

(c) know-how that COMPANY can show was in COMPANY's possession at the time of disclosure and was not acquired directly or indirectly from PSRC; or

(d) know-how that has been, is now, or is hereinafter furnished or made known to COMPANY by a third party as a matter of right and who did not derive the information from PSRC.

Public Sector Technology License

9.2 Confidential information shall not be deemed to fall within the exceptions of Article 9.1 if (1) the confidential information is specific and is merely embraced by more general information in the public domain or in COMPANY's possession or (2) the confidential information is a combination that can be pieced together or reconstructed from multiple sources, none of which shows the whole combination, its principle of operation, and method of use.

9.3 Nothing contained in this Article shall be construed as restricting COMPANY from disclosing the know-how to others as a necessary adjunct to the manufacture, use, and sale of the Licensed Products in accordance with this Agreement.

9.4 COMPANY further agrees to take all reasonable precautions to prevent any of its personnel from divulging to any other party, except as is otherwise provided herein, any know-how furnished to COMPANY by PSRC under this Agreement.

9.5 The provisions of this Article shall survive the term of this Agreement.

Article X: HOLD HARMLESS

10.1 Except for willful misconduct by PSRC, its employees, and its agents, COMPANY agrees to hold harmless and indemnify PSRC, its employees, and its agents from the following:

(a) all product liability claims arising from the manufacture, use, or sale of the Licensed Products made by COMPANY;

(b) all medical malpractice claims in the use of Licensed Products manufactured, distributed, or sold by COMPANY; and

(c) all claims against COMPANY or PSRC for infringement arising from the manufacture, use, or sale by COMPANY of Licensed Products.

10.2 COMPANY shall, at its sole expense, assume the defense of such suit or claim. PSRC shall promptly notify COMPANY of any suit or claim against PSRC, its employees, or its agents of which PSRC is aware. COMPANY may elect to cease to manufacture, distribute, or sell the Licensed Product or obtain a license for the Licensed Product rather than defend a patent infringement suit, provided that COMPANY shall continue to be responsible with respect to any claim of infringement arising from any action of COMPANY occurring before the date of ceasing manufacture, distribution, or sale or before the date of any license.

10.3 Insofar as COMPANY has liability insurance, COMPANY agrees to instruct its insurance carrier to name PSRC as an additional insured; to issue PSRC a certificate of coverage; and to give PSRC notice of any cancellation, renewal, or changes in coverage.

10.4 The provisions of this Article shall survive the term of this Agreement, whether or not such cause of action or claim had accrued at the time of termination.

Public Sector Technology License

Article XI: TERM

11.1 This Agreement shall commence on the Effective Date and may be terminated under the following circumstances and on the following terms and conditions:

(a) If COMPANY has not successfully manufactured pilot-scale batches of at least one (1) Licensed Product in its facility within one (1) year from the Effective Date, PSRC may terminate by giving COMPANY thirty (30) days' written notice;

(b) If COMPANY has not obtained permission, if such permission is required under [COUNTRY] law, from the Government of [COUNTRY] to sell at least one (1) Licensed Product in [COUNTRY] within _______ (__) months from the Effective Date, PSRC may terminate this Agreement by giving COMPANY thirty (30) days' written notice;

(c) If COMPANY has not successfully completed the milestones established in Exhibit A, PSRC may terminate by giving COMPANY sixty (60) days' written notice;

(d) COMPANY may terminate at any time by giving PSRC ninety (90) days' written notice;

(e) If either party should at any time default or commit any breach of any covenant or any obligation of the license and should fail to remedy any default or breach within thirty (30) days after written notice thereof by the other party, the injured party may, at its sole option, terminate this license by notice in writing to such effect;

(f) If an order be made or an effective resolution be passed for the winding up of COMPANY; if there is a failure, distress, execution, or other legal process levied or enforced upon or against a substantial portion of the chattels or property of COMPANY; if a receiver is appointed for the undertaking of the property and assets of COMPANY or a substantial portion thereof; or if COMPANY shall make any assignment or composition for the benefit of its creditors or shall cease to carry on business, then PSRC may terminate by giving COMPANY thirty (30) days' written notice;

(g) If COMPANY has not obtained permission to sell a Licensed Product from the government of any country requiring such permission within two (2) years from the date a grant was first made by PSRC to COMPANY for sales anywhere in the world for such Licensed Product, PSRC may terminate that portion of the grant giving COMPANY permission to sell said Licensed Product in those countries for which a complete application for said government permission has not been filed. This partial termination shall occur by PSRC giving COMPANY thirty (30) days' written notice; and

(h) By mutual agreement of the parties.

11.2 UPON TERMINATION OF THIS AGREEMENT:

(a) All rights and licenses created by this Agreement shall expire, except that any sums due as royalty shall remain payable and Articles X and XI shall remain in effect. COMPANY shall return to PSRC all technical documents given to COMPANY pursuant to this Agreement, shall destroy any copies of the technical documentation supplied, shall refrain from further manufacture or distribution of the Licensed Products anywhere in the world, and shall assign and transfer to PSRC or to PSRC's designee all governmental approvals and/or registrations to make, use, and/or sell Licensed Products that are owned or controlled by COMPANY to the extent permitted by local law and/or government regulation. Within thirty (30) days after termination, COMPANY shall return to PSRC all know-how; and

(b) The relationships of PSRC and COMPANY to any third parties thereafter shall be determined by mutual agreement and both parties agree to cooperate in executing and delivering such instruments of assignment or other instruments or documents as may be required to effectuate such assignments.

11.3 Unless terminated sooner, this Agreement shall remain in force and effect for _______ (__) years from the Effective Date. Thereafter, COMPANY shall retain the right to make, use, and sell Licensed Products without further obligation to PSRC hereunder except as provided under Articles IX and X.

Article XII: THIRD PARTY LICENSES

COMPANY agrees to cooperate fully with PSRC in assisting other companies or organizations selected by PSRC, and to whom COMPANY has no reasonable objection, to manufacture the Licensed Products in a country outside of the Territory and to obtain any requisite governmental approval for the manufacture, use, and sale of the Licensed Products in such country, provided that such other company or organization covenants not to manufacture, use, and/or sell Licensed Products in the Territory. In providing manufacturing and registration assistance hereunder, COMPANY shall be reimbursed its expenses, including employee time, plus ten percent (10%).

Article XIII: NOTICE

Any notice required or provided for by the terms of this Agreement shall be in writing, and all notices, reports, and payments (other than royalties) provided for hereunder shall be sent by registered mail, prepaid, or facsimile to the business address of the party to be served therewith. It is agreed that the business addresses of the parties shall be as follows:

If to COMPANY:
Facsimile:
With copies to: ______________________________
[ADDRESS]______________________________
____
____

If to PSRC: Facsimile:
With copies to: The PSRC Organization
[ADDRESS]______________________________
____
____

or such other addresses as either party shall have notified the other party. Any such notice, royalty, or payment shall be deemed to have been given or made on the date such letter was registered or delivered for transmission to the sender's facsimile operator, but any assumption of actual notice or payment shall be subject to rebuttal to show that it has not actually been received.

Public Sector Technology License

Article XIV: ASSIGNABILITY
This Agreement shall not be assignable or transferable by COMPANY without the written consent of PSRC, but may be assigned or transferred by PSRC without the consent of COMPANY. COMPANY may, however, assign this Agreement to any wholly owned subsidiary of COMPANY, but COMPANY assumes full responsibility to PSRC for the fulfillment of all the terms of this Agreement by the assignee.

Article XV: ENTIRE AGREEMENT
This Agreement contains the entire agreement and understanding between the parties with respect to the subject matter herein and supersedes all prior discussions and other writings, except for those specifically referred to herein. This Agreement shall not be modified except in writing.

Article XVI: ARBITRATION
This Agreement shall be governed by the law of __________. The parties hereto undertake to settle any dispute, controversy, or claim arising under, out of, or in connection with this Agreement, including, without limitation, its formation, validity, binding effect, interpretation, performance, breach, or termination, as well as noncontractual claims, in an amicable manner. If an amicable settlement cannot be reached within 30 days for any reason, the dispute shall be referred to and finally settled by arbitration in accordance with the UNCITRAL Arbitration Rules then obtaining. The appointing authority shall be the Secretary-General of the Permanent Court of Arbitration, the number of arbitrators shall be three, and the language to be used in the arbitral proceedings shall be English. The place of arbitration shall be determined by mutual agreement, but if agreement cannot be reached the proceedings shall take place in ____________.

Either party to this Agreement may request any judicial authority to order any interim measures of protection for the preservation of its rights and interests to the extent permitted by law, including, without limitation, injunctions and measures for the conservation of such property and information that form part of the subject matter in dispute. Such requests shall not be deemed incompatible with, or as a waiver of, this agreement to arbitrate. In respect of any requests for interim measures of protection, and without limitation to proceeding in any other forum, the parties hereby consent to the exercise of jurisdiction by the judicial authorities of ________________.

In the event a party fails to proceed with arbitration, unsuccessfully challenges the arbitrator's award, fails to comply with the arbitrator's award, or fails to comply with any interim measure of protection issued by any competent authority, the other party shall be entitled to costs of suit, including reasonable attorney's fees, for having to compel arbitration or defend or enforce the award or interim measure.

Article XVII: WAIVER
No waiver by either party of any provision hereof shall be deemed a waiver of any other provision hereof or of any subsequent breach by either party of the same or any other provision. None of the terms of this Agreement will be held to have been waived or altered unless such waiver or alteration is in writing and signed by both the parties hereto.

PUBLIC SECTOR TECHNOLOGY LICENSE

Article XVIII: WARRANTY

PSRC warrants that the Technology and the know-how associated therewith were developed by PSRC; that the Technology and know-how are owned by PSRC; and that, subject to [LIST HERE ANY RESTRICTIONS], PSRC has a worldwide, exclusive license for such Technology and know-how together with the right to sublicense others thereunder to make, use, and sell Licensed Products anywhere in the world. Nothing in this Agreement will be construed as a warranty or representation that anything made, used, sold, or otherwise disposed of under any license granted in this Agreement is or will be free from infringement of patents of third parties. Except as specifically set forth herein, PSRC makes no representations or warranties, either express or implied, arising by law or otherwise, including, but not limited to, implied warranties of merchantability or fitness for a particular purpose. In no event will PSRC, its employees, and its agents have any obligation or liability arising from tort or for loss of revenue or profit or for incidental or consequential damages.

It is the sole responsibility of COMPANY to undertake a thorough search for third party patents before selling any Licensed Product in any national market, and COMPANY agrees that PSRC, its employees, and its agents shall not have any liability for infringement of any patent as a consequence of the manufacture, use, or sale of any Licensed Product or by reason of any other use of the Technology by COMPANY.

IN WITNESS WHEREOF the parties hereto have caused these presents to be executed in duplicate by their respective duly authorized officers.

For PSRC:	For COMPANY:
By:____________________	By:____________________
Date:____________________	Date:____________________

Exhibit A

Milestones (in months from signing agreement)	Target Date

1. Establish a validated assembly or manufacturing process
2. Conduct clinical trials of product produced in [COUNTRY]
3. etc.

Public Sector Technology License

Exhibit B: Guidelines for Appropriate Use of Product Based on the Technology

Exhibit C: Good Manufacturing Practices for the General Release of Product Based on Technology

The following items must be completed or in place before products are released for general use:

[The Institute may specify a number of conditions depending on local considerations. Examples are listed here:

1. Review and approval by PSRC of package inserts and packaging materials

2. Inspection of manufacturing and control procedures, facilities, records, inventories, and reference samples by an PSRC consultant

3. Establishment of a system of post-market surveillance that [specify requirements] ______ ______________________________

4. Establishment of an introduction program to product users

5. Notification/discussion with PSRC of any change in manufacturing procedure. Pilot batches using new manufacturing procedures must pass stability testing before going into industrial production. The first two industrial batches using new manufacturing procedures must go into ongoing stability testing at room temperature (manufacturer's warehouse). No less than one batch per year must be placed in the same program. Results are to be used to support registration and/or modify information about shelf life.

6. PSRC reserves the right to audit the manufacturer's installations, process, and inventory, and to retain samples on a yearly basis. At any point in the distribution system, product may be tested by PSRC as part of this audit.]

Sample Agreement 3: Public Sector Patent License (Medical Research Council of South Africa)

Source: MRC Innovation Centre, PO Box 19070, Tygerberg, 7505, Cape Town, South Africa. Reproduced with permission. Visit innovation.mrc.ac.za to download many additional template and sample agreements.

MRC SOUTH AFRICA
MEDICAL
RESEARCH
COUNCIL

Standard License Agreement
Between
The South African Medical Research Council

and

____

For

____

This License Agreement (the "Agreement") is made by and between ______________________, a________________________, duly established under the Companies Act having an address at _____________________ ("LICENSEE") and the South African Medical Research Council, a non-profit organization having an address at Fransie van Zijl Drive, Parowvallei, Cape Town ("MRC").

This Agreement is effective on the date of the last signature ("Effective Date") the _____ day of ____________ 20_____.

WITNESSETH

RECITALS

WHEREAS the MRC is the sole owner of the Technology titled _____________________, which is covered by Patent Rights as defined below;

WHEREAS the MRC warrants that it possesses the right to license the Technology;

WHEREAS the MRC is desirous that the Technology be developed and utilized to the fullest possible extent to produce commercially marketable Products so that its benefits can be enjoyed by the general public;

WHEREAS LICENSEE entered into a Confidential Disclosure Agreement with the MRC, effective ______, for the purpose of evaluating the Invention;

WHEREAS LICENSEE entered into a Letter of Intent with the MRC, effective ________, for the purpose of negotiating this Agreement;

WHEREAS LICENSEE desires to acquire an exclusive (or non-exclusive) license, under the terms and conditions hereinafter set forth, in the Territory for commercial development, use, and sale of the Invention.

Public Sector Patent License

NOW THEREFORE, For these and other valuable considerations, the receipt of which is hereby acknowledged, the parties agree as follows:

1. DEFINITIONS

The terms, as defined herein, shall have the same meanings in both their singular and plural forms.

1.1 "Affiliate" shall mean any corporation or other business entity controlled by, controlling or under common control with LICENSEE. For this purpose, "control" shall mean direct or indirect beneficial ownership of at least a fifty percent (50%) of the voting stock of, or at least a fifty percent (50%) interest in the income of such corporation or other business entity, or such other relationship as in fact, constitutes actual control.

1.2 "Combination Product" means any product which is a Licensed Product and contains other products(s) or product component(s) that
(i) does not use Invention, Technology or Patent Rights;
(ii) the sale, use or import by itself does not contribute to the infringement of Patent Rights;
(iii) can be sold separately by LICENSEE, its Sublicensee or an Affiliate; and
(iv) enhances the market price of the final product(s) sold used or imported by LICENSEE, its Sublicensee, or an Affiliate.

1.3 "Field of use" means .... [*specific field of use for the license*].

1.4 "Know-How" means the ideas, methods, characterization and techniques developed by the MRC before the Effective Date, which are necessary for practicing the Patent Rights.

1.5 "Licensed Method" means any method that uses Technology or any part thereof, or is covered by Patent Rights the use of which would constitute, but for the license granted to LICENSEE under this Agreement, an infringement of any pending or issued and unexpired claim within Patent Rights.

1.6 "Licensed Product" shall mean any composition or product or part thereof which:
(i) is covered in whole or in part by an issued, unexpired claim or a pending claim contained in the Patent Rights in the country in which any Licensed Product is made, used or sold;
(ii) is manufactured by the Licensed Method or using a process which is covered in whole or in part by an issued, unexpired claim or a pending claim contained in the Patent Rights in the country in which any Licensed Process is used or in which such product or part thereof is used or sold.

1.7 "Net Sales" means the total of the gross invoice prices of Licensed Products sold by LICENSEE, its Sublicensee, an Affiliate, or any combination thereof, less the sum of the following actual and customary deductions where applicable and separately listed: cash, trade, or quantity discounts; sales, use, tariff, import/export duties or other excise taxes imposed on particular sales (except for value-added and income taxes imposed on the sales of Product in foreign countries); transportation or shipping charges to purchasers; or credits to LICENSEE because of rejections, allowances or returns. For purposes of calculating Net Sales, transfers to a Sublicensee or an Affiliate of Licensed Product under this Agreement for
(i) end use (but not resale) by the Sublicensee or Affiliate shall be treated by LICENSEE at list price of LICENSEE, or
(ii) resale by a Sublicensee or an Affiliate shall be treated as sales at the list price of the Sublicensee or Affiliate.

PUBLIC SECTOR PATENT LICENSE

1.8 "Patent Costs" means all out-of-pocket expenses for the preparation, filing, prosecution, and maintenance of all South African and foreign patents included in Patent Rights. Patent Costs shall also include reasonable out-of-pocket expenses for patentability opinions, inventorship determination, preparation and prosecution of patent application, re-examination, re-issue, interference, and opposition activities related to patents or applications in Patent Rights.

1.9 "Patent Rights" shall mean all of the following MRC intellectual property:
(i) SA Patent Application Number ______________,
(ii) PCT Patent Application Number ______________,
and any new claim, reissues, reexaminations, improvements or extensions, continuations, divisionals, and the foreign counterpart patents, patent applications and patents issuing therefrom relating to the product. The MRC shall be the assignee and owner of all such Patents and Patent Applications.

1.10 "Process" shall mean any process which is covered in whole or in part by an issued, unexpired claim or pending claim contained in the Patent Rights.

1.11 "Sponsor Rights" means all the applicable provisions of any license to the South African Government executed by the MRC and the overriding obligations to the Government and the overriding obligations to .... under the sponsorship agreement with the same.

1.12 "Sublicensee" as used in this Agreement shall mean any third party to whom LICENSEE has granted a license to make, have made, use and/or sell the Product under the Patent Rights, provided said third party has agreed in writing with LICENSEE to accept the conditions and restrictions agreed to by LICENSEE in this Agreement.

1.13 "Technology" means the following MRC intellectual property:
(i) Patent Rights, as defined in 1.8 above.
(ii) Any and all copyrights, mask works, trademarks, service marks, trade dress, trade secrets, confidential information, proprietary information or know-how pertaining to Invention.

1.14 "Term" means the period of time beginning on the Effective Date and ending on the later of
(i) the expiration date of the longest-lived Patent Rights; or
(ii) the twenty-first (21st) anniversary of Effective Date.

1.15 "Territory" means .... [*areas in which the license is valid*].

2. GRANTS

2.1 **License**. In consideration for payment of royalties and subject to the limitations set forth in this Agreement and Sponsor's Rights, the MRC hereby grants to LICENSEE, and LICENSEE hereby accepts, a license under Patent Rights to make, have made for its own use and sale, use, sell, offer for sale, and import Licensed Products and to practice Licensed Methods and to use Technology, in the Field within the Territory and during the Term.

The license granted herein is exclusive and the MRC shall not grant to third parties a further license under Patent Rights or to use Technology in the Field, within the Territory and during the Term. (or The license granted herein is non-exclusive and the MRC may grant to third parties further licenses under Patent Rights or to use Technology in the Field, within the Territory and during the Term.)

PUBLIC SECTOR PATENT LICENSE

The MRC grants to the LICENSEE the authority to make application for Patents, in the name of the MRC; all expenses of obtaining and maintaining said patents shall be paid by LICENSEE.

LICENSEE agrees that the Technology constitutes a trade secret. LICENSEE agrees to include a clause in any and all agreements for sale or transfer to third parties that use of Technology or Products derived therefrom shall be for non-commercial research and development purposes only.

LICENSEE agrees that if commercially useful derivatives or developments from the Technology are made by LICENSEE or sublicensees, such derivatives or developments shall belong to the MRC. The MRC shall negotiate with LICENSEE or sublicensees for appropriate compensation which will be reasonable and customary in the field of the developments.

2.2 **Sublicense.**

(i) The license granted in Paragraph 2.1 includes the right of LICENSEE to grant sublicenses to third parties during the Term but only for as long the license is exclusive.

(ii) With respect to sublicense granted pursuant to Paragraph 2.2(a), LICENSEE shall:

(1) not receive, or agree to receive, anything of value in lieu of cash as considerations from a third party under a sublicense granted pursuant to Paragraph 2.2(a) without the express written consent of the MRC;

(2) to the extent applicable, include all of the rights of and obligations due to the MRC (and, if applicable, the Sponsor's Rights) and contained in this Agreement;

(3) promptly provide the MRC with a copy of each sublicense issued; and

(4) collect and guarantee payment of all payments due, directly or indirectly, to the MRC from Sublicensees and summarize and deliver all reports due, directly or indirectly, to the MRC from Sublicensees.

(iii) Upon termination of this Agreement for any reason, the MRC, at its sole discretion, shall determine whether LICENSEE shall cancel or assign to the MRC any and all sublicenses.

2.3 **Reservation of Rights.** The MRC reserves the right to:

(i) practice the Invention, Technology and Patent Rights for its own use

(ii) use the Invention, Technology and Patent Rights for educational and research purposes;

(iii) publish or otherwise disseminate any information about the Invention and Technology at any time;

(iv) allow other nonprofit institutions to use Invention, Technology and Patent Rights for educational and non-commercial research purposes in their facilities; and

(v) require LICENSEE to sublicense to a third party the Invention, Technology and Patent Rights in any field of no commercial interest to LICENSEE.

3. **CONSIDERATIONS**

3.1 **Fees and Royalties.** The parties hereto understand that the fees and royalties payable by LICENSEE to the MRC under this Agreement are partial considerations for the license granted herein to LICENSEE under Technology, and Patent Rights. LICENSEE shall pay the MRC:

(i) a **license issue fee** of ___ Rands (R_______) upon execution of this Agreement; or

(i) a **license issue fee** of ___ Rands (R_______), within thirty (30) days after the Effective Date; or

Public Sector Patent License

(i) in recognition of LICENSEE being a startup business and partially in lieu of cash, a **license issue fee** in the form of ___ % [*or* ___ *shares*] of the LICENSEE'S common stock authorized in the Shareholder's Agreement of the LICENSEE (*or authorized in the Article of Incorporation of the LICENSEE*) dated ________, and a copy of which is attached to this Agreement as Exhibit A; or

(i) partially in lieu of cash, a **license issue fee** in the form of an option granted to the MRC to purchase for one Rand (R1.00) ___ % [*or* ___ *shares*] of the LICENSEE'S common stock authorized in the Shareholder's Agreement of the LICENSEE (*or authorized in the Article of Incorporation of the LICENSEE*) dated _________, and a copy of which is attached to this Agreement as Exhibit A. The option period commences on Effective Date and shall terminate _____ years thereafter. This option, in whole or in part, can be exercised by the MRC or transferred by the MRC to the inventors any time during the option period.

(ii) **license maintenance fees** of ___ Rand (R_______) per year and payable on the first anniversary of the Effective Date and annually thereafter on each anniversary; provided however, that LICENSEE's obligation to pay this fee shall end on the date when LICENSEE is commercially selling a Licensed Product;

(iii) **milestone payments** in the amounts payable according to the following schedule or events:

Amount Date or Event
(1) ....
(2) ....

(iv) an **Earned Royalty** of ___ percent (___%) on Net Sales of Licensed Products by LICENSEE and/or its Affiliate(s); or

(iv) an **Earned Royalty**
(1) of ___ percent (___%) on Net Sales of Licensed Products for diagnostic uses by LICENSEE and/or its Affiliate(s); and
(2) of ___ percent (___%) on Net Sales of Licensed Products for therapeutic uses by LICENSEE and/or its Affiliate(s); provided, however, that the earned royalty due on Net Sales of Combination Product by LICENSEE and/or its Affiliate(s) shall be calculated as below:
Earned Royalties due the MRC = A/(A+B+C . . .) x Royalty Rate on Net Sales of the Licensed Products applicable in (i) or (ii) x Net Sales of Combination Product, where: A is the separately listed sale price of the Licensed Product or Licensed Product components; and B and C . . . are the separately listed sale prices of the individual products or product components. If LICENSEE does not separately sell any of the B, C . . . products or product components used in Combination Product, the purchase price paid by LICENSEE in the procurement of said products or product components shall be used.

(v) fifty percent (50%) of all **sublicense fees** received by LICENSEE from its Sublicensees that are not earned royalties;

(vi) on each and every **sublicense royalty** payment received by LICENSEE from its Sublicensees on sales of Licensed Product by Sublicensee, the higher of
(1) fifty percent (50%) of the royalties received by LICENSEE; or
(2) royalties based on the royalty rate in Paragraph 3.1(iv) as applied to Net Sales of Sublicensee;

(vii) beginning the calendar year of commercial sales of the first License Product by LICENSEE, its Sublicensee, or an Affiliate and if the total earned royalties paid by LICENSEE under Paragraphs 3.1(iv) and (vi) to the MRC in any such year cumulatively amounts to less than _______ Rand (R_______) ("**Minimum Annual Royalty**"), LICENSEE shall pay to the MRC a Minimum Annual Royalty on or before February 28 following the last quarter of such year the difference between amount noted above and the total earned royalty paid by LICENSEE for such year under Paragraphs 3.1(iv) and (vi); provided, however, that for the

PUBLIC SECTOR PATENT LICENSE

year of commercial sales of the first Licensed Product, the amount of Minimum Annual Royalty payable shall be pro-rated for the number of months remaining in that calendar year; or

(vii) beginning the calendar year of commercial sales of the first License Product by LICENSEE, its Sublicensee, or an Affiliate, LICENSEE will pay _______ Rand (R_______) ("**Minimum Annual Royalty**"). Such Minimum Annual Royalties will be considered as a credit toward earned royalties due for the applicable calendar year and the royalty reports shall reflect the use of such credit.

a) The provisions for Minimum Annual Royalties shall be construed as an annual minimum payment requirement and none of the Minimum Annual Royalty payments are refundable or applicable to succeeding years.

b) If the aggregate earned royalties for any given calendar year for which a minimum annual royalty is payable do not exceed the Minimum Annual Royalty, LICENSEE may pay the MRC, with its payment for the last calendar year quarter, a balancing payment in an amount equal to the difference between the minimum annual royalty and the earned royalties for that calendar year.

c) If LICENSEE does not make the balancing payment described in Section 5.3.b, the MRC shall have the option on sixty (60) days notice to LICENSEE to convert this Agreement from an Exclusive License into a non-exclusive license, in which event, LICENSEE shall not have any further Minimum Annual Royalty obligation and LICENSEE will also be granted any more favorable term or terms granted by the MRC to another licensee.

3.2 All fees and royalty payments specified in Paragraphs 3.1(i) through 3.1(vii) above shall be paid by LICENSEE pursuant to Paragraph 4.3 and shall be delivered by LICENSEE to the MRC as noted in Paragraph 10.1. Earned Royalties will be reduced by ____ % upon expiration of Patent Rights.

3.3 **Patent Costs.** LICENSEE shall reimburse the MRC all past (prior to the Effective Date) and future (on or after the Effective Date) Patent Costs plus a fifteen percent (15%) patent service fee within thirty (30) days following receipt by LICENSEE of an itemized invoice from the MRC. Past Patent Costs are _______ (R_______)

3.4 **Due Diligence.**

(a) LICENSEE shall:

(1) adhere to the following milestones: (a) 1st milestone within ____ (_) years from the Effective Date of this Agreement; (b) 2nd milestone within ____ (_) years from the Effective Date;

(2) use its best efforts to develop, manufacture, market and sell the Products in the Territory and will exert its best efforts to create a demand for the Licensed Products;

(3) maintain satisfactory standards in respect to the nature of the Product manufactured and/or sold by LICENSEE. LICENSEE, agrees that all Product manufactured and/or sold by it shall be of a quality which is appropriate to products of the type here involved. LICENSEE agrees that similar provisions shall be included by sublicenses of all tiers;

(4) annually spend not less than _______ Rand (R_______) for the development of Licensed Products during the first __ years of this Agreement. LICENSEE may, at its sole option, fund the research of any one of the Inventors and credit the amount of such funding actually paid to the MRC against its obligation under this paragraph; or

(3) annually spend not less than _______ Rand (R_______) for the development of Licensed Products during the first __ years of this Agreement. LICENSEE recognizes the expertise of the Inventors in Invention and is committed to contract the Inventors to further develop Invention at the MRC at R__ a year for a total of ___ years. LICENSEE may credit the amount actually paid to the MRC under such contract against its obligation under this paragraph;

Public Sector Patent License

(4) market Licensed Products in South Africa within six (6) months of receiving regulatory approval to market such Licensed Products;
(5) reasonably fill the market demand for Licensed Products following commencement of marketing at any time during the term of this Agreement; and
(6) obtain all necessary governmental approvals for the manufacture, use and sale of Licensed Products.

(b) If LICENSEE fails to perform any of its obligations specified in Paragraphs 3.4(a)(1)-(6), then the MRC shall have the right and option upon thirty (30) days written notice to either terminate this Agreement or change LICENSEE's exclusive license to a nonexclusive license unless LICENSEE begins to diligently cure any breach and such breach is cured within thirty (30) days of receipt of the notice. This right, if exercised by the MRC, supersedes the rights granted in Article 2.

4. REPORTS, RECORDS AND PAYMENTS

4.1 **Reports.**

(a) **Progress Reports.**

(1) Beginning _____ 200_ and ending on the date of first commercial sale of a Licensed Product in South Africa, LICENSEE shall submit to the MRC semi-annual progress reports covering LICENSEE's (and Affiliate's and Sublicensee's) activities to develop and test all Licensed Products and obtain governmental approvals necessary for marketing the same. Such reports shall include a summary of work completed; summary of work in progress; current schedule of anticipated events or milestones; market plans for introduction of Licensed Products; and summary of resources (Rand value) spent in the reporting period.
(2) LICENSEE shall also report to the MRC, in its immediately subsequent progress report, the date of first commercial sale of a Licensed Product in each country.

(b) **Royalty Reports.** After the first commercial sale of a Licensed Product anywhere in the world, LICENSEE shall submit to the MRC quarterly royalty reports on or before each February 28, May 31, August 31 and November 30 of each year. Each royalty report shall cover LICENSEE's (and each Affiliate's and Sublicensee's) most recently completed calendar quarter and shall show:
(1) the gross sales, deductions and Net Sales during the most recently completed calendar quarter and the royalties, in Rands, payable with respect thereto;
(2) the number of each type of Licensed Product sold;
(3) sublicense fees and royalties received during the most recently completed calendar quarter in Rands, payable with respect thereto;
(4) the method used to calculate the royalties; and
(5) the exchange rates used.

If no sales of Licensed Products have been made and no sublicense revenues have been received by LICENSEE during any reporting period, LICENSEE shall so report.

4.2 **Records & Audits.**

(a) LICENSEE shall keep, and shall require its Affiliates and Sublicensees to keep, accurate and correct records of all Licensed Products manufactured, used, and sold, and sublicense fees received under this Agreement. Such records shall be retained by LICENSEE for at least five (5) years following a given reporting period.

Public Sector Patent License

(b) All records shall be available during normal business hours for inspection at the expense of the MRC by the MRC's Internal Audit Department or by a Certified Public Accountant selected by the MRC and in compliance with the other terms of this Agreement for the sole purpose of verifying reports and payments. Such inspector shall not disclose to the MRC any information other than information relating to the accuracy of reports and payments made under this Agreement or other compliance issues. In the event that any such inspection shows an under reporting and underpayment in excess of five percent (5%) for any twelve (12) month period, then LICENSEE shall pay the cost of the audit as well as any additional sum that would have been payable to the MRC had the LICENSEE reported correctly, plus an interest charge at a rate of ten percent (10%) per year. Such interest shall be calculated from the date the correct payment was due to the MRC up to the date when such payment is actually made by LICENSEE. For underpayment not in excess of five percent (5%) for any twelve (12) month period, LICENSEE shall pay the difference within thirty (30) days without interest charge or inspection cost.

4.3 **Payments.**

(a) All fees and royalties due the MRC shall be paid in Rands and all cheques shall be made payable to the MRC. When Licensed Products are sold in currencies other than Rands, LICENSEE shall first determine the earned royalty in the currency of the country in which Licensed Products were sold and then convert the amount into equivalent Rands, using the exchange rate quoted in the Wall Street Journal, or the exchange rate fixed for such date by the appropriate South African governmental agency, on the last business day of the applicable reporting period.

(b) **Royalty Payments.**

(1) Royalties shall accrue when Licensed Products are invoiced, or if not invoiced, when delivered to a third party or Affiliate.

(2) LICENSEE shall pay earned royalties quarterly on or before February 28, May 31, August 31 and November 30 of each calendar year. Each such payment shall be for earned royalties accrued within LICENSEE's most recently completed calendar quarter.

(3) Royalties earned on sales occurring or under sublicense granted pursuant to this Agreement in any country outside South Africa shall not be reduced by LICENSEE for any taxes, fees, or other charges imposed by the government of such country on the payment of royalty income, except that all payments made by LICENSEE in fulfillment of the MRC tax liability in any particular country may be credited against earned royalties or fees due the MRC for that country. LICENSEE shall pay all bank charges resulting from the transfer of such royalty payments.

(4) If at any time legal restrictions prevent the prompt remittance of part or all royalties by LICENSEE with respect to any country where a Licensed Product is sold or a sublicense is granted pursuant to this Agreement, LICENSEE shall convert the amount owed to the MRC into Rands and shall pay the MRC directly from its SA sources of fund for as long as the legal restrictions apply.

(5) LICENSEE shall not collect royalties from, or cause to be paid on Licensed Products sold to the account of the SA Government or any agency thereof as provided for in the license to the SA Government.

(6) In the event that any patent or patent claim within Patent Rights is held invalid in a final decision by a patent office from which no appeal or additional patent prosecution has been or can be taken, or by a court of competent jurisdiction and last resort and from which no appeal has or can be taken, all obligation to pay royalties based solely on that patent or claim or any claim patentably indistinct therefrom shall cease as of the date of such final decision. LICENSEE shall not, however, be relieved from paying

any royalties that accrued before the date of such final decision, that are based on another patent or claim not involved in such final decision, or that are based on the use of Technology.

(c) **Late Payments.** In the event royalty, reimbursement and/or fee payments are not received by the MRC when due, LICENSEE shall pay to the MRC interest charges at a rate of ten percent (10%) per year on the amount due. Such interest shall be calculated from the date payment was due until actually received by the MRC.

5. PATENT MATTERS

5.1 **Patent Prosecution and Maintenance.**

(a) Provided that LICENSEE has reimbursed the MRC for Patent Costs pursuant to Paragraph 3.2, the MRC shall diligently prosecute and maintain the South African and, if available, foreign patents, and applications in Patent Rights using counsel of its choice. the MRC shall provide LICENSEE with copies of all relevant documentation relating to such prosecution and LICENSEE shall keep this documentation confidential. The counsel shall take instructions only from the MRC, and all patents and patent applications in Patent Rights shall be assigned solely to the MRC.

(b) The MRC shall consider amending any patent application in Patent Rights to include claims reasonably requested by LICENSEE to protect the products contemplated to be sold by LICENSEE under this Agreement at LICENSEE'S cost.

(c) LICENSEE shall apply for an extension of the term of any patent in Patent Rights if appropriate. LICENSEE shall prepare all documents for such application, and the MRC shall execute such documents and take any other additional action as LICENSEE reasonably requests in connection therewith.

d) LICENSEE may elect to terminate its reimbursement obligations with respect to any patent application or patent in Patent Rights upon three (3) months' written notice to the MRC. The MRC shall use reasonable efforts to curtail further Patent Costs for such application or patent when such notice of termination is received from LICENSEE. The MRC, in its sole discretion and at its sole expense, may continue prosecution and maintenance of said application or patent, and LICENSEE shall then have no further license with respect thereto. Non-payment of any portion of Patent Costs with respect to any application or patent may be deemed by the MRC as an election by LICENSEE to terminate its reimbursement obligations with respect to such application or patent. The failure of LICENSEE to pay any such fee or costs within thirty (30) days of receipt of an invoice for same shall cause LICENSEE to automatically, without further action by the MRC, lose all rights in the jurisdiction for which fees or costs were due.

5.2 **Patent Infringement.**

(a) If LICENSEE learns of any substantial infringement of Patent Rights, LICENSEE shall so inform the MRC and provide the MRC with reasonable evidence of the infringement. Neither party shall notify a third party of the infringement of Patent Rights without the consent of the other party. Both parties shall use reasonable efforts and cooperation to terminate infringement without litigation.

(b) LICENSEE may request the MRC to take legal action against such third party for the infringement of Patent Rights. Such request shall be made in writing and shall include reasonable evidence of such infringement and damages to LICENSEE. If the infringing activity has not abated ninety (90) days following LICENSEE's request, the MRC shall elect to or not to commence suit on its own account. The MRC shall give notice of its election in writing to LICENSEE by the end of the one-hundredth (100th) day after receiving notice of such request from LICENSEE. LICENSEE may thereafter bring suit for patent infringement at its own expense, if and only if the MRC elects not to commence suit and the infringement occurred in a jurisdiction where LICENSEE has an exclusive license under this Agreement. If LICENSEE elects to bring suit, the MRC may join that suit at its own expense.

(c) Recoveries from actions brought pursuant to Paragraph 5.2(b) shall belong to the party bringing suit. Legal actions brought jointly by the MRC and LICENSEE and fully participated in by both shall be at the joint expense of the parties and all recoveries shall be shared jointly by them in proportion to the share of expense paid by each party.

(d) Each party shall cooperate with the other in litigation proceedings at the expense of the party bringing suit. Litigation shall be controlled by the party bringing the suit, except that the MRC may be represented by counsel of its choice in any suit brought by LICENSEE.

5.3 **Patent Marking.** LICENSEE shall mark all Licensed Products made, used or sold under the terms of this Agreement, or their containers, in accordance with the applicable patent marking laws.

6. GOVERNMENTAL MATTERS

6.1 **Governmental Approval or Registration.** If this Agreement or any associated transaction is required by the law of any nation to be either approved or registered with any governmental agency, LICENSEE shall assume all legal obligations to do so. LICENSEE shall notify the MRC if it becomes aware that this Agreement is subject to a South African or foreign government reporting or approval requirement. LICENSEE shall make all necessary filings and pay all costs including fees, penalties, and all other out-of-pocket costs associated with such reporting or approval process.

6.2 **Export Control Laws.** LICENSEE shall observe all applicable South African and foreign laws with respect to the transfer of Licensed Products and related technical data to foreign countries, including, without limitation, the International Traffic in Arms Regulations and the Export Administration Regulations.

6.3 **Preference for South African Industry.** If LICENSEE sells a Licensed Product or Combination Product in SA, LICENSEE shall manufacture said product substantially in SA.

6.4 **Rights of SA Government**. If the technology was developed with funds provided by South Africa, this agreement is subject to all applicable laws, regulations, and the terms of any agreements under which funds were provided. This includes any rights of South Africa to use the Technology for governmental purposes, any limits on the place of manufacture of products using the Technology, and any obligations to make products based on the technology available with in a reasonable time.

7. TERMINATION OF THE AGREEMENT

7.1 **Termination by the MRC.** If LICENSEE fails to perform or violates any term of this Agreement, then the MRC may give written notice of default ("Notice of Default") to LICENSEE. If LICENSEE fails to cure the default within thirty (30) days of the Notice of Default, the MRC

PUBLIC SECTOR PATENT LICENSE

may terminate this Agreement and the license granted herein by a second written notice ("Notice of Termination") to LICENSEE. If a Notice of Termination is sent to LICENSEE, this Agreement shall automatically terminate on the effective date of that notice. Termination shall not relieve LICENSEE of its obligation to pay any fees owed at the time of termination and shall not impair any accrued right of the MRC.

7.2 **Termination by Licensee.**

(a) LICENSEE shall have the right at any time and for any reason to terminate this Agreement upon a ninety (90) day written notice to the MRC. Said notice shall state LICENSEE's reason for terminating this Agreement.

(b) Any termination under Paragraph 7.2(a) shall not relieve LICENSEE of any obligation or liability accrued under this Agreement prior to termination or rescind any payment made to the MRC or action by LICENSEE prior to the time termination becomes effective. Termination shall not affect in any manner any rights of the MRC arising under this Agreement prior to termination.

7.3 **Survival on Termination.** The following Paragraphs and Articles shall survive the termination of this Agreement:

(a) Article 4 (REPORTS, RECORDS AND PAYMENTS);

(b) Paragraph 7.4 (Disposition of Licensed Products on Hand);

(c) Paragraph 8.2 (Indemnification);

(d) Article 9 (USE OF NAMES AND TRADEMARKS);

(e) Paragraph 10.2 hereof (Secrecy); and

(f) Paragraph 10.5 (Failure to Perform).

7.4 **Disposition of Licensed Products on Hand.** Upon termination of this Agreement, LICENSEE may dispose of all previously made or partially made Licensed Product within a period of one hundred and twenty (120) days of the effective date of such termination provided that the sale of such Licensed Product by LICENSEE, its Sublicensees, or Affiliates shall be subject to the terms of this Agreement, including but not limited to the rendering of reports and payment of royalties required under this Agreement.

8. LIMITED WARRANTY AND INDEMNIFICATION

8.1 **Limited Warranty.**

(a) The MRC warrants that it has the lawful right to grant this license.

(b) The license granted herein and the associated Technology are provided "AS IS WITH ALL FAULTS", AND THE ENTIRE RISK AS TO SATISFACTORY QUALITY, PERFORMANCE, ACCURACY, AND EFFORT IS WITH THE LICENSEE. THE MRC MAKES NO WARRANTIES, EXPRESS OR IMPLIED, AND HEREBY DISCLAIMS ALL SUCH WARRANTIES, AS TO ANY MATTER WHATSOEVER, INCLUDING, WITHOUT LIMITATION, THE CONDITION, INCLUDING PURITY, OF ANY INVENTION(S), TECHNOLOGY OR PRODUCT, WHETHER TANGIBLE OR INTANGIBLE, LICENSED UNDER THIS AGREEMENT; OR OF MERCHANTABILITY, OR FITNESS FOR A PARTICULAR PURPOSE OF THE INVENTION, TECHNOLOGY OR PRODUCT; OR OWNERSHIP;

PUBLIC SECTOR PATENT LICENSE

OR THAT THE USE OF THE LICENSED TECHNOLOGY OR PRODUCT WILL NOT INFRINGE ANY PATENT, COPYRIGHTS, TRADEMARKS, OR OTHER RIGHTS. LICENSOR SHALL NOT BE LIABLE FOR ANY DIRECT, CONSEQUENTIAL, OR OTHER DAMAGES SUFFERED BY ANY LICENSEE OR ANY THIRD PARTIES RESULTING FROM THE USE, PRODUCTION, MANUFACTURE, SALE, LEASE, CONSUMPTION, OR ADVERTISEMENT OF THE TECHNOLOGY OR PRODUCT.

(c) In no event shall the MRC be liable for any consequential, special, exemplary, punitive or incidental damages even if it has been advised of the possibility of such damages resulting from this Agreement or the exercise of the license granted herein or the use of the Invention, Licensed Product, Licensed Method or Technology.

(d) Nothing in this Agreement shall be construed as:
   (1) a warranty or representation by the MRC as to the validity or scope of any Patent Rights;
   (2) a warranty or representation that anything made, used, sold or otherwise disposed of under any license granted in this Agreement is or shall be free from infringement of patents of third parties;
   (3) an obligation to bring or prosecute actions or suits against third parties for patent infringement except as provided in Paragraph 5.2 hereof;
   (4) conferring by implication, estoppel or otherwise any license or rights under any patents of the MRC other than Patent Rights as defined in this Agreement, regardless of whether those patents are dominant or subordinate to Patent Rights;
   (5) an obligation to furnish any know-how not provided in Patent Rights and Technology; or
   (6) an obligation to update Technology.

(e) LICENSEE agrees that the MRC's liability in connection with the Invention, Technology and Patent Rights, whether arising in contract, negligence, strict liability, tort or otherwise shall not exceed the lesser of (i) the amount paid by LICENSEE to the MRC for the license, or (ii) R_____.

(f) LICENSEE understands and agrees that the MRC is not engaged, and does not purport to be engaged, in LICENSEE'S business and LICENSEE assumes all responsibilities and obligations with respect to any decision LICENSEE makes or action LICENSEE may take as a result of LICENSEE'S use of the Invention, Technology and Patent Rights. The limitations of warranties, liabilities, and remedies under this Agreement are a reflection of the risks assumed by the parties in order to obtain the Invention, Technology or Patent Rights at the specified license fee. LICENSEE agrees to assume the risk for: (i) all liabilities disclaimed by the MRC contained herein, and (ii) all alleged damages in excess of the amount, if any, of the remedy provided hereunder.

8.2 **Indemnification.**

(a) LICENSEE shall indemnify, hold harmless and defend the MRC, its officers, employees, and agents; the sponsors of the research that led to the Invention; and the Inventors of the patents and patent applications in Patent Rights and their employers against any and all claims, suits, losses, damage, costs, fees, and expenses resulting from or arising out of exercise of this license or any sublicense or which may be brought against LICENSOR, its Trustees, officers, faculty, employees or students as a result of or arising out of any negligent act or omission of LICENSEE, its agents, or employees, or arising out of use, production, manufacture, sale, lease, consumption or advertisement by LICENSEE or any third party of any licensed Product, Invention or Technology licensed under this Agreement. This indemnification shall include, but not be limited to, any product liability.

PUBLIC SECTOR PATENT LICENSE

(b) LICENSEE, at its sole cost and expense, shall insure its activities in connection with the work under this Agreement and obtain, keep in force and maintain insurance or an equivalent program of self insurance as follows:
  (1) comprehensive or commercial general liability insurance (contractual liability included) with limits of at least: (i) each occurrence, R1,000,000; (ii) products/completed operations aggregate, R5,000,000; (iii) personal and advertising injury, R1,000,000; and (iv) general aggregate (commercial form only), R5,000,000; and
  (2) the coverage and limits referred to above shall not in any way limit the liability of LICENSEE.

(c) LICENSEE shall furnish the MRC with certificates of insurance showing compliance with all requirements. Such certificates shall: (i) provide for thirty (30) day advance written notice to the MRC of any modification; (ii) indicate that the MRC has been endorsed as an additional insured under the coverage referred to above; and (iii) include a provision that the coverage shall be primary and shall not participate with nor shall be excess over any valid and collectable insurance or program of self-insurance carried or maintained by the MRC.

(d) The MRC shall notify LICENSEE in writing of any claim or suit brought against the MRC in respect of which the MRC intends to invoke the provisions of this Article. LICENSEE shall keep the MRC informed on a current basis of its defense of any claims under this Article.

9. USE OF NAMES AND TRADEMARKS

9.1 Nothing contained in this Agreement confers any right to use in advertising, publicity, or other promotional activities any name, trade name, trademark, or other designation of either party hereto (including contraction, abbreviation or simulation of any of the foregoing). Unless required by law, the use by LICENSEE of the name, Medical Research Council is prohibited, without the express written consent of the MRC.

9.2 The MRC may disclose to the Inventors the terms and conditions of this Agreement upon their request. If such disclosure is made, the MRC shall request the Inventors not disclose such terms and conditions to others.

9.3 The MRC may acknowledge the existence of this Agreement and the extent of the grant in Article 2 to third parties, but the MRC shall not disclose the financial terms of this Agreement to third parties, except where the MRC is required by law to do so.

10. MISCELLANEOUS PROVISIONS

10.1 **Correspondence.** Any notice or payment required to be given to either party under this Agreement shall be deemed to have been properly given and effective:
(a) on the date of delivery if delivered in person, or
(b) five (5) days after mailing if mailed by first-class or certified mail, postage paid, to the respective addresses given below, or to such other address as is designated by written notice given to the other party.

PUBLIC SECTOR PATENT LICENSE

If sent to LICENSEE:
[Name and address of licensee]
Attention: ________________

If sent to the MRC:
MRC Innovation Centre
Fransie van Zijl Drive, Parowvallei
Cape Town
Attention: Prof Bunn

10.2 **Secrecy.**

(a) "Confidential Information" shall mean information, including Technology, relating to the Invention and disclosed by the MRC to LICENSEE during the term of this Agreement, which if disclosed in writing shall be marked "Confidential", or if first disclosed otherwise, shall within thirty (30) days of such disclosure be reduced to writing by the MRC and sent to LICENSEE:

(b) Licensee shall:

(1) use the Confidential Information for the sole purpose of performing under the terms of this Agreement;

(2) safeguard Confidential Information against disclosure to others with the same degree of care as it exercises with its own data of a similar nature;

(3) not disclose Confidential Information to others (except to its employees, agents or consultants who are bound to LICENSEE by a like obligation of confidentiality) without the express written permission of the MRC, except that LICENSEE shall not be prevented from using or disclosing any of the Confidential Information that:

(i) LICENSEE can demonstrate by written records was previously known to it;

(ii) is now, or becomes in the future, public knowledge other than through acts or omissions of LICENSEE; or

(iii) is lawfully obtained by LICENSEE from sources independent of the MRC; and

(c) The secrecy obligations of LICENSEE with respect to Confidential Information shall continue for a period ending five (5) years from the termination date of this Agreement.

10.3 **Assignability.** This Agreement may be assigned by the MRC. This Agreement may not be assigned by LICENSEE except in connection with the sale or other transfer of LICENSEE's entire business or that part of LICENSEE's business to which the license granted hereby relates. LICENSEE shall give the MRC thirty (30) days' prior notice of such assignment or transfer. Any other assignment of this License Agreement without the prior written consent of the MRC shall be void. Such written consent shall not be unreasonably withheld or delayed.

10.4 **No Waiver.** No waiver by either party of any breach or default of any covenant or agreement set forth in this Agreement shall be deemed a waiver as to any subsequent and/or similar breach or default.

10.5 **Failure to Perform.** In the event of a failure of performance due under this Agreement and if it becomes necessary for either party to undertake legal action against the other on account thereof, then the prevailing party shall be entitled to reasonable attorney's fees in addition to costs and necessary disbursements.

Public Sector Patent License

10.6 **Governing Laws.** THIS AGREEMENT SHALL BE INTERPRETED AND CONSTRUED IN ACCORDANCE WITH THE LAWS OF THE REPUBLIC OF SOUTH AFRICA, but the scope and validity of any patent or patent application shall be governed by the applicable laws of the country of the patent or patent application. Any action in connection with this Agreement shall be commenced and maintained only in the South African Court and LICENSEE consent to personal jurisdiction and venue in any such court.

10.7 **Force Majeure.** A party to this Agreement may be excused from any performance required herein if such performance is rendered impossible or unfeasible due to any catastrophe or other major event beyond its reasonable control, including, without limitation, war, riot, and insurrection; laws, proclamations, edicts, ordinances, or regulations; strikes, lockouts, or other serious labor disputes; and floods, fires, explosions, or other natural disasters. When such events have abated, the non-performing party's obligations herein shall resume.

10.8 **Headings.** The headings of the several sections are inserted for convenience of reference only and are not intended to be a part of or to affect the meaning or interpretation of this Agreement.

10.9 **Entire Agreement.** This Agreement embodies the entire understanding of the parties and supersedes all previous communications, representations or understandings, either oral or written, between the parties relating to the subject matter hereof.

10.10 **Amendments.** No amendment or modification of this Agreement shall be valid or binding on the parties unless made in writing and signed on behalf of each party.

10.11 **Severability.** In the event that any of the provisions contained in this Agreement is held to be invalid, illegal, or unenforceable in any respect, such invalidity, illegality or unenforceability shall not affect any other provisions of this Agreement, and this Agreement shall be construed as if the invalid, illegal, or unenforceable provisions had never been contained in it.

10.12 **Time Limitation on Initiation of actions.** No action, regardless of form, arising out of the subject matter of this Agreement may be brought by LICENSEE more than one (1) year after the cause of action has arisen.

IN WITNESS WHEREOF, both the MRC and LICENSEE have executed this Agreement, in duplicate originals, by their respective and duly authorized officers on the day and year written.

[LICENSEE]:	**THE MRC:**
By: ______________________ (Signature)	By: ______________________ (Signature)
Name: ______________________	Name: ______________________
Title: ______________________	Title: ______________________
Date: ______________________	Date: ______________________

Public Sector Patent License

WITNESS:	WITNESS:
By: ____________________ (Signature)	By: ____________________ (Signature)
Name: ____________________	Name: ____________________
Title: ____________________	Title: ____________________
Date: ____________________	Date: ____________________

Sample Agreement 4: Plant Variety and Trademark License

Source: Based on an actual agreement. The names of the parties are purely fictional and have been inserted for illustration purposes only. Certain terms of the license have been modified or left blank.

COMMERCIAL VARIETY and TRADE MARK LICENCE AGREEMENT

by and between

BETTER SEED COMPANY, a company organised and existing under the laws of __________ [country] in the name of BETTER SEED CO Ltd and having its registered office in __________ [address] (hereinafter called "BETTER SEED CO.")

and

GOLDEN HYBRIDS, a company organised and existing under the laws of __________ [country] in the name of Golden Hybrids S.A. and having its registered office in __________ [address] (hereinafter called "GOLDEN HYBRIDS").

INTRODUCTION

GOLDEN HYBRIDS is a company breeding, trialling, producing, marketing and selling new varieties of various agricultural species.

BETTER SEED CO. is a breeder of corn and other species and the owner of the varieties provided under this agreement, or has the right to organize for introduction and marketing of varieties in certain territories.

Under this agreement, GOLDEN HYBRIDS will trial certain varieties owned or represented by BETTER SEED CO., with the intention of becoming an exclusive distributor and exclusive producer of the varieties in Southeast Asia.

1. DEFINITIONS

GROSS SALES means income at invoice values received for LICENSED PRODUCTS over a given period of time.

INTELLECTUAL PROPERTY means the patents, copyrights, trademarks, design rights, data protection rights, PVP and any other statutory rights for inventions, improvements, designs, and any other intellectual property rights.

INVENTION means the invention, which is the subject matter of PATENTS, PLANT VARIETY PROTECTION or any other form of INTELLECTUAL PROPERTY protection and licensed hereunder to GOLDEN HYBRIDS.

LICENSED MATERIAL means all forms of living and non-living biological material including without limitation strains, clones, plants, parts of plants, cultivars, germplasm, and genetic material provided by BETTER SEED CO. to GOLDEN HYBRIDS under this agreement.

LICENSED PRODUCTS means the crop or crops or any parts of the VARIETIES or TRIAL VARIETIES listed in Schedule 1 and 2.

Plant Variety and Trademark License

LICENSED TRADEMARK means the trademark BETTER SEED CO.-Tropical Genetics as defined in Schedule 3.

NET SALES means GROSS SALES reduced by customer discounts, returns, freight out, and allowances.

PVP means Plant Variety Protection; the protection of varieties as a form of exclusive ownership and use rights determined based on distinctness, uniformity and stability of the Plant Material.

RIGHTS shall mean plant variety rights to the Varieties in the Territory.

SUBSIDIARY of BETTER SEED CO. means any corporation over 50% of the voting stock of which is directly or indirectly owned or controlled by such GOLDEN HYBRIDS.

TERRITORY shall mean Country X and any such countries listed in Schedule 4 as amended from time to time by mutual written agreement.

TRIAL VARIETIES shall mean all the varieties of corn listed in Schedule 1, as amended from time to time.

VARIETY (and VARIETIES) shall mean the varieties listed in Schedule 2, as amended from time to time, whether or not relevant certificate of Plant Variety Protection have been obtained and marketed under the tradename given by GOLDEN HYBRIDS.

2. TRIAL VARIETIES

2.1 BETTER SEED CO. grants to GOLDEN HYBRIDS and its SUBSIDIARIES the exclusive right to grow the Trial Varieties in the Territory for a period of 7 (seven) growing seasons in order to evaluate the material.

2.2 GOLDEN HYBRIDS shall at its own expense in relation to each of the Trial Varieties carry out the necessary trials to determine whether they are suitable for use in the Territory. GOLDEN HYBRIDS shall keep full and accurate records and provide BETTER SEED CO. with the results of the trials in the form provided. GOLDEN HYBRIDS shall notify BETTER SEED CO. of the location of the trials. GOLDEN HYBRIDS shall allow BETTER SEED CO. and its agents access to the trials on reasonable notice.

2.3 BETTER SEED CO. shall at GOLDEN HYBRIDS's expense provide GOLDEN HYBRIDS with sufficient seed to carry out trials, and supply such technical and other relevant information in its possession as will assist GOLDEN HYBRIDS effectively to evaluate the Trial Varieties.

2.4 BETTER SEED CO. shall supply GOLDEN HYBRIDS with up to 20 (twenty) new Trial Varieties per year, if available.

2.5 At any time before and for 6 (six) months after the end of the trials of a variety, GOLDEN HYBRIDS may request that the Trial Variety becomes one of the Varieties, and on consent from BETTER SEED CO. it becomes one of the Varieties. If a variety is not selected for commercialisation, the Agreement shall terminate with regard to that variety.

3. GRANT OF RIGHTS

3.1 GOLDEN HYBRIDS is granted the exclusive right to produce Varieties.

Plant Variety and Trademark License

3.2 GOLDEN HYBRIDS is granted the exclusive right to authorise third parties to exercise Rights for the marketing and sale in the Territory of seed of the Varieties.

3.3 BETTER SEED CO. undertakes not to produce or market seed of the Licensed Products in the Territory.

3.4 GOLDEN HYBRIDS agrees to permit BETTER SEED CO. or its representatives to inspect the facilities where Licensed Products are being produced and packaged.

4. DURATION

4.1 Notwithstanding clause 10 below, this agreement shall become effective upon signature by both parties hereto and will remain in force for 5 (five) years (EXPIRATION).

4.2 Unless terminated with at least 6 months notice and 6 months prior to Expiration, this agreement shall automatically be renewed for periods of 2 (two) years.

5. GOLDEN HYBRIDS'S OBLIGATIONS

5.1 GOLDEN HYBRIDS shall at its own expense ensure (so far as possible) that the Varieties are entered on the National Plant Variety List in the Territory and shall maintain the Varieties on the National Plant Variety List and the Rights for the Varieties in the Territory and shall not allow such entry or Rights to lapse unless the variety is withdrawn from the market.

5.2 GOLDEN HYBRIDS shall at its own expense apply for Rights on the single-cross hybrids in the Territory. If a variety is not granted Rights by the authorities within a reasonable time frame in the Territory, the Agreement shall terminate with regard to that variety.

5.3 Before selling LICENSED PRODUCTS, GOLDEN HYBRIDS shall submit to BETTER SEED CO., at no cost to BETTER SEED CO. and for approval as to quality, at least one complete set of all promotional and advertising material associated therewith. Failure of BETTER SEED CO. to approve such samples within 10 (ten) working days after receipt hereof will be deemed approval. If BETTER SEED CO. should disapprove any such sample, it shall provide specific reasons for such disapproval. Once such samples have been approved by BETTER SEED CO., GOLDEN HYBRIDS shall not materially depart therefrom without BETTER SEED CO.'s prior express written consent, which shall not be unreasonably withheld.

5.4 GOLDEN HYBRIDS shall advise BETTER SEED CO. annually of its marketing plans for the subsequent year for the Varieties.

5.5 GOLDEN HYBRIDS shall purchase all inbreds for the production of the Varieties from BETTER SEED CO. unless BETTER SEED CO. licenses to GOLDEN HYBRIDS the right to produce inbreds. Such license, if agreed, shall be granted through an amendment to this Agreement.

5.6 GOLDEN HYBRIDS shall maintain clear records of all amounts of hybrids and varieties produced and inform BETTER SEED CO. on a yearly basis of such quantities produced, sold, distributed as promotional materials, and stock levels.

5.7 GOLDEN HYBRIDS shall use in its promotion of the Varieties BETTER SEED CO. trademarks, logos and distinctive BETTER SEED CO. business marks (Schedule 3).

PLANT VARIETY AND TRADEMARK LICENSE

5.8 On all advertising and technical material GOLDEN HYBRIDS shall refer to BETTER SEED CO. as breeder of the Varieties.

5.9 GOLDEN HYBRIDS shall maintain ISO 9001 quality standards. If ISO 9001 standard certification were revoked or suspended, GOLDEN HYBRIDS shall notify BETTER SEED CO. immediately and GOLDEN HYBRIDS shall use its best efforts to restore such a quality in a timely manner. In the event that GOLDEN HYBRIDS has not taken appropriate steps to restore such a quality within 24 months after notification by BETTER SEED CO., BETTER SEED CO. has the right to terminate this Agreement.

6. BETTER SEED CO.'s OBLIGATIONS

6.1 Subject to the receipt of reasonable notice from GOLDEN HYBRIDS BETTER SEED CO. shall sell to GOLDEN HYBRIDS such quantities of pre-basic and basic seed of the parental lines as GOLDEN HYBRIDS may from time to time request. Seed shall meet the standard of certification required in the Territory. The prices for such seed shall be agreed in writing prior to delivery and may be reviewed from time to time.

6.2 BETTER SEED CO. shall provide all the supporting information reasonably required by GOLDEN HYBRIDS for the purposes of any application, entry or maintenance.

6.3 BETTER SEED CO. shall notify GOLDEN HYBRIDS should it become aware of any changes that materially affect the execution of this Agreement.

6.4 BETTER SEED CO. agrees to supply GOLDEN HYBRIDS, at BETTER SEED CO. expense, at least 5kg (five kg) per F1 hybrid seed per growing season (where 1 calendar year has 2-3 seasons) for the exclusive purpose of GOLDEN HYBRIDS registering the hybrid in national field trials. Such national field trials require 3 (three) growing seasons.

6.5 BETTER SEED CO. agrees to supply GOLDEN HYBRIDS, at BETTER SEED CO. expense, at least 50kg (fifty kg) per F1 hybrid seed per growing season (where 1 calendar year has 2-3 seasons) for the exclusive purpose of GOLDEN HYBRIDS conducting demonstration trials prior to commercialization. Such demonstration trials require 3 (three) growing seasons.

6.6 At the request of GOLDEN HYBRIDS, BETTER SEED CO. shall provide adequate samples of all certification lots of seed of the Varieties in order that GOLDEN HYBRIDS can use such samples for check plots.

7. ROYALTIES

7.1 BETTER SEED CO. shall, at its own discretion, supply GOLDEN HYBRIDS with inbreds for the production of hybrids. If such inbreds are supplied, BETTER SEED CO. shall sell them to GOLDEN HYBRIDSS at a price not to exceed $10/kg per inbred line.

7.2 All seed of the Varieties, whether certified or not (and sold) as seed within the Territory under the licence granted under Clause 3 above shall be subject to the payment of a royalty.

7.3 The royalty shall be on a sliding scale basis where royalties are due on the sale price on net volume sold, for each calendar year, for each hybrid:
- first 100 metric tons of seed sold: calculated at 6.5 % royalties
- 101-250 metric tons of seed sold: calculated at 5.5 % royalties

PLANT VARIETY AND TRADEMARK LICENSE

- 251-500 metric tons of seed sold: calculated at 4.5 % royalties
- 501+ metric tons of seed sold: calculated at 4.0 % royalties

7.4 GOLDEN HYBRIDS shall keep accurate accounts and records of all Gross and Net sales of seed of the Varieties upon which royalties are payable. BETTER SEED CO. or an independent accountant authorised by BETTER SEED CO. shall be permitted to inspect such accounts and records at least once in each year solely for the purpose of verifying the volume and type of sales upon which royalties are payable.

7.5 If on investigation GOLDEN HYBRIDS has underpaid royalties by more than 2.5% of the amount due, the costs of the investigation shall be paid by GOLDEN HYBRIDS.

7.6 GOLDEN HYBRIDS agrees to negotiate in good faith with BETTER SEED CO. on minimum aggregate inbred and royalty payments. Such negotiations shall commence as soon as field trial data and sales projections permit, but must be concluded prior to the first commercial sales. Such clause shall be incorporated in Schedule 5 and duly signed.

8. PAYMENT OF ROYALTIES

8.1 Not later than 15 January in each year GOLDEN HYBRIDS shall send a report to BETTER SEED CO. giving details of seed sales of the Varieties upon which royalty is payable in respect of the period of 12 months ended the previous year. GOLDEN HYBRIDS shall pay to BETTER SEED CO. by bank transfer on or around 1 February of every year.

8.2 Royalty shall be paid in US Dollars converted at the exchange rate in effect with BANK on the close of business on the business day before the due date for payment. In the event such bank rates differ, the average of the two shall be applied.

8.3 BETTER SEED CO. shall deduct the mandatory 10% taxes (effective deduction of 11.11%) on royalty payments prior to making the transfer. Such tax shall not be applied to payments for the supply of inbred seed. GOLDEN HYBRIDS shall notify BETTER SEED CO. immediately if and when the mandatory tax rate changes.

9. BREACH OF RIGHTS

9.1 If GOLDEN HYBRIDS becomes aware of any third party breach of rights in the Varieties it shall promptly notify BETTER SEED CO., which shall take such steps as it considers appropriate to remedy the breach and take such steps as are necessary to enforce the Rights.

10. TERMINATION

10.1 Notwithstanding Clause 4 above either party may terminate this Agreement at any time without incurring any liability thereby and without prejudice to any other remedies it may have if:

i) the other commits a breach of this Agreement and if capable of remedy that other fails to remedy the breach within one month having been required to do so by written notice; or

ii) the other enters into liquidation whether compulsorily or voluntarily (except for the purposes of reconstruction or amalgamation) or if a receiver or administrative receiver is appointed over the assets of the other of the equivalent of any of these events in the Territory.

PLANT VARIETY AND TRADEMARK LICENSE

iii) If by reason of any circumstances as described in Clause 18 the performance of this Agreement becomes impossible for more than 12 (twelve) consecutive months either party shall be entitled to terminate this Agreement.

iv) GOLDEN HYBRIDS shall, notwithstanding termination of this Agreement, remain permanently as non-exclusive license of each of the Varieties under the terms of this Agreement where Rights have been applied for and granted or seed has been marketed for more than 3 (three) growing seasons by GOLDEN HYBRIDS.

v) Sales and invoicing of seeds remaining at the termination of this Agreement have to be made as "certified seed". BETTER SEED CO. has the right to take over seed of higher qualities from GOLDEN HYBRIDS at cost price.

11. SURRENDER

GOLDEN HYBRIDS may surrender its rights to any of the Trial Varieties or Varieties at any time, on giving notice to BETTER SEED CO.. Surrender will not affect the obligations of GOLDEN HYBRIDS under Clauses 5, 6 and 7 above.

12. ASSIGNMENT

12.1 This Agreement is personal to the parties and may not be assigned by either party without the prior written consent of the other, although each party may perform the obligations undertaken by it and exercise the rights granted to it under this Agreement either itself or through any one or more of its subsidies or associated companies.

13. CONFIDENTIALITY

The parties shall not disclose during the validity of this Agreement or thereafter to any third party any commercial, technical or other information of a confidential nature received of obtained by either party from the other except information provided as a marketing aid.

14. DISPUTE RESOLUTION

14.1 In event of dispute shall arise between the PARTIES to this Agreement, the PARTIES agree to participate in at least 4 (four) hours mediation in accordance with the mediation rules of F.I.S. Arbitration procedures for the International Seed Trade.

14.2 In case the PARTIES are unable to resolve the dispute in mediation they agree to submit the dispute to final and binding arbitration under the arbitration rules of F.I.S. Arbitration procedures for the International Seed Trade, and the judgment upon the award rendered by the Arbitrator(s) may be entered into any court having judgment thereof.

14.3 The PARTIES agree to share equally the costs of mediation and arbitration.

15. GOODWILL

In the event of termination of the Agreement, neither party shall be entitled under law or otherwise to receive any payment from the other for actual, consequential, indirect, special or incidental damages, costs or expenses, whether foreseeable or unforeseeable (including but not limited to labour claims and loss of profits, investments or goodwill), any right to which the parties hereby waive and disclaim to the fullest extent permitted by law.

PLANT VARIETY AND TRADEMARK LICENSE

16. IMPROVEMENTS AND DISCOVERIES

16.1 If during the term of this Agreement GOLDEN HYBRIDS generates any improvement or discovery which improves the Licensed Material, it shall notify BETTER SEED CO. immediately and the Parties shall meet to discuss the ownership and intellectual property protection of such improvement or discovery , and if appropriate, the Territory and Countries in which such intellectual property protection should be sought.

16.2 Should such improvement or discovery be protectable under intellectual property statutes, BETTER SEED CO. will be granted a royalty-free, worldwide, non-exclusive commercial license thereunder including the right to sublicense for all applications and an exclusive first option for a period of 12 months from the date the BETTER SEED CO. received notification of such improvement or discovery to negotiate worldwide exclusive access or all uses.

16.3 In any event, GOLDEN HYBRIDS shall retain royalty bearing non-exclusive licenses for use in Territory. The terms of any license to be granted under this section shall be negotiated between the PARTIES and reduced to writing.

17. WARRANTY AND LIABILITY

17.1 Except as expressly provided herein, BETTER SEED CO. makes no representation and extends no warranties. BETTER SEED CO. disclaims any responsibility with respect to performance, merchantability, fitness for a particular purpose or freedom from infringement of third party patent rights (other than as of the date of the execution of this Agreement BETTER SEED CO. is not aware of any such infringement).

17.2 BETTER SEED CO. shall in no event be liable for damages, whether direct or otherwise, arising out of the use by GOLDEN HYBRIDS or any third party of information or materials supplied hereunder.

17.3 In no event shall BETTER SEED CO. be liable for lost or prospective profits or special or consequential damages, whether or not BETTER SEED CO. has been advised of the possibility of the damages, nor for any claim by a third party against GOLDEN HYBRIDS.

17.4 BETTER SEED CO. warrants that it is the sole owner of the LICENSED INTELLECTUAL PROPERTY and Materials and that it has the right to grant licenses.

18. DISCLAIMER

18.1 The seed supplied under this Agreement is from a conventional breeding program in which genetically modified material has never been deliberately introduced. The methods used in the breeding, development and production of these plant progenies, lines and varieties include procedures aimed at minimizing the adventitious presence of Genetically Modified Organisms ("GMO").

18.2 BETTER SEED CO. shall not be liable, in any circumstances or for any reason, for incidental or consequential losses or special or punitive damages or any losses or damages of a similar nature. As a condition of liability, BETTER SEED CO. must receive notice by Registered Post within 30 (thirty) days after any defect becomes apparent.

PLANT VARIETY AND TRADEMARK LICENSE

19. FORCE MAJEURE

Neither party shall be in default hereunder by reason of its delay in performance of, or failure to perform, any of its obligations hereunder, if such delay or failure is caused by strikes or other labour disturbance, acts of God, acts of the public enemy, riots or other civil disturbances, fire, flood, interference by civil or military authorities, compliance with government laws, rules or regulations, delays in transportation, failure of suppliers, inability to secure necessary governmental priorities for materials, or any other circumstances beyond its control and without its fault or negligence.

20. GOVERNMENT AND REGULATORY APPROVALS

GOLDEN HYBRIDS shall be responsible for adhering to all laws and regulations and for obtaining and complying with all government and regulatory approvals, licenses, clearances and consents pertinent to or required to cover its activities under this Agreement.

21. MISCELLANEOUS

21.1 If any part of this Agreement is declared illegal or invalid then it shall be severed from the Agreement without affecting the remainder.

21.2 Each PARTY is acting as an independent entity. Nothing in this Agreement shall be construed so as to constitute a partnership or joint venture of any kind between BETTER SEED CO. and GOLDEN HYBRIDS.

21.3 Any notices given by either party to the other shall be made in writing by International Courier and addressed to them at their addresses set out above.

21.4 This Agreement shall be construed and interpreted according to the laws of Lucerne, Switzerland.

In witness whereof, the duly authorised representatives for and on behalf of the parties hereto have executed this Agreement in duplicate, each party taking 1 (one) copy, as of the day and year written below.

Date ____________________

[BETTER SEED CO.]	[GOLDEN HYBRIDS]
By: ________________________	By: ________________________
Title: ________________________	Title: ________________________

SCHEDULE 1

TRIAL VARIETIES

[description and listing of varieties]

Plant Variety and Trademark License

Schedule amended on [Date] ______________________________

By: ______________________ By: ______________________
(Signature) (Signature)

By signature of this amended Schedule, all earlier Schedules under this Agreement are automatically cancelled.

SCHEDULE 2

COMMERCIAL VARIETIES

[description and listing of varieties]

Schedule amended on [Date] ______________________________

By: ______________________ By: ______________________
(Signature) (Signature)

By signature of this amended Schedule, all earlier Schedules under this Agreement are automatically cancelled.

SCHEDULE 3

LICENSED TRADEMARK

[insert trademark here]

Schedule amended on [Date] ______________________________

By: ______________________ By: ______________________
(Signature) (Signature)

By signature of this amended Schedule, all earlier Schedules under this Agreement are automatically cancelled.

SCHEDULE 4

TERRITORY

Exclusive
[list countries here]

Non-exclusive
[list countries here]

PLANT VARIETY AND TRADEMARK LICENSE

Schedule amended on [Date] ______________________________

By: ______________________ By: ______________________
(Signature) (Signature)

By signature of this amended Schedule, all earlier Schedules under this Agreement are automatically cancelled.

SCHEDULE 5

MINIMUM PAYMENTS

In case the royalties paid for a given variety do not aggregate a minimum of _________ US dollars for the year ending December 31, ______, increasing by ____% annually for _____ succeeding calendar years, and continuing at such level ______ subsequently, BETTER SEED CO. shall be entitled to revert to a non-exclusive license for each such variety for which payments do not meet the minimum payment.

Schedule amended on [Date] ______________________________

By: ______________________ By: ______________________
(Signature) (Signature)

By signature of this amended Schedule, all earlier Schedules under this Agreement are automatically cancelled.

SAMPLE AGREEMENT 5: INTELLECTUAL PROPERTY AND TRADEMARK LICENSE (STANFORD)

Source: Office of Technology Licensing, Stanford University, 1705 El Camino Real, Palo Alto, CA, 94306, U.S.A. Reproduced with permission. Visit http://otl.stanford.edu and www.stanford.edu/group/ICO/ to download many additional template and sample agreements.

THE COLLEGIATE LICENSING COMPANY STANDARD RETAIL PRODUCT LICENSE AGREEMENT

This is an Agreement between ________________, a ____________ organized under the laws of the state of ________, having a principal place of business at ______________________ ________________ ("Licensee"), and the Collegiate Licensing Company, a Georgia corporation, having a principal place of business at 290 Interstate North, Suite 200, Atlanta, Georgia 30339 ("CLC"), as agent on behalf of the Collegiate Institutions (as defined below).

WHEREAS, the individual Collegiate Institutions have authorized CLC as agent to administer their respective trademark licensing programs; and

WHEREAS, certain Collegiate Institutions have authorized CLC to enter into this Agreement on their behalf to license the use of certain Licensed Indicia (as defined below); and

WHEREAS, Licensee desires to manufacture, advertise, distribute and sell certain Licensed Articles (as defined below) containing the Licensed Indicia, and certain Collegiate Institutions, through CLC, are willing, subject to certain conditions, to grant this license.

NOW, THEREFORE, in consideration of the parties' mutual covenants and undertakings, and other good and valuable consideration the receipt and sufficiency of which are acknowledged, the parties agree as follows:

1. DEFINITIONS

In addition to the terms defined elsewhere in this Agreement, as used in this Agreement, the following terms shall have the following respective meanings:

(a) "Collegiate Institutions" means the individual colleges, universities and other institutions represented by CLC, including any additions or deletions that may be made from time-to-time by CLC.

(b) "Licensed Indicia" means the names and identifying indicia of the Collegiate Institutions including, without limitation, the trademarks, service marks, trade dress, team names, nicknames, abbreviations, city/state names in the appropriate context, slogans, designs, colors, uniform and helmet designs, distinctive landmarks, logographics, mascots, seals and other symbols associated with or referring to the respective Collegiate Institutions. Licensed Indicia includes those shown in Appendix B, modifications of the Licensed Indicia approved for use by the Collegiate Institutions, and any other names or identifying indicia adopted and approved for use by the Collegiate Institutions.

(c) "Licensed Articles" means the products listed in Appendix C which contain Licensed Indicia.

(d) "Authorized Brands" means any additional brand names or labels Licensee may use in

Intellectual Property and Trademark License

association with the Licensed Articles. Authorized Brands are listed in Appendix D.

(e) "Distribution Channels" means the channels of trade in which Licensee may advertise, distribute and sell the Licensed Articles in the Territory. The Distribution Channels authorized herein are indicated in Appendix D, which may also identify Distribution Channels that are not authorized in this Agreement. Licensee shall not advertise, distribute or sell Licensed Articles to any third party that Licensee knows or should reasonably know intends or is likely to advertise, redistribute or resell Licensed Articles outside the authorized Distribution Channels.

(f) "Territory" means the United States of America, its territories and possessions, and United States military bases abroad. Licensee shall not advertise, distribute or sell Licensed Articles outside the Territory, or to any person or entity that Licensee knows or should reasonably know intends or is likely to advertise, redistribute or resell Licensed Articles outside the Territory.

(g) "Net Sales" means the total gross sales of all Licensed Articles distributed or sold at the greater of Licensee's invoiced selling price or Licensee's regular domestic wholesale warehouse price, including the royalty amount, less lawful quantity trade discounts actually allowed and taken as such by customers and shown on the invoice, less any credits for returns actually made as supported by credit memoranda issued to customers, less sales taxes, and less prepaid transportation charges on Licensed Articles shipped by Licensee from its facilities to the purchaser. There shall be no other deductions allowed including, without limitation, deductions for direct or indirect costs incurred in the manufacturing, distributing, selling, importing or advertising (including cooperative and other advertising and promotional allowances) of the Licensed Articles, nor shall any deductions be allowed for non-collected or uncollectable accounts, commissions, cash or early payment discounts, close-out sales, distress sales, sales to employees, or any other costs.

(h) "Premiums" means any products, including Licensed Articles, bearing any Licensed Indicia featured alone or in combination with the indicia of any third party, that Licensee sells or gives away for the purposes of (i) promoting, publicizing or increasing the sale of its own products or services; or (ii) promoting, publicizing or increasing the sale of the products or services of any third party. Premiums include, without limitation, combination sales, incentives for sales force, and trade or consumer promotions such as sweepstakes.

2. GRANT OF LICENSE

(a) Grant: Upon execution of this Agreement, and subject to its terms and conditions, the Collegiate Institutions listed in Appendix A, through CLC, grant Licensee the nonexclusive, revocable, nontransferable rights to manufacture, advertise, distribute and sell the Licensed Articles listed in Appendix C, containing the Licensed Indicia shown in Appendix B, under the applicable Authorized Brands and in the Distribution Channels indicated in Appendix D, in the Territory, during the Term. Licensee shall exercise such rights in accordance with all CLC and Collegiate Institution guidelines, policies and requirements provided to Licensee, which shall be deemed part of the Agreement.

(b) Rights Reserved: Nothing in this Agreement shall be construed to prevent CLC or any Collegiate Institution from granting any other licenses or rights for use of the Licensed Indicia. The Collegiate Institutions retain all rights to use and license their respective Licensed Indicia.

Intellectual Property and Trademark License

(c) Term: This Agreement shall begin effective as of last date of signature below and shall expire ________________, unless terminated sooner or renewed in the manner provided in this Agreement.

(d) Renewal: Upon expiration, if Licensee has complied with all terms and conditions of this Agreement during the preceding Term or renewal period, Licensee shall be considered for renewal of this Agreement. Renewal is at the discretion of the individual Collegiate Institutions in consultation with CLC. Licensee recognizes and agrees that CLC and the Collegiate Institutions have no express or implied obligation to renew the Agreement. CLC and the Collegiate Institutions will have no liability to Licensee for any expenses incurred by Licensee in anticipation of any renewal of the Agreement.

(e) Limitations on License: This license is subject to the following limitations and obligations, as well as other limitations and obligations set forth in the Agreement:

(1) Licensee shall not use the Licensed Indicia for any purpose other than as authorized in this Agreement. Any proposed additions to the Licensed Articles and/or new designs shall be submitted in writing or via iCLC to CLC and samples shall be submitted to CLC for prior approval, as provided in Section 10. Licensee shall, upon notice by CLC, immediately recall any unauthorized products or designs from the marketplace, and destroy them or submit them to CLC, at CLC's option and at Licensee's expense.
(2) Licensee shall not use any brand names other than Authorized Brands in connection with the manufacture, advertising, distribution and sale of the Licensed Articles. CLC and the Collegiate Institutions shall have the right to remove or change any of the Authorized Brands during the Term.
(3) Licensee shall advertise, distribute and sell Licensed Articles only in the authorized Distribution Channels. CLC and the Collegiate Institutions shall have the right to determine whether a particular retail account falls within a particular Distribution Channel. Unless specified in Appendix D, Licensee shall have no right to advertise, distribute or sell Licensed Articles directly to consumers.
(4) Licensee must receive CLC's prior written authorization to use any Distributor of any Licensed Article. A "Distributor" shall mean any party whose business includes purchasing manufactured products from any other third party and shipping such products to retailers without changing such products. Licensee will remain primarily obligated to CLC and the Collegiate Institutions under this Agreement notwithstanding CLC's approval of a Distributor and Licensee shall ensure that any approved Distributor complies with all applicable terms and conditions of the Agreement including, without limitation, providing such Distributor with instructions relating to the distribution of the Licensed Articles and the Distribution Channels for the Licensed Articles. If an approved Distributor engages in conduct that would be a default under the Agreement if Licensee engaged in such conduct, Licensee shall be deemed in default and shall fully cooperate with CLC to ensure that such conduct ceases promptly.
(5) Licensee shall not provide any method of application of Licensed Indicia for any third party unless CLC authorizes Licensee to provide said application under the terms of an authorized manufacturer's or supplier's agreement.
(6) Licensee shall not contract with any domestic or foreign third party for the production of Licensed Articles or application of Licensed Indicia by that party ("Manufacturer") without CLC's prior written authorization. In the event that Licensee desires to have a Manufacturer produce one or more Licensed Article, or any component thereof, Licensee shall provide CLC with the name, address, telephone number and principal contact of the proposed Manufacturer. CLC must approve any Manufacturer, and the Manufacturer must execute an authorized manufacturer's or supplier's agreement

Intellectual Property and Trademark License

provided by CLC prior to use of the Licensed Indicia. In addition, Licensee shall take the steps necessary to ensure the following: Manufacturer shall produce the Licensed Articles only as and when directed by Licensee, which remains fully responsible for ensuring that the Licensed Articles are manufactured in accordance with the terms herein including approval, labor code requirements and royalty payment; Manufacturer shall not advertise, distribute or sell Licensed Articles to any person or entity other than Licensee; and Manufacturer shall not delegate in any manner whatsoever its obligations with respect to the Licensed Articles. Licensee's failure to comply with this Section may result in termination of this Agreement and/or confiscation and seizure of Licensed Articles. CLC and the individual Collegiate Institutions hereby reserve the right to terminate the engagement of any Manufacturer at any time.

(7) Licensee shall comply, and ensure that all Manufacturers comply, with labor code and monitoring requirements as established by the respective Collegiate Institutions and as set forth in The Collegiate Licensing Company Special Agreement Regarding Labor Codes of Conduct, which is incorporated herein by reference. CLC shall give Licensee reasonable written notice of any changes in labor code requirements. Licensee, upon receipt of the notice, is responsible for complying with the new labor code requirements.

(8) Any Licensed Articles manufactured at a location outside of the United States shall be taken into the possession of Licensee prior to being distributed or sold in the Territory.

(9) Licensee shall have no right to delegate any responsibility to any Sublicensee of any Licensed Article without the prior written approval of CLC. A "Sublicensee" shall mean any third party that manufactures any Licensed Article, ships such product to retailers, and invoices retailers directly.

(10) Licensee shall not use any of the Licensed Articles as Premiums unless Licensee receives prior written authorization through CLC pursuant to a separate agreement with CLC. Licensee shall not provide Licensed Articles as Premiums to any third party whom Licensee knows or should reasonably know intends to use the Licensed Articles as Premiums.

(11) Licensee is not permitted, without the applicable Collegiate Institution's prior written authorization, to promote or market a Licensed Article by means of a direct mailing or any other direct solicitation to a list of alumni, students, parents, athletic contributors, faculty or staff, or other group associated with the Collegiate Institution, regardless of how Licensee acquires such list.

(12) The National Collegiate Athletic Association (NCAA) rules prohibit the use of the name or likeness of any person who has current or remaining collegiate athletic eligibility on or in connection with the sale or promotion of any commercial product or service. In conducting activity under this Agreement, Licensee shall not encourage or participate in any activity that would cause an athlete or a Collegiate Institution to violate any such rule of the NCAA or other governing body of any intercollegiate athletic conference.

3. MARKETING EFFORTS / PERFORMANCE

(a) Marketing Efforts: Licensee recognizes that marketing efforts for Licensed Articles are important to the success of this program and Licensee, if requested, will assist CLC with such efforts by its participation.

(b) Performance: With respect to each of the Collegiate Institutions listed in Appendix A, Licensee shall manufacture, distribute, sell and maintain inventory of sufficient quantities of Licensed Articles to meet the reasonable market demand in the Distribution Channels.

INTELLECTUAL PROPERTY AND TRADEMARK LICENSE

4. SELECTION OF COLLEGIATE INSTITUTIONS

Prior to execution of this Agreement, Licensee requested a license for certain Collegiate Institutions. Appendix A lists those Collegiate Institutions that have approved Licensee's request for a license. Licensee may from time-to-time request the addition of Collegiate Institutions to this Agreement, as provided in Section 5(d).

5. MODIFICATION OF APPENDICES

(a) The Collegiate Institutions and their royalty charges listed in Appendix A, the Licensed Indicia shown in Appendix B, the Collegiate Institution policies including those in Appendix B-1, the Licensed Articles listed in Appendix C, the Authorized Brands and Distribution Channels indicated in Appendix D, and labor code requirements may be changed by CLC when and if such changes are directed by CLC and the Collegiate Institutions.

(b) Through periodic advisory bulletins or notices, including, without limitation, notification through online publications (e.g., iCLC) or via email, CLC will give Licensee written reasonable notice of any changes to appendices or policies. Licensee, upon receipt of the bulletins or notices, is responsible for distributing them promptly to the appropriate party(s) and complying with the modified appendices and policies.

(c) Licensee recognizes and agrees that certain changes to Appendices A, B, B-1, C, or D may affect Licensee's rights regarding certain Collegiate Institutions, Licensed Indicia, Licensed Articles, Authorized Brands or Distribution Channels. Licensee agrees that such rights shall cease on the effective date of the notice of such changes, in accordance with the terms of the notice. In such event, those provisions of Section 17 regarding disposal of inventory shall become effective for the affected Collegiate Institutions, Licensed Indicia, Licensed Articles, Authorized Brands or Distribution Channels unless Licensee obtains written permission from the affected Collegiate Institutions concerned to continue to use the Licensed Indicia, or to manufacture, advertise, distribute or sell the Licensed Articles.

(d) Upon notification by CLC of the addition of a Collegiate Institution to the CLC program, or at any other time, Licensee may request in writing or through iCLC the addition of Collegiate Institutions to the Agreement. Any such addition will require an addendum to Appendix A. Such addendum will be fully executed only upon Licensee's completion of product and design approval requirements, as provided in Section 10.

6. PAYMENTS

(a) Rate: Licensee agrees that it shall pay to CLC the applicable royalty charges set forth adjacent to the respective Collegiate Institutions listed in Appendix A. Unless otherwise specified, the royalties paid ("Royalty Payments") shall be based upon Net Sales, as defined in Section 1(g), of all Licensed Articles sold during the Term and any renewal, and during any period allowed pursuant to Section 17.

(b) For purposes of determining the Royalty Payments, sales shall be deemed to have been made when Licensed Articles are billed, invoiced, shipped, or paid for, whichever occurs first.

(c) Advance Payments: Upon execution of this Agreement by Licensee, and upon any renewal, Licensee shall pay CLC, as a nonrefundable payment, the Advance Payments set forth in Appendix A. Upon renewal, the Advance Payments will be prorated, where applicable, as per CLC's written instructions. Licensee may apply the Advance Payments as credits against Royalty Payments and Minimum Guarantee payments (if applicable) due for the

Intellectual Property and Trademark License

specific Collegiate Institutions, which credits shall expire no later than twenty (20) days after the expiration of the Term and any renewal period.

(d) Minimum Guarantee: Licensee shall pay CLC the Minimum Guarantee amounts (if applicable) set forth in Appendix A by no later than twenty (20) days after the end of the Term and any renewal period, unless specified otherwise in Appendix A.

(e) Administrative Fee: Upon execution of this Agreement by Licensee, and upon any renewal, Licensee shall pay CLC, as a non-refundable payment, the Administrative Fee set forth in Appendix A.

(f) Royalty Payments shall be paid by Licensee to CLC on all Licensed Articles (including, without limitation, any seconds, irregulars, etc. permitted pursuant to the provisions of Section 10(b) of this Agreement) distributed or sold by Licensee or any of its affiliated or subsidiary companies even if not billed or billed at less than the regular Net Sales price for such Licensed Articles, and payment shall be computed based upon the regular Net Sales price for such Licensed Articles distributed or sold to the trade by Licensee or, if such regular Net Sales pricing is not available, as determined by CLC's evaluation of comparable prices charged the trade for similar products.

(g) Distribution: In the event Licensee distributes or sells Licensed Articles at a special price directly or indirectly to itself, including without limitation, any affiliate or subsidiary of Licensee, to any other person, firm or corporation related in any manner to Licensee or its officers, directors or major stockholders, or through a Distributor (such distribution arrangements being subject to prior written approval by CLC), Licensee shall pay royalties with respect to such distribution or sales based upon the regular Net Sales price for such Licensed Articles distributed or sold to the trade by Licensee or, if such regular Net Sales pricing is not available, as determined by CLC's evaluation of comparable prices charged the trade for similar products.

(h) FOB Sales: If a customer of Licensee purchases Licensed Articles FOB the manufacturing source or participates in other arrangements which result in such customer paying less for the Licensed Articles than Licensee's regular selling price to the trade (such FOB Sales or other arrangements being subject to prior written approval by CLC), Licensee shall pay royalties with respect to such distribution or sales based upon the regular Net Sales price for such Licensed Articles distributed or sold to the trade by Licensee or, if such regular Net Sales pricing is not available, as determined by CLC's evaluation of comparable prices charged the trade for similar products.

(i) Multiple Royalties: CLC recognizes that Licensee may be a party to other license agreements which, together with this Agreement, would subject certain Licensed Articles to one or more additional royalty payments above and beyond the Royalty Payments. Royalty Payments required to be paid to CLC for Licensed Articles may be reduced only by mutually agreed upon amounts set forth in writing.

(j) Exempt Area: On or around certain Collegiate Institution campuses, certain accounts or areas may be exempt from the obligation to pay Royalty Payments for sales made and delivered by Licensee to customers located within the exempt area. If, however, Licensee charges royalties for such sales, then Royalty Payments are due and payable on such sales. Appendix B-1 lists those exemptions. CLC and the Collegiate Institutions reserve the right to add to or delete from Appendix B-1, and will notify Licensee of these changes in writing as provided in Section 5(b). Licensee shall be responsible for obtaining and documenting

Intellectual Property and Trademark License

confirmation from CLC or a Collegiate Institution licensing official that a particular account is exempt.

7. ROYALTY STATEMENT AND PENALTIES

(a) On or before the twentieth (20th) day of each month, Licensee shall submit to CLC, in a format provided or approved by CLC, a full and complete statement, certified by an officer of the Licensee to be true and accurate, showing the quantity, description, and Net Sales (including itemization of any permitted deductions and/or exemptions) of the Licensed Articles distributed and/or sold during the preceding month, listed (i) by Collegiate Institution, (ii) by Licensed Article, (iii) by applicable Authorized Brand, and (iv) by Distribution Channel. Such report shall include any additional information kept in the normal course of business by the Licensee which is appropriate to enable an independent determination of the amount due hereunder with respect to each Collegiate Institution. All Royalty Payments then due CLC shall be made simultaneously with the submission of the statements. If no sales or use of the Licensed Articles were made during any reporting period for one or more Collegiate Institutions, Licensee shall provide CLC a written statement to that effect as part of the report.

(b) Licensee shall pay CLC an additional charge of one and one-half percent (1.5%) per month, compounded on a monthly basis, or the maximum rate allowed by law, if lower, on any payment due under the Agreement that remains unpaid after such payment becomes due.

(c) CLC's receipt or acceptance of any statements or Royalty Payments, or the cashing of any royalty checks, shall not preclude CLC from questioning the correctness thereof at any time. Upon discovery of any verifiable inconsistency or mistake in such statements or payments, Licensee shall immediately rectify such inconsistency or mistake.

(d) Licensee shall, unless otherwise directed in writing by CLC, send all payments and statements to CLC at the address set forth in the heading of this Agreement, or transmit the same via electronic format approved by CLC.

8. OWNERSHIP OF LICENSED INDICIA AND PROTECTION OF RIGHTS

(a) Licensee acknowledges and agrees that the respective Collegiate Institutions own each of their respective Licensed Indicia, modifications of the Licensed Indicia, as well as any other Licensed Indicia adopted for use by the Collegiate Institutions, that each of the Licensed Indicia is valid, and that each Collegiate Institution has the exclusive right to use each of its Licensed Indicia subject only to limited permission granted to Licensee to use the Licensed Indicia pursuant to this Agreement. Licensee acknowledges the validity of the state and federal registrations each Collegiate Institution owns, obtains or acquires for its Licensed Indicia. Licensee shall not, at any time, file any trademark application with the United States Patent and Trademark Office, or with any other governmental entity for the Licensed Indicia, regardless of whether such Licensed Indicia is shown in Appendix B. Licensee shall not use any of the Licensed Indicia or any similar mark as, or as part of, a trademark, service mark, trade name, fictitious name, company or corporate name anywhere in the world. Any trademark or service mark registration obtained or applied for that contains the Licensed Indicia or any similar mark shall be immediately transferred to the applicable Collegiate Institution without compensation.

Intellectual Property and Trademark License

(b) Licensee shall not oppose or seek to cancel or challenge, in any forum, including, but not limited to, the United States Patent and Trademark Office, any application or registration of the Licensed Indicia of any Collegiate Institution. Licensee shall not object to, or file any action or lawsuit because of, any use by the Collegiate Institutions of their Licensed Indicia for any goods or services, whether such use is by the Collegiate Institutions directly or through licensees or authorized users.

(c) Licensee recognizes the great value of the good will associated with the Licensed Indicia and acknowledges that such good will belongs to the Collegiate Institutions, and that such Licensed Indicia have inherent and/or acquired distinctiveness. Licensee shall not, during the term of this Agreement or thereafter, dispute or contest the property rights of the Collegiate Institutions, dispute or contest the validity of this Agreement, or use the Licensed Indicia or any similar mark in any manner other than as licensed hereunder.

(d) Licensee agrees to assist CLC in the protection of the rights of the Collegiate Institutions in and to the Licensed Indicia and shall provide, at reasonable cost to be borne by CLC and/or the Collegiate Institutions, any evidence, documents, and testimony concerning the use by Licensee of the Licensed Indicia, which CLC may request for use in obtaining, defending, or enforcing rights in any Licensed Indicia or related application or registration. Licensee shall notify CLC in writing of any infringements by others of the Licensed Indicia of which it is aware. CLC and the applicable Collegiate Institution shall have the right to determine whether any action shall be taken on account of any such alleged infringements. Licensee shall not institute any suit or take any action on account of any such alleged infringements without first obtaining the written authorization of CLC and the Collegiate Institutions. Licensee agrees that it is not entitled to share in any proceeds received by CLC or any Collegiate Institution (by settlement or otherwise) in connection with any formal or informal action brought by CLC, Collegiate Institutions or other entity.

(e) Nothing in this Agreement gives Licensee any right, title, or interest in the Licensed Indicia except the right to use the Licensed Indicia in accordance with the terms of this Agreement. Licensee's use of the Licensed Indicia shall inure to the benefit of the respective Collegiate Institutions.

(f-1) Acknowledgment: Licensee acknowledges that any original designs, artwork or other compilations ("Works") created by it pursuant to this Agreement that contain the Licensed Indicia are "compilations" or "supplementary works" as those terms are used in Section 101 of the Copyright Act, and that the Works will be, and will be treated as having been, specially ordered or commissioned for use as a compilation or supplementary work rendered for, at the instigation and under the overall direction of the Collegiate Institutions; and therefore that all the work on and contributions to the Works by Licensee, as well as the Works themselves, are and at all times shall be regarded as "work made for hire" by the Licensee for the Collegiate Institutions. Without limiting the foregoing acknowledgment or subsequent assignment, Licensee further acknowledges that any rights that Licensee might have under this Agreement do not in any way dilute or affect the interests of the Collegiate Institutions in the Licensed Indicia or any derivatives thereof; nor permit Licensee to copy or use the Works or the Licensed Indicia, except as expressly permitted under this Agreement; nor to affix a copyright or trademark notice to any product bearing the Works or the Licensed Indicia, except as expressly permitted under this Agreement.

INTELLECTUAL PROPERTY AND TRADEMARK LICENSE

(f-2) Assignment: Without curtailing or limiting the foregoing acknowledgment, Licensee assigns, grants and delivers (and agrees further to assign, grant and deliver) exclusively to the respective Collegiate Institutions, all rights, titles and interests of every kind and nature whatsoever in and to the Works, and all copies and versions, including all copyrights and all renewals. Licensee further agrees to execute and deliver to CLC and the Collegiate Institutions such other and further instruments and documents as CLC or the particular Collegiate Institutions from time-to-time reasonably may request for the purpose of establishing, evidencing and enforcing or defending the complete, exclusive, perpetual and worldwide ownership by such respective Collegiate Institutions of all rights, titles and interests of every kind and nature whatsoever, including all copyrights, in and to the Works, and Licensee appoints CLC as agent and attorney-in-fact, with full power of substitution, to execute and deliver such documents or instruments as Licensee may fail or refuse promptly to execute and deliver, this power and agency being coupled with an interest and being irrevocable.

(g) Licensee acknowledges that its breach or threatened breach of this Agreement will result in immediate and irremediable damage to CLC and/or the Collegiate Institutions and that money damages alone would be inadequate to compensate CLC and/or the Collegiate Institutions. Therefore, in the event of a breach or threatened breach of this Agreement by Licensee, CLC and/or the Collegiate Institutions may, in addition to other remedies, immediately obtain and enforce injunctive relief prohibiting the breach or threatened breach or compelling specific performance. In the event of any breach or threatened breach of this Agreement by Licensee or infringement of any rights of the Collegiate Institutions, if CLC and/or the Collegiate Institutions employ attorneys or incur other expenses, Licensee shall reimburse CLC and/or the Collegiate Institutions for their reasonable attorney's fees and other expenses.

9. DISPLAY AND APPROVAL OF LICENSED INDICIA

(a) Licensee shall use the Licensed Indicia properly on all Licensed Articles, as well as labels, containers, packages, tags and displays (collectively "Packaging"), and in all print and online advertisements and promotional literature, and television and radio commercials promoting Licensed Articles (collectively "Advertising Materials"). On all visible Packaging and Advertising Materials, the Licensed Indicia shall be emphasized in relation to surrounding material by using a distinctive typeface, color, underlining, or other technique approved by CLC and the Collegiate Institutions. Any use of any Licensed Indicia shall conform to the requirements as specified in Appendix B. Wherever appropriate, the Licensed Indicia shall be used as a proper adjective, and the common noun for the product shall be used in conjunction with the Licensed Indicia. The proper symbol to identify the Licensed Indicia as a trademark (i.e., the ® symbol if the Licensed Indicia is registered in the United States Patent and Trademark Office or the ™ symbol if not so registered) and/or copyright legend (i.e., © [Date][Collegiate Institution]) shall be placed adjacent to each Licensed Indicia. Except when otherwise expressly authorized in writing by CLC, Licensee shall not use on any one Licensed Article or its Packaging the Licensed Indicia of more than one Collegiate Institution.

(b) CLC will provide to Licensee guidance on the proper use of the Licensed Indicia. A true representation or example of any proposed use by Licensee of any of the Licensed Indicia listed, in any visible or audible medium, and all proposed Licensed Articles, Packaging and Advertising Materials containing or referring to any Licensed Indicia, shall be submitted at Licensee's expense to CLC for written approval prior to such use, as provided in Section 10.

INTELLECTUAL PROPERTY AND TRADEMARK LICENSE

Licensee shall not use any Licensed Indicia in any form or in any material disapproved or not approved by CLC.

(c) Licensee shall display on each Licensed Article or its Packaging and Advertising Materials the trademark and license notices required by CLC's written instructions in effect as of the date of manufacture.

10. PROCEDURE FOR APPROVAL

(a) Licensee understands and agrees that it is an essential condition of this Agreement to protect the standards and good reputations of the Collegiate Institutions, and agrees that the Licensed Articles, Packaging, Advertising Materials and/or designs containing the Licensed Indicia shall be of high and consistent quality, subject to the prior written approval and continuing supervision and control of CLC and the Collegiate Institutions. Licensee shall submit all Licensed Articles, Packaging, Advertising Materials and/or designs containing the Licensed Indicia to CLC in a timely fashion to ensure that CLC and the Collegiate Institutions have adequate time to review such materials prior to the date of their proposed use by Licensee, and Licensee must receive prior written quality control approval by CLC as provided herein.

(b) Prior to the manufacture, use, distribution or sale of any Licensed Article, Packaging, Advertising Materials and/or designs containing the Licensed Indicia, Licensee shall submit to CLC for approval, at Licensee's expense and in the format required by CLC, at least one sample of each proposed Licensed Article, Packaging, Advertising Materials and/or design for each Collegiate Institution and one sample for CLC as the same would be manufactured, used, distributed or sold. If CLC approves in writing or via iCLC the proposed Licensed Article, Packaging, Advertising Materials and/or design, the same shall be accepted to serve as an example of quality for that Licensed Article, Packaging, Advertising Materials and/or design, and production quantities may be manufactured by Licensee in strict conformity with the approved sample. All approvals provided herein are effective only for the Term or renewal period in which Licensee has submitted and CLC has approved the Licensed Articles, Packaging, Advertising Materials and/or designs, unless Licensee is otherwise notified in writing by CLC. Licensee shall not depart from the approved quality standards in any material respect without the prior written approval of CLC. Licensed Articles, Packaging, Advertising Materials and/or designs not meeting those standards, including seconds, irregulars, etc., shall not be distributed or sold under any circumstances without CLC's prior written authorization.

(c) Licensee may only use the Licensed Indicia as shown in Appendix B and approved in the manner set forth herein. Licensee may not modify the Licensed Indicia without the prior written approval of CLC as provided in Section 10(b) above. The use of the Licensed Indicia in conjunction with original artwork supplied by the Licensee requires the express approval of CLC as provided in Section 10(b) above. Licensee may submit sketches of proposed artwork for preliminary approval before submitting finished samples.

(d) The descriptions of the Licensed Articles are set out in Appendix C. Licensee agrees to adhere strictly to the description of each Licensed Article.

(e) At time of renewal, or upon request by CLC at any other time, in addition to any other requirement, Licensee shall submit to CLC such number of each Licensed Article, Packaging, Advertising Materials and/or design manufactured, used, distributed or sold under the

INTELLECTUAL PROPERTY AND TRADEMARK LICENSE

Licensed Indicia as may be necessary for CLC to examine and test to assure compliance with the quality and standards for Licensed Articles, Packaging, Advertising Materials and/or designs approved herein. Each item shall be shipped in its usual container or wrapper, together with all labels, tags, and other materials usually accompanying the item. Licensee shall bear the expense of manufacturing and shipping the required number of Licensed Articles, Packaging, Advertising Materials and/or designs to the destination(s) designated by CLC.

(f) If CLC notifies Licensee of any defect in any Licensed Article, Packaging, Advertising Materials and/or designs or of any deviation from the approved use of any of the Licensed Indicia, Licensee shall have fifteen (15) days from the date of notification from CLC to correct every noted defect or deviation. Defective Licensed Articles, Packaging, Advertising Materials and/or designs in Licensee's inventory shall not be used, distributed or sold and shall, upon request by CLC, be immediately recalled from the marketplace and destroyed or submitted to CLC, at CLC's option and at Licensee's expense. However, if it is possible to correct all defects in the Licensed Articles, Packaging, Advertising Materials and/or designs in Licensee's inventory, said items may be distributed or sold after all defects are corrected to the satisfaction of CLC, which shall be indicated in writing. CLC and/or its authorized representatives shall have the right at reasonable times without notice to inspect Licensee's plants, warehouses, storage facilities and operations related to the production of Licensed Articles.

(g) Licensee shall comply with all applicable laws, regulations, standards and procedures relating or pertaining to the manufacture, use, advertising, distribution or sale of the Licensed Articles. Licensee shall comply with the requirements, including reporting requirements, of any regulatory agencies (including, without limitation, the United States Consumer Product Safety Commission, Federal Trade Commission, or Food and Drug Administration) which shall have jurisdiction over the Licensed Articles. Both before and after Licensed Articles are put on the market, Licensee shall follow reasonable and proper procedures for testing Licensed Articles for compliance with laws, regulations, standards and procedures, and shall permit CLC and/or its authorized representatives, upon reasonable notice, to inspect its and its Manufacturer's testing, manufacturing and quality control records, procedures and facilities and to test or sample Licensed Articles for compliance with this Section. Licensed Articles found by CLC at any time not to comply with applicable laws, regulations, standards and procedures shall be deemed disapproved, even if previously approved by CLC, and shall not be shipped and/or shall be subject to recall unless and until Licensee can demonstrate to CLC's satisfaction that such Licensed Articles have been brought into full compliance.

(h) Licensee shall inform CLC in writing of any complaint regarding the Licensed Articles promptly upon Licensee's receipt of such complaint.

(i) Any unauthorized or unapproved use by Licensee of any Licensed Indicia of any Collegiate Institution shall constitute grounds for immediate termination of this Agreement and also may result in action against Licensee for trademark infringement and/or unfair competition, other applicable claims, and collection of monetary damages.

11. DISPLAY OF OFFICIAL LABEL

(a) Licensee shall, prior to advertising, distribution or sale of any Licensed Article, affix to each Licensed Article, its Packaging and Advertising Materials an "Officially Licensed Collegiate Products" tag or label in the form prescribed by CLC ("Official Label"). In addition, Licensee shall affix Licensee's Authorized Brand(s) to each Licensed Article, its Packaging and Advertising Materials. It is acceptable for Licensee's Authorized Brand(s) to appear on the Official Label subject to prior written approval by CLC. Licensee shall obtain Official Labels from the supplier(s) authorized by CLC to provide those labels.

(b) Licensee is responsible for affixing the Official Label to each Licensed Article, its Packaging and Advertising Materials. Licensee shall not provide Official Labels to any third party for any purpose whatsoever, without prior written approval by CLC.

(c) Licensee agrees to defend, indemnify and hold harmless CLC, the Collegiate Institutions, and those Indemnified Parties set forth in Section 14(a) from all liability claims, costs or damages, including but not limited to any liability for the conversion or seizure of any of the Licensed Articles not containing the Official Label and/or Licensee's Authorized Brand(s) as required by this Section. This provision is in addition to and in no way limits Section 14.

(d) Licensee's purchase and use of the Official Label is contingent upon the Licensee maintaining its rights under this Agreement. Upon termination or expiration of this Agreement, subject to those provisions of Section 17 regarding disposal of inventory, Licensee must return all Official Labels to CLC for destruction. Licensee agrees that there will be no financial reimbursement to the Licensee by CLC, its agents, employees, or business partners for any unused Official Labels.

12. NO JOINT VENTURE OR ENDORSEMENT OF LICENSEE

Nothing in this Agreement shall be construed to place the parties in the relationship of partners, joint venturers or agents, and Licensee shall have no power to obligate or bind CLC or any Collegiate Institution in any manner whatsoever. Neither CLC nor any Collegiate Institution is in any way a guarantor of the quality of any product produced by Licensee. Licensee shall neither state nor imply, directly or indirectly, that the Licensee or its activities, other than under this license, are supported, endorsed or sponsored by CLC or by any Collegiate Institution and, upon the direction of CLC, shall issue express disclaimers to that effect.

13. REPRESENTATIONS

Licensee represents, warrants and agrees that the Licensed Articles, Packaging, Advertising Materials and/or designs shall (i) be of good quality in design, material and workmanship and suitable for their intended purpose, (ii) not cause harm when used with ordinary care, and (iii) not infringe or violate the rights of any third party. Licensee further represents, warrants and agrees that all work on and contribution to the Works shall be by bona fide "employees" of Licensee working "within the scope of employment" as those terms are used in 17 U.S.C. § 101, et. seq. Each party represents and warrants that it has the right and authority to enter into and perform under this Agreement.

14. INDEMNIFICATION AND INSURANCE

(a) Licensee is solely responsible for, and will defend, indemnify and hold harmless CLC, the Collegiate Institutions, and their respective officers, agents, and employees (collectively "Indemnified Parties") from any claims, demands, causes of action or damages, including reasonable attorney's fees, arising out of (i) any unauthorized use of or infringement of any patent, copyright, trademark or other proprietary right of a third party by Licensee in

Intellectual Property and Trademark License

connection with the Licensed Articles, Packaging, Advertising Materials and/or designs covered by this Agreement, (ii) defects or alleged defects or deficiencies in said Licensed Articles, Packaging, Advertising Materials and/or designs or the use thereof, (iii) false advertising, fraud, misrepresentation or other claims related to the Licensed Articles, Packaging, Advertising Materials and/or designs not involving a claim of right to the Licensed Indicia, (iv) the unauthorized use of the Licensed Indicia or any breach or alleged breach by Licensee of any of its representations, warranties, covenants or obligations contained in this Agreement, (v) libel or slander against, or invasion of the right of privacy, publicity or property of, or violation or misappropriation of any other right of any third party, and/or (vi) agreements or alleged agreements made or entered into by Licensee to effectuate the terms of this Agreement. The indemnifications hereunder shall survive the expiration or termination of this Agreement.

(b) Prior to the first sale or distribution of any Licensed Article, or use of the Licensed Indicia, Licensee shall obtain from an insurance carrier having a rating of at least A-7 by the A.M. Best & Co. or other rating satisfactory to CLC, and thereafter maintain, Commercial General Liability insurance, including product, advertising and contractual liability insurance. Licensee's insurance coverage shall provide adequate protection for the Indemnified Parties as additional insured parties on Licensee's policy against any claims, demands, or causes of action and damages, including reasonable attorney's fees, arising out of any of the circumstances described in Section 14(a) above. Such insurance policy shall not be canceled or materially changed in form without at least thirty (30) days written notice to CLC. Prior to the first sale or distribution of any Licensed Article, or use of the Licensed Indicia, Licensee shall furnish CLC a certificate of such insurance and endorsements in the form prescribed by CLC. Licensee agrees that such insurance policy or policies shall provide coverage of one million dollars ($1,000,000) for personal and advertising injury, bodily injury and property damage arising out of each occurrence, or Licensee's standard insurance policy limits, whichever is greater. However, recognizing that the aforesaid amounts may be inappropriate with regard to specific classes of goods, it is contemplated that CLC may require reasonable adjustment to the foregoing amounts. Any adjustment must be confirmed in writing by CLC.

15. RECORDS AND RIGHT TO AUDIT

(a) Licensee shall keep, maintain and preserve at its principal place of business during the Term, any renewal periods and at least three (3) years following termination or expiration, complete and accurate books, accounts, records and other materials covering all transactions related to this Agreement in a manner such that the information contained in the statements referred to in Section 7 can be readily determined including, without limitation, customer records, invoices, correspondence and banking, financial and other records in Licensee's possession or under its control. CLC and/or its authorized representatives shall have the right to inspect and audit all materials related to this Agreement regarding any Collegiate Institution represented by CLC, which right to inspect and audit shall include the conduct of normal audit tests of additional Licensee records including those covering "non-licensed" sales to verify that they are not sales covered by this Agreement. In addition to the materials required by normal accounting practices, Licensee must retain detail of Licensed Article sales to the invoice number level for audit purposes, and invoices must indicate the Collegiate Institution name beside each Licensed Article. Licensee will provide CLC and/or its authorized representatives the above-referenced invoice detail information in an Excel CD-ROM or disk format.

INTELLECTUAL PROPERTY AND TRADEMARK LICENSE

(b) Such materials shall be available for inspection and audit (including photocopying) at any time during the Term, any renewal periods and at least three (3) years following termination or expiration during reasonable business hours and upon at least five (5) days notice by CLC and/or its representatives. Licensee will cooperate and will not cause or permit any interference with CLC and/or its representatives in the performance of their duties of inspection and audit. CLC and/or its representatives shall have free and full access to said materials for inspection and audit purposes. Licensee shall pay CLC the amount of any additional costs beyond the cost of the originally scheduled audit incurred by CLC (i) due to a change in a scheduled audit date, which change is made at Licensee's request and approved by CLC, or (ii) if Licensee's books and records are not organized and/or available for audit.

(c) Following the conduct of the audit, Licensee shall take immediate steps to timely resolve all issues raised therein, including payment of any monies owing and due. Should an audit indicate either (i) an underpayment of five percent (5%) or more, or (ii) an underpayment of $5,000 or more, of the monies due CLC, the cost of the audit shall be paid by Licensee. Payment of any audit costs is in addition to the full amount of any underpayment including late payment charges as provided in Section 7(b). Without prejudice to the rights set forth in Section 16 below, Licensee must cure any contract breaches discovered during the audit, provide amended reports if required, and submit the amount of any underpayment including late payment charges and, if applicable, the cost of the audit and/or cancellation fees within fifteen (15) days from the date Licensee is notified of the audit result.

16. DEFAULT; CORRECTIVE ACTIONS; TERMINATION

(a) Licensee's failure to fully comply with each provision of the Agreement, including but not limited to Licensee's failure to perform as required or breach of any provision, shall be deemed a default under the Agreement. Upon default, CLC and the individual Collegiate Institutions may require the Licensee to take action to correct such default for such Collegiate Institutions. In the event that Licensee is required to take corrective action, CLC and the Collegiate Institutions shall determine the corrective action that Licensee will be required to take for such failure to perform or breach commensurate with the scope and history of Licensee's past performance. Such action may include, without limitation, requiring Licensee to adopt remedial accounting and reporting measures; requiring Licensee to conduct an internal audit; requiring Licensee to train its personnel or permitting CLC to assist therein at Licensee's expense; and requiring Licensee to discontinue the manufacture, advertising, distribution and sale of certain products bearing the Licensed Indicia. Additionally, in the event any default by Licensee results in damages to CLC or the Collegiate Institutions in an amount that would be difficult or impossible to ascertain (including, without limitation, sales of products bearing the Licensed Indicia that have not been approved pursuant to Section 10, sales of Licensed Articles without labeling as required in Section 11, etc.), then CLC and the Collegiate Institutions shall be entitled to receive compensation for damages in an amount to be determined by CLC in consultation with the Collegiate Institutions. The amount of such compensation payable pursuant to this provision shall not be less than an amount equivalent to the greater of the Advance Payment or $100, per occurrence, for each affected Collegiate Institution; provided, however, that nothing contained herein shall limit CLC's or the Collegiate Institutions' rights under this Agreement, in law, in equity or otherwise, including, without limitation, the amount of damages CLC or the Collegiate Institutions may be entitled to. If damages are assessed against the Licensee pursuant to this provision, then Licensee's ability to continue to operate under this Agreement shall be contingent upon payment of such damages in the time allowed by CLC and the Collegiate Institutions.

Intellectual Property and Trademark License

(b) In addition to the right to require corrective action for default as set forth in Section 16(a), CLC and the individual Collegiate Institutions shall have the right to terminate this Agreement without prejudice to any other rights under this Agreement, in law, in equity or otherwise, upon written notice to Licensee at any time should any of the following occur, which shall also be deemed defaults under the Agreement:

(1) Licensee has not begun the bona fide manufacture, distribution, and sale of Licensed Articles within one (1) month of the date of approval of the samples of Licensed Articles.
(2) Licensee fails to continue the bona fide manufacture, distribution, and sale of Licensed Articles during the Term. If, during any calendar quarter of the Term, Licensee fails to sell any of the Licensed Articles or fails to sell any Licensed Articles for a particular Collegiate Institution, CLC may terminate this Agreement with respect to said Licensed Article or Collegiate Institution.
(3) Licensee fails to make any payment due or fails to deliver any required statement.
(4) The amounts stated in the periodic statements furnished pursuant to Section 7 are significantly or consistently understated.
(5) Licensee fails to generate royalties during the Term or any renewal period that meet or exceed the amount of the Advance Payments and Minimum Guarantee amounts as provided in Section 6 and Appendix A.
(6) Licensee fails to make available its premises, records or other business information for any audit or to resolve any issue raised in connection with any audit, as required in Section 15.
(7) Licensee fails to pay its liabilities when due, or makes any assignment for the benefit of creditors, or files any petition under any federal or state bankruptcy statute, or is adjudicated bankrupt or insolvent, or if any receiver is appointed for its business or property, or if any trustee in bankruptcy shall be appointed under the laws of the United States government or the several states.
(8) Licensee attempts to grant or grants a sublicense or attempts to assign or assigns any right or duty under this Agreement to any person or entity without the prior written authorization of CLC.
(9) Licensee distributes or sells any Licensed Articles outside the authorized Distribution Channels for such Licensed Articles, or distributes or sells any Licensed Articles to any third party that Licensee knows or should reasonably know intends to distribute or sell such Licensed Articles outside the authorized Distribution Channels for such Licensed Articles.
(10) Licensee distributes or sells any Licensed Articles outside the Territory or distributes or sells any Licensed Articles to a third party that Licensee knows or should reasonably know intends to distribute or sell such Licensed Articles outside the Territory.
(11) If an entity acquires in a single transaction or through a series of transactions more than fifty percent (50%) ownership or controlling interest in Licensee.
(12) Licensee or any related entity manufactures, distributes or sells any product infringing or diluting the trademark, property or any other right of any Collegiate Institution or any other party.
(13) Licensee fails to deliver to CLC and maintain in full force and effect the insurance referred to in Section 14(b).
(14) CLC, a Collegiate Institution, or any governmental agency or court of competent jurisdiction finds that the Licensed Articles are defective in any way, manner or form.
(15) Any monitoring agency authorized by a Collegiate Institution determines that Licensee is in violation of the labor code adopted by that Collegiate Institution, and Licensee fails to effectively remediate said violation for that Collegiate Institution within a time

period that is reasonable with respect to the nature and extent of the violation.

(16) Licensee commits any act or omission that damages or reflects unfavorably, embarrasses or otherwise detracts from the good reputation of any Collegiate Institution.

(17) Licensee manufactures, distributes or sells Licensed Articles of quality lower than the samples approved, or manufactures, distributes, sells or uses Licensed Articles or Licensed Indicia in a manner not approved or disapproved by CLC.

(18) Licensee fails to affix to each Licensed Article, its Packaging and Advertising Materials an Official Label and Authorized Brand in the manner provided in Section 11.

(19) Licensee commits a default under any other provision of this Agreement, and fails to cure such default within fifteen (15) days of written notice from CLC.

(c) CLC shall have the right to terminate this Agreement upon written notice to Licensee without cause with respect to a particular Collegiate Institution in the event that said Collegiate Institution directs CLC to terminate this Agreement. This termination shall be without prejudice to any other rights CLC may have, whether under the provisions of this Agreement, in law, in equity or otherwise.

(d) The entire unpaid balance of all Royalty Payments and other amounts owing and due under this Agreement shall immediately become due and payable upon termination.

17. EFFECT OF EXPIRATION OR TERMINATION; DISPOSAL OF INVENTORY

(a) Effect of Expiration or Termination: After expiration or termination of this Agreement for any reason, Licensee shall immediately discontinue the manufacture, advertising, use, distribution and sale of all Licensed Articles, Packaging and Advertising Materials, the use of all Licensed Indicia, and all similar marks, except as provided in Section 17(b), or unless expressly authorized in writing by CLC or the applicable Collegiate Institution. Until payment to CLC of any monies due it, CLC shall have a lien on any units of Licensed Articles not then disposed of by Licensee and on any monies due Licensee from any jobber, wholesaler, distributor, or other third parties with respect to sales of Licensed Articles.

(b) Disposal of Inventory: After expiration or termination of this Agreement for any reason, Licensee shall have no further right to manufacture, advertise, use, distribute or sell Licensed Articles, Packaging or Advertising Materials utilizing the Licensed Indicia, but may continue to distribute its remaining inventory of Licensed Articles in existence at the time of expiration or termination for a period of sixty (60) days; provided, however, that Licensee has delivered all statements (including Final Statement) and payments then due, that during the disposal period Licensee shall deliver all statements and payments due in accordance with Section 7, that Licensed Articles are sold at Licensee's regular Net Sales price and within the Distribution Channels, and that Licensee shall comply with all other terms and conditions of this Agreement. Notwithstanding the foregoing, Licensee shall not manufacture, advertise, use, distribute or sell any Licensed Articles, Packaging or Advertising Materials after the expiration or termination of this Agreement because of: (i) departure of Licensee from the quality and style approved by CLC under this Agreement, (ii) failure of Licensee to obtain product or design approval, or (iii) a default under Section 16.

18. FINAL STATEMENT

Upon expiration or termination of this Agreement for any reason, or at any other time upon request by CLC or the Collegiate Institutions, Licensee shall furnish to CLC a statement showing the number and description of Licensed Articles on hand or in process. Following such expiration

Intellectual Property and Trademark License

or termination, including inventory disposal period, if allowed, CLC may request Licensee to either (i) surrender unsold Licensed Articles, Packaging and Advertising Materials, as well as dies, molds and screens used to manufacture such Licensed Articles and Packaging, or (ii) destroy all such remaining unsold materials, certifying their destruction to CLC and specifying the number of each destroyed. CLC and/or its authorized representatives reserve the right to conduct physical inventories to ascertain or verify Licensee's compliance with the foregoing.

19. SURVIVAL OF RIGHTS

The terms and conditions of this Agreement necessary to protect the rights and interests of CLC and the Collegiate Institutions, including, without limitation, Licensee's obligations under Sections 8, 13, 14 and 15, shall survive the termination or expiration of this Agreement. The terms and conditions of this Agreement providing for any other activity following the effective date of termination or expiration of this Agreement shall survive until such time as those terms and conditions have been fulfilled or satisfied.

20. NOTICES

All notices and statements to be given and all payments to be made, shall be given or made to the parties at their respective addresses set forth herein, unless notification of a change of address is given in writing. Unless otherwise provided in the Agreement, all notices shall be sent by certified mail, return receipt requested; facsimile, the receipt of which is confirmed by confirmation document; email, confirmed by email receipt confirmation notice; or nationally recognized overnight delivery service that provides evidence of delivery, and shall be deemed to have been given at the time they are sent.

21. CONFORMITY TO LAW AND POLICY

(a) Licensee shall comply with such guidelines, policies, and requirements as CLC may give written notice from time-to-time including, without limitation, guidelines, policies and/or requirements contained in periodic CLC bulletins or notices.

(b) Licensee undertakes and agrees to obtain and maintain all applicable permits and licenses at Licensee's expense.

(c) Licensee shall pay all federal, state and local taxes due on or by reason of the manufacture, distribution or sale of the Licensed Articles.

22. SEVERABILITY

The determination that any provision of this Agreement is invalid or unenforceable shall not invalidate this Agreement, and the remainder of this Agreement shall be valid and enforceable to the fullest extent permitted by law.

23. NON-ASSIGNABILITY

This Agreement is personal to Licensee. Neither this Agreement nor any of Licensee's rights shall be sold, transferred or assigned by Licensee without CLC's prior written approval, and no rights shall devolve by operation of law or otherwise upon any assignee, receiver, liquidator, trustee or other party. Subject to the foregoing, this Agreement shall be binding upon any approved assignee or successor of Licensee and shall inure to the benefit of CLC, its successors and assigns.

Intellectual Property and Trademark License

24. ENTIRE AGREEMENT / NO WAIVER

Unless otherwise specified herein, this Agreement or any renewal, including appendices, constitutes the entire agreement and understanding between the parties and cancels, terminates, and supersedes any prior agreement or understanding, written or oral, relating to the subject matter hereof between Licensee, CLC and the Collegiate Institutions. There are no representations, promises, agreements, warranties, covenants or understandings other than those contained herein. None of the provisions of this Agreement may be waived or modified, except expressly in writing signed by both parties. However, failure of either party to require the performance of any term in this Agreement or the waiver by either party of any breach shall not prevent subsequent enforcement of such term nor be deemed a waiver of any subsequent breach.

25. COLLEGIATE INSTITUTION RIGHT TO ENFORCE

Each Collegiate Institution is entitled to enforce its rights in the Licensed Indicia and the terms of this Agreement directly against the Licensee; and each Collegiate Institution is entitled to all the rights and remedies available under this Agreement.

26. MISCELLANEOUS

When necessary for appropriate meaning, a plural shall be deemed to be the singular and singular shall be deemed to be the plural. The attached appendices are an integral part of this Agreement. Section headings are for convenience only and shall not add to or detract from any of the terms or provisions of this Agreement. This Agreement shall be governed by and construed in accordance with the laws of the state of Georgia, which shall be the sole jurisdiction for any disputes. This Agreement shall not be binding on CLC until signed by CLC as agent on behalf of the Collegiate Institutions.

IN WITNESS WHEREOF, the parties hereto have caused this Agreement to be executed effective as of the last date of signature below.

LICENSEE:

By: ______________________________ [Seal]
(Signature of officer, partner, or person duly authorized to sign)

Title: ______________________________

Date: ______________________________

THE COLLEGIATE LICENSING COMPANY, as agent on behalf of the Collegiate Institutions

By: ______________________________
(Signature of person duly authorized to sign)

Title: ______________________________

Date: ______________________________

Agreement 6: Distributorship Agreement

Source: Mahoney RT (ed.). 2004. *Handbook of Best Practices for Management of Intellectual Property in Health Research and Development.* MIHR: Oxford, U.K. Reproduced with permission.

A Public Sector Research Center intends neither to be a manufacturer or distributor. This sample agreement would be used in circumstances where the public sector institution is assisting a manufacturer to obtain one or more distributors or vice versa.

This Agreement is effective this _______ day of __________, 20__ ("Effective Date"), by and between [MANUFACTURER], a corporation organized and existing under the laws of [COUNTRY] ("Manufacturer"), and [DISTRIBUTOR], a corporation organized and existing under the laws of [COUNTRY] ("Distributor").

WITNESSETH

WHEREAS the intellectual property related to product (as hereafter defined), including Trademarks, is owned or controlled by [TMOWNER] ("Trademark Owner");

WHEREAS Trademark Owner has granted to Manufacturer the right to use and sell the Product in [COUNTRY] under the Trademarks using the Distributor; and

WHEREAS Manufacturer wishes to appoint Distributor to sell the Product in [COUNTRY], and Distributor is willing and able to import, promote, distribute, and sell Product under the Trademarks in [COUNTRY].

NOW, THEREFORE, the parties intending to be legally bound agree as follows:

ARTICLE I: DEFINITIONS

Wherever used in this Agreement, the following terms have the following meanings:

1.1 Product means [PRODUCT DEFINITION/COMPOSITION]

1.2 Public Sector [CURRENT DEFINITION]

1.3 Private Sector means all markets not defined as Public Sector.

1.4 Trademark means all trademarks, service marks, logotypes, commercial symbols, insignias, and designs pertaining thereto, including, but not limited to, the trademark [TM] _____ and the logotype associated therewith, now owned by Trademark Owner and licensed to Manufacturer, as the same may be amended, modified, revised, or improved hereafter that are associated and identified with the manufacture and sale of the Product.

ARTICLE II: APPOINTMENT OF DISTRIBUTOR

2.1 Subject to the terms and conditions of this Agreement, Manufacturer appoints Distributor as its nonexclusive agent for the importation, promotion, distribution, and sale of Product under the Trademarks in [COUNTRY].

2.2 This Agreement grants the Distributor the right to package the Product but does not grant the Distributor the right to manufacture the Product or to have it manufactured by a third party.

ARTICLE III: USE OF TRADEMARK

3.1 Distributor recognizes the substantial value of the goodwill associated with the Trademark and acknowledges that the Trademark and all rights therein and the goodwill pertaining thereto belong exclusively to Trademark Owner. Distributor agrees not to commit any act or omission adverse or injurious to said rights.

3.2 Distributor agrees that every use of the Trademark by Distributor shall inure to the benefit of Trademark Owner, and that Distributor shall not at any time acquire any rights in the Trademark by virtue of any use Distributor may make of the Trademark.

3.3 Distributor agrees to cooperate fully and in good faith with Trademark Owner for the purpose of securing, preserving, and protecting Trademark Owner's rights in and to the Trademarks, including executing a trademark license with Trademark Owner and/or with the Manufacturer which license may be registered with the Patent and Trademark Office (or its equivalent) in [COUNTRY].

3.4 Distributor acknowledges that Distributor's failure to cease the use of the Trademark on the termination or expiration of this Agreement will result in immediate and irreparable damage to Trademark Owner and to the rights of any subsequent licensee. Distributor acknowledges and admits that there is no adequate remedy at law for such failure and agrees that, in the event of such failure, Trademark Owner shall be entitled to equitable relief by way of temporary and permanent injunctions and such other and further relief as any court with jurisdiction may deem just and proper.

3.5 Distributor shall report to Trademark Owner and Manufacturer, in writing, any infringement or imitation of the Trademarks of which Distributor becomes aware. Trademark Owner shall have the sole right to determine whether to institute litigation upon such infringements as well as the selection of counsel. Trademark Owner may commence or prosecute any claims or suits for infringement of the Trademarks in its own name or in the name of Manufacturer or the Distributor or may join Distributor and/or Manufacturer as a party thereto. If Trademark Owner brings an action against any infringer of the Trademark, Distributor and Manufacturer shall cooperate with Trademark Owner and lend whatever assistance is necessary in the prosecution of such litigation. If Trademark Owner decides not to institute such litigation, it may authorize, within its sole discretion, in writing, Distributor or Manufacturer to institute such litigation.

3.6 Distributor shall not contest or deny the validity or enforceability of the Trademark or oppose or seek to cancel any registration thereof by Trademark Owner, or aid or abet others in doing so, either during the term of this Agreement or at any time thereafter.

3.7 Distributor acknowledges that any use of the Trademark in violation of the provisions of this Article will cause irreparable damage to Trademark Owner and its licensees, constitutes an incurable default of this Agreement, and is grounds for immediate termination of this Agreement.

Distributorship Agreement

ARTICLE IV: REGISTRATION OF PRODUCT

4.1 Distributor shall register the Product with the regulatory authorities of [COUNTRY].

4.2 Manufacturer will provide a technical dossier for registration and assist in responding to specific questions that may arise during registration. Registration documents supplied to Distributor shall be in English. Any translation of the registration documents shall be the responsibility of Distributor.

4.3 If the law and/or regulations of [COUNTRY] require that Distributor be named as sole or joint owner of the subject registration, Distributor agrees that upon termination or expiration of this Agreement, Distributor will promptly assign to Trademark Owner all right, title, and interest that the Distributor may have in the subject registration and will terminate Distributor's own interest therein.

4.4 The technical dossier provided hereunder contains technical and proprietary information supplied by the Manufacturer and shall be deemed to have been provided in confidence for the sole purpose herein set forth. Distributor undertakes not to use any of the information for any purpose other than the registration in [COUNTRY] of the Product manufactured by Manufacturer and not to disclose any information to any third party, other than government regulatory authorities, without the written consent of Manufacturer.

4.5 Distributor will use all possible care and diligence to obtain the prompt issuance of the registration for the Product.

4.6 All expenses incurred relating to the registration of Product, including but not limited to taxes, official fees, and clinical trials that might be required by the government authorities of [COUNTRY], shall be borne by the Distributor.

ARTICLE V: SUPPLY TERMS

5.1 Distributor agrees that Manufacturer shall be Distributor's sole supplier of Product and agrees that it will distribute only Product purchased from manufacturer except that Distributor may receive and distribute in the Public Sector, Product supplied to Distributor by international donor agencies for such distribution regardless of where the donor agency obtained such Product. All Product sold to Distributor hereunder shall be manufactured by Manufacturer in accordance with Good Manufacturing Practices.

5.2 Since Manufacturer has made, or is making, distribution arrangements for Product with representatives in other countries, Distributor agrees that it shall not knowingly allow Product to be distributed for use in countries outside of [COUNTRY] without prior approval from Manufacturer.

5.3 All orders submitted by Distributor to Manufacturer are subject to acceptance by Manufacturer, to government restrictions and approval, and to allocations that may be necessary due to production capacity restrictions.

5.4 To assure a constant supply of the Product, Distributor shall stock a sufficient quantity of the Product to satisfy without delay the demands for it, and Distributor undertakes to keep at all times _____-months' (__) stock of the Product for the Private and Public Sectors in [COUNTRY]. However, Distributor is not required to carry a stock of Product in the expectation of sales to entities of the Public Sector who purchase through public bidding. To this end, Distributor shall place with Manufacturer timely and sufficient orders for the Product, taking into account the market demand, shipping time, and filling of the order by Manufacturer. Manufacturer shall supply and ship to Distributor as quickly as possible and always within ninety (90) days of receiving its purchase order with the amount of Product specified therein.

5.5 Distributor shall at all times remain in close contact with those entities of the Public Sector that purchase through public bidding and with those of their officers whose responsibilities have a bearing on these purchases; assure that it is immediately advised whenever a tender for [PRODUCT TYPE] is being solicited; and whenever such is the case, relay the information to the Manufacturer within forty-eight (48) hours.

5.6 Distributor shall clear the Products from the airport or other port of entry at its own expense within fifteen (15) days after their arrival in [COUNTRY]. Distributor shall be responsible for the clearance of customs of Product and local transport to its facilities.

5.7 Distributor agrees to inspect the Product immediately upon delivery and to give notice by fax to the manufacturer within fifteen (15) days of such delivery of any matter of thing by reason whereof it alleges that the Product is not in good condition. If no such notice is served by the Distributor upon the Manufacturer, the Product shall be deemed to be in accordance with this Agreement in all respects and the Distributor shall be deemed to have accepted the Product. If Distributor, having served notice on the Manufacturer, demonstrates that the Product is not in good condition, Manufacturer shall at its option either replace the defective goods with Product complying with this Agreement or refund to the Distributor the price paid for defective Product.

5.8 Distributor agrees to inform Manufacturer in writing three (3) months before the end of each calendar year of its estimated requirements of the Product for the following year.

5.9 Manufacturer shall be free to accept or not the return of expired Product.

ARTICLE VI: PRICING AND PAYMENTS

6.1 Prices charged to Distributor will, unless otherwise negotiated, be in accord with the Manufacturer's prices for Public Sector and Private Sector distribution prevailing at the time of shipment. After the initial order, two (2) months' notice will be given for any price increase. Manufacturer will endeavor to keep Distributor supplied with current information regarding pricing. The price for the Public Sector shall be preferential and set at the lowest possible reasonable level permitting a commercially reasonable return; however, nothing herein shall be interpreted as requiring the sale of Product below fully allocable costs plus a mark up of _______ percent (__%). Price shall be FOB [PLACE].

6.2 Distributor guarantees payment of all orders placed or approved by Distributor. Orders will provide for payment terms of thirty (30) days from date of order in [CURRENCY], except that in the event of unsatisfactory payment history, Manufacturer reserves the right to provide Product to Distributor on a COD basis.

Distributorship Agreement

ARTICLE VII: PACKAGING

7.1 Distributor shall package the Product under the supervision of its own technicians, in its own factory or in another qualified factory, and shall oversee the process with all necessary care, strictly following good pharmaceutical manufacturing practices.

7.2 Should Distributor need to have the Product packaged by a third party, said party must first have been approved by the Manufacturer and must commit itself in writing to comply with the articles of this Agreement relevant to the Product.

7.3 Distributor agrees to mark all packaging for Product in accordance with the applicable laws in [COUNTRY]. Said packaging shall be submitted to Manufacturer for approval before it is made up or printed.

7.4 All packaging costs and expenses shall be borne entirely by the Distributor.

ARTICLE VIII: ADVERTISING AND PROMOTION

8.1 Distributor undertakes at its own expense to actively promote the Product in [COUNTRY] by the best legal and appropriate means and to retain a trained sales force of [TYPE] representatives and detailers to assure an effective promotion of the Product with the [TYPE] community and with other professional [TYPE] personnel. Distributor shall further place, at its own expense, promotional advertisement and writings on the Product in [TYPE] and other suitable publications covering [COUNTRY].

8.2 Each year by September 30, Distributor shall submit to Manufacturer for approval Distributor's promotional plan for the following year detailing promotional visits to health professionals and to others, distribution of samples of the Products and promotional materials and to whom, advertisements and writings of the Product to be placed in publications in [COUNTRY], and planned participation and contributions to [TYPE] reunions and function. Said promotional plan to be reviewed jointly by the Distributor and the Manufacturer each six (6) months for the eventual modifications by mutual agreement. Special educational or promotional activities not included in the promotional plan require the approval of the Manufacturer, and the apportionment of their cost, if there be any, will be decided by mutual consent of the parties.

8.3 Distributor shall submit to Manufacturer quarterly marketing reports listing the promotional activities carried out during each month of the last period.

8.4 Manufacturer shall provide to Distributor free of charge, save customs duties, a certain quantity of free samples of the Product and dummies of the scientific, technical, commercial, and training materials required to carry out promotional programs for the Product. Should the laws of [COUNTRY] not recognize the distribution of free samples, in no event shall Distributor be allowed to claim any discount on the price of Product. Distributor shall provide, at its own cost, to the Public Sector agencies that use or may use Product, the needed scientific, technical, educative, and training material for this purpose and previously approved by the Manufacturer.

8.5 In promoting the Product, Distributor shall refrain from making any claims regarding its therapeutic action or effectiveness different or greater than those specified by the Manufacturer and by the Sanitary Authorities of [COUNTRY].

Distributorship Agreement

8.6 All advertisements and promotional materials, including text and graphics, used by Distributor shall be subject to prior written approval of Manufacturer, which approval shall not be unreasonably withheld.

ARTICLE IX: REPORTING

9.1 Distributor will provide to Manufacturer monthly sales reports containing such information retarding sales of Product as Manufacturer shall specify, and including nonbinding, good-faith forecasts of its anticipated requirements and shipping dates for the three (3) month periods following such reports.

9.2 On or before February 1 of each year, Distributor shall supply Manufacturer with a report for the preceding calendar year or part thereof showing separately the quantity of Product purchased from Manufacturer and sold to the Private Sector and Public Sector in [COUNTRY], the average selling price of the Product purchased from Manufacturer, the quantity of Product supplied to Distributor by international donor agencies, and the selling price, if any, of the Product supplied by the international donor agencies. Manufacturer shall promptly provide a copy of the report to Trademark Owner.

9.3 Distributor shall, upon request by Manufacturer or Trademark Owner provide supporting documentation adequately justifying the pricing structure for the Public Sector. The rights created by this paragraph are directly enforceable by Manufacturer on behalf of any Public Sector agency wishing to purchase Product. To the extent possible under [COUNTRY] law, the rights created by this paragraph are directly enforceable by any Public-Sector agency on its own behalf. In addition to any other rights possessed by Manufacturer, a breach of any of the provisions of this paragraph shall be sufficient basis for termination of this Agreement by Manufacturer upon thirty (30) days' written notice.

ARTICLE X: COVENANTS AND REPRESENTATIONS OF DISTRIBUTOR

10.1 Distributor is a corporation duly formed, validly existing, and in good standing under the laws of [COUNTRY] and is duly qualified to transact business.

10.2 Distributor agrees that it shall not use or distribute Product in any manner inconsistent with the terms and intent of this Agreement.

10.3 Distributor agrees to use its best efforts to successfully market and distribute Product from Manufacturer in [COUNTRY] on a continuing basis during the term of this Agreement and to comply with good business practices and all laws and regulations relevant to this Agreement or the subject matter hereof.

10.4 Distributor agrees to keep Manufacturer informed as to any problems encountered with the Products and any resolutions arrived at for those problems and to communicate promptly to Manufacturer any and all suggested modifications, design changes or improvements of the Products. Manufacturer agrees to promptly pass this information on to Trademark Owner. Distributor and Manufacturer further agree that Trademark Owner shall have all right, title, and interest in and to any such suggested modifications, design changes, or improvements of the Products, without the payment of any additional consideration thereof.

DISTRIBUTORSHIP AGREEMENT

ARTICLE XI: CONFIDENTIALITY
All technical, corporate, business, and other proprietary information furnished by Trademark Owner of Manufacturer hereunder, or which results from the joint efforts of Trademark Owner and/or Manufacturer's and Distributor's personnel, shall be deemed to have been furnished to Distributor in confidence for the sole purposes herein set forth, and Distributor undertakes not to use any of this information for any purpose not connected with the orders accepted under this Agreement. Distributor shall also take all reasonable precautions to prevent communication, without the written consent of Trademark Owner or Manufacturer, of any such technical or other proprietary information to any third party, except as may be necessary to carry out the purposes of this Agreement.

ARTICLE XII: TERM AND TERMINATION

12.1 This Agreement is effective as of the Effective Date and will expire ________(__) years thereafter, provided the following minimum volumes have been purchased by Distributor:

Year	Amount

The above minimum volume for [YEAR] assumes approval for product registration by the government in [COUNTRY] by [DATE], and will be reduced or expanded pro rata using the quantity specified for [YEAR] in case of delayed or expedited approval. If Distributor does not purchase the minimum volume specified for any year, Manufacturer may terminate this Agreement by giving thirty (30) days' written notice, provided that if Distributor during the thirty (30)-day period orders sufficient Product for immediate delivery to make up the deficiency, the notice will be revoked.

12.2 This Agreement may also be terminated in the event Manufacturer determines that a change in management or effective financial control of Distributor has or will adversely affect the distribution of Product in accordance with this Agreement.

12.3 Distributor may terminate this Agreement any time by giving six (6) months' written notice to Manufacturer. During this six (6)-month period, Distributor will continue to use its best efforts to promote the sale and use of Product.

12.4 Extensions and renewals of this Agreement will be subject to agreement between the parties made at least six (6) months prior to its expiration.

ARTICLE XIII: RIGHTS AND OBLIGATIONS ON TERMINATION
Upon termination or expiration of this Agreement, Distributor shall return unused inventory to Manufacturer. Distributor shall dispose of all advertising material relating to the Product or the Trademark and shall discontinue immediately any use of the Trademark. Distributor shall maintain as confidential all proprietary information supplied to Distributor hereunder.

ARTICLE XIV: RELATIONSHIP OF PARTIES
The parties hereto expressly understand and agree that Distributor is an independent contractor in the performance of each and every part of this Agreement and is solely responsible for the actions of all of its employees and agents. Neither Trademark Owner nor Manufacturer shall be obligated by any agreements, representations, or warranties made by Distributor, its employees, or its agents nor with respect to any other action of Distributor, its employees, or its agents, nor

shall Trademark Owner or Manufacturer be obligated for any claims, liabilities, damages, debts, settlements, costs, expenses, and liabilities that my arise on account of Distributor's activities, or those of its employees or its agents.

ARTICLE XV: HOLD HARMLESS

15.1 Distributor agrees to use its best efforts to ensure that Product is transported, stored, and distributed in accordance with handling instructions provided by Manufacturer. Distributor further agrees to use its best efforts to ensure that Product is provided to customers in a manner which facilitates its safe and proper use. Manufacturer shall have the right to enter and inspect any premises or facilities used by Distributor for or in connection with the preparation, promotion, marketing, and distribution of the Product, at any time during normal business hours and shall further have the right to take a reasonable number of samples of the Product at no charge in order to determine Distributor's compliance with the terms and condition of this Agreement.

The Distributor shall sell the Product on its own account and in no event shall the Manufacturer be deemed liable for credits the Distributor may grant or for any other obligations the Distributor may have to fulfill for its sales or other types of transaction in [COUNTRY]. It is understood and agreed that the Distributor has no right or authority whatsoever to accept any financial obligation on the Manufacturer's name or account without the Manufacturer's prior written approval.

15.2 Distributor shall, in respect of Product distributed by it, indemnify and hold harmless Trademark Owner, and its employees and agents against any and all claims that might arise, and liabilities and related fees and expenses that might be incurred, on account of any injury, illness, suffering, disease, or death to any person or unborn offspring of any such person by reason of the distribution, sale, or use of the Product distributed by Distributor.

ARTICLE XVI: NOTICES

Any report, accounting, objection, notice, or consent required or provided for by the terms of this Agreement shall be in writing, and all accounting, obligations, notices, consents, and reports provided for hereunder shall be sent by registered mail, prepaid, or by facsimile to the business address of the party to be served therewith. It is agreed that the business addresses of the parties shall be as follows:

If to Manufacturer: ____

If to Distributor: ____

ARTICLE XVII: PROHIBITION AGAINST ASSIGNMENT

This Agreement is entered into in reliance upon and in consideration of the experience, knowledge, skills, and qualifications of and trust and confidence placed in Distributor by Manufacturer. Therefore, neither Distributor's interest in this Agreement nor any of its rights or privileges hereunder shall be assigned, transferred, shared, or divided voluntarily or involuntarily, by operation of law or otherwise, in any manner, without the prior written consent of Manufacturer and Trademark Owner. In the event of any change in management or effective financial control of Distributor, Distributor shall inform Manufacturer immediately. If, in the opinion of Manufacturer, this change adversely affects the management of Distributor or the business or general best

Distributorship Agreement

interest of either party, Manufacturer may, within sixty (60) days of Distributor's notice, terminate this Agreement and cancel any or all pending orders by giving Distributor ninety (90) days' written notice, such termination and/or cancellation to be effective at the end of such ninety (90)-day period.

ARTICLE XVIII: FORCE MAJEURE

No failure or omission by any party in the performance of any obligation of this Agreement shall be deemed a breach of this Agreement nor create any liability if the same shall arise from any cause or causes beyond the control of such party, including, but not restricted to, the following, which for the purposes of this Agreement shall be regarded as beyond the control of the party concerned:

Government regulations, acts of God, strikes or other acts of workers, fire, storm, explosions, riots, war, rebellion, transportation embargoes, or failures or delays in transportation.

ARTICLE XIV: AMENDMENTS

No amendment or other modification of this Agreement shall be valid or binding on any party hereto unless reduced to writing and executed by the parties hereto.

ARTICLE XX: WAIVER

No waiver by any party of any provision hereof shall be deemed a waiver of any other provision hereof or of any subsequent breach by any party of the same or any other provision. None of the terms of this Agreement will be held to have been waived or altered unless such waiver or alteration is in writing and signed by all of the parties hereto.

ARTICLE XXI: GOVERNING LAW AND ARBITRATION

21.1 This Agreement shall be governed by the law of ________________.

21.2 The parties hereto undertake to settle any dispute, controversy, or claim arising under, out of, or in connection with this Agreement, including, without limitation, its formation, validity, binding effect, interpretation, performance, breach, or termination, as well as noncontractual claims, in an amicable manner. If an amicable settlement cannot be reached within 30 days for any reason, the dispute shall be referred to and finally settled by arbitration in accordance with the UNCITRAL Arbitration Rules then obtaining. The appointing authority shall be the Secretary-General of the Permanent Court of Arbitration, the number of arbitrators shall be three, and the language to be used in the arbitral proceedings shall be English. The place of arbitration shall be determined by mutual agreement, but if agreement cannot be reached the proceedings shall take place in ______________.

21.3 Either party to this Agreement may request any judicial authority to order any interim measures of protection for the preservation of its rights and interests to the extent permitted by law, including, without limitation, injunctions and measures for the conservation of such property and information that form part of the subject matter in dispute. Such requests shall not be deemed incompatible with, or as a waiver of, this agreement to arbitrate. In respect of any requests for interim measures of protection, and without limitation to proceeding in any other forum, the parties hereby consent to the exercise of jurisdiction by the judicial authorities of ________________.

21.4 In the event a party fails to proceed with arbitration, unsuccessfully challenges the arbitrator's award, fails to comply with the arbitrator's award, or fails to comply with any

DISTRIBUTORSHIP AGREEMENT

interim measure of protection issued by any competent authority, the other party shall be entitled to costs of suit, including reasonable attorney's fees, for having to compel arbitration or defend or enforce the award or interim measure.

IN WITNESS WHEREOF the parties hereto have caused these presents to be executed in triplicate by their duly authorized officers.

For Manufacturer:	For Distributor:
By: ____________________	By: ____________________
Date: ____________________	Date: ____________________

Biographical Sketches of Authors and Members of the Board of Patrons*

AMMANN, Klaus
Klaus is Emeritus Professor of Biodiversity at the University of Bern, Switzerland. He has worked in vegetation and glacial history, vegetation ecology, urban ecology, lichen chemistry, biomonitoring air pollution and plant taxonomy. He also served as director of the Botanical Garden at the University of Bern. He has been the leader of numerous research projects supported by the Swiss Government and the E.U. on research in ecological monitoring in Bulgaria, biomonitoring air pollution, European plant conservation, risk assessment of gene flow of transgenic crops, and communication strategies. He is a member of numerous scientific committees and organizations, such as chair of the section of biodiversity of the European Federation of Biotechnology, and the Swiss committee on biosafety. He is member of the board of directors of Africa Harvest and involved in biosafety research on sorghum in Africa. He also is active in the field of philosophy and methodology of science communication, together with his wife Dr. Biljana Papazov Ammann. Presently he is guest professor at the Delft University of Technology, Department of Biotechnology, Holland.

ANDERSON, Carolyn
Ms. Anderson was a member of the founding management team of Diversa Corp. where she served as Vice President of Intellectual Property and Licensing. Ms. Anderson led the company's IP group from inception in 1994 to her departure in 2005. In 2000, the Diversa management team achieved the most successful biotechnology IPO at that time, raising over $200 million in gross proceeds. In addition, during her tenure as the head of intellectual property, Ms. Anderson was the IP lead in business negotiations that raised $300 million in committed funding from corporate partners. Diversa's patents were cited by *MIT Magazine* as being in the top 10 in the world both in 2003 and 2004 across all industries based on citation frequency. Before joining Diversa Corp., Ms. Anderson served in multiple roles at Stratagene Cloning Systems, a molecular biology company based in La Jolla, California, including in sales, marketing, product management, and business development.

Ms. Anderson earned her undergraduate degree at the University of California, San Diego, in Biochemistry and Cell Biology, is a registered patent agent with the U.S. Patent Office, and has published in the area of intellectual property concerning biodiversity access. In 2003, she was a nominee in the *T Sector Magazine* and BIOCOM BioFUSION award for the "Life Sciences In-House Legal Counsel of the Year." Currently she is the President of Capia IP, which provides business-based IP advice and services to the life-sciences industry, and co-owner of BioAtla, a U.S.-based biotechnology company with operations in China, which offers protein engineering and evolution services.

ANDERSON, Mark
Mark Anderson is a U.K.-qualified solicitor who specializes in intellectual property and commercial transactions. He founded Anderson & Company, the Technology Law Practice™, in 1994, after having spent seven years with Bristows, a specialist IP law firm in London. He works from offices overlooking the River Thames in the Oxfordshire countryside. Most of his clients are biotech and IT companies and universities in the U.K. and continental Europe. He is a member of the U.K. Intellectual Property Lawyers Association and a member of the IP Working Party of the Law Society of England and Wales. He has written and co-authored several books on IP-related subjects, including *Technology Transfer: Law, Practice and Precedents*, 2nd ed. (Haywards Heath: Tottel, 2003) and *Modern Law of Patents*: Butterworths, 2005).

BALLANTYNE, Zoë
Zoë Ballantyne is currently responsible for the preparation and negotiation of legal agreements and IP advice for the technology transfer division of the Wellcome Trust and the Wellcome Trust Sanger Institute. Her work covers a broad spectrum of corporate, commercial, and intellectual property law, as well as intellectual property policy. Ms. Ballantyne graduated from the University of Cambridge with a degree in natural sciences, specializing in genetics. She obtained a post-graduate diploma in law and completed the Legal Practice Course at Nottingham Law School. She subsequently obtained a diploma in intellectual property law from the University of Bristol. Ms. Ballantyne qualified as a solicitor in 2001 and practiced in the IP department and the life-sciences group of the international law firm Ashurst before joining the Wellcome Trust in 2004.

* Several Members of the Board of Patrons joined after the first printing of this Handbook. Their biographical sketches can be found on www.iphandbook.org/handbook/authors/.

BANERJI, Jaya

Jaya Banerji has spent over 15 years in communications. She has worked in India, the Middle East, Switzerland, and the U.K. She acquired her expertise in writing, editing, scripting, publishing, and advocacy from both the not-for-profit and commercial sectors. Ms. Banerji has been a freelance writer, editor, and reviewer for the print media and a number of publishing houses. She has worked at Kali for Women, Delhi, Asia's first feminist publishing house, McKinsey & Company India, Médecins Sans Frontières' Campaign for Access to Essential Medicines, and the Drugs for Neglected Diseases Initiative. She has recently joined the Medicines for Malaria venture. Throughout her career, Ms. Banerji has retained her strong belief in the rights of the most vulnerable and impoverished, especially women and children. She continues to do what she loves best—writing, editing, and scripting communications that she hopes will contribute in some small way to making the world a more just and equitable place.

BARBOUR, Eric

Eric Barbour joined Syngenta in November of 2005. He currently heads a team of licensing managers supporting biotech and seeds organizations. In this capacity, he manages a range of activities, from licensing university technologies to making deals with major competitors. Eric's background is in IP licensing and valuation. He also has a strong research background in the area of insect control and herbicide tolerance traits and gene expression. Eric has worked in both the public and private sectors, with ten years management experience in licensing and intellectual property in the agricultural seed business.

Prior to joining Syngenta, Eric was employed by Pioneer Hi-Bred International Inc. in the Intellectual Property Licensing and Management group. Before Pioneer, he worked as a research scientist at Allelix Crop Technologies, working on transgenic traits in canola and other crops.

Eric is an active member of the Licensing Executive Society, the Biotechnology Industry Organization and, until recently, an affiliate member of the Association of University Technology Managers. He graduated with a BS.c. and M.Sc. degrees from the University of Guelph, Canada, and has an M.B.A. from Drake University.

BATES, Tony

Tony Bates joined the National Health Service (NHS) in England in 1992, when the NHS R&D initiative began. He was a member of the team in the Department of Health that introduced management of R&D into the NHS. Dr. Bates was the Intellectual Property Adviser to the National Health Service in England beginning in 1998. His main responsibility was to advise NHS Trusts and Primary Care Trusts on how NHS policy should be implemented in order to have the greatest benefit for the NHS and its patients. He helped develop policies and produced the Department of Health Framework and Guidance (for the management of intellectual property), which was published in 2002. He was instrumental in changing the law to allow NHS bodies to take shareholdings in spinout companies. He also created the IP management network of NHS Innovations Hubs.

He is a physicist by training and was in the Department of Physics at the University of Cardiff for about 18 years, where he was a researcher in solid-state physics. During that time, two of his inventions reached the marketplace. He moved from academia into IP management. After ten more years' experience in Cardiff University, he became Director of Planning and Marketing and was responsible for research and intellectual property.

Dr. Bates retired in 2004 but continues to work as a consultant in intellectual property.

BEACHY, Roger

Roger Beachy is president of the Donald Danforth Plant Science Center in St. Louis, Missouri. He previously held academic positions at Washington University, St. Louis, and the Scripps Research Institute, La Jolla, California. His research includes projects to reduce virus infection in plants via biotechnology, and in studies of the control of gene expression in plants. Beachy is a member of the U.S. National Academy of Sciences and a Fellow of the Academy of Microbiology; he has received several awards for his work, including the Wolf Prize in Agriculture. The Danforth Center has committed significant efforts to research in developing countries, including through private-public partnerships, and Beachy is involved in a variety of efforts with regard to rationalizing regulations that control commercialization of agricultural biotechnology.

Beachy is President of the International Association of Plant Biotechnology. He belongs to numerous institutional boards, including the PNAS Editorial Board, the NRC Governing Board, the Board on Agriculture and Natural Resources (National Research Council of the National Academy of Sciences), Malaysia's International Advisory Panel, and the Governing Board of Directors of the International Crops Research Institute for the Semi-Arid Tropics (ICRISAT), and the Burrill and Company Board of Advisors.

BENNETT, Alan B.

Alan Bennett currently serves as the Associate Vice Chancellor for Research at U.C. Davis. He is responsible for technology transfer, strengthening research-based alliances with industry, and supporting technology-based economic development in the Sacramento/Davis region. He is the founding Executive Director of the Public Intellectual Property Resource for Agriculture (PIPRA), an organization consisting of 37 universities in nine countries that is dedicated to the collective management of intellectual property and supports broad commercial innovation and humanitarian uses of technology in agriculture. From 2000 to 2004, Dr. Bennett served as the Executive Director of the University of California Systemwide Office of Technology Transfer and Research Administration, where he was responsible for IP management and research policy for the University of California system; this task involved managing a portfolio of more than 5,000 cases, 700 active licenses, and revenue in excess of US$350 million for the four-year period. He earned B.S. and Ph.D. degrees in Plant Biology at U.C. Davis and Cornell University, respectively. He joined the U.C. Davis faculty in 1983. His research in plant molecular genetics has focused on cell-wall disassembly and fruit development. Dr. Bennett has published over 130 research papers in leading scientific journals, holds several utility patents related to crop quality traits, and is a regular speaker at universities, international symposia, and private companies. He is a Fellow of the American Association for the Advancement of Science (AAAS) and of the California Council for Science and Technology (CCST).

BICH, T. Q.
Mr. Truong Quang Bich is Director of the Cuc Phuong National Park, Nho Quan, Ninh Binh, Vietnam. He is a forestry specialist with special expertise in forest management and biodiversity conservation. His research has focused on natural regeneration of forest following shifting cultivation, especially in some areas within Cuc Phuong National Park formerly settled by ethnic communities, before the establishment of this national park in 1962. More recently, he has been Project Director of community-based environmental education and visitor interpretation of the Cuc Phuong National Park, and, since 1998, he has served as a Co-Project Leader of an ICBG program with responsibility of implementing biotic survey and biodiversity conservation at the Cuc Phuong National Park.

BINH, Le Tran
A doctoral graduate from the University of Greifswald, Germany, Professor Le Tan Binh is a plant biotechnologist with expertise in plant tissue culture. His research for the past 25 years has focused on the use of molecular biology techniques for the improvement of crop plants, especially rice, and for improved utilization of Vietnam's biological resources. He has served as Principal Investigator of numerous projects, and since 1999 he has served as the Secretary General of the Vietnamese Association of Biotechnologists. He has also been the focal person in the Subcommittee for Biotechnology for the Vietnamese Biotechnology Committee of Science and Technology. He is the Director of Institute of Biotechnology, Vietnamese Academy of Science and Technology, Hanoi.

BJÖRK, Lars
Lars Björk studied biotechnology and chemistry at the Royal Institute of Technology in Stockholm, where he received a M.Sc. in 1965 for research on the production of secondary metabolites and a Licenciate degree in 1970 for research on secondary metabolite accumulation in plant tissue cultures. From 1984 to 1986, he was director of a Nordic project for introduction of this art in Sweden, Norway, Denmark, and Finland. During this time, he became an advisor to pharmaceutical and food companies, advising them on the usage of raw materials and the development of new products. He became Associate Professor of Pharmacognosy at Uppsala University in 1986. He received a Ph.D. in Pharmaceutical Sciences in 1990 with a thesis on techniques to improve secondary metabolism in plant tissues. He was responsible for the creation of a new unit within the Swedish University of Agriculture, the Phytochemical Centre, Balsgård. From 1991 to 2001, he was the research director of the Centre and investigated the selection of plant species and the domestication of wild species. From 1995 to 2005, he was a member of the board of the European Federation of Medicinal Plant Producers and contributed to the creation of new rules for Good Agricultural Practices for the production of raw materials to be used in the manufacture of pharmaceuticals; these rules were later adopted by the European Drug Administration (EMEA). In 2001, he became Senior Research Officer at Lund University, directing projects that determine the biological activity of natural products, a post he still holds. His institute is involved in bioprospecting in Bolivia, Morocco, Egypt, India, and other countries. He instituted Ethnobotany as an academic subject at Uppsala University in the spring of 2005. His current research concentrates on Laos but also includes other countries in Southeast Asia, as well as Morocco. He works on bioprospecting projects with Ph.D. students at both Lund and Uppsala. As a member of the Swedish Scientific Council for Biological Diversity (an advisory committee to the Swedish government on questions concerning the Convention on Biological Diversity), he has been especially involved in questions concerning access and benefit sharing.

BLAKENEY, Michael
Michael Blakeney is Herchel Smith Professor of Intellectual Property Law at Queen Mary, University of London and Director of the Queen Mary Intellectual Property Research Institute. He has held academic positions at a number of universities in Australia and the U.K. and worked in the Asia Pacific Bureau of the World Intellectual Property Organization. He is an arbitrator for the International Court of Arbitration. Professor Blakeney has acted as an intellectual property management advisor for the Asian Development Bank, the Consulting Group for International Agricultural Research, the European Commission (EC), the European Patent Office, the Food and Agricultural Organization, the World Intellectual Property Organization, and a number of universities and public research institutes.

He has directed E.C. projects to create intellectual property infrastructures in a number of new E.U. Member States and E.U. Applicant States. He has written and edited a number of books in the fields of intellectual property, media, and competition law. His most recent publications are: *Trade Related Aspects of Intellectual Property Rights. A Concise Guide to the TRIPS Agreement* (London: Sweet & Maxwell, 1996); *Intellectual Property Aspects of Ethnobiology* (Editor) (London: Sweet & Maxwell, 1999); *Border Control of Intellectual Property Rights* (Editor) (London: Sweet & Maxwell, 2001); *IP in Biodiversity and Agriculture: Regulating the Biosphere* (Editor with P. Drahos) (London: Sweet & Maxwell, 2001); *Enforcement Handbook* (Brussels: EC, 2003), and *International Encyclopaedia of Intellectual Property Treaties* (with A. Ilardi) (Oxford: Oxford University Press, 2004).

BLAYA ALGARRA, Alicia
Alicia Blaya earned the title Magister Lvcentinvs in Intellectual Property and Information Society Law from Universidad de Alicante, Spain. She earned an LLM in International Commercial Law, University of Westminster, London, U.K., in 2000 and a law degree from the Universidad de Alicante in 1996.

Since 1997, Blaya has been a registered member of the Professional Association of Lawyers of Alicante (Spain), where she was born. During her early years of working for a law firm, she provided legal assistance and trial work on civil, administrative and commercial cases.

She became Legal Advisor of the IPR-Helpdesk Project in 2002 and Senior Legal Advisor in 2003. Since 2005 she has served as Coordinator of the legal team and has been responsible for the research and technological development content and training actions of the project. She has worked inside the IP field, especially in the area of the IP-related aspects relevant for E.U.-funded research projects. She has contributed to several publications in various fields of IP and has taught courses on innovation in Spain, both to postgraduate students and Latin American professionals. Blaya has given many seminars in a wide range of IP-related issues relevant for research projects as well as in IP law and innovation, both in European countries and other countries, such as Cuba, Ukraine, Russia, and Egypt.

BOADI, Richard Y.

Richard Boadi is a national of Ghana and a member of the bars of the State of New York and the Republic of Ghana. He is currently legal counsel to the African Agricultural Technology Foundation (AATF). In this capacity, he advises the Board, management, and staff about current technology transfer policy and legal developments at national, regional, and international levels; he drafts, reviews, and negotiates agreements to which AATF is a party; he creates and fosters networks with licensees; and he handles other in-house legal needs of AATF. Before joining AATF, Mr. Boadi worked in the following capacities: as a senior attorney with the New York State Office for Technology; a contracts and commercial lawyer with the New York City Human Resources Administration; a Teaching Assistant for the Faculty of Law at the University of Ghana; a Junior Barrister with Reindorf Chambers in Accra, Ghana; and an Assistant Legal Officer with the Ghana Copyright Office. He is a graduate of Cornell University (LL.M.), the Ghana School of Law (B.L.), and the University of Ghana (LL.B.).

BOBROWICZ, Donna

Donna Bobrowicz has a B.S. in Medical Laboratory Sciences, an M.B.A., and a J.D. She is a U.S. patent attorney with over 20 years of experience, specializing in chemical and biochemical technologies in both human health and agriculture. She has been the in-house counsel in licensing matters for the technology acquisition group of the seed producer Pioneer Hi-Bred Inc. She has also been counsel for licensing and patents at the human diagnostics divisions at Abbott Laboratories and Akzo Pharma and for the cellulose and plastic casings manufacturer Viskase Companies, Inc. She has been affiliated with SWIFTT, the Strategic World Initiative for Technology Transfer at Cornell University, and has been part of an IP audit team at agricultural institutes in Kenya and Colombia. Currently, she is setting up technology transfer processes at the Stritch School of Medicine, Loyola University Chicago. She has her own IP practice in the metropolitan Chicago area.

BOETTIGER, Sara

Sara Boettiger is an agricultural economist with a background in intellectual property (IP) law. She works as Director of Strategic Planning and Development at The Public Intellectual Property Resource for Agriculture (PIPRA) and is a consultant for the Bill & Melinda Gates Foundation. She publishes in the field of IP law and policy and is a member of the Board of Directors for the Institute of Forest Biotechnology. Her professional interests are the design and implementation of practical services that support innovation and improve livelihoods in developing countries. Her research interests are the legal and economic ramifications of IP rights and developing countries, collaborative innovation systems, open source in copyright and patents, university technology transfer systems, and the strategic use of patents in developed countries.

Dr. Boettiger holds a B.A. from the University of Arizona, an M.S. from the University of California, Berkeley, and a Ph.D. in Agricultural and Resource Economics from the University of California, Berkeley.

BOKANGA, Mpoko

Mpoko Bokanga is a food scientist with a Master's degree from the Massachusetts Institute of Technology and a doctorate from Cornell University. He has been involved in agricultural research and development in Africa for the past 17 years. Before becoming the first Executive Director of AATF, Dr. Bokanga worked as an Industrial Development Officer of Agro-industries with the United Nations Industrial Development Organization (UNIDO) in Abuja, Nigeria. From 1989 to 2002, he was a Research Scientist with the International Institute of Tropical Agriculture (IITA). He has also been a Visiting Professor of Food Science at Alabama A & M University and a Research Associate for Westreco Inc., a Nestlé Research Company. At Westreco, Dr. Bokanga developed processes based on immobilized microbial and enzyme systems; and at IITA, he developed technologies for processing cassava and yams into new products that were subsequently introduced into more than a dozen African countries. He has co-authored or edited three books and published several papers on the biochemistry and health implications of cyanogenesis in cassava and on the processing of root and tuber crops. He is the coordinator of the Working Group on Cassava Safety (WOCAS), a subcommittee of the International Society for Tropical Root Crops (ISTRC), whose main function is to monitor the progress of and encourage research on cyanogenesis in cassava and its implications for food safety. Dr. Bokanga is the current chair of ISTRC-AB, the African branch of the ISTRC, and holds a visiting professorship at the University of Greenwich in England (2005-2008).

BORLAUG, Norman E.

In 1970, Norman Borlaug won the Nobel Peace Prize for his lifelong work to feed a hungry world. His work, more than that of any other person, is credited with saving lives.

In 1944, Dr. Borlaug joined the Rockefeller Foundation's pioneering technical-assistance program in Mexico, at which he was a research scientist in charge of wheat improvement. For the next two decades, he worked to solve a series of wheat production problems in Mexico and to train a generation of young scientists.

With the establishment of the International Maize and Wheat Improvement Center (CIMMYT) in Mexico in 1966, Borlaug assumed leadership of the wheat program; he continues to serve as a consultant for it. The high-yielding, disease-resistant wheat cultivars he developed, along with improved management practices, transformed agricultural production in Mexico during the 1950s and in Asia and Latin America in the 1960s and 1970s. This transformation has come to be known as the Green Revolution.

In 1984, Dr. Borlaug joined Texas A&M University and was named Distinguished Professor of International Agriculture. Since 1986, he has also served as president of the Sasakawa Africa Association and leader of the Sasakawa-Global 2000 agricultural program in Sub-Saharan Africa, in partnership with former U.S. President Jimmy Carter and Yohei Sasakawa.

Borlaug has been awarded 58 honorary doctorate degrees, and is a member or fellow of the academies of science in 12 nations. The U.S. National Academies of Science awarded him the National Service Medal in 2002 and in 2004 President Bush bestowed upon Borlaug the U.S. National Medal of Science. He was the driving force behind the establishment of the World Food Prize in 1985 and serves as Chairman of its Council of Advisors.

BREMER, Howard

Howard Bremer holds degrees in Chemical Engineering and Law from the University of Wisconsin-Madison. He has been admitted to membership in the bars of the U.S. Supreme Court, the Court of Appeals for the Federal Circuit, the District Court for the Southern District of Ohio, and the State of Wisconsin, and has practiced before the Patent and Trademark Office. He has twice been Chairman of the Patent Law Section of the State Bar of Wisconsin, President and Trustee of the Association of University Technology Managers (AUTM), and President of the Wisconsin Intellectual Property Law Association. He has been active in the American Bar Association Section on Intellectual Property Law and the American Intellectual Property Law Association. He recently was awarded the Jefferson Medal by the New Jersey Intellectual Property Law Association. He has engaged in legislative activities involving questions of intellectual property, and served on the National Advisory Commission on Patent Law Reform. He was employed by the Procter and Gamble Company for 12 years and was Patent Counsel for the Wisconsin Alumni Research Foundation for 28 years.

BREWSTER, Amanda L.

Amanda Brewster serves as a Policy Officer for the Wellcome Trust in London. Previously, she worked as a Program Associate of the Science and Intellectual Property in the Public Interest project of the American Association for the Advancement of Science (AAAS) in Washington, D.C. She has also worked on science policy for the National Council for Science and the Environment. Although she has conducted field research on the vector ecology of Lyme disease and on insect pollination, in recent years she has used her training as a biologist to study the intersections of scientific research, innovation, human health, and the environment. She holds a master's degree in international health policy from the London School of Economics and Political Science and a bachelor's degree in molecular, cellular, and developmental biology from Yale University.

BROOKE, Steve

Steve Brooke, an advisor for commercialization and corporate partnerships in PATH's Technology Solutions Strategic Program, plays a lead role in public-private product development collaboration, product commercialization strategy, and intellectual property management. Mr. Brooke conceptualizes, develops, negotiates, and implements complex collaborations and strategies, and negotiates co-development, licensing issues, and agreements. He also provides guidance and advice to staff working in these areas and manages projects that have a high degree of commercialization and/or private partner focus. Prior to receiving a master of business administration degree in marketing and finance from Northwestern University, he served as marketing manager for a medical products company. Mr. Brooke has lived in both Europe and Asia.

BROWN, Alfred (Buz)

Alfred Brown has served as the President of BCM Technologies (BCMT) since 2003. He is its chief visionary and was responsible for transforming BCM Technologies into BCM Ventures. He has over 30 years of experience in biotechnology commercialization and venture creation, including large pharmaceutical, biotech, academic and venture capital roles, with particular expertise in the areas of cancer, immunology, regenerative medicine, molecular diagnostics, and predictive medicine. Before joining BCMT, Dr. Brown was Director of the Office of Cooperative Research at the Yale School of Medicine, where he co-developed the "Yale Model" of academic venture creation with Drs. Gardiner, Soderstrom, and Swartley. The Yale Model is now recognized as one of the leading academic technology commercialization and venture creation programs. Dr. Brown has helped to create many companies, including Achillion Pharmaceuticals (in registration), Applied Spine Technologies, HistoRx, Kemia, RibX Pharmaceuticals, and VaxInnate. Before working at Yale, Dr. Brown served as founder and CEO of Penn Technology Group, Knowledge Express Data System and Ontyx, Inc. (now Apelon, Inc.). He started his professional career at SmithKline & French Labs in immunology and cancer drug research and later assumed responsibility for strategic planning and biotechnology business development. Dr. Brown has a B.A. in biology from Colby College, a Ph.D. in pharmacology and toxicology from the University of Rochester School of Medicine; in addition, he has completed a post-doctoral fellowship in the Pharmacology Department at the Yale School of Medicine. He serves on the boards of Oncovance Technologies, Molecular LogiX, Kardia Therapeutics, Progression Therapeutics, and EnVivo Pharmaceuticals (observer). Dr. Brown will serve as the managing director of BCMV and focus his efforts on investments in the areas of therapeutics, diagnostics, and devices.

BUBELA, Tania
Tania Bubela (B.Sc. Ph.D. LL.B.) is an Assistant Professor in the Department of Marketing, Business Economics and Law in the School of Business at the University of Alberta. She is also a Research Associate at the Health Law Institute at the University of Alberta and a Member of the Centre for Intellectual Property Policy in the Faculty of Law at McGill University. Her doctoral research was in the biological sciences; she taught biology and genetics as a faculty member at the University of Toronto at Mississauga. After gaining a law degree in 2003, Dr. Bubela clerked for The Honourable Louise Arbour at the Supreme Court of Canada. Dr. Bubela's research focuses on intellectual property systems in biotechnology, as well as questions of health law, ethics, and policy as they relate to emerging technologies such as functional genomics and stem cell research.

BURDON, Jeremy
Dr. Burdon is currently Director of IP Assets at Arizona Technology Enterprises, LLC (AzTE) in Tempe, Arizona, with responsibility for the Health Sciences portfolio. Prior to moving to AzTE, Dr. Burdon was with Medtronic, Inc. There he was responsible for managing the Intellectual Property portfolio and patent liaison activities for implantable medical device technologies, in both research and advanced development environments.

Dr. Burdon spent more than nine years at Motorola, Inc., initially with its Component Products Division in New Mexico, researching polymer thin-film technologies and oxide semiconductor thin-films for RF/Microwave applications, where he moved several technologies into the advanced development stage. Dr. Burdon then worked at Motorola Corporate Research in Tempe, Arizona, developing material technologies for micro-devices, and on advanced development of micro fluidic devices for analytical and on-chip bio-analysis.

Dr. Burdon holds a B.Sc. in chemistry and a Ph.D. in Polymer Science from the University Of Sussex, U.K., where his research focused on oxidative degradation of organic/polymer materials and the polymerization behavior of bisphenol-A epoxy systems for graphite-based composites using chemiluminescence and ion-recom luminescence techniques. Dr. Burdon holds 14 issued patents in the areas of materials, micro-devices, microfluidic systems and implantable medical devices.

CAHOON, Richard S.
Richard Cahoon received his undergraduate degrees in biology and political science from the University of Utah. His Master's degree is from Montana State University in the field of bioprocess engineering. He founded and was president of a biotechnology device company and was later a managing partner of a bioprocess engineering consulting firm.

In 1990, Dr. Cahoon joined the Cornell Center for Technology, Enterprise & Commercialization as Assistant Director for Technology Marketing. Previously, he was Associate Director of the Center for Biofilm Engineering, an NSF Engineering Research Center at Montana State University. In 1992, he was appointed Cornell's Associate Director for Patents and Technology Marketing; a year later, he was promoted to Vice President of the Cornell Research Foundation (CRF), Cornell's intellectual property subsidiary. In January 2003, Richard became Senior Vice President of CRF; he has been serving as Acting Executive Director of the Cornell Center for Technology, Enterprise, and Commercialization since 2002.

Dr. Cahoon has more than 25 years of experience in various aspects of technology commercialization, including R&D management, inventing, project engineering, product development, marketing and sales, process engineering, entrepreneurship, collaboration management, intellectual property, and licensing. He also holds a patent for a bioprocess system. His Ph.D. is from Cornell in Natural Resource Policy with a dissertation that focused on the relationship between intellectual property and biological resource conservation law and policy.

CAMPBELL, Alison F.
Alison Campbell is Director of KCL Enterprises Ltd., the commercialization and research support company of King's College, London. KCLE manages all aspects of the College's external partnering activities, from business development to IP management, licensing, start-up company formation, and the administration and negotiation of research grants and contracts. KCLE also supports training in enterprise within the College. Dr. Campbell has worked in technology transfer and business development for 15 years. Before joining KCLE, she was Acting CEO of MRC Technology. She has experience in the biotechnology industry and worked for a number of years at Celltech Ltd. A biochemist, Dr. Campbell is a graduate of University College London and earned a Ph.D. in chemical biology from Imperial College. She currently serves as a nonexecutive director on the boards of a number of spinout companies and two London enterprise initiatives (Simfonec and the London Technology Network). She is a nonexecutive director of the university seed fund, Kinetique. She is a member of the UNICO committee (the U.K. university commercialization organization), and is Chair of Praxis (the U.K. Technology Transfer Training Programme).

CARRIERE, Michael
Michael D. Carriere, manager of the strawberry licensing program at the University of California, Davis, received his B.S. degree in Agricultural Science and Management from the University of California, Davis in 1988 and his Ph.D. degree in Plant Biology also from the University of California, Davis in 2000. His Ph.D. work focused on functional genomics and the physiology of submergence tolerance in rice. Prior to graduate school, Dr. Carriere spent four years in a private sector rice cultivar improvement program. In his current role he is charged with building the global licensing presence of UC Davis strawberry cultivars within the public sector framework of a land-grant institution. Dr. Carriere has given invited presentations at conferences and universities on the topic of university plant licensing. He lives in Davis, California with his wife and two children.

CARVALHO, Sergio M. Paulino de
Sergio M. Paulino de Carvalho is General Coordinator of Institutional Partnership and Regional Diffusion of the Brazilian National Institute of Industrial Property.

He is also Researcher at the Agricultural Research Enterprise of the State of Rio de Janeiro (PESAGRO-RIO) and Associate Researcher with the Study Group on the Organization of Research and Innovation (GEOPI), State University of Campinas (UNICAMP), Campinas, São Paulo, Brazil.

He is an economist from the Federal Fluminense University (UFF), with a Master's and Ph.D. Degree in Scientific and Technological Policy at the State University of Campinas (UNICAMP). He is author of several publications and articles on intellectual property policies and the organization of research.

CHAMAS, Claudia Inês

Claudia Chamas is a researcher at the Oswaldo Cruz Institute (FIOCRUZ, Ministry of Health, Ministry of Health). She works on intellectual property issues and is the author of several journal articles on the topic that focus on biotechnology and pharmaceuticals. She received a bachelor's degree in Chemical Engineering and a MSc and DSc in Production Engineering at the Federal University of Rio de Janeiro. In the years 2000 and 2002, she was visiting researcher at the Max-Planck-Institut für Geistiges Eigentum, Wettbewerbs- und Steuerrecht in Munich. She has organised seminars and coordinated research projects that were funded by Brazilian funding agencies (CNPq, Faperj, etc). She is a regular guest speaker at various universities and is a member of the Brazilian Association of Intellectual Property, the Brazilian Ministry of Health's Intellectual Property Committee, and the Brazilian Association of Fine Chemicals, Biotechnology, and Specialties Industries' Industrial Property Committee.

CHAPMAN, Audrey R.

Audrey Chapman holds the Healey Endowed Chair in Medical Humanities, Law, and Ethics at the University of Connecticut Health Center. She formerly served as the Director of the Science and Human Rights Program at the American Association for the Advancement of Science (AAAS) and as the Co-Director of the AAAS initiative on Science and Intellectual Property in the Public Interest. She is the author, coauthor, or editor of sixteen books and numerous articles and reports dealing with ethical, human rights, and intellectual property issues related to health, pharmaceuticals, and genetic developments. She received a Ph.D. in public law and government from Columbia University and graduate degrees in theology and ethics from New York Theological Seminary and Union Theological Seminary. She has worked closely with the United Nations Committee on Economic, Social and Cultural Rights and the UN Special Rapporteur on the Right to Health. She is currently a member of the University of Connecticut Embryonic Stem Cell Oversight Committee, the John Dempsey Hospital Ethics Committee, and the Expert Genomics Advisory Panel of the Connecticut Department of Public Health.

She has worked on a wide range of ethical, human rights, and intellectual property issues related to health and pharmaceuticals.

CHAPMAN, Junko

Junko has been a Research Associate at MIHR (Centre for the Management of Intellectual Property in Health Research and Development) since April 2005. Before MIHR, she spent ten years at RIKEN (The Institute for Physical and Chemical Research), a semi-governmental research institute in Japan. At RIKEN, Junko's responsibilities included conclusion of MTAs (Material Transfer Agreements) and collaborative research agreements and management of RIKEN's intellectual property portfolio. Junko was also heavily involved in establishing a system for disseminating RIKEN's key inventions internationally. These inventions included RIKEN's mouse cDNA clones, which have since become globally recognized and widely used.

Junko graduated from SPRU (Science and Technology Policy Research), University of Sussex, in 2001, where she received her M.Sc. in Science and Technology Policy, during which time she focused, in part, on the impact of harmonization of patent systems. Junko is also a graduate of GRIPS (National Graduate Institute for Policy Studies, Japan), where she received a Master in International Development Studies in 2005.

CHEN, Zhang Liang

Zhang Liang Chen was born on February 3, 1961, in Fujian, China. He received his Ph.D. in 1987 from Washington University for his research in the Division of Biology and Biomedical Sciences in the field of plant molecular biology and his work in early transgenic plant research. He then returned to China as an associate professor. Two years later, he was a full professor at Beijing University. He has continued his research in transgenic plants and biosafety. He served as director of National Key Laboratory of Protein Engineering and Plant Genetic Engineering. In 1995, he became vice-president of research at Peking University. In 2002, he became the president of China Agricultural University. He and his research group have published over 190 international papers and seven book, and hold over eight patents.

Dr. Chen is also Chair of the Plant Biotech Committee of UNESCO, Consultant for the International Society for Plant Molecular Biology (ISPMB), and member of the Sino-Euro Administration Committee for Biotechnology Cooperation. He also serves as a member and Vice-Chairman of the Council of Scientific Advisers to the International Center for Genetic Engineering and Biotechnology (ICGEB) in Italy and India.

CHI-HAM, Cecilia

Cecilia Chi-Ham, a native of Honduras, earned a B.S. degree in Chemistry and Environmental Sciences at the University of the Ozarks and a Ph.D. in Chemistry and Biochemistry at the University of Southern Mississippi. In 2004, upon completing her post-doctoral work at Michigan State University in the field of plant biology, Dr. Chi-Ham joined the Public Intellectual Property Resource for Agriculture (PIPRA). Dr. Chi-Ham is a plant biologist interested in facilitating agricultural innovations, particularly in developing countries, and leads PIPRA's Biotechnology Resources Program. The Biotechnology Resources Program's activities include the following: developing research tools with maximum freedom-to-operate that can support a wide array of agricultural applications for humanitarian and commercial purposes; facilitating technology transfer; building new partnerships and research collaborations; and providing legal information on biotechnology tools. The program's multi-disciplinary activities straddle the delicate junction between the scientific, legal, business development, and regulatory affairs that are an integral part of research and development of new agricultural innovations in developed and developing countries.

CHRISTOFFERSEN, Leif
Leif Christoffersen is a Founder at the E.O. Wilson Biodiversity Foundation, a nonprofit environmental organization focused on linking businesses to sustainable resources and conservation efforts, as well as educating them about the importance of biodiversity. From 2000 to 2005, Mr. Christoffersen served as the Biodiversity Manager at the Diversa Corporation and managed biodiversity collaborations and bioprospecting efforts in Alaska, Antarctica, Australia, Bermuda, Costa Rica, Ghana, Hawaii, Indonesia, Kenya, Puerto Rico, Russia, and South Africa; he also managed Diversa's biodiversity collaboration with the Center for Reproduction of Endangered Species (CRES). From 1995 to 2000, he served as the Vice President for the World Foundation for Environment and Development (WFED), working with Yellowstone National Park and the National Park Service; he also worked with the National Institute for Biodiversity in Costa Rica and the Center for Ecological Research and BioResources Development in Russia, developing bioprospecting programs and facilitating negotiations for benefit-sharing arrangements with biotech companies. While at WFED, Mr. Christoffersen also served as a Climate Change consultant to the United Nations Environment Programme. Previously, Mr. Christoffersen had worked for CARE International in marketing, public relations, and environmental monitoring and evaluation in Costa Rica, Kenya, and Norway. Mr. Christoffersen received his B.A. in Economics from Hobart College in Geneva, New York, where he won the Elizabeth and Ruth Young Peace Prize. He is scheduled to graduate with an M.B.A. from the Rady School of Management at the University of California, San Diego, in August 2007.

CLIFT, Charles
Charles Clift has had a great deal of experience in the U.K. Department for International Development (DFID), where he works principally as an economist on all aspects of DFID's work. He began his career as an agricultural economist, advising DFID on its agricultural research priorities. He has lived and worked in Africa, the Caribbean, and India. He has also been responsible for the management of DFID's economic and social research, and the coordination of all of DFID's research programmes, including those concerned with health and agriculture. From 2001 to 2002, he acted as Head of the Secretariat of the U.K. Commission on Intellectual Property Rights (www.iprcommission.org). From 2004 to 2006, he was employed in a similar capacity by the WHO Commission on Intellectual Property Rights, Innovation and Public Health (www.who.int/intellectualproperty).

COLAIANNI, Alessandra
Alessandra Colaianni is a member of the undergraduate class of 2007 at Duke University, where she is double-majoring in Biology and Philosophy. While at Duke, she has worked as a research assistant to Dr. Robert Cook-Deegan, director of the Center for Genome Ethics, Law, and Policy, which is a part of the Institute for Genome Sciences and Policy at Duke. Her research there has focused mainly on compiling histories of the Cohen-Boyer patents, which she has written about with Dr. Maryann Feldman, and the Axel patents. After she graduates in May, she will continue to work with Dr. Cook-Deegan for a year while applying to law school.

CONWAY, Gordon
Gordon Conway took up his appointment as Chief Scientific Adviser for the Department of International Development (DFID) in January 2005. He was educated at the University of Wales, Bangor, the University of Cambridge, the University of Trinidad, and the University of California, Davis. His discipline is agricultural ecology. In the early 1960's, he worked in Sabah, North Borneo, and became one of the pioneers of sustainable agriculture. From 1970 to 1986, he was Professor of Environmental Technology at the Imperial College of Science and Technology in London. During this period, he lived and worked in many countries in Asia and the Middle East. He then directed the sustainable agriculture program of the International Institute for Environment and Development in London. From 1988 to 1992, he was Representative of the Ford Foundation in New Delhi; from 1992 to 1998, he was Vice-Chancellor of the University of Sussex and Chair of the Institute for Development Studies. He was President of The Rockefeller Foundation from 1998 to 2004. He has honorary degrees from the Universities of Sussex, Brighton, Wales, and the West Indies; he is an honorary fellow of the Institute of Biology, and a fellow of Imperial College, the American Academy of Arts and Sciences, and the Royal Society. He authored *Unwelcome Harvest: Agriculture and Pollution* (London: Earthscan), *The Doubly Green Revolution: Food for All in the 21st Century* (London: Penguin; Ithaca, New York: Cornell University Press), and *Islamophobia: A Challenge for Us All* (London: The Runnymede Trust).

COOK, Tim
After being awarded a doctorate in cryogenic engineering at Oxford, Tim Cook joined the Oxford Instruments Group (a spinout company from the University of Oxford) in 1975. During his 12 years with them, the Group's turnover grew from £1million to £100 million. In 1983, Dr. Cook was appointed Managing Director of the Group's subsidiary, Oxford Analytical Instruments. After two more appointments as Managing Director, Dr. Cook became a private investor in 1990 and the founding Managing Director of Oxford Semiconductor and Oxford Asymmetry (a spinout from the University of Oxford, floated in 1998 and recently sold for over £300 million).

In 1997, Dr. Cook was appointed Managing Director of Isis Innovation, the technology transfer company of the University of Oxford. Since then, Isis has recruited 37 staff members and negotiated over 100 option and licence agreements. In the last nine years, Isis has established 54 new spinout companies from the University, which have collectively raised over £300 million in investment capital. In January 2006, Dr. Cook became Visiting Professor in Science Entrepreneurship;

he became Deputy Chairman of Isis Innovation in April of that year.

Is addition to his work in Oxford, Dr. Cook is also working on technology transfer with other universities. He has given invited lectures in many countries and visited University Technology Transfer Organizations in the U.S., Australia, Europe, and Japan.

COOK, Trevor
Trevor Cook joined Bird & Bird in 1974 with a degree in chemistry from Southampton University. He was admitted as a Solicitor in 1977 and then joined the Intellectual Property Department of Bird & Bird, where since 1981 he has been a partner. He specializes in intellectual property and regulatory law. He is Treasurer of the U.K. Group of The International Association for the Protection of Industrial Property (AIPPI), Secretary to the British Copyright Council Standing Committee on Copyright and Technology, and a member of the Council of the Intellectual Property Institute. In recent years, he has acted in many of the leading patent infringement cases that have come before the English courts, most of which have concerned pharmaceuticals and biotechnology, and also in many of the leading cases regarding the protection of regulatory data that have come before the European Court of Justice.

CORREA, Carlos M.
Carlos M. Correa is Director of the Center for Interdisciplinary Studies of Industrial Property Law and Economics at the University of Buenos Aires, as well as Director of the Post-graduate Courses on Intellectual Property at the same University. He has been a Visiting Professor and taught post-graduate courses at several universities. He has also been a consultant in different areas of law and economics (including investment, science and technology, and intellectual property) to UNCTAD, UNIDO, UNDP, WHO, FAO, the Inter-American Development Bank, INTAL, the World Bank, SELA, ECLA, UNDP, and other regional and international organizations, as well as several governments. He was a member of the U.K. International Commission on Intellectual Property, which was established in 2001, and of the Commission on Intellectual Property, Innovation, and Public Health, which was established by the World Health Assembly in 2004. He is the author of several books and numerous articles on law and economics, particularly on the topics of investment, technology, and intellectual property.

COSTANZA, Charles
Charles (Chuck) Costanza is a consultant to the biotechnology industry. Since 2002, he has consulted for Diversa Corporation, a San Diego-based corporation that has pioneered the development of high-performance specialty enzymes. Mr. Costanza advises the company on business development in emerging markets, and specializes in biodiversity access agreements. He has presented papers around the world and to many different audiences (business, NGO, academic, and government) on the subject of biodiversity access agreements. Prior to working with Diversa, Chuck worked in the financial and technology industries, managing projects, consulting, and developing business across the U.S., Europe, the former Soviet Union, and Asia. He worked for ICF Consulting, Inc. as a project manager, and developed and oversaw projects for clients such as The World Bank, British Petroleum, and the U.S. Environmental Protection Agency. Prior to working for ICF, Mr. Costanza worked as a project manager for the International Finance Corporation; in this capacity, he led a multinational team of 70 economists and lawyers who helped to privatize collective farms in Ukraine in 1996, the first privatizations of their kind. Fluent in Russian and German, Mr. Costanza received a Bachelor's Degree from the College of the Holy Cross and a Master's Degree from Harvard University.

CROW, Michael
Michael Crow—educator, knowledge enterprise architect, and science and technology policy scholar—became president of Arizona State University on July 1, 2002. He is currently helping to transform ASU into one of the nation's leading public metropolitan research universities. Under his direction, the university's faculty pursue teaching, research, and creative work that is focused on the major challenges and questions of our time and, especially, the challenges related to Arizona's environment and economy. He has committed the university to global engagement and to setting a new standard for public service. During his tenure, ASU has marked a number of important milestones: the establishment of major global interdisciplinary research initiatives such as the Biodesign Institute, the Global Institute for Sustainability, and the Flexible Display Center; an unprecedented expansion of research infrastructure that added more than one million square feet of new research space; a dramatic increase in federal research awards; and the four largest gifts in the history of the university. Prior to joining ASU, Dr. Crow was Professor of Public and International Affairs and Executive Vice Provost of Columbia University. He is the author of books and articles on the analysis of knowledge organizations, knowledge transfer, science and technology policy, and the practice and theory of public policy.

CROWELL, W. Mark
W. Mark Crowell is Associate Vice Chancellor for Economic Development and Technology Transfer at the University of North Carolina at Chapel Hill (UNC). Prior to joining UNC, he held similar positions at North Carolina State University and at Duke University. He has extensive experience in technology transfer, new company development, seed capital formation, and research park development and marketing. Mr. Crowell also leads UNC's efforts to connect its research enterprises with economic and business development opportunities in the region, state, nation, and world. Mr. Crowell, as a representative of UNC, sits on the boards of major statewide and regional economic development and entrepreneurial support agencies in North Carolina.

Mr. Crowell is Past President (2005) of the Association of University Technology Managers (AUTM). AUTM is the pre-eminent international organization in the field of academic technology transfer. AUTM has a membership of nearly 3,500 professionals, almost 12% of whom are from outside of North America. AUTM's mission is to promote and enhance the global technology transfer profession through education, training, networking, and advocacy, and through the identification and dissemination of best practices in academic technology transfer.

Mr. Crowell's speaking and consulting experience in the past two years includes keynote addresses at international conferences in at least 15 countries outside the U.S., as well as advisory roles with many major national and international organizations, including the AAAS and the National Academies of Sciences. He has extensive management experience in organizations and initiatives related to technology transfer and innovation-based economic development.

CRUZ, Richard L.

Richard L. Cruz focuses his practice on securing, licensing, and enforcing intellectual property rights, primarily in the electrical, electro-mechanical, and electronic arts. His practice includes domestic and foreign patent prosecution, patent validity and infringement analysis, state-of-the-art and patentability opinions, licensing, due diligence, copyrights, and trademarks, and the litigation circuits, software, and electronics fields.

Mr. Cruz earned his law degree, with honors, from Widener University School of Law. While at Widener, Mr. Cruz was a member of the Moot Court and Trial Advocacy Honor Societies. Mr. Cruz also earned a Certificate in Trial Advocacy, with honors, while in law school. He is a member of the Pennsylvania bar and is admitted to practice before the U.S. Patent and Trademark Office and the Eastern District of Pennsylvania. Prior to attending law school, Mr. Cruz earned a degree in Engineering from the University of Pittsburgh.

CUNNINGHAM, Sean

Sean Cunningham is a partner at the firm of DLA Piper U.S. L.L.P. in San Diego, California. Mr. Cunningham is a trial lawyer who specializes in patent litigation, with an emphasis on litigation in the International Trade Commission, multi-jurisdictional litigation, and litigation involving conduct in standard-setting organizations. He received his law degree from the University of Kansas School of Law, where he was Order of the Coif and Editor of the Kansas Law Review. Sean has spent his entire legal career with DLA Piper (formerly known as Gray Cary Ware & Freidenrich). He is a member of the Federal Circuit Bar Association, the American Intellectual Property Law Association, and the Intellectual Property Section of the California Bar Association. Mr. Cunningham frequently works with companies such as Hewlett-Packard, Agilent Technologies, and Qualcomm Incorporated. His full biography can be found at www.dlapiper.com/sean_cunningham/.

DI GIORGIO, Rosana Ceron

Rosana Di Giorgio currently serves as Intellectual Property & Partnership Development Director of Inova Unicamp, the Technology Licensing Office of the State University of Campinas, Brazil (Unicamp). She is responsible for business development between the university and the marketplace. In the three years since she joined Unicamp, she has signed 150 technology transfer agreements involving IP development and licensing between the university and the commercial sector, a new record for both Brazil and Latin America; as a result, Unicamp is the biggest licensor in the country. Some of these technology transfer agreements have already resulted in commercialized products, such as BiPhor and Aglicon-Soy. She is also responsible for defining policies, practices, legal affairs, team building, and management of the university's IP portfolio, which, since 1999, has been considered the country's largest.

In the past three years, she has been invited to speak at about 40 national and international conferences, symposia, and workshops on the subject of transferring academic technology to the commercial sector. These talks have attracted investors, industry representatives, and academics. Some of her publications include: articles for business magazines, such as *Líderes Empresariais* and *Les Novelles*, and business newspapers, such as *Gazeta Mercantil*; books like *Propriedade Intelectual: O Caminho para o Desenvolvimento*, chapter 7, a book sponsored and launched by Microsoft; television interviews for such channels as Globo News, EPTV, and TV Bandeirantes; radio interviews for such channels as CBN and Eldorado; and electronic reports, such as a WIPO report on best practices.

Her previous experience includes eight years managing people, projects, accounts, and business development in Brazil and abroad; creating innovative solutions for several market sectors (financial, energy, IT, and pharmaceutical); strategic planning; market research; and business plan and business viability analysis. She previously served as Executive Director of facTI, a private foundation concerned with IT; under her guidance, the foundation became financially viable in one year. She was also Corporate Business Development Manager and Semiconductor Division Manager at CPqD, the biggest research and development center in Brazil for telecommunications and IT.

DI SANTE, Anne C.

Anne C. Di Sante (University of Michigan M.B.A., Marketing; University of Michigan M.S., Microbiology/Immunology; University of Michigan B.S., Medical Technology) is the Director of the Technology Transfer Office at Wayne State University. She is responsible for the daily operations related to traditional invention management activities undertaken by the TTO staff, including resource allocation, commercialization strategy development and implementation, and agreement negotiation and maintenance. Prior to joining WSU in 1998, she held various positions within the Technology Management Office (now the Office of Technology Transfer) at the University of Michigan, ranging from student intern to Acting Director. She is experienced in managing inventions from all scientific disciplines; however, she specializes in medical, pharmaceutical, and biotechnology inventions. She has had the pleasure of managing several key biomedical inventions, including the cystic fibrosis gene, several gene therapy technologies, anti-viral compounds, anti-cancer compounds, methods for antibiotic development, a nasal vaccine for the prevention of influenza, and several immunotherapy technologies. Currently, she is a member of the Intellectual Property Commercialization Committee, the governing body for the Michigan Universities Commercialization Initiative, a $9.0 million project funded by the State of Michigan; she is also a member of the MUCI Finance Committee. Ms. Di Sante is also an active member of the Association of University Technology Managers, currently serving as editor of the AUTM Newsletter.

She has also served on AUTM's Board of Trustees as Vice President of the Central Region and as an Executive Committee participant. She participated in the development of the BioMed Expo (now the MichBio Expo), serving on the planning committee for four years. She is a member of the Licensing Executives Society.

DODDS, John

John Dodds was born in the U.K. but has lived and worked in the Middle East, Latin America, and the United States. He became a U.S. citizen in the 1980s. Originally trained as a biochemist, he earned both a Bachelor's and Doctoral degree at the University of London. Dr. Dodds also earned a law degree in the U.S. and founded the law firm Dodds and Associates in 1999. His early research career focused on plant biochemistry and plant tissue culture. In the early 1980s, Dr. Dodds co-authored a standard and well-used textbook on techniques for culturing plant tissue. His work focused on plant genetic conservation *in vitro* and plant transformation systems. He then moved into the area of agricultural development and has worked in, and traveled to, more than 100 countries. Dr. Dodds has also been a teacher in many capacities, from a professor in both the U.S. and the U.K. to the coordinator of regular IP training programs that have been offered through his law firm in several countries. In his few hours of spare time, Dr. Dodds also writes novels and assists his wife in restoring historic buildings.

DUNN, Martha

Dr. Dunn joined Ciba-Geigy in 1994 as a Research Scientist, working in the areas of protein structure/function and molecular recombination for the discovery of novel traits, in support of the agribusiness sector of Ciba-Geigy.

Martha joined the Licensing Department in 1998, where she helped negotiate agreements to support R&D and commercialization activities for Syngenta Biotechnology and its affiliated companies.

Prior to this, Martha worked at Genetics Institute in Cambridge, Massachusetts, as a research scientist in the Immunology Department and was responsible for assay development for GI's ongoing projects.

Dr. Dunn is a registered as a Patent Agent with the U.S. Patent and Trademark Office, an active member of the Licensing Executive Society, the Biotechnology Industry Organization and an affiliate member of the Association of University Technology Managers.

Martha graduated with a B.S. from Boston College and has received her Ph.D. in Biochemistry and Biophysics from the University of North Carolina at Chapel Hill.

EDWARDS, Mark G.

Mark G. Edwards is the Managing Director of Recombinant Capital, Inc. (Recap), a consulting firm based in Walnut Creek, California. More than 500 biotechnology, pharmaceutical, and service companies subscribe to Recap's databases (Recap.com & rDNA.com) or retain Recap to advise them on biotech alliances and valuations.

Mr. Edwards has been invited to speak to many trade and industry groups about structuring alliances and other business relationships that are related to the development and commercialization of new technologies, compounds, or products. These groups include the Institute of Medicine, the Licensing Executives Society, the Association of University Technology Managers, Sigma Xi (the Scientific Research Society of the National Academy of Sciences), and the Biotechnology Industry Organization. He has also provided expert testimony at deposition or trial in lawsuits dealing with either reasonable royalties or normal custom and practice in the biotechnology and pharmaceutical industries.

Mr. Edwards is on the Board of Directors of Allos Therapeutics, Inc. Prior to founding Recap in 1988, Mr. Edwards was Manager of Business Development at Chiron Corporation. He received his B.A. and M.B.A. degrees from Stanford University.

EISS, Robert

Robert Eiss presently serves as the CEO of MIHR. He has held senior management positions at the U.S. National Institutes of Health (NIH) and the White House Office of National Drug Control Policy. He has more than twenty years of experience in the planning and management of NIH-supported global health programs. He helped initiate the Multilateral Initiative on Malaria, a consortium of international investors that supports research for improved control and prevention of malaria in Sub-Saharan Africa. He has also helped start a cooperative venture between NIH-supported institutions and the World Bank's Global Development Network, whose goal is to assess the effects of health on economic productivity. He also conceived of the "Global Forum on Bioethics," an informal partnership among multiple organizations that addresses issues of equity and social justice in North-South research enterprises, including allocation of intellectual property rights.

He has been responsible for the analysis and implementation of policies related to IP allocation in cooperative programs involving NIH and its international counterparts. As a representative of NIH to the White House Committee on International Science, Engineering and Technology, he was lead author on reports that established policy frameworks for cooperative programs between the U.S. and both the European Union and Russia. He also served as Associate Director of the White House Office of National Drug Control Policy, where he helped institute a national initiative to improve access to treatment in the public and criminal justice settings through block and discretionary grants.

A native of Washington, D.C., Mr. Eiss graduated from the University of Maryland at College Park and received his M.A. from Oxford University.

FATHALLA, Mahmoud F.

Dr. Fathalla is a professor of Obstetrics and Gynecology and former Dean of the Medical School at Assiut University in Egypt and is currently the chairman of the World Health Organization (WHO) Global Advisory Committee on Health Research. He has served as the Director of the UNDP, UNFPA, World Bank, WHO Special Programme of Research, Development and Research Training in Human Reproduction and has served as a consultant to various international bodies such as the WHO, UNPF, IPPF, Population Council, and the Ford and Rockefeller foundations. He is the author of more than 150 scientific publications. Professor

Fathalla has been an international campaigner for Safe Motherhood and a founder of the Safer Motherhood Initiative. His scientific interests include women's health, safe motherhood, reproductive health, ethics and human rights, and contraceptive research and development.

FEINDT, Hans H.
Hans Feindt currently supervises the Monitoring and Enforcement Branch of the Division of Technology Development and Transfer in the NIH Office of Technology Transfer (OTT). He has been with OTT since November 2002. Before joining NIH, he worked in various roles as a scientist, project leader, and research director in the medical diagnostics industry at a number of large and small U.S. companies. During his 20-year industry career, he was employed by Bethesda Research Laboratories, Becton Dickinson & Co., Quidel Corporation, and OraSure Technologies. He contributed to the development and commercialization of rapid, antibody-based tests for a variety of important infectious disease agents. Some of these tests are still in commercial use. In addition to developing new products, he also transferred and established new technologies that were obtained through licensing or acquisition deals by the companies where he was employed. He is listed as a co-inventor on numerous patents and a co-author on a number of scientific publications. Dr. Feindt earned a Ph.D. in biochemistry from Brandeis University and a B.S. in chemistry from the University of Delaware. He and his family currently reside in Baltimore, Maryland.

FELDMAN, Maryann P.
Maryann P. Feldman is the inaugural Zell Miller Distinguished Professor of Higher Education at the Institute of Higher Education of The University of Georgia. Previously, she was Professor of Business Economics at the University of Toronto, where she also held the Jeffrey S. Skoll Chair in Technical Innovation and Entrepreneurship at the Rotman School of Management. Dr. Feldman's work focuses on the ways in which universities transfer technology and the implications of those transfers for economic development. She explores the means by which geographic clusters produce economic growth and has special expertise in university-generated technologies and the commercialization of academic research. Prior to her appointment at Toronto, Dr. Feldman was at Johns Hopkins University, where she was a faculty member at the Institute for Policy Studies. At Johns Hopkins, Dr. Feldman was the founding policy director at the Information Security Institute (JHUISI) at the Whiting School of Engineering. Her most recent book, co-edited with P. Braunerhjelm, is *Cluster Genesis: the Origins and Emergence of Technology-Based Economic Development* (Oxford: Oxford University Press, 2006).

FENTON, Gillian M.
Gillian M. Fenton is a patent and intellectual property attorney who specializes in the fields of biotechnology and pharmaceutical patents and licensing. She has over fourteen years' experience as an attorney and is a member of the bar in Massachusetts, Maryland, and the District of Columbia and is also registered to practice before the U.S. Patent and Trademark Office. Ms. Fenton currently is Chief Intellectual Property Counsel at Emergent BioSolutions Inc., where she is responsible for all patent matters and licensing-related intellectual property matters. She previously served as in-house counsel at Biogen, Inc. and at several Boston-based law firms, including Foley Hoag LLP. Ms. Fenton is a 1992 graduate of Suffolk University Law School in Boston. Before she entered the field of law, she was a scientist in the immunology laboratory at Genetics Institute, Inc. She is a 1984 graduate of Trinity College in Hartford, Connecticut, where she received a bachelor's degree in biochemistry.

FERNÁNDEZ, Carlos
Carlos Fernández studied agronomy at Universidad de Chile. After working as an Assistant Professor at the Agronomy Faculty of the same university, he received a Ph.D. in Plant Physiology at the University of California, Davis. Upon graduation, he joined Monsanto Company, where he held various management positions that gave him responsibilities in several countries. He led the development of agricultural technologies in Latin American countries, first from the company headquarters in St. Louis and later from Sao Paulo, Brazil. Among other things, he contributed to the development of new applications for Roundup, the most successful herbicide in the world, and the development of nontillage systems for various crops. In Europe, he developed new products and actively participated in the design of the Roundup post-patent policy for Europe and Africa.

While working for Monsanto in California, he evaluated and contributed to the development and introduction of transgenic crops to the market. During his stay in California, he returned to the University of California, Davis, and earned an M.B.A. In 1999, he returned to Santiago, Chile, and began working at Fundación Chile, where he coordinated programs related to technology transfer, intellectual property, regulatory matters, and the development of transgenic crops. He contributed to the Cooperative Agreement between the University of California, Davis and Fundación Chile. In addition to his work at Fundación Chile, he serves as a consultant to the Food and Agricultural Organization of the UN and the Chilean Ministry of Economy. Some of his latest contributions as a consultant include two studies sponsored by the Ministry of Economy of Chile: "Comparative Analysis of Biotechnology Policies in N. Zealand, Canada, United States, Australia, Japan, China, Argentina, Brazil, Spain and Chile" and "Formulation of a Model for a Technology Transfer Office for Chile." He also contributed to a recent study, sponsored by UNDP, titled "Commercialization Impact on Agricultural Export Products Caused by the Introduction of GMO in Chile."

As of July 2006, Dr. Fernández is the Head of Strategic Studies and the technology transfer unit of the Foundation for Agriculture Innovation.

FERNANDEZ, Dennis S.
Dennis Fernandez has over 20 years' experience in Silicon Valley and high-tech industry as a patent prosecutor and intellectual property litigator, a venture capitalist, and an engineering manager. He specializes in developing offensive and defensive patent strategies for start-up electronics, software, and biotech companies and their investors. Mr. Fernandez serves as strategic

advisor to leading venture capital firms, including Sevin Rosen, Venrock, Charles River Ventures, and Walden International. Some of his clients include Marvell Technology, SiRF Technology, Ayala Corporation, Stanford University, and Northwestern University, as well as various start-up companies acquired by Cisco, Broadcom, Ciena, and Cadence Design Systems. He also serves on the Editorial Board of the *Nanotechnology Law & Business Journal*, the Board of Directors of the Association of Patent Law Firms, and the Science and Technology Advisory Council. Previously, Dennis served on a consultancy with the United Nations Development Programme on Asian economic development.

Mr. Fernandez also holds several U.S. and international patents in the areas of digital television, sensor networks, and bioinformatics. He has an electrical engineering degree from Northwestern University, a law degree from Suffolk University Law School, and is a Registered U.S. Patent Attorney.

FINSTON, Susan K.

Susan K. Finston has more than 20 years of experience in the management of international legal and public policy issues. In June 2005, Ms. Finston founded Finston Consulting, LLC. Her company provides a range of services to the biotechnology industry, including business development, strategic marketing, technology transfer, policy analysis and advocacy, ally development, and education and awareness programs for start-ups and multinational companies.

Ms. Finston is a board member of BayhDole25, a technology transfer NGO that was established in 2005 to study the social and economic impact of the Bayh-Dole Act of 1980 and related international technology transfer legislation. She also serves as Executive Director of the American BioIndustry Alliance (ABIA), an advocacy organization that seeks enabling conditions for biotechnology through sustainable, mutually beneficial Access and Benefit Sharing (ABS) policies. She was recently elected to the Alumni Board of the Ford School of Public Policy at the University of Michigan, where she received a Joint JD/MPP degree in 1986. She also served on the Board of Governors of the Washington Foreign Law Society and was a member of the National Advisory Board of the International Society of Environmental Biotechnology. For publications, presentations, and upcoming events, see www.finstonconsulting.com.

FONG, H. H. S.

Professor Harry Fong is a Professor Emeritus of Pharmacognosy and Associate Director, WHO Collaborating Centre on Traditional Medicine at the College of Pharmacy, the University of Illinois at Chicago, as well as Adjunct Professor at RMIT University in Melbourne, Australia. He is an internationally known pharmacognosist/natural products chemist with more than 48 years of experience centering on the search for anti-tumor, cancer chemopreventive, antimalarial, anti-HIV, and anti-TB agents from plants and on quality-control standardization and clinical evaluation of herbal medicine/botanical-dietary supplements. His research has resulted in more than 255 research papers, a number of books, book chapters, and review articles. He was an Associate Editor of the *Journal of Natural Products* and was a primary writer of the *WHO Monographs on Selected Medicinal Plants* Vol. 1-4 and the *WHO Guidelines on GACP*. His work and services led to many honors. He served as President of the American Society of Pharmacognosy from 1978 to 1979, President of the Society for Economic Botany from 1981 to 1982, Visiting Professor at Guangzhou University of Traditional Chinese Medicine University, and Honorary Member of the American Society of Pharmacognosy beginning in 2004. He is a recipient of the Jack L. Beal Post-Baccalaureate Alumnae Award, the College of Pharmacy, Ohio State University; Member, WHO Traditional Medicine Expert Panel (1997 to present); and Member, International Advisory Board on Hong Kong Chinese Materia Medica Standards (2002 to present).

FRASER, John A.

John A. Fraser has been the Executive Director of the Office of IP Development and Commercialization at Florida State University, Tallahassee, since 1996. Previously, he was Director of the University/Industry Liaison Office at Simon Fraser University, Vancouver. Mr. Fraser has substantial corporate and university experience. He has also held the following positions: Executive Vice President and co-founder of UTC, Inc., a venture-capital-backed, North Carolina-based university licensing/technology transfer firm; President and CEO of UTI, a University of Calgary-based for-profit technology transfer company; Vice President of TDC, Inc., a Toronto and Vancouver-based venture capital firm; and President of Burnside Development, a technology commercialization consulting firm. He has co-founded three companies and assisted in the launching of another 12 technology-based firms.

In 2006, he became President of AUTM, the global, academic professional technology transfer association, and served a two-year term as VP Membership (2001–2003). He is a Founding Board Director of TalTech Alliance, the technology association of the Tallahassee region, and its Executive Committee. He is also a Founding Board member of the Florida Research Consortium and its Executive Committee; he was appointed by the governor to increase university/company interactions to better the Florida economy. Through the Johns Hopkins University technology transfer program, he has helped scientists and engineers create business plans for new start-up companies. In 2006, he joined the Board of BioFlorida, the statewide biotechnology trade association.

Mr. Fraser holds a Master's Degree in Biochemistry from the University of California, Berkeley.

FREEMAN, John W.

John Freeman is a Principal in the law firm of Fish & Richardson P.C., with 35 years of experience. He has a diverse practice emphasizing patent licensing and patent opinions. He specializes in biotechnology, chemistry, bioinformatics and biology. He has extensive experience in academic-industry collaborations, diligence involving intellectual property, and all aspects of patent counseling and prosecution. He also has experience in pharmaceutical patent counselling, including pre-suit investigation and strategic issues under Hatch-Waxman provisions.

Prior to joining Fish & Richardson, Mr. Freeman served in the office of general counsel at the Civil

Aeronautics Board, where he was responsible for litigated cases involving administrative law. He also served as a law clerk to Justice Robert N.C. Nix of the Pennsylvania Supreme Court. He received a B.A. from Williams College in organic chemistry and a J.D. from the University of Pennsylvania Law School.

FREIRE, Maria

Dr. Maria C. Freire is CEO and President of The Global Alliance for TB Drug Development, a position she has held since 2001. During her service, the Alliance has built the largest pipeline of TB drugs in the world, advanced compounds into clinical testing, and pioneered precedent-setting agreements with industry.

From 1995 to 2001, Dr. Freire directed the Office of Technology Transfer at the NIH, where she was responsible of technology transfer policies and procedures for the Department of Health and Human Services and for patenting and licensing activities at the NIH and the FDA.

Dr. Freire is an internationally recognized expert in technology commercialization. She is a member of the NIH Advisory Board for Clinical Research, a Governor of the New York Academy of Sciences, and the Chair of the Working Group for New TB Drugs for the global Stop TB Partnership. Dr. Freire was selected as one of ten Commissioners of the World Health Organization's Commission on Intellectual Property Rights, Innovation and Public Health (CIPIH) and a member of Time magazine's Global Health Summit Board of Advisors.

Born in Lima, Peru, Dr. Freire trained at the Universidad Peruana Cayetano Heredia. She holds a Ph.D. in biophysics and completed post-graduate studies in immunology and virology at the University of Virginia and the University of Tennessee, respectively, and at the John F. Kennedy School of Government at Harvard University. She has received numerous national and international awards, including the Arthur S. Flemming Award, DHHS Secretary's Award for Distinguished Service, and the Bayh-Dole Award.

GACEL, Rafael A.

Rafael A. Gacel is an Associate Director of Technology Transfer Services at the University of California, Davis. Since joining U.C. Davis in January 2000, he has also served as an intellectual property officer and a material transfer analyst; he has worked on thousands of MTAs, confidentiality agreements, research agreements, and licenses; and he has given presentations and classes on MTA-related topics. From 1998 to 2000, he was an administrative analyst at UC Berkeley; from 1996 to 1998, he was a deputy director of financial management with the U.S. Senate (assigned to the U.S. Capital Police); from 1976 to 1996, he was a financial management officer, an information systems officer, a commanding officer, and an engineer equipment operator in the U. S. Marine Corps. Mr. Gacel earned a B.S. in Civil Engineering from the University of Washington in 1983, an M.B.A from National University in 1986, and an M.S. in Information Systems from the Naval Postgraduate School in 1991. He immigrated to the United States of America as a Cuban refugee in 1964, and has been living in Davis, California, since 1998. Mr. Gacel has also lived in the Dominican Republic, Guatemala, and Okinawa. He has three children: Enrique, Maria, and Emmanuel.

GAO, Wangsheng

Wangsheng Gao is the Director of the Regional Agricultural Development Center at China Agriculture University (CAU). He is also a Professor of Farming Systems and Ecology at the College of Agronomy and Biotechnology at CAU and Chairman of the China Farming-system Research Society. From 1979 to 1983, he was an undergraduate in the Department of Agronomy of Gansu Agriculture University (GAU), where he earned his B.A. From 1984 to 1990, he worked at GAU and earned his M.S. degree there in 1989. From 1991 to 1994, he returned to CAU and earned a Ph.D. there. Since 1995, he has taught at CAU. Between 1995 and 2006, he finished more than 10 research programs supported by national science & technology project. His areas of research interest are farming systems and regional rural development, conservation tillage and sustainable agriculture, agro-ecosystem eco-economical analyses, and agricultural high-technology assessment and developmental policy.

GARNER, Cathy

Cathy Garner is currently Chief Executive of Manchester: Knowledge Capital. Dr. Garner has a background in university-business links and technology transfer, as well as extensive experience in the fields of urban regeneration, education, and knowledge-based business development.

She is a Trustee of the U.K. registered charity MIHR and was its founding CEO until 2004. Dr. Garner established and ran the Research and Enterprise Office at the University of Glasgow in Scotland. She helped establish the Scottish Institute for Enterprise and was a founder-director of the Scottish North American Business Council. She is a member of the Association of University Technology Managers (AUTM) in the U.S. and has served as their inaugural Vice President for International Relations.

Her career includes eight years of policy and research management in the public sector and ten years of academic research in education and urban regeneration. She has acted as an intellectual property advisor to the U.K., Canada, Japan, and South Africa and served as a Non-Executive Director on numerous Boards. She is a Fellow of the Royal Statistical Society.

GHAFELE, Roya

Roya Ghafele works as an economist with the World Intellectual Property Organization. She concentrates on questions related to value creation and intellectual property in the area of life sciences. Dr. Ghafele has published widely in the field of IP management in the life sciences and has advised the governments of several developing countries on how to better align intellectual property with overall innovation and health policies. Previously, Dr. Ghafele worked with the OECD Trade Directorate, McKinsey & Company, and as a professional ballet dancer. Dr. Ghafele was trained at Johns Hopkins University, the Sorbonne, and Vienna University. Her doctoral dissertation, "Globalization, Francophone Africa and the WTO – a Historical Discourse Analysis" was awarded the Theodor Körner Research Prize by the president of Austria.

GOLD, E. Richard
Richard Gold is the Director of the Centre for Intellectual Property Policy. He teaches courses on intellectual property and innovation at McGill University's Faculty of Law. His research centers on the nexus between innovation and development, particularly with respect to biotechnology in the international context. He is the Principal Investigator of the Intellectual Property Modelling Group, a transdisciplinary research team investigating intellectual property regimes and their links to innovation, financing, public opinion, and development. Dr. Gold has consulted with the World Intellectual Property Organization, the Organization for Economic Cooperation and Development, the World Health Organization, and various Canadian federal and provincial governments and institutions. Dr. Gold holds an S.J.D. and LL.M. from the University of Michigan, a LL.B. (Honors) from the University of Toronto and a B.Sc. from McGill University.

GOLDMAN, Michael L.
Michael L. Goldman practices law at Nixon Peabody LLP. He has extensive experience in patent licensing and intellectual property agreements, particularly with sponsored research agreements and license agreements. He deals regularly with pharmaceutical, genomics, and biotechnology companies, as well as universities, agricultural cooperatives, and international entities. His work has focused on serving large and small companies, universities, and other institutions in the biotechnology and chemical fields. He served as a law clerk for Hon. Jack R. Miller, Circuit Judge, U.S. Court of Appeals for the Federal Circuit in Washington, D.C., and served as a patent examiner for the U.S. Patent and Trademark Office. He is a chemical engineer and was previously employed by the Gulf Oil Corporation and the Bendix Corporation. Mr. Goldman has authored articles and given presentations on various licensing projects around the world. He is admitted to practice in the District of Columbia Court of Appeals, the New York Court of Appeals, and the U.S. Court of Appeals for the Federal Circuit. Mr. Goldman is a registered attorney in the U.S. Patent and Trademark Office and a member of the American Intellectual Property Law Association, the American Bar Association, the District of Columbia Bar Association, the New York State Bar Association, and the Federal Circuit Bar Association.

GRACEN, Vernon
Vernon Gracen is currently a Visiting Professor in the department of Plant Breeding and Genetics at Cornell University. He teaches an introductory course in Plant Breeding and manages curriculum development for the Cornell Transnational Learning Program. Raised in Savannah, Georgia, he now lives in Ithaca, New York. He earned a B.S. in Education from Georgia Southern College and a Ph.D. in Agronomy from the University of Florida. He joined Cornell University as an Assistant Professor in Plant Breeding and Biometry in 1970 and moved through the ranks of Associate and Full Professor. His research interests were in the areas of breeding for disease and insect resistance in maize and cassava. He joined Cargill Hybrid Seed in 1987 as Vice President, Director of Research and Development for the North American Seed Division. He became Director of Research for Cargill's Worldwide Seed Business in 1992. He returned to Cornell as a Visiting Professor in 2001.

GRAFF, Gregory D.
Gregory D. Graff is an applied economist with expertise in the economics of innovation, entrepreneurship, intellectual property, and technology transfer, especially as they apply to the agricultural life sciences and biotechnology. He applies microeconomic and econometric tools to scientific, patent, regulatory, and commercial data, building uniquely thorough industry-level datasets to analyze the impacts of innovation and technology transactions on markets, industrial organization, and the political economy of science policy.

Dr. Graff currently manages research projects for the Public Sector Intellectual Property Resource for Agriculture (PIPRA), a consortium of 37 agricultural research universities and institutes that is hosted by the University of California. PIPRA uses an innovative model of collaborative intellectual property management to mobilize its members' technologies for the purpose of genetically improving "orphan" crops. Dr. Graff has taught as a university lecturer at both U.C. Berkeley and U.C. Davis and has recently published articles in *The Review of Economics and Statistics, World Development, California Management Review,* and *Nature Biotechnology* as well as chapters in several books. Dr. Graff has a Ph.D. in agricultural and resource economics from U.C. Berkeley (2002), an M.A. in economics from Ohio State University (1995), and a B.S. in biology from Cornell University (1992).

GUO, Hua
Hua Guo focuses her practice on patent prosecution, opinions, and agreements dealing with biotechnology, pharmaceuticals, chemicals, and medical devices.

Prior to joining Jones Day, Dr. Guo worked as an intern in the Corporation Sponsored Research and Licensing Office of Massachusetts General Hospital, where she was actively involved in technology licensing, marketing, prior art searches, and patent prosecution before the USPTO. Previously, she was an associate at King & Wood, PRC Lawyers, where she prepared, filed, and prosecuted domestic and foreign patent applications pertaining to biotechnology, pharmaceuticals, chemicals, and medical devices. She also consulted with clients on invention patentability and patent validity.

Dr. Guo got her Master's degree in Intellectual Property at Franklin Pierce Law Center. Before attending law school, she earned an M.D. and a Ph.D. in molecular pathology. She is a member of Association of Attorneys specializing in the practice of Intellectual Property Law (AIPLA), the All China Patent Agent Association (ACPAA), and the Chinese Bar Association.

GURRY, Francis
Francis Gurry, a national of Australia, is Deputy Director General of the World Intellectual Property Organization (WIPO) in Geneva. He is responsible for WIPO's activities in the area of patents, which include patent policy questions and the administration of the Patent Cooperation Treaty (PCT), under which some 145,000 international patent applications were filed in 2006, biotechnology and genetic resource policy questions, traditional knowledge, and the WIPO Arbitration

and Mediation Center, which has administered over 26,000 disputes over Internet domain names since 2000.

Dr. Gurry holds law degrees from the University of Melbourne and a Doctor of Philosophy from the University of Cambridge in the United Kingdom. He is a Professorial Fellow of the Faculty of Law of the University of Melbourne.

He is the author of a textbook on the law of trade secrets and confidential information, entitled *Breach of Confidence*, published by Oxford University Press in the United Kingdom in 1984, and co-author, with Frederick Abbott and Thomas Cottier, of *The International Intellectual Property System: Commentary and Materials*, published by Kluwer in July 1999.

GYLLENHAAL, C.

Dr. Gyllenhaal is an ethnobotanist by training. She is a Research Assistant Professor at the University of Illinois at Chicago and Research Program Manager at the Block Center for Integrative Cancer Treatment located in Evanston, Illinois. She has been active in a variety of research projects that includes herbal supplements and, especially, cancer therapy, traditional medicines, biological activities of natural products, and IP issues related to indigenous traditional medicinal plant knowledge. Her involvement with a project led by D. D. Soejarto (funded through the NIH/FIC ICBG) has given Gyllenhaal substantial experience in administering and subcontracting for international collaborative projects. She has been an Associate Editor of *Economic Botany and Taxonomy* and Editor of *Journal of Natural Products*. Since 2001, she has served as Associate Editor of *Integrative Cancer Therapies*.

HAEUSSLER, H. Walter

H. Walter Haeussler is a member of the Board of Directors at HemoBioTech. He has served as the Director of Technology Transfer at Texas Tech University, as General Counsel to Advisys Inc., an animal biotechnology company, and as the former President of the Cornell Research Foundation at Cornell University. Before that, he was with Jones, Tullar & Cooper P.C., Arlington, Virginia, rising to the position of managing partner. From 1963 to 1972, he was an intellectual property attorney with PPG Industries, Pittsburgh. Haeussler earned a B.S. in chemistry from Bowling Green University, Bowling Green, Ohio, and a J.D. from Duquesne University School of Law, Pittsburgh.

HAMZAOUI, Amina

Dr. Amina Hamzaoui is the Associate Director of Intellectual Property at the Whitehead Institute of Biomedical Research in Cambridge, Massachusetts, U.S.A. She manages and oversees activities related to the intellectual property that is produced from research and other work conducted at the Institute. Prior to joining the Whitehead Institute in April 2005, Dr. Hamzaoui was an Assistant Professor at St. Thomas University, in Miami, Florida. During her tenure at St. Thomas, she served as the Chairperson of the Institutional Review Board/Research Ethics Committee for research with human participants. She did her postdoctoral fellowship in 1998 at the University of Miami School of Medicine. She has over seven years of comprehensive research experience in analysis, theory development, and practical application in the area of cardiovascular disease and in the development and improvement of the biosynthetic materials used in arterial and vascular reconstructive surgery. Dr. Hamzaoui received her Ph.D. in Chemistry of Materials & Polymer Science in 1997 from Université des Sciences Montpellier II – Montpellier, France, her Master of Science in Biomedical Engineering in 1993 from Université Paris XIII/Galileo Institute, Paris, France and a Bachelor of Science in Biochemistry from Université des Sciences Montpellier II – Montpellier, France.

HANDLER, Philana S.

Philana S. Handler is an associate at the firm of Whitham, Curtis, Christofferson & Cook. Ms. Handler has worked in the field of intellectual property for nearly a decade since joining the firm of Whitham, Curtis, Whitham & McGinn in 1997. She received the degree of Bachelor of Science in Psychology from the George Mason University in 2000. While an undergraduate, Ms. Handler was a member of the Honor Committee, as well as president of the Judicial Board of the George Mason University. Ms. Handler received her law degree, with honors, from the David A. Clarke School of Law in 2004. During law school, Ms. Handler assisted in the development of small businesses and nonprofit organizations; her duties included filing government forms, preparing contracts, and trademark work. In addition, during law school, Ms. Handler advocated on behalf of children in need of special education services.

Currently, Ms. Handler primarily handles trademark, copyright, and unfair competition matters, as well as litigation matters. She also heads up the firm's patent annuity program. Ms. Handler is involved in a supporting role in patent matters concerning mechanical and biological technologies and also works on various contract and licensing matters.

HANNA, Kathi E.

Kathi E. Hanna has over 25 years of experience in science and health policy as an analyst, writer, and editor specializing in biomedical research policy and bioethics. She served as Research Director and Editorial Consultant to President Clinton's National Bioethics Advisory Commission (NBAC) and directed the completion of NBAC's reports. In the mid-1990s, Dr. Hanna was Senior Advisor on Reproductive Health to the Advisory Committee on Gulf War Veterans Illnesses. More recently, she served as the lead author and editor of President Bush's Task Force to Improve Health Care Delivery for Our Nation's Veterans. In the 1980s and 1990s, Dr. Hanna was a Senior Analyst at the congressional Office of Technology Assessment, where she contributed to numerous science policy studies requested by congressional committees on science education, research funding, biotechnology, women's health, human genetics, bioethics, and reproductive technologies. In the past two decades, she has served as an analyst and editorial consultant to the Howard Hughes Medical Institute, the National Institutes of Health, the U.S. National Academies, the U.S. Office for Human Research Protections, and various charitable foundations, voluntary health organizations, and biotechnology companies. Before moving to the Washington, D.C., area, she was the Genetics Coordinator at Children's Memorial Hospital in Chicago, where she directed clinical counseling and coordinated an international research program in prenatal diagnosis. Dr. Hanna received an A.B. in Biology from Lafayette College, an M.S. in Human Genetics from Sarah Lawrence College, and a doctorate from George Washington University. She is currently Senior Vice President at Styllus, LLC, a medical and scientific writing company based in the Boston and Washington, D.C., areas.

HANSEN, Stephen A.
Stephen A. Hansen is Project Director with the Science & Human Rights Program. His work currently focuses on projects that relate to the effects of intellectual property rights on science, particularly those that relate to traditional knowledge and human rights. He serves as the Project Manager for an AAAS project: Science & Intellectual Property in the Public Interest (SIPPI). He is co-author of the handbook *Traditional Knowledge and Intellectual Property*. He also designed the Traditional Ecological Knowledge Prior Art Database (T.E.K.*P. A.D), an online digital archive of traditional practices from local communities throughout the world that are already in the public domain. Mr. Hansen's other main area of work is in economic, social, and cultural rights (ESCR); he has worked with the United Nations Committee on Economic, Social and Cultural Rights and UNESCO in this capacity. He is the author of a chapter on cultural rights in the AAAS publication *Core Obligations: Building a Framework for Economic, Social and Cultural Rights*. He has also been involved in violations monitoring and documentation and has authored *The Thesaurus of Economic, Social and Cultural Rights*. He has directed projects with the National Commission for Human Rights in Honduras, as well as the Centro de Estudios Legales y Sociales (Center for Legal and Social Research, CELS) in Buenos Aires, Argentina. Mr. Hansen holds a Bachelor's degree in Anthropology from Oberlin College and an M.A. in Anthropology from The George Washington University in Washington, D.C.

HARNER-JAY, Claudia
Claudia Harner-Jay is a program officer with PATH's Technology Solutions Strategic Program. Her major responsibilities include creating and implementing commercialization strategies for health technologies; managing intellectual property issues; identifying potential partners and performing due diligence; and negotiating collaboration agreements. She also serves as team leader or project manager for select health technology initiatives. While at PATH, Ms. Harner-Jay has managed numerous market research studies to help refine product development activities, identify potential partners, and inform introduction strategies. Before joining PATH, Ms. Harner-Jay was a business development manager at Monsanto Life Sciences Company and helped develop the market for agricultural products produced by small farmers in low-resource settings in Latin America. In addition, she has worked with coffee farmers in Central America, where her findings and recommendations led to a US$50 million World Bank loan for the coffee industry in El Salvador. Ms. Harner-Jay also worked for UBS in Zurich, where she earned a Swiss Banking Diploma. She holds an M.B.A. and an M.S. in environmental policy from the University of Michigan and a B.A. in international affairs from the University of Puget Sound. She is fluent in both Spanish and German.

HARNEY, Dennis J.
Dennis J. Harney is an attorney at Sonnenschein Nath & Rosenthal LLP, where he is a member of the Intellectual Property and Technology Practice Group. His practice encompasses all areas of intellectual property law, including preparation and prosecution of patent applications in the United States and foreign countries. Dr. Harney's work focuses primarily on biotechnology and biochemical patent preparation and prosecution, including patents for transgenic plants and bacteria; novel DNA and protein sequences; and biological, chemical, and pharmaceutical therapeutics. Dr. Harney's practice also encompasses validity/invalidity and infringement opinions and counseling related to patentability and freedom to operate.

In 2003, Dr. Harney earned a J.D. from University of Dayton School of Law, where he graduated first in his class; in 2003, he earned a Ph.D. in Botany from Miami University, where he specialized in plant stress physiology; and in 1996, he earned an M.S. in Botany from Miami University, where he studied plant secondary product chemistry.

Dr. Harney is a member of the American Intellectual Property Law Association (AIPLA); the Intellectual Property Owner's Association (IPO); the Missouri Bar Association; and the American Society of Plant Biologists. Dr. Harney has served on the AIPLA Biotechnology Committee and currently serves on the IPO Committee for Genetic Resources and Traditional Knowledge. Dr. Harney is also an adjunct professor at Saint Louis University School of Law, where he teaches a course titled "Biotechnology and the Law."

HEHER, Anthony D.
Anthony D. Heher has a longstanding interest in research and innovation and the contribution that they can make to economic and social development. His experience includes 12 years at South Africa's Council for Scientific and Industrial Research (CSIR); 16 years as founder and CEO of a high-tech spinout company based on his doctoral research; and ten years working in economic development, including two years at the South African Department of Trade & Industry, where he was Chief Director for Industrial Promotion in 1997 and head of the national economic cluster program. From 2000 to 2005, he was Director of UCT Innovation at the University of Cape Town. In 2006, he rejoined AfED, an economic and business consultancy that he founded in 1998. He is actively involved in a projects ranging from entrepreneurship development to economic development through public-private partnerships. He was instrumental in the establishment of the Southern African Research & Innovation Management Association (SARIMA) in 2001 and was its founding President. SARIMA is focused on capacity building and linking the various players in the research and innovation spectrum so that they can cooperate more effectively. He has a Ph.D. in engineering. He has had a varied academic career: he is a graduate of the Universities of Natal, Pretoria, and California in Physics, Mathematics,

Electrical Engineering, Mechanical Engineering and Computer Science; he also holds an executive M.B.A. from Wits Business School. He has remained an active researcher his whole working life, although his research output has taken rather varied forms!

HENNESSEY, William O.
William Hennessey is Professor of Law at the Franklin Pierce Law Center. He directed Pierce Law's graduate programs in intellectual property and summer from 1986 until 2003. A noted IP expert, author, and lecturer, he recently directed the fourth annual Pierce Law Intellectual Property Summer Institute at Tsinghua University School of Law in Beijing, China. He co-authored a legal casebook on international IP law and policy.

Professor Hennessey has served as a legal advisor to the governments of Indonesia and the People's Republic of China and has served as a consultant to the World Bank, Asian Development Bank, United Nations Development Programme, U.S. Agency for International Development, U.S. Department of State, and the U.S. Patent and Trademark Office. He has also served as consultant for the World Intellectual Property Organization in many countries on various issues concerning IP protection and economic development.

HERSEY, Karen
Karen Hersey is Visiting Professor of Law at the Franklin Pierce Law Center in Concord, New Hampshire. Professor Hersey recently retired as Senior Counsel for intellectual property at Massachusetts Institute of Technology where she represented M.I.T.'s interests on intellectual property matters with a variety of constituencies including industrial research partners and both U.S. and foreign governments. In 1992, she served as the academic community's representative to a Congressionally mandated Department of Defense Government-Industry Advisory Committee on Rights in Technical Data and Computer Software to study and recommend changes in the Department of Defense Procurement Regulations in the areas of technical data and computer software. She publishes widely in the area of intellectual property law as it impacts institutions of higher education. Professor Hersey is a past President of the Association of University Technology Managers (AUTM). In addition to offering courses dealing with technology transfer for nonprofit organizations and intellectual property management in universities, she also teaches U.S. Copyright Law. She received her B.A. from Goucher College and her LL.B. from Boston University School of Law.

HIEP, N. T.
A doctoral graduate from Komarov Botanical Institue of the Russian Academy of Sciences, St. Petersburg, Russia, Dr. Nguyen Tien Hiep's expertise is in plant taxonomy, evolutionary biology, and ecology, specializing on the studies of the flora of Vietnam. As a senior scientist at the Institute of Ecology and Biological Resources of the Vietnamese Academy of Science and Technology, Hanoi, he is best known for his research output in the study of the cycads and other gymnosperms of Vietnam. He has served as coordinator of many international collaborative projects, especially projects between the Institute of Ecology and Biological Resources and Museum National D'Histoire Naturelle, Paris, France, and the Missouri Botanical Garden, St. Louis, USA. Since 1998, he has served as a Project Co-Leader of an ICBG Program, with responsibility in biotic survey of the Cuc Phuong National Park.

HINES, Sally
Sally Hines is the Administrative Services Manager of the Office of Technology Licensing (OTL) at Stanford University. She has been with the Stanford OTL since its beginning in 1970 and was instrumental in setting up many office procedures that are still in use. The Stanford OTL has increased its staff from two people in 1970 to approximately 30 people today. Ms. Hines handles office management, facilities management, equity management, human resources, and the $4M operating budget for the office. She has also been involved with the Association of University Technology Managers (AUTM) since its inception and has served on its Board. She currently serves as Chairperson of the AUTM TOOLS course for administrative staff.

HOPE, Janet
Janet Hope is a qualified biochemist and molecular biologist, as well as a former practicing lawyer. She has published in the fields of constitutional, criminal, administrative, environmental, human rights, and intellectual property law.

In January 2003, she published the first substantial treatise on open-source biotechnology (available at rsss.anu.edu.au/~janeth). Under the supervision of Professor Peter Drahos, she completed her doctoral dissertation on Open Source Biotechnology at the Australian National University in 2004.

Together with colleagues Dianne Nicol and John Braithwaite, Dr. Hope is the recipient of an Australian Research Council grant to investigate a range of collaborative intellectual property mechanisms in the Australian biotechnology industry. Her book *Bio Bazaar: The Open Source Revolution and Biotechnology* will be published by Harvard University Press in late 2007.

HSU, Justin
Justin Hsu is currently a third year undergraduate Industrial Engineering and Operations Research student at the University of California, Berkeley. Prior to his summer tenure at Fernandez & Associates L.L.P., he was a mechanical engineering intern at Lawrence Berkeley National Laboratory, where he worked on magnetic measurement systems at the Advanced Light Source, a synchrotron radiation facility. His professional and academic interests include intellectual property law, venture capital, stochastic processes, and optimization with applications in finance and operations research. In his spare time, he enjoys football and playing the guitar, among other things. Upon graduation, Justin hopes to attend graduate school and work several years in the Silicon Valley high-tech industry before taking the U.S. patent bar examination.

HUIE, James T.
James T. Huie earned his bachelor of arts degree in Molecular and Cell Biology, with an emphasis in Biochemistry, at the University of California, Berkeley. Subsequently, Mr. Huie obtained his law degree at Santa

Clara University School of Law. During and after law school, Mr. Huie developed skills in patent prosecution at Fernandez & Associates, LLP. After some time in the patent field, Mr. Huie changed his focus to corporate practice within the venture capital industry. He has practiced in the Silicon Valley at such law firms as Greenberg Traurig, LLP and Wilson, Sonsini, Goodrich & Rosati, PC. Currently an associate at Wilson, Sonsini, Goodrich & Rosati, PC. he represents early-stage and public medical device companies, as well as institutional and venture capital investors, in transactions that involve equity and debt financings, mergers and acquisitions, public offerings, and public filings.

HUNG, N. V.

Dr. Nguyen Van Hung is Deputy Director of the Institute of Chemistry and Director of Bioorganic Division, Vietnamese Academy of Science and Technology, Hanoi, Vietnam. He is a phytochemist with special expertise in the isolation, structure elucidation, and synthesis of medicinally important compounds, such as the antimalarial drug artemisinin. Early in 2006, he and his chemistry team at the Institute of Chemistry successfully isolated shikimic acid from the starting raw material Chinese star anise fruit, *Illicium verum Hook. f. (Illiciaceae)*, and using this molecule as a precursor, they successfully synthesized the drug oseltamivir (Tamiflu). Since 1998, he has served as a Co-Project Leader in chemistry of an International Cooperative Biodiversity Groups (ICBG) program with responsibility in the isolation and structure elucidation of biologically active compounds.

IDRIS, Kamil

Dr. Kamil Idris has been Director General of the World Intellectual Property Organization (WIPO) since November 1997. He is head of the International Union for the Protection of New Varieties of Plants (UPOV). He was formally re-appointed to a second six-year term as Director General of WIPO on May 27, 2003. His mandate will end on November 30, 2009. Formerly, Kamil Idris was a member of the International Law Commission from 1992 to 1996 and from 2000 to 2001.

Kamil Idris holds a Bachelor of Law (LL.B.) from Khartoum University, Sudan; a Bachelor of Arts in Philosophy, Political Science and Economic Theories from Cairo University, Egypt; a master's in International Law and International Affairs from Ohio University, United States; and a Doctorate in International Law from the Graduate Institute of International Studies, University of Geneva, Switzerland.

JACOBY, Erica

Erica Jacoby, a senior commercialization associate in PATH's Technology Solutions Strategic Program, serves as a resource in the general areas of business planning, market development, and commercialization activities. Her primary responsibilities include identifying, conducting due diligence, and selecting partners; conducting market assessments; developing commercialization strategies; and writing and negotiating legal agreements. In these capacities, she is involved with several different technology development projects at PATH, including needle-free injections, vitamin-fortified rice, neonatal resuscitators, and HIV/STI prevention technologies such as the female condom. Prior to joining PATH, Ms. Jacoby worked in marketing and research at several organizations, including a diagnostic test manufacturer, the Institute of the Americas, and the Graduate School of International Relations/Pacific Studies at the University of California, San Diego (UCSD). Her overseas experience includes work and study in Mexico, Slovakia, Spain, and Venezuela. She holds a Master's degree in Pacific international affairs, with a concentration in international management, from the University of California, San Diego.

JAHN, Molly

Molly Jahn holds degrees from Swarthmore College, M.I.T., and Cornell University, and pursued postdoctoral work at U.C. Berkeley. At Cornell, Molly focused her research on plant breeding, genetics, genomics and molecular biology and on the development of improved crop germplasm. Her group at Cornell has produced a number of globally successful crop varieties currently grown commercially on six continents. Molly has worked extensively internationally in Latin America, Asia and Africa to link crop breeding objectives to outcomes that improve human welfare, such as nutritional status and income. Molly was recently named a Fellow of the AAAS and was elected to the Board of Directors of The World Vegetable Center, the international research center for vegetables. On August 1, 2006, she was named the twelfth dean of the College of Agricultural and Life Sciences at the University of Wisconsin - Madison.

JONES, Keith J.

Keith J. Jones is the Director of the Office of Intellectual Property Administration and the Executive Director of the Washington State University Research Foundation.

Dr. Jones, assisted by a staff of seven, is responsible for the evaluation, patenting, marketing, and licensing of the approximately 70 invention disclosures per year submitted to the Office of Intellectual Property at WSU. WSU is a large research university that produces intellectual property ranging from medical applications to new wheat varieties. WSU intellectual property results in about 15 licenses per year, and last year produced an income of over US$2 million. Dr. Jones is also responsible for initiating and managing other university technology commercialization activities, including the WSU Research and Technology Park, the Cougar Gap Fund, and the Venture Partner Program. Dr. Jones is frequently invited to speak nationally and internationally on university technology commercialization.

Previously, Dr. Jones was Director of Commercialization-Life Sciences at Virginia Tech for six years. He has experience in international business development at a San Diego Ag-Biotechnology company, Mycogen Corp., where he developed new markets for biotech products in the Middle and Far East. He worked for two years as a scientific advisor at one of the most prestigious intellectual property litigation law firms in the U.S. He was a USAID contractor for two years, during which time he was involved with university development based in Sumatra, Indonesia.

He has a Ph.D. from North Carolina State University in Plant Pathology and holds three issued patents and one pending patent.

JORDA, Karl F.
Karl F. Jorda is the David Rines Professor of Intellectual Property Law as well as the Director of the Germeshausen Center for the Law of Innovation and Entrepreneurship at Franklin Pierce Law Center, where he primarily teaches Technology Licensing and IP Management. From 1995 to 2003, he also taught International IP Law as Adjunct Professor at the Fletcher School of Law and Diplomacy, Tufts University. Before joining Pierce Law in 1989, he was Chief IP Counsel for 26 years at CIBA-GEIGY Corporation (now Novartis, Syngenta, and Ciba Speciality Chemicals).

Dr. Jorda was President of the Pacific Intellectual Property Association (PIPA) and the New York Intellectual Property Law Association. He served on the Boards of Directors of AIPLA, ABA-IPL Section, INTA, IPO, ACPC and AIPPI-American Group. Dr. Jorda is also the recipient of several rewards: the 1996 Jefferson Medal of the NJIPLA, "the United States' highest honor in intellectual property," for "extraordinary contributions to the U.S. intellectual property law system"; the 1989 PIPA medal for "Outstanding Contributions to International Cooperation in the Intellectual Property Field"; and the 1998 Distinguished Alumni Award of the University of Great Falls. In 1990 and 1991, he served as a consultant to the Indonesian and Bulgarian IP offices. From 1999 to 2005, he was the U.S. Representative to the Confidentiality Commission of the Organization for the Prohibition of Chemical Weapons in The Hague, Netherlands.

Dr. Jorda received his undergraduate degree (summa cum laude) from the University of Great Falls, and an M.A. and a J.D. from Notre Dame University. He is admitted to the bars of Illinois, Indiana, and New York as well as to practice before the U.S. Supreme Court, the Court of Appeals for Federal Circuit, and the U.S. and Canadian Patent and Trademark Offices. He is a frequent speaker in IP programs in foreign countries under the auspices of WIPO, USAID, USIA, IESC, etc. He has lectured in 41 countries, 27 of them developing countries, including Madagascar and Mongolia.

KAMPF, Roger
Roger Kampf is from Hamburg, Germany. He joined the World Trade Organization in May 2004 and works as Counsellor in the Intellectual Property Division. He is responsible for the Secretariat's work in the area of TRIPS and public health and enforcement, as well as for providing technical assistance in relation to intellectual property. Mr. Kampf previously worked for the European Commission, both at its headquarters in Brussels and at its permanent representation in Geneva; from 1998 to 2004, he was responsible for intellectual property issues in WTO and WIPO, as well as for government procurement. Previously, he was involved in negotiating financial services under the GATS Agreement (the General Agreement on Trade in Services), and also worked as an assistant in public law and European Communities law at the University of Hamburg. Mr. Kampf holds a law degree from the University of Hamburg and a degree in public administration from the Ecole Nationale d'Administration in Paris. He has published on various aspects of EC and WTO law.

KARJALA, Dennis S.
Dennis S. Karjala was an engineering/physics major at Princeton University and holds a Ph.D. in electrical engineering from the University of Illinois (1965). He received his J.D. from the University of California, Berkeley, in 1972, where he was editor-in-chief of the California Law Review and Order of the Coif. After five years of private practice in San Francisco, Dr. Karjala joined the College of Law at Arizona State University in January 1978 as Associate Professor. He has been a Professor of Law since the fall of 1981 and currently holds the Jack E. Brown Chair. His teaching and research are primarily in the area of intellectual property law, especially copyright and the application of intellectual property law to digital technologies. He has also taught and written on the subjects of corporate and securities law and federal income taxation. Dr. Karjala has also done some comparative work on Japanese copyright and corporate law. Dr. Karjala was active in the opposition to the 1998 Sonny Bono Copyright Term Extension Act and in the *Eldred* case, which unsuccessfully challenged the constitutionality of the Copyright Term Extension Act.

KEEVEY-KHOTARI, Simon
Simon Keevey-Khotari is an English-qualified barrister who specializes in intellectual property and commercial transactions. He spent more than five years with Anderson & Company, most of which was spent on a long-term secondment with Imperial Innovations Limited, the technology transfer company of Imperial College, London. He has also co-authored a book with Mark Anderson titled *Drafting Confidentiality Agreements*, 2nd ed. (The Law Society of England and Wales 2004).

KEILLER, Todd S.
Todd S. Keiller has more than 30 years of licensing, business development, and marketing experience. He has worked for 16 years in the industrial sector in a variety of sales, marketing, and business development roles, ten of which were in the Science and Medical Products Divisions of Corning Glass Works. He has over 16 years of academic licensing experience and is the former Vice President, Ventures of the Brigham and Women's Hospital in Boston. In 1998, Todd joined the University of Vermont and assisted the College of Medicine in technology affairs. In 1999, he was appointed Director of Technology Transfer for the entire University. He also handles technology transfer for Maine Medical Center, Caritas St. Elizabeth's Medical Center of Boston, and Boston Biomedical Research Institute. Mr. Keiller has contributed to the foundation of seven companies, the most recent of which are Nephromics (a company focused on the diagnosis of preeclampsia), Vascular Genetics, Inc. (a gene therapy company, now traded publicly under the name of CorAutus), and Tolerance Pharmaceuticals (a diagnostic and therapeutic company devoted to transplantation, which was purchased by Roche). Mr. Keiller holds an A.B. from Dartmouth College and an M.B.A. from the Tuck School of Business Administration.

KESAN, Jay P.
Jay P. Kesan is Professor and Director of the Program in Intellectual Property and Technology Law at the

University of Illinois at Urbana-Champaign. His academic interests and writings are in the area of patent law, cyberlaw, and law and technology. Some of his recent scholarly work is focused on computer software and agricultural biotechnology. He is a registered patent attorney and received his J.D. *summa cum laude* from Georgetown University. He also has a Ph.D. in Electrical and Computer Engineering from the University of Texas at Austin and worked for several years as a research scientist at the IBM T.J. Watson Research Center in New York. For a more complete biography, please see www.jaykesan.com.

KEUSCH, Gerald T.
Professor Jerry Keusch is Associate Provost and Associate Dean for Global Health at Boston University and Director of the university's Global Health Initiative. He is a physician-scientist, whose career has focused on the study of infectious diseases of developing countries at the laboratory and field levels. He has received all three of the major recognition awards of the Infectious Diseases Society of America for this work, including the Squibb, Finland and Bristol Awards. Professor Keusch is a member of Institute of Medicine at the U.S. National Academies, where he serves on its Board on Global Health. He is also a member of the Roundtable on Science and Technology for Sustainability at the National Research Council and its Task Force on Linking Knowledge with Action for Sustainable Development, both at the U.S. National Academies.

Prior to joining Boston University Professor Keusch was the Associate Director for International Research and Director of the Fogarty International Center at the U.S. National Institutes of Health. During his tenure funding for global health research and capacity building dramatically increased. Among the innovations he initiated were explorations into the creative use of intellectual property rights deemed to NIH grantees to insure that developing countries could benefit from discovery funded by public resources. His work in this area, together with the Rockefeller Foundation, led to the formation of the Center for the Management of Intellectual Property in Health Research (MIHR). He served on the Founding Board of MIHR and has subsequently been Vice-Chair of the MIHR Board.

KHUSH, Gurdev Singh
Dr. Khush was born in a small village in Punjab. After receiving his education at the Punjab Agricultural University and the University of California, Davis, Dr. Khush, in 1967, joined the International Rice Research Institute in the Philippines where he served as the Head of Plant Breeding, Genetics, and Biochemistry Division until 2002. As a result of wide-scale adoption of his high-yielding varieties, rice production increased 135% between 1967 and 2000, to feed an estimated one billion additional consumers. His contributions to rice genetics and biotechnology are equally well recognized. He has written three books, more than 80 book chapters and 160 research papers.

Dr. Khush has served as consultant to rice breeding programs of 15 countries as well as The Rockefeller Foundation, the Third World Academy of Sciences, Italy, and the International Science Foundation, Sweden. He is now serving as a member of Scientific Advisory Committee (overseas) to the Department of Biotechnology, Government of India and member of Science Council, an advisory body to Chinese Academy of Agricultural Sciences, Beijing.

For his monumental contributions to the World Food Security, Dr. Khush has been honored with numerous awards and honors such as the Japan Prize (1987), World Food Prize (1996), Rank Prize (1998), Wolf Prize (2000), International Scientific and Technological Cooperation Award from the Government of China (2001), and Padma Shriaward from the president of India. He is one of five Indian scientists who have been elected to membership of Royal Society (FRS) as well as the U.S. National Academy of Sciences. Dr Khush has received Doctor of Science, *honoris causa*, degrees from nine universities including from University of Cambridge in England and Ohio State University.

Commenting on his life work, Dr. Cantrell, Director of the International Rice Research Institute said, "While Dr. Khush's name may have passed the lips of many, his life's work has passed the lips of almost half of humanity."

KOWALSKI, Stanley P.
Stanley P. Kowalski was born and grew up in a working-class neighborhood in Pittsburgh, Pennsylvania, where he attended Catholic primary and public high school. He matriculated at the Pennsylvania State University, and later at the University of Pittsburgh, earning B.S. degrees in horticulture and biology, with emphases in genetics and biochemistry. Later, he earned a Ph.D. in plant breeding from Cornell University. Dr. Kowalski's experience as a research scientist has included studies of plant nutrition at the Pennsylvania State University, wheat breeding at the University of Nebraska, purification and characterization of DNA polymerases at the University of Rochester, biochemical characterization of insect resistance in potatoes at Cornell University, lipid-mediated signal transduction at the National University of Singapore, plant genome mapping at Texas A&M University, glycolipid biosynthesis at Cornell University, and a study of the biochemical/genetic basis of plant/insect interactions at the U.S.D.A. Beltsville Agricultural Research Center. He has been long interested in international development, due both to his exposure to the dynamic international programs at Cornell and the influence of Professor Norman Borlaug, whose office was located directly across the hall from Dr. Kowalski's laboratory at Texas A&M University.

The second phase of Dr. Kowalski's career has been defined by a transition from research to international work. He received a foreign language area studies scholarship and completed Cornell's one-year intensive Chinese-language program (Chinese FALCON). Subsequently, he worked for the International Service for the Acquisition of Agri-Biotech Applications (ISAAA) in the intellectual property/technology transfer initiative, during which time he conducted the preliminary freedom-to-operate analysis of *Golden*Rice. After working at ISAAA, he earned a J.D. with an emphasis in intellectual property at the Franklin Pierce Law center. He has published numerous research and legal articles.

KRATTIGER, Anatole
Anatole Krattiger, a Swiss citizen, began his career as a farmer, lived in many parts of the world, and is currently a research professor at the Biodesign Institute at Arizona State University (ASU). As adjunct professor at the Sandra Day O'Connor College of Law at ASU, he co-teaches a course on innovation management and controversies in health and agri-biotechnology. He is an Adjunct Professor at Cornell University where he co-teaches a course on IP management in the life sciences. He founded, and serves as Chairman of, *bio*Developments-International Institute, a nonprofit organization that brings people together to jointly develop solutions to problems that extend beyond geographic and cultural frontiers. He recently served as Executive to the Humanitarian Board for *Golden*Rice, a position that required him to work on licensing, technology transfer, and regulatory issues; he also served as Director of Research at MIHR in the U.K. during its formative years. In the early 1990s, he contributed to the international establishment of ISAAA, a global agri-biotechnology broker developing public-private partnerships in agriculture; he served as executive director of ISAAA until 2000. He also briefly worked on biodiversity-policy issues at the International Academy of the Environment in Geneva, Switzerland, and as a scientist in biotechnology at CIMMYT in Mexico.

Dr. Krattiger is a member of the Advisory Council on Intellectual Property of the Franklin Pierce Law Center in Concord, New Hampshire, and a member of the board of the Black Sea Biotechnology Association. He is editor-in-chief of *Innovation Strategy Today* and a member of the editorial boards of the *International Journal of Biotechnology* and the *International Journal of Technology Transfer and Commercialization.* He was a Distinguished Advisor to the Council for Biotechnology Information in Washington, D.C., until the Council merged with BIO. He holds a diploma in farming, a bachelor's degree in agronomy from the Swiss Agricultural College, a master's degree in plant breeding, and a Ph.D. in biochemistry and genetics from the University of Cambridge, U.K.

LASHER, Heidi
Heidi Lasher is a freelance writer and communications consultant. Her work spans a range of health topics from immunization and vaccines to HIV/AIDS, reproductive health, and financial sustainability. Formerly a communications officer for the Children's Vaccine Program at PATH, Ms. Lasher worked extensively in Andhra Pradesh, India on an immunization-strengthening and hepatitis B vaccine introduction project. Before working at PATH, Ms. Lasher was a project manager for a Seattle-based strategic communications consulting firm. Ms. Lasher now provides freelance services to various international public health organizations in training, facilitation, writing and editing, media relations, and advocacy. Ms. Lasher and her family live in Bozeman, Montana.

LESSER, William H.
William Lesser has been in the Department of Applied Economics and Management at Cornell University since receiving his Ph.D. in agricultural economics, with a specialization in marketing, from the University of Wisconsin in 1978. Early on at Cornell he was an innovator in the application of PCs to food distribution, writing some of the earliest specialized software. Much of the time has, however, been focused on the farm and consumer level effects of biotechnology on agriculture. A particular specialization is the ramifications of patents and Plant Breeders Rights. In a related area, he has examined ownership of and access to genetic resources. Work has involved advising the governments of Brazil, Bangladesh, Switzerland and Indonesia, among others. He has written three books and numerous articles and chapters on the subject of agricultural biotechnology. His teaching has included export marketing, international marketing, and futures and options. In 2003, he was appointed as Chair of the Department of Applied Economics and Management at Cornell.

LIU, Connie Kang
Connie Liu has a B.Eng. in biomedical engineering from the Huazhong University of Science and Technology in China and a Master's degree in biotechnology from the University of Toronto. She was a bioprocess engineer at a large biotech company in China before joining the academic research community as a researcher at Hong Kong Polytechnic University. She has industrial experience in project management and technology transfer from academia to industry. She has also received training in Chinese and Canadian intellectual property law, including intellectual property policy and strategies. Her interests revolve around innovation and intellectual property in the biotech and pharmaceutical industries; she has special interest in the commercialization of academic research, licensing technologies, firms' research strategies, and intellectual property management. She was a Research Associate to Dr. Maryann Feldman from 2004 to 2006. Currently, Ms Liu is an M.B.A. candidate at the Rotman School of Management at the University of Toronto.

LIVNE, Oren
Oren Livne is currently the Associate Director for Licensing at the University of California, Santa Barbara (UCSB). He is a founding member of UCSB's first on-campus technology transfer office, the Office of Technology and Industry Alliances. He oversees the patenting and licensing of all inventions developed at UCSB. He has the pleasure of directly managing a significant number of inventions, including those produced by UCSB's world-renowned Solid State Lighting & Display Center. Prior to joining UCSB, he was based at the University of California's Office of The President. He serves as a mentor to UCSB's chapter of Engineers Without Borders, whose mission is to improve quality of life through environmentally and economically sustainable engineering projects. He is also an advisor to students in UCSB's Technology Management Program, which aims to educate the next generation of technology-based entrepreneurs. He is a registered patent agent with several pending and issued patents of his own.

LUND, Brett
Brett Lund has represented and counselled dozens of life science and technology companies in the areas of technology transfer, company formation, venture capital financings, licensing, mergers and acquisitions, and

initial public offerings. Mr. Lund served as a corporate attorney specializing in emerging growth companies at the San Diego office of the law firm Cooley Godward LLP, where he represented a wide range of companies including Maxim Pharmaceuticals, Qualcomm, Acadia Pharmaceuticals, Vertex Pharmaceuticals, AMCC, and the Titan Corporation. Mr. Lund left Cooley Godward to become the Associate General Counsel for Ford Motor Company's telematics division. At Ford, he helped to develop a nationwide wireless voice/data network, in-vehicle hardware, and associated services.

Mr. Lund has also worked in Marketing and Product Development for Johnson & Johnson's diabetes group, and served as Business Development Manager for Incyte Genomics. He is currently the Licensing Manager for Syngenta Biotechnology and is primarily responsible for developing alliances in their BioFuels business. Mr. Lund holds a Masters in Business Administration from the Fuqua School of Business at Duke University, a Juris Doctorate from Duke Law School, and a Bachelors degree in political science from the University of California, San Diego. He is a member of the California Bar Association, North Carolina Bar Association, Association of University Technology Managers, and the Licensing Executives Society.

MACWRIGHT, Robert S.

Robert S. MacWright is Executive Director and CEO of the University of Virginia Patent Foundation, which evaluates, protects, and licenses inventions that originate from research performed at the University of Virginia (UVA). In addition, Dr. MacWright is the founder and President of Spinner Technologies, Inc., a UVA Patent Foundation subsidiary that encourages and assists faculty entrepreneurs and their start-up companies. Dr. MacWright also led Spinner's efforts to create the Jefferson Corner Group, an angel investment fund focused on UVA start-ups. He has been Executive Director of the UVA Patent Foundation since 1997.

Dr. MacWright joined the university licensing profession in 1985, when he became Assistant Director of the Rutgers University Office of Corporate and Industrial Research Services. He initiated the first independent patent and licensing program at Rutgers, and served as Director of the Rutgers Office of Corporate Liaison and Technology Transfer from 1988 to 1992.

Dr. MacWright holds a Ph.D. in biochemistry from Rutgers University and the University of Medicine and Dentistry of New Jersey, and has carried out postdoctoral and industrial research in molecular genetics and protein chemistry. He also holds a law degree from Rutgers Law-Newark. He is a Registered U .S. Patent Attorney, and is admitted to practice in New York, New Jersey, and Virginia. He formerly practiced law in the New York City IP firm of Kenyon & Kenyon, and also in the intellectual property department of Skadden, Arps, Slate, Meagher & Flom, LLP, a well-known New York mergers and acquisitions firm.

MAHONEY, Richard T.

Richard T. Mahoney is Director, Vaccine Access, for the Pediatric Dengue Vaccine Initiative, a program of the International Vaccine Institute (IVI) in Korea. Previously, he was Research Professor in the School of Life Sciences and in the Biodesign Institute of Arizona State University. As a consultant to the Rockefeller Foundation, he played a lead role in the consultative process that led to the formation of MIHR. Previously, he was responsible for institutional development in the establishment and launching of the IVI in Seoul, Korea. In this role, he was responsible for cultivating relations with vaccine manufacturers and managing intellectual property, among other things. Dr. Mahoney has had a long career in public health and is known for his work with the International Task Force on Hepatitis B Immunization, accomplished while he was with the Program for Appropriate Technology in Health (PATH). Before co-founding and joining PATH, he was a Program Officer in Population with the Ford Foundation. He oversaw the development and implementation of IP management policies for the Ford Foundation, PATH, and IVI. Prof. Mahoney continues to write on policy and economic research.

MANGENA, Mosibudi

Mosibudi Mangena is the Minister of Science and Technology in South Africa and President of the Azanian People's Organisation (AZAPO). He was born in Tzaneen, matriculated from Hebron Training College in 1969, and received an M.Sc. degree in Applied Mathematics from the University of South Africa (called the University of Azania on the AZAPO website). He joined the South African Students' Organisation (SASO) and was elected to the Student's Representative Council at the University of Zululand in 1971. Moving back to Pretoria, he became chairperson of the SASO Pretoria branch in 1972. He chaired the Botswana region of the Black Consciousness Movement of Azania (BCMA) in 1981 and the BCMA central committee from 1982 to 1994. He returned from exile in 1994 and became leader of Azapo. He was appointed Deputy Minister of Education in South Africa by Nelson Mandela in 2001, and became Minister of Science and Technology in 2004.

MARCHANT, Gary

Gary Marchant is the Lincoln Professor Emerging Technologies, Law and Ethics at the Sandra Day O'Connor College of Law at Arizona State University. He is also a Professor of Life Sciences at ASU and Executive Director of the ASU Center for the Study of Law, Science and Technology. Professor Marchant has a Ph.D. in Genetics from the University of British Columbia, a Masters of Public Policy degree from the Kennedy School of Government, and a law degree from Harvard. Prior to joining the ASU faculty in 1999, he was a partner in a Washington, D.C. law firm, where his practice focused on environmental and administrative law. Professor Marchant teaches and researches in the subject areas of environmental law, risk assessment and risk management, genetics and the law, biotechnology law, food and drug law, legal aspects of nanotechnology, and law, science, and technology.

MASHELKAR, R.A.

Dr. R.A. Mashelkar is presently the President of the Indian National Science Academy (INSA) and President of Global Research Alliance (GRA), a network of publicly funded R&D institutes from five continents with over 60,000 scientists. Prior to this, for over eleven years Dr. Mashelkar served as the Director General of Council of Scientific and Industrial Research (CSIR), an organization with thirty-eight laboratories and about 20,000 employees. His leadership transformed CSIR into a user-focused, performance-driven organization, a process of transformation that has been recently heralded as one of the ten most significant achievements of Indian Science and Technology in the 20th century.

Dr. Mashelkar is only the third Indian engineer to have been elected as a Fellow to the Royal Society (FRS), London, in the 20th century. He was elected Foreign Associate of the National Academy of Science (U.S.) in 2005, and was only the eighth Indian since 1863 to be elected. He was elected a Foreign Fellow of the U.S. National Academy of Engineering (2003), Fellow of the Royal Academy of Engineering (U.K.) in 1996, and Fellow of the World Academy of Art & Science (U.S.) in 2000. Twenty-six universities have honored him with honorary doctorates, including the universities of London, Salford, Pretoria, Wisconsin, and Delhi. He is currently the President of the Materials Research Society of India.

In post-liberalized India, Dr. Mashelkar has played a critical role in shaping the country's S&T policies. He was a member of the Scientific Advisory Council to the Prime Minister and also of the Scientific Advisory Committee to the Cabinet set up by successive governments.

Dr. Mashelkar has won more than 50 awards and medals, including the S.S. Bhatnagar Prize (1982), the Pandit Jawaharlal Nehru Technology Award (1991), the G.D. Birla Scientific Research Award (1993), the Material Scientist of Year Award (2000), the IMC Juran Quality Medal (2002), the HRD Excellence Award (2002), the Lal Bahadur Shastri National Award for Excellence in Public Administration and Management Sciences (2002), the World Federation of Engineering Organizations (WFEO) Medal of Engineering Excellence by WFEO, Paris (2003), the Lifetime Achievement Award by the Indian Science Congress (2004), the Science Medal by the Academy of Science for the Developing World (2005), and the Ashutosh Mookherjee Memorial Award by the Indian Science Congress (2005), among others.

The President of India honored Dr. Mashelkar with the Padmashri (1991) and with the Padmabhushan (2000), which are two of the highest civilian honors in India, in recognition of his contribution to nation building.

MATTHEWS, Duncan

Duncan Matthews is Senior Lecturer in Intellectual Property Law. He has acted as a consultant to the Directorate General Trade of the European Commission, the ECAP II EC-ASEAN Intellectual Property Rights Co-operation Programme, and the Science and Intellectual Property in the Public Interest Program (SIPPI) of the American Association for the Advancement of Science (AAAS). He holds an Economic and Social Research Council (ESRC) research grant on NGOs, Intellectual Property Rights, and Multilateral Institutions. He is the author of the following publications: "From the August 30, 2003 WTO Decision to the December 6, 2005 Agreement on an Amendment to TRIPS: Improving Access to Medicines in Developing Countries?" [2006] 10 *Intellectual Property Quarterly* 91-130, ISSN: 1364-906X; "TRIPS Flexibilities and Access to Medicines in Developing Countries: The Problem with Technical Assistance and Free Trade Agreements" [2005] 27 *European Intellectual Property Review* 420-427, ISSN: 0142-0461; "Is History Repeating Itself? The Outcome of Negotiations on Access to Medicines, the HIV/AIDS Pandemic and Intellectual Property Rights in the World Trade Organisation" [2004] 1 *Law, Social Justice and Global Development Journal* (LGD), ISSN: 1467-0437; "The WTO Decision on Implementation of Paragraph 6 of the Doha Declaration on the TRIPs Agreement and Public Health: A Solution to the Access to Essential Medicines Problem?" [2004] 7(1) *Journal of International Economic Law* 73-107, ISSN: 1369-3034; "A Strategic Approach to Managing Intellectual Property" (Co-author with J. Pickering and J. Kirkland) in R. Blackburn (Editor), *Intellectual Property and Innovation Management in Small Firms*, London: Routledge, 2003, 35-54, ISBN: 0415228840; *Globalising Intellectual Property Rights: The TRIPs Agreement*, London: Routledge, 2002 ISBN: 041522327X.

MCBRIDE, Timothy B.

Timothy B. McBride is an attorney for the intellectual property law firm of Senniger Powers. Mr. McBride's practice includes all areas of intellectual property law, including the preparation and prosecution of patent applications in the United States and abroad. His work is focused in the following areas: preparation and prosecution of patents in the fields of biotechnology, molecular biology, immunology, and animal science, including patents for transgenic plants and bacteria, novel DNA and protein sequences, vaccines, gene therapeutics, and diagnostics; validity/invalidity and infringement opinions; and counseling related to patentability and freedom to operate.

Mr. McBride received his Bachelor of Science degree in Neurobiology and Animal Physiology from Purdue University in 1996. He received his Bachelor of Arts degree in Psychology from Purdue University in 1997 and his Juris Doctorate from Indiana University School of Law-Indianapolis in 2001. Mr. McBride is also an adjunct professor in the School of Engineering and Applied Science at Washington University, where he teaches a course on intellectual property for engineers and scientists.

MCCALLA, Alex F.

Alex is Professor of Agricultural and Resource Economics, Emeritus, at the University of California, Davis. He was born in Alberta, Canada, and received his first two degrees from the University of Alberta before moving on to the University of Minnesota where he received his doctorate in Agricultural Economics in 1966. Throughout his academic career he was associated with the University of California-Davis where he served as Dean of the College of Agricultural and Environmental Sciences and Associate Director of the California Agricultural Experiment Station (1970–1975) and Founding Dean, Graduate School of Management (1979–1981).

Dr. McCalla is best known for his research in international trade where he has published extensively.

The quality of his research and communication skills has been recognized by the American Agricultural Economics Association, which presented him with its Quality of Communication Award in 1979 and its Quality of Research Discovery Award in 1982. He was elected Fellow of the American Agricultural Economics Association in 1988, Fellow of the Canadian Agricultural Economics Society in 2000, and a Distinguished Scholar of the Western Agricultural Economics Association in 2004. He was a founding member and co-convener of the International Agricultural Trade Research Consortium. He served as the Chair of the Technical Advisory Committee (TAC) of the Consultative Group on International Agricultural Research (CGIAR) from 1988 to 1994.

He elected early retirement from the University of California in June 1994 and was appointed Director of the Agriculture and Natural Resources Department of the World Bank in Washington, D.C., effective September 12, 1994. During his tenure he led a major effort to revitalize the World Bank's commitment to Rural Development. He was appointed Director of Rural Development in July 1997, following a Bank reorganization. He retired from the World Bank December 31, 1999.

In June 1998 he was awarded the Degree of Doctor of Science, *honoris causa,* by McGill University in Montreal, Canada. On December 28, 1999, he was awarded the Doctor's Degree of Honor by the Georgian State Agrarian University. In September of 2004 he received the Distinguished Alumni Award from the University of Alberta.

He served as Chair of the Board of Trustees of CIMMYT, the International Maize and Wheat Improvement Center with Headquarters in Mexico, (2001–2005) and is a member of the Board of Directors of the Danforth Plant Science Center in St. Louis.

MCCRACKIN, Ann M.

Ann M. McCrackin is a registered patent attorney and a shareholder of Schwegman, Lundberg, Woessner & Kluth. Her practice focuses on computer architecture, software, and business methods. She also specializes in reexamination practice and international patent protection. Ms. McCrackin holds a Bachelor's degree from Iowa State University, with a major in speech communication and minors in computer science and English (B.S., 1992). She also completed graduate coursework in computer engineering at Iowa State University. Ms McCrackin received her J.D. from Franklin Pierce Law Center in Concord, New Hampshire, in 1997. At Franklin Pierce, she was a senior editor of *IDEA: The Journal of Law and Technology*. Her legal curriculum concentrated on intellectual property law. Prior to attending law school, she worked for the Center for Advanced Technology Development (CATD) at Iowa State University. At CATD, she analyzed the suitability of commercialization software and negotiated and prepared license agreements for scientific software. Ms. McCrackin is frequently invited to speak on various patent prosecution topics and is a co-editor of *Electronic and Software Patents: Law and Practice*, a treatise published by BNA Books.

MCGEE, David R.

David R. McGee joined the University of California, Davis in January 2004. He is the Executive Director for Technology and Industry Alliances (TIA), an organization within the Office of Research at U.C. Davis. TIA provides patent and copyright protection and licensing services for the University's intellectual property, and helps faculty and industry collaborate on research projects and spinout companies.

Prior to joining U.C. Davis, Dr. McGee was a consultant in biotechnology and intellectual property strategy. In 1987, Dr. McGee founded Large Scale Biology Corporation (LSBC), a biotechnology healthcare firm that develops new biopharmaceuticals, including patient-specific vaccines in plants using viral vectors. He served as the corporate Executive Vice President and the President of the Biomanufacturing Business Unit at LSBC until June 2003. From 1982 to 1987, Dr. McGee was a founding member and Vice President of Operations at Sungene Technologies Corporation, a plant biotechnology company that improved major commercial crop species using genetic engineering and tissue culture. Dr. McGee received his Ph.D. in genetics from Louisiana State University and served as a faculty instructor of zoology and genetics. Dr. McGee currently serves on the boards of a number of private and nonprofit companies.

MEDAKKER, Akshat

Akshat Medakker is a technology manager at Sathguru Management consultants, where he focuses on a variety of projects relating to intellectual property and technology transfer. Currently, he assists clients in the Indian government as well as the private sector in both India and the U.S. in matters of patenting, negotiation for licensing, and drafting of licensing and other IP agreements. Prior to joining Sathguru, Mr. Medakker oversaw strategic business development for a biotech company in India. He graduated from the University of Sydney with a degree in Molecular Biotechnology and has research experience as a molecular biologist in both Australia and India.

Mr. Medakker is also actively involved in the Society for Technology Management. He coordinates the activities of the association, including course design and the delivery of intensive training for technology managers and researchers in IP rights and technology management.

MIN, Eun-Joo

Eun-Joo Min is Senior Legal Officer at the Arbitration and Mediation Center of the World Intellectual Property Organization (WIPO). She holds a Ph.D. in law from Yonsei University in Seoul, Republic of Korea, a certificate degree in international law from the Graduate Institute of International Affairs in Geneva, Switzerland, and was a Fulbright Scholar at the University of Michigan Law School. Prior to joining WIPO in February 2000, Dr. Min taught international law and international economic law at the International Division and College of Law at Yonsei University.

At the WIPO Arbitration and Mediation Center, Dr. Min manages arbitration and mediation cases, develops and implements new alternative dispute resolution

(ADR) procedures, develops educational programs on ADR, collaborates in the development of new information technology applications in support of WIPO case administration, and manages relations between the Center and other organizations and users of the WIPO Arbitration and Mediation Center's services. Dr. Min is the co-editor of *Collection of WIPO Domain Name Panel Decisions* (The Hague: Kluwer Law International, 2003) and has written and spoken extensively on ADR and intellectual property.

MONGEON, Marcel D.

Marcel Mongeon is an Intellectual Property Coach who assists companies and institutions in devising and implementing strategies to help them profit from their intangible assets. He is an experienced international speaker and seminar leader in many fields, including business strategy, understanding legal issues (such as intellectual property management), and negotiations.

Mr. Mongeon is a lawyer qualified to practice in the Canadian provinces of Ontario and Québec as well as in New York. He is a Canadian Registered Patent and Trade-mark Agent. He holds business, law, and science degrees from McGill University (B.Com., LL.B., B.C.L.), McMaster University (M.B.A.), and Swinburne University (M.Sc.). He is a Fellow of the Intellectual Property Institute of Canada. He oversaw all sponsored research, patenting, and commercialization activities of the technology transfer office at McMaster University from 1997 to 2006. He has also had experience in the manufacturing, hospitality, and information technology industries. He has practiced law with major firms in Montréal and Toronto.

Mr. Mongeon is an active Rotarian and has served on the boards of his local chamber of commerce, the Canadian Council of Better Business Bureaus, the international Association of University Technology Managers, and the Canadian University Intellectual Property Group, among others.

MOREL, Carlos

Dr. Carlos M. Morel is currently the Director of the Centre for Technological Development in Health (CDTS), a new unit being implemented at the Oswaldo Cruz Foundation (FIOCRUZ, Rio de Janeiro, Brazil) to stimulate health product innovation.

A molecular biologist and medical doctor by training, Dr. Morel received his M.D. from the Medical Faculty of the Federal University of Pernambuco. He completed his graduate studies at the Biophysics Institute of the Federal University of Rio de Janeiro and at the Molecular Biology Department of the Swiss Cancer Institute in Lausanne (ISREC), Switzerland. His research has been in the field of molecular parasitology, and he has collaborated with various international organizations and research programs working on neglected diseases and capacity building.

Dr. Morel was previously a Professor at Brasilia University (UnB, Brasilia, Brazil) and President of FIOCRU. He was also Director of the UNICEF/UNDP/World Bank/WHO Special Programme for Research and Training in Tropical Diseases (TDR) at the World Health Organization in Geneva, where he established close working relationships with product-development public private partnerships and global ventures committed to public health. He participated actively in the establishment of the Medicines for Malaria Venture (MMV), the Global Alliance for TB Drug Development (GATB), the Drugs for Neglected Diseases Initiative (DNDi), and the Foundation for Innovative New Diagnostics (FIND).

A member of the Brazilian Academy of Sciences and an Honorary Fellow of the Royal Society of Tropical Medicine and Hygiene in London, Dr. Morel holds the National Order of Scientific Merit (Brazil) and Doctor *Honoris Causa* from the Federal University of Pernambuco (Brazil). He has been a member of the MIHR Board of Trustees since its founding.

MOYNIHAN, Michael R.

Michael R. Moynihan has more than 20 years' experience in plant biotechnology. For the past ten years, he has primarily been involved in technology transfer and the development of public-private research networks in his capacities as a Senior Project Director at InterLink Biotechnologies LLC and Director of Biotechnology Development at Fundación Chile. Dr. Moynihan earned an Sc.B. in Biology from Brown University in 1974 and M.A. and Ph.D. degrees in Biology from Harvard University in 1979 and 1982. He was a Visiting Research Fellow at the Institute for Molecular and Cellular Biology at Osaka University, a Postdoctoral Associate in the Section of Plant Biology at Cornell University, a Postdoctoral Associate at the Center for Agricultural Molecular Biology at Rutgers University, and a Principal Scientist in the plant biotechnology laboratory of EniChem Americas.

MUNOŻ TELLEZ, Viviana

Viviana Munoz is a Programme Officer at the South Centre, an intergovernmental organization of developing countries based in Geneva, Switzerland. Ms. Munoz assists the research, policy analysis, policy advice, capacity building, and training activities of the Centre's Innovation and Access to Knowledge Programme. Her efforts support the development, coordinated use, and improvement of the capacities of developing countries and their institutions. The Centre aims to integrate the development dimension into their policies on innovation, access to knowledge, and intellectual property.

Previously, Ms. Munoz worked at Queen Mary Intellectual Property Research Institute, University of London, as Research Assistant for an Economic and Social Research Council (ESRC) project (www.ipngos.org/). The project examined the role that nongovernmental organizations (NGOs) play in supporting the positions of countries on intellectual property, public health, and biodiversity at multilateral institutions. She has also worked as an independent consultant. Some of her recent works include: Munoz V. (with Matthews D.), *Bilateral technical assistance and the TRIPS agreement: the United States, Japan and the European Communities in comparative perspective,* Journal of World Intellectual Property, Vol. 9, No. 6. (November 2006), pp. 629-653, and Munoz V. (with Waitara C.), *An Analysis of the Impact of the Treaty on the Protection of Broadcasting Organisations and Cablecasting Organisations on Developing Countries,* Research Paper 9, South Centre, (December 2006).

Ms. Munoz holds a B.A. in International Relations from the U. Rosario, Colombia, and an M.Sc. in Development Management from the London School of Economics.

MUNYI, Peter

Peter Munyi is the Chief Legal Officer of the International Centre of Insect Physiology and Ecology (ICIPE). He offers in-house legal advice to ICIPE in many areas; intellectual property law is his speciality. Through SEAPRI, an ICIPE initiative, Mr. Munyi also offers legal and policy advice on agriculture, genetic resources, and environmental issues to the governments of developing countries, international organizations, and nongovernmental organizations. He has published widely on many topics, including intellectual property-related issues, genetic resources, and biodiversity. He holds a Master's degree in European Intellectual Property Law from Stockholm University, Sweden, and a Bachelor's Degree in Law from Moi University, Kenya. Prior to joining ICIPE, Mr. Munyi was a commercial lawyer in a private practice based in Nairobi.

MUTSCHLER, Martha

Martha Mutschler is a professor in the Department of Plant Breeding, College of Agriculture and Life Sciences, Cornell University. She directs a research program in tomato and onion breeding and genetics. Her work deals with plant genetics and breeding projects concerning the genetic control of novel traits derived from wild species, the genetic control/physiological mechanisms underlying these traits, and the use of these mechanisms in vegetable improvement. This work has resulted in several U.S. patents, as well as the release of elite breeding lines with novel forms of disease resistance or insect resistance, or modified production traits such as extended shelf life or early maturity. Dr. Mutschler has served on the board of directors for the Cornell Research Foundation (the patent and licensing unit for Cornell University) for over a decade. This service led to her interest in plant intellectual property, a subject on which she has published. It also led her to develop a computer-assisted instruction module for training undergraduate and graduate students in IP issues.

NEAGLEY, Clinton H.

Clinton H. Neagley is Associate Director of Technology Transfer Services at the University of California, Davis. His responsibilities include patenting and licensing inventions in the fields of agriculture, biotechnology, chemistry, and physical science, as well as assisting academic researchers on intellectual property matters. Before joining U.C. Davis, Dr. Neagley was Chief Patent Counsel and Director of Licensing for DNA Plant Technology Corporation (DNAP), where he was responsible for intellectual property and technology contracts, managing the patent portfolio, and licensing and freedom-to-operate assessment. Prior to his position at DNAP, Dr. Neagley spent ten years with the New York City intellectual property law firm of Davis Hoxie Faithfull & Hapgood, where he worked on litigation, licensing, and patent prosecution. As a partner at Davis Hoxie, he played a lead role in the biotechnology group, representing clients from large established companies, start-up companies, and universities. He has lectured and published on many topics of patent law. He is a member of the California and New York Bars and is a registered Patent Attorney. Dr. Neagley has a J.D. from Cornell Law School and a Ph.D. in Chemistry from the University of California, Davis.

NEEDLE, William H.

William H. Needle is the founder of Needle & Rosenberg, P.C., one of the largest intellectual property law firms in the Southeast. He has exclusively practiced patent, trademark, copyright, and trade secret law over his entire 36-year career. Mr. Needle is an adjunct professor of Licensing Law at Emory University School of Law and adjunct professor of Patent Law at Georgia State University College of Law. He serves as a mediator or arbitrator in complex disputes involving intellectual property issues and has been an expert witness in patent, trademark, and copyright infringement actions. On several occasions, he has been appointed to serve as a Special Master by U.S. District Court Judges in patent infringement cases wherein his recommendations on validity, infringement, and damages in two actions were affirmed by the Federal Circuit Court of Appeals.

Mr. Needle has served as a Special Assistant Attorney General for the State of Georgia for intellectual property law issues for over 30 years and is a Fellow of the Lawyers Foundation of the State Bar of Georgia. He is a member of several organizations: the Advisory Board of The Technological Innovation: Generating Economic Results (TI: GER) program, a collaboration between Georgia Tech and Emory Law School that prepares students to commercialize new technologies; the Advisory Board of the School of Biomedical Engineering at the University of Alabama at Birmingham; and the Advisory Board of Georgia Tech's College of Sciences. In addition, he served on the committee that was tasked with formulating Local Rules regarding patent litigation for the U.S. District Court, Northern District of Georgia.

Mr. Needle's peers in the legal community voted him one of *Georgia Trend* magazine's "Legal Elite" every year from 2003 to 2007 and named him one of the "Top 100 Georgia Super Lawyers" every year from 2004 to 2007. He has also been listed as one of the best intellectual property lawyers in Atlanta for over ten years in *The Best Lawyers in America®*. Additionally, he is a certified "Memphis in May" barbecue judge.

Mr. Needle graduated with a B.S. in Chemistry from the Georgia Institute of Technology in 1967. He received his J.D. from the Emory University School of Law in 1970.

NELKI, Daniel

Daniel Nelki is Head of Legal and Operations in Technology Transfer at the Wellcome Trust. Dr. Nelki obtained his Ph.D. in 1987 from the University of London and went on to conduct postdoctoral research in mouse molecular genetics at King's College, London. He obtained a Diploma in Law in 1991 and, following completion of his professional solicitor's qualifications, practiced with the international law firm Baker & McKenzie, eventually specializing as an intellectual property lawyer. He joined the Wellcome Trust in 1995 and played a key role in planning and structuring the Technology Transfer group, thanks to his uniquely multi-faceted perspectives. He is particularly knowledgeable on the topics of intellectual property and charity law, as well as the commercial exploitation of fundamental research. He has earned a Master's in Law (LLM) for his thesis on the ownership of human genes and tissue.

NELSEN, Lita
Lita Nelsen is the Director of the Technology Licensing Office at the Massachusetts Institute of Technology, where she has been since 1986. Every year, the office manages over 400 new inventions originating from M.I.T., the Whitehead Institute, and Lincoln Laboratory. Typically, the office negotiates over 100 licenses and starts up over 20 new companies each year. Ms. Nelsen earned her B.S. and M.S. degrees in Chemical Engineering from M.I.T., as well as an M.S. in Management from M.I.T. as a Sloan Fellow. Prior to joining the M.I.T. Technology Licensing Office, Ms. Nelsen spent 20 years in industry, primarily in the fields of membrane separations, medical devices, and biotechnology; she worked at such companies as Amicon, Millipore, Arthur D. Little, Inc., and Applied Biotechnology. Ms. Nelsen was the 1992 President of the Association of University Technology Managers. She serves on the board of the Mount Auburn Hospital and the Scientific Advisory Board of the Children's Hospital Oakland Research Foundation. She also serves as the intellectual property advisor to the International AIDS Vaccine Initiative and is a founding and current board member of MIHR. Ms. Nelsen is widely published in the fields of technology transfer and university/industry collaborations. She was a CMI Fellow at Cambridge MIT Institute (at the University of Cambridge), where she studied the role of university/industry/government partnerships in technology transfer and local economic development. She is a co-founder of Praxis, the U.K. University Technology Transfer Training Programme.

NEWMAN, Pauline
Pauline Newman is a judge of the U.S. Court of Appeals for the Federal Circuit. She received a B.A. from Vassar College, an M.A. in Pure Science from Columbia University, a Ph.D. in Chemistry from Yale University, and an LL.B. from New York University School of Law. Before her appointment as circuit judge in 1984, she was Director of Patents and Licensing at FMC Corporation in Philadelphia. She worked as Science Policy Specialist at UNESCO, Paris, and as a research chemist at American Cyanamid Company. She has served as adviser to various governmental programs, and has been an officer and director of several bar and scientific associations. She has been awarded the Wilbur Cross Medal by Yale University, the Vanderbilt Medal by New York University School of Law, the Jefferson Medal by the New Jersey Patent Law Association, and the Award for Outstanding Contributions to International Cooperation by the Pacific Industrial Property Association. She is a distinguished Professor of Law at George Mason University School of Law and the author of articles in the fields of innovation and science and the law.

NILSSON, Malin
Malin Nilsson graduated in 2001 from the Swedish University of Agricultural Sciences, where she received her M.Sc. in Horticulture. She studied crop production, plant protection, and developmental studies, but finally got hooked on plant breeding and genetics; it was an interest that brought her to the Research Institute Geisenheim, Germany, where she spent part of her time as a student. Ms. Nilsson has worked in the International Division of the Swedish plant breeding company Svalöf Weibull since 2001. Initially, her responsibilities were mainly within the field of plant variety licensing, ranging from establishing agreements to practical follow-up of the obligations and rights under such agreements. During this period, she became interested in intellectual property rights, an interest that was further developed when she moved on to work with the complete product portfolio as Product Manager in Oilcrops. In 2006, after a short period spent sharpening her skills in the role of Sales Manager Sweden, she took the position of Marketing Manager in Cereals and Oilseeds. Ms. Nilsson's current position allows her to use and further develop her competence in the field of intellectual property, especially in plant variety rights, throughout the value chain.

NOTTENBURG, Carol
Carol Nottenburg is a patent lawyer specializing in biotechnology. She migrated to law after a career in science. Her career path took her from undergraduate days at Caltech (B.S. Biology) to graduate work at Stanford University (Ph.D. Genetics), to a postdoctoral fellowship at the University of California, San Francisco, where she worked in the laboratory of Dr. Harold Varmus, Nobel Laureate, and ultimately to the Fred Hutchinson Cancer Research Center (FHCRC) in Seattle, Washington, where she joined the faculty of the Clinical Division. For five years, her laboratory studied the causes of poor immune system development in patients who have had bone marrow transplants. Dr. Nottenburg then turned to the pursuit of a career in patent law. After graduating *magna cum laude* from the University of Puget Sound Law School (now Seattle University Law School), she joined the law firm of Seed and Berry in Seattle, Washington. She took primary responsibility for a number of small biotech company clients, and her practice focused on integrating patent strategies with business strategies. A move to CAMBIA, a nonprofit, private research institute in Australia, allowed her to become involved in policy issues surrounding intellectual property. At the same time that she was responsible for intellectual property matters at CAMBIA, she spearheaded the development of the internet-based Patent Lens Resource (www.patentlens.net). Now returned to private practice, Dr. Nottenburg assists clients with integrating patent and business strategies and with strengthening the clients' capacity to deal with both ordinary and more esoteric patent matters.

NUGENT, Rachel
Rachel is a senior associate in CGD's Global Health Programs. She provides economic and policy expertise to support HPRN Working Groups, manages CGD programs on Population and Economic Development, and conducts research on other global health topics. She has 25 years of experience as a development economist, managing and carrying out research and policy analysis in the fields of health, agriculture, and the environment.

Prior to joining CGD, Rachel worked at the Population Reference Bureau, the Fogarty International Center of the U.S. National Institutes of Health, and the United Nations Food and Agriculture Organization.

She received her Ph.D. in Economics from George Washington University, and served as associate professor and chair of the economics department at Pacific Lutheran University in Tacoma, Washington. Rachel's publications include a range of topics, from the cost-effectiveness of noncommunicable disease interventions and health impacts of fiscal policies, to impacts of microcredit on the environment in developing countries and the economic impacts of transboundary diseases and pests.

NYAGAH, Ruth Ruguru

Ruth is a Kenyan citizen and has been working in the horticultural industry for the last 12 years, in work ranging from managing export farms to quality control. She currently works as an auditor technical and social auditor for various consumer standards, including EUREPGAP, ETP, HACCP, and BRC. She is also the managing director of Africert Ltd, the first local certification company to be accredited to ISO 65/EN45011 in East and Central Africa, which she set up through funding from GTZ in 2004. Africert's entry in the certification arena in East and Central Africa provides producers, who would have otherwise been marginalized in accessing the lucrative export markets due to the high costs of importing certification services, with an internationally accredited certification company, right within their region.

Ruth is a member of the EUREPGAP Certification body Committee and is a private consultant for both UNCTAD and FAO on issues related to the uptake of private standards and their effect on smallholder farmers in export horticulture in Kenya.

Ruth holds a Bachelor of Science degree in General Agriculture from the University of Eastern Africa, Baraton (Kenya), and a Master of Science in Post Harvest Horticulture from the University of Greenwich, Natural Resources Institute (NRI), U.K.

OEHLER, Joachim

Joachim Oehler is CEO of Concept Foundation, an internationally operating, independent not-for-profit organization that is active in IP management for health products; its goal is to create the best public sector benefits through the out-licensing of intellectual property. Dr. Oehler initiated and established the value-based approach of Concept Foundation to the out-licensing model of its products, and developed an authoritative framework for the performance orientation of license agreements through detailed milestones that maximize public sector benefits. Before joining Concept Foundation, he managed a fully integrated pharmaceutical company out of Tokyo, Japan. His company's capacities spanned the entire spectrum of pharmaceutical business: product development, application laboratories, clinical research, local manufacturing, drug regulatory affairs and registration, clinical market development, marketing, and sales, as well as administration, logistics management, warehousing, and so on. Previously, Dr. Oehler had held several senior divisional management positions in Japanese subsidiaries of multinational pharmaceutical companies. Before working with Japanese companies, he held several marketing positions in the pharmaceutical industry in Europe and North America. Dr. Oehler continues to advise a select group of companies and organizations on cross-cultural management issues.

OLSON, Arne M.

A native of the Chicago area, Arne M. Olson has practiced intellectual property law for over 25 years. He has handled complex litigation involving patent, trademark, copyright, trade secret, Internet domain name protection, cybersquatting, and licensing matters. Mr. Olson advises clients on licensing, intellectual property protection, and litigation involving a wide range of technologies. His clients include several universities and research institutions, as well as publicly traded companies.

Mr. Olson has a bachelor's degree in physics from the University of Chicago and a J.D. with honors from the DePaul University School of Law. He is admitted to practice in Illinois before the U. S. Court of Appeals for the Seventh Circuit and the U. S. Court of Appeals for the Federal Circuit. He is also a member of the Trial Bar of the U. S. District Court for the Northern District of Illinois and the U.S. District Court for the Western District of Michigan. He is registered to practice before the U.S. Patent and Trademark Office.

His professional memberships include the International Trademark Association, the American Bar Association, the American Intellectual Property Law Association, and the American Association for the Advancement of Science. He was designated a "Super Lawyer" by *Law & Politics* magazine in 2005, 2006, and 2007, and was selected as a Leading Lawyer in intellectual property law by the Illinois Leading Lawyers Network, Law Bulletin Publishing Company.

PAGE, Nigel

Nigel is the Finance Editor of *IAM* magazine (www.iam-magazine.com). He has worked for more than 15 years as a journalist and editor in print media. After qualifying as a barrister in the U.K., Nigel switched careers to work in journalism and helped to launch various legal and business publications, including *Legal Business* magazine, the U.K.'s leading monthly magazine for the legal market. He continues to contribute to various newspapers and magazines, including the *Financial Times*, *Euromoney*, and *Financial News*. He was one of the founders of *IAM* magazine in 2004 and became its Finance Editor in 2005.

PARDEE, William

William Pardee received his PhD from Cornell University, and has been professor of Agronomy and Plant Breeding at Cornell since 1966. Earlier, he served as assistant, then associate professor of Agronomy, at the University of Illinois and as visiting professor at Oregon State University. Professor Pardee's research focuses on seed production, seed policies, and crop varietal improvement. He has participated in numerous national committees and symposia, and in international seed programs. He has been elected as fellow in the American Society of Agronomy, the Crop Science Society of America, and the American Society of the Advancement of Science. From 1978 to 1987, he served as Chairman of the Department of Plant Breeding and Biometry at Cornell.

His extension goal is to help farmers and seed growers improve their competitive positions and income through the use of improved seeds of superior varieties. To achieve this, he works closely with seed growers and dealers to encourage the production and distribution of high quality seed, and writes and speaks regularly for grower, dealer, and farmer audiences, providing information designed to help them in their choice of seed. He has participated in policy decisions related to seed at state and national levels, seeking to maintain practical yet effective seed policies and regulations.

PÉCOUL, Bernard

Bernard Pécoul has been Executive Officer of the Geneva-based Drugs for Neglected Diseases Initiative (DNDi) since its inception in 2003.

Dr. Pécoul earned a medical degree from the French University of Clermont Ferrand, France, and a Master's of Public Health from Tulane University in the U.S. He joined Médecins Sans Frontières (MSF) as a volunteer physician in 1983; in Honduras, he provided healthcare to refugees from El Salvador, Nicaragua, and Guatemala. In 1985, still with MSF, he moved to Thailand and Malaysia, managing public health projects for refugees from Vietnam, Burma, and Laos. He was a co-founder and director of research and training from 1988-1991 at Epicentre, an epidemiological research organization in Paris, France. Then, from 1991-1998, he was the Executive Director of the French section of MSF, where he oversaw 100 field projects in 40 countries. From 1998 to 2003, Dr. Pécoul was Executive Director of Médecins Sans Frontières' Campaign for Access to Essential Medicines, whose goal is to increase access to essential medicines in developing countries by advocating for a combination of policies: lower drug prices on a sustainable basis, increased research on neglected diseases, and production of unprofitable but medically necessary drugs.

While at MSF, Dr. Pécoul had been active in the creation of the Drugs for Neglected Diseases Initiative (DNDi), which was finally launched as a foundation in July 2003. In October 2003, he was selected as Executive Director of the fledgling Initiative. DNDi is a not-for-profit organization that seeks to develop and make available drugs that treat neglected diseases (such as sleeping sickness, leishmaniasis, and Chagas disease) that afflict the poor in developing countries. As executive director, Bernard is coordinating the entire research and development initiative and managing a team of project managers and scientists who are located in various parts of the world, particularly in Asia, Africa, and Latin America.

PEFILE, Sibongile

Prior to her appointment as Council of Scientific and Industrial Research (CSIR) Group Manager for R&D Outcomes, Sibongile Pefile was the CSIR Intellectual Property and Innovation Manager. With an academic background in pharmacy, which includes an M.Sc. in Pharmaceutics and a Ph.D. in Pharmacology, Dr. Pefile moved into the field of intellectual property when she became Programme Director at the Centre for the Management of Intellectual Property in Health Research and Development (MIHR). In this capacity, she was responsible for the strategic planning, implementation, and coordination of MIHR capacity development programs in intellectual property management.

Her work as a consultant for the Rockefeller Foundation led to the formation of MIHR. Prior to consulting for the Rockefeller Foundation, she spent several years working for the Technology and Business Development Directorate at the Medical Research Council, where she was responsible for establishing the Indigenous Knowledge systems office. The office addressed policy, ethical, and intellectual property issues relating to health research and indigenous knowledge systems and technologies. She is a member of several international bodies concerned with intellectual property and its management, and has published and presented numerous papers on this topic. In 2006, she attended the Mastering Technology Enterprises programme, IMD, and completed the UNISA/WIPO IP Law Specialization course.

PEZZUTO, J. M.

A biochemist with research interests in the areas of biology-driven natural-product drug discovery and characterization, in particular cancer chemotherapy, cancer chemoprevention, malaria, and AIDS, Professor John Pezzuto is Dean of the College of Pharmacy, University of Hawaii. He has also served as Dean of the College of Pharmacy, Nursing, and Health Sciences at Purdue University in West Lafayette, Indiana. Previously, he held the rank of Full Professor in both the College of Pharmacy and the College of Medicine of the University of Illinois at Chicago, where he was a Distinguished University Professor. Since 1977, he has continuously received support from the U.S. National Institutes of Health and currently serves as the Principal Investigator of a program project grant in cancer chemoprevention, while also a Co-Investigator for an International Collaborative Biodiversity Group. He wrote and coauthored 400 publications and co-invented several patented technologies. He has edited three books, serves on the editorial boards of 11 international journals. He is the former editor-in-chief of the *International Journal of Pharmacognosy,* and of *Combinatorial Chemistry and High Throughput Screening.* He is the current editor-in-chief of *Pharmaceutical Biology.*

PHILLIPS, Peter W. B.

Peter W.B. Phillips is a professor in the department of Political Studies at the University of Saskatchewan. He holds concurrent faculty appointments in both Agricultural Economics and Management at the U. of S. and is a Professor-at-Large in the Institute of Advanced Studies at the University of Western Australia. Dr. Phillips' research concentrates on issues related to governing transformative innovations, a topic that involves examining intellectual property rights for agricultural biotechnology, the economics and management of innovation and trade, and marketing issues related to new technologies. He has done theoretical, empirical, institutional, and policy analysis of technological change, has published a variety of books and journal articles on governing innovation, and has consulted on innovation policy with industry and governments in Canada, the U.S., the E.U., and Australia, as well as with the Organization for Economic Co-operation and Development (OECD). He is either a principal investigator or investigator of seven internationally peer-reviewed research programs that have a combined budget of CAD$52 million. He is currently a member of the Canadian Biotechnology Advisory Committee, a senior research associate with the Estey Centre for Law and Economics in International Trade, a member of the editorial boards of AgBioForum and IP Strategy, a fellow at The Centre for Innovation Studies (THECIS), and a member of the Canadian Association of Business Economists.

PITKETHLY, Robert

Robert Pitkethly is a university lecturer in management studies (intellectual property) at the Said Business School, University of Oxford. In addition, he is a fellow and tutor in management at St. Peter's College, where he is also a senior research associate of the Oxford Intellectual Property Research Centre. His teaching and research interests are centered on strategic management and the management of intellectual property. He has worked as a management consultant in connection with a wide variety of technology-based and general management issues and has also worked as a qualified U.K. and European Patent Attorney in both private practice and industry.

He holds degrees in chemistry, business administration, and Japanese studies. Prior to moving to Oxford, he was a Research Fellow at the Judge Institute at Cambridge University. He has also been a Visiting Research Fellow at the Institute of Intellectual Property and the National Institute of Science and Technology in Tokyo.

POTRYKUS, Ingo

Ingo Potrykus is the engine behind the *Golden*Rice Project and the Humanitarian Board. Together with Peter Beyer, he was one of the inventors of the *Golden*Rice technology. Since his retirement as a professor in 1999, far from settling down, he has devoted enormous efforts to bringing biofortified *Golden*Rice to those who need it.

Prof. Potrykus was born in 1933 in Hirschberg, Silesia, Germany. He has been married since 1960, and has three children and eight grandchildren. In 1968, he earned a Ph.D. in Plant Genetics at the Max-Planck-Institute for Plant Breeding Research, Cologne, Germany.

He conducted research in botany at the University of Basel, Switzerland, and was an Assistant Professor at the Institute of Plant Physiology, Stuttgart-Hohenheim from 1970 to 1974. From 1974 to 1976, he was Research Group Leader at the Max-Planck-Institute for Genetics, Ladenburg-Heidelberg, and then, until 1986, at the Friedrich Miescher-Institute, Basel, Switzerland. From 1986 until his academic retirement in 1999, he was Full Professor in Plant Sciences at the Swiss Federal Institute of Technology (ETH), Zurich.

Since 1974, his research has focused on plant-science-based contributions to food security in developing countries, where he was involved in the development and application of genetic engineering technology for "food security" crops such as rice (*Oryza sativa*), wheat (*Triticum aestivum*), sorghum (*Sorghum bicolor*), and cassava (*Manihot esculenta*). Focusing on problems in the areas of disease and pest resistance that were difficult to solve with traditional techniques, he worked to improve food quality and yield, improved exploitation of natural resources, and improved biosafety. This work was performed by an international team of 60 coworkers, on average, that was financed from competitive grants and core funding. The *Golden*Rice project, initiated in 1991 as Ph.D. project, was possible only because of that core funding. Details of the *Golden*Rice project can be found in approximately 340 publications in refereed journals and 30 international patents.

Professor Potrykus's teaching activities have included lectures and courses in basic and advanced plant biology and plant biotechnology in Biology, Agronomy, Pharmacy, Forestry, and Environmental Sciences departments, as well as International Training Courses such as EMBO. His numerous awards include: the KUMHO (ISPMB) Science International Award in Plant Molecular Biology and Biotechnology in 2000, the American Society of Plant Biologists (ASPB) Leadership in Science Public Service Award in 2001, the Crop Science of America (CSSA) Klepper Endowment Lectureship in 2001, the CSSA President's Award in 2002, and the European Culture Award in Science in 2002. He received an Honorary Doctorate from the Swedish University of Agricultural Sciences in 2002. He is a member of *Academia Europaea*, the World Technology Network, the Swiss Academy of Technical Sciences, and the Hungarian Academy of Sciences.

POTTER, Robert H.

As a Senior Associate at AGBIOS, Robert H. Potter provides biotechnology regulatory, intellectual property rights, and risk assessment expertise to a variety of capacity-building and commercial projects. Before joining AGBIOS in 2005, Dr. Potter was the Technology Coordinator for the U.S. Agency for International Development (USAID) Agricultural Biotechnology Support Project II at Cornell University, where he was responsible for technology evaluation and product delivery planning. Dr. Potter was previously employed as an intellectual property specialist for Cornell University's Strategic World Initiative for Technology Transfer program. He has extensive experience in the preparation and presentation of workshops on intellectual property issues related to agricultural biotechnology and plant genomics. Dr. Potter's scientific training is in plant molecular biology. He holds a Ph.D. from Rothamsted Experimental Station in the U.K., and has postdoctoral experience at the Agricultural University of Norway and Murdoch University, West Australia. His research has included gene expression studies in the developing barley grain, investigations of the molecular basis of host plant response to attack by root-knot nematodes (*Meloidogyne* spp.), and the use of molecular markers in wheat and barley breeding.

RAZGAITIS, Richard

Dr. Richard Razgaitis has 40 years of experience working with development, commercialization, and technology management. He began his professional career as a "rocket scientist" on the Saturn/Apollo lunar launch team, and worked on every launch from Apollo 1 through the first lunar landing. He was a faculty member for 10 years and taught more than 20 different undergraduate and graduate level courses. He was also a research scientist and inventor at a billion-dollar private institute. For 10 years, he was vice president of commercial development/licensing at two different billion-dollar companies. Since 1998, he has been a consultant in IP/technology management, opportunity discovery, valuation, and dealmaking.

He is the author of three books on valuation and dealmaking: *Early-Stage Technologies: Valuation and Pricing*; *Valuation and Pricing of Technology-Based Intellectual Property*; *Dealmaking Using Real Options and Monte Carlo Analysis* (all published by John Wiley). He has also authored two book chapters: "Technology Valuation," published in *The LESI Guide to Licensing Best Practices* (Wiley, 2002), and "Pricing the Intellectual Property Rights to Early-Stage Technologies: A Primer of Basic Tools," *AUTM Technology Transfer Practice Manual* (2003). For more than 10 years, he taught technology valuation and pricing courses for AUTM to more than 2,000 students.

For more than 10 years he held a variety of senior positions in the Licensing Executives Society, including VP and Treasurer. Since 2000, he has been on the Board of the Licensing Foundation, the past three years as its President. He has served on the Board of the National Inventors Hall of Fame Foundation.

He has B.Sc., M.Sc., and Ph.D. degrees in engineering, and an M.B.A. Dr. Razgaitis has been a registered Professional Engineer in Texas, Oregon and Ohio, and is an inventor on four patents. He and his wife have been married 40 years and have five children.

RILEY, M. C.

Mary Riley received her doctorate in cultural anthropology from Tulane University and her law degree from Northern Illinois University. She is a co-investigator with the UIC-based International Cooperative Biodiversity Group (ICBG) and is a visiting senior research specialist at the University of Illinois at Chicago, Program for Collaborative Research in the Pharmaceutical Sciences (PCRPS). In addition, she is an attorney with Merritt, Flebotte, Wilson, Webb & Caruso, PLLC, in Columbia, South Carolina. She edited the volume *Indigenous Intellectual Property Rights: Legal Obstacles and Innovative Solutions* (Altamira Press).

RITTER, John F.

John F. Ritter is the Director of the Office of Technology Licensing and Intellectual Property at Princeton University. Mr. Ritter is a Registered Patent Attorney and has been at Princeton University since September 1996. Prior to joining Princeton, Mr. Ritter was a senior member of the technology licensing team at Rutgers University and held several marketing positions in industry. Mr. Ritter has a B.S. in Engineering and an M.B.A. in Marketing, in addition to a law degree. He is a member of the New Jersey and Pennsylvania bars.

RODIN, Judith

Judith Rodin has served as president of the Rockefeller Foundation since March 2005. Trained as a research psychologist, Dr. Rodin was previously the president of the University of Pennsylvania, and earlier the provost of Yale University. The Rockefeller Foundation was established in 1913 by John D. Rockefeller, Sr. to "promote the well-being" of humanity by addressing the root causes of serious problems. The Foundation works globally to expand opportunities for poor and vulnerable people and to help ensure that the benefits of globalization are shared more equitably.

Judith Rodin was born and raised in Philadelphia, Pennsylvania. She graduated from the University of Pennsylvania, and received her Ph.D. from Columbia University. A pioneer in the behavioral medicine movement, she taught at New York University before embarking on 22 years on the faculty at Yale, where she ultimately held appointments in both the School of Arts and Sciences and the School of Medicine. Named president at Penn in 1994, she was the first woman to serve as president of an Ivy League institution.

Dr. Rodin serves on a number of leading nonprofit boards, as well as on the boards of AMR Corporation, Citigroup, and Comcast Corporation. She is the author of more than 200 academic articles and chapters and has written or co-written 11 books. She served on President Clinton's Committee of Advisors on Science and Technology. A member of a number of leading academic societies, including the Institute of Medicine of the National Academy of Sciences, she has received nine honorary doctorate degrees.

ROHRBAUGH, Mark L.

Mark L. Rohrbaugh, Ph.D., J.D., has served since 1991 as the Director of the Office of Technology Transfer (OTT), National Institutes of Health (NIH), Department of Health and Human Services (HHS). OTT manages the patenting and commercial licensing of a large portfolio of NIH and FDA intramural inventions and contributes to the HHS's intramural and extramural technology transfer policy. OTT licensees have brought to market well over 100 products, 25 of which are FDA-approved; in 2005, the licensee sales generated by these products approached US$5 billion. OTT also advises NIH on the terms and conditions of funding agreements with respect to intellectual property, material transfer, and data rights. Dr. Rohrbaugh serves as Vice-Chair of the Public Health Service Technology Transfer Policy Board and represents the HHS on the National Science and Technology Council Technology Committee. He has represented the HHS at meetings of the World Health Organization (WHO), the Organisation for Economic Co-operation and Development (OECD), and the United Nations Industrial Development Organization (UNIDO).

Dr. Rohrbaugh previously served as Director of the Office of Technology Development at the National Institute of Allergy and Infectious Diseases (NIAID), where he managed a staff that was responsible for the negotiation of technology transfer agreements between industry and academic institutions for the conduct of NIAID intramural basic and clinical research and extramural cooperative networks.

Prior to joining the NIH, Dr. Rohrbaugh conducted molecular and cell biology research in academic and industrial laboratories. He received his Ph.D. in biochemistry from The Pennsylvania State University and a degree in law with honors from The George Washington University Law School, where he served as an Articles Editor for *American Intellectual Property Law Association Quarterly*.

ROSS, Gavin S.
Gavin Ross is Vice President of Business Development for HortResearch (USA), a U.S. subsidiary of HortResearch, a New Zealand fruit research company. He has held a number of positions within HortResearch. In the 1990s, he established a laboratory that focused on the genes and enzymes involved in apple fruit ripening. In 1997, he took over the leadership of the postharvest science group of HortResearch, which was one of the largest research groups in this area in the world. Three years later, he was part of the team that raised the funds to launch HortResearch's fruit genomics program, and he led the program in its early phases. In recent years, he has made the transition from science to business development. His experience includes several years working at a publicly listed New Zealand biotechnology company. In his current role, he represents HortResearch in the U.S. market and is building a genuine presence for the company in North America. He is actively involved in product development and support, as well as technology licensing.

RYAN, Camille D.
A self-professed "late bloomer," Camille D. Ryan began her academic career after working for several years in the local agricultural biotechnology industry. She first worked with a small plant biotechnology company that specialized in developing proprietary technologies and cloned plant varieties for mine reclamation work and site remediation. Subsequently, she moved into a position with the biotechnology department of a large multinational corporation that specializes in crop production, and worked on cross-functional team efforts to bring the first genetically modified canola varieties to the market.

Throughout her academic career, Ms. Ryan has been involved in a number of research projects, including the Innovation Systems Research Network's "Cluster Initiative," which examines intellectual property structures and innovation in the Saskatoon agricultural biotechnology cluster. Her links with local industry have been a natural segué for her collaborations with the National Research Council's Plant Biotechnology Institute (NRC-PBI) and the Industrial Research Assistance Program (IRAP), as well as with the Canadian Light Source Synchrotron (CLS). Currently, Ms. Ryan works as a research assistant for Genome Canada's GE3LS project. She conducts theoretical, empirical, and policy analysis that explores the management of intellectual property in genomics-based research projects. She is currently in the process of finalizing her Ph.D. dissertation in Interdisciplinary Studies at the University of Saskatchewan.

RYGNESTAD, Hild
Hild Rygnestad provides consulting services in the area of project controls, as well as economic and financial analyses, to a range of private and public sector clients with operations in the international marketplace. In particular, she focuses on cost controls, contract management, and risk analysis. Before becoming an independent consultant, Dr. Rygnestad worked as a Program Associate with the Strategic World Initiative for Technology Transfer (SWIFTT) at Cornell University, where she was responsible for the development, design, and maintenance of a contract management database and a Web-based course in intellectual property management. Previously, Dr. Rygnestad had worked as a researcher with the Danish Research Institute of Food Economics, where she conducted research in the area of agricultural economics, especially analyses of environmental economics and policy. Dr. Rygnestad's scientific education is in agricultural economics. She holds a B.Sc. from the Agricultural University of Norway and a Ph.D. from the University of Western Australia.

SALICRUP, Luis A.
Luis A. Salicrup serves as Senior Advisor for International Technology Transfer Activities at the Office of Technology Transfer (OTT) in the Office of the Director of the National Institutes of Health (NIH). He leads OTT's efforts to transfer public health service technologies from NIH and the Food & Drug Administration (FDA) to institutions in developing countries in order to solve global health problems. Dr. Salicrup also developed and implemented The International Training Program for NIH OTT. The goal of this program is to provide practical experience in relevant areas of global health to staff from public and private institutions located in developing countries. Dr. Salicrup is also responsible for teaching the courses "Technology Transfer," "Biomedical Business Development and International Strategic Partnering," and "Biotechnology Business Leadership" at NIH's Foundation for Advanced Studies.

Before joining OTT, Dr. Salicrup was International Health Research Scientist/Program Director at NIH's Fogarty International Center. Before working at NIH, Dr. Salicrup was CEO and President of Techno-Sur and Associates, a consulting firm that provides international health, technology management, and university-industry alliances services to international and regional organizations, as well as to government agencies and universities worldwide. Dr. Salicrup received his Ph.D. in microbiology and molecular genetics from Rutgers University. He also holds a Master's in Technology Management. After completing postdoctoral training at Princeton University and NIH, he served as Manager of the Divisions of Quality Control and Technical Support at Baxter Diagnostics International Inc. and was Associate Professor of Microbiology and Immunology at the Inter American University and the University of Puerto Rico. Dr. Salicrup is a member of numerous professional organizations and has published in several scientific journals.

SALIM, Emil
Emil Salim is on the faculty of economics at the University of Indonesia. Previously, he was the State Minister for Population and Environment from 1978 to 1993. He currently serves as a member of many international and national committees, including the United Nations High Level Advisory Board on Sustainable Development. He serves as Chairman of the National Economic Board, an economic expert team to President Abdurachman Wahid. He was a member of the economic expert team to President Suharto on debt and development issues of the nonaligned countries, and a member of the Indonesian Peoples' Assembly. In addition, he was Co-chairman of the World Commission on Forestry and Sustainable Development.

Dr. Salim also serves as Chairman of the Board of Trustees for a number of leading Indonesian environmental organizations, including the Indonesian Biodiversity Foundation, the Foundation for Sustainable Development, and the Indonesian Ecolabelling Institute. He received his master's degree and his doctorate in economics from the University of California, Berkeley, in the United States.

SALLES-FILHO, Sergio
Sergio Salles-Filho is a full professor at the Department of Science and Technology Policy at the State University of Campinas, SP, Brazil. He earned a B.S. in Agronomic Engineering at the Rural Federal University of Rio de Janeiro in 1980. He earned his Master's degree at the State University of São Paulo in 1985 in the biological treatment of agro-industrial wastes. Since then, he has dedicated himself to studies of the social ramifications of science, technology, and innovation. His first studies were technological assessments, with a specialization in the impact of modern biotechnology on less developed countries.

In 1995, he founded, with a group of colleagues, the Study Group on Organization of Research and Innovation (GEOPI), which is dedicated to the development of theoretical and empirical studies of the management of technology and innovation. Nowadays, his main areas of research are the organization of areas and institutions of ST&I, technological assessment and institutional evaluation, technological prospective, and intellectual property and financing in ST&I.

SANDELIN, Jon
Jon Sandelin graduated from the University of Washington with a degree in Chemistry in 1962, served four years as a Naval Officer on the U.S. submarine Ronquil, and then earned an M.B.A. from Stanford University in 1968. He returned to Stanford University in 1970 as the Financial Officer of the Stanford Computer Center; he later became the Associate Director of the Center. He joined Stanford's Office of Technology Licensing (OTL) in 1984. At the OTL, he was responsible for licensing all forms of intellectual property, including inventions, computer software, and university trademarks. Mr. Sandelin has served as a consultant for the licensing of research-related inventions to other universities, nonprofit research organizations, and governments. He is the author of many articles on technology transfer through licensing, and has given numerous workshops and presentations on this topic in the United States and overseas. Mr. Sandelin served two terms as a vice president of the Association of University Technology Managers (AUTM), where he was responsible for developing AUTM's overseas relationships. He is also past president of the Association of Collegiate Licensing Administrators (ACLA). On July 1, 2002, he was selected to serve a three-year term on the Public Advisory Committee for the U.S. Patent and Trademark Office (USPTO). This Committee prepares an annual report for the U.S. President and Congress on the operations of the USPTO. He was granted emeritus status in March 2003, and now devotes most of his time to consulting projects, primarily for overseas clients.

SASSON, Albert
Professor Albert Sasson, a Moroccan, is a world-renowned international consultant in biotechnology. He has authored more than 200 publications concerning his research and popularization activities in soil microbiology, algology, and agrobiology. He has published books and contributed to publications on biology teaching, environment and development issues, biotechnologies, and food and nutrition. *Biotechnologies in Developing Countries* is one of his outstanding publications.

Professor Sasson is a prolific speaker, with invaluable information and insight in the areas of cloning, genetically modified foods, the use of biotechnology in agriculture and its possible impact on man and the environment, and ethical and legal issues related to biotechnology. He has expert knowledge of how biotechnology can reduce poverty and the successes and failures of its application worldwide.

After a career as a university dean, he joined UNESCO in 1974, where he served as Special Advisor to the UN for over 27 years. Since January 2000, Prof. Sasson has been senior consultant to UNESCO, Moroccan institutions, and the company Publicis Dialog (Paris). He provides special advice to governments worldwide on the development of national policies on biotechnology, and is an advocate for the adaptation of technologies by the third world for their social and economic development.

Professor Albert Sasson is a man with a passion for science, especially for discoveries in the life sciences. He is truly fascinated with the application of science to food, agriculture, medicine, pharmaceuticals, energy, the environment, and bio-remediation.

SATYANARAYANA, Kanikaram
Kanikaram Satyanarayana holds a doctorate degree in biosciences. After a brief postdoctoral stint, he joined the Council of Scientific and Industrial Research in New Delhi. In 1980, he moved to the Indian Council of Medical Research (ICMR). He is involved in science and technology policy and evaluation, and is Chief of the Intellectual Property Rights Unit. For over twenty years, he has worked extensively in the areas of science and technology evaluation and science policy issues; he was instrumental in the formulation of Indian national policies in these areas.

In 1996, Dr. Satyanarayana published the first guidelines for promoting industry-academia partnerships in medical research in *Contract Research, Consultancy and Technology Transfer policy of the ICMR*. These guidelines are currently being revised to be in agreement with the new WTO and IPR regimes. He has organized several training workshops on WTO and IP rights issues for the benefit of scientists at ICMR institutes, medical colleges, and other institutes. Some of these training workshops were conducted with international funding (WHO). He set up the Intellectual Property Rights Unit at the ICMR in 1999 and brought out the *Intellectual Property Rights Policy of ICMR* in 2002. He is a member of several national committees on intellectual property and has participated in several national and international conferences on such topics as globalization, the impact of TRIPS on public health, access to health care in developing countries, and so on. An active researcher, he has obtained competitive grants from various agencies in

India and the World Health Organization. He has also published several papers in national and international journals. He is closely associated with the U.K.-based Centre for the Management of intellectual property in Health R&D (MIHR) and has contributed to their Manual for Technology Transfer Managers. Currently, he is the only member of the International Editorial Board of the second edition of this *Handbook* who is from a developing country. He is a founder and Secretary of the Society for Technology Management, India, and is currently a Senior Deputy Director-General and Chief of the Intellectual Property Rights Unit at the ICMR.

SCHNEIDERMAN, Anne M.

Anne M. Schneiderman is an intellectual property lawyer in private practice in Ithaca, New York. A scientist and registered patent attorney, Dr. Schneiderman counsels clients in a wide range of high-technology industries, including biotechnology, pharmaceuticals, medical devices, agroscience, and mechanical and electrical engineering. Her law practice is involved in the following activities: worldwide patent procurement; conducting due diligence reviews for financings, collaborations, and partnering deals; the preparation of patentability, freedom-to-operate, noninfringement, and validity opinions; and the analysis, development, and establishment of intellectual property portfolios.

Before establishing her law offices, Dr. Schneiderman served as in-house counsel and director of intellectual property for a high-technology start-up company. She then practiced for six years with Pennie & Edmonds LLP, a leading U.S. intellectual property law firm (now dissolved) in their Palo Alto and Manhattan offices.

Dr. Schneiderman is a graduate of Stanford University, with degrees in biological sciences (B.S. with distinction) and law (J.D.). She also holds a Ph.D. in neurobiology from Harvard University. Before becoming a lawyer, she was a neurobiologist with academic appointments at Cornell and Yale Universities. Dr. Schneiderman's training in both science and law has allowed her to assist scientists, inventors, and management teams in transforming their ideas into patentable inventions.

SCHUBERT, Karel R.

Dr. Karel R. Schubert is internationally recognized for his academic and industrial work on plant and microbial biochemistry, molecular biotechnology, metabolic engineering, and for his discovery of natural products and genes to control pests, pathogens, and parasites.

Dr. Schubert received his B.S. degree in chemistry (Magna Cum Laude), from West Virginia University in 1971, and M. S. and Ph.D. degrees in biochemistry from the University of Illinois, Urbana-Champaign, in 1973 and 1975. After completing his doctoral degree, Dr. Schubert was a Research Fellow in the Department of Botany and Plant Pathology at Oregon State University. He received additional postgraduate training in nematology at the University of California, Davis. He was an Assistant and Associate Professor of Biochemistry at Michigan State University, a Research Manager with Monsanto, Assistant Director of the Center for Plant Science and Biotechnology at Washington University, and Director of the Plant Genetic Resources Center at the Missouri Botanical Garden.

From 1990 to 2000, Dr. Schubert held The George Lynn Cross Endowed Chair of Botany and Microbiology and OCAST Most Eminent Scholar at the University of Oklahoma. While in Oklahoma, Dr. Schubert also founded ProTech, Inc., an Oklahoma-based start-up company, and served as Chief Executive Officer and Director of Research. In August 2000, Dr. Schubert joined the scientific and administrative staff of the Donald Danforth Plant Science Center, as the Vice President for Technology Management and Science Administration. As VP for Technology Management and Science Administration, Dr. Schubert has been involved in technology transfer activities, patenting, and licensing within the agricultural and healthcare sectors. In addition to his administrative responsibilities, Dr. Schubert has focused on humanitarian projects, including the nutritional biofortification of cereal and root crops. Dr. Schubert has been involved in the formation of PIPRA and served as the Chairman of its Executive Committee.

SEKI, Akinori

Akinori Seki is president of the Sasakawa Peace Foundation (SPF), an organization committed to fostering international understanding, exchange, and cooperation. Seki studied at the Gakushuuin University of Economics and received his Ph.D. from the London School of Business.

He worked for many years for the Marubeni Corporation, where he became General Manager (Strategies and Coordination) and Deputy Executive Officer (Corporate Strategies Department). He also lived in Africa briefly as President of Gambia Fisheries' Co. Ltd. He joined the SPF in 1999, initially as Program Director, before becoming Chief Operating Officer, then Executive Director, and now President.

He has served as an advisor to many organizations, including the Myanmar Economic and Management Institute, the United Nations Industrial Development Organization (UNIDO), and the University of Cambodia, and he was a committee member of KEIDANREN and of the study group for Indo-China. He serves on the Board of Directors of the Bellagio Forum and is Member of the Advisory Committee, UNIDO (Tokyo Office). He is an Honorary Professor of Tafaccur University, in the Republic of Azerbaijan.

SERAGELDIN, Ismail

Ismail Serageldin is Director of the Library of Alexandria and also chairs the Boards of Directors for each of the Biblioteca Alexandria's affiliated research institutes and museums. He is also a Distinguished Professor at Wageningen University in the Netherlands. He serves as Chair and Member of a number of advisory committees for academic, research, scientific and international institutions and civil society efforts, including the Institut d'Egypte (Egyptian Academy of Science), TWAS (Third World Academy of Sciences), the Indian National Academy of Agricultural Sciences, and the European Academy of Sciences and Arts. He is former Chairman of the Consultative Group on International Agricultural Research (CGIAR, 1994-2000), Founder and former Chairman of the Global Water Partnership (GWP, 1996-2000) and the Consultative Group to Assist the Poorest (CGAP), a microfinance program (1995-2000).

Serageldin has also served in a number of capacities at the World Bank, including as Vice President for Environmentally and Socially Sustainable Development (1992-1998), and for Special Programs (1998-2000). He has published over 50 books and monographs and over 200 papers on a variety of topics, including biotechnology, rural development, sustainability, and the value of science to society. He holds a Bachelor of Science degree in engineering from Cairo University and a Master's and Ph.D. from Harvard University. He has received 19 honorary doctorates.

SHEVELUKHA, Victor S.

Victor Shevelukha was born in 1929, currently lives in Moscow, and is head of the Agricultural Biotechnology Department, Russian State Agrarian University, Moscow. He is a member of the V.I. Lenin All-Union Academy of Agricultural Sciences (VASKhNIL), the Russian academy of Agricultural Sciences, the International Academy of Agrarian Education, the Slavonic Academy, the Agrarian Academy of the Belarus Republic, the International Academy of Informational Sciences, and the Academy of Natural Sciences, among other public academies.

Victor has authored more than 400 scientific works, including 10 monographs and manuals on plant production, plant breeding, seed production, agricultural biotechnology, plant physiology, and agricultural economic policy. He has advised 45 Ph.D. students and 12 doctors of sciences and is currently Chairman of the Scientific Council in RSAU-MAAS, which confers doctorate degrees in the fields of genetics, biotechnology, plant breeding, and seed production.

He worked as a senior agronomist at MAAT's training farm, Druzhba, in the Yaroslavl region (1955-1957); as a secretary of the Ryazantcev CPSU district committee, Yaroslavl region (1957-1959); as the head of agricultural department, Yaroslavl CPSU regional committee; as the first vice-chairman of Yaroslavl regional executive committee (1959-1964); as senior lecturer, associate professor, professor, and head of Crop Science Department at the Belarus Agricultural Academy (1964-1973); the director of Belarus Research Institute for Arable Farming (1973-1974); a secretary of the Central Committee, Belarussia Communist Party (1974-1979); a Deputy Minister of Agriculture of the USSR; a member of Collegium in the USSR Ministry of Agriculture (1979-1983); academic-secretary of Plant Production and Breeding Department, V.I. Lenin All-Union Academy of Agricultural Sciences and Russian Academy of Agricultural Sciences (1983-1994); a deputy of the State Duma, Federal Assembly of the Russian Federation; and vice-chairman of the Committee for Education & Science, the State Duma (1994-2000).

Prof. Shevelukha is also a member of both the Russian Federation Union of Writers and the Russian Federation Union of journalists. He has written and published 10 volumes of fiction and sociopolitical journalism.

Finally, Victor has been awarded the K.A. Timiryazev and V.I. Vernadsky gold medals, orders and medals of the USSR, Russia, and foreign countries, and honorary deeds and titles from the State Duma (Russian Parliament), the Federal Assembly of the Russian Federation, the Ministry of Agriculture, and the Ministry of Education and Science.

SHORT, Jay M.

Jay M. Short is the Founder, President, and Chairman of the E.O. Wilson Biodiversity Foundation. He has more than 20 years' experience working in biotechnology-based businesses, environmentally compatible development, and the commercialization of products derived from biodiversity. He is a Founder of Diversa and has also served as its CEO, President, and CTO. During his tenure, the company established the first agreement ever negotiated with a National Park to access biodiversity for commercial development. Under Dr. Short's stewardship, Diversa established similar pioneering agreements with Russia, Ghana, Kenya, Mexico, Bermuda, and Indonesia. These agreements were in alignment with and even exceeded the recommendations of the Convention on Biological Diversity. The company was also the first to generate biodiversity access royalties for Costa Rica.

Dr. Short led the company's highly successful initial public offering (IPO), which raised over $200 million in gross proceeds; at the time, it was the largest biotechnology IPO ever completed. His team also raised $300 million in committed funding from corporate partners, including Novartis, Syngenta, Dow, Merck, Dupont, Danisco, Givaudan, and Cargill. Dr. Short invented key genomic technologies for the discovery and optimization of products from microbial genes and gene pathways; these technologies are used in industrial, chemical, agricultural, and pharmaceutical applications. Under his leadership, Diversa was one of Deloitte and Touche's "Technology Fast 50" for every year following the company's IPO. Dr. Short also directed industry-leading efforts in bioethics through Diversa's pioneering practice of establishing equitable benefit-sharing relationships with countries that provided genetic materials.

Before joining Diversa, Dr. Short served as President of Stratacyte and V.P. of R&D and Operations at Stratagene. Dr. Short earned his B.A. in chemistry at Taylor University and his Ph.D. in biochemistry at CWRU. He is the author of more than 100 publications and is named as inventor on more than 100 issued patents. His patents were cited by *MIT Magazine* as among the top 10 in the world both in 2003 and 2004 across all industries. He received San Diego's 2001 E&Y Entrepreneur of the Year Award and was the recipient of two first-place awards granted by the UCSD-Connect Program, which recognizes innovation in biotechnology. In 2003, he received the ABL Innovations in HealthCare Gold Award; in 2004, he received the Henry F. Whalen, Jr. Award for Business Development from the ACS. Dr. Short has served on the National Research Council (NRC) panel for the National Institute of Standards and Technology (NIST) and numerous other governmental committees. He currently serves as a Director for Invitrogen, Senomyx, and Anaptys. He is Entrepreneur-in-Residence for UCSD-Connect. In addition, he is an advisor for City National Bank, a fellow of the Explorer's Club, and is founder and co-owner of Capia IP.

SHOTWELL, Sandra L.

Sandra L. Shotwell is a Founder and Managing Partner of Alta Biomedical Group, a consulting firm specializing in technology commercialization. Her current clients at Alta Biomedical Group include both early-stage and established companies, as well as research institutions and nonprofit organizations.

Dr. Shotwell has over 20 years' experience in managing technologies on behalf of U.S. and international research organizations and corporations. She has extensive experience in license negotiation, the development of business strategies, and the management of research administration. Her technology management experience includes positions at Stanford University and the European Union's Joint Research Center. She established the National Institutes of Health (NIH) Technology Licensing Branch, where she directed licensing for the NIH, the Centers for Disease Control, and the Food and Drug Administration. She served as Director of Technology and Research Collaborations at Oregon Health Sciences University (OHSU), where she was responsible for technology licensing, company spinouts, research grants and contracts, and company-sponsored research. While at OHSU, she assisted in the creation of six new spinout companies.

Dr. Shotwell currently serves on the Board of Directors of the Oregon Bioscience Association and the International Sustainable Development Foundation. She has previously served on the Boards of Virogenomics, Inc., the Stanford OTL Gap Fund, and the Association of University Technology Managers. She is active in the Licensing Executives Society and the Association of University Technology Managers. Dr. Shotwell did her undergraduate work at Princeton University, earned a Ph.D. in Biology from the California Institute of Technology, and did postdoctoral research in Neurobiology at Stanford University School of Medicine.

SLATE, Peter J.

Peter J. Slate is the founding Chief Executive Officer of Arizona Technology Enterprises, the technology licensing and venturing arm of Arizona State University. Mr. Slate has extensive experience as an entrepreneur and advisor to emerging and start-up venture companies. He has also held senior business development and strategy positions with public and private companies, including Baxter International, where he was Director of Corporate Strategy and the founder of Baxter's Global Technology Outlicensing Group, and Zenith Electronics, where he played a key role in the company's operational and financial restructuring. Prior to joining Zenith, Mr. Slate was the Vice President and Associate General Counsel of Primecare International, Inc., a leading physician practice management company, where he oversaw acquisitions and financing transactions. Mr. Slate began his career as a corporate attorney with the law firm of Katten, Muchin, and Zavis in Chicago, specializing in mergers and acquisitions, securities, private equity, and technology development transactions.

Mr. Slate has a B.A. from the University of Michigan, a Juris Doctorate from George Washington University, and a Master's in Business Administration from the Kellogg Graduate School of Management at Northwestern University. He has served on a number of corporate and philanthropic Boards of Directors and is a past Chairman of the Chicago Chapter of the Licensing Executives Society (LES). Mr. Slate lectures regularly on the subjects of technology and investment due diligence, venture capital, strategic alliances, and licensing.

SLOMAN, Robert G.

In May 1992, Robert G. Sloman acquired exclusive rights to develop and license technology transfer management software from his former employer, Washington Research Foundation. The software was originally designed by Mr. Sloman and his WRF team. In October 1992, he founded Inteum Co. and began to develop and license the software. Inteum Co. has grown to be the leader in its field, with software installations across North America and around the world. The head office of Inteum is in Kirkland, Washington, and there is a branch office in Akron, Ohio. Inteum has long-term business relationships with several companies that contribute to its services and products. In 2000, the originally licensed software was superseded by a completely new system, designed in-house and called Inteum C/S. Mr. Sloman has hired and trained a highly successful team of individuals with diverse backgrounds and established Inteum Co. LLC as an internationally recognized, vital, and successful organization. The team members are also company shareholders.

Mr. Sloman had prior professional experience with Flow Systems, Inc. in Kent, Washington and Monsanto Australia Ltd. in Melbourne, Australia. He has developed an effective management style and a structured, managed approach to growth, focusing on managing talented people effectively and providing a rigorous infrastructure to support them.

SODERSTROM, Jon

Jon Soderstrom is currently the Managing Director of the Office of Cooperative Research at Yale University. The Office manages the intellectual assets created at Yale in order to achieve the maximum benefit for the public and provide a financial return that will support the university's research efforts. He is responsible for (1) developing and managing the intellectual property portfolio, (2) defining and executing commercialization strategies, including the negotiation of licenses and corporate-sponsored research agreements, and (3) developing and marketing business concepts for new spin-off ventures to the investment community. Since joining the Office in 1996, he has participated in the formation of more than 25 new ventures, including polyGenomics, Molecular Staging (acquired by Qiagen), Agilix, Asilas Genomic Systems, Achillion Pharmaceuticals (NASQ: ACHN), PhytoCeutica, Protometrix (acquired by Invitrogen), Iconic Therapeutics, Applied Spine Technologies, HistoRx, and VaxInnate. Collectively, these companies have raised over $350 million in professional venture capital.

Prior to holding this position, Dr. Soderstrom was the Director of Program Development for Oak Ridge National Laboratory (ORNL); previously, he had served for ten years as Director of Technology Licensing for Martin Marietta Energy Systems. In the Office of Technology Transfer, he directed a group of ten professionals responsible for negotiating licenses and Cooperative Research and Development Agreements (CRADAs). Dr. Soderstrom was a founding board member and past president of the Association of Federal Technology Transfer Executives. He is also a member of the Licensing Executive Society and the Association of University Technology Managers (AUTM). He is the President-Elect of AUTM, has served as Vice President for Public Policy, and is a member of the Board of Directors and Executive Committee. He is frequently asked to lecture and teach seminars on various aspects of the technology transfer process and economic development both within the United States and abroad. He has testified before Congress on technology transfer issues and served as an expert witness in patent infringement litigation.

In addition to his professional accomplishments, Dr. Soderstrom was honored as the 87th "Point of Light" by President George H. W. Bush in March of 1990 for his volunteer work in constructing and rehabilitating low-income housing in East Tennessee. Dr. Soderstrom received his Ph.D. from Northwestern University in 1980 and his A.B. from Hope College in 1976.

SOEJARTO, D. D.

A plant taxonomist and economic botanist by training, Professor Soejarto is best known for his more than 40 years of plant exploration work, which covers more than 20 countries. He has established three Herbarium Research Institutions (in Colombia, the Philippines, and Vietnam). He was honored as Founder of the Herbarium of the University of Antioquia (Medellin, Colombia), during a 2004 national conference on medicinal plants that celebrated the founding of the herbarium and the deposit of its 100,000th specimen. During his exploration program in Southeast Asia under the U.S. National Cancer Institute's funding (1986-2004), the anti-HIV calanolides were discovered from a species of *Calophyllum* trees of Malaysia.

He is a co-author, together with the NCI scientists, of the calanolides patent, and was elected University of Illinois Senior University Scholar, 1996-1999. He developed an expertise in pharmacognosy and IP issues as a result of a long period of association with chemist and biologist colleagues, as well as with the Office of Intellectual Property/Technology Management at the University of Illinois, Chicago. He has served as Editor of *Journal of Ethnopharmacology* (1988-2004), and is currently a member of the editorial board of five scientific journals. He has been a recipient of an NIH/FIC ICBG grant as Principal Investigator for two cycles (1998-2003; 2003-2008), and is author and co-author of more than 200 scientific papers and book chapters, and three books.

SOMERSALO, Susanne

Susanne Somersalo has her Ph.D. in Plant Physiology from the University of Turku, Finland. After an academic career at the University of Turku and University of Helsinki, Dr. Somersalo served several years in Helsinki University's licensing office as project manager and research evaluator. She then earned a master's degree in Intellectual Property Law at Franklin Pierce Law Center, New Hampshire, U.S.A. She is a registered patent agent with the U.S. Patent and Trademark Office and currently works as an IP specialist in the Washington, D.C.-based law firm Dodds and Associates.

SOUTHAVONG, B. H.

Professor Bounhong Southavong is a pharmacist and pharmacognosist with special expertise in the study of Lao Traditional Medicines, especially medicinal plants. Since 1997, he has been Director of the Traditional Medicine Research Center, Ministry of Health of Laos, Vientiane, Lao P.D.R. He is also President of the Council of Medical Sciences, and has been Vice President of the National Ethic Committee on Medical Sciences Research, and of the Council of Medical Care Professionals of the Ministry of Health of Laos. The Project Leader of an ICBG program since 1998, he has the responsibility of implementing research on the studies of medicinal plants of Laos. One outcome of this ICBG program is the Lao Biodiversity Fund (LBF), established in 2004, for which he serves as Vice President. The mission of LBF is to promote the sustainable utilization of Lao plant resources, the conservation of biodiversity (in particular, medicinal plants), community development, and the protection of Lao Traditional Medicines.

STEINBOCK, Martha Bair

Martha Bair Steinbock serves as the Deputy Assistant Administrator, Office of Technology Transfer, Agricultural Research Service (ARS), U.S. Department of Agriculture (USDA), in Beltsville, Maryland. In this capacity, she helps oversee the national technology transfer efforts of the USDA, including the development of cooperative research agreements. She also serves as the U.S. Executive Secretary for the US-EC Task Force on Biotechnology Research. Prior to becoming Deputy Assistant Administrator, Ms. Steinbock was the Technology Transfer Coordinator for the Pacific West Area of ARS. She has also worked as an international affairs specialist for the USDA Office of Agricultural Biotechnology and the USDA Foreign Agricultural Service. Prior to joining USDA, she worked as a consulting economist for the Food and Agriculture Organization of the United Nations in Rome, Italy. Ms. Steinbock received a Masters degree in International Affairs from the Johns Hopkins University, School of Advanced International Studies, and did her undergraduate studies at Portland State University and Reed College, in Portland, Oregon. Ms. Steinbock is a native of northern California, where she was raised on a family farm.

STEVENS, Ashley J.

Dr. Stevens has been Director of the Office of Technology Transfer at Boston University since 1995 and is also Director for Research Programs in the Institute for Technology Entrepreneurship and Commercialization in the School of Management, where he teaches a graduate level, inter-disciplinary course on Technology Commercialization. Before joining Boston University, he was Director of the Office of Technology Transfer at the Dana-Farber Cancer Institute, a teaching affiliate of the Harvard Medical School.

Prior to entering the technology transfer profession, Dr. Stevens worked in the biotechnology industry for nearly ten years. He was a co-founder of Kytogenics, Inc., where he is still a Director. He was also co-founder and General Manager of Genmap, Inc., and Vice President of Business Development for BioTechnica International. He started his career with The Procter & Gamble Company, where he held a number of positions in sales, marketing, strategic planning, and acquisitions.

Dr. Stevens is very active with the Association of University Technology Managers, most recently as Vice President, Annual Meeting and Surveys, and publishes and lectures frequently on many aspects of technology transfer, including the Bayh-Dole Act, the economic impact of technology transfer and its role in economic development, and the role of technology transfer in global health and technology valuation. AUTM presented him with the Bayh-Dole Award for 2007. Dr. Stevens holds a Bachelor of Arts in Natural Sciences, and a Master of Arts and a Doctor of Philosophy in Physical Chemistry from Oxford University.

STREITZ, Wendy D.

Wendy D. Streitz is the Director of Policy, Analysis, and Campus Services (PACS) in the University of California's central Office of Technology Transfer. The PACS unit coordinates the system-wide technology transfer program and has wide-ranging responsibilities, including developing and implementing policies, providing guidance for campuses and external entities regarding the University's policies and practices, training, and legislative analysis.

Prior to joining the University of California, Ms. Streitz was Associate Director of Intellectual Property and Technology Transfer at Auburn University in Alabama, where she was directly involved in the broad spectrum of technology transfer. Her caseload included technologies from both the physical and life sciences. Previously, she had spent twelve years as an electrical engineer and engineering manager at Westinghouse Electric Corporation, holding leadership positions in radar signal processing.

Ms. Streitz received a BS in Engineering from Harvey Mudd College and an MSEE from Johns Hopkins University.

SWAMINATHAN, M. S.

Professor M. S. Swaminathan has been acclaimed by *TIME* magazine as one of the twenty most influential Asians of the 20th century, one of the only three from India, the other two being Mahatma Gandhi and Rabindranath Tagore. He has been described by the United Nations Environment Programme as "the Father of Economic Ecology," and by Javier Perez de Cuellar, Secretary General of the United Nations, as "a living legend who will go into the annals of history as a world scientist of rare distinction." He was Chairman of the UN Science Advisory Committee, set up in 1980 to take follow-up action on the Vienna Plan of Action. He has also served as Independent Chairman of the FAO Council and President of the International Union for the Conservation of Nature and Natural Resources. He is the current President of the Pugwash Conferences on Science and World Affairs.

A plant geneticist by training, Professor Swaminathan's contributions to the agricultural renaissance of India have led to his being widely referred to as the scientific leader of the green revolution movement. His advocacy of sustainable agriculture leading to an "evergreen revolution" has made him an acknowledged world leader in the field of sustainable food security. The International Association of Women and Development conferred on him their first international award for his significant contributions to promoting the knowledge, skill, and technological empowerment of women in agriculture, and for his pioneering role in mainstreaming gender considerations in agriculture and rural development. Professor Swaminathan was awarded the Ramon Magsaysay Award for Community Leadership in 1971, the Albert Einstein World Science Award in 1986, and the first World Food Prize in 1987.

Professor Swaminathan is a Fellow of many of the leading scientific academies of India and the world, including the Royal Society of London and the U.S. National Academy of Sciences. He has received 55 honorary doctorate degrees from universities around the world. He currently holds the UNESCO Chair in Ecotechnology at the M. S. Swaminathan Research Foundation in Chennai (Madras), India, and was Chairman of the National Commission on Agriculture, Food, and Nutrition Security of India until October 2006.

SYDARA, K.

Associate Professor Kongmany Sydara is a natural product chemist with special expertise in the isolation of biologically active compounds from Lao plants and in the standardization of Lao Traditional Medicines. He is the Deputy Director of the Traditional Medicine Research Center, Ministry of Health, Vientiane, Lao P.D.R. Since 1998, he has served as a Co-Project Leader of an ICBG program, with responsibility for implementing the studies of medicinal plants of Laos. He is co-author of more than 10 scientific papers and books.

TARZIAN SORENSEN, J. A.

Dr. Jill Tarzian Sorensen is Associate Provost and Director of the Office of Licensing and Technology Development at Johns Hopkins University. She is an IP lawyer with nearly 20 years of law and business experience. She has spent 18 of those years at the University of Illinois at Chicago, where she served as Assistant University Counsel and then Associate University Counsel beginning in 1987. In 1998, she became Director of Technology Management and Assistant Vice Chancellor for Research, reorganizing the office according to a decentralized model responsive to the needs of the university's schools and faculty. In five years, her office nearly doubled the university's invention disclosures and the number of licenses it executed. The office also promoted new models of technology transfer, including leveraging intellectual property, particularly in global health, for sustainable economic development in developing countries. Last year, she assumed a new position as Director of Health Initiatives, building international partnerships focused on global health.

TAUBMAN, Antony

Antony Taubman is currently Acting Director and Head of the Global Intellectual Property Issues Division (including the Traditional Knowledge Division and Life Sciences Program) of WIPO, a position he assumed in May 2002, with responsibility for programs on intellectual property and genetic resources, traditional knowledge and folklore, the life sciences, and related global issues. After a diplomatic career, he left the Australian Department of Foreign Affairs and Trade (DFAT) in 2001 to join the newly formed Australian Centre for Intellectual Property in Agriculture, at the Australian National University, teaching and researching on international IP law. From 1998 to 2001, he was Director of the International Intellectual Property Section of DFAT, and in that capacity was engaged in multilateral and bilateral negotiations on intellectual property issues, domestic policy development, regional cooperation, and TRIPS dispute settlement. He has taken part in many training and capacity building programs on intellectual property law and TRIPS in Australia and a number of Asian countries. He has authored a training handbook on intellectual property and biotechnology, a comprehensive study on the implementation of the TRIPS Agreement, and a range of academic and general publications on international intellectual property law and policy. He has held a teaching appointment at the School of Law at the University of Melbourne, delivering a specialist postgraduate course on TRIPS Law and Practice.

He joined DFAT in 1988 as a career diplomat, and his service included disarmament policy and participation in the negotiations on the Chemical Weapons Convention, a posting in the Australian Embassy in Tehran as Deputy Head of Mission, and a posting to the Hague as Alternate Representative to the Preparatory Commission for the Organisation for the Prohibition of Chemical Weapons and Chair of the Expert Group on Confidentiality. He previously worked for WIPO from 1995 to 1998; his duties then included development cooperation in Asia and the Pacific, the development of the revised WIPO program and budget, and associated policy development. A registered patent attorney, he worked in private practice in the law of patents, trademarks, and designs in Melbourne in the 1980s. His tertiary education has included computer science, mathematics, engineering, classical languages, philosophy, international relations and law, and he has taught ancient Greek philosophy at Melbourne University.

TERNOUTH, Philip

Philip Ternouth is the Associate Director for R&D and Knowledge Transfer at the Council for Industry and Higher Education (CIHE). He is also a Regional Advisor for Knowledge Transfer Partnerships, a U.K. government-sponsored program that employs graduates who are supervised by researchers and work with companies on business transformation projects. He originally read Natural Sciences at Sidney Sussex College at Cambridge University. He then spent the next 11 years in research, urban regeneration projects, and management in the public sector. In 1985, he went into the IT industry, specializing in pre-sales consultancy. He then moved into marketing. In 1995, after five years of board-level appointments in small companies and management and marketing consultancies, he joined Vuman Limited, the technology exploitation company of the University of Manchester, where he became Business Development Director.

Since 2000, he has been active in a number of knowledge transfer activities, including researching and writing "Knowledge Transfer, Towards a Strategic Framework" and "The Business of Knowledge Transfer" for the Council for Industry and Higher Education. He has chaired the board of a start-up company, Manchester Geomatics Limited. Recently, he researched and co-authored "International Competitiveness: Businesses Working with U.K. Universities." He is currently engaged in editing for publication a set of case studies that illustrate the support of entrepreneurship in and around Oxford.

Mr. Ternouth is an active member of national and international organizations that provide mutual support and professional development in the transfer of university technology to commercial application. He has presented numerous papers on aspects of technology and knowledge transfer at conferences in the U.K., U.S., and Europe. He has also undertaken a number of consultancy assignments in the U.K. and overseas for research institutions and governments, assisting them in developing their knowledge transfer agendas. He holds professional and postgraduate qualifications in marketing and is a member of the Institute of Directors.

THANGARAJ, Harry

Harry Thangaraj is Director of Research at MIHR. He began his career as a doctor after graduating from the Christian Medical College, Vellore, India. After completing a Ph.D. in Microbial Genetics in London in 1991, he worked until 2004 on various research projects involving pathogenic mycobacteria; he has several publications on this topic.

In 2005, Dr. Thangaraj completed a Master's degree in Intellectual Property Management at Queen Mary University, London (QMUL), during which he was awarded a prize by Glaxo Smith Kline for highest scores in the Patent Law examinations. He worked as an intern both at the Enterprise and Innovation Office of St. George's University London and at Bristows, one of the U.K.'s largest and oldest firms specializing in IP law. In 2006, he obtained a Certificate in Intellectual Property Law from QMUL, a foundational exam (part-qualification) for U.K. Patent Attorneys. He is currently involved with "Global Access Strategies" and the creative uses of IP management for the Pharma-Planta Consortium, an E.U.-funded initiative to develop plant-derived biopharmaceuticals for the treatment and diagnosis of diseases that disproportionately affect poorer populations of the world.

THOMSON, Jennifer A.

Jennifer Ann Thomson is Professor of Microbiology in the Department of Molecular and Cell Biology at the University of Cape Town, South Africa (UCT). Previously, she had held the positions of Head of the Department of Microbiology at UCT, the Director of the Laboratory for Molecular and Cell Biology at the Council for Scientific and Industrial Research, and Associate Professor in the Department of Genetics at the University of the Witwatersrand in Johannesburg. Her research involves the development of genetically modified maize that is resistant to the Maize streak virus (endemic to Africa) and tolerant to drought. She received an honorary doctorate from the Sorbonne University, Paris in 2005, and the UNESCO/L'Oreal award for Women in Science in 2004. She is Chair of the Board of the African Agricultural Technology Foundation, based in Nairobi, Kenya. She is a Director of the South African Pebble Bed Modular Reactor (Pty) Ltd. She has published a book, *Genes for Africa: Genetically Modified Crops in the Developing World.*

THORNSTRÖM, Carl-Gustaf

Carl-Gustaf Thornström is Associate Professor in social and economic geography. His teaching and research emphasize agricultural issues, natural resources management, and geopolitics. He is also a guest researcher in genetic policies at the Swedish Biodiversity Centre, an adviser to the Swedish University of Agricultural Sciences (Sida), and a referee to the Swedish Government office regarding genetic policy issues. For more than 20 years, Dr. Thornström has worked with policy issues related to international agricultural research at SAREC, a department for research cooperation within Sida (Swedish International Development Cooperation Agency). Dr. Thornström's research focuses mainly on policy: genetic resources, intellectual property rights, and coherence issues that affect international agreements and processes; specific topics include life patents, GMOs, protection of traditional knowledge, enclosure of the biological/genetic commons, access to genetic resources, and proprietary science. Dr. Thornström was born June 18, 1946. He is married, with two children.

TUCKER, William T.

William T. Tucker was born in the U.K. and educated in Australia. He holds a B.Sc. (Hons) and a Ph.D. in Microbiology from the University of Queensland. Dr. Tucker has held postdoctoral research fellowships at Stanford University (with Professor Stanley Cohen) and at the Research School of Biological Sciences at the Australian National University in Canberra, Australia. He also holds an M.B.A. degree from St. Mary's College in Moraga, California.

Dr. Tucker's career has focused on agricultural biotechnology. During his ten-year tenure with Advanced Genetic Sciences, and later its successor organization, DNA Plant Technology, he worked first as a research scientist, and later in technology management and business development. He then joined Applera Corporation (Applied Biosystems in Foster City, California), where he was part of the team that licensed PCR technology for commercial applications. Dr. Tucker then joined the business development team at the agricultural genomics unit of Celera Genomics, where he sought out agricultural applications of molecular marker technology, high throughput sequencing, and related genomics platforms. He continued this work when Paradigm Genetics (based in North Carolina) acquired the plant-related part of Celera's agricultural genomics business.

In 2003, Dr. Tucker joined the Office of Technology Transfer at the University of California, Office of the President (UCOP), in Oakland, California, where he focused on the licensing of plant varieties developed by scientists at U.C. Davis and U.C. Riverside. Since 2004, Dr. Tucker has been the Executive Director of Research Administration and Technology Transfer at UCOP.

VAN FLEET, Justin W.

Justin W. van Fleet is the Founder and Principal educational consultant for The Advance Associates, an international education and development consulting company based in Washington, D.C. He specializes in the development of online and offline curricula, capacity-building and training programs, and evaluation methodologies for a variety of audiences. His most recent intellectual property and indigenous knowledge project involved the development and facilitation of a training program for South African small-scale farmers on intellectual property rights to promote and protect indigenous knowledge and resources.

Previously, he has held staff positions in educational capacities for NetAid (New York City), the Harvard University François-Xavier Bagnoud Center for Health and Human Rights (Boston), and the American Association for the Advance of Science's Science and Human Rights program (Washington, DC). He holds a Master of Education from the Harvard University Graduate School of Education and is currently pursuing his Ph.D. in International Education Policy at the University of Maryland.

VAN MONTAGU, Baron Marc

Baron Marc Van Montagu is an Emeritus Professor at Ghent University, and founder and Chairman of the Board of IPBO, the Institute for Plant Biotechnology for Developing Countries. He received a Ph.D. in organic chemistry/biochemistry from Ghent University in 1965, and served as the Director of the Department of Genetics at the Flanders Interuniversity Institute for Biotechnology, before joining the faculty at Ghent University in 1999.

Dr. Van Montagu has made pioneering contributions to plant gene discovery, including the discovery of the gene transfer mechanism between *Agrobacterium* and plants, which was central to the development of transgenic plants. His work at the Lab of Genetics, Ghent University, produced two spin-off biotech companies, Plant Genetic Systems (PGS) and Crop Design. His research at PGS led to the construction of the first herbicide tolerant plants, as well as the construction of the first plants producing the Bt (*Bacillus thuringensis*) insecticide. His was listed among the top 100 living contributors to biotechnology by *The Scientist* magazine and, until 2004, was the most cited scientist in the field of Plant and Animal Science.

He is currently the President of the European Federation of Biotechnology (EFB) and the Chairman of the Public Research and Regulation Initiative (PRRI), and is a member of several other scientific advisory committees. He has been granted numerous prizes and awards in recognition of his pioneering research, including the Japan Prize and the Theodor Bucher medal. In 1990, he was granted the title of Baron by King Baudouin of the Belgians.

VIJAYARAGHAVAN, K.

K. Vijayaraghavan (Vijay) is a Certified Management Consultant (CMC) and a Fellow of the International Council of Management Consulting Institutes (ICMCI). He holds a Master's degree and Fellowship in public accounting and management consulting, with a focus on strategic and technology management consulting. He is the Chief Executive of Sathguru Management Consultants Pvt Ltd, a large consulting firm based in Hyderabad, India. Sathguru advises government organizations, multilateral and bilateral development institutions, private enterprises, and NGOs in several countries across the Asian region. Mr. Vijay is engaged in shaping a number of Indian policy initiatives in the life sciences, and is a member of selected national committees constituted for this purpose. Sathguru is also an Associate of the program Cornell-in-India. Mr. Vijay co-directs Cornell's program in India, which also extends to several other countries in Asia. He is a Regional Coordinator for ABSPII South Asian activities (in India and Bangladesh).

VIKSNINS, Ann S.

Ann S. Viksnins is an acknowledged expert in PCT law. She prepares and prosecutes patent applications in the biological arts, including molecular biology, immunology, plant sciences, and cellular biology. Ms. Viksnins graduated *cum laude* from St. Olaf College with a double major in biology and religion. At the University of Minnesota Law School, Ms. Viksnins earned a Juris Doctor degree *cum laude* and was an Editor of the *Minnesota Law Review*. Ms. Viksnins is also a graduate of the College of Biological Sciences at the University of Minnesota, where she earned a Master of Science degree in genetics. Ms. Viksnins has been practicing intellectual property law since 1992. She is a frequent writer and speaker on patent law issues, and has served in leadership roles in various professional organizations at the national level. Ms. Viksnins is a member of the Minnesota bar, a registered U.S. patent attorney, registered to practice before the U.S. Court of Appeals for the Federal Circuit, and the managing partner of the law firm of Viksnins Harris & Padys PLLP.

VILJAMAA, Kimmo

Kimmo Viljamaa is consultant at Advansis Ltd, Finland. He also works as a part-time researcher at the Research Unit for Urban and Regional Development Studies at the University of Tampere. His current work deals with various domestic and international development projects related to national and regional innovation policies. Recently, he has worked actively with these issues, particularly in several Eastern European countries. Before taking his current positions, he worked for seven years at the University of Tampere on various research projects related to regional innovation policy and regional development, including the Local Innovation Systems (LIS) project coordinated by the Industrial Performance Center at MIT. His recent research focuses on the dynamics of regional innovation systems, the interplay of technological development and regional development policy, regional industrial clustering, the role of universities in regional innovation systems, and the role of knowledge management in regional innovation policy.

VU, Bui Minh

Professor Bui Minh Vu is an economist with special expertise in forest economics. He is Director of the Institute for International Business Management and Training, Vietnam, and Chairman of the Forest Economics branch, Vietnamese Forestry Science Association. He is best known for his research on Agro-forestry economics. His work has been honored by the Vietnamese Agriculture and Rural Development Ministry, which presented him with its Hung King Award in 1981, and an Award of Science and Technology in 1986. He also received the Award for Forestry Development of Vietnam. Throughout his academic career, Professor Bui Minh Vu was associated with Hanoi University of Economics, Hanoi University of Agriculture, and the Institute for Agricultural Science Research, Hanoi University of Forestry. At the moment, he is a special expert in community development through a micro-loan program.

WATAL, Jayashree

Jayashree Watal has been a Counsellor in the Intellectual Property Division of the WTO since February 2001. Ms. Watal has more than 22 years of experience working in the Indian government; for ten of those years, she was devoted to policy, diplomacy, research, and administration on intellectual property rights. She worked in the Indian Ministry of Commerce as Director of the Trade Policy Division in New Delhi from 1995 to 1998, representing India at a crucial stage in the Uruguay Round TRIPS negotiations in 1989-90.

She has researched and published articles on issues related to intellectual property rights, including a book, *Intellectual Property Rights in the WTO and Developing Countries* (Oxford University Press, India and Kluwer Law International, 2001). She was a Visiting Scholar at the Center for International Development at Harvard University (2000), the Institute for International Economics, Washington, D.C. (Oct. 1998-August 2000), and the George Washington University Law School, Washington, D.C. (1997- 2000).

WATANABE, Kazuo N.

Kazuo N. Watanabe is a research professor who studies plant genetic resources, IP rights issues, biosafety, and bioethics research at the Gene Research Center of the University of Tsukuba, Japan (2001–present). He has a part-time service to the International Plant Genetics Resource Institute (IPGRI) for providing his scientific expertise on plant genetics and biotechnology, including biosafety and IP rights aspects for developing countries.

After receiving his Ph.D. in plant breeding and plant genetics from the University of Wisconsin-Madison, Dr. Watanabe worked from 1988 to 1996 on the germplasm enhancement of tuber-bearing *Solanum* species for the Centro Internacional de la Papa (CIP) in Lima, Peru. CIP gives scientific and technical support regarding potato production and breeding to developing countries. From the end of 1991 to 1996, he was seconded to the Department of Plant Breeding at Cornell University as an adjunct assistant professor as part of a shuttle research program on *Solanaceae* molecular genetics. He was also an advisor to ISAAA for issues of biotechnology transfer in developing countries. Dr. Watanabe returned to Japan and was an associate professor at the Institute of Biology-oriented Science and Technology at Kinki University from 1996-2001. His work dealt with transgenic crops and biosafety issues associated with environmental risks to plant genetic diversity. Since his return to Japan, Dr. Watanabe has been serving the Japanese government in negotiation sessions at the Cartagena Protocol on Biosafety for the Convention on Biological Diversity (CBD). He also serves developing countries on plant genetic resources associated with issues related to the CBD and FAO-IT. Dr. Watanabe has served as a technical expert for projects involving plant genetic resources in several developing countries, as well as the global PGR program that has been run by the Japan International Cooperation Agency since 1996. Currently, he serves as a research review board member for the National Genebank Project, the Rice Genome Project, and the STAFF research institute of the Japan Bioindustry Association. Dr. Watanabe is also an adjunct professor at Cornell University and the Institute of Advanced Studies at the United Nations University. He teaches courses on multi-disciplinary research activities that use plant genetic resources to build capacity and create policies in developing countries.

WEIDEMIER, B. Jean
B. Jean Weidemier is the principal of Cambridge Licensing Law, LLC. Ms. Weidemier's legal practice concentrates on technology transfer, and she has extensive experience in the following areas: patent and software licensing in life sciences and high-technology industries, university and nonprofit technology transfer, strategic collaboration agreements, research and development contracts, and related agreements. Prior to founding her own firm in 2005, Ms. Weidemier was employed in the following capacities: at Testa, Hurwitz & Thibeault as Counsel in the Licensing Group; at the Massachusetts Institute of Technology as Counsel and Technology Licensing Officer; and at Hershey Foods Corporation as Counsel. Ms. Weidemier is a graduate of Dickinson College (B.A., Psychology) and the University of Richmond School of Law.

WHITHAM, Michael E.
Michael E. Whitham is a principal in the law firm of Whitham, Curtis, Christofferson & Cook. He has worked in the field of intellectual property law for over 20 years and has been frequently invited to lecture on topics such as inventorship, licensing, technology audits, and patent practice. His law firm is focused on intellectual property law. It handles the following issues: intellectual property licenses; litigation; contractual matters, including employee/consulting agreements, joint development agreements, material transfer agreements, supply agreements, and so on; counseling, including technology audits, patent evaluations, strategic acquisitions, patent development strategies, trademark branding strategies, and copyrighted material acquisition and distribution; patent preparation and prosecution; trademark and copyright registration; and antitrust. Mr. Whitham represents large and small companies, as well as several nonprofit research organizations and universities throughout the world (particularly in the U.S., Germany, Japan, Korea, and Brazil). Mr. Whitham is a licensed attorney and holds degrees in biochemistry, chemistry, and law. Mr. Whitham also serves on the Board of the Albert B. Sabin Vaccine Institute, a nonprofit organization dedicated to the development, supply, and use of vaccines by people throughout the world, particularly those in most need.

WOLSON, Rosemary
Rosemary Wolson is Intellectual Property Manager at the Council for Scientific and Industrial Research (CSIR) in Pretoria, South Africa. She has a B.Sc. (Hons) degree in Microbiology and an LL.B., both from the University of Cape Town. In her previous position as Intellectual Property Manager at the University of Cape Town (UCT), she participated in establishing UCT Innovation, the division responsible for UCT's technology transfer and research contract management functions. Her experience as an early technology transfer practitioner in a developing country sparked her interest in broader policy issues related to the roles of innovation and intellectual property rights in promoting development; she takes on selected applied-research projects in these areas from time to time. She is also involved in various capacity-building and information-sharing initiatives in South Africa, other countries in Africa, and other parts of the world. She is a member of the International Advisory Committee of Public Interest Intellectual Property Advisors (PIIPA) and sits on the Executive Board of the Southern African Research and Innovation Management Association (SARIMA), a regional network of stakeholders.

WYSE, Roger E.
Roger E. Wyse is Managing Director and General Partner of Burrill & Company, a life sciences merchant bank and leading life sciences venture capital firm located in San Francisco, California. Dr. Wyse joined Burrill & Company in 1998 and has overseen venture capital investing, partnering, and the spinout of technology from large companies in the agricultural, nutraceutical, health and wellness, and industrial biotechnology fields. The firm has over $850 million under management.

Dr. Wyse chairs or serves on the boards of 11 private companies. He is Co-Chairman of the newly formed $150 million Malaysian Life Capital Fund. He is also a member of the International Advisory Panel for Biotechnology (BioIAP) for the Prime Minister of Malaysia. He was founder and Chairman of the Alliance for Animal Genome Research.

He has over 27 years of experience as an internationally recognized scientist and as a dean at two major research universities, Rutgers and the University of Wisconsin-Madison. Before joining Burrill & Company, Dr. Wyse served for five years as Dean of the College of Agricultural and Life Sciences at the University of Wisconsin-Madison. From 1986 to 1992, he was Dean of Research at Rutgers University.

Dr. Wyse earned international recognition for his basic studies in plant biochemistry. He has published over 150 scientific papers. In 1982, he received the prestigious Arthur Flemming Award for the Outstanding Young Scientist in the U.S. Federal Service. He was elected a Fellow of both the Crop Science Society of America and The American Society of Agronomy. He has also served as a consultant to numerous Fortune 500 companies.

XU, Ji
Ji XU, graduated from China Agricultural University, received post-graduate education in the United States (University of Maryland and California State University), and on-the-job training by United Nations Specialized Organizations in Italy/Rome, Thailand/Bangkok, and the Philippines/Cebu. He served the Chinese Ministry of Agriculture as a government officer and was then appointed by the Central Government as the Alternate Permanent Representative of P. R. China to the United Nations Agencies on Food and Agriculture in Rome. He has also served the Food and Agriculture Organization of the United Nations and was assigned by FAO as the Assistant FAO Representative in China. Currently, he is a Professor and Advisor of International Relations in China Agricultural University.

XUAN, L. T.
A plant physiologist, biochemist, and biotechnologist by training, Dr. Le Thi Xuan has dedicated more than 20 years of her research to plant tissue culture technology. She has been a major contributor in the establishment of plant biotechnology facilities and training in many research institutes throughout Vietnam, from 1975 to the present. From 1994 through 1999, she led a MacArthur Foundation-funded project as its Principal Investigator on the discovery of bioactive compounds in plants from the Cuc Phuong National Park, Vietnam. Since 1998, she has been Project Leader of an International Cooperative Biodiversity Groups (ICBG) program, responsible for implementing the bio-conservation of bioactive and threatened plants of Cuc Phuong National Park.

XUAN, Vo-Tong
Dr. Vo-Tong Xuan is a distinguished agricultural scientist, an outstanding educator, a low-profile institution builder, and a national and international leader in agricultural development.

As a scientist, he is widely recognized for his expertise in the management of saline and acid-sulphate soils and other problem soils in Vietnam. He is an expert in rice production and in rice-based farming systems, as well as in agricultural diversification in the Mekong Delta. His technical expertise and strong farmer-focused leadership in the Mekong Delta greatly increased rice productivity and contributed to the emergence of Vietnam as the third-largest rice exporting country in the world. Xuan has authored and co-authored six books and more than 100 technical papers about agricultural, rural development, and sustainable food security.

As an educator, he emphasized scientific as well as down-to-earth hands-on-training in the University of Cantho, at which he served as Chairman of the Departments of Bio-Agronomy and Agronomy, and Assistant Dean of Agriculture. He rose to the rank of Vice Rector of the University of Cantho and, in 2000, was elected President of Angiang University, a position he still holds.

As an institution builder, Xuan developed and strengthened the Mekong Delta Farming Systems Research and Development Institute and served as its Director from 1983 to 2001. He also served as FAO Project Coordinator for the establishment of Agricultural Service Centers for Small Farmers. He organized the Vietnam Farming Systems R & D Network and has been serving as its Coordinator since 1991.

As a national leader in agriculture, Dr. Xuan was appointed member of the National Council on Science and Technology, the National Council on Education, the National Council on Professorial Titles Advisory Council of the Vietnam Chamber of Commerce and Industry, the Steering Committee of the Vietnam-Holland Research Program on Rural Development, and the Consultants' Group to the Prime Minister.

As an international leader in agriculture, he is widely recognized for his integrated approaches to agricultural development and deep concern for efficient and effective use of natural resources, sustainability, and environmental issues, as well as, for food security problems of developing countries. He is a strong advocate of the farming system approach in agricultural development.

He has served in key positions in the following international organizations: Member, Board of Governors, Asian Institute of Management in Manila; Member, Board of Trustees of IRRI; Member, Board of Trustees of The Rockefeller Foundation; Member, Board of Trustees of the International Potato Center at Lima, Peru; Member, FAO's Advisory Committee on Farmer-Centered Agricultural Resource Management Program; Member, Technical Advisory Committee of the CGIAR; Member, Policy Advisory Council, Australian Centre for International Agricultural Research; Member, Advisory Council of the Asian Development Research Forum.

Dr. Xuan served as international consultant, lecturer of IFAD, FAO, DANIDA, SIDA, and IDRC-Singapore since the 1980's.

He received from the Prime Minister of Canada a certificate of recognition for his "dedication and contribution to the world of sciences." The Ministry of Agriculture, Fisheries and Forestry of the Republic of France, awarded him the "*Chevalier de l'Ordre du Merite Agricole Medal*." He was elected the 2002 Nikkei Asia Prize for Regional Growth; Most Distinguished Alumnus of the University of the Philippines College of Agriculture Alumni Association; Ramon Magsaysay Award for Government Service; and the 2005 ASTD Derek Tribe Award. Other awards include: the People's Teacher Award, Vietnam Farmers' Federation Medal "For the Cause of the Farmers' State Award as "Hero of the Working Class," Outstanding Scientific Achievement Award from the Prime Minister, Most Distinguished Alumnus Award from the University of the Philippines at Los Banos.

YIN, Ronald
Ronald Yin is a partner with DLA Piper US LLP. He specializes in patent law and has over 30 years of experience in this field. He received a B.S. in Physics from M.I.T., an M.S. in Applied Physics from Cornell University, and a J.D. from Georgetown University. He began practicing patent law at RCA Corporation in Princeton, New Jersey. He later moved to California, where he joined a small, multi-national company, Measurex Corp., where he was the sole in-house counsel. At Measurex, he had responsibility for general matters, as well as patent matters. After Measurex, he practiced for 20 years with the firm of Limbach, Limbach & Sutton, headquartered in the historic Ferry Building in San Francisco. After the firm dissolved, he joined the firm of Gray Cary Ware & Freidenrich, which merged with Piper Rudnick and DLA in 2004 and subsequently became DLA Piper. He is a member of the bar of the states of California and New Jersey, and is registered to practice before the U.S. Patent and Trademark Office. He is also a former member of the Executive Committee of the Intellectual Property Section of the State Bar of California.

YOUNG, Terry A.
Terry A. Young has more than 20 years experience in IP rights, innovation management, and technology commercialization. He was Assistant Vice Chancellor for Technology Transfer for The Texas A&M University System and Executive Director of its Technology Licensing Office. He currently serves as the Director of Research Development at the University of South Dakota.

He has started three technology transfer offices (TTOs) from scratch and led each office to success. He has also started two companies, one of them a university spinout that licensed university technology. From 2001-2002, he served as President of the International Association of University Technology Transfer Managers, which has nearly 4,000 members. He has made more than 55 trips abroad in the capacity of an expert consultant on IP rights and technology transfer issues. He has authored more than ten book chapters or journal articles on IP rights and technology commercialization, including a chapter entitled National Innovation Systems to be published in *Innovation and Business Partnering in Japan, Europe and the United States*, (London: Rutledge, September 2006). In 2002, he was appointed a member of the National Academy of Engineering of the Czech Republic, in recognition of his work establishing an intellectual property and innovation commercialization regime in that country. In 2004, he was recognized by Nigerian academicians as the country's intellectual property Man of the Year. In 2005, he received a U.S. National Service Award for his contributions to economic growth in Eurasia. Also in 2005, he was appointed by U.S. President George W. Bush as one of only five members of the U.S.-Russian Innovation Council on High Technologies, whose goal was to improve scientific cooperation between the two countries.

ZABLOCKI, Edward M.

Mr. Edward M. Zablocki, M.S., C.I.P. has worked for 24 years in the area of research administration at the State University of New York at Buffalo. He graduated from Williams College *magna cum laude* with induction into the Phi Beta Kappa national honor society. Mr. Zablocki has a master's degree from the University at Buffalo, with a concentration in the area of public administration. He has been involved in a range of research administration activities, including developing a technology incubator, fostering university/industry relations, creating research centers, promoting the University's contribution to local economic development, and promoting research compliance. He is presently the Research Subjects Protection Administrator, overseeing compliance with the ethical imperatives and regulatory mandates that apply to research involving humans and animals. Mr. Zablocki is a Certified IRB Professional (C.I.P.), which demonstrates his knowledge in the field of human research subject protection.

ZUCKER, Howard A.

Dr. Howard Zucker, M.D., J.D., is the Assistant Director General of the WHO for Health Technology and Pharmaceuticals and also the Representative of the WPO Director General for Intellectual Property, Innovation, and Public Health. He received his B.S. degree from McGill University and his M.D. from George Washington University School of Medicine. He trained in pediatrics at Johns Hopkins Hospital, anaesthesiology at The Hospital of the University of Pennsylvania, pediatric critical care medicine and paediatric anaesthesiology at The Children's Hospital of Philadelphia, and pediatric cardiology at Children's Hospital Boston at Harvard Medical School. He was an Assistant Professor at Yale University School of Medicine, an Associate Professor at Columbia University College of Physicians and Surgeons, Adjunct Associate Professor at Cornell University Medical School, research affiliate at M.I.T. and on the faculty at the National Institutes of Health. He received his J.D. from Fordham University School of Law and his Masters in Law from Columbia Law School. Dr Zucker served as a White House Fellow and most recently was Deputy Assistant Secretary of Health at the U.S. Department of Health and Human Services. He completed the National Preparedness Leadership Initiative Executive Education program at the Kennedy School of Government/Harvard School of Public Health and is admitted to the Bar of the Supreme Court of the United States. Dr Zucker is also a member of the Council on Foreign Relations and listed in Best Doctors in America and Who's Who in the World.

Glossary

In order to develop a coherent system for best practices in intellectual property management, the various *terms of art* commonly used must be clearly and unambiguously defined, such that they are standardized and universally understood to have the same meaning. In this glossary we have attempted to present precise, accurate definitions for important, commonly used terms in the fields of technology transfer and IP management. We hope that providing such definitions will make possible clear, transparent communication and thereby lead to increased mutual understanding between technology transfer professionals, IP managers, researchers, investors, and entrepreneurs involved in the business of promoting innovation. Clear communication that promotes increased understanding will be particularly important in international contexts. For the use of specific definitions in agreements, readers should also consult their institution's legal advisors. While some areas of law are fairly uniform throughout the world, language differs significantly from country to country, and even within countries, for some areas of law.

The definitions contained in the glossary are derived, in part, from McCarthy's Desk Encyclopedia of Intellectual Property.[1] In addition to this glossary, the reader is encouraged to refer, for expanded definitions and additional terms, to online intellectual property glossaries, including those found on the following Web sites:

- World Intellectual Property Organization: www.wipo.int/tk/en/glossary/index.html.
- U.S. Patent and Trademark Office: www.uspto.gov/main/glossary/index.html.
- U.S. Department of State: usinfo.state.gov/products/pubs/intelprp/glossary.htm.

assignment
A transfer of intellectual property (IP) rights. An assignment of a patent, for example, is a transfer of sufficient rights so that the recipient has title to the patent. An assignment can be a transfer of all rights of exclusivity in the patent, a transfer of an undivided portion (for example, a 50 percent interest), or a transfer of all rights within a specified location (for example, a certain area of the United States). Anything less is considered to be a license transfer, rather than a patent transfer.

Bayh-Dole Act of 1980
The U.S. Bayh-Dole Act (P.L. 96-517, 94 Stat. 3015, codified at 35 U.S.C. §§ 200–211) allows universities, not-for-profit organizations, and small businesses to retain certain IP rights related to inventions made via federally supported R&D. Serving as the statutory foundation facilitating federally supported R&D technology transfer, the Act was designed to promote commercialization of innovations arising from such R&D through cooperation between the research community, industry, and state and local governments.

MIHR/PIPRA. 2007. Glossary. In *Intellectual Property Management in Health and Agricultural Innovation: A Handbook of Best Practices* (eds. A Krattiger, RT Mahoney, L Nelsen, et al.). MIHR: Oxford, U.K., and PIPRA: Davis, U.S.A. Available online at www.ipHandbook.org.

Berne Convention
A major multinational copyright treaty, with nearly 150 members. There are five main points to the Berne Convention: (1) national treatment, that is, nondiscrimination with respect to foreign authors and copyright owners; (2) no formalities, that is, copyright is automatically granted and is not conditioned on formalities such as registration or notice; (3) minimum duration of copyright; (4) moral rights provided to authors under the national laws of member nations; and (5) copyright protection independent of whether such protection exists in the country of origin.

best mode
A condition for the grant of a patent, found in the patent specification. An inventor must describe and disclose the best method he or she knows for carrying out the invention.

biotechnology
The use of biological methods (often genetic engineering and related advanced-molecular-biology applications) to produce products, processes, and related services. Generally, these are patentable under U.S. patent law.

claims
The section of the patent that defines an invention (the technology that is the exclusive property of the patentee for the duration of the patent) and is legally enforceable; that is, the claims set the metes and bounds of the patent rights. The patent specification must conclude with a claim, particularly pointing out and distinctly claiming the subject matter that the applicant regards as the invention or discovery. The claim or claims are interpreted as set forth in the specification: the terms and phrases used in the claims must be sufficiently described in the specification, that is, patent claims must read in the light of the specification. The specification discloses and the claims define the invention.

commercialization
The process of taking an invention or discovery to the marketplace. It involves working the idea into a business plan, consideration of protection options, and determining how to market and distribute the finished product.

compulsory license
A license granted by the state upon request to a third party that, through the license, is permitted to exploit a patented invention after the owner of the patent has refused to provide a voluntary license under acceptable conditions.

confidential disclosure agreement
See **confidentiality agreement**.

confidentiality agreement (nondisclosure agreement, confidential disclosure agreement)
A legal document through which intellectual property can be disclosed by one party to another wherein the latter party is permitted to use the information for certain purposes, and only those purposes, that are stated in the agreement and agrees not to disclose the information to others.

continuation
A second patent application containing a disclosure identical to one in a previous (parent or grandparent) application filed by the same applicant as the original application, while the original application is still pending, and that is entitled to the filing (priority) date of the original application.

continuation-in-part
A second patent application containing a disclosure identical to one in a previous (parent or grandparent) application filed by the same applicant as the original application, but, in contrast to the continuation application, also contains new matter not found in the original application. Hence, whereas claims that rely on matter in the parent application are entitled to the original filing date priority, claims that rely on any new matter are entitled only to the later continuation-in-part application filing date.

contributory infringement
An indirect infringement of IP rights in which people, or organizations, contribute to a direct act of infringement by another (in order to aid or abet the act of infringement), for example, knowingly selling an article that is used solely to practice a patented process or to manufacture a patented product.

copyright
An exclusive right conferred by the government on the creator of a work to bar others from reproducing, adapting, distributing to the public, performing in public, or publicly displaying said work. Copyright does not protect an abstract idea; it protects only the concrete expression of an idea. In order to obtain copyright protection, a work must have originality and some modicum of creativity.

Convention on Biological Diversity
An international agreement articulated at the 1992 Earth Summit in Rio de Janeiro, the Convention seeks to establish a comprehensive strategy for sustainable development, setting out commitments for maintaining the world's ecological underpinnings in light of increasing business and economic development. The Convention established three main goals: the conservation of biological diversity, the sustainable use of its components, and the fair and equitable sharing of the benefits from the use of genetic resources.

cross licensing
A legal agreement in which two or more parties that have potentially conflicting patent claims, or other conflicting IP rights, reach an agreement to share the IP rights in question through a reciprocal licensing arrangement.

dependent claim
A claim in a patent that refers back to a previous claim and defines an invention that is narrower in scope than that in the previous claim. A dependent claim is written in such a way as to be more restricting than the technology defined in the previous claim (often an independent claim).

descriptive mark
A word, picture, or other symbol that describes some quality or trait of a product or service, such as the purpose, size, color, class of users, or end effect on users. A descriptive term is not considered to be inherently distinctive; to establish validity of a descriptive mark for registration or protection in court, proof of acquired distinctiveness of the mark is needed. This acquired distinctiveness confers secondary meaning. For example, "Kentucky Fried Chicken" a mark that originally was descriptive, subsequently acquired secondary meaning as a trademark for a distinctive type of commercial food product.

design patent
A government grant of exclusive rights in a novel, nonobvious, and ornamental industrial design. A design patent confers the right to exclude others from making, using, or selling designs that closely resemble the patented design. A design patent covers the ornamental aspects of a design; its functional aspects are covered by a utility patent. A design patent and a utility patent can cover different aspects of the same article.

differential pricing (tiered pricing)
The practice of setting different prices for different markets—typically higher prices in richer markets and lower prices in poorer markets.

disclosure of origin
A requirement imposed on patent applicants to disclose in patent applications the geographic origin of biological material on which the invention (subject of the patent application) is based.

divisional patent application
A patent application that is carved out of a parent application, such that the parent application is divided into one or more divisional patent applications. Divisional applications are entitled to the original filing-date priority of the parent application.

due diligence
Investigations undertaken to assess the ownership and scope of one or more IP rights that are being sold, licensed or used as collateral in a transaction. This is done in order to identify business and legal risks associated with the IP rights being analyzed.

duration
The term, or length of time that an IP right lasts. A U.S. utility patent on an invention, for example, has a duration of 20 years from the date on which the patent application was filed, as does a plant patent. The duration of a U.S. copyright is usually the life of the author plus 70 years (for works created after January 1, 1978). Protection of information as a trade secret lasts as long as the information remains secret. Duration of a trademark continues as long as it is used (as a source indicator) and properly maintained/protected.

examination. *See patent examination.*

exclusive license agreement
A legal document licensing intellectual property to another party for its exclusive use. Exclusively licensed patent rights cannot, within the scope or field of the exclusive license, be subsequently or simultaneously licensed to any other party.

field-of-use restriction
A provision in an IP license that restricts use of the licensed intellectual property by the licensee to only in a defined product or service market.

first to file
A rule under which patent priority is determined. The rule gives priority to the party that first files a patent application for an invention, rather than to the party that is first to invent. First to file is followed by almost every nation in the world except the United States. For trademarks, priority between conflicting applications to register a trademark is handled by publishing the application with the earliest filing date for possible opposition by the applicant with a later filing date. In the United States, ownership of a trademark is determined by who was first to use it, not by who was first to file an application for registration. However, under the intent-to-use system, an application for registration can be filed prior to actual use of a mark.

first to invent
A rule under which patent priority is determined by which inventor was the first to actually invent, rather than by who was the first to file a patent application. This is the rule followed in the United States. *Compare to first to file.*

freedom to operate
The ability to undertake research and/or commercial development of a product without infringing the unlicensed intellectual or tangible property rights of others.

functionality
That aspect of design that makes a product work better for its intended purpose, as opposed to making the product look better or to identify its commercial source.

Indigenous Cultural and IP Rights
Indigenous Cultural and IP Rights refers to the rights to a heritage, that its, to the objects, sites, knowledge, and methods of transmission of communities that have traditionally been defined by the social ownership of knowledge. This right privileges customary law over modern law. Heritage includes all aspects of culture (art, music, dance, literature, and so on), indigenous knowledge (medicinal, nutritional), and land management practices. There are numerous attempts today to give legal substance and scientific validity to indigenous knowledge. Article 29 of the Draft Declaration of the Rights of World Indigenous People states that "*[i]ndigenous people are entitled to the recognition of full ownership, control and protection of their cultural and intellectual property.*"

industrial property
Industrial property is a subset of intellectual property, referring to those types of intellectual property that have an industrial application. Specifically, it refers to patents, trademarks, designs, mask works, and plant breeders' rights.

infringement
An invasion of an exclusive right of intellectual property. Infringement of a utility patent includes making, using, or selling a patented product or process without permission. Infringement of a design patent involves fabrication of a design that, to the ordinary observer, is substantially the same as an existing design, where the resemblance is intended to induce the observer to purchase one thing supposing it to be another. Infringement of a trademark consists of the unauthorized use or imitation of a mark that is the property of another in order to deceive, confuse, or mislead others. Infringement of a copyright involves reproducing, adapting, distributing, performing in public, or displaying in public the copyrighted work of someone else.

intellectual property (IP)
Creative ideas and expressions of the human mind that have commercial value and are entitled to the legal protection of a property right. The major legal mechanisms for protecting intellectual property are copyrights, patents, and trademarks. IP rights enable owners to select who may access and use their intellectual property and to protect it from unauthorized use.

international patent application
*Refer to **Patent Cooperation Treaty (PCT)**.*

intellectual property management
The means by which an institutionally owned IP portfolio is managed with regard to marketing, patenting, licensing, and administration.

invention
The creation of a new technical idea and of the physical embodiment of the idea or the means to accomplish it. To be patentable, an invention must be novel, must have utility, and would not have been obvious to those possessing ordinary skill in the particular art of the invention.

inventive step (nonobviousness)
A condition for patentability, which means that the invention would not be obvious to someone with knowledge and experience in the technological field of the invention. According to the European Patent Convention, "*An invention shall be considered as involving an inventive step if, having regard to the state of the art, it is not obvious to a person skilled in the art.*"

joint inventors
Two or more inventors of a single invention who work together during the inventive process.

know-how
Information that enables a person to accomplish a particular task or to operate a particular device or process. *Refer to **trade secret**.*

license
A grant of permission to use an IP right within a defined time, context, market line, or territory. There are important distinctions between exclusive licenses and nonexclusive licenses. An exclusive license is "exclusive" as to a defined scope, that is, the license might not be the only license granted for a particular IP asset, as there might be many possible fields and scopes of use that can also be subject to exclusive licensing. In giving an exclusive license, the licensor promises that he or she will not grant other licenses of the same rights within the same scope or field covered by the exclusive license. The owner of IP rights may also grant any number of nonexclusive licenses covering rights within a defined scope. A patent license is a transfer of rights that does not amount to an assignment of the patent. A trademark or service mark can be validly licensed only if the licensor controls the nature and quality of the goods or services sold by the licensee under the licensed mark. Under copyright law, an exclusive licensee is the owner of a particular right of copyright, and he or she may sue for infringement of the licensed right. There is never more than a single copyright in a work regardless of the owner's exclusive license of various rights to different persons.

licensee
A party obtaining rights under a license agreement.
licensor
A party granting rights under a license agreement.

license out
The process by which one person, company, or institution extends to another person, company, or institution permission to use the former's intellectual property.

license in
The process by which a person, company, or institution obtains permission to use the intellectual property owned by someone else.

material transfer agreement (MTA)
A contract between the owner of a tangible material and a party seeking the right to use the material for research or other assessment purposes. The material may be either patented or unpatented. Material transfer agreements tend to be shorter than license agreements. The purpose of an MTA is to document the transfer the material and outline the terms of use, including identification of the research or assessment project, terms of confidentiality, publication, and liability.

maintenance fees
Fees for maintaining in force a patent. The fees typically have to be paid at irregular intervals, depending on the jurisdiction, and significantly increase over time.

notice
A formal sign or notification attached to items that embody or reproduce an intellectual property assset—for example, the presence of the word *patent* or its abbreviation, *pat.*, together with the patent number, on a patented article made by a patent holder or his/her licensees. The formal statutory notice of U.S. trademark registration is the letter *R* inside a circle: ®, *Reg. U.S. Pat. & Tm. Off.*, or *Registered in U.S. Patent and Trademark Office*. Many firms use informal trademark notices, such as *Brand*, *TM*, *Trademark*, *SM*, or *Service Mark*, adjacent to words or other symbols considered to be protectable marks. Notice of copyright consists of the letter *C* in a circle symbol: © or the word *Copr.* or *Copyright*, the copyright owner's name, and the year of first publication.

nonassignable
A condition whereby a licensing agreement and/or the rights, obligations, and terms thereof may not be assigned to any party who is not a signatory to the agreement.

nondisclosure agreement
See **confidentiality agreement**.

nonexclusive license
A license under which rights are granted to the licensee but not exclusively to that licensee; the licensor reserves the right to give the same or similar rights to use the licensed materials to other parties.

nonobviousness
One of three conditions an invention must meet to be patentable. *See also* ***inventive step***.

nontransferable
The licensing agreement and/or the rights, obligations, and terms thereof that may not be sold, given, assigned, or otherwise conveyed to any party who is not a signatory to the agreement.

novelty
One of three conditions an invention must meet to be patentable.

obviousness
A condition of an invention that makes it ineligible to receive a valid patent; the condition of an invention whereby a person with ordinary skill in a field of technology can readily deduce it from publicly available information (prior art). *See also* ***ordinary skill in the art***.

ordinary skill in the art
The level of technical knowledge, experience, and expertise possessed by the ordinary engineer, scientist, or designer in a technology that is relevant to an invention.

Paris Convention
The main international treaty governing patents, trademarks, and unfair competition. The Convention is administered by the World Intellectual Property Organization (WIPO) and has four principal provisions: (1) national treatment for all seeking protection of IP rights, whether foreign or nationals; (2) minimum level of protection; (3) Convention priority, with a specified time (12 months for patents, six months for trademarks) for applications to be filed in other member nations; and (4) administrative framework within the Paris Union.

patent (U.S.)
A grant by the federal government to an inventor of the right to exclude others from making, using, or selling his or her invention. There are three kinds of patents in the United States: a standard utility patent on the functional aspects of products and processes; a design patent on the ornamental design of useful objects; and a plant patent on a new variety of a living plant. Patents do not protect ideas, only structures and methods that apply technological concepts. Each type of patent confers the right to exclude others from a precisely defined scope of technology, industrial design, or plant variety. In return for the right to exclude, an inventor must fully disclose

the details of the invention to the public so that others can understand it and use it to further develop the technology. Once the patent expires, the public is entitled to make and use the invention and is entitled to a full and complete disclosure of how to do so.

patent application
A technical document that describes in detail an invention for which a patent is sought.

patent examination
A process of review of a patent application, undertaken by a patent examiner, to determine whether the application complies with all statutory requirements for patentability. The examination process reviews prior art to ensure novelty, along with determining compliance with other statutory requirements, rules, and matters of procedure and form.

Patent Cooperation Treaty (PCT)
An international treaty that provides a mechanism through which an applicant can file a single application that, when certain requirements have been fulfilled, may then be pursued as a regular national filing in any of the PCT member nations. There are currently more than 120 PCT member nations.

patent pooling
A patent pool is an agreement between two or more patent owners to license one or more of their patents to one another or to third parties. A patent pool allows interested parties to gather all the necessary tools to practice a certain technology.

patent searching
A process carried out by the patent examiner for checking the novelty of a patent application. The subsequent patent research report lists published items comprising both patent and nonpatent literature relevant to the subject of the invention.

plant breeders' rights
Plant breeder's rights are used to protect new varieties of plants by giving exclusive commercial rights to market a new variety or its reproductive material.

plant patent
In the United States, the Plant Patent Act of 1930 provides a grant of exclusive IP rights to applicants who have invented or discovered a new asexually propagated variety of plant. Tuberous plants are not covered by plant patents.

plant variety protection (PVP)
A form of patent-like protection for sexually propagated plants, as well as hybrids, tubers, and harvested plant parts. The Plant Variety Protection Act of 1970 is administered by the U.S. Department of Agriculture and not the U.S. Patent and Trademark Office (which does issue plant patents).

prior art
The existing body of technological information against which an invention is judged in order to determine whether it is novel and nonobvious and can thus be patented.

prior informed consent
The consent given by a party with respect to an activity after being fully informed of all material facts relating to that activity. The Convention for Biological Diversity requires that access to genetic resources shall be subject to the prior informed consent of the country providing the resources.

priority date
The date of the first filing of a patent application that describes an invention in detail. Priority date, as well as patentability, with respect to novelty of invention, is determined in light of any relevant prior art existing at the time of filing. In other words, depending on the specific jurisdiction, if the invention was known or published previous to the priority date, the applicant will be unable to obtain a patent.

provisional application
A provisional application is a document in patent actions that serves to establish an early priority date of an invention. A provisional application will not mature into a regular application, and does not form the basis of a grant of a patent. It is a document that precedes the complete application upon which the grant is based. A provisional application establishes a priority date for disclosure of the details of an invention and allows a period of up to 12 months for development and refinement of the invention before the patent claims take their final form in a complete, regular patent application.

process claim
A claim of a patent that covers the method by which an invention is performed by defining the steps to be followed. This differs from a product claim or an apparatus claim, which covers the structure of a product.

product-by-process claim
A patent claim through which a product is claimed by defining the process by which it is made. The product-by-process form of claim is most often used to define new chemical compounds, since many new chemicals, drugs, and pharmaceuticals can practicably be defined only by describing the process of making them.

public domain
The status of an invention, creative work, commercial symbol, or any other creation that is not protected by some form of IP right. Items that have been determined

to be in the public domain are available for copying and use by anyone.

reduction to practice
The physical part of the inventive process that completes and ends the process of invention by demonstrating that the invention has a practical application. Reduction to practice can be carried out either by the actual construction of an apparatus, by performing the steps in a process, or by formally filing a patent application (constructive reduction to practice).

research tools
The term *research tool* includes the full range of tools that scientists may use in the laboratory, including cell lines, monoclonal antibodies, reagents, animal models, growth factors, combinatorial chemistry and DNA libraries, clones and cloning tools (such as PCR), methods, and laboratory equipment and machines.[2] There is concern about the patenting of research tools, because such patents may inhibit the free undertaking of research.

royalty
Income derived from the sale or use of a licensed product or process.

tiered pricing
See ***differential pricing.***

trademark
(1) A word, slogan, design, picture, or other symbol used to identify and distinguish goods. (2) Any identifying symbol, including a word, design, or shape of a product or container, that qualifies for legal status as a trademark, service mark, collective mark, certification mark, trade name, or trade dress. Trademarks identify one seller's goods and distinguish them from goods sold by others. They signify that all goods bearing the mark come from, or are controlled by, a single source and are of an equal level of quality. And they advertise, promote, and generally assist in selling goods. A trademark is infringed by another if the second use causes confusion of source, affiliation, connection, or sponsorship.

trade secret
Business information that is the subject of reasonable efforts to preserve confidentiality and has value because it is not generally known in the corresponding trade. Such confidential information is protected against those who gain access to it through improper methods or by a breach of confidence. Misappropriation of a trade secret is a type of unfair competition.

traditional knowledge
Tradition-based creations, innovations, literary, artistic or scientific works, performances and designs originating from or associated with a particular people or territory.

Trade-Related Aspects of Intellectual Property Rights (TRIPS)
An international agreement that was initiated under the forerunner of the World Trade Organization (WTO), the General Agreement on Tariffs and Trade (GATT), under the Uruguay round of trade negotiations. The TRIPS Agreement is the most comprehensive multilateral agreement on Intellectual Property covering all IP instruments. It was the first IP rights accord to legitimize the patenting of living organisms. TRIPS provides the guidelines for the harmonization of IP rights laws under the WTO. All WTO member countries have substantive TRIPS obligations.

unfair competition
Commercial conduct that the law views as unjust, providing a civil claim against a person who has been injured by the conduct. Trademark infringement has long been considered to be unfair competition. Other recognized legal categories of unfair competition are false advertising, trade libel, misappropriation of a trade secret, infringement of the right of publicity, and misappropriation.

UPOV (the Convention of the International Union for the Protection of New Varieties of Plants)
An international treaty that guarantees to plant breeders in member nations national treatment and a right of priority. National plant variety protection statutes of member nations are brought into harmonization with the various UPOV provisions, for example, the requirements of distinctness, uniformity, stability, and novelty for new crop varieties.

utility
The usefulness of a patented invention. To be patentable an invention must operate and be capable of use, and it must perform some "useful" function for society. ■

---

1 McCarthy JT, RE Schechter and DJ Franklyn. 1995 and 2004. *McCarthy's Desk Encyclopedia of Intellectual Property*, 2nd and 3rd editions. The Bureau of National Affairs: Washington, DC.

2 From NIH Research Tools Guidelines. ott.od.nih.gov/policy/rt_guide.html.

Readers are encouraged to consult the online version of this *Handbook* (www.iphandbook.org) for fuller indexing, a robust search engine, and a "Web log" presentation of key content areas with comments.

Index

C

I

K

L

M

U